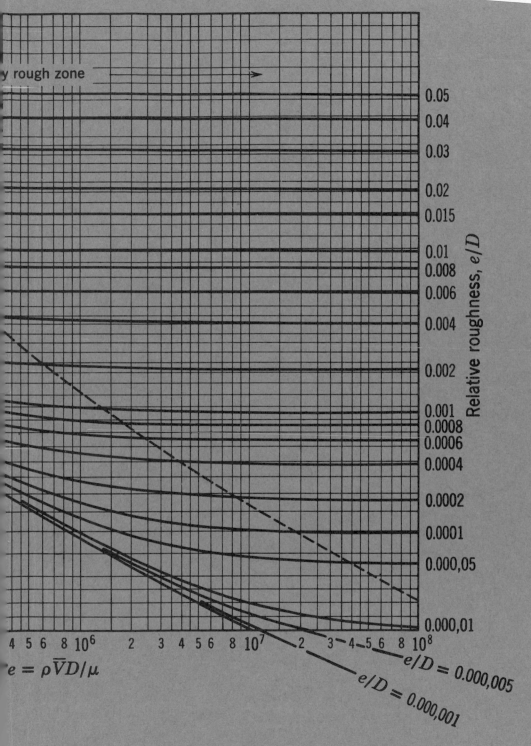

y rough zone

0.05
0.04
0.03
0.02
0.015
0.01
0.008
0.006
0.004
0.002
0.001
0.0008
0.0006
0.0004
0.0002
0.0001
0.000,05
0.000,01

Relative roughness, e/D

4 5 6 8 10^6 2 3 4 5 6 8 10^7 2 3 4 5 6 8 10^8

$e = \rho \overline{V} D / \mu$

$e/D = 0.000,005$

$e/D = 0.000,001$

INTRODUCTION
TO FLUID
MECHANICS

INTRODUCTION TO FLUID MECHANICS
Fourth Edition

ROBERT W. FOX
ALAN T. McDONALD

School of Mechanical Engineering
Purdue University

JOHN WILEY & SONS, INC.

New York · Chichester · Brisbane · Toronto · Singapore

On the Cover: *A Large Wind Tunnel and Model*

The cover photo shows a full-size "modcl" of a short takeoff, vertical landing, supersonic aircraft. The model is mounted 40 ft above the floor of the world's largest wind tunnel—the new 80×120 ft test section—at NASA Ames Research Center. (The man on the tunnel floor shows the scale.)

This aircraft uses the ejector concept, in which relatively cool fan air is directed to the ejector system—the inlet louvers visible along the aircraft fuselage—to produce lift. This creates a relatively cool flow field on the ground beneath the vehicle. The remaining air is exhausted out the rear nozzle to provide thrust for forward flight.

A key test objective was to measure the thrust augmentation of the ejector system. Full-scale testing allows duplication of flight Reynolds number, which is important in the transition from hover to wing-borne flight following vertical takeoff. Large-scale testing also provides more reliable powerplant and ejector performance data than small-scale tests. The very large test section minimizes extraneous effects on vertical force data.

(Reprinted with the permission of Aviation Week & Space Technology.)

Acquisition Editor:	Charity Robey
Production Manager:	Joe Ford
Production Supervisor:	Lucille Buonocore
Manufacturing Manager:	Lorraine Fumoso
Designer:	Laura Nicholls
Photo Researcher:	Hilary Newman
Illustrations:	Sigmund Malinowski

Recognizing the importance of preserving what has been written, it is a policy of John Wiley & Sons, Inc. to have books of enduring value published in the United States printed on acid-free paper, and we exert our best efforts to that end.

Library of Congress Cataloging in Publication Data:

Fox, Robert W.
    Introduction to fluid mechanics / Robert W. Fox, Alan T. McDonald.
    – 4th ed.
        p.   cm.
    Includes index.
    ISBN 0-471-54852-9
    1. Fluid mechanics.   I. McDonald, Alan T.   II. Title.
TA357.F69   1992
620.1'06–dc20   91-29150
                                                                  CIP

Printed in the United States of America

10 9 8 7 6 5 4 3 2 1

PREFACE

This text was written for an introductory course in fluid mechanics. Our approach to the subject in the Fourth Edition is unchanged. The physical concepts of fluid mechanics and analysis methods, beginning from basic principles, are emphasized throughout. The primary objective of this book is to help students develop an orderly approach to problem solving. Thus we start from basic equations, state assumptions clearly, and relate results to expected physical behavior. The approach is illustrated by the 123 example problems in the text. Solutions to these examples have been prepared to demonstrate good solution techniques and to explain troublesome points of theory. The example problems are set apart from the text in format, so they are particularly easy to follow.

In the Fourth Edition, the international system of units (SI) again is used in approximately 70 percent of the example problems and end-of-chapter problems. English engineering units are retained in the remaining problems to provide experience with this traditional system and to highlight methods of conversion among unit systems.

Complete explanations in the text, together with numerous detailed examples, make this book understandable for students. This allows the instructor freedom to depart from conventional lecture teaching methods. Classroom time can be used to bring in outside material, expand on special topic areas (such as non-Newtonian flow, boundary-layer flow, lift and drag, or measurement methods), solve example problems, or explain any difficult points of the assigned homework. Thus each class period can be used most appropriately to satisfy student needs.

The material in this book has been selected carefully. There is a detailed presentation of a broad range of topics suitable for a one- or two-semester course in fluid mechanics at the junior or senior level. Introductory courses in rigid-body dynamics and mathematics through integral calculus are necessary prerequisites. Some background in thermodynamics is desirable for the study of one-dimensional compressible flow.

The presentation is organized into broad topic areas:

- Introductory concepts, scope of fluid mechanics, and fluid statics (Chapters 1, 2, and 3).
- Development and application of control volume forms of basic equations (Chapter 4).
- Development and application of differential forms of basic equations (Chapters 5 and 6).
- Dimensional analysis and correlation of experimental data (Chapter 7).
- Applications for incompressible flow (internal flows in Chapter 8 and external flows in Chapter 9).
- Analysis and applications of flow in open channels (Chapter 10).

- Analysis and applications of fluid machinery (Chapter 11).
- Analysis and applications of one-dimensional compressible flow (Chapters 12 and 13).

Summary Objectives listed at the end of each chapter indicate specific concepts that should be understood and tasks students should be able to do after they have studied the material.

Major additions in the Fourth edition include a new chapter (Chapter 11) on fluid machinery, a new section on supersonic channel flow, and a large number of new homework problems. A *Software Supplement* is being developed independently for use with the text.

The new chapter on fluid machinery emphasizes actual machinery and applications. Pump curves are included in a separate appendix to allow treatment of a variety of realistic fluid system applications. The new section on supersonic channel flow with shocks permits a meaningful discussion of supersonic wind tunnel flows. With the addition of 440 new end of chapter problems, the Fourth Edition now contains 1500 problems for homework assignment and student exercises so that as many as eight semesters may be covered without repeating problem assignments. Some end of chapter problems are best solved by writing simple calculator or computer programs (these are identified by the double dagger symbol). Selected computer programs are contained in the *Software Supplement*.

In addition to the major changes, subtle changes and improvements have been made throughout the text. These have emphasized: improved pedagogy, more current data for real situations, more applications, design problems, and computer applications.

The *Solutions Manual* for the Fourth Edition continues the tradition established for this text. The *Solutions Manual*—available from the publisher when the text is adopted—contains a complete, detailed, full-size solution for each of the 1500 homework problems. Each solution has been prepared using the format of the example problems. Solutions may be photocopied for classroom or library use, eliminating the labor of problem solving for the instructor using the text.

The new *Instructor's Guide* in the *Solutions Manual* gives a difficulty rating for each problem and keys each problem to the relevant text section. This makes it simple to assign homework at the desired difficulty range for each section of the book.

We have been using open-ended *design problems* several times each semester in place of traditional laboratory experiments. These problems give students more time to explore the application of fluid mechanics principles to the design of devices and systems. Selected design problems have been placed in the *Instructor's Guide*, along with suggestions for their use.

The *Software Supplement* is designed to allow students to play the *What if?* game. Parameters are easy to vary so their effects on system behavior can be identified readily. The *Supplement* includes software for analysis of fluid properties, standard atmosphere, accelerating control volumes, pipe flow head loss, surface depth in open-channel flow, and one-dimensional compressible flow. For each compressible flow case, the temperature-entropy diagram may be plotted if desired. Hints for using the software also are presented in the *Instructor's Guide*.

Many fine instructional videos and films are available to further clarify and demonstrate basic principles in fluid mechanics. We refer to them in the text where their use is appropriate; a complete list of suppliers and titles is included in Appendix C.

When our students have finished this course, we expect them to be able to apply the basic equations to a variety of problems, including new problems that they have not encountered previously We emphasize physical understanding throughout to make students aware of the variety of phenomena that occur in real fluid flow situations. By minimizing the number of "magic formulas" and emphasizing the fundamental approach, we believe students will develop confidence in their ability to apply the material and will be able to reason out solutions to challenging problems.

The book also is well suited for independent study by students or practicing engineers. Its readability and clear examples help to build student confidence. The summary objectives at the end of each chapter may be used for review or to assess the achievement of educational goals.

We recognize that no single approach can satisfy all needs. We are grateful to the many students and faculty whose comments have helped us improve the Fourth Edition. We especially thank the reviewers for the Fourth Edition, who were: Seppo Korpela of The Ohio State University, Darryl Alofs of University of Missouri—Rolla, Jim Liburdy of Clemson University, Stanley Berger of University of California—Berkeley, Willem Brutsaert of University of Maine, Charles Merkle of The Pennsylvania State University, Frank Champagne of University of Arizona, Chris Rogers of Tufts University, Eugene Kordyban of University of Detroit, Ed Shaughnessy of Duke University, and Edgar O'Rear of University of Oklahoma. Professors Merkle and O'Rear provided particularly detailed and insightful reviews.

Thanks also are due to our wives, Beryl and Tania, who from behind the scenes supported the long hours of preparation that went into this effort. As always, we welcome criticisms and suggestions from interested readers or users of this book.

Robert W. Fox
Alan T. McDonald

CONTENTS

Chapter 1

INTRODUCTION

The goal of this texbook is to provide a clear, concise introduction to the subject of fluid mechanics. In beginning the study of any subject, a number of questions may come to mind. Students in the first course in fluid mechanics might ask:

What is fluid mechanics all about?

Why do I have to study it?

Why should I want to study it?

How does it relate to subject areas with which I am already familiar?

In this chapter we shall try to present at least qualitative answers to these and similar questions. This should serve to establish a base and a perspective for our study of fluid mechanics. Before proceeding with the definition of a fluid, we digress for a moment with a few pointed comments to students.

1-1 NOTE TO STUDENTS

In writing this book we have kept you, the student, uppermost in our minds; the book is written for you. It is our strong feeling that classroom time should not be devoted to a regurgitation of textbook material by the instructor. Instead, the time should be used to amplify the textbook material by discussing related material and applying basic principles to the solution of problems. The necessary conditions for accomplishing this goal are: (1) a clear, concise presentation of the fundamentals that you, the student, can read and understand, and (2) your willingness to read the text material before going to class. We have assumed responsibility for meeting the first condition. You must assume responsibility for satisfying the second condition. There probably will be times when we fall short of satisfying these objectives. If so, we would appreciate hearing of these shortcomings either directly or through your instructor.

It goes without saying that an introductory text is not all-inclusive. Your instructor undoubtedly will expand on the material presented, suggest alternative approaches to topics, and introduce additional new material. We encourage you to refer to the many other fluid mechanics textbooks and references available in the library; where another text presents a particularly good discussion of a given topic, we shall refer to it directly. We also encourage you to learn from your fellow students and from the graduate assistant(s) assigned to the course as well as from your instructor. We assume that you have had an introduction to thermodynamics (either in a basic physics course or an introductory course in thermodynamics) and prior courses in statics, dynamics, and differential and integral calculus. No attempt will be made to restate this subject material; however, the pertinent aspects of this previous study will be reviewed briefly when appropriate.

It is our strong belief that one learns best by *doing*. This is true whether the subject under study is fluid mechanics, thermodynamics, or golf. The fundamentals in any of these cases are few, and mastery of them comes through practice. *Thus it is extremely important, in fact essential, that you solve problems.* The numerous problems included at the end of each chapter provide the opportunity to gain facility in applying fundamentals to the solution of problems. You should avoid the temptation to adopt a "plug and chug" approach to solving problems. Most of the problems are such that this approach simply will not work. In solving problems we strongly recommend that you proceed using the following logical steps:

1. State briefly and concisely (in your own words) the information given.
2. State the information to be found.
3. Draw a schematic of the system or control volume to be used in the analysis. Be sure to label the boundaries of the system or control volume and label appropriate coordinate directions.
4. Give the appropriate mathematical formulation of the *basic* laws that you consider necessary to solve the problem.
5. List the simplifying assumptions that you feel are appropriate in the problem.
6. Complete the analysis algebraically before substituting numerical values.
7. Substitute numerical values (using a consistent set of units) to obtain a numerical answer.
   a. Reference the source of values for any physical properties.
   b. Be sure the significant figures in the answer are consistent with the given data.
8. Check the answer and review the assumptions made in the solution to make sure they are reasonable.
9. Label the answer.

In your initial work this problem format may seen unnecessary. However, such an orderly approach to the solution of problems will reduce errors, save time, and permit a clearer understanding of the limitations of a particular solution. This approach also prepares you for communicating your solution method and results to others, as will often be necessary in your career. This format is used in all example problems presented in this text; answers to example problems are given to three significant figures.

Most engineering calculations involve measured values or physical property data. Every measured value has associated with it an experimental uncertainty. The uncertainty in a measurement can be reduced with care and by applying more precise measurement techniques, but cost and time needed to obtain data rise sharply as measurement precision is increased. Consequently, few engineering data are sufficiently precise to justify the use of more than three significant figures.

The principles of specifying the experimental uncertainty of a measurement and of estimating the uncertainty of a calculated result are reviewed in Appendix F. These should be understood thoroughly by anyone who performs laboratory work. We suggest you take time to review Appendix F before performing laboratory work or solving the homework problems at the end of this chapter.

1-2 DEFINITION OF A FLUID

Fluid mechanics deals with the behavior of fluids at rest and in motion. It is logical to begin with a definition of a *fluid*: a fluid is a substance that deforms continuously under the application of a shear (tangential) stress no matter how small the shear stress may be.

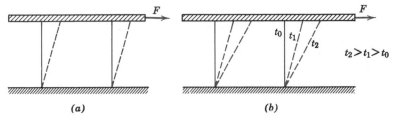

Fig. 1.1 Behavior of (a) solid and (b) fluid, under the action of a constant shear force.

Thus fluids comprise the liquid and gas (or vapor) phases of the physical forms in which matter exists. The distinction between a fluid and the solid state of matter is clear if you compare fluid and solid behavior. A solid deforms when a shear stress is applied, but it does not deform continuously.

In Fig. 1.1 the behavior of a solid (Fig. 1.1*a*) and a fluid (Fig. 1.1*b*) under the action of a constant shear force are contrasted. In Fig. 1.1*a* the shear force is applied to the solid through the upper of two plates to which the solid has been bonded. When the shear force is applied to the plate, the block is deformed as shown. From our previous work in mechanics, we know that, provided the elastic limit of the solid material is not exceeded, the deformation is proportional to the applied shear stress, $\tau = F/A$, where A is the area of the surface in contact with the plate.

To repeat the experiment with a fluid between the plates, use a dye marker to outline a fluid element as shown by the solid lines (Fig. 1.1*b*). When the force, F, is applied to the upper plate, the fluid element continues to deform as long as the force is applied. The fluid in direct contact with the solid boundary has the same velocity as the boundary itself; there is no slip at the boundary. This is an experimental fact based on numerous observations of fluid behavior.[1] The shape of the fluid element, at successive instants of time $t_2 > t_1 > t_0$, is shown (Fig. 1.1*b*) by the dashed lines, which represent the positions of the dye markers at successive times. Because the fluid motion continues under the application of a shear stress, we may alternatively define a fluid as a substance that cannot sustain a shear stress when at rest.

1-3 SCOPE OF FLUID MECHANICS

Having defined a fluid and noted the characteristics that distinguish it from a solid, we might ask the question: "Why study fluid mechanics?"

Knowledge and understanding of the basic principles and concepts of fluid mechanics are essential to analyze any system in which a fluid is the working medium. The design of virtually all means of transportation requires application of the principles of fluid mechanics. Included are aircraft for both subsonic and supersonic flight, ground effect machines, hovercraft, vertical takeoff and landing aircraft requiring minimum runway length, surface ships, submarines, and automobiles. In recent years automobile manufacturers have given more consideration to aerodynamic design. This has been true for some time for the designers of both racing cars and boats. The design of propulsion systems for space flight as well as for toy rockets is based on

---

[1] The no-slip condition is demonstrated in the NCFMF film, *Fundamentals of Boundary Layers*, F. H. Abernathy, principal. A complete list of fluid mechanics film titles and sources is given in Appendix C.

the principles of fluid mechanics. The collapse of the Tacoma Narrows Bridge in 1940 is evidence of the possible consequences of neglecting the basic principles of fluid mechanics.[2] It is commonplace today to perform model studies to determine the aerodynamic forces on, and flow fields around, buildings and structures. These include studies of skyscrapers, baseball stadiums, smokestacks, and shopping plazas.

The design of all types of fluid machinery including pumps, fans, blowers, compressors, and turbines clearly requires knowledge of the basic principles of fluid mechanics. Lubrication is an application of considerable importance in fluid mechanics. Heating and ventilating systems for private homes, large office buildings, and underground tunnels, and the design of pipeline systems are further examples of technical problem areas requiring knowledge of fluid mechanics. The circulatory system of the body is essentially a fluid system. It is not surprising that the design of blood substitutes, artificial hearts, heart-lung machines, breathing aids, and other such devices must rely on the basic principles of fluid mechanics.

Even some of our recreational endeavors are directly related to fluid mechanics. The slicing and hooking of golf balls can be explained by the principles of fluid mechanics (although they can be corrected only by a golf pro!).

The list of applications of the principles of fluid mechanics could be extended considerably. Our main point here is that fluid mechanics is not a subject studied for purely academic interest; rather, it is a subject with widespread importance both in our everyday experiences and in modern technology.

Clearly, we cannot hope to consider in detail even a small percentage of these and other specific problems of fluid mechanics. Instead, the purpose of this text is to present the basic laws and associated physical concepts that provide the basis or starting point in the analysis of any problem in fluid mechanics.

1-4   BASIC EQUATIONS

Analysis of any problem in fluid mechanics necessarily begins, either directly or indirectly, with statements of the basic laws governing the fluid motion. The basic laws, which are applicable to any fluid, are:

1. The conservation of mass.
2. Newton's second law of motion.
3. The principle of angular momentum.
4. The first law of thermodynamics.
5. The second law of thermodynamics.

Clearly, not all basic laws always are required to solve any one problem. On the other hand, in many problems it is necessary to bring into the analysis additional relations, in the form of equations of state or constitutive equations, that describe the behavior of physical properties of fluids under given conditions.

You probably recall studying properties of gases in basic physics or thermodynamics. The *ideal gas* equation of state

$$p = \rho RT \tag{1.1}$$

is a model that relates density to pressure and temperature for many gases for calculations of engineering accuracy under normal conditions. In Eq. 1.1, R is the gas

---

[2] For dramatic evidence of aerodynamic forces in action, see the film, *Collapse of the Tacoma Narrows Bridge*.

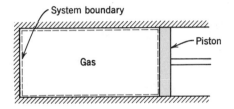

Fig. 1.2   Piston-cylinder assembly.

constant. Values of R are given in Appendix A for several common gases; p and T in Eq. 1.1 are the absolute pressure and absolute temperature, respectively. Example Problem 1.1 illustrates use of the ideal gas equation of state.

It is obvious that the basic laws with which we shall deal are the same as those used in mechanics and thermodynamics. Our task will be to formulate these laws in suitable forms to solve fluid flow problems and to apply them to a wide variety of problems.

We must emphasize that there are, as we shall see, many apparently simple problems in fluid mechanics that cannot be solved analytically. In such cases we must resort to more complicated numerical solutions and/or results of experimental tests.

Not all measurements can be made to the same degree of accuracy and not all data are equally good; the validity of data should be documented before test results are used for design. A statement of the probable uncertainty of data is an important part of reporting experimental results completely and clearly. Analysis of uncertainty also is useful during experiment design. Careful study may indicate potential sources of unacceptable error and suggest improved measurement methods.

1-5 METHODS OF ANALYSIS

The first step in solving a problem is to define the system that you are attempting to analyze. In basic mechanics, extensive use was made of the free-body diagram. In thermodynamics closed or open systems were considered. In this text we use the terms *system* and *control volume*. The importance of defining the system or control volume before applying the basic equations in the analysis of a problem cannot be overemphasized. At this point we review the definitions of systems and control volumes.

1-5.1 System and Control Volume

A system is defined as a fixed, identifiable quantity of mass; the system boundaries separate the system from the surroundings. The boundaries of the system may be fixed or movable; however, there is no mass transfer across the system boundaries.

In the familiar piston-cylinder assembly from thermodynamics, Fig. 1.2, the gas in the cylinder is the system. If a high-temperature source is brought in contact with the left end of the cylinder, the piston will move to the right; the boundary of the system thus moves. Heat and work may cross the boundaries of the system, but the quantity of matter within the system boundaries remains fixed. There is no mass transfer across the system boundaries.

EXAMPLE 1.1—First Law Application to Closed System

A piston-cylinder device contains 0.95 kg of oxygen initially at a temperature of 27 C and a pressure of 150 kPa. Heat is added to the gas and it expands at constant pressure to a temperature of 627 C. Determine the amount of heat added during the process.

EXAMPLE PROBLEM 1.1

GIVEN: Piston-cylinder containing O_2, $m = 0.95$ kg.

$T_1 = 27$ C $T_2 = 627$ C
p = constant = 150 kPa (abs)

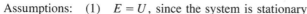

FIND: $Q_{1 \rightarrow 2}$.

SOLUTION:
We are dealing with a system, $m = 0.95$ kg.
Basic equation: First law for the system, $Q_{12} - W_{12} = E_2 - E_1$

Assumptions: (1) $E = U$, since the system is stationary
              (2)   Ideal gas with constant specific heats

Under the above assumptions,

$$E_2 - E_1 = U_2 - U_1 = m(u_2 - u_1) = m c_v (T_2 - T_1)$$

The work done during the process is moving boundary work

$$W_{12} = \int_{V_1}^{V_2} p \, dV = p(V_2 - V_1)$$

For an ideal gas, $pV = mRT$. Hence $W_{12} = mR(T_2 - T_1)$. Then from the first law equation,

$$Q_{12} = E_2 - E_1 + W_{12} = m c_v (T_2 - T_1) + mR(T_2 - T_1)$$

$$Q_{12} = m(T_2 - T_1)(c_v + R)$$

$$Q_{12} = m c_p (T_2 - T_1) \qquad \{R = c_p - c_v\}$$

From the Appendix, Table A.6, for O_2, $c_p = 909.4$ J/kg · K. Solving for Q_{12}, we obtain

$$Q_{12} = \frac{0.95 \text{ kg}}{} \times 909.4 \ \frac{\text{J}}{\text{kg} \cdot \text{K}} \times 600 \text{ K}$$

$$Q_{12} = 518 \ \text{kJ} \qquad\qquad\qquad\qquad\qquad\qquad Q_{12}$$

$\left\{\begin{array}{l} \text{The purpose of this problem was to review the use of:} \\ \text{(i) the first law of thermodynamics for a system, and} \\ \text{(ii) the equation of state for an ideal gas.} \end{array}\right\}$

In mechanics courses you made extensive use of the free-body diagram (system approach). This was logical because you were dealing with an easily identifiable rigid body. However, in fluid mechanics we normally are concerned with the flow of fluids through devices such as compressors, turbines, pipelines, nozzles, and so on. In these cases it is difficult to focus attention on a fixed identifiable quantity of mass. It is much more convenient, for analysis, to focus attention on a volume in space through which the fluid flows. Consequently, we use the control volume approach.

A control volume is an arbitrary volume in space through which fluid flows. The geometric boundary of the control volume is called the control surface. The control surface may be real or imaginary; it may be at rest or in motion. Figure 1.3

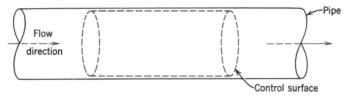

Fig. 1.3 Fluid flow through a pipe.

shows a possible control surface for analysis of flow through a pipe. Here the inside surface of the pipe, a real physical boundary, comprises part of the control surface. However, the vertical portions of the control surface are imaginary. There is no corresponding physical surface; these imaginary boundaries are selected arbitrarily for accounting purposes. Since the location of the control surface has a direct effect on the accounting procedure in applying the basic laws, it is extremely important that the control surface be carefully chosen and clearly defined before beginning any analysis.

1-5.2 Differential versus Integral Approach

The basic laws that we apply in our study of fluid mechanics can be formulated in terms of infinitesimal or finite systems and control volumes. As you might suspect, the equations will look different in each case. Both approaches are important in the study of fluid mechanics and both will be developed in the course of our work.

In the first case the resulting equations are differential equations. Solution of the differential equations of motion provides a means of determining the detailed (point by point) behavior of the flow.

Frequently, in the problems under study, the information sought does not require a detailed knowledge of the flow. We often are interested in the gross behavior of a device; in such cases it is more appropriate to use the integral formulation of the basic laws. The integral formulation, using finite systems or control volumes, usually is easier to treat analytically. The basic laws of mechanics and thermodynamics, formulated in terms of finite systems, are the basis for deriving the control volume equations in Chapter 4.

1-5.3 Methods of Description

Mechanics deals almost exclusively with systems; you have made extensive use of the basic equations applied to a fixed, identifiable quantity of mass. In attempting to analyze thermodynamic devices, you often found it necessary to use a control volume (open system) analysis. Clearly, the type of analysis depends on the problem. Where it is easy to keep track of identifiable elements of mass (e.g., in particle mechanics), we utilize a method of description that follows the particle. This sometimes is referred to as the *Lagrangian* method of description.

Consider, for example, the application of Newton's second law to a particle of fixed mass, m. Mathematically, we can write Newton's second law for a system of mass, m, as

$$\sum \vec{F} = m\vec{a} = m\frac{d\vec{V}}{dt} = m\frac{d^2\vec{r}}{dt^2} \qquad (1.2)$$

In Eq. 1.2, $\sum \vec{F}$ is the sum of all external forces acting on the system, $\vec{a}$ is the acceleration of the center of mass of the system, $\vec{V}$ is the velocity of the center of mass of the system, and $\vec{r}$ is the position vector of the center of mass of the system relative to a fixed coordinate system.

EXAMPLE 1.2—Free-Fall of Ball in Air

The air resistance on a 200 g ball in free flight is given by $f = 2 \times 10^{-4} v^2$, where f is in newtons and v is in meters per second. If the ball is dropped from rest 500 m above the ground, determine the speed at which it hits the ground. What percentage of the terminal speed is the result?

EXAMPLE PROBLEM 1.2

GIVEN: Ball, $m = 0.2$ kg, released from rest at $y_0 = 500$ m
          Air resistance, $f = kv^2$, where $k = 2 \times 10^{-4}$ N · sec²/m²

        Units: f(N), v(m/sec)

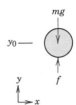

FIND: (a)   The speed at which the ball hits the ground.
       (b)   Ratio of speed to terminal speed.

SOLUTION:
Basic equation: $\sum \vec{F} = m\vec{a}$
The motion of the ball is governed by the equation

$$\sum F_y = ma_y = m\frac{dv}{dt}$$

Since $v = v(y)$, we write $\sum F_y = m\dfrac{dv}{dy}\dfrac{dy}{dt} = mv\dfrac{dv}{dy}$. Then,

$$\sum F_y = f - mg = kv^2 - mg = mv\frac{dv}{dy}$$

Separating variables and integrating,

$$\int_{y_0}^{y} dy = \int_{0}^{v} \frac{mv\,dv}{kv^2 - mg}$$

$$y - y_0 = \left[\frac{m}{2k}\ln(kv^2 - mg)\right]_0^v = \frac{m}{2k}\ln\frac{kv^2 - mg}{-mg}$$

Taking antilogarithms, we obtain

$$kv^2 - mg = -mg\, e^{\left[\frac{2k}{m}(y - y_0)\right]}$$

Solving for v gives

$$v = \left\{\frac{1}{k}mg\left(1 - e^{\left[\frac{2k}{m}(y - y_0)\right]}\right)\right\}^{1/2}$$

Substituting numerical values with $y = 0$ yields

$$v = \left\{0.2\text{ kg} \times 9.81\,\frac{\text{m}}{\text{sec}^2} \times \frac{\text{m}^2}{2 \times 10^{-4}\text{ N} \cdot \text{sec}^2} \times \frac{\text{N} \cdot \text{sec}^2}{\text{kg} \cdot \text{m}}\left(1 - e^{\left[\frac{2 \times 2 \times 10^{-4}}{0.2}(-500)\right]}\right)\right\}^{1/2}$$

$$v = 78.7 \text{ m/sec}$$ v

At terminal speed, $a_y = 0$ and $\sum F_y = 0 = kv_t^2 - mg$

Then, $v_t = \left[\dfrac{mg}{k}\right]^{1/2} = \left[0.2 \text{ kg} \times 9.81 \dfrac{\text{m}}{\text{sec}^2} \times \dfrac{\text{m}^2}{2\times 10^{-4} \text{ N}\cdot\text{sec}^2} \times \dfrac{\text{N}\cdot\text{sec}^2}{\text{kg}\cdot\text{m}}\right]^{1/2}$

$$v_t = 99.0 \text{ m/sec} \quad \text{and} \quad \frac{v}{v_t} = \frac{78.7}{99.0} = 0.795, \text{ or } 79.5\%$$

$$\frac{v}{v_t}$$

$\left\{\begin{array}{l}\text{This problem is included as a reminder of the method of description used in particle}\\ \text{mechanics.}\end{array}\right\}$

We may consider a fluid to be composed of a very large number of particles whose motion must be described; keeping track of the motion of each fluid particle separately would become a horrendous bookkeeping problem. Consequently, a particle description becomes unmanageable. Often we find it convenient to use a different type of description. Particularly with control volume analyses, it is convenient to use the field, or *Eulerian*, method of description, which focuses attention on the properties of a flow at a given point in space as a function of time. In the Eulerian method of description, the properties of a flow field are described as functions of space coordinates and time. We shall see in Chapter 2 that this method of description is a logical outgrowth of the assumption that fluids may be treated as continuous media.

1-6 DIMENSIONS AND UNITS

Engineering problems are solved to answer specific questions. It goes without saying that the answer must include units. (It makes a difference whether a pipe diameter required is 1 meter or 1 foot!) Consequently, it is appropriate to present a brief review of dimensions and units. We say "review" because the topic is familiar from your earlier work in mechanics.

We refer to physical quantities such as length, time, mass, and temperature as *dimensions*. In terms of a particular system of dimensions all measurable quantities can be subdivided into two groups—primary quantities and secondary quantities. We refer to a small group of dimensions from which all others can be formed as primary quantities. Primary quantities are those for which we set up arbitrary scales of measure; secondary quantities are those quantities whose dimensions are expressible in terms of the dimensions of the primary quantities.

Units are the arbitrary names (and magnitudes) assigned to the primary dimensions adopted as standards for measurement. For example, the primary dimension of length may be measured in units of meters, feet, yards, or miles. These units of length are related to each other through unit conversion factors (1 mile = 5280 feet = 1609 meters).

1-6.1 Systems of Dimensions

Any valid equation that relates physical quantities must be dimensionally homogeneous; each term in the equation must have the same dimensions. We recognize that Newton's second law ($\vec{F} \propto m\vec{a}$) relates the four dimensions, F, M, L, and t. Thus force and mass cannot both be selected as primary dimensions without introducing a constant of proportionality that has dimensions (and units).

Length and time are primary dimensions in all dimensional systems in common use. In some systems, mass is taken as a primary dimension. In others, force is selected as a primary dimension; a third system chooses both force and mass as primary dimensions. Thus we have three basic systems of dimensions, corresponding to the different ways of specifying the primary dimensions.

a. Mass $[M]$, length $[L]$, time $[t]$, temperature $[T]$.
b. Force $[F]$, length $[L]$, time $[t]$, temperature $[T]$.
c. Force $[F]$, mass $[M]$, length $[L]$, time $[t]$, temperature $[T]$.

In system a, force $[F]$ is a secondary dimension and the constant of proportionality in Newton's second law is dimensionless. In system b, mass $[M]$ is a secondary dimension, and again the constant of proportionality in Newton's second law is dimensionless. In system c, both force $[F]$ and mass $[M]$ have been selected as primary dimensions. In this case the constant of proportionality, g_c, in Newton's second law (written $\vec{F} = m\vec{a}/g_c$) is not dimensionless. The dimensions of g_c must in fact be $[ML/Ft^2]$ for the equation to be dimensionally homogeneous. The numerical value of the constant of proportionality depends on the units of measure chosen for each of the primary quantities.

1-6.2 Systems of Units

There is more than one way to select the unit of measure for each primary dimension. We shall present only the more common engineering systems of units for each of the basic systems of dimensions.

a. MLtT

SI, which is the official abbreviation in all languages for the Système International d'Unités,[3] is an extension and refinement of the traditional metric system. More than 30 countries have declared it to be the only legally accepted system.

In the SI system of units, the unit of mass is the kilogram (kg), the unit of length is the meter (m), the unit of time is the second (sec), and the unit of temperature is the kelvin (K). Force is a secondary dimension, and its unit, the newton (N), is defined from Newton's second law as

$$1 \text{ N} \equiv 1 \text{ kg} \cdot \text{m/sec}^2$$

In the Absolute Metric system of units, the unit of mass is the gram, the unit of length is the centimeter, the unit of time is the second, and the unit of temperature is the kelvin. Since force is a secondary dimension, the unit of force, the dyne, is defined in terms of Newton's second law as

$$1 \text{ dyne} \equiv 1 \text{ g} \cdot \text{cm/sec}^2$$

b. FLtT

In the British Gravitational system of units, the unit of force is the pound (lbf), the unit of length is the foot (ft), the unit of time is the second, and the unit of temperature is the Rankine (R). Since mass is a secondary dimension, the unit of mass,

---

[3] American Society for Testing and Materials, *ASTM Standard for Metric Practice*, E380–89. Philadelphia: ASTM, 1989.

the slug, is defined in terms of Newton's second law as

$$1 \text{ slug} \equiv 1 \text{ lbf} \cdot \text{sec}^2/\text{ft}$$

c. FMLtT

In the English Engineering system of units, the unit of force is the pound force (lbf), the unit of mass is the pound mass (lbm), the unit of length is the foot, the unit of time is the second, and the unit of temperature is the Rankine. Since both force and mass are chosen as primary dimensions, Newton's second law is written as

$$\vec{F} = \frac{m \vec{a}}{g_c}$$

A force of one pound (1 lbf) is the force that gives a pound mass (1 lbm) an acceleration equal to the standard acceleration of gravity on Earth, 32.17 ft/sec^2. From Newton's second law we see that (to three significant figures)

$$1 \text{ lbf} \equiv \frac{1 \text{ lbm} \times 32.2 \text{ ft/sec}^2}{g_c}$$

or

$$g_c \equiv 32.2 \text{ ft} \cdot \text{lbm/lbf} \cdot \text{sec}^2$$

The constant of proportionality, g_c, has both dimensions and units. The dimensions arose because we selected both force and mass as primary dimensions; the units (and the numerical value) are a consequence of our choices for the standards of measurement.

Since a force of 1 lbf accelerates 1 lbm at 32.2 ft/sec^2, it would accelerate 32.2 lbm at 1 ft/sec^2. A slug also is accelerated at 1 ft/sec^2 by a force of 1 lbf. Therefore,

$$1 \text{ slug} \equiv 32.2 \text{ lbm}$$

1-6.3 Preferred Systems of Units

In this text we shall use both the SI and the British Gravitational systems of units. In either case, the constant of proportionality in Newton's second law is dimensionless and has a value of unity. Consequently, Newton's second law is written as $\vec{F} = m\vec{a}$. In these systems, it follows that the gravitational force (the "weight"[4]) on an object of mass, m, is given by $W = mg$.

SI units and prefixes, together with other defined units and useful conversion factors, are summarized in Appendix G.

1-7 SUMMARY OBJECTIVES

After completing study of Chapter 1, you should be able to do the following:

1. Give operational definitions of:

fluid	Eulerian method of description
no-slip condition	dimensions
system	units
control volume	dimensional homogeneity
Lagrangian method of description	weight

---

[4] Note that in the English Engineering system, the weight of an object is given by $W = mg/g_c$.

2. Give examples in which fluid mechanics is important to an understanding of phenomena from everyday experience and modern technology.
3. List the five basic laws governing the motion of fluids.
4. State the three basic systems of dimensions.
5. Give typical units of physical quantities in the SI, British Gravitational, and English Engineering systems of units.
6. Solve the problems at the end of the chapter that relate to the material you have studied.

PROBLEMS

1.1 A number of common substances are

Tar	Sand
"Silly Putty"	Jello
Modeling clay	Toothpaste
Wax	Shaving cream

Some of these materials exhibit characteristics of both solid and fluid behavior under different conditions. Explain and give examples.

1.2 A tank of compressed oxygen for flame cutting is to contain 10 kg of oxygen at a pressure of 14 MPa (the temperature is 35 C). How large must be the tank volume? What is the diameter of a sphere with this volume?

1.3 Air at a pressure of 40 psia and a temperature of 70 F is moving with a speed of 100 ft/sec. Calculate: (a) the kinetic energy per unit mass of the air and (b) the kinetic energy per unit volume of the air.

1.4 Air at an absolute pressure of 300 kPa and a temperature of 20 C is moving at a speed of 30 m/sec. Calculate: (a) the kinetic energy per unit mass of the air and (b) the kinetic energy per unit volume of the air.

1.5 Give a word statement of each of the five basic conservation laws stated in Section 1-4, as they apply to a system.

1.6 Moist air is a mixture of water vapor and air. At 100 percent relative humidity, the water vapor is at its saturation pressure. The atmospheric pressure given by a barometer equals the sum of the partial pressures of (dry) air and of the water vapor. Thus

$$p_{atm} = p_{air} + p_{vapor}$$

The saturation pressure of water at 15 C is 1.70 kPa. Use Dalton's law of partial pressures to calculate the density of moist air with 100 percent relative humidity at standard temperature and pressure (STP conditions are $T = 15$ C and $p = 101.3$ kPa absolute). Compare with the density of dry air at the same conditions.

1.7 A can of pet food has the following internal dimensions: 102 mm height and 73 mm diameter (each $\pm$ 1 mm at odds of 20 to 1). The label lists the mass of the contents as 397 g. Evaluate the magnitude and estimated uncertainty of the density of the pet food if the mass value is accurate to $\pm$ 1 g at the same odds.

1.8 The mass of the standard American golf ball is 1.62 ± 0.01 oz and its mean diameter is 1.68 ± 0.01 in. Determine the density and specific gravity of the American golf ball. Estimate the uncertainties in the calculated values.

1.9 The mass of the standard British golf ball is 1.62 ± 0.01 oz and its mean diameter is 1.62 ± 0.01 in. Determine the density and specific gravity of the British golf ball. Estimate the uncertainties in the calculated values.

1.10 The estimated dimensions of a soda can are $D = 66.0 \pm 0.5$ mm and $H = 110 \pm 0.5$ mm. Measure the mass of a full can and an empty can using a kitchen scale or postal scale. Estimate the volume of soda contained in the can. From your measurements estimate the depth to which the can is filled and the uncertainty in the estimate. Assume the value of SG = 1.055, as supplied by the bottler.

1.11 Calculate the density of standard air in a laboratory from the ideal gas equation of state. Estimate the experimental uncertainty in the air density calculated for standard conditions (29.9 in. of mercury and 59 F) if the uncertainty in measuring the barometer height is ± 0.1 in. of mercury and the uncertainty in measuring temperature is ±0.5 F. (Note that 29.9 in. of mercury corresponds to 14.7 psia.)

1.12 Repeat the calculation of uncertainty described in Problem 1.11 for air in a freezer. Assume the measured barometer height is 759 ± 1 mm of mercury and the temperature is − 20 ± 0.5 C. [Note that 759 mm of mercury corresponds to 101 kPa (abs).]

1.13 The mass flow rate in a water flow system determined by collecting the discharge over a timed interval is 0.3 kg/sec. The scales used can be read to the nearest 0.05 kg and the stopwatch is accurate to 0.2 sec. Estimate the precision with which the flow rate can be calculated for time intervals of (a) 10 sec and (b) 1 min.

1.14 The mass flow rate of water in a tube is measured using a beaker to catch water during a timed interval. The nominal mass flow rate is 100 g/sec. Assume that mass is measured using a balance with a least count of 1 g and a maximum capacity of 1 kg, and that the timer has a least count of 0.1 sec. Estimate the time intervals and uncertainties in measured mass flow rate that would result from using 100, 500, and 1000 mL beakers. Would there be any advantage in using the largest beaker? Assume the tare mass of the empty 1000 mL beaker is 500 g.

1.15 Measured data for pressure drop of air flow in a smooth pipe are to be reduced to friction factors. The friction factor is defined as

$$f = \frac{\Delta p}{\frac{L}{D}\frac{1}{2}\rho \bar{V}^2}$$

where Δp = pressure drop, L = pipe length, D = pipe diameter, ρ = air density, and $\bar{V}$ = average flow velocity. The measured data and their estimated experimental uncertainties are:

$$\Delta p = 8.5 \pm 0.1 \text{ mm } H_2O \qquad \rho = 1.23 \pm 0.01 \text{ kg/m}^3$$
$$L = 750 \pm 1 \text{ mm} \qquad \bar{V} = 25 \pm 0.5 \text{ m/sec}$$
$$D = 62.5 \pm 0.1 \text{ mm}$$

(Note that 1 mm of H_2O corresponds to 9.80 N/m^2.)

Estimate the uncertainty in the calculated friction factor. Which variable contributes the most uncertainty to the friction factor?

1.16 An enthusiast magazine publishes data from its road tests on the lateral acceleration capability of cars. The measurements are made using a 200 ft diameter skid pad. Assume the vehicle path deviates from the circle by ± 2 ft and that the vehicle speed is read from a fifth-wheel speed-measuring system to ± 0.5 mph. Estimate the experimental uncertainty in a reported lateral acceleration of 0.823 g. How would you improve the experimental procedure to reduce the uncertainty?

1.17 Using the nominal dimensions of the soda can given in Problem 1.10, determine the precision with which the diameter and height must be measured to estimate the volume of the can within an uncertainty of ± 0.5 percent.

1.18 An American golf ball is described in Problem 1.8. Assuming the measured mass and its uncertainty as given, determine the precision to which the diameter of the ball must be measured so the density of the ball may be estimated within an uncertainty of ± 1 percent.

1.19 Many measurements today are made using digital instruments, which convert analog signals from transducers to digital form via A-D (analog-to-digital) converters. The resolution of an A-D converter is specified by the maximum number of bits of information it produces; the resolution is ± 0.5 in the least significant bit. Estimate the uncertainty in the output from an 8-bit and a 12-bit converter as a percentage of the full-scale readings. (Assume the analog input signal exactly represents the measured value.) Comment on the uncertainties in the readings at 1/10 of full scale.

‡1.20   The height of a building may be estimated by measuring the horizontal distance to a point on the ground and the angle from this point to the top of the building. Assuming these measurements are $L = 100 \pm 0.5$ ft and $\theta = 30 \pm 0.2$ degrees, estimate the height of the building and the uncertainty in the estimate. For the same building height and measurement uncertainties, determine the distance from the building at which measurements should be made to minimize the uncertainty in estimated height. Evaluate and plot the optimum measurement angle as a function of building height.

‡1.21   In the design of a medical instrument it is desired to dispense 1 cubic millimeter of liquid using a piston-cylinder syringe made from molded plastic. The molding operation produces plastic parts with estimated dimensional uncertainties of $\pm\,0.002$ in. Estimate the uncertainty in dispensed volume that results from the uncertainties in the dimensions of the device. Determine the ratio of stroke length to bore diameter that gives a design with minimum uncertainty in volume dispensed. Is the result influenced by the magnitude of the dimensional uncertainty?

1.22   A tank for scuba diving is designed to contain 50 standard cubic feet (SCF) of air when filled to a pressure of 3000 pounds per square inch (gage) at an ambient temperature of 80 F. Calculate the interior volume of the tank and its length if the inside diameter is 6 in. A standard cubic foot of gas occupies one cubic foot at standard temperature and pressure ($T = 15$ C and $p = 101.3$ kPa absolute).

1.23   A compressed air tank in a service station holds 0.2 m^3 of compressed air at 800 kPa (gage). Determine the amount of energy required to compress this much air isothermally from atmospheric pressure, assuming a frictionless process. (Note that the release of this much energy would be catastrophic if the tank were ruptured.)

1.24   Air trapped in a bicycle tire pump is compressed suddenly to $\frac{1}{5}$ of its original volume. Both heat transfer and friction may be neglected as a first approximation. Determine the final temperature of the air in the pump if the initial temperature is 20 C.

1.25   The catapult on an aircraft carrier has a piston diameter of 2 ft and a stroke of 100 ft. Initially, the cylinder contains steam at 250 psia, 600 F in a volume of 100 ft^3. Assume that heat transfer and friction are negligible as the piston moves through its stroke. Determine (a) the temperature and pressure of the steam at the end of the stroke and (b) the work done by the steam during the expansion process.

1.26   A small particle moving in water experiences drag force, $F_D = kV$, where the dimensions of k are force per unit speed. A particle, of mass m, is set in motion horizontally with initial speed V_0. Show that the horizontal distance traveled before the particle stops is equal to mV_0/k. Assume there is no vertical motion.

1.27   The aerodynamic drag force on a tractor-trailer rig moving in still air is given by $F_D = kV^2$, where $k = 0.135$ lbf $\cdot$ sec^2/ft^2. Calculate the force required to overcome aerodynamic drag at a speed of 55 mph. Evaluate the power saving if the aerodynamic drag were reduced 6 percent by installing a fairing on the cab roof.

1.28   A projectile is fired with velocity $\vec{V}_0$ and elevation angle θ, above the horizon. Air resistance may be neglected. Express the range of the projectile in terms of V_0 and θ. Determine the angle that gives the maximum range.

1.29   Very small particles moving in fluids experience a drag force proportional to speed. Consider a particle of net weight, W, dropped in a fluid. The particle experiences a drag force, kV, where V is the particle speed. Determine the time required for the particle to accelerate from rest to 95 percent of its terminal speed, V_t, in terms of k, W, and g.

1.30   Consider again the small particle of Problem 1.29. Express the distance required to reach 95 percent of its terminal speed in terms of g, k, and W.

‡ You may wish to use simple computer programs to help solve problems marked with daggers.

1.31   A skydiver with a mass of 75 kg jumps from an aircraft. The aerodynamic drag force acting on the sky diver is known to be $F_D = kV^2$, where $k = 0.228$ N · sec^2/m^2. Determine the maximum speed of free fall for the sky diver and the speed reached after 100 m of fall.

1.32   A spear, of mass $m = 0.3$ kg, is propelled horizontally from a spear gun by a scuba diver. The initial speed of the spear is $V_0 = 30$ m/sec. The force that resists its motion through the water is given by $F_D = kV^2$, where $k = 0.033$ N · sec^2/m^2. The spear is effective against sharks when its speed is above $V = 10$ m/sec. Estimate the effective range of the spear.

1.33   A swimmer in freshwater can move at a maximum steady speed of $V_s = 1.5$ m/sec in still water. To do so, the swimmer must produce enough power to overcome water resistance. The drag force acting on the swimmer is estimated to be $F_D = kV^2$, where $k = 30$ N · sec^2/m^2, and V is the speed of the swimmer *relative* to the water. Evaluate the power produced by the swimmer in still water. If the swimmer can maintain the same power output in a river current that moves at 3 km/hr, estimate the maximum speeds the swimmer can reach (a) swimming upstream and (b) swimming downstream.

1.34   The English perfected the longbow as a weapon after the Medieval period. In the hands of a skilled archer, the longbow was reputed to be accurate at ranges to 100 meters or more. If the maximum altitude of an arrow is less than 10 m while traveling to a target 100 m away from the archer, and neglecting air resistance, estimate the speed and angle at which the arrow must leave the bow.

1.35   A sky diver, with mass $m = 80$ kg, drops from a slow-moving aircraft and falls straight down. The aerodynamic drag force acting on the diver is $F_D = kV^2$, where $k = 0.27$ N · sec^2/m^2, and V is the speed relative to the air. Evaluate the terminal speed of the sky diver. Estimate the vertical distance required for the sky diver to reach 95 percent of terminal speed. Compare with the distance required to reach the same speed if air resistance were neglected.

1.36   A droplet of lubricating oil with $D = 0.7$ mm is placed in water at 20 C. Since the droplet is small, its resistance to motion may be characterized approximately by $F_D = kV$, where $k = 6.60 \times 10^{-6}$ N · sec/m. Calculate the magnitude and direction of the terminal speed of the droplet.

1.37   A ball is thrown vertically upward with initial speed V_0. Air resistance on the ball is proportional to the square of its speed, $F_D = kV^2$. Analyze the motion to obtain expressions for the velocity of the ball and its height as functions of time. Express the time for the ball to reach maximum height in terms of V_0, m, k, and g. Compare with the value for no air resistance.

1.38   For each quantity listed, indicate dimensions using the *MLtT* system of dimensions, and give typical SI and English units:

(a)	Power	(b)	Pressure
(c)	Modulus of elasticity	(d)	Angular velocity
(e)	Energy	(f)	Momentum
(g)	Shear stress	(h)	Specific heat
(i)	Thermal expansion coefficient		

1.39   For each quantity listed, indicate dimensions using the *FLtT* system of dimensions, and give typical SI and English units:

(a)	Power	(b)	Pressure
(c)	Modulus of elasticity	(d)	Angular velocity
(e)	Energy	(f)	Moment of a force
(g)	Momentum	(h)	Shear stress
(i)	Strain		

1.40   The unit of pressure in the SI system is the pascal (Pa). How many pounds force per square inch (psi) correspond to 1 Pa?

1.41 A U.S. gallon, by definition, contains a volume of 231 cubic inches. Determine (a) the number of gallons in a cubic foot and (b) the number of liters in a gallon.

1.42 Obtain the conversion factor for converting the viscosity, μ, from units of newtons, seconds, and meters to units of pounds force, seconds, and feet. Check your answer using Appendix A.

1.43 Truck weight laws in Michigan allow a gross combination weight of 130,000 lbf. A tractor-trailer tank truck weighs 36,000 lbf empty. Calculate the number of gallons of gasoline that it can carry legally. Use data for gasoline from Appendix A.

1.44 A super-tanker carries a cargo of 400,000 long tons (1 long ton = 2240 lbm) of crude oil. Assume that the specific gravity of the oil is SG = 0.85. Calculate the number of barrels of oil in the tanker's cargo (the petroleum industry defines 1 bbl as 42 U.S. gallons).

1.45 The density of mercury is given as 26.3 slug/ft^3. Calculate the specific gravity and the specific volume in m^3/kg of the mercury. Calculate the specific weight in lbf/ft^3 on Earth and on the moon. Acceleration of gravity on the moon is 5.47 ft/sec^2.

1.46 Derive the following conversion factors:
  (a) Convert a volume flow rate in cubic meters per second to cubic feet per second.
  (b) Convert a volume flow rate in cubic feet per second to gallons per minute.
  (c) Convert a volume flow rate of water in gallons per hour to kg per minute.
  (d) Convert a volume flow rate of air in standard cubic feet per minute (SCFM) to pounds per hour. A standard cubic foot of gas occupies one cubic foot at standard temperature and pressure ($T = 15$ C and $p = 101.3$ kPa absolute).

1.47 At an institution known for its basketball prowess, a new set of physical units has been suggested. The basic unit of length is to be the "three-pointer," which is 21 ft; the basic unit of time is the "shot clock," which is 45 sec; the basic unit of force is the "basketball," which is 21 oz. Determine the conversion factors among these units and their SI equivalents. How are the units of mass in the two systems related?

1.48 At an institution known for its baseball prowess, it has been suggested that force, velocity, and length be considered as basic dimensions. The basic unit of force is the "baseball," which is 5.1 oz; the basic unit of velocity is the "fastball," which is 90 mph; the basic unit of length is the "homerun," which is 385 ft. Determine the conversion factors among these units and their SI equivalents. What is the unit of mass in the new system? What is the conversion factor between this unit of mass and the SI unit?

1.49 A container weighs 2.9 lbf when empty. When filled with water at 90 F, the mass of the container and its contents is 1.95 slug. Find the weight of water in the container, and its volume in cubic feet, using data from Appendix A.

1.50 Evaluate the change in specific weight of mercury, in lbf/ft^3, as its temperature changes from 70 to 90 F. Use the data in Appendix A.

Chapter 2

FUNDAMENTAL CONCEPTS

In Chapter 1 we indicated that our study of fluid mechanics will build on earlier studies in mechanics and thermodynamics. To develop a unified approach, we review some familiar topics and introduce some new concepts and definitions. The purpose of this chapter is to develop these fundamental concepts.

2-1 FLUID AS A CONTINUUM

In our definition of a fluid, no mention was made of the molecular structure of matter. All fluids are composed of molecules in constant motion. However, in most engineering applications we are interested in the average or macroscopic effects of many molecules. It is these macroscopic effects that we ordinarily perceive and measure. We thus treat a fluid as an infinitely divisible substance, a *continuum*, and do not concern ourselves with the behavior of individual molecules.

The concept of a continuum is the basis of classical fluid mechanics. The continuum assumption is valid in treating the behavior of fluids under normal conditions. However, it breaks down whenever the mean free path of the molecules (approximately 10^{-7} m for gas molecules that show ideal behavior at STP)[1] becomes the same order of magnitude as the smallest significant characteristic dimension of the problem. In problems such as rarefied gas flow (e.g., as encountered in flights into the upper reaches of the atmosphere), we must abandon the concept of a continuum in favor of the microscopic and statistical points of view.

As a consequence of the continuum assumption, each fluid property is assumed to have a definite value at every point in space. Thus fluid properties such as density, temperature, velocity, and so on, are considered to be continuous functions of position and time.

To illustrate the concept of a property at a point, consider the manner in which we determine the density at a point. A region of fluid is shown in Fig. 2.1. We are interested in determining the density at the point C, whose coordinates are x_0, y_0, and z_0. The density is defined as mass per unit volume. Thus the mean density within the volume, $\forall$, would be given by $\rho = m/\forall$. In general, this will not be equal to the value of the density at point C. To determine the density at point C, we must select a small volume, $\delta\forall$, surrounding point C and determine the ratio $\delta m/\delta\forall$. The question is, how small can we make the volume $\delta\forall$? Let us answer this question by plotting the ratio, $\delta m/\delta\forall$, and allowing the volume to shrink continuously in size. Assuming that the volume, $\delta\forall$, is initially relatively large (but still small compared to the volume, $\forall$) a typical plot of $\delta m/\delta\forall$ might appear as in Fig. 2.1*b*. In other

---

[1] STP (Standard Temperature and Pressure) for air are 15 C (59 F) and 101.3 kPa absolute (14.696 psia), respectively.

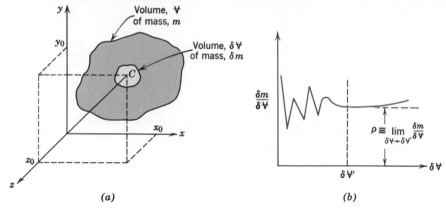

Fig. 2.1 Definition of density at a point.

words, $\delta \mathsf{V}$ must be sufficiently large to yield a meaningful, reproducible value for the density at a location and yet small enough to be able to resolve spatial variations in density. The average density tends to approach an asymptotic value as the volume is shrunk to enclose only homogeneous fluid in the immediate neighborhood of point C. When $\delta \mathsf{V}$ becomes so small that it contains only a small number of molecules, it becomes impossible to fix a definite value for $\delta m / \delta \mathsf{V}$; the value will vary erratically as molecules cross into and out of the volume. Thus there is a lower limiting value of $\delta \mathsf{V}$, designated $\delta \mathsf{V}'$ in Fig. 2.1b, allowable for use in defining fluid density at a point.[2] The density at a "point" is then defined as

$$\rho \equiv \lim_{\delta \mathsf{V} \to \delta \mathsf{V}'} \frac{\delta m}{\delta \mathsf{V}} \tag{2.1}$$

Since point C was arbitrary, the density at any point in the fluid could be determined in a like manner. If density determinations were made simultaneously at an infinite number of points in the fluid, we would obtain an expression for the density distribution as a function of the space coordinates, $\rho - \rho(x, y, z)$, at the given instant in time. Clearly, the density at a point may vary with time as a result of work done on or by the fluid and/or heat transfer to the fluid. Thus the complete representation of density (the *field* representation) is given by

$$\rho = \rho(x, y, z, t) \tag{2.2}$$

Since density is a scalar quantity, requiring only the specification of a magnitude for a complete description, the field represented by Eq. 2.2 is a scalar field.

2-2 VELOCITY FIELD

In the previous section we saw that the continuum assumption led directly to the notion of the density field. Other fluid properties also may be described by fields.

In dealing with fluids in motion, we shall necessarily be concerned with the description of a velocity field. Refer again to Fig. 2.1a. Define the fluid velocity at point C as the instantaneous velocity of the center of gravity of the volume, $\delta \mathsf{V}'$,

---

[2] The size of $\delta \mathsf{V}'$ is extremely small. For example, 1 m^3 of air at STP contains approximately 2.5×10^{25} molecules. Thus the number of molecules in a volume of 10^{-12} m^3 (about the size of a grain of sand) would be 2.5×10^{13}. This number is certainly large enough to ensure that the average mass within $\delta \mathsf{V}'$ will be constant.

instantaneously surrounding point C. If we define a *fluid particle* as a small mass of fluid of fixed identity of volume $\delta V'$, then the velocity at point C is defined as the instantaneous velocity of the fluid particle which, at a given instant, is passing though point C. The velocity at any point in the flow field is defined similarly. At a given instant the velocity field, $\vec{V}$, is a function of the space coordinates x, y, z. The velocity at any point in the flow field might vary from one instant to another. Thus the complete representation of velocity (the velocity field) is given by

$$\vec{V} = \vec{V}(x, y, z, t) \qquad (2.3)$$

The velocity vector, $\vec{V}$, can be written in terms of its three scalar components. Denoting the components in the x, y, and z directions by u, v, and w, then

$$\vec{V} = u\hat{i} + v\hat{j} + w\hat{k} \qquad (2.4)$$

In general, each of the components, u, v, and w will be a function of x, y, z, and t.

If properties at every point in a flow field do not change with time, the flow is termed *steady*. Stated mathematically, the definition of steady flow is

$$\frac{\partial \eta}{\partial t} = 0$$

where η represents any fluid property. For steady flow,

$$\frac{\partial \rho}{\partial t} = 0 \qquad \text{or} \qquad \rho = \rho(x, y, z)$$

and

$$\frac{\partial \vec{V}}{\partial t} = 0 \qquad \text{or} \qquad \vec{V} = \vec{V}(x, y, z)$$

Thus, in steady flow, any property may vary from point to point in the field, but all properties remain constant with time at every point.

2-2.1 One-, Two-, and Three-Dimensional Flows

A flow is classified as one-, two-, or three-dimensional depending on the number of space coordinates required to specify the velocity field.[3] Equation 2.3 indicates that the velocity field may be a function of three space coordinates and time. Such a flow field is termed *three-dimensional* (it is also *unsteady*) because the velocity at any point in the flow field depends on the three coordinates required to locate the point in space.

Although most flow fields are inherently three-dimensional, analysis based on fewer dimensions is frequently meaningful. Consider, for example, the steady flow through a long straight pipe of constant cross section. Far from the entrance to the pipe the velocity distribution may be described by

$$u = u_{\max} \left[1 - \left(\frac{r}{R} \right)^2 \right] \qquad (2.5)$$

---

[3] Some authors choose to classify a flow as one-, two-, or three-dimensional on the basis of the number of space coordinates required to specify all fluid properties. In this text, classification of flow fields will be based on the number of space coordinates required to specify the velocity field only.

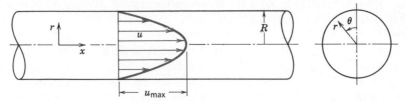

Fig. 2.2 Example of one-dimensional flow.

This profile is shown in Fig. 2.2, where cylindrical coordinates r, θ, and x are used to locate any point in the flow field. The velocity field is a function of r only; it is independent of the coordinates x and θ. Thus this is a one-dimensional flow.

An example of a two-dimensional flow is illustrated in Fig. 2.3; the velocity distribution is depicted for a flow between diverging straight walls that are imagined to be infinite in extent (in the z direction). Since the channel is considered to be infinite in the z direction, the velocity field will be identical in all planes perpendicular to the z axis. Consequently, the velocity field is a function only of the space coordinates x and y; the flow field is classified as two-dimensional.

As you might suspect, the complexity of analysis increases considerably with the number of dimensions of the flow field. For many problems encountered in engineering, a one-dimensional analysis is adequate to provide approximate solutions of engineering accuracy.

Since all fluids satisfying the continuum assumption must have a zero relative velocity at a solid surface (to satisfy the no-slip condition), most flows are inherently two- or three-dimensional. For purposes of analysis it often is convenient to introduce the notion of *uniform flow* at a given cross section. In a flow that is uniform at a given cross section, the velocity is constant across any section normal to the flow. Under this assumption,[4] the two-dimensional flow of Fig. 2.3 is modeled as the flow shown in Fig. 2.4. In the flow of Fig. 2.4, the velocity field is a function of x alone, and thus the flow model is one-dimensional. (Other properties, such as density or pressure, also may be assumed uniform at a section, if appropriate.)

The term *uniform flow field* (as opposed to uniform flow at a cross section) is used to describe a flow in which the magnitude and direction of the velocity vector are constant, i.e., independent of all space coordinates, throughout the entire flow field.

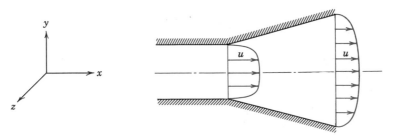

Fig. 2.3 Example of two-dimensional flow.

---

[4] Convenience alone does not justify this assumption; often results of acceptable accuracy are obtained. Sweeping assumptions such as uniform flow at a cross section should always be reviewed carefully to be sure they provide a reasonable analytical model of the real flow.

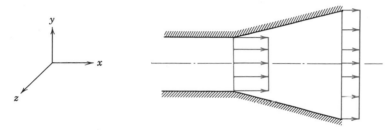

Fig. 2.4 Example of uniform flow at a section.

2-2.2 Timelines, Pathlines, Streaklines, and Streamlines

In the analysis of problems in fluid mechanics, frequently it is advantageous to obtain a visual representation of a flow field. Such a representation is provided by timelines, pathlines, streaklines, and streamlines.[5]

If a number of adjacent fluid particles in a flow field are marked at a given instant, they form a line in the fluid at that instant; this line is called a *timeline*. Subsequent observations of the line may provide information about the flow field. For example, in discussing the behavior of a fluid under the action of a constant shear force (Section 1-2) timelines were introduced to demonstrate the deformation of a fluid at successive instants.

A *pathline* is the path or trajectory traced out by a moving fluid particle. To make a pathline visible, we might identify a fluid particle at a given instant, e.g., by the use of dye, and then take a long exposure photograph of its subsequent motion. The line traced out by the particle is a pathline.

On the other hand, we might choose to focus our attention on a fixed location in space and identify, again by the use of dye, all fluid particles passing through this point. After a short period of time we would have a number of identifiable fluid particles in the flow, all of which had, at some time, passed through one fixed location in space. The line joining these fluid particles is defined as a *streakline*.

Streamlines are lines drawn in the flow field so that at a given instant they are tangent to the direction of flow at every point in the flow field. Since the streamlines are tangent to the velocity vector at every point in the flow field, there can be no flow across a streamline. The procedure used to obtain the equation for a streamline in two-dimensional flow is illustrated in Example Problem 2.1.

In steady flow, the velocity at each point in the flow field remains constant with time and, consequently, the streamlines do not vary from one instant to the next. This implies that a particle located on a given streamline will remain on the same streamline. Furthermore, consecutive particles passing through a fixed point in space will be on the same streamline and, subsequently, will remain on this streamline. Thus in a steady flow, pathlines, streaklines, and streamlines are identical lines in the flow field.

The shape of the streamlines may vary from instant to instant if the flow is unsteady. In the case of unsteady flow, pathlines, streaklines, and streamlines do not coincide.

EXAMPLE 2.1—Streamlines and Pathlines in Two-Dimensional Flow
A velocity field is given by $\vec{V} = ax\hat{i} - ay\hat{j}$; the units of velocity are m/sec; x and y are given in meters; $a = 0.1$ sec^{-1}.

---

[5] Timelines, pathlines, streaklines, and streamlines are demonstrated in the NCFMF film, *Flow Visualization*, S. J. Kline, principal.

(a)   Obtain an equation for the streamlines in the xy plane.

(b)   Plot the streamline passing through the point $(x_0, y_0, 0) = (2, 8, 0)$.

(c)   Determine the velocity of a particle at the point $(2, 8, 0)$.

(d)   If the particle passing through the point $(x_0, y_0, 0)$ is marked at time $t_0 = 0$, determine the location of the particle at time $t = 20$ sec.

(e)   What is the velocity of the particle at $t = 20$ sec?

(f)   Show that the equation of the particle path (the pathline) is the same as the equation of the streamline.

EXAMPLE PROBLEM 2.1

GIVEN: Velocity field, $\vec{V} = ax\hat{i} - ay\hat{j}$; x and y in meters; $a = 0.1$ sec^{-1}

FIND: (a)   Equation of the streamlines in the xy plane.
   (b)   Plot the streamline through point $(2, 8, 0)$.
   (c)   Velocity of particle at point $(2, 8, 0)$.
   (d)   Position at $t = 20$ sec of particle located at $(2, 8, 0)$ at $t = 0$.
   (e)   Velocity of particle at position found in (d).
   (f)   Equation of pathline of particle located at $(2, 8, 0)$ at $t = 0$.

SOLUTION:

(a)   Streamlines are lines drawn in the flow field such that, at a given instant, they are tangent to the direction of flow at every point.

Consequently,

$$\frac{dy}{dx}\bigg)_{\text{streamline}} = \frac{v}{u} = \frac{-ay}{ax} = \frac{-y}{x}$$

Separating variables and integrating, we obtain

$$\int \frac{dy}{y} = -\int \frac{dx}{x}$$

or

$$\ln y = -\ln x + c_1$$

This can be written as $xy = c$.

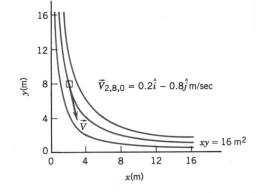

$$\vec{V}_{2,8,0} = 0.2\hat{i} - 0.8\hat{j} \text{ m/sec}$$

$$xy = 16 \text{ m}^2$$

(b)   For the streamline passing through the point $(x_0, y_0, 0) = (2, 8, 0)$ the constant, c, has a value of 16 and the equation of the streamline through the point $(2, 8, 0)$ is

$$xy = x_0 y_0 = 16 \text{ m}^2$$

(c)   The velocity field is $\vec{V} = ax\hat{i} - ay\hat{j}$. At the point $(2, 8, 0)$

$$\vec{V} = a(x\hat{i} - y\hat{j}) = 0.1 \text{ sec}^{-1}(2\hat{i} - 8\hat{j}) \text{ m} = 0.2\hat{i} - 0.8\hat{j} \text{ m/sec}$$

(d)   A particle moving in the flow field will have velocity given by

$$\vec{V} = ax\hat{i} - ay\hat{j}$$

Thus

$$u_p = \frac{dx}{dt} = ax \quad \text{and} \quad v_p = \frac{dy}{dt} = -ay$$

Separating variables and integrating (in each equation) gives

$$\int_{x_0}^{x} \frac{dx}{x} = \int_{0}^{t} a\, dt \qquad \text{and} \qquad \int_{y_0}^{y} \frac{dy}{y} = \int_{0}^{t} -a\, dt$$

Then

$$\ln \frac{x}{x_0} = at \qquad \text{and} \qquad \ln \frac{y}{y_0} = -at$$

or

$$x = x_0 e^{at} \qquad \text{and} \qquad y = y_0 e^{-at}$$

At $t = 20$ sec,

$$x = 2 \text{ m } e^{(0.1)20} = 14.8 \text{ m} \qquad \text{and} \qquad y = 8 \text{ m } e^{-(0.1)20} = 1.08 \text{ m}$$

At $t = 20$ sec, particle is at $(14.8, 1.08, 0)$ m

(e) At the point $(14.8, 1.08, 0)$ m,

$$\vec{V} = a(x\hat{i} - y\hat{j}) = 0.1 \text{ sec}^{-1}(14.8\hat{i} - 1.08\hat{j}) \text{ m} = 1.48\hat{i} - 0.108\hat{j} \text{ m/sec}$$

(f) To determine the equation of the pathline, we use the parametric equations

$$x = x_0 e^{at} \qquad \text{and} \qquad y = y_0 e^{-at}$$

and eliminate t. Solving for e^{at} from both equations

$$e^{at} = \frac{y_0}{y} = \frac{x}{x_0} \qquad \text{therefore } xy = x_0 y_0 = 16 \text{ m}^2$$

Note: (i)   the equation of the streamline through $(x_0, y_0, 0)$ and the equation of the pathline traced out by the particle passing through $(x_0, y_0, 0)$ are the same for this steady flow.

(ii)   in following a particle (Lagrangian method of description), both the coordinates of the particle (x, y) and the components of the particle velocity $(u_p = dx/dt$ and $v_p = dy/dt)$ are functions of time.

2-3   STRESS FIELD

Surface and body forces are encountered in the study of continuum fluid mechanics. *Surface forces* act on the boundaries of a medium through direct contact. Forces developed without physical contact, and distributed over the volume of the fluid, are termed *body forces*. Gravitational and electromagnetic forces are examples of body forces.

The gravitational body force acting on an element of volume, $d\forall$, is given by $\rho \vec{g}\, d\forall$, where ρ is the density (mass per unit volume) and $\vec{g}$ is the local gravitational acceleration. Thus the gravitational body force per unit volume is $\rho \vec{g}$ and the gravitational body force per unit mass is $\vec{g}$.

Stresses in a medium result from forces acting on some portion of the medium. The concept of stress provides a convenient means to describe the manner in which forces acting on the boundaries of the medium are transmitted through the medium. Since force and area are both vector quantities, we might anticipate that the stress

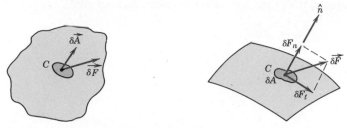

Fig. 2.5 The concept of stress in a continuum.

field will not be a vector field. We shall show that, in general, nine quantities are required to specify the state of stress in a fluid. (Stress is a tensor quantity of second order.)

In a flowing fluid, consider a portion, $\delta \vec{A}$, of the surface passing through the point C. The orientation of $\delta \vec{A}$ is given by the unit vector, $\hat{n}$, shown in Fig. 2.5. The direction of $\hat{n}$ is normal to the surface.

The force, $\delta \vec{F}$, acting on $\delta \vec{A}$ may be resolved into two components, one normal to and the other tangent to the area. A normal stress σ_n and a shear stress τ_n are then defined as

$$\sigma_n = \lim_{\delta A_n \to 0} \frac{\delta F_n}{\delta A_n} \qquad (2.6)$$

and

$$\tau_n = \lim_{\delta A_n \to 0} \frac{\delta F_t}{\delta A_n} \qquad (2.7)$$

Subscript n on the stress is included as a reminder that the stresses are associated with the surface $\delta \vec{A}$ through C, having an outer normal in the $\hat{n}$ direction. For any other surface through C the values of the stresses could be different.

In dealing with vector quantities such as force, it is customary to consider components in an orthogonal coordinate system. In rectangular coordinates we might consider the stresses acting on planes whose outward drawn normals are in the x, y, or z directions. In Fig. 2.6 we consider the stress on the element δA_x, whose outward drawn normal is in the x direction. The force, $\delta \vec{F}$, has been resolved into components along each of the coordinate directions. Dividing the magnitude of each force component by the area, δA_x, and taking the limit as δA_x approaches zero, we

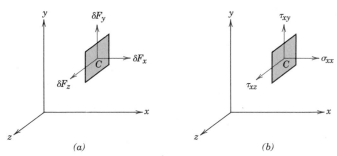

Fig. 2.6 (a) Force components and (b) stress components, on the element of area δA_x.

define the three stress components shown in Fig. 2.6*b*:

$$\sigma_{xx} = \lim_{\delta A_x \to 0} \frac{\delta F_x}{\delta A_x}$$ (2.8)

$$\tau_{xy} = \lim_{\delta A_x \to 0} \frac{\delta F_y}{\delta A_x} \qquad \tau_{xz} = \lim_{\delta A_x \to 0} \frac{\delta F_z}{\delta A_x}$$

We have used a double subscript notation to label the stresses. The first subscript (in this case, x) indicates the plane on which the stress acts (in this case, a surface perpendicular to the x axis). The second subscript indicates the direction in which the stress acts.

Consideration of area element δA_y would lead to the definitions of the stresses, σ_{yy}, τ_{yx}, and τ_{yz}; use of area element δA_z would similarly lead to the definitions of σ_{zz}, τ_{zx}, τ_{zy}.

An infinite number of planes can be passed through point C, resulting in an infinite number of stresses associated with that point. Fortunately, the state of stress at a point can be described completely by specifying the stresses acting on three mutually perpendicular planes through the point. The stress at a point is specified by the nine components

$$\begin{bmatrix} \sigma_{xx} & \tau_{xy} & \tau_{xz} \\ \tau_{yx} & \sigma_{yy} & \tau_{yz} \\ \tau_{zx} & \tau_{zy} & \sigma_{zz} \end{bmatrix}$$

where σ has been used to denote a normal stress, and shear stresses are denoted by τ. The notation for designating stress is shown in Fig. 2.7.

Referring to the infinitesimal element shown in Fig. 2.7, we see that there are six planes (two x planes, two y planes, and two z planes) on which stresses may act. In order to designate the plane of interest, we could use terms like front and back, top and bottom, or left and right. However, it is more logical to name the planes in terms of the coordinate axes. The planes are named and denoted as positive or negative according to the direction of the outward drawn normal to the plane. Thus the top plane, for example, is a positive y plane and the back plane is a negative z plane.

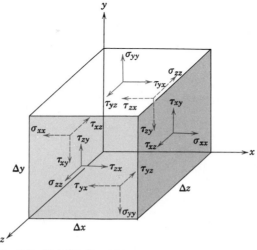

Fig. 2.7 Notation for stress.

It also is necessary to adopt a sign convention for the stress. A stress component is positive when the direction of the stress component and the plane on which it acts are both positive or both negative. Thus $\tau_{yx} = 5$ lbf/in.^{2} represents a shear stress on a positive y plane in the positive x direction or a shear stress on a negative y plane in the negative x direction. In Fig. 2.7 all stresses have been drawn as positive stresses. Stress components are negative when the direction of the stress component and the plane on which it acts are of opposite sign.

2-4 VISCOSITY

We have defined a fluid as a substance that deforms continuously under the action of a shear stress. In the absence of a shear stress, there will be no deformation. Fluids may be broadly classified according to the relation between the applied shear stress and the rate of deformation.

Consider the behavior of a fluid element between the two infinite plates shown in Fig. 2.8. The upper plate moves at constant velocity, δu, under the influence of a constant applied force, δF_x. The shear stress, τ_{yx}, applied to the fluid element is given by

$$\tau_{yx} = \lim_{\delta A_y \to 0} \frac{\delta F_x}{\delta A_y} = \frac{dF_x}{dA_y}$$

where δA_y is the area of the fluid element in contact with the plate. During time interval δt, the fluid element is deformed from position $MNOP$ to position $M'NOP'$. The rate of deformation of the fluid is given by

$$deformation\ rate = \lim_{\delta t \to 0} \frac{\delta \alpha}{\delta t} = \frac{d\alpha}{dt}$$

To calculate the shear stress, τ_{yx}, it is desirable to express $d\alpha/dt$ in terms of readily measurable quantities. This can be done easily. The distance, δl, between the points M and M' is given by

$$\delta l = \delta u\, \delta t$$

or alternatively, for small angles,

$$\delta l = \delta y\, \delta \alpha$$

Equating these two expressions for δl gives

$$\frac{\delta \alpha}{\delta t} = \frac{\delta u}{\delta y}$$

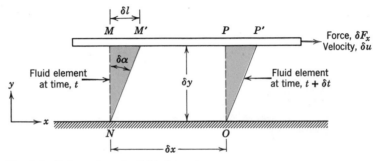

Fig. 2.8 Deformation of a fluid element.

Taking the limit of both sides of the equality, we obtain

$$\frac{d\alpha}{dt} = \frac{du}{dy}$$

Thus, the fluid element of Fig. 2.8, when subjected to shear stress, τ_{yx}, experiences a rate of deformation *(shear rate)* given by du/dy. Fluids in which shear stress is directly proportional to rate of deformation are *Newtonian fluids*. The term *non-Newtonian* is used to classify all fluids in which shear stress is not directly proportional to shear rate.

2-4.1 Newtonian Fluid

Most common fluids such as water, air, and gasoline are Newtonian under normal conditions. If the fluid of Fig. 2.8 is Newtonian, then

$$\tau_{yx} \propto \frac{du}{dy} \tag{2.9}$$

If one considers the deformation of two different Newtonian fluids, say glycerin and water, one recognizes that they will deform at different rates under the action of the same applied shear stress. Glycerin exhibits a much larger resistance to deformation than water. Thus we say it is much more viscous. The constant of proportionality in Eq. 2.9 is the *absolute* (or *dynamic*) *viscosity*, μ. Thus in terms of the coordinates of Fig. 2.8, Newton's law of viscosity is given for one-dimensional flow by

$$\tau_{yx} = \mu \frac{du}{dy} \tag{2.10}$$

Note that since the dimensions of τ are $[F/L^2]$ and the dimensions of du/dy are $[1/t]$, then μ has dimensions $[Ft/L^2]$. Since the dimensions of force, F, mass, M, length, L, and time, t, are related by Newton's second law of motion, the dimensions of μ can also be expressed as $[M/Lt]$. In the British Gravitational system, the units of viscosity are lbf · sec/ft^2 or slug/ft · sec. In the Absolute Metric System, the basic unit of viscosity is called a poise (poise $\equiv$ g/cm · sec); in the SI system the units of viscosity are kg/m · sec or Pa · sec (= N · sec/m^2). The calculation of viscous shear stress is illustrated in Example Problem 2.2.

In fluid mechanics the ratio of absolute viscosity, μ, to density, ρ, often arises. This ratio is given the name *kinematic viscosity* and is represented by the symbol ν. Since density has dimensions $[M/L^3]$, the dimensions of ν are $[L^2/t]$. In the Absolute Metric system of units, the unit for ν is a stoke (stoke $\equiv$ cm^2/sec).

Viscosity data for a number of common Newtonian fluids are given in Appendix A. Note that for gases, viscosity increases with temperature, whereas for liquids, viscosity decreases with increasing temperature.

EXAMPLE 2.2—Viscosity and Shear Stress in Newtonian Fluid

An infinite plate is moved over a second plate on a layer of liquid as shown. For small gap width, d, we assume a linear velocity distribution in the liquid. The liquid viscosity is 0.65 centipoise and its specific gravity is 0.88. Calculate:

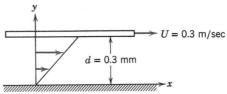

(a)   The absolute viscosity of the liquid, in lbf · sec/ft².
(b)   The kinematic viscosity of the liquid, in m²/sec.
(c)   The shear stress on the upper plate, in lbf/ft².
(d)   The shear stress on the lower plate, in Pa.
(e)   Indicate the direction of each shear stress calculated in parts (c) and (d).

EXAMPLE PROBLEM 2.2

GIVEN: Linear velocity profile in the liquid between infinite parallel plates as shown

$$\mu = 0.65 \text{ cp}$$

$$SG = 0.88$$

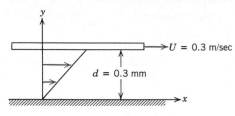

FIND: (a) μ in units of lbf · sec/ft².   (b) ν in units of m²/sec.
(c) τ on upper plate in units of lbf/ft².   (d) τ on lower plate in units of
(e)   Direction of stress in parts (c) and (d).   Pa.

SOLUTION:

Basic equation: $\tau_{yx} = \mu \dfrac{du}{dy}$ Definition: $\nu = \dfrac{\mu}{\rho}$

Assumptions: (1)   Linear velocity distribution
(2)   Steady flow
(3) μ = constant

(a) $\mu = \dfrac{0.65 \text{ cp}}{} \times \dfrac{\text{poise}}{100 \text{ cp}} \times \dfrac{g}{\text{cm} \cdot \text{sec} \cdot \text{poise}} \times \dfrac{\text{lbm}}{453.6 \text{ g}} \times \dfrac{\text{slug}}{32.3 \text{ lbm}}$

$\times \dfrac{30.48 \text{ cm}}{\text{ft}} \times \dfrac{\text{lbf} \cdot \text{sec}^2}{\text{slug} \cdot \text{ft}}$

$\mu = 1.36 \times 10^{-5} \text{ lbf} \cdot \text{sec/ft}^2$ ←————————————— μ

(b) $\nu = \dfrac{\mu}{\rho} = \dfrac{\mu}{SG\rho_{H_2O}}$

$= \dfrac{1.36 \times 10^{-5} \text{ lbf} \cdot \text{sec}}{\text{ft}^2} \times \dfrac{\text{ft}^3}{(0.88)1.94 \text{ slug}} \times \dfrac{\text{slug} \cdot \text{ft}}{\text{lbf} \cdot \text{sec}^2} \times \dfrac{(0.3048)^2 \text{ m}^2}{\text{ft}^2}$

$\nu = 7.40 \times 10^{-7} \text{ m}^2/\text{sec}$ ←————————————— ν

(c) $\tau_{\text{upper}} = \tau_{yx,\text{upper}} = \mu \left. \dfrac{du}{dy} \right)_{y=d}$.

Since u varies linearly with y,

$$\dfrac{du}{dy} = \dfrac{\Delta u}{\Delta y} = \dfrac{U-0}{d-0} = \dfrac{U}{d} = \dfrac{0.3 \text{ m}}{\text{sec}} \times \dfrac{1}{0.3 \text{ mm}} \times \dfrac{1000 \text{ mm}}{\text{m}} = 1000 \text{ sec}^{-1}$$

$\tau_{\text{upper}} = \mu \dfrac{U}{d} = \dfrac{1.36 \times 10^{-5} \text{ lbf} \cdot \text{sec}}{\text{ft}^2} \times \dfrac{1000}{\text{sec}} = 0.0136 \text{ lbf/ft}^2$ ←——— τ_{upper}

(d) $\quad \tau_{\text{lower}} = \mu \dfrac{U}{d} = \dfrac{0.0136 \text{ lbf}}{\text{ft}^2} \times \dfrac{4.448 \text{ N}}{\text{lbf}} \times \dfrac{\text{ft}^2}{(0.3048)^2 \text{ m}^2} \times \dfrac{\text{Pa} \cdot \text{m}^2}{\text{N}} = 0.651 \text{ Pa} \qquad \tau_{\text{lower}}$

(e) Direction of shear stress on upper and lower plates.

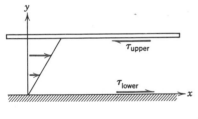

$\left\{ \begin{array}{l} \text{The upper plate is a negative } y \text{ surface, so} \\ \text{positive } \tau_{yx} \text{ acts in the negative } x \text{ direction.} \end{array} \right\}$

$\left\{ \begin{array}{l} \text{The lower plate is a positive } y \text{ surface, so} \\ \text{positive } \tau_{yx} \text{ acts in the positive } x \text{ direction.} \end{array} \right\}$

(e)

2-4.2 Non-Newtonian Fluids

Fluids in which shear stress is not directly proportional to deformation rate are non-Newtonian. Many common fluids exhibit non-Newtonian behavior. Two familiar examples are toothpaste and Lucite[6] paint. The latter is very "thick" when in the can, but becomes "thin" when sheared by brushing. Toothpaste behaves as a "fluid" when squeezed from the tube. However, it does not run out by itself when the cap is removed. There is a threshold or yield stress below which toothpaste behaves as a solid. Strictly speaking, our definition of a fluid is valid only for materials that have zero yield stress. Non-Newtonian fluids commonly are classified as having time-independent or time-dependent behavior. Examples of time-independent behavior are shown in the rheological diagram of Fig. 2.9.

Numerous empirical equations have been proposed to model the observed relations between τ_{yx} and du/dy for time-independent fluids. They may be adequately represented for many engineering applications by the power law model, which for one-dimensional flow becomes

$$\tau_{yx} = k \left(\frac{du}{dy} \right)^n \tag{2.11}$$

where the exponent, n, is called the flow behavior index and k, the consistency index. This equation reduces to Newton's law of viscosity for $n = 1$ with $k = \mu$.

If Eq. 2.11 is rewritten in the form

$$\tau_{yx} = k \left| \frac{du}{dy} \right|^{n-1} \frac{du}{dy} = \eta \frac{du}{dy} \tag{2.12}$$

then $\eta = k|du/dy|^{n-1}$ is referred to as the *apparent viscosity*. Most non-Newtonian fluids have apparent viscosities that are relatively high compared to the viscosity of water.

Fluids in which the apparent viscosity decreases with increasing deformation rate ($n < 1$) are called *pseudoplastic* (or shear thinning) fluids. Most non-Newtonian fluids fall into this group; examples include polymer solutions, colloidal suspensions, and paper pulp in water. If the apparent viscosity increases with increasing deforma-

---

[6] Trademark, E. I. du Pont de Nemours & Company.

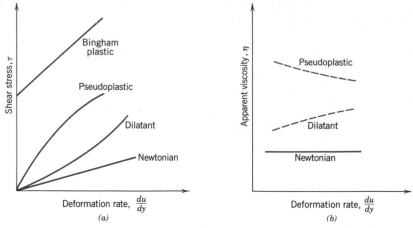

Fig. 2.9   (a) Shear stress, τ, and (b) apparent viscosity, η, as a function of deformation rate for one-dimensional flow of various non-Newtonian fluids.

tion rate ($n > 1$) the fluid is termed *dilatant* (or shear thickening). Suspensions of starch and of sand are examples of dilatant fluids.

A "fluid" that behaves as a solid until a minimum yield stress, τ_y, is exceeded and subsequently exhibits a linear relation between stress and rate of deformation is referred to as an ideal or *Bingham plastic*. The appropriate shear stress model is

$$\tau_{yx} = \tau_y + \mu_p \frac{du}{dy} \qquad (2.13)$$

Clay suspensions, drilling muds, and toothpaste are examples of substances exhibiting this behavior.

The study of non-Newtonian fluids is further complicated by the fact that the apparent viscosity may be time-dependent. *Thixotropic* fluids show a decrease in η with time under a constant applied shear stress; many paints are thixotropic. *Rheopectic* fluids show an increase in η with time. After deformation some fluids partially return to their original shape when the applied stress is released; such fluids are called *viscoelastic*.[7]

2-5   DESCRIPTION AND CLASSIFICATION OF FLUID MOTIONS

In Chapter 1 we listed a wide variety of typical problems encountered in fluid mechanics and outlined our method of approach to the subject. Before proceeding with our detailed study, we shall attempt a broad classification of fluid mechanics on the basis of observable physical characteristics of flow fields. Since there is much overlap in the types of flow fields encountered, there is no universally accepted classification scheme. One possible classification is shown in Fig. 2.10.

2-5.1   Viscous and Inviscid Flows

The main subdivision indicated is between inviscid and viscous flows. Flows in which the effects of viscosity are neglected are termed *inviscid flows*. In an inviscid flow

---

[7] Examples of time-dependent and viscoelastic fluids are illustrated in the NCFMF film, *Rheological Behavior of Fluids*, H. Markowitz, principal.

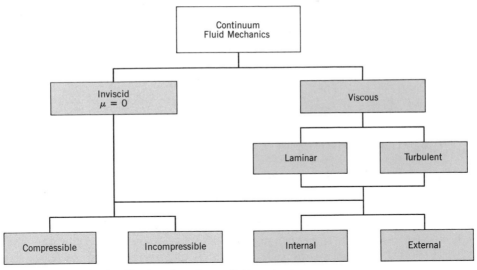

Fig. 2.10 Possible classification of continuum fluid mechanics.

the fluid viscosity, μ, is assumed to be zero. Fluids with zero viscosity do not exist; however, there are many problems where neglecting the viscous forces will simplify the analysis and, at the same time, lead to meaningful results. (Simplification of the analysis is always desirable, but the results must be reasonably accurate if the solution is to be of value.)

All fluids possess viscosity and, consequently, viscous flows are of paramount importance in the study of continuum fluid mechanics. We shall study viscous flows in some detail later; here we consider a few examples of viscous flow phenomena.

In our discussion following the definition of a fluid (Section 1-2), we noted that in any viscous flow, the fluid in direct contact with a solid boundary has the same velocity as the boundary itself; there is no slip at the boundary. For the one-dimensional viscous flow of Fig. 2.8, the shear stress[8] was given by Eq. 2.10,

$$\tau_{yx} = \mu \frac{du}{dy} \tag{2.10}$$

The fluid velocity at a stationary solid surface in a moving fluid is zero. Since the bulk fluid is in motion, velocity gradients and hence shear stresses must be present in the flow. These stresses in turn affect the fluid motion.

As a practical case, consider the fluid motion around a thin wing or ship hull. Such a flow might be represented by the crude approximation of flow over a flat plate, as shown in Fig. 2.11. The flow approaching the plate is of uniform velocity, U_∞. We are interested in providing a qualitative picture of the velocity distribution at various locations along the plate. Two such locations are denoted by x_1 and x_2. Consider the first location, x_1. In order to arrive at a qualitative picture of the velocity distribution, we start by labeling the y coordinates at which the velocity is known. (For clarity, distances in the y direction have been exaggerated greatly in Fig. 2.11.)

From the no-slip condition, we know the velocity at point A must be zero; we have one point on the velocity profile. Can we locate any other points on the profile?

---

[8] In general, $\tau_{yx} = \mu \left(\dfrac{\partial u}{\partial y} + \dfrac{\partial v}{\partial x} \right)$ for flows that are not one-dimensional; see Chapter 5.

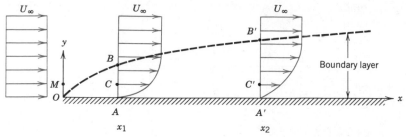

Fig. 2.11 Incompressible laminar viscous flow over a semi-infinite flat plate.

Let us stop for a minute and ask ourselves, "What is the effect of the plate on the flow?" The plate is stationary and, therefore, exerts a retarding force on the flow; it slows the fluid in the neighborhood of the surface. At a y location sufficiently far from the plate, say point B, the flow will not be influenced by the presence of the plate. If the pressure does not vary in the x direction (as is the case for flow over a semi-infinite flat plate) the velocity at point B will be U_∞. It seems reasonable to expect the velocity to increase smoothly and monotonically from the value $u = 0$ at $y = 0$ to $u = U_\infty$ at $y = y_B$. The profile has been so drawn; thus at some point, C, intermediate between points A and B, the velocity lies between zero and U_∞. For $0 \le y \le y_B$, then $0 \le u \le U_\infty$. From these characteristics of the velocity profile and our definition of the shear stress,[9] we note that shear stresses are present within the region $0 \le y \le y_B$; for $y > y_B$, the velocity gradient is zero and hence no shear stresses are present.

What about the velocity profile at location x_2? Is it exactly the same as the profile at x_1? A look at Fig. 2.11 suggests that it is not. At least it has not been drawn that way! Even though it is qualitatively the same, why is it not exactly the same? We might guess that the plate would influence a greater region of the flow field as we move farther down the plate. Looking again at the profile at location x_1, we see that the slower-moving fluid adjacent to the plate exerts a retarding force on the faster-moving fluid above it. We can see this by considering the shear stress on the y plane through point C. Since we are interested in the stress exerted on the faster-moving fluid above the plane, we are looking for the direction of the shear stress on a negative y plane through point C. Since $\partial u / \partial y > 0$, τ_{yx} on the plane through point C has a positive numerical value; consequently the shear stress must be in the negative x direction.

To establish the qualitative picture of the velocity profile at x_2, we recognize that the no-slip condition requires the velocity at the wall to be zero; this fixes the velocity at A' as zero. Since at location x_1, the slower-moving fluid exerts a retarding force on the fluid above it, we would expect the distance out to the point where the velocity is U_∞ to be increased at location, x_2; i.e., $y_{B'} > y_B$. Furthermore, it is reasonable to expect that $u_{C'} < u_C$.

From our qualitative picture of the flow field, we see that we can divide the flow into two general regions. In the region adjacent to the boundary, shear stresses are present; this region is called the boundary layer.[10] Outside the boundary layer

---

[9] For the two-dimensional boundary-layer flow of Fig. 2.11, the shear stress is given closely by $\tau_{yx} = \mu \dfrac{\partial u}{\partial y}$.

[10] The formation of a boundary layer is illustrated in the NCFMF film, *Fundamentals of Boundary Layers*, F. H. Abernathy, principal.

the velocity gradient is zero and hence the shear stresses are zero. In this region we may use inviscid flow theory to analyze the flow.

Before leaving our discussion of the viscous flow over a semi-infinite flat plate, we should stop and reflect on two points. In our qualitative description of the flow field, we were only concerned about the behavior of the x component of velocity, the component u. What about the y component of velocity, the component v? Is it zero throughout the flow field? We also might ask if the edge of the boundary layer is a streamline.

To answer these questions, consider the streamlines of the flow. Rather than consider all possible streamlines, let us consider the streamline through the point M. Recalling that a streamline is defined as a line drawn tangent to the velocity vector at every point in the flow, our first inclination might be to depict the streamline through M as a straight line parallel to the x axis. However, this would violate the requirement that there can be no flow across a streamline. Because there can be no flow across a streamline, the mass flow between adjacent streamlines (or between a streamline and a solid boundary) must be constant. For the incompressible viscous flow of Fig. 2.11, we recognize that the streamline through point M cannot be a straight line parallel to the x axis. To maintain a constant mass flow between the streamline through point M and the x axis, the spacing between the streamline and the x axis must increase continuously as we move along the plate. Therefore, although small, the y component of velocity is not zero.

The streamline through M crosses the dashed line we have used to denote the edge of the boundary layer. Consequently, we conclude that the edge of the boundary layer is not a streamline and that there is flow into the boundary layer as we move down the plate. Indeed, if the boundary layer is to grow, there must be flow across the edge of the boundary layer.

For a given freestream velocity, U_∞, the size of the boundary layer will depend on the properties of the fluid. Since the shear stress is directly proportional to the viscosity, we expect the size of the boundary layer to depend on the viscosity of the fluid. In Chapter 9, we shall develop expressions for determining the rate of boundary-layer growth.

We have used incompressible flow over a semi-infinite flat plate to establish a qualitative picture of the viscous flow over a solid boundary. In that example, we had to consider only the effect of shear forces; the pressure was constant throughout the flow field. Now let us consider a steady flow field (the incompressible flow over a cylinder) where both pressure forces and viscous forces are important. For steady flow, pathlines, streaklines, and streamlines all are identical. If we were to use some means of flow visualization, we would find the flow field to be of the general character shown in Fig. 2.12a.[11]

We see that the streamlines are symmetric about the x axis. The fluid along the central streamline impinges on the cylinder at point A, divides, and flows around the cylinder. Point A on the cylinder is called a *stagnation point*. As in flow over a flat plate, a boundary layer develops in the neighborhood of the solid surface. The velocity distribution outside the boundary layer can be determined qualitatively from the spacing of the streamlines. Since there can be no flow across a streamline, we would expect the flow velocity to increase in regions where the spacing between streamlines

---

[11] The details of the flow will depend on the various flow properties. For all but very low-speed flows, the qualitative picture will be as shown.

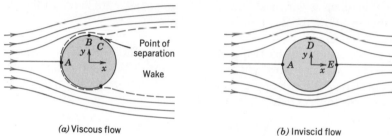

(a) Viscous flow   (b) Inviscid flow

Fig. 2.12 Qualitative picture of incompressible flow over a cylinder.

decreases. Conversely, an increase in streamline spacing implies a decrease in flow velocity.

Consider for a moment the incompressible flow field around a cylinder calculated assuming an inviscid flow, as shown in Fig. 2.12b; this flow is symmetric about both the x and y axes. The velocity around the cylinder increases to a maximum at point D and then decreases as we move further around the cylinder. For inviscid flow, an increase in velocity is accompanied by a decrease in pressure; conversely, a decrease in velocity is accompanied by an increase in pressure. Thus in the case of an incompressible inviscid flow, the pressure along the surface of the cylinder decreases as we move from point A to point D and then increases again from point D to point E. Since the flow is symmetric with respect to both the x and y axes, we would also expect the pressure distribution to be symmetric with respect to these axes. This is indeed the case for inviscid flow.

Since no shear stresses are present in an inviscid flow, the pressure forces are the only forces we need consider in determining the net force on the cylinder. The symmetry of the pressure distribution leads to the conclusion that for an inviscid flow, there is no net force on the cylinder in either the x or y directions. The net force in the x direction is termed the *drag*. Thus for an inviscid flow over a cylinder, we are led to the conclusion that the drag is zero; this conclusion is contrary to experience, for we know that all bodies experience some drag when placed in a real flow. In treating the inviscid flow over a body we have, by the definition of inviscid flow, neglected the presence of the boundary layer. Let us go back and look again at the real flow situation.

In the real flow, Fig. 2.12a, experiments show the boundary layer to be thin between points A and C. Since the boundary layer is thin, it is reasonable to assume that the pressure field is qualitatively the same as in the inviscid flow case. Since the pressure decreases continuously between points A and B, a fluid element inside the boundary layer experiences a net pressure force in the direction of flow. In the region between A and B, this net pressure force is sufficient to overcome the resisting shear force and motion of the element in the flow direction is maintained.

Now consider an element of fluid inside the boundary layer on the back of the cylinder beyond point B. Since the pressure increases in the direction of flow, the fluid element experiences a net pressure force opposite to its direction of motion. At some point the momentum of the fluid in the boundary layer is insufficient to carry the element further into the region of increasing pressure. The fluid layers adjacent to the solid surface are brought to rest and the flow *separates* from the surface.[12] Boundary-layer separation results in the formation of a relatively low-pressure region

---

[12] The flow over a variety of models, illustrating flow separation, is demonstrated in the University of Iowa film, *Form Drag, Lift, and Propulsion*, H. Rouse, principal.

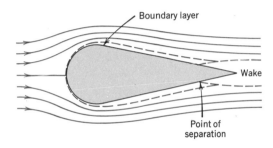

Fig. 2.13   Flow over a streamlined object.

behind a body; this region, which is deficient in momentum, is called the *wake*. Thus, for separated flow over a body, there is a net unbalance of pressure forces in the direction of flow; this results in a pressure drag on the body. The greater the size of the wake behind a body, the greater is the pressure drag.

It is logical to ask how one might reduce the size of the wake and thus reduce the pressure drag. Since a large wake results from boundary-layer separation, which in turn is related to the presence of an adverse pressure gradient (increase of pressure in the direction of flow), reducing the adverse pressure gradient should delay the onset of separation and, hence, reduce the drag.

Streamlining a body reduces the adverse pressure gradient by spreading a given pressure rise over a greater distance. For example, if a gradually tapered rear section were added to the cylinder of Fig. 2.12, the flow field would appear qualitatively as shown in Fig. 2.13. Streamlining the body delays the onset of separation; although the increased surface area of the body causes the total shear force acting on the body to increase, the drag is reduced significantly.[13]

Flow separation also may occur in internal flows (flows through ducts) as a result of rapid or abrupt changes in duct geometry.

2-5.2 Laminar and Turbulent Flows

Viscous flow regimes are classified as laminar or turbulent on the basis of flow structure. In the laminar regime, flow structure is characterized by smooth motion in laminae, or layers. Flow structure in the turbulent regime is characterized by random, three-dimensional motions of fluid particles in addition to the mean motion.

In laminar flow there is no macroscopic mixing of adjacent fluid layers. A thin filament of dye injected into a laminar flow appears as a single line; there is no dispersion of dye throughout the flow, except the slow dispersion due to molecular motion. On the other hand, a dye filament injected into a turbulent flow disperses quickly throughout the flow field; the line of dye breaks up into myriad entangled threads of dye. This behavior of turbulent flow is due to the velocity fluctuations present; the macroscopic mixing of fluid particles from adjacent layers of fluid results in rapid dispersion of the dye. The straight filament of smoke rising from a cigarette in still surroundings gives a clear picture of laminar flow. As the smoke continues to rise, it breaks up into random, haphazard motions; this is an example of turbulent flow.[14]

---

[13] The effect of streamlining a body is demonstrated in the NCFMF film, *Fluid Dynamics of Drag*, A. H. Shapiro, principal.

[14] Several examples illustrating the nature of laminar and turbulent flows are shown in the NCFMF film, *Turbulence*, R. W. Stewart, principal and in the University of Iowa film, *Characteristics of Laminar and Turbulent Flow*, H. Rouse, principal.

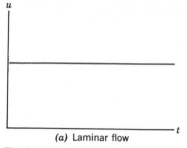
(a) Laminar flow

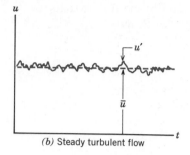
(b) Steady turbulent flow

Fig. 2.14 Variation of axial velocity with time.

One can obtain a more quantitative picture of the difference between laminar and turbulent flow by examining the output from a sensitive velocity-measuring device immersed in the flow. If one measures the x component of velocity at a fixed location in a pipe for both laminar and turbulent steady flow, the traces of velocity versus time appear as shown in Fig. 2.14. For steady laminar flow, the velocity at a point remains constant with time. In turbulent flow the velocity trace indicates random fluctuations of the instantaneous velocity, u, about the time mean velocity, $\bar{u}$. We can consider the instantaneous velocity, u, as the sum of the time mean velocity, $\bar{u}$, and the fluctuating component, u',

$$u = \bar{u} + u'$$

Because the flow is steady, the mean velocity, $\bar{u}$, does not vary with time.

Although many turbulent flows of interest are steady in the mean ($\bar{u}$ is not a function of time), the presence of the random, high-frequency velocity fluctuations makes the analysis of turbulent flows extremely difficult. In a one-dimensional laminar flow, the shear stress is related to the velocity gradient by the simple relation

$$\tau_{yx} = \mu \frac{du}{dy} \tag{2.10}$$

For a turbulent flow in which the mean velocity field is one-dimensional, no such simple relation is valid. Random, three-dimensional velocity fluctuations (u', v', and w') transport momentum across the mean flow streamlines, increasing the effective shear stress. Consequently, in turbulent flow there is no universal relationship between the stress field and the mean-velocity field. Thus in turbulent flows we must rely heavily on semi-empirical theories and on experimental data.

2-5.3 Compressible and Incompressible Flows

Flows in which variations in density are negligible are termed *incompressible*; when density variations within a flow are not negligible, the flow is called *compressible*. The most common example of compressible flow concerns the flow of gases, while the flow of liquids may frequently be treated as incompressible. However, water hammer and cavitation[15] are examples of the importance of compressibility effects in liquid flows. Gas flows with negligible heat transfer also may be considered incompressible provided that the flow speeds are small relative to the speed of sound;

---

[15] Examples of cavitation are illustrated in the NCFMF film, *Cavitation*, P. Eisenberg, principal.

the ratio of the flow speed, V, to the local speed of sound, c, in the gas is defined as the Mach number,

$$M \equiv \frac{V}{c}$$

For $M < 0.3$, the maximum density variation is less than 5 percent. Thus gas flows with $M < 0.3$ can be treated as incompressible; a value of $M = 0.3$ in air at standard conditions corresponds to a speed of approximately 100 m/sec.

Compressible flows occur frequently in engineering applications. Common examples include compressed air systems used to power shop tools and dental drills, transmission of gases in pipelines at high pressure, and pneumatic or fluidic control and sensing systems. Compressibility effects are very important in the design of modern high-speed aircraft and missiles, power plants, fans, and compressors.

2-5.4   Internal and External Flows

Flows completely bounded by solid surfaces are called internal or duct flows. Flows over bodies immersed in an unbounded fluid are termed external flows. Both internal and external flows may be laminar or turbulent, compressible or incompressible.

In the case of incompressible flow through a pipe, the nature of the flow (laminar or turbulent) is determined by the value of a dimensionless parameter, the Reynolds number, $Re = \rho \bar{V} D / \mu$, where ρ is the density of the fluid, $\bar{V}$ the average flow velocity, D the pipe diameter, and μ the viscosity of the fluid. Pipe flow is laminar when $Re \leq 2300$; it may be turbulent for larger values. (The Reynolds number and other important dimensionless parameters encountered in fluid mechanics will be discussed in Chapter 7.) Chapter 8 will be devoted to a study of internal incompressible flow.

External flows occur over bodies immersed in an unbounded fluid. The flow over a semi-infinite flat plate (Fig. 2.11) and the flow over a cylinder (Fig. 2.12) are examples of external flows. Boundary-layer flows also may be laminar or turbulent; the definitions of laminar and turbulent flows given earlier also apply to boundary-layer flows; the details of a flow field may be significantly different depending on whether the boundary layer is laminar or turbulent. For the flat plate flow of Fig. 2.11, the nature of the boundary layer (laminar or turbulent) is determined by the Reynolds number, $Re_x = \rho U_\infty x / \mu$ where x is the distance downstream from the leading edge of the plate. The flow is generally laminar for $Re_x \leq 5 \times 10^5$; it may be turbulent for larger values. In Chapter 9, boundary-layer flows and flow over immersed bodies will be discussed in detail.

The internal flow of liquids in which the duct does not flow full—where there is a free surface subject to a constant pressure—is termed *open-channel* flow. Common examples of open-channel flow include flow in rivers, irrigation ditches, and aqueducts. Open-channel flow will be treated in Chapter 10.

The internal flow through fluid machines is considered in Chapter 11. The principle of angular momentum is applied to develop fundamental equations for fluid machines. Pumps, fans, blowers, compressors, and propellers that add energy to fluid streams are considered, as are turbines and windmills that extract energy. The chapter features detailed discussion of operation of fluid systems.

In the case of internal compressible flows, proper duct design is necessary to attain supersonic flow. The variation of fluid properties within a variable-area flow passage is not the same for supersonic flow ($M > 1$) as it is for subsonic flow ($M < 1$). Likewise the boundary conditions on the flow at the exit of an internal flow (e.g., the discharge from a nozzle) are different in the two cases. For subsonic flow discharge, the pressure in the exit plane of the nozzle is ambient pressure. For sonic flow, the nozzle exit pressure may be greater than ambient. For a supersonic jet, the pressure in the exit plane of the nozzle may be greater than, equal to, or less than ambient pressure. One-dimensional, steady compressible flow will be treated in Chapters 12 and 13.

2-6 SUMMARY OBJECTIVES

After completing study of Chapter 2, you should be able to do the following:

1. Give operational definitions of:

continuum	pseudoplastic fluid
property at a point	thixotropic fluid
scalar field	rheopectic fluid
vector field	Bingham plastic
steady flow	viscous flow
uniform flow at a section	inviscid flow
timeline	boundary layer
pathline	stagnation point
streakline	drag
streamline	separation
body force	wake
surface force	laminar flow
shear stress	turbulent flow
normal stress	Reynolds number
Newtonian fluid	compressible flow
non-Newtonian fluid	incompressible flow
viscosity	Mach number
kinematic viscosity	internal flow
apparent viscosity	external flow
dilatant fluid	open-channel flow

2. Give examples of one-, two-, and three-dimensional flows.
3. Calculate and plot streamlines, pathlines, and streaklines for specified velocity fields.
4. State the convention for designating the nine components of the stress field.
5. Write Newton's law of viscosity and determine the shear stress and shear force that correspond to a given one-dimensional velocity profile.
6. Solve the problems at the end of the chapter that relate to the material you have studied.

PROBLEMS

2.1 For the velocity fields given below, determine:
   (a) whether the flow field is one-, two-, or three-dimensional, and why.
   (b) whether the flow is steady or unsteady, and why.
   (The quantities a and b are constants.)
   (1) $\vec{V} = [ae^{-bx}]\hat{i}$ (2) $\vec{V} = ax^2\hat{i} + bx\hat{j}$
   (3) $\vec{V} = [ax^2e^{-bt}]\hat{i}$ (4) $\vec{V} = ax\hat{i} - by\hat{j}$
   (5) $\vec{V} = (ax + t)\hat{i} - by^2\hat{j}$ (6) $\vec{V} = ax^2\hat{i} + bxz\hat{j}$
   (7) $\vec{V} = a(x^2 + y^2)^{1/2}(1/z^3)\hat{k}$ (8) $\vec{V} = axy\hat{i} - byzt\hat{j}$

2.2 For the velocity fields given below, determine:
   (a) whether the flow field is one-, two-, or three-dimensional, and why.
   (b) whether the flow is steady or unsteady, and why.

(The quantities a, b, and c are constants.)

(1) $\vec{V} = [ae^{-by}]\hat{i}$ (2) $\vec{V} = by^2\hat{j} + cy\hat{k}$

(3) $\vec{V} = [ay^2 e^{-bt}]\hat{j}$ (4) $\vec{V} = by\hat{i} - ax\hat{j}$

(5) $\vec{V} = ax\hat{i} + (t - by)\hat{j}$ (6) $\vec{V} = ax^2\hat{i} + bxy\hat{k}$

(7) $\vec{V} = ax^2\hat{i} + by\hat{j} + cxz\hat{k}$ (8) $\vec{V} = ax\hat{i} - by\hat{j} + (t - cz)\hat{k}$

2.3 For the velocity fields given below, determine:

(a) whether the flow field is one-, two-, or three-dimensional, and why.

(b) whether the flow is steady or unsteady, and why.

(The quantities a, b, and c are constants.)

(1) $\vec{V} = [ae^{-bz}]\hat{k}$ (2) $\vec{V} = az^2\hat{i} + bz\hat{k}$

(3) $\vec{V} = ax\hat{i} + bx^2 e^{-ct}\hat{j}$ (4) $\vec{V} = ax\hat{i} + by\hat{j} + cx^2\hat{k}$

(5) $\vec{V} = ax\hat{i} + by^2\hat{j} + cyt\hat{k}$ (6) $\vec{V} = ax\hat{i} + bx^2\hat{j} - cx^2\hat{k}$

(7) $\vec{V} = ay^2\hat{i} + bx\hat{j} + cxy\hat{k}$ (8) $\vec{V} = ax\hat{i} + by^2\hat{j} + czt\hat{k}$

2.4 A viscous liquid is sheared between two parallel disks; the upper disk rotates and the lower one is fixed. The velocity field between the disks is given by $\vec{V} = \hat{e}_\theta r\omega z/h$. (The origin of coordinates is located at the center of the lower disk; the upper disk is located at $z = h$.) What are the dimensions of this velocity field? Does this velocity field satisfy appropriate physical boundary conditions? What are they?

2.5 A velocity field is given by the expression

$$\vec{V} = U\cos\theta\left[1 - \left(\frac{a}{r}\right)^2\right]\hat{e}_r - U\sin\theta\left[1 + \left(\frac{a}{r}\right)^2\right]\hat{e}_\theta$$

Find all points in the $r\theta$ plane where: (a) $V_r = 0$, (b) $V_\theta = 0$, and (c) $V_r = V_\theta = 0$.

2.6 A streamline is a line drawn in the flow field so that at each instant it is tangent to the velocity vector at each point. Consider a two-dimensional, steady flow in the xy plane with velocity field given by $\vec{V} = u\hat{i} + v\hat{j}$. Use the vector cross product, $\vec{V} \times d\vec{s}$, where $d\vec{s}$ is an element of distance along a streamline, to show that $dx/u = dy/v$ for the streamline.

2.7 The two-dimensional steady flow described by the velocity field $\vec{V} = ax\hat{i} - ay\hat{j}$ was analyzed in Example Problem 2.1, with x and y given in meters, and $a = 0.1$ sec^{-1}. Consider this flow field again. Assume that a timeline is marked in the flow at $t = 0$, connecting points along the line $y = $ constant, from $(x, y) = (1, 8)$ to $(3, 8)$. Calculate the position of the timeline at $t = 10$ sec. What general conclusion can you draw about the motion of any timeline that initially is horizontal in this flow?

2.8 The velocity field $\vec{V} = ax\hat{i} - by\hat{j}$, where $a = b = 1$ sec^{-1}, can be interpreted to represent flow in a corner. Find an equation for the flow streamlines. Plot several streamlines in the first quadrant, including the one that passes through the point $(x, y) = (0, 0)$.

2.9 The velocity field $\vec{V} = -ax\hat{i} + by\hat{j}$ can be interpreted to represent flow in a corner. Obtain an equation for the flow streamlines. Plot several streamlines in the first quadrant for $a = b = 2$ sec^{-1}, including the one that passes through the point $(x, y) = (1, 1)$.

2.10 A flow is described by the velocity field $\vec{V} = (Ax + B)\hat{i} + (-Ay)\hat{j}$, where $A = 10$ ft/sec/ft and $B = 3$ ft/sec. Plot a few streamlines in the xy plane, including the one that passes through the point $(x, y) = (1, 2)$.

2.11 A velocity field is specified as $\vec{V} = ax^2\hat{i} + bxy\hat{j}$, where $a = 2/$m·sec, $b = -4/$m·sec, and the coordinates are measured in meters. Is the flow field one-, two-, or three-dimensional? Why? Calculate the velocity components at the point $(2, \frac{1}{2}, 0)$. Develop an equation for the streamline passing through this point.

2.12 For the velocity field $\vec{V} = Axy\hat{i} + By^2\hat{j}$, where $A = 1$ m^{-1}sec^{-1}, $B = -\frac{1}{2}$ m^{-1}sec^{-1}, and the coordinates are measured in meters, plot streamlines for positive y.

2.13 A velocity field is represented by $\vec{V} = (Ax - B)\hat{i} + Cy\hat{j} + Dt\hat{k}$, where $A = 2$ sec^{-1}, $B = 4$ m/sec, $C = -2$ sec^{-1}, $D = 5$ m/sec^2, and the coordinates are measured in meters. Does this expression describe a one-, two-, or three-dimensional flow field?

Why? Obtain the equation for streamlines in the xy plane. Sketch a few streamlines in the upper half plane.

2.14 The velocity for a steady, incompressible flow in the xy plane is given by $\vec{V} - \hat{i}A/x + \hat{j}Ay/x^2$, where $A = 2$ m²/sec, and the coordinates are measured in meters. Obtain an equation for the streamline that passes through the point $(x, y) = (1, 3)$. Calculate the time required for a fluid particle to move from $x = 1$ m to $x = 3$ m in this flow field.

2.15 Beginning with the velocity field of Problem 2.8, verify the parametric equations for particle motion given in Problem 2.18. Obtain the equation for the pathline of the particle located at the point $(x, y) = (1, 2)$ at the instant $t = 0$. Compare this pathline with the streamline through the same point.

2.16 Consider again the velocity field of Problem 2.9. Obtain parametric equations for particle motion and the equation for the pathline of the particle located at the point $(x, y) = (2, 1)$ at the instant $t = 0$. Compare this pathline with the streamline through the same point.

2.17 Consider the flow field given in Eulerian description by the expression $\vec{V} = A\hat{i} + Bt\hat{j}$, where $A = 2$ m/sec, $B = 0.3$ m/sec², and the coordinates are measured in meters. Derive the Lagrangian position functions for the fluid particle that was located at the point $(x, y) = (1, 1)$ at the instant $t = 0$. Obtain an algebraic expression for the pathline followed by this particle. Plot the pathline and compare with the streamlines through the same point at the instants $t = 0$, 1, and 2 sec.

2.18 Parametric equations for the position of a particle in a flow field are given as $x_p = c_1 e^{at}$ and $y_p = c_2 e^{-bt}$. Find the equation of the pathline for a particle located at $(x, y) = (1, 2)$ at $t = 0$. Show that for this flow field, $\vec{V} = ax\hat{i} - by\hat{j}$. Compare the pathline with a streamline through the same point.

2.19 Consider the flow described by the velocity field $\vec{V} = x(1 + At)\hat{i} + y\hat{j}$, with $A = 0.5$ sec^{-1}. For the point $(1, 1, 0)$, calculate and plot (a) the streamline through the point at $t = 0$ and (b) the pathline traced out by the particle that passes through the point at this instant.

2.20 Consider the flow described by the velocity field $\vec{V} = x(1 + At)\hat{i} + y\hat{j}$, with $A = 0.5$ sec^{-1}. For the point $(1, 1, 0)$, calculate and plot (a) the streamline through the point at $t = 0$ and (b) the streakline formed by particles that passed through the point at earlier times.

2.21 Consider the flow field $\vec{V} = ax(1 + bt)\hat{i} + cy\hat{j}$, where $a = c = 1$ sec^{-1} and $b = 0.2$ sec^{-1}. For the particle that passes through the point $(x, y) = (1, 1)$ at the instant $t = 0$, plot the pathline during the interval from $t = 0$ to $t = 3$ sec. Compare with the streamline through the same point at the instant $t = 0$.

2.22 Streaklines are traced out by neutrally buoyant marker fluid injected into a flow field from a fixed point in space. A particle of the marker fluid that is at point (x, y) at time t, must have passed through the injection point (x_0, y_0) at some earlier instant $t = \tau$. The time history of a marker particle may be found by solving the pathline equations for the initial conditions that $x = x_0$, $y = y_0$ when $t = \tau$. The present locations of particles on the streakline are obtained by setting τ equal to values in the range $0 \leq \tau \leq t$. Consider the flow field $\vec{V} = ax(1 + bt)\hat{i} + cy\hat{j}$, where $a = c = 1$ sec^{-1} and $b = 0.2$ sec^{-1}. Evaluate the streakline that passes through the initial point $(x_0, y_0) = (1, 1)$, during the interval from $t = 0$ to $t = 3$ sec. Compare with the streamline through the same point at the instant $t = 0$.

‡2.23 Consider the flow field $\vec{V} = axt\hat{i} + b\hat{j}$, where $a = 2$ sec^{-2} and $b = 3$ m/sec. For the particle that passes through the point $(x, y) = (3, 1)$ at the instant $t = 0$, plot the pathline during the interval from $t = 0$ to 3 sec. Compare this pathline with the streamlines through the same point at the instants $t = 1$, 2, and 3 sec.

2.24 Tiny hydrogen bubbles are being used as tracers to visualize a flow. All the bubbles are generated at the origin $(x = 0, y = 0)$. The velocity field is unsteady and obeys the

‡ You may wish to use simple computer programs to help solve problems marked with daggers.

equations:

$$u = 1 \text{ m/sec} \qquad v = 1 \text{ m/sec} \qquad 0 \le t < 2 \text{ sec}$$
$$u = 0 \qquad v = 1.5 \text{ m/sec} \qquad 2 \le t \le 4 \text{ sec}$$

Plot the pathlines of bubbles that leave the origin at $t = 0, 1, 2, 3,$ and 4 sec. Mark the locations of these five bubbles at $t = 4$ sec. Use a dashed line to indicate the position of a streakline at $t - 4$ sec.

‡2.25 Consider again the velocity field given in Problem 2.21. Locate the positions at $t = 3$ sec of particles that passed the point $(x, y) = (1, 1)$ at times $t = 0, 1,$ and 2 sec. Join the particle positions with a dashed line. Compare the resulting streakline with the streamline through the same point at the instant $t = 0$.

‡2.26 Consider the velocity field $\vec{V} = ax\hat{i} + by(1 + ct)\hat{j}$, where $a = b = 2 \text{ sec}^{-1}$, and $c = 0.4$ sec^{-1}. For the particle that passes through the point $(x, y) = (1, 1)$ at the instant $t = 0$, plot the pathline during the interval from $t = 0$ to 1.5 sec. Compare this pathline with the streakline through the same point at the instant $t = 0$.

‡2.27 Consider the flow field $\vec{V} = axt\hat{i} + b\hat{j}$, where $a = 3 \text{ sec}^{-2}$ and $b = 2 \text{ m/sec}$. For the particle that passes through the point $(x, y) = (1, 2)$ at the instant $t = 0$, plot the pathline during the time interval from $t = 0$ to 3 sec. Compare this pathline with the streakline through the same point at the instant $t = 3$ sec.

2.28 A tornado can be represented in polar coordinates by the velocity field

$$\vec{V} = -\frac{a}{r}\hat{e}_r + \frac{b}{r}\hat{e}_\theta$$

where $\hat{e}_r$ and $\hat{e}_\theta$ are unit vectors in the r and θ directions, respectively. Use the vector cross product, $\vec{V} \times d\vec{s}$, where $d\vec{s}$ is an element of distance along a streamline, to show that streamlines form logarithmic spirals,

$$r = ce^{-(a/b)\theta}$$

2.29 Use double index notation to label the six shear stresses shown.

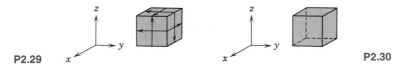

P2.29   P2.30

2.30 On the element shown, indicate all possible stresses represented by

$$\tau_{zx} = -10 \text{ lbf/ft}^2 \qquad \text{and} \qquad \sigma_{yy} = 15 \text{ lbf/ft}^2$$

2.31 The density distribution in the column of saline solution shown is given by $\rho = \rho_0(1 + ky)$, where $\rho_0 = 1.94 \text{ slug/ft}^3$ and $k = 0.10 \text{ ft}^{-1}$. Compute the body force acting on the volume.

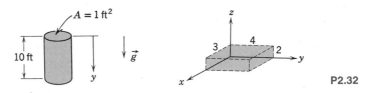

P2.31   P2.32

2.32 A body force distribution is given as $\vec{B} = ax\hat{i} + b\hat{j} + cz\hat{k}$ per unit mass of the material acted on. The density of the material is given as $\rho = lx^2 + ry + nz$. All coordinates are measured in meters. Determine the resultant body force in the region shown when: $a = 0$, $b = 0.1 \text{ N/kg}$, $c = 0.5 \text{ N/kg} \cdot \text{m}$, $l = 2.0 \text{ kg/m}^5$, $r = 0$, and $n = 1.0 \text{ kg/m}^4$.

‡ You may wish to use simple computer programs to help solve problems marked with daggers.

2.33 For the region shown in Problem 2.32, determine the resultant body force when: $a = 5.0$ N/kg $\cdot$ m, $b = 0$, $c = 10.0$ N/kg $\cdot$ m, $l = 0.1$ kg/m^5, $r = 0.5$ kg/m^4, and $n = 0$.

2.34 For the region shown in Problem 2.32, determine the resultant body force when: $a = 1.0$ N/kg $\cdot$ m, $b = 2.0$ N/kg, $c = 0$, $l = 0$, $r = 1.0$ kg/m^4, and $n = 2.0$ kg/m^4.

2.35 The following stress levels are known to exist at a point in a continuous medium:

$\sigma_{xx} = 200$ psi
$\sigma_{yy} = -100$ psi
$\tau_{xy} = \tau_{yx} = 30$ psi

P2.35

Find the magnitudes of the normal and shear stresses acting on plane AA.

2.36 The temperature dependence of the viscosity of water is represented well by the empirical equation

$$\mu = Ae^{B/T}$$

where T is absolute temperature. The following data were measured in a laboratory viscometer: at 10 C, $\mu = 0.0013$ kg/m $\cdot$ sec, and at 20 C, $\mu = 0.0010$ kg/m $\cdot$ sec. Use the method suggested in Appendix A to evaluate constants A and B in the correlating equation. Check against data from Fig. A.2.

2.37 The variation with temperature of the viscosity of air is correlated well by the empirical Sutherland equation

$$\mu = \frac{bT^{1/2}}{1 + S/T}$$

Best-fit values of b and S are given in Appendix A for use with SI units. Use these values to develop an equation for calculating air viscosity in British Gravitational units as a function of absolute temperature in degrees Rankine. Check your result using data from Appendix A.

2.38 The variation with temperature of the viscosity of air is represented well by the empirical Sutherland correlation

$$\mu = \frac{bT^{1/2}}{1 + S/T}$$

Best-fit values of b and S are given in Appendix A. Develop an equation in SI units for kinematic viscosity versus temperature for air at atmospheric pressure. Assume ideal gas behavior. Check using data from Appendix A.

2.39 Data for the temperature variation of the viscosity of water are correlated well by the equation

$$\mu = 2.414 \times 10^{-5} \exp\left(\frac{570.6}{T - 140}\right)$$

where μ is dynamic viscosity in N $\cdot$ sec/m^2 and T is absolute temperature in kelvins (Reference 9 of Appendix A). Calculate the viscosity of water at 0, 20, and 40 C. Compare with values obtained from the plots in Appendix A.

2.40 Capillary-tube viscometers are used widely for industrial purposes, particularly for petroleum products and lubricants. The American Society for Testing and Materials (ASTM) has standardized dimensions and test procedures for Saybolt Universal and Saybolt Furol (contraction for "fuel and motor oils") capillary viscometers. These instruments are simple to use: the liquid to be tested is placed in a cylinder that has a short small-bore tube and stopper at its lower end. The cylinder is surrounded by a constant-temperature bath. After the sample temperature stabilizes, the stopper is removed. The time is measured for 60 cubic centimeters of liquid to flow out. The Saybolt reading is the flow time in seconds. These "viscosity" readings may be converted to engineering units using the empirical equations (Streeter, V. L., *Handbook of Fluid Dynamics*. New York: McGraw-Hill, 1961, Section 1.11):

Viscometer	Time Range (sec)	Kinematic Viscosity ν(St)
Saybolt Universal:	$32 \le t \le 100$	$0.00226t - \dfrac{1.95}{t}$
	$100 < t < 1000$	$0.00220t - \dfrac{1.35}{t}$
Saybolt Furol:	$25 \le t \le 40$	$0.0224t - \dfrac{1.84}{t}$
	$40 < t$	$0.0216t - \dfrac{0.60}{t}$

Calculate the ranges of kinematic viscosity (in SI units) for which each instrument may be used.

2.41 The velocity distribution for laminar flow between parallel plates is given by

$$\frac{u}{u_{max}} = 1 - \left(\frac{2y}{h}\right)^2$$

where h is the distance separating the plates and the origin is placed midway between the plates. Consider a flow of water at 15 C, with $u_{max} = 0.30$ m/sec and $h = 0.50$ mm. Calculate the shear stress on the upper plate and give its direction.

2.42 The velocity distribution for laminar flow between parallel plates is given by

$$\frac{u}{u_{max}} = 1 - \left(\frac{2y}{h}\right)^2$$

where h is the distance separating the plates and the origin is placed midway between the plates. Consider flow of water at 15 C with maximum speed of 0.05 m/sec and $h = 5$ mm. Calculate the force on a 0.3 m^2 section of the lower plate and give its direction.

2.43 Two infinite parallel horizontal surfaces are separated by a film of SAE 10W oil at $T = 10$ C that is $d = 0.0144$ in. thick. The upper surface moves with speed V. The lower surface is stationary and contains a spring-mounted plate of area $A = 0.725$ ft^2. The horizontal force on the plate is $F = 0.155$ lbf. Calculate the speed of the upper plate.

2.44 A small flat-bottomed sled used for demonstrations is supported on a film of air; the air film is $h = 0.00135$ in. thick and the contact area is $A = 12.5$ in.^{2} At one instant, the speed of the sled is $V = 6.25$ ft/sec. Determine the force that opposes the motion of the sled at this instant.

2.45 A female freestyle ice skater, weighing 112 lbf, glides on one skate at speed $V = 20.5$ ft/sec. Her weight is supported by a thin film of liquid water melted from the ice by the pressure of the skate blade. Assume the blade is $L = 11.5$ in. long and $w = 0.125$ in. wide, and that the water film is $h = 0.0000575$ in. thick. Estimate the deceleration of the skater that results from viscous shear in the water film, if end effects are neglected.

2.46 Crude oil, with specific gravity SG = 0.85 and viscosity $\mu = 2.15 \times 10^{-3}$ lbf · sec/ft^2, flows steadily down a surface inclined $\theta = 30$ degrees below the horizontal in a film of thickness $h = 0.125$ in. The velocity profile is given by

$$u = \frac{\rho g}{\mu}\left(hy - \frac{y^2}{2}\right)\sin\theta$$

(Coordinate x is along the surface and y is normal to the surface.) Plot the velocity profile. Determine the magnitude and direction of the shear stress that acts on the surface.

2.47 A block weighing 10 lbf and having dimensions 10 in. on each edge is pulled up an inclined surface on which there is a film of SAE 10W oil at 100 F. If the speed of the block is 5 ft/sec and the oil film is 0.001 in. thick, find the force required to pull the block. Assume the velocity distribution in the oil film is linear. The surface is inclined at an angle of 15° from the horizontal.

2.48 A block has a mass of 2 kg and is 0.2 m square. It slides down a smooth incline on a thin film of oil. The slope is 30° from the horizontal. The oil is SAE 30 at 20 C, the film is 0.02 mm thick, and the velocity profile may be assumed linear. Calculate the terminal speed of the block.

2.49 A slab of mass $M = 125$ kg slides down a large smooth plane surface inclined at $\theta = 2.15°$ from the horizontal and covered with a film of water $h = 0.0250$ mm thick. Estimate the terminal speed of the slab if its dimensions are $L = 515$ mm by $W = 525$ mm.

2.50 Recording tape is to be coated on both sides with lubricant by drawing it through a narrow gap. The tape is 0.015 in. thick and 1.00 in. wide. It is centered in the gap with a clearance of 0.012 in. on each side. The lubricant, of viscosity $\mu = 0.021$ slug/ft · sec, completely fills the space between the tape and gap for a length of 0.75 in. along the tape. If the tape can withstand a maximum tensile force of 7.5 lbf, determine the maximum speed with which it can be pulled through the gap.

2.51 The air hockey game at Rocky's Rec Room has pucks of mass 30 g, with a diameter of 100 mm. The air film under a puck is 0.1 mm thick. Calculate the time required after impact for a puck to lose 10 percent of its initial speed.

2.52 A block of mass M slides on a thin film of oil. The film thickness is h and the area of the block is A. When released, mass m exerts tension on the cord, causing the block to accelerate. Neglect friction in the pulley and air resistance. Develop an algebraic expression for the viscous force that acts on the block when it moves at speed V. Derive a differential equation for the block speed as a function of time. Obtain an algebraic expression for the maximum speed of the block.

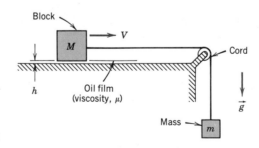

P2.52

2.53 A rectangular steel block with length $L = 101$ mm, width $W = 50.5$ mm, and height $H = 20.2$ mm slides along a smooth horizontal surface on a film of SAE 30 oil at $T = 20$ C, with depth $d = 0.505$ mm. At time $t = 0$, the initial speed of the block is $U_0 = 0.525$ m/sec. Calculate the magnitude and state the direction of the viscous shear force that acts on the block at time $t = 0$. Develop a differential equation for the speed of the block and find its speed at $t = 5$ sec. Will the speed of the block ever reach zero? Determine the maximum displacement of the block.

2.54 The air hockey table of Problem 2.51 has been installed poorly; its surface slopes at 2° from the horizontal. A player fires a puck directly uphill at an initial speed of 10 m/sec. Determine its speed at the instant it passes through the goal, 2 m away.

2.55 Magnet wire is to be coated with varnish for insulation by drawing it through a circular die of 0.9 mm diameter. The wire diameter is 0.8 mm and it is centered in the die. The varnish ($\mu = 20$ centipoise) completely fills the space between the wire and the die for a length of 20 mm. The wire is drawn through the die at a speed of 50 m/sec. Determine the force required to pull the wire.

2.56 A concentric cylinder viscometer may be formed by rotating the inner member of a pair of closely fitting cylinders (see Fig. P2.60). The annular gap must be made small so that a linear velocity profile will exist in the liquid sample. Consider a viscometer with an inner cylinder of 3 in. diameter and 6 in. height, and a clearance gap width of 0.001 in., filled with castor oil at 90 F. Determine the torque required to turn the inner cylinder at 250 rpm.

2.57 A concentric cylinder viscometer may be formed by rotating the inner member of a pair of closely fitting cylinders (see Fig. P2.60). For small clearances, a linear velocity profile may be assumed in the liquid filling the annular clearance gap. A viscometer has an inner cylinder of 75 mm diameter and 150 mm height, with a clearance gap width of 0.02 mm. A torque of 0.021 N · m is required to turn the inner cylinder at 100 rpm. Determine the viscosity of the liquid in the clearance gap of the viscometer.

2.58 A shaft with outside diameter of 18 mm turns at 20 revolutions per second inside a stationary journal bearing 60 mm long. A thin film of oil 0.2 mm thick fills the concentric annulus between the shaft and journal. The torque needed to turn the shaft is 0.0036 N · m. Estimate the viscosity of the oil that fills the gap.

2.59 A cylinder with mass $M = 0.225$ kg slides down a long vertical tube. The tube is lubricated with a thin layer of glycerin at $T = 30$ C in the gap between the cylinder and the tube. The tube diameter is $D = 30.1$ mm, the cylinder is $H = 20.5$ mm high, and the gap width is estimated at $h = 0.00125$ mm. At a certain instant the cylinder is traveling downward at $V = 12.5$ mm/sec. Assume that the air pressure on both sides of the cylinder is the same. Calculate the acceleration of the cylinder at this instant. Estimate the ultimate speed the cylinder would reach after sliding a long way down the tube.

2.60 A concentric cylinder viscometer is driven by a falling mass M connected by a cord and pulley to the inner cylinder, as shown. The liquid to be tested fills the annular gap of width a and height H. After a brief starting transient, the mass falls at constant speed V_m. Develop an algebraic expression for the viscosity of the liquid in the device in terms of M, g, V_m, r, R, a, and H. Evaluate the viscosity of the liquid using:

$$M = 0.10 \text{ kg} \qquad r = 25 \text{ mm}$$
$$R = 50 \text{ mm} \qquad a = 0.20 \text{ mm}$$
$$H = 50 \text{ mm} \qquad V_m = 40 \text{ mm/sec}$$

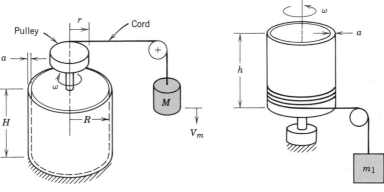

P2.60   P2.61

2.61 The thin outer cylinder (mass m_2 and radius R) of a small portable concentric cylinder viscometer is driven by a falling mass, m_1, attached to a cord. The inner cylinder is stationary. The clearance between the cylinders is a. Neglect bearing friction, air resistance, and the mass of liquid in the viscometer. Obtain an algebraic expression for the torque due to viscous shear that acts on the cylinder at angular speed ω. Derive and solve a differential equation for the angular speed of the outer cylinder as a function of time. Obtain an expression for the maximum angular speed of the cylinder.

2.62 Estimates place the range of shear rates encountered in the lubricating oil during winter cranking of a cold engine between 10^4 and 10^5 sec^{-1} (*SAE Handbook*, 1976). A concentric-cylinder viscometer is to be designed to obtain shear rates in the necessary range. Constraints on the viscometer include a maximum speed for the outer cylinder of 12,000 rpm and a maximum Reynolds number for the lubricant in the clearance gap of 1200, based on the linear speed of the outer cylinder and the gap width. Using viscosity data for SAE 10W-30 motor oil at −20 C (from Appendix A), calculate the gap width required. Assess the feasibility of the design.

2.63 A circular aluminum shaft mounted in a journal is shown. The symmetric clearance gap between the shaft and journal is filled with SAE 10W-30 oil at $T = 30$ C. The shaft is caused to turn by the attached mass and cord. Develop and solve a differential equation for the angular speed of the shaft as a function of time. Calculate the maximum angular speed of the shaft and the time required to reach 95 percent of this speed.

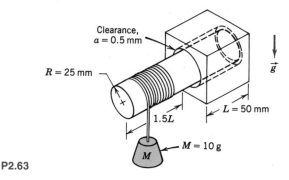

P2.63

2.64 A shock-free coupling for a low-power mechanical drive is to be made from a pair of concentric cylinders. The annular space between the cylinders is to be filled with oil. The drive must transmit power, $P = 5$ W. Other dimensions and properties are as shown. Neglect any bearing friction and end effects. Assume the minimum practical gap clearance δ for the device is $\delta = 0.5$ mm. Dow manufactures silicone fluids with viscosities as high as 10^6 centipoise. Determine the viscosity that should be specified to satisfy the requirement for this device.

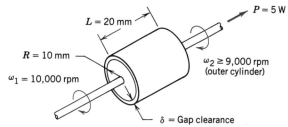

P2.64

2.65 A proposal has been made to use a pair of parallel disks to measure the viscosity of a liquid sample. The upper disk rotates at height h above the lower disk. The viscosity of the liquid in the gap is to be calculated from measurements of the torque needed to turn the upper disk steadily. Obtain an algebraic expression for the torque needed to turn the disk.

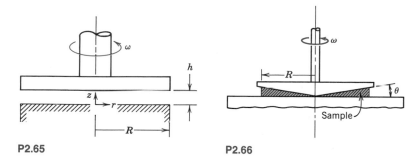

P2.65　　　　　　　　　　　**P2.66**

2.66 The cone and plate viscometer shown is an instrument used frequently to characterize non-Newtonian fluids. It consists of a flat plate and a rotating cone with a very obtuse angle (typically θ is less than 0.5 degrees). The apex of the cone just touches the plate surface and the liquid to be tested fills the narrow gap formed by the cone and plate. Derive an expression for the shear rate in the liquid that fills the gap in terms of the geometry of the system. Evaluate the torque on the driven cone in terms of the shear stress and geometry of the system.

2.67 A viscous clutch is to be made from a pair of closely spaced parallel disks enclosing a thin layer of viscous liquid. Develop algebraic expressions for the torque and the power transmitted by the disk pair, in terms of liquid viscosity, μ, disk radius, R, disk spacing, a, and the angular speeds: ω_i of the input disk and ω_o of the output disk. Also develop expressions for the slip ratio, $s = \Delta\omega/\omega_i$, in terms of ω_i and the torque transmitted. Determine the efficiency, η, in terms of the slip ratio.

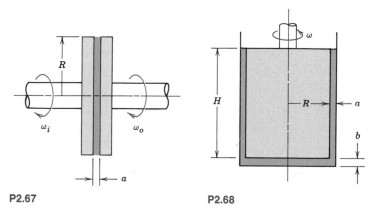

P2.67　　　　　　　　　　　**P2.68**

2.68 A concentric-cylinder viscometer is shown. Viscous torque is produced by the annular gap around the inner cylinder. Additional viscous torque is produced by the flat bottom of the inner cylinder as it rotates above the flat bottom of the stationary outer cylinder. Obtain an algebraic expression for the viscous torque due to flow in the annular gap of width a. Obtain an algebraic expression for the viscous torque due to flow in the bottom clearance gap of height b. Prepare a plot showing the ratio, b/a, required to hold the bottom torque to 1 percent or less of the annulus torque, versus the other geometric variables. What are the design implications? What modifications to the design can you recommend?

2.69 A conical pointed shaft turns in a conical bearing. The gap between shaft and bearing is filled with heavy oil having the viscosity of SAE 30 at 30 C. Obtain an algebraic expression for the shear stress that acts on the surface of the conical shaft. Calculate the viscous torque that acts on the shaft.

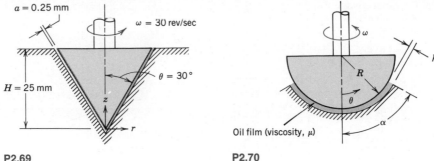

P2.69   **P2.70**

‡2.70   A spherical thrust bearing is shown. The gap between the spherical member and the housing is of constant width h. Obtain and plot an algebraic expression for the torque on the spherical member, as a function of angle α.

‡2.71   A cross section of a rotating bearing is shown. The spherical member rotates with angular speed ω, a small distance, a, above the plane surface. The narrow gap is filled with viscous oil, having $\mu = 1250$ cp. Obtain an algebraic expression for the shear stress acting on the spherical member. Evaluate the maximum shear stress that acts on the spherical member for the conditions shown. (Is the maximum necessarily located at the maximum radius?) Develop an algebraic expression (in the form of an integral) for the total viscous shear torque that acts on the spherical member. Calculate the torque using the dimensions shown.

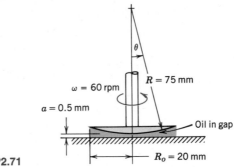

P2.71

2.72   Pulverized coal mixed with water to form a slurry often is transported over long distances in pipelines. When the weight percent of solids exceeds approximately 45 percent, the slurry behaves as a Bingham plastic. Data from a coal slurry with 50 percent solids by weight showed a yield stress value of 3.1 N/m² and a plastic viscosity of 0.003 N · sec/m². Use the Bingham plastic model to evaluate the shear stress that would be expected in this suspension at a shear rate of 500 sec⁻¹. Compare to the shear stress for pure water at the same shear rate.

‡2.73   An analysis developed in Chapter 8 for the shear stress and velocity profile for laminar flow in a circular tube may be extended. The results for a power-law non-Newtonian fluid may be written in the form

$$\frac{u}{u_{\text{mean}}} = \left(\frac{3n+1}{n+1}\right)\left[1 - \left(\frac{r}{R}\right)^{\frac{n+1}{n}}\right]$$

Plot velocity profiles, u/u_{mean} versus r/R, for a Newtonian fluid with $n = 1$, a dilatant fluid with $n = 3$, and a pseudoplastic fluid with $n = \frac{1}{3}$. Compare features of the curves.

‡ You may wish to use simple computer programs to help solve problems marked with daggers.

‡2.74 Consider again the velocity profile for laminar flow of a power-law non-Newtonian fluid in a long circular tube given in Problem 2.73. Investigate the limiting cases of infinite dilatancy and infinite pseudoplasticity, represented by $n = \infty$ and $n = 0$, respectively. Plot u/u_{mean} versus r/R for these cases and compare to the Newtonian case for which $n = 1$.

2.75 Viscometric data for heavy cream show it to behave as a pseudoplastic fluid that can be modeled well by a power-law relationship between shear stress and rate of shearing strain at low shear rates. Assume the following data:

Shear stress, τ_{yx} (dyne/cm^2)	0.1	1.0
Shear rate, $\dfrac{du}{dy}$ (sec^{-1})	0.023	0.75

Use these data to fit a power-law model. Evaluate the flow behavior and consistency indexes using SI units. Estimate the shear stress that would be expected at a shear rate of 0.1 sec^{-1}. Compare the apparent viscosity of cream at this shear rate to that of water.

2.76 Dilatant behavior sometimes is found when dilute suspensions are tested at high shear rates. The following data were measured in a test of a suspension containing 12 percent solids by volume:

Shear stress, τ (N/m^2)	6.5	4.8	2.7	1.7
Shear rate, $\dfrac{du}{dy}$ (sec^{-1})	600	470	300	200

Evaluate the consistency index and flow behavior index for this suspension by fitting the data to a power-law model.

2.77 The Reynolds number, which is an important parameter in viscous flow phenomena, is defined by the equation

$$Re = \frac{\rho V L}{\mu}$$

where ρ and μ are the fluid density and viscosity, and V and L are characteristic velocity and length, respectively. Express each variable in terms of its basic dimensions. Show that the Reynolds number is dimensionless.

2.78 Air at standard conditions flows through a pipe of 1 in. diameter. The average velocity is 1 ft/sec. Is the flow laminar or turbulent?

2.79 Water at 15 C flows in a tube with an inside diameter of 50 mm. Determine the maximum value of average velocity for which the flow would be laminar.

2.80 The fluid mechanics of the human arterial system is of vital importance to our health and safety. The specific gravity and viscosity of blood are about 1.06 and 3.3 centipoise (cp), respectively. The mean flow speed in the aorta (30 mm i.d.) of a large human is about 0.15 m/sec. Calculate the flow Reynolds number using these properties. Would you expect this flow to be laminar or turbulent?

2.81 Consider the incompressible laminar viscous flow over a semi-infinite flat plate shown in Fig. 2.11. Sketch the variation in shear stress, τ_{yx}, as a function of y at the locations x_1 and x_2. Also sketch the variation in shear stress along the plate surface ($y = 0$) as a function of x.

‡ You may wish to use the simple computer programs to help solve problems with daggers.

Chapter 3

FLUID STATICS

By definition, a fluid must deform continuously when a shear stress of any magnitude is applied. The absence of relative motion (and thus, angular deformation) implies the absence of shear stresses. Therefore, fluids either at rest or in "rigid-body" motion are able to sustain only normal stresses. Analysis of hydrostatic cases is thus appreciably simpler than for fluids undergoing angular deformation (see Section 5-3.3).

   Mere simplicity does not justify our study of a subject. Normal forces transmitted by fluids are important in many practical situations. Using the principles of hydrostatics, we can compute forces on submerged objects, develop instruments for measuring pressures, and deduce properties of the atmosphere and oceans. The principles of hydrostatics also may be used to determine forces developed by hydraulic systems in applications such as industrial presses or automobile brakes.

   In a static, homogeneous fluid, or in a fluid undergoing rigid-body motion, a fluid particle retains its identity for all time. Since there is no relative motion within the fluid, a fluid element does not deform. We may apply Newton's second law of motion to evaluate the reaction of the particle to the applied forces.

3-1   THE BASIC EQUATION OF FLUID STATICS

Our primary objective is to obtain an equation that will enable us to determine the pressure field within a static fluid. To do this, we apply Newton's second law to a differential fluid element of mass $dm = \rho \, d\Psi$, with sides dx, dy, and dz, as shown in Fig. 3.1. The fluid element is stationary relative to the stationary rectangular coordinate system shown. (Fluids in rigid-body motion will be treated in Section 3-7.)

   From our previous discussion, recall that two general types of forces may be applied to a fluid: body forces and surface forces. The only body force that must be considered in most engineering problems is due to gravity. In some situations body forces due to electric or magnetic fields might be present; they will not be considered in this text.

   For a differential fluid element, the body force, $d\vec{F}_B$, is

$$d\vec{F}_B = \vec{g} \, dm = \vec{g}\rho \, d\Psi$$

where $\vec{g}$ is the local gravity vector, ρ is the density, and $d\Psi$ is the volume of the element. In Cartesian coordinates $d\Psi = dx \, dy \, dz$, so

$$d\vec{F}_B = \rho\vec{g} \, dx \, dy \, dz$$

   In a static fluid no shear stresses can be present. Thus the only surface force is the pressure force. Pressure is a field quantity, $p = p(x, y, z)$; the pressure varies with position within the fluid. The net pressure force that results from this variation can be evaluated by summing the forces that act on the six faces of the fluid element.

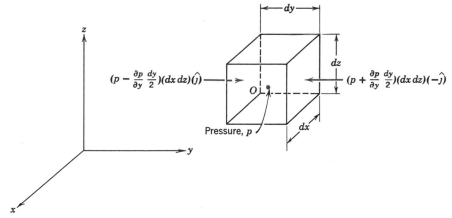

Fig. 3.1 Differential fluid element and pressure forces in the y direction.

Let the pressure at the center, O, of the element be p. To determine the pressure at each of the six faces of the element, we use a Taylor series expansion of the pressure about the point O. The pressure at the left face of the differential element is

$$p_L = p + \frac{\partial p}{\partial y}(y_L - y) = p + \frac{\partial p}{\partial y}\left(-\frac{dy}{2}\right) = p - \frac{\partial p}{\partial y}\frac{dy}{2}$$

(Terms of higher order are omitted because they will vanish in the subsequent limiting process.) The pressure on the right face of the differential element is

$$p_R = p + \frac{\partial p}{\partial y}(y_R - y) = p + \frac{\partial p}{\partial y}\frac{dy}{2}$$

The pressure *forces* acting on the two y surfaces of the differential element are shown in Fig. 3.1. Each pressure force is a product of three terms. The first is the magnitude of the pressure. The magnitude is multiplied by the area of the face to give the pressure force, and a unit vector is introduced to indicate direction. Note also in Fig. 3.1 that the pressure force on each face acts *against* the face. A positive pressure corresponds to a *compressive* normal stress.

Pressure forces on the other faces of the element are obtained in the same way. Combining all such forces gives the net surface force acting on the element. Thus

$$d\vec{F_S} = \left(p - \frac{\partial p}{\partial x}\frac{dx}{2}\right)(dy\ dz)(\hat{i}) + \left(p + \frac{\partial p}{\partial x}\frac{dx}{2}\right)(dy\ dz)(-\hat{i})$$

$$+ \left(p - \frac{\partial p}{\partial y}\frac{dy}{2}\right)(dx\ dz)(\hat{j}) + \left(p + \frac{\partial p}{\partial y}\frac{dy}{2}\right)(dx\ dz)(-\hat{j})$$

$$+ \left(p - \frac{\partial p}{\partial z}\frac{dz}{2}\right)(dx\ dy)(\hat{k}) + \left(p + \frac{\partial p}{\partial z}\frac{dz}{2}\right)(dx\ dy)(-\hat{k})$$

Collecting and canceling terms, we obtain

$$d\vec{F_S} = \left(-\frac{\partial p}{\partial x}\hat{i} - \frac{\partial p}{\partial y}\hat{j} - \frac{\partial p}{\partial z}\hat{k}\right)dx\ dy\ dz$$

or

$$d\vec{F}_S = -\left(\frac{\partial p}{\partial x}\hat{i} + \frac{\partial p}{\partial y}\hat{j} + \frac{\partial p}{\partial z}\hat{k}\right) dx\ dy\ dz \qquad (3.1a)$$

The term in parentheses is called the gradient of the pressure or simply the pressure gradient and may be written grad p or ∇p. In rectangular coordinates

$$\text{grad}\ p \equiv \nabla p \equiv \left(\hat{i}\frac{\partial p}{\partial x} + \hat{j}\frac{\partial p}{\partial y} + \hat{k}\frac{\partial p}{\partial z}\right) \equiv \left(\hat{i}\frac{\partial}{\partial x} + \hat{j}\frac{\partial}{\partial y} + \hat{k}\frac{\partial}{\partial z}\right) p$$

The gradient can be viewed as a vector operator; taking the gradient of a scalar field gives a vector field. Using the gradient designation, Eq. 3.1a can be written as

$$d\vec{F}_S = -\text{grad}\ p(dx\ dy\ dz) = -\nabla p\ dx\ dy\ dz \qquad (3.1b)$$

Physically the gradient of pressure is the negative of the surface force per unit volume due to pressure. We note that the level of pressure is not important in evaluating the net pressure force. Instead, what matters is the rate at which pressure changes occur with distance, the *pressure gradient*. We shall find this term very useful throughout our study of fluid mechanics.

We combine the formulations for surface and body forces that we have developed to obtain the total force acting on a fluid element. Thus

$$d\vec{F} = d\vec{F}_S + d\vec{F}_B = (-\text{grad}\ p + \rho\vec{g})\,dx\ dy\ dz = (-\text{grad}\ p + \rho\vec{g})\,d\forall$$

or on a per unit volume basis

$$\frac{d\vec{F}}{d\forall} = -\text{grad}\ p + \rho\vec{g} \qquad (3.2)$$

For a fluid particle, Newton's second law gives $d\vec{F} = \vec{a}\,dm = \vec{a}\rho\,d\forall$. For a static fluid, $\vec{a} = 0$. Thus

$$\frac{d\vec{F}}{d\forall} = \rho\vec{a} = 0$$

Substituting for $d\vec{F}/d\forall$ from Eq. 3.2, we obtain

$$-\text{grad}\ p + \rho\vec{g} = 0 \qquad (3.3)$$

Let us review briefly our derivation of this equation. The physical significance of each term is

$$-\text{grad}\ p \qquad + \qquad \rho\vec{g} \qquad = 0$$

$$\left\{\begin{array}{c}\text{net pressure force}\\\text{per unit volume}\\\text{at a point}\end{array}\right\} + \left\{\begin{array}{c}\text{body force per}\\\text{unit volume}\\\text{at a point}\end{array}\right\} = 0$$

This is a vector equation, which means that it consists of three component equations that must be satisfied individually. The components are

$$\begin{array}{ll} -\dfrac{\partial p}{\partial x} + \rho g_x = 0 & x\ \text{direction} \\[2mm] -\dfrac{\partial p}{\partial y} + \rho g_y = 0 & y\ \text{direction} \\[2mm] -\dfrac{\partial p}{\partial z} + \rho g_z = 0 & z\ \text{direction} \end{array} \right\} \qquad (3.4)$$

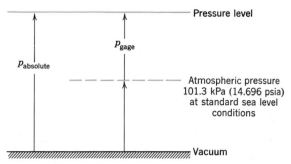

Fig. 3.2 Absolute and gage pressures, showing reference levels.

Equations 3.4 describe the pressure variation in each of the three coordinate directions in a static fluid. To simplify further, it is logical to choose a coordinate system such that the gravity vector is aligned with one of the coordinate axes. If the coordinate system is chosen such that the z axis is directed vertically, then $g_x = 0$, $g_y = 0$, and $g_z = -g$. Under these conditions, the component equations become

$$\frac{\partial p}{\partial x} = 0 \qquad \frac{\partial p}{\partial y} = 0 \qquad \frac{\partial p}{\partial z} = -\rho g \qquad (3.5)$$

Equations 3.5 indicate that under the assumptions made, the pressure is independent of coordinates x and y; it depends on z alone. Thus since p is a function of a single variable, a total derivative may be used instead of a partial derivative. With these simplifications, Eqs. 3.5 finally reduce to

$$\frac{dp}{dz} = -\rho g \equiv -\gamma \qquad (3.6)$$

Restrictions: (1) Static fluid
                  (2) Gravity is the only body force
                  (3) The z axis is vertical

This equation is the basic pressure-height relation of fluid statics. It is subject to the restrictions noted. Therefore it must be applied only where these restrictions are reasonable for the physical situation. To determine the pressure distribution in a static fluid, Eq. 3.6 may be integrated and appropriate boundary conditions applied.

Before considering specific cases that are readily treated analytically, it is important to note that pressure values must be stated with respect to a reference level. If the reference level is a vacuum, pressures are termed *absolute*, as shown in Fig. 3.2.

Most pressure gages read a pressure *difference*—the difference between the measured pressure and the ambient level (usually atmospheric pressure). Pressure levels measured with respect to atmospheric pressure are termed *gage* pressures. Thus

$$p_{\text{absolute}} = p_{\text{gage}} + p_{\text{atmosphere}}$$

Absolute pressures must be used in all calculations with the ideal gas or other equations of state.

3-2 PRESSURE VARIATION IN A STATIC FLUID

We have seen that pressure variation in any static fluid is described by the basic pressure-height relation

$$\frac{dp}{dz} = -\rho g \qquad (3.6)$$

Although ρg may be defined as the *specific weight*, γ, it has been written as ρg in Eq. 3.6 to emphasize that *both* ρ and g must be considered variables. In order to integrate Eq. 3.6 to find the pressure distribution, assumptions must be made about variations in both ρ and g.

For most practical engineering situations, the variation in g is negligible. Only for a situation such as computing very precisely the pressure change over a large elevation difference would the variation in g need to be included. Unless stated otherwise, we shall assume g to be constant with elevation at any given location.

In many practical engineering problems the variation in ρ will be appreciable, and accurate results will require that it be accounted for. Several types of variation are easy to treat analytically.

Pressure variation in a compressible fluid can be evaluated by integrating Eq. 3.6. Before this can be done, density must be expressed as a function of one of the other variables in the equation. Property information or an equation of state may be used to obtain the required relation for density.

The density of gases generally depends on pressure and temperature. The ideal gas equation of state

$$p = \rho R T \qquad (1.1)$$

where R is the gas constant (see Appendix A) and T the absolute temperature, accurately models the behavior of most gases under engineering conditions. However, the use of Eq. 1.1 introduces the gas temperature as an additional variable. Therefore, an additional assumption must be made about temperature variation before Eq. 3.6 can be integrated.

For an incompressible fluid, $\rho =$ constant. Then for constant gravity,

$$\frac{dp}{dz} = -\rho g = \text{constant}$$

To determine the pressure variation, we must integrate and apply appropriate boundary conditions. If the pressure at the reference level, z_0, is designated as p_0, then the pressure, p, at location z is found by integration

$$\int_{p_0}^{p} dp = -\int_{z_0}^{z} \rho g \, dz$$

or

$$p - p_0 = -\rho g(z - z_0) = \rho g(z_0 - z)$$

For liquids, it is often convenient to take the origin of the coordinate system at the free surface (reference level) and to measure distances as positive downward from the free surface. With h measured positive downward, then

$$z_0 - z = h$$

and

$$p - p_0 = \rho g h \qquad (3.7)$$

Equation 3.7 indicates that the pressure difference between two points in a static fluid can be determined by measuring the elevation difference between the two points. Devices used for this purpose are called *manometers*.

Atmospheric pressure may be obtained from a *barometer*, in which the height of a mercury column is measured. The measured height may be converted to engineering units using Eq. 3.7 and the data for specific gravity of mercury given in Appendix A. Although the vapor pressure of mercury may be neglected (see Problem 3.8), for precise work, temperature and altitude corrections must be applied to the measured level and the effects of surface tension must be considered.

A simple U-tube manometer is shown in Fig. 3.3. Since the right leg is open to the atmosphere, measurements of h_1 and h_2 will allow the determination of the gage pressure at A. Using the notation of Fig. 3.3, and applying Eq. 3.7 between A and B and between B and C, gives

$$p_A - p_B = \rho_1 g(z_B - z_A) = -\rho_1 g h_1$$

and

$$p_B - p_C = \rho_2 g(z_C - z_B) = \rho_2 g h_2$$

Adding the two equations

$$p_A - p_C = \rho_2 g h_2 - \rho_1 g h_1$$

Since $p_C = p_{atm}$, then $p_A - p_C = p_{Agage}$.

If the density ρ_1 is negligible compared to ρ_2, then $p_{Agage} = \rho_2 g h_2$. Note that the pressures at B' and B are equal because they are at the same elevation in a continuous length of the same fluid.

Manometers are simple and inexpensive devices used frequently for pressure measurements. Students sometimes have trouble analyzing multiple tube manometer situations. The following rules of thumb are useful:

1. Any two points at the same elevation in a continuous length of the same liquid are at the same pressure.

2. Pressure increases as one goes *down* a liquid column (remember the pressure change on diving into a swimming pool).

Example Problem 3.1 illustrates the use of a multiple-liquid manometer for measuring a pressure difference. Because the liquid level change is small at low pressure differential, a U-tube manometer may be difficult to read accurately. The level change can be increased by changing the manometer design or by using two immiscible liquids of slightly different density. Analysis of a typical reservoir manometer design is illustrated in Example Problem 3.2.

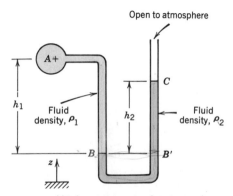

U-tube manometer for measuring gage pressure at A.

EXAMPLE 3.1—Multiple-Liquid Manometer

Water flows through pipes A and B. Oil, with specific gravity 0.8, is in the upper portion of the inverted U. Mercury (specific gravity 13.6) is in the bottom of the manometer bends. Determine the pressure difference, $p_A - p_B$, in units of lbf/in.^{2}

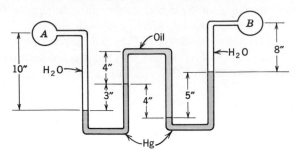

EXAMPLE PROBLEM 3.1

GIVEN: Multiple-tube manometer as shown. Specific gravity of oil is 0.8; specific gravity of mercury is 13.6.

FIND: The pressure difference, $p_A - p_B$, in lbf/in.^{2}

SOLUTION:

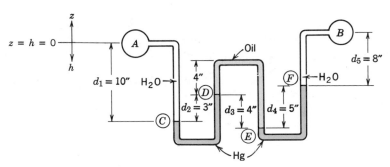

Basic equations:
$$\frac{dp}{dz} = -\frac{dp}{dh} = -\rho g \qquad SG = \frac{\rho}{\rho_{H_2O}}$$

Assumptions: (1) Static fluid
(2) Incompressible

Then
$$dp = \rho g \; dh \qquad \text{and} \qquad \int_{p_1}^{p_2} dp = \int_{h_1}^{h_2} \rho g \; dh$$

For $\rho = $ constant
$$p_2 - p_1 = \rho g (h_2 - h_1)$$

Beginning at point A and applying the equation between successive points around the manometer gives

$$p_C - p_A = +\rho_{H_2O} g \, d_1$$
$$p_D - p_C = -\rho_{Hg} g \, d_2$$

$$p_E - p_D = +\rho_{\text{oil}} g\, d_3$$
$$p_F - p_E = -\rho_{\text{Hg}} g\, d_4$$
$$p_B - p_F = -\rho_{\text{H}_2\text{O}} g\, d_5$$

Adding, we obtain

$$p_A - p_B = (p_A - p_C) + (p_C - p_D) + (p_D - p_E) + (p_E - p_F) + (p_F - p_B)$$
$$= -\rho_{\text{H}_2\text{O}} g\, d_1 + \rho_{\text{Hg}} g\, d_2 - \rho_{\text{oil}} g\, d_3 + \rho_{\text{Hg}} g\, d_4 + \rho_{\text{H}_2\text{O}} g\, d_5$$

Substituting $\rho = \text{SG}\,\rho_{\text{H}_2\text{O}}$ yields

$$p_A - p_B = g(-\rho_{\text{H}_2\text{O}}d_1 + 13.6\rho_{\text{H}_2\text{O}}d_2 - 0.8\rho_{\text{H}_2\text{O}}d_3 + 13.6\rho_{\text{H}_2\text{O}}d_4 + \rho_{\text{H}_2\text{O}}d_5)$$
$$= g\,\rho_{\text{H}_2\text{O}}(-d_1 + 13.6d_2 - 0.8d_3 + 13.6d_4 + d_5)$$
$$= g\,\rho_{\text{H}_2\text{O}}(-10 + 40.8 - 3.2 + 68 + 8) \text{ in.}$$
$$= g\,\rho_{\text{H}_2\text{O}} \times 103.6 \text{ in.}$$
$$= \frac{32.2 \text{ ft}}{\text{sec}^2} \times \frac{1.94 \text{ slug}}{\text{ft}^3} \times 103.6 \text{ in.} \times \frac{\text{ft}}{12 \text{ in.}} \times \frac{\text{ft}^2}{144 \text{ in.}^2} \times \frac{\text{lbf} \cdot \text{sec}^2}{\text{slug} \cdot \text{ft}}$$

$$p_A - p_B = 3.74 \text{ lbf/in.}^2 \qquad\qquad\qquad\qquad\qquad p_A - p_B$$

EXAMPLE 3.2—Reservoir Manometer

A reservoir manometer is built with a tube diameter of 10 mm and a reservoir diameter of 30 mm. The manometer liquid is Meriam red oil. Determine the manometer sensitivity, i.e., the deflection in millimeters per millimeter of water applied pressure differential.

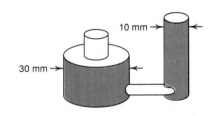

EXAMPLE PROBLEM 3.2

GIVEN: Reservoir manometer as shown.

$d = 10$ mm
$D = 30$ mm

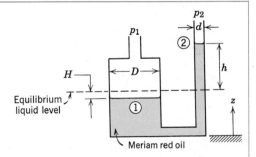

FIND: Liquid deflection, h, in millimeters per millimeter of water applied pressure differential.

SOLUTION:

Basic equations: $\qquad \dfrac{dp}{dz} = -\rho g \qquad \text{SG} = \dfrac{\rho}{\rho_{\text{H}_2\text{O}}}$

Assumptions: (1) Static fluid
(2) Incompressible

Then

$$dp = -\rho g \ dz \qquad \text{and} \qquad \int_{p_1}^{p_2} dp = -\int_{z_1}^{z_2} \rho g \ dz$$

For $\rho = $ constant,

$$p_2 - p_1 = -\rho g (z_2 - z_1)$$

or

$$p_1 - p_2 = \rho g (z_2 - z_1) = \rho_{oil} g (h + H)$$

To eliminate H, note that the *volume* of manometer liquid must remain constant. Thus the volume displaced from the reservoir must be the same as that which rises into the tube,

$$\frac{\pi}{4} D^2 H = \frac{\pi}{4} d^2 h \qquad \text{or} \qquad H = \left(\frac{d}{D}\right)^2 h$$

Substituting gives

$$p_1 - p_2 = \rho_{oil} g h \left[1 + \left(\frac{d}{D}\right)^2 \right]$$

This equation can be simplified by expressing the applied pressure differential as an equivalent water column of height Δh_e,

$$p_1 - p_2 = \rho_{H_2O} g \Delta h_e,$$

and noting that $\rho_{oil} = SG_{oil} \rho_{H_2O}$. Then

$$\rho_{H_2O} g \Delta h_e = SG_{oil} \rho_{H_2O} g h \left[1 + \left(\frac{d}{D}\right)^2 \right]$$

or

$$\frac{h}{\Delta h_e} = \frac{1}{SG_{oil}[1 + (d/D)^2]}$$

For Meriam red oil, $SG = 0.827$ (Table A.1). Thus, the sensitivity is

$$\frac{h}{\Delta h_e} = \frac{1}{0.827[1 + (10/30)^2]} = 1.09 \qquad \qquad \underleftarrow{\hspace{2cm}} \quad \frac{h}{\Delta h_e}$$

$\left\{ \begin{array}{l} \text{This problem illustrates the effects of manometer design and choice of gage liquid on} \\ \text{sensitivity.} \end{array} \right\}$

For many liquids, density is only a weak function of temperature. At modest pressures, liquids may be considered incompressible. However, at high pressures, compressibility effects in liquids can be important. Pressure and density changes in liquids are related by the *bulk compressibility modulus*, or modulus of elasticity,

$$E_v \equiv \frac{dp}{(d\rho/\rho)} \tag{3.8}$$

If the bulk modulus is assumed constant, then density is only a function of pressure (the fluid is *barotropic*) and Eq. 3.8 provides the additional density relation needed to integrate the basic pressure-height relation. Bulk modulus data for some common liquids are given in Appendix A.

3-3 THE STANDARD ATMOSPHERE

Several International Congresses for Aeronautics have been held so that aviation experts around the world might communicate better. The outcome of one such Congress was an internationally accepted definition of the Standard Atmosphere. The sea level conditions of the U.S. Standard Atmosphere are summarized in Table 3.1.

Table 3.1 Sea Level Conditions of the U.S. Standard Atmosphere

Property	Symbol	SI	English
Temperature	T	288 K	59 F
Pressure	p	101.3 kPa (abs)	14.696 psia
Density	ρ	1.225 kg/m^3	0.002377 slug/ft^3
Specific weight	γ	—	0.07651 lbf/ft^3
Viscosity	μ	1.781×10^{-5} kg/m · sec (Pa · sec)	3.719×10^{-7} lbf · sec/ft^2

The temperature profile of the U.S. Standard Atmosphere is shown in Fig. 3.4. Additional property values are tabulated as functions of elevation in Appendix A.

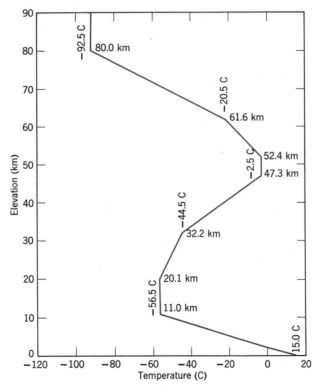

Fig. 3.4 Temperature variation with altitude in the U.S. Standard Atmosphere.

EXAMPLE 3.3—Pressure and Density Variation in the Atmosphere

The maximum power output capability of an internal combustion engine decreases with altitude because the air density and hence the mass flow rate of air decrease. A truck leaves Denver (elevation 5280 ft) on a day when the local temperature and barometric pressure are 80 F and 24.8 in. of mercury, respectively. It travels through

Vail Pass (elevation 10,600 ft) where the temperature is 62 F. Determine the local barometric pressure at Vail Pass and the percent change in density if the temperature is assumed to be a linear function of altitude.

EXAMPLE PROBLEM 3.3

GIVEN: Truck travels from Denver to Vail Pass.

$$
\begin{array}{ll}
\text{Denver: } z = 5{,}280 \text{ ft} & \text{Vail Pass: } \quad z = 10{,}600 \text{ ft} \\
\qquad\quad p = 24.8 \text{ in. Hg} & \qquad\qquad\quad T = 62 \text{ F} \\
\qquad\quad T = 80 \text{ F} &
\end{array}
$$

FIND: Atmospheric pressure at Vail Pass.
Percent change in air density between Denver and Vail.

SOLUTION:

Basic equations:
$$
\frac{dp}{dz} = -\rho g \qquad p = \rho R T
$$

Assumptions: (1) Static fluid
(2) Air behaves as an ideal gas
(3) Temperature varies linearly with altitude

Substituting into the basic pressure-height relation yields

$$
\frac{dp}{dz} = -\frac{p}{RT} g \qquad \text{or} \qquad \frac{dp}{p} = -\frac{g\,dz}{RT}
$$

But temperature varies linearly with elevation, so $T = T_0 + m(z - z_0)$. Thus

$$
\frac{dp}{p} = -\frac{g\,dz}{R[T_0 + m(z - z_0)]} = -\frac{g\,m\,d(z - z_0)}{mR[T_0 + m(z - z_0)]}
$$

Integrating from p_0 in Denver to p at Vail, we obtain

$$
\ln\left(\frac{p}{p_0}\right) = -\frac{g}{mR} \ln\left[\frac{T_0 + m(z - z_0)}{T_0}\right] = -\frac{g}{mR} \ln\left(\frac{T}{T_0}\right)
$$

or

$$
\frac{p}{p_0} = \left(\frac{T}{T_0}\right)^{-g/mR}
$$

Evaluating gives

$$
m = \frac{T - T_0}{z - z_0} = \frac{(62 - 80)\text{ F}}{(10.6 - 5.28)10^3 \text{ ft}} = -3.38 \times 10^{-3} \text{ F/ft}
$$

and

$$
\frac{g}{mR} = \frac{32.2}{\sec^2} \frac{\text{ft}}{} \times \frac{(-1)\text{ ft}}{3.38 \times 10^{-3}\,\text{F}} \times \frac{\text{lbm}\cdot\text{R}}{53.3\ \text{ft}\cdot\text{lbf}} \times \frac{\text{slug}}{32.2\ \text{lbm}} \times \frac{\text{lbf}\cdot\sec^2}{\text{slug}\cdot\text{ft}} = -5.55
$$

Thus

$$
\frac{p}{p_0} = \left(\frac{T}{T_0}\right)^{-g/mR} = \left(\frac{460 + 62}{460 + 80}\right)^{5.55} = (0.967)^{5.55} = 0.830
$$

and

$$
p = 0.830\,p_0 = (0.830)24.8 \text{ in. Hg} = 20.6 \text{ in. Hg} \qquad\qquad\qquad p
$$

$\Big\{$ Note that temperature must be expressed as an absolute temperature because it came from the ideal gas equation of state. $\Big\}$

The percent change in density is given by

$$\frac{\rho-\rho_0}{\rho_0}=\frac{\rho}{\rho_0}-1=\frac{p}{p_0}\frac{T_0}{T}-1=\frac{0.830}{0.967}-1=-0.142 \text{ or } -14.2\%$$

$$\overset{\leftarrow}{\frac{\Delta\rho}{\rho_0}}$$

$\left\{\begin{array}{l}\text{This problem is included to illustrate use of the ideal gas equation of state with the}\\ \text{basic pressure-height relation to evaluate the pressure distribution in the atmosphere.}\end{array}\right\}$

3-4 HYDRAULIC SYSTEMS

Hydraulic systems are characterized by very high pressures. As a consequence of these high system pressures, hydrostatic pressure variations often may be neglected. Automobile hydraulic brakes develop pressures up to 10 MPa (1500 psi); aircraft and machinery hydraulic actuation systems frequently are designed for pressures up to 30 MPa (4500 psi), and jacks use pressures to 70 MPa (10,000 psi). Special-purpose laboratory test equipment is commercially available for use at pressures to 1000 MPa (150,000 psi)!

Although liquids generally are considered incompressible at ordinary pressures, density changes may be appreciable at high pressures. Compressibility moduli of hydraulic fluids also may vary sharply at high pressures. In problems involving unsteady flow, both compressibility of the fluid and elasticity of the boundary structure must be considered. Analysis of problems such as noise and vibration in hydraulic systems, actuators, and shock absorbers quickly becomes complex and is beyond the scope of this book.

3-5 HYDROSTATIC FORCE ON SUBMERGED SURFACES

Now that we have determined the manner in which the pressure varies in a static fluid, we can examine the force on a surface submerged in a liquid.

In order to determine completely the force acting on a submerged surface, we must specify:

1. The magnitude of the force.
2. The direction of the force.
3. The line of action of the resultant force.

We shall consider both plane and curved submerged surfaces.

3-5.1 Hydrostatic Force on a Plane Submerged Surface

A plane submerged surface, on whose upper face we wish to determine the resultant hydrostatic force, is shown in Fig. 3.5. The coordinates have been chosen so that the surface lies in the xy plane.

Since there can be no shear stresses in a static fluid, the hydrostatic force on any element of the surface must act normal to the surface. The pressure force acting on an element of the upper surface, $d\vec{A}=dx\,dy\,\hat{k}$, is given by

$$d\vec{F}=-p\,d\vec{A} \tag{3.9}$$

The positive direction of the vector $d\vec{A}$ is the outward drawn normal to the area; the negative sign in Eq. 3.9 indicates that the force, $d\vec{F}$, acts *against* the surface in a direction opposite to that of $d\vec{A}$. The *resultant* force acting on the surface is found by summing the contributions of the infinitesimal forces over the entire area. Thus

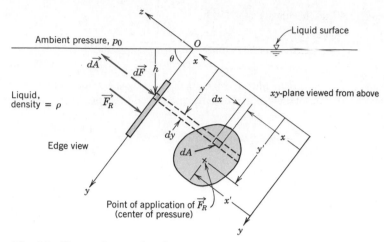

Fig. 3.5 Plane submerged surface.

$$\vec{F}_R = \int_A -p\, d\vec{A} \qquad (3.10)$$

In order to evaluate the integral in Eq. 3.10, both the pressure, p, and the element of area, $d\vec{A}$, must be expressed in terms of the same variables. The basic pressure-height relation for a static fluid can be written as

$$\frac{dp}{dh} = \rho g$$

where h is measured positive downward from the liquid free surface. Then, if the pressure at the free surface ($h = 0$) is p_0, we may integrate the pressure-height relation to obtain an expression for the pressure, p, at any depth, h. Thus, since $\rho =$ constant,

$$p = p_0 + \int_0^h \rho g\, dh = p_0 + \rho g h$$

This expression for p then can be substituted into Eq. 3.10. The geometry of the surface is expressed in terms of x and y; since the depth, h, is expressible in terms of y, i.e., $h = y \sin\theta$, the equation can be integrated to determine the resultant force.

The point of application of the resultant force must be such that the moment of the resultant force about any axis is equal to the moment of the distributed force about the same axis. If the position vector, from an arbitrary origin of coordinates to the point of application of the resultant force, is designated as $\vec{r}\,'$, then

$$\vec{r}\,' \times \vec{F}_R = \int \vec{r} \times d\vec{F} = -\int_A \vec{r} \times p\, d\vec{A} \qquad (3.11)$$

Referring to Fig. 3.5, we see that $\vec{r}\,' = \hat{i}x' + \hat{j}y'$, $\vec{r} = \hat{i}x + \hat{j}y$, and $d\vec{A} = dA\,\hat{k}$. Since the resultant force, $\vec{F}_R$, acts against the surface (in a direction opposite to that of $d\vec{A}$), then $\vec{F}_R = -F_R\hat{k}$. Substituting into Eq. 3.11 gives

$$(\hat{i}x' + \hat{j}y') \times -F_R\hat{k} = \int (\hat{i}x + \hat{j}y) \times d\vec{F} = -\int_A (\hat{i}x + \hat{j}y) \times p\, dA\,\hat{k}$$

Evaluating the cross product, we obtain

$$\hat{j}x'F_R - \hat{i}y'F_R = \int_A (\hat{j}xp - \hat{i}yp)dA$$

This is a vector equation, so the components must be equal. Thus

$$y'F_R = \int_A yp\,dA \quad \text{and} \quad x'F_R = \int_A xp\,dA \qquad (3.12)$$

where x' and y' are the coordinates of the point of application of the resultant force. Note that Eqs. 3.10 and 3.11 can be used to determine the magnitude of the resultant force and its point of application on any plane submerged surface. They do not require that the density be a constant or that the free surface of the liquid be at atmospheric pressure.

Equations 3.10 and 3.12 are mathematical statements of basic principles familiar to you from previous courses in physics and statics:

1. The resultant force is the sum of the infinitesimal forces (Eq. 3.10).
2. The moment of the resultant force about any axis is equal to the moment of the distributed force about the same axis (Eq. 3.12).

In evaluating the hydrostatic force acting on a plane submerged surface, we have used vector notation to emphasize that forces and moments are vector quantities. Since all elements of the force are parallel, the use of vectors is not essential. Summarizing:

1. The magnitude of $\vec{F}_R$ is given by

$$F_R = |\vec{F}_R| = \int p\,dA$$

2. The direction of $\vec{F}_R$ is normal to the surface.
3. For a surface in the xy plane, the line of action of $\vec{F}_R$ passes through the point x', y' (the center of pressure), where

$$y'F_R = \int_A yp\,dA \quad \text{and} \quad x'F_R = \int_A xp\,dA$$

◄EXAMPLE 3.4—Computing Equations for Pressure Force and Point of Application on Plane Submerged Surface

Consider a plane submerged surface with free surface at atmospheric pressure. Using the notation of Fig. 3.5, (a) show that the hydrostatic force on the upper face of any plane submerged surface is equal to the pressure at the centroid times the area of the surface and (b) derive expressions for the coordinates of the center of pressure in terms of the geometric parameters of the surface.

EXAMPLE PROBLEM 3.4

GIVEN: Plane submerged surface as shown, with centroid of area at x_c, y_c. Free surface at ambient pressure (zero gage pressure).

FIND: (a) Show that $F_R = p_c A$.
(b) Determine expressions for coordinates of center of pressure.

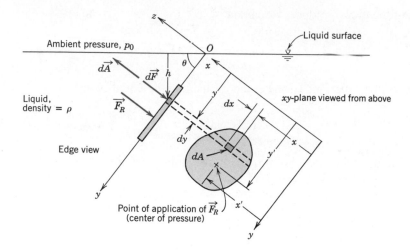

SOLUTION:

Basic equations:
$$F_R = \int p\, dA \qquad \frac{dp}{dh} = \rho g$$

For an incompressible fluid, integrating the pressure-height relation from the free surface $(h = 0, p = p_0)$ gives

$$p = p_0 + \rho g h$$

The force, F_R, is then

$$F_R = \int_A p\, dA = \int_A (p_0 + \rho g h) dA = \int_A (p_0 + \rho g y \sin\theta) dA$$

$$F_R = p_0 \int_A dA + \rho g \sin\theta \int_A y\, dA = p_0 A + \rho g \sin\theta \int_A y\, d\vec{A}$$

The integral is the first moment of the surface area about the x axis, which may be written

$$\int_A y\, dA = y_c A$$

where y_c is the y coordinate of the *centroid* of the area, A. Thus

$$F_R = p_0 A + \rho g \sin\theta y_c A = (p_0 + \rho g h_c) A = p_c A \qquad\qquad F_R$$

where p_c is the pressure in the liquid at the location of the centroid of area A. This result is valid for any pressure, p_0, at the free surface of the liquid. When p_0 is atmospheric pressure (zero gage pressure) and it acts on both sides of the surface, then p_0 makes no contribution to the *net* hydrostatic force, and it may be dropped.

To find expressions for the coordinates of the center of pressure we recognize that the moment of the resultant force about any axis must be equal to the moment of the distributed force about the same axis. Taking moments about the x axis gives

$$y' F_R = \int_A y p\, dA$$

Substituting $F_R = \rho g \sin\theta y_c A$, $p = \rho g h$, and $h = y \sin\theta$, we obtain

$$y' \rho g \sin\theta y_c A = \int_A y \rho g h\, dA = \int_A y^2 \rho g \sin\theta\, dA = \rho g \sin\theta \int_A y^2\, dA$$

Recognizing that $\int_A y^2\,dA = I_{xx}$, the second moment of the area about the x axis, we find that

$$y' = I_{xx}/Ay_c$$

From the parallel axis theorem, $I_{xx} = I_{\hat{x}\hat{x}} + Ay_c^2$, where $I_{\hat{x}\hat{x}}$ is the second moment of the area about the centroidal $\hat{x}$ axis,

$$y' = y_c + \frac{I_{\hat{x}\hat{x}}}{Ay_c} \qquad\qquad y'$$

Taking moments about the y axis gives $x'F_R = \int_A xp\,dA$.

Substituting for F_R, p, and h as above results in

$$x'\rho g \sin\theta\, y_c A = \int_A x\rho g h\,dA = \int_A xy\rho g \sin\theta\,dA = \rho g \sin\theta \int_A xy\,dA$$

Recognizing that $\int_A xy\,dA = I_{xy}$, the area product of inertia, we obtain

$$x' = I_{xy}/Ay_c$$

From the parallel axis theorem, $I_{xy} = I_{\hat{x}\hat{y}} + Ax_c y_c$, where $I_{\hat{x}\hat{y}}$ is the area product of inertia with respect to the centroidal $\hat{x}\hat{y}$ axis. Then

$$x' = x_c + \frac{I_{\hat{x}\hat{y}}}{Ay_c} \qquad\qquad x'$$

Note: The equations derived for x' and y' are valid only when the pressure at the free surface is atmospheric.

$\left\{\begin{array}{l}\text{This problem is included to illustrate the derivation of computing equations that would}\\\text{be convenient to use if a number of such problems were to be solved.}\end{array}\right\}$

EXAMPLE 3.5—Resultant Force on Inclined Plane Submerged Surface

The inclined surface shown, hinged along A, is 5 m wide. Determine the resultant force, $\vec{F}_R$, of the water on the inclined surface.

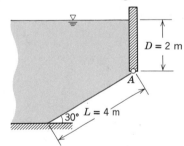

EXAMPLE PROBLEM 3.5

GIVEN: Rectangular gate, hinged along A, $w = 5$ m.

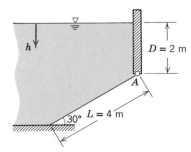
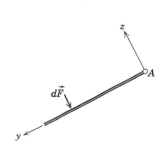

FIND: Resultant force, $\vec{F}_R$, of the water on the gate.

SOLUTION:
In order to completely determine $\vec{F}_R$, we must specify: (a) the magnitude, (b) the direction, and (c) the line of action, of the resultant force.

Basic equations:
$$\vec{F}_R = -\int p \, d\vec{A} \qquad \frac{dp}{dh} = \rho g$$

Consider the gate, hinged along A, lying in the xy plane, with coordinates as shown.

$$\vec{F}_R = -\int_A p \, d\vec{A} = -\int_A pw \, dy \, \hat{k} \qquad (d\vec{A} = w \, dy \, \hat{k})$$

We now need p as a function of y to perform the integration. From the basic pressure-height relation,

$$\frac{dp}{dh} = \rho g \qquad \text{so} \qquad dp = \rho g \, dh \qquad \text{and} \qquad \int_{p_a}^{p} dp = \int_0^h \rho g \, dh$$

By assuming $\rho = $ constant,

$$p = p_a + \rho g h \qquad \{\text{This gives } p = p(h). \text{We need } p = p(y).\}$$

From the diagram

$$h = D + y \sin 30° \qquad \text{where } D = 2 \text{ m}$$

Since we are interested in the resultant force from the water on the gate, then we drop p_a and obtain

$$p = \rho g (D + y \sin 30°)$$

Note the bottom face of the gate is open to the atmosphere and subject to p_a as well. Thus

$$\vec{F}_R = -\int_A p \, d\vec{A} = -\int_0^L \rho g (D + y \sin 30°) w \, dy \, \hat{k}$$

$$= -\rho g w \left[Dy + \frac{y^2}{2} \sin 30° \right]_0^L \hat{k} = -\rho g w \left[DL + \frac{L^2}{2} \sin 30° \right] \hat{k}$$

$$= \frac{-999 \text{ kg}}{\text{m}^3} \times \frac{9.81 \text{ m}}{\text{sec}^2} \times 5 \text{ m} \left[2 \text{ m} \times 4 \text{ m} + \frac{16 \text{ m}^2}{2} \times \frac{1}{2} \right] \frac{\text{N} \cdot \text{sec}^2}{\text{kg} \cdot \text{m}} \hat{k}$$

$$\vec{F}_R = -588 \hat{k} \text{ kN} \qquad\qquad \{\text{Force acts in negative } z \text{ direction}\} \; \vec{F}_R$$

To find the line of action of the resultant force, $\vec{F}_R$, we recognize that the line of action of the resultant force must be such that the moment of the resultant force about any axis must be equal to the moment of the distributed force about the same axis. Considering moments about the x axis through point A $(0, 0, 0)$, we obtain

$$F_R y' = \int_A yp \, dA$$

Then

$$y' = \frac{1}{F_R} \int_A yp \, dA = \frac{1}{F_R} \int_0^L ypw \, dy = \frac{\rho g w}{F_R} \int_0^L y(D + y \sin 30°) \, dy$$

$$= \frac{\rho g w}{F_R} \left[\frac{Dy^2}{2} + \frac{y^3}{3} \sin 30° \right]_0^L = \frac{\rho g w}{F_R} \left[\frac{DL^2}{2} + \frac{L^3}{3} \sin 30° \right]$$

$$= \frac{999 \text{ kg}}{\text{m}^3} \times \frac{9.81 \text{ m}}{\text{sec}^2} \times \frac{5 \text{ m}}{5.88 \times 10^5 \text{ N}} \left[\frac{2 \text{ m} \times 16 \text{ m}^2}{2} + \frac{64 \text{ m}^3}{3} \times \frac{1}{2} \right] \frac{\text{N} \cdot \text{sec}^2}{\text{kg} \cdot \text{m}}$$

$y' = 2.22$ m

Also, from consideration of moments about the y axis through point A,

$$x' = \frac{1}{F_R} \int_A xp \, dA$$

In calculating the moment of the distributed force (right side), recall, from your earlier courses in statics, that the centroid of the area element must be used for "x." Since the area element is of constant width, then $x = w/2$, and

$$x' = \frac{1}{F_R} \int_A \frac{w}{2} p \, dA = \frac{w}{2F_R} \int_A p \, dA = \frac{w}{2} = 2.5 \text{ m}$$

$$\vec{r}' = 2.5\hat{i} + 2.22\hat{j} \text{ m} \qquad \left\{ \begin{array}{l} \text{Line of action of } \vec{F}_R \text{ is along} \\ \text{negative } z \text{ axis through } \vec{r}' \end{array} \right\} \qquad \vec{r}'$$

⟵

$\left\{ \begin{array}{l} \text{This problem illustrates the procedure used to determine the resultant force, } \vec{F}_R\text{, equiv-} \\ \text{alent to a distributed force on a plane submerged surface.} \end{array} \right.$

EXAMPLE 3.6—Force on Vertical Plane Submerged Surface with Nonzero Gage Pressure at Free Surface

The door shown in the side of the tank is hinged along its bottom edge. A pressure of 100 psfg is applied to the liquid free surface. Find the force, F_t, required to keep the door closed.

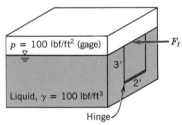

EXAMPLE PROBLEM 3.6

GIVEN: Door as shown in the figure; x axis is along the hinge.

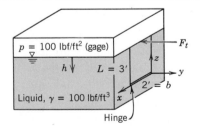

FIND: Force required to keep door shut.

SOLUTION:

Basic equations: $\qquad F_R = \int p \, dA \qquad \frac{dp}{dh} = \rho g \qquad \sum \vec{M} = 0$

Summing moments about the hinge axis yields

$$\sum M_x = 0 = LF_t - \int z \, dF = 0$$

$$\therefore F_t = \frac{1}{L} \int z \, dF = \frac{1}{L} \int z \, p \, dA = \frac{1}{L} \int_0^L z \, pb \, dz$$

To solve for F_t we need to know p as a function of z:

$$\frac{dp}{dh} = \rho g = \gamma \quad \text{and} \quad dp = \gamma \, dh$$

Then

$$p - p_0 = \int_{p_0}^{p} dp = \int_0^h \gamma \, dh \quad \text{so} \quad p = p_0 + \gamma h$$

Since atmospheric pressure acts on the outside of the door, the pressure p_0 in the above expression should be gage pressure. With $p = p_0 + \gamma h$ and $h = L - z$,

$$F_t = \frac{1}{L} \int_0^L z[p_0 + \gamma(L-z)]b \, dz = \frac{b}{L} \int_0^L p_0 z \, dz + \frac{\gamma b}{L} \int_0^L (Lz - z^2) \, dz$$

$$= \frac{p_0 b z^2}{2L} \bigg|_0^L + \frac{\gamma b}{L} \left[\frac{Lz^2}{2} - \frac{z^3}{3} \right]_0^L$$

$$= \frac{p_0 bL}{2} + \gamma bL^2 \left[\frac{1}{2} - \frac{1}{3} \right] = \frac{p_0 bL}{2} + \frac{\gamma bL^2}{6}$$

$$= \frac{100 \text{ lbf}}{\text{ft}^2} \times 2 \text{ ft} \times 3 \text{ ft} \times \frac{1}{2} + \frac{100 \text{ lbf}}{\text{ft}^3} \times 2 \text{ ft} \times 9 \text{ ft}^2 \times \frac{1}{6}$$

$$F_t = 600 \text{ lbf} \qquad\qquad\qquad\qquad\qquad F_t$$

This problem illustrates:
  (i) inclusion of a nonzero gage pressure at the free surface of the liquid.
  (ii) direct use of the distributed moment without evaluating the resultant force and line of application separately.

3-5.2 Hydrostatic Force on a Curved Submerged Surface

Determining the hydrostatic force on a curved submerged surface is slightly more involved than calculating the force on a plane surface. The hydrostatic force on an infinitesimal element of a curved surface, $d\vec{A}$, acts normal to the surface. However, the differential pressure force on each element of the surface acts in a different direction because of the surface curvature. Accounting for this change in direction makes the problem a little more involved.

What do we normally do when we wish to sum a series of force vectors acting in different directions? The usual procedure is to sum the components of the vectors relative to a convenient coordinate system.

Consider the curved surface shown in Fig. 3.6. The pressure force acting on the element of area, $d\vec{A}$, is given by

$$d\vec{F} = -p \, d\vec{A} \tag{3.9}$$

The resultant force is again given by

$$\vec{F}_R = -\int_A p \, d\vec{A} \tag{3.10}$$

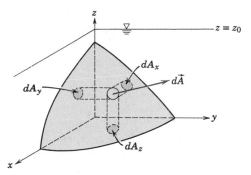

Fig. 3.6 Curved submerged surface.

We can write

$$\vec{F}_R = \hat{i} F_{R_x} + \hat{j} F_{R_y} + \hat{k} F_{R_z} \tag{3.13}$$

where F_{R_x}, F_{R_y}, and F_{R_z} are the components of $\vec{F}_R$ in the positive x, y, and z directions, respectively.

To evaluate the component of the force in a given direction, we take the dot product of the force with the unit vector in the given direction. For example, taking the dot product of each side of Eq. 3.10 with unit vector $\hat{i}$ gives

$$F_{R_x} = \int dF_x = \vec{F}_R \cdot \hat{i} = \int d\vec{F} \cdot \hat{i} = -\int_A p \, d\vec{A} \cdot \hat{i} = -\int_{A_x} p \, dA_x$$

where dA_x is the projection of $d\vec{A}$ on a plane perpendicular to the x axis (see Fig. 3.6).

Since, in any problem, the direction of the force component can be determined by inspection, the use of vectors is not necessary. In general, the magnitude of the component of the resultant force in the l direction is given by

$$F_{R_l} = \int_{A_l} p \, dA_l \tag{3.14}$$

where dA_l is the projection of the area element dA on a plane perpendicular to the l direction. The line of action of each component of the resultant force is found by recognizing that the moment of the resultant force component about a given axis must be equal to the moment of the corresponding distributed force component about the same axis.

The vertical component of the resultant hydrostatic force on a curved submerged surface is equal to the total weight of the liquid directly above the surface. This can be seen by taking the dot product of Eq. 3.9 with unit vector $\hat{k}$ to obtain

$$dF_z = -p \, dA_z$$

With the free surface at atmospheric pressure, then $p = \rho g h$, and

$$dF_z = -\rho g h \, dA_z = -\rho g \, d\forall$$

where $\rho g h \, dA_z = \rho g \, d\forall$ is the weight of a differential cylinder of liquid above the element of surface area, dA_z, extending a distance h from the curved surface to the free surface. The vertical component of the resultant force is obtained by integrating over the entire submerged surface. Thus

$$F_z = -\int_{A_z} \rho g h \, dA_z = -\int_{\forall} \rho g \, d\forall = -\rho g \forall$$

The minus sign indicates that a curved surface with a positive dA_z projection is subjected to a force in the negative z direction. The line of action of the vertical force component passes through the center of gravity of the volume of liquid between the submerged surface and the free surface of the liquid.

We have shown that the resultant hydrostatic force on a curved submerged surface is specified in terms of its components. To determine the components and their corresponding lines of action, we proceed for each component just as we did for plane submerged surfaces. When all three components are present, the lines of action of the components of the resultant force will not necessarily coincide; the complete resultant may not be expressed as a single force. In most problems, it is the components parallel and perpendicular to the liquid free surface that are of interest.

EXAMPLE 3.7—Force Components on a Curved Submerged Surface

The gate shown has constant width, $w = 5$ m. The equation of the surface is $x = y^2/a$, where $a = 4$ m. The depth of water to the right of the gate is $D = 4$ m. Find the horizontal and vertical components of the resultant force due to the water and the line of action of each.

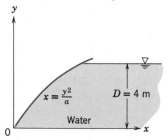

EXAMPLE PROBLEM 3.7

GIVEN: Gate of constant width, $w = 5$ m.
          Equation of surface in xy plane is $x = y^2/a$, where $a = 4$ m.
          Water stands at depth $D = 4$ m to the right of the gate.

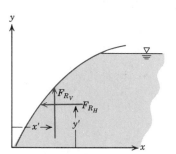

FIND: F_{R_H}, F_{R_V}, and line of action of each.

SOLUTION:

Basic equations: $\vec{F_R} = -\int p \, d\vec{A}$ $\dfrac{dp}{dh} = \rho g$

$$F_{R_H} = \int_0^D p w \, dy \qquad F_{R_V} = \int_0^{D^2/a} p w \, dx$$

In order to integrate, we need expressions for $p(y)$ and $p(x)$ along the surface of the gate.

$$\frac{dp}{dh} = \rho g, \quad \text{so} \quad dp = \rho g \, dh \quad \text{and} \quad \int_{p_a}^{p} dp = \int_0^h \rho g \, dh$$

If we assume ρ = constant, then

$$p = p_a + \rho g h$$

Since atmospheric pressure acts on both the top of the gate and the free surface of the liquid, there is no net contribution of the atmospheric pressure force. Thus, in determining the force due to the liquid, we take $p = \rho g h$.

We now need an expression for $h = h(y)$ and $h = h(x)$ along the surface of the gate. Along the surface of the gate, $h = D - y$. Since the equation of the gate surface is $x = y^2/a$, then along the gate $y = \sqrt{a}\,x^{1/2}$ and thus h can also be written as $h = D - \sqrt{a}\,x^{1/2}$. Substituting the appropriate equations for h into the expressions for F_{R_H} and F_{R_V} gives

$$F_{R_H} = \int_0^D pw\,dy = \int_0^D \rho g h w\,dy = \rho g w \int_0^D h\,dy = \rho g w \int_0^D (D - y)\,dy$$

$$= \rho g w \left[Dy - \frac{y^2}{2} \right]_0^D = \rho g w \left[D^2 - \frac{D^2}{2} \right] = \frac{\rho g w D^2}{2}$$

$$F_{R_H} = \frac{999 \text{ kg}}{\text{m}^3} \times \frac{9.81 \text{ m}}{\text{sec}^2} \times 5 \text{ m} \times \frac{(4)^2 \text{ m}^2}{2} \times \frac{N \cdot \text{sec}^2}{\text{kg} \cdot \text{m}} = 392 \text{ kN} \qquad \overset{F_{R_H}}{\longleftarrow}$$

$$F_{R_V} = \int_0^{D^2/a} pw\,dx = \int_0^{D^2/a} \rho g h w\,dx = \rho g w \int_0^{D^2/a} h\,dx$$

$$= \rho g w \int_0^{D^2/a} (D - \sqrt{a}\,x^{1/2})\,dx$$

$$= \rho g w \left[Dx - \frac{2}{3}\sqrt{a}\,x^{3/2} \right]_0^{D^2/a} = \rho g w \left[\frac{D^3}{a} - \frac{2}{3}\sqrt{a}\,\frac{D^3}{a^{3/2}} \right] = \frac{\rho g w D^3}{3a}$$

$$F_{R_V} = \frac{999 \text{ kg}}{\text{m}^3} \times \frac{9.81 \text{ m}}{\text{sec}^2} \times 5 \text{ m} \times \frac{(4)^3 \text{ m}^3}{3} \times \frac{1}{4 \text{ m}} \times \frac{N \cdot \text{sec}^2}{\text{kg} \cdot \text{m}} = 261 \text{ kN} \qquad \overset{F_{R_V}}{\longleftarrow}$$

To find the line of action of F_{R_H}, the moment of F_{R_H} about the z axis through O must be equal to the sum of the moments of dF_H about the same axis.

$$y'F_{R_H} = \int_{A_x} yp\,dA_x \qquad \text{and} \qquad y' = \frac{1}{F_{R_H}} \int_{A_x} yp\,dA_x$$

$$y' = \frac{1}{F_{R_H}} \int_0^D ypw\,dy = \frac{1}{F_{R_H}} \int_0^D y\rho g h w\,dy = \frac{\rho g w}{F_{R_H}} \int_0^D y(D - y)\,dy$$

$$= \frac{\rho g w}{F_{R_H}} \left[\frac{Dy^2}{2} - \frac{y^3}{3} \right]_0^D$$

$$y' = \frac{\rho g w D^3}{6 F_{R_H}} = \frac{\rho g w D^3}{6} \left[\frac{2}{\rho g w D^2} \right] = \frac{D}{3} = \frac{4 \text{ m}}{3} = 1.33 \text{ m} \qquad \overset{y'}{\longleftarrow}$$

To find the line of action of F_{R_V}, the moment of F_{R_V} about the z axis through O must be equal to the sum of the moments of dF_V about the same axis.

$$x'F_{R_V} = \int_{A_y} xp\,dA_y \qquad \text{and} \qquad x' = \frac{1}{F_{R_V}} \int_{A_y} xp\,dA_y$$

$$x' = \frac{1}{F_{R_V}} \int_0^{D^2/a} xpw\,dx = \frac{1}{F_{R_V}} \int_0^{D^2/a} x\rho g h w\,dx = \frac{\rho g w}{F_{R_V}} \int_0^{D^2/a} x(D - \sqrt{a}\,x^{1/2})\,dx$$

$$= \frac{\rho g w}{F_{R_V}} \left[\frac{D}{2} x^2 - \frac{2}{5} \sqrt{a} x^{5/2} \right]_0^{D^2/a} = \frac{\rho g w}{F_{R_V}} \left[\frac{D^5}{2a^2} - \frac{2}{5} \sqrt{a} \frac{D^5}{a^{5/2}} \right] = \frac{\rho g w D^5}{10 F_{R_V} a^2}$$

$$= \frac{\rho g w D^5}{10 a^2} \left[\frac{3a}{\rho g w D^3} \right]$$

$$x' = \frac{3 D^2}{10 a} = \frac{3}{10} \times \frac{(4)^2 \ \text{m}^2}{} \times \frac{1}{4 \ \text{m}} = 1.2 \ \text{m}$$

x'

{ This problem illustrates the calculation of resultant force components on a curved submerged surface. }

****3-6 BUOYANCY AND STABILITY**

If an object is immersed in, or floating on the surface of, a liquid, the net vertical force acting on it due to liquid pressure is termed *buoyancy*. Consider an object totally immersed in static liquid, as shown in Fig. 3.7.

The vertical force on the body due to hydrostatic pressure may be found most easily by considering cylindrical volume elements similar to the one shown in Fig. 3.7. For a static fluid

$$\frac{dp}{dh} = \rho g$$

Integrating for constant ρ gives

$$p = p_0 + \rho g h$$

The net vertical force on the element is

$$dF_z = (p_0 + \rho g h_2) \, dA - (p_0 + \rho g h_1) \, dA = \rho g (h_2 - h_1) \, dA$$

But $(h_2 - h_1) dA = d\mathbb{V}$, the volume of the element. Thus

$$F_z = \int dF_z = \int_{\mathbb{V}} \rho g \, d\mathbb{V} = \rho g \mathbb{V} \tag{3.15}$$

where $\mathbb{V}$ is the volume of the object. Thus the net vertical pressure force, or buoyancy, on the object, equals the force of gravity on the liquid displaced by the object. This relation reportedly was used by Archimedes in 220 B.C. to determine the gold content in the crown of King Hiero II. Consequently, it is often called "Archimedes' Principle." In more current technical applications, Eq. 3.15 is used to design displacement vessels, flotation gear, and bathyscaphes.

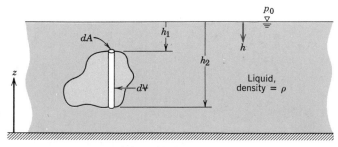

Fig. 3.7 Immersed body in static liquid.

** This section may be omitted without loss of continuity in the text material.

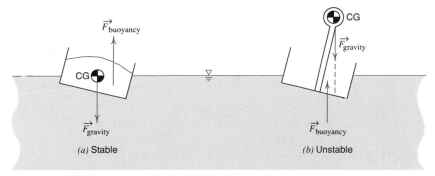

Fig. 3.8 Stability of floating bodies.

It should be emphasized that Eq. 3.15 predicts the net vertical force on a body that is totally submerged in a single fluid. In cases of partial immersion, one should consider the surface forces directly rather than attempting to deal with displaced volumes.

The line of action of the buoyancy force may be found using the methods of Section 3-5.2. Since floating bodies are in equilibrium under body and buoyancy forces, the location of the line of action of the buoyancy force determines stability, as shown in Fig. 3.8.

The body force due to gravity on an object acts through its center of gravity, CG. In Fig. 3.8a, the buoyant force is offset and produces a couple that tends to right the craft. In Fig. 3.8b, the couple tends to capsize the craft. In sailing, wind loads bring additional forces onto a boat; such additional forces must be considered in analyzing stability.

**3-7 FLUIDS IN RIGID-BODY MOTION

A fluid in rigid-body motion moves without deformation as though it were a solid body. Since there is no deformation, there can be no shear stress. Consequently, the only surface stress on each element of fluid is that due to pressure.

A fluid particle retains its identity in rigid-body motion because the fluid does not deform. As in the case of the static fluid, we may apply Newton's second law of motion to determine the pressure field that results from a specified rigid-body motion.

In Section 3-1 we derived an expression for the total force due to pressure and gravity acting on a fluid particle of volume $d\forall$. We obtained

$$d\vec{F} = (-\operatorname{grad}\, p + \rho\vec{g})d\forall$$

or

$$\frac{d\vec{F}}{d\forall} = -\operatorname{grad}\, p + \rho\vec{g} \tag{3.2}$$

Newton's second law was written

$$d\vec{F} = \vec{a}\, dm = \vec{a}\rho\, d\forall \quad \text{or} \quad \frac{d\vec{F}}{d\forall} = \rho\vec{a}$$

Substituting from Eq. 3.2, we obtain

$$-\operatorname{grad}\, p + \rho\vec{g} = \rho\vec{a} \tag{3.16}$$

** This section may be omitted without loss of continuity in the text material.

The physical significance of each term in this equation is

$$-\,\text{grad}\ p \qquad + \qquad \rho\vec{g} \qquad = \qquad \rho\vec{a}$$

$$\left\{\begin{array}{l}\text{net pressure force}\\\text{per unit volume}\\\text{at a point}\end{array}\right\} + \left\{\begin{array}{l}\text{body force}\\\text{per unit volume}\\\text{at a point}\end{array}\right\} = \left\{\begin{array}{l}\text{mass per}\\\text{unit}\\\text{volume}\end{array}\right\} \times \left\{\begin{array}{l}\text{acceleration}\\\text{of fluid}\\\text{particle}\end{array}\right\}$$

This vector equation consists of three component equations that must be satisfied individually. In rectangular coordinates the component equations are

$$\left.\begin{array}{ll} -\dfrac{\partial p}{\partial x} + \rho g_x = \rho a_x & x \text{ direction} \\[2mm] -\dfrac{\partial p}{\partial y} + \rho g_y = \rho a_y & y \text{ direction} \\[2mm] -\dfrac{\partial p}{\partial z} + \rho g_z = \rho a_z & z \text{ direction} \end{array}\right\} \qquad (3.17)$$

Component equations for other coordinate systems can be written using the appropriate expression for grad p. In cylindrical coordinates the vector operator, ∇, is given by

$$\nabla = \hat{e}_r \frac{\partial}{\partial r} + \hat{e}_\theta \frac{1}{r}\frac{\partial}{\partial \theta} + \hat{k}\frac{\partial}{\partial z} \qquad (3.18)$$

where $\hat{e}_r$ and $\hat{e}_\theta$ are unit vectors in the r and θ directions, respectively. Thus

$$\text{grad}\ p = \nabla p = \hat{e}_r \frac{\partial p}{\partial r} + \hat{e}_\theta \frac{1}{r}\frac{\partial p}{\partial \theta} + \hat{k}\frac{\partial p}{\partial z} \qquad (3.19)$$

EXAMPLE 3.8—Liquid in Rigid-Body Motion with Linear Acceleration

As a result of a promotion, you are transferred from your present location. You must transport a fish tank in the back of your station wagon. The tank is 12 in. × 24 in. × 12 in. How much water should you leave in the tank to be reasonably sure that it will not spill over during the trip?

EXAMPLE PROBLEM 3.8

GIVEN: Fish tank 12 in. × 24 in. × 12 in. partially filled with water to be transported in an automobile.

FIND: Allowable depth of water for reasonable assurance that it will not spill during the trip.

SOLUTION:

The first step in the solution is to formulate the problem by translating the general problem into a more specific one.

We recognize that there will be motion of the water surface as a result of the car's traveling over bumps in the road, going around corners, etc. However, we shall assume that the main effect on the water surface is due to linear accelerations (and decelerations) of the car; we shall neglect sloshing.

Thus we have reduced the problem to one of determining the effect of a linear acceleration on the free surface. We have not yet decided on the orientation of the tank relative to the direction of motion. Choosing the x coordinate in the direction of motion, should we align the tank with the long side parallel, or perpendicular, to the direction of motion?

If there will be no relative motion in the water, we must assume we are dealing with a constant acceleration, a_x. What is the shape of the free surface under these conditions?

Let us restate the problem to answer the original questions without making any restrictive assumptions at the outset.

GIVEN: Tank partially filled with water (to a depth d in.) subject to constant linear acceleration, a_x. Tank height is 12 in.; length parallel to direction of motion is b in. Width perpendicular to direction of motion is c in.

FIND: (a)   Shape of free surface under constant a_x.
          (b)   Allowable water height, d, to avoid spilling as a function of a_x and tank orientation.
          (c)   Optimum tank orientation and allowable depth.

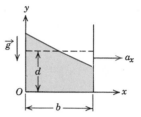

SOLUTION:

Basic equation: $$-\nabla p + \rho \vec{g} = \rho \vec{a}$$

$$-\left(\hat{i}\frac{\partial p}{\partial x} + \hat{j}\frac{\partial p}{\partial y} + \hat{k}\frac{\partial p}{\partial z}\right) + \rho(\hat{i}g_x + \hat{j}g_y + \hat{k}g_z) = \rho(\hat{i}a_x + \hat{j}a_y + \hat{k}a_z)$$

Since p is not a function of z, $\partial p/\partial z = 0$. Also, $g_x = 0$, $g_y = -g$, $g_z = 0$, and $a_y = a_z = 0$.

$$\therefore -\hat{i}\frac{\partial p}{\partial x} - \hat{j}\frac{\partial p}{\partial y} - \hat{j}\rho g = \hat{i}\rho a_x$$

The component equations are:

$$\frac{\partial p}{\partial x} = -\rho a_x$$
$$\frac{\partial p}{\partial y} = -\rho g$$

{ Recall that a partial derivative means that all other independent variables are held constant in the differentiation. }

The problem now is to find an expression for $p = p(x, y)$. This would enable us to find the equation of the free surface. But perhaps we do not have to do that.

Since the pressure is $p = p(x, y)$, then the difference in pressure between two points (x, y) and $(x + dx, y + dy)$ is

$$dp = \frac{\partial p}{\partial x}\,dx + \frac{\partial p}{\partial y}\,dy$$

Since the free surface is a line of constant pressure, then along the free surface $p = $ constant, so $dp = 0$ and

$$0 = \frac{\partial p}{\partial x}\,dx + \frac{\partial p}{\partial y}\,dy = -\rho a_x\,dx - \rho g\,dy$$

Therefore,

$$\left.\frac{dy}{dx}\right)_{\text{free surface}} = -\frac{a_x}{g} \longleftarrow \qquad\qquad \{\text{The free surface is a straight line.}\}$$

In the diagram,

d = original depth

e = height above the original depth

b = tank length parallel to direction of motion

$$e = \frac{b}{2}\tan\theta = \frac{b}{2}\left(-\frac{dy}{dx}\right)_{\text{free surface}} = \frac{b}{2}\frac{a_x}{g} \qquad \left\{\text{Only valid for } d \le \frac{b}{2}\right\}$$

Since we want e to be smallest for a given a_x, the tank should be aligned with b as small as possible. We should align the tank with the long side perpendicular to the direction of motion, that is, choose $b = 12$ in.

With $b = 12$ in.

$$e = 6\frac{a_x}{g} \text{ in.}$$

The maximum allowable value of $e = 12 - d$ in. Thus

$$12 - d = 6\frac{a_x}{g} \qquad \text{and} \qquad d_{\max} = 12 - 6\frac{a_x}{g}$$

If the maximum a_x is assumed to be $\frac{2}{3}g$, then allowable $d = 8$ in.

To allow a margin of safety, perhaps we should select $d = 6$ in.

Recall that a steady acceleration was assumed in this problem. The car would have to be driven *very* carefully.

$$\left\{\begin{array}{l} \text{This problem has been included to demonstrate:} \\ \text{(i) \quad not all problems are clearly defined, nor do they have unique answers, and} \\ \text{(ii) \quad the application of the equation, } -\nabla p + \rho\vec{g} = \rho\vec{a}. \end{array}\right\}$$

EXAMPLE 3.9—Liquid in Rigid-Body Motion with Constant Angular Velocity

A cylindrical container, partially filled with liquid, is rotated at a constant angular velocity, ω, about its axis as shown in the diagram. After a short time there is no relative motion; the liquid rotates with the cylinder as if the system were a rigid body. Determine the shape of the free surface.

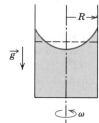

EXAMPLE PROBLEM 3.9

GIVEN: A cylinder of liquid in rigid-body rotation with angular velocity ω about its axis.

FIND: The shape of the free surface.

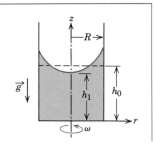

SOLUTION:

Basic equation:
$$-\nabla p + \rho \vec{g} = \rho \vec{a}$$

It is convenient to use a cylindrical coordinate system, r, θ, z. Since $g_r = g_\theta = 0$ and $g_z = -g$, then

$$-\left(\hat{e}_r \frac{\partial p}{\partial r} + \hat{e}_\theta \frac{1}{r}\frac{\partial p}{\partial \theta} + \hat{k}\frac{\partial p}{\partial z}\right) - \hat{k}\rho g = \rho(\hat{e}_r a_r + \hat{e}_\theta a_\theta + \hat{k}a_z)$$

Also $a_\theta = a_z = 0$ and $a_r = -\omega^2 r$.

$$\therefore -(\hat{e}_r \frac{\partial p}{\partial r} + \hat{e}_\theta \frac{1}{r}\frac{\partial p}{\partial \theta} + \hat{k}\frac{\partial p}{\partial z}) = -\hat{e}_r \rho \omega^2 r + \hat{k}\rho g$$

The component equations are: $\dfrac{\partial p}{\partial r} = \rho \omega^2 r$, $\dfrac{\partial p}{\partial \theta} = 0$, and $\dfrac{\partial p}{\partial z} = -\rho g$

From the component equations we see that the pressure is not a function of θ; it is a function of r and z only.

Since $p = p(r, z)$, the differential change, dp, in pressure between two points with coordinates (r, θ, z) and $(r + dr, \theta, z + dz)$ is given by

$$dp = \frac{\partial p}{\partial r}\bigg)_z dr + \frac{\partial p}{\partial z}\bigg)_r dz$$

Then

$$dp = \rho \omega^2 r \, dr - \rho g \, dz$$

To obtain the pressure difference between a reference point (r_1, z_1), where the pressure is p_1, and the arbitrary point (r, z), where the pressure is p, we must integrate

$$\int_{p_1}^{p} dp = \int_{r_1}^{r} \rho \omega^2 r \, dr - \int_{z_1}^{z} \rho g \, dz$$

$$p - p_1 = \frac{\rho \omega^2}{2}(r^2 - r_1^2) - \rho g(z - z_1)$$

Taking the reference point on the cylinder axis at the free surface gives

$$p_1 = p_{\text{atm}} \qquad r_1 = 0 \qquad z_1 = h_1$$

Then

$$p - p_{\text{atm}} = \frac{\rho \omega^2 r^2}{2} - \rho g(z - h_1)$$

Since the free surface is a surface of constant pressure ($p = p_{\text{atm}}$), the equation of the free surface is given by

$$0 = \frac{\rho \omega^2 r^2}{2} - \rho g(z - h_1)$$

or

$$z = h_1 + \frac{(\omega r)^2}{2g}$$

The equation of the free surface is a parabola with vertex on the axis at $z = h_1$.

We can solve for the height h_1 under conditions of rotation in terms of the original surface height, h_0, in the absence of rotation. To do this, we use the fact that the volume of liquid must remain constant. With no rotation

$$V = \pi R^2 h_0$$

With rotation

$$V = \int_0^R \int_0^z 2\pi r \, dz \, dr = \int_0^R 2\pi z r \, dr = \int_0^R 2\pi \left(h_1 + \frac{\omega^2 r^2}{2g} \right) r \, dr$$

$$V = 2\pi \left[h_1 \frac{r^2}{2} + \frac{\omega^2 r^4}{8g} \right]_0^R = \pi \left[h_1 R^2 + \frac{\omega^2 R^4}{4g} \right]$$

Then

$$\pi R^2 h_0 = \pi \left[h_1 R^2 + \frac{\omega^2 R^4}{4g} \right]$$

and

$$h_1 = h_0 - \frac{(\omega R)^2}{4g}$$

Finally,

$$z = h_0 - \frac{(\omega R)^2}{4g} + \frac{(\omega r)^2}{2g}$$

$$z = h_0 - \frac{(\omega R)^2}{2g} \left[\frac{1}{2} - \left(\frac{r}{R} \right)^2 \right] \qquad\qquad z(r)$$

$\left\{ \begin{array}{l} \text{This problem illustrates the application of Newton's second law in cylindrical coordi-} \\ \text{nates and the physical behavior of a liquid with a free surface undergoing solid-body} \\ \text{rotation.} \end{array} \right\}$

3-8 SUMMARY OBJECTIVES

After completing study of Chapter 3, you should be able to do the following:

1. Write the basic equation of fluid statics in vector form and indicate the physical significance of each term.
2. Write the basic pressure-height relation for a static fluid and integrate it to determine the pressure variation for any given fluid property variation.
3. State the relation between absolute and gage pressures.
4. Determine the pressure difference indicated by readings from a variety of manometers.
5. Define temperature and pressure conditions for the standard atmosphere.
6. For a plane submerged surface:
    (a) Determine the resultant force due to the fluid acting on the surface and its line of action.
    (b) Determine the external force(s) required to maintain the surface in equilibrium.
7. For a submerged surface with curvature in one plane:
    (a) Determine the components of the resultant force due to the fluid acting on the surface and their lines of action.
    (b) Determine the external force(s) required to maintain the surface in equilibrium.
**8. Determine the buoyancy force on a body immersed in, or floating on the surface of, a liquid; determine the stability of a floating object.

** This objective applies to a section that may be omitted without loss of continuity in the text material.

**9. Apply the basic hydrostatic equation to determine the pressure field and/or free surface shape for any body of fluid in rigid-body motion.

10. Solve the problems at the end of the chapter that relate to the material you have studied.

PROBLEMS

3.1 A device known as a deadweight tester can be used as a standard to calibrate mechanical pressure gages (the useful range is about 30 kPa to 35 MPa). Known pressures are generated by loading weights on a vertical piston-cylinder arrangement. The weighted piston is rotated to minimize frictional effects. The maximum convenient load is 100 kg. Determine an appropriate piston size to cover the pressure range given.

3.2 The pressure-height equation for a static incompressible fluid was integrated assuming that the gravitational acceleration, g, was constant. The Law of Gravitational Attraction is

$$g = g_0 \left(\frac{R}{R+h} \right)^2$$

where R is the radius of the Earth and h is altitude above the surface. Find the percent variation in g for the following two cases (take $R = 4000$ miles): (a) $h = 6$ mile altitude and (b) $h = -4$ mile altitude.

3.3 A pneumatic hoist is to be designed for a service station. Shop air is available at a gage pressure of 600 kPa. The hoist must lift automobiles of up to 3000 kg. Friction in the piston-cylinder mechanism and seals causes a force of 980 N opposing the piston motion. Determine the piston diameter necessary to provide the lift force. What pressure should be maintained in the lift cylinder to lower smoothly a car with a mass of 895 kg?

3.4 Pipe for the Alaskan pipeline has an internal diameter of 1.22 m. Wall thicknesses of 11 and 14 mm are used. Lengths of pipe were capped and tested hydrostatically to a pressure of 10 MPa. Calculate the maximum tensile stress in the pipe wall. Will the direction of the maximum stress in the pipe wall be axial or circumferential?

3.5 Compressed nitrogen is shipped in a cylindrical tank of diameter $D = 0.25$ m and length $L = 1.3$ m. The gas in the tank is at an absolute pressure of 20 MPa and a temperature of 20 C. Calculate the mass of gas in the tank. If the maximum allowable stress in the tank wall is 210 MPa, determine the theoretical minimum thickness of the cylinder wall.

3.6 A CO_2 cartridge for an air rifle is 60 mm long and has 16.5 mm inside diameter. The wall thickness is 0.5 mm. The label states it contains 12 g of CO_2. Estimate the maximum pressure inside a fully charged cartridge. Assuming a biaxial stress state in the cylinder wall, compute the maximum axial, circumferential, and shear stresses in the cylinder wall.

3.7 A mechanical pressure gage attached to the closed reservoir tank of an air compressor indicates a pressure of 827 kPa on a day when the barometer height is 750 mm of mercury. Calculate the absolute pressure in the tank. What pressure would the gage indicate if the barometer reading changed to 775 mm of mercury?

3.8 A closed container contains water at a 5 m depth. The absolute pressure above the water surface is 0.3 atm. Calculate the absolute pressure on the inside of the bottom surface of the container.

3.9 The vapor pressure of mercury is $p_v = 2.5 \times 10^{-5}$ psia at 70 F. Calculate the error in barometer height due to neglecting the vapor pressure of mercury. Would this be detectable for engineering calculations?

3.10 The actual column height of a mercury barometer is $h = 29.5$ in. when $T = 70$ F. Find the atmospheric pressure in lbf/ft². Express the barometer height in millimeters of mercury at 0 C.

3.11 Many recreation facilities use inflatable "bubble" structures. A tennis bubble to enclose four courts is shaped roughly as a circular semicylinder with a diameter of
30 m and a length of 60 m. The blowers used to inflate the structure can maintain
the air pressure inside the bubble at 10 mm of water above ambient pressure. The
fabric "skin" of the bubble is of uniform thickness. Determine the maximum material
density, in mass per unit area, that can be used to fabricate a pressure-supported bubble.

3.12 The tube shown is filled with mercury at 20 C. Calculate the force applied to the
piston.

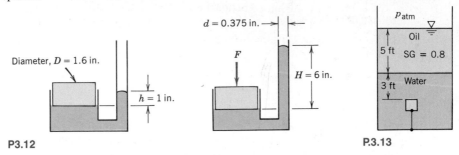

P3.12   P.3.13

3.13 A 1 ft cube of solid oak is held submerged by a tether as shown. Calculate the
actual force of the water on the bottom surface of the cube and the tension in the
tether.

3.14 A cube with 4 in. sides is submerged in a liquid and suspended by a cord so its top
is horizontal and 6 in. below the surface. The mass of the cube is $M = 0.569$ slug;
the tension in the cord is $T = 11.5$ lbf. Calculate the density and specific gravity of
the liquid.

3.15 A power plant exhaust stack is $H = 75$ m tall. The average gas temperature in the stack
is $T_g = 210$ C. Treat the stack gas as an ideal gas with the thermodynamic properties
of air. The pressure at the stack outlet is equal to the pressure of the surrounding
air at the same elevation. Assume standard sea-level air conditions outside the stack
base. Calculate the pressure in the stack gas at the bottom of the stack. Evaluate the
pressure difference (expressed in millimeters of water) between the air and the stack
gas at the base.

3.16 A hand-operated piston pump for a home well is shown. Water lifted by the piston
is carried to the pump discharge through a tube of 0.75 in. inside diameter. The
pump rod and piston weigh 30 lbf. The maximum comfortable pumping speed is 50
strokes per minute. The mechanical advantage of the pump handle is 7:1. Evaluate
the pressure on the piston when the entire discharge line is full of water. Calculate the

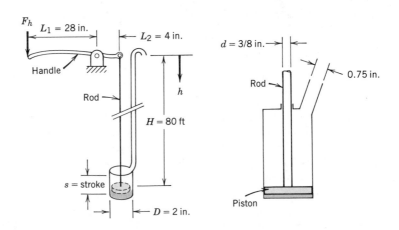

P3.16

force on the pump handle needed to begin moving water (neglect friction between the piston and housing). Neglecting leakage, determine the piston stroke required if a flow of 2.5 gpm is to be delivered by the pump.

3.17 On a certain calm day, a mild inversion causes the atmospheric temperature to remain constant at 30 C between sea level and 5 km altitude. Under these conditions, calculate the elevation change for which a 1 percent reduction in air pressure occurs.

3.18 Determine the change of elevation necessary to effect a 15 percent reduction in density for an isothermal atmosphere at 20 C.

3.19 Water usually is assumed to be incompressible when evaluating static pressure variations. Actually, it is about 100 times more compressible than steel. Assume the bulk modulus is constant. Compute the percent change in density for water raised to a gage pressure of 100 atm.

3.20 High-speed jets of water are used to cut concrete and other composite materials, e.g., for aircraft components. The maximum pressures are in the vicinity of 50,000 psi. Would you expect the assumption of constant density to be reasonable for engineering calculations?

3.21 Lubricating oil is used as the working fluid in a high-pressure hydraulic system. Estimate the percent change in oil density as its pressure is raised from ambient conditions to 300 atm (gage). Is constant density a reasonable model for the oil?

3.22 The Martian atmosphere behaves as an ideal gas with mean molecular mass of 32.0 and constant temperature of 200 K. The atmospheric density at the planet surface is $\rho = 0.015$ kg/m^3 and Martian gravity is 3.92 m/sec^2. Calculate the density of the Martian atmosphere at height $z = 20$ km above the surface.

3.23 At ground level in Denver, Colorado, the atmospheric pressure and temperature are 83.2 kPa and 25 C. Calculate the pressure on Pike's Peak at an elevation of 2690 m above the city assuming (a) an incompressible and (b) an adiabatic atmosphere.

3.24 An inverted cylindrical container is lowered slowly beneath the surface of a pool of water. Air trapped in the container is compressed isothermally as the hydrostatic pressure increases. Develop an expression for the water height, y, inside the container in terms of the container height, H, and depth of submersion, h.

3.25 Water usually is assumed to be incompressible when evaluating static pressure variations. Actually, its compressibility can be important in the design of submersible vehicles. Assume that the bulk modulus of water is constant. Compute the pressure and density at a depth of 4 miles in seawater. Express the density change as a percentage of the sea-level density.

3.26 Oceanographic research vessels have descended to 10 km below sea level. At these extreme depths, the compressibility of seawater can be significant. One may model the behavior of seawater by assuming that its bulk compressibility modulus remains constant. Using this assumption, evaluate the deviations in density and pressure compared to values computed using the incompressible assumption at a depth of 10 km in seawater. Express your answers in percent.

3.27 If air is assumed to be an ideal gas, knowledge of the temperature variation with altitude allows the determination of pressure at any elevation when conditions are known at a reference elevation, z_0.
(a) For $T = T_0(1+mz)$, derive the equation for the variation of pressure as a function of altitude if the pressure at the reference elevation is p_0.
(b) Using the results of part (a), show that the variation of pressure for the isothermal case ($m \to 0$) is given by

$$\frac{p}{p_0} = e^{-(g/RT_0)(z-z_0)}$$

3.28 Assuming the bulk modulus is constant for seawater, derive an expression for the density variation with depth, h, below the surface. Show that the result may be written

$$\rho \approx \rho_0 + bh$$

where ρ_0 is the density at the surface. Evaluate the constant b. Then using the approximation, obtain an equation for the variation of pressure with depth below the surface. Determine the percent error in pressure predicted by the approximate solution at a depth of 1000 m.

3.29 The pressure in a water main is measured using the dual-fluid manometer shown. Evaluate the gage pressure in the water main.

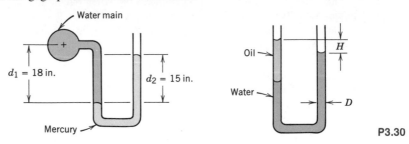

P3.29

P3.30

3.30 A manometer is formed from glass tubing with uniform inside diameter, $D = 6.35$ mm, as shown. The U-tube is partially filled with water. Then $\forall = 3.25$ cm^3 of Meriam red oil is added to the left side, as shown. Calculate the equilibrium height, H, when both legs of the U-tube are open to the atmosphere.

3.31 The manometer shown contains water and kerosine. With both tubes open to the atmosphere, the free-surface elevations differ by $H_0 = 20.0$ mm. Determine the elevation difference when a pressure of 98.0 Pa (gage) is applied to the right tube.

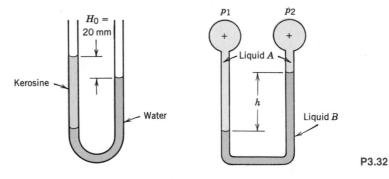

P3.31

P3.32

3.32 The manometer shown contains two liquids. Liquid A has SG = 0.88 and liquid B has SG = 2.95. Calculate the deflection, h, when the applied pressure difference is $p_1 - p_2 = 870$ Pa.

3.33 Consider the two-fluid manometer shown. Calculate the applied pressure difference.

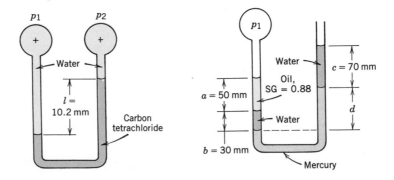

P3.33

P3.34

3.34 The manometer shown contains three liquids. When $p_1 = 10.0$ kPa (gage), determine the deflection distance d.

3.35 Determine the gage pressure in psig at point a, if liquid A has SG $= 0.75$ and liquid B has SG $= 1.20$. The liquid surrounding point a is water and the tank on the left is open to the atmosphere.

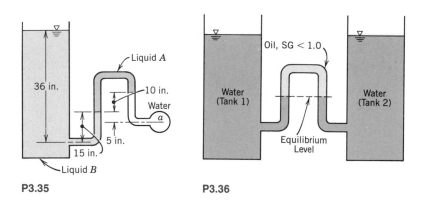

P3.35   P3.36

3.36 The NIH Corporation's engineering department is evaluating a sophisticated $80,000 laser system to measure the difference in water level between two large water storage tanks. It is important that small differences be measured accurately. You suggest that the job can be done with a $200 manometer arrangement. An oil less dense than water can be used to give a 10 : 1 amplification of meniscus movement; a small difference in level between the tanks will cause 10 times as much deflection in the oil levels in the manometer. Determine the specific gravity of the oil required for 10 : 1 amplification.

3.37 Consider a manometer connected as shown. Calculate the pressure difference.

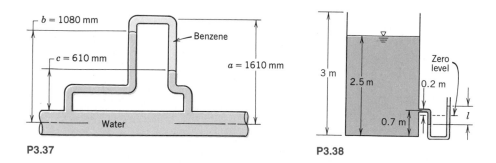

P3.37   P3.38

3.38 A rectangular tank, open to the atmosphere, is filled with water to a depth of 2.5 m as shown. A U-tube manometer is connected to the tank at a location 0.7 m above the tank bottom. If the zero level of the Meriam blue manometer fluid is 0.2 m below the connection, determine the deflection l after the manometer is connected and all the air has been removed from the connecting leg.

3.39 The manometer fluid of Problem 3.38 is replaced with mercury (same zero level). The tank is sealed and the air pressure is increased to a gage pressure of 0.5 atm. Determine the deflection l.

3.40 A reservoir manometer is calibrated for use with a liquid of specific gravity 0.827. The reservoir diameter is $\frac{5}{8}$ in. and the (vertical) tube diameter is $\frac{3}{16}$ in. Calculate the required distance between marks on the vertical scale for 1 in. of water pressure difference.

3.41 A student wishes to design a manometer with better sensitivity than a water-filled U-tube of constant diameter. The student's concept involves using tubes with different diameters and two liquids, as shown. Evaluate the deflection h of this manometer, if the applied pressure difference is $\Delta p = 250$ N/m^2. Determine the sensitivity of this manometer.

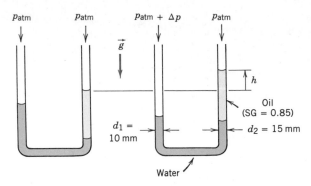

P3.41

3.42 For the inclined-tube reservoir manometer shown, obtain a general expression for the liquid deflection, L, in the inclined leg, in terms of the applied pressure difference, Δp. Also obtain a general expression for the sensitivity of the manometer.

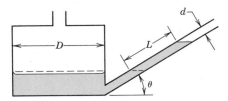

P3.42 through 3.48

3.43 Consider the inclined-tube reservoir manometer shown. Assume $\theta = 15.0°$, $D = 72.0$ mm, $d = 6.35$ mm, and the liquid is Meriam blue. The liquid deflection in the inclined tube is $L = 230$ mm when the manometer is connected to pressure differential Δp. Evaluate Δp. Determine the sensitivity of this manometer.

3.44 The inclined-tube reservoir manometer shown is used to measure pressures in a wind tunnel. The liquid is Meriam red oil. Assume $\theta = 15.0°$, $D = 19.1$ mm, and $d = 6.35$ mm. Calculate the pressure difference across the manometer when the liquid deflection along the inclined tube is $L = 32.7$ mm. Evaluate the sensitivity of this manometer.

3.45 The inclined-tube manometer shown has $D = 90$ mm and $d = 6$ mm; the liquid is Meriam red oil. The length of the measuring tube is 0.6 m; $\theta = 30°$. Determine the maximum pressure, in Pa, that can be measured with this manometer. Evaluate the sensitivity of the manometer.

3.46 The inclined-tube manometer shown has $D = 3$ in. and $d = 0.25$ in., and is filled with Meriam red oil. Compute the angle, θ, that will give a 5 in. oil deflection along the inclined tube for an applied pressure of 1 in. of water (gage). Determine the sensitivity of this manometer.

3.47 The inclined-tube manometer shown has $D = 96$ mm and $d = 8$ mm. Determine the angle, θ, required to provide a 5 : 1 increase in liquid deflection, L, compared to the total deflection in a regular U-tube manometer. Evaluate the sensitivity of this inclined-tube manometer.

3.48 Basic dimensions of an inclined-tube reservoir manometer in a fluid mechanics laboratory are shown. Assume: the area of the reservoir is $A_r = 277$ mm^2, $d = 3.64$ mm, and $\theta = 10.5°$. A gage pressure equivalent to 10.2 mm of water is applied to the reservoir; the inclined tube is open to atmosphere. Find the deflection distance, L, of the manometer liquid along the tube, if the liquid is Meriam red oil. Determine the sensitivity of this manometer.

3.49 If the tank of Problem 3.38 is sealed tightly and water drains slowly from the bottom of the tank, determine the deflection, l, after the system has attained equilibrium.

3.50 Surface tension causes the meniscus to rise in a water-filled manometer. This phenomenon of *capillary rise* becomes significant in small-diameter tubes. Develop an expression for the capillary rise of water in a vertical tube. Show that the result is

$$\Delta h = \frac{2\sigma \cos\theta}{\rho g R}$$

where R is the tube radius and θ is the contact angle. Evaluate and plot the results for water in tubes from 1 to 10 mm inside diameter.

3.51 Solve Problem 3.50 for the capillary depression of mercury in a tube. Evaluate and plot for tubes between 1 and 10 mm inside diameter.

3.52 Pressure variations that result from altitude changes can cause ear "popping" and discomfort to airplane passengers or those driving in the mountains. Each individual is affected differently, but one ear "pop" per 75 m of elevation change might be a reasonable average figure. Determine the pressure change, expressed in millimeters of water, that corresponds to this elevation difference on a standard day at an altitude of 2000 m.

3.53 Because the pressure falls, water boils at a lower temperature with increasing altitude. Consequently, cake mixes and boiled eggs, among other foods, must be cooked different lengths of time. Determine the boiling temperature of water at 1000 and 2000 m elevation on a standard day, and compare with the sea-level value.

3.54 A section of vertical wall is to be constructed from ready-mix concrete poured between forms. The wall is to be 3 m high, 0.25 m thick, and 5 m wide. Calculate the force exerted by the ready-mix concrete on each form. Determine the line of application of the force.

3.55 A door 1 m wide and 1.5 m high is located in a plane vertical wall of a water tank. The door is hinged along its upper edge, which is 1 m below the water surface. Atmospheric pressure acts on the outer surface of the door and at the water surface. Determine the magnitude and line of action of the total resultant force from all fluids acting on the door.

3.56 If, in Problem 3.55, the water surface gage pressure is raised to 0.3 atm, determine the magnitude and line of action of the total resultant force from all fluids acting on the door.

3.57 A door 1 m wide and 1.5 m high is located in a plane vertical wall of a water tank. The door is hinged along its upper edge, which is 1 m below the water surface. Atmospheric pressure acts on the outer surface of the door. If the pressure at the water surface is atmospheric, what force must be applied at the lower edge of the door in order to keep the door from opening?

3.58 If, in Problem 3.57, the gage pressure at the water surface is 0.5 atm, what force must be applied at the lower edge of the door in order to keep the door from opening?

3.59 A tank with a center partition has a small "door" 0.5 m wide by 1 m high at the bottom. This door is hinged along the top edge. The left side has 0.6 m of water and the right side contains 1 m of nitric acid (SG = 1.5). What force (magnitude and direction) is required at the lower edge of the door to hold it closed?

3.60 The door shown is 5 ft wide and 10 ft high. Find the resultant force from all fluids acting on the door.

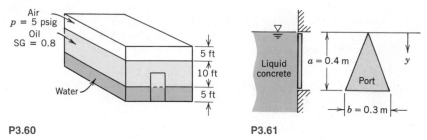

P3.60   P3.61

3.61 A triangular access port must be provided in the side of a form containing liquid concrete. Using the coordinates and dimensions shown, determine the resultant force that acts on the port and its point of application.

3.62 The circular access port in the side of a water standpipe has a diameter of 0.6 m and is held in place by eight bolts evenly spaced around the circumference. If the standpipe diameter is 7 m and the center of the port is located 12 m below the free surface of the water, determine (a) the total force on the port and (b) the appropriate bolt diameter.

3.63 The gate shown is hinged at H. The gate is 2 m wide normal to the plane of the diagram. Calculate the force required at A to hold the gate closed.

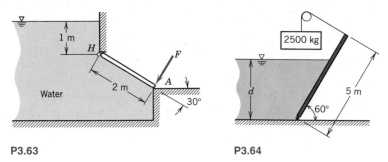

P3.63   P3.64

3.64 The gate shown is 3 m wide and for analysis can be considered massless. For what depth of water will this rectangular gate be in equilibrium as shown?

3.65 A plane gate is held in equilibrium by the uniformly distributed force per unit width, F, as shown. The gate weighs 600 lbf/ft of width and its center of gravity is 6 ft from the hinge at O. Find F when $D = 5$ ft and $\theta = 30°$.

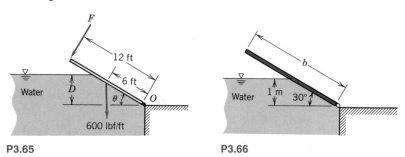

P3.65   P3.66

3.66 A gate of mass 2000 kg is mounted on a frictionless hinge along the lower edge. The length of the reservoir and gate (perpendicular to the plane of view) is 8 m. For the equilibrium conditions shown, compute the width, b, of the gate.

3.67 A plane gate of uniform thickness holds back a depth of water as shown. Find the minimum weight needed to keep the gate closed.

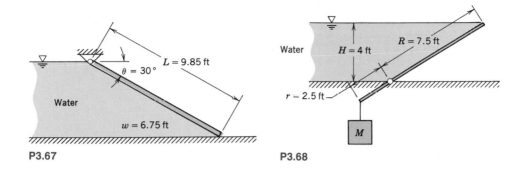

P3.67   **P3.68**

3.68 The water level is controlled by a plane gate of uniform thickness as shown. The width of the gate normal to the diagram is $w = 10$ ft. Determine the mass, M, needed to maintain the water level at depth H or less, if the mass of the gate is negligible.

3.69 The rectangular gate AB, as shown, is 2 m wide. Find the force per unit width exerted against the stop at A. Assume that the gate mass is negligible.

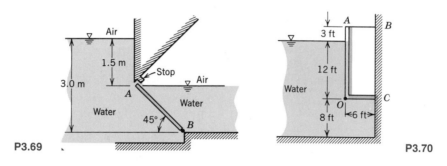

P3.69   **P3.70**

3.70 The gate AOC shown is 6 ft wide and is hinged along O. Neglecting the weight of the gate, determine the force in bar AB.

3.71 As water rises on the left side of the rectangular gate, the gate will open automatically. At what depth above the hinge will this occur? Neglect the mass of the gate.

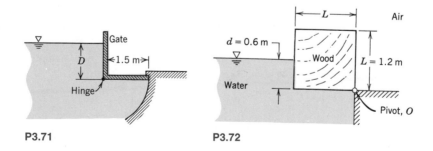

P3.71   **P3.72**

3.72 A long, square wooden block is pivoted along one edge. The block is in equilibrium when immersed in water to the depth shown. Evaluate the specific gravity of the wood, if friction in the pivot is negligible.

3.73 Consider a semicylindrical trough of radius R and length L. Develop general expressions for the magnitude and line of action of the hydrostatic force on one end, if the trough is full of water and open to atmosphere.

3.74 A submarine is 100 ft below the sea surface as shown. Find the net force, F, required to open the circular hatch when applied as shown. The pressure inside the submarine is equal to atmospheric pressure.

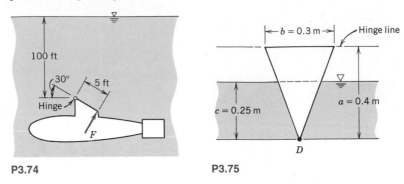

P3.74   P3.75

3.75 A window in the shape of an isosceles triangle and hinged at the top is placed in the vertical wall of a form that contains liquid concrete. Determine the minimum force that must be applied at point D to keep the window closed for the configuration of form and concrete shown.

3.76 Gates in the Poe Lock at Sault Ste. Marie, Michigan, close a channel $W = 110$ ft wide, $L = 1200$ ft long, and $D = 32$ ft deep. The geometry of one pair of gates is shown; each gate is hinged at the channel wall. When closed, the gate edges are forced together at the center of the channel by water pressure. Evaluate the force exerted by the water on gate A. Determine the magnitude and direction of the force components exerted by the gate on the hinge. (Neglect the weight of the gate.)

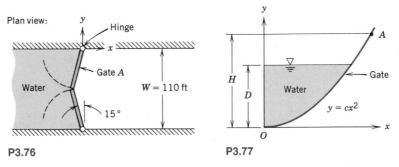

P3.76   P3.77

3.77 The parabolic gate shown is 2 m wide. Determine the magnitude and line of action of the vertical force on the gate due to the water; $c = 0.25$ m^{-1}, $D = 2$ m, and $H = 3$ m.

3.78 The gate shown is 1.5 m wide. Determine the magnitude and moment of the vertical component of the force about O; $a = 1.0$ m^{-2}, $D = 1.20$ m, and $H = 1.40$ m.

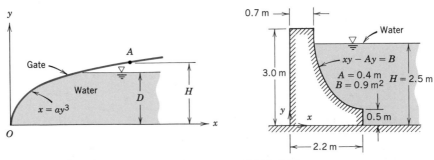

P3.78, 3.87, 3.88, 3.90   P3.79, 3.89

3.79 A dam is to be constructed across the Wabash River using the cross section shown. For water height $H = 2.5$ m, calculate the magnitude and line of action of the vertical force of water on the dam face. Assume the dam width is $w = 50$ m.

3.80 A spillway gate formed in the shape of a circular arc is w m wide. Find the magnitude and line of action of the vertical component of the force due to all fluids acting on the gate.

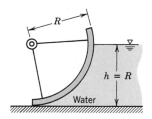

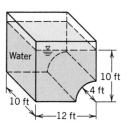

P3.80

P3.81

3.81 An open tank is filled with water to the depth indicated. Atmospheric pressure acts on all outer surfaces of the tank. Determine the magnitude and line of action of the vertical component of the force of the water on the curved part of the tank bottom.

3.82 Determine the magnitude and line of action of the vertical force on the curved section AB; section AB is 1 ft wide. Atmospheric pressure acts at the free surface; $k = 1.0$ ft^{-1}, $D = 4$ ft, and $H = 6$ ft.

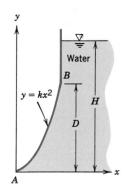

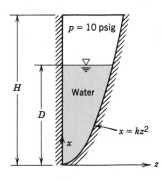

P3.82 A

P3.83, P3.84

3.83 The tank shown is 2 ft wide (perpendicular to the xz plane). It is filled with water to a depth of 8 ft. The air between the top of the tank and the water is pressurized to 10 psig. Determine the magnitude and the line of action of the vertical force on the curved portion of the tank; $k = 0.5$ ft^{-1}, $D = 8$ ft, and $H = 12$ ft.

3.84 If the water depth, D, in the tank of Problem 3.83 is reduced to 4 ft and the air pressure is maintained at 10 psig, determine the magnitude and line of action of the vertical force on the curved portion of the tank.

3.85 For the conditions of Problem 3.77, determine the horizontal force applied at A required to maintain the gate in equilibrium.

3.86 For the parabolic gate of Problem 3.77, assume no moment is applied at the origin, where the gate is hinged. Evaluate the vertical force that must be applied at point A to hold the gate in position.

3.87 For the conditions of Problem 3.78, determine the reaction at O required for equilibrium.

3.88 For the cubic gate of Problem 3.78, assume no moment is applied at the origin, where the gate is hinged. Evaluate the horizontal force that must be applied at point A to hold the gate in position.

3.89 Consider again the dam of Problem 3.79. Is it possible for water forces to overturn this dam? Under what circumstances?

3.90 If water stands at a depth of 0.5 m to the left of the gate of Problem 3.78, determine the total moment about O.

3.91 The depth of water to the right of the gate of Problem 3.77 is increased from zero to L m. Determine the depth, L, required to reduce the moment about O to 50 percent of the value for $L = 0$.

3.92 A gate, in the form of a quarter-cylinder, hinged at A and sealed at B, is 2 m wide. The bottom of the gate is 3 m below the water surface. Determine the force on the stop at B if the gate is made of concrete; $R = 2$ m and $D = 3$ m.

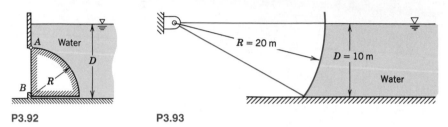

P3.92   P3.93

3.93 A Tainter gate used to control water flow from the Uniontown Dam on the Ohio River is shown; the gate width is $w = 35$ m. Determine the magnitude, direction, and line of action of the force from the water acting on the gate.

3.94 Ready-mix concrete is to be poured into the form shown. You are asked to make calculations on which to base the design of the structure needed to hold the cylindrical form in place (the form is $w = 6$ ft wide). Calculate the vertical force exerted by the concrete on the cylindrical portion of the form and find its line of action.

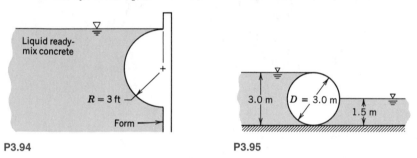

P3.94   P3.95

3.95 A cylindrical weir has a diameter of 3 m and a length of 6 m. Find the magnitude and direction of the resultant force acting on the weir from the water.

3.96 A cylindrical log of diameter D rests against the top of a dam. The water is level with the top of the log and the center of the log is level with the top of the dam. Obtain expressions for (a) the mass of the log per unit length and (b) the contact force per unit length between the log and dam.

3.97 A curved submerged surface, in the form of a quarter cylinder, with radius $R = 0.3$ m is shown. The form is filled to depth $H = 0.24$ m, with liquid concrete. The width is $w = 1.25$ m. Calculate the magnitude of the vertical hydrostatic force on the form from the concrete. Find the line of action of the force.

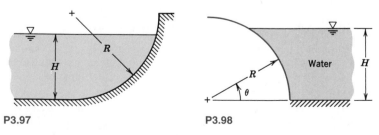

P3.97   P3.98

3.98 A curved surface is formed as a circular arc with $R = 0.750$ m as shown. The surface is $w = 3.55$ m wide. Water stands to the right of the curved surface to depth $H = 0.650$ m. Calculate the vertical hydrostatic force on the curved surface. Evaluate the line of action of this force. Find the magnitude and line of action of the horizontal force on the surface.

3.99 A canoe is represented by a right circular semicylinder, with $R = 0.35$ m and $L = 5.25$ m. The canoe floats in water that is $d = 0.245$ m deep. Set up a general algebraic expression for the maximum total mass (canoe and contents) that can be floated, as a function of depth. Evaluate for the given conditions.

‡**3.100** The cylinder shown is supported by an incompressible liquid of density ρ, and is hinged along its length. The cylinder, of mass M, length L, and radius R, is immersed in liquid to depth H. Obtain a general expression for the cylinder specific gravity versus the ratio of liquid depth to cylinder radius, $\alpha = H/R$, needed to hold the cylinder in equilibrium for $0 \leq \alpha < 1$. Plot the results.

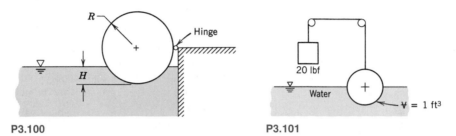

P3.100 P3.101

3.101 Find the specific weight of the sphere shown if its volume is 1 ft^3. State all assumptions. Is the weight necessary to float the sphere?

3.102 A hemispherical viewing port, of radius $R = 0.75$ m, is installed at depth $H = 2.5$ m, in the side of an aquarium filled with seawater, as shown. Evaluate the magnitudes of the vertical and horizontal forces of the water acting on the viewing port.

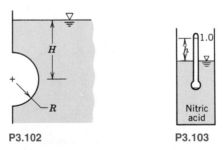

P3.102 P3.103

3.103 A hydrometer is a specific gravity indicator, the value being indicated by the level at which the free surface intersects the stem when floating in a liquid. The 1.0 mark is the level when in distilled water. For the unit shown, the immersed volume in distilled water is 15 cm^3. The stem is 6 mm in diameter. Find the distance, h, from the 1.0 mark to the surface when the hydrometer is placed in a nitric acid solution of specific gravity 1.5.

3.104 The fat-to-muscle ratio of a person may be determined from a specific gravity measurement. The measurement is made by immersing the body in a tank of water and measuring the net weight. Develop an expression for the specific gravity of a person in terms of their weight in air, net weight in water, and SG $= f(T)$ for water.

3.105 Quantify the statement, "Only the tip of an iceberg shows (in seawater)."

3.106 Hydrogen bubbles are used to visualize water flow streaklines in the film, *Flow Visualization*. A typical hydrogen bubble diameter is $d = 0.025$ mm. The bubbles

‡ You may wish to use simple computer programs to help solve problems marked with daggers.

tend to rise slowly in water because of buoyancy; eventually they reach terminal speed. The drag force of the water on a bubble is given by $F_D = 3\pi\mu V d$, where μ is the viscosity of water and V is the bubble speed relative to the water. Find the buoyancy force that acts on a hydrogen bubble immersed in water. Estimate the terminal speed of a bubble rising in water.

3.107 A manufacturer's catalog lists a buoyancy compensator, BC (similar to a life vest), for scuba diving. The BC claims a lift up to 40 lbf, obtained from an inflation cartridge that contains 25 g of carbon dioxide. Evaluate the manufacturer's claim if the BC mass is negligible. To what depth in seawater can the BC produce the lift claimed?

3.108 A modern supertanker has a tank capacity of a half-million metric tons of Arabian crude oil with SG = 0.86. The ship is essentially rectangular with a length of 400 m and a beam (width) of 65 m. The mass of the ship is approximately 230,000 metric tons. When the ship is unloaded, it is necessary to take on seawater ballast to maintain sufficient draft for stability and to keep the propeller submerged. A minimum draft of 20 m is required. Determine the maximum draft of the fully loaded tanker in seawater. Also determine what fraction of the tanks must be filled with seawater ballast when traveling unloaded.

3.109 Hot-air ballooning is a popular sport. According to a recent article, "hot-air volumes must be large because air heated to 150 F over ambient lifts only 0.018 lbf/ft^3 compared to 0.066 and 0.071 for helium and hydrogen, respectively." Check these statements for sea-level conditions. Calculate the effect of increasing the hot-air maximum temperature to 250 F above ambient.

3.110 One cubic foot of material weighing 67 lbf is allowed to sink in water as shown. A circular wooden rod 10 ft long and 3 in.^{2} in cross section is attached to the weight and also to the wall. If the rod weighs 3 lbf, what will be the angle, θ, for equilibrium?

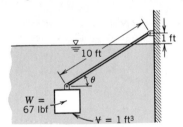

P3.110

3.111 A manufacturer's catalog of aluminum air tanks for diving claims that the tanks are neutrally buoyant when empty. A tank holds 50 standard cubic feet of air when filled to 3000 psig, is 6.9 in. outside diameter, has a wall thickness of 0.467 in., and is 19 in. long. Evaluate the claim.

3.112 A helium balloon is to lift a payload to an altitude of 40 km, where the atmospheric pressure and temperature are 3.0 mbar and -25 C, respectively. The balloon skin is polyester with specific gravity of 1.28 and thickness of 0.015 mm. To maintain a spherical shape, the balloon is pressurized to a gage pressure of 0.45 mbar. Determine the maximum balloon diameter if the allowable tensile stress in the skin is limited to 62 MN/m^2. What payload can be carried?

3.113 Scientific balloons operating at pressure equilibrium with the surroundings have been used to lift instrument packages to extremely high altitudes. One such balloon, constructed of polyester with a skin thickness of 0.013 mm, lifted a payload of 230 kg to an altitude of approximately 49 km, where atmospheric conditions are 0.95 mbar and -20 C. The helium gas in the balloon was at a temperature of approximately -10 C. The specific gravity of the skin material is 1.28. Determine the diameter and mass of the balloon. Assume that the balloon is spherical.

3.114 A sphere, of radius R, is partially immersed, to depth d, in a liquid of specific gravity SG. Obtain an algebraic expression for the buoyancy force acting on the sphere as a function of submersion depth d, for $0 \le d \le 2R$.

3.115 A sphere of radius R, made from material of specific gravity SG, is submerged in a tank of water. The sphere is placed over a hole, of radius a, in the tank bottom. Develop a general expression for the range of specific gravities for which the sphere will float to the surface. For the dimensions given, determine the minimum SG required for the sphere to remain in the position shown.

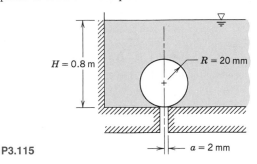

P3.115

3.116 A soda straw is made of plastic, with specific gravity SG = 1.1. The straw is 5 mm inside diameter and the thickness of the plastic is 0.4 mm. Its length is 250 mm. Experiments have shown that when placed in a glass of soft drink (SG = 1.055), the straw remains submerged. Estimate the external force required to support a straw submerged vertically in soft drink to a depth of 100 mm. Assume surface tension, σ, for the soft drink is similar to that of water.

3.117 A square oak timber, with sides W and length L, floats in seawater. Calculate the equilibrium depth d, at which the timber floats in calm water. Estimate the torque needed to hold the timber in a position rotated 15° clockwise from its undisturbed equilibrium position.

3.118 Consider a long object with square cross section (a m on a side) floating horizontally on the surface of a liquid. Assume that the object initially is one-fourth submerged. Considering small angular displacements only, develop an expression for the torque that tends to return the object to a horizontal position. Evaluate the range of positions of the symmetric center of gravity (CG) for which the object will remain stable.

3.119 A cylindrical timber, with $D = 0.3$ m and $L = 4$ m, is weighted on its lower end so that it floats vertically with 3 m submerged in seawater. When displaced vertically from its equilibrium position, the timber oscillates or "heaves" in a vertical direction upon release. Estimate the frequency of oscillation in this heave mode. Neglect viscous effects and water motion.

3.120 A cylindrical container, similar to that analyzed in Example Problem 3.9, is rotated at constant angular velocity about its axis. The cylinder is 1 ft in diameter, and initially contains water that is 4 in. deep. Determine the maximum rate at which the container can be rotated before the liquid free surface just touches the bottom of the tank. Does your answer depend on the density of the liquid? Explain.

3.121 A crude accelerometer can be made from a liquid-filled U-tube as shown. Derive an expression for the acceleration $\vec{a}$, in terms of liquid level difference h, tube geometry, and fluid properties.

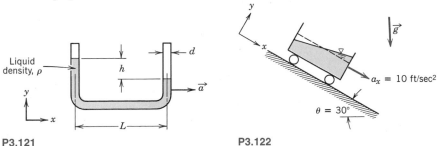

P3.121   P3.122

3.122 A rectangular container of water undergoes constant acceleration down an incline as shown. Determine the slope of the free surface using the coordinate system shown.

3.123 A runner training for a marathon works out on an indoor running track. He decides to carry a half cup of water to quench his thirst. Assume he runs at steady speed, $V = 5.3$ m/sec, around the turns of radius $R = 17.5$ m. Calculate the radial acceleration of the runner and the water. Estimate the average inclination angle of the water surface, making the direction clear.

3.124 A sealed chamber, which contains manometer oil (SG = 0.8), rotates about its axis with angular velocity ω. Derive an expression for the radial pressure gradient in the oil, $\partial p/\partial r$, in terms of radius r, and angular velocity ω.

3.125 A test tube is spun in a centrifuge. The tube support is mounted on a pivot so that the tube swings outward as rotation speed increases. At high speeds, the tube is nearly horizontal. Find (a) an expression for the radial component of acceleration of a liquid element located at radius r, (b) the radial pressure gradient $\partial p/\partial r$, and (c) the maximum pressure on the bottom of the test tube if it contains water. (The free surface and bottom radii are 50 and 130 mm, respectively.)

3.126 A centrifugal micromanometer can be used to create small and accurate differential pressures in air for precise measurement work. The device consists of a pair of parallel disks that rotate to develop a radial pressure difference. There is no flow between the disks. Obtain an expression for pressure difference in terms of rotation speed, radius, and air density. Evaluate the speed of rotation required to develop a differential pressure of 8 μm of water using a device with 50 mm radius.

3.127 The U-tube shown is filled with water at $T = 20$ C. It is sealed at A and open to the atmosphere at D. The tube is rotated about vertical axis AB. For the dimensions shown, compute the maximum angular speed if there is to be no cavitation.

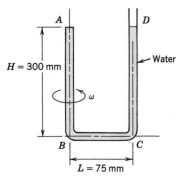

P3.127

3.128 A cubical box, 1 m on a side, half-filled with oil (SG = 0.80), is given a constant horizontal acceleration of $0.2g$. Determine the slope of the free surface and the pressure along the bottom of the box.

3.129 A rectangular container, of base dimensions 0.4 m $\times$ 0.2 m and height 0.4 m, is filled with water to a depth of 0.2 m; the mass of the empty container is 10 kg. The container is placed on a plane inclined at 30° to the horizontal. If the coefficient of sliding friction between the container and the plane is 0.3, determine the angle of the water surface relative to the horizontal.

3.130 If the container of Problem 3.129 slides without friction, determine the angle of the water surface relative to the horizontal. What is the slope of the free surface for the same acceleration up the plane?

3.131 A partially full can of soft drink is placed at the outer edge of a child's merry-go-round, located $R = 1.5$ m from the axis of rotation. The can diameter and height are $D = 65$ mm and $H = 120$ mm. The can is half-full of soda, with specific gravity SG = 1.06. Evaluate the slope of the liquid surface in the can if the merry-go-round spins at 0.3 revolution per second. Calculate the spin rate at which the can would spill, assuming no slippage between the can bottom and the merry-go-round. Would the can most likely spill or slide off the merry-go-round?

3.132 A pail, 1 ft in diameter and 1 ft deep, weighs 3 lbf and contains 8 in. of water. The pail is swung in a vertical circle of 3 ft radius at a speed of 15 ft/sec. Assume the water moves as a rigid body. At the instant when the pail is at the top of its trajectory, compute the tension in the string and the pressure on the bottom of the pail from the water.

3.133 Gas centrifuges are used in one process to produce enriched uranium for nuclear fuel rods. The maximum peripheral speed of a gas centrifuge is limited by stress considerations to about 300 m/sec. Assume a gas centrifuge containing uranium hexafluoride gas, with molecular mass $M_m = 352$, and ideal gas behavior. Develop an expression for the ratio of maximum pressure to pressure at the centrifuge axis. Evaluate the pressure ratio for a gas temperature of 325 C.

3.134 A rectangular container, of base dimensions 0.4 m × 0.2 m and height 0.5 m, is filled with water to a depth of 0.2 m; the mass of the empty container is 10 kg. The container is placed on a horizontal surface and is subjected to a constant horizontal force of 150 N. If the coefficient of sliding friction between the container and the surface is 0.25 and the tank is aligned with the short dimension along the direction of motion, determine (a) the force of the water on each end of the tank and (b) the force of the water on the bottom of the tank.

3.135 An automobile traveling at 90 km/hr rounds a long sweeping curve of radius 250 m. The air conditioner is on and the windows are rolled up so that the air within the car moves essentially as a rigid body. A child in the back seat holds the string of a balloon filled with helium. On a straight road the string is vertical, but in the curve it is not. Determine the magnitude and direction of the string angle as measured from the vertical.

3.136 Cast iron or steel molds are used in a horizontal-spindle machine to make tubular castings such as liners and tubes. A charge of molten metal is poured into the spinning mold. The radial acceleration permits nearly uniformly thick wall sections to form. A steel liner, of length $L = 2$ m, outer radius $r_o = 0.15$ m, and inner radius $r_i = 0.10$ m, is to be formed by this process. To attain nearly uniform thickness, the minimum radial acceleration should be $10g$. Determine (a) the required angular velocity and (b) the maximum and minimum pressures on the surface of the mold.

Chapter 4

BASIC EQUATIONS
IN INTEGRAL FORM
FOR A CONTROL VOLUME

We shall begin our study of fluids in motion by developing the basic equations in integral form for application to control volumes. Why the control volume formulation rather than the system formulation? There are two basic reasons. First, since fluid media are capable of continuous distortion and deformation, often it is extremely difficult to identify and follow the same mass of fluid at all times (as must be done to apply the system formulation). Second, we often are interested, not in the motion of a given mass of fluid, but rather in the effect of the fluid motion on some device or structure. Thus it is more convenient to apply the basic laws to a defined volume in space, using a control volume analysis.

The basic laws for a system should be familiar from your earlier studies in physics, mechanics, and thermodynamics. Our approach to developing the mathematical formulation of these laws for a control volume will be to develop a general formulation that will allow us to convert from a system analysis to a control volume analysis.

4-1 BASIC LAWS FOR A SYSTEM

The basic laws for a system are summarized briefly; for reasons that will become apparent in the next section, each of the basic equations for a system is written as a rate equation.

4-1.1 Conservation of Mass

Since a system is, by definition, an arbitrary collection of matter of fixed identity, a system is composed of the same quantity of matter at all times. The conservation of mass states that the mass, M, of the system is contant. On a rate basis, we have

$$\left. \frac{dM}{dt} \right)_{\text{system}} = 0 \tag{4.1a}$$

where

$$M_{\text{system}} = \int_{\text{mass(system)}} dm = \int_{\Psi(\text{system})} \rho \; d\Psi \tag{4.1b}$$

4-1.2 Newton's Second Law

For a system moving relative to an inertial reference frame, Newton's second law states that the sum of all external forces acting on the system is equal to the time rate of change of linear momentum of the system,

$$\vec{F} = \frac{d\vec{P}}{dt}\bigg)_{\text{system}} \tag{4.2a}$$

where the linear momentum, $\vec{P}$, of the system is given by

$$\vec{P}_{\text{system}} = \int_{\text{mass(system)}} \vec{V}\, dm = \int_{\Psi(\text{system})} \vec{V} \rho \, d\Psi \tag{4.2b}$$

4-1.3 The Angular Momentum Principle

The angular momentum principle for a system states that the rate of change of angular momentum is equal to the sum of all torques acting on the system,

$$\vec{T} = \frac{d\vec{H}}{dt}\bigg)_{\text{system}} \tag{4.3a}$$

where the angular momentum of the system is given by

$$\vec{H}_{\text{system}} = \int_{\text{mass(system)}} \vec{r} \times \vec{V}\, dm = \int_{\Psi(\text{system})} \vec{r} \times \vec{V} \rho \, d\Psi \tag{4.3b}$$

Torque can be produced by surface and body forces, and also by shafts that cross the system boundary,

$$\vec{T} = \vec{r} \times \vec{F}_s + \int_{\text{mass(system)}} \vec{r} \times \vec{g}\, dm + \vec{T}_{\text{shaft}} \tag{4.3c}$$

4-1.4 The First Law of Thermodynamics

The first law of thermodynamics is a statement of conservation of energy for a system,

$$\delta Q - \delta W = dE$$

In rate form the equation can be written as

$$\dot{Q} - \dot{W} = \frac{dE}{dt}\bigg)_{\text{system}} \tag{4.4a}$$

where the total energy of the system is given by

$$E_{\text{system}} = \int_{\text{mass(system)}} e\, dm = \int_{\Psi(\text{system})} e \rho \, d\Psi \tag{4.4b}$$

and

$$e = u + \frac{V^2}{2} + gz \tag{4.4c}$$

In Eq. 4.4a the rate of heat transfer, $\dot{Q}$, is positive when heat is added to the system from the surroundings; the rate of work, $\dot{W}$, is positive when work is done by the system on its surroundings. In Eq. 4.4c, u is the specific internal energy, V the speed, and z the height relative to a convenient datum of a particle of substance having mass dm.

4-1.5   The Second Law of Thermodynamics

If an amount of heat, δQ, is transferred to a system at temperature T, the second law of thermodynamics states that the change in entropy, dS, of the system is given by

$$dS \geq \frac{\delta Q}{T}$$

On a rate basis we can write

$$\left.\frac{dS}{dt}\right)_{\text{system}} \geq \frac{1}{T}\dot{Q} \tag{4.5a}$$

where the total entropy of the system is given by

$$S_{\text{system}} = \int_{\text{mass(system)}} s\, dm = \int_{\Psi(\text{system})} s\rho\, d\Psi \tag{4.5b}$$

4-2   RELATION OF SYSTEM DERIVATIVES TO THE CONTROL VOLUME FORMULATION

In the previous section we summarized the basic equations for a system. We found that when written on a rate basis, each equation involved the time derivative of an extensive property of the system (the total mass, momentum, angular momentum, energy, or entropy of the system). To develop the control volume formulation of each basic law from the system formulation, we shall use the symbol N to designate any arbitrary extensive property of the system. The corresponding intensive property (extensive property per unit mass) will be designated by η. Thus

$$N_{\text{system}} = \int_{\text{mass(system)}} \eta\, dm = \int_{\Psi(\text{system})} \eta\rho\, d\Psi \tag{4.6}$$

Comparing Eq. 4.6 with Eqs. 4.1b, 4.2b, 4.3b, 4.4b, and 4.5b, we see that if:

$$\begin{aligned} N &= M, & \text{then } \eta &= 1 \\ N &= \vec{P}, & \text{then } \eta &= \vec{V} \\ N &= \vec{H}, & \text{then } \eta &= \vec{r} \times \vec{V} \\ N &= E, & \text{then } \eta &= e \\ N &= S, & \text{then } \eta &= s \end{aligned}$$

The major task in going from the system to the control volume formulation of the basic laws is to express the rate of change of the arbitrary extensive property, N, for a system, in terms of variations of this property associated with a control volume. Since mass crosses the boundary of a control volume, time variations of the property N associated with the control volume involve the mass flux and the properties convected with it. A convenient way to account for mass flux is to use a limiting process involving a system and a control volume that coincide at a certain instant. Flux quantities in regions of overlap and regions surrounding the control volume are then formulated approximately, and the limiting process is applied to obtain exact results. The final equation relates the rate of change of the arbitrary extensive property, N, for a system to the time variations of this property associated with a control volume.

4-2.1 Derivation

The system and control volume to be used in the analysis are shown in Fig. 4.1. The flow field, $\vec{V}(x, y, z, t)$, is arbitrary relative to coordinates x, y, and z. The control volume is fixed in space relative to coordinate system xyz; by definition, the system always must consist of the same fluid particles, and consequently it must move with the flow field. In Fig. 4.1 the boundaries of the system are shown at two different instants, t_0 and $t_0 + \Delta t$. At t_0, the boundaries of the system and the control volume coincide; at $t_0 + \Delta t$, the system occupies regions II and III. The system has been chosen so that the mass within region I enters the control volume during the interval Δt, and the mass in region III leaves the control volume during the same interval.

Recall that our objective is to relate the rate of change of any arbitrary extensive property, N, of the system to the time variations of this property associated with the control volume. From the definition of a derivative, the rate of change of N_{system} is given by

$$\left. \frac{dN}{dt} \right)_{\text{system}} \equiv \lim_{\Delta t \to 0} \frac{N_s)_{t_0 + \Delta t} - N_s)_{t_0}}{\Delta t} \tag{4.7}$$

For convenience, subscript s has been used to denote the system in the definition of a derivative in Eq. 4.7.

At $t_0 + \Delta t$, the system occupies regions II and III; at t_0, the system and the control volume coincide. Thus,

$$N_s)_{t_0 + \Delta t} = (N_{\text{II}} + N_{\text{III}})_{t_0 + \Delta t} = (N_{\text{CV}} - N_{\text{I}} + N_{\text{III}})_{t_0 + \Delta t}$$

and

$$N_s)_{t_0} = (N_{\text{CV}})_{t_0}$$

Since

$$N_{\text{system}} = \int_{\text{mass(system)}} \eta \, dm = \int_{\Psi(\text{system})} \eta \rho \, d\Psi \tag{4.6}$$

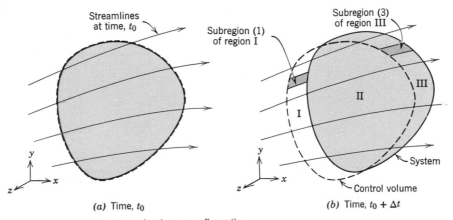

(a) Time, t_0 *(b)* Time, $t_0 + \Delta t$

Fig. 4.1 System and control volume configuration.

we can write

$$N_s)_{t_0 + \Delta t} = \left[\int_{CV} \eta \rho \, d\Psi\right]_{t_0 + \Delta t} - \left[\int_{I} \eta \rho \, d\Psi\right]_{t_0 + \Delta t} + \left[\int_{III} \eta \rho \, d\Psi\right]_{t_0 + \Delta t}$$

and

$$N_s)_{t_0} = (N_{CV})_{t_0} = \left[\int_{CV} \eta \rho \, d\Psi\right]_{t_0}$$

Substituting into the defintion of the system derivative, Eq. 4.7, we obtain

$$\frac{dN}{dt}\bigg]_s$$

$$= \lim_{\Delta t \to 0} \frac{\left[\int_{CV} \eta \rho \, d\Psi\right]_{t_0 + \Delta t} + \left[\int_{III} \eta \rho \, d\Psi\right]_{t_0 + \Delta t} - \left[\int_{I} \eta \rho \, d\Psi\right]_{t_0 + \Delta t} - \left[\int_{CV} \eta \rho \, d\Psi\right]_{t_0}}{\Delta t}$$

(4.8)

Since the limit of a sum is equal to the sum of the limits, we can write

$$\frac{dN}{dt}\bigg]_s = \underbrace{\lim_{\Delta t \to 0} \frac{\left[\int_{CV} \eta \rho \, d\Psi\right]_{t_0 + \Delta t} - \left[\int_{CV} \eta \rho \, d\Psi\right]_{t_0}}{\Delta t}}_{①}$$

$$+ \underbrace{\lim_{\Delta t \to 0} \frac{\left[\int_{III} \eta \rho \, d\Psi\right]_{t_0 + \Delta t}}{\Delta t}}_{②} - \underbrace{\lim_{\Delta t \to 0} \frac{\left[\int_{I} \eta \rho \, d\Psi\right]_{t_0 + \Delta t}}{\Delta t}}_{③} \qquad (4.9)$$

Our task now is to evaluate each of the three terms in Eq. 4.9.

Term ① in Eq. 4.9 simplifies to

$$\lim_{\Delta t \to 0} \frac{\left[\int_{CV} \eta \rho \, d\Psi\right]_{t_0 + \Delta t} - \left[\int_{CV} \eta \rho \, d\Psi\right]_{t_0}}{\Delta t} = \lim_{\Delta t \to 0} \frac{N_{CV})_{t_0 + \Delta t} - N_{CV})_{t_0}}{\Delta t}$$

$$= \frac{\partial N_{CV}}{\partial t} = \frac{\partial}{\partial t} \int_{CV} \eta \rho \, d\Psi$$

Term ② in Eq. 4.9 simplifies to

$$\lim_{\Delta t \to 0} \frac{\left[\int_{III} \eta \rho \, d\Psi\right]_{t_0 + \Delta t}}{\Delta t} = \lim_{\Delta t \to 0} \frac{N_{III})_{t_0 + \Delta t}}{\Delta t}$$

To evaluate $N_{III})_{t_0 + \Delta t}$, let us look at the enlarged view of a typical subregion of region III shown in Fig. 4.2. Vector $d\vec{A}$ has magnitude equal to the element of area, dA, of the control surface; the direction of $d\vec{A}$ is that of the normal drawn outward from the element of control surface area. Angle α is the angle between $d\vec{A}$ and the velocity vector, $\vec{V}$. Since the mass in region III is that which flows *out* of the control

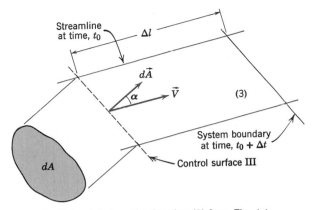

Fig. 4.2 Enlarged view of subregion (3) from Fig. 4.1.

volume during the interval Δt, angle α will always be less than $\pi/2$ over the entire area of the control surface bounding region III.

For subregion (3) we can write

$$dN_{\mathrm{III}})_{t_0+\Delta t} = (\eta\rho\, d\mathbf{V})_{t_0+\Delta t} = [\eta\rho(\Delta l\cos\alpha\, dA)]_{t_0+\Delta t}$$

since $d\mathbf{V} = \Delta l\cos\alpha\, dA$. Then for the entire region III,

$$N_{\mathrm{III}})_{t_0+\Delta t} = \left[\iint_{\mathrm{CS_{III}}} \eta\rho\,\Delta l\cos\alpha\, dA\right]_{t_0+\Delta t}$$

where $\mathrm{CS_{III}}$ is the surface common to region III and the control volume. In this expression, Δl is the distance traveled by a particle on the system surface during the interval Δt, along a streamline that existed at t_0.

Now that we have an expression for $N_{\mathrm{III}})_{t_0+\Delta t}$, we can evaluate term ② in Eq. 4.9:

$$\lim_{\Delta t\to 0}\frac{\left[\iint_{\mathrm{III}}\eta\rho\, d\mathbf{V}\right]_{t_0+\Delta t}}{\Delta t} = \lim_{\Delta t\to 0}\frac{N_{\mathrm{III}})_{t_0+\Delta t}}{\Delta t} = \lim_{\Delta t\to 0}\frac{\int_{\mathrm{CS_{III}}}\eta\rho\,\Delta l\cos\alpha\, dA}{\Delta t}$$

$$= \lim_{\Delta t\to 0}\int_{\mathrm{CS_{III}}}\eta\rho\frac{\Delta l}{\Delta t}\cos\alpha\, dA = \int_{\mathrm{CS_{III}}}\eta\rho|\vec{V}|\cos\alpha|d\vec{A}|$$

The last equality follows from the fact that

$$\lim_{\Delta t\to 0}\frac{\Delta l}{\Delta t} = |\vec{V}| \qquad \text{and} \qquad dA = |d\vec{A}|$$

Term ③ in Eq. 4.9 simplifies to

$$-\lim_{\Delta t\to 0}\frac{\left[\iint_{\mathrm{I}}\eta\rho\, d\mathbf{V}\right]_{t_0+\Delta t}}{\Delta t} = -\lim_{\Delta t\to 0}\frac{N_{\mathrm{I}})_{t_0+\Delta t}}{\Delta t}$$

To evaluate $N_{\mathrm{I}})_{t_0+\Delta t}$, look at the enlarged view of a typical subregion of region I shown in Fig. 4.3. Vector $d\vec{A}$ has magnitude equal to the area element, dA, of the control surface; the direction of $d\vec{A}$ is that of the outward drawn normal from the element of control surface area. Angle α is the angle between $d\vec{A}$ and the velocity vector, $\vec{V}$. Since the mass in region I flows *into* the control volume during the time

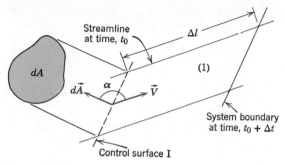

Fig. 4.3   Enlarged view of subregion (1) from Fig. 4.1.

interval Δt, angle α will always be greater than $\pi/2$ over the entire area of the control surface bounding region I.

For subregion (1) we can write

$$dN_{\mathrm{I}})_{t_0+\Delta t} = (\eta\rho\, d\mathbf{\Psi})_{t_0+\Delta t} = [\eta\rho\,\Delta l(-\cos\alpha)\,dA]_{t_0+\Delta t}$$

since $d\mathbf{\Psi} = \Delta l(-\cos\alpha)\,dA$. Why the minus sign? Recall that volume is a scalar quantity that must have a positive numerical value. Since $\alpha > \pi/2$, then $\cos\alpha$ will be negative.

Then, for the entire region I,

$$N_{\mathrm{I}})_{t_0+\Delta t} = \left[\iint_{\mathrm{CS_I}} -\eta\rho\,\Delta l\cos\alpha\, dA\right]_{t_0+\Delta t}$$

where $\mathrm{CS_I}$ is the surface common to region I and the control volume. In this expression, Δl is the distance traveled by a particle on the system surface during interval Δt, along a streamline that existed at t_0.

Now that we have an expression for $N_{\mathrm{I}})_{t_0+\Delta t}$, we can evaluate term ③ in Eq. 4.9:

$$-\lim_{\Delta t\to 0}\frac{\left[\displaystyle\int_{\mathrm{I}}\eta\rho\, d\mathbf{\Psi}\right]_{t_0+\Delta t}}{\Delta t} = -\lim_{\Delta t\to 0}\frac{N_{\mathrm{I}})_{t_0+\Delta t}}{\Delta t} = -\lim_{\Delta t\to 0}\frac{\displaystyle\int_{\mathrm{CS_I}} -\eta\rho\,\Delta l\cos\alpha\, dA}{\Delta t}$$

$$= \lim_{\Delta t\to 0}\int_{\mathrm{CS_I}} \eta\rho\frac{\Delta l}{\Delta t}\cos\alpha\, dA = \int_{\mathrm{CS_I}} \eta\rho|\vec{V}|\cos\alpha|d\vec{A}|$$

The last equality follows from the fact that

$$\lim_{\Delta t\to 0}\frac{\Delta l}{\Delta t} = |\vec{V}| \qquad \text{and} \qquad dA = |d\vec{A}|$$

Now that we have obtained expressions for each of the three terms on the right side, Eq. 4.9 can be written

$$\frac{dN}{dt}\bigg)_{\mathrm{system}} = \frac{\partial}{\partial t}\int_{\mathrm{CV}}\eta\rho\, d\mathbf{\Psi} + \int_{\mathrm{CS_I}}\eta\rho|\vec{V}|\cos\alpha|d\vec{A}| + \int_{\mathrm{CS_{III}}}\eta\rho|\vec{V}|\cos\alpha|d\vec{A}|$$

Referring to Fig. 4.1, we see that the entire control surface, CS, consists of three surfaces,

$$\mathrm{CS} = \mathrm{CS_I} + \mathrm{CS_{III}} + \mathrm{CS}_p$$

where CS_p is characterized by no flow across the surface, where either $\alpha = \pi/2$ or $\vec{V} = 0$.

Consequently, we can write

$$\left.\frac{dN}{dt}\right)_{system} = \frac{\partial}{\partial t}\int_{CV}\eta\rho\,d\Psi + \int_{CS}\eta\rho|\vec{V}|\cos\alpha|d\vec{A}| \tag{4.10}$$

Recognizing that $|\vec{V}|\cos\alpha|d\vec{A}| = \vec{V}\cdot d\vec{A}$, Eq. 4.10 becomes

$$\left.\frac{dN}{dt}\right)_{system} = \frac{\partial}{\partial t}\int_{CV}\eta\rho\,d\Psi + \int_{CS}\eta\rho\vec{V}\cdot d\vec{A} \tag{4.11}$$

Equation 4.11 is the relation we set out to obtain.

4-2.2 Physical Interpretation

We have taken several pages to derive Eq. 4.11. Recall that our objective was to obtain a general relation between the rate of change of any arbitrary extensive property, N, of a system and variations of this property associated with the control volume. The main reason for deriving it was to reduce the algebra required to obtain the control volume formulations of the basic equations. Because the working form of each basic equation for application to control volumes is developed from Eq. 4.11, we consider the equation itself to be "basic" and rewrite it to emphasize its importance:

$$\left.\frac{dN}{dt}\right)_{system} = \frac{\partial}{\partial t}\int_{CV}\eta\rho\,d\Psi + \int_{CS}\eta\rho\vec{V}\cdot d\vec{A} \tag{4.11}$$

It is important to recall that in deriving Eq. 4.11, the limiting process (taking the limit as $\Delta t \to 0$) ensured that the relation is valid at the instant when the system and the control volume coincide. In using Eq. 4.11 to go from the system formulations of the basic laws to the control volume formulations, we recognize that Eq. 4.11 relates the rate of change of any extensive property, N, of a system to variations of this property associated with a control volume at the instant when the system and the control volume coincide; this is true since, as $\Delta t \to 0$, the system and the control volume occupy the same volume and have the same boundaries.

Before using Eq. 4.11 to develop control volume formulations of the basic laws, let us make sure we understand each of the terms and symbols in the equation:

$\left.\dfrac{dN}{dt}\right)_{system}$ is the total rate of change of any arbitrary extensive property of the system.

$\dfrac{\partial}{\partial t}\displaystyle\int_{CV}\eta\rho\,d\Psi$ is the time rate of change of the arbitrary extensive property, N, within the control volume.

: η is the intensive property corresponding to N; $\eta = N$ per unit mass.

: $\rho\,d\Psi$ is an element of mass contained in the control volume.

: $\int_{CV}\eta\rho\,d\Psi$ is the total amount of the extensive property, N, contained within the control volume.

$\int_{\text{CS}} \eta \rho \vec{V} \cdot d\vec{A}$ is the net rate of flux of the extensive property, N, out through the control surface.

: $\rho \vec{V} \cdot d\vec{A}$ is the rate of mass flux through area element $d\vec{A}$ per unit time (we recognize that the dot product is a scalar product; the sign of $\rho \vec{V} \cdot d\vec{A}$ depends on the direction of the velocity vector, $\vec{V}$, relative to the area vector, $d\vec{A}$).

: $\eta \rho \vec{V} \cdot d\vec{A}$ is the rate of flux of the extensive property, N, through the area, $d\vec{A}$.

Two additional points about Eq. 4.11 should be made. First, velocity $\vec{V}$ is measured relative to the surface of the control volume. In developing Eq. 4.11, we considered a control volume fixed relative to the reference coordinates, x, y, and z. Since the velocity field was specified relative to the same reference coordinates, it follows that velocity $\vec{V}$ is measured relative to the control volume. Second, in our development, the system moved in the specified velocity field; thus the time rate of change of the arbitrary extensive property, N, within the control volume must be evaluated by an observer fixed in the control volume.

We shall further emphasize these points in deriving the control volume formulation of each of the basic laws. In each case, we begin with the familiar system formulation and use Eq. 4.11 to relate system derivatives to time variations associated with a fixed control volume at the instant when the system and the control volume coincide.[1]

4-3 CONSERVATION OF MASS

The first physical principle to which we apply the relation between system and control volume formulations is conservation of mass. It is intuitive that mass neither can be created nor destroyed; if the flow rate of mass into a control volume exceeds the rate of flow out, mass will accumulate within the CV.

Recall that conservation of mass states simply that the mass of a system is constant,

$$\frac{dM}{dt}\bigg)_{\text{system}} = 0 \tag{4.1a}$$

where

$$M_{\text{system}} = \int_{\text{mass(system)}} dm = \int_{\Psi(\text{system})} \rho\, d\Psi \tag{4.1b}$$

The system and control volume formulations are related by Eq. 4.11,

$$\frac{dN}{dt}\bigg)_{\text{system}} = \frac{\partial}{\partial t} \int_{\text{CV}} \eta \rho\, d\Psi + \int_{\text{CS}} \eta \rho \vec{V} \cdot d\vec{A} \tag{4.11}$$

where

$$N_{\text{system}} = \int_{\text{mass(system)}} \eta\, dm = \int_{\Psi(\text{system})} \eta \rho\, d\Psi \tag{4.6}$$

---

[1] Equation 4.11 has been derived for a control volume fixed in space relative to coordinates xyz. For the case of a *deformable* control volume, whose shape varies with time, Eq. 4.11 may be applied provided that the velocity, $\vec{V}$, in the flux integral is measured relative to the local control surface through which the flux occurs.

To derive the control volume formulation of the conservation of mass, we set

$$N = M \quad \text{and} \quad \eta = 1$$

With this substitution, we obtain

$$\left.\frac{dM}{dt}\right)_{\text{system}} = \frac{\partial}{\partial t}\int_{CV} \rho\, d\mathbf{V} + \int_{CS} \rho\vec{V}\cdot d\vec{A} \tag{4.12}$$

Comparing Eqs. 4.1a and 4.12, we arrive at the control volume formulation of the conservation of mass:

$$0 = \frac{\partial}{\partial t}\int_{CV} \rho\, d\mathbf{V} + \int_{CS} \rho\vec{V}\cdot d\vec{A} \tag{4.13}$$

In Eq. 4.13 the first term represents the rate of change of mass within the control volume; the second term represents the net rate of mass flux out through the control surface. Conservation of mass requires that the sum of the rate of change of mass within the control volume and the net rate of mass outflow through the control surface be zero.

We emphasize that the velocity, $\vec{V}$, in Eq. 4.13 is measured relative to the control surface. Furthermore, the dot product, $\rho\vec{V}\cdot d\vec{A}$, is a scalar product. The sign depends on the direction of the velocity vector, $\vec{V}$, relative to the area vector, $d\vec{A}$. Referring back to the derivation of Eq. 4.11, we see that the dot product, $\rho\vec{V}\cdot d\vec{A}$, is positive where flow is out through the control surface, negative where flow is in through the control surface, and zero where flow is tangent to the control surface.

4-3.1 Special Cases

In special cases it is possible to simplify Eq. 4.13. Consider first the case of incompressible flow, in which the density remains constant. When ρ is a constant, it is not a function of space or time. Consequently, for incompressible flow, Eq. 4.13 may be written as

$$0 = \rho\frac{\partial}{\partial t}\int_{CV} d\mathbf{V} + \rho\int_{CS} \vec{V}\cdot d\vec{A} \tag{4.14a}$$

The integral of $d\mathbf{V}$ over the control volume is simply the volume of the control volume. Thus, on dividing through by ρ, we write Eq. 4.14a as

$$0 = \frac{\partial\mathbf{V}}{\partial t} + \int_{CS} \vec{V}\cdot d\vec{A} \tag{4.14b}$$

For a nondeformable control volume, i.e., a control volume of fixed size and shape, $\mathbf{V}$ = constant. The conservation of mass for incompressible flow through a fixed control volume becomes

$$0 = \int_{CS} \vec{V}\cdot d\vec{A} \tag{4.14c}$$

Note that we have not assumed the flow to be steady in reducing Eq. 4.13 to the form 4.14c. We have only imposed the restriction of incompressible flow. Thus Eq. 4.14c is a statement of the conservation of mass for an incompressible flow that may be steady or unsteady.

The dimensions of the integrand in Eq. 4.14c are L^3/t. The integral of $\vec{V} \cdot d\vec{A}$ over a section of the control surface is commonly called the *volume flow rate* or *volume rate of flow*. Thus, for incompressible flow, the volume flow rate into a fixed control volume must be equal to the volume flow rate out of the control volume. The volume flow rate, Q, through a section of a control surface of area A, is given by

$$Q = \int_A \vec{V} \cdot d\vec{A} \qquad (4.15a)$$

The average velocity, $\bar{V}$, at a section is defined as

$$\bar{V} = \frac{Q}{A} = \frac{1}{A} \int_A \vec{V} \cdot d\vec{A} \qquad (4.15b)$$

Consider now the general case of steady flow that is not incompressible. Since the flow is steady, this means that at most $\rho = \rho(x, y, z)$. By definition, none of the fluid properties varies with time in a steady flow. Consequently, the first term of Eq. 4.13 must be zero and, hence, for steady flow, the statement of conservation of mass reduces to

$$0 = \int_{CS} \rho \vec{V} \cdot d\vec{A} \qquad (4.16)$$

Thus, for steady flow, the mass flow rate into a control volume must be equal to the mass flow rate out of the control volume.

As we noted in our previous discussion of velocity fields in Section 2-2, the idealization of *uniform flow at a section* frequently provides an adequate flow model. Uniform flow at a section implies the velocity is constant across the entire area at a section. When the density also is constant at a section, the flux integral in Eq. 4.13 may be replaced by a product. Thus, when uniform flow at section n is assumed,

$$\int_{A_n} \rho \vec{V} \cdot d\vec{A} = \rho_n \vec{V}_n \cdot \vec{A}_n$$

or using scalar magnitudes

$$\int_{A_n} \rho \vec{V} \cdot d\vec{A} = \pm |\rho_n V_n A_n|$$

Again note that when $\rho \vec{V} \cdot d\vec{A}$ is negative, mass flows in through the control surface. Mass flows out through the control surface in regions where $\rho \vec{V} \cdot d\vec{A}$ is positive. This fact provides a quick check of the signs on the various flux terms in an analysis.

EXAMPLE 4.1—Mass Flow through Multiport Device

Consider steady flow of water through the device shown in the diagram. The areas are: $A_1 = 0.2$ ft^2, $A_2 = 0.5$ ft^2, and $A_3 = A_4 = 0.4$ ft^2. The mass flow rate out through section ③ is given as 3.88 slug/sec. The volume flow rate in through section ④ is given as 1 ft^3/sec, and $\vec{V}_1 = 10\hat{i}$ ft/sec. If properties are assumed uniform across all inlet and outlet flow sections, determine the flow velocity at section ②.

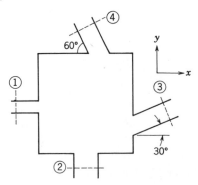

EXAMPLE PROBLEM 4.1

GIVEN: Steady flow of water through the device. Properties uniform at all ports.

$A_1 = 0.2 \text{ ft}^2 \qquad A_2 = 0.5 \text{ ft}^2$

$A_3 = A_4 = 0.4 \text{ ft}^2 \qquad \rho = 1.94 \text{ slug/ft}^3$

$\dot{m}_3 = 3.88 \text{ slug/sec (outflow)}$

$\vec{V}_1 = 10\hat{i} \text{ ft/sec}$

Volume flow rate in at ④ $= 1.0 \text{ ft}^3/\text{sec}$

FIND: Velocity at section ②.

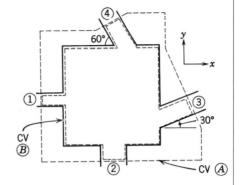

SOLUTION:
Choose a fixed control volume. Two possibilities are shown by dashed lines.

Basic equation:
$$0 = \frac{\partial}{\partial t} \int_{CV} \rho \, d\Psi + \int_{CS} \rho \vec{V} \cdot d\vec{A}$$

Assumptions: (1) Steady flow
(2) Incompressible flow
(3) Uniform properties at each section where fluid crosses the CV boundaries

For steady flow, the first term is zero by definition, so
$$0 = \int_{CS} \rho \vec{V} \cdot d\vec{A}$$

In looking at either control volume, we see that there are four sections where mass flows across the control surface. Thus we write

$$\int_{CS} \rho \vec{V} \cdot d\vec{A} = \int_{A_1} \rho \vec{V} \cdot d\vec{A} + \int_{A_2} \rho \vec{V} \cdot d\vec{A} + \int_{A_3} \rho \vec{V} \cdot d\vec{A} + \int_{A_4} \rho \vec{V} \cdot d\vec{A} = 0 \qquad (1)$$

Let us look at these integrals one at a time, recognizing that properties are uniform over each area and assuming that $\rho = $ constant.

$$\int_{A_1} \rho \vec{V} \cdot d\vec{A} = -\int_{A_1} |\rho V \, dA| = -|\rho V_1 A_1|$$

$\left\{ \begin{array}{l} \text{Sign of } \vec{V} \cdot d\vec{A} \text{ is negative at} \\ \text{surface } ①. \end{array} \right\}$

$\left\{ \begin{array}{l} \text{With the absolute value signs indicated, we have accounted for the directions of } \vec{V} \text{ and} \\ d\vec{A} \text{ in taking the dot product.} \end{array} \right\}$

Since we do not know the direction of $\vec{V}_2$, we shall leave section ② for the moment.

$$\int_{A_3} \rho \vec{V} \cdot d\vec{A} = \int_{A_3} |\rho V \, dA| = |\rho V_3 A_3| = \dot{m}_3$$

$\left\{ \begin{array}{l} \text{Sign of } \vec{V} \cdot d\vec{A} \text{ is positive at} \\ \text{surface } ③, \text{ since flow is out.} \end{array} \right\}$

$$\int_{A_4} \rho \vec{V} \cdot d\vec{A} = -\int_{A_4} |\rho V \, dA| = -|\rho V_4 A_4|$$

$$= -\rho |V_4 A_4| = -\rho |Q_4|$$

$\left\{ \begin{array}{l} \text{Sign of } \vec{V} \cdot d\vec{A} \text{ is negative at} \\ \text{surface } ④. \end{array} \right\}$

where Q is the volume flow rate.

From Eq. 1 above,

$$\int_{A_2} \rho \vec{V} \cdot d\vec{A} = -\int_{A_1} \rho \vec{V} \cdot d\vec{A} - \int_{A_3} \rho \vec{V} \cdot d\vec{A} - \int_{A_4} \rho \vec{V} \cdot d\vec{A}$$

$$= +|\rho V_1 A_1| - \dot{m}_3 + \rho |Q_4|$$

$$= \left| 1.94 \frac{\text{slug}}{\text{ft}^3} \times 10 \frac{\text{ft}}{\text{sec}} \times 0.2 \text{ ft}^2 \right| - 3.88 \frac{\text{slug}}{\text{sec}} + 1.94 \frac{\text{slug}}{\text{ft}^3} \left| 1.0 \frac{\text{ft}^3}{\text{sec}} \right|$$

$$\int_{A_2} \rho \vec{V} \cdot d\vec{A} = 1.94 \text{ slug/sec}$$

Since this is positive, $\vec{V} \cdot d\vec{A}$ at section ② is positive. Flow is out, as shown in the sketch:

$$\int_{A_2} \rho \vec{V} \cdot d\vec{A} = \int_{A_2} |\rho V \, dA| = |\rho V_2 A_2| = 1.94 \text{ slug/sec}$$

$$|V_2| = \frac{1.94 \text{ slug/sec}}{\rho A_2} = \frac{1.94 \text{ slug}}{\text{sec}} \times \frac{\text{ft}^3}{1.94 \text{ slug}} \times \frac{1}{0.5 \text{ ft}^2} = 2 \text{ ft/sec}$$

Since V_2 is in the negative y direction, then

$$\vec{V}_2 = -2\hat{j} \text{ ft/sec}$$

$\left\{ \text{This problem illustrates the procedure recommended for evaluating } \int_{CS} \rho \vec{V} \cdot d\vec{A}. \right\}$

EXAMPLE 4.2—Mass Flow Rate in Boundary Layer

The fluid in direct contact with a stationary solid boundary has zero velocity; there is no slip at the boundary. Thus the flow over a flat plate adheres to the plate surface and forms a boundary layer, as depicted below. The flow ahead of the plate is uniform with velocity, $\vec{V} = U\hat{i}$; $U = 30$ m/sec. The velocity distribution within the boundary layer $(0 \le y \le \delta)$ along cd is approximated as $u/U = 2(y/\delta) - (y/\delta)^2$.

The boundary-layer thickness at this location is 5 mm. The fluid is air with density $\rho = 1.24$ kg/m^3. Assuming the plate width to be 0.6 m, calculate the mass flow rate across surface bc of control volume $abcd$.

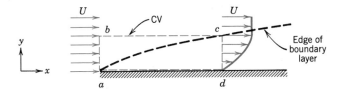

EXAMPLE PROBLEM 4.2

GIVEN: Steady, incompressible flow over a flat plate, $\rho = 1.24$ kg/m^3.
Width of plate, $w = 0.6$ m.
Velocity ahead of plate is uniform: $\vec{V} = U\hat{i}$, $U = 30$ m/sec.

At $x = x_d$:

$\delta = 5$ mm

$\dfrac{u}{U} = 2\left(\dfrac{y}{\delta}\right) - \left(\dfrac{y}{\delta}\right)^2$

FIND: The mass flow rate across surface bc.

SOLUTION:
The fixed control volume is selected as shown by the dashed lines.

Basic equation:
$$0 = \frac{\partial}{\partial t} \int_{CV} \rho \, d\mathcal{V} + \int_{CS} \rho \vec{V} \cdot d\vec{A}$$

Assumptions: (1)   Steady flow
(2)   Incompressible flow
(3)   Two-dimensional flow, properties are independent of z

For steady flow,
$$\frac{\partial}{\partial t} \int_{CV} \rho \, d\mathcal{V} = 0 \qquad \text{and hence} \qquad \int_{CS} \rho \vec{V} \cdot d\vec{A} = 0$$

If it is assumed that there is no flow in the z direction, then

$$0 = \int_{CS} \rho \vec{V} \cdot d\vec{A} \qquad\qquad \left(\begin{array}{c}\text{no flow}\\\text{across } da\end{array}\right)\nearrow$$

$$0 = \int_{A_{ab}} \rho \vec{V} \cdot d\vec{A} + \int_{A_{bc}} \rho \vec{V} \cdot d\vec{A} + \int_{A_{cd}} \rho \vec{V} \cdot d\vec{A} + \int_{A_{da}} \rho \vec{V} \cdot d\vec{A}$$

$$\therefore \dot{m}_{bc} = \int_{A_{bc}} \rho \vec{V} \cdot d\vec{A} = -\int_{A_{ab}} \rho \vec{V} \cdot d\vec{A} - \int_{A_{cd}} \rho \vec{V} \cdot d\vec{A}$$

{We need to evaluate the integrals on the right side of the equation.}

For depth w in the z direction, we obtain

$$\int_{A_{ab}} \rho \vec{V} \cdot d\vec{A} = -\int_{A_{ab}} |\rho u \, dA| = -\int_{y_a}^{y_b} |\rho u w \, dy| \qquad \left\{\begin{array}{c}\vec{V} \cdot d\vec{A} \text{ is negative}\\ dA = w \, dy\end{array}\right\}$$

$$= -\int_0^\delta |\rho u w \, dy| = -\left|\int_0^\delta \rho U w \, dy\right| \qquad \{u = U \text{ over area } ab\}$$

$$\int_{A_{ab}} \rho \vec{V} \cdot d\vec{A} = -|[\rho U w y]_0^\delta| = -\rho U w \delta$$

$$\int_{A_{cd}} \rho \vec{V} \cdot d\vec{A} = \int_{A_{cd}} |\rho u \, dA| = \int_{y_d}^{y_c} |\rho u w \, dy| \qquad \left\{\begin{array}{c}\vec{V} \cdot d\vec{A} \text{ is positive}\\ dA = w \, dy\end{array}\right\}$$

$$= \int_0^\delta |\rho u w \, dy| = \int_0^\delta \left|\rho w U\left[2\left(\frac{y}{\delta}\right) - \left(\frac{y}{\delta}\right)^2\right] dy\right|$$

$$\int_{A_{cd}} \rho \vec{V} \cdot d\vec{A} = \left|\rho w U\left[\frac{y^2}{\delta} - \frac{y^3}{3\delta^2}\right]_0^\delta\right| = \left|\rho w U \delta\left[1 - \frac{1}{3}\right]\right| = \frac{2\rho U w \delta}{3}$$

$$\therefore \dot{m}_{bc} = \int_{A_{bc}} \rho \vec{V} \cdot d\vec{A} = -\int_{A_{ab}} \rho \vec{V} \cdot d\vec{A} - \int_{A_{cd}} \rho \vec{V} \cdot d\vec{A} = \rho U w \delta - \frac{2\rho U w \delta}{3}$$

$$= \frac{\rho U w \delta}{3} = \frac{1}{3} \times \frac{1.24 \text{ kg}}{\text{m}^3} \times \frac{30 \text{ m}}{\text{sec}} \times 0.6 \text{ m} \times 5 \text{ mm} \times \frac{\text{m}}{1000 \text{ mm}}$$

$$\left. \int_{A_{bc}} \rho \vec{V} \cdot d\vec{A} = 0.0372 \text{ kg/sec} \right.$$ $\left\{ \begin{array}{l} \text{Positive sign indicates flow out} \\ \text{across surface } bc. \end{array} \right\} \dot{m}_{bc}$

$\left\{ \begin{array}{l} \text{This problem illustrates the application of the control volume formulation of conser-} \\ \text{vation of mass to the case of nonuniform flow at a section.} \end{array} \right\}$

EXAMPLE 4.3—Density Change in Venting Tank

A tank of 0.05 m^3 volume contains air at 800 kPa (absolute) and 15 C. At $t = 0$, air escapes from the tank through a valve with a flow area of 65 mm^2. The air passing through the valve has a speed of 311 m/sec and a density of 6.13 kg/m^3. Properties in the rest of the tank may be assumed uniform at each instant. Determine the instantaneous rate of change of density in the tank at $t = 0$.

EXAMPLE PROBLEM 4.3

GIVEN: Tank of volume $\Psi = 0.05$ m^3 contains
air at $p = 800$ kPa (absolute), $T = 15$ C.
At $t = 0$, air escapes through a valve.
Air leaves with speed $V = 311$ m/sec and
density $\rho = 6.13$ kg/m^3 through area $A = 65$ mm^2.

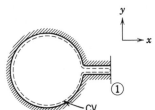

FIND: Rate of change of air density in the tank at $t = 0$.

SOLUTION:
Choose a fixed control volume as shown by the dashed line.

Basic equation: $0 = \dfrac{\partial}{\partial t} \int_{CV} \rho \, d\Psi + \int_{CS} \rho \vec{V} \cdot d\vec{A}$

Assumptions: (1) Properties in the tank are uniform, but time-dependent
              (2) Uniform flow at section ①

Since properties are assumed uniform in the tank at any instant, we can take ρ out from within the integral of the first term,

$$\frac{\partial}{\partial t}\left[\rho_{CV} \int_{CV} d\Psi \right] + \int_{CS} \rho \vec{V} \cdot d\vec{A} = 0$$

Now, $\int_{CV} d\Psi = \Psi$, and hence

$$\frac{\partial}{\partial t}(\rho \Psi)_{CV} + \int_{CS} \rho \vec{V} \cdot d\vec{A} = 0$$

The only place where mass crosses the boundary of the control volume is at surface ①. Hence

$$\int_{CS} \rho \vec{V} \cdot d\vec{A} = \int_{A_1} \rho \vec{V} \cdot d\vec{A} \qquad \text{and} \qquad \frac{\partial}{\partial t}(\rho \Psi) + \int_{A_1} \rho \vec{V} \cdot d\vec{A} = 0$$

At surface ① the sign of $\rho \vec{V} \cdot d\vec{A}$ is positive, so

$$\frac{\partial}{\partial t}(\rho \Psi) + \int_{A_1} |\rho V \, dA| = 0$$

If it is assumed that properties are uniform over surface ①, then

$$\frac{\partial}{\partial t}(\rho \Psi) + |\rho_1 V_1 A_1| = 0 \qquad \text{or} \qquad \frac{\partial}{\partial t}(\rho \Psi) = -|\rho_1 V_1 A_1|$$

Since the volume, V, of the tank is not a function of time

$$\mathrm{V}\frac{\partial\rho}{\partial t}=-|\rho_1 V_1 A_1| \qquad \text{and} \qquad \frac{\partial\rho}{\partial t}=-\frac{|\rho_1 V_1 A_1|}{\mathrm{V}}$$

At $t=0$,

$$\frac{\partial\rho}{\partial t}=\frac{-6.13 \text{ kg}}{\text{m}^3}\times\frac{311 \text{ m}}{\text{sec}}\times 65 \text{ mm}^2 \times\frac{1}{0.05 \text{ m}^3}\times\frac{\text{m}^2}{10^6 \text{ mm}^2}$$

$$\frac{\partial\rho}{\partial t}=-2.48 \text{ kg/m}^3\text{/sec} \qquad\qquad \{\text{the density is decreasing}\} \quad \frac{\partial\rho}{\partial t}$$

$\left\{\begin{array}{l}\text{This problem illustrates the application of the control volume formulation of conser-}\\\text{vation of mass to an unsteady flow.}\end{array}\right\}$

4-4 MOMENTUM EQUATION FOR INERTIAL CONTROL VOLUME

We wish to develop a mathematical formulation of Newton's second law suitable for application to a control volume. In this section our derivation will be restricted to an inertial control volume fixed in space relative to coordinate system xyz that is not accelerating relative to stationary reference frame XYZ.

In deriving the control volume formulation of Newton's second law, the procedure is analogous to the procedure followed in deriving the mathematical formulation for the conservation of mass applied to a control volume. We begin with the mathematical formulation for a system and then use Eq. 4.11 to go from the system to the control volume formulation.

Recall that Newton's second law for a system moving relative to an inertial coordinate system was given by Eq. 4.2a as

$$\vec{F}=\frac{d\vec{P}}{dt}\bigg)_{\text{system}} \tag{4.2a}$$

where the linear momentum, $\vec{P}$, of the system is given by

$$\vec{P}_{\text{system}}=\int_{\text{mass(system)}}\vec{V}\,dm=\int_{\mathrm{V}\text{(system)}}\vec{V}\rho\,d\mathrm{V} \tag{4.2b}$$

and the resultant force, $\vec{F}$, includes all surface and body forces acting on the system,

$$\vec{F}=\vec{F}_S+\vec{F}_B$$

The system and control volume formulations are related by Eq. 4.11,

$$\frac{dN}{dt}\bigg)_{\text{system}}=\frac{\partial}{\partial t}\int_{\text{CV}}\eta\rho\,d\mathrm{V}+\int_{\text{CS}}\eta\rho\vec{V}\cdot d\vec{A} \tag{4.11}$$

where

$$N_{\text{system}}=\int_{\text{mass(system)}}\eta\,dm=\int_{\mathrm{V}\text{(system)}}\eta\rho\,d\mathrm{V} \tag{4.6}$$

To derive the control volume formulation of Newton's second law, we set

$$N=\vec{P} \qquad \text{and} \qquad \eta=\vec{V}$$

From Eq. 4.11, with this substitution, we obtain

$$\frac{d\vec{P}}{dt}\bigg)_{\text{system}}=\frac{\partial}{\partial t}\int_{\text{CV}}\vec{V}\rho\,d\mathrm{V}+\int_{\text{CS}}\vec{V}\rho\vec{V}\cdot d\vec{A} \tag{4.17}$$

From Eq. 4.2a

$$\left. \frac{d\vec{P}}{dt} \right)_{\text{system}} = \vec{F})_{\text{on system}} \tag{4.2a}$$

Since, in deriving Eq. 4.11, the system and the control volume coincided at t_0, then

$$\vec{F}]_{\text{on system}} = \vec{F}]_{\text{on control volume}}$$

In light of this, Eqs. 4.2a and 4.17 may be combined to yield the control volume formulation of Newton's second law for a nonaccelerating control volume

$$\vec{F} = \vec{F}_S + \vec{F}_B = \frac{\partial}{\partial t} \int_{\text{CV}} \vec{V} \rho \, d\Psi + \int_{\text{CS}} \vec{V} \rho \vec{V} \cdot d\vec{A} \tag{4.18}$$

This equation states that the sum of all forces (surface and body forces) acting on a nonaccelerating control volume is equal to the sum of the rate of change of momentum inside the control volume and the net rate of flux of momentum out through the control surface.

The derivation of the momentum equation for a control volume was straightforward. Application of this basic equation to the solution of problems will not be difficult if you exercise care in using the equation.

In using any basic equation for a control volume analysis, the first step must be to draw the boundaries of the control volume and label appropriate coordinate directions. In Eq. 4.18, the force, $\vec{F}$, represents all forces acting on the control volume. It includes both surface forces and body forces. If we denote the body force per unit mass as $\vec{B}$, then

$$\vec{F}_B = \int \vec{B} \, dm = \int_{\text{CV}} \vec{B} \rho \, d\Psi$$

When the force of gravity is the only body force, then the body force per unit mass is $\vec{g}$. The surface force due to pressure is given by

$$\vec{F}_S = \int_A -p \, d\vec{A}$$

The nature of the forces acting on the control volume undoubtedly will influence the choice of the control volume boundaries.

All velocities, $\vec{V}$, in Eq. 4.18 are measured relative to the control volume. The momentum flux, $\vec{V} \rho \vec{V} \cdot d\vec{A}$, through an element of the control surface area, $d\vec{A}$, is a vector. The sign of the scalar product, $\rho \vec{V} \cdot d\vec{A}$, depends on the direction of the velocity vector, $\vec{V}$, relative to the area vector, $d\vec{A}$. The sign of the vector velocity, $\vec{V}$, depends on the coordinate system chosen.

The momentum equation is a vector equation. As with all vector equations, it may be written as three scalar component equations. Relative to an xyz coordinate system, the scalar components of Eq. 4.18 are

$$F_x = F_{S_x} + F_{B_x} = \frac{\partial}{\partial t} \int_{\text{CV}} u \rho \, d\Psi + \int_{\text{CS}} u \rho \vec{V} \cdot d\vec{A} \tag{4.19a}$$

$$F_y = F_{S_y} + F_{B_y} = \frac{\partial}{\partial t} \int_{\text{CV}} v \rho \, d\Psi + \int_{\text{CS}} v \rho \vec{V} \cdot d\vec{A} \tag{4.19b}$$

$$F_z = F_{S_z} + F_{B_z} = \frac{\partial}{\partial t} \int_{CV} w \rho \, d\Psi + \int_{CS} w \rho \vec{V} \cdot d\vec{A} \qquad (4.19c)$$

To use the scalar equations, it again is necessary to select a coordinate system at the outset. The positive directions of the velocity components, u, v, and w, and the force components, F_x, F_y, F_z, are then established relative to the selected coordinate system. As we have previously pointed out, the sign of the scalar product, $\rho \vec{V} \cdot d\vec{A}$, depends on the direction of the velocity vector, $\vec{V}$, relative to the area vector, $d\vec{A}$. Thus the flux term in either Eq. 4.18 or Eqs. 4.19 is a product of two quantities, both of which have algebraic signs. We suggest that you proceed in two steps to determine the momentum flux through any portion of a control surface:

1. The first step is to determine the sign of $\rho \vec{V} \cdot d\vec{A}$,

$$\rho \vec{V} \cdot d\vec{A} = \rho |V \, dA| \cos\alpha = \pm |\rho V \, dA \cos\alpha|$$

2. The second step is to determine the sign for each velocity component, u, v, and w. The sign, which depends on the choice of coordinate system, should be accounted for when substituting numerical values into the terms $u \rho \vec{V} \cdot d\vec{A} = u\{\pm |\rho V \, dA \cos\alpha|\}$, and so on.

EXAMPLE 4.4—Choice of Control Volume for Momentum Analysis

Water from a stationary nozzle strikes a flat plate as shown. The velocity of the water leaving the nozzle is 15 m/sec; the nozzle area is 0.01 m². Assuming the water is directed normal to the plate, and flows along the plate, determine the horizontal force on the support.

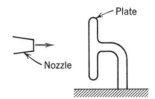

EXAMPLE PROBLEM 4.4

GIVEN: Water from a stationary nozzle is directed normal to the plate; subsequent flow is parallel to plate.

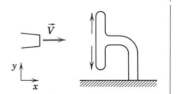

Jet velocity, $\vec{V} = 15\hat{i}$ m/sec

Nozzle area, $A_n = 0.01$ m²

FIND: Horizontal force on the support.

SOLUTION:
We chose a coordinate system in defining the problem above. We must now choose a suitable control volume. A number of possible choices are shown by the dashed lines below.

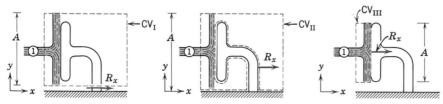

In all three cases, the water from the nozzle crosses the control surface through area A_1 (assumed equal to the nozzle area) and the water is assumed to leave the control volume tangent to the plate surface in the $+y$ or $-y$ direction. Before trying to decide which is the "best" control volume to use, let us write the basic equations.

$$\vec{F} = \vec{F}_S + \vec{F}_B = \frac{\partial}{\partial t} \int_{CV} \vec{V} \rho \, d\Psi + \int_{CS} \vec{V} \rho \vec{V} \cdot d\vec{A} \qquad \text{and} \qquad \frac{\partial}{\partial t} \int_{CV} \rho \, d\Psi + \int_{CS} \rho \vec{V} \cdot d\vec{A} = 0$$

Assumptions: (1)   Steady flow
                 (2)   Incompressible flow
                 (3)   Uniform flow at each section where fluid crosses the CV boundaries

Regardless of our choice of control volume, the flow is steady and the basic equations become

$$\vec{F} = \vec{F}_S + \vec{F}_B = \int_{CS} \vec{V} \rho \vec{V} \cdot d\vec{A} \quad \text{and} \quad \int_{CS} \rho \vec{V} \cdot d\vec{A} = 0$$

Evaluating the momentum flux term will lead to the same result for all of the control volumes. We should choose the control volume that allows the most straightforward evaluation of the forces.

Remember in applying the momentum equation that the force, $\vec{F}$, represents all forces acting *on* the control volume.

Let us solve the problem using each of the three control volumes.

CV_I

The control volume has been selected so that the area of the left surface is equal to the area of the right surface. Denote this area by A.

The control volume cuts through the support. We denote the force of the support on the control volume as R_x and assume it to be positive. (The force of the control volume on the support is equal and opposite to R_x.)

Since we are looking for the horizontal force, we write the x component of the steady flow momentum equation

$$F_{S_x} + F_{B_x} = \int_{CS} u \rho \vec{V} \cdot d\vec{A}$$

There are no body forces in the x direction, so $F_{B_x} = 0$, and

$$F_{S_x} = \int_{CS} u \rho \vec{V} \cdot d\vec{A}$$

To evaluate F_{S_x}, we must include all surface forces acting on the control volume

$$F_{S_x} \quad = \qquad\qquad p_a A \qquad\qquad\qquad - \qquad\qquad p_a A \qquad\qquad\qquad + \qquad\qquad R_x$$

force due to atmospheric pressure acts to right (positive direction) on left surface	force due to atmospheric pressure acts to left (negative direction) on right surface	force of support on control volume (assumed positive)

Consequently, $F_{S_x} = R_x$, and

$$R_x = \int_{CS} u \rho \vec{V} \cdot d\vec{A} = \int_{A_1} u \rho \vec{V} \cdot d\vec{A} \qquad \left\{ \begin{array}{l} \text{For mass crossing top and bottom} \\ \text{surfaces, } u = 0. \end{array} \right\}$$

$$= \int_{A_1} u \{-|\rho V_1 \, dA|\} \qquad\qquad \left\{ \begin{array}{l} \text{At } ①, \rho \vec{V} \cdot d\vec{A} = -|\rho V_1 \, dA|, \text{ since direction} \\ \text{of } \vec{V}_1 \text{ and } d\vec{A}_1 \text{ are } 180° \text{ apart.} \end{array} \right\}$$

$$= -u_1 |\rho V_1 A_1| \qquad\qquad\qquad\qquad \text{\{properties uniform over } A_1\}$$

$$= -\frac{15}{} \frac{m}{sec} \left| \frac{999 \text{ kg}}{m^3} \times \frac{15}{} \frac{m}{sec} \times \frac{0.01 \text{ m}^2}{} \right| \frac{N \cdot sec^2}{kg \cdot m} \qquad\qquad \text{\{}u_1 = 15 \text{ m/sec\}}$$

$$R_x = -2.25 \text{ kN} \qquad\qquad\qquad \text{\{}R_x \text{ acts opposite to positive direction assumed.\}}$$

The force on the support is

$$K_x = -R_x = 2.25 \text{ kN} \qquad\qquad\qquad \text{\{force on support acts to the right\} } K_x$$

CV$_{\mathrm{II}}$

The control volume has been selected so the area of the left surface is equal to the area of the right surface. Denote this area by A.

The control volume does not cut through the support. However, the control volume is in contact with the support over several portions of the area of the control surface. There is a force exerted by the support on the control surface. We denote the x component of this force as R_x.

Then for this control volume the x component of the momentum equation leads directly (in the same step by step manner) to the same solution as for CV$_{\mathrm{I}}$ above.

CV$_{\mathrm{III}}$

The control volume has been selected so the areas of the left surface and of the right surface are equal to the area of the plate. Denote this area by A.

As in the case of CV$_{\mathrm{II}}$ the x component of the force of the support on the CV is denoted by R_x.

Then the x component of the momentum equation,

$$F_{S_x} = \int_{CS} u \rho \vec{V} \cdot d\vec{A}$$

yields

$$F_{S_x} = p_a A + R_x = \int_{A_1} u \rho \vec{V} \cdot d\vec{A} = \int_{A_1} u\{-|\rho V_1\, dA|\} = -2.25 \text{ kN}$$

Then

$$R_x = -p_a A - 2.25 \text{ kN} \qquad \text{and} \qquad K_x = -R_x = p_a A + 2.25 \text{ kN}$$

To determine the net force on the plate, we need a free-body diagram of the plate:

$$F_{\text{net}} = K_x - p_a A \qquad \{\text{since atmospheric pressure acts on the back of the plate}\}$$

$$F_{\text{net}} = p_a A + 2.25 \text{ kN} - p_a A = 2.25 \text{ kN}$$

$\left\{\begin{array}{l}\text{This problem illustrates the application of the momentum equation to an inertial control}\\ \text{volume, with emphasis on choosing a suitable control volume.}\end{array}\right\}$

EXAMPLE 4.5—Tank on Scale: Body Force

A metal container 2 ft high, with an inside cross-sectional area of 1 ft^2, weighs 5 lbf when empty. The container is placed on a scale and water flows in through an opening in the top and out through the two equal area openings in the sides, as shown in the diagram. Under steady flow conditions, the height of the water in the tank is 1.9 ft. Determine the reading on the scale.

$A_1 = 0.1 \text{ ft}^2$

$\vec{V}_1 = -5\hat{j} \text{ ft/sec}$

$A_2 = A_3 = 0.1 \text{ ft}^2$

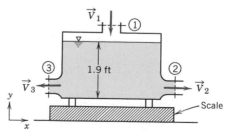

EXAMPLE PROBLEM 4.5

GIVEN: Metal container, of height 2 ft and cross-sectional area $A = 1$ ft^2, weighs 5 lbf when empty. Container rests on scale. Under steady flow conditions water depth is 1.9 ft. Water enters vertically at section ① and leaves horizontally through sections ② and ③.

$$A_1 = 0.1 \text{ ft}^2$$

$$\vec{V}_1 = -5\hat{j} \text{ ft/sec}$$

$$A_2 = A_3 = 0.1 \text{ ft}^2$$

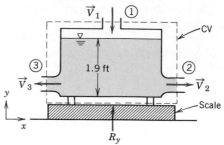

FIND: The reading on the scale.

SOLUTION:
Choose a control volume as shown; R_y is the force of the scale on the control volume and is assumed positive.

Basic equations:

$$\vec{F}_S + \vec{F}_B = \overset{\overset{=0(1)}{\cancel{\nearrow}}}{\frac{\partial}{\partial t}} \int_{CV} \vec{V} \rho \, d\forall + \int_{CS} \vec{V} \rho \vec{V} \cdot d\vec{A}$$

$$0 = \overset{\overset{=0(1)}{\cancel{\nearrow}}}{\frac{\partial}{\partial t}} \int_{CV} \rho \, d\forall + \int_{CS} \rho \vec{V} \cdot d\vec{A}$$

Assumptions: (1) Steady flow
(2) Incompressible flow
(3) Uniform flow at each section where fluid crosses the CV boundaries

We write the y component of the momentum equation

$$F_{S_y} + F_{B_y} = \int_{CS} v \rho \vec{V} \cdot d\vec{A} \tag{1}$$

$F_{S_y} = R_y$ {There is no net force due to atmospheric pressure.}
$F_{B_y} = -W_{\text{tank}} - W_{H_2O}$ {Both body forces act in negative y direction.}

$$W_{H_2O} = \rho g \forall = \gamma A h$$

$$\int_{CS} v \rho \vec{V} \cdot d\vec{A} = \int_{A_1} v \rho \vec{V} \cdot d\vec{A} = \int_{A_1} v\{-|\rho_1 V_1 dA|\} \quad \left\{ \begin{array}{l} \vec{V} \cdot d\vec{A} \quad \text{is negative at } ①. \\ v = 0 \text{ at sections } ② \text{ and } ③. \end{array} \right\}$$

$$= -v_1 |\rho_1 V_1 A_1| \qquad \text{{We are assuming uniform properties at } ①.}$$

Substituting into Eq. 1 gives

$$R_y - W_{\text{tank}} - \gamma A h = -v_1 |\rho_1 V_1 A_1|$$

or

$$R_y = W_{\text{tank}} + \gamma A h - v_1 |\rho_1 V_1 A_1|$$

Substituting numbers with $v_1 = -5$ ft/sec gives

$$R_y = 5 \text{ lbf} + \frac{62.4 \text{ lbf}}{\text{ft}^3} \times 1 \text{ ft}^2 \times 1.9 \text{ ft} - \left(\frac{-5 \text{ ft}}{\text{sec}}\right) \left| 1.94 \frac{\text{slug}}{\text{ft}^3} \left(\frac{-5 \text{ ft}}{\text{sec}}\right) 0.1 \text{ ft}^2 \right| \frac{\text{lbf} \cdot \text{sec}^2}{\text{slug} \cdot \text{ft}}$$

$$R_y = 128 \text{ lbf} \qquad\qquad\qquad \text{{Force of scale on CV is upward.}}$$

The force of the control volume on the scale is $K_y = -R_y = -128$ lbf.

The minus sign indicates that the force on the scale is downward in our coordinate system. Therefore, the scale reading is 128 lbf.

{ This problem illustrates the application of the momentum equation to an inertial control volume with body forces included. }

EXAMPLE 4.6—Flow under Sluice Gate: Hydrostatic Pressure Force

Water in an open channel flows under a sluice gate as shown in the sketch. The flow is incompressible and uniform at sections ① and ②. Hydrostatic pressure distributions may be assumed at sections ① and ② because the flow streamlines are essentially straight there. Determine the magnitude and direction of the force per unit width exerted on the gate by the flow.

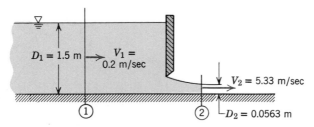

EXAMPLE PROBLEM 4.6

GIVEN: Flow under sluice gate. Width $= w$.

FIND: Force exerted (per unit width) on the gate.

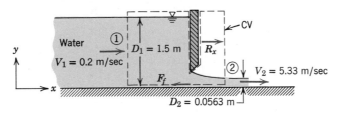

SOLUTION:
Choose the CV and coordinate system shown for analysis. Apply the x component of the momentum equation.

$$= 0(2) \quad = 0(3)$$

Basic equation

$$F_{S_x} + F_{B_x} = \frac{\partial}{\partial t} \int_{CV} u \rho \, d\forall + \int_{CS} u \rho \vec{V} \cdot d\vec{A}$$

Assumptions: (1) F_f negligible (neglect friction on channel bottom)
(2) $F_{B_x} = 0$
(3)   Steady flow
(4)   Incompressible flow
(5)   Uniform flow at each section
(6)   Hydrostatic pressure distributions at ① and ②

Then

$$F_{S_x} = u_1 \{ -|\rho V_1 w D_1| \} + u_2 \{ |\rho V_2 w D_2| \}$$

The surface forces acting on the CV are due to pressure and the unknown force, R_x. From assumption (6),

$$\frac{dp}{dy} = -\rho g \qquad p = p_0 + \rho g(y_0 - y) = p_{\text{atm}} + \rho g(D - y)$$

Evaluating F_{S_x} gives

$$F_{S_x} = \int_0^{D_1} p_1\, dA_1 - \int_0^{D_2} p_2\, dA_2 - p_{\text{atm}}(D_1 - D_2)w + R_x$$

$$= \int_0^{D_1} [p_{\text{atm}} + \rho g(D_1 - y)]w\, dy$$

$$- \int_0^{D_2} [p_{\text{atm}} + \rho g(D_2 - y)]w\, dy - p_{\text{atm}}(D_1 - D_2)w + R_x$$

$$F_{S_x} = \cancel{p_{\text{atm}} D_1 w} + \frac{\rho g D_1^2}{2}w - \cancel{p_{\text{atm}} D_2 w} - \frac{\rho g D_2^2}{2}w - \cancel{p_{\text{atm}} D_1 w} + \cancel{p_{\text{atm}} D_2 w} + R_x$$

or

$$F_{S_x} = R_x + \frac{\rho g w}{2}(D_1^2 - D_2^2)$$

Substituting into the momentum equation, with $u_1 = V_1$ and $u_2 = V_2$, gives

$$R_x + \frac{\rho g w}{2}(D_1^2 - D_2^2) = -V_1|\rho V_1 w D_1| + V_2|\rho V_2 w D_2|$$

or

$$R_x = \rho w(V_2^2 D_2 - V_1^2 D_1) - \frac{\rho g w}{2}(D_1^2 - D_2^2)$$

and

$$\frac{R_x}{w} = \rho(V_2^2 D_2 - V_1^2 D_1) - \frac{\rho g}{2}(D_1^2 - D_2^2)$$

$$= \frac{999\ \text{kg}}{\text{m}^3}\left[(5.33)^2(0.0563) - (0.2)^2(1.5)\right]\frac{\text{m}^2}{\text{sec}^2}\,\text{m} \times \frac{\text{N}\cdot\text{sec}^2}{\text{kg}\cdot\text{m}}$$

$$- \frac{1}{2} \times \frac{999\ \text{kg}}{\text{m}^3} \times \frac{9.81\ \text{m}}{\text{sec}^2}\left[(1.5)^2 - (0.0563)^2\right]\text{m}^2 \times \frac{\text{N}\cdot\text{sec}^2}{\text{kg}\cdot\text{m}}$$

$$\frac{R_x}{w} = -9.47\ \text{kN/m}$$

R_x is the unknown external force acting *on* the control volume. It is applied to the CV by the gate. Therefore, the force from all fluids *on* the gate is K_x, where $K_x = -R_x$. Thus

$$\frac{K_x}{w} = -\frac{R_x}{w} = 9.47\ \text{kN/m} \qquad \{\text{applied to the right}\} \qquad \overset{K_x}{\underset{w}{\xleftarrow{\hspace{1cm}}}}$$

$$\left\{\begin{array}{l}\text{This problem illustrates application of the momentum equation to a control volume in}\\ \text{which the pressure is not uniform over the entire control surface.}\end{array}\right\}$$

EXAMPLE 4.7—Flow through Elbow: Use of Gage Pressures

Water flows steadily through the 90° reducing elbow shown in the diagram. At the inlet to the elbow, the absolute pressure is 221 kPa and the cross-sectional area is 0.01 m². At the outlet, the cross-sectional area is 0.0025 m² and the velocity is 16 m/sec. The pressure at the outlet is atmospheric. Determine the force required to hold the elbow in place.

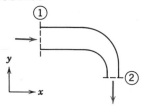

EXAMPLE PROBLEM 4.7

GIVEN: Steady flow of water through 90° reducing elbow.

$$p_1 = 221 \text{ kPa (abs)} \qquad A_1 = 0.01 \text{ m}^2$$

$$\vec{V}_2 = -16\hat{j} \text{ m/sec} \qquad A_2 = 0.0025 \text{ m}^2$$

FIND: Force required to hold elbow in place.

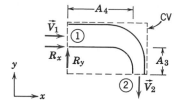

SOLUTION:

Choose control volume as shown by the dashed line.

R_x, R_y are components of force required to hold the elbow in place. (They are assumed positive and are forces acting on CV.)

A_3 is the area of vertical sides of CV excluding A_1; $A_{\text{vertical sides}} = A_1 + A_3$.

A_4 is the area of horizontal sides of CV excluding A_2; $A_{\text{horizontal sides}} = A_2 + A_4$.

Basic equations:

$$\vec{F} = \vec{F}_S + \vec{F}_B = \overset{\cancel{=0(4)}}{\frac{\partial}{\partial t}\int_{\text{CV}} \vec{V}\rho \, d\forall} + \int_{\text{CS}} \vec{V}\rho\vec{V}\cdot d\vec{A}$$

$$0 = \overset{\cancel{=0(4)}}{\frac{\partial}{\partial t}\int_{\text{CV}} \rho \, d\forall} + \int_{\text{CS}} \rho\vec{V}\cdot d\vec{A}$$

Assumptions: (1) Uniform flow at each section
(2) Atmospheric pressure, $p_a = 101$ kPa
(3) Incompressible flow
(4) Steady flow

Writing the x component of the momentum equation results in

$$F_{S_x} = \int_{\text{CS}} u\rho\vec{V}\cdot d\vec{A} = \int_{A_1} u\rho\vec{V}\cdot d\vec{A} \qquad\qquad \{F_{B_x} = 0 \text{ and } u_2 = 0\}$$

$$p_1 A_1 + p_a A_3 - p_a(A_1 + A_3) + R_x = \int_{A_1} u\rho\vec{V}\cdot d\vec{A} \qquad \left\{\begin{array}{l}\text{Pressure over right side is}\\ p_a. \text{ Pressure over left side}\\ \text{is } p_1 \text{ on } A_1 \text{ and } p_a \text{ on } A_3.\end{array}\right\}$$

$$(p_1 - p_a)A_1 + R_x = \int_{A_1} u\{-|\rho V_1 \, dA|\} \qquad\qquad \{\vec{V}\cdot d\vec{A} \text{ is negative at } A_1.\}$$

$$R_x = -p_{1_g}A_1 - u_1|\rho V_1 A_1| \qquad\qquad \{p_1 - p_a = p_{1_{\text{gage}}}\}$$

To find V_1, use the continuity equation:

$$\int_{\text{CS}} \rho\vec{V}\cdot d\vec{A} = 0 = \int_{A_1} \rho\vec{V}\cdot d\vec{A} + \int_{A_2} \rho\vec{V}\cdot d\vec{A}$$

$$\therefore 0 = -\int_{A_1} |\rho V \, dA| + \int_{A_2} |\rho V \, dA| = -|\rho V_1 A_1| + |\rho V_2 A_2|$$

and

$$|V_1| = |V_2|\frac{A_2}{A_1} = 16 \, \frac{\text{m}}{\text{sec}} \times \frac{0.0025}{0.01} = 4 \text{ m/sec} \qquad \therefore \vec{V}_1 = 4\hat{i} \text{ m/sec}$$

$$R_x = -p_{1_g}A_1 - u_1|\rho V_1 A_1|$$

$$= -\frac{1.20\times 10^5 \text{ N}}{\text{m}^2} \times 0.01 \text{ m}^2 - 4 \, \frac{\text{m}}{\text{sec}}\left|999 \, \frac{\text{kg}}{\text{m}^3} \times 4 \, \frac{\text{m}}{\text{sec}} \times 0.01 \text{ m}^2 \times \frac{\text{N}\cdot\text{sec}^2}{\text{kg}\cdot\text{m}}\right|$$

$$R_x = -1.36 \text{ kN} \qquad\qquad \{R_x \text{ to hold elbow acts to left}\} \; R_x$$

Writing the y component of the momentum equation gives

$$F_{S_y} + F_{B_y} = \int_{CS} v\rho \vec{V} \cdot d\vec{A} = \int_{A_2} v\rho \vec{V} \cdot d\vec{A} \qquad \{v_1 = 0\}$$

$$p_a A_4 + p_a A_2 - p_a A_4 - p_a A_2 + F_{B_y} + R_y = \int_{A_2} v\{|\rho V \, dA|\} \qquad \left\{ \begin{array}{l} \text{Pressure is } p_a \text{ over} \\ \text{top and bottom of CV.} \\ \vec{V} \cdot d\vec{A} \text{ is positive at } ②. \end{array} \right\}$$

$$F_{B_y} + R_y = v_2 |\rho V_2 A_2|$$

$$R_y = -F_{B_y} + v_2 |\rho V_2 A_2| \qquad \left\{ \begin{array}{l} \text{Since we do not know the volume or mass of} \\ \text{the elbow, we cannot evaluate } F_{B_y}. \end{array} \right\}$$

Substituting numbers, recognizing $\vec{V}_2 = -16\hat{j}$ m/sec, so $v_2 = -16$ m/sec

$$R_y = -F_{B_y} + \left(\frac{-16 \text{ m}}{\text{sec}} \right) \left| 999 \frac{\text{kg}}{\text{m}^3} \left(\frac{-16 \text{ m}}{\text{sec}} \right) 0.0025 \text{ m}^2 \times \frac{\text{N} \cdot \text{sec}^2}{\text{kg} \cdot \text{m}} \right|$$

$$R_y = -F_{B_y} - 639 \text{ N}$$

Neglecting F_{B_y} gives

$$R_y = -639 \text{ N} \qquad\qquad \{R_y \text{ to hold elbow acts down}\} \qquad R_y$$

$$\left\{ \begin{array}{l} \text{This problem illustrates the application of the momentum equation to an inertial control} \\ \text{volume in which the pressure is not atmospheric across the entire control surface.} \\[6pt] \text{Since the pressure forces over the entire control surface must be included in the analysis,} \\ \text{use of gage pressures on all surfaces gives correct (and often more direct) results.} \end{array} \right.$$

EXAMPLE 4.8—Conveyor Belt Filling: Rate of Change of Momentum in Control Volume

A horizontal conveyor belt moving at 3 ft/sec receives sand from a hopper. The sand falls vertically from the hopper to the belt at a speed of 5 ft/sec and a flow rate of 500 lbm/sec (the density of sand is approximately 2700 lbm/cubic yard). The conveyor belt is initially empty but begins to fill with sand. If friction in the drive system and rollers is negligible, find the tension required to pull the belt while the conveyor is filling.

EXAMPLE PROBLEM 4.8

GIVEN: Conveyor and hopper shown in sketch.

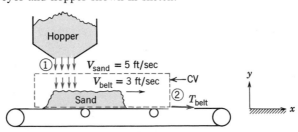

FIND: T_{belt} at the instant shown.

SOLUTION:
Use the control volume and coordinates shown. Apply the x component of the momentum equation.

Basic equations:

$$F_{S_x} + \cancel{F_{B_x}}^{= 0(2)} = \frac{\partial}{\partial t} \int_{CV} u \rho \, d\forall + \int_{CS} u \rho \vec{V} \cdot d\vec{A} \qquad 0 = \frac{\partial}{\partial t} \int_{CV} \rho \, d\forall + \int_{CS} \rho \vec{V} \cdot d\vec{A}$$

Assumptions: (1) $F_{S_x} = T_{\text{belt}} = T$
(2) $F_{B_x} = 0$
(3) Uniform flow at section ①
(4) All sand on belt moves with $V_{\text{belt}} = V_b$

Then

$$T = \frac{\partial}{\partial t} \int_{CV} u \rho \, d\forall + u_1\{-|\rho V_1 A_1|\} + u_2\{|\rho V_2 A_2|\}$$

Since $u_1 = 0$, and there is no flow at section ②, then $T = \dfrac{\partial}{\partial t} \displaystyle\int_{CV} u \rho \, d\forall$

From assumption (4), inside the CV, $u = V_b =$ constant, and hence

$$T = V_b \frac{\partial}{\partial t} \int_{CV} \rho \, d\forall = V_b \frac{\partial M_s}{\partial t}$$

where M_s is the mass of sand on the belt (inside the control volume).
From the continuity equation,

$$\frac{\partial}{\partial t} \int_{CV} \rho \, d\forall = \frac{\partial}{\partial t} M_s = -\int_{CS} \rho \vec{V} \cdot d\vec{A} = \dot{m}_s = 500 \text{ lbm/sec}$$

Then

$$T = V_b \dot{m}_s = \frac{3}{\text{sec}} \frac{\text{ft}}{} \times \frac{500}{\text{sec}} \frac{\text{lbm}}{} \times \frac{\text{slug}}{32.2 \text{ lbm}} \times \frac{\text{lbf} \cdot \text{sec}^2}{\text{slug} \cdot \text{ft}}$$

$$T = 46.6 \text{ lbf} \hspace{8cm} T$$

$\left\{ \begin{array}{l} \text{This problem illustrates an application of the momentum equation to a problem in} \\ \text{which the rate of change of momentum within the control volume is not equal to zero.} \end{array} \right\}$

**4-4.1 Differential Control Volume Analysis

We have considered a number of examples in which conservation of mass and the momentum equation have been applied to finite control volumes. The control volume chosen for analysis need not be finite in size.

Application of the basic equations to a differential control volume leads to differential equations describing the relationships among properties in the flow field. In some cases, the differential equation can be solved to give detailed information about property variations in the flow field. For the case of steady, incompressible, frictionless flow along a streamline, integration of one such differential equation leads to a useful relationship among velocity, pressure, and elevation in a flow field. This case is presented to illustrate the use of differential control volumes.

Let us apply the continuity and momentum equations to a steady incompressible flow without friction, as shown in Fig. 4.4. The control volume chosen is fixed in space and bounded by flow streamlines, and is thus an element of a streamtube. The length of the control volume is ds.

** This section may be omitted without loss of continuity in the text material.

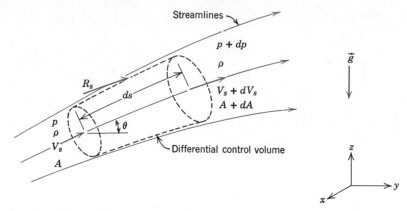

Fig. 4.4   Differential control volume for momentum analysis of flow through a streamtube.

Because the control volume is bounded by streamlines, the only flow across the bounding surfaces occurs at the end sections. These are located at coordinates s and $s + ds$, measured along the central streamline.

Properties at the inlet section are assigned arbitrary symbolic values. Properties at the outlet section are assumed to increase by a differential amount. Thus at $s + ds$, the flow speed is assumed to be $V_s + dV_s$, and so forth. The differential changes, dp, dV_s, and dA, all are assumed to be positive in formulating the problem. (As in a free-body analysis in statics or dynamics, the actual algebraic sign of each differential change will be determined from the results of the analysis.)

Now let us apply the continuity equation and the s component of the momentum equation to the control volume of Fig. 4.4.

a. Continuity Equation

Basic equation:
$$0 = \frac{\partial}{\partial t} \int_{CV} \rho \, dV + \int_{CS} \rho \vec{V} \cdot d\vec{A} \tag{4.13}$$

$$= 0(1)$$

Assumptions: (1)   Steady flow
(2)   No flow across bounding streamlines
(3)   Incompressible flow, $\rho = $ constant

Then
$$0 = \{-|\rho V_s A|\} + \{|\rho(V_s + dV_s)(A + dA)|\}$$

and
$$\rho V_s A = \rho(V_s + dV_s)(A + dA) \tag{4.20a}$$

On expanding the right side and simplifying, we obtain
$$0 = V_s \, dA + A \, dV_s + dA \, dV_s$$

But $dA \, dV_s$ is a product of differentials, which may be neglected compared to $V_s \, dA$ or $A \, dV_s$. Thus
$$0 = V_s \, dA + A \, dV_s \tag{4.20b}$$

b. Streamwise Component of the Momentum Equation

Basic equation:
$$F_{S_s} + F_{B_s} = \overbrace{\frac{\partial}{\partial t} \int_{CV} u_s \rho \, dV}^{=0(1)} + \int_{CS} u_s \rho \vec{V} \cdot d\vec{A} \tag{4.21}$$

Assumption: (4) No friction: $R_s = 0$ and F_{S_s} is due to pressure forces only

The pressure force will have three terms:

$$F_{S_s} = pA - (p + dp)(A + dA) + \left(p + \frac{dp}{2}\right) dA \tag{4.22a}$$

The first and second terms in Eq. 4.22a are the pressure forces on the end faces of the control surface. The third term is the pressure force acting in the s direction on the bounding stream surface of the control volume. Its magnitude is the product of the average pressure acting on the stream surface, $p + \frac{1}{2}dp$, times the area component of the stream surface in the s direction, dA. Equation 4.22a simplifies to

$$F_{S_s} = -A \, dp - \tfrac{1}{2}dp \, dA \tag{4.22b}$$

The body force component in the s direction is

$$F_{B_s} = \rho g_s \, dV = \rho(-g \sin\theta)\left(A + \frac{dA}{2}\right) ds$$

But $\sin\theta \, ds = dz$, so that

$$F_{B_s} = -\rho g \left(A + \frac{dA}{2}\right) dz \tag{4.22c}$$

The momentum flux will be

$$\int_{CS} u_s \rho \vec{V} \cdot d\vec{A} = V_s\{-|\rho V_s A|\} + (V_s + d V_s)\{|\rho(V_s + d V_s)(A + dA)|\}$$

since there is no mass flux across the stream surfaces. The terms in braces are equal from continuity, Eq. 4.20a, so

$$\int_{CS} u_s \rho \vec{V} \cdot d\vec{A} = V_s(-\rho V_s A) + (V_s + d V_s)(\rho V_s A) = \rho V_s A \, d V_s \tag{4.23}$$

Substitution of Eqs. 4.22b, 4.22c, and 4.23 into the momentum equation gives

$$-A \, dp - \tfrac{1}{2}dp \, dA - \rho g A \, dz - \tfrac{1}{2}\rho g \, dA \, dz = \rho V_s A \, d V_s$$

Dividing by ρA and noting that products of differentials are negligible compared to the remaining terms, we obtain

$$-\frac{dp}{\rho} - g \, dz = V_s d V_s = d\left(\frac{V_s^2}{2}\right)$$

or

$$\frac{dp}{\rho} + d\left(\frac{V_s^2}{2}\right) + g \, dz = 0 \tag{4.24}$$

For incompressible flow, this equation may be integrated to obtain

$$\frac{p}{\rho} + \frac{V_s^2}{2} + gz = \text{constant}$$

or dropping subscript s,

$$\frac{p}{\rho} + \frac{V^2}{2} + gz = \text{constant} \tag{4.25}$$

This equation is subject to the restrictions:

1. Steady flow.
2. No friction.
3. Flow along a streamline.
4. Incompressible flow.

By applying the momentum equation to an infinitesimal streamtube control volume, for steady incompressible flow without friction, we have derived a relation among the pressure, velocity, and elevation. This relationship is very powerful and useful. For example, it could have been used to evaluate the pressure at the inlet of the reducing elbow analyzed in Example Problem 4.7 or to determine the velocity of water leaving the sluice gate of Example Problem 4.6. In both of these flow situations the restrictions required to derive Eq. 4.25 are reasonable idealizations of the actual flow behavior. The restrictions must be emphasized heavily because they do not always form a realistic model for flow behavior; consequently, they must be justified carefully each time Eq. 4.25 is applied.

Equation 4.25 is a form of the Bernoulli equation. It will be derived again in detail in Chapter 6 because it is such a useful tool for flow analysis and because an alternative derivation will give added insight into the need for care in applying the equation.

EXAMPLE 4.9—Nozzle Flow: Application of Bernoulli Equation

Water flows steadily through a horizontal nozzle, discharging to the atmosphere. At the nozzle inlet the diameter is D_1; at the nozzle outlet the diameter is D_2. Derive an expression for the minimum gage pressure required at the nozzle inlet to produce a given volume flow rate, Q. Evaluate the inlet gage pressure if $D_1 = 3.0$ in., $D_2 = 1.0$ in., and the desired flow rate is 0.7 ft³/sec.

EXAMPLE PROBLEM 4.9

GIVEN: Steady flow of water through a horizontal nozzle, discharging to the atmosphere.

$$D_1 = 3.0 \text{ in.} \qquad D_2 = 1.0 \text{ in.} \qquad p_2 = p_{\text{atm}}$$

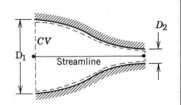

FIND: (a) p_{1_g} as a function of volume flow rate, Q.
(b) Evaluate for $Q = 0.7$ ft³/sec.

SOLUTION:

Basic equations:

$$\frac{p_1}{\rho} + \frac{V_1^2}{2} + gz_1 = \frac{p_2}{\rho} + \frac{V_2^2}{2} + gz_2$$
$$= 0(1)$$

$$0 = \frac{\partial}{\partial t} \int_{CV} \rho \, d\Psi + \int_{CS} \rho \vec{V} \cdot d\vec{A}$$

Assumptions: (1) Steady flow
(2) Incompressible flow

    (3)   Frictionless flow

    (4)   Flow along a streamline

    (5) $z_1 = z_2$

    (6)   Uniform flow at sections ① and ②

Apply the Bernoulli equation along a streamline between points ① and ② to evaluate p_1. Then

$$p_{1_g} = p_1 - p_{atm} = p_1 - p_2 = \frac{\rho}{2}(V_2^2 - V_1^2) = \frac{\rho}{2}V_1^2\left[\left(\frac{V_2}{V_1}\right)^2 - 1\right]$$

Apply the continuity equation

$$0 = \{-|\rho V_1 A_1|\} + \{|\rho V_2 A_2|\} \qquad \text{or} \qquad V_1 A_1 = V_2 A_2 = Q$$

so that

$$\frac{V_2}{V_1} = \frac{A_1}{A_2} \qquad \text{and} \qquad V_1 = \frac{Q}{A_1}$$

Then

$$p_{1_g} = \frac{\rho Q^2}{2A_1^2}\left[\left(\frac{A_1}{A_2}\right)^2 - 1\right]$$

Since $A = \pi D^2/4$, then

$$p_{1_g} = \frac{8\rho Q^2}{\pi^2 D_1^4}\left[\left(\frac{D_1}{D_2}\right)^4 - 1\right]$$

With $D_1 = 3.0$ in., $D_2 = 1.0$ in., and $\rho = 1.94$ slug/ft^3,

$$p_{1_g} = \frac{8}{\pi^2} \times \frac{1.94 \text{ slug}}{\text{ft}^3} \times \frac{1}{(3)^4 \text{ in.}^4} \times Q^2\left[(3.0)^4 - 1\right]\frac{\text{lbf} \cdot \text{sec}^2}{\text{slug} \cdot \text{ft}} \times \frac{144 \text{ in.}^2}{\text{ft}^2}$$

$$p_{1_g} = 224Q^2 \frac{\text{lbf} \cdot \text{sec}^2}{\text{in.}^2 \cdot \text{ft}^6} \qquad\qquad\qquad\qquad\qquad\qquad\quad p_{1_g}$$

With $Q = 0.7$ ft^3/sec, then $p_{1_g} = 110$ lbf/in.^{2}

$\left[\begin{array}{l}\text{This problem illustrates the application of the Bernoulli equation to a flow where}\\\text{the restrictions of steady, incompressible, frictionless flow along a streamline are a}\\\text{reasonable flow model.}\end{array}\right\}$

4-4.2 Control Volume Moving with Constant Velocity

In the preceding problems, which illustrate applications of the momentum equation to inertial control volumes, we have considered only stationary control volumes. A control volume (fixed relative to reference frame xyz) moving with a constant velocity, $\vec{V}_{rf}$, relative to a fixed (inertial) reference frame XYZ, is also inertial, since it has no acceleration with respect to XYZ.

Equation 4.11, which expresses system derivatives in terms of control volume variables, is valid for any motion of coordinate system xyz (fixed to the control volume), provided that:

1. All velocities are measured *relative* to the control volume.

2. All time derivatives are measured *relative* to the control volume.

To emphasize this point, we rewrite Eq. 4.11 as

$$\left.\frac{dN}{dt}\right)_{\text{system}} = \frac{\partial}{\partial t}\int_{\text{CV}} \eta\rho\,d\Psi + \int_{\text{CS}} \eta\rho\vec{V}_{xyz}\cdot d\vec{A} \qquad (4.26)$$

Since all time derivatives must be measured relative to the control volume, in using this equation to obtain the momentum equation for an inertial control volume from the system formulation, we must set

$$N = \vec{P}_{xyz} \qquad \text{and} \qquad \eta = \vec{V}_{xyz}$$

The control volume equation is then written as

$$\vec{F} = \vec{F}_S + \vec{F}_B = \frac{\partial}{\partial t}\int_{\text{CV}} \vec{V}_{xyz}\rho\,d\Psi + \int_{\text{CS}} \vec{V}_{xyz}\rho\vec{V}_{xyz}\cdot d\vec{A} \qquad (4.27)$$

Equation 4.27 is the formulation of Newton's second law applied to any inertial control volume (stationary or moving with a constant velocity). It is identical to Eq. 4.18 except that we have included the subscript xyz to emphasize that quantities must be measured relative to the control volume. (It is helpful to imagine that the velocities are those that would be seen by an observer moving at constant speed with the control volume.) The momentum equation is applied to an inertial control volume moving with constant velocity in Example Problem 4.10.

EXAMPLE 4.10—Vane Moving with Constant Velocity

The sketch shows a vane with a turning angle of 60°. The vane moves at constant speed, $U = 10$ m/sec, and receives a jet of water that leaves a stationary nozzle with speed $V = 30$ m/sec. The nozzle has an exit area of 0.003 m². Determine the force that must be applied to maintain the vane speed constant.

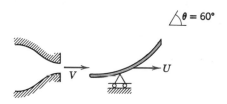

EXAMPLE PROBLEM 4.10

GIVEN: Vane, with turning angle $\theta = 60°$, moves with constant velocity, $\vec{U} = 10\hat{i}$ m/sec. Water from a constant-area nozzle, $A = 0.003$ m², with velocity $\vec{V} = 30\hat{i}$ m/sec, flows over the vane as shown.

FIND: The force that must be applied to maintain the vane speed constant.

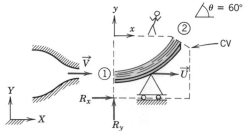

SOLUTION:
Select a control volume moving with the vane at constant velocity, $\vec{U}$, as shown by the dashed lines. R_x and R_y are the components of force required to maintain the velocity of the control volume at $10\hat{i}$ m/sec.

The control volume is inertial, since it is not accelerating (U = constant). Remember that all velocities must be measured relative to the control volume in applying the basic equations.

Basic equations:
$$\vec{F_S} + \vec{F_B} = \frac{\partial}{\partial t} \int_{CV} \vec{V}_{xyz}\, \rho\, d\Psi + \int_{CS} \vec{V}_{xyz}\, \rho \vec{V}_{xyz} \cdot d\vec{A}$$

$$0 = \frac{\partial}{\partial t} \int_{CV} \rho\, d\Psi + \int_{CS} \rho \vec{V}_{xyz} \cdot d\vec{A}$$

Assumptions: (1) Flow is steady relative to the vane
(2) Magnitude of relative velocity along the vane is constant:
$|\vec{V_1}| = |\vec{V_2}| = V - U$
(3) Properties are uniform at sections ① and ②
(4) $F_{B_x} = 0$
(5) Incompressible flow

The x component of the momentum equation is

$$F_{S_x} + \overset{=\,0(4)}{\cancel{F_{B_x}}} = \overset{=\,0(1)}{\cancel{\frac{\partial}{\partial t} \int_{CV}}} u_{xyz}\, \rho\, d\Psi + \int_{CS} u_{xyz}\, \rho \vec{V}_{xyz} \cdot d\vec{A}$$

There is no net pressure force, since p_{atm} acts on all sides of the CV. Thus

$$R_x = \int_{A_1} u\{-|\rho V\, dA|\} + \int_{A_2} u\{|\rho V\, dA|\} = -u_1|\rho V_1 A_1| + u_2|\rho V_2 A_2|$$

(All velocities are measured relative to xyz.) From the continuity equation

$$0 = \int_{A_1} \{-|\rho V\, dA|\} + \int_{A_2} |\rho V\, dA| = -|\rho V_1 A_1| + |\rho V_2 A_2|$$

or

$$|\rho V_1 A_1| = |\rho V_2 A_2|$$

Therefore,

$$R_x = (u_2 - u_1)|\rho V_1 A_1|$$

All velocities must be measured relative to the CV, so we note that

$$V_1 = V - U \qquad V_2 = V - U$$

$$u_1 = V - U \qquad u_2 = (V - U)\cos\theta$$

Substituting yields

$$R_x = [(V - U)\cos\theta - (V - U)]|\rho(V - U)A_1| = (V - U)(\cos\theta - 1)|\rho(V - U)A_1|$$

$$= \frac{(30 - 10)}{} \frac{m}{\sec} (0.50 - 1) \left| \frac{999}{} \frac{kg}{m^3} \frac{(30 - 10)}{} \frac{m}{\sec} \times 0.003\ m^2 \right| \frac{N \cdot \sec^2}{kg \cdot m}$$

$$R_x = -599\ N \qquad \{\text{to the left}\}$$

Writing the y component of the momentum equation, we obtain

$$F_{S_y} + F_{B_y} = \overset{=\,0(1)}{\cancel{\frac{\partial}{\partial t} \int_{CV}}} v_{xyz}\, \rho\, d\Psi + \int_{CS} v_{xyz}\, \rho \vec{V}_{xyz} \cdot d\vec{A}$$

Denoting the mass of the CV as M, then

$$R_y - Mg = \int_{CS} v\rho\vec{V} \cdot d\vec{A} = \int_{A_2} v\rho\vec{V} \cdot d\vec{A} \qquad \{v_1 = 0\} \qquad \left\{ \begin{array}{l} \text{All velocities are} \\ \text{measured relative to} \\ xyz. \end{array} \right\}$$

$$= \int_{A_2} v|\rho V \, dA| = v_2|\rho V_2 A_2| = v_2|\rho V_1 A_1| \qquad \{\text{Recall } |\rho V_2 A_2| = |\rho V_1 A_1|.\}$$

$$= (V - U)\sin\theta|\rho(V - U)A_1|$$

$$= \frac{(30 - 10)}{\text{sec}} \frac{\text{m}}{} (0.866) \left| 999 \frac{\text{kg}}{\text{m}^3} \frac{(30 - 10)}{\text{sec}} \frac{\text{m}}{} \times 0.003 \text{ m}^2 \right| \frac{\text{N} \cdot \text{sec}^2}{\text{kg} \cdot \text{m}}$$

$$R_y - Mg = 1.04 \text{ kN} \qquad \{\text{upward}\}$$

Thus the vertical force (in addition to the weight of the vane and water within the CV) required to maintain the motion is

$$R_y = 1.04 \text{ kN} \qquad \{\text{upward}\}$$

Then the net force required to maintain the constant vane speed is

$$\vec{R} = -0.599\hat{i} + 1.04\hat{j} \text{ kN} \qquad\qquad \vec{R}$$

{ This problem illustrates the fact that in applying the momentum equation to an inertial control volume all velocities must be measured relative to the control volume. }

4-5 MOMENTUM EQUATION FOR CONTROL VOLUME WITH RECTILINEAR ACCELERATION

For an inertial control volume (having no acceleration relative to a stationary frame of reference), the appropriate formulation of Newton's second law is given by Eq. 4.27,

$$\vec{F} = \vec{F}_S + \vec{F}_B = \frac{\partial}{\partial t} \int_{CV} \vec{V}_{xyz}\rho \, d\Psi + \int_{CS} \vec{V}_{xyz}\rho\vec{V}_{xyz} \cdot d\vec{A} \qquad (4.27)$$

Not all control volumes are inertial; a rocket must accelerate if it is to get off the ground. Since we are interested in analyzing control volumes that may accelerate relative to inertial coordinates, it is logical to ask whether Eq. 4.27 can be used for an accelerating control volume. To answer this question, let us briefly review the two major elements used in developing Eq. 4.27.

First, in relating the system derivatives to the control volume formulation (Eq. 4.26 or 4.11), the control volume was fixed relative to xyz; the flow field, $\vec{V}(x, y, z, t)$, was specified relative to the coordinates x, y, and z. No restriction was placed on the motion of the xyz reference frame. Consequently, Eq. 4.26 is valid at any instant for any arbitrary motion of the coordinates x, y, and z provided that all time derivatives and velocities in the equation are measured relative to the control volume.

Second, the system equation

$$\vec{F} = \frac{d\vec{P}}{dt}\bigg)_{\text{system}} \qquad (4.2a)$$

where the linear momentum, $\vec{P}$, of the system is given by

$$\vec{P}_{\text{system}} = \int_{\text{mass(system)}} \vec{V} \, dm = \int_{\Psi \text{ (system)}} \vec{V}\rho \, d\Psi \qquad (4.2b)$$

is valid only for velocities measured relative to an inertial reference frame. Thus, if we denote the inertial reference frame by XYZ, then Newton's second law states that

$$\vec{F} = \frac{d\vec{P}_{XYZ}}{dt}\bigg)_{\text{system}} \tag{4.28}$$

Since the time derivatives of $\vec{P}_{XYZ}$ and $\vec{P}_{xyz}$ are not equal for a system accelerating relative to an inertial reference frame, Eq. 4.27 is not valid for an accelerating control volume.

To develop the momentum equation for a linearly accelerating control volume, it is necessary to relate $\vec{P}_{XYZ}$ of the system to $\vec{P}_{xyz}$ of the system. The system derivative $d\vec{P}_{xyz}/dt$ can be related to control volume variables through Eq. 4.26. We begin by writing Newton's second law for a system, remembering that the acceleration must be measured relative to an inertial reference frame that we have designated XYZ. We write

$$\vec{F} = \frac{d\vec{P}_{XYZ}}{dt}\bigg)_{\text{system}} = \frac{d}{dt}\int_{\text{mass(system)}} \vec{V}_{XYZ}\, dm = \int_{\text{mass(system)}} \frac{d\vec{V}_{XYZ}}{dt}\, dm$$

$$\vec{F} = \int_{\text{mass(system)}} \vec{a}_{XYZ}\, dm \tag{4.29}$$

The only problem now is to obtain a suitable expression for $\vec{a}_{XYZ}$, for the special case in which coordinate frame xyz undergoes pure translation, without rotation, relative to inertial frame XYZ.

Since the motion of xyz is pure translation, without rotation, relative to inertial reference frame XYZ, then

$$\vec{a}_{XYZ} = \vec{a}_{xyz} + \vec{a}_{rf} \tag{4.30}$$

where

$\vec{a}_{XYZ}$ is the rectilinear acceleration of the system relative to inertial reference frame XYZ,

$\vec{a}_{xyz}$ is the rectilinear acceleration of the system relative to noninertial reference frame xyz, and

$\vec{a}_{rf}$ is the rectilinear acceleration of noninertial reference frame xyz relative to inertial frame XYZ.

For this case, we write the system equation as

$$\vec{F} = \int_{\text{mass(system)}} \vec{a}_{XYZ}\, dm = \int_{\text{mass(system)}} (\vec{a}_{xyz} + \vec{a}_{rf})\, dm$$

Alternately,

$$\vec{F} - \int_{\text{mass(system)}} \vec{a}_{rf}\, dm = \int_{\text{mass(system)}} \vec{a}_{xyz}\, dm$$

Since

$$\vec{a}_{xyz} = \frac{d\vec{V}_{xyz}}{dt}$$

then

$$\vec{F} - \int_{\text{mass(system)}} \vec{a}_{rf}\, dm = \int_{\text{mass(system)}} \frac{d\vec{V}_{xyz}}{dt}\, dm$$

$$= \frac{d}{dt}\int_{\text{mass(system)}} \vec{V}_{xyz}\, dm = \frac{d\vec{P}_{xyz}}{dt}\bigg)_{\text{system}}$$

Since $dm = \rho \, d\forall$, the system equation can be written as

$$\vec{F} - \int_{\forall(\text{system})} \vec{a}_{rf} \rho \, d\forall = \left.\frac{d\vec{P}_{xyz}}{dt}\right)_{\text{system}} \tag{4.31a}$$

where the linear momentum, $\vec{P}_{xyz}$, of the system is given by

$$\left.\vec{P}_{xyz}\right)_{\text{system}} = \int_{\text{mass(system)}} \vec{V}_{xyz} \, dm = \int_{\forall(\text{system})} \vec{V}_{xyz} \rho \, d\forall \tag{4.31b}$$

and the force, $\vec{F}$, includes all surface and body forces acting on the system.

The system formulation is related to the formulation for a moving control volume by Eq. 4.26,

$$\left.\frac{dN}{dt}\right)_{\text{system}} = \frac{\partial}{\partial t} \int_{\text{CV}} \eta \rho \, d\forall + \int_{\text{CS}} \eta \rho \vec{V}_{xyz} \cdot d\vec{A} \tag{4.26}$$

where

$$N_{\text{system}} = \int_{\text{mass(system)}} \eta \, dm = \int_{\forall(\text{system})} \eta \rho \, d\forall \tag{4.6}$$

To derive the control volume formulation of Newton's second law, we set

$$N = \vec{P}_{xyz} \qquad \text{and} \qquad \eta = \vec{V}_{xyz}$$

From Eq. 4.26, with this substitution, we obtain

$$\left.\frac{d\vec{P}_{xyz}}{dt}\right)_{\text{system}} = \frac{\partial}{\partial t} \int_{\text{CV}} \vec{V}_{xyz} \rho \, d\forall + \int_{\text{CS}} \vec{V}_{xyz} \rho \vec{V}_{xyz} \cdot d\vec{A} \tag{4.32}$$

From the system equation,

$$\left.\frac{d\vec{P}_{xyz}}{dt}\right)_{\text{system}} = \vec{F}_{\text{on system}} - \int_{\forall(\text{system})} \vec{a}_{rf} \rho \, d\forall \tag{4.31c}$$

Since the system and the control volume coincided at t_0, then,

$$\vec{F}_{\text{on system}} - \int_{\forall(\text{system})} \vec{a}_{rf} \rho \, d\forall = \vec{F}_{\text{on CV}} - \int_{\text{CV}} \vec{a}_{rf} \rho \, d\forall$$

In light of this, Eqs. 4.31c and 4.32 may be combined to yield the formulation of Newton's second law for a control volume accelerating, without rotation, relative to an inertial reference frame:

$$\vec{F} - \int_{\text{CV}} \vec{a}_{rf} \rho \, d\forall = \frac{\partial}{\partial t} \int_{\text{CV}} \vec{V}_{xyz} \rho \, d\forall + \int_{\text{CS}} \vec{V}_{xyz} \rho \vec{V}_{xyz} \cdot d\vec{A} \tag{4.33}$$

Since $\vec{F} = \vec{F}_S + \vec{F}_B$, Eq. 4.33 becomes

$$\vec{F}_S + \vec{F}_B - \int_{\text{CV}} \vec{a}_{rf} \rho \, d\forall = \frac{\partial}{\partial t} \int_{\text{CV}} \vec{V}_{xyz} \rho \, d\forall + \int_{\text{CS}} \vec{V}_{xyz} \rho \vec{V}_{xyz} \cdot d\vec{A} \tag{4.34}$$

Comparing the momentum equation for a control volume with rectilinear acceleration, Eq. 4.34, to that for a nonaccelerating control volume, Eq. 4.27, we see that the only difference is the presence of one additional term in Eq. 4.34. When the control volume is not accelerating relative to inertial reference frame XYZ, then $\vec{a}_{rf} = 0$, and Eq. 4.34 reduces to Eq. 4.27.

The precautions concerning the use of Eq. 4.27 also apply to the use of Eq. 4.34. Before attempting to apply either equation, one must draw the boundaries of the control volume and label appropriate coordinate directions. For an accelerating control volume, one must label two coordinate systems: one (xyz) on the control volume and the other (XYZ) an inertial reference frame.

In Eq. 4.34, $\vec{F}_S$ represents all surface forces acting on the control volume. Since the mass within the control volume may vary with time, both the remaining terms on the left side of the equation may be functions of time. Futhermore, the acceleration, $\vec{a}_{rf}$, of the reference frame xyz relative to an inertial frame, in general will be a function of time.

All velocities in Eq. 4.34 are measured relative to the control volume. The momentum flux, $\vec{V}_{xyz} \rho \vec{V}_{xyz} \cdot d\vec{A}$, through an element of the control surface area, $d\vec{A}$, is a vector. The sign of the scalar product, $\rho \vec{V}_{xyz} \cdot d\vec{A}$, depends on the direction of the velocity vector, $\vec{V}_{xyz}$, relative to the area vector, $d\vec{A}$. The sign of the vector velocity, $\vec{V}_{xyz}$, depends on the coordinate system chosen.

The momentum equation is a vector equation. As with all vector equations, it may be written as three scalar component equations. The scalar components of Eq. 4.34 are

$$F_{S_x} + F_{B_x} - \int_{CV} a_{rf_x} \rho \, d\forall = \frac{\partial}{\partial t} \int_{CV} u_{xyz} \rho \, d\forall + \int_{CS} u_{xyz} \rho \vec{V}_{xyz} \cdot d\vec{A} \qquad (4.35a)$$

$$F_{S_y} + F_{B_y} - \int_{CV} a_{rf_y} \rho \, d\forall = \frac{\partial}{\partial t} \int_{CV} v_{xyz} \rho \, d\forall + \int_{CS} v_{xyz} \rho \vec{V}_{xyz} \cdot d\vec{A} \qquad (4.35b)$$

$$F_{S_z} + F_{B_z} - \int_{CV} a_{rf_z} \rho \, d\forall = \frac{\partial}{\partial t} \int_{CV} w_{xyz} \rho \, d\forall + \int_{CS} w_{xyz} \rho \vec{V}_{xyz} \cdot d\vec{A} \qquad (4.35c)$$

EXAMPLE 4.11—Vane Moving with Rectilinear Acceleration

A vane, with turning angle $\theta = 60°$, is attached to a cart. The cart and vane, of mass $M = 75$ kg, roll on a level track. Friction and air resistance may be neglected. The vane receives a jet of water, which leaves a stationary nozzle horizontally at $V = 35$ m/sec. The nozzle exit area is $A = 0.003$ m². Determine the velocity of the cart as a function of time and plot the results.

EXAMPLE PROBLEM 4.11

GIVEN: Vane and cart as sketched, with $M = 75$ kg.

FIND: (a) $U(t)$.
   (b) Plot results.

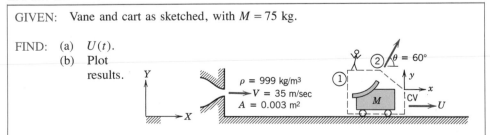

SOLUTION:
Choose the control volume and coordinate systems shown for the analysis. Note that XY is a fixed frame, while frame xy moves with the cart. Apply the x component of the momentum equation.

Basic equation:
$$F_{S_x} + F_{B_x} - \int_{CV} a_{rf_x}\rho\, d\forall = \frac{\partial}{\partial t}\int_{CV} u_{xyz}\rho\, d\forall + \int_{CS} u_{xyz}\rho\vec{V}_{xyz}\cdot d\vec{A}$$

with $= 0(1)$, $= 0(2)$ over first two terms and $\simeq 0(3)$ over first right term.

Assumptions:
(1) $F_{S_x} = 0$, since no resistance is present
(2) $F_{B_x} = 0$
(3) Neglect the mass and rate of change of u_{xyz} for water in contact with the vane compared to the cart mass

$$\frac{\partial}{\partial t}\int_{CV} u_{xyz}\rho\, d\forall \simeq 0$$

(4) Uniform flow at sections ① and ②
(5) Water stream is not slowed by friction on the vane, so $|\vec{V}_{xyz_1}| = |\vec{V}_{xyz_2}|$
(6) $A_2 = A_1 = A$

Then

$$-\int_{CV} a_{rf_x}\rho\, d\forall = u_{xyz_1}\{-|\rho V_{xyz_1}A_1|\} + u_{xyz_2}\{|\rho V_{xyz_2}A_2|\}$$

where all velocities must be measured relative to the xyz frame. Dropping subscripts rf and xyz, we obtain

$$-\int_{CV} a_x\rho\, d\forall = u_1\{-|\rho V_1 A_1|\} + u_2\{|\rho V_2 A_2|\} \tag{1}$$

Evaluating these terms separately gives

$$-\int_{CV} a_x\rho\, d\forall = -a_x M_{CV} = -a_x M = -\frac{dU}{dt}M$$

$$u_1\{-|\rho V_1 A_1|\} = (V - U)\{-|\rho(V - U)A|\} = -\rho(V - U)^2 A$$

$$u_2\{|\rho V_2 A_2|\} = (V - U)\cos\theta\{|\rho(V - U)A|\} = \rho(V - U)^2 A\cos\theta$$

Absolute value signs have been dropped from the flux terms, since $V \geq U$. Substitution into Eq. 1 gives

$$-M\frac{dU}{dt} = -\rho(V - U)^2 A + \rho(V - U)^2 A\cos\theta$$

or

$$-M\frac{dU}{dt} = (\cos\theta - 1)\rho(V - U)^2 A$$

Separating variables, we obtain

$$\frac{dU}{(V - U)^2} = \frac{(1 - \cos\theta)\rho A}{M}dt = b\, dt \quad \text{where } b = \frac{(1 - \cos\theta)\rho A}{M}$$

Note that since $V = $ constant, $dU = -d(V - U)$. Integrating between limits $U = 0$ at $t = 0$, and $U = U$ at $t = t$,

$$\int_0^U \frac{dU}{(V - U)^2} = \int_0^U \frac{-d(V - U)}{(V - U)^2} = \frac{1}{(V - U)}\Big]_0^U = \int_0^t b\, dt = bt$$

or

$$\frac{1}{(V - U)} - \frac{1}{V} = \frac{U}{V(V - U)} = bt$$

Solving for U, we obtain

$$\frac{U}{V} = \frac{Vbt}{1+Vbt}$$

Evaluating the term Vb gives

$$Vb = V\frac{(1-\cos\theta)\rho A}{M}$$

$$Vb = \frac{35}{\text{sec}} \frac{m}{\text{sec}} \times \frac{(1-0.5)}{75 \text{ kg}} \times \frac{999 \text{ kg}}{m^3} \times 0.003 \text{ m}^2 = 0.699 \text{ sec}^{-1}$$

Thus

$$\frac{U}{V} = \frac{0.699t}{1+0.699t} \qquad\qquad (t \text{ in sec}) \qquad\qquad U(t)$$

Plot:

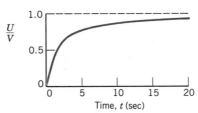

As the vane speed, U, nears the jet speed, V, the mass flow rate crossing the control surface decreases toward zero. The plot shows the corresponding decrease in vane acceleration.

EXAMPLE 4.12—Rocket Directed Vertically

A small rocket, with an initial mass of 400 kg, is to be launched vertically. Upon ignition the rocket consumes fuel at the rate of 5 kg/sec and ejects gas at atmospheric pressure with a speed of 3500 m/sec relative to the rocket. Determine the initial acceleration of the rocket and the rocket speed after 10 sec, if air resistance is neglected.

EXAMPLE PROBLEM 4.12

GIVEN: Small rocket accelerates vertically from rest.
            Air resistance may be neglected.
            Rate of fuel consumption, $\dot{m}_{\text{out}} = 5$ kg/sec.
            Exhaust velocity, $V_e = 3500$ m/sec, relative to rocket,
            leaving at atmospheric pressure.

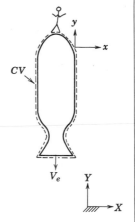

FIND: (a)   Initial acceleration of the rocket.
         (b)   Rocket velocity after 10 sec.

SOLUTION:
Choose a control volume as shown by dashed lines. Because the control volume is accelerating, define inertial coordinate system XY and coordinate system xy attached to the CV. Apply the y component of the momentum equation.

Basic equation:

$$F_{S_y} + F_{B_y} - \int_{CV} a_{rf_y} \rho \, d\forall = \frac{\partial}{\partial t} \int_{CV} v_{xyz} \rho \, d\forall + \int_{CS} v_{xyz} \rho \vec{V}_{xyz} \cdot d\vec{A}$$

Assumptions: (1) Atmospheric pressure acts on all surfaces of the CV; since air resistance is neglected, then $F_{S_y} = 0$
 (2) Gravity is the only body force; g is constant
 (3) Flow leaving the rocket is uniform and V_e is constant

Under these assumptions the momentum equation reduces to

$$F_{B_y} - \int_{CV} a_{rf_y} \rho \, d\forall = \frac{\partial}{\partial t} \int_{CV} v_{xyz} \rho \, d\forall + \int_{CS} v_{xyz} \rho \vec{V}_{xyz} \cdot d\vec{A} \qquad (1)$$
$$\quad\quad\text{Ⓐ} \qquad\qquad\quad\text{Ⓑ} \qquad\qquad\quad\text{©} \qquad\qquad\quad\text{Ⓓ}$$

Let us look at the equation term by term:

Ⓐ $\quad F_{B_y} = -\int_{CV} g \rho \, d\forall = -g \int_{CV} \rho \, d\forall = -g M_{CV} \qquad$ {since g is constant}

The mass of the CV will be a function of time because mass is leaving the CV at rate $\dot{m}_e$. To determine M_{CV} as a function of time, we use the conservation of mass equation

$$\frac{\partial}{\partial t} \int_{CV} \rho \, d\forall + \int_{CS} \rho \vec{V} \cdot d\vec{A} = 0$$

Then

$$\frac{\partial}{\partial t} \int_{CV} \rho \, d\forall = -\int_{CS} \rho \vec{V} \cdot d\vec{A} = -\int_{A_e} \rho \vec{V} \cdot d\vec{A} = -\int_{A_e} \{|\rho V \, dA|\} = -|\dot{m}_e|$$

The minus sign indicates that the mass of the CV is decreasing with time. Since the mass of the CV is only a function of time, we can write

$$\frac{d M_{CV}}{dt} = -|\dot{m}_e|$$

To find the mass of the CV at any time, t, we integrate

$$\int_{M_0}^{M} d M_{CV} = -\int_0^t |\dot{m}_e| dt \qquad \text{where at } t=0, M_{CV} = M_0, \text{ and at } t=t, M_{CV} = M$$

Then, $M - M_0 = -|\dot{m}_e| t$, or $M = M_0 - \dot{m}_e t$.
{Since $\dot{m}_e$ is positive, we have dropped the absolute value sign.}
 Substituting the expression for M into term Ⓐ, we obtain

$$F_{B_y} = -\int_{CV} g \rho \, d\forall = -g M_{CV} = -g(M_0 - \dot{m}_e t)$$

Ⓑ $\quad -\int_{CV} a_{rf_y} \rho \, d\forall$

The acceleration, a_{rf_y}, of the CV is that seen by an observer in the XY coordinate system. Thus a_{rf_y} is not a function of the coordinates xyz, and

$$-\int_{CV} a_{rf_y} \rho \, d\forall = -a_{rf_y} \int_{CV} \rho \, d\forall = -a_{rf_y} M_{CV} = -a_{rf_y}(M_0 - \dot{m}_e t)$$

© $\quad \dfrac{\partial}{\partial t} \displaystyle\int_{CV} v_{xyz} \rho \, d\forall$

is the time rate of change of the y momentum of the fluid in the control volume measured relative to the control volume.

Even though the y momentum of the fluid inside the CV, measured relative to the CV, is a large number, it does not change appreciably with time. To see this, we must recognize that:

(1)   The unburned fuel and the rocket structure have zero momentum relative to the rocket.
(2)   The velocity of the gas at the nozzle exit remains constant with time as does the velocity at various points in the nozzle.

Consequently, it is reasonable to assume that

$$\frac{\partial}{\partial t} \int_{CV} v_{xyz} \rho \, d\Psi \approx 0$$

ⓓ $$\int_{CS} v_{xyz} \rho \vec{V}_{xyz} \cdot d\vec{A} = \int_{A_e} v_{xyz} |\rho V_{xyz} \, dA| = v_{xyz} |\dot{m}_e|$$

Since $\vec{V}_e = -V_e \hat{j}$, then

$$v_{xyz} |\dot{m}_e| = -V_e |\dot{m}_e| = -V_e \dot{m}_e$$

Substituting terms ⓐ through ⓓ into Eq. 1, we obtain

$$-g(M_0 - \dot{m}_e t) - a_{rf_y}(M_0 - \dot{m}_e t) = -V_e \dot{m}_e$$

or

$$a_{rf_y} = \frac{V_e \dot{m}_e}{M_0 - \dot{m}_e t} - g \qquad (2)$$

At time $t = 0$,

$$a_{rf_y}\Big)_{t=0} = \frac{V_e \dot{m}_e}{M_0} - g = 3500 \frac{m}{sec} \times 5 \frac{kg}{sec} \times \frac{1}{400 \ kg} - 9.81 \frac{m}{sec^2}$$

$$a_{rf_y}\Big)_{t=0} = 33.9 \ m/sec^2 \qquad\qquad\qquad\qquad a_{rf_y})_{t=0}$$

The acceleration of the CV is by definition

$$a_{rf_y} = \frac{dV_{CV}}{dt}$$

Substituting from Eq. 2,

$$\frac{dV_{CV}}{dt} = \frac{V_e \dot{m}_e}{M_0 - \dot{m}_e t} - g$$

Separating variables and integrating gives

$$V_{CV} = \int_0^{V_{CV}} dV_{CV} = \int_0^t \frac{V_e \dot{m}_e \, dt}{M_0 - \dot{m}_e t} - \int_0^t g \, dt = -V_e \ln\left[\frac{M_0 - \dot{m}_e t}{M_0}\right] - g t$$

At $t = 10$ sec,

$$V_{CV} = -3500 \frac{m}{sec} \times \ln\left[\frac{350 \ kg}{400 \ kg}\right] - 9.81 \frac{m}{sec^2} \times 10 \ sec$$

$$V_{CV} = 369 \ m/sec \qquad\qquad\qquad\qquad\qquad V_{CV})_{t=10 \ sec}$$

{ This problem illustrates the application of the momentum equation to a linearly accelerating control volume. }

****4-6 MOMENTUM EQUATION FOR CONTROL VOLUME WITH ARBITRARY ACCELERATION**

In Section 4-5 we formulated the momentum equation for a control volume with rectilinear acceleration. The purpose of this section is to extend the formulation to include rotation and angular acceleration of the control volume, in addtion to translation and rectilinear acceleration.

First, we develop an expression for Newton's second law in an arbitrary, noninertial coordinate system. Then we use Eq. 4.26 to complete the formulation for a control volume. Newton's second law for a system moving relative to an inertial coordinate system is given by

$$\vec{F} = \frac{d\vec{P}_{XYZ}}{dt}\bigg)_{\text{system}}$$

Since

$$\vec{P}_{XYZ}\big)_{\text{system}} = \int_{M(\text{system})} d\vec{V}_{XYZ}\, dm$$

and $M(\text{system})$ is constant, then

$$\vec{F} = \frac{d}{dt}\int_{M(\text{system})} \vec{V}_{XYZ}\, dm = \int_{M(\text{system})} \frac{d\vec{V}_{XYZ}}{dt}\, dm$$

or

$$\vec{F} = \int_{M(\text{system})} \vec{a}_{XYZ}\, dm \tag{4.36}$$

The basic problem is to relate $\vec{a}_{XYZ}$ to the acceleration $\vec{a}_{xyz}$, measured relative to a noninertial coordinate system. For this purpose, consider the noninertial reference frame, xyz, shown in Fig. 4.5.

The noninertial frame, xyz, is located by position vector $\vec{R}$ relative to the fixed frame. The noninertial frame rotates with angular velocity $\vec{\omega}$.[2] The particle is located relative to the moving frame by position vector $\vec{r} = \hat{i}x + \hat{j}y + \hat{k}z$. Relative to the inertial reference frame, XYZ, the position of the particle is denoted by position vector $\vec{X}$. From the geometry of the figure, $\vec{X} = \vec{R} + \vec{r}$.

The velocity of the particle, relative to an observer in the XYZ system, is

$$\vec{V}_{XYZ} = \frac{d\vec{X}}{dt} = \frac{d\vec{R}}{dt} + \frac{d\vec{r}}{dt} = \vec{V}_{rf} + \frac{d\vec{r}}{dt} \tag{4.37}$$

We must be careful in evaluating $d\vec{r}/dt$ because both the magnitude, $|\vec{r}|$, and the orientation of the unit vectors, $\hat{i}$, $\hat{j}$, and $\hat{k}$, are functions of time. Thus

$$\frac{d\vec{r}}{dt} = \frac{d}{dt}(x\hat{i} + y\hat{j} + z\hat{k}) = \hat{i}\frac{dx}{dt} + x\frac{d\hat{i}}{dt} + \hat{j}\frac{dy}{dt} + y\frac{d\hat{j}}{dt} + \hat{k}\frac{dz}{dt} + z\frac{d\hat{k}}{dt} \tag{4.38a}$$

The terms dx/dt, dy/dt, and dz/dt are the velocity components of the particle relative to xyz. Thus

$$\vec{V}_{xyz} = \hat{i}\frac{dx}{dt} + \hat{j}\frac{dy}{dt} + \hat{k}\frac{dz}{dt} \tag{4.38b}$$

** This section may be omitted without loss of continuity in the text material.

[2] Note that any arbitrary motion can be decomposed into a translation plus a rotation.

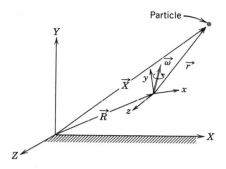

Fig. 4.5   Location of a particle in inertial (XYZ) and noninertial (xyz) reference frames.

You may recall from dynamics (see Example Problem 4.13) that for a rotating coordinate system,

$$\vec{\omega} \times \vec{r} = x\frac{d\hat{i}}{dt} + y\frac{d\hat{j}}{dt} + z\frac{d\hat{k}}{dt} \tag{4.38c}$$

Combining Eqs. 4.38a, 4.38b, and 4.38c, we obtain

$$\frac{d\vec{r}}{dt} = \vec{V}_{xyz} + \vec{\omega} \times \vec{r} \tag{4.38d}$$

Substituting into Eq. 4.37 gives

$$\vec{V}_{XYZ} = \vec{V}_{rf} + \vec{V}_{xyz} + \vec{\omega} \times \vec{r} \tag{4.39}$$

The acceleration of the particle relative to an observer in the XYZ system is

$$\vec{a}_{XYZ} = \frac{d\vec{V}_{XYZ}}{dt} = \frac{d\vec{V}_{rf}}{dt} + \frac{d\vec{V}_{xyz}}{dt} + \frac{d}{dt}(\vec{\omega} \times \vec{r})$$

or

$$\vec{a}_{XYZ} = \vec{a}_{rf} + \frac{d\vec{V}_{xyz}}{dt} + \frac{d}{dt}(\vec{\omega} \times \vec{r}) \tag{4.40}$$

Both $\vec{V}_{xyz}$ and $\vec{r}$ are measured relative to xyz, so the same caution observed in developing Eq. 4.38d applies. Thus

$$\frac{d\vec{V}_{xyz}}{dt} = \vec{a}_{xyz} + \vec{\omega} \times \vec{V}_{xyz} \tag{4.41a}$$

and

$$\frac{d}{dt}(\vec{\omega} \times \vec{r}) = \frac{d\vec{\omega}}{dt} \times \vec{r} + \vec{\omega} \times \frac{d\vec{r}}{dt}$$
$$= \dot{\vec{\omega}} \times \vec{r} + \vec{\omega} \times (\vec{V}_{xyz} + \vec{\omega} \times \vec{r})$$

or

$$\frac{d}{dt}(\vec{\omega} \times \vec{r}) = \dot{\vec{\omega}} \times \vec{r} + \vec{\omega} \times \vec{V}_{xyz} + \vec{\omega} \times (\vec{\omega} \times \vec{r}) \tag{4.41b}$$

Substituting Eqs. 4.41a and 4.41b into Eq. 4.40, we obtain

$$\vec{a}_{XYZ} = \vec{a}_{rf} + \vec{a}_{xyz} + 2\vec{\omega} \times \vec{V}_{xyz} + \vec{\omega} \times (\vec{\omega} \times \vec{r}) + \dot{\vec{\omega}} \times \vec{r} \tag{4.42}$$

The physical meaning of each term in Eq. 4.42 is

$\vec{a}_{XYZ}$:   Absolute rectilinear acceleration of a particle relative to fixed reference frame XYZ

$\vec{a}_{rf}$:   Absolute rectilinear acceleration of moving reference frame xyz relative to fixed frame XYZ

$\vec{a}_{xyz}$:   Rectilinear acceleration of a particle *relative* to moving reference frame xyz (this acceleration would be that seen by an observer on moving frame xyz)

$2\vec{\omega} \times \vec{V}_{xyz}$:   Coriolis acceleration due to motion of the particle *within* moving frame xyz

$\vec{\omega} \times (\vec{\omega} \times \vec{r})$:   Centripetal acceleration due to rotation of moving frame xyz

$\dot{\vec{\omega}} \times \vec{r}$:   Tangential acceleration due to angular acceleration of moving reference frame xyz

Substituting $\vec{a}_{XYZ}$, as given by Eq. 4.42, into Eq. 4.36, we obtain

$$\vec{F}_{\text{system}} = \int_{M(\text{system})} [\vec{a}_{rf} + \vec{a}_{xyz} + 2\vec{\omega} \times \vec{V}_{xyz} + \vec{\omega} \times (\vec{\omega} \times \vec{r}) + \dot{\vec{\omega}} \times \vec{r}]\, dm$$

or

$$\vec{F} - \int_{M(\text{system})} [\vec{a}_{rf} + 2\vec{\omega} \times \vec{V}_{xyz} + \vec{\omega} \times (\vec{\omega} \times \vec{r}) + \dot{\vec{\omega}} \times \vec{r}]\, dm = \int_{M(\text{system})} \vec{a}_{xyz}\, dm \quad (4.43a)$$

But

$$\int_{M(\text{system})} \vec{a}_{xyz}\, dm = \int_{M(\text{system})} \frac{d\vec{V}_{xyz}}{dt}\, dm = \frac{d}{dt} \int_{M(\text{system})} \vec{V}_{xyz}\, dm = \frac{d\vec{P}_{xyz}}{dt} \Bigg)_{\text{system}}$$

$$(4.43b)$$

where all time derivatives are those seen by an observer fixed in noninertial frame xyz. Combining Eqs. 4.43a and 4.43b, we obtain

$$\vec{F} - \int_{M(\text{system})} [\vec{a}_{rf} + 2\vec{\omega} \times \vec{V}_{xyz} + \vec{\omega} \times (\vec{\omega} \times \vec{r}) + \dot{\vec{\omega}} \times \vec{r}]\, dm = \frac{d\vec{P}_{xyz}}{dt} \Bigg)_{\text{system}}$$

or

$$\vec{F}_S + \vec{F}_B - \int_{\Psi(\text{system})} [\vec{a}_{rf} + 2\vec{\omega} \times \vec{V}_{xyz} + \vec{\omega} \times (\vec{\omega} \times \vec{r}) + \dot{\vec{\omega}} \times \vec{r}]\rho\, d\Psi = \frac{d\vec{P}_{xyz}}{dt} \Bigg)_{\text{system}} \quad (4.44)$$

Equation 4.44 is a statement of Newton's second law for a system. The system derivative, $d\vec{P}_{xyz}/dt$, represents the rate of change of momentum, $\vec{P}_{xyz}$, of the system measured relative to xyz, as seen by an observer in xyz. This system derivative can be related to control volume variables through Eq. 4.26,

$$\frac{dN}{dt} \Bigg)_{\text{system}} = \frac{\partial}{\partial t} \int_{\text{CV}} \eta \rho\, d\Psi + \int_{\text{CS}} \eta \rho \vec{V}_{xyz} \cdot d\vec{A} \quad (4.26)$$

To obtain the control volume formulation, we set $N = \vec{P}_{xyz}$, and $\eta = \vec{V}_{xyz}$. Then Eqs. 4.26 and 4.44 may be combined to give

$$\vec{F}_S + \vec{F}_B - \int_{\text{CV}} [\vec{a}_{rf} + 2\vec{\omega} \times \vec{V}_{xyz} + \vec{\omega} \times (\vec{\omega} \times \vec{r}) + \dot{\vec{\omega}} \times \vec{r}]\rho\, d\Psi$$

$$= \frac{\partial}{\partial t} \int_{\text{CV}} \vec{V}_{xyz} \rho\, d\Psi + \int_{\text{CS}} \vec{V}_{xyz} \rho \vec{V}_{xyz} \cdot d\vec{A} \quad (4.45)$$

Equation 4.45 is the most general control volume formulation of Newton's second law. Comparing the momentum equation for a control volume moving with arbitrary acceleration, Eq. 4.45, to that for a control volume moving with rectilinear acceleration, Eq. 4.34, we see that the only difference is the presence of three additional terms on the left side of Eq. 4.45. These terms result from the angular motion of noninertial reference frame xyz. Note that Eq. 4.45 reduces to Eq. 4.34, when the angular terms are zero, and to Eq. 4.27 for an inertial control volume.

The precautions concerning the use of Eqs. 4.27 and 4.34 also apply to the use of Eq. 4.45. Before attempting to apply this equation, one must draw the boundaries of the control volume and label appropriate coordinate directions. For a control volume moving with arbitrary acceleration, one must label a coordinate system (xyz) on the control volume and an inertial reference frame (XYZ).

EXAMPLE 4.13—Velocity in Fixed and Noninertial Reference Frames

A reference frame, xyz, moves arbitrarily with respect to a fixed frame, XYZ. A particle moves with velocity $\vec{V}_{xyz} = (dx/dt)\hat{i} + (dy/dt)\hat{j} + (dz/dt)\hat{k}$, relative to frame xyz. Show that the absolute velocity of the particle is given by

$$\vec{V}_{XYZ} = \vec{V}_{rf} + \vec{V}_{xyz} + \vec{\omega} \times \vec{r}$$

EXAMPLE PROBLEM 4.13

GIVEN: Fixed and noninertial frames as shown.

FIND: $\vec{V}_{XYZ}$ in terms of $\vec{V}_{xyz}$, $\vec{\omega}$, $\vec{r}$, and $\vec{V}_{rf}$.

SOLUTION:

From the geometry of the sketch, $\vec{X} = \vec{R} + \vec{r}$, so

$$\vec{V}_{XYZ} = \frac{d\vec{X}}{dt} = \frac{d\vec{R}}{dt} + \frac{d\vec{r}}{dt} = \vec{V}_{rf} + \frac{d\vec{r}}{dt}$$

Since

$$\vec{r} = x\hat{i} + y\hat{j} + z\hat{k}$$

then

$$\frac{d\vec{r}}{dt} = \frac{dx}{dt}\hat{i} + \frac{dy}{dt}\hat{j} + \frac{dz}{dt}\hat{k} + x\frac{d\hat{i}}{dt} + y\frac{d\hat{j}}{dt} + z\frac{d\hat{k}}{dt}$$

or

$$\frac{d\vec{r}}{dt} = \vec{V}_{xyz} + x\frac{d\hat{i}}{dt} + y\frac{d\hat{j}}{dt} + z\frac{d\hat{k}}{dt}$$

The problem now is to evaluate $d\hat{i}/dt$, $d\hat{j}/dt$, and $d\hat{k}/dt$ due to the angular motion of frame xyz. To evaluate these derivatives, we must consider the rotation of each unit vector due to the three components of the angular velocity, $\vec{\omega}$, of frame xyz.

Consider the unit vector $\hat{i}$. It will rotate in the xy plane due to ω_z, as follows:

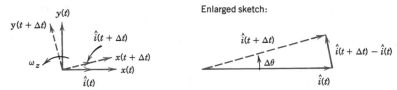

Now

$$\left.\frac{d\,\hat{i}}{dt}\right]_{\text{due to }\omega_z} = \lim_{\Delta t \to 0}\left[\frac{\hat{i}(t+\Delta t)-\hat{i}(t)}{\Delta t}\right] = \lim_{\Delta t \to 0}\left[\frac{(1)\Delta\theta\,\hat{j}}{\Delta t}\right] = \lim_{\Delta t \to 0}\left[\frac{(1)\omega_z\Delta t\,\hat{j}}{\Delta t}\right] = \hat{j}\omega_z$$

Similarly, $\hat{i}$ will rotate in the xz plane due to ω_y.

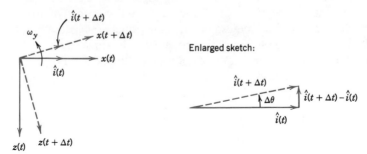

Then

$$\left.\frac{d\,\hat{i}}{dt}\right]_{\text{due to }\omega_y} = \lim_{\Delta t \to 0}\left[\frac{\hat{i}(t+\Delta t)-\hat{i}(t)}{\Delta t}\right] = \lim_{\Delta t \to 0}\left[\frac{(1)\Delta\theta(-\hat{k})}{\Delta t}\right] = \lim_{\Delta t \to 0}\left[\frac{(1)\omega_y\Delta t(-\hat{k})}{\Delta t}\right]$$

$$\left.\frac{d\,\hat{i}}{dt}\right]_{\text{due to }\omega_y} = -\hat{k}\omega_y$$

Rotation in the yz plane due to ω_x does not affect $\hat{i}$. Combining terms,

$$\frac{d\,\hat{i}}{dt} = \omega_z\hat{j} - \omega_y\hat{k}$$

By similar reasoning,

$$\frac{d\,\hat{j}}{dt} = \omega_x\hat{k} - \omega_z\hat{i} \qquad \text{and} \qquad \frac{d\,\hat{k}}{dt} = \omega_y\hat{i} - \omega_x\hat{j}$$

Thus

$$x\frac{d\,\hat{i}}{dt} + y\frac{d\,\hat{j}}{dt} + z\frac{d\,\hat{k}}{dt} = (z\omega_y - y\omega_z)\hat{i} + (x\omega_z - z\omega_x)\hat{j} + (y\omega_x - x\omega_y)\hat{k}$$

But

$$\vec{\omega}\times\vec{r} = \begin{vmatrix} \hat{i} & \hat{j} & \hat{k} \\ \omega_x & \omega_y & \omega_z \\ x & y & z \end{vmatrix} = (z\omega_y - y\omega_z)\hat{i} + (x\omega_z - z\omega_x)\hat{j} + (y\omega_x - x\omega_y)\hat{k}$$

Combining these results, we obtain

$$\vec{V}_{XYZ} = \vec{V}_{rf} + \vec{V}_{xyz} + \vec{\omega}\times\vec{r} \qquad\qquad \vec{V}_{XYZ}$$

**4-7 THE ANGULAR MOMENTUM PRINCIPLE

Next we develop a control volume expression for the angular momentum principle. We begin with the mathematical statement for a system and use Eq. 4.11 to complete the formulation for a fixed (inertial) control volume (Section 4-7.1). To obtain the control volume formulation for a rotating (noninertial) control volume

---

** This section may be omitted without loss of continuity in the text material.

(Section 4-7.2), we first develop a suitable expression for the angular momentum principle applied to a system in general motion. We then use Eq. 4.26 to complete the formulation for a control volume.

4-7.1 Equation for Fixed Control Volume

The angular momentum principle for a system is

$$\vec{T} = \frac{d\vec{H}}{dt}\bigg)_{\text{system}} \qquad (4.3a)$$

where $\vec{T}$ = total torque exerted on the system by its surroundings, and
$\vec{H}$ = angular momentum of the system,

$$\vec{H} = \int_{M(\text{system})} \vec{r} \times \vec{V} \, dm = \int_{\Psi(\text{system})} \vec{r} \times \vec{V} \rho \, d\Psi \qquad (4.3b)$$

All quantities in the system equation must be formulated with respect to an inertial reference frame. Reference frames at rest, or moving with constant linear velocity, are inertial, and Eq. 4.3b can be used directly to develop the control volume form of the angular momentum principle. (Rotating reference frames are noninertial and will be treated in Section 4-7.2.)

The position vector, $\vec{r}$, locates each mass or volume element of the system with respect to the coordinate system. The torque, $\vec{T}$, applied to a system may be written

$$\vec{T} = \vec{r} \times \vec{F}_s + \int_{M(\text{system})} \vec{r} \times \vec{g} \, dm + \vec{T}_{\text{shaft}} \qquad (4.3c)$$

The relation between the system and fixed control volume formulations is

$$\frac{dN}{dt}\bigg)_{\text{system}} = \frac{\partial}{\partial t} \int_{\text{CV}} \eta \rho \, d\Psi + \int_{\text{CS}} \eta \rho \vec{V} \cdot d\vec{A} \qquad (4.11)$$

where

$$N_{\text{system}} = \int_{M(\text{system})} \eta \, dm$$

If we set $N = \vec{H}$, then $\eta = \vec{r} \times \vec{V}$, and

$$\frac{d\vec{H}}{dt}\bigg)_{\text{system}} = \frac{\partial}{\partial t} \int_{\text{CV}} \vec{r} \times \vec{V} \rho \, d\Psi + \int_{\text{CS}} \vec{r} \times \vec{V} \rho \vec{V} \cdot d\vec{A} \qquad (4.46)$$

Combining Eqs. 4.3a, 4.3c, and 4.46, we obtain

$$\vec{r} \times \vec{F}_s + \int_{M(\text{system})} \vec{r} \times \vec{g} \, dm + \vec{T}_{\text{shaft}} = \frac{\partial}{\partial t} \int_{\text{CV}} \vec{r} \times \vec{V} \rho \, d\Psi + \int_{\text{CS}} \vec{r} \times \vec{V} \rho \vec{V} \cdot d\vec{A}$$

Since the system and control volume coincided at time t_0,

$$\vec{T}_{\text{system}} = \vec{T}_{\text{CV}}$$

and

$$\vec{r} \times \vec{F}_s + \int_{\text{CV}} \vec{r} \times \vec{g} \rho \, d\Psi + \vec{T}_{\text{shaft}} = \frac{\partial}{\partial t} \int_{\text{CV}} \vec{r} \times \vec{V} \rho \, d\Psi + \int_{\text{CS}} \vec{r} \times \vec{V} \rho \vec{V} \cdot d\vec{A} \qquad (4.47)$$

Equation 4.47 is a general formulation of the angular momentum principle for an inertial control volume. The left side of the equation is an expression for all the torques that act on the control volume. Terms on the right express the rate of change of angular momentum within the control volume and the net rate of flux of angular momentum from the control volume. All velocities in Eq. 4.47 are measured relative to the fixed control volume.

For analysis of rotating machinery, Eq. 4.47 is often used in scalar form by considering only the component directed along the axis of rotation. This application is illustrated in Chapter 11.

The application of Eq. 4.47 to the analysis of a simple lawn sprinkler is illustrated in Example Problem 4.14. This same problem is considered in Example Problem 4.15 using the formulation of the angular momentum principle for a rotating control volume.

EXAMPLE 4.14—Lawn Sprinkler: Fixed Control Volume Analysis

A small lawn sprinkler is shown in the sketch below. At an inlet gage pressure of 20 kPa, the total volume flow rate of water through the sprinkler is 7.5 liters per minute and it rotates at 30 rpm. The diameter of each jet is 4 mm. Calculate the jet speed relative to each sprinkler nozzle. Evaluate the friction torque at the sprinkler pivot.

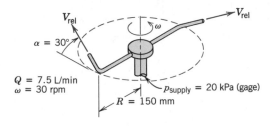

EXAMPLE PROBLEM 4.14

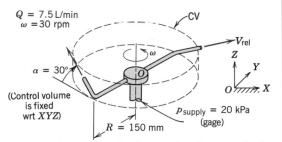

GIVEN: Small lawn sprinker as shown.

FIND: (a) Jet speed relative to each nozzle.
  (b) Friction torque at pivot.

SOLUTION:
Apply continuity and angular momentum equations using fixed control volume enclosing sprinkler arms.

Basic equations:

$$0 = \cancelto{0(1)}{\frac{\partial}{\partial t}} \int_{CV} \rho \, d\forall + \int_{CS} \rho \vec{V} \cdot d\vec{A}$$

$$\vec{r} \times \vec{F}_s + \int_{CV} \vec{r} \times \vec{g} \rho \, d\forall + \vec{T}_{shaft} = \frac{\partial}{\partial t} \int_{CV} \vec{r} \times \vec{V} \rho \, d\forall + \int_{CS} \vec{r} \times \vec{V} \rho \vec{V} \cdot d\vec{A} \qquad (1)$$

where all velocities are measured relative to the inertial coordinates XYZ.

Assumptions: (1) Incompressible flow
  (2) Uniform flow at each section
  (3) $\vec{\omega}$ = constant

From continuity, the jet speed relative to the nozzle is given by

$$V_{rel} = \frac{Q}{2A_{jet}} = \frac{Q}{2}\frac{4}{\pi D_{jet}^2}$$

$$= \frac{1}{2} \times 7.5\,\frac{L}{min} \times \frac{4}{\pi}\frac{1}{(4)^2 mm^2} \times \frac{m^3}{1000\,L} \times \frac{10^6\,mm^2}{m^2} \times \frac{min}{60\,sec}$$

$$V_{rel} = 4.97\,m/sec \qquad\qquad\qquad\qquad\qquad V_{rel}$$

Consider terms in the angular momentum equation separately. Since atmospheric pressure acts on the entire control surface, and the pressure force at the inlet causes no moment about O, then $\vec{r} \times \vec{F}_s = 0$. The moment of the body force in each of the two arms is equal and opposite and hence the second term on the left side of the equation is zero. The only external torque acting on the CV is friction in the pivot. It opposes the motion, so

$$\vec{T}_{shaft} = -T_f \hat{K} \qquad\qquad (2)$$

Before we can evaluate the control volume integral on the right side of Eq. 1, we need to develop expressions for the position vector, $\vec{r}$, and velocity vector, $\vec{V}$, (measured relative to the fixed coordinate system XYZ) of each element of fluid in the control volume.

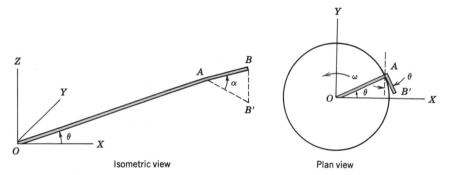

Isometric view   Plan view

OA lies in the XY plane; AB is inclined at angle α to the XY plane; point B' is the projection of point B on the XY plane.

We assume that the length, L, of the tip AB is small compared to the length, R, of the horizontal arm OA. Consequently we neglect the angular momentum of the fluid in the tips compared to the angular momentum in the horizontal arms.

Consider flow in the horizontal tube OA of length R. Denote the radial position from O by the distance r. At any point in the tube the fluid velocity relative to fixed coordinates XYZ is the sum of the velocity in the tube and the angular velocity $\vec{\omega} \times \vec{r}$. Thus

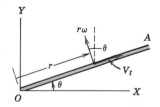

$$\vec{V} = \hat{I}(V_t \cos\theta - r\omega \sin\theta) + \hat{J}(V_t \sin\theta + r\omega \cos\theta)$$

The position vector $\vec{r} = \hat{I} r \cos\theta + \hat{J} r \sin\theta$, and

$$\vec{r} \times \vec{V} = \hat{K}(r^2\omega \cos^2\theta + r^2\omega \sin^2\theta) = \hat{K} r^2\omega$$

Then

$$\int_{\mathbf{V}_{OA}} \vec{r} \times \vec{V} \rho\, d\mathbf{V} = \int_O^R \hat{K} r^2\omega\, \rho A\, dr = \hat{K}\frac{R^3\omega}{3}\rho A$$

and

$$\frac{\partial}{\partial t}\int_{\Psi_{OA}}\vec{r}\times\vec{V}\rho\,d\Psi=\frac{\partial}{\partial t}\left[\hat{K}\frac{R^3\omega}{3}\rho A\right]=0 \tag{3}$$

where A is the cross-sectional area of the horizontal tube. Identical results are obtained for the other horizontal tube in the control volume.

To evaluate the flux of angular momentum through the control surface we need an expression for $\vec{r}_{\text{jet}}=\vec{r}_B$ and the jet velocity, $\vec{V}_j$, measured relative to the fixed coordinate system XYZ. From the geometry of arm OAB,

$$\vec{r}_B=\hat{I}(R\cos\theta+L\cos\alpha\,\sin\theta)+\hat{J}(R\sin\theta-L\cos\alpha\,\cos\theta)+\hat{K}L\sin\alpha$$

For $L\ll R$, then

$$\vec{r}_B=\hat{I}R\cos\theta+\hat{J}R\sin\theta$$

$$\vec{V}_j=\vec{V}_{\text{rel}}+\vec{V}_{\text{tip}}=\hat{I}\,V_{\text{rel}}\cos\alpha\,\sin\theta-\hat{J}\,V_{\text{rel}}\cos\alpha\,\cos\theta+\hat{K}\,V_{\text{rel}}\sin\alpha-\hat{I}\omega R\sin\theta+\hat{J}\omega R\cos\theta$$

$$\vec{V}_j=\hat{I}(V_{\text{rel}}\cos\alpha-\omega R)\sin\theta-\hat{J}(V_{\text{rel}}\cos\alpha-\omega R)\cos\theta+\hat{K}\,V_{\text{rel}}\sin\alpha$$

$$\vec{r}_B\times\vec{V}_j=\hat{I}RV_{\text{rel}}\sin\alpha\,\sin\theta-\hat{J}RV_{\text{rel}}\sin\alpha\,\cos\theta-\hat{K}R(V_{\text{rel}}\cos\alpha-\omega R)(\sin^2\theta+\cos^2\theta)$$

$$\vec{r}_B\times\vec{V}_j=\hat{I}RV_{\text{rel}}\sin\alpha\,\sin\theta-\hat{J}RV_{\text{rel}}\sin\alpha\,\cos\theta-\hat{K}R(V_{\text{rel}}\cos\alpha-\omega R)$$

The flux integral is evaluated for flow crossing the control surface. For arm OAB,

$$\int_{\text{CS}}\vec{r}\times\vec{V}_j\rho\vec{V}\cdot d\vec{A}=\left[\hat{I}RV_{\text{rel}}\sin\alpha\,\sin\theta-\hat{J}RV_{\text{rel}}\sin\alpha\,\cos\theta-\hat{K}R(V_{\text{rel}}\cos\alpha-\omega R)\right]\rho\frac{Q}{2}$$

The velocity and radius vectors for flow in the left arm must be described in terms of the same unit vectors used for the right arm. In the left arm the $\hat{I}$ and $\hat{J}$ components of the cross product are of opposite sign, since $\sin(\theta+\pi)=-\sin(\theta)$ and $\cos(\theta+\pi)=-\cos(\theta)$. Thus for the complete CV,

$$\int_{\text{CS}}\vec{r}\times\vec{V}_j\rho\vec{V}\cdot d\vec{A}=-\hat{K}R(V_{\text{rel}}\cos\alpha-\omega R)\rho Q \tag{4}$$

In the supply line, $\vec{r}\times\vec{V}=0$ and hence term (4) represents the total angular momentum flux through the control surface.

Substituting terms (2), (3), and (4) into Eq. 1, we obtain

$$-T_f\hat{K}=-\hat{K}R(V_{\text{rel}}\cos\alpha-\omega R)\rho Q$$

or

$$T_f=R(V_{\text{rel}}\cos\alpha-\omega R)\rho Q$$

From the data given,

$$\omega R=\frac{30\ \text{rev}}{\text{min}}\times 150\ \text{mm}\times\frac{2\pi\ \text{rad}}{\text{rev}}\times\frac{\text{min}}{60\ \text{sec}}\times\frac{\text{m}}{1000\ \text{mm}}=\frac{0.471\ \text{m}}{\text{sec}}$$

Substituting gives

$$T_f=150\ \text{mm}\left(\frac{4.97\ \text{m}}{\text{sec}}\times\frac{\cos 30°}{}-\frac{0.471\ \text{m}}{\text{sec}}\right)\frac{999\ \text{kg}}{\text{m}^3}\times\frac{7.5\ \text{L}}{\text{min}}$$

$$\times\frac{\text{m}^3}{1000\ \text{L}}\times\frac{\text{min}}{60\ \text{sec}}\times\frac{\text{N}\cdot\text{sec}^2}{\text{kg}\cdot\text{m}}\times\frac{\text{m}}{1000\ \text{mm}}$$

$$T_f=0.0718\ \text{N}\cdot\text{m} \qquad\qquad T_f$$

$\left[\begin{array}{l}\text{This problem has been included to illustrate use of the angular momentum principle for} \\ \text{an inertial control volume. Note that in using Eq. 4.47, the angular momentum, } \vec{r} \times \vec{V}, \\ \text{must be measured relative to an inertial reference frame. The problem is reworked} \\ \text{using a noninertial control volume as Example Problem 4.15.}\end{array}\right\}$

4-7.2   Equation for Rotating Control Volume

In problems involving rotating elements, such as the rotating sprinkler of Example Problem 4.14, it is often convenient to express all fluid velocities relative to the rotating component. The most convenient control volume is a noninertial one that rotates with the component. In this section we develop a formulation of the angular momentum principle for a noninertial control volume rotating about an axis fixed in space.

Inertial and noninertial reference frames were related in Section 4-6. Figure 4.5 showed the notation used. For a system,

$$\vec{T}_{\text{system}} = \frac{d\vec{H}}{dt}\bigg)_{\text{system}} \tag{4.3a}$$

The angular momentum of a system in general motion must be specified relative to an inertial reference frame. Using the notation of Fig. 4.5,

$$\vec{H}_{\text{system}} = \int_{M(\text{system})} (\vec{R} + \vec{r}) \times \vec{V}_{XYZ}\, dm = \int_{\forall(\text{system})} (\vec{R} + \vec{r}) \times \vec{V}_{XYZ}\, \rho\, d\forall$$

With $\vec{R} = 0$ and the xyz frame restricted to rotation about XYZ, the equation becomes

$$\vec{H}_{\text{system}} = \int_{M(\text{system})} \vec{r} \times \vec{V}_{XYZ}\, dm = \int_{\forall(\text{system})} \vec{r} \times \vec{V}_{XYZ}\, \rho\, d\forall$$

so that

$$\vec{T}_{\text{system}} = \frac{d}{dt} \int_{M(\text{system})} \vec{r} \times \vec{V}_{XYZ}\, dm$$

Since the mass of a system is constant,

$$\vec{T}_{\text{system}} = \int_{M(\text{system})} \frac{d}{dt} (\vec{r} \times \vec{V}_{XYZ})\, dm$$

or

$$\vec{T}_{\text{system}} = \int_{M(\text{system})} \left(\frac{d\vec{r}}{dt} \times \vec{V}_{XYZ} + \vec{r} \times \frac{d\vec{V}_{XYZ}}{dt} \right) dm \tag{4.48}$$

From the analysis of Section 4-6,

$$\vec{V}_{XYZ} = \vec{V}_{rf} + \frac{d\vec{r}}{dt} \tag{4.37}$$

With xyz restricted to pure rotation, $\vec{V}_{rf} = 0$. The first term under the integral on the right side of Eq. 4.48 is then

$$\frac{d\vec{r}}{dt} \times \frac{d\vec{r}}{dt} = 0$$

Thus Eq. 4.48 reduces to

$$\vec{T}_{\text{system}} = \int_{M(\text{system})} \vec{r} \times \frac{d\vec{V}_{XYZ}}{dt} \, dm = \int_{M(\text{system})} \vec{r} \times \vec{a}_{XYZ} \, dm \qquad (4.49)$$

From Eq. 4.42 with $\vec{a}_{rf} = 0$ (since xyz does not translate),

$$\vec{a}_{XYZ} = \vec{a}_{xyz} + 2\vec{\omega} \times \vec{V}_{xyz} + \vec{\omega} \times (\vec{\omega} \times \vec{r}) + \dot{\vec{\omega}} \times \vec{r}$$

Substituting into Eq. 4.49, we obtain

$$\vec{T}_{\text{system}} = \int_{M(\text{system})} \vec{r} \times [\vec{a}_{xyz} + 2\vec{\omega} \times \vec{V}_{xyz} + \vec{\omega} \times (\vec{\omega} \times \vec{r}) + \dot{\vec{\omega}} \times \vec{r}] \, dm$$

or

$$\vec{T}_{\text{system}} - \int_{M(\text{system})} \vec{r} \times [2\vec{\omega} \times \vec{V}_{xyz} + \vec{\omega} \times (\vec{\omega} \times \vec{r}) + \dot{\vec{\omega}} \times \vec{r}] \, dm$$

$$= \int_{M(\text{system})} \vec{r} \times \vec{a}_{xyz} \, dm = \int_{M(\text{system})} \vec{r} \times \frac{d\vec{V}_{xyz}}{dt} \, dm \qquad (4.50)$$

Using the time rate of change as observed from the system, we can write the last term as

$$\int_{M(\text{system})} \vec{r} \times \frac{d\vec{V}_{xyz}}{dt} \, dm = \frac{d}{dt} \int_{M(\text{system})} \vec{r} \times \vec{V}_{xyz} \, dm = \frac{d\vec{H}_{xyz}}{dt}\bigg)_{\text{system}} \qquad (4.51)$$

The torque on the system is given by

$$\vec{T}_{\text{system}} = \vec{r} \times \vec{F}_S + \int_{M(\text{system})} \vec{r} \times \vec{g} \, dm + \vec{T}_{\text{shaft}} \qquad (4.3c)$$

The relation between the system and control volume formulations is

$$\frac{dN}{dt}\bigg)_{\text{system}} = \frac{\partial}{\partial t} \int_{\text{CV}} \eta\rho \, d\Psi + \int_{\text{CS}} \eta\rho \vec{V}_{xyz} \cdot d\vec{A} \qquad (4.26)$$

where

$$N_{\text{system}} = \int_{M(\text{system})} \eta \, dm$$

Setting N equal to $\vec{H}_{xyz})_{\text{system}}$ and $\eta = \vec{r} \times \vec{V}_{xyz}$ yields

$$\frac{d\vec{H}_{xyz}}{dt}\bigg)_{\text{system}} = \frac{\partial}{\partial t} \int_{\text{CV}} \vec{r} \times \vec{V}_{xyz} \rho \, d\Psi + \int_{\text{CS}} \vec{r} \times \vec{V}_{xyz} \rho \vec{V}_{xyz} \cdot d\vec{A} \qquad (4.52)$$

Combining Eqs. 4.50, 4.51, 4.52, and 4.3c, we obtain

$$\vec{r} \times \vec{F}_S + \int_{M(\text{system})} \vec{r} \times \vec{g} \, dm + \vec{T}_{\text{shaft}}$$

$$- \int_{M(\text{system})} \vec{r} \times [2\vec{\omega} \times \vec{V}_{xyz} + \vec{\omega} \times (\vec{\omega} \times \vec{r}) + \dot{\vec{\omega}} \times \vec{r}] dm$$

$$= \frac{\partial}{\partial t} \int_{CV} \vec{r} \times \vec{V}_{xyz}\, \rho\, d\forall + \int_{CS} \vec{r} \times \vec{V}_{xyz}\, \rho \vec{V}_{xyz} \cdot d\vec{A}$$

Since the system and control volume coincided at t_0,

$$\vec{r} \times \vec{F}_S + \int_{CV} \vec{r} \times \vec{g}\, \rho\, d\forall + \vec{T}_{\text{shaft}}$$

$$- \int_{CV} \vec{r} \times [2\vec{\omega} \times \vec{V}_{xyz} + \vec{\omega} \times (\vec{\omega} \times \vec{r}) + \dot{\vec{\omega}} \times \vec{r}] \rho\, d\forall$$

$$= \frac{\partial}{\partial t} \int_{CV} \vec{r} \times \vec{V}_{xyz}\, \rho\, d\forall + \int_{CS} \vec{r} \times \vec{V}_{xyz}\, \rho \vec{V}_{xyz} \cdot d\vec{A} \qquad (4.53)$$

Equation 4.53 is the formulation of the angular momentum principle for a (noninertial) control volume rotating about an axis fixed in space. All fluid velocities and rates of change in Eq. 4.53 are evaluated relative to the control volume. Application of the equation to a rotating sprinkler is illustrated in Example Problem 4.15.

EXAMPLE 4.15—Lawn Sprinkler: Rotating Control Volume Analysis

A small lawn sprinkler is shown in the sketch below. At an inlet gage pressure of 20 kPa, the total volume flow rate of water through the sprinkler is 7.5 liters per minute and it rotates at 30 rpm. The diameter of each jet is 4 mm. Calculate the jet speed relative to each sprinkler nozzle. Evaluate the friction torque at the sprinkler pivot.

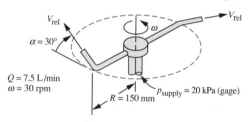

EXAMPLE PROBLEM 4.15

GIVEN: Small lawn sprinkler as shown.

FIND: (a) Jet speed relative to each nozzle.
   (b) Friction torque at pivot.

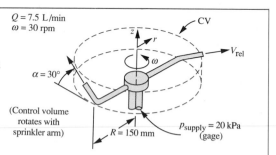

(Control volume rotates with sprinkler arm)

SOLUTION:
Apply continuity and angular momentum equations using rotating control volume enclosing sprinkler arms.

Basic equations:

$$0 = \frac{\cancel{\partial}}{\cancel{\partial t}} \int_{CV} \rho\, d\forall + \int_{CS} \rho \vec{V}_{xyz} \cdot d\vec{A} \qquad {}^{=\,0(1)}$$

$${}^{=\,0(3)}$$

$$\vec{r} \times \vec{F}_S + \int_{CV} \vec{r} \times \vec{g}\, \rho\, d\forall + \vec{T}_{\text{shaft}} - \int_{CV} \vec{r} \times [2\vec{\omega} \times \vec{V}_{xyz} + \vec{\omega} \times (\vec{\omega} \times \vec{r}) + \cancel{\dot{\vec{\omega}} \times \vec{r}}] \rho\, d\forall$$

$$= \underset{= 0(1)}{\overset{\displaystyle \frac{\partial}{\partial t}}{\Big/}} \int_{CV} \vec{r} \times \vec{V}_{xyz}\, \rho\, d\Psi + \int_{CS} \vec{r} \times \vec{V}_{xyz}\, \rho \vec{V}_{xyz} \cdot d\vec{A}$$

Assumptions: (1)   Steady flow relative to the rotating CV
               (2)   Uniform flow at each section
               (3) $\vec{\omega}$ = constant

From continuity

$$V_{\text{rel}} = \frac{Q}{2A_{\text{jet}}} = \frac{Q}{2}\frac{4}{\pi D_{\text{jet}}^2}$$

$$= \frac{1}{2} \times \frac{7.5}{\text{min}}\frac{L}{\text{min}} \times \frac{4}{\pi}\frac{1}{(4)^2\ \text{mm}^2} \times \frac{\text{m}^3}{1000\ \text{L}} \times \frac{10^6\ \text{mm}^2}{\text{m}^2} \times \frac{\text{min}}{60\ \text{sec}}$$

$$V_{\text{rel}} = 4.97\ \text{m/sec} \qquad\qquad\qquad\qquad\qquad\qquad\qquad V_{\text{rel}}$$

Consider terms in the angular momentum equation separately. As in Example Problem 4.14, the only external torque acting on the CV is friction in the pivot. It opposes the motion, so

$$\vec{T}_{\text{shaft}} = -T_f\hat{k} \tag{1}$$

The integral on the left is evaluated for flow *within* the CV. Let the velocity and area within the sprinkler tubes be V_{CV} and A_{CV}, respectively. Then, for *one side*, the first term is

$$\int_{CV} \vec{r} \times [2\vec{\omega} \times \vec{V}_{xyz}]\rho\, d\Psi = \int_0^R r\hat{e}_r \times [2\omega\hat{k} \times V_{CV}\hat{e}_r]\rho A_{CV}\, dr$$

$$= \int_0^R r\hat{e}_r \times 2\omega V_{CV}\hat{e}_\theta\, \rho A_{CV}\, dr$$

$$= \int_0^R 2\omega V_{CV}\rho A_{CV} r\, dr\,\hat{k} = \omega R^2 \rho V_{CV} A_{CV}\hat{k} \qquad \text{\{one side\}}$$

(The flow in the bent portion of the tube has no r component of velocity, so it does not contribute to the integral.)

From continuity, $Q = 2V_{CV}A_{CV}$, so for *both sides* the integral becomes

$$\int_{CV} \vec{r} \times [2\vec{\omega} \times \vec{V}_{xyz}]\rho\, d\Psi = \omega R^2 \rho Q\hat{k} \tag{2}$$

The second term in the integral is evaluated as

$$\int_{CV} \vec{r} \times [\vec{\omega} \times (\vec{\omega} \times \vec{r})]\rho\, d\Psi = \int_{CV} r\hat{e}_r \times [\omega\hat{k} \times (\omega\hat{k} \times r\hat{e}_r)]\rho\, d\Psi$$

$$= \int_{CV} r\hat{e}_r \times [\omega\hat{k} \times \omega r\hat{e}_\theta]\rho\, d\Psi = \int_{CV} r\hat{e}_r \times \omega^2 r(-\hat{e}_r)\rho\, d\Psi = 0$$

so it contributes no torque.

    The integral on the right is evaluated for flow crossing the control surface. For the right side,

$$\int_{CS} \vec{r} \times \vec{V}_{xyz}\, \rho \vec{V}_{xyz} \cdot d\vec{A} = R\hat{e}_r \times V_{\text{rel}}[\cos\alpha(-\hat{e}_\theta) + \sin\alpha\hat{k}]\{+\rho V_{\text{rel}} A_{\text{jet}}\}$$

$$= RV_{\text{rel}}[\cos\alpha(-\hat{k}) + \sin\alpha(-\hat{e}_\theta)]\rho\frac{Q}{2}$$

The velocity and radius vectors for flow in the left arm must be described in terms of the same unit vectors used for the right arm. In the left sprinkler arm, the θ component has the same magnitude but opposite sign, so it cancels. For the complete CV,

$$\int_{CS} \vec{r} \times \vec{V}_{xyz}\, \rho \vec{V}_{xyz} \cdot d\vec{A} = -RV_{\text{rel}} \cos\alpha\, \rho Q \hat{k} \tag{3}$$

Combining terms (1), (2), and (3), we obtain

$$-T_f \hat{k} - \omega R^2 \rho Q \hat{k} = -R V_{\text{rel}} \cos\alpha\, \rho Q \hat{k}$$

or

$$T_f = R(V_{\text{rel}} \cos\alpha - \omega R)\rho Q$$

From the data given,

$$\omega R = \frac{30 \text{ rev}}{\text{min}} \times \frac{150 \text{ mm}}{} \times \frac{2\pi \text{ rad}}{\text{rev}} \times \frac{\text{min}}{60 \text{ sec}} \times \frac{\text{m}}{1000 \text{ mm}} = \frac{0.471 \text{ m}}{\text{sec}}$$

Substituting gives

$$T_f = 150 \text{ mm} \left(\frac{4.97 \text{ m}}{\text{sec}} \times \cos 30° - \frac{0.471 \text{ m}}{\text{sec}} \right) \frac{999 \text{ kg}}{\text{m}^3} \times \frac{7.5 \text{ L}}{\text{min}}$$

$$\times \frac{\text{m}^3}{1000 \text{ L}} \times \frac{\text{min}}{60 \text{ sec}} \times \frac{\text{N} \cdot \text{sec}^2}{\text{kg} \cdot \text{m}} \times \frac{\text{m}}{1000 \text{ mm}}$$

$$\underline{T_f = 0.0718 \text{ N} \cdot \text{m}} \qquad\qquad\qquad\qquad T_f$$

$$\left\{ \begin{array}{l} \text{This problem has been included to illustrate use of the angular momentum equation} \\ \text{for a (noninertial) rotating control volume. The result is identical to the result obtained} \\ \text{using the fixed control volume analysis of Example Problem 4.14.} \end{array} \right.$$

4-8   THE FIRST LAW OF THERMODYNAMICS

The first law of thermodynamics is a statement of conservation of energy. Recall that the system formulation of the first law was

$$\dot{Q} - \dot{W} = \left. \frac{dE}{dt} \right)_{\text{system}} \tag{4.4a}$$

where the total energy of the system is given by

$$E_{\text{system}} = \int_{M(\text{system})} e\, dm = \int_{\Psi(\text{system})} e\rho\, d\Psi \tag{4.4b}$$

and

$$e = u + \frac{V^2}{2} + gz$$

In Eq. 4.4a, the rate of heat transfer, $\dot{Q}$, is positive when heat is added to the system from the surroundings; the rate of work, $\dot{W}$, is positive when work is done by the system on its surroundings.

The system and control volume formulations are related by

$$\left. \frac{dN}{dt} \right)_{\text{system}} = \frac{\partial}{\partial t} \int_{CV} \eta \rho\, d\Psi + \int_{CS} \eta \rho \vec{V} \cdot d\vec{A} \tag{4.11}$$

where

$$N_{\text{system}} = \int_{M(\text{system})} \eta \, dm = \int_{\Psi(\text{system})} \eta \rho \, d\Psi \qquad (4.6)$$

To derive the control volume formulation of the first law of thermodynamics, we set

$$N = E \qquad \text{and} \qquad \eta = e$$

From Eq. 4.11, with this substitution, we obtain

$$\left.\frac{dE}{dt}\right)_{\text{system}} = \frac{\partial}{\partial t} \int_{\text{CV}} e\rho \, d\Psi + \int_{\text{CS}} e\rho \vec{V} \cdot d\vec{A} \qquad (4.54)$$

In deriving Eq. 4.11, the system and the control volume coincided at t_0, so

$$[\dot{Q} - \dot{W}]_{\text{system}} = [\dot{Q} - \dot{W}]_{\text{control volume}}$$

In light of this, Eqs. 4.4a and 4.54 yield the control volume formulation of the first law of thermodynamics,

$$\dot{Q} - \dot{W} = \frac{\partial}{\partial t} \int_{\text{CV}} e\rho \, d\Psi + \int_{\text{CS}} e\rho \vec{V} \cdot d\vec{A} \qquad (4.55)$$

where

$$e = u + \frac{V^2}{2} + gz$$

Note that for steady flow the first term on the right side of Eq. 4.55 is zero. Is Eq. 4.55 the form of the first law used in thermodynamics? Even for steady flow, Eq. 4.55 is not quite the same form used in applying the first law to control volume problems. To obtain a formulation suitable and convenient for problem solutions, let us take a closer look at the work term, $\dot{W}$.

4-8.1 Rate of Work Done by a Control Volume

The term $\dot{W}$ in Eq. 4.55 has a positive numerical value when work is done by the control volume on the surroundings. The rate of work done on the control volume is of opposite sign to the work done by the control volume.

The rate of work done by the control volume is conveniently subdivided into four classifications,

$$\dot{W} = \dot{W}_s + \dot{W}_{\text{normal}} + \dot{W}_{\text{shear}} + \dot{W}_{\text{other}}$$

Let us consider these separately:

1. Shaft Work

We shall designate shaft work W_s and hence the rate of work transferred out through the control surface by shaft work is designated $\dot{W}_s$.

2. Work Done by Normal Stresses at the Control Surface

Recall that work requires a force to be moved through a distance. Thus, in moving a force, $\vec{F}$, through an infinitesimal distance, $d\vec{s}$, the work done is given by

$$\delta W = \vec{F} \cdot d\vec{s}$$

To obtain the rate at which work is done by the force, divide by the time increment, Δt, and take the limit as $\Delta t \to 0$. Thus the rate of work done by the force, $\vec{F}$, is given by

$$\dot{W} = \lim_{\Delta t \to 0} \frac{\delta W}{\Delta t} = \lim_{\Delta t \to 0} \frac{\vec{F} \cdot d\vec{s}}{\Delta t} \qquad \text{or} \qquad \dot{W} = \vec{F} \cdot \vec{V}$$

The rate of work done on an element of area, $d\vec{A}$, of the control surface by normal stresses is given by

$$d\vec{F} \cdot \vec{V} = \sigma_{nn} d\vec{A} \cdot \vec{V}$$

Since the work out across the boundaries of the control volume is the negative of the work done on the control volume, the total rate of work out of the control volume due to normal stresses is given by

$$\dot{W}_{\text{normal}} = -\int_{\text{CS}} \sigma_{nn} \, d\vec{A} \cdot \vec{V} = -\int_{\text{CS}} \sigma_{nn} \vec{V} \cdot d\vec{A}$$

3.   Work Done by Shear Stresses at the Control Surface

Just as work is done by the normal stresses at the boundaries of the control volume, so may work be done by the shear stresses.

The shear force acting on an element of area of the control surface is given by

$$d\vec{F} = \vec{\tau} \, dA$$

where the shear stress vector, $\vec{\tau}$, is the shear stress acting in the plane of dA.

The rate of work done on the entire control surface by shear stresses is given by

$$\dot{W}_{\text{shear}} = \int_{\text{CS}} \vec{\tau} \, dA \cdot \vec{V} = \int_{\text{CS}} \vec{\tau} \cdot \vec{V} \, dA$$

Since the work out across the boundaries of the control volume is the negative of the work done on the control volume, then the rate of work out of the control volume due to shear stresses is given by

$$\dot{W}_{\text{shear}} = -\int_{\text{CS}} \vec{\tau} \cdot \vec{V} \, dA$$

This integral is better expressed as three terms

$$\dot{W}_{\text{shear}} = -\int_{\text{CS}} \vec{\tau} \cdot \vec{V} \, dA$$

$$= -\int_{A(\text{shafts})} \vec{\tau} \cdot \vec{V} \, dA - \int_{A(\text{solid surface})} \vec{\tau} \cdot \vec{V} \, dA - \int_{A(\text{ports})} \vec{\tau} \cdot \vec{V} \, dA$$

We have already accounted for the first term, since we included $\dot{W}_{\text{shaft}}$ previously. At solid surfaces, $\vec{V} = 0$, so the second term is zero (for a fixed control volume). Thus

$$\dot{W}_{\text{shear}} = -\int_{A(\text{ports})} \vec{\tau} \cdot \vec{V} \, dA$$

This last term can be made zero by proper choice of control surfaces. If we choose a control surface that cuts across each port perpendicular to the flow, then $d\vec{A}$ is parallel to $\vec{V}$. Since $\vec{\tau}$ is in the plane of dA, then $\vec{\tau}$ is perpendicular to $\vec{V}$. Thus, for a control surface perpendicular to $\vec{V}$,

$$\vec{\tau} \cdot \vec{V} = 0 \qquad \text{and} \qquad \dot{W}_{\text{shear}} = 0$$

4. Other Work

Electrical energy could be added to the control volume. Also electromagnetic energy, e.g., in radar or laser beams, could be absorbed. In most problems, such contributions will be absent, but we should note them in our general formulation.

With all of the terms in $\dot{W}$ evaluated, we obtain

$$\dot{W} = \dot{W}_s - \int_{CS} \sigma_{nn} \vec{V} \cdot d\vec{A} + \dot{W}_{shear} + \dot{W}_{other} \qquad (4.56)$$

4-8.2 Control Volume Equation

Substituting the expression for $\dot{W}$ from Eq. 4.56 into Eq. 4.55 gives

$$\dot{Q} - \dot{W}_s + \int_{CS} \sigma_{nn} \vec{V} \cdot d\vec{A} - \dot{W}_{shear} - \dot{W}_{other} = \frac{\partial}{\partial t} \int_{CV} e\rho \, d\Psi + \int_{CS} e\rho \vec{V} \cdot d\vec{A}$$

Rearranging this equation, we obtain

$$\dot{Q} - \dot{W}_s - \dot{W}_{shear} - \dot{W}_{other} = \frac{\partial}{\partial t} \int_{CV} e\rho \, d\Psi + \int_{CS} e\rho \vec{V} \cdot d\vec{A} - \int_{CS} \sigma_{nn} \vec{V} \cdot d\vec{A}$$

Since $\rho = 1/v$, where v is *specific volume*, then

$$\int_{CS} \sigma_{nn} \vec{V} \cdot d\vec{A} = \int_{CS} \sigma_{nn} v\rho \vec{V} \cdot d\vec{A}$$

Hence

$$\dot{Q} - \dot{W}_s - \dot{W}_{shear} - \dot{W}_{other} = \frac{\partial}{\partial t} \int_{CV} e\rho \, d\Psi + \int_{CS} (e - \sigma_{nn} v)\rho \vec{V} \cdot d\vec{A}$$

Viscous effects can make the normal stress, σ_{nn}, different from the negative of the thermodynamic pressure, $-p$. However, for most flows of common engineering interest, $\sigma_{nn} \simeq -p$. Then

$$\dot{Q} - \dot{W}_s - \dot{W}_{shear} - \dot{W}_{other} = \frac{\partial}{\partial t} \int_{CV} e\rho \, d\Psi + \int_{CS} (e + pv)\rho \vec{V} \cdot d\vec{A}$$

Finally, substituting $e = u + V^2/2 + gz$ into the last term, we obtain the familiar form of the first law formulation for a control volume,

$$\dot{Q} - \dot{W}_s - \dot{W}_{shear} - \dot{W}_{other} = \frac{\partial}{\partial t} \int_{CV} e\rho d\Psi + \int_{CS} \left(u + pv + \frac{V^2}{2} + gz \right) \rho \vec{V} \cdot d\vec{A} \quad (4.57)$$

Each work term in Eq. 4.57 represents the rate of work done by the control volume on the surroundings.

EXAMPLE 4.16—Compressor: First Law Analysis

Air at 14.7 psia, 70 F, enters a compressor with negligible velocity and is discharged at 50 psia, 100 F through a pipe with 1 ft^2 area. The flow rate is 20 lbm/sec. The power input to the compressor is 600 hp. Determine the rate of heat transfer.

EXAMPLE PROBLEM 4.16

GIVEN: Air enters a compressor at ① and leaves at ② with conditions as shown. The air flow rate is 20 lbm/sec and the power input to the compressor is 600 hp.

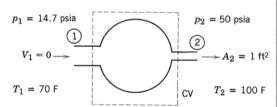

$p_1 = 14.7$ psia $p_2 = 50$ psia

①   ②

$V_1 \approx 0 \longrightarrow$ $\longrightarrow A_2 = 1$ ft^2

FIND: The rate of heat transfer.

$T_1 = 70$ F   CV $T_2 = 100$ F

SOLUTION:

Basic equations:

$$\overset{=0(1)}{0 = \frac{\partial}{\partial t}\int_{CV} \rho\, d\Psi} + \int_{CS} \rho \vec{V} \cdot d\vec{A}$$

$$\overset{=0(4)\ \ =0(1)}{\dot{Q} - \dot{W_s} - \dot{W}_{shear} = \frac{\partial}{\partial t}\int_{CV} e\rho\, d\Psi} + \int_{CS}\left(u + pv + \frac{V^2}{2} + gz\right)\rho\vec{V}\cdot d\vec{A}$$

Assumptions: (1) Steady flow
(2) Properties uniform over inlet and outlet sections
(3) Treat air as an ideal gas, $p = \rho RT$
(4) Area of CV at ① and ② perpendicular to velocity, thus $\dot{W}_{shear} = 0$
(5) $z_1 = z_2$
(6) Inlet kinetic energy is negligible

Under the assumptions listed, the first law becomes

$$\dot{Q} - \dot{W_s} = \int_{CS}\left(u + pv + \frac{V^2}{2} + gz\right)\rho\vec{V}\cdot d\vec{A}$$

$$\dot{Q} - \dot{W_s} = \int_{CS}\left(h + \frac{V^2}{2} + gz\right)\rho\vec{V}\cdot d\vec{A} \qquad \{h \equiv u + pv\}$$

or

$$\dot{Q} = \dot{W_s} + \int_{CS}\left(h + \frac{V^2}{2} + gz\right)\rho\vec{V}\cdot d\vec{A}$$

For uniform properties (assumption 2) we can write

$$\dot{Q} = \dot{W_s} + \left(h_1 + \overset{\approx 0(6)}{\frac{V_1^2}{2}} + gz_1\right)\{-|\rho_1 V_1 A_1|\} + \left(h_2 + \frac{V_2^2}{2} + gz_2\right)\{|\rho_2 V_2 A_2|\}$$

For steady flow, from conservation of mass,

$$\int_{CS} \rho\vec{V}\cdot d\vec{A} = 0$$

Therefore, $-|\rho_1 V_1 A_1| + |\rho_2 V_2 A_2| = 0$, or $|\rho_1 V_1 A_1| = |\rho_2 V_2 A_2| = \dot{m}$. Hence we can write

$$\dot{Q} = \dot{W_s} + \dot{m}\left[(h_2 - h_1) + \frac{V_2^2}{2} + g\overset{=0(5)}{(z_2 - z_1)}\right]$$

Assume that air behaves as an ideal gas with constant c_p. Then $h_2 - h_1 = c_p(T_2 - T_1)$, and

$$\dot{Q} = \dot{W}_s + \dot{m}\left[c_p(T_2 - T_1) + \frac{V_2^2}{2}\right]$$

From continuity $|V_2| = \dot{m}/\rho_2 A_2$. Since $p_2 = \rho_2 R T_2$, then

$$|V_2| = \frac{\dot{m}}{A_2}\frac{RT_2}{p_2} = 20\ \frac{\text{lbm}}{\text{sec}} \times \frac{1}{1\ \text{ft}^2} \times 53.3\ \frac{\text{ft}\cdot\text{lbf}}{\text{lbm}\cdot\text{R}} \times 560\ \text{R} \times \frac{\text{in.}^2}{50\ \text{lbf}} \times \frac{\text{ft}^2}{144\ \text{in.}^2}$$

$$|V_2| = 82.9\ \text{ft/sec}$$

$$\dot{Q} = \dot{W}_s + \dot{m}\,c_p(T_2 - T_1) + \dot{m}\,\frac{V_2^2}{2}$$

Note that power input is *to* the CV, so $\dot{W}_s = -600$ hp, and

$$\dot{Q} = -\ 600\ \text{hp} \times 550\ \frac{\text{ft}\cdot\text{lbf}}{\text{hp}\cdot\text{sec}} \times \frac{\text{Btu}}{778\ \text{ft}\cdot\text{lbf}} + 20\ \frac{\text{lbm}}{\text{sec}} \times 0.24\ \frac{\text{Btu}}{\text{lbm}\cdot\text{R}} \times 30\ \text{R}$$

$$+\ 20\ \frac{\text{lbm}}{\text{sec}} \times \frac{(82.9)^2}{2}\ \frac{\text{ft}^2}{\text{sec}^2} \times \frac{\text{slug}}{32.2\ \text{lbm}} \times \frac{\text{Btu}}{778\ \text{ft}\cdot\text{lbf}} \times \frac{\text{lbf}\cdot\text{sec}^2}{\text{slug}\cdot\text{ft}}$$

$$\dot{Q} = -277\ \text{Btu/sec} \qquad\qquad \{\text{heat rejection}\}\ \dot{Q}$$

$\left[\begin{array}{l}\text{In addition to demonstrating a straightforward application of the first law, this problem}\\ \text{illustrates the need for keeping units straight.}\end{array}\right]$

EXAMPLE 4.17—Tank Filling: First Law Analysis

A tank of 0.1 m³ volume is connected to a high-pressure air line; both line and tank are initially at a uniform temperature of 20 C. The initial tank gage pressure is 100 kPa. The absolute line pressure is 2.0 MPa; the line is large enough so that its temperature and pressure may be assumed constant. The tank temperature is monitored by a fast-response thermocouple. At the instant after the valve is opened, the tank temperature rises at the rate of 0.05 C/sec. Determine the instantaneous flow rate of air into the tank if heat transfer is neglected.

EXAMPLE PROBLEM 4.17

GIVEN: Air supply pipe and tank as shown. At $t = 0^+$, $\partial T/\partial t = 0.05$ C/sec.

FIND: $\dot{m}$ at $t = 0^+$.

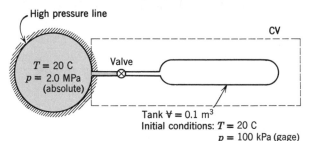

SOLUTION:
Choose CV shown, apply energy equation.

$$\overset{=\,0(1)}{}\ \overset{=\,0(2)}{}\ \overset{=\,0(3)}{}\ \overset{=\,0(4)}{}$$

Basic equation: $\displaystyle \not{\dot{Q}} - \not{\dot{W}}_s - \not{\dot{W}}_{\text{shear}} - \not{\dot{W}}_{\text{other}} = \frac{\partial}{\partial t}\int_{\text{CV}} e\rho\,d\forall + \int_{\text{CS}} (e + pv)\rho\vec{V}\cdot d\vec{A}$

$$\begin{array}{cc} \approx 0(5) & \approx 0(6) \\ \nearrow & \nearrow \end{array}$$
$$e = u + \frac{V^2}{2} + gz$$

Assumptions:
(1) $\dot{Q} = 0$ (given)
(2) $\dot{W}_s = 0$
(3) $\dot{W}_{shear} = 0$
(4) $\dot{W}_{other} = 0$
(5) Velocities in line and tank are small
(6) Neglect potential energy
(7) Uniform flow at tank inlet
(8) Properties uniform in tank
(9) Ideal gas, $p = \rho R T$, $du = c_v dT$

Then

$$0 = \frac{\partial}{\partial t} \int_{CV} u_{tank} \rho \, d\Psi + (u_{line} + pv)\{-|\rho VA|\}$$

But initially, T is uniform, so $u_{tank} = u_{line} = u$, and

$$0 = \frac{\partial}{\partial t} \int_{CV} u \rho \, d\Psi + (u + pv)\{-|\rho VA|\}$$

Since tank properties are uniform, $\partial/\partial t$ may be replaced by d/dt, and

$$0 = \frac{d}{dt}[uM] - (u + pv)\dot{m}$$

or

$$0 = u \frac{dM}{dt} + M \frac{du}{dt} - u\dot{m} - pv\dot{m} \tag{1}$$

The term dM/dt may be evaluated from continuity:

Basic equation: $\quad 0 = \frac{\partial}{\partial t} \int_{CV} \rho \, d\Psi + \int_{CS} \rho \vec{V} \cdot d\vec{A}$

$$0 = \frac{dM}{dt} + \{-|\rho VA|\} \quad \text{or} \quad \frac{dM}{dt} = \dot{m}$$

Substituting into Eq. 1 gives

$$0 = \cancel{u\dot{m}} + M \frac{du}{dt} \cancel{- u\dot{m}} - pv\dot{m} = M c_v \frac{dT}{dt} - pv\dot{m}$$

or

$$\dot{m} = \frac{M c_v (dT/dt)}{pv} = \frac{\rho \Psi c_v (dT/dt)}{pv} = \frac{\rho \Psi c_v (dT/dt)}{RT} \tag{2}$$

But at $t = 0$, $p_{tank} = 100$ kPa (gage), and

$$\rho = \rho_{tank} = \frac{p_{tank}}{RT} = \frac{(1.00 + 1.01)10^5 \frac{N}{m^2} \times \frac{kg \cdot K}{287 \, N \cdot m} \times \frac{1}{293 \, K}}{} = 2.39 \text{ kg/m}^3$$

Substituting into Eq. 2, we obtain

$$\dot{m} = 2.39 \frac{kg}{m^3} \times 0.1 \text{ m}^3 \times 717 \frac{N \cdot m}{kg \cdot K} \times 0.05 \frac{K}{sec} \times \frac{kg \cdot K}{287 \, N \cdot m} \times \frac{1}{293 \, K} \times 1000 \frac{g}{kg}$$

$\dot{m} = 0.102$ g/sec $\dot{m}$

$\left\{ \begin{array}{l} \text{This problem illustrates the application of the energy equation to an unsteady flow} \\ \text{situation.} \end{array} \right\}$

4–9 THE SECOND LAW OF THERMODYNAMICS

Recall that the system formulation of the second law is

$$\left.\frac{dS}{dt}\right)_{\text{system}} \geq \frac{1}{T}\dot{Q} \tag{4.5a}$$

where the total entropy of the system is given by

$$S_{\text{system}} = \int_{M(\text{system})} s\,dm = \int_{\mathbf{V}(\text{system})} s\rho\,d\mathbf{V} \tag{4.5b}$$

The relation between system and control volume formulations is

$$\left.\frac{dN}{dt}\right)_{\text{system}} = \frac{\partial}{\partial t}\int_{\text{CV}} \eta\rho\,d\mathbf{V} + \int_{\text{CS}} \eta\rho\vec{V}\cdot d\vec{A} \tag{4.11}$$

where

$$N_{\text{system}} = \int_{M(\text{system})} \eta\,dm = \int_{\mathbf{V}(\text{system})} \eta\rho\,d\mathbf{V} \tag{4.6}$$

To derive the control volume formulation of the second law of thermodynamics, we set

$$N = S \qquad \text{and} \qquad \eta = s$$

From Eq. 4.11, with this substitution, we obtain

$$\left.\frac{dS}{dt}\right)_{\text{system}} = \frac{\partial}{\partial t}\int_{\text{CV}} s\rho\,d\mathbf{V} + \int_{\text{CS}} s\rho\vec{V}\cdot d\vec{A} \tag{4.58}$$

From Eq. 4.5a,

$$\left.\frac{dS}{dt}\right)_{\text{system}} \geq \frac{1}{T}\dot{Q}$$

The system and the control volume coincided at t_0; thus

$$\left.\frac{1}{T}\dot{Q}\right)_{\text{system}} = \left.\frac{1}{T}\dot{Q}\right)_{\text{CV}} = \int_{\text{CS}}\frac{1}{T}\left(\frac{\dot{Q}}{A}\right)dA$$

In light of this, Eqs. 4.5a and 4.58 yield the control volume formulation of the second law of thermodynamics

$$\frac{\partial}{\partial t}\int_{\text{CV}} s\rho\,d\mathbf{V} + \int_{\text{CS}} s\rho\vec{V}\cdot d\vec{A} \geq \int_{\text{CS}}\frac{1}{T}\left(\frac{\dot{Q}}{A}\right)dA \tag{4.59}$$

In Eq. 4.59, the term $(\dot{Q}/A)$ represents the heat flux per unit area into the control volume through the area element dA. To evaluate the term

$$\int_{\text{CS}}\frac{1}{T}\left(\frac{\dot{Q}}{A}\right)dA$$

both the local heat flux, $(\dot{Q}/A)$, and local temperature, T, must be known for each area element of the control surface.

4–10 SUMMARY OBJECTIVES

After completing study of Chapter 4, you should be able to do the following:

1. Write each of the five basic laws (conservation of mass, Newton's second law, angular momentum principle, the first law of thermodynamics, and the second law of thermodynamics) for a system as a rate equation.

2. If the extensive property in the rate equations of Summary Objective 1 is designated N, define the corresponding intensive property, designated η, in each of the basic equations.

3. Write the equation that relates the rate of change of any arbitrary extensive property, N, of a system, to the variations of the property associated with a control volume. Give the physical significance of each quantity in the equation.

4. Write the control volume formulation of the conservation of mass and state the physical meaning of each term in the equation. Apply the equation to the solution of flow problems.

5. Write the control volume formulation of the momentum equation for an inertial control volume and state the physical meaning of each term in the equation. Apply the equation to the solution of flow problems.

** 6. State the relationship among fluid properties (the Bernoulli equation) that results from applying the momentum equation to a differential control volume. List the restrictions on the use of the Bernoulli equation.

7. Write the control volume formulation of the momentum equation for a control volume with rectilinear acceleration and state the physical meaning of each term in the equation. Apply the equation to the solution of flow problems.

** 8. Write the control volume formulation of the momentum equation for a control volume with arbitrary acceleration and state the physical meaning of each term in the equation. Apply the equation to the solution of flow problems.

** 9. Write the control volume formulation of the angular momentum principle for (a) a fixed and (b) a rotating control volume and state the physical meaning of each term in the equation. Apply the equation to the solution of flow problems.

10. Write the control volume formulation of the first law of thermodynamics and state the physical meaning of each term in the equation. Apply the equation to the solution of flow problems.

11. Write the control volume formulation of the second law of thermodynamics and state the physical meaning of each term in the equation. Apply the equation to the solution of flow problems.

12. Solve the problems at the end of the chapter that relate to the material you have studied.

PROBLEMS

4.1 A person with 75 kg total mass rides an elevator in a tall building. Calculate the force exerted by the person on the elevator floor during steady accelerations of (a) 3 m/sec^2 upward and (b) 2 m/sec^2 downward.

4.2 Calculate the minimum amount of work required to accelerate a 1500 kg automobile from rest to a speed of 25 m/sec on a level highway, if air resistance and mechanical losses are neglected.

** These objectives apply to sections that may be omitted without loss of continuity in the text material.

4.3   Air is expanded isothermally in a nonflow process from initial conditions of 170 C, 276 kPa (abs), and 0.05 m^3 to a final volume of 0.1 m^3. Determine the heat transferred.

4.4   In order to cool a six-pack as quickly as possible, it is placed in a freezer for a period of 1 hr. If the room temperature is 25 C and the cooled beverage is at a final temperature of 5 C, determine the change in specific entropy of the beverage.

4.5   A body of mass M is at rest on a horizontal plane. At $t = 0$, the body is acted upon by a constant force, F_1, of 10 lbf. The force is applied for a period of 5 sec. Immediately upon removal of this force, the body is acted upon by a constant force, F_2, of 3 lbf in the opposite direction. There is no friction. Determine the length of time of action by F_2 required to bring the body to rest.

4.6   A police investigation of tire marks showed that a car traveling along a straight level street had skidded to a stop for a total distance of 50 m after the brakes were applied. The coefficient of friction between tires and pavement is estimated to be $\mu = 0.6$. What was the probable minimum speed of the car when the brakes were applied?

4.7   A fully loaded Boeing 727-200 jet transport aircraft weighs 190,000 lbf. The pilot brings the 3 engines to full takeoff thrust of 15,500 lbf each before releasing the brakes. Neglecting aerodynamic and rolling resistance, estimate the minimum runway length and time needed to reach a takeoff speed of 140 mph. Assume engine thrust remains constant during ground roll.

4.8   A pistol bullet fired horizontally in air experiences an aerodynamic drag force proportional to the square of its speed, $F_D = kV^2$. Ballistic data, measured at the muzzle and at a range of 50 m, show that $V_0 = 260$ m/sec at the muzzle and $V = 240$ m/sec at 50 m. Estimate k if the bullet mass is 15.6 g.

4.9   A small steel ball of radius r, placed atop a much larger sphere of radius R, begins to roll under the influence of gravity. Rolling and air resistance are negligible. As the speed of the ball increases, it leaves the surface of the sphere and becomes a projectile. Determine the location at which the ball loses contact with the sphere.

4.10 The average rate of heat loss from a person to the surroundings when not actively working is about 300 Btu/hr. Suppose that in an auditorium with volume of approximately 1.2×10^7 ft^3, containing 6000 people, the ventilation system fails. How much does the internal energy of the air in the auditorium increase during the first 15 min after the ventilation system fails? Considering the auditorium and people as a system, and assuming no heat transfer to the surroundings, how much does the internal energy of the system change? How do you account for the fact that the temperature of the air increases? Estimate the rate of temperature rise under these conditions.

4.11 Air at 20 C and an absolute pressure of 1 atm is compressed adiabatically, without friction, to an absolute pressure of 3 atm. Determine the internal energy change.

4.12 The cooling system of an automobile contains 20 liters of coolant. The auto is started on a cold winter morning when the temperature is -10 C. During warmup, the engine rejects heat to the coolant at the rate of 5 kW. Estimate the minimum time needed for the coolant to reach 15 C, if its heat capacity is that of water.

4.13 An aluminum can of soft drink is to be cooled in a refrigerator where the temperature is $T_r = 5$ C. The rate of heat transfer from the can is $\dot{Q} = -k(T - T_r)$, where $k = 0.25$ W/C. Calculate the energy that must be removed if the mass of the can and its contents is equivalent to 390 g of water and its initial temperature is 25 C. Estimate the time required to cool the can to 7 C.

4.14 A 1500 kg automobile has a drivetrain that supplies 50 kW to the drive wheels. Determine the minimum time and distance required to accelerate the vehicle from rest to 25 m/sec on level road neglecting air and rolling resistance.

4.15 Most modern racing cars depend heavily on aerodynamic design features to generate downforce, which increases the loading on the tires and therefore the cornering speeds. A race car running at the Indianapolis Motor Speedway has a mass of 800 kg. Its aerodynamic devices produce a downforce of 8 kN at racing speeds. Each turn at Indy has an effective radius of 250 m and is banked at 9.2°. For these conditions, estimate the maximum radial acceleration and the corresponding speed that an Indy car can achieve in the turns. Assume the coefficient of friction for racing tires is 1.2.

4.16 The mass of an aluminum beverage can is 20 g. Its diameter and height are 65 and 120 mm, respectively. When full, the can contains 354 milliliters of soft drink with $SG = 1.05$. Evaluate the height of the center of gravity of the can as a function of liquid level. At what level would the can be least likely to tip over when subjected to a steady lateral acceleration? Calculate the minimum coefficient of static friction for which the full can would tip rather than slide on a horizontal surface.

4.17 A flywheel, consisting of a steel disk 20 mm thick and 0.6 m in diameter, rotates at 20,000 rpm. The flywheel is mounted in a subway car with its axis of rotation placed lengthwise. The car, traveling at a speed of 15 m/sec, rounds a curve of 80 m radius. Determine the reaction torque exerted by the flywheel assembly on the subway car.

4.18 A fluid mechanics laboratory experiment consists of a cylindrical tank containing water and mounted on a turntable, as in Example 3.9. The cylinder diameter is 0.2 m; the initial water depth, h_0, is 40 mm. Measurements show that when the turntable is switched off from a speed of 78 rpm, all motion of the liquid ceases 180 sec later. Determine the average torque applied to the water during this interval.

4.19 The velocity field in the region shown is given by $\vec{V} = az\,\hat{j} + b\hat{k}$, where $a = 10$ sec^{-1} and $b = 5$ m/sec. For depth w perpendicular to the diagram, an element of area ① may be represented by $w\,dz(-\hat{j})$ and an element of area ② by $w\,dy(-\hat{k})$. (Note that both are drawn *outward* from the control volume, hence the minus signs.)
  (a) Find an expression for $\vec{V}\cdot d\vec{A}_1$.　(b) Evaluate $\int_{A_1}\vec{V}\cdot d\vec{A}_1$.
  (c) Find an expression for $\vec{V}\cdot d\vec{A}_2$.　(d) Find an expression for $\vec{V}(\vec{V}\cdot d\vec{A}_2)$.
  (e) Evaluate $\int_{A_2}\vec{V}(\vec{V}\cdot d\vec{A}_2)$.

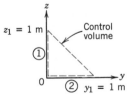

P4.19

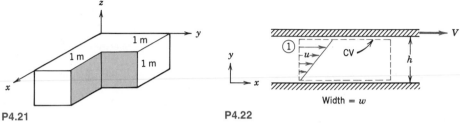

P4.20

4.20 A flow field is given by $\vec{V} = ay\hat{i} + b\hat{k}$, where $a = 2$ sec^{-1} and $b = 1$ ft/sec. Evaluate the volume flow rate through the shaded surface. All dimensions are in feet.

4.21 The shaded area shown is in a flow where the velocity field is given by $\vec{V} = ax\hat{i} - by\hat{j}$; $a = b = 1$ sec^{-1}. Evaluate the integrals $\int \vec{V}\cdot d\vec{A}$ and $\int \vec{V}(\vec{V}\cdot d\vec{A})$ over the shaded area.

P4.21　　　　　　　　　　P4.22

4.22 Evaluate the x direction momentum flux, $\int u(\rho\vec{V}\cdot d\vec{A})$, for cross section ① of the control volume shown in the diagram.

4.23 The velocity distribution for laminar flow in a long circular tube of radius R is given by the one-dimensional expression,

$$\vec{V} = u\hat{i} = u_{max}\left[1 - \left(\frac{r}{R}\right)^2\right]\hat{i}$$

For this profile, evaluate (a) $\int \vec{V} \cdot d\vec{A}$ and (b) $\int \vec{V}(\vec{V} \cdot d\vec{A})$ for the tube cross section.

4.24 The area shown shaded is in a flow where the velocity field is given by $\vec{V} = -ax\hat{i} + by\hat{j} + c\hat{k}$; $a = b = 1 \text{ sec}^{-1}$ and $c = 1$ m/sec. Write a vector expression for an element of the shaded area. Evaluate the integrals $\int \vec{V} \cdot d\vec{A}$ and $\int \vec{V}(\vec{V} \cdot d\vec{A})$ over the shaded area.

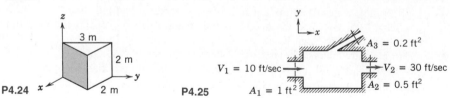

P4.24

P4.25

4.25 Consider steady, incompressible flow through the device shown. Determine the magnitude and direction of the volume flow rate through port 3.

4.26 Air at standard atmospheric conditions enters a compressor at a rate of 20 m³/min. The air is discharged at 800 kPa (abs) and 60 C. If the velocity in the discharge line must be limited to 20 m/sec, calculate the required diameter of the line.

4.27 In the incompressible flow through the device shown, velocities may be considered uniform over the inlet and outlet sections. If the fluid flowing is water, obtain an expression for the mass flow rate at section ③. The following conditions are known: $A_1 = 0.1$ m², $A_2 = 0.2$ m², $A_3 = 0.15$ m², $V_1 = 5$ m/sec, and $V_2 = 10 + 5\cos(4\pi t)$ m/sec.

P4.27

P4.28

4.28 Fluid with a 1050 kg/m³ density is flowing steadily through the rectangular box shown. Given $A_1 = 0.05$ m², $A_2 = 0.01$ m², $A_3 = 0.06$ m², $\vec{V}_1 = 4\hat{i}$ m/sec, and $\vec{V}_2 = -8\hat{j}$ m/sec, determine velocity $\vec{V}_3$.

4.29 Oil flows steadily in a thin layer down an inclined plane. The velocity profile is

$$u = \frac{\rho g \sin\theta}{2\mu}\left[hy - \frac{y^2}{2}\right]$$

Express the mass flow rate per unit width in terms of ρ, μ, g, θ, and h.

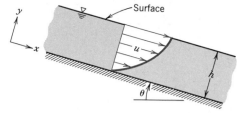

P4.29

4.30 Incompressible fluid flows through the device shown. The inlet flow is uniform with $V_1 = 2.0$ ft/sec. The outlet profile is linear, $V_2 = ky$. The device is $w = 1.25$ ft wide. Find k if the flow is steady.

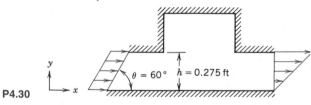

P4.30

4.31 Water enters a wide, flat channel of height $2h$ with a uniform velocity of 5 m/sec. At the channel outlet the velocity distribution is given by

$$\frac{u}{u_{max}} = 1 - \left(\frac{y}{h}\right)^2$$

where y is measured from the centerline of the channel. Determine the exit centerline velocity, u_{max}.

4.32 Incompressible fluid flows steadily through a plane diverging channel. At the inlet, of height H, the flow is uniform with magnitude V_1. At the outlet, of height $2H$, the velocity profile is

$$V_2 = V_m \cos\left(\frac{\pi y}{2H}\right)$$

where y is measured from the channel centerline. Express V_m in terms of V_1.

4.33 Water flows steadily through a pipe of length L and radius $R = 3$ in. Calculate the uniform inlet velocity, U, if the velocity distribution across the outlet is given by

$$u = u_{max}\left[1 - \frac{r^2}{R^2}\right]$$

and $u_{max} = 10$ ft/sec.

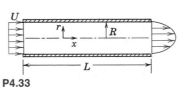

P4.33

4.34 A two-dimensional reducing bend has a linear velocity profile at section ①. The flow is uniform at sections ② and ③. The fluid is incompressible and the flow is steady. Find the magnitude and direction of the uniform velocity at section ③.

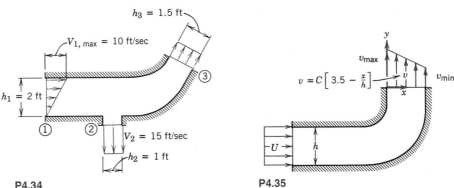

P4.34   **P4.35**

4.35 Water enters a two-dimensional channel of constant width, h, with uniform velocity, U. The channel makes a 90° bend that distorts the flow to produce the linear velocity profile shown at the exit, with $v_{max} = 2\,v_{min}$. Evaluate v_{min}.

4.36 A porous round tube with $D = 60$ mm carries water. The inlet velocity is uniform with $V_1 = 7.0$ m/sec. Water flows radially and axisymmetrically through the porous walls with velocity distribution

$$v = V_0\left[1 - \left(\frac{x}{L}\right)^2\right]$$

where $V_0 = 0.03$ m/sec and $L = 0.950$ m. Calculate the mass flow rate inside the tube at $x = L$.

4.37 A *hydraulic accumulator* is designed to reduce pressure pulsations in a machine tool hydraulic system. For the instant shown, determine the rate at which the accumulator gains or loses hydraulic oil.

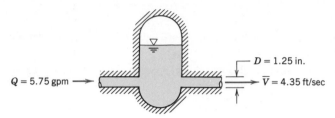

$Q = 5.75$ gpm

$D = 1.25$ in.

$\bar{V} = 4.35$ ft/sec

P4.37

4.38 A section of pipe carrying water contains an expansion chamber with a free surface whose area is 2 m^2. The inlet and outlet pipes are both 1 m^2 in area. At a given instant, the velocity at section ① is 3 m/sec into the chamber. Water flows out at section ② at 4 m^3/sec. Both flows are uniform. Find the rate of change of free surface level at the given instant. Indicate whether the level rises or falls.

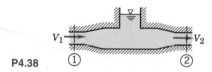

V_1 V_2

P4.38   ①   ②

4.39 A rectangular tank used to supply water for a Reynolds flow experiment is 230 mm deep. Its width and length are $W = 150$ mm and $L = 230$ mm. Water flows from the outlet tube (inside diameter $D = 6.35$ mm) at Reynolds number $Re = 2000$, when the tank is half full. The supply valve is closed. Find the rate of change of water level in the tank at this instant.

4.40 Viscous liquid from a circular tank, $D = 300$ mm in diameter, drains through a long circular tube of radius $R = 50$ mm. The velocity profile at the tube discharge is

$$u = u_{max}\left[1 - \left(\frac{r}{R}\right)^2\right]$$

Show that the average speed of flow in the drain tube is $\bar{V} = \frac{1}{2}u_{max}$. Evaluate the rate of change of liquid level in the tank at the instant when $u_{max} = 1.5$ m/sec.

4.41 A tank of 0.5 m^3 volume contains compressed air. A valve is opened and air escapes with a velocity of 300 m/sec through an opening of 130 mm^2 area. Air temperature passing through the opening is -15 C and the absolute pressure is 350 kPa. Find the rate of change of density of the air in the tank at this moment.

4.42 Air enters a tank through an area of 0.2 ft^2 with a velocity of 15 ft/sec and a density of 0.03 slug/ft^3. Air leaves with a velocity of 5 ft/sec and a density equal to that in the tank. The initial density of the air in the tank is 0.02 slug/ft^3. The total tank volume is 20 ft^3 and the exit area is 0.4 ft^2. Find the initial rate of change of density in the tank.

4.43 A cylindrical tank, 0.3 m in diameter, drains through a hole in its bottom. At the instant when the water depth is 0.6 m, the flow rate from the tank is observed to be 4 kg/sec. Determine the rate of change of water level at this instant.

4.44 A cylindrical tank, of diameter $D = 50$ mm, drains through an opening, $d = 5$ mm, in the bottom of the tank. The speed of the liquid leaving the tank is approximately $V = \sqrt{2gy}$, where y is the height from the tank bottom to the free surface. If the tank is initially filled with water to $y_0 = 0.4$ m, determine the water depth at $t = 12$ sec.

4.45 For the conditions of Problem 4.44, estimate the time required to drain the tank to depth $y = 20$ mm.

4.46 A conical funnel of half-angle $\theta = 15°$, with maximum diameter $D = 70$ mm, drains through a hole (diameter $d = 3.12$ mm) in its bottom. The speed of the liquid leaving the funnel is approximately $V = (2gy)^{1/2}$, where y is the height of the liquid free surface above the hole. Find the rate of change of surface level in the funnel at the instant when $y = H/2$.

4.47 A conical flask contains water to height $H = 36.8$ mm, where the flask diameter is $D = 29.4$ mm. Water drains out through a smoothly rounded hole of diameter $d = 7.35$ mm at the apex of the cone. The flow speed at the exit is approximately $V = (2gy)^{1/2}$, where y is the height of the liquid free surface above the hole. A stream of water flows into the top of the flask at constant volume flow rate, $Q = 3.75 \times 10^{-7}$ m^3/hr. Find the volume flow rate from the bottom of the flask. Evaluate the direction and rate of change of water surface level in the flask at this instant.

4.48 Water enters the tank through a pipe at rate $Q = 0.35$ ft^3/sec. Water also leaves the tank through a smoothly rounded nozzle of 2 in. diameter. The exit velocity through this hole, which depends on the height, h, of the water level above the hole, is $V_{\text{exit}} = \sqrt{2gh}$. At $t = 0$, $h = 9$ ft. Derive an equation that could be solved for $h(t)$ when $h > 4$ ft. For $4 < h < 9$ ft, is dh/dt greater than, less than, or equal to zero?

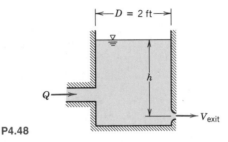

P4.48

4.49 Water flows steadily past a porous flat plate. Constant suction is applied along the porous section. The velocity profile at section cd is

$$\frac{u}{U_\infty} = 3\left[\frac{y}{\delta}\right] - 2\left[\frac{y}{\delta}\right]^{1.5}$$

Evaluate the mass flow rate across section bc.

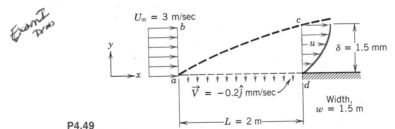

P4.49

4.50 A conical funnel of half-angle θ drains through a small hole of area A at the vertex. The speed of the liquid leaving the funnel is approximately $V = \sqrt{2gy}$, where y is the height of the liquid free surface above the hole. The funnel initially is filled to height y_0. Obtain an expression for the time, t, required to drain the funnel. Express the result in terms of the initial volume, V_0, of liquid in the funnel and the initial volume flow rate, $Q_0 = A\sqrt{2gy_0} = AV_0$.

4.51 A tank of fixed volume contains brine with initial density, ρ_i, greater than water. Pure water enters the tank steadily and mixes thoroughly with the brine in the tank. The liquid level in the tank remains constant. Derive expressions for (a) the rate of change of density of the liquid mixture in the tank and (b) the time required for the density to reach the value ρ_f, where $\rho_i > \rho_f > \rho_{H_2O}$.

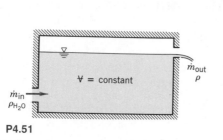

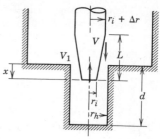

P4.51   P4.52

4.52   Motion of a hydraulic cylinder is cushioned at the end of its stroke by a piston that enters a hole as shown. The cavity and cylinder are filled with hydraulic fluid of uniform density, ρ. Obtain an expression for the velocity, V_1, at which hydraulic fluid escapes from the cylindrical hole as a function of piston displacement, x.

‡**4.53**   Over time, air seeps through pores in the rubber of high-pressure bicycle tires. The saying is that a tire loses pressure at the rate of "a pound [1 psi] a day." The true rate of pressure loss is not constant; instead, the instantaneous leakage mass flow rate is proportional to the air density in the tire and to the gage pressure in the tire, $\dot{m} \propto \rho p$. Because the leakage rate is slow, air in the tire is nearly isothermal. Consider a tire that initially is inflated to 0.6 MPa (gage). Assume the initial rate of pressure loss is 1 psi per day. Estimate the pressure that remains in the tire at the end of 30 days. How accurate is "a pound a day" over the entire 30 day period?

4.54   Evaluate the net rate of flux of momentum out through the control surface of Problem 4.28.

4.55   For the conditions of Problem 4.31, evaluate the ratio of the x-direction momentum flux at the channel outlet to that at the inlet.

4.56   For the conditions of Problem 4.33, evaluate the ratio of the x-direction momentum flux at the pipe outlet to that at the inlet.

4.57   A large tank is affixed to a cart as shown. Water issues from the tank through a 600 mm² nozzle at a speed of 10 m/sec. The water level in the tank is maintained constant by adding water through a vertical pipe. Determine the tension in the wire holding the cart stationary.

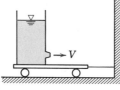

P4.57   P4.58, 4.59

4.58   A jet of water issuing from a stationary nozzle at 15 m/sec ($A_j = 0.05$ m²) strikes a turning vane mounted on a cart as shown. The vane turns the jet through angle $\theta = 50°$. Determine the value of M required to hold the cart stationary.

‡**4.59**   If the vane angle, θ, of Problem 4.58 is adjustable, plot the mass, M, needed to hold the cart stationary versus θ for $0 \le \theta \le 180°$.

4.60   A circular cylinder inserted across a stream of flowing water deflects the stream through angle θ, as shown. (This is termed the "Coanda effect.") For $a = 0.5$ in., $b = 0.1$ in., $V = 10$ ft/sec, and $\theta = 20°$, determine the horizontal component of the force on the cylinder due to the flowing water.

4.61   A vertical plate has a sharp-edged orifice at its center. A water jet of speed V strikes the plate concentrically. Obtain an expression for the external force needed to hold

‡ You may wish to use simple computer programs to help solve problems marked with daggers.

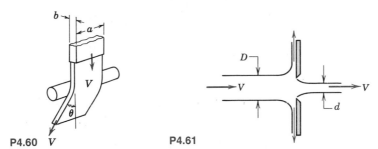

P4.60 *V*   P4.61

the plate in place, if the jet leaving the orifice also has speed V. Evaluate the force for $V = 5$ m/sec, $D = 100$ mm, and $d = 25$ mm.

4.62 Dry ready-mix concrete is packaged in 90 lb sacks; the density of the dry mix is 100 lbm/ft^3. The sacks are supported by pallets as they move through the automatic bagging machine. If the filling time is 4 sec and the vertical velocity of the mix is 10 ft/sec, determine the maximum vertical force on the pallet.

4.63 A farmer purchases 675 kg of bulk grain from the local co-op. The grain is loaded into his pickup truck from a hopper with an outlet diameter of 0.3 m. The loading operator determines the payload by observing the indicated gross mass of the truck as a function of time. The grain flow from the hopper ($\dot{m} = 40$ kg/sec) is terminated when the indicated scale reading reaches the desired gross mass. If the grain density is 600 kg/m^3, determine the true payload.

4.64 A large "weigh tank" is used to calibrate a flow meter. Measurements of weight as a function of time are made. Water enters the tank vertically from the flow-metering system at a speed of 20 ft/sec through a 1.5 in. diameter pipe. If the weight of the empty tank is 50 lbf, determine the scale reading at $t = 10$ sec. What is the true weight of the water and tank at $t = 10$ sec?

4.65 Water flows steadily through a fire hose and nozzle. The hose is 75 mm inside diameter, and the nozzle tip is 25 mm i.d.; water gage pressure in the hose is 510 kPa, and the stream leaving the nozzle is uniform. The exit speed and pressure are 32 m/sec and atmospheric, respectively. Find the force transmitted by the coupling between the nozzle and hose. Indicate whether the coupling is in tension or compression.

4.66 Water flows through a well-made nozzle at the end of a 50 mm diameter pipe. The nozzle exit diameter is 25 mm. The water flow rate is 53.0 m^3/hr, and the pressure immediately upstream from the nozzle is 522 kPa (abs). Find the force transmitted by the coupling between the nozzle and pipe. Indicate whether the coupling is in tension or compression.

4.67 A shallow circular dish has a sharp-edged orifice at its center. A water jet, of speed V, strikes the dish concentrically. Obtain an expression for the external force needed to hold the dish in place if the jet issuing from the orifice also has speed V. Evaluate the force for $V = 5$ m/sec, $D = 100$ mm, and $d = 20$ mm.

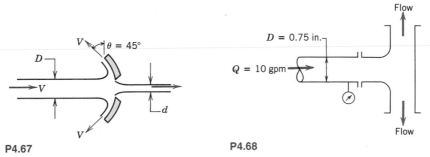

P4.67   P4.68

4.68 Water flows steadily at the rate of 10 gpm through a horizontal tee and discharges to atmosphere. The pressure just upstream from the tee is 15.0 psia. Determine the force exerted on the line by the tee.

4.69 Water is flowing steadily through the 180° elbow shown. At the inlet to the elbow the gage pressure is 96 kPa. The water discharges to atmospheric pressure. Assume properties are uniform over the inlet and outlet areas; $A_1 = 2600$ mm², $A_2 = 650$ mm², and $V_1 = 3.05$ m/sec. Find the horizontal component of force required to hold the elbow in place.

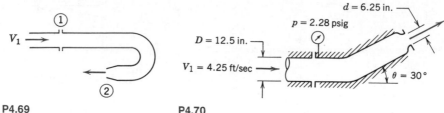

P4.69   P4.70

4.70 Water flows steadily through the nozzle shown, discharging to atmosphere. Calculate the horizontal component of force in the flanged joint. Indicate whether the joint is in tension or compression.

4.71 A flat plate orifice of 50 mm diameter is located at the end of a 100 mm diameter pipe. Water flows through the pipe and orifice at 0.05 m³/sec. The diameter of the water jet downstream from the orifice is 35 mm. Calculate the external force required to hold the orifice in place. Neglect friction on the pipe wall.

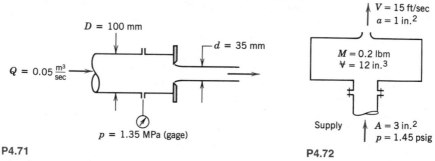

P4.71   P4.72

4.72 A spray system is shown in the diagram. Water is supplied at $p = 1.45$ psig, through the flanged opening of area $A = 3$ in.² The water leaves in a steady free jet at atmospheric pressure. The jet area and speed are $a = 1.0$ in.² and $V = 15$ ft/sec. The mass of the spray system is 0.2 lbm and it contains $V\!\!\!\!- = 12$ in.³ of water. Find the force exerted on the supply pipe by the spray system.

4.73 The nozzle shown discharges a sheet of water through a 180° arc. The water speed is 15 m/sec and the jet thickness is 30 mm at a radial distance of 0.3 m from the centerline of the supply pipe. Find (a) the volume flow rate of water in the jet sheet and (b) the y component of force required to hold the nozzle in place.

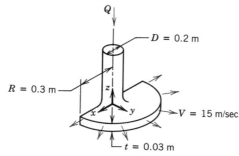

P4.73

4.74 A typical jet engine test stand installation is shown, together with some test data. Fuel enters the top of the engine vertically at a rate equal to 2 percent of the mass flow rate of the inlet air. For the given conditions, compute the air flow rate through the engine and estimate the thrust.

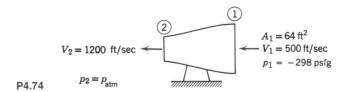

P4.74 $p_2 = p_{atm}$

$A_1 = 64 \text{ ft}^2$
$V_1 = 500 \text{ ft/sec}$
$p_1 = -298 \text{ psig}$

$V_2 = 1200 \text{ ft/sec}$

4.75 At rated thrust, a liquid-fueled rocket motor consumes 180 lbm/sec of nitric acid as oxidizer and 70 lbm/sec of aniline as fuel. Flow leaves axially at 6000 ft/sec relative to the nozzle and at 16.5 psia. The nozzle exit diameter is $D = 2$ ft. Calculate the thrust produced by the motor on a test stand at standard sea-level pressure.

4.76 The designer of a fluidic device wishes to deflect an air jet, with mass flow rate $\dot{m}_1$ and cross-sectional area A_1, by 30°. The designer proposes doing this with a smaller air jet of cross-sectional area A_2, oriented perpendicular to the original jet. What mass flow rate, $\dot{m}_2$, is needed in the smaller jet to accomplish this task? Assume incompressible flow of standard air and neglect gravity.

4.77 Consider flow through the sudden expansion shown. If the flow is incompressible and friction is neglected, show that the pressure rise, $\Delta p = p_2 - p_1$, is given by

$$\frac{\Delta p}{\frac{1}{2}\rho \bar{V}_1^2} = 2\left(\frac{d}{D}\right)^2\left[1-\left(\frac{d}{D}\right)^2\right]$$

Hint: Assume the pressure is uniform and equal to p_1 on the vertical surface of the expansion.

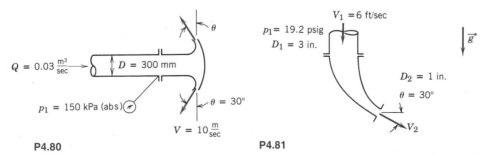

P4.77   **P4.78, 4.79**

4.78 A turning vane, which deflects the water through 60°, is attached to the cart under the conditions of Problem 4.57. Determine the tension in the wire holding the cart stationary and the force of the vane on the cart.

4.79 If the turning vane of Problem 4.78 deflects the water through 90°, determine the tension in the wire holding the cart stationary and the force of the vane on the cart.

4.80 A conical spray head is shown. The fluid is water and the exit stream is uniform. Evaluate (a) the thickness of the spray sheet at 400 mm radius and (b) the axial force exerted by the spray head on the supply pipe.

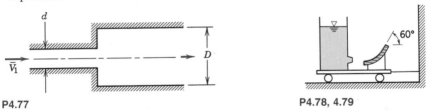

P4.80   **P4.81**

4.81 A curved nozzle assembly that discharges to the atmosphere is shown. The nozzle weighs 10 lbf and its internal volume is 150 in.[3] The fluid is water. Determine the reaction force exerted by the nozzle on the coupling to the inlet pipe.

4.82 A reducer in a piping system is shown. The internal volume of the reducer is $0.2 \ m^3$ and its mass is 25 kg. Evaluate the total force that must be provided by the surrounding pipes to support the reducer. The fluid is gasoline.

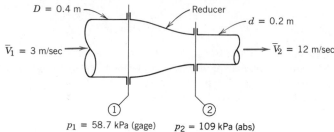

P4.82 $p_1 = 58.7 \ kPa \ (gage)$ $p_2 = 109 \ kPa \ (abs)$

4.83 A water jet pump has jet area $0.01 \ m^2$ and jet speed 30 m/sec. The jet is within a secondary stream of water having speed $V_s = 3$ m/sec. The total area of the duct (the sum of the jet and secondary stream areas) is $0.075 \ m^2$. The water is thoroughly mixed and leaves the jet pump in a uniform stream. The pressures of the jet and secondary stream are the same at the pump inlet. Determine the speed at the pump exit and the pressure rise, $p_2 - p_1$.

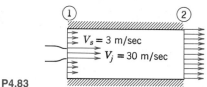

P4.83

4.84 A 30° reducing elbow is shown. The fluid is water. Evaluate the components of force that must be provided by the adjacent pipes to keep the elbow from moving.

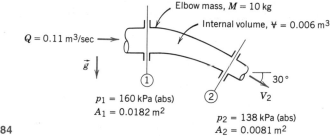

P4.84

4.85 A jet pump operating on water is shown. The primary jet discharges into a housing with constant inside diameter; water leaves at section ③. The pressures of the jet and secondary stream at sections ① and ② are the same. Neglect friction between the water and the housing. Calculate the magnitude and direction of the volume flow rate in the secondary stream in the annular region between the primary jet and the housing. Evaluate the pressure at the outlet from the jet pump.

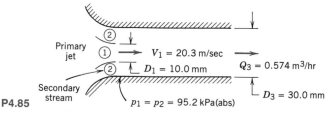

P4.85

4.86 Consider the steady adiabatic flow of air through a long straight pipe with $0.5 \ ft^2$ cross-sectional area. At the inlet, the air is at 30 psia, 140 F, and has a velocity of 500 ft/sec. At the exit, the air is at 11.3 psia and has a velocity of 985 ft/sec. Calculate the axial force of the air on the pipe. (Be sure to make the direction clear.)

4.87 A monotube boiler consists of a 20 ft length of tubing with 0.375 in. inside diameter. Water enters at the rate of 0.3 lbm/sec at 500 psia. Steam leaves at 400 psig with 0.024 slug/ft^3 density. Find the magnitude and direction of the force exerted by the flowing fluid on the tube.

4.88 A gas flows steadily through a heated porous pipe of constant 0.2 m^2 cross-sectional area. At the pipe inlet, the absolute pressure is 340 kPa, the density is 5.1 kg/m^3, and the mean velocity is 152 m/sec. The fluid passing through the porous wall leaves in a direction normal to the pipe axis, and the total flow rate through the porous wall is 29.2 kg/sec. At the pipe outlet, the absolute pressure is 280 kPa and the density is 2.6 kg/m^3. Determine the axial force of the fluid on the pipe.

4.89 Water is discharged from a narrow slot in a 150 mm diameter pipe. The resulting horizontal two-dimensional jet is 1 m long and 15 mm thick, but of nonuniform velocity. The pressure at the inlet section is 30 kPa (gage). Calculate (a) the volume flow rate at the inlet section and (b) the forces required at the coupling to hold the spray pipe in place. Neglect the mass of the pipe and the water it contains.

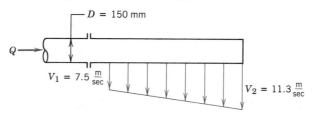

P4.89 Thickness, t = 15 mm

4.90 Water flows steadily through the flat bend of Problem 4.35. Flow at the inlet is uniform and horizontal. Flow at the exit is nonuniform, vertical, and at atmospheric pressure. The mass of the channel structure is $M_c = 2.05$ kg; the internal volume of the channel is $\Psi = 0.00355$ m^3. The channel width is constant and equal to $h = 75.5$ mm. Evaluate the force exerted by the channel assembly on the supply duct.

4.91 A nozzle for a spray system is designed to produce a flat radial sheet of water. The sheet leaves the nozzle at $V_2 = 10$ m/sec, covers 180° of arc, and has thickness $t = 1.5$ mm. The nozzle discharge radius is $R = 50$ mm. The water supply pipe is 35 mm in diameter and the inlet pressure is $p_1 = 150$ kPa (abs). Evaluate the axial force exerted by the spray nozzle on the coupling.

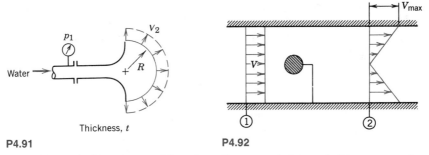

P4.91 Thickness, t **P4.92**

4.92 A small round object is tested in a 1 m diameter wind tunnel. The pressure is uniform across sections ① and ②. The upstream pressure is 20 mm H$_2$O (gage), the downstream pressure is 10 mm H$_2$O (gage), and the mean air speed is 10 m/sec. The velocity profile at section ② is linear; it varies from zero at the tunnel centerline to a maximum at the tunnel wall. Calculate (a) the mass flow rate in the wind tunnel, (b) the maximum velocity at section ②, and (c) the drag of the object and its supporting vane. Neglect viscous resistance at the tunnel wall.

4.93 An incompressible fluid flows steadily in the entrance region of a two-dimensional channel of height $2h$. The uniform velocity at the channel entrance is $U_1 = 20$ ft/sec. The velocity distribution at a section downstream is

$$\frac{u}{u_{max}} = 1 - \left[\frac{y}{h}\right]^2$$

Evaluate the maximum velocity at the downstream section. Calculate the pressure drop that would exist in the channel if viscous friction at the walls could be neglected.

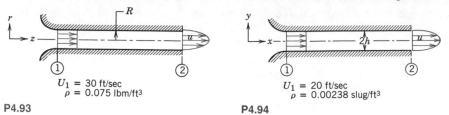

$U_1 = 30$ ft/sec
$\rho = 0.075$ lbm/ft3

$U_1 = 20$ ft/sec
$\rho = 0.00238$ slug/ft3

P4.93

P4.94

4.94 An incompressible fluid flows steadily in the entrance region of a circular tube of radius R. The uniform velocity at the tube entrance is $U_1 = 30$ ft/sec. The velocity distribution at a section downstream is

$$\frac{u}{u_{max}} = 1 - \left[\frac{r}{R}\right]^2$$

Evaluate the maximum velocity at the downstream section. Calculate the pressure drop that would exist in the tube if viscous friction at the walls could be neglected.

4.95 A fluid of constant density ρ enters a pipe of radius R with uniform velocity U. At a downstream section the velocity varies with radius, r, according to the equation

$$u = 2U\left(1 - \frac{r^2}{R^2}\right)$$

The pressure at sections ① (inlet) and ② (downstream) are p_1 and p_2, respectively. Show that the frictional force, F, of the pipe walls on the fluid between sections ① and ② is

$$F = \pi R^2\left[-(p_1 - p_2) + \tfrac{1}{3}\rho U^2\right]$$

in a direction opposing the flow.

4.96 Air enters a duct, of diameter $D = 25.0$ mm, through a well-rounded inlet with uniform speed, $U_1 = 0.870$ m/sec. At a downstream section where $L = 2.25$ m, the fully developed velocity profile is

$$\frac{u(r)}{U_c} = 1 - \left(\frac{r}{R}\right)^2$$

The pressure drop between these sections is $p_1 - p_2 = 1.92$ N/m². Find the total force of friction exerted by the tube on the air.

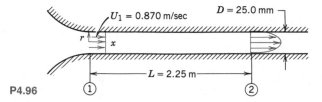

$U_1 = 0.870$ m/sec $D = 25.0$ mm

$L = 2.25$ m

P4.96   ①   ②

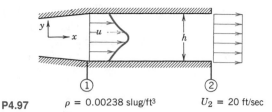

P4.97 $\rho = 0.00238$ slug/ft³ $U_2 = 20$ ft/sec

4.97 Air flows steadily through a plane-wall diffuser and duct system. The channel width normal to the plane of the diagram is b. At section ①, the velocity profile is

$$\frac{u}{u_{max}} = e^{-2\frac{|y|}{h}}$$

The velocity is uniform at section ②. The pressure is uniform across sections ① and ②. Assume incompressible flow and neglect friction at the channel walls. Calculate (a) the value of u_{max} and (b) the pressure difference, $p_2 - p_1$.

4.98 Air at standard conditions flows along a flat plate. The undisturbed freestream speed is $U_0 = 30$ m/sec. At $L = 0.3$ m downstream from the leading edge of the plate, the boundary-layer thickness is $\delta = 1.5$ mm. The velocity profile at this location is approximated as $u/U_0 = y/\delta$. Calculate the horizontal component of force per unit width required to hold the plate stationary.

4.99 Consider the incompressible flow of fluid in a boundary layer as depicted in Example Problem 4.2. Show that the friction drag force of the fluid on the surface is given by

$$F_f = \int_0^\delta \rho u(U - u)w \, dy$$

Evaluate the drag force for the conditions of Example Problem 4.2.

4.100 Air at standard conditions flows along a flat plate. The undisturbed freestream speed is $U_0 = 10$ m/sec. At $L = 145$ mm downstream from the leading edge of the plate, the boundary-layer thickness is $\delta = 2.3$ mm. The velocity profile at this location is

$$\frac{u}{U_0} = \frac{3}{2}\frac{y}{\delta} - \frac{1}{2}\left[\frac{y}{\delta}\right]^3$$

Calculate the horizontal component of force per unit width required to hold the plate stationary.

4.101 Consider again the flow over a flat plate with suction depicted in Problem 4.49. For the given conditions, evaluate the force per unit width required to hold the plate stationary.

4.102 A windmill operating in a uniform air stream is shown. The Rankine propeller theory (Problem 4.103) predicts that half the velocity change occurs upstream and half downstream from the windmill. Analyze the flow to develop expressions for V_2, V_3, and D_3, and the thrust on the windmill. Evaluate the thrust on the windmill if $V_1 = 10$ m/sec, $D = 4$ m, and the atmospheric pressure streamline diameter is 3 m upstream from the windmill. Assume uniform velocity distributions and standard air.

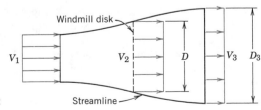

P4.102

4.103 A propeller operating in still air is shown. A theory for uniform flow without losses developed and published by Rankine in 1885 predicts that the air speed through the propeller disk is half that of the slipstream behind the propeller. Pressure on the streamlines bounding the slipstream is atmospheric. Develop an expression for the thrust developed by a propeller for the conditions shown. Evaluate for standard air with $V = 30$ m/sec and $D = 3$ m.

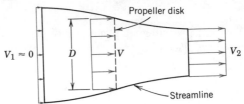

P4.103, 4.114, 4.150

4.104 A sharp-edged splitter plate inserted part way into a flat stream of flowing water produces the flow pattern shown. Analyze the situation to evaluate θ as a function of α, where $0 \leq \alpha < 0.5$. Evaluate the force needed to hold the splitter plate in place. (Neglect any friction force between the water stream and the splitter plate.)

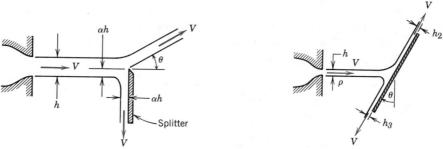

P4.104   P4.105

‡**4.105** When a plane liquid jet strikes an inclined flat plate, it splits into two streams of equal speed but unequal thickness. For frictionless flow there can be no tangential force on the plate surface. Use this assumption to develop an expression for h_2/h as a function of plate angle, θ. Plot your results and comment on the limiting cases, $\theta = 0$ and $\theta = 90°$.

4.106 Experimental measurements are made in a low-speed air jet to determine the drag force on a circular cylinder. Velocity measurements at two sections, where the pressure is uniform and equal, give the results shown. Evaluate the drag force on the cylinder, per unit width.

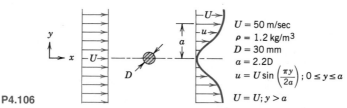

$U = 50$ m/sec
$\rho = 1.2$ kg/m^3
$D = 30$ mm
$a = 2.2D$
$u = U \sin\left(\dfrac{\pi y}{2a}\right); 0 \leq y \leq a$
$U = U; y > a$

P4.106

‡**4.107** The constant-area mixing section of a water jet pump is shown. The primary jet is within a secondary stream that is uniform and at the same pressure as the jet exit. Assume the two streams mix thoroughly and leave in a uniform stream. Develop an algebraic expression for the dimensionless pressure rise, $\Delta p/q = (p_2 - p_1)/\frac{1}{2}\rho\bar{V}_j^2$, in terms of the area ratio, $a = A_j/A_2$, and the flow rate ratio, $r = Q_2/Q_j$. Plot curves of $\Delta p/q$ versus r for $a = 0.05$, 0.1, and 0.2.

‡ You may wish to use simple computer programs to help solve problems marked with daggers.

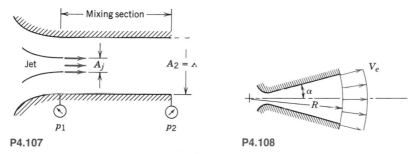

P4.107   **P4.108**

4.108 Gases leaving the propulsion nozzle of a rocket are modeled as flowing radially outward from a point upstream from the nozzle throat. Assume the speed of the exit flow, V_e, has constant magnitude. Develop an expression for the axial thrust, T_a, developed by flow leaving the nozzle exit plane. Compare your result to the one-dimensional approximation, $T = \dot{m} V_e$. Evaluate the percent error for $\alpha = 15°$.

**4.109 Consider a cylindrical tank, of inside area A_t and mass M, placed on a scale. Assume the tank is filled to level h, with water, which drains from a well-rounded orifice of area A in the tank bottom. (The jet does not hit the scale platform.) Develop an expression for the percentage reduction in scale reading due to the momentum efflux through the orifice.

**4.110 Two large tanks containing water have small smoothly contoured orifices of equal area. A jet of liquid issues from the left tank. Assume the flow is uniform and unaffected by friction. The jet impinges on a flat plate covering the opening of the right tank. Obtain an expression for the height, h, required to balance the hydrostatic force on the plate from water in the right tank.

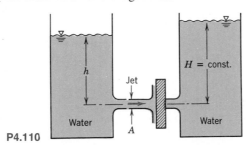

P4.110

**4.111 A horizontal axisymmetric jet of air with 10 mm diameter strikes a stationary vertical disk of 200 mm diameter. The jet speed is 50 m/sec at the nozzle exit. A manometer is connected to the center of the disk. Calculate (a) the deflection, h, if the manometer liquid has SG = 1.75 and (b) the force exerted by the jet on the disk.

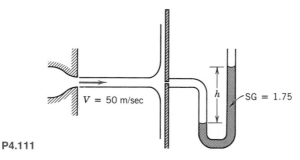

P4.111

** These problems require material from sections that may be omitted without loss of continuity in the text material.

**4.112 Consider a cylindrical tank, of inside area A_t and mass M, placed on a smooth plane surface. Assume the tank is filled to level h with water. A small rounded orifice of area A is placed in the side near the tank bottom so that a jet discharges horizontally. Develop an expression for the coefficient of friction that would allow the tank to slide as a result of momentum flux from the jet. Evaluate for $A/A_t = 0.1$. Comment on your result.

**4.113 Water flows at the rate of 0.019 m³/sec into a tall cylindrical tank that is 1 m inside diameter. The tank bottom has a well-rounded orifice 55 mm in diameter. Water flows vertically from the orifice due to gravity. When the water in the tank reaches a certain level, the rate of outflow equals the rate of inflow. Neglecting friction, calculate the equilibrium water level.

**4.114 Consider again the statement and diagram of Problem 4.103. Apply the Bernoulli equation twice to obtain expressions for the pressure just in front of and just behind the propeller disk. (Note that the thrust on the propeller may be written as the pressure difference across the disk times the disk area.) Use the momentum equation for a control volume to obtain an expression for the thrust exerted by the propeller. Equate the two expressions for thrust to show that half the velocity increase occurs ahead of and half behind the propeller.

**4.115 A uniform jet of water leaves a 15 mm diameter nozzle and flows directly downward. The jet speed at the nozzle exit plane is 1.5 m/sec. The jet impinges on a horizontal disk and flows radially outward in a flat sheet. Obtain a general expression for the velocity the liquid stream would reach at the level of the disk. Develop an expression for the force required to hold the disk stationary, neglecting the mass of the disk and water sheet. Evaluate for $h = 1.5$ m.

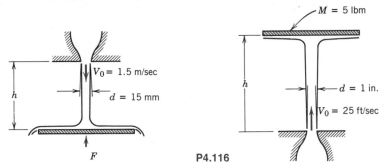

P4.115   P4.116

**4.116 A 5 lbm disk is constrained horizontally but is free to move vertically. The disk is struck from below by a vertical jet of water. The speed and diameter of the water jet are 25 ft/sec and 1 in. at the nozzle exit. Obtain a general expression for the speed of the water jet as a function of height, h. Find the height to which the disk will rise and remain stationary.

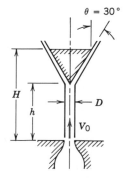

P4.117

** These problems require material from sections that may be omitted without loss of continuity in the text material.

**4.117 Water from a jet of diameter D is used to support the cone-shaped object shown. Derive an expression for the *combined* mass of the cone and water, M, that can be supported by the jet, in terms of parameters associated with a suitably chosen control volume. Use your expression to calculate M when $V_0 = 10$ m/sec, $H = 1$ m, $h = 0.8$ m, $D = 50$ mm, and $\theta = 30°$. Estimate the mass of water in the control volume.

‡**4.118 The tank of Problem 4.113 initially is empty. At $t = 0$ the water supply is turned on at a constant rate of 0.019 m³/sec. Develop a differential equation for the rate of change of water level. Use numerical integration to estimate the time required for water in the tank to reach 95 percent of its equilibrium level.

**4.119 In ancient Egypt, circular vessels filled with water sometimes were used as crude clocks. The vessels were shaped in such a way that, as water drained from the bottom, the surface level dropped at constant rate, s. Assume that water drains from a small hole of area A. Find an expression for the radius of the vessel, r, as a function of the water level, h. Determine the volume of water needed so that the clock will operate for n hours.

**4.120 A stream of incompressible liquid moving at low speed leaves a nozzle pointed directly downward. Assume the velocity at any cross section is uniform and neglect viscous effects. The velocity and area of the jet at the nozzle exit are V_0 and A_0, respectively. Apply conservation of mass and the momentum equation to a differential control volume of length dz in the flow direction. Derive expressions for the variations of jet velocity and area as functions of z. Evaluate the distance at which the jet area is half its original value. (Take the origin of coordinates at the nozzle exit.)

**4.121 A stream of incompressible liquid moving at low speed leaves a nozzle pointed directly upward. Assume the velocity at any cross section is uniform and neglect viscous effects. The velocity and area of the jet at the nozzle exit are V_0 and A_0, respectively. Apply conservation of mass and the momentum equation to a differential control volume of length dz in the flow direction. Derive expressions for the variations of jet velocity and area as functions of z. Evaluate the vertical distance required to reduce the jet speed to zero. (Take the origin of coordinates at the nozzle exit.)

**4.122 Incompressible fluid of negligible viscosity is pumped, at total volume flow rate Q, through a porous surface into the small gap between closely spaced parallel plates as shown. The fluid has only horizontal motion in the gap. Assume uniform flow across any vertical section. Obtain an expression for the pressure variation as a function of x. *Hint:* Apply conservation of mass and the momentum equation to a differential control volume of thickness dx, located at position x.

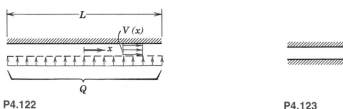

P4.122   P4.123

**4.123 Incompressible liquid of negligible viscosity is pumped, at total volume flow rate Q, through two small holes into the narrow gap between closely spaced parallel plates as shown. The liquid flowing away from the holes has only radial motion. Assume uniform flow across any vertical section and discharge to atmospheric pressure at $r = R$. Obtain an expression for the pressure variation as a function of radius. *Hint:* Apply conservation of mass and the momentum equation to a differential control volume of thickness dr, located at radius r.

** These problems require material from sections that may be omitted without loss of continuity in the text material.

‡ You may wish to use simple computer programs to help solve problems marked with daggers.

**4.124 Incompressible liquid flows steadily through a pipe of constant diameter. The pipe contains a section of porous wall of length L. There liquid is removed at constant rate q, expressed as volume flow rate per unit length. The liquid velocity in the pipe at the entrance to the porous section is V_0. The liquid removed in the porous section has no axial component of velocity. Evaluate the velocity and pressure distributions for flow through the porous section. Neglect viscous effects. *Hint:* Apply conservation of mass and the momentum equation to a differential control volume of length dx.

**4.125 Incompressible liquid flows steadily through a pipe of constant diameter. The pipe contains a section of porous wall of length L. There liquid is removed at rate

$$q(x) = q_{max}\frac{x}{L}$$

with q_{max} expressed as volume flow rate per unit length. The liquid velocity in the pipe at the entrance to the porous section is V_0. The liquid removed in the porous section has no axial component of velocity. Evaluate the velocity and pressure distributions for flow through the porous section. Neglect viscous effects. *Hint:* Apply conservation of mass and the momentum equation to a differential control volume of length dx.

**4.126 Incompressible liquid flows steadily through a pipe of constant diameter. The pipe contains a section of porous wall of length L, where liquid is supplied at constant rate q, expressed as volume flow rate per unit length. The liquid velocity in the pipe at the entrance to the porous section is V_0. The liquid supplied in the porous section has no axial component of velocity. Evaluate the velocity and pressure distributions for flow through the porous section. Neglect viscous effects. *Hint:* Apply conservation of mass and the momentum equation to a differential control volume of length dx.

**4.127 Incompressible liquid flows steadily through a pipe of constant diameter. The pipe contains a section of porous wall of length L, where liquid is supplied at the rate

$$q(x) = q_{max}\frac{x}{L}$$

with q_{max} expressed as volume flow rate per unit length. The liquid velocity in the pipe at the entrance to the porous section is V_0. The liquid supplied in the porous section has no axial component of velocity. Evaluate the velocity and pressure distributions for flow through the porous section. Neglect viscous effects. *Hint:* Apply conservation of mass and the momentum equation to a differential control volume of length dx.

**4.128 Propagation of small waves on a liquid free surface may be analyzed using a differential control volume. Consider a small solitary wave moving with speed c from right to left. Assume a small change in water surface elevation across the wave. To make the flow appear steady, choose a differential control volume that encloses the wave and moves with it. Apply conservation of mass and the momentum equation to derive an expression for wave speed. Be sure to include in your analysis the hydrostatic pressure forces on the control surface. Neglect friction on the channel bed.

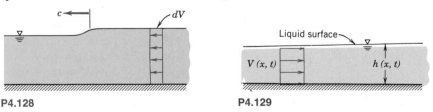

P4.128   P4.129

** These problems require material from sections that may be omitted without loss of continuity in the text material.

**4.129 Consider unsteady, one-dimensional flow of an incompressible liquid in a horizontal open channel. The free surface elevation is not constant, but the velocity distribution at any section is uniform. Depth h and mean velocity V are functions of both x and t. Assume the channel width, b, is large. Derive appropriate forms of conservation of mass and the linear momentum equation for this flow. *Hint:* Use a differential control volume of length dx in the flow direction. Be sure to include the effects of hydrostatic pressure forces and friction on the channel bed.

**4.130 Liquid falls vertically into a short horizontal rectangular open channel of width b. The total volume flow rate, Q, is distributed uniformly over area bL. Neglect viscous effects. Obtain an expression for h_1 in terms of h_2, Q, and b. *Hint:* Choose a control volume with outer boundary located at $x = L$. Sketch the surface profile, $h(x)$. *Hint:* Use a differential control volume of width, dx.

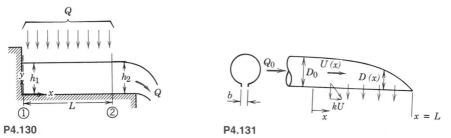

P4.130   P4.131

**4.131 As part of an industrial process, a pipe is arranged with a small slot along its length and liquid is allowed to drain out in a continuous sheet. The pipe is tapered so the pressure and vertical flow velocity are constant along its length. The horizontal velocity component of the fluid leaving the pipe is a constant fraction, k, of the local average velocity in the pipe. Neglect friction and gravity effects as a first approximation. Show that k must be 1.0 and the velocity within the pipe must be constant. Obtain an expression for the required diameter variation, $D(x)$. *Hint:* Note that mechanical energy is conserved along a streamline.

**4.132 The small gap between two long narrow parallel plates initially is filled with incompressible liquid. At $t = 0$ the upper plate begins to move downward toward the lower plate with constant velocity, V_0, causing the liquid to be squeezed from the narrow gap. Neglecting viscous effects, and assuming uniform flow in the horizontal direction, develop an expression for the velocity field between the parallel plates. *Hint:* Apply conservation of mass to a control volume with outer surface located at position x. Note that even though the velocity of the upper plate is constant, the flow is unsteady.

**4.133 The narrow gap between two closely spaced circular plates initially is filled with incompressible liquid. At $t = 0$ the upper plate begins to move downward toward the lower plate with constant velocity, V_0, causing the liquid to be squeezed from the narrow gap. Neglecting viscous effects and assuming uniform flow in the radial direction, develop an expression for the velocity field between the parallel plates. *Hint:* Apply conservation of mass to a control volume with outer surface located at radius r. Note that even though the velocity of the upper plate is constant, the flow is unsteady.

4.134 A jet of water is directed against a vane, which could be a blade in a turbine or in any other piece of hydraulic machinery. The water leaves the stationary 50 mm diameter nozzle with a speed of 20 m/sec and enters the vane tangent to the surface at A. The inside surface of the vane at B makes angle $\theta = 150°$ with the x direction. Compute the force that must be applied to maintain the vane speed constant at $U = 5$ m/sec.

** These problems require material from sections that may be omitted without loss of continuity in the text material.

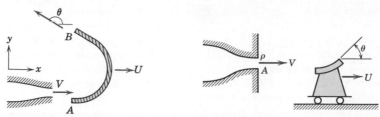

P4.134   **P4.135, 4.136, 4.137, 4.144, 4.157, 4.188, 4.189**

4.135 Water from a stationary nozzle impinges on a moving vane with a turning angle of 120°. The vane moves away from the nozzle with constant speed, $U = 30$ ft/sec, and receives a jet that leaves the nozzle with speed $V = 100$ ft/sec. The nozzle has an exit area of 0.04 ft². Find the force that must be applied to maintain the vane speed constant.

4.136 A water jet, issuing from a stationary nozzle, encounters a vane curved through an angle of 90° that is moving away from the nozzle at a constant speed of 15 m/sec. The jet has a cross-sectional area of 600 mm² and a speed of 30 m/sec. Determine the force that must be applied to maintain the vane speed constant.

4.137 A jet of oil (SG = 0.8) strikes a curved blade that turns the fluid through an angle of 180°. The jet area is 1200 mm² and its speed relative to the stationary nozzle is 20 m/sec. The blade moves toward the nozzle at 10 m/sec. Determine the force that must be applied to maintain the blade speed constant.

4.138 A snow plow mounted on a truck clears a path 12 ft wide through heavy wet snow. The snow is 8 in. deep and its density is 10 lbm/ft³. The truck travels at 20 mph. Snow is discharged from the plow at an angle of 45° from the travel direction and 45° above the horizontal. Evaluate the force required to push the plow.

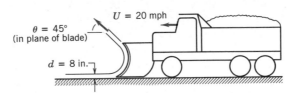

P4.138

4.139 A boat propelled by thrust from a water jet is to achieve a speed of 30 mph. Towing tests have shown that 200 lbf is needed to move the boat at this speed. Assume water for the pump system enters the inlet at a speed (relative to the boat) equal to the boat speed, and that inlet and outlet pressures are equal to that of undisturbed water at the same level. The pump flow rate is 900 gpm. Determine the exit jet speed (relative to the boat) required to achieve the desired boat speed.

4.140 The circular dish, whose cross section is shown, has an outside diameter of 0.20 m. A water jet with speed of 30 m/sec strikes the dish concentrically. The dish moves to the left at 10 m/sec. The jet diameter is 20 mm. The dish has a hole at its center that allows a stream of water 10 mm in diameter to pass through without resistance. The remainder of the jet is deflected and flows along the dish. Calculate the force required to maintain the dish motion.

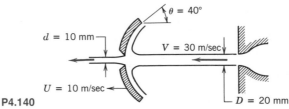

P4.140

4.141 The Canadair CL-215T amphibious aircraft is specially designed to fight fires. It is the only production aircraft that can scoop water—1620 gallons in 12 seconds—from any lake, river, or ocean. Determine the added thrust required during water scooping, as a function of aircraft speed, for a reasonable range of speeds.

4.142 Beginning in the 1900s, water troughs called "track pans" were built between the rails to allow steam locomotives to scoop water into their tenders without stopping. Each tender equipped with a scoop could pick up 2.5 gallons of water per foot of scooping distance while traveling at 50 mph. Estimate the added drawbar force required to pull a tender at constant speed while scooping water.

4.143 Police attempt to control demonstrators by using a "water cannon" mounted on a moving truck. The cannon shoots a steady stream of water, with diameter $D = 25$ mm and speed $V = 10$ m/sec (relative to the truck). The truck moves at speed $U = 3$ m/sec *toward* the demonstrators. Calculate the maximum force that could be exerted on a demonstrator by the water stream. Describe the manner in which the force would vary if the stream were to impinge at an angle other than perpendicular.

4.144 Consider a single vane, with turning angle θ, moving horizontally at constant speed, U, under the influence of an impinging jet as in Problem 4.135. The absolute speed of the jet is V. Obtain general expressions for the resultant force and power that the vane could produce. Show that the power is maximized when $U = V/3$.

4.145 The circular dish, whose cross section is shown, has an outside diameter of 0.15 m. A water jet strikes the dish concentrically and then flows outward along the surface of the dish. The jet speed is 45 m/sec and the dish moves to the left at 10 m/sec. Find the thickness of the jet sheet at a radius of 75 mm from the jet axis. What horizontal force on the dish is required to maintain this motion?

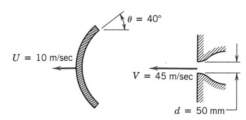

P4.145

4.146 Water, in a 100 mm diameter jet with speed of 30 m/sec to the right, is deflected by a cone that moves to the left at 15 m/sec. Determine (a) the thickness of the jet sheet at a radius of 200 mm and (b) the external horizontal force needed to move the cone.

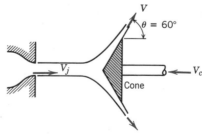

P4.146

4.147 Consider a series of turning vanes struck by a continuous jet of water that leaves a 50 mm diameter nozzle at constant speed, $V = 86.6$ m/sec. The vanes move with constant speed, $U = 50$ m/sec. Note that all the mass flow leaving the jet crosses the vanes. The curvature of the vanes is described by angles $\theta_1 = 30°$ and $\theta_2 = 45°$, as shown. Evaluate the nozzle angle, α, required to ensure that the jet enters tangent to the leading edge of each vane. Calculate the force that must be applied to maintain the vane speed constant.

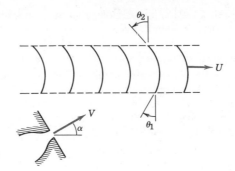

4.148 Consider again the moving multiple-vane system described in Problem 4.147. Assuming that a way could be found to make α nearly zero (and thus, θ_1 nearly 90°), evaluate the vane speed, U, that would result in maximum power output from the moving vane system.

4.149 A plane jet of water strikes a splitter vane and divides into two flat streams as shown. Find the mass flow rate ratio, $\dot{m}_2/\dot{m}_3$, required to produce zero net vertical force on the splitter vane. Determine the horizontal force that must be applied under these conditions to maintain the vane motion at steady speed.

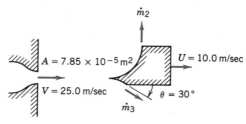

**4.150 Consider again the statement and diagram for Problem 4.114. Solve the problem for the case in which the propeller advances with speed V_1 into stationary fluid.

**4.151 The propeller on an airboat used in the Florida Everglades moves air at the rate of 40 kg/sec. When at rest, the speed of the slipstream behind the propeller is 40 m/sec at a location where the pressure is atmospheric. Calculate (a) the propeller diameter, (b) the thrust produced at rest, and (c) the thrust produced when the airboat is moving ahead at 10 m/sec, if the mass flow rate through the propeller remains constant.

4.152 A jet of lye solution (SG = 1.10) 50 mm in diameter has an absolute speed of 15 m/sec. It strikes a single flat plate that is moving away from the nozzle with an absolute speed of 5 m/sec. The plate makes an angle of 60° with the horizontal. Calculate the force on the plate from the jet. Assume no friction along the plate surface.

4.153 A steady jet of water is used to propel a small cart along a horizontal track as shown. Total resistance to motion of the cart assembly is given by $F_D = kU^2$, where $k = 0.92$ N·sec^2/m^2. Evaluate the acceleration of the cart at the instant when its speed is $U = 10$ m/sec.

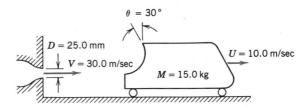

P4.153

** These problems require material from sections that may be omitted without loss of continuity in the text material.

4.154 A vehicle is moving at a speed of 50 ft/sec along level ground under the action of constant force $F = 100$ lbf. At $t = 0$, mass begins leaving the vehicle through a hole in the bottom. Assuming the mass leaves the vehicle vertically at a rate of 10 lbm/sec and the vehicle continues to move under the action of the constant force, determine the vehicle speed after 20 sec. The initial mass of the vehicle is 2000 lbm.

4.155 A vane/slider assembly moves under the influence of a liquid jet as shown. The coefficient of kinetic friction for motion of the slider along the surface is $\mu_k = 0.30$. Calculate (a) the acceleration of the slider at the instant when $U = 10$ m/sec and (b) the terminal speed of the slider.

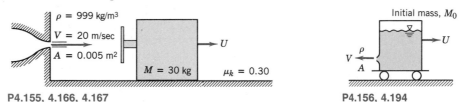

P4.155, 4.166, 4.167   P4.156, 4.194

4.156 A cart is propelled by a liquid jet issuing horizontally from a tank as shown. The track is horizontal; resistance to motion may be neglected. The tank is pressurized so that the jet speed may be considered constant. Obtain a general expression for the speed of the cart as it accelerates from rest.

4.157 The acceleration of the vane/cart assembly of Problem 4.135 is to be controlled as it accelerates from rest by changing the vane angle, θ. A constant acceleration, $a = 1.5$ m/sec^2, is desired. The water jet leaves the nozzle of area $A = 0.025$ m^2, with speed $V = 15$ m/sec. The vane/cart assembly has a mass of 55 kg; neglect friction. Determine θ at $t = 5$ sec.

4.158 The wheeled cart shown rolls with negligible resistance. The cart is to accelerate to the right at a *constant rate* of 2 m/sec^2. This is to be accomplished by "programming" the water jet area, $A(t)$, that reaches the cart. The jet speed remains constant at 10 m/sec. Obtain an expression for $A(t)$ required to produce the motion. Sketch the area variation for $t \le 4$ sec. Evaluate the jet area at $t = 2$ sec.

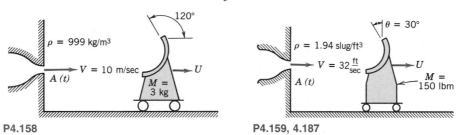

P4.158   P4.159, 4.187

4.159 A small vaned cart, of mass $M = 150$ lbm, is to be accelerated by a water jet. The jet speed is constant at $V = 32$ ft/sec, but its area is controlled to vary the mass flow rate. The cart is to start from rest at $t = 0$ and undergo *constant acceleration* at 4 ft/sec^2 until it reaches $U = 16$ ft/sec. Then it is to travel at constant speed. Neglect resistance to motion. Obtain a general expression for the jet cross-sectional area, $A(t)$, that must reach the cart during acceleration. Evaluate the jet area required at the instant the cart starts to move. Determine the time at which the jet flow from the nozzle must be cut off.

4.160 A rocket sled, weighing 10,000 lbf and traveling 600 mph, is to be braked by lowering a scoop into a water trough. The scoop is 6 in. wide. Determine the time required (after lowering the scoop to a depth of 3 in. into the water) to bring the sled to a speed of 20 mph. (See diagram on next page.)

4.161 A rocket sled is to be slowed from an initial speed of 300 m/sec by lowering a scoop into a water trough. The scoop is 0.3 m wide; it deflects the water through

150°. The trough is 800 m in length. The mass of the sled is 8000 kg. At the initial speed it experiences an aerodynamic drag force of 90 kN. The aerodynamic force is proportional to the square of the sled speed. It is desired to slow the sled to 100 m/sec. Determine the depth to which the scoop must be lowered into the water.

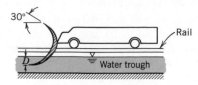

P4.160, 4.161

4.162 Starting from rest, the cart shown is propelled by a hydraulic catapult (liquid jet). The jet strikes the curved surface and makes a 180° turn, leaving horizontally. Air and rolling resistance may be neglected. If the mass of the cart is 100 kg and the jet of water leaves the nozzle (area of 0.001 m²) with a speed of 30 m/sec, determine the speed of the cart 5 sec after the jet is directed against the cart.

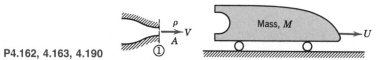

P4.162, 4.163, 4.190

4.163 Consider the jet and cart of Problem 4.162 again, but include an aerodynamic drag force proportional to the square of cart speed, $F_D = kU^2$, with $k = 2.0$ N · sec²/m². Derive an expression for the cart acceleration as a function of cart speed and other given parameters. Evaluate the acceleration of the cart at $U = 10$ m/sec. What fraction is this speed of the terminal speed of the cart?

4.164 A small cart that carries a single turning vane rolls on a level track. The cart mass is $M = 10.5$ kg and its initial speed is $U_0 = 12.5$ m/sec. At $t = 0$, the vane is struck by an opposing jet of water, as shown. Neglect any external forces due to air or rolling resistance. Determine the time and distance needed for the liquid jet to bring the cart to rest.

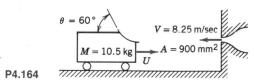

P4.164

4.165 A curved vane and cart assembly moves horizontally toward a water jet, under the conditions shown. The mass of the assembly is $M = 20$ kg and its initial speed is $U_0 = 5.75$ m/sec. Neglect rolling resistance and aerodynamic drag. Evaluate the time and distance needed for the cart assembly to slow to a stop.

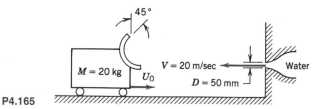

P4.165

4.166 Solve Problem 4.155 if the vane and slider ride on a film of oil instead of sliding in contact with the surface. Assume motion resistance is proportional to speed, $F_R = kU$, with $k = 7.5$ N · sec/m.

4.167 Consider again the statement and diagram of Problem 4.155. Obtain general expressions for the acceleration and speed of the slider as functions of time. Evaluate the terminal speed for the conditions given in Problem 4.155.

4.168 A rectangular block of mass M, with vertical faces, rolls without resistance along a smooth horizontal plane as shown. The block travels initially at speed U_0. At $t = 0$ the block is struck by a liquid jet and its speed begins to slow. Obtain an algebraic expression for the acceleration of the block for $t > 0$. Solve the equation to determine the time at which $U = 0$.

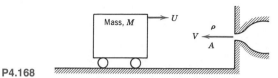

P4.168

4.169 A rectangular block of mass M, with vertical faces, rolls on a horizontal surface between two opposing jets as shown. At $t = 0$ the block is set into motion at speed U_0. Subsequently, it moves without friction parallel to the jet axes with speed $U(t)$. Neglect the mass of any liquid adhering to the block compared to M. Obtain general expressions for the acceleration of the block, $a(t)$, and the block speed, $U(t)$.

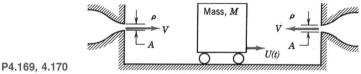

P4.169, 4.170

4.170 Consider the statement and diagram of Problem 4.169. Assume that at $t = 0$, when the block is at $x = 0$, it is set into motion at speed $U_0 = 10$ m/sec, to the right. Calculate the time required to reduce the block speed to $U = 0.5$ m/sec, and the block position at that instant.

**4.171 The cart shown is supplied with water from nozzles above it. The flow rates are adjusted to maintain the water level in the cart at height $h = 3$ ft. Water from each jet falls straight downward. The weight of the tank and water is 300 lbf. Compute the maximum speed, U, that will be attained by the cart if it rolls without friction on a level surface. Find the time required to reach 90 percent of terminal speed.

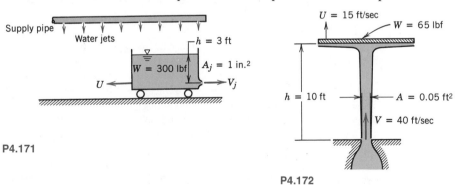

P4.171

P4.172

**4.172 A vertical jet of water impinges on a horizontal disk as shown. The disk assembly weighs 65 lbf. When the disk is 10 ft above the nozzle exit, it is moving upward at $U = 15$ ft/sec. Compute the vertical acceleration of the disk at this instant.

4.173 A container rolls along a horizontal surface with negligible friction. A restraining cord applies a constant retarding force, F_R, to the container. The initial mass and speed of the container are M_0 and U_0. Mass leaves vertically at constant rate $\dot{m}$. Obtain algebraic expressions for acceleration, velocity, and position of the container as functions of time.

** These problems require material from sections that may be omitted without loss of continuity in the text material.

4.174 An open container rolls with negligible friction along a smooth horizontal surface. Its initial mass and speed are M_0 and U_0. At $t = 0$, water begins to enter the container from overhead nozzles pointed straight down, at constant mass flow rate, $\dot{m}$. A restraining cord applies a constant retarding force, F_R. Obtain algebraic expressions for the speed and displacement of the container as functions of time.

4.175 A manned space capsule travels in level flight above the Earth's atmosphere at initial speed $U_0 = 8.05$ km/sec. The capsule is to be slowed by a retro-rocket to $U = 5.00$ km/sec in preparation for a reentry maneuver. The initial mass of the capsule is $M_0 = 1600$ kg. The rocket consumes fuel at $\dot{m} = 8.0$ kg/sec, and exhaust gases leave at $V_e = 2940$ m/sec relative to the capsule. Evaluate the duration of the retro-rocket firing needed to accomplish this.

4.176 A rocket sled accelerates from rest on a level track with negligible air and rolling resistances. The initial mass of the sled is $M_0 = 600$ kg. The rocket initially contains 150 kg of fuel. The rocket motor burns fuel at constant rate $\dot{m} = 15$ kg/sec. Exhaust gases leave the rocket nozzle uniformly and axially at $V_e = 2900$ m/sec relative to the nozzle. Find the maximum speed reached by the rocket sled. Calculate the maximum acceleration of the sled during the run.

4.177 A rocket sled with initial mass of 3 metric tons, including 1 ton of fuel, rests on a level section of track. At $t = 0$, the solid fuel of the rocket is ignited and the rocket burns fuel at the rate of 75 kg/sec. The exit speed of the exhaust gas relative to the rocket is 2500 m/sec. Neglecting friction and air resistance, calculate the acceleration and speed of the sled at $t = 10$ sec.

4.178 A rocket sled with initial mass of 2000 lbm is to be accelerated on a level track. The rocket motor burns fuel at constant rate $\dot{m} = 30$ lbm/sec. The rocket exhaust flow is uniform and axial. Gases leave the nozzle at 9000 ft/sec relative to the nozzle. Determine the minimum mass of rocket fuel needed to propel the sled to a speed of 600 mph before burnout occurs. As a first approximation, neglect resistance forces.

4.179 A rocket sled has an initial mass of 4 metric tons, including 1 ton of fuel. The motion resistance in the track on which the sled rides and that of the air total kU, where k is 75 N·sec/m, and U is the speed of the sled in m/sec. The exit speed of the exhaust gas relative to the rocket is 1500 m/sec, and it burns fuel at the rate of 90 kg/sec. Compute the speed of the sled after 10 sec.

4.180 A rocket motor is used to accelerate a kinetic energy weapon to a speed of 1.8 km/sec in horizontal flight. The exit stream leaves the nozzle axially and at atmospheric pressure with a speed of 3000 m/sec relative to the rocket. The rocket motor ignites upon release of the weapon from an aircraft flying horizontally at $U_0 = 300$ m/sec. Neglecting air resistance, obtain an algebraic expression for the speed reached by the weapon in level flight. Determine the minimum fraction of the initial mass of the weapon that must be fuel to accomplish the desired acceleration.

4.181 A large two-stage liquid rocket with mass of 30,000 kg is to be launched from a sea-level launch pad. The main engine burns liquid hydrogen and liquid oxygen in a stoichiometric mixture at 2450 kg/sec. The thrust nozzle has an exit diameter of 2.6 m. The exhaust gases exit the nozzle at 2270 m/sec and an exit plane pressure of 66 kPa absolute. Calculate the acceleration of the rocket at liftoff. Obtain an expression for speed as a function of time, neglecting air resistance.

4.182 Neglecting air resistance, what speed would a vertically directed rocket attain in 10 sec if it starts from rest, has initial mass of 200 kg, burns 10 kg/sec, and ejects gas to atmospheric pressure at 2900 m/sec relative to the rocket?

4.183 A "home-made" solid propellant rocket has an initial mass of 20 lbm; 15 lbm of this is fuel. The rocket is directed vertically upward from rest, burns fuel at a constant rate of 0.5 lbm/sec, and ejects exhaust gas at a speed of 6500 ft/sec relative to the rocket. Assume that the pressure at the exit is atmospheric and that air resistance may be neglected. Calculate the rocket speed after 20 sec and the distance traveled by the rocket in 20 sec.

4.184 A rocket cart that is initially at rest and weighs 1610 lbf is to be fired and is to have a constant acceleration of 20 ft/sec². To accomplish this, the exhaust gases will be deflected through angle θ, which varies as a function of time. This will account for a frictional force that is proportional to the speed squared, $F = 0.002U^2$, where F is in lbf and U is in ft/sec. The rocket exhausts gases at the rate of 1 slug/sec at constant speed, $V_e = 7500$ ft/sec, relative to the vehicle. Find an expression for $\cos\theta$ as a function of time, t. Determine t at which $\cos\theta$ is a minimum, and find θ_{max}.

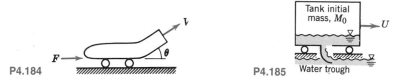

P4.184

P4.185   Water trough

4.185 The moving tank shown is to be slowed by lowering a scoop to pick up water from a trough. The initial mass and speed of the cart and its contents are M_0 and U_0, respectively. Neglect external forces due to pressure or friction and assume that the track is horizontal. Apply the continuity and momentum equations to show that at any instant $U = U_0 M_0/M$. Obtain a general expression for U/U_0 as a function of time.

4.186 The tank shown rolls along a level track. Water received from a jet is retained in the tank. The tank is to accelerate from rest toward the right with *constant* acceleration, a. Neglect wind and rolling resistance. Find an algebraic expression for the force (as a function of time) required to maintain the tank acceleration at constant a.

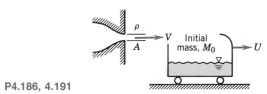

P4.186, 4.191

4.187 Consider again the statement and diagram of Problem 4.159. Determine the cross-sectional area variation *at the nozzle*, $A_n(t)$, required to accelerate the cart. Specify the time at which flow from the nozzle must be cut off.

‡4.188 The vane/cart assembly of mass $M = 30$ kg, shown in Problem 4.135, is driven by a water jet. The water leaves the stationary nozzle of area $A = 0.02$ m², with a speed of 20 m/sec. The coefficient of kinetic friction between the assembly and the surface is 0.10. Plot the terminal speed of the assembly as a function of vane turning angle, θ, for $0 \le \theta \le \pi/2$. At what angle does the assembly begin to move if the coefficient of static friction is 0.15?

4.189 The vane angle, θ, of the vane/cart assembly in Problem 4.188 is set at 60°. Determine the cart speed at $t = 1$ sec.

4.190 Consider the vehicle shown in Problem 4.162. Starting from rest, it is propelled by a hydraulic catapult (liquid jet). The jet strikes the curved surface and makes a 180° turn, leaving horizontally. Air and rolling resistance may be neglected. Using the notation shown, obtain an equation for the acceleration of the vehicle at any time and determine the time required for the vehicle to reach $U = V/2$.

4.191 The tank of Problem 4.186 rolls with negligible resistance along a horizontal track. It is to be accelerated from rest by a liquid jet that strikes the vane and is deflected

‡ You may wish to use simple computer problems to help solve problems marked with daggers.

into the tank. The initial mass of the tank is M_0. Use the continuity and momentum equations to show that at any instant the mass of the vehicle and liquid contents is $M = M_0 V/(V - U)$. Obtain a general expression for U/V as a function of time.

4.192 A small rocket motor is used to power a "jet pack" device to lift a single astronaut above the Earth's surface. The rocket motor produces a uniform exhaust jet with *constant* speed, $V_e = 2940$ m/sec. The total initial mass of the astronaut and the jet pack is $M_0 = 130$ kg. Of this, 40 kg is fuel for the rocket motor. Develop an algebraic expression for the *variable* fuel mass flow rate required to keep the jet pack and astronaut hovering in a fixed position above the ground. Calculate the maximum hover time aloft before the fuel supply is expended.

4.193 A model solid propellant rocket has a mass of 69.6 g, of which 12.5 g is fuel. The rocket produces 1.3 lbf of thrust for a duration of 1.7 sec. For these conditions, calculate (a) the total impulse of the rocket motor, (b) the specific impulse of the rocket motor, and (c) the maximum speed and height attainable in the absence of air resistance. (*Specific impulse* is defined as the ratio of thrust to weight flow rate of propellant, expressed in seconds.)

4.194 The small cart of Problem 4.156 is to be accelerated horizontally by a liquid jet. The acceleration is to be *constant* at 0.5 g for 5 sec. The jet leaves the nozzle in a uniform stream at 200 ft/sec relative to the cart. The only resistance to motion is frictional; the kinetic friction coefficient is $\mu_k = 0.15$. The initial mass of the cart and its contents is 3000 lbm. Obtain an algebraic expression for the mass flow rate of liquid needed to produce the desired acceleration. Obtain an algebraic expression for the vehicle mass as a function of time. Determine the flow rate at $t = 5$ sec.

4.195 A rocket sled, of initial mass M_0, is fired at $t = 0$ along a horizontal track. Fuel is burned at the rate $\dot{m}$; the speed of the exhaust gas relative to the rocket is V. If the total motion resistance is given by kU^2, where U is the sled speed, obtain an algebraic expression for the speed, U, of the rocket as a function of time.

4.196 A plastic toy rocket is shown. The rocket is propelled by a jet of water forced out the nozzle by compressed air. As a first approximation, assume that the water speed in the rocket chamber is given by $V = V_0 - kt$. The chamber and exit areas are A_c and A_e, respectively. The area ratio is approximately $A_e/A_c = 0.10$. Assume the initial mass of the rocket is M_0 and neglect the mass of the air. Find (a) the velocity of the water at the nozzle exit, (b) the mass of the rocket, M, and (c) the acceleration of the rocket, as functions of time. *Hint:* Use different control volumes for each calculation.

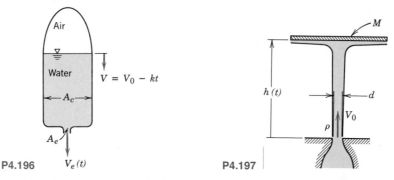

P4.196 $V_e(t)$ P4.197

**4.197 A disk, of mass M, is constrained horizontally but is free to move vertically. A jet of water strikes the disk from below. The jet leaves the nozzle at initial speed V_0. Obtain a differential equation for the disk height, $h(t)$, above the jet exit plane if the disk

---

** These problems require material from sections that may be omitted without loss of continuity in the text material.

is released from large height, H. Assume that when the disk reaches equilibrium, its height above the jet exit plane is h_0. Sketch $h(t)$ for the disk released at $t = 0$ from $H > h_0$. Explain why the sketch is as you show it.

4.198 A small solid-fuel rocket motor is fired on a test stand. The combustion chamber is circular, with 100 mm diameter. Fuel, of density 1660 kg/m³, burns uniformly at the rate of 12.7 mm/sec. Measurements show that the exhaust gases leave the rocket at ambient pressure, at a speed of 2750 m/sec. The absolute pressure and temperature in the combustion chamber are 7.0 MPa and 3610 K, respectively. Treat the combustion products as an ideal gas with molecular mass of 25.8. Evaluate the rate of change of mass and of linear momentum within the rocket motor. Express the rate of change of linear momentum within the motor as a percentage of the motor thrust.

**4.199 A fluid particle moves through the impeller of a centrifugal pump with radial speed (relative to the impeller) $V_r = V_0 r_0 / r$, while the impeller turns at constant angular speed, ω. Determine the total acceleration of the particle just prior to leaving the impeller at radius R.

**4.200 The rocket shown has initial mass M_0. It is attached to a rigid horizontal rod that pivots about the origin. Assume exhaust gases leave axially at atmospheric pressure, at mass flow rate $\dot{m}$, and with speed V_e, relative to the rocket. Develop a differential equation for the motion of the rocket and the rod. Neglect aerodynamic drag and the mass of the rod. Solve for angular speed as a function of time.

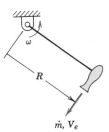

P4.200, 4.201, 4.202, 4.203 $\dot{m}, V_e$

**4.201 Repeat Problem 4.200, using as numerical data $R = 2.2$ m, $M_0 = 3.5$ kg, $\dot{m} = 0.075$ kg/sec, and $V_e = 2900$ m/sec. Solve for angular speed as a function of time after the rocket motor is ignited. Calculate the angular speed and the tension in the support rod after 5 seconds.

**4.202 Repeat Problem 4.200, using as numerical data $R = 4.5$ ft, $M_0 = 0.16$ slug, $\dot{m} = 0.25$ lbm/sec, and $V_e = 7100$ ft/sec. Find the time after ignition when the system attains angular speed $\omega = 95$ rad/sec. Calculate the angular acceleration and tension in the support rod at this instant.

**4.203 Repeat Problem 4.202, with the same numerical data, but including the effect of a constant aerodynamic drag force, $F_D = 5.0$ lbf, on the rocket motor.

**4.204 A small "pod" propelled by a jet of air rotates on a strut about a vertical axis at a radius of 1 m from a fixed center. The pod mass is 2 kg; the jet velocity, density, and area are 200 m/sec, 1.5 kg/m³, and 75 mm², respectively. At one instant, the pod moves at a speed of 50 m/sec, and the air drag force is 10 N. Determine the angular acceleration of the pod at this instant. Neglect friction and the mass of the strut. (See diagram on next page.)

**4.205 Solve Problem 4.204 for the pod angular speed as a function of time starting from rest. Assume the drag force on the pod is given by $F_D = kV^2$. Evaluate k and find the time required for the pod to reach $V = 30$ m/sec.

** These problems require material from sections that may be omitted without loss of continuity in the text material.

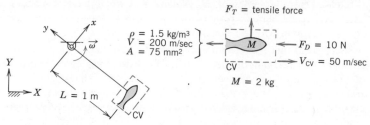

P4.204, 4.205

**4.206 A large irrigation sprinkler unit, mounted on a cart, discharges water with a speed of 40 m/sec at an angle of 30° to the horizontal. The 50 mm diameter nozzle is 3 m above the ground. Calculate the magnitude of the moment that tends to overturn the cart.

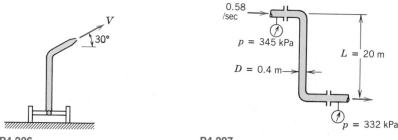

P4.206   **P4.207**

**4.207 Crude oil (SG = 0.95) from a tanker dock flows through a pipe of 0.4 m diameter in the configuration shown. The flow rate is 0.58 m³/sec, and the gage pressures are shown in the diagram. Determine the force and torque that are exerted by the pipe assembly on its supports.

**4.208 The simplified lawn sprinkler shown rotates in the horizontal plane. At the center pivot, $Q = 4.5$ gpm of water enters vertically. Water discharges in the horizontal plane from each jet. If the pivot is frictionless, calculate the torque needed to keep the sprinkler from rotating. Neglecting the inertia of the sprinkler itself, calculate the angular acceleration that results when the torque is removed.

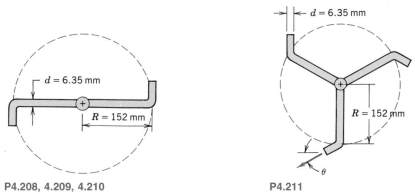

P4.208, 4.209, 4.210   **P4.211**

**4.209 Consider the sprinkler of Problem 4.208 again. Derive a differential equation for the angular speed of the sprinkler as a function of time. Evaluate its steady-state speed of rotation, if there is no friction in the pivot.

**4.210 Repeat Problem 4.209, but assume a constant retarding torque in the pivot of 0.045 ft· lbf.

** These problems require material from sections that may be omitted without loss of continuity in the text material.

**4.211 The lawn sprinkler shown is supplied with water at a rate of 68 L/min. Neglecting friction in the pivot, calculate and plot the steady-state angular speed of the sprinkler for $0 \le \theta < 90°$.

**4.212 A small lawn sprinkler is shown. The sprinkler operates at a gage pressure of 140 kPa. The total flow rate of water through the sprinkler is 4 L/min. Each jet discharges at 17 m/sec (relative to the sprinkler arm) in a direction inclined 30° above the horizontal. The sprinkler rotates about a vertical axis. Friction in the bearing causes a torque of 0.18 N·m opposing rotation. Evaluate the torque required to hold the sprinkler stationary.

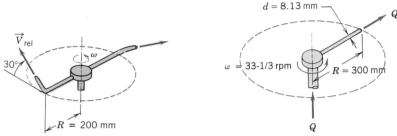

P4.212, 4.214, 4.217   P4.213

**4.213 A single tube carrying water rotates at constant angular speed, as shown. Water is pumped through the tube at volume flow rate $Q = 13.8$ L/min. Find the torque that must be applied to maintain the steady rotation of the tube using *two* methods of analysis: (a) a *rotating* control volume and (b) a *fixed* control volume.

**4.214 In Problem 4.212, calculate the initial acceleration of the sprinkler from rest if no external torque is applied and its moment of inertia is 0.1 kg·m² when filled with water.

**4.215 A Pelton wheel is a form of water turbine well adapted to situations of high head and low flow rate. The wheel consists of a series of vanes mounted on a rotor, as shown. One or more jets are arranged to strike the buckets tangentially. In practice it is possible to deflect the jet stream through angles, θ, of up to 165°. Consider the Pelton wheel and single jet arrangement shown. Obtain an expression for the torque exerted by the water stream on the wheel and the corresponding power output. (Let the bucket speed be $U = \omega R$.) Determine the value of U/V required to maximize the power produced by the wheel.

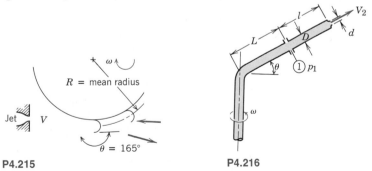

P4.215   P4.216

**4.216 The large irrigation sprinkler is fabricated such that a "nozzle section" of length l can be mounted to the boom arm of length L; both the boom and nozzle section have inside diameter D; the opening at the exit of the nozzle section is of diameter d. The boom, inclined at angle θ to the horizontal, rotates with angular speed ω about the

** These problems require material from sections that may be omitted without loss of continuity in the text material.

vertical axis. Determine the force of the nozzle section on the boom given that the weight of the nozzle section is $W - 50$ lbf and the following operating conditions: $D = 4$ in., $d = 1$ in., $\theta = 30°$, $L = 10$ ft, $l = 3$ ft, $\omega = 3$ rpm, $V_2 = 120$ ft/sec, and $p_1 = 100$ psig.

**4.217 A small lawn sprinkler is shown (P4.212). The sprinkler operates at an inlet gage pressure of 140 kPa. The total flow rate of water through the sprinkler is 4.0 L/min. Each jet discharges at 17 m/sec (relative to the sprinkler arm) in a direction inclined 30° above the horizontal. The sprinkler rotates about a vertical axis. Friction in the bearing causes a torque of 0.18 N·m opposing rotation. Determine the steady speed of rotation of the sprinkler and the approximate area covered by the spray.

**4.218 Water flows at the rate of 0.15 m³/sec through a nozzle assembly that rotates steadily at 30 rpm. The arm and nozzle masses are negligible compared to the water inside. Determine the torque required to drive the device and the reaction torques at the flange.

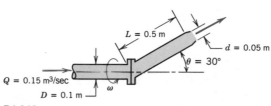

P4.218

**4.219 A pipe branches symmetrically into two legs of length L, and the whole system rotates with angular speed ω around its axis. Each branch is inclined at angle α to the axis of rotation. Liquid enters the pipe steadily, with zero angular momentum, at volume flow rate Q. The pipe diameter, D, is much smaller than L. Obtain an expression for the external torque required to turn the pipe. What additional torque would be required to impart angular acceleration $\dot{\omega}$?

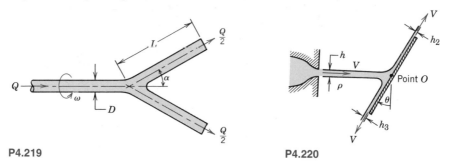

P4.219   P4.220

**4.220 Liquid in a thin sheet, of width w and thickness h, flows from a slot and strikes a stationary inclined flat plate, as shown. Experiments show that the resultant force of the liquid jet on the plate does not act through point O, where the jet centerline intersects the plate. Determine the magnitude and line of application of the resultant force as functions of θ. Evaluate the equilibrium angle of the plate if the resultant force is applied at point O. Neglect any viscous effects.

‡**4.221 Repeat Example Problem 4.15 for the case when the spray arm starts from rest and accelerates. Assume the moment of inertia of the empty spray arm is $I_0 = 0.8 \times 10^{-4}$ kg·m² and its inside diameter is $D = 8.10$ mm. Derive the differential equation that

**** These problems require material from sections that may be omitted without loss of continuity in the text material.

‡ You may wish to use simple computer programs to help solve problems marked with daggers.

describes the angular speed of the sprinkler. Solve it numerically to calculate the time needed to reach 95 percent of terminal speed.

4.222 Air at standard conditions enters a compressor at 75 m/sec and leaves at an absolute pressure and temperature of 200 kPa and 345 K, and speed $V = 125$ m/sec. The flow rate is 1 kg/sec. The cooling water circulating around the compressor casing removes 18 kJ/kg of air. Determine the power required by the compressor.

4.223 Air enters a compressor at 14 psia, 80 F with negligible speed and is discharged at 70 psia, 500 F with a speed of 500 ft/sec. If the power input is 3200 hp and the flow rate is 20 lbm/sec, determine the rate of heat transfer.

4.224 Air is drawn from the atmosphere into a turbomachine. At the exit, conditions are 500 kPa (gage) and 130 C. The exit speed is 100 m/sec and the mass flow rate is 0.8 kg/sec. Flow is steady and there is no heat transfer. Compute the shaft work interaction with the surroundings.

4.225 A turbine is supplied with 0.6 m³/sec of water from a 0.3 m diameter pipe; the discharge pipe has a 0.4 m diameter. Determine the pressure drop across the turbine if it delivers 60 kW.

4.226 A pump system in a dishwasher circulates water at 55 gpm. The toal head, $p/\rho g + V^2/2g$, leaving the pump is 120 in. of water. The head at the pump inlet may be neglected. If the power supplied to the pump is 0.4 hp, determine its efficiency.

4.227 Compressed air is stored in a pressure bottle with a volume of 10 ft³, at 3000 psia and 140 F. At a certain instant a valve is opened and mass flows from the bottle at $\dot{m} = 0.105$ lbm/sec. Find the rate of change of temperature in the bottle at this instant.

4.228 A pump draws water from a reservoir through a 150 mm diameter suction pipe and delivers it to a 75 mm diameter discharge pipe. The end of the suction pipe is 2 m below the free surface of the reservoir. The pressure gage on the discharge pipe (2 m above the reservoir surface) reads 170 kPa. The average speed in the discharge pipe is 3 m/sec. If the pump efficiency is 75 percent, determine the power required to drive it.

4.229 A centrifugal water pump with a 4 in. diameter inlet and a 4 in. diameter discharge pipe has a flow rate of 300 gpm. The inlet pressure is 8 in. Hg vacuum and the exit pressure is 35 psig. The inlet and outlet sections are located at the same elevation. The measured power input is 9.1 hp. Determine the pump efficiency.

4.230 A hydraulic turbine, located 150 ft below the level of a reservoir, develops 500 hp at a water flow rate of 45 slug/sec. The inlet and outlet of the turbine are at the same elevation, and each has a flow area of 1.0 ft². The outlet discharges directly to atmosphere as a free jet. Neglect friction in the pipes connecting the reservoir and turbine. Calculate the pressure drop across the turbine. Estimate the turbine efficiency.

4.231 A jet-propelled aircraft traveling at 200 m/sec takes in 40 kg/sec of air and discharges it at 500 m/sec relative to the aircraft. Determine the *propulsive efficiency* (defined as the ratio of the useful work output to the mechanical energy input to the fluid) of the aircraft.

4.232 A fire truck draws water from an open constant-level reservoir through a hose of 0.2 ft² cross section. The water is discharged from the pump through a 0.2 ft² hose to a surge tank pressurized to 120 psig with air. The pump adds 970 ft · lbf/slug of energy to the fluid; the water level in the surge tank is 3 ft above the pump centerline. From the surge tank the water flows steadily through a hose of 0.04 ft² area to a nozzle of 0.01 ft² area. The nozzle is held at the same level as the water free surface in the surge tank. Neglect all frictional losses and sketch the system. Find the
(a) speed of the water leaving the nozzle.
(b) elevation of the pump relative to the reservoir.
(c) power required to drive the pump if it has an efficiency of 80 percent.
(d) maximum height above the reservoir that the stream of water can reach.

4.233 All major harbors are equipped with fireboats for extinguishing ship fires. A 75 mm diameter hose is attached to the discharge of a 10 kW pump on such a boat. The nozzle attached to the end of the hose has a diameter of 25 mm. If the nozzle discharge is held 3 m above the surface of the water, determine the volume flow rate through the nozzle, the maximum height to which the water will rise, and the force on the boat if the water jet is directed horizontally over the stern.

4.234 The propulsive efficiency, η, of a jet-propelled device is defined as the ratio of the useful work output to the mechanical energy input to the fluid. Determine the propulsive efficiency for the jet boat of Problem 4.139.

**4.235 The propulsive efficiency, η, of a propeller is defined as the ratio of the useful work produced to the mechanical energy input to the fluid. Determine the propulsive efficiency of the moving airboat of Problem 4.151. What would be the efficiency of the boat if it were not moving?

4.236 The efficiency, η, of a windmill is defined as the ratio of the useful work output to the kinetic energy flux contained in a streamtube of undisturbed wind the same diameter as the windmill blades. Evaluate the efficiency of the windmill of Problem 4.102.

‡4.237 Express the efficiency of the water jet pump of Problem 4.107 as

$$\eta = \frac{\text{Fluid energy rise}}{\text{Energy flux input}} = \frac{\Delta p Q_2}{q_j Q_j}$$

where $q_j = \frac{1}{2}\rho V_j^2$. Write an algebraic expression for efficiency in terms of the area ratio, $a = A_j/A_2$, and the flow rate ratio, $r = Q_2/Q_1$. Find the combination of a and r that produces maximum efficiency. Plot the optimum efficiency and dimensionless pressure rise versus area ratio.

**4.238 The total mass of the helicopter-type craft shown is 1500 kg. The pressure of the air is atmospheric at the outlet. Assume the flow is steady and one-dimensional. Treat the air as incompressible at standard conditions and calculate, for a hovering position, the speed of air leaving the craft and the minimum power that must be delivered to the air by the propeller.

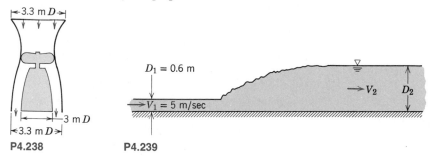

P4.238   P4.239

4.239 Liquid flowing at high speed in a wide, horizontal open channel under some conditions can undergo a hydraulic jump, as shown. For a suitably chosen control volume, the flows entering and leaving the jump may be considered uniform with hydrostatic pressure distributions (see Example Problem 4.6). Consider a channel of width w, with water flow at $D_1 = 0.6$ m and $V_1 = 5$ m/sec. Show that in general, $D_2 = D_1 [\sqrt{1 + 8V_1^2/gD_1} - 1]/2$. Evaluate the change in mechanical energy through the hydraulic jump. If heat transfer to the surroundings is negligible, determine the change in water temperature through the jump.

** These problems require material from sections that may be omitted without loss of continuity in the text material.

‡ You may wish to use simple computer programs to help solve problems marked with daggers.

Chapter 5

INTRODUCTION TO DIFFERENTIAL ANALYSIS OF FLUID MOTION

In Chapter 4, we developed the basic equations in integral form for a control volume. The integral equations are particularly useful when we are interested in the gross behavior of a flow field and its effect on various devices. However, the integral approach does not enable us to obtain detailed point by point knowledge of the flow field.

To obtain this detailed knowledge, we must apply the equations of fluid motion in differential form. In this chapter we shall develop differential equations for the conservation of mass and Newton's second law of motion. Since we are interested in formulating differential equations, our analysis will be in terms of infinitesimal systems and control volumes.

5-1 CONSERVATION OF MASS

In Chapter 2, we found that the continuum assumption—the assumption that a fluid could be treated as a continuous distribution of matter—led directly to a field representation of fluid properties. The property fields are defined by continuous functions of the space coordinates and time. The density and velocity fields are related through conservation of mass. We shall derive the differential equation for conservation of mass in rectangular and in cylindrical coordinates. In both cases the derivation is carried out by applying conservation of mass to a differential control volume.

5-1.1 Rectangular Coordinate System

In rectangular coordinates, the control volume chosen is an infinitesimal cube with sides of length dx, dy, dz as shown in Fig. 5.1. The density at the center, O, of the control volume is ρ and the velocity there is $\vec{V} = \hat{i}u + \hat{j}v + \hat{k}w$.

To evaluate the properties at each of the six faces of the control surface, we use a Taylor series expansion about point O. For example, at the right face,

$$\rho\Big)_{x+dx/2} = \rho + \left(\frac{\partial \rho}{\partial x}\right)\frac{dx}{2} + \left(\frac{\partial^2 \rho}{\partial x^2}\right)\frac{1}{2!}\left(\frac{dx}{2}\right)^2 + \cdots$$

Neglecting higher order terms, we can write

$$\rho\Big)_{x+dx/2} = \rho + \left(\frac{\partial \rho}{\partial x}\right)\frac{dx}{2}$$

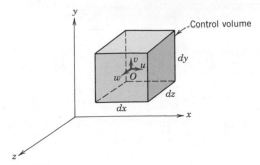

Fig. 5.1 Differential control volume in rectangular coordinates.

and

$$u\Big)_{x+dx/2} = u + \left(\frac{\partial u}{\partial x}\right)\frac{dx}{2}$$

where ρ, u, $\dfrac{\partial \rho}{\partial x}$, and $\dfrac{\partial u}{\partial x}$ are all evaluated at point O. The corresponding terms at the left face are

$$\rho\Big)_{x-dx/2} = \rho + \left(\frac{\partial \rho}{\partial x}\right)\left(-\frac{dx}{2}\right) = \rho - \left(\frac{\partial \rho}{\partial x}\right)\frac{dx}{2}$$

$$u\Big)_{x-dx/2} = u + \left(\frac{\partial u}{\partial x}\right)\left(-\frac{dx}{2}\right) = u - \left(\frac{\partial u}{\partial x}\right)\frac{dx}{2}$$

A word statement of conservation of mass is

$$\left[\begin{array}{c}\text{Net rate of mass flux out} \\ \text{through the control surface}\end{array}\right] + \left[\begin{array}{c}\text{Rate of change of mass} \\ \text{inside the control volume}\end{array}\right] = 0$$

To evaluate the first term in this equation, we must evaluate $\int_{CS} \rho\vec{V}\cdot d\vec{A}$; we must consider the mass flux through each of the six surfaces of the control surface. The details of this evaluation are shown in Table 5.1. Velocities across each of the six faces have been assumed to be in the positive coordinate directions. Higher-order terms [e.g., $(dx)^2$] have been neglected.

We see that the net rate of mass flux out through the control surface is given by

$$\left[\frac{\partial \rho u}{\partial x} + \frac{\partial \rho v}{\partial y} + \frac{\partial \rho w}{\partial z}\right] dx\, dy\, dz$$

The mass inside the control volume at any instant is the product of the mass per unit volume, ρ, and the volume, $dx\, dy\, dz$. Thus the rate of change of mass inside the control volume is given by

$$\frac{\partial \rho}{\partial t} dx\, dy\, dz$$

In rectangular coordinates the differential equation for conservation of mass is then

$$\frac{\partial \rho u}{\partial x} + \frac{\partial \rho v}{\partial y} + \frac{\partial \rho w}{\partial z} + \frac{\partial \rho}{\partial t} = 0 \tag{5.1a}$$

Table 5.1 Mass Flux through the Control Surface of a Rectangular Differential Control Volume

Surface	$\int \rho \vec{V} \cdot d\vec{A}$

Left $(-x)$
$$= -\left[\rho - \left(\frac{\partial \rho}{\partial x}\right)\frac{dx}{2}\right]\left[u - \left(\frac{\partial u}{\partial x}\right)\frac{dx}{2}\right]dy\,dz = -\rho u\,dy\,dz + \frac{1}{2}\left[u\left(\frac{\partial \rho}{\partial x}\right) + \rho\left(\frac{\partial u}{\partial x}\right)\right]dx\,dy\,dz$$

Right $(+x)$
$$= \left[\rho + \left(\frac{\partial \rho}{\partial x}\right)\frac{dx}{2}\right]\left[u + \left(\frac{\partial u}{\partial x}\right)\frac{dx}{2}\right]dy\,dz = \rho u\,dy\,dz + \frac{1}{2}\left[u\left(\frac{\partial \rho}{\partial x}\right) + \rho\left(\frac{\partial u}{\partial x}\right)\right]dx\,dy\,dz$$

Bottom $(-y)$
$$= -\left[\rho - \left(\frac{\partial \rho}{\partial y}\right)\frac{dy}{2}\right]\left[v - \left(\frac{\partial v}{\partial y}\right)\frac{dy}{2}\right]dx\,dz = -\rho v\,dx\,dz + \frac{1}{2}\left[v\left(\frac{\partial \rho}{\partial y}\right) + \rho\left(\frac{\partial v}{\partial y}\right)\right]dx\,dy\,dz$$

Top $(+y)$
$$= \left[\rho + \left(\frac{\partial \rho}{\partial y}\right)\frac{dy}{2}\right]\left[v + \left(\frac{\partial v}{\partial y}\right)\frac{dy}{2}\right]dx\,dz = \rho v\,dx\,dz + \frac{1}{2}\left[v\left(\frac{\partial \rho}{\partial y}\right) + \rho\left(\frac{\partial v}{\partial y}\right)\right]dx\,dy\,dz$$

Back $(-z)$
$$= -\left[\rho - \left(\frac{\partial \rho}{\partial z}\right)\frac{dz}{2}\right]\left[w - \left(\frac{\partial w}{\partial z}\right)\frac{dz}{2}\right]dx\,dy = -\rho w\,dx\,dy + \frac{1}{2}\left[w\left(\frac{\partial \rho}{\partial z}\right) + \rho\left(\frac{\partial w}{\partial z}\right)\right]dx\,dy\,dz$$

Front $(+z)$
$$= \left[\rho + \left(\frac{\partial \rho}{\partial z}\right)\frac{dz}{2}\right]\left[w + \left(\frac{\partial w}{\partial z}\right)\frac{dz}{2}\right]dx\,dy = \rho w\,dx\,dy + \frac{1}{2}\left[w\left(\frac{\partial \rho}{\partial z}\right) + \rho\left(\frac{\partial w}{\partial z}\right)\right]dx\,dy\,dz$$

Then,

$$\int_{CS} \rho \vec{V} \cdot d\vec{A} = \left[\left\{u\left(\frac{\partial \rho}{\partial x}\right) + \rho\left(\frac{\partial u}{\partial x}\right)\right\} + \left\{v\left(\frac{\partial \rho}{\partial y}\right) + \rho\left(\frac{\partial v}{\partial y}\right)\right\} + \left\{w\left(\frac{\partial \rho}{\partial z}\right) + \rho\left(\frac{\partial w}{\partial z}\right)\right\}\right]dx\,dy\,dz$$

or

$$\int_{CS} \rho \vec{V} \cdot d\vec{A} = \left[\frac{\partial \rho u}{\partial x} + \frac{\partial \rho v}{\partial y} + \frac{\partial \rho w}{\partial z}\right]dx\,dy\,dz$$

Since the vector operator, ∇, in rectangular coordinates, is given by

$$\nabla = \hat{i}\frac{\partial}{\partial x} + \hat{j}\frac{\partial}{\partial y} + \hat{k}\frac{\partial}{\partial z}$$

then

$$\frac{\partial \rho u}{\partial x} + \frac{\partial \rho v}{\partial y} + \frac{\partial \rho w}{\partial z} = \nabla \cdot \rho \vec{V}$$

and the conservation of mass may be written as

$$\nabla \cdot \rho \vec{V} + \frac{\partial \rho}{\partial t} = 0 \qquad (5.1b)$$

Two flow cases for which the differential continuity equation may be simplified are worthy of note. For incompressible flow, $\rho =$ constant; density is neither a function of space coordinates nor time. For incompressible flow, the continuity equation simplifies to

$$\frac{\partial u}{\partial x} + \frac{\partial v}{\partial y} + \frac{\partial w}{\partial z} = 0$$

Thus the velocity field, $\vec{V}(x, y, z, t)$, for incompressible flow must satisfy $\nabla \cdot \vec{V} = 0$.

For steady flow, all fluid properties are, by definition, independent of time. Thus at most $\rho = \rho(x, y, z)$; for steady flow, the continuity equation can be written as

$$\frac{\partial \rho u}{\partial x} + \frac{\partial \rho v}{\partial y} + \frac{\partial \rho w}{\partial z} = 0$$

or

$$\nabla \cdot \rho \vec{V} = 0$$

EXAMPLE 5.1—Integration of Two-Dimensional Differential Continuity Equation

For a two-dimensional flow in the xy plane, the x component of velocity is given by $u = Ax$. Determine a possible y component for steady, incompressible flow. How many possible y components are there?

EXAMPLE PROBLEM 5.1

GIVEN: Two-dimensional flow in the xy plane for which $u = Ax$.

FIND: (a) Possible y component for steady, incompressible flow.
(b) How many y components are possible?

SOLUTION:

Basic equation: $\nabla \cdot \rho \vec{V} + \dfrac{\partial \rho}{\partial t} = 0$ ~Continuity eq.~

For steady, incompressible flow, $\dfrac{\partial \rho}{\partial t} = 0$, and $\rho =$ constant; thus $\nabla \cdot \vec{V} = 0$. In rectangular coordinates

$$\frac{\partial u}{\partial x} + \frac{\partial v}{\partial y} + \frac{\partial w}{\partial z} = 0$$

For two-dimensional flow in the xy plane, $\vec{V} = \vec{V}(x, y)$. Then partial derivatives with respect to z are zero, and

$$\frac{\partial u}{\partial x} + \frac{\partial v}{\partial y} = 0$$

Then

$$\frac{\partial v}{\partial y} = -\frac{\partial u}{\partial x} = -A$$

which gives an expression for the rate of change of v holding x constant. This equation can be integrated to obtain an expression for v. The result is

$$v = \int \frac{\partial v}{\partial y} dy + f(x) = -Ay + f(x) \qquad \qquad v \; \underleftarrow{\hspace{3cm}}$$

{The function of x appears because we had a partial derivative of v with respect to y.}

Any function $f(x)$ is allowable, since $\dfrac{\partial}{\partial y} f(x) = 0$. Thus any number of expressions for v could satisfy the differential continuity equation under the given conditions. The simplest expression for v would be obtained by setting $f(x) = 0$. Then $v = -Ay$, and

$$\vec{V} = Ax\,\hat{i} - Ay\,\hat{j} \qquad \qquad \vec{V} \; \underleftarrow{\hspace{3cm}}$$

⎧ This problem illustrates use of the differential continuity equation for steady, incom-
⎨ pressible flow to evaluate a possible velocity component and introduces the integration
⎩ of a partial derivative. ⎭

EXAMPLE 5.2—Unsteady Differential Continuity Equation

A gas-filled pneumatic strut in an automobile suspension system behaves like a piston-cylinder apparatus. At one instant when the piston is $L = 0.15$ m away from the closed end of the cylinder, the gas density is uniform at $\rho = 18$ kg/m^3 and the piston begins to move away from the closed end at $V = 12$ m/sec. The gas motion is one-dimensional and proportional to distance from the closed end; it varies linearly from zero at the end to $u = V$ at the piston. Evaluate the rate of change of gas density at this instant. Obtain an expression for the average density as a function of time.

EXAMPLE PROBLEM 5.2

GIVEN: Piston-cylinder as shown.

FIND: (a) Rate of change of density.
      (b) $\rho(t)$.

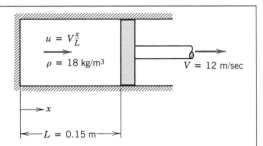

SOLUTION:

Basic equation: $\nabla \cdot \rho \vec{V} + \dfrac{\partial \rho}{\partial t} = 0$

In rectangular coordinates, $\dfrac{\partial \rho u}{\partial x} + \dfrac{\partial \rho v}{\partial y} + \dfrac{\partial \rho w}{\partial z} + \dfrac{\partial \rho}{\partial t} = 0$

Since $u = u(x)$, then partial derivatives with respect to y and z are zero, and

$$\frac{\partial \rho u}{\partial x} + \frac{\partial \rho}{\partial t} = 0$$

Then

$$\frac{\partial \rho}{\partial t} = -\frac{\partial \rho u}{\partial x} = -\rho \frac{\partial u}{\partial x} - u \frac{\partial \rho}{\partial x}$$

Since ρ is assumed uniform in the volume, then $\frac{\partial \rho}{\partial x} = 0$, and $\frac{d\rho}{dt} = -\rho \frac{\partial u}{\partial x}$.

Since $u = V\frac{x}{L}$, then $\frac{\partial u}{\partial x} = \frac{V}{L}$, and $\frac{d\rho}{dt} = -\rho \frac{V}{L}$. However, note that $L = L_0 + Vt$.

Separate variables and integrate,

$$\int_{\rho_0}^{\rho} \frac{d\rho}{\rho} = -\int_0^t \frac{V}{L} dt = -\int_0^t \frac{V \, dt}{L_0 + Vt}$$

$$\ln \frac{\rho}{\rho_0} = \ln \frac{L_0}{L_0 + Vt} \qquad \text{and} \qquad \rho(t) = \rho_0 \left[\frac{1}{1 + Vt/L_0} \right] \qquad\qquad \rho(t) \; \longleftarrow$$

At $t = 0$,

$$\frac{\partial \rho}{\partial t} = -\rho_0 \frac{V}{L} = -\frac{18 \text{ kg}}{\text{m}^3} \times \frac{12 \text{ m}}{\text{sec}} \times \frac{1}{0.15 \text{ m}} = -1440 \text{ kg/m}^3 \cdot \text{sec} \qquad \frac{\partial \rho}{\partial t} \; \longleftarrow$$

This problem illustrates use of the differential continuity equation to evaluate a density variation.

5-1.2 Cylindrical Coordinate System

In cylindrical coordinates, a suitable differential control volume is shown in Fig. 5.2. The density at the center, O, of the control volume is ρ and the velocity there is $\vec{V} = \hat{e}_r V_r + \hat{e}_\theta V_\theta + \hat{k} V_z$, where $\hat{e}_r$, $\hat{e}_\theta$, and $\hat{k}$ are unit vectors in the r, θ, and z directions, respectively, and V_r, V_θ, and V_z are the velocity components in the r, θ, and z directions, respectively. To evaluate $\int_{CS} \rho \vec{V} \cdot d\vec{A}$, we must consider the mass flux through each of the six faces of the control surface. The properties at each of the six faces of the control surface are obtained from a Taylor series expansion about point O. The details of the mass flux evaluation are shown in Table 5.2. Velocity components V_r, V_θ, and V_z are all assumed to be in the positive direction, and higher-order terms have been neglected.

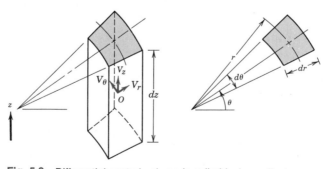

Fig. 5.2 Differential control volume in cylindrical coordinates.

Table 5.2 Mass Flux through the Control Surface of a Cylindrical Differential Control Volume

Surface	$\displaystyle\int \rho \vec{V} \cdot d\vec{A}$
Inside $(-r)$	$\displaystyle = -\left[\rho - \left(\frac{\partial \rho}{\partial r}\right)\frac{dr}{2}\right]\left[V_r - \left(\frac{\partial V_r}{\partial r}\right)\frac{dr}{2}\right]\left(r - \frac{dr}{2}\right)d\theta\,dz = -\rho V_r r\,d\theta\,dz + \rho V_r\frac{dr}{2}d\theta\,dz + \rho\left(\frac{\partial V_r}{\partial r}\right)r\frac{dr}{2}d\theta\,dz + V_r\left(\frac{\partial \rho}{\partial r}\right)r\frac{dr}{2}d\theta\,dz$
Outside $(+r)$	$\displaystyle = \left[\rho + \left(\frac{\partial \rho}{\partial r}\right)\frac{dr}{2}\right]\left[V_r + \left(\frac{\partial V_r}{\partial r}\right)\frac{dr}{2}\right]\left(r + \frac{dr}{2}\right)d\theta\,dz = \rho V_r r\,d\theta\,dz + \rho V_r\frac{dr}{2}d\theta\,dz + \rho\left(\frac{\partial V_r}{\partial r}\right)r\frac{dr}{2}d\theta\,dz + V_r\left(\frac{\partial \rho}{\partial r}\right)r\frac{dr}{2}d\theta\,dz$
Front $(-\theta)$	$\displaystyle = -\left[\rho - \left(\frac{\partial \rho}{\partial \theta}\right)\frac{d\theta}{2}\right]\left[V_\theta - \left(\frac{\partial V_\theta}{\partial \theta}\right)\frac{d\theta}{2}\right]dr\,dz = -\rho V_\theta dr\,dz + \rho\left(\frac{\partial V_\theta}{\partial \theta}\right)\frac{d\theta}{2}dr\,dz + V_\theta\left(\frac{\partial \rho}{\partial \theta}\right)\frac{d\theta}{2}dr\,dz$
Back $(+\theta)$	$\displaystyle = \left[\rho + \left(\frac{\partial \rho}{\partial \theta}\right)\frac{d\theta}{2}\right]\left[V_\theta + \left(\frac{\partial V_\theta}{\partial \theta}\right)\frac{d\theta}{2}\right]dr\,dz = \rho V_\theta dr\,dz + \rho\left(\frac{\partial V_\theta}{\partial \theta}\right)\frac{d\theta}{2}dr\,dz + V_\theta\left(\frac{\partial \rho}{\partial \theta}\right)\frac{d\theta}{2}dr\,dz$
Bottom $(-z)$	$\displaystyle = -\left[\rho - \left(\frac{\partial \rho}{\partial z}\right)\frac{dz}{2}\right]\left[V_z - \left(\frac{\partial V_z}{\partial z}\right)\frac{dz}{2}\right]r\,d\theta\,dr = -\rho V_z r\,d\theta\,dr + \rho\left(\frac{\partial V_z}{\partial z}\right)\frac{dz}{2}r\,d\theta\,dr + V_z\left(\frac{\partial \rho}{\partial z}\right)\frac{dz}{2}r\,d\theta\,dr$
Top $(+z)$	$\displaystyle = \left[\rho + \left(\frac{\partial \rho}{\partial z}\right)\frac{dz}{2}\right]\left[V_z + \left(\frac{\partial V_z}{\partial z}\right)\frac{dz}{2}\right]r\,d\theta\,dr = \rho V_z r\,d\theta\,dr + \rho\left(\frac{\partial V_z}{\partial z}\right)\frac{dz}{2}r\,d\theta\,dr + V_z\left(\frac{\partial \rho}{\partial z}\right)\frac{dz}{2}r\,d\theta\,dr$

Then,

$$\int_{CS} \rho \vec{V}\cdot d\vec{A} = \left[\rho V_r + r\left\{\rho\left(\frac{\partial V_r}{\partial r}\right) + V_r\left(\frac{\partial \rho}{\partial r}\right)\right\} + \left\{\rho\left(\frac{\partial V_\theta}{\partial \theta}\right) + V_\theta\left(\frac{\partial \rho}{\partial \theta}\right)\right\} + r\left\{\rho\left(\frac{\partial V_z}{\partial z}\right) + V_z\left(\frac{\partial \rho}{\partial z}\right)\right\}\right]dr\,d\theta\,dz$$

or

$$\int_{CS}\rho\vec{V}\cdot d\vec{A} = \left[\rho V_r + r\frac{\partial \rho V_r}{\partial r} + \frac{\partial \rho V_\theta}{\partial \theta} + r\frac{\partial \rho V_z}{\partial z}\right]dr\,d\theta\,dz$$

We see that the net rate of mass flux out through the control surface is given by

$$\left[\rho V_r + r \frac{\partial \rho V_r}{\partial r} + \frac{\partial \rho V_\theta}{\partial \theta} + r \frac{\partial \rho V_z}{\partial z} \right] dr \, d\theta \, dz$$

The mass inside the control volume at any instant is the product of the mass per unit volume, ρ, and the volume, $r \, d\theta \, dr \, dz$. Thus the rate of change of mass inside the control volume is given by

$$\frac{\partial \rho}{\partial t} r \, d\theta \, dr \, dz$$

In cylindrical coordinates the differential equation for conservation of mass is then

$$\rho V_r + r \frac{\partial \rho V_r}{\partial r} + \frac{\partial \rho V_\theta}{\partial \theta} + r \frac{\partial \rho V_z}{\partial z} + r \frac{\partial \rho}{\partial t} = 0$$

Dividing by r gives

$$\frac{\rho V_r}{r} + \frac{\partial \rho V_r}{\partial r} + \frac{1}{r} \frac{\partial \rho V_\theta}{\partial \theta} + \frac{\partial \rho V_z}{\partial z} + \frac{\partial \rho}{\partial t} = 0$$

or

$$\frac{1}{r} \frac{\partial (r \rho V_r)}{\partial r} + \frac{1}{r} \frac{\partial (\rho V_\theta)}{\partial \theta} + \frac{\partial (\rho V_z)}{\partial z} + \frac{\partial \rho}{\partial t} = 0 \tag{5.2}$$

In cylindrical coordinates the vector operator ∇ is given by

$$\nabla = \hat{e}_r \frac{\partial}{\partial r} + \hat{e}_\theta \frac{1}{r} \frac{\partial}{\partial \theta} + \hat{k} \frac{\partial}{\partial z} \tag{3.18}$$

Equation 5.2 also may be written[1] in vector notation as

$$\nabla \cdot \rho \vec{V} + \frac{\partial \rho}{\partial t} = 0 \tag{5.1b}$$

For incompressible flow, $\rho = $ constant, and Eq. 5.2 reduces to

$$\frac{1}{r} \frac{\partial (r V_r)}{\partial r} + \frac{1}{r} \frac{\partial V_\theta}{\partial \theta} + \frac{\partial V_z}{\partial z} = 0$$

Thus the velocity field, $\vec{V}(x, y, z, t)$, for incompressible flow must satisfy $\nabla \cdot \vec{V} = 0$. For steady flow, Eq. 5.2 reduces to

$$\frac{1}{r} \frac{\partial (r \rho V_r)}{\partial r} + \frac{1}{r} \frac{\partial (\rho V_\theta)}{\partial \theta} + \frac{\partial (\rho V_z)}{\partial z} = 0$$

or

$$\nabla \cdot \rho \vec{V} = 0$$

---

[1] To evaluate $\nabla \cdot \rho \vec{V}$ in cylindrical coordinates, we must remember that

$$\frac{\partial \hat{e}_r}{\partial \theta} = \hat{e}_\theta \qquad \text{and} \qquad \frac{\partial \hat{e}_\theta}{\partial \theta} = -\hat{e}_r$$

as shown in Example Problem 4.13.

When written in vector form, the differential continuity equation (the mathematical statement of conservation of mass), Eq. 5.1b, may be applied in any coordinate system. One simply substitutes the appropriate expression for the vector operator ∇. In retrospect, this result is not surprising since conservation of mass applies regardless of our choice of coordinate system.

EXAMPLE 5.3—Differential Continuity Equation in Cylindrical Coordinates

Consider a one-dimensional radial flow in the $r\theta$ plane, characterized by $V_r = f(r)$ and $V_\theta = 0$. Determine the conditions on $f(r)$ required for incompressible flow.

EXAMPLE PROBLEM 5.3

GIVEN: One-dimensional radial flow in the $r\theta$ plane: $V_r = f(r)$ and $V_\theta = 0$.

FIND: Requirements on $f(r)$ for incompressible flow.

SOLUTION:

Basic equation:
$$\nabla \cdot \rho\vec{V} + \frac{\partial \rho}{\partial t} = 0$$

For incompressible flow, $\rho = $ constant, so $\dfrac{\partial \rho}{\partial t} = 0$. In cylindrical coordinates

$$\frac{1}{r}\frac{\partial}{\partial r}(rV_r) + \frac{1}{r}\frac{\partial}{\partial \theta}V_\theta + \frac{\partial V_z}{\partial z} = 0$$

For the given velocity field, $\vec{V} = \vec{V}(r)$. $V_\theta = 0$ and partial derivatives with respect to z are zero, so

$$\frac{1}{r}\frac{\partial}{\partial r}(rV_r) = 0$$

Integrating with respect to r gives

$$rV_r = \text{constant}$$

Thus the continuity equation shows that the radial velocity must be $V_r = f(r) = C/r$ for one-dimensional radial flow of an incompressible fluid.

**5-2   STREAM FUNCTION FOR TWO-DIMENSIONAL INCOMPRESSIBLE FLOW

It is convenient to have a means of describing mathematically any particular pattern of flow. An adequate description should portray the notion of the shape of the streamlines (including the boundaries) and the scale of the velocity at representative points in the flow. A mathematical device that serves this purpose is the stream function, ψ. The stream function is formulated as a relation between the streamlines and the statement of conservation of mass. The stream function is a single mathematical function, $\psi(x, y, t)$, that replaces the two velocity components, $u(x, y, t)$ and $v(x, y, t)$.

For a two-dimensional incompressible flow in the xy plane, conservation of mass, Eq. 5.1a, can be written

$$\frac{\partial u}{\partial x} + \frac{\partial v}{\partial y} = 0 \tag{5.3}$$

** This section may be omitted without loss of continuity in the text material.

If a continuous function, $\psi(x, y, t)$, called the stream function, is defined such that

$$u \equiv \frac{\partial \psi}{\partial y} \quad \text{and} \quad v \equiv -\frac{\partial \psi}{\partial x} \tag{5.4}$$

then the continuity equation, Eq. 5.3, is satisfied exactly, since

$$\frac{\partial u}{\partial x} + \frac{\partial v}{\partial y} = \frac{\partial^2 \psi}{\partial x \, \partial y} - \frac{\partial^2 \psi}{\partial y \, \partial x} = 0$$

Recall that streamlines are lines drawn in the flow field, such that, at a given instant, they are tangent to the direction of flow at every point in the flow field. Thus, if $d\vec{r}$ is an element of length along a streamline, the equation of the streamline is given by

$$\vec{V} \times d\vec{r} = 0 = (\hat{i}u + \hat{j}v) \times (\hat{i} \, dx + \hat{j} \, dy)$$
$$= \hat{k}(u \, dy - v \, dx)$$

The equation of a streamline in a two-dimensional flow is

$$u \, dy - v \, dx = 0$$

Substituting for the velocity components, u and v, in terms of the stream function, ψ, from Eq. 5.4, then along a streamline

$$\frac{\partial \psi}{\partial x} dx + \frac{\partial \psi}{\partial y} dy = 0 \tag{5.5}$$

Since $\psi = \psi(x, y, t)$, then at an instant, t_0, $\psi = \psi(x, y, t_0)$; at this instant, a change in ψ may be evaluated as though $\psi = \psi(x, y)$. Thus, at any instant,

$$d\psi = \frac{\partial \psi}{\partial x} dx + \frac{\partial \psi}{\partial y} dy \tag{5.6}$$

Comparing Eqs. 5.5 and 5.6, we see that along an instantaneous streamline, $d\psi = 0$; ψ is a constant along a streamline. Since the differential of ψ is exact, the integral of $d\psi$ between any two points in a flow field, $\psi_2 - \psi_1$, depends only on the end points of integration.

From the definition of a streamline, we recognize that there can be no flow across a streamline. Thus, if the streamlines in a two-dimensional, incompressible flow field, at a given instant are as shown in Fig. 5.3, the rate of flow between streamlines ψ_1 and ψ_2 across the lines AB, BC, DE, and DF must be equal.

The volume flow rate, Q, between streamlines ψ_1 and ψ_2 can be evaluated by considering the flow across AB or across BC. For a unit depth, the flow rate across AB is

$$Q = \int_{y_1}^{y_2} u \, dy = \int_{y_1}^{y_2} \frac{\partial \psi}{\partial y} dy$$

Along AB, $x = $ constant, and $d\psi = \partial \psi / \partial y \, dy$. Therefore,

$$Q = \int_{y_1}^{y_2} \frac{\partial \psi}{\partial y} dy = \int_{\psi_1}^{\psi_2} d\psi = \psi_2 - \psi_1$$

For a unit depth, the flow rate across BC is

$$Q = \int_{x_1}^{x_2} v \, dx = -\int_{x_1}^{x_2} \frac{\partial \psi}{\partial x} dx$$

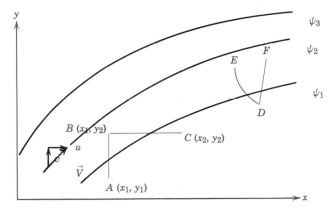

Fig. 5.3   Instantaneous streamlines in a two-dimensional flow.

Along BC, $y = $ constant, and $d\psi = \partial\psi/\partial x \, dx$. Therefore,

$$Q = -\int_{x_1}^{x_2} \frac{\partial\psi}{\partial x}dx = -\int_{\psi_2}^{\psi_1} d\psi = \psi_2 - \psi_1$$

Thus the volume flow rate (per unit depth) between any two streamlines can be written as the difference between the constant values of ψ defining the two streamlines.[2] The sign convention used to define the stream function, ψ, is such that flow is positive from left to right. If the streamline through the origin is designated $\psi = 0$, then the ψ value for any other streamline represents the flow from left to right between the origin and that streamline. This is illustrated in Example Problem 5.4.

For a two-dimensional, incompressible flow in the $r\theta$ plane, conservation of mass, Eq. 5.2, can be written as

$$\frac{\partial(rV_r)}{\partial r} + \frac{\partial V_\theta}{\partial \theta} = 0 \tag{5.7}$$

The stream function, $\psi(r, \theta, t)$, then is defined such that

$$V_r \equiv \frac{1}{r}\frac{\partial\psi}{\partial\theta} \qquad \text{and} \qquad V_\theta \equiv -\frac{\partial\psi}{\partial r} \tag{5.8}$$

With ψ defined according to Eq. 5.8, the continuity equation, Eq. 5.7, is satisfied exactly.

EXAMPLE 5.4—Stream Function for Flow In a Corner

Given the velocity field for the steady, incompressible flow of Example 5.1, $\vec{V} = Ax\hat{i} - Ay\hat{j}$, with $A = 2 \text{ sec}^{-1}$, determine the stream function that will yield

---

[2] For two-dimensional steady compressible flow in the xy plane, the stream function, ψ, is defined such that

$$\rho u \equiv \frac{\partial\psi}{\partial y} \qquad \text{and} \qquad \rho v \equiv -\frac{\partial\psi}{\partial x}$$

The difference between the constant values of ψ defining two streamlines is the mass flow rate (per unit depth) between the two streamlines.

this velocity field. Plot and interpret the streamline pattern in the first and second quadrants of the xy plane.

EXAMPLE PROBLEM 5.4

GIVEN: Velocity field, $\vec{V} = Ax\hat{i} - Ay\hat{j}$, with $A = 2$ sec^{-1}.

FIND: (a) Stream function, ψ.
(b) Plot in first and second quadrants and interpret.

SOLUTION:
The flow is incompressible, so the stream function satisfies Eq. 5.4.

From Eq. 5.4, $u = \dfrac{\partial \psi}{\partial y}$ and $v = -\dfrac{\partial \psi}{\partial x}$. From the given velocity field,

$$u = Ax = \frac{\partial \psi}{\partial y}$$

Integrating with respect to y gives

$$\psi = \int \frac{\partial \psi}{\partial y} dy + f(x) = Axy + f(x) \qquad (1)$$

where $f(x)$ is arbitrary. The function $f(x)$ may be evaluated using the equation for v. Thus, from Eq. 1,

$$v = -\frac{\partial \psi}{\partial x} = -Ay - \frac{df}{dx} \qquad (2)$$

From the given velocity field, $v = -Ay$. Comparing this with Eq. 2 shows that $\dfrac{df}{dx} = 0$, or $f(x) = $ constant. Therefore, Eq. 1 becomes

$$\psi = Axy + c \qquad\qquad\qquad \psi$$

Lines of constant ψ represent streamlines in the flow field. The constant c may be chosen as any convenient value for plotting purposes. The constant is chosen as zero in order that the streamline through the origin be designated as $\psi = \psi_1 = 0$. Then the value for any other streamline represents the flow from left to right between the origin and that streamline. With $c = 0$ and $A = 2$ sec^{-1}, then

$$\psi = 2xy \qquad \text{(m}^3\text{/sec/m)}$$

Separate plots of the streamlines in the first and second quadrants are presented below.

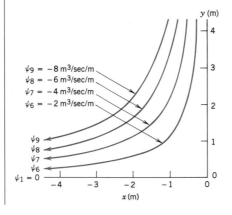

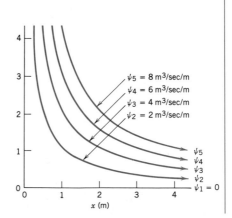

In the first quadrant, since $u > 0$ and $v < 0$, the flow is from left to right and down. The volume flow rate between the streamline, $\psi = \psi_1$, through the origin and the streamline $\psi = \psi_2$, is

$$Q_{12} = \psi_2 - \psi_1 = 2 \text{ m}^3/\text{sec/m}$$

In the second quadrant, since $u < 0$ and $v < 0$, the flow is from right to left and down. The volume flow rate between streamlines ψ_7 and ψ_9 is

$$Q_{79} = \psi_9 - \psi_7 = [-8 - (-4)] \text{ m}^3/\text{sec/m} = -4 \text{ m}^3/\text{sec/m}$$

The negative sign is consistent with flow from right to left.

Regions of high-speed flow occur where the streamlines are close together. Lower-speed flow occurs near the origin, where streamline spacing is wider. In either quadrant, the flow qualitatively looks like flow in a "corner," formed by a pair of walls.

5-3 MOTION OF A FLUID ELEMENT (KINEMATICS)

Before formulating the effects of forces on fluid motion (dynamics), let us consider first the motion (kinematics) of a fluid element in a flow field. For convenience, we follow an infinitesimal element of fixed identity (mass), as shown in Fig. 5.4.

As the infinitesimal element of mass, dm, moves in a flow field, several things may happen to it. Obviously the element translates; it undergoes a linear displacement from a location x, y, z to a different location x_1, y_1, z_1. The element may also rotate; the orientation of the element as shown in Fig. 5.4, where the sides of the element are parallel to the coordinate axes x, y, z may change as a result of pure rotation about any one (or all three) of the coordinate axes. In addition the element may deform. The deformation may be subdivided into two parts—linear and angular deformation. Linear deformation involves a change in shape without a change in orientation of the element: a deformation in which planes of the element that were originally perpendicular (e.g., the top and side of the element) remain perpendicular. Angular deformation involves a distortion of the element in which planes that were originally perpendicular are no longer perpendicular. In general, a fluid element may undergo a combination of translation, rotation, and linear and angular deformation during the course of its motion.

These four components of fluid motion are illustrated in Fig. 5.5 for motion in the xy plane. For a general three-dimensional flow, similar motions of the particle would be depicted in the yz and xz planes. For pure translation or rotation, the fluid element retains its shape; there is no deformation. Thus shear stresses do not arise as

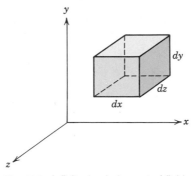

Fig. 5.4 Infinitesimal element of fluid.

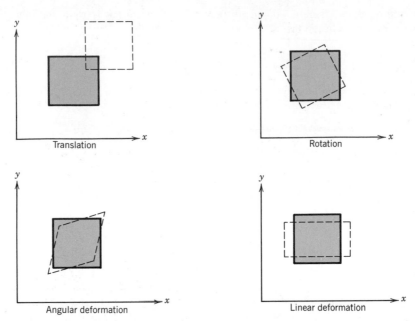

Fig. 5.5 Pictorial representation of the components of fluid motion.

a result of pure translation or rotation. (Recall from Chapter 2 that in a Newtonian fluid the shear stress is directly proportional to the rate of angular deformation.)

5-3.1 Acceleration of a Fluid Particle in a Velocity Field

Let us remember first that we are dealing with an element of fixed mass, dm. As discussed in Section 1-5.3, one may obtain the equation of motion for a particle by applying Newton's second law to that particle. The disadvantage of this approach is that a separate equation is required for each particle. Thus the bookkeeping for many particles becomes a problem.

A more general description of acceleration can be obtained by considering a particle moving in a velocity field. The basic hypothesis of continuum fluid mechanics has led us to a field description of fluid flow in which the properties of a flow field are defined by continuous functions of the space coordinates and time. In particular, the velocity field is given by $\vec{V} = \vec{V}(x, y, z, t)$. The field description is very powerful, since information for the entire flow is given by one equation.

The problem, then, is to retain the field description for fluid properties and obtain an expression for the acceleration of a fluid particle as it translates in a flow field. Stated simply, the problem is:

> Given the velocity field, $\vec{V} = \vec{V}(x, y, z, t)$, find the acceleration of a fluid particle, $\vec{a}_p$.

Consider a particle moving in a velocity field. At time t, the particle is at the position x, y, z and has a velocity corresponding to the velocity at that point in space at t,

$$\vec{V}_p]_t = \vec{V}(x, y, z, t)$$

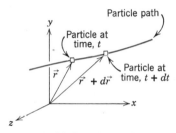

Fig. 5.6 Motion of a particle in a flow field.

At $t + dt$, the particle has moved to a new position, with coordinates $x + dx, y + dy, z + dz$, and has a velocity given by

$$\vec{V}_p]_{t+dt} = \vec{V}(x + dx, y + dy, z + dz, t + dt)$$

This is shown pictorially in Fig. 5.6.

The particle velocity at t (position $\vec{r}$) is given by $\vec{V}_p = \vec{V}(x, y, z, t)$. Then $d\vec{V}_p$, the change in the velocity of the particle, in moving from location $\vec{r}$ to $\vec{r} + d\vec{r}$, is given by

$$d\vec{V}_p = \frac{\partial \vec{V}}{\partial x} dx_p + \frac{\partial \vec{V}}{\partial y} dy_p + \frac{\partial \vec{V}}{\partial z} dz_p + \frac{\partial \vec{V}}{\partial t} dt$$

The total acceleration of the particle is given by

$$\vec{a}_p = \frac{d\vec{V}_p}{dt} = \frac{\partial \vec{V}}{\partial x} \frac{dx_p}{dt} + \frac{\partial \vec{V}}{\partial y} \frac{dy_p}{dt} + \frac{\partial \vec{V}}{\partial z} \frac{dz_p}{dt} + \frac{\partial \vec{V}}{\partial t}$$

Since

$$\frac{dx_p}{dt} = u \qquad \frac{dy_p}{dt} = v \qquad \text{and} \qquad \frac{dz_p}{dt} = w$$

then

$$\vec{a}_p = \frac{d\vec{V}_p}{dt} = u\frac{\partial \vec{V}}{\partial x} + v\frac{\partial \vec{V}}{\partial y} + w\frac{\partial \vec{V}}{\partial z} + \frac{\partial \vec{V}}{\partial t}$$

To remind us that calculation of the acceleration of a fluid particle in a velocity field requires a special derivative, it is given the symbol $D\vec{V}/Dt$. Thus

$$\frac{D\vec{V}}{Dt} \equiv \vec{a}_p = u\frac{\partial \vec{V}}{\partial x} + v\frac{\partial \vec{V}}{\partial y} + w\frac{\partial \vec{V}}{\partial z} + \frac{\partial \vec{V}}{\partial t} \qquad (5.9)$$

The derivative, $D\vec{V}/Dt$, defined by Eq. 5.9, is commonly called the *substantial derivative* to remind us that it is computed for a particle of "substance." It often is called the material or particle derivative.

From Eq. 5.9 we recognize that a fluid particle moving in a flow field may undergo acceleration for either of two reasons. It may be accelerated because it is convected into a region of higher (or lower) velocity. For example, in the steady flow through a nozzle, in which, by definition, the velocity field is not a function of time, a fluid particle will accelerate as it moves through the nozzle. The particle is convected into a region of higher velocity.[3] If a flow field is unsteady a fluid particle will undergo an additional "local" acceleration, because the velocity field is a function of time.

The physical significance of the terms in Eq. 5.9 is

$$\vec{a}_p = \underbrace{\frac{D\vec{V}}{Dt}}_{\substack{\text{total}\\ \text{acceleration}\\ \text{of a particle}}} = \underbrace{u\frac{\partial\vec{V}}{\partial x} + v\frac{\partial\vec{V}}{\partial y} + w\frac{\partial\vec{V}}{\partial z}}_{\substack{\text{convective}\\ \text{acceleration}}} + \underbrace{\frac{\partial\vec{V}}{\partial t}}_{\substack{\text{local}\\ \text{acceleration}}}$$

The convective acceleration may be written as a single vector expression using the vector gradient operator, ∇. Thus

$$u\frac{\partial\vec{V}}{\partial x} + v\frac{\partial\vec{V}}{\partial y} + w\frac{\partial\vec{V}}{\partial z} = (\vec{V}\cdot\nabla)\vec{V}$$

(We suggest that you check this equality by expanding the right side of the equation using the familiar dot product operation.) Thus Eq. 5.9 may be written as

$$\frac{D\vec{V}}{Dt} \equiv \vec{a}_p = (\vec{V}\cdot\nabla)\vec{V} + \frac{\partial\vec{V}}{\partial t} \tag{5.10}$$

For a two-dimensional flow, say $\vec{V} = \vec{V}(x, y, t)$, Eq. 5.9 reduces to

$$\frac{D\vec{V}}{Dt} = u\frac{\partial\vec{V}}{\partial x} + v\frac{\partial\vec{V}}{\partial y} + \frac{\partial\vec{V}}{\partial t}$$

For a one-dimensional flow, say $\vec{V} = \vec{V}(x, t)$, Eq. 5.9 becomes

$$\frac{D\vec{V}}{Dt} = u\frac{\partial\vec{V}}{\partial x} + \frac{\partial\vec{V}}{\partial t}$$

Finally, for a steady flow in three dimensions, Eq. 5.9 becomes

$$\frac{D\vec{V}}{Dt} = u\frac{\partial\vec{V}}{\partial x} + v\frac{\partial\vec{V}}{\partial y} + w\frac{\partial\vec{V}}{\partial z}$$

which is not necessarily zero. Thus a fluid particle may undergo a convective acceleration due to its motion, even in a steady velocity field.

Equation 5.9 is a vector equation. As with all vector equations, it may be written in scalar component equations. Relative to an xyz coordinate system, the scalar components of Eq. 5.9 are written

$$a_{x_p} = \frac{Du}{Dt} = u\frac{\partial u}{\partial x} + v\frac{\partial u}{\partial y} + w\frac{\partial u}{\partial z} + \frac{\partial u}{\partial t} \tag{5.11a}$$

$$a_{y_p} = \frac{Dv}{Dt} = u\frac{\partial v}{\partial x} + v\frac{\partial v}{\partial y} + w\frac{\partial v}{\partial z} + \frac{\partial v}{\partial t} \tag{5.11b}$$

$$a_{z_p} = \frac{Dw}{Dt} = u\frac{\partial w}{\partial x} + v\frac{\partial w}{\partial y} + w\frac{\partial w}{\partial z} + \frac{\partial w}{\partial t} \tag{5.11c}$$

[3] Convective accelerations are demonstrated and calculation of total acceleration of a fluid particle is illustrated in the NCFMF film, *Eulerian and Lagrangian Descriptions in Fluid Mechanics*, J. L. Lumley, principal.

The components of acceleration in cylindrical coordinates may be obtained from Eq. 5.10 by expressing the velocity, $\vec{V}$, in cylindrical coordinates (Section 5-1.2) and utilizing the appropriate expression (Eq. 3.18) for the vector operator ∇. Thus[4]

$$a_{r_p} = V_r \frac{\partial V_r}{\partial r} + \frac{V_\theta}{r} \frac{\partial V_r}{\partial \theta} - \frac{V_\theta^2}{r} + V_z \frac{\partial V_r}{\partial z} + \frac{\partial V_r}{\partial t} \tag{5.12a}$$

$$a_{\theta_p} = V_r \frac{\partial V_\theta}{\partial r} + \frac{V_\theta}{r} \frac{\partial V_\theta}{\partial \theta} + \frac{V_r V_\theta}{r} + V_z \frac{\partial V_\theta}{\partial z} + \frac{\partial V_\theta}{\partial t} \tag{5.12b}$$

$$a_{z_p} = V_r \frac{\partial V_z}{\partial r} + \frac{V_\theta}{r} \frac{\partial V_z}{\partial \theta} + V_z \frac{\partial V_z}{\partial z} + \frac{\partial V_z}{\partial t} \tag{5.12c}$$

We have obtained an expression for the acceleration of a particle anywhere in the flow field; this is the Eulerian method of description. To determine the acceleration of a particle at a particular point in the flow field, one substitutes the coordinates of the point into the field expression for acceleration. In the Lagrangian method of description, the motion (position, velocity, and acceleration) of the particle is described as a function of time. The Eulerian and Lagrangian methods of description are illustrated in Example Problem 5.5.

EXAMPLE 5.5—Particle Acceleration in Eulerian and Lagrangian Descriptions

Consider one-dimensional, steady, incompressible flow through the plane converging channel shown. The velocity field is given by $\vec{V} = V_1[1 + (x/L)]\hat{i}$. Find the x component of acceleration for a particle moving in the flow field. If we use the method of description of particle mechanics, the position of the particle, located at $x = 0$ at time $t = 0$, will be a function of time, $x_p = f(t)$. Obtain the expression for $f(t)$ and then, by taking the second derivative of the function with respect to time, obtain an expression for the x component of the particle acceleration.

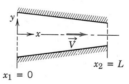

EXAMPLE PROBLEM 5.5

GIVEN: Steady, one-dimensional, incompressible flow through the converging channel shown.

$$\vec{V} = V_1 \left(1 + \frac{x}{L}\right)\hat{i}$$

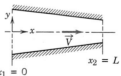

FIND: (a) The x component of acceleration of a particle moving in the flow field.
 (b) For the particle located at $x = 0$ at $t = 0$, obtain an expression for its
 (1) position, x_p, as a function of time.
 (2) x component of acceleration, a_{x_p}, as a function of time.

[4] In evaluating $(\vec{V} \cdot \nabla)\vec{V}$, recall that $\hat{e}_r$ and $\hat{e}_\theta$ are functions of θ (see Footnote 1 on p. 200).

SOLUTION:
The acceleration of a particle moving in a velocity field is given by

$$\frac{D\vec{V}}{Dt} = u\frac{\partial \vec{V}}{\partial x} + v\frac{\partial \vec{V}}{\partial y} + w\frac{\partial \vec{V}}{\partial z} + \frac{\partial \vec{V}}{\partial t}$$

The x component of acceleration is given by

$$\frac{Du}{Dt} = u\frac{\partial u}{\partial x} + v\frac{\partial u}{\partial y} + w\frac{\partial u}{\partial z} + \frac{\partial u}{\partial t}$$

For the flow field given, $v = w = 0$, and $u = V_1\left(1 + \frac{x}{L}\right)$.

Therefore, $\dfrac{Du}{Dt} = u\dfrac{\partial u}{\partial x} = V_1\left(1 + \dfrac{x}{L}\right)\dfrac{V_1}{L} = \dfrac{V_1^2}{L}\left(1 + \dfrac{x}{L}\right)$ $\dfrac{Du}{Dt}$

$\left\{\begin{array}{l}\text{To determine the acceleration of a particle at any point in the flow field, we merely} \\ \text{substitute the present location of the particle into the above result.}\end{array}\right\}$

In the second part of this problem we are interested in following a particular particle, namely the one located at $x = 0$ at $t = 0$, as it flows through the channel.

The x coordinate that locates this particle will be a function of time, $x_p = f(t)$. Furthermore, $u_p = df/dt$ will be a function of time. The particle will have the velocity corresponding to its location in the velocity field. At $t = 0$, the particle is at $x = 0$, and its velocity is $u_p = V_1$. At some later time, t, the particle will reach the exit, $x = L$; at that time it will have velocity $u_p = 2V_1$. To find the expression for $x_p = f(t)$, we write

$$u_p = \frac{dx_p}{dt} = \frac{df}{dt} = V_1\left(1 + \frac{x}{L}\right) = V_1\left(1 + \frac{f}{L}\right)$$

Separating variables gives

$$\frac{df}{(1 + f/L)} = V_1\,dt$$

Since at $t = 0$, the particle in question was located at $x = 0$, and at t, this particle is located at $x_p = f$, then

$$\int_0^f \frac{df}{(1 + f/L)} = \int_0^t V_1\,dt \quad \text{and} \quad L\ln\left(1 + \frac{f}{L}\right) = V_1 t$$

Then, $\ln\left(1 + \dfrac{f}{L}\right) = \dfrac{V_1 t}{L}$, or $1 + \dfrac{f}{L} = e^{V_1 t/L}$

and

$$f = L[e^{V_1 t/L} - 1]$$

Then the position of the particle, located at $x = 0$ at $t = 0$, is given as a function of time by

$$x_p = f(t) = L[e^{V_1 t/L} - 1] \qquad\qquad x_p$$

The x component of acceleration of this particle is given by

$$a_{x_p} = \frac{d^2 x_p}{dt^2} = \frac{d^2 f}{dt^2} = \frac{V_1^2}{L}e^{V_1 t/L} \qquad\qquad a_{x_p}$$

We now have two different ways of expressing the acceleration of the particle that was located at $x = 0$ at $t = 0$. Note that although the flow field is steady, when we follow a particular particle, its position and acceleration (and velocity) are functions of time.

We check to see that both expressions for the acceleration give identical results:

$$a_{x_p} = \frac{V_1^2}{L} e^{V_1 t/L} \qquad\qquad a_{x_p} = \frac{Du}{Dt} = \frac{V_1^2}{L}\left(1 + \frac{x}{L}\right)$$

(a) At $t = 0$, $x_p = 0$ \qquad\qquad At $t = 0$, the particle is at $x = 0$

$$a_{x_p} = \frac{V_1^2}{L} e^0 = \frac{V_1^2}{L} \qquad (a) \qquad\qquad \frac{Du}{Dt} = \frac{V_1^2}{L}(1+0) = \frac{V_1^2}{L} \qquad (a)$$

Check.

(b) When $x_p = \dfrac{L}{2}$, $t = t_1$, \qquad\qquad At $x = 0.5L$

$$x_p = \frac{L}{2} = L[e^{V_1 t_1/L} - 1] \qquad\qquad \frac{Du}{Dt} = \frac{V_1^2}{L}(1+0.5)$$

Therefore, $e^{V_1 t_1/L} = 1.5$, and \qquad\qquad $\dfrac{Du}{Dt} = \dfrac{1.5 V_1^2}{L}$ \qquad (b)

$$a_{x_p} = \frac{V_1^2}{L} e^{V_1 t_1/L}$$

Check.

$$a_{x_p} = \frac{V_1^2}{L}(1.5) = \frac{1.5 V_1^2}{L} \qquad (b)$$

(c) When $x_p = L$, $t = t_2$, \qquad\qquad At $x = L$

$$x_p = L = L[e^{V_1 t_2/L} - 1] \qquad\qquad \frac{Du}{Dt} = \frac{V_1^2}{L}(1+1)$$

Therefore, $e^{V_1 t_2/L} = 2$, and

$$a_{x_p} = \frac{V_1^2}{L} e^{V_1 t_2/L} \qquad\qquad \frac{Du}{Dt} = \frac{2V_1^2}{L} \qquad (c)$$

Check.

$$a_{x_p} = \frac{V_1^2}{L}(2) = \frac{2V_1^2}{L} \qquad (c)$$

$\left\{\begin{array}{l}\text{This problem illustrates the Eulerian and Lagrangian methods of describing the motion}\\ \text{of a particle.}\end{array}\right\}$

5-3.2 Fluid Rotation

The rotation, $\vec{\omega}$, of a fluid particle is defined as the average angular velocity of any two mutually perpendicular line elements of the particle. Rotation is a vector quantity. A particle moving in a general three-dimensional flow field may rotate about all three coordinate axes. Thus, in general,

$$\vec{\omega} = \hat{i}\omega_x + \hat{j}\omega_y + \hat{k}\omega_z$$

where ω_x is the rotation about the x axis, ω_y is the rotation about the y axis, and ω_z is the rotation about the z axis. The positive sense of rotation is given by the right-hand rule.

To obtain a mathematical expression for fluid rotation, consider motion of a fluid element in the xy plane. The components of the velocity at every point in the flow field are given by $u(x, y)$ and $v(x, y)$. The rotation of a fluid element in such a flow field is illustrated in Fig. 5.7. The two mutually perpendicular lines, oa and ob, will rotate to the positions shown during the interval Δt, only if the velocities at points a and b are different from the velocity at o.

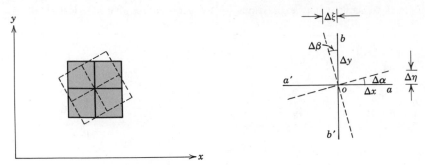

Fig. 5.7 Rotation of a fluid element in a two-dimensional flow field.

Consider first the rotation of line oa, of length Δx. Rotation of this line is due to variations of the y component of velocity. If the y component of velocity at the point o is taken as v_o, then the y component of velocity at point a can be written, using a Taylor series expansion, as

$$v = v_o + \frac{\partial v}{\partial x}\Delta x$$

The angular velocity of line oa is given by

$$\omega_{oa} = \lim_{\Delta t \to 0} \frac{\Delta \alpha}{\Delta t} = \lim_{\Delta t \to 0} \frac{\Delta \eta/\Delta x}{\Delta t}$$

Since

$$\Delta \eta = \frac{\partial v}{\partial x}\Delta x\,\Delta t$$

$$\omega_{oa} = \lim_{\Delta t \to 0} \frac{(\partial v/\partial x)\Delta x\,\Delta t/\Delta x}{\Delta t} = \frac{\partial v}{\partial x}$$

Rotation of line ob, of length Δy, results from variations in the x component of velocity. If the x component of velocity at point o is taken as u_o, then the x component of velocity at point b can be written, using a Taylor series expansion, as

$$u = u_o + \frac{\partial u}{\partial y}\Delta y$$

The angular velocity of line ob is given by

$$\omega_{ob} = \lim_{\Delta t \to 0} \frac{\Delta \beta}{\Delta t} = \lim_{\Delta t \to 0} \frac{\Delta \xi/\Delta y}{\Delta t}$$

Since

$$\Delta \xi = -\frac{\partial u}{\partial y}\Delta y\,\Delta t$$

$$\omega_{ob} = \lim_{\Delta t \to 0} \frac{-(\partial u/\partial y)\Delta y\,\Delta t/\Delta y}{\Delta t} = -\frac{\partial u}{\partial y}$$

(The negative sign is introduced to give a positive value of ω_{ob}. According to our sign convention, counterclockwise rotation is positive.)

The rotation of the fluid element about the z axis is the average angular velocity of the two mutually perpendicular line elements, oa and ob, in the xy plane.

Thus

$$\omega_z = \frac{1}{2}\left(\frac{\partial v}{\partial x} - \frac{\partial u}{\partial y}\right)$$

By considering the rotation of two mutually perpendicular lines in the yz and xz planes, one can show that

$$\omega_x = \frac{1}{2}\left(\frac{\partial w}{\partial y} - \frac{\partial v}{\partial z}\right) \quad \text{and} \quad \omega_y = \frac{1}{2}\left(\frac{\partial u}{\partial z} - \frac{\partial w}{\partial x}\right)$$

Then

$$\vec{\omega} = \hat{i}\omega_x + \hat{j}\omega_y + \hat{k}\omega_z = \frac{1}{2}\left[\hat{i}\left(\frac{\partial w}{\partial y} - \frac{\partial v}{\partial z}\right) + \hat{j}\left(\frac{\partial u}{\partial z} - \frac{\partial w}{\partial x}\right) + \hat{k}\left(\frac{\partial v}{\partial x} - \frac{\partial u}{\partial y}\right)\right] \quad (5.13)$$

We recognize the term in the square brackets as

$$\text{curl } \vec{V} = \nabla \times \vec{V}$$

Then, in vector notation, we can write

$$\vec{\omega} = \tfrac{1}{2}\nabla \times \vec{V} \qquad (5.14)$$

Under what conditions might we expect to have an irrotational flow? A fluid particle moving, without rotation, in a flow field cannot develop a rotation under the action of a body force or normal surface (pressure) forces. Development of rotation in a fluid particle, initially without rotation, requires the action of a shear stress on the surface of the particle. Since shear stress is proportional to the rate of angular deformation, then a particle that is initially without rotation will not develop a rotation without a simultaneous angular deformation. The shear stress is related to the rate of angular deformation through the viscosity. The presence of viscous forces means the flow is rotational.[5]

The condition of irrotationality may be a valid assumption for those regions of a flow in which viscous forces are negligible.[6] (For example, such a region exists outside the boundary layer in the flow over a solid surface.) The factor of $\frac{1}{2}$ can be eliminated in Eq. 5.12 by defining a quantity called the *vorticity*, $\vec{\zeta}$, to be twice the rotation.

$$\vec{\zeta} \equiv 2\vec{\omega} = \nabla \times \vec{V} \qquad (5.15)$$

The vorticity is a measure of the rotation of a fluid element as it moves in the flow field. In cylindrical coordinates the vorticity is [7]

$$\nabla \times \vec{V} = \hat{e}_r\left(\frac{1}{r}\frac{\partial V_z}{\partial \theta} - \frac{\partial V_\theta}{\partial z}\right) + \hat{e}_\theta\left(\frac{\partial V_r}{\partial z} - \frac{\partial V_z}{\partial r}\right) + \hat{k}\left(\frac{1}{r}\frac{\partial r V_\theta}{\partial r} - \frac{1}{r}\frac{\partial V_r}{\partial \theta}\right) \qquad (5.16)$$

The *circulation*, Γ, is defined as the line integral of the tangential velocity component about a closed curve fixed in the flow,

$$\Gamma = \oint_C \vec{V} \cdot d\vec{s} \qquad (5.17)$$

---

[5] A rigorous proof using the complete equations of motion for a fluid particle is given in [1], pp. 142–145. (The references are located just after the objectives, see p. 222.)

[6] Examples of rotational and irrotational motion are shown in the NCFMF film, *Vorticity*, A. H. Shapiro, principal.

[7] In carrying out the curl operation, recall that $\hat{e}_r$ and $\hat{e}_\theta$ are functions of θ (see Footnote 1 on p. 200).

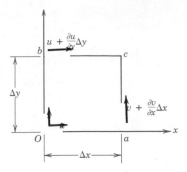

Fig. 5.8 Velocity components on the boundaries of a fluid element.

where $d\vec{s}$ is an elemental vector, of length ds, tangent to the curve; a positive sense corresponds to a counterclockwise path of integration around the curve. A relationship between circulation and vorticity can be obtained by considering the fluid element of Fig. 5.7. The element has been redrawn in Fig. 5.8; the velocity variations shown are consistent with those used in determining the fluid rotation.

For the closed curve $oacb$,

$$d\Gamma = u\,\Delta x + \left(v + \frac{\partial v}{\partial x}\Delta x\right)\Delta y - \left(u + \frac{\partial u}{\partial y}\Delta y\right)\Delta x - v\,\Delta y$$

$$d\Gamma = \left(\frac{\partial v}{\partial x} - \frac{\partial u}{\partial y}\right)\Delta x\,\Delta y$$

$$d\Gamma = 2\omega_z\,\Delta x\,\Delta y$$

Then,

$$\Gamma = \oint_C \vec{V}\cdot d\vec{s} = \int_A 2\omega_z\,dA = \int_A (\nabla\times\vec{V})_z\,dA \qquad (5.18)$$

Equation 5.18 is a statement of the Stokes Theorem in two dimensions. Thus the circulation around a closed contour is the sum of the vorticity enclosed within it.

EXAMPLE 5.6—Free and Forced Vortex Flows

Consider flow fields with purely tangential motion (circular streamlines): $V_r = 0$ and $V_\theta = f(r)$. Evaluate the rotation, vorticity, and circulation for rigid-body rotation, a *forced vortex*. Show that it is possible to choose $f(r)$ so that flow is irrotational, i.e., to produce a *free vortex*.

EXAMPLE PROBLEM 5.6

GIVEN: Flow field with tangential motion, $V_r = 0$ and $V_\theta = f(r)$.

FIND: (a) Rotation, vorticity, and circulation for rigid-body motion (a *forced vortex*).
　　　 (b) Evaluate $f(r)$ for irrotational motion (a *free vortex*).

SOLUTION:
Basic equation: $\qquad\qquad\qquad\qquad \vec{\zeta} = 2\vec{\omega} = \nabla\times\vec{V} \qquad\qquad\qquad (5.15)$

For motion in the $r\theta$ plane, the only components of rotation and vorticity are in the z direction,

$$\zeta_z = 2\omega_z = \frac{1}{r}\frac{\partial r V_\theta}{\partial r} - \frac{1}{r}\frac{\partial V_r}{\partial \theta}$$

Because $V_r = 0$ everywhere in this field, this reduces to $\zeta_z = 2\omega_z = \dfrac{1}{r}\dfrac{\partial r V_\theta}{\partial r}$.

(a) For rigid-body rotation, $V_\theta = \omega r$.

Then $\omega_z = \dfrac{1}{2}\dfrac{1}{r}\dfrac{\partial r V_\theta}{\partial r} = \dfrac{1}{2}\dfrac{1}{r}\dfrac{\partial}{\partial r}\omega r^2 = \dfrac{1}{2r}(2\omega r) = \omega$ \qquad and \qquad $\zeta_z = 2\omega$.

The circulation is $\Gamma = \displaystyle\oint_C \vec{V}\cdot d\vec{s} = \int_A 2\omega_z\, dA.$ \hfill (5.18)

Since $\omega_z = \omega = $ constant, the circulation about any closed contour is given by $\Gamma = 2\omega A$, where A is the area enclosed by the contour. Thus for rigid-body motion (a forced vortex), the rotation and vorticity are constants; the circulation depends on the area enclosed by the contour.

(b) For irrotational flow, $\dfrac{1}{r}\dfrac{\partial}{\partial r} r V_\theta = 0$. Integrating, we find

$$r V_\theta = \text{constant} \qquad \text{or} \qquad V_\theta = f(r) = \frac{C}{r}$$

For this flow, the origin is a singular point where $V_\theta \to \infty$. The circulation for any contour enclosing the origin is

$$\Gamma = \oint_C \vec{V}\cdot d\vec{s} = \int_0^{2\pi} \frac{C}{r} r\, d\theta = 2\pi C$$

The circulation around any contour not enclosing the singular point at the origin is zero.

5-3.3 Fluid Deformation

Angular deformation of a fluid element involves changes in the angle between two mutually perpendicular lines in the fluid. Referring to Fig. 5.9, we see that the rate of angular deformation of the fluid element is the rate of decrease of the angle between lines oa and ob. The rate of angular deformation is given by

$$-\frac{d\gamma}{dt} = \frac{d\alpha}{dt} + \frac{d\beta}{dt}$$

Now,

$$\frac{d\alpha}{dt} = \lim_{\Delta t \to 0}\frac{\Delta\alpha}{\Delta t} = \lim_{\Delta t \to 0}\frac{\Delta\eta/\Delta x}{\Delta t} = \lim_{\Delta t \to 0}\frac{(\partial v/\partial x)\Delta x\,\Delta t/\Delta x}{\Delta t} = \frac{\partial v}{\partial x}$$

and

$$\frac{d\beta}{dt} = \lim_{\Delta t \to 0}\frac{\Delta\beta}{\Delta t} = \lim_{\Delta t \to 0}\frac{\Delta\xi/\Delta y}{\Delta t} = \lim_{\Delta t \to 0}\frac{(\partial u/\partial y)\Delta y\,\Delta t/\Delta y}{\Delta t} = \frac{\partial u}{\partial y}$$

Consequently, the rate of angular deformation in the xy plane is

$$\frac{d\alpha}{dt} + \frac{d\beta}{dt} = -\frac{d\gamma}{dt} = \frac{\partial v}{\partial x} + \frac{\partial u}{\partial y}$$ \hfill (5.19)

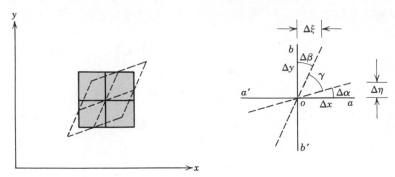

Fig. 5.9 Angular deformation of a fluid element in a two-dimensional flow field.

The shear stress is related to the rate of angular deformation through the fluid viscosity. In a viscous flow (where velocity gradients are present) it is highly unlikely that $\partial v/\partial x$ will be equal and opposite to $\partial u/\partial y$ throughout the flow field (e.g., consider the boundary-layer flow of Fig. 2.11 and the flow over a cylinder, shown in Fig. 2.12). The presence of viscous forces means the flow is rotational.

Calculation of angular deformation is illustrated for a simple flow field in Example Problem 5.7.

EXAMPLE 5.7—Rotation in Viscometric Flow

A viscometric flow in the narrow gap between large parallel plates is shown. The velocity field in the narrow gap is given by $\vec{V} = U(y/h)\hat{i}$, where $U = 4$ mm/sec and $h = 4$ mm. At $t = 0$ two lines, ac and bd, are marked in the fluid as shown. Evaluate the positions of the marked points at $t = 1.5$ sec and sketch for comparison. Calculate the rate of angular deformation and the rate of rotation of a fluid particle in this velocity field. Comment on your results.

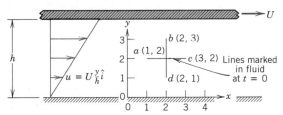

EXAMPLE PROBLEM 5.7

GIVEN: Velocity field, $\vec{V} = U\dfrac{y}{h}\hat{i}$; $U = 4$ mm/sec, and $h = 4$ mm. Fluid particles marked at $t = 0$ to form cross as shown.

FIND: (a) Positions of points a', b', c', and d' at $t = 1.5$ sec; plot.
  (b) Rate of angular deformation.
  (c) Rate of rotation of a fluid particle.
  (d) Comment on the significance of these results.

SOLUTION:
For the given flow field $v = 0$, so there is no vertical motion. The velocity of each point stays constant, so $\Delta x = u\Delta t$ for each point. At point b, $u = 3$ mm/sec, so

$$\Delta x_b = \frac{3}{} \frac{\text{mm}}{\text{sec}} \times \frac{1.5 \text{ sec}}{} = 4.5 \text{ mm}$$

Points a and c each move 3 mm, and point d moves 1.5 mm. The plot at $t = 1.5$ sec is

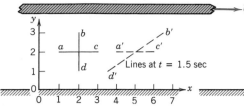

The rate of angular deformation is

$$-\dot{\gamma} = \frac{\partial u}{\partial y} + \frac{\partial v}{\partial x} = U\frac{1}{h} + 0 = \frac{U}{h} = \frac{4}{\text{sec}}\frac{\text{mm}}{\text{sec}} \times \frac{1}{4 \text{ mm}} = 1 \text{ sec}^{-1} \qquad \underleftarrow{\quad -\dot{\gamma}\quad}$$

The rate of rotation is

$$\omega_z = \frac{1}{2}\left(\frac{\partial v}{\partial x} - \frac{\partial u}{\partial y}\right) = \frac{1}{2}\left(0 - \frac{U}{h}\right) = -\frac{1}{2} \times \frac{4}{\text{sec}}\frac{\text{mm}}{\text{sec}} \times \frac{1}{4 \text{ mm}} = -0.5 \text{ sec}^{-1} \qquad \underleftarrow{\quad \omega_z \quad}$$

This flow is viscous, so we expect to have both angular deformation and rotation; shape and orientation of a fluid particle both change. The concepts of rotation and deformation are treated at length in the NCFMF film, *Deformation of Continuous Media*, J. L. Lumley, principal.

During linear deformation, the shape of the fluid element, described by the angles at its vertices, remains unchanged, since all right angles continue to be right angles (see Fig. 5.5). The element will change length in the x direction only if $\partial u/\partial x$ is other than zero. Similarly, a change in the y dimension requires a nonzero value of $\partial v/\partial y$ and a change in the z dimension requires a nonzero value of $\partial w/\partial z$. These quantities represent the components of longitudinal rates of strain in the x, y, and z directions, respectively. Changes in length of the sides may produce changes in volume of the element. The rate of local instantaneous *volume dilation* is given by

$$\text{Volume dilation rate} = \frac{\partial u}{\partial x} + \frac{\partial v}{\partial y} + \frac{\partial w}{\partial z} = \nabla \cdot \vec{V} \tag{5.20}$$

For incompressible flow, the rate of volume dilation is zero.

EXAMPLE 5.8—Deformation Rates for Flow in a Corner
The velocity field $\vec{V} = Ax\hat{i} - Ay\hat{j}$ represents flow in a "corner," as shown in Example Problem 5.4. Consider the case where $A = 0.3$ sec^{-1} and the coordinates are measured in meters. A square is marked in the fluid as shown at $t = 0$. Evaluate the new positions of the four corner points when point a has moved to $x = \frac{3}{2}$ m after τ seconds. Evaluate the rates of linear deformation in the x and y directions. Compare area $a'b'c'd'$ at $t = \tau$ with area $abcd$ at $t = 0$. Comment on this result.

EXAMPLE PROBLEM 5.8

GIVEN: $\vec{V} = Ax\hat{i} - Ay\hat{j}$; $A = 0.3 \text{ sec}^{-1}$,
      x and y in meters.

FIND: (a) Position of square at $t = \tau$
          when a is at a' at $x = \frac{3}{2}$ m.
     (b) Rates of linear deformation.
     (c) Compare area $a'b'c'd'$ with $abcd$.
     (d) Comment on the results.

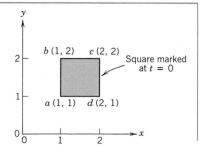

SOLUTION:
First we must find τ, so we must follow a fluid particle using Lagrangian description.
Thus

$$u = \frac{dx_p}{dt} = Ax_p \qquad \frac{dx}{x} = A\,dt \qquad \int_{x_0}^{x} \frac{dx}{x} = \int_{0}^{\tau} A\,dt \qquad \ln\frac{x}{x_0} = A\tau$$

$$\tau = \frac{\ln x/x_0}{A} = \frac{\ln(\frac{3}{2})}{0.3\ \text{sec}^{-1}} = 1.35\ \text{sec}$$

In the y direction

$$v = \frac{dy_p}{dt} = -Ay_p \qquad \frac{dy}{y} = -A\,dt \qquad \frac{y}{y_0} = e^{-A\tau}$$

The point coordinates at τ are:

The plot is:

Point	$t = 0$	$t = \tau$
a	$(1, 1)$	$(\frac{3}{2}, \frac{2}{3})$
b	$(1, 2)$	$(\frac{3}{2}, \frac{4}{3})$
c	$(2, 2)$	$(3, \frac{4}{3})$
d	$(2, 1)$	$(3, \frac{2}{3})$

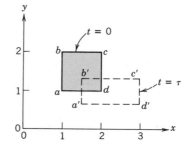

The rates of linear deformation are:

$$\frac{\partial u}{\partial x} = \frac{\partial}{\partial x}Ax = A = 0.3\ \text{sec}^{-1} \qquad \text{in the } x \text{ direction}$$

$$\frac{\partial v}{\partial y} = \frac{\partial}{\partial y}(-Ay) = -A = -0.3\ \text{sec}^{-1} \qquad \text{in the } y \text{ direction}$$

The rate of volume dilation is

$$\nabla \cdot \vec{V} = \frac{\partial u}{\partial x} + \frac{\partial v}{\partial y} = A - A = 0$$

Area $abcd = 1\ \text{m}^2$ and area $a'b'c'd' = (3 - \frac{3}{2})(\frac{4}{3} - \frac{2}{3}) = 1\ \text{m}^2$.

$$\left\{ \begin{array}{l} \text{Note that parallel planes remain parallel; there is linear deformation but no angular} \\ \text{deformation. The rates of linear deformation are equal and opposite, so the area of the} \\ \text{marked region is conserved.} \end{array} \right.$$

In the NCFMF film, *Film Visualization*, S. J. Kline, principal, hydrogen bubble time-streak markers are used to demonstrate experimentally that the area of a marked fluid square is conserved in two-dimensional, incompressible flow.

We have shown in this section that the velocity field contains all information needed to determine the acceleration, rotation, and deformation of a particle in a flow.

5-4 MOMENTUM EQUATION

A dynamic equation describing fluid motion may be obtained by applying Newton's second law to a particle. To derive the differential form of the momentum equation, we shall apply Newton's second law to an infinitesimal fluid particle of mass dm.

Recall that Newton's second law for a finite system is given by

$$\vec{F} = \frac{d\vec{P}}{dt}\bigg)_{system}$$

where the linear momentum, $\vec{P}$, of the system is given by

$$\vec{P}_{system} = \int_{mass\,(system)} \vec{V}\,dm \qquad (4.2b)$$

Then, for an infinitesimal system of mass dm, Newton's second law can be written

$$d\vec{F} = dm\,\frac{d\vec{V}}{dt}\bigg)_{system} \qquad (5.21)$$

Having obtained an expression for the acceleration of a fluid element of mass dm, moving in a velocity field (Eq. 5.9), we can write Newton's second law as the vector equation

$$d\vec{F} = dm\,\frac{D\vec{V}}{Dt} = dm\left[u\frac{\partial\vec{V}}{\partial x} + v\frac{\partial\vec{V}}{\partial y} + w\frac{\partial\vec{V}}{\partial z} + \frac{\partial\vec{V}}{\partial t}\right] \qquad (5.22)$$

We need now to obtain a suitable formulation for the force, $d\vec{F}$, or its components, dF_x, dF_y, and dF_z, acting on the element.

5-4.1   Forces Acting on a Fluid Particle

Recall that the forces acting on a fluid element may be classified as body forces and surface forces; surface forces include both normal forces and tangential (shear) forces.

We shall consider the x component of the force, acting on a differential element of mass dm and volume $d\Psi = dx\,dy\,dz$. Only those stresses that act in the x direction will give rise to surface forces in the x direction. If the stresses at the center of the differential element are taken to be σ_{xx}, τ_{yx}, and τ_{zx}, then the stresses acting in the x direction on each face of the element (obtained by a Taylor series expansion about the center of the element) are as shown in Fig. 5.10.

To obtain the net surface force in the x direction, dF_{S_x}, we must sum the forces in the x direction. Thus

$$dF_{S_x} = \left(\sigma_{xx} + \frac{\partial\sigma_{xx}}{\partial x}\frac{dx}{2}\right)dy\,dz - \left(\sigma_{xx} - \frac{\partial\sigma_{xx}}{\partial x}\frac{dx}{2}\right)dy\,dz$$

$$+ \left(\tau_{yx} + \frac{\partial\tau_{yx}}{\partial y}\frac{dy}{2}\right)dx\,dz - \left(\tau_{yx} - \frac{\partial\tau_{yx}}{\partial y}\frac{dy}{2}\right)dx\,dz$$

$$+ \left(\tau_{zx} + \frac{\partial\tau_{zx}}{\partial z}\frac{dz}{2}\right)dx\,dy - \left(\tau_{zx} - \frac{\partial\tau_{zx}}{\partial z}\frac{dz}{2}\right)dx\,dy$$

On simplifying, we obtain

$$dF_{S_x} = \left(\frac{\partial\sigma_{xx}}{\partial x} + \frac{\partial\tau_{yx}}{\partial y} + \frac{\partial\tau_{zx}}{\partial z}\right)dx\,dy\,dz$$

When the force of gravity is the only body force acting, then the body force per unit mass is $\vec{g}$. The net force in the x direction, dF_x, is given by

$$dF_x = dF_{B_x} + dF_{S_x} = \left(\rho g_x + \frac{\partial\sigma_{xx}}{\partial x} + \frac{\partial\tau_{yx}}{\partial y} + \frac{\partial\tau_{zx}}{\partial z}\right)dx\,dy\,dz \qquad (5.23a)$$

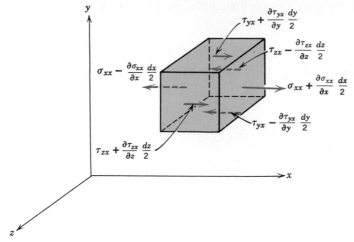

Fig. 5.10 Stresses in the x direction on an element of fluid.

One can derive similar expressions for the force components in the y and z directions:

$$dF_y = dF_{B_y} + dF_{S_y} = \left(\rho g_y + \frac{\partial \tau_{xy}}{\partial x} + \frac{\partial \sigma_{yy}}{\partial y} + \frac{\partial \tau_{zy}}{\partial z}\right) dx\, dy\, dz \qquad (5.23b)$$

$$dF_z = dF_{B_z} + dF_{S_z} = \left(\rho g_z + \frac{\partial \tau_{xz}}{\partial x} + \frac{\partial \tau_{yz}}{\partial y} + \frac{\partial \sigma_{zz}}{\partial z}\right) dx\, dy\, dz \qquad (5.23c)$$

5-4.2 Differential Momentum Equation

We have now formulated expressions for the components, dF_x, dF_y, and dF_z, of the force, $d\vec{F}$, acting on the element of mass dm. If we substitute these expressions (Eqs. 5.23) for the force components into the x, y, and z components of Eq. 5.22, we obtain the differential equations of motion,

$$\rho g_x + \frac{\partial \sigma_{xx}}{\partial x} + \frac{\partial \tau_{yx}}{\partial y} + \frac{\partial \tau_{zx}}{\partial z} = \rho\left(\frac{\partial u}{\partial t} + u\frac{\partial u}{\partial x} + v\frac{\partial u}{\partial y} + w\frac{\partial u}{\partial z}\right) \qquad (5.24a)$$

$$\rho g_y + \frac{\partial \tau_{xy}}{\partial x} + \frac{\partial \sigma_{yy}}{\partial y} + \frac{\partial \tau_{zy}}{\partial z} = \rho\left(\frac{\partial v}{\partial t} + u\frac{\partial v}{\partial x} + v\frac{\partial v}{\partial y} + w\frac{\partial v}{\partial z}\right) \qquad (5.24b)$$

$$\rho g_z + \frac{\partial \tau_{xz}}{\partial x} + \frac{\partial \tau_{yz}}{\partial y} + \frac{\partial \sigma_{zz}}{\partial z} = \rho\left(\frac{\partial w}{\partial t} + u\frac{\partial w}{\partial x} + v\frac{\partial w}{\partial y} + w\frac{\partial w}{\partial z}\right) \qquad (5.24c)$$

Equations 5.24 are the differential equations of motion for any fluid satisfying the continuum assumption. Before the equations can be used to solve problems, suitable expressions for the stresses must be obtained in terms of the velocity and pressure fields.

5-4.3 Newtonian Fluid: Navier–Stokes Equations

For a Newtonian fluid the viscous stress is proportional to the rate of shearing strain (angular deformation rate). The stresses may be expressed in terms of velocity gradients and fluid properties in rectangular coordinates as follows:[8]

$$\tau_{xy} = \tau_{yx} = \mu\left(\frac{\partial v}{\partial x} + \frac{\partial u}{\partial y}\right) \tag{5.25a}$$

$$\tau_{yz} = \tau_{zy} = \mu\left(\frac{\partial w}{\partial y} + \frac{\partial v}{\partial z}\right) \tag{5.25b}$$

$$\tau_{zx} = \tau_{xz} = \mu\left(\frac{\partial u}{\partial z} + \frac{\partial w}{\partial x}\right) \tag{5.25c}$$

$$\sigma_{xx} = -p - \frac{2}{3}\mu\nabla\cdot\vec{V} + 2\mu\frac{\partial u}{\partial x} \tag{5.25d}$$

$$\sigma_{yy} = -p - \frac{2}{3}\mu\nabla\cdot\vec{V} + 2\mu\frac{\partial v}{\partial y} \tag{5.25e}$$

$$\sigma_{zz} = -p - \frac{2}{3}\mu\nabla\cdot\vec{V} + 2\mu\frac{\partial w}{\partial z} \tag{5.25f}$$

where p is the local thermodynamic pressure.

If these expressions are introduced into the differential equations of motion (Eqs. 5.24), we obtain

$$\rho\frac{Du}{Dt} = \rho g_x - \frac{\partial p}{\partial x} + \frac{\partial}{\partial x}\left[\mu\left(2\frac{\partial u}{\partial x} - \frac{2}{3}\nabla\cdot\vec{V}\right)\right] + \frac{\partial}{\partial y}\left[\mu\left(\frac{\partial u}{\partial y} + \frac{\partial v}{\partial x}\right)\right]$$
$$+ \frac{\partial}{\partial z}\left[\mu\left(\frac{\partial w}{\partial x} + \frac{\partial u}{\partial z}\right)\right] \tag{5.26a}$$

$$\rho\frac{Dv}{Dt} = \rho g_y - \frac{\partial p}{\partial y} + \frac{\partial}{\partial x}\left[\mu\left(\frac{\partial u}{\partial y} + \frac{\partial v}{\partial x}\right)\right] + \frac{\partial}{\partial y}\left[\mu\left(2\frac{\partial v}{\partial y} - \frac{2}{3}\nabla\cdot\vec{V}\right)\right]$$
$$+ \frac{\partial}{\partial z}\left[\mu\left(\frac{\partial v}{\partial z} + \frac{\partial w}{\partial y}\right)\right] \tag{5.26b}$$

$$\rho\frac{Dw}{Dt} = \rho g_z - \frac{\partial p}{\partial z} + \frac{\partial}{\partial x}\left[\mu\left(\frac{\partial w}{\partial x} + \frac{\partial u}{\partial z}\right)\right] + \frac{\partial}{\partial y}\left[\mu\left(\frac{\partial v}{\partial z} + \frac{\partial w}{\partial y}\right)\right]$$
$$+ \frac{\partial}{\partial z}\left[\mu\left(2\frac{\partial w}{\partial z} - \frac{2}{3}\nabla\cdot\vec{V}\right)\right] \tag{5.26c}$$

These equations of motion are called the Navier–Stokes equations. The equations are greatly simplified when applied to incompressible flow with constant viscosity. Under these conditions the equations reduce to

$$\rho\left(\frac{\partial u}{\partial t} + u\frac{\partial u}{\partial x} + v\frac{\partial u}{\partial y} + w\frac{\partial u}{\partial z}\right) = \rho g_x - \frac{\partial p}{\partial x} + \mu\left(\frac{\partial^2 u}{\partial x^2} + \frac{\partial^2 u}{\partial y^2} + \frac{\partial^2 u}{\partial z^2}\right) \tag{5.27a}$$

$$\rho\left(\frac{\partial v}{\partial t} + u\frac{\partial v}{\partial x} + v\frac{\partial v}{\partial y} + w\frac{\partial v}{\partial z}\right) = \rho g_y - \frac{\partial p}{\partial y} + \mu\left(\frac{\partial^2 v}{\partial x^2} + \frac{\partial^2 v}{\partial y^2} + \frac{\partial^2 v}{\partial z^2}\right) \tag{5.27b}$$

$$\rho\left(\frac{\partial w}{\partial t} + u\frac{\partial w}{\partial x} + v\frac{\partial w}{\partial y} + w\frac{\partial w}{\partial z}\right) = \rho g_z - \frac{\partial p}{\partial z} + \mu\left(\frac{\partial^2 w}{\partial x^2} + \frac{\partial^2 w}{\partial y^2} + \frac{\partial^2 w}{\partial z^2}\right) \tag{5.27c}$$

---

[8] The derivation of these results is beyond the scope of this book. Detailed derivations may be found in References 2, 3, and 4.

The Navier–Stokes equations in cylindrical coordinates, for constant density and viscosity, are given in Appendix B.

For the case of frictionless flow ($\mu = 0$) the equations of motion (Eqs. 5.26 or Eqs. 5.27) reduce to Euler's equation,

$$\rho \frac{D\vec{V}}{Dt} = \rho \vec{g} - \nabla p$$

We shall consider the case of frictionless flow in Chapter 6.

5-5 SUMMARY OBJECTIVES

After completing study of Chapter 5, you should be able to do the following:

1. Write the differential form of the conservation of mass in (a) vector form, (b) rectangular coordinates, and (c) cylindrical coordinates.

2. Given an algebraic expression for a velocity field, determine if the field represents a possible incompressible flow.

3. Given one component of velocity in a two-dimensional flow field, evaluate another component for steady, incompressible flow.

**4. For a two-dimensional incompressible flow field, define the stream function, ψ; given the velocity field, determine the stream function; given the stream function, determine the velocity field.

5. For a fluid particle moving in a given velocity field, determine the total, convective, and local accelerations.

6. For a fluid particle moving in a flow field, illustrate: translation, rotation, linear deformation, and angular deformation.

7. Define: fluid rotation, vorticity, and circulation.

8. Write the differential form of the momentum equation for viscous flow and state the physical meaning of each term in the equation.

9. Solve the problems at the end of the chapter that relate to the material you have studied.

REFERENCES

1. Li, W. H., and S. H. Lam, *Principles of Fluid Mechanics*. Reading, MA: Addison-Wesley, 1964.

2. Daily, J. W., and D. R. F. Harleman, *Fluid Dynamics*. Reading, MA: Addison-Wesley, 1966.

3. Schlichting, H., *Boundary-Layer Theory*, 7th ed. New York: McGraw-Hill, 1979.

4. White, F. M., *Viscous Fluid Flow,* 2nd ed. New York: McGraw-Hill, 1991.

PROBLEMS

5.1 Which of the following sets of equations represent possible two-dimensional incompressible flow cases?

(a) $u = x + y$; $v = x - y$ (b) $u = x + 2y$; $v = x^2 - y^2$

(c) $u = 4x + y$; $v = x - y^2$ (d) $u = xt + 2y$; $v = x^2 - yt^2$

(e) $u = xt^2$; $v = xyt + y^2$

** This objective applies to a section that may be omitted without loss of continuity in the text material.

5.2 Which of the following sets of equations represent possible two-dimensional incompressible flow cases?

(a) $u = 2x^2 + y^2$;
    $v = x^3 - x(y^2 - 2y)$

(b) $u = 2xy - x^2 + y$; $v = 2xy - y^2 + x^2$

(c) $u = xt + 2y$; $v = xt^2 - yt$

(d) $u = (x + 2y)xt$; $v = (2x - y)yt$

5.3 Which of the following sets of equations represent possible three-dimensional incompressible flow cases?

(a) $u = x + y + z^2$; $v = x - y + z$; $w = 2xy + y^2 + 4$

(b) $u = xyzt$; $v = -xyzt^2$; $w = (z^2/2)(xt^2 - yt)$

(c) $u = y^2 + 2xz$; $v = -2yz + x^2yz$; $w = \frac{1}{2}x^2z^2 + x^3y^4$

5.4 The three components of velocity in a velocity field are given by $u = Ax + By + Cz$, $v = Dx + Ey + Fz$, and $w = Gx + Hy + Jz$. Determine the relationship among the coefficients A through J that is necessary if this is to be a possible incompressible flow field.

5.5 For a flow in the xy plane, the y component of velocity is given by $v = y^2 - 2x + 2y$. Determine a possible x component for steady, incompressible flow. Is it also valid for unsteady, incompressible flow? Why? How many possible x components are there?

5.6 The x component of velocity in a steady, incompressible flow field in the xy plane is $u = A/x$, where $A = 2$ m²/sec, and x is measured in meters. Find the simplest y component of velocity for this flow field.

5.7 A crude approximation for the x component of velocity in an incompressible laminar boundary layer is a linear variation from $u = 0$ at the surface ($y = 0$) to the freestream velocity, U, at the boundary-layer edge ($y = \delta$). The equation for the profile is $u = Uy/\delta$, where $\delta = cx^{1/2}$ and c is a constant. Show that the simplest expression for the y component of velocity is $v = Uy/4x$. Evaluate the maximum value of the ratio v/U, at a location where $x = 0.5$ m and $\delta = 5$ mm.

5.8 A useful approximation for the x component of velocity in an incompressible laminar boundary layer is a parabolic variation from $u = 0$ at the surface ($y = 0$) to the freestream velocity, U, at the edge of the boundary layer ($y = \delta$). The equation for the profile is $u/U = 2(y/\delta) - (y/\delta)^2$, where $\delta = cx^{1/2}$ and c is a constant. Show that the simplest expression for the y component of velocity is

$$\frac{v}{U} = \frac{\delta}{x}\left[\frac{1}{2}\left(\frac{y}{\delta}\right)^2 - \frac{1}{3}\left(\frac{y}{\delta}\right)^3\right]$$

Obtain an expression for the maximum value of the ratio v/U. Evaluate at a location where $\delta = 5$ mm and $x = 0.5$ m.

5.9 A useful approximation for the x component of velocity in an incompressible laminar boundary layer is a sinusoidal variation from $u = 0$ at the surface ($y = 0$) to the freestream velocity, U, at the edge of the boundary layer ($y = \delta$). The equation for the profile is $u = U \sin(\pi y/2\delta)$, where $\delta = cx^{1/2}$ and c is a constant. Show that the simplest expression for the y component of velocity is

$$\frac{v}{U} = \frac{1}{\pi}\frac{\delta}{x}\left[\cos\left(\frac{\pi}{2}\frac{y}{\delta}\right) + \left(\frac{\pi}{2}\frac{y}{\delta}\right)\sin\left(\frac{\pi}{2}\frac{y}{\delta}\right) - 1\right]$$

Obtain an expression for the maximum value of the ratio v/U. Evaluate at a location where $x = 0.5$ m and $\delta = 5$ mm.

5.10 Consider the one-dimensional, unsteady velocity field in which the only component of velocity is $u = Axt$, where u is in ft/sec, x is in ft, t is in sec, and $A = 5$ sec^{-2}. Assuming density is a function of time only, obtain a general expression for ρ. Show that $\rho/\rho_0 = e^{-At^2/2}$ is a possible solution.

5.11 Which of the following sets of equations represent possible incompressible flow cases?

(a) $V_r = U \cos\theta$; $V_\theta = -U \sin\theta$

(b) $V_r = -q/2\pi r$; $V_\theta = K/2\pi r$

(c) $V_r = U \cos\theta\,[1 - (a/r)^2]$; $V_\theta = -U \sin\theta\,[1 + (a/r)^2]$

5.12 A velocity field in cylindrical coordinates is given as

$$\vec{V} = \left(\frac{q}{2\pi r} + U \cos\theta\right)\hat{e}_r - U \sin\theta\,\hat{e}_\theta$$

where $q = 200$ m²/sec and $U = 10$ m/sec. Show that this is a possible incompressible flow case. Locate the stagnation points where $|\vec{V}| = 0$.

5.13 For an incompressible flow in the $r\theta$ plane, the r component of velocity is given as $V_r = -\Lambda \cos\theta/r^2$. Determine a possible θ component of velocity. How many possible θ components are there?

5.14 A viscous liquid is sheared between two parallel disks of radius R, one of which rotates while the other is fixed. The velocity field is purely tangential, and the velocity varies linearly with z from $V_\theta = 0$ at $z = 0$ (the fixed disk) to the velocity of the rotating disk at its surface ($z = h$). Derive an expression for the velocity field between the disks.

5.15 A velocity field in cylindrical coordinates is given as $\vec{V} = (A/r)(\hat{e}_r + \hat{e}_\theta)$, where $A = 0.25\,\text{m}^2/\text{sec}$. Does this represent a possible incompressible flow case? Obtain the equation for a streamline passing through the point $r_0 = 1$ m, $\theta = 0$. Compare with the pathline through the same point.

**5.16 A uniform flow field is inclined at angle α above the x axis. Evaluate the velocity components u and v. Determine the stream function for this flow field.

**5.17 The stream function for a certain incompressible flow is given as $\psi = Axy$. Plot several streamlines, including $\psi = 0$. Obtain an expression for the velocity field.

**5.18 The velocity field for the viscometric flow of Example Problem 5.7 is $\vec{V} = U(y/h)\hat{i}$. Find the stream function for this flow. Locate the streamline that divides the total flow rate into two equal parts.

**5.19 Determine the family of ψ functions that will yield the velocity field $\vec{V} = (x^2 - y^2)\hat{i} - 2xy\,\hat{j}$.

**5.20 Does the velocity field of Problem 5.15 represent a possible incompressible flow case? If so, evaluate the stream function for the flow. If not, evaluate the rate of change of density in the flow field.

**5.21 The stream function for a certain incompressible flow field is given by the expression $\psi = -Ur\sin\theta + q\theta/2\pi$. Obtain an expression for the velocity field. Find the stagnation point(s) where $|\vec{V}| = 0$, and show that $\psi = 0$ there.

**5.22 Incompressible flow around a circular cylinder of radius a is represented by the stream function $\psi = -Ur\sin\theta + Ua^2\sin\theta/r$, where U represents the freestream velocity. Obtain an expression for the velocity field. Show that $V_r = 0$ along the circle $r = a$. Locate the points along $r = a$ where $|\vec{V}| = U$.

**5.23 Consider a flow with velocity components $u = 0$, $v = -y^3 - 4z$, and $w = 3y^2z$.

(a) Is this a one-, two-, or three-dimensional flow?

(b) Demonstrate whether this is an incompressible or compressible flow.

(c) If possible, derive a stream function for this flow.

**5.24 An incompressible frictionless flow field is specified by the stream function $\psi = -6Ax - 8Ay$, where $A = 1$ m/sec, and x and y are coordinates in meters. Sketch the streamlines $\psi = 0$ and $\psi = 8$. Indicate the direction of the velocity vector at the

---

** These problems require material from sections that may be omitted without loss of continuity in the text material.

point $(0, 0)$ on the sketch. Determine the magnitude of the flow rate between the streamlines passing through the points $(2, 2)$ and $(4, 1)$.

**5.25 In a parallel one-dimensional flow in the positive x direction, the velocity varies linearly from zero at $y = 0$ to 100 ft/sec at $y = 4$ ft. Determine an expression for the stream function, ψ. Also determine the y coordinate above which the volume flow rate is half the total between $y = 0$ and $y = 4$ ft.

**5.26 A linear velocity profile was used to model flow in a laminar incompressible boundary layer in Problem 5.7. Derive the stream function for this flow field. Locate streamlines at one-quarter and one-half the total volume flow rate in the boundary layer.

**5.27 A parabolic velocity profile was used to model flow in a laminar incompressible boundary layer in Problem 5.8. Derive the stream function for this flow field. Locate streamlines at one-quarter and one-half the total volume flow rate in the boundary layer.

**5.28 Derive the stream function that represents the sinusoidal approximation used to model the x component of velocity for the boundary layer of Problem 5.9. Locate streamlines at one-quarter and one-half the total volume flow rate in the boundary layer.

**5.29 A rigid-body motion was modeled in Example Problem 5.6 by the velocity field $\vec{V} = r\omega \hat{e}_\theta$. Find the stream function for this flow. Evaluate the volume flow rate per unit depth between $r_1 = 0.05$ m and $r_2 = 0.07$ m, if $\omega = 0.5$ rad/sec. Sketch the velocity profile along a line of constant θ. Check the flow rate calculated from the stream function by integrating the velocity profile along this line.

**5.30 Example Problem 5.6 showed that the velocity field for a free vortex in the $r\theta$ plane is $\vec{V} = \hat{e}_\theta C/r$. Find the stream function for this flow. Evaluate the volume flow rate per unit depth between $r_1 = 0.05$ m and $r_2 = 0.07$ m, if $C = 0.5$ m^2/sec. Sketch the velocity profile along a line of constant θ. Check the flow rate calculated from the stream function by integrating the velocity profile along this line.

**5.31 Flow in a "sector" of included angle $\alpha = \pi/n$ is represented by the stream function $\psi = Ur^n \sin(n\theta)$. Show that for $n = 2$, this stream function reduces to the form found in Example Problem 5.4 for flow in a square corner ($\alpha = \pi/2$).

5.32 Consider the velocity field $\vec{V} = Axy\hat{i} - \frac{1}{2}Ay^2\hat{j}$ in the xy plane, where $A = 0.25$ m$^{-1} \cdot$ sec^{-1}, and the coordinates are measured in meters. Is this a possible incompressible flow field? Calculate the acceleration of a fluid particle at point $(x, y) = (2, 1)$.

5.33 Consider the flow field given by $\vec{V} = xy^2\hat{i} - \frac{1}{3}y^3\hat{j} + xy\hat{k}$. Determine (a) the number of dimensions of the flow, (b) if it is a possible incompressible flow, and (c) the acceleration of a fluid particle at point $(x, y, z) = (1, 2, 3)$.

5.34 Consider the flow field given by $\vec{V} = ax^2y\hat{i} - by\hat{j} + cz^2\hat{k}$, where $a = 1$ m$^{-2} \cdot$ sec^{-1}, $b = 3$ sec^{-1}, and $c = 2$ m$^{-1} \cdot$ sec^{-1}. Determine (a) the number of dimensions of the flow, (b) if it is a possible incompressible flow, and (c) the acceleration of a fluid particle at point $(x, y, z) = (3, 1, 2)$.

5.35 The velocity field within a laminar boundary layer is given by the expression

$$\vec{V} = \frac{AUy}{x^{1/2}}\hat{i} + \frac{AUy^2}{4x^{3/2}}\hat{j}$$

In this expression, $A = 141$ m$^{-1/2}$, and $U = 0.240$ m/sec is the freestream velocity. Show that this velocity field represents a possible incompressible flow. Calculate the acceleration of a fluid particle at point $(x, y) = (0.5$ m, 5 mm$)$.

5.36 The y component of velocity in a two-dimensional, incompressible flow field is given by $v = -Axy$, where v is in m/sec, x and y are in meters, and A is a dimensional constant. There is no velocity component or variation in the z direction. Determine

** These problems require material from sections that may be omitted without loss of continuity in the text material.

the dimensions of the constant, A. Find the simplest x component of velocity in this flow field. Calculate the acceleration of a fluid particle at point $(x, y) = (1, 2)$.

5.37 The x component of velocity in a steady, incompressible flow field in the xy plane is $u = A/x$, where $A = 2$ m²/sec and x is measured in meters. Show that the simplest y component of velocity for this flow field is $v = Ay/x^2$. Evaluate the acceleration of a fluid particle at point $(x, y) = (1, 3)$.

5.38 The variation in cross-sectional area with distance along a diffuser may be expressed as $A = A_1 e^{ax}$, where A_1 is the cross-sectional area at the diffuser inlet. Assume that flow is incompressible and uniform at any cross section. Sketch the variations of area and velocity along the diffuser as functions of x. Develop an algebraic expression for the acceleration of a fluid particle in the diffuser, in terms of V_1 and x.

5.39 Solve Problem 4.123 to show that the radial velocity in the narrow gap is $V_r = Q/2\pi rh$. Derive an expression for the acceleration of a fluid particle in the gap.

5.40 Consider the low-speed flow of air between parallel disks as shown. Assume that the flow is incompressible and inviscid, and that the velocity is purely radial and uniform at any section. The flow speed is $V = 15$ m/sec at $R = 75$ mm. Simplify the continuity equation to a form applicable to this flow field. Show that a general expression for the velocity field is $\vec{V} = V(R/r)\hat{e}_r$ for $r_i \leq r \leq R$. Calculate the acceleration of a fluid particle at the locations $r = r_i$ and $r = R$.

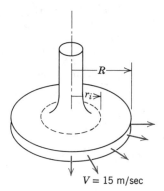

P5.40 $V = 15$ m/sec

5.41 An "air hockey" puck may be modeled as a circular disk supported by a layer of air that issues from multiple tiny holes in the game table. Assume that a puck floats distance $h = 1$ mm above the table, through which air flows vertically at average speed $v = 0.08$ m/sec. Obtain an expression for radial flow speed under the puck, if the flow is considered to be uniform and incompressible. If the puck diameter is 75 mm, determine the magnitude and location of the maximum radial acceleration experienced by a fluid particle under the puck.

5.42 A cylindrical tank, of radius $R = 4$ in., is filled with water to a depth of 6 in. The tank is rotated about its vertical axis. During start up, $0 \leq t \leq \tau$, the rate of rotation is given by $\omega = \omega_0 t/\tau$, where $\tau = 2$ sec, and the steady-state rotational speed is $\omega_0 = 78$ rpm. The no-slip condition requires that fluid particles at the tank wall have zero velocity relative to the wall. For a particle at the wall, determine the acceleration at time $t = 1$ sec and the steady-state acceleration.

5.43 The temperature, T, in a long tunnel is known to vary approximately as $T = T_0 - \alpha e^{-x/L} \sin(2\pi t/\tau)$, where T_0, α, L, and τ are constants, and x is measured from the entrance. A particle moves into the tunnel with a constant speed, U. Obtain an expression for the rate of change of temperature experienced by the particle. What are the dimensions of this expression?

5.44 An aircraft flies due North at 300 mph ground speed. Its rate of climb is 3000 ft/min. The vertical temperature gradient is -3 F per 1000 ft of altitude. The ground temperature varies with position through a cold front, falling at the rate of 1 F per mile. Compute the rate of temperature change shown by a recorder on board the aircraft.

5.45 As an aircraft flies through a cold front, an on-board instrument indicates that ambient temperature drops at the rate of 0.5 F per minute. Other instruments show an air speed of 300 knots and a 3500 ft/min rate of climb. The front is stationary and vertically uniform. Compute the rate of change of temperature with respect to horizontal distance through the cold front.

5.46 After a rainfall the sediment concentration at a certain point in a river increases at the rate of 100 parts per million (ppm) per hour. In addition, the sediment concentration increases with distance downstream as a result of influx from tributary streams; this rate of increase is 50 ppm per mile. At this point the stream flows at 0.5 mph. A boat is used to survey the sediment concentration. The operator is amazed to find three different apparent rates of change of sediment concentration when the boat travels upstream, drifts with the current, or travels downstream. Explain physically why the different rates are observed. If the speed of the boat is 2.5 mph, compute the three rates of change.

5.47 A steady, two-dimensional velocity field is given by $\vec{V} = Ax\hat{i} - Ay\hat{j}$, where $A = 1 \text{ sec}^{-1}$. Show that the streamlines for this flow are rectangular hyperbolas, $xy = C$. Plot streamlines that correspond to $C = 0$, 1, and 2 m^2. Obtain a general expression for the acceleration of a fluid particle in this velocity field. Calculate the acceleration of fluid particles at the points $(x, y) = (\frac{1}{2}, 2)$, $(1, 1)$, and $(2, \frac{1}{2})$, where x and y are measured in meters. Show the acceleration vectors on the streamline plot.

5.48 A flow is described by the velocity field $\vec{V} = (Ax + B)\hat{i} - Ay\hat{j}$, where $A = 10 \text{ sec}^{-1}$, $B = 3 \text{ ft} \cdot \text{sec}^{-1}$, and the coordinates are measured in feet. Plot a few streamlines in the xy plane. Is this a possible incompressible flow? Compute the acceleration of a fluid particle at point $(x, y) = (3, 2)$. Show the acceleration vector on the sketch.

5.49 A velocity field is represented by the expression $\vec{V} = (Ax - B)\hat{i} + Cy\hat{j} + Dt\hat{k}$, where $A = 2 \text{ sec}^{-1}$, $B = 4 \text{ m} \cdot \text{sec}^{-1}$, $D = 5 \text{ m} \cdot \text{sec}^{-2}$, and the coordinates are measured in meters. Determine the proper value for C if the flow field is to be incompressible. Calculate the acceleration of a fluid particle located at point $(x, y) = (3, 2)$. Sketch the flow streamlines in the xy plane.

‡5.50 A linear approximate velocity profile was used in Problem 5.7 to model a laminar incompressible boundary layer on a flat plate. For this profile, obtain expressions for the x and y components of acceleration of a fluid particle in the boundary layer. Locate the maximum magnitudes of the x and y accelerations. Compute the ratio of the maximum x magnitude to the maximum y magnitude for the flow conditions of Problem 5.7.

‡5.51 A parabolic approximate velocity profile was used in Problem 5.8 to model flow in a laminar incompressible boundary layer on a flat plate. For this profile, show that the x component of acceleration of a fluid particle within the boundary layer may be written

$$a_x = -\frac{U^2}{x}\left[\left(\frac{y}{\delta}\right)^2 - \frac{4}{3}\left(\frac{y}{\delta}\right)^3 + \frac{1}{3}\left(\frac{y}{\delta}\right)^4\right]$$

For a given x location, find the value of y/δ for which a_x is a maximum. Calculate the maximum a_x, if $U = 8.7 \text{ m/sec}$ and $x = 0.5 \text{ m}$.

‡5.52 A sinusoidal approximate velocity profile was used in Problem 5.9 to model flow in a laminar incompressible boundary layer on a flat plate. For this profile, obtain an expression for the x component of acceleration of a fluid particle in the boundary layer. Locate the maximum x component of acceleration within the boundary layer. Calculate the maximum a_x, if $U = 8.7 \text{ m/sec}$ and $x = 0.5 \text{ m}$.

---

‡You may wish to use simple computer programs to help solve problems marked with daggers.

5.53 Air flows into the narrow gap, of height h, between closely spaced parallel plates through a porous surface as shown. Use a control volume, with outer surface located at position x, to show that the uniform velocity in the x direction is $u = v_0 x/h$. Find an expression for the velocity component in the y direction. Evaluate the acceleration of a fluid particle in the gap.

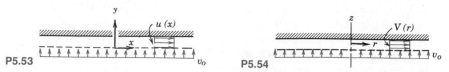

P5.53   P5.54

5.54 Air flows into the narrow gap, of height h, between closely spaced parallel disks through a porous surface as shown. Use a control volume, with outer surface located at position r, to show that the uniform velocity in the r direction is $V = v_0 r/2h$. Find an expression for the velocity component in the z direction ($v_0 \ll V$). Evaluate the components of acceleration for a fluid particle in the gap.

5.55 The velocity field for steady inviscid flow from left to right over a circular cylinder, of radius a, is given by

$$\vec{V} = U\cos\theta\left[1 - \left(\frac{a}{r}\right)^2\right]\hat{e}_r - U\sin\theta\left[1 + \left(\frac{a}{r}\right)^2\right]\hat{e}_\theta$$

Obtain expressions for the acceleration of a fluid particle moving along the stagnation streamline ($\theta = \pi$) and for the acceleration along the cylinder surface ($r = a$). Determine the locations at which these accelerations reach maximum and minimum values.

5.56 Consider the incompressible flow of a fluid through a nozzle as shown. The area of the nozzle is given by $A = A_0(1 - bx)$ and the inlet velocity varies according to $U = C(1 + at)$, where $A_0 = 1$ ft^2, $L = 4$ ft, $b = 0.1$ ft^{-1}, $a = 2$ sec^{-1}, and $C = 10$ ft/sec. The flow may be assumed one-dimensional. Find the acceleration of a fluid particle at $x = L/2$ for $t = 0$ and 0.5 sec.

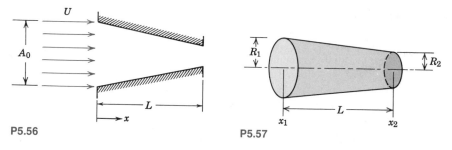

P5.56   P5.57

‡5.57 Consider the one-dimensional, incompressible flow through the circular channel shown. The velocity at section ① is given by $U = U_0 + U_1 \sin\omega t$, where $U_0 = 20$ m/sec, $U_1 = 2$ m/sec, and $\omega = 0.3$ rad/sec. The channel dimensions are $L = 1$ m, $R_1 = 0.2$ m, and $R_2 = 0.1$ m. Determine the particle acceleration at the channel exit. Plot the results as a function of time over a complete cycle.

5.58 Consider steady incompressible flow of an inviscid fluid, with the density of standard air, through the horizontal nozzle shown in the diagram for Problem 5.56. The x component of velocity in the nozzle is given by $u = U_0/(1 - bx)$, where $U_0 = 18.7$ m/sec, $b = 0.3$ m^{-1}, and $L = 0.65$ m. By symmetry, the y component of velocity is $v = 0$ along the nozzle centerline ($y = 0$). There is no flow or variation in the z direction. Derive the simplest possible form of the y component of velocity in the nozzle. Obtain an algebraic expression for the acceleration of a fluid particle

‡You may wish to use simple computer programs to help solve problems marked with daggers.

moving in the nozzle. Evaluate the maximum x component of acceleration of a fluid particle within the nozzle.

‡**5.59** The circular channel of Problem 5.57 is replaced by a plane channel of uniform width with the same linear dimensions. For the conditions given, determine the particle acceleration at the channel exit and plot the results as a function of time over a complete cycle.

5.60 Consider again the steady, two-dimensional velocity field of Problem 5.47. Obtain expressions for the particle coordinates, $x_p = f_1(t)$ and $y_p = f_2(t)$, as functions of time and the initial particle position, (x_0, y_0) at $t = 0$. Determine the time required for a particle to travel from initial position, $(x_0, y_0) = (\frac{1}{2}, 2)$, to positions $(x, y) = (1, 1)$ and $(2, \frac{1}{2})$. Compare the particle accelerations determined by differentiating $f_1(t)$ and $f_2(t)$ with those obtained in Problem 5.47.

5.61 In cylindrical coordinates, the velocity field for a two-dimensional flow is given by $\vec{V} = \vec{V}(r, \theta, t)$. Show, by direct substitution into Eq. 5.10, that the radial and tangential components of the acceleration of a particle are given by Eqs. 5.12a and 5.12b.

5.62 The differential form of the equation for conservation of mass can be used to evaluate the relative rate of change of density of a fluid particle as it moves through a flow. Show that

$$\frac{1}{\rho}\frac{D\rho}{Dt} = -\nabla \cdot \vec{V}$$

Explain the physical significance of $\nabla \cdot \vec{V}$.

5.63 A flow is represented by the velocity field $\vec{V} = 10x\hat{i} - 10y\hat{j} + 30\hat{k}$. Determine if the field is (a) a possible incompressible flow and (b) irrotational.

5.64 A flow is represented by the velocity field $\vec{V} = (4x^2 + 3y)\hat{i} + (3x - 2y)\hat{j}$. Determine if the field is (a) a possible incompressible flow and (b) irrotational.

5.65 Consider flow in the $r\theta$ plane with velocity field given by $\vec{V} = \hat{e}_r V_r + \hat{e}_\theta V_\theta$. By direct substitution into Eq. 5.15, verify that the vorticity is given by Eq. 5.16.

5.66 Consider again the sinusoidal velocity profile used to model the x component of velocity for a boundary layer in Problem 5.9. Neglect the vertical component of velocity. Evaluate the circulation around the contour bounded by $x = 0.4$ m, $x = 0.6$ m, $y = 0$, and $y = 8$ mm. What would be the results of this evaluation if it were performed 0.2 m further downstream? Assume $U = 0.5$ m/sec.

5.67 Consider the velocity field for flow in a rectangular "corner," $\vec{V} = Ax\hat{i} - Ay\hat{j}$, with $A = 0.3$ sec^{-1}, as in Example Problem 5.8. Evaluate the circulation about the unit square of Example Problem 5.8.

5.68 Consider the two-dimensional incompressible flow field in which $u = Axy$ and $v = By^2$, where $A = 1$ m$^{-1}\cdot$sec^{-1}, $B = -\frac{1}{2}$ m$^{-1}\cdot$sec^{-1}, and the coordinates are measured in meters. Determine the rotation at point $(x, y) = (1, 1)$. Evaluate the circulation about the "curve" bounded by $y = 0$, $x = 1$, $y = 1$, and $x = 0$.

**5.69 Consider the flow field represented by the stream function $\psi = 10xy + 17$. Is this a possible two-dimensional, incompressible flow? Is the flow irrotational?

**5.70 Consider a velocity field for motion parallel to the x axis with constant shear. The shear rate is $du/dy = A$, where $A = 0.1$ sec^{-1}. Obtain an expression for the velocity field, $\vec{V}$. Calculate the rate of rotation. Evaluate the stream function for this flow field.

**5.71 Consider the velocity field given by $\vec{V} = Axy\hat{i} + By^2\hat{j}$, where $A = 4$ m$^{-1}\cdot$sec^{-1}, $B = -2$ m$^{-1}\cdot$sec^{-1}, and the coordinates are measured in meters. Determine the fluid

‡ You may wish to use simple computer programs to help solve problems marked with daggers.
** These problems require material from sections that may be omitted without loss of continuity in the text material.

rotation. Evaluate the circulation about the "curve" bounded by $y = 0$, $x = 1$, $y = 1$, and $x = 0$. Obtain an expression for the stream function.

5.72 Consider the flow represented by the velocity field $\vec{V} = (Ay + B)\hat{i} + Ax\hat{j}$, where $A = 6\ \text{sec}^{-1}$, $B = 3\ \text{m} \cdot \text{sec}^{-1}$, and the coordinates are measured in meters. Obtain an expression for the stream function. Evaluate the circulation about the "curve" bounded by $y = 0$, $x = 1$, $y = 1$, and $x = 0$.

5.73 Consider again the viscometric flow of Example Problem 5.7. Evaluate the average rate of rotation of a pair of perpendicular line segments oriented at $\pm 45°$ from the x axis. Show that the rates of rotation for these segments are the same as in the example.

5.74 Consider the pressure-driven flow between stationary parallel plates separated by distance b. Coordinate y is measured from the bottom plate. The velocity field is given by $u = U(y/b)[1 - (y/b)]$. Obtain an expression for the circulation about a closed contour of height h and length L. Evaluate when $h = b/2$ and when $h = b$. Show that the same result is obtained from the area integral of the Stokes Theorem (Eq. 5.18).

5.75 A flow field is represented by the stream function $\psi = x^2 - y^2$. Find the corresponding velocity field. Plot several streamlines and illustrate the velocity field. Show that this flow field is irrotational.

5.76 The velocity field near the core of a tornado can be approximated as

$$\vec{V} = -\frac{q}{2\pi r}\hat{e}_r + \frac{K}{2\pi r}\hat{e}_\theta$$

Is this an irrotational flow field? Obtain the stream function for this flow.

5.77 The velocity profile for fully developed flow in a circular tube is $V_z = V_{\max}[1 - (r/R)^2]$. Evaluate the rates of linear and angular deformation for this flow. Obtain an expression for the vorticity vector, $\vec{\zeta}$.

5.78 Consider the velocity field given by $\vec{V} = Ax^3yt\hat{i} + By^2z\hat{j} + Cxyt^2\hat{k}$, where $A = 1\ \text{m}^{-3} \cdot \text{sec}^{-2}$, $B = 1\ \text{m}^{-2} \cdot \text{sec}^{-1}$, $C = 1\ \text{m}^{-1} \cdot \text{sec}^{-3}$, and the coordinates are measured in meters. Obtain an expression for the y component of acceleration of a fluid particle. Develop an algebraic expression for the time rate of linear strain in the z direction.

5.79 Consider the pressure-driven flow between stationary parallel plates separated by distance $2b$. Coordinate y is measured from the channel centerline. The velocity field is given by $u = u_{\max}[1 - (y/b)^2]$. Evaluate the rates of linear and angular deformation. Obtain an expression for the vorticity vector, $\vec{\zeta}$. Find the location where the vorticity is a maximum.

5.80 A linear velocity profile was used to model flow in a laminar incompressible boundary layer in Problem 5.7. Express the rotation of a fluid particle. Locate the maximum rate of rotation. Express the rate of angular deformation for a fluid particle. Locate the maximum rate of angular deformation. Express the rates of linear deformation for a fluid particle. Locate the maximum rates of linear deformation. Express the shear force per unit volume in the x direction. Locate the maximum shear force per unit volume; interpret this result.

5.81 Solve Problem 5.80 using the parabolic velocity profile of Problem 5.8.

5.82 Solve Problem 5.80 using the sinusoidal velocity profile of Problem 5.9.

5.83 Problem 4.31 gave the velocity profile for fully developed laminar flow between stationary parallel plates separated by distance $2h$, as $u = u_{\max}[1 - (y/h)^2]$. Coordinate y is measured from the channel centerline. Obtain an expression for the shear

---

** These problems require material from sections that may be omitted without loss of continuity in the text material.

force per unit volume in the x direction for this flow. Calculate its maximum value if $u_{max} = 0.4$ m/sec and $h = 1.5$ mm.

5.84 The x component of velocity in a laminar boundary layer in water is approximated as $u = U \sin(\pi y / 2\delta)$, where $U = 3$ m/sec and $\delta = 2$ mm. The y component of velocity is much smaller than u. Obtain an expression for the net shear force per unit volume in the x direction on a fluid element. Calculate its maximum value for this flow.

5.85 Problem 4.23 gave the velocity profile for fully developed laminar flow in a circular tube as $u = u_{max}[1 - (r/R)^2]$. Obtain an expression for the shear force per unit volume in the x direction for this flow. Evaluate its maximum value for the conditions of Problem 4.33.

Chapter 6

INCOMPRESSIBLE INVISCID FLOW

All real fluids possess viscosity. However, there are many flow cases in which it is reasonable to neglect the effects of viscosity. (This is analogous to neglecting friction forces in the analysis of solid systems for some situations.) Consequently, it is useful to investigate the dynamics of an *ideal fluid* that is incompressible and has zero viscosity. The analysis of ideal fluid motions is simpler than for viscous flows because no shear stresses are present in inviscid flow. Normal stresses are the only stresses that must be considered in the analysis. For a nonviscous fluid in motion, the normal stress at a point is the same in all directions (it is a scalar quantity). The normal stress in an inviscid flow is the negative of the thermodynamic pressure, $\sigma_{nn} = -p$. (This result is consistent with Eqs. 5.25 for $\mu = 0$.)

6-1 MOMENTUM EQUATION FOR FRICTIONLESS FLOW: EULER'S EQUATIONS

The equations of motion for frictionless flow, called Euler's equations, can be obtained from the general equations of motion (Eqs. 5.24). Since, in a frictionless flow, there can be no shear stresses and the normal stress is the negative of the thermodynamic pressure, then the equations of motion for a frictionless flow are

$$\rho g_x - \frac{\partial p}{\partial x} = \rho \left(\frac{\partial u}{\partial t} + u \frac{\partial u}{\partial x} + v \frac{\partial u}{\partial y} + w \frac{\partial u}{\partial z} \right) \tag{6.1a}$$

$$\rho g_y - \frac{\partial p}{\partial y} = \rho \left(\frac{\partial v}{\partial t} + u \frac{\partial v}{\partial x} + v \frac{\partial v}{\partial y} + w \frac{\partial v}{\partial z} \right) \tag{6.1b}$$

$$\rho g_z - \frac{\partial p}{\partial z} = \rho \left(\frac{\partial w}{\partial t} + u \frac{\partial w}{\partial x} + v \frac{\partial w}{\partial y} + w \frac{\partial w}{\partial z} \right) \tag{6.1c}$$

We can also write the above equations as a single vector equation

$$\rho \vec{g} - \nabla p = \rho \left(\frac{\partial \vec{V}}{\partial t} + u \frac{\partial \vec{V}}{\partial x} + v \frac{\partial \vec{V}}{\partial y} + w \frac{\partial \vec{V}}{\partial z} \right)$$

or

$$\rho \vec{g} - \nabla p = \rho \frac{D\vec{V}}{Dt} \tag{6.2}$$

If the z coordinate is directed vertically, then, since $\nabla z = \hat{k}$,

$$\rho\vec{g} = -\rho g\hat{k} = -\rho g\nabla z$$

and Euler's equation can be written as

$$-g\nabla z - \frac{1}{\rho}\nabla p = \frac{D\vec{V}}{Dt} = \frac{\partial\vec{V}}{\partial t} + (\vec{V}\cdot\nabla)\vec{V} \qquad (6.3)$$

In cylindrical coordinates, the equations in component form, with gravity the only body force, are

$$g_r - \frac{1}{\rho}\frac{\partial p}{\partial r} = a_r = \frac{\partial V_r}{\partial t} + V_r\frac{\partial V_r}{\partial r} + \frac{V_\theta}{r}\frac{\partial V_r}{\partial\theta} + V_z\frac{\partial V_r}{\partial z} - \frac{V_\theta^2}{r} \qquad (6.4a)$$

$$g_\theta - \frac{1}{\rho r}\frac{\partial p}{\partial\theta} = a_\theta = \frac{\partial V_\theta}{\partial t} + V_r\frac{\partial V_\theta}{\partial r} + \frac{V_\theta}{r}\frac{\partial V_\theta}{\partial\theta} + V_z\frac{\partial V_\theta}{\partial z} + \frac{V_r V_\theta}{r} \qquad (6.4b)$$

$$g_z - \frac{1}{\rho}\frac{\partial p}{\partial z} = a_z = \frac{\partial V_z}{\partial t} + V_r\frac{\partial V_z}{\partial r} + \frac{V_\theta}{r}\frac{\partial V_z}{\partial\theta} + V_z\frac{\partial V_z}{\partial z} \qquad (6.4c)$$

If the z axis is directed vertically upward, then $g_r = g_\theta = 0$ and $g_z = -g$.

In Chapter 3 we found that if a fluid is accelerated so that there is no relative motion between adjacent layers, the fluid moves without deformation and no shear stresses occur. We were able to determine the pressure variation within the fluid by applying the equations of motion to an appropriate free body. We considered two specific cases. For the case of rectilinear acceleration, we derived the differential equation of motion, Eq. 3.16; in Example Problem 3.8 we applied the equation to a tank of water moving as a rigid body. In Example Problem 3.9 we considered the case of a liquid undergoing steady rotation about a vertical axis. It is left to you as an exercise to show that the use of Euler's equations to solve Example Problems 3.8 and 3.9 leads to results identical to those obtained previously.

6-2 EULER'S EQUATIONS IN STREAMLINE COORDINATES

In Chapter 2 we pointed out that streamlines, drawn tangent to the velocity vectors at every point in the flow field, provide a convenient graphical representation. In steady flow a fluid particle will move along a streamline because, for steady flow, pathlines and streamlines coincide. Thus, in describing the motion of a fluid particle in a steady flow, the distance along a streamline is a logical coordinate to use in writing the equations of motion. "Streamline coordinates" also may be used to describe unsteady flow. Streamlines in unsteady flow give a graphical representation of the instantaneous velocity field.

For simplicity, consider the flow in the yz plane shown in Fig. 6.1. The equations of motion are to be written in terms of the coordinate s, distance along a streamline, and the coordinate n, distance normal to the streamline. Since the velocity vector must be tangent to the streamline, the velocity field is given by $\vec{V} = \vec{V}(s, t)$. The pressure at the center of the fluid element is p. If we apply Newton's second law in the streamwise (the s) direction to the fluid element of volume $ds\,dn\,dx$, then neglecting viscous forces we obtain

$$\left(p - \frac{\partial p}{\partial s}\frac{ds}{2}\right)dn\,dx - \left(p + \frac{\partial p}{\partial s}\frac{ds}{2}\right)dn\,dx - \rho g\sin\beta\,ds\,dn\,dx = \rho a_s\,ds\,dn\,dx$$

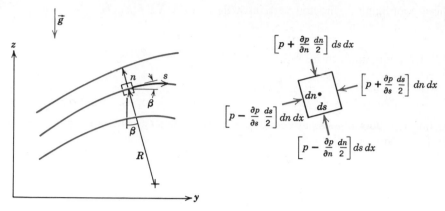

Fig. 6.1   Fluid particle moving along a streamline.

where β is the angle between the tangent to the streamline and the horizontal, and a_s is the acceleration of the fluid particle along the streamline. Simplifying the equation, we obtain

$$-\frac{\partial p}{\partial s} - \rho g \sin\beta = \rho a_s$$

Since $\sin\beta = \partial z/\partial s$, we can write

$$-\frac{1}{\rho}\frac{\partial p}{\partial s} - g\frac{\partial z}{\partial s} = a_s$$

Along any streamline $V_s = V_s(s, t)$, and the total acceleration of a fluid particle in the streamwise direction is given by

$$a_s = \frac{DV_s}{Dt} = \frac{\partial V_s}{\partial t} + V_s\frac{\partial V_s}{\partial s}$$

The velocity is tangent to the streamline, so the subscript, s, on V_s is redundant and can be dropped. Euler's equation in the streamwise direction with the z axis directed vertically is then

$$-\frac{1}{\rho}\frac{\partial p}{\partial s} - g\frac{\partial z}{\partial s} = \frac{\partial V}{\partial t} + V\frac{\partial V}{\partial s} \qquad (6.5a)$$

For steady flow, and neglecting body forces, Euler's equation in the streamwise direction reduces to

$$\frac{1}{\rho}\frac{\partial p}{\partial s} = -V\frac{\partial V}{\partial s} \qquad (6.5b)$$

which indicates that a decrease in velocity is accompanied by an increase in pressure and conversely.[1]

To obtain Euler's equation in a direction normal to the streamlines, we apply Newton's second law in the n direction to the fluid element. Again, neglecting

[1] The relationship between variations in pressure and velocity in the streamwise direction for steady incompressible inviscid flow is illustrated in the NCFMF film, *Pressure Fields and Fluid Acceleration*, A. H. Shapiro, principal.

viscous forces, we obtain

$$\left(p - \frac{\partial p}{\partial n}\frac{dn}{2}\right)ds\,dx - \left(p + \frac{\partial p}{\partial n}\frac{dn}{2}\right)ds\,dx - \rho g\cos\beta\,dn\,dx\,ds = \rho a_n\,dn\,dx\,ds$$

where β is the angle between the n direction and the vertical, and a_n is the acceleration of the fluid particle in the n direction. Simplifying the equation, we obtain

$$-\frac{\partial p}{\partial n} - \rho g\cos\beta = \rho a_n$$

Since $\cos\beta = \partial z/\partial n$, we write

$$-\frac{1}{\rho}\frac{\partial p}{\partial n} - g\frac{\partial z}{\partial n} = a_n$$

The normal acceleration of the fluid element is toward the center of curvature of the streamline, in the minus n direction; thus in the coordinate system of Fig. 6.1, the familiar centripetal acceleration is written

$$a_n = \frac{-V^2}{R}$$

for steady flow,[2] where R is the radius of curvature of the streamline. Then, Euler's equation normal to the streamline is written for steady flow as

$$\frac{1}{\rho}\frac{\partial p}{\partial n} + g\frac{\partial z}{\partial n} = \frac{V^2}{R} \tag{6.6a}$$

For steady flow in a horizontal plane, Euler's equation normal to a streamline becomes

$$\frac{1}{\rho}\frac{\partial p}{\partial n} = \frac{V^2}{R} \tag{6.6b}$$

Equation 6.6b indicates that pressure increases in the direction outward from the center of curvature of the streamlines.[3] In regions where the streamlines are straight, the radius of curvature, R, is infinite and there is no pressure variation normal to the streamlines.

EXAMPLE 6.1—Flow in a Bend

The flow rate of air at standard conditions in a flat duct is to be determined by installing pressure taps across a bend. The duct is 0.3 m deep and 0.1 m wide. The inner radius of the bend is 0.25 m. If the measured pressure difference between the taps is 40 mm of water, compute the approximate flow rate, assuming the velocity is uniform across the bend.

---

[2] If the flow were not steady, the streamline pattern would change with time. In that case,

$$a_n = -\frac{V_s^2}{R} + \frac{\partial V_n}{\partial t}$$

[3] The effect of streamline curvature on the pressure gradient normal to a streamline is illustrated in the NCFMF film, *Pressure Fields and Fluid Acceleration*, A.H. Shapiro, principal.

EXAMPLE PROBLEM 6.1

GIVEN: Flow through duct bend as shown.

$$p_2 - p_1 = \rho_{\text{H}_2\text{O}} g \, \Delta h,$$

where $\Delta h = 40$ mm H_2O.
Flow uniform. Air at STP.

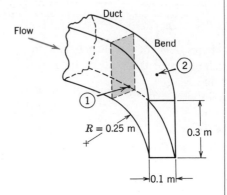

FIND: Q.

SOLUTION:
Apply Euler's n component equation across flow streamlines.

Basic equation: $\dfrac{\partial p}{\partial r} = \dfrac{\rho V^2}{r}$

Assumptions: (1) Frictionless flow
(2) Incompressible flow
(3) Uniform flow at measurement section

For this flow, $p = p(r)$, so

$$\frac{\partial p}{\partial r} = \frac{dp}{dr} = \frac{\rho V^2}{r} \qquad \text{or} \qquad dp = \rho V^2 \frac{dr}{r}$$

Integrating gives

$$p_2 - p_1 = \rho V^2 \ln r \bigg]_{r_1}^{r_2} = \rho V^2 \ln \frac{r_2}{r_1}$$

Then

$$V = \left[\frac{p_2 - p_1}{\rho \ln(r_2/r_1)} \right]^{1/2}$$

But $\Delta p = p_2 - p_1 = \rho_{\text{H}_2\text{O}} g \, \Delta h$, so

$$V = \left[\frac{\rho_{\text{H}_2\text{O}} g \, \Delta h}{\rho \ln(r_2/r_1)} \right]^{1/2}$$

$$= \left[999 \, \frac{\text{kg}}{\text{m}^3} \times 9.81 \, \frac{\text{m}}{\text{sec}^2} \times 0.04 \, \text{m} \times \frac{\text{m}^3}{1.23 \, \text{kg}} \times \frac{1}{\ln(0.35 \, \text{m}/0.25 \, \text{m})} \right]^{1/2}$$

$$V = 30.8 \text{ m/sec}$$

For uniform flow

$$Q = VA = 30.8 \, \frac{\text{m}}{\text{sec}} \times 0.1 \, \text{m} \times 0.3 \, \text{m} = 0.924 \text{ m}^3/\text{sec} \qquad\qquad Q$$

$\left\{\begin{array}{l}\text{In actual applications, the velocity profile in a channel bend will not be uniform. The} \\ \text{velocity profile in a bend tends toward a free vortex (irrotational) profile in which the} \\ \text{velocity varies inversely with radius (Section 6-7.4).}\end{array}\right\}$

6-3 BERNOULLI EQUATION—INTEGRATION OF EULER'S EQUATION ALONG A STREAMLINE FOR STEADY FLOW

We have written the momentum equations and the continuity equation in differential form. Theoretically, for an incompressible inviscid flow, these equations can be solved to give the complete velocity and pressure fields. (If density is not constant, an additional thermodynamic relation for density is required.) Although, in theory, the equations can be solved, the solution for a particular flow field may be very involved. However, we can integrate Euler's equation readily for steady flow along a streamline. The differential control volume analysis of Section 4-4.1 led to a differential equation (Eq. 4.24), which when integrated, led to a form of the Bernoulli equation. In order to give added physical insight about restrictions on the results, two additional derivations of the Bernoulli equation are presented.

6-3.1 Derivation Using Streamline Coordinates

Euler's equation for steady flow along a streamline is given by

$$-\frac{1}{\rho}\frac{\partial p}{\partial s} - g\frac{\partial z}{\partial s} = V\frac{\partial V}{\partial s} \qquad (6.7)$$

If a fluid particle moves a distance, ds, along a streamline, then

$$\frac{\partial p}{\partial s}ds = dp \qquad \text{(the change in pressure along } s\text{)}$$

$$\frac{\partial z}{\partial s}ds = dz \qquad \text{(the change in elevation along } s\text{)}$$

$$\frac{\partial V}{\partial s}ds = dV \qquad \text{(the change in velocity along } s\text{)}$$

Thus, after multiplying Eq. 6.7 by ds, we can write

$$-\frac{dp}{\rho} - g\,dz = V\,dV \qquad \text{(along } s\text{)}$$

or

$$\frac{dp}{\rho} + V\,dV + g\,dz = 0 \qquad \text{(along } s\text{)}$$

Integration of this equation gives

$$\int \frac{dp}{\rho} + \frac{V^2}{2} + gz = \text{constant} \qquad \text{(along } s\text{)} \qquad (6.8)$$

Before Eq. 6.8 can be applied, we must specify the relation between the pressure, p, and the density, ρ. For the special case of incompressible flow, $\rho = \text{constant}$, and Eq. 6.8 becomes the Bernoulli equation,

$$\frac{p}{\rho} + \frac{V^2}{2} + gz = \text{constant} \qquad (6.9)$$

Restrictions: (1) Steady flow
                (2) Incompressible flow
                (3) Frictionless flow
                (4) Flow along a streamline

The Bernoulli equation is a powerful and useful equation because it relates pressure changes to velocity and elevation changes along a streamline. However, it gives correct results only when applied to a flow situation where all four of the restrictions are reasonable. Keep the restrictions firmly in mind whenever you consider using the Bernoulli equation. (In general, the Bernoulli constant in Eq. 6.9 has different values along different streamlines.[4])

**6-3.2 Derivation Using Rectangular Coordinates

The vector form of Euler's equation, Eq. 6.3, also can be integrated along a stream-line. Since the velocity field, $\vec{V}$, is specified in terms of the rectangular coordinates, x, y, z, it is convenient to use vector notation. We shall restrict the derivation to steady flow; thus, the end result of our effort should be Eq. 6.8.

For steady flow, Euler's equation in rectangular coordinates can be expressed as

$$-\frac{1}{\rho}\nabla p - g\,\nabla z = \frac{D\vec{V}}{Dt} = u\frac{\partial \vec{V}}{\partial x} + v\frac{\partial \vec{V}}{\partial y} + w\frac{\partial \vec{V}}{\partial z} = (\vec{V}\cdot\nabla)\vec{V} \qquad (6.10)$$

For steady flow the velocity field is given by $\vec{V} = \vec{V}(x, y, z)$. The streamlines are lines drawn in the flow field tangent to the velocity vector at every point. Recall again that for steady flow, streamlines, pathlines, and streaklines coincide. The motion of a particle along a streamline is governed by Eq. 6.10. In the time interval dt, the particle moves a distance $d\vec{s}$, along the streamline.

If we take the dot product of the terms in Eq. 6.10 with the distance, $d\vec{s}$, along the streamline, we obtain a scalar equation relating the pressure, p, speed, V, and elevation, z, along the streamline. Taking the dot product of $d\vec{s}$ with Eq. 6.10 gives

$$-\frac{1}{\rho}\nabla p \cdot d\vec{s} - g\nabla z \cdot d\vec{s} = (\vec{V}\cdot\nabla)\vec{V}\cdot d\vec{s} \qquad (6.11)$$

where

$$d\vec{s} = dx\,\hat{i} + dy\,\hat{j} + dz\,\hat{k} \qquad \text{(along } s)$$

Now we evaluate each of the three terms in Eq. 6.11

$$-\frac{1}{\rho}\nabla p \cdot d\vec{s} = -\frac{1}{\rho}\left[\hat{i}\frac{\partial p}{\partial x} + \hat{j}\frac{\partial p}{\partial y} + \hat{k}\frac{\partial p}{\partial z}\right] \cdot [dx\,\hat{i} + dy\,\hat{j} + dz\,\hat{k}]$$

$$= -\frac{1}{\rho}\left[\frac{\partial p}{\partial x}dx + \frac{\partial p}{\partial y}dy + \frac{\partial p}{\partial z}dz\right] \qquad \text{(along } s)$$

---

[4] For the case of irrotational flow, the constant has a single value throughout the entire flow field (Section 6-7.1).

** This section may be omitted without loss of continuity in the text material.

$$-\frac{1}{\rho}\nabla p \cdot d\vec{s} = -\frac{1}{\rho}dp \qquad \text{(along } s)$$

and

$$-g\,\nabla z \cdot d\vec{s} = -g\hat{k} \cdot [dx\hat{i} + dy\hat{j} + dz\hat{k}]$$

$$-g\,\nabla z \cdot d\vec{s} = -g\,dz \qquad \text{(along } s)$$

Using a vector identity,[5] we can write the third term as

$$(\vec{V} \cdot \nabla)\vec{V} \cdot d\vec{s} = [\tfrac{1}{2}\nabla(\vec{V} \cdot \vec{V}) - \vec{V} \times (\nabla \times \vec{V})] \cdot d\vec{s}$$

$$= \{\tfrac{1}{2}\nabla(\vec{V} \cdot \vec{V})\} \cdot d\vec{s} - \{\vec{V} \times (\nabla \times \vec{V})\} \cdot d\vec{s}$$

The last term on the right side of this equation is zero, since $\vec{V}$ is parallel to $d\vec{s}$. Consequently,

$$(\vec{V} \cdot \nabla)\vec{V} \cdot d\vec{s} = \tfrac{1}{2}\nabla(\vec{V} \cdot \vec{V}) \cdot d\vec{s} = \tfrac{1}{2}\nabla(V^2) \cdot d\vec{s} \qquad \text{(along } s)$$

$$= \frac{1}{2}\left[\hat{i}\frac{\partial V^2}{\partial x} + \hat{j}\frac{\partial V^2}{\partial y} + \hat{k}\frac{\partial V^2}{\partial z}\right] \cdot [dx\hat{i} + dy\hat{j} + dz\hat{k}]$$

$$= \frac{1}{2}\left[\frac{\partial V^2}{\partial x}dx + \frac{\partial V^2}{\partial y}dy + \frac{\partial V^2}{\partial z}dz\right]$$

$$(\vec{V} \cdot \nabla)\vec{V} \cdot d\vec{s} = \tfrac{1}{2}d(V^2) \qquad \text{(along } s)$$

Substituting these three terms into Eq. 6.11 yields

$$\frac{dp}{\rho} + \frac{1}{2}d(V^2) + g\,dz = 0 \qquad \text{(along } s)$$

Integrating this equation, we obtain

$$\int \frac{dp}{\rho} + \frac{V^2}{2} + gz = \text{constant} \qquad \text{(along } s)$$

If the density is constant, we obtain the Bernoulli equation

$$\frac{p}{\rho} + \frac{V^2}{2} + gz = \text{constant}$$

As expected, we see that the last two equations are identical to Eqs. 6.8 and 6.9 derived previously using streamline coordinates. The Bernoulli equation, derived using rectangular coordinates, is still subject to the restrictions: (1) steady flow, (2) incompressible flow, (3) frictionless flow, and (4) flow along a streamline.

6-3.3 Static, Stagnation, and Dynamic Pressures

The pressure, p, which we have used in deriving the Bernoulli equation, Eq. 6.9, is the thermodynamic pressure; it is commonly called the static pressure. The static pressure is that pressure which would be measured by an instrument moving with the

---

[5] The vector identity

$$(\vec{V} \cdot \nabla)\vec{V} = \tfrac{1}{2}\nabla(\vec{V} \cdot \vec{V}) - \vec{V} \times (\nabla \times \vec{V})$$

may be verified by expanding each side into components.

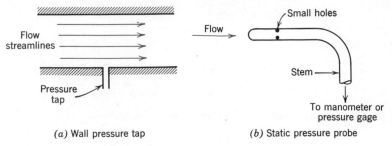

(a) Wall pressure tap   (b) Static pressure probe

Fig. 6.2 Measurement of static pressure.

flow. However, such a measurement is rather difficult to make in a practical situation! How do we measure static pressure experimentally?

In Section 6-2 we showed that there was no pressure variation normal to straight streamlines. This fact makes it possible to measure the static pressure in a flowing fluid using a wall pressure "tap," placed in a region where the flow streamlines are straight, as shown in Fig. 6.2a. The pressure tap is a small hole, drilled carefully in the wall, with its axis perpendicular to the surface. If the hole is perpendicular to the duct wall and free from burrs, accurate measurements of static pressure can be made by connecting the tap to a suitable measuring instrument.

In a fluid stream far from a wall, or where streamlines are curved, accurate static pressure measurements can be made by careful use of a static pressure probe, shown in Fig. 6.2b. Such probes must be designed so that the measuring holes are placed correctly with respect to the probe tip and stem to avoid erroneous results. In use, the measuring section must be aligned with the local flow direction.

Static pressure probes, such as that shown in Fig 6.2b, and in a variety of other forms, are available commercially in sizes as small as 1.5 mm ($\frac{1}{16}$ in.) in diameter.

The stagnation pressure is obtained when a flowing fluid is decelerated to zero speed by a frictionless process. In incompressible flow, the Bernoulli equation can be used to relate changes in speed and pressure along a streamline for such a process. Neglecting elevation differences, Eq. 6.9 becomes

$$\frac{p}{\rho} + \frac{V^2}{2} = \text{constant}$$

If the static pressure is p at a point in the flow where the speed is V, then the stagnation pressure, p_0, where the stagnation speed, V_0, is zero, may be computed from

$$\frac{p_0}{\rho} + \frac{\overset{=0}{\cancel{V_0^2}}}{\cancel{2}} = \frac{p}{\rho} + \frac{V^2}{2}$$

or

$$p_0 = p + \frac{1}{2}\rho V^2 \tag{6.12}$$

Equation 6.12 is a mathematical statement of the definition of stagnation pressure, valid for incompressible flow. The term $\frac{1}{2}\rho V^2$ generally is called the *dynamic pressure*. Solving for the dynamic pressure gives

$$\frac{1}{2}\rho V^2 = p_0 - p$$

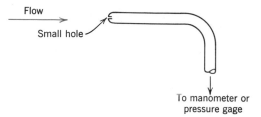

Fig. 6.3   Measurement of stagnation pressure.

and for the speed

$$V = \sqrt{\frac{2(p_0 - p)}{\rho}} \tag{6.13}$$

Thus, if the stagnation pressure and the static pressure could be measured at a point, Eq. 6.13 would give the local flow speed.

Stagnation pressure is measured in the laboratory using a probe with a hole that faces directly upstream as shown in Fig. 6.3. Such a probe is called a stagnation pressure probe, or pitot (pronounced *pea-toe*) tube. Again, the measuring section must be aligned with the local flow direction.

We have seen that static pressure at a point can be measured with a static pressure tap or probe (Fig. 6.2). If we knew the stagnation pressure at the same point, then the flow speed could be computed from Eq. 6.13. Two possible experimental setups are shown in Fig. 6.4.

In Fig 6.4a, the static pressure corresponding to point A is read from the wall static pressure tap. The stagnation pressure is measured directly at A by the total head tube, as shown. (The stem of the total head tube is placed downstream from the measurement location to minimize disturbance of the local flow.)

Two probes often are combined, as in the pitot-static tube shown in Fig. 6.4b. The inner tube is used to measure the stagnation pressure at point B, while the static pressure at C is sensed by the small holes in the outer tube. In flow fields where the static pressure variation in the streamwise direction is small, the pitot-static tube may be used to infer the speed at point B in the flow, by assuming $p_B = p_C$, and using Eq. 6.13. (Note that when $p_B \neq p_C$, this procedure will give erroneous results.)

Remember that the Bernoulli equation applies only for incompressible flow (Mach number, $M \le 0.3$). The definition and calculation of the stagnation pressure for compressible flow will be discussed in Section 12-3.1.

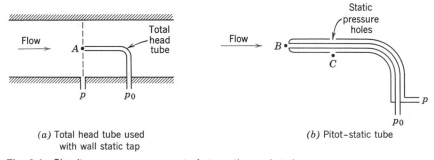

(a) Total head tube used with wall static tap

(b) Pitot–static tube

Fig. 6.4   Simultaneous measurement of stagnation and static pressures.

EXAMPLE 6.2—Pitot Tube

A pitot probe is inserted in an air flow (at STP) to measure the flow speed. The tube is inserted so that it points upstream into the flow and the pressure sensed by the probe is the stagnation pressure. The static pressure is measured at the same location in the flow, using a wall pressure tap. If the pressure difference is 30 mm of mercury, determine the flow speed.

EXAMPLE PROBLEM 6.2

GIVEN: A pitot tube inserted in a flow as shown. The flowing fluid is air and the manometer liquid is mercury.

FIND: The flow speed.

SOLUTION:

Basic equation: $\dfrac{p}{\rho} + \dfrac{V^2}{2} + gz = \text{constant}$

Assumptions: (1) Steady flow
(2) Incompressible flow
(3) Flow along a streamline
(4) Frictionless deceleration along stagnation streamline

Writing Bernoulli's equation along the stagnation streamline, taking $\Delta z = 0$, yields

$$\frac{p_0}{\rho} = \frac{p}{\rho} + \frac{V^2}{2}$$

p_0 is the stagnation pressure at the tube opening where the speed has been reduced, without friction, to zero. Solving for V gives

$$V = \sqrt{\frac{2(p_0 - p)}{\rho_{\text{air}}}}$$

From the diagram,

$$p_0 - p = \rho_{\text{Hg}} g h = \rho_{\text{H}_2\text{O}} g h (\text{SG}_{\text{Hg}})$$

and

$$V = \sqrt{\frac{2\rho_{\text{H}_2\text{O}} g h (\text{SG}_{\text{Hg}})}{\rho_{\text{air}}}}$$

$$= \sqrt{2 \times \frac{1000 \text{ kg}}{\text{m}^3} \times \frac{9.81 \text{ m}}{\text{sec}^2} \times 30 \text{ mm} \times \frac{\text{m}}{1000 \text{ mm}} \times 13.6 \times \frac{\text{m}^3}{1.23 \text{ kg}}}$$

$V = 80.8 \text{ m/sec}$ \hfill V

At $T = 20$ C, the speed of sound in air is 343 m/sec. Hence, $M = 0.236$ and the assumption of incompressible flow is valid.

{This problem illustrates the use of a pitot tube to determine the speed at a point.}

6-3.4 Applications

The Bernoulli equation can be applied between any two points on a streamline provided that the other three restrictions are satisfied. The result is

$$\frac{p_1}{\rho} + \frac{V_1^2}{2} + g z_1 = \frac{p_2}{\rho} + \frac{V_2^2}{2} + g z_2 \qquad (6.14)$$

where subscripts 1 and 2 represent any two points on a streamline. Applications of Eqs. 6.9 and 6.14 to typical flow problems are illustrated in Example Problems 6.3 through 6.5.

In some situations, the flow appears unsteady from one reference frame, but steady from another, which translates in the flow. Since the Bernoulli equation was derived by integrating Newton's second law for a fluid particle, it can be applied in any inertial reference frame (see the discussion of translating frames in Section 4-4.2). The procedure is illustrated in Example Problem 6.6.

EXAMPLE 6.3—Nozzle Flow
Air flows steadily and at low speed through a horizontal nozzle, discharging to the atmosphere. At the nozzle inlet, the area is 0.1 m². At the nozzle exit, the area is 0.02 m². The flow is essentially incompressible, and frictional effects are negligible. Determine the gage pressure required at the nozzle inlet to produce an outlet speed of 50 m/sec.

EXAMPLE PROBLEM 6.3

GIVEN: Flow through a nozzle, as shown. The air flow is steady, incompressible, and frictionless.

FIND: $p_1 - p_{atm}$.

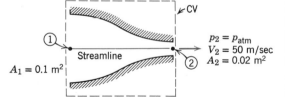

$A_1 = 0.1$ m²

$p_2 = p_{atm}$
$V_2 = 50$ m/sec
$A_2 = 0.02$ m²

SOLUTION:
Basic equations:
$$\frac{p_1}{\rho} + \frac{V_1^2}{2} + g z_1 = \frac{p_2}{\rho} + \frac{V_2^2}{2} + g z_2$$

$$= 0(1)$$

$$0 = \frac{\partial}{\partial t} \int_{CV} \rho \, d\Psi + \int_{CS} \rho \vec{V} \cdot d\vec{A}$$

Assumptions: (1) Steady flow
           (2) Incompressible flow
           (3) Frictionless flow
           (4) Flow along a streamline
           (5) $z_1 = z_2$
           (6) Uniform flow at sections ① and ②

The maximum speed of 50 m/sec is well below the value of 100 m/sec, which corresponds to a Mach number $M \approx 0.3$ in standard air. Hence, the flow may be treated as incompressible.

Apply the Bernoulli equation along a streamline between points ① and ② to evaluate p_1. Then

$$p_1 - p_{atm} = p_1 - p_2 = \frac{\rho}{2}(V_2^2 - V_1^2)$$

Apply the continuity equation to determine V_1,

$$0 = \{-|\rho V_1 A_1|\} + \{|\rho V_2 A_2|\} \qquad \text{or} \qquad V_1 A_1 = V_2 A_2$$

so that

$$V_1 = V_2 \frac{A_2}{A_1} = 50 \ \frac{\text{m}}{\text{sec}} \times \frac{0.02 \ \text{m}^2}{0.1 \ \text{m}^2} = 10 \ \text{m/sec}$$

For air at standard conditions, $\rho = 1.23 \ \text{kg/m}^3$. Then

$$p_1 - p_{atm} = \frac{\rho}{2}(V_2^2 - V_1^2)$$

$$= \frac{1}{2} \times 1.23 \ \frac{\text{kg}}{\text{m}^3} \left[(50)^2 \ \frac{\text{m}^2}{\text{sec}^2} - (10)^2 \ \frac{\text{m}^2}{\text{sec}^2} \right] \frac{\text{N} \cdot \text{sec}^2}{\text{kg} \cdot \text{m}}$$

$$p_1 - p_{atm} = 1.48 \ \text{kPa} \qquad\qquad\qquad p_1 - p_{atm}$$

{ This problem illustrates a typical application of the Bernoulli equation. Note that if the flow streamlines are straight at the nozzle inlet and exit, the pressure will be uniform at those sections. }

EXAMPLE 6.4—Flow through a Siphon

A U-tube acts as a water siphon. The bend in the tube is 1 m above the water surface; the tube outlet is 7 m below the water surface. The fluid issues from the bottom of the siphon as a free jet at atmospheric pressure. If the flow is frictionless as a first approximation, determine (after listing the necessary assumptions) the speed of the free jet and the absolute pressure of the fluid in the bend.

EXAMPLE PROBLEM 6.4

GIVEN: Water flowing through a siphon as shown.

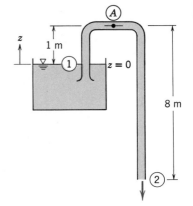

FIND: (a) Speed of water leaving as a free jet.
(b) Pressure at point Ⓐ in the flow.

SOLUTION:

Basic equation: $\dfrac{p}{\rho} + \dfrac{V^2}{2} + gz = \text{constant}$

Assumptions: (1) Neglect friction
(2) Steady flow
(3) Incompressible flow
(4) Flow along a streamline
(5) Reservoir is large compared to pipe

Apply the Bernoulli equation between points ① and ②.

$$\frac{p_1}{\rho} + \frac{V_1^2}{2} + gz_1 = \frac{p_2}{\rho} + \frac{V_2^2}{2} + gz_2$$

Since $\text{area}_{\text{reservoir}} \gg \text{area}_{\text{pipe}}$, $V_1 \approx 0$. Also $p_1 = p_2 = p_{\text{atm}}$, so

$$g z_1 = \frac{V_2^2}{2} + g z_2 \quad \text{and} \quad V_2^2 = 2g(z_1 - z_2)$$

$$V_2 = \sqrt{2g(z_1 - z_2)} = \sqrt{2 \times 9.81 \, \frac{\text{m}}{\text{sec}^2} \times 7 \, \text{m}} = 11.7 \text{ m/sec} \qquad\qquad V_2$$

To determine the pressure at location Ⓐ, we write the Bernoulli equation between ① and Ⓐ.

$$\frac{p_1}{\rho} + \frac{V_1^2}{2} + g z_1 = \frac{p_A}{\rho} + \frac{V_A^2}{2} + g z_A$$

Again $V_1 \approx 0$ and from conservation of mass $V_A = V_2$. Hence

$$\frac{p_A}{\rho} = \frac{p_1}{\rho} + g z_1 - \frac{V_2^2}{2} - g z_A = \frac{p_1}{\rho} + g(z_1 - z_A) - \frac{V_2^2}{2}$$

$$p_A = p_1 + \rho g(z_1 - z_A) - \rho \frac{V_2^2}{2}$$

$$= \frac{1.01 \times 10^5 \, \text{N}}{\text{m}^2} + \frac{999 \, \text{kg}}{\text{m}^3} \times 9.81 \, \frac{\text{m}}{\text{sec}^2} \times (-1 \, \text{m}) \times \frac{\text{N} \cdot \text{sec}^2}{\text{kg} \cdot \text{m}}$$

$$- \frac{1}{2} \times \frac{999 \, \text{kg}}{\text{m}^3} \times \frac{(11.7)^2 \, \text{m}^2}{\text{sec}^2} \times \frac{\text{N} \cdot \text{sec}^2}{\text{kg} \cdot \text{m}}$$

$$p_A = 22.8 \text{ kPa (abs) or } -78.5 \text{ kPa (gage)} \qquad\qquad p_A$$

$\left\{ \begin{array}{l} \text{This problem illustrates a straightforward application of the Bernoulli equation with} \\ \text{elevation changes included.} \end{array} \right\}$

EXAMPLE 6.5—Flow under a Sluice Gate

Water flows under a sluice gate on a horizontal bed at the inlet to a flume. Upstream of the gate, the water depth is 1.5 ft and the speed is negligible. At the vena contracta downstream of the gate, the flow streamlines are straight and the depth is 2 in. Hydrostatic pressure distributions and uniform flow may be assumed at each section; friction is negligible. Determine the flow speed downstream of the gate, and the discharge in cubic feet per second per foot of width.

EXAMPLE PROBLEM 6.5

GIVEN: Flow of water under a sluice gate. Flow is frictionless, uniform at each section, and the pressure distribution is hydrostatic at sections ① and ②.

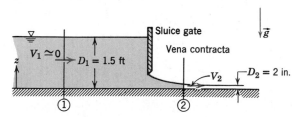

FIND: (a) V_2.

(b) Q in ft³/sec/ft of width.

SOLUTION:
The flow satisfies all conditions necessary to apply the Bernoulli equation. The question is, what streamline do we use?

Basic equation: $\dfrac{p_1}{\rho} + \dfrac{V_1^2}{2} + gz_1 = \dfrac{p_2}{\rho} + \dfrac{V_2^2}{2} + gz_2$

Assumptions: (1) Steady flow
(2) Incompressible flow
(3) Frictionless flow
(4) Flow along a streamline
(5) Uniform flow at each section
(6) Hydrostatic pressure distribution

From assumption 6,

$$\frac{dp}{dz} = -\rho g \quad \text{so that} \quad p = p_{atm} + \rho g(D - z) \quad \text{or} \quad \frac{p}{\rho} = \frac{p_{atm}}{\rho} + g(D - z)$$

Substituting this relation into the Bernoulli equation gives

$$\frac{p_{atm}}{\rho} + g(D_1 - z_1) + \frac{V_1^2}{2} + gz_1 = \frac{p_{atm}}{\rho} + g(D_2 - z_2) + \frac{V_2^2}{2} + gz_2$$

or

$$\frac{V_1^2}{2} + gD_1 = \frac{V_2^2}{2} + gD_2$$

This result implies that $V^2/2 + gD = $ constant, and the constant has the same value along *any* streamline for this flow. Solving for V_2 yields

$$V_2 = \sqrt{2g(D_1 - D_2) + V_1^2}$$

But $V_1^2 \approx 0$, so

$$V_2 = \sqrt{2g(D_1 - D_2)} = \sqrt{2 \times 32.2 \frac{\text{ft}}{\text{sec}^2} \left(1.5 \text{ ft} - 2 \text{ in.} \times \frac{\text{ft}}{12 \text{ in.}}\right)}$$

$$V_2 = 9.27 \text{ ft/sec} \quad\longleftarrow\quad V_2$$

For uniform flow, $Q = VA = VDw$, or

$$\frac{Q}{w} = VD = V_2 D_2 = 9.27 \frac{\text{ft}}{\text{sec}} \times 2 \text{ in.} \times \frac{\text{ft}}{12 \text{ in.}} = 1.55 \text{ ft}^2/\text{sec}$$

$$\frac{Q}{w} = 1.55 \text{ ft}^3/\text{sec per foot of width} \quad\longleftarrow\quad \frac{Q}{w}$$

EXAMPLE 6.6—Bernoulli Equation in Translating Reference Frame
A light plane flies at 150 km/hr in standard air at an altitude of 1000 m. Determine the stagnation pressure at the leading edge of the wing. At a certain point close to the wing, the air speed *relative* to the wing is 60 m/sec. Compute the pressure at this point.

EXAMPLE PROBLEM 6.6

GIVEN: Aircraft in flight at 150 km/hr at 1000 m altitude in standard air.

$V_{air} = 0$

$V_B = 60$ m/sec
(relative to wing)

A   B

$V_w = 150$ km/hr

Observer

FIND: Stagnation pressure, p_{0_A}, at point A and static pressure, p_B, at point B.

SOLUTION:
Flow is unsteady when observed from a fixed frame, that is, by an observer on the ground. However, an observer *on* the wing sees the following steady flow:

Observer

B $V_B = 60$ m/sec
A

$V_{air} = V_w = 150$ km/hr

At $z = 1000$ m in standard air, the temperature is 281 K and the speed of sound is 336 m/sec. Hence at point B, $M_B = V_B/c = 0.178$. This is less than 0.3, so the flow may be treated as incompressible. Thus the Bernoulli equation can be applied along a streamline in the moving observer's inertial reference frame.

Basic equation: $\dfrac{p_{air}}{\rho} + \dfrac{V_{air}^2}{2} + g z_{air} = \dfrac{p_A}{\rho} + \dfrac{V_A^2}{2} + g z_A = \dfrac{p_B}{\rho} + \dfrac{V_B^2}{2} + g z_B$

Assumptions: (1) Steady flow
(2) Incompressible flow ($V < 100$ m/sec)
(3) Frictionless flow
(4) Flow along a streamline
(5) Neglect Δz

Values for pressure and density may be found from Table A.3. Thus, at 1000 m, $p/p_{SL} = 0.8870$ and $\rho/\rho_{SL} = 0.9075$. Consequently,

$$p = 0.8870\, p_{SL} = \overset{0.8870}{} \times 1.01 \times 10^5\ \frac{N}{m^2} = 8.96 \times 10^4\ N/m^2$$

and

$$\rho = 0.9075 \rho_{SL} = \overset{0.9075}{} \times 1.23\ \frac{kg}{m^3} = 1.12\ kg/m^3$$

Since the speed is $V_A = 0$ at the stagnation point, then

$$p_{0_A} = p_{air} + \frac{1}{2}\rho V_{air}^2$$

$$= \frac{8.96 \times 10^4\ N}{m^2} + \frac{1}{2} \times 1.12\ \frac{kg}{m^3} \left(\frac{150\ km}{hr} \times \frac{1000\ m}{km} \times \frac{hr}{3600\ sec}\right)^2 \times \frac{N \cdot sec^2}{kg \cdot m}$$

$$p_{0_A} = 90.6\ kPa\ (abs) \hspace{3cm} p_{0_A}$$

Solving for the static pressure at B, we obtain

$$p_B = p_{air} + \frac{\rho}{2}(V_{air}^2 - V_B^2)$$

$$p_B = \frac{8.96 \times 10^4}{} \frac{N}{m^2}$$

$$+ \frac{1}{2} \times 1.12 \frac{kg}{m^3} \left[\left(150 \frac{km}{hr} \times 1000 \frac{m}{km} \times \frac{hr}{3600 \; sec} \right)^2 - (60)^2 \frac{m^2}{sec^2} \right] \frac{N \cdot sec^2}{kg \cdot m}$$

$$p_B = 88.6 \; kPa \; (abs) \hspace{8cm} p_B$$

6-3.5 Cautions on Use of the Bernoulli Equation

We have seen, in Example Problems 6.3 through 6.6, several situations where the Bernoulli equation may be applied because the restrictions on its use led to a reasonable flow model. However, in some situations you might be tempted to apply the Bernoulli equation where the restrictions are not satisfied. Some subtle cases that violate the restrictions are discussed briefly in this section.

Flow through the nozzle of Example Problem 6.3 was modeled well by the Bernoulli equation. Because the pressure gradient in a nozzle is favorable, there is no separation and the boundary layers on the walls remain thin. Friction has a negligible effect on the flow velocity profile, so one-dimensional flow is a good model. The velocity at any section can be calculated from the corresponding flow area.

A diverging passage or sudden expansion should not be modeled using the Bernoulli equation. Adverse pressure gradients cause rapid growth of boundary layers, severely distorted velocity profiles, and possible flow separation.[6] One-dimensional flow is a poor model for such flows. Because of area blockage due to boundary-layer growth, pressure rise in actual diffusers always is less than that predicted for inviscid one-dimensional flow.

The Bernoulli equation was a reasonable model for the siphon of Example Problem 6.4 because the entrance was well rounded, the bends were gentle, and the overall length was short. Flow separation, that can occur at inlets with sharp corners and in abrupt bends, causes the flow to depart from that predicted by a one-dimensional model and the Bernoulli equation. Frictional effects would not be negligible if the tube were long.

Example Problem 6.5 presented an open-channel flow situation analogous to a nozzle, for which the Bernoulli equation is a good flow model. The hydraulic jump[7] is an example of an open-channel flow with adverse pressure gradient. Flow through a hydraulic jump is mixed violently, making it impossible to identify streamlines. Thus the Bernoulli equation cannot be used to model flow through a hydraulic jump.

The Bernoulli equation cannot be applied *through* a machine such as a propeller, pump, or windmill. The equation was derived by integrating along a streamtube (Section 4-4.1) or a streamline (Section 6-3) in the absence of moving surfaces such as blades or vanes. It is impossible to have locally steady flow or to identify streamlines during flow through a machine. As suggested in Problem 4.114, the Bernoulli equation may be applied before and following a machine if the restrictions on its use are satisfied. However, it may not be applied through a machine or across an impeller.

Finally, compressibility must be considered for flow of gases. Density changes caused by dynamic compression due to motion may be neglected for engineering purposes if the local Mach number remains below about $M \approx 0.3$, as noted in Ex-

---

[6] See the NCFMF film, *Flow Visualization,* S. J. Kline, principal.
[7] See the NCFMF films, *Waves in Fluids,* A. E. Bryson, principal, and *Stratified Flow,* R. R. Long, principal, for examples of this behavior.

ample Problems 6.3 and 6.6. Temperature changes can cause significant changes in density of a gas, even for low-speed flow. Thus the Bernoulli equation could not be applied to air flow through a heating element (e.g., of a hand-held hair dryer) where significant changes in temperature occur.

6-4   RELATION BETWEEN THE FIRST LAW OF THERMODYNAMICS AND THE BERNOULLI EQUATION

The Bernoulli equation, Eq. 6.9, was obtained by integrating Euler's equation along a streamline for steady, incompressible, frictionless flow. Thus Eq. 6.9 was derived from the momentum equation for a fluid particle.

An equation identical in form to Eq. 6.9 (although requiring very different restrictions) may be obtained from the first law of thermodynamics. Our objective in this section is to reduce the energy equation to the form of the Bernoulli equation given by Eq. 6.9. Having arrived at this form, we then compare the restrictions on the two equations to help us understand more clearly the restrictions on the use of Eq. 6.9.

Consider steady flow in the absence of shear forces. We choose a control volume bounded by streamlines along its periphery. Such a control volume, shown in Fig. 6.5, often is called a *streamtube*.

Basic equation:

$$\overset{=\,0(1)}{\dot{Q}} - \overset{=\,0(2)}{\dot{W}_s} - \overset{=\,0(3)}{\dot{W}_{\text{shear}}} - \overset{=\,0(4)}{\dot{W}_{\text{other}}} = \frac{\partial}{\partial t}\int_{\text{CV}} e\rho\,d\forall + \int_{\text{CS}} (e + pv)\rho\vec{V}\cdot d\vec{A} \qquad (4.57)$$

$$e = u + \frac{V^2}{2} + gz$$

Restrictions: (1) $\dot{W}_s = 0$
 (2) $\dot{W}_{\text{shear}} = 0$
 (3) $\dot{W}_{\text{other}} = 0$
 (4) Steady flow
 (5) Uniform flow and properties at each section

Under these restrictions, Eq. 4.57 becomes

$$0 = \left(u_1 + p_1 v_1 + \frac{V_1^2}{2} + gz_1\right)\{-|\rho_1 V_1 A_1|\}$$

$$+ \left(u_2 + p_2 v_2 + \frac{V_2^2}{2} + gz_2\right)\{|\rho_2 V_2 A_2|\} - \dot{Q}$$

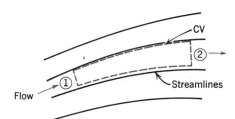

Fig. 6.5 Flow through a streamtube.

But from continuity under these restrictions,

$$0 = \underset{}{\overset{=0(4)}{\cancel{\frac{\partial}{\partial t} \int_{CV} \rho \, d\forall}}} + \int_{CS} \rho \vec{V} \cdot d\vec{A}$$

or

$$0 = \{-|\rho_1 V_1 A_1|\} + \{|\rho_2 V_2 A_2|\}$$

That is,

$$\dot{m} = \rho_1 V_1 A_1 = \rho_2 V_2 A_2$$

Also

$$\dot{Q} = \frac{\delta Q}{dt} = \frac{\delta Q}{dm} \frac{dm}{dt} = \frac{\delta Q}{dm} \dot{m}$$

Thus, from the energy equation,

$$0 = \left[\left(p_2 v_2 + \frac{V_2^2}{2} + g z_2 \right) - \left(p_1 v_1 + \frac{V_1^2}{2} + g z_1 \right) \right] \dot{m} + \left(u_2 - u_1 - \frac{\delta Q}{dm} \right) \dot{m}$$

or

$$p_1 v_1 + \frac{V_1^2}{2} + g z_1 = p_2 v_2 + \frac{V_2^2}{2} + g z_2 + \left(u_2 - u_1 - \frac{\delta Q}{dm} \right)$$

Under the restriction of incompressible flow, $v_1 = v_2 = 1/\rho$ and hence

$$\frac{p_1}{\rho} + \frac{V_1^2}{2} + g z_1 = \frac{p_2}{\rho} + \frac{V_2^2}{2} + g z_2 + \left(u_2 - u_1 - \frac{\delta Q}{dm} \right) \qquad (6.15)$$

Equation 6.15 would reduce to the Bernoulli equation if the term in parentheses were zero. Thus, under the additional restrictions,

(6)   Incompressible flow, $v_1 = v_2 = 1/\rho = \text{constant}$
(7) $(u_2 - u_1 - \delta Q/dm) = 0$

the energy equation reduces to

$$\frac{p_1}{\rho} + \frac{V_1^2}{2} + g z_1 = \frac{p_2}{\rho} + \frac{V_2^2}{2} + g z_2$$

or

$$\frac{p}{\rho} + \frac{V^2}{2} + g z = \text{constant} \qquad (6.16)$$

Equation 6.16 is identical in form to the Bernoulli equation, Eq. 6.9. The Bernoulli equation was derived from momentum considerations (Newton's second law), and is valid for steady, incompressible, frictionless flow along a streamline. Equation 6.16 was obtained by applying the first law of thermodynamics to a streamtube control volume, subject to restrictions 1 through 7 above. Thus the Bernoulli equation (Eq. 6.9) and the identical form of the energy equation (Eq. 6.16) were developed from entirely different models, coming from entirely different basic concepts, and involving different restrictions.

Note that restriction 7,

$$(u_2 - u_1) - \frac{\delta Q}{dm} = 0$$

was necessary to obtain the Bernoulli equation from the first law of thermodynamics. This restriction can be satisfied if $\delta Q/dm$ is zero (there is no heat transfer to the fluid) and $u_2 = u_1$ (there is no change in the internal thermal energy of the fluid). The restriction also is satisfied if $(u_2 - u_1)$ and $\delta Q/dm$ are nonzero provided that the two terms are equal. That this is true for incompressible frictionless flow is shown in Example Problem 6.7.

For the special case considered in this section it is true that the first law of thermodynamics reduces to the Bernoulli equation. It is important to emphasize that the Bernoulli equation was obtained by integrating the differential form of Newton's second law (Euler's equation) for steady, incompressible, frictionless flow along a streamline. Each term in the Bernoulli equation has dimensions of energy per unit mass. Thus the Bernoulli equation is obtained in a manner similar to the way that the energy method is introduced in particle mechanics. The Bernoulli equation may be viewed as a mechanical energy balance. In those cases wherein there is no conversion of mechanical to thermal energy, then mechanical energy and thermal energy are separately conserved. For these cases, the first law of thermodynamics and Newton's second law do not yield separate information. However, in general, the first law of thermodynamics and Newton's second law are independent equations that must be satisfied separately.

EXAMPLE 6.7—Internal Energy and Heat Transfer
in Frictionless Incompressible Flow

Consider frictionless, incompressible flow with heat transfer. Show that

$$u_2 - u_1 = \frac{\delta Q}{dm}$$

EXAMPLE PROBLEM 6.7

GIVEN: Frictionless, incompressible flow with heat transfer.

SHOW: $u_2 - u_1 = \dfrac{\delta Q}{dm}$.

SOLUTION:
In general, the internal energy, u, can be expressed as $u = u(T, v)$. For incompressible flow, $v = $ constant, and $u = u(T)$. Thus the thermodynamic state of the fluid is determined by the single thermodynamic property, T. The internal energy change for any process, $u_2 - u_1$, depends only on the temperatures at the end states.

From the Gibbs equation, $T\,ds = du + p\,dv$, valid for a pure substance undergoing any process, we obtain

$$T\,ds = du$$

for incompressible flow, since $dv = 0$. Since the internal energy change, du, between specified end states, is independent of the process, we take a reversible process, for which $T\,ds = d(\delta Q/dm) = du$. Therefore,

$$u_2 - u_1 = \frac{\delta Q}{dm}$$

EXAMPLE 6.8—Frictionless Flow with Heat Transfer

Water flows steadily from a large open reservoir through a short length of pipe and a nozzle with cross-sectional area $A = 0.864$ in.^{2} A well-insulated 10 kW heater surrounds the pipe. The flow is assumed to be steady, frictionless, and incompressible. Find the temperature rise of the fluid.

EXAMPLE PROBLEM 6.8

GIVEN: Water flows from a large reservoir through the system shown and discharges to atmospheric pressure. The heater is 10 kW; $A_4 = 0.864$ in.^{2}

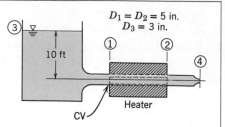

$D_1 = D_2 = 5$ in.
$D_3 = 3$ in.

FIND: The temperature rise of the fluid between points ① and ②.

SOLUTION:

Basic equations:

$$\frac{p}{\rho} + \frac{V^2}{2} + gz = \text{constant}$$

$$0 = \overset{=\,0(1)}{\overbrace{\frac{\partial}{\partial t} \int_{CV} \rho\, d\forall}} + \int_{CS} \rho \vec{V} \cdot d\vec{A}$$

$$\dot{Q} - \overset{=\,0(4)}{\overbrace{\dot{W}_s}} - \overset{=\,0(4)}{\overbrace{\dot{W}_{shear}}} = \overset{=\,0(1)}{\overbrace{\frac{\partial}{\partial t} \int_{CV} e\rho\, d\forall}} + \int_{CS} \left(u + pv + \frac{V^2}{2} + gz \right)\rho\vec{V} \cdot d\vec{A}$$

Assumptions: (1) Steady flow
(2) Frictionless flow
(3) Incompressible flow
(4) No shaft work, no shear work
(5) Flow along a streamline

Under the assumptions listed, the first law of thermodynamics for the CV shown becomes

$$\dot{Q} = \int_{CS} \left(u + pv + \frac{V^2}{2} + gz \right)\rho\vec{V} \cdot d\vec{A}$$

$$= \int_{A_1} \left(u + pv + \frac{V^2}{2} + gz \right)\rho\vec{V} \cdot d\vec{A} + \int_{A_2} \left(u + pv + \frac{V^2}{2} + gz \right)\rho\vec{V} \cdot d\vec{A}$$

For uniform properties at ① and ②

$$\dot{Q} = -|\rho V_1 A_1| \left(u_1 + p_1 v + \frac{V_1^2}{2} + gz_1 \right) + |\rho V_2 A_2| \left(u_2 + p_2 v + \frac{V_2^2}{2} + gz_2 \right)$$

From conservation of mass, $|\rho V_1 A_1| = |\rho V_2 A_2| = \dot{m}$, so

$$\dot{Q} = \dot{m} \left[u_2 - u_1 + \left(\frac{p_2}{\rho} + \frac{V_2^2}{2} + gz_2 \right) - \left(\frac{p_1}{\rho} + \frac{V_1^2}{2} + gz_1 \right) \right]$$

For frictionless, incompressible, steady flow, along a streamline,

$$\frac{p}{\rho} + \frac{V^2}{2} + gz = \text{constant}$$

Therefore,

$$\dot{Q} = \dot{m}(u_2 - u_1)$$

Since, for an incompressible fluid, $u_2 - u_1 = c(T_2 - T_1)$, then

$$T_2 - T_1 = \frac{\dot{Q}}{\dot{m}c}$$

From continuity,

$$\dot{m} - \rho V_4 A_4$$

To find V_4, write the Bernoulli equation between the free surface at ③ and point ④.

$$\frac{p_3}{\rho} + \frac{V_3^2}{2} + g z_3 = \frac{p_4}{\rho} + \frac{V_4^2}{2} + g z_4$$

Since $p_3 = p_4$ and $V_3 \approx 0$, then

$$V_4 = \sqrt{2g(z_3 - z_4)} = \sqrt{2 \times \frac{32.2 \ \text{ft}}{\text{sec}^2} \times 10 \ \text{ft}} = 25.4 \ \text{ft/sec}$$

and

$$\dot{m} = \rho V_4 A_4 = \frac{1.94 \ \text{slug}}{\text{ft}^3} \times \frac{25.4 \ \text{ft}}{\text{sec}} \times 0.864 \ \text{in.}^2 \times \frac{\text{ft}^2}{144 \ \text{in.}^2}$$

$$\dot{m} = 0.296 \ \text{slug/sec}$$

Assuming no heat loss to the surroundings, we obtain

$$T_2 - T_1 = \frac{\dot{Q}}{\dot{m}c} = 10 \ \text{kW} \times \frac{3413 \ \text{Btu}}{\text{kW} \cdot \text{hr}} \times \frac{\text{hr}}{3600 \ \text{sec}} \times \frac{\text{sec}}{0.296 \ \text{slug}} \times \frac{\text{slug}}{32.2 \ \text{lbm}} \times \frac{\text{lbm} \cdot \text{R}}{1 \ \text{Btu}}$$

$$T_2 - T_1 = 0.995 \ \text{R} \qquad\qquad\qquad\qquad\qquad\qquad T_2 - T_1$$

{ This problem illustrates that, in general, the first law of thermodynamics and the
Bernoulli equation are independent equations. }

For steady, frictionless, incompressible flow along a streamline, we have shown
that the first law of thermodynamics reduces to the Bernoulli equation. From Eq.
6.16 we conclude that there is no loss of mechanical energy in such a flow.

Often it is convenient to represent the mechanical energy level of a flow graphically. The energy equation in the form of Eq. 6.16 suggests such a representation.
Dividing Eq. 6.16 by g, we obtain

$$\frac{p}{\rho g} + \frac{V^2}{2g} + z = H = \text{constant} \qquad\qquad (6.17)$$

Each term in Eq. 6.17 has dimensions of length, or "head" of flowing fluid. The
individual terms are

$\dfrac{p}{\rho g}$ is the head due to local static pressure

$\dfrac{V^2}{2g}$ is the head due to local dynamic pressure
(kinetic energy per unit mg of flowing fluid)

z is the elevation head

H is the total head for the flow

The *energy grade line* (EGL) represents the total head height. As shown by Eq.
6.17, the EGL height remains constant for frictionless flow when no work is done

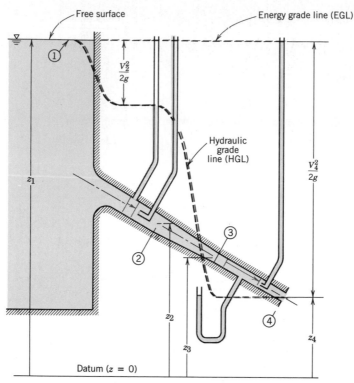

Energy and hydraulic grade lines for frictionless flow.

on or by the flowing fluid. Liquid would rise to the EGL height in a total head tube placed in the flow.

The *hydraulic grade line* (HGL) height represents the sum of the elevation and static pressure heads, $z + p/\rho g$. In a static pressure tap attached to the flow conduit, liquid would rise to the HGL height. For open-channel flow, the HGL is at the liquid free surface.

The difference in heights between the EGL and the HGL represents the dynamic (velocity) head, $V^2/2g$. The relationship among the EGL, HGL, and velocity head is illustrated schematically in Fig. 6.6 for frictionless flow from a tank through a pipe with a reducer.

The total head of the flow shown in Fig. 6.6 is obtained by applying Eq. 6.17 at point ①, the free surface in the large reservoir. There the velocity is negligible and the pressure is atmospheric (zero gage). Thus total head is equal to z_1. This defines the height of the energy grade line, which remains constant for this flow, since there is no friction or work.

The velocity head increases from zero to $V_2^2/2g$ as the fluid accelerates into the first section of constant-diameter tube. Since the EGL height is constant, the HGL must decrease in height. When the velocity becomes constant, then the HGL height stays constant.

The velocity increases again in the reducer between sections ② and ③. As the velocity head increases, the HGL height drops. When the velocity becomes constant between sections ③ and ④, the HGL stays constant at a lower height.

At the free discharge at section ④, the static head is zero (gage). There the HGL height is equal to z_4. As shown, the velocity head is $V_4^2/2g$. The sum of the HGL height and velocity head equals the EGL height. (The static head is negative between sections ③ and ④, because the pipe centerline is above the HGL.)

Static taps and total head tubes connected to manometers are shown schematically in Fig. 6.6. The static taps give readings equal to the HGL height. The total head tubes give readings equal to the EGL height.

The effects of friction, and work interactions with a flow, will be discussed in detail in Chapter 8. The effect of friction is to convert mechanical energy to internal thermal energy. Thus friction reduces the total head of the flowing fluid, causing a gradual reduction in the EGL height. Work addition to the fluid, as delivered by a pump for example, increases the EGL height.

**6-5 UNSTEADY BERNOULLI EQUATION—INTEGRATION OF EULER'S EQUATION ALONG A STREAMLINE

It is not necessary to restrict the development of the Bernoulli equation to steady flow. The purpose of this section is to develop the corresponding equation for unsteady flow along a streamline and to illustrate its use.

The momentum equation for frictionless flow was found in Section 6-1 to be

$$-\frac{1}{\rho}\nabla p - g\nabla z = \frac{D\vec{V}}{Dt} \tag{6.3}$$

Equation 6.3 is a vector equation. It can be converted to a scalar equation by taking the dot product with $d\vec{s}$, where $d\vec{s}$ is an element of distance along a streamline. Thus

$$-\frac{1}{\rho}\nabla p \cdot d\vec{s} - g\nabla z \cdot d\vec{s} = \frac{D\vec{V}}{Dt}\cdot d\vec{s} = \frac{DV_s}{Dt}ds = V_s\frac{\partial V_s}{\partial s}ds + \frac{\partial V_s}{\partial t}ds \tag{6.18}$$

Examining the terms in Eq. 6.18, we note that

$$\nabla p \cdot d\vec{s} = dp \qquad \text{(the change in pressure along } s)$$

$$\nabla z \cdot d\vec{s} = dz \qquad \text{(the change in } z \text{ along } s)$$

$$\frac{\partial V_s}{\partial s}ds = dV_s \qquad \text{(the change in } V_s \text{ along } s)$$

Substituting into Eq. 6.18, we obtain

$$-\frac{dp}{\rho} - g\,dz = V_s\,dV_s + \frac{\partial V_s}{\partial t}ds \tag{6.19}$$

Integrating along a streamline from point 1 to point 2 yields

$$\int_1^2 \frac{dp}{\rho} + \frac{V_2^2 - V_1^2}{2} + g(z_2 - z_1) + \int_1^2 \frac{\partial V_s}{\partial t}ds = 0 \tag{6.20}$$

For incompressible flow, the density is constant. For this special case, Eq. 6.20 becomes

$$\frac{p_1}{\rho} + \frac{V_1^2}{2} + gz_1 = \frac{p_2}{\rho} + \frac{V_2^2}{2} + gz_2 + \int_1^2 \frac{\partial V_s}{\partial t}ds \tag{6.21}$$

Restrictions: (1) Incompressible flow
               (2) Frictionless flow
               (3) Flow along a streamline

**This section may be omitted without loss of continuity in the text material.

To evaluate the integral term in Eq. 6.21, the variation in $\partial V_s/\partial t$ must be known as a function of s, the distance along the streamline measured from point 1. (For steady flow, $\partial V_s/\partial t = 0$, and Eq. 6.21 reduces to Eq. 6.9.) Equation 6.21 may be applied to any flow in which the restrictions are compatible with the physical situation.

Application of Eq. 6.21 is illustrated in Example Problem 6.9.

EXAMPLE 6.9—Unsteady Bernoulli Equation

A long pipe is connected to a large reservoir that initially is filled with water to a depth of 3 m. The pipe is 150 mm in diameter and 6 m long. As a first approximation, friction may be neglected. Determine the flow velocity leaving the pipe as a function of time after a cap is removed from its free end. The reservoir is large enough so that the change in its level may be neglected.

EXAMPLE PROBLEM 6.9

GIVEN: Pipe and large reservoir as shown.

FIND: $V_2(t)$.

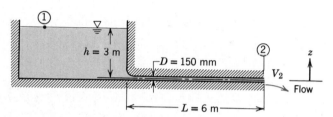

SOLUTION:
Apply the Bernoulli equation to the unsteady flow along a streamline from point ① to point ②.

Basic equation:
$$\frac{\cancel{p_1}}{\rho}^{\approx\,0(5)} + \frac{\cancel{V_1^2}}{2} + gz_1 = \frac{\cancel{p_2}}{\rho}^{=\,0(6)} + \frac{V_2^2}{2} + g\cancel{z_2} + \int_1^2 \frac{\partial V_s}{\partial t}\,ds$$

Assumptions: (1) Incompressible flow
(2) Frictionless flow
(3) Flow along a streamline from ① to ②
(4) $p_1 = p_2 = p_{atm}$
(5) $V_1^2 \approx 0$
(6) $z_2 = 0$
(7) $z_1 = h = $ constant
(8) Neglect velocity in reservoir, except for small region near the inlet to the tube

Then
$$gz_1 = gh = \frac{V_2^2}{2} + \int_1^2 \frac{\partial V_s}{\partial t}\,ds$$

In view of assumption (8), the integral becomes
$$\int_1^2 \frac{\partial V_s}{\partial t}\,ds \approx \int_0^L \frac{\partial V_s}{\partial t}\,ds$$

In the tube, $V_s = V_2$ everywhere, so that
$$\int_0^L \frac{\partial V_s}{\partial t}\,ds = \int_0^L \frac{dV_2}{dt}\,ds = L\frac{dV_2}{dt}$$

Substituting gives
$$gh = \frac{V_2^2}{2} + L\frac{dV_2}{dt}$$

Separating variables, we obtain

$$\frac{d V_2}{2gh - V_2^2} = \frac{dt}{2L}$$

Integrating between limits $V = 0$ at $t = 0$ and $V = V_2$ at $t = t$,

$$\int_0^{V_2} \frac{d V}{2gh - V^2} = \left[\frac{1}{\sqrt{2gh}} \tanh^{-1}\left(\frac{V}{\sqrt{2gh}}\right) \right]_0^{V_2} = \frac{t}{2L}$$

Since $\tanh^{-1}(0) = 0$, we obtain

$$\frac{1}{\sqrt{2gh}} \tanh^{-1}\left(\frac{V_2}{\sqrt{2gh}}\right) = \frac{t}{2L}$$

or

$$\frac{V_2}{\sqrt{2gh}} = \tanh\left(\frac{t}{2L}\sqrt{2gh}\right) \qquad\qquad V_2(t)$$

For the given conditions,

$$\sqrt{2gh} = \sqrt{2 \times 9.81 \frac{m}{sec^2} \times 3\ m} = 7.67\ m/sec$$

and

$$\frac{t}{2L}\sqrt{2gh} = \frac{t}{2} \times \frac{1}{6\ m} \times 7.67 \frac{m}{sec} = 0.639t$$

The result is then $V_2 = 7.67\ \tanh\,(0.639t)$ m/sec.

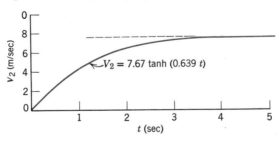

**6-6 IRROTATIONAL FLOW

An *irrotational flow* is one in which fluid elements moving in the flow field do not undergo any rotation. For $\vec{\omega} = 0$, $\nabla \times \vec{V} = 0$, and from Eq. 5.13,

$$\frac{\partial w}{\partial y} - \frac{\partial v}{\partial z} = \frac{\partial u}{\partial z} - \frac{\partial w}{\partial x} = \frac{\partial v}{\partial x} - \frac{\partial u}{\partial y} = 0 \qquad (6.22)$$

In cylindrical coordinates, from Eq. 5.16, the irrotationality condition requires that

$$\frac{1}{r}\frac{\partial V_z}{\partial \theta} - \frac{\partial V_\theta}{\partial z} = \frac{\partial V_r}{\partial z} - \frac{\partial V_z}{\partial r} = \frac{1}{r}\frac{\partial r V_\theta}{\partial r} - \frac{1}{r}\frac{\partial V_r}{\partial \theta} = 0 \qquad (6.23)$$

** This section may be omitted without loss of continuity in the text material. (Note that Section 5-2 contains background material needed for study of Sections 6-6.3, 6-6.4, and 6-6.5.)

6-6.1 Bernoulli Equation Applied to Irrotational Flow

In Section 6-3.1, we integrated Euler's equation along a streamline for steady, incompressible, inviscid flow to obtain the Bernoulli equation

$$\frac{p}{\rho} + \frac{V^2}{2} + gz = \text{constant} \tag{6.9}$$

Equation 6.9 can be applied between any two points on the same streamline. The value of the constant will vary, in general, from streamline to streamline.

If, in addition to being inviscid, steady and incompressible, the flow field is also irrotational (the velocity field is such that $2\vec{\omega} = \nabla \times \vec{V} = 0$), we can show that Bernoulli's equation can be applied between any two points in the flow. Then the value of the constant in Eq. 6.9 is the same for all streamlines. To illustrate this, we start with Euler's equation in vector form,

$$-\frac{1}{\rho}\nabla p - g\nabla z = (\vec{V} \cdot \nabla)\vec{V} \tag{6.10}$$

Using the vector identity

$$(\vec{V} \cdot \nabla)\vec{V} = \frac{1}{2}\nabla(\vec{V} \cdot \vec{V}) - \vec{V} \times (\nabla \times \vec{V})$$

we see that for irrotational flow, since $\nabla \times \vec{V} = 0$, then

$$(\vec{V} \cdot \nabla)\vec{V} = \frac{1}{2}\nabla(\vec{V} \cdot \vec{V})$$

and Euler's equation for irrotational flow can be written as

$$-\frac{1}{\rho}\nabla p - g\nabla_z = \frac{1}{2}\nabla(\vec{V} \cdot \vec{V}) = \frac{1}{2}\nabla(V^2) \tag{6.24}$$

During the interval dt, a fluid particle moves from the vector position $\vec{r}$, to the position $\vec{r} + d\vec{r}$; the displacement $d\vec{r}$ is an arbitrary infinitesimal displacement in any direction. Taking the dot product of $d\vec{r} = dx\hat{i} + dy\hat{j} + dz\hat{k}$ with each of the terms in Eq. 6.24, we have

$$-\frac{1}{\rho}\nabla p \cdot d\vec{r} - g\nabla z \cdot d\vec{r} = \frac{1}{2}\nabla(V^2) \cdot d\vec{r}$$

and hence

$$-\frac{dp}{\rho} - g\,dz = \frac{1}{2}d(V^2)$$

or

$$\frac{dp}{\rho} + \frac{1}{2}d(V^2) + g\,dz = 0$$

Integrating this equation gives

$$\int \frac{dp}{\rho} + \frac{V^2}{2} + gz = \text{constant} \tag{6.25}$$

For incompressible flow, $\rho = $ constant, and

$$\frac{p}{\rho} + \frac{V^2}{2} + gz = \text{constant} \tag{6.26}$$

Since $d\vec{r}$ was an arbitrary displacement, then for a steady, incompressible, inviscid flow that is also irrotational, Eq. 6.26 is valid between any two points in the flow field.

6-6.2 Velocity Potential

In Section 5-2 we formulated the stream function, ψ, a relation between the streamlines and conservation of mass for two-dimensional, incompressible flow.

We can formulate a relation called the potential function, ϕ, for a velocity field that is irrotational. To do so, we must use the fundamental vector identity[8]

$$\text{curl}\,(\text{grad}\,\phi) = \nabla \times \nabla \phi = 0 \tag{6.27}$$

which is valid if ϕ is a scalar function (of the space coordinates and time) having continuous first and second derivatives.

Then, for an irrotational flow in which $\nabla \times \vec{V} = 0$, a scalar function, ϕ, must exist such that the gradient of ϕ is equal to the velocity vector, $\vec{V}$. In order that the positive direction of flow be in the direction of decreasing ϕ (analogous to the positive direction of heat transfer being defined in the direction of decreasing temperature), we define ϕ such that

$$\vec{V} \equiv -\nabla \phi \tag{6.28}$$

Thus

$$u = -\frac{\partial \phi}{\partial x} \qquad v = -\frac{\partial \phi}{\partial y} \qquad w = -\frac{\partial \phi}{\partial z} \tag{6.29}$$

With the potential function defined in this way, the irrotationality condition, Eq. 6.22, is satisfied identically.

In cylindrical coordinates,

$$\nabla = \hat{e}_r \frac{\partial}{\partial r} + \hat{e}_\theta \frac{1}{r} \frac{\partial}{\partial \theta} + \hat{k} \frac{\partial}{\partial z} \tag{3.18}$$

From Eq. 6.28, then, in cylindrical coordinates

$$V_r = -\frac{\partial \phi}{\partial r} \qquad V_\theta = -\frac{1}{r} \frac{\partial \phi}{\partial \theta} \qquad V_z = -\frac{\partial \phi}{\partial z} \tag{6.30}$$

The velocity potential, ϕ, exists only for irrotational flow. The stream function, ψ, satisfies the continuity equation for incompressible flow; the stream function is not subject to the restriction of irrotational flow.

Irrotationality may be a valid assumption for those regions of a flow in which viscous forces are negligible.[9] (For example, such a region exists outside the boundary layer in the flow over a solid surface.) The theory for irrotational flow is developed in terms of an imaginary ideal fluid whose viscosity is identically zero. Since, in an

---

[8] The proof of this identity may be found in any book treating vector analysis, or it may be proved by expanding Eq. 6.27 into components.

[9] Examples of rotational and irrotational motion are shown in the NCFMF film, *Vorticity*, A. H. Shapiro, principal.

irrotational flow, the velocity field may be defined by the potential function, ϕ, the theory is often referred to as potential flow theory.

All real fluids possess viscosity, but there are many situations in which the assumption of inviscid flow considerably simplifies the analysis and, at the same time, gives meaningful results. Because of its utility and mathematical appeal, potential flow has been studied extensively.[10]

6-6.3 Stream Function and Velocity Potential for Two-Dimensional, Irrotational, Incompressible Flow; Laplace's Equation

For a two-dimensional, incompressible, irrotational flow we have expressions for the velocity components, u and v, in terms of both the stream function, ψ, and the velocity potential, ϕ,

$$u = \frac{\partial \psi}{\partial y} \qquad v = -\frac{\partial \psi}{\partial x} \tag{5.4}$$

$$u = -\frac{\partial \phi}{\partial x} \qquad v = -\frac{\partial \phi}{\partial y} \tag{6.29}$$

Substituting for u and v from Eq. 5.4 into the irrotationality condition,

$$\frac{\partial v}{\partial x} - \frac{\partial u}{\partial y} = 0 \tag{6.22}$$

we obtain

$$\frac{\partial^2 \psi}{\partial x^2} + \frac{\partial^2 \psi}{\partial y^2} = 0 \tag{6.31a}$$

Substituting for u and v from Eq. 6.29 into the continuity equation,

$$\frac{\partial u}{\partial x} + \frac{\partial v}{\partial y} = 0 \tag{5.3}$$

we obtain

$$\frac{\partial^2 \phi}{\partial x^2} + \frac{\partial^2 \phi}{\partial y^2} = 0 \tag{6.31b}$$

Equations 6.31a and 6.31b are forms of Laplace's equation—an equation that arises in many areas of the physical sciences and engineering. Any function ψ or ϕ that satisfies Laplace's equation represents a possible two-dimensional, incompressible, irrotational flow field.

In Section 5-2 we showed that the stream function, ψ, is constant along a streamline. For $\psi = $ constant, $d\psi = 0$ and

$$d\psi = \frac{\partial \psi}{\partial x} dx + \frac{\partial \psi}{\partial y} dy = 0$$

The slope of a streamline—a line of constant ψ—is given by

$$\frac{dy}{dx}\bigg)_\psi = -\frac{\partial \psi/\partial x}{\partial \psi/\partial y} = -\frac{-v}{u} = \frac{v}{u} \tag{6.32}$$

---

[10] Anyone interested in a detailed study of potential flow theory may find [1–3] of interest.

Along a line of constant ϕ, $d\phi = 0$ and

$$d\phi = \frac{\partial \phi}{\partial x}dx + \frac{\partial \phi}{\partial y}dy = 0$$

Consequently, the slope of a potential line—a line of constant ϕ—is given by

$$\frac{dy}{dx}\bigg)_\phi = -\frac{\partial \phi/\partial x}{\partial \phi/\partial y} = -\frac{u}{v} \tag{6.33}$$

The last equality of Eq. 6.33 follows from use of Eq. 6.29.

Comparing Eqs. 6.32 and 6.33, we see that the slope of a constant ψ line at any point is the negative reciprocal of the slope of the constant ϕ line at that point: lines of constant ψ and constant ϕ are orthogonal. This property of potential lines and streamlines is useful in graphical analyses of flow fields.

EXAMPLE 6.10—Velocity Potential

Consider the flow field given by $\psi = ax^2 - ay^2$, where $a = 3$ sec^{-1}. Show that the flow is irrotational. Determine the velocity potential for this flow.

EXAMPLE PROBLEM 6.10

GIVEN: Incompressible flow field with $\psi = ax^2 - ay^2$, where $a = 3$ sec^{-1}.

FIND: (a)   Show that the flow is irrotational.
(b)   Determine the velocity potential for this flow.

SOLUTION:
If the flow is irrotational, then $\omega_z = 0$. Since

$$2\omega_z = \frac{\partial v}{\partial x} - \frac{\partial u}{\partial y} \quad \text{and} \quad u = \frac{\partial \psi}{\partial y} \quad v = -\frac{\partial \psi}{\partial x}$$

then

$$u = \frac{\partial}{\partial y}(ax^2 - ay^2) = -2ay \quad \text{and} \quad v = -\frac{\partial}{\partial x}(ax^2 - ay^2) = -2ax$$

so

$$2\omega_z = \frac{\partial v}{\partial x} - \frac{\partial u}{\partial y} = \frac{\partial}{\partial x}(-2ax) - \frac{\partial}{\partial y}(-2ay) = -2a + 2a = 0 \qquad \underleftarrow{\hspace{2cm}} 2\omega_z$$

Therefore, the flow is irrotational.

The velocity components can be written in terms of the velocity potential as

$$u = -\frac{\partial \phi}{\partial x} \quad \text{and} \quad v = -\frac{\partial \phi}{\partial y}$$

Consequently, $u = -\dfrac{\partial \phi}{\partial x} = -2ay$ and $\dfrac{\partial \phi}{\partial x} = 2ay$. Integrating with respect to x gives $\phi = 2axy + f(y)$, where $f(y)$ is an arbitrary function of y. Then

$$v = -2ax = -\frac{\partial \phi}{\partial y} = -\frac{\partial}{\partial y}[2axy + f(y)]$$

Therefore, $-2ax = -2ax - \dfrac{\partial f(y)}{\partial y} = -2ax - \dfrac{df}{dy}$, so $\dfrac{df}{dy} = 0$ and $f = $ constant. Thus

$\phi = 2axy + $ constant $\qquad\qquad\qquad\qquad\qquad\qquad\qquad\qquad\qquad\qquad \underleftarrow{\hspace{1cm}} \phi$

We also can show that lines of constant ψ and ϕ are orthogonal.

$$\psi = ax^2 - ay^2 \qquad \text{and} \qquad \phi = 2axy + c$$

For $\psi = $ constant, $d\psi = 0 = 2ax\,dx - 2ay\,dy$; hence $\left.\dfrac{dy}{dx}\right)_{\psi=c} = \dfrac{x}{y}$

For $\phi = $ constant, $d\phi = 0 = 2ay\,dx + 2ax\,dy$; hence $\left.\dfrac{dy}{dx}\right)_{\phi=c} = -\dfrac{y}{x}$

The slopes of lines of constant ϕ and constant ψ are negative reciprocals. Therefore lines of constant ϕ are orthogonal to lines of constant ψ.

$\left\{\begin{array}{l}\text{This problem illustrates the relation among the stream function, the velocity field, and} \\ \text{the velocity potential.}\end{array}\right\}$

6-6.4 Elementary Plane Flows

A variety of potential flows can be constructed by superposing elementary flow patterns. The ψ and ϕ functions for five elementary two-dimensional flows—a uniform flow, a source, a sink, a vortex, and a doublet—are summarized in Table 6.1. The ψ and ϕ functions can be obtained from the velocity field for each elementary flow.

A *uniform flow* of constant velocity parallel to the x axis satisfies the continuity equation and the irrotationality condition identically. In Table 6.1 we have shown the ψ and ϕ functions for a uniform flow in the positive x direction.

For a uniform flow of constant magnitude, V, inclined at angle α to the x axis,

$$\psi = (V\cos\alpha)y - (V\sin\alpha)x$$
$$\phi = -(V\sin\alpha)y - (V\cos\alpha)x$$

A simple *source* is a flow pattern in the xy plane in which flow is radially outward from the z axis and symmetrical in all directions. The strength, q, of the source is the volume flow rate per unit depth. At any radius, r, from a source, the tangential velocity, V_θ, is zero; the radial velocity, V_r, is the volume flow rate per unit depth, q, divided by the flow area per unit depth, $2\pi r$. Thus $V_r = q/2\pi r$ for a source. The ψ and ϕ functions for a source are shown in Table 6.1.

In a simple *sink,* flow is radially inward; a sink is a negative source. The ψ and ϕ functions for a sink shown in Table 6.1 are the negatives of the corresponding functions for a source flow.

The origin of either a sink or a source is a singular point, since the radial velocity approaches infinity as the radius approaches zero. Thus, while an actual flow may resemble a source or a sink for some values of r, sources and sinks have no exact physical counterparts. The primary value of the concept of sources and sinks is that, when combined with other elementary flows, they produce flow patterns that adequately represent realistic flows.

A flow pattern in which the streamlines are concentric circles is a vortex; in a *free (irrotational) vortex,* fluid particles do not rotate as they move around the vortex center. The velocity distribution in an irrotational vortex can be determined from Euler's equation and the Bernoulli equation. For irrotational flow, the Bernoulli equation is valid between any two points in the flow field. For flow in a horizontal plane,

$$\frac{1}{\rho}dp = -V_\theta\,dV_\theta$$

Table 6.1 Elementary Plane Flows

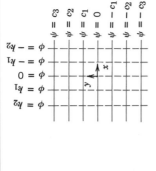

Uniform Flow (positive x direction)

$$u = U \qquad \psi = Uy$$

$$v = 0 \qquad \phi = -Ux$$

$\Gamma = 0$ around any closed curve

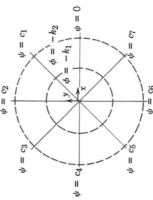

Source Flow (from origin)

$$V_r = \frac{q}{2\pi r} \qquad \psi = \frac{q}{2\pi}\theta$$

$$V_\theta = 0 \qquad \phi = -\frac{q}{2\pi}\ln r$$

Origin is singular point
q is volume flow rate per unit depth
$\Gamma = 0$ around any closed curve

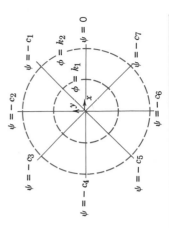

Sink Flow (toward origin)

$$V_r = -\frac{q}{2\pi r} \qquad \psi = -\frac{q}{2\pi}\theta$$

$$V_\theta = 0 \qquad \phi = \frac{q}{2\pi}\ln r$$

Origin is singular point
q is volume flow rate per unit depth
$\Gamma = 0$ around any closed curve

Table 6.1 Elementary Plane Flows (*cont'd*)

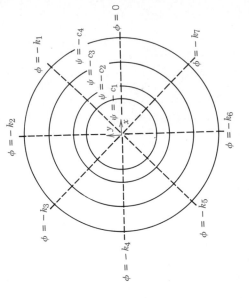

Irrotational Vortex (counter-clockwise, center at origin)

$$V_r = 0 \qquad \psi = -\frac{K}{2\pi}\ln r$$

$$V_\theta = \frac{K}{2\pi r} \qquad \phi = -\frac{K}{2\pi}\theta$$

Origin is singular point
K is strength of the vortex
$\Gamma = K$ around any closed
    curve enclosing origin
$\Gamma = 0$ around any closed
    curve not enclosing origin

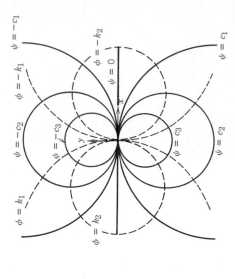

Doublet (center at origin)

$$V_r = -\frac{\Lambda}{r^2}\cos\theta \qquad \psi = -\frac{\Lambda\sin\theta}{r}$$

$$V_\theta = -\frac{\Lambda}{r^2}\sin\theta \qquad \phi = -\frac{\Lambda\cos\theta}{r}$$

Origin is singular point
Λ is strength of the doublet
$\Gamma = 0$ around any closed curve

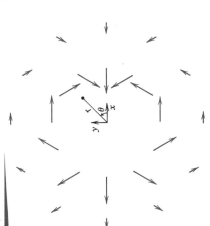

Euler's equation normal to the streamline is

$$\frac{1}{\rho}\frac{dp}{dr} = \frac{V_\theta^2}{r}$$

Combining these equations yields

$$\frac{dp}{\rho} = -V_\theta\, dV_\theta = \frac{V_\theta^2}{r}\, dr$$

From the last equality

$$V_\theta\, dr + r\, dV_\theta = 0$$

Integrating this equation gives

$$V_\theta r = \text{constant}$$

The strength, K, of the vortex is defined as $K = 2\pi r V_\theta$; the dimensions of K are L^2/t (volume flow rate per unit depth). The irrotational vortex is a reasonable approximation to the flow field in a tornado (except in the region of the origin; the origin is a singular point).

The final "elementary" flow listed in Table 6.1 is the *doublet*. This flow is produced mathematically by allowing a source and a sink of numerically equal strengths to merge. In the limit, as the distance, δs, between them approaches zero, their strengths increase so the product $q\,\delta s/2\pi$ tends to a finite value, Λ, which is termed the strength of the doublet.

6-6.5 Superposition of Elementary Plane Flows

We showed in Section 6-6.3 that both ϕ and ψ satisfy Laplace's equation for flow that is both incompressible and irrotational. Since Laplace's equation is a linear, homogeneous partial differential equation, solutions may be superposed (added together) to develop more complex and interesting patterns of flow. Thus if ψ_1 and ψ_2 satisfy Laplace's equation, then so does $\psi_3 = \psi_1 + \psi_2$. The elementary plane flows are the building blocks in this superposition process.

The objective of superposition of elementary flows is to produce flow patterns similar to those of practical interest. The ideal flow model postulates an ideal fluid with zero viscosity and therefore zero shear stresses. Since there is no flow across a streamline, any streamline contour can be imagined to represent a solid surface.

The combination of mathematical elegance and utility of potential flow attracted many workers to its study. Some of history's most famous applied mathematicians studied the theory and application of "hydrodynamics," as potential flow was termed before 1900. The list of names includes Bernoulli, Lagrange, d'Alembert, Cauchy, Rankine, and Euler [4].

Through the end of the nineteenth century, workers in pure hydrodynamics failed to produce results that agreed with experiment. Potential flows produced body shapes with lift but predicted zero drag (the "d'Alembert paradox"). Two influences changed this situation: first Prandtl introduced the boundary-layer concept and began to develop the theory, and second, interest in aeronautics increased dramatically in the early 1900s.

Prandtl showed by mathematical analysis and through elegantly simple experiments that viscous effects are confined to a thin boundary layer on the surface of a body. Even for real fluids, flow outside the boundary layer behaves as though the fluid

Table 6.2 Superposition of Elementary Plane Flows

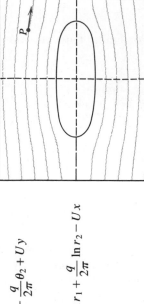

Source and Uniform Flow (flow past a half-body)

$$\psi = \psi_{so} + \psi_{uf} = \psi_1 + \psi_2 = \frac{q}{2\pi}\theta + Uy = \frac{q}{2\pi}\theta + Ur\sin\theta$$

$$\phi = \phi_{so} + \phi_{uf} = \phi_1 + \phi_2 = -\frac{q}{2\pi}\ln r - Ux = -\frac{q}{2\pi}\ln r - Ur\cos\theta$$

Source and Sink (equal strength, separation distance on x axis $= 2a$)

$$\psi = \psi_{so} + \psi_{si} = \psi_1 + \psi_2 = \frac{q}{2\pi}\theta_1 - \frac{q}{2\pi}\theta_2 = \frac{q}{2\pi}(\theta_1 - \theta_2)$$

$$\phi = \phi_{so} + \phi_{si} = \phi_1 + \phi_2 = -\frac{q}{2\pi}\ln r_1 + \frac{q}{2\pi}\ln r_2 = \frac{q}{2\pi}\ln\frac{r_2}{r_1}$$

Source, Sink, and Uniform Flow (flow past a Rankine body)

$$\psi = \psi_{so} + \psi_{si} + \psi_{uf} = \psi_1 + \psi_2 + \psi_3 = \frac{q}{2\pi}\theta_1 - \frac{q}{2\pi}\theta_2 + Uy$$

$$\psi = \frac{q}{2\pi}(\theta_1 - \theta_2) + Ur\sin\theta$$

$$\phi = \phi_{so} + \phi_{si} + \phi_{uf} = \phi_1 + \phi_2 + \phi_3 = -\frac{q}{2\pi}\ln r_1 + \frac{q}{2\pi}\ln r_2 - Ux$$

$$\phi = \frac{q}{2\pi}\ln\frac{r_2}{r_1} - Ur\cos\theta$$

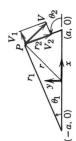

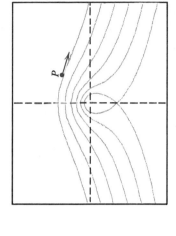

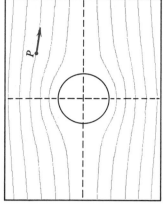

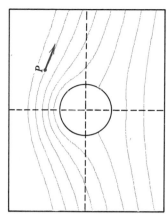

Vortex (clockwise) and Uniform Flow

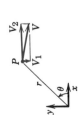

$$\psi = \psi_v + \psi_{uf} = \psi_1 + \psi_2 = \frac{K}{2\pi}\ln r + Uy = \frac{K}{2\pi}\ln r + Ur\sin\theta$$

$$\phi = \phi_v + \phi_{uf} = \phi_1 + \phi_2 = \frac{K}{2\pi}\theta - Ux = \frac{K}{2\pi}\theta - Ur\cos\theta$$

Doublet and Uniform Flow (flow past a cylinder)

$$\psi = \psi_d + \psi_{uf} = \psi_1 + \psi_2 = -\frac{\Lambda\sin\theta}{r} + Uy = -\frac{\Lambda\sin\theta}{r} + Ur\sin\theta \qquad a = \sqrt{\frac{\Lambda}{U}}$$

$$\psi = U\left(r - \frac{\Lambda}{Ur}\right)\sin\theta = Ur\left(1 - \frac{a^2}{r^2}\right)\sin\theta$$

$$\phi = \phi_d + \phi_{uf} = \phi_1 + \phi_2 = -\frac{\Lambda\cos\theta}{r} - Ux = -\frac{\Lambda\cos\theta}{r} - Ur\cos\theta$$

$$\phi = U\left(r + \frac{\Lambda}{Ur}\right)\cos\theta = -Ur\left(1 + \frac{a^2}{r^2}\right)\cos\theta$$

Doublet, Vortex (clockwise), and Uniform Flow (flow past a cylinder with circulation)

$$\psi = \psi_d + \psi_v + \psi_{uf} = \psi_1 + \psi_2 + \psi_3 = -\frac{\Lambda\sin\theta}{r} + \frac{K}{2\pi}\ln r + Uy$$

$$\psi = -\frac{\Lambda\sin\theta}{r} + \frac{K}{2\pi}\ln r + Ur\sin\theta = Ur\left(1 - \frac{a^2}{r^2}\right)\sin\theta + \frac{K}{2\pi}\ln r$$

$$\phi = \phi_d + \phi_v + \phi_{uf} = \phi_1 + \phi_2 + \phi_3 = -\frac{\Lambda\cos\theta}{r} + \frac{K}{2\pi}\theta - Ux$$

$$\phi = -\frac{\Lambda\cos\theta}{r} + \frac{K}{2\pi}\theta - Ur\cos\theta = -Ur\left(1 + \frac{a^2}{r^2}\right)\cos\theta + \frac{K}{2\pi}\theta$$

$$a = \sqrt{\frac{\Lambda}{U}}; \quad K < 4\pi aU$$

Table 6.2 Superposition of Elementary Plane Flows (cont'd.)

Source and Vortex (spiral vortex)

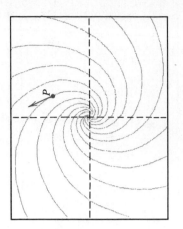

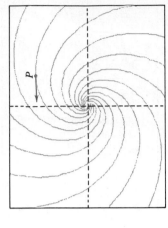

$$\psi = \psi_{so} + \psi_v = \psi_1 + \psi_2 = \frac{q}{2\pi}\theta - \frac{K}{2\pi}\ln r$$

$$\phi = \phi_{so} + \phi_v = \phi_1 + \phi_2 = -\frac{q}{2\pi}\ln r - \frac{K}{2\pi}\theta$$

Sink and Vortex

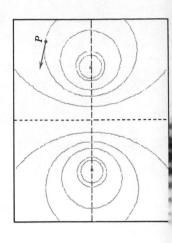

$$\psi = \psi_{si} + \psi_v = \psi_1 + \psi_2 = -\frac{q}{2\pi}\theta - \frac{K}{2\pi}\ln r$$

$$\phi = \phi_{si} + \phi_v = \phi_1 + \phi_2 = \frac{q}{2\pi}\ln r - \frac{K}{2\pi}\theta$$

Vortex Pair (equal strength, opposite rotation, separation distance on x axis $= 2a$)

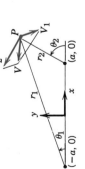

$$\psi = \psi_{v1} + \psi_{v2} = \psi_1 + \psi_2 = -\frac{K}{2\pi}\ln r_1 + \frac{K}{2\pi}\ln r_2 = \frac{K}{2\pi}\ln\frac{r_2}{r_1}$$

$$\phi = \phi_{v1} + \phi_{v2} = \phi_1 + \phi_2 = -\frac{K}{2\pi}\theta_1 + \frac{K}{2\pi}\theta_2 = \frac{K}{2\pi}(\theta_2 - \theta_1)$$

had zero viscosity. Pressure gradients from the external flow are impressed on the boundary layer.

The most important input to a calculation of the real fluid flow in the boundary layer (and thus the drag on a body) is the pressure distribution. Once the velocity field is known from the potential flow solution, the pressure distribution may be calculated. Knowledge of the boundary layer behavior makes it possible to predict the drag, and in some cases the lift, on an object.

Two methods of combining elementary flows may be used. The direct method consists of combining elementary flows, then calculating the streamline pattern, body shape, velocity field, stagnation point, and pressure distribution directly. Several examples of flow patterns produced by the direct method are given in Table 6.2.

Example Problem 6.11 demonstrates the method of superposition. The combination of a doublet with a uniform flow gives the steady flow of an ideal fluid past a cylinder. This combination was used to develop the flow pattern for ideal flow around a cylinder shown in Fig. 2.12. The potential flow solution gives the velocity field. The Bernoulli equation then can be used to obtain the pressure field. Several additional examples of superposition are included in the problems at the end of the chapter.

Complex flow patterns can be developed by combining elementary flows. Distributed line sources, sinks, and "images" may be used to create bodies of arbitrary shape. Powerful mathematical techniques, such as complex variables and conformal transformations, can be used to obtain flow fields for interesting geometries [5–8].

The inverse method of superposition calculates the body shape to produce a desired pressure distribution. Distributed singularities—vortices, sources, and sinks located on the axis or body surface—are used to model the body [9]. A large computer code must be used to evaluate the distribution of singularities required to produce the desired pressure distribution [10].

The use of ever larger computers and development of faster codes today is allowing solution of complex three-dimensional problems using direct or inverse methods [11].

EXAMPLE 6.11—Flow over a Cylinder: Superposition of Doublet and Uniform Flow

For two-dimensional, incompressible, irrotational flow, the superposition of a doublet and a uniform flow represents flow around a circular cylinder. Obtain the stream function and velocity potential for this flow pattern. Find the velocity field, locate the stagnation points and the cylinder surface, and obtain the surface pressure distribution.

EXAMPLE PROBLEM 6.11

GIVEN: Two-dimensional, incompressible, irrotational flow formed from superposition of a doublet and a uniform flow.

FIND: (a) Stream function and velocity potential.
(b) Velocity field.
(c) Stagnation points.
(d) Cylinder surface.
(e) Surface pressure distribution.

SOLUTION:
Stream functions may be added because the flow field is incompressible and irrotational. Thus from Table 6.1, the stream function for the combination is

$$\psi = \psi_d + \psi_{uf} = -\frac{\Lambda \sin\theta}{r} + Ur\sin\theta \qquad \psi$$

The velocity potential is

$$\phi = \phi_d + \phi_{uf} = -\frac{\Lambda \cos\theta}{r} - Ur\cos\theta \qquad \longleftarrow \qquad \phi$$

The corresponding velocity components are obtained using Eqs. 6.29 as

$$V_r = -\frac{\partial \phi}{\partial r} = -\frac{\Lambda \cos\theta}{r^2} + U\cos\theta$$

$$V_\theta = -\frac{1}{r}\frac{\partial \phi}{\partial \theta} = -\frac{\Lambda \sin\theta}{r^2} - U\sin\theta$$

The velocity field is

$$\vec{V} = V_r\hat{e}_r + V_\theta\hat{e}_\theta = \left(-\frac{\Lambda \cos\theta}{r^2} + U\cos\theta\right)\hat{e}_r + \left(-\frac{\Lambda \sin\theta}{r^2} - U\sin\theta\right)\hat{e}_\theta \qquad \longleftarrow \qquad \vec{V}$$

Stagnation points are where $\vec{V} = V_r\hat{e}_r + V_\theta\hat{e}_\theta = 0$, so

$$V_r = 0 = -\frac{\Lambda \cos\theta}{r^2} + U\cos\theta = \cos\theta\left(U - \frac{\Lambda}{r^2}\right)$$

Thus $V_r = 0$ when $r = \sqrt{\dfrac{\Lambda}{U}} = a$. Also,

$$V_\theta = 0 = -\frac{\Lambda \sin\theta}{r^2} - U\sin\theta = -\sin\theta\left(U + \frac{\Lambda}{r^2}\right)$$

Thus $V_\theta = 0$ when $\theta = 0, \pi$.

Stagnation points are $(r, \theta) = (a, 0), (a, \pi)$. $\qquad \longleftarrow \qquad$ Stagnation Points

Note that $V_r = 0$ along $r = a$, so this represents flow around a circular cylinder, as shown in Table 6.2.

Flow is irrotational, so the Bernoulli equation may be applied between any two points. Applying the equation between a point far upstream and a point on the surface of the cylinder (neglecting elevation differences), we obtain

$$\frac{p_\infty}{\rho} + \frac{U^2}{2} + gz = \frac{p}{\rho} + \frac{V^2}{2} + gz$$

Thus

$$p - p_\infty = \tfrac{1}{2}\rho(U^2 - V^2)$$

Along the surface, $r = a$, and

$$V^2 = V_\theta^2 = \left(-\frac{\Lambda}{a^2} - U\right)^2 \sin^2\theta = 4U^2\sin^2\theta$$

since $\Lambda = Ua^2$. Substituting yields

$$p - p_\infty = \tfrac{1}{2}\rho(U^2 - 4U^2\sin^2\theta) = \tfrac{1}{2}\rho U^2(1 - 4\sin^2\theta)$$

or

$$\frac{p - p_\infty}{\tfrac{1}{2}\rho U^2} = 1 - 4\sin^2\theta \qquad \longleftarrow \qquad \text{Pressure Distribution}$$

$\left\{\begin{array}{l}\text{This problem illustrates the techniques used to combine elementary plane flows to form} \\ \text{a flow pattern of practical interest.}\end{array}\right\}$

6-8 SUMMARY OBJECTIVES

After completing study of Chapter 6, you should be able to do the following:

1. Write Euler's equations in (a) vector form, (b) rectangular coordinates, (c) cylindrical coordinates, and (d) streamline coordinates.
2. Integrate Euler's equation along a streamline in steady flow to obtain the Bernoulli equation. State the restrictions on use of the Bernoulli equation.
3. Define static pressure, stagnation pressure, and dynamic pressure.
4. For steady, incompressible, inviscid flow through a streamtube, state the conditions under which the first law of thermodynamics reduces to the Bernoulli equation.
**5. Write the unsteady Bernoulli equation for flow along a streamline. State the restrictions on use of the equation.
**6. For a steady, inviscid, incompressible flow that is irrotational, show that the Bernoulli equation may be applied between any two points in the flow field.
**7. For a two-dimensional, irrotational, incompressible flow field:
    (a) Given the velocity field, determine the velocity potential, ϕ.
    (b) Given the velocity potential, determine the velocity field.
    (c) Show that lines of constant ψ and constant ϕ are orthogonal.
    (d) Given ψ, determine ϕ (and vice versa).
**8. Given the stream function, ψ, for a two-dimensional, irrotational flow about a body, determine:
    (a) the equation of the body.
    (b) the pressure distribution along the body surface.
9. Solve the problems at the end of the chapter that relate to the material you have studied.

REFERENCES

1. Robertson, J. M., *Hydrodynamics in Theory and Application*. Englewood Cliffs, NJ: Prentice-Hall, 1965.
2. Streeter, V. L., *Fluid Dynamics*. New York: McGraw-Hill, 1948.
3. Vallentine, H. R., *Applied Hydrodynamics*. London: Butterworths, 1959.
4. Rouse, H., and S. Ince, *History of Hydraulics*. New York: Dover, 1957.
5. Lamb, H., *Hydrodynamics*. New York: Dover, 1945.
6. Milne-Thomson, L. M., *Theoretical Hydrodynamics*, 4th ed. New York: Macmillan, 1960.
7. Karamcheti, K., *Principles of Ideal-Fluid Aerodynamics*. New York: Wiley, 1966.
8. Kirchhoff, R. H., *Potential Flows: Computer Graphic Solutions*. New York: Marcel Dekker, 1985.
9. Kuethe, A. M., and C.-Y. Chow, *Foundations of Aerodynamics: Bases of Aerodynamic Design*, 4th ed. New York: Wiley, 1986.
10. Hess, J. L., and A. M. O. Smith, "Calculation of Potential Flow about Arbitrary Bodies," in *Progress in Aeronautical Sciences*, Volume 8, ed. by D. Kuchemann. Oxford: Pergamon Press, 1966.
11. Shang, J. S., "An Assessment of Numerical Solutions of the Complete Navier-Stokes Equations," *AIAA J. Aircraft, 22,* 5, May 1985, pp. 353–370.

PROBLEMS

6.1 Consider the flow field with velocity given by $\vec{V} = (Axy - Bx^2)\hat{i} + (Axy - By^2)\hat{j}$, where $A = 2/\text{ft} \cdot \text{sec}$, $B = 1/\text{ft} \cdot \text{sec}$, and the coordinates are measured in feet. The density

** These objectives apply to sections that may be omitted without loss of continuity in the text material.

is 2 slug/ft³ and gravity acts in the negative y direction. Determine the acceleration of a fluid particle and the pressure gradient at point $(x, y) = (1, 1)$.

6.2 An incompressible flow field is given by $\vec{V} = (Ax + By)\hat{i} - Ay\hat{j}$, where $A = 1$ sec⁻¹, $B = 2$ sec⁻¹, and the coordinates are measured in meters. Find the magnitude and direction of the acceleration of a fluid particle at point $(x, y) = (1, 2)$. Find the pressure gradient at the same point, if $\vec{g} = -g\hat{j}$ and the fluid is water.

6.3 A horizontal flow of water is described by the velocity field $\vec{V} = (Ax + Bt)\hat{i} + (-Ay + Bt)\hat{j}$, where $A = 10$ sec⁻¹, $B = 5$ ft·sec⁻², x and y are in feet, and t is in seconds. Compute the acceleration of a fluid particle at point $(x, y) = (1, 5)$ at $t = 10$ sec. Evaluate $\partial p/\partial x$ under the same conditions.

6.4 A velocity field in a fluid with density of 1500 kg/m³ is given by $\vec{V} = (Ax - By)t\hat{i} - (Ay + Bx)t\hat{j}$, where $A = 1$ sec⁻², $B = 2$ sec⁻², x and y are in meters, and t is in seconds. Body forces are negligible. Evaluate ∇p at point $(x, y) = (1, 2)$ at $t = 1$ sec.

6.5 The x component of velocity in an incompressible flow field is given by $u = Ax$, where $A = 2$ sec⁻¹ and the coordinates are measured in meters. The pressure at point $(x, y) = (0, 0)$ is $p_0 = 190$ kPa (gage). The density is $\rho = 1.50$ kg/m³ and the z axis is vertical. Evaluate the simplest possible y component of velocity. Calculate the fluid acceleration and determine the pressure gradient at point $(x, y) = (2, 1)$. Find the pressure distribution along the positive x axis.

6.6 In a frictionless, incompressible flow, the velocity field in m/sec and the body force are given by $\vec{V} = Ax\hat{i} - Ay\hat{j}$ and $\vec{g} = -g\hat{k}$; the coordinates are measured in meters. The pressure is p_0 at point $(x, y, z) = (0, 0, 0)$. Obtain an expression for the pressure field, $p(x, y, z)$.

6.7 The velocity distribution in a two-dimensional steady flow field in the xy plane is $\vec{V} = (Ax - B)\hat{i} + (C - Ay)\hat{j}$, where $A = 2$ sec⁻¹, $B = 5$ m·sec⁻¹, and $C = 3$ m·sec⁻¹; the coordinates are measured in meters, and the body force distribution is $\vec{g} = -g\hat{k}$. Does the velocity field represent the flow of an incompressible fluid? Find the stagnation point of the flow field. Obtain an expression for the pressure gradient in the flow field. Evaluate the difference in pressure between point $(x, y) = (1, 3)$ and the origin, if the density is 1.2 kg/m³.

6.8 Consider a steady fluid motion with rigid-body swirl about the z axis. Assume $\vec{g} = -g\hat{k}$. Show that Eqs. 6.4 reduce to

$$\frac{\partial p}{\partial r} = \rho r \omega^2 \qquad \frac{\partial p}{\partial \theta} = 0 \qquad \frac{\partial p}{\partial z} = -\rho g$$

6.9 For the flow of Problem 4.123 show that the variation of uniform radial velocity is $V_r = Q/2\pi r h$. Obtain expressions for the r component of acceleration of a fluid particle in the gap and for the pressure variation as a function of radial distance from the central holes.

6.10 An incompressible, inviscid fluid flows into a horizontal round tube through its porous wall. The tube is closed at the left end and the flow discharges from the tube to the atmosphere at the right end. For simplicity, consider the x component of velocity in the tube uniform across any cross section. The density of the fluid is ρ, the tube diameter and length are D and L respectively, and the uniform inflow velocity is v_0. The flow is steady. Obtain an algebraic expression for the x component of acceleration of a fluid particle located at position x, in terms of v_0, x, and D. Find an expression for the pressure gradient, $\partial p/\partial x$, at position x. Integrate to obtain an expression for the gage pressure at $x = 0$.

6.11 Consider the flow field $\vec{V} = Ax\hat{i} - Ay\hat{j}$, where $A = 3$ sec⁻¹ and the coordinates are measured in feet. Gravity is in the negative z direction, $\vec{g} = -g\hat{k}$. Find the force of the fluid on the yz plane in the region bounded by $y = 1$, $y = 2$, $z = 0$, and $z = 2$, if the pressure at $(x, y, z) = (0, \frac{1}{2}, 0)$ is 250 pounds per square foot and the fluid has the density of water.

6.12 Consider the flow of Problem 5.40. Evaluate the magnitude and direction of the net pressure force that acts on the upper plate between r_i and R, if $r_i = R/2$.

6.13 Consider steady, incompressible flow without friction in the narrow gap between solid parallel disks. Assume that flow is uniform across each plane, $r = $ constant. Show that the Euler equations reduce to $dp = -\rho V dV$ for this situation. Show that the same result is obtained by applying the continuity and momentum equations to a segment, $d\theta$, of an annular differential control volume of thickness dr.

6.14 Consider again the flow field of Problem 5.53. Assume the flow is incompressible with $\rho = 1.23$ kg/m^3 and friction is negligible. Further assume the vertical air flow velocity is $v_0 = 15$ mm/sec, the half-width of the cavity is $L = 22$ mm, and its height is $h = 1.2$ mm. Calculate the pressure gradient at $(x, y) = (L, h)$. Obtain an equation for the flow streamlines in the cavity.

6.15 A rectangular microcircuit "chip" floats on a thin layer of air, $h = 0.5$ mm thick, above a porous surface. The chip width is $b = 20$ mm, as shown. Its length, L, is very long in the direction perpendicular to the diagram. There is no flow in the z direction. Assume flow in the x direction in the gap under the chip is uniform. Flow is incompressible and frictional effects may be neglected. Use a suitably chosen control volume to show that $U(x) = qx/h$ in the gap. Find a general expression for the acceleration of a fluid particle in the gap. Evaluate the maximum acceleration. Obtain an expression for the pressure gradient, $\partial p/\partial x$, and sketch the pressure distribution under the chip. Show p_{atm} on your sketch. Is the net pressure force on the chip directed upward or downward? Explain. For the conditions shown, with $q = 0.06$ m^3/sec/m, estimate the mass per unit length of the chip.

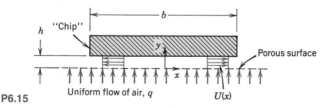

P6.15

6.16 A liquid layer separates two plane surfaces as shown. The lower surface is stationary; the upper surface moves downward at constant speed, V. The moving surface has width w, perpendicular to the plane of the diagram, and $w \gg L$. The incompressible liquid layer, of density ρ, is squeezed from between the surfaces. Assume the flow is uniform at any cross section and neglect viscosity as a first approximation. Use a suitably chosen control volume to show that $u = Vx/b$ within the gap, where $b = b_0 - Vt$. Obtain an algebraic expression for the acceleration of a fluid particle located at x. Determine the pressure gradient, $\partial p/\partial x$, in the liquid layer. Find the pressure distribution, $p(x)$. Obtain an expression for the net pressure force that acts on the upper (moving) flat surface.

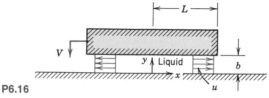

P6.16

6.17 Heavy weights can be moved with relative ease on air cushions by using a load pallet as shown. Air is supplied from the plenum through porous surface AB. It enters the gap vertically at uniform speed, q. Once in the gap, all air flows in the positive x

direction (there is no flow across the plane at $x = 0$). Assume air flow in the gap is incompressible and uniform at each cross section, with speed $u(x)$, as shown in the enlarged view. Although the gap is narrow ($h \ll L$), neglect frictional effects as a first approximation. Use a suitably chosen control volume to show that $u(x) = qx/h$ in the gap. Calculate the acceleration of a fluid particle in the gap. Evaluate the pressure gradient, $\partial p/\partial x$, and sketch the pressure distribution within the gap. Be sure to indicate the pressure at $x = L$.

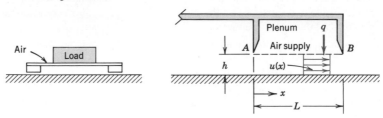

P6.17

6.18 The flow field in a forced vortex (rigid-body motion) is given by $\vec{V} = \omega r \hat{e}_\theta$, where $\omega = 10 \text{ sec}^{-1}$ and r is in meters. Assume a frictionless fluid with $\rho = 1000 \text{ kg/m}^3$. Express the radial pressure gradient, $\partial p/\partial r$, as a function of r. Evaluate the pressure change between $r_1 = 1$ m and $r_2 = 2$ m.

6.19 The flow field in a free (irrotational) vortex is given by $\vec{V} = \hat{e}_\theta K/2\pi r$, for $r > 0$, where r is in meters. Assume a frictionless fluid with $\rho = 1000 \text{ kg/m}^3$, and $K = 20\pi \text{ m}^2/\text{sec}$. Express the radial pressure gradient, $\partial p/\partial r$, as a function of r, and evaluate the pressure change between $r_1 = 1$ m and $r_2 = 2$ m.

6.20 Air at 20 psia and 100 F flows around a smooth corner at the inlet to a diffuser. The air speed is 150 ft/sec, and the radius of curvature of the streamlines is 3 in. Determine the magnitude of the centripetal acceleration experienced by a fluid particle rounding the corner. Express your answer in gs. Evaluate the pressure gradient, $\partial p/\partial r$.

6.21 Steady, frictionless, and incompressible flow from right to left over a stationary circular cylinder of radius a is given by the velocity field

$$\vec{V} = U\left[\left(\frac{a}{r}\right)^2 - 1\right]\cos\theta\, \hat{e}_r + U\left[\left(\frac{a}{r}\right)^2 + 1\right]\sin\theta\, \hat{e}_\theta$$

Consider flow along the streamline forming the cylinder surface, $r = a$. Express the components of the pressure gradient in terms of angle θ. Plot speed, V, as a function of r along the radial line $\theta = \pi/2$ for $r > a$.

6.22 To model the velocity distribution in the curved inlet section of a wind tunnel, the radius of curvature of the streamlines is expressed as $R = LR_0/2y$. As a first approximation, assume the air speed along each streamline is $V = 20$ m/sec. Evaluate the pressure change from $y = 0$ to the tunnel wall at $y = L/2$, if $L = 150$ mm and $R_0 = 0.6$ m.

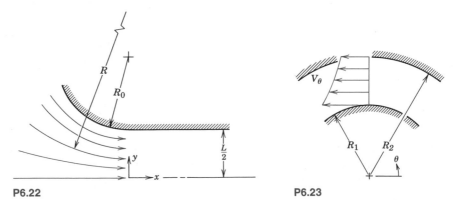

P6.22   **P6.23**

6.23 The radial variation of velocity at the midsection of the 180° bend shown is given by $r V_\theta$ = constant. The cross section of the bend is square. Assume that the velocity is not a function of z. Derive an equation for the pressure difference between the outside and the inside of the bend. Express your answer in terms of the mass flow rate, the fluid density, the geometric parameters, R_1 and R_2, and the depth of the bend, $h = R_2 - R_1$.

6.24 The y component of velocity in a two-dimensional incompressible flow field is given by $v = -Axy$, where v is in m/sec, the coordinates are measured in meters, and A is a dimensional constant. There is no velocity component or variation in the z direction. Calculate the acceleration of a fluid particle at point $(x, y) = (1, 2)$. Estimate the radius of curvature of the streamline passing through this point.

6.25 A pitot-static tube is immersed in a flowing stream of atmospheric air. A manometer indicates a dynamic pressure of 1.05 inches of water. Calculate the air speed.

6.26 A pitot-static tube is used to measure the centerline speed in a duct carrying room air at a pressure of 101 kPa (abs) and temperature of 32 C. Determine the differential pressure reading in millimeters of water that corresponds to an air speed of 28.5 m/sec.

6.27 A pitot-static tube is used to measure the speed of air flowing at standard conditions in a wind tunnel. The stagnation pressure and static pressure in the tunnel are found to be − 0.56 and −1.54 inches of water (gage), respectively. Calculate the speed of the flowing air.

6.28 An aircraft flies at an air speed of 315 kilometers per hour at 2500 m elevation through a standard atmosphere. Evaluate the stagnation pressure on the nose of the aircraft.

6.29 A pitot-static tube is used to measure the speed of air at standard conditions at a point in a flow. To ensure that the flow may be assumed incompressible for calculations of engineering accuracy, the speed is to be maintained at 100 m/sec or less. Determine the manometer deflection, in millimeters of water, that corresponds to the maximum desirable speed.

6.30 A jet of air from a nozzle is blown at right angles against a wall in which a pressure tap is located. A manometer connected to the tap shows a head of 0.14 in. of mercury above atmospheric. Determine the approximate speed of the air leaving the nozzle if it is at 40 F and 14.7 psia.

6.31 Maintenance work on high-pressure hydraulic systems requires special precautions. A small leak can result in a high-speed jet of hydraulic fluid that can penetrate the skin and cause serious injury (therefore troubleshooters are cautioned to use a piece of paper or cardboard, *not a finger,* to search for leaks). Calculate and plot the jet speed of a leak versus system pressure, for pressures up to 40 MPa (gage). Explain how a high-speed jet of hydraulic fluid can cause injury.

6.32 An open-circuit wind tunnel draws in air from the atmosphere through a well-contoured nozzle. In the test section, where the flow is straight and nearly uniform, a static pressure tap is drilled into the tunnel wall. A manometer connected to the tap shows that static pressure within the tunnel is 45 mm of water below atmospheric. Assume that the air is incompressible, and at 25 C, 100 kPa (abs). Calculate the air speed in the wind-tunnel test section.

6.33 The inlet contraction and test section of a laboratory wind tunnel are shown. The air speed in the test section is $U = 22.5$ m/sec. A total-head tube pointed upstream indicates that the stagnation pressure on the test section centerline is 6.0 mm of water below atmospheric. The corrected barometric pressure and temperature in the laboratory are 99.1 kPa (abs) and 23 C. Evaluate the dynamic pressure on the centerline of the wind tunnel test section. Compute the static pressure at the same point. Qualitatively compare the static pressure at the tunnel wall with that at the centerline. Explain why the two may not be identical.

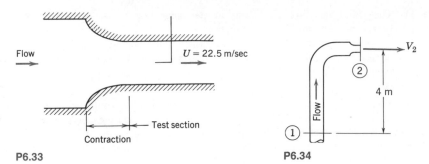

P6.33   **P6.34**

6.34 Water flows steadily up the vertical 0.1 m diameter pipe and out the nozzle, which is 0.05 m in diameter, discharging to atmospheric pressure. The stream velocity at the nozzle exit must be 20 m/sec. Calculate the gage pressure required at section ①, assuming frictionless flow.

6.35 Water flows in a circular pipe. At one section the diameter is 0.3 m, the static pressure is 260 kPa (gage), the velocity is 3 m/sec, and the elevation is 10 m above ground level. At a section downstream at ground level, the pipe diameter is 0.15 m. Find the gage pressure at the downstream section if frictional effects may be neglected.

6.36 Water may be considered to flow without friction through the siphon. The water flow rate is 0.03 m^3/sec, its temperature is 20 C, and the pipe diameter is 75 mm. Compute the maximum allowable height, h, so that the pressure at point A is above the vapor pressure of the water.

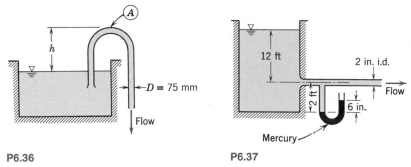

P6.36   **P6.37**

6.37 Water flows from a very large tank through a 2 in. diameter tube. The dark liquid in the manometer is mercury. Estimate the velocity in the pipe and the rate of discharge from the tank.

6.38 A stream of liquid moving at low speed leaves a nozzle pointed directly downward. The velocity may be considered uniform across the nozzle exit and the effects of friction may be ignored. At the nozzle exit, located at elevation z_0, the jet velocity and area are V_0 and A_0, respectively. Determine the variation of jet area with elevation.

6.39 In a laboratory experiment, water flows radially outward at moderate speed through the space between circular plane parallel disks. The perimeter of the disks is open to the atmosphere. The disks have diameter $D = 150$ mm and the spacing between the disks is $h = 0.8$ mm. The measured mass flow rate of water is $\dot{m} = 305$ g/sec. Assuming frictionless flow in the space between the disks, estimate the theoretical static pressure between the disks at radius $r = 50$ mm. In the laboratory situation, where *some* friction is present, would the pressure measured at the same location be above or below the theoretical value? Why?

6.40 Consider again the flat spray nozzle of Problem 4.91. Assume the thickness of the water sheet leaving the nozzle is $t = 0.5$ mm and the inside diameter of the inlet pipe is $D_1 = 20$ mm. Evaluate the minimum upstream pressure, p_1, required to produce a volume flow rate of $Q = 3.0$ m^3/hr through the nozzle assembly.

6.41 A fire nozzle is coupled to the end of a hose with inside diameter $D = 75$ mm. The nozzle is contoured smoothly and has outlet diameter $d = 25$ mm. The design inlet pressure for the nozzle is $p_1 = 689$ kPa (gage). Evaluate the maximum flow rate the nozzle could deliver.

6.42 Consider steady, frictionless, incompressible flow of air over the wing of an airplane. The air approaching the wing is at 10 psia, 40 F, and has a speed of 200 ft/sec relative to the wing. At a certain point in the flow, the pressure is −0.40 psi (gage). Calculate the speed of the air relative to the wing at this point.

6.43 A seawater inlet for reactor cooling is to be located on the outer hull of a nuclear submarine. The maximum submerged speed of the sub is 35 knots. At the location of the inlet, the water speed parallel to the hull is 20 knots when the sub moves at maximum speed. Determine the maximum static pressure that might be expected at the inlet at 100 m depth below the free surface.

6.44 A mercury barometer is carried in a car on a day when there is no wind. The temperature is 20 C and the corrected barometer height is 761 mm of mercury. One window is open slightly as the car travels at 105 km/hr. The barometer reading in the moving car is 5 mm lower than when the car is stationary. Explain what is happening. Calculate the local speed of the air flowing past the window, *relative to the automobile*.

6.45 An Indianapolis racing car travels at a maximum speed of 350 km/hr. A pressure gage attached to the airfoil reads −50 mm of water (gage). Estimate the air speed relative to the car at that location.

6.46 A person riding in an automobile at 60 mph holds a pitot-static tube far enough out the window that it faces upstream in undisturbed flow. The tube is connected to a differential manometer inside the car. Assuming no wind, determine the manometer reading that would be observed. What would be the manometer reading if the car traveled into a headwind of 20 mph?

6.47 An Indianapolis racing car travels at 98.3 m/sec along a straightaway. The team engineer wishes to locate an air inlet on the body of the car to obtain cooling air for the driver's suit. The plan is to place the inlet at a location where the air speed is 25.5 m/sec along the surface of the car. Calculate the static pressure at the proposed inlet location. Express the pressure rise above ambient as a fraction of the freestream dynamic pressure.

6.48 A miniature submarine is propelled by a water jet system. The sub "swims" upstream in a river at 10.5 knots relative to the water. The inlet duct centerline is located 8.5 ft below the surface of the river. At section ① in the duct, the water speed is 23.5 ft/sec relative to the sub. Compute the static and stagnation gage pressures at section ①.

6.49 A fire nozzle is coupled to the end of a hose with inside diameter $D = 3.0$ in. The nozzle is smoothly contoured and its outlet diameter is $d = 1.0$ in. The nozzle is designed to operate at an inlet water pressure of 100 psig. Determine the design flow rate of the nozzle. (Express your answer in gpm.) Evaluate the axial force required to hold the nozzle in place. Indicate whether the hose coupling is in tension or compression.

6.50 Water flows steadily through a 3.25 in. diameter pipe and discharges through a 1.25 in. diameter nozzle to atmospheric pressure. The flow rate is 24.5 gpm. Calculate the minimum static pressure required in the pipe to produce this flow rate. Evaluate the axial force of the nozzle assembly on the pipe flange.

6.51 A smoothly contoured nozzle, with outlet diameter $d = 20$ mm, is coupled to a straight pipe by means of flanges. Water flows in the pipe, of diameter $D = 50$ mm, and the nozzle discharges to the atmosphere. For steady flow and neglecting the effects of viscosity, find the volume flow rate in the pipe corresponding to a calculated axial force of 45.5 N needed to keep the nozzle attached to the pipe.

6.52 Water flows steadily through the reducing elbow shown. The elbow is smooth and short, and the flow accelerates, so the effect of friction is small. The volume flow rate is $Q = 1.27$ L/sec. The elbow is in a horizontal plane. Estimate the gage pressure at section ①. Calculate the x component of the force exerted *by* the reducing elbow *on* the supply pipe.

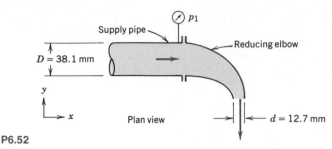

P6.52

6.53 The flow system of parallel disks shown contains water. As a first approximation, friction may be neglected. Determine the volume flow rate and the pressure at point ©. ($R = 300$ mm and $r_C = 150$ mm.)

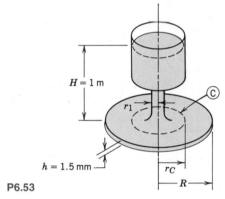

P6.53

6.54 A water jet is directed upward from a well-designed nozzle of area $A_1 = 600$ mm²; the exit jet speed is $V_1 = 6.3$ m/sec. The flow is steady and the liquid stream does not break up. Point 2 is located $H = 1.55$ m above the nozzle exit plane. Determine the velocity in the undisturbed jet at point 2. Calculate the pressure that would be sensed by a stagnation tube located there. Evaluate the force that would be exerted on a flat plate placed normal to the stream at point 2. Sketch the pressure distribution on the plate.

6.55 An object, with a flat horizontal lower surface, moves downward into the jet of the spray system of Problem 4.72 with speed U = 5 ft/sec. Determine the minimum supply pressure needed to produce the jet leaving the spray system at V = 15 ft/sec. Calculate the maximum pressure exerted by the liquid jet on the flat object at the instant when the object is h = 1.5 ft above the jet exit. Estimate the force of the water jet on the flat object.

6.56 According to best-selling author Tom Clancy, the Soviet nuclear submarine *Red October* is propelled by the water jet propulsion system shown. When operating at depth $H = 30.5$ m below the free surface, estimate the static pressure at section ①. Calculate the pressure required at the inlet to the propulsion nozzle, section ②. Compute the minimum pumping power needed to propel the vessel.

6.57 The tank shown has a well-rounded orifice with area A_j. At $t = 0$, the water level is at height h_0. Develop an expression for dimensionless water height, h/h_0, at any later time.

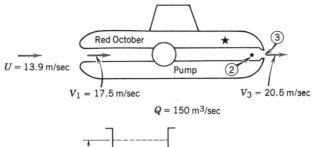

$U = 13.9$ m/sec

$V_1 = 17.5$ m/sec

$V_3 = 20.5$ m/sec

$Q = 150$ m³/sec

P6.56

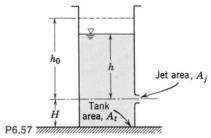

P6.57

6.58 The water level in a large tank is maintained at height H above the surrounding level terrain. A rounded orifice placed in the side of the tank discharges a horizontal jet. Neglecting friction, determine the height at which the orifice should be placed so the water strikes the ground at the maximum horizontal distance from the tank.

6.59 A horizontal axisymmetric jet of air with 10 mm diameter strikes a stationary vertical disk of 200 mm diameter. The jet speed is 50 m/sec at the nozzle exit. A manometer is connected to the center of the disk. Calculate (a) the deflection, if the manometer liquid has SG = 1.75, and (b) the force exerted by the jet on the disk. Sketch the streamline pattern and the distribution of pressure on the face of the disk.

6.60 Obtain an expression for the pressure distribution as a function of radius for the "air hockey" puck of Problem 5.41. Estimate the mass of the puck.

6.61 The tennis "bubble" of Problem 3.11 is subjected to a wind that blows at 50 km/hr in a direction perpendicular to the axis of the semicylindrical shape. By using polar coordinates, with angle θ measured from the ground on the upwind side of the structure, the resulting pressure distribution may be expressed as

$$\frac{p - p_\infty}{\frac{1}{2}\rho V_w^2} = 1 - 4\sin^2\theta$$

where p is the pressure at the surface, p_∞ the atmospheric pressure, and V_w the wind speed. Determine the net vertical force exerted on the structure.

6.62 Steady, frictionless, and incompressible flow from left to right over a stationary circular cylinder, of radius a, is represented by the velocity field

$$\vec{V} = U\left[1 - \left(\frac{a}{r}\right)^2\right]\cos\theta\,\hat{e}_r - U\left[1 + \left(\frac{a}{r}\right)^2\right]\sin\theta\,\hat{e}_\theta$$

Obtain an expression for the pressure distribution along the streamline forming the cylinder surface, $r = a$. Determine the locations where the static pressure on the cylinder is equal to the freestream static pressure. Evaluate the net pressure force on the cylinder.

6.63 The flow over a Quonset hut may be approximated by the velocity distribution of Problem 6.62 with $0 \le \theta \le \pi$ (see next page). During a storm the wind speed reaches 100 km/hr; the outside temperature is 5 C. A barometer inside the hut reads 720 mm of mercury; pressure p_∞ is also 720 mm Hg. The hut has a diameter of 6 m and a length of 18 m. Determine the net force tending to lift the hut off its foundation.

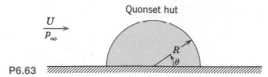

P6.63

6.64 Water flows at low speed through a circular tube with inside diameter of 50 mm. A smoothly contoured plug of 40 mm diameter is held in the end of the tube where the water discharges to atmosphere. Neglect frictional effects and assume uniform velocity profiles at each section. Determine the pressure measured by the gage and the force required to hold the plug.

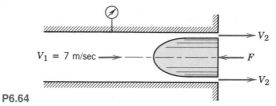

P6.64

6.65 The pressure delivered by the turbocharger on an engine used for the Indianapolis 500 mile race is controlled by a pressure relief valve. The valve is set to release when the pressure in the intake manifold reaches 75 in. of mercury above the pressure outside the valve. The pressure relief valve is located at a point on the car where the local air speed is 1.3 times the air speed in the freestream far from the car. For a standard day and a car speed of 220 mph, estimate the static pressure near the pressure relief valve. Specify clearly whether this is a gage or absolute pressure. What is the effect on the boost pressure delivered to the engine?

6.66 High-pressure air forces a stream of water from a tiny, rounded orifice, of area A, in a tank. The pressure is high enough that gravity may be neglected. The air expands slowly, so that the expansion may be considered isothermal. The initial volume of air in the tank is $\mathbb{V}_0$. At later instants the volume of air is $\mathbb{V}(t)$; the total volume of the tank is $\mathbb{V}_t$. Obtain an algebraic expression for the mass flow rate of water leaving the tank. Find an algebraic expression for the rate of change in mass of the water inside the tank. Develop an ordinary differential equation and solve for the water mass in the tank at any instant. Sketch a graph of water mass in the tank versus time.

6.67 A tank with a *reentrant* orifice called a *Borda mouthpiece* is shown. The fluid is inviscid and incompressible. The reentrant orifice essentially eliminates flow along the tank walls, so the pressure there is nearly hydrostatic. Calculate the *contraction coefficient*, $C_c = A_j/A_0$. *Hint:* Equate the unbalanced hydrostatic pressure force and momentum flux from the jet.

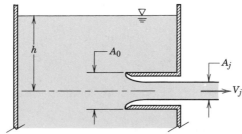

P6.67

**6.68 Compressed air is used to accelerate water from a tube. Neglect the velocity in the reservoir and assume the flow in the tube is uniform at any section. At a particular instant, it is known that $V = 2$ m/sec and $dV/dt = 2.50$ m/sec^2. The cross-sectional area of the tube is $A = 0.02$ m^2. Determine the pressure in the tank at this instant.

** These problems require material from sections that may be omitted without loss of continuity in the text material.

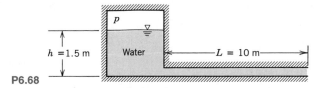

P6.68

**6.69 The air speed in the inlet manifold of an internal-combustion engine is approximated as $V = V_0(1 + \sin\omega t)$. Assume $V_0 = 30$ m/sec, $\omega = 50$ Hz, and that the density is half that of standard air. If the length of the inlet passage is $L = 0.3$ m, calculate the pressure variation required to cause the motion. Neglect frictional effects.

**6.70 A frictionless, incompressible fluid flows through a nozzle whose area varies linearly with distance according to $A = A_1(1 - ax/L)$. The volume flow rate is unsteady and described by $Q = Q_0(1 + qt)$. Section ② is located distance L downstream from section ①. Develop an expression for the pressure difference, $p_2 - p_1$.

**6.71 Consider the reservoir and disk flow system of Problem 6.53, with the reservoir level maintained constant. Flow between the disks is started from rest at $t = 0$. Evaluate the rate of change of volume flow rate at $t = 0$, if $r_1 = 50$ mm.

**6.72 Apply the unsteady Bernoulli equation to the U-tube manometer of constant diameter shown. Assume that the manometer is initially deflected and then released. Obtain a differential equation for l as a function of time.

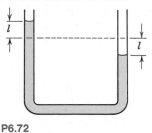

P6.72

**6.73 Two circular disks, of radius R, are separated by distance b. The upper disk moves toward the lower one at constant speed V. The space between the disks is filled with a frictionless, incompressible fluid, which is squeezed out as the disks come together. Assume that, at any radial section, the velocity is uniform across the gap width, b. However, note that b is a function of time. The pressure surrounding the disks is atmospheric. Determine the gage pressure at $r = 0$.

**6.74 Consider the tank of Problem 4.44. Using the Bernoulli equation for unsteady flow along a streamline, evaluate the minimum diameter ratio, D/d, required to justify the assumption that flow from the tank is quasi-steady.

**6.75 Consider incompressible flow of standard air, with velocity field given by $\vec{V} = Ax\hat{i} - Ay\hat{j} + B\hat{k}$, where $A = 10$ sec^{-1}, $B = 30$ m/sec, and the coordinates are measured in meters. Neglect gravity. Is it possible to calculate the pressure change between points (0, 0, 0) and (3, 1, 0)? If possible, do so.

**6.76 Consider the flow of water represented by the velocity field $\vec{V} = Ay\hat{i} + Ax\hat{j}$, where $A = 3$ sec^{-1} and the coordinates are measured in meters. Is it possible to calculate the pressure change between points (0, 0, 0) and (1, 1, 1)? If possible, do so.

**6.77 An incompressible flow field is given as $\vec{V} = Ax^2y\hat{i} - Axy^2\hat{j}$, where $A = 6$/m$^2 \cdot$ sec and the coordinates are measured in meters. Evaluate the rotation of the flow. What can be said about the use of the Bernoulli equation for this flow?

** These problems require material from sections that may be omitted without loss of continuity in the text material.

**6.78 Determine whether the Bernoulli equation can be applied between different radii for the vortex flow fields (a) $\vec{V} = \omega r \hat{e}_\theta$ and (b) $\vec{V} = \hat{e}_\theta \dfrac{K}{2\pi r}$.

**6.79 Consider the flow represented by the stream function $\psi = Ax^2y$, where A is a dimensional constant equal to 2.5 m$^{-1}\cdot$sec^{-1}. The density is 2.45 slug/ft^3. Is the flow rotational? Can the pressure difference between points $(x, y) = (1, 4)$ and $(2, 1)$ be evaluated? If so, calculate it, and if not, explain why.

**6.80 Consider the flow field represented by the stream function $\psi = Axy + Ay^2$, where $A = 1$ sec^{-1}. Show that this represents a possible incompressible flow field. Evaluate the rotation of the flow. Plot a few streamlines in the upper half plane.

**6.81 The velocity field for a two-dimensional flow is $\vec{V} = (Ax - By)t\hat{i} - (Bx + Ay)t\hat{j}$, where $A = 1$ sec^{-1}, $B = 2$ sec^{-2}, t is in seconds, and the coordinates are measured in meters. Is this a possible incompressible flow? Is the flow steady or unsteady? Show that the flow is irrotational and derive an expression for the velocity potential.

**6.82 The stream function of a flow field is $\psi = Ax^2y - By^3$, where $A = 1$ m$^{-1}\cdot$sec^{-1}, $B = \frac{1}{3}$ m$^{-1}\cdot$sec^{-1}, and the coordinates are measured in meters. Find an expression for the velocity potential.

**6.83 A flow field is represented by the stream function $\psi = x^2 - y^2$. Find the corresponding velocity field. Show that this flow field is irrotational and obtain the potential function.

**6.84 Consider the flow field represented by the potential function $\phi = x^2 - y^2$. Verify that this is an incompressible flow and obtain the corresponding stream function.

**6.85 Consider the flow field represented by the potential function $\phi = Ax^2 + Bxy - Ay^2$. Verify that this is an incompressible flow and determine the corresponding stream function.

**6.86 Consider the flow field represented by the velocity potential $\phi = Ax + Bx^2 - By^2$, where $A = 1$ sec^{-1}, $B = 1$ m$^{-1}\cdot$sec^{-1}, and the coordinates are measured in meters. Obtain expressions for the velocity field and the stream function. Calculate the pressure difference between the origin and point $(xy) = (1, 2)$.

**6.87 A flow field is represented by the potential function $\phi = Ay^3 - Bx^2y$, where $A = \frac{1}{3}$ m$^{-1}\cdot$sec^{-1}, $B = 1$ m$^{-1}\cdot$sec^{-1}, and the coordinates are measured in meters. Obtain an expression for the magnitude of the velocity vector. Find the stream function for the flow. Plot streamlines $\psi = 0$ and $\psi = -4$ m^2/sec for $x \geq 0$.

**6.88 An incompressible flow field is characterized by the stream function $\psi = 3Ax^2y - Ay^3$, where $A = 1$ m$^{-1}\cdot$sec^{-1}. Sketch a few streamlines in the first quadrant. Show that this flow field is irrotational. Derive the velocity potential for the flow.

**6.89 The velocity distribution in a two-dimensional, steady, inviscid flow field in the xy plane is $\vec{V} = (Ax + B)\hat{i} + (C - Ay)\hat{j}$, where $A = 3$ sec^{-1}, $B = 6$ m/sec, $C = 4$ m/sec, and the coordinates are measured in meters. The body force distribution is $\vec{B} = -g\hat{k}$ and the density is 825 kg/m^3. Does this represent a possible incompressible flow field? Sketch the streamlines in the upper half plane. Find the stagnation point(s) of the flow field. Is the flow irrotational? If so, obtain the potential function. Evaluate the pressure difference between the origin and point $(x, y, z) = (2, 2, 2)$.

**6.90 A certain irrotational flow field in the xy plane has the stream function $\psi = Bxy$, where $B = 0.25$ sec^{-1}, and the coordinates are measured in meters. Sketch several streamlines in the first quadrant. Label the direction of increasing ψ. Indicate the direction of flow. Determine the rate of flow between points $(x, y) = (2, 2)$ and $(3, 3)$. Find the velocity potential for this flow. Sketch some lines of constant velocity potential on the streamline plot. Label the direction of increasing ϕ.

** These problems require material from sections that may be omitted without loss of continuity in the text material.

**6.91   Consider the flow field of water given by $\vec{V} = Ax^2y^2\hat{i} - Bxy^3\hat{j}$, where $A = 3/m^3 \cdot \sec$ and $B = 2/m^3 \cdot \sec$. Determine the stream function for this flow. Determine the fluid rotation. Neglecting gravity, is it possible to calculate the pressure difference between points $(0, 0, 0)$ and $(1, 1, 1)$? If so, calculate it, and if not, explain why.

**6.92   Beginning with the velocity components for a simple source shown in Table 6.1, verify the expressions for the stream function and velocity potential.

**6.93   Beginning with the velocity components for a simple sink shown in Table 6.1, verify the expressions for the stream function and velocity potential.

**6.94   Beginning with the velocity components for a free vortex shown in Table 6.1, verify the expressions for the stream function and velocity potential.

**6.95   Consider the superposition of two flows: (1) velocity in the positive x direction varying linearly from zero at $y = 0$ to 100 ft/sec at $y = 4$ and (2) velocity in the positive y direction with speed of 20 ft/sec. Find the stream function for the combined flow. Evaluate the volume flow rate between points $(x, y) = (1, 1)$ and $(2, 2)$. Find the velocity potential for the combined flow if one exists.

**6.96   Consider the flow past a circular cylinder of Example Problem 6.11. Plot the pressure coefficient

$$C_p = \frac{p - p_\infty}{\frac{1}{2}\rho U^2}$$

along the surface of the cylinder versus θ for $0 \le \theta \le \pi$. Find the maximum and minimum values of C_p. Locate the points on the cylinder surface where the static pressure is equal to the freestream pressure. Evaluate the angular sensitivity where $C_p = 0$.

**6.97   A spiral vortex is formed by combining a source and a vortex. Show that the streamlines of this combined flow are logarithmic spirals such that $r = ce^{\pm q\theta/K}$.

**6.98   Consider the flow past a circular cylinder, of radius a, used in Example Problem 6.11. Show that $V_r = 0$ along the lines $(r, \theta) = (r, \pm\pi/2)$. Plot V_θ/U versus radius for $r \ge a$, along the line $(r, \theta) = (r, \pi/2)$. Find the distance beyond which the influence of the cylinder is less than 1 percent of U.

**6.99   Consider the flow past a circular cylinder, of radius a, used in Example Problem 6.11. Integrate the pressure distribution around the cylinder surface to show that the drag and lift forces predicted by this ideal flow model are zero.

**6.100   To model flow around a circular cylinder with circulation, add a clockwise vortex to the flow considered in Example Problem 6.11. Find the stream function, velocity potential, and velocity field for this combined flow. Locate the stagnation points on the cylinder surface. Evaluate the range of vortex strengths for which two stagnation points exist.

**6.101   Consider again the ideal fluid flow around a circular cylinder with circulation given in Problem 6.100. Show that the circulation may be expressed as $\Gamma = -K$ by integrating around the cylinder surface. Integrate the pressure distribution to evaluate the aerodynamic lift (force in the y direction) on the cylinder. Show that the lift force is $L = -\rho U\Gamma$.

**6.102   A crude model of a tornado is formed by combining a sink, of strength $q = 2800$ m^2/sec, and a free vortex, of strength $K = 5600$ m^2/sec. Obtain the stream function and velocity potential for this flow field. Estimate the radius beyond which the flow may be treated as incompressible. Find the gage pressure at that radius.

**6.103   Consider the flow obtained by superposing a source, of strength q, and a free vortex, of strength K. The velocity components at point A, where $(r_A, \theta_A) = (2, \pi/2)$, are

** These problems require material from sections that may be omitted without loss of continuity in the text material.

$V_r = 2.5$ ft/sec and $V_\theta = 3.75$ ft/sec. Determine the source strength, q, and the vortex strength, K. Neglecting body forces, find the pressure difference between points A and B, where $(r_B, \theta_B) = (1.25, \pi/2)$, if the density is that of water. Find the equation of the line representing $\psi = 0$ (i.e., for $\psi = 0$, express r as a function of θ).

**6.104 A source and a sink of equal strength are placed on the x axis at $x = -a$ and $x = a$, respectively. Obtain the stream function and velocity potential for the resulting flow field. Find points on the y axis between which half the total volume flow rate passes.

**6.105 Consider incompressible, inviscid flow in the xy plane. Assume that two sources of equal strength, q, are located along the x axis at $x = \pm a$. Sketch streamlines for the flow field produced by the source pair. Indicate the location of the stagnation point(s) of the flow. Evaluate the stream function for the combined flow field at point $(x, y) = (a, a)$. Determine the volume flow rate that passes between the y axis and point $(x, y) = (a, a)$. If pressure at the stagnation point is p_0, obtain expressions for the static and dynamic pressures at point $(x, y) = (a, a)$.

**6.106 The flow pattern of a vortex pair is to be investigated. A clockwise vortex is located at $(x, y) = (-a, 0)$, and a counterclockwise one of equal strength is at $(x, y) = (a, 0)$. Find the stream function, velocity potential, and velocity field for this combined flow. Evaluate the pressure distribution along the y axis.

**6.107 A source and a sink with strengths of equal magnitude, $q = 3\pi$ m²/sec, are placed on the x axis at $x = -a$ and $x = a$, respectively. A uniform flow, with speed $U = 20$ m/sec, in the positive x direction, is added to obtain the flow past a Rankine body. Obtain the stream function, velocity potential, and velocity field for the combined flow. Find the value of $\psi = $ constant on the stagnation streamline. Locate the stagnation points if $a = 0.3$ m.

**6.108 Consider again the flow past a Rankine body of Problem 6.107. The half-width, h, of the body in the y direction is given by the transcendental equation

$$\frac{h}{a} = \cot\left(\frac{\pi U h}{q}\right)$$

Evaluate the half-width, h. Find the local velocity and the pressure at points $(x, y) = (0, \pm h)$. Assume the fluid density is that of standard air.

**6.109 A flow field is formed by combining a uniform flow in the positive x direction, with $U = 10$ m/sec, and a counterclockwise vortex, with strength $K = 4\pi$ m²/sec, located at the origin. Obtain the stream function, velocity potential, and velocity field for the combined flow. Locate the stagnation point(s) for the flow. Sketch the streamline that passes through point $(r, \theta) = (5K/\pi U, \pi)$.

**6.110 Consider the flow field formed by combining a uniform flow in the positive x direction and a source located at the origin. Obtain expressions for the stream function, velocity potential, and velocity field for the combined flow. If $U = 25$ m/sec, determine the source strength if the stagnation point is located at $x = -0.5$ m. Sketch the stagnation streamline. Evaluate the locations of the branches of the stagnation streamline far downstream.

‡**6.111 Consider the flow field formed by combining a uniform flow in the positive x direction and a source located at the origin. Let $U = 30$ m/sec and $q = 150$ m²/sec. Use a calculator or simple computer program to locate the points on the stagnation streamline where the velocity reaches its maximum value. Find the gage pressure there if the fluid density is 1.2 kg/m³.

** These problems require material from sections that may be omitted without loss of continuity in the text material.

‡ You may wish to use simple computer programs to help solve problems marked with daggers.

**6.112 Consider the flow field formed by combining a uniform flow in the positive x direction with a sink located at the origin. Let $U = 50$ m/sec and $q = 90$ m^2/sec. Use a suitably chosen control volume to evaluate the net force per unit depth needed to hold in place (in standard air) the surface shape formed by the stagnation streamline.

**6.113 A flow field is to be constructed by combining a vortex pair and a uniform flow. A counterclockwise vortex is placed at $(x, y) = (0, a)$, and a clockwise one of equal strength is at $(0, -a)$. The uniform flow is in the positive y direction. Obtain the stream function, velocity potential, and velocity field for this combined flow. Evaluate the uniform flow speed needed to form a single closed streamline with stagnation point at $(0, \pm a/2)$. Sketch this streamline.

Chapter 7

DIMENSIONAL ANALYSIS AND SIMILITUDE

Because so few real flows can be solved exactly by analytical methods alone, the development of fluid mechanics has depended heavily on experimental results. Solutions of real problems usually involve a combination of analysis and experimental information. First, the real physical flow situation is approximated with a mathematical model that is simple enough to yield a solution. Then experimental measurements are made to check the analytical results. Based on the measurements, refinements in the analysis are made. The experimental results are an essential link in this iterative process. Empirical designs, developed without analysis or careful review of available experimental data, are often high in cost and poor or inadequate in performance.

However, experimental work in the laboratory is both time-consuming and expensive. One obvious goal is to obtain the most information from the fewest experiments. Dimensional analysis is an important tool that often helps us to achieve this goal. The dimensionless parameters that we obtain also can be used to correlate data for presentation using the minimum possible number of plots.

When experimental testing of a full-size prototype is either impossible or prohibitively expensive (as happens so often), the only feasible way of attacking the problem is through model testing in the laboratory. If we are to predict the prototype behavior from measurements on the model, it is obvious that we cannot run just any test on any model. The model flow and the prototype flow must be related by known scaling laws. We shall investigate the conditions necessary to obtain this similarity of model and prototype flows following the discussion of dimensional analysis.

7-1 NATURE OF DIMENSIONAL ANALYSIS

Most phenomena in fluid mechanics depend in a complex way on geometric and flow parameters. For example, consider the drag force on a stationary smooth sphere immersed in a uniform stream. What experiments must be conducted to determine the drag force on the sphere? To answer this question, we must specify the parameters that are important in determining the drag force. Clearly, we would expect the drag force to depend on the size of the sphere (characterized by the diameter, D), the fluid velocity, V, and the fluid viscosity, μ. In addition, the density of the fluid, ρ, also might be important. Representing the drag force by F, we can write the symbolic equation

$$F = f(D, V, \rho, \mu)$$

Although we may have neglected parameters on which the drag force depends, such as surface roughness (or have included parameters on which it does not depend), we have formulated the problem of determining the drag force for a stationary sphere in terms of quantities that are both controllable and measurable in the laboratory.

Let us imagine a series of experiments to determine the dependence of F on the variables, D, V, ρ, and μ. After building a suitable experimental facility, the work could begin. To obtain a curve of F versus V for fixed values of ρ, μ, and D, we might need tests at 10 values of V. To explore the diameter effect, each test would be repeated for spheres of 10 different diameters. If the procedure were repeated for 10 values of ρ and μ in turn, simple arithmetic shows that 10^4 separate tests would be needed. Assuming each test takes $\frac{1}{2}$ hour and we work 8 hours per day, the testing will require $2\frac{1}{2}$ years to complete. Needless to say, there also would be some difficulty in presenting the data. By plotting F versus V, with D as a parameter, for each combination of density and viscosity values, all of the data could be presented on a total of 100 sheets of graph paper. The utility of such results would be limited at best.

Fortunately, we can obtain meaningful results with significantly less effort through the use of dimensional analysis. As shown in Example Problem 7.1, all the data for the drag force on a smooth sphere can be plotted as a functional relation between two nondimensional parameters in the form

$$\frac{F}{\rho V^2 D^2} = f_1\left(\frac{\rho V D}{\mu}\right)$$

The form of the function still must be determined experimentally. However, rather than needing to conduct 10^4 experiments, we could establish the nature of the function as accurately with only 10 tests. The time saved in performing only 10 rather than 10^4 tests is obvious. Even more important is the greater experimental convenience. No longer must we find fluids with 10 different values of density and viscosity. Nor must we make 10 spheres of different diameters. Instead, only the ratio, $\rho V D/\mu$, must be varied. This can be accomplished simply by changing the velocity, for example.

The Buckingham Pi theorem is a statement of the relation between a function expressed in terms of dimensional parameters and a related function expressed in terms of nondimensional parameters. Use of the Buckingham Pi theorem allows us to develop the important nondimensional parameters quickly and easily.

7-2 BUCKINGHAM PI THEOREM

Given a physical problem in which the dependent parameter is a function of $n-1$ independent parameters, we may express the relationship among the variables in functional form as

$$q_1 = f(q_2, q_3, \ldots, q_n)$$

where q_1 is the dependent parameter, and $q_2, q_3, \ldots, q_n$ are the $n-1$ independent parameters. Mathematically, we can express the functional relationship in the equivalent form

$$g(q_1, q_2, \ldots, q_n) = 0$$

where g is an unspecified function, different from f. For the drag on a sphere we wrote the symbolic equation

$$F = f(D, V, \rho, \mu)$$

We could just as well have written

$$g(F, D, V, \rho, \mu) = 0$$

The Buckingham Pi theorem [1] states that: Given a relation among n parameters of the form

$$g(q_1, q_2, \ldots, q_n) = 0$$

then the n parameters may be grouped into $n - m$ independent dimensionless ratios, or Π parameters, expressible in functional form by

$$G(\Pi_1, \Pi_2, \ldots, \Pi_{n-m}) = 0$$

or

$$\Pi_1 = G_1(\Pi_2, \Pi_3, \ldots, \Pi_{n-m})$$

The number m is usually,[1] but not always, equal to the minimum number of independent dimensions required to specify the dimensions of all the parameters, $q_1, q_2, \ldots, q_n$.

The theorem does not predict the functional form of G or G_1. The functional relation among the independent dimensionless Π parameters must be determined experimentally.

The $n - m$ Π parameters are not unique. A Π parameter is not independent if it can be formed from a product or quotient of the other parameters of the problem. For example, if

$$\Pi_5 = \frac{2\Pi_1}{\Pi_2 \Pi_3} \qquad \text{or} \qquad \Pi_6 = \frac{\Pi_1^{3/4}}{\Pi_3^2}$$

then neither Π_5 nor Π_6 is independent of the other dimensionless parameters.

Several methods for determining the dimensionless parameters are available. A detailed procedure is presented in the next section.

7-3 DETERMINING THE Π GROUPS

Regardless of the method to be used to determine the dimensionless parameters, one begins by listing all parameters that are known (or believed) to affect the given flow phenomenon. Some experience admittedly is helpful in compiling the list. Students, who do not have this experience, often are troubled by the need to apply engineering judgment in an apparent massive dose. However, it is difficult to go wrong if a generous selection of parameters is made.

If you suspect that a phenomenon depends on a given parameter, include it. If your suspicion is correct, experiments will show that the parameter must be included to get consistent results. If the parameter is extraneous, an extra Π parameter may result, but experiments will show that it may be eliminated from consideration. Therefore, do not be afraid to include *all* the parameters that you feel are important.

The six steps listed below outline a recommended procedure for determining the Π parameters:

Step 1. *List all the parameters involved.*   (Let n be the number of parameters.) If all the pertinent parameters are not included, a relation may be obtained, but it will not give the complete story. If parameters that actually have

[1] See Example Problem 7.3.

no effect on the physical phenomenon are included, either the
dimensional analysis will show that these do not enter the relation
or one or more dimensionless groups will be obtained that expe.
will show to be extraneous.

Step 2. *Select a set of fundamental (primary) dimensions, e.g., MLt or*
(Note that for heat transfer problems you may also need T for tempc
ature, and in electrical systems, q for charge.)

Step 3. *List the dimensions of all parameters in terms of primary dimensions.*
(Let r be the number of primary dimensions.) Either force or mass may
be selected as a primary dimension.

Step 4. *Select from the list of parameters a number of repeating parameters equal
to the number of primary dimensions, r, and including all the primary
dimensions.* No two repeating parameters may have the same net dimensions differing by only a single exponent; for example, do not include both
a length (L) and a moment of inertia of an area (L^4) as repeating parameters. The repeating parameters chosen may appear in all the dimensionless
groups obtained; consequently, do *not* include the dependent parameter
among those selected in this step.

Step 5. *Set up dimensional equations, combining the parameters selected in Step
4 with each of the other parameters in turn, to form dimensionless groups.*
(There will be $n-m$ equations.) Solve the dimensional equations to obtain
the $n-m$ dimensionless groups.

Step 6. *Check to see that each group obtained is dimensionless.* If mass was
initially selected as a primary dimension, it is wise to check the groups
using force as a primary dimension, or vice versa.

The functional relationship among the Π parameters must be determined experimentally. The detailed procedure for determining the dimensionless Π parameters is
illustrated in Example Problems 7.1 and 7.2.

EXAMPLE 7.1—Drag Force on a Smooth Sphere

As noted in Section 7-1, the drag force, F, on a smooth sphere depends on the relative
velocity, V, the sphere diameter, D, the fluid density, ρ, and the fluid viscosity, μ.
Obtain a set of dimensionless groups that can be used to correlate experimental data.

EXAMPLE PROBLEM 7.1

GIVEN: $F = f(\rho, V, D, \mu)$ for a smooth sphere.

FIND: An appropriate set of dimensionless groups.

SOLUTION:
(Circled numbers refer to steps in the procedure for determining dimensionless Π parameters.)

① F V D ρ μ $n = 5$ parameters

②   Select primary dimensions M, L, and t

③ F V D ρ μ

$\dfrac{ML}{t^2}$ $\dfrac{L}{t}$ L $\dfrac{M}{L^3}$ $\dfrac{M}{Lt}$ $r = 3$ primary dimensions

④ Select repeating parameters ρ, V, D

⑤ Then $n - m = 2$ dimensionless groups will result. Setting up dimensional equations, we obtain

$$\Pi_1 = \rho^a V^b D^c F = \left(\frac{M}{L^3}\right)^a \left(\frac{L}{t}\right)^b (L)^c \left(\frac{ML}{t^2}\right) = M^0 L^0 t^0$$

Equating the exponents of M, L, and t results in

$$\begin{matrix} M: & a+1=0 & a=-1 \\ L: & -3a+b+c+1=0 & c=-2 \\ t: & -b-2=0 & b=-2 \end{matrix} \right\} \quad \text{Therefore, } \Pi_1 = \frac{F}{\rho V^2 D^2}$$

Similarly,

$$\Pi_2 = \rho^d V^e D^f \mu = \left(\frac{M}{L^3}\right)^d \left(\frac{L}{t}\right)^e (L)^f \left(\frac{M}{Lt}\right) = M^0 L^0 t^0$$

$$\begin{matrix} M: & d+1=0 & d=-1 \\ L: & -3d+e+f-1=0 & f=-1 \\ t: & -e-1=0 & e=-1 \end{matrix} \right\} \quad \text{Therefore, } \Pi_2 = \frac{\mu}{\rho V D}$$

⑥ Check using F, L, t dimensions

$$\Pi_1 = \frac{F}{\rho V^2 D^2} : F \frac{L^4}{Ft^2}\left(\frac{t}{L}\right)^2 \frac{1}{L^2} = [1]$$

where [ ] means "has dimensions of," and

$$\Pi_2 = \frac{\mu}{\rho V D} : \frac{Ft}{L^2}\frac{L^4}{Ft^2}\frac{t}{L}\frac{1}{L} = [1]$$

The functional relationship is $\Pi_1 = f(\Pi_2)$, or

$$\frac{F}{\rho V^2 D^2} = f\left(\frac{\mu}{\rho V D}\right)$$

as noted before. The form of the function, f, must be determined experimentally.

EXAMPLE 7.2—Pressure Drop in Pipe Flow

The pressure drop, Δp, for steady, incompressible viscous flow through a straight horizontal pipe depends on the pipe length, l, the average velocity, $\bar{V}$, the fluid viscosity, μ, the pipe diameter, D, the fluid density, ρ, and the average variation, e, of the inside radius (the average "roughness" height). Determine a set of dimensionless groups that can be used to correlate data.

EXAMPLE PROBLEM 7.2

GIVEN: $\Delta p = f(\rho, \bar{V}, D, l, \mu, e)$ for flow in a circular pipe.

FIND: A suitable set of dimensionless groups.

SOLUTION:
(Circled numbers refer to steps in the procedure for determining dimensionless Π parameters.)

① Δp ρ μ $\bar{V}$ l D e $n = 7$ parameters

②   Choose primary dimensions M, L, and t

③ Δp ρ μ $\bar{V}$ l D e

     $\dfrac{M}{Lt^2}$ $\dfrac{M}{L^3}$ $\dfrac{M}{Lt}$ $\dfrac{L}{t}$ L L L $r = 3$ primary dimensions

④   Select repeating parameters $\rho, \bar{V}, D$ $m = r = 3$ repeating parameters

⑤   Then $n - m = 4$ dimensionless groups will result. Setting up dimensional equations we have:

$$\Pi_1 = \rho^a \bar{V}^b D^c \Delta p$$

$$= \left(\frac{M}{L^3}\right)^a \left(\frac{L}{t}\right)^b (L)^c \left(\frac{M}{Lt^2}\right) = M^0 L^0 t^0$$

$$\Pi_2 = \rho^d \bar{V}^e D^f \mu$$

$$= \left(\frac{M}{L^3}\right)^d \left(\frac{L}{t}\right)^e (L)^f \frac{M}{Lt} = M^0 L^0 t^0$$

$M:$ $0 = a + 1$ $a = -1$

$L:$ $0 = -3a + b + c - 1$ $b = -2$

$t:$ $0 = -b - 2$ $c = 0$

$M:$ $0 = d + 1$ $d = -1$

$L:$ $0 = -3d + e + f - 1$ $e = -1$

$t:$ $0 = -e - 1$ $f = -1$

Therefore, $\Pi_1 = \rho^{-1} \bar{V}^{-2} D^0 \Delta p = \dfrac{\Delta p}{\rho \bar{V}^2}$

Therefore, $\Pi_2 = \dfrac{\mu}{\rho \bar{V} D}$

$$\Pi_3 = \rho^g \bar{V}^h D^i l$$

$$= \left(\frac{M}{L^3}\right)^g \left(\frac{L}{t}\right)^h (L)^i L = M^0 L^0 t^0$$

$$\Pi_4 = \rho^j \bar{V}^k D^l e$$

$$= \left(\frac{M}{L^3}\right)^j \left(\frac{L}{t}\right)^k (L)^l L = M^0 L^0 t^0$$

$M:$ $0 = g$ $g = 0$

$L:$ $0 = -3g + h + i + 1$ $h = 0$

$t:$ $0 = -h$ $i = -1$

$M:$ $0 = j$ $j = 0$

$L:$ $0 = -3j + k + l + 1$ $k = 0$

$t:$ $0 = -k$ $l = -1$

Therefore, $\Pi_3 = \dfrac{l}{D}$

Therefore, $\Pi_4 = \dfrac{e}{D}$

⑥   Check, using F, L, t dimensions

$$\Pi_1 = \frac{\Delta p}{\rho \bar{V}^2} : \frac{F}{L^2} \frac{L^4}{Ft^2} \frac{t^2}{L^2} = [1] \qquad \Pi_3 = \frac{l}{D} : \frac{L}{L} = [1]$$

$$\Pi_2 = \frac{\mu}{\rho \bar{V} D} : \frac{Ft}{L^2} \frac{L^4}{Ft^2} \frac{t}{L} \frac{1}{L} = [1] \qquad \Pi_4 = \frac{e}{D} : \frac{L}{L} = [1]$$

Finally, the functional relationship is

$$\Pi_1 = f(\Pi_2, \Pi_3, \Pi_4)$$

or

$$\frac{\Delta p}{\rho \bar{V}^2} = f\left(\frac{\mu}{\rho \bar{V} D}, \frac{l}{D}, \frac{e}{D}\right)$$

$\left\{\begin{array}{l}\text{Experiments in many laboratories have shown that this relationship correlates the data} \\ \text{well. We shall discuss this result in greater detail in Section 8-7.1.}\end{array}\right\}$

The procedure outlined above, where m is taken equal to r (the fewest independent dimensions required to specify the dimensions of all parameters involved) almost always produces the correct number of dimensionless Π parameters. In a few cases, trouble arises because the number of primary dimensions differs when variables are expressed in terms of different systems of dimensions. The value of m can be established with certainty by determining the rank of the dimensional matrix; m is equal to the rank of the dimensional matrix. The procedure is illustrated in Example Problem 7.3.

The $n - m$ dimensionless groups obtained from the procedure are independent but not unique. If a different set of repeating parameters is chosen, different groups result. Since the repeating parameters chosen may appear in all of the dimensionless groups obtained, the viscosity, μ, is not an appropriate choice for a repeating variable. The choice of $\rho[M/L^3]$, $V[L/t]$, and a characteristic length $[L]$ as repeating variables generally leads to a set of dimensionless parameters that have been found to be most suitable for correlating a wide range of experimental data. This is not surprising if one recognizes that inertia forces are important in most fluid mechanics problems. From Newton's second law, $F = ma$; the mass can be written as $m = \rho \mathbb{V}$ and, since volume has dimensions of L^3, then qualitatively $m \propto \rho L^3$. The acceleration can be written as $a = dv/dt = v\,dv/ds$, and hence qualitatively $a \propto V^2/L$. Thus the inertia force, F, is proportional to $\rho V^2 L^2$.

If $n-m = 1$, then a single dimensionless Π parameter is obtained. In this case, the Buckingham Pi theorem indicates that the single Π parameter must be a constant.

EXAMPLE 7.3—Capillary Rise: Use of Dimensional Matrix

When a small tube is dipped into a pool of liquid, surface tension causes a *meniscus* to form at the free surface, which is elevated or depressed depending on the contact angle at the liquid-solid-gas interface. Experiments indicate that the magnitude of this *capillary effect*, Δh, is a function of the tube diameter, D, liquid specific weight, γ, and surface tension, σ. Determine the number of independent Π parameters that can be formed and obtain a set.

EXAMPLE PROBLEM 7.3

GIVEN: $\Delta h = f(D, \gamma, \sigma)$

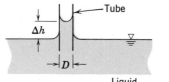

FIND: (a)   Number of independent Π parameters.
        (b)   Evaluate one set.

SOLUTION:
(Circled numbers refer to steps in the procedure for determining dimensionless Π parameters.)

Liquid
(Specific weight = γ
Surface tension = σ)

① Δh D γ σ $n = 4$ parameters

②   Choose primary dimensions (use both M, L, t and F, L, t dimensions to illustrate the problem in determining m)

③   (a) M, L, t (b) F, L, t

Δh	D	γ	σ		Δh	D	γ	σ
L	L	$\dfrac{M}{L^2 t^2}$	$\dfrac{M}{t^2}$		L	L	$\dfrac{F}{L^3}$	$\dfrac{F}{L}$

$r = 3$ primary dimensions $r = 2$ primary dimensions

Thus we ask, "Is m equal to r?" Let us check the dimensional matrix to find out. The dimensional matrix is

	Δh	D	γ	σ
M	0	0	1	1
L	1	1	-2	0
t	0	0	-2	-2

	Δh	D	γ	σ
F	0	0	1	1
L	1	1	-3	-1

The rank of a matrix is equal to the order of its largest nonzero determinant.

$$\begin{vmatrix} 0 & 1 & 1 \\ 1 & -2 & 0 \\ 0 & -2 & -2 \end{vmatrix} = 0 - (1)(-2) \\ \qquad\qquad +(1)(-2) = 0$$

$$\begin{vmatrix} -2 & 0 \\ -2 & -2 \end{vmatrix} = 4 \neq 0 \qquad \begin{array}{l} \therefore\, m = 2 \\ m \neq r \end{array}$$

$$\begin{vmatrix} 1 & 1 \\ -3 & -1 \end{vmatrix} = -1 + 3 = 2 \neq 0$$

$$\begin{array}{l} \therefore\, m = 2 \\ m = r \end{array}$$

④ $m = 2$. Choose D, γ as repeating parameters.

$m = 2$. Choose D, γ as repeating parameters.

⑤ $n - m = 2$ dimensionless groups will result.

$n - m = 2$ dimensionless groups will result.

$$\Pi_1 = D^a \gamma^b \Delta h$$

$$= (L)^a \left(\frac{M}{L^2 t^2}\right)^b (L) = M^0 L^0 t^0$$

$$\Pi_1 = D^e \gamma^f \Delta h$$

$$= (L)^e \left(\frac{F}{L^3}\right)^f L = F^0 L^0 t^0$$

$$\begin{array}{ll} M: & b + 0 = 0 \\ L: & a - 2b + 1 = 0 \\ t: & -2b + 0 = 0 \end{array} \right\} \begin{array}{l} b = 0 \\ a = -1 \end{array}$$

$$\begin{array}{ll} F: & f = 0 \\ L: & e - 3f + 1 = 0 \end{array} \right\} \; e = -1$$

Therefore, $\Pi_1 = \dfrac{\Delta h}{D}$

Therefore, $\Pi_1 = \dfrac{\Delta h}{D}$

$$\Pi_2 = D^c \gamma^d \sigma$$

$$= (L)^c \left(\frac{M}{L^2 t^2}\right)^d \frac{M}{t^2} = M^0 L^0 t^0$$

$$\Pi_2 = D^g \gamma^h \sigma$$

$$= (L)^g \left(\frac{F}{L^3}\right)^h \frac{F}{L} = F^0 L^0 t^0$$

$$\begin{array}{ll} M: & d + 1 = 0 \\ L: & c - 2d = 0 \\ t: & -2d - 2 = 0 \end{array} \right\} \begin{array}{l} d = -1 \\ c = -2 \end{array}$$

$$\begin{array}{ll} F: & h + 1 = 0 \\ L: & g - 3h - 1 = 0 \end{array} \right\} \begin{array}{l} h = -1 \\ g = -2 \end{array}$$

Therefore, $\Pi_2 = \dfrac{\sigma}{D^2 \gamma}$

Therefore, $\Pi_2 = \dfrac{\sigma}{D^2 \gamma}$

⑥ Check, using F, L, t dimensions

Check, using M, L, t dimensions

$$\Pi_1 = \frac{\Delta h}{D} : \frac{L}{L} = [1]$$

$$\Pi_1 = \frac{\Delta h}{D} : \frac{L}{L} = [1]$$

$$\Pi_2 = \frac{\sigma}{D^2 \gamma} : \frac{F}{L} \frac{1}{L^2} \frac{L^3}{F} = [1]$$

$$\Pi_2 = \frac{\sigma}{D^2 \gamma} : \frac{M}{t^2} \frac{1}{L^2} \frac{L^2 t^2}{M} = [1]$$

Therefore, both systems of dimensions yield the same dimensionless Π parameters. The predicted functional relationship is

$$\Pi_1 = f(\Pi_2) \qquad \text{or} \qquad \frac{\Delta h}{D} = f\left(\frac{\sigma}{D^2 \gamma}\right)$$

> This result is reasonable on physical grounds. The fluid is static; one would not expect time to be an important dimension.
>
> The purpose of this problem is to illustrate use of the dimensional matrix to determine the required number of repeating parameters.

7-4 DIMENSIONLESS GROUPS OF SIGNIFICANCE IN FLUID MECHANICS

Over the years, several hundred different dimensionless groups that are important in engineering have been identified. Following tradition, each such group has been given the name of a prominent scientist or engineer, usually the one who pioneered its use. Several are so fundamental and occur so frequently in fluid mechanics that we should take time to learn their definitions. Understanding their physical significance also gives insight into the phenomena we study.

Forces encountered in flowing fluids include those due to inertia, viscosity, pressure, gravity, surface tension, and compressibility. The ratio of any two forces will be dimensionless. We have previously shown that the inertia force is proportional to $\rho V^2 L^2$. To facilitate forming ratios of forces, we can express each of the remaining forces as follows:

$$\text{Viscous force} = \tau A \propto \mu \frac{du}{dy} A \propto \mu \frac{V}{L} L^2 \propto \mu VL$$

$$\text{Pressure force} = (\Delta p)A \propto (\Delta p)L^2$$

$$\text{Gravity force} = mg \propto g\rho L^3$$

$$\text{Surface tension force} = \sigma L$$

$$\text{Compressibility force} = E_v A \propto E_v L^2$$

Inertia forces are important in most fluid mechanics problems. The ratio of the inertia force to each of the other forces listed above leads to five fundamental dimensionless groups encountered in fluid mechanics.

In the 1880s, Osborne Reynolds, the British engineer, studied the transition between laminar and turbulent flow regimes in a tube. He discovered that the parameter (later named after him)

$$Re = \frac{\rho \bar{V} D}{\mu} = \frac{\bar{V} D}{\nu}$$

is a criterion by which the flow regime may be determined. Later experiments have shown that the *Reynolds number* is a key parameter for other flow cases as well. Thus, in general,

$$Re = \frac{\rho VL}{\mu} = \frac{VL}{\nu}$$

where L is a characteristic length descriptive of the flow field geometry. The Reynolds number is the ratio of inertia forces to viscous forces. "Large" Reynolds number flows generally are turbulent. Flows in which the inertia forces are "small" compared to viscous forces are characteristically laminar flows.

In aerodynamic and other model testing, it is convenient to present pressure data in dimensionless form. The ratio

$$Eu = \frac{\Delta p}{\frac{1}{2}\rho V^2}$$

is formed, where Δp is the local pressure minus the freestream pressure, and ρ and V are properties of the freestream flow. This ratio has been named after Leonhard Euler, the Swiss mathematician who did much of the early analytical work in fluid mechanics. Euler is credited with being the first to recognize the role of pressure in fluid motion; the Euler equations of Chapter 6 demonstrate this role. The *Euler number* is the ratio of pressure forces to inertia forces. (The factor $\frac{1}{2}$ is introduced into the denominator to give the dynamic pressure.) The Euler number is often called the *pressure coefficient*, C_p.

In the study of cavitation phenomena, the pressure difference, Δp, is taken as $\Delta p = p - p_v$, where p, ρ, and V are conditions in the liquid stream, and p_v is the liquid vapor pressure at the test temperature. The resulting dimensionless parameter is referred to as the *cavitation number*,

$$Ca = \frac{p - p_v}{\frac{1}{2}\rho V^2}$$

$\Delta\ Pressure \over Viscous\ forces$

William Froude was a British naval architect. Together with his son, Robert Edmund Froude, he discovered that the parameter

$$Fr = \frac{V}{\sqrt{gL}}$$

was significant for flows with free surface effects. Squaring the *Froude number* gives

$$Fr^2 = \frac{V^2}{gL} = \frac{\rho V^2 L^2}{\rho g L^3}$$

$inertia\ forces \over gravity\ force$

which may be interpreted as the ratio of inertia forces to gravity forces. The length, L, is a characteristic length descriptive of the flow field. In the case of open-channel flow, the characteristic length is the water depth; Froude numbers less than one indicate subcritical flow and values greater than one indicate supercritical flow.

The *Weber number* is the ratio of inertia forces to surface tension forces. It may be written

$$We = \frac{\rho V^2 L}{\sigma}$$

$inertia\ forces \over surface\ tension\ forces$

In the 1870s, the Austrian physicist Ernst Mach introduced the parameter

$$M = \frac{V}{c}$$

where V is the flow speed and c is the local sonic speed. Analysis and experiments have shown that the *Mach number* is a key parameter that characterizes compressibility effects in a flow. The Mach number may be written

$$M = \frac{V}{c} = \frac{V}{\sqrt{\dfrac{dp}{d\rho}}} = \frac{V}{\sqrt{\dfrac{E_v}{\rho}}} \quad \text{or} \quad M^2 = \frac{\rho V^2 L^2}{E_v L^2}$$

$inertia\ forces \over forces\ ..\ compressible$

which may be interpreted as a ratio of inertia forces to forces due to compressibility. For truly incompressible flow (under some conditions even liquids are quite compressible), $c = \infty$ so that $M = 0$.

7-5 FLOW SIMILARITY AND MODEL STUDIES

To be useful, a model test must yield data that can be scaled to obtain the forces, moments, and dynamic loads that would exist on the full-scale prototype. What conditions must be met to ensure the similarity of model and prototype flows?

Perhaps the most obvious requirement is that the model and prototype must be geometrically similar. *Geometric similarity* requires that the model and prototype be the same shape, and that all linear dimensions of the model be related to corresponding dimensions of the prototype by a constant scale factor.

A second requirement is that the model and prototype flows must be *kinematically similar*. Two flows are kinematically similar when the velocities at corresponding points are in the same direction and are related in magnitude by a constant scale factor. Thus two flows that are kinematically similar also have streamline patterns related by a constant scale factor. Since the boundaries form the bounding streamlines, flows that are kinematically similar must be geometrically similar.

In principle, kinematic similarity would require that a wind tunnel of infinite cross section be used to obtain data for drag on an object, in order to model correctly the performance in an infinite flow field. In practice, this restriction may be relaxed considerably, permitting use of equipment of reasonable size.

Kinematic similarity requires that the regimes of flow be the same for model and prototype. If compressibility or cavitation effects, which may change even the qualitative patterns of flow, are not present in the prototype flow, they must be avoided in the model flow.

When two flows have force distributions such that identical types of forces are parallel and are related in magnitude by a constant scale factor at all corresponding points, the flows are *dynamically similar*.

The requirements for dynamic similarity are the most restrictive: Two flows must possess both geometric and kinematic similarity to be dynamically similar.

To establish the conditions required for complete dynamic similarity, all forces that are important in the flow situation must be considered. Thus the effects of viscous forces, of pressure forces, of surface tension forces, and so on, must be considered. Test conditions must be established such that all important forces are related by the same scale factor between model and prototype flows. When dynamic similarity exists, data measured in a model flow may be related quantitatively to conditions in the prototype flow. What, then, are the conditions for ensuring dynamic similarity between model and prototype flows?

The Buckingham Pi theorem may be used to obtain the governing dimensionless groups for a flow phenomenon; to achieve dynamic similarity between geometrically similar flows, we must duplicate all but one of these dimensionless groups.

For example, in considering the drag force on a sphere in Example Problem 7.1, we began with

$$F = f(D, V, \rho, \mu)$$

The Buckingham Pi theorem predicted the functional relation

$$\frac{F}{\rho V^2 D^2} = f_1\left(\frac{\rho V D}{\mu}\right)$$

In Section 7-4 we showed that the dimensionless parameters can be viewed as ratios of forces. Thus, in considering a model flow and a prototype flow about a sphere (the flows are geometrically similar), the flows also will be dynamically similar if

$$\left(\frac{\rho V D}{\mu}\right)_{model} = \left(\frac{\rho V D}{\mu}\right)_{prototype}$$

Furthermore, if

$$Re_{model} = Re_{prototype}$$

then

$$\left(\frac{F}{\rho V^2 D^2}\right)_{model} = \left(\frac{F}{\rho V^2 D^2}\right)_{prototype}$$

and the results determined from the model study can be used to predict the drag on the full-scale prototype.

The actual force on the object due to the fluid is not the same in both cases, but its dimensionless value is. The two tests can be run using different fluids, if desired, as long as the Reynolds numbers are matched. For experimental convenience, test data can be measured in a wind tunnel in air and the results used to predict drag in water, as illustrated in Example Problem 7.4.

EXAMPLE 7.4—Similarity: Drag of a Sonar Transducer

The drag of a sonar transducer is to be predicted, based on wind tunnel test data. The prototype, a 1 ft diameter sphere, is to be towed at 5 knots (nautical miles per hour) in seawater at 5 C. The model is 6 in. in diameter. Determine the required test speed in air. If the drag of the model at test conditions is 5.58 lbf, estimate the drag of the prototype.

EXAMPLE PROBLEM 7.4

GIVEN: Sonar transducer to be tested in a wind tunnel.

FIND: (a) V_m.
　　　　(b) F_p.

SOLUTION:
Since the prototype operates in water and the model test is to be performed in air, useful results can be expected only if cavitation effects are absent in the prototype flow and compressibility effects are absent from the model test. Under these conditions,

$$\frac{F}{\rho V^2 D^2} = f\left(\frac{\rho V D}{\mu}\right)$$

and the test should be run at

$$Re_{model} = Re_{prototype}$$

to ensure dynamic similarity. For seawater at 5 C, $\rho = 1.99$ slug/ft^3 and $\nu = 1.68 \times 10^{-5}$ ft^2/sec. At prototype conditions,

$$V_p = \frac{5 \text{ nmi}}{\text{hr}} \times \frac{6080 \text{ ft}}{\text{nmi}} \times \frac{\text{hr}}{3600 \text{ sec}} = 8.44 \text{ ft/sec}$$

$$Re_p = \frac{V_p D_p}{\nu_p} = \frac{8.44}{\text{sec}} \frac{\text{ft}}{} \times 1 \text{ ft} \times \frac{\text{sec}}{1.68 \times 10^{-5} \text{ ft}^2} = 5.02 \times 10^5$$

The model test conditions must duplicate this Reynolds number. Thus

$$Re_m = \frac{V_m D_m}{\nu_m} = 5.02 \times 10^5$$

For air at STP, $\rho = 0.00238$ slug/ft^3 and $\nu = 1.56 \times 10^{-4}$ ft^2/sec. The wind tunnel must be operated at

$$V_m = Re_m \frac{\nu_m}{D_m} = 5.02 \times 10^5 \times 1.56 \times 10^{-4} \frac{\text{ft}^2}{\text{sec}} \times \frac{1}{0.5 \text{ ft}}$$

$$V_m = 156 \text{ ft/sec} \overset{\longleftarrow}{} V_m$$

This speed is low enough to neglect compressibility effects.

At these test conditions, the model and prototype flows are dynamically similar. Hence

$$\left. \frac{F}{\rho V^2 D^2} \right)_m = \left. \frac{F}{\rho V^2 D^2} \right)_p$$

and

$$F_p = F_m \frac{\rho_p}{\rho_m} \frac{V_p^2}{V_m^2} \frac{D_p^2}{D_m^2} = 5.58 \text{ lbf} \times \frac{1.99}{0.00238} \times \frac{(8.44)^2}{(156)^2} \times \frac{1}{(0.5)^2}$$

$$F_p = 54.6 \text{ lbf} \overset{\longleftarrow}{} F_p$$

If cavitation were expected—if the sonar probe were operated at high speed near the free surface of the seawater—then useful results could not be obtained from a model test in air.

{This problem demonstrates the calculation of prototype values from model test data.}

7-5.1 Incomplete Similarity

We have shown that it is necessary to duplicate all but one of the significant dimensionless groups to achieve complete dynamic similarity between geometrically similiar flows.

In the simplified situation of Example Problem 7.4, duplicating the Reynolds number between model and prototype ensured dynamically similar flows. Testing in air allowed the Reynolds number to be duplicated exactly (this also could have been accomplished in a water tunnel for this situation). The drag force on a sphere actually depends on the nature of the boundary-layer flow. Therefore, geometric similarity requires that the relative surface roughness of the model and prototype be the same. This means that relative roughness also is a parameter that must be duplicated between model and prototype situations. If we assume that the model was constructed carefully, measured values of drag from model tests could be scaled to predict drag for the operating conditions of the prototype.

In many model studies, to achieve dynamic similarity requries duplication of several dimensionless groups. In some cases, complete dynamic similarity between model and prototype may not be attainable. Determining the drag force (resistance) of a surface ship is an example of such a situation. Resistance on a surface ship arises from skin friction on the hull (viscous forces) and surface wave resistance (gravity forces). Complete dynamic similarity requires that both Reynolds and Froude numbers be duplicated between model and prototype.

In general it is not possible to predict wave resistance analytically, so it must be modeled. This requires that

$$Fr_m = \frac{V_m}{(gL_m)^{1/2}} = Fr_p = \frac{V_p}{(gL_p)^{1/2}}$$

To match Froude numbers between model and prototype requires a velocity ratio of

$$\frac{V_m}{V_p} = \left(\frac{L_m}{L_p}\right)^{1/2}$$

to ensure dynamically similar surface wave patterns.

For any model length scale, matching the Froude numbers determines the velocity ratio. Only the kinematic viscosity can be varied to match Reynolds numbers. Thus

$$Re_m = \frac{V_m L_m}{\nu_m} = Re_p = \frac{V_p L_p}{\nu_p}$$

leads to the condition that

$$\frac{\nu_m}{\nu_p} = \frac{V_m}{V_p}\frac{L_m}{L_p}$$

If we use the velocity ratio obtained from matching Froude numbers, equality of Reynolds numbers leads to a kinematic viscosity ratio of

$$\frac{\nu_m}{\nu_p} = \left(\frac{L_m}{L_p}\right)^{1/2}\frac{L_m}{L_p} = \left(\frac{L_m}{L_p}\right)^{3/2}$$

If L_m/L_p equals $\frac{1}{100}$ (a typical length scale for ship model tests), then ν_m/ν_p must be $\frac{1}{1000}$. Fig. A.3 shows that mercury is the only liquid with kinematic viscosity less than water. However, it is only about an order of magnitude less, so the kinematic viscosity ratio required to duplicate Reynolds numbers cannot be attained.

Water is the only practical fluid for most model tests of free-surface flows. To obtain complete dynamic similarity then would require a full-scale test. However, model studies do provide useful information even though complete similarity cannot be obtained. Fig. 7.1 shows data from a test of a 1:80 scale model of a ship conducted at the U.S. Naval Academy Hydromechanics Laboratory. The plot displays resistance coefficient data versus Froude number. The square points are calculated from values of total resistance measured in the test.

Resistance of the full-scale ship can be calculated from model test results using the following procedure. The pattern of surface waves, and thus the wave resistance, is matched between model and prototype at corresponding Froude numbers. The viscous drag on the model is estimated using the analytical methods of Chapter 9 (the estimated frictional resistance coefficients are plotted in Fig. 7.1 as diamonds). The model wave resistance is calculated as the difference between total drag and estimated friction drag (estimated wave resistance coefficients for the model are plotted as circles).

The prototype wave resistance is calculated using Froude number scaling by equating wave resistance coefficients for model and prototype. The points plotted as circles in Fig. 7.2 for the prototype are identical to model coefficients at corresponding Froude numbers. The skin friction drag calculated analytically for the prototype, shown in Fig. 7.2 by the diamonds, is added to the scaled wave drag coefficients, to predict the total prototype drag coefficients.

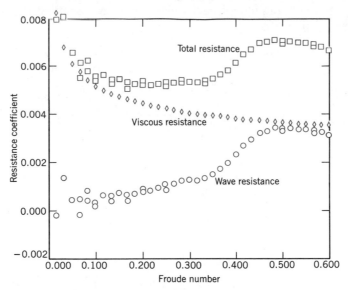

Fig. 7.1 Data from test of 1:80 scale model of U.S. Navy guided missile frigate *Oliver Hazard Perry* (FFG-7). (Data from U.S. Naval Academy Hydromechanics Laboratory, courtesy of Professor Bruce Johnson.)

Because the Reynolds number cannot be matched for model tests of surface ships, the boundary-layer behavior is not the same for model and prototype. The model Reynolds number is only $(L_m/L_p)^{3/2}$ as large as the prototype value, so the extent of laminar flow in the boundary layer on the model is too large by a corresponding factor. The method just described assumes that boundary-layer behavior can be scaled. To make this possible, the model boundary layer is "tripped" or "stimulated" to become turbulent at a location that corresponds to the behavior on the full-scale vessel. "Studs" were used to stimulate the boundary layer for the model test results shown in Fig. 7.1.

A correction factor sometimes is added to the full-scale coefficients calculated from model test data. This factor accounts for roughness, waviness, and unevenness that inevitably are more pronounced on the full-scale ship than on the model. Comparisons between predictions from model tests and measurements made in full-scale trials suggest an overall accuracy within ±5 percent [2].

The Froude number is an important parameter in the modeling of rivers and harbors. In these situations it is not practical to obtain complete similarity. Use of a reasonable model scale would lead to extremely small water depths. Viscous forces and surface tension forces would have much larger relative effects on the model flow than in the prototype. Consequently, different length scales are used for the vertical and horizontal directions. Viscous forces in the deeper model flow are increased using artificial roughness elements.

Emphasis on fuel economy has made reduction of aerodynamic drag important for automobiles, trucks, and buses. Most work on development of low-drag configurations is done using model tests. Traditionally, automobile models have been built to $\frac{3}{8}$ scale, at which a model of a full-size automobile has a frontal area of about 0.3 m². Thus testing can be done in a wind tunnel with test section area of 6 m² or larger. At $\frac{3}{8}$ scale, a wind speed of about 150 mph is needed to model a

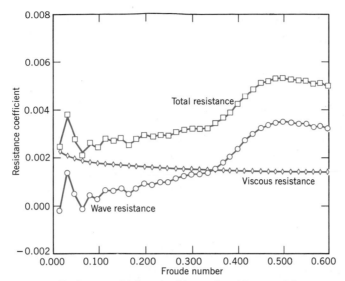

Fig. 7.2   Resistance of full-scale ship predicted from model test
results. (Data from U.S. Naval Academy Hydromechanics
Laboratory, courtesy of Professor Bruce Johnson.)

prototype automobile traveling at the legal speed limit. Thus there is no problem
with compressibility effects, but the scale models are expensive and time-consuming
to build.

Figure 7.3 shows a large wind tunnel (test section dimensions are 5.4 m high,
10.4 m wide, and 21.3 m long; maximum air speed is 250 km/hr with the tunnel
empty) used by General Motors to test full-scale automobiles at highway speeds.
The large test section allows use of production autos or of full-scale clay mockups
of proposed auto body styles. The relatively low speed permits flow visualization
using tufts or "smoke" streams.[2] Using full-size "models," stylists and engineers can
work together to achieve optimum results.

It is harder to achieve dynamic similarity in tests of trucks and buses; models must
be made to smaller scale than those of automobiles.[3] A large scale for truck and bus
testing is 1:8. To achieve complete dynamic similarity by matching Reynolds numbers
at this scale would require a test speed of 440 mph. This would introduce unwanted
compressibility effects, and model and prototype flows would not be kinematically
similar. Fortunately, trucks and buses are "bluff" objects. Experiments show that
above a certain Reynolds number, their nondimensional drag becomes independent
of Reynolds number [5]. Although similarity is not complete, measured test data can
be scaled to predict prototype drag forces. The procedure is illustrated in Example
Problem 7.5.

For additional details on techniques and applications of dimensional analysis consult [6–9].

---

[2] A mixture of liquid nitrogen and steam is used to produce "smoke" streaklines that evaporate and do
not clog the fine mesh screens used to reduce the turbulence level in the tunnel. Streaklines may be
made to appear "colored" in photos by placing a filter over the camera lens. This and other techniques
for flow visualization are detailed in [3] and [4].

[3] The vehicle length is particularly important when testing at large yaw angles to simulate crosswind
behavior. Tunnel blockage considerations limit the acceptable model size. See [5] for recommended
practices.

Fig. 7.3 Full-scale automobile under test in 5.4 × 10.4 m wind tunnel test section at the General Motors Technical Center, Warren, Michigan. (Photograph courtesy of General Motors.)

EXAMPLE 7.5—Incomplete Similarity: Aerodynamic Drag on a Bus

The following wind tunnel test data from a 1:16 scale model of a bus are available:

Air Speed (m/sec)	18.0	21.8	26.0	30.1	35.0	38.5	40.9	44.1	46.7
Drag Force (N)	3.10	4.41	6.09	7.97	10.7	12.9	14.7	16.9	18.9

Using the properties of standard air, calculate and plot the dimensionless aerodynamic drag coefficient,

$$C_D = \frac{F_D}{\frac{1}{2}\rho V^2 A}$$

versus Reynolds number, $Re = \rho V w/\mu$, where w is model width. Find the minimum test speed above which C_D remains constant. Estimate the aerodynamic drag force and power requirement for the prototype vehicle at 100 km/hr. (The width and frontal area of the prototype are 8 ft and 84 ft^2, respectively.)

EXAMPLE PROBLEM 7.5

GIVEN: Data from a wind tunnel test of a model bus. Prototype dimensions are width of 8 ft and frontal area of 84 ft^2. Model scale is 1:16. Standard air is the test fluid.

FIND: (a) Calculate and plot aerodynamic drag coefficient, $C_D = F_D/\frac{1}{2}\rho V^2 A$, versus Reynolds number, $Re = \rho V w/\mu$.
   (b) Determine the speed above which C_D is constant.
   (c) Estimate the aerodynamic drag force and power required for the full-scale vehicle at 100 km/hr.

SOLUTION:
The model width is

$$w_m = \frac{1}{16} w_p = \frac{1}{16} \times \frac{8 \text{ ft}}{} \times \frac{0.3048 \text{ m}}{\text{ft}} = 0.152 \text{ m}$$

The model area is

$$A_m = \left(\frac{1}{16}\right)^2 A_p = \left(\frac{1}{16}\right)^2 \times \frac{84 \text{ ft}^2}{} \times \frac{(0.3048)^2 \text{ m}^2}{\text{ft}^2} = 0.0305 \text{ m}^2$$

The aerodynamic drag coefficient may be calculated as

$$C_D = \frac{F_D}{\frac{1}{2}\rho V^2 A}$$

$$= \frac{2}{} \times F_D \text{ (N)} \times \frac{\text{m}^3}{1.23 \text{ kg}} \times \frac{\text{sec}^2}{(V)^2 \text{ m}^2} \times \frac{1}{0.0305 \text{ m}^2} \times \frac{\text{kg} \cdot \text{m}}{\text{N} \cdot \text{sec}^2}$$

$$C_D = \frac{53.3 \, F_D \text{ (N)}}{[V(\text{m/sec})]^2}$$

The Reynolds number may be calculated as

$$Re = \frac{\rho V w}{\mu} = \frac{V w}{\nu} = \frac{V}{} \frac{\text{m}}{\text{sec}} \times 0.152 \text{ m} \times \frac{\text{sec}}{1.45 \times 10^{-5} \text{ m}^2}$$

$$Re = 1.05 \times 10^4 \, V(\text{m/sec})$$

The calculated values are plotted in the following figure:

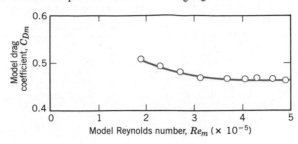

C_{Dm} versus Re_m

The plot shows that the model drag coefficient becomes constant at $C_{Dm} = 0.463$ above $Re_m = 4 \times 10^5$, which corresponds to an air speed of approximately 40 m/sec. Since the drag coefficient is independent of Reynolds number above $Re \approx 4 \times 10^5$, then for the prototype vehicle ($Re \approx 4.5 \times 10^6$), $C_D = 0.463$. The drag force on the full-scale vehicle is

$$F_{Dp} = C_D \tfrac{1}{2} \rho V_p^2 A_p$$

$$= \frac{0.463}{2} \times 1.23 \; \frac{\text{kg}}{\text{m}^3} \left(100 \; \frac{\text{km}}{\text{hr}} \times 1000 \; \frac{\text{m}}{\text{km}} \times \frac{\text{hr}}{3600 \; \text{sec}} \right)^2$$

$$\times \; 84 \; \text{ft}^2 \times (0.3048)^2 \; \frac{\text{m}^2}{\text{ft}^2} \times \frac{\text{N} \cdot \text{sec}^2}{\text{kg} \cdot \text{m}}$$

$$F_{Dp} = 1.71 \; \text{kN} \qquad\qquad\qquad\qquad\qquad\qquad\qquad\qquad F_{Dp}$$

The corresponding power required to overcome aerodynamic drag is

$$\mathcal{P}_p = F_{Dp} V_p$$

$$= \frac{1.71 \times 10^3 \; \text{N}}{} \times 100 \; \frac{\text{km}}{\text{hr}} \times 1000 \; \frac{\text{m}}{\text{km}} \times \frac{\text{hr}}{3600 \; \text{sec}} \times \frac{\text{W} \cdot \text{sec}}{\text{N} \cdot \text{m}}$$

$$\mathcal{P}_p = 47.6 \; \text{kW} \qquad\qquad\qquad\qquad\qquad\qquad\qquad\qquad\qquad \mathcal{P}_p$$

This example illustrates application of model test data in a situation where nondimensional drag is constant above a certain minimum Reynolds number (the SAE *Recommended Practice* [5] suggests $Re \geq 2 \times 10^6$ for truck and bus testing). In this situation, it is not necessary to duplicate the prototype Reynolds number to obtain useful model test data.

7-5.2 Scaling with Multiple Dependent Parameters

In some situations of practical importance there may be more than one dependent parameter. In such cases, dimensionless groups must be formed separately for each dependent parameter.

As an example, consider a typical centrifugal pump. The detailed flow pattern within a pump changes with volume flow rate and speed; these changes affect the pump's performance. Performance parameters of interest include the pressure rise (or head) developed, the power input required, and the machine efficiency measured under specific operating conditions.[4] Performance curves are generated by varying

---

[4] Efficiency is defined as the ratio of power delivered to the fluid divided by input power, $\eta = \mathcal{P}/\mathcal{P}_{\text{in}}$. For incompressible flow, the energy equation reduces to $\mathcal{P} = \rho Q H$ (when head is expressed as energy per unit mass) or to $\mathcal{P} = \rho g Q H$ (when head is expressed as energy per unit weight).

an independent parameter such as the volume flow rate. Thus the independent variables are volume flow rate, angular speed, impeller diameter, and fluid properties. Dependent variables are the several performance quantities of interest.

Finding dimensionless parameters begins from the symbolic equations for the dependence of head, H (energy per unit mass, L^2/t^2), and power, $\mathscr{P}$, on the independent parameters, given by

$$H = g_1(Q, \rho, \omega, D, \mu)$$

and

$$\mathscr{P} = g_2(Q, \rho, \omega, D, \mu)$$

Straightforward use of the Pi theorem gives the dimensionless *head coefficient* and *power coefficient* as

$$\frac{H}{\omega^2 D^2} = f_1\left(\frac{Q}{\omega D^3}, \frac{\rho \omega D^2}{\mu}\right) \tag{7.1}$$

and

$$\frac{\mathscr{P}}{\rho \omega^3 D^5} = f_2\left(\frac{Q}{\omega D^3}, \frac{\rho \omega D^2}{\mu}\right) \tag{7.2}$$

The dimensionless parameter $Q/\omega D^3$ in these equations is the *flow coefficient*. The dimensionless parameter $\rho \omega D^2/\mu$ ($\approx \rho V D/\mu$) is a form of Reynolds number.

Head and power in a pump are developed by inertia forces. Both the flow pattern within a pump and the pump performance change with volume flow rate and speed of rotation. Performance is difficult to predict analytically except at the design point of the pump, so it is measured experimentally. Typical characteristic curves plotted from experimental data for a centrifugal pump tested at constant speed are shown in Fig. 7.4 as functions of volume flow rate. The head and power curves in Fig. 7.4 are faired through points plotted from measured data. The efficiency curve is faired through points calculated from the data. Maximum efficiency usually occurs at the design point.

Complete similarity in pump performance tests would require identical flow coefficients and Reynolds numbers. In practice, it has been found that viscous effects are relatively unimportant when two geometrically similar machines operate under "similar" flow conditions. Thus, from Eqs. 7.1 and 7.2, when

$$\frac{Q_1}{\omega_1 D_1^3} = \frac{Q_2}{\omega_2 D_2^3} \tag{7.3}$$

it follows that

$$\frac{H_1}{\omega_1^2 D_1^2} = \frac{H_2}{\omega_2^2 D_2^2} \tag{7.4}$$

and

$$\frac{\mathscr{P}_1}{\rho_1 \omega_1^3 D_1^5} = \frac{\mathscr{P}_2}{\rho_2 \omega_2^3 D_2^5} \tag{7.5}$$

The empirical observation that viscous effects are unimportant under similar flow conditions allows use of Eqs. 7.3 through 7.5 to scale the performance characteristics of machines to different operating conditions, as either the speed or diameter are

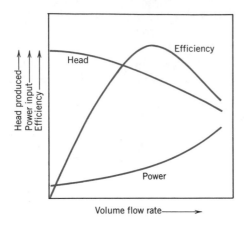

Fig 7.4 Typical characteristic curves for centrifugal pump tested at constant speed.

changed. These useful scaling relationships[5] are known as pump or fan "laws." If operating conditions for one machine are known, operating conditions for any geometrically similar machine can be found by changing D and ω according to Eqs. 7.3 through 7.5.

Another useful pump parameter can be obtained by eliminating the machine diameter from Eqs. 7.3 and 7.4. Designating $\Pi_1 = Q/\omega D^3$ and $\Pi_2 = H/\omega^2 D^2$, then the ratio $\Pi_1^{1/2}/\Pi_2^{3/4}$ is a dimensionless parameter; this parameter is the *specific speed*, N_s,

$$N_s = \frac{\omega Q^{1/2}}{H^{3/4}} \tag{7.6}$$

The specific speed, as defined in Eq. 7.6, is a dimensionless parameter (provided that the head, H, is expressed as energy per unit mass). You may think of specific speed as the speed required for a machine to produce unit head at unit volume flow rate. A constant specific speed describes all operating conditions of geometrically similar machines with similar flow conditions.

Although specific speed is a dimensionless parameter, it is common practice to use a convenient but inconsistent set of units in specifying the variables, ω, Q, and H. When this is done the specific speed is not a unitless parameter and its magnitude depends on the units used to calculate it. Typical units used in U.S. engineering practice for pumps are rpm for ω, gpm for Q, and feet (energy per unit weight) for H. In these units, "low" specific speed means $500 < N_s < 4000$ and "high" means $10,000 < N_s < 15,000$. Example Problem 7.6 illustrates use of the pump scaling laws and specific speed parameter. More details of specific speed calculations and additional examples of applications to fluid machinery are presented in Chapter 11.

EXAMPLE 7.6—Pump Laws

A centrifugal pump has an efficiency of 80 percent at its design point specific speed of 2000 (units of rpm, gpm, and feet). The impeller diameter is 8 in. At design point flow conditions, the volume flow rate is 300 gpm of water at 1170 rpm. To obtain a larger flow rate, the pump is to be fitted with a 1750 rpm motor. Use the pump

[5] More details on dimensional analysis, design, and performance curves for fluid machinery are presented in Chapter 11.

"laws" to find the design point performance characteristics of the pump at the higher speed. Show that the specific speed remains constant for the higher operating speed. Determine the motor size required.

EXAMPLE PROBLEM 7.6

GIVEN: Centrifugal pump with design specific speed of 2000 (in rpm, gpm, and feet units). Impeller diameter is $D = 8$ in. At the pump's design point flow conditions, $\omega = 1170$ rpm, and $Q = 300$ gpm, with water.

FIND: Use pump "laws" to find for similar flow conditions at 1750 rpm:
  (a) Performance characteristics.
  (b) Specific speed.
  (c) Motor size required.

SOLUTION:
From pump "laws," $Q/\omega D^3 = $ constant, so

$$Q_2 = Q_1 \frac{\omega_2}{\omega_1} \left(\frac{D_2}{D_1}\right)^3 = 300 \text{ gpm} \left(\frac{1750}{1170}\right)(1)^3 = 449 \text{ gpm} \qquad\qquad \underline{Q_2}$$

The pump head is not specified at $\omega_1 = 1170$ rpm, but it can be calculated from the specific speed, $N_s = 2000$. Using the given units and the definition of N_s,

$$N_s = \frac{\omega Q^{1/2}}{H^{3/4}} \qquad \text{so} \qquad H_1 = \left(\frac{\omega_1 Q_1^{1/2}}{N_s}\right)^{4/3} = 21.9 \text{ ft}$$

Then $H/\omega^2 D^2 = $ constant, so

$$H_2 = H_1 \left(\frac{\omega_2}{\omega_1}\right)^2 \left(\frac{D_2}{D_1}\right)^2 = 21.9 \text{ ft} \left(\frac{1750}{1170}\right)^2 (1)^2 = 49.0 \text{ ft} \qquad\qquad \underline{H_2}$$

The pump ouput power is $\mathcal{P}_1 = \rho g Q_1 H_1$, so at $\omega_1 = 1170$ rpm,

$$\mathcal{P}_1 = \frac{1.94 \text{ slug}}{\text{ft}^3} \times \frac{32.2 \text{ ft}}{\text{sec}^2} \times \frac{300 \text{ gal}}{\text{min}} \times \frac{21.9 \text{ ft}}{} \times \frac{\text{ft}^3}{7.48 \text{ gal}}$$

$$\times \frac{\text{min}}{60 \text{ sec}} \times \frac{\text{lbf} \cdot \text{sec}^2}{\text{slug} \cdot \text{ft}} \times \frac{\text{hp} \cdot \text{sec}}{550 \text{ ft} \cdot \text{lbf}}$$

$$\mathcal{P}_1 = 1.66 \text{ hp}$$

But $\mathcal{P}/\rho \omega^3 D^5 = $ constant, so

$$\mathcal{P}_2 = \mathcal{P}_1 \left(\frac{\rho_2}{\rho_1}\right)\left(\frac{\omega_2}{\omega_1}\right)^3 \left(\frac{D_2}{D_1}\right)^5 = 1.66 \text{ hp} \,(1)\left(\frac{1750}{1170}\right)^3 (1)^5 = 5.55 \text{ hp} \qquad \underline{\mathcal{P}_2}$$

The required input power may be calculated as

$$\mathcal{P}_{\text{in}} = \frac{\mathcal{P}_2}{\eta} = \frac{5.55 \text{ hp}}{0.80} = 6.94 \text{ hp} \qquad\qquad \underline{\mathcal{P}_{\text{in}}}$$

Thus a 7.5 hp motor (the next largest standard size) probably would be specified. The specific speed at $\omega_2 = 1750$ rpm is

$$N_s = \frac{\omega Q^{1/2}}{H^{3/4}} = \frac{1750 \,(449)^{1/2}}{(49.0)^{3/4}} = 2000 \qquad\qquad \underline{N_s}$$

> This example illustrates application of the pump "laws" and specific speed to scaling of performance data. Pump and fan "laws" are used widely in industry to scale performance for families of machines from a single performance curve, and to specify drive speed and power in machine applications.

7-5.3 Comments on Model Testing

While outlining the procedures involved in model testing, we have tried not to imply that testing is a simple task which automatically gives results that are easily interpretable, accurate, and complete. As in all experimental work, careful planning and execution are needed to obtain valid results. Models must be constructed carefully and accurately, and they must include sufficient detail in areas critical to the phenomenon being measured. Aerodynamic balances or other force measuring systems must be aligned carefully and calibrated correctly. Mounting methods must be devised that offer adequate rigidity and model motion, yet do not interfere with the phenomenon being measured. References [10–12] are considered the standard sources for details of wind tunnel test techniques. More specialized techniques for water impact testing are described in [13].

Experimental facilities must be designed and constructed carefully. The quality of flow in a wind tunnel must be documented. Flow in the test section should be as nearly uniform as possible (unless the desire is to simulate a special profile such as an atmospheric boundary layer), free from angularity, and with little swirl. If they interfere with measurements, boundary layers on tunnel walls must be removed by suction or energized by blowing. Pressure gradients in a wind tunnel test section may cause erroneous drag force readings due to pressure variations in the flow direction.

Special facilities are needed for unusual conditions or for special test requirements, especially to achieve large Reynolds numbers. Many facilities are so large or specialized that they cannot be supported by university laboratories or private industry. A few examples include [14–16]:

- National Full-Scale Aerodynamics Complex, NASA, Ames Research Center, Moffett Field, California
  Two wind tunnel test sections, powered by a 125,000 hp electric drive system:
  - 40 ft high and 80 ft wide (12 × 24 m) test section, maximum wind speed of 300 knots.
  - 80 ft high and 120 ft wide (24 × 36 m) test section, maximum wind speed of 137 knots (see cover illustration).
- U.S. Navy, David Taylor Research Center, Carderock, Maryland
  - High-Speed Towing Basin 2968 ft long, 21 ft wide, and 16 ft deep. Towing carriage can travel at up to 100 knots while measuring drag loads to 8000 lbf and side loads to 2000 lbf.
  - 36 in. variable-pressure water tunnel with 50 knot maximum test speed at pressures between 2 and 60 psia.
  - Anechoic Flow Facility with quiet, low turbulence air flow in 8 ft square by 21 ft long open-jet test section. Flow noise at maximum speed of 200 ft/sec is less than that of conversational speech.
- U.S. Army Corps of Engineers, Sausalito, California
  - San Francisco Bay and Delta Model with slightly more than 1 acre in area, 1:1000 horizontal scale and 1:100 vertical scale, 13,500 gpm of pumping capacity, use of fresh- and saltwater, and tide simulation.

- NASA, Langley Research Center, Hampton, Virginia
  - National Transonic Facility (NTF) with cryogenic technology (temperatures as low as −300 F) to reduce gas viscosity, raising Reynolds number by a factor of 6, while halving drive power.

7-6 NONDIMENSIONALIZING THE BASIC DIFFERENTIAL EQUATIONS

Ultimate success in use of the Buckingham Pi theorem is determined by the insight used to select the parameters included. If a complete set is chosen, the results will be complete. If an important variable is omitted, the results will be without meaning. Additional variables can be included if there is any uncertainty. As more experience is gained with fluid flow phenomena, the selection process becomes easier. Experience also gives more insight into the physical significance of each dimensionless group.

A more rigorous and broader approach to determine the conditions under which two flows are similar is to use the governing differential equations and boundary conditions. Similitude may be obtained when two physical phenomena are governed by differential equations and boundary conditions that have the same dimensionless forms. Dynamic similarity is guaranteed by duplicating the dimensionless coefficients of the equations and boundary conditions between prototype and model.

As an example of nondimensionalizing the basic differential equations, consider steady incompressible two-dimensional flow in the xy plane. Assume gravity acts in the negative y direction.

The equation for conservation of mass is

$$\frac{\partial u}{\partial x} + \frac{\partial v}{\partial y} = 0 \tag{7.7}$$

and the Navier–Stokes equations (Eqs. 5.27) reduce to

$$\rho\left(u\frac{\partial u}{\partial x} + v\frac{\partial u}{\partial y}\right) = -\frac{\partial p}{\partial x} + \mu\left(\frac{\partial^2 u}{\partial x^2} + \frac{\partial^2 u}{\partial y^2}\right) \tag{7.8}$$

$$\rho\left(u\frac{\partial v}{\partial x} + v\frac{\partial v}{\partial y}\right) = -\rho g - \frac{\partial p}{\partial y} + \mu\left(\frac{\partial^2 v}{\partial x^2} + \frac{\partial^2 v}{\partial y^2}\right) \tag{7.9}$$

To nondimensionalize these equations, divide all lengths by a reference length, L, and all velocities by a reference velocity, V_∞, which usually is taken as the freestream velocity. Make the pressure nondimensional by dividing by ρV_∞^2 (twice the freestream dynamic pressure). Denoting nondimensional quantities by asterisks, we obtain

$$x^* = \frac{x}{L} \quad y^* = \frac{y}{L} \quad u^* = \frac{u}{V_\infty} \quad v^* = \frac{v}{V_\infty} \quad \text{and} \quad p^* = \frac{p}{\rho V_\infty^2} \tag{7.10}$$

To illustrate the procedure for nondimensionalizing the equations, consider two typical terms in the equations,

$$u\frac{\partial u}{\partial x} = V_\infty\left(\frac{u}{V_\infty}\right)\frac{\partial(u/V_\infty)V_\infty}{\partial(x/L)L} = \frac{V_\infty^2}{L}u^*\frac{\partial u^*}{\partial x^*}$$

and

$$\frac{\partial^2 u}{\partial y^2} = \frac{\partial}{\partial y}\left(\frac{\partial u}{\partial y}\right) = \frac{\partial}{\partial (y/L)L}\left[\frac{\partial (u/V_\infty)V_\infty}{\partial (y/L)L}\right] = \frac{V_\infty}{L^2}\frac{\partial^2 u^*}{\partial y^{*2}}$$

By following this procedure, Eqs. 7.7, 7.8, and 7.9 can be written

$$\frac{V_\infty}{L}\frac{\partial u^*}{\partial x^*} + \frac{V_\infty}{L}\frac{\partial v^*}{\partial y^*} = 0 \tag{7.11}$$

$$\frac{\rho V_\infty^2}{L}\left(u^*\frac{\partial u^*}{\partial x^*} + v^*\frac{\partial u^*}{\partial y^*}\right) = -\frac{\rho V_\infty^2}{L}\frac{\partial p^*}{\partial x^*} + \frac{\mu V_\infty}{L^2}\left(\frac{\partial^2 u^*}{\partial x^{*2}} + \frac{\partial^2 u^*}{\partial y^{*2}}\right) \tag{7.12}$$

$$\frac{\rho V_\infty^2}{L}\left(u^*\frac{\partial v^*}{\partial x^*} + v^*\frac{\partial v^*}{\partial y^*}\right) = -\rho g - \frac{\rho V_\infty^2}{L}\frac{\partial p^*}{\partial y^*} + \frac{\mu V_\infty}{L^2}\left(\frac{\partial^2 v^*}{\partial x^{*2}} + \frac{\partial^2 v^*}{\partial y^{*2}}\right) \tag{7.13}$$

Dividing Eq. 7.11 by V_∞/L and Eqs. 7.12 and 7.13 by $\rho V_\infty^2/L$ gives

$$\frac{\partial u^*}{\partial x^*} + \frac{\partial v^*}{\partial y^*} = 0 \tag{7.14}$$

$$u^*\frac{\partial u^*}{\partial x^*} + v^*\frac{\partial u^*}{\partial y^*} = -\frac{\partial p^*}{\partial x^*} + \frac{\mu}{\rho V_\infty L}\left(\frac{\partial^2 u^*}{\partial x^{*2}} + \frac{\partial^2 u^*}{\partial y^{*2}}\right) \tag{7.15}$$

$$u^*\frac{\partial v^*}{\partial x^*} + v^*\frac{\partial v^*}{\partial y^*} = -\frac{gL}{V_\infty^2} - \frac{\partial p^*}{\partial y^*} + \frac{\mu}{\rho V_\infty L}\left(\frac{\partial^2 v^*}{\partial x^{*2}} + \frac{\partial^2 v^*}{\partial y^{*2}}\right) \tag{7.16}$$

From the nondimensional equations (Eqs. 7.14, 7.15, 7.16), we conclude that the differential equations for two flow systems will be identical, if and only if, the quantities $\mu/\rho V_\infty L$ and gL/V_∞^2 are the same for both flows. Thus, model studies to determine the drag force on a surface ship require duplication of both the Froude number and the Reynolds number to ensure dynamically similar flows.

For flows around submerged bodies far below the free surface, as for the sphere of Example Problem 7.4, body forces are not important. The governing equations do not include the body force term, ρg. Nondimensionalizing the governing equations for this case shows that the nondimensional equations governing two flows will be identical if the Reynolds number is the same for both flows.

So far we have concentrated on the differential equations governing the flow. It is important to emphasize that in addition to identical nondimensional equations, the nondimensional boundary conditions also must be identical if the two flows are to be kinematically similar. This leads to the requirement of geometric similarity between the flows. Nondimensionalizing the boundary conditions may lead to additional requirements that must be satisfied between the two flows. For example, consider the case where the velocity at a specified location is periodic. The boundary condition (*bc*) then specifies

$$u_{bc} = V_\infty \sin \omega t$$

If we nondimensionalize time using the ratio of reference velocity to reference length, then

$$t^* = \frac{t V_\infty}{L}$$

The nondimensional boundary condition becomes

$$u_{bc}^* = \frac{u_{bc}}{V_\infty} = \sin\left(\frac{\omega L}{V_\infty} t^*\right)$$

Duplication of the boundary condition requires that the parameter $\omega L/V_\infty$ be the same between the two flows. This parameter is the *Strouhal number*

$$St = \frac{\omega L}{V_\infty}$$

which is named after the German physicist who first discovered its importance while investigating the self-excited "singing" of wires in the wind.

Establishing similitude from the differential equations and boundary conditions that describe the flow is a rigorous procedure. If one begins with the correct equations, and performs each step correctly, one can be sure that all appropriate variables have been included. Additional derivations and examples of establishing similitude from the governing equations are presented in [17] and [18].

The governing differential equations often are written in nondimensional form for numerical solution. Scaling is simplified and unit conversion problems are reduced when dimensionless forms of the equations are used. Use of dimensionless equations frequently permits solutions to be presented in generalized form.

7-7 SUMMARY OBJECTIVES

After completing study of Chapter 7, you should be able to do the following:

1. Define:

Reynolds number	Mach number
Euler number (pressure coefficient)	Strouhal number
Cavitation number	geometric similarity
Froude number	kinematic similarity
Weber number	dynamic similarity

2. State the Buckingham Pi theorem.
3. Given a physical problem in which the dependent parameter is a function of specified independent parameters, determine a set of independent dimensionless ratios that characterize the problem.
4. State the conditions under which prototype behavior can be predicted from model tests.
5. Predict results for a prototype from model test data.
6. Obtain dimensionless coefficients by nondimensionalizing the governing differential equations.
7. Solve the problems at the end of the chapter that relate to the material you have studied.

REFERENCES

1. Buckingham, E., "On Physically Similar Systems: Illustrations of the Use of Dimensional Equations," *Physical Review, 4*, 4, 1914, pp. 345–376.
2. Todd, L. H., "Resistance and Propulsion," Chapter VII in *Principles of Naval Architecture*, J. P. Comstock, ed. New York: Society of Naval Architects and Marine Engineers, 1967.
3. "Aerodynamic Flow Visualization Techniques and Procedures," Warrendale, PA: Society of Automotive Engineers, SAE Information Report, HS J1566, January 1986.

4. Merzkirch, W., *Flow Visualization*, 2nd ed. New York: Academic Press, 1987.

5. "SAE Wind Tunnel Test Procedure for Trucks and Buses," *Recommended Practice* SAE J1252, Warrendale, PA: Society of Automotive Engineers, 1981.

6. Sedov, L. I., *Similarity and Dimensional Methods in Mechanics*. New York: Academic Press, 1959.

7. Birkhoff, G., *Hydrodynamics—A Study in Logic, Fact, and Similitude,* 2nd ed. Princeton, NJ: Princeton University Press, 1960.

8. Ipsen, D. C., *Units, Dimensions, and Dimensionless Numbers*. New York: McGraw-Hill, 1960.

9. Yalin, M. S., *Theory of Hydraulic Models*. New York: Macmillan, 1971.

10. Pankhurst, R. C., and D. W. Holder, *Wind-Tunnel Technique*. London: Pitman, 1965.

11. Rae, W. H., and A. Pope, *Low-Speed Wind Tunnel Testing,* 2nd ed. New York: Wiley-Interscience, 1984.

12. Pope, A., and K. L. Goin, *High-Speed Wind Tunnel Testing*. New York: Krieger, 1978.

13. Waugh, J. G., and G. W. Stubstad, *Hydroballistics Modeling*. San Diego, CA: U.S. Naval Undersea Center, *ca*. 1965.

14. Baals, D. W., and W. R. Corliss, *Wind Tunnels of NASA*. Washington, D.C.: National Aeronautics and Space Administration, SP-440, 1981.

15. Vincent, M., "The Naval Ship Research and Development Center," Carderock, MD: Naval Ship Research and Development Center, Report 3039 (Revised), November 1971.

16. Smith, B. E., P. T. Zell, and P. M. Shinoda, "Comparison of Model- and Full-Scale Wind-Tunnel Performance," *Journal of Aircraft, 27,* 3, March 1990, pp. 232–238.

17. Kline, S. J., *Similitude and Approximation Theory*. New York: McGraw-Hill, 1965.

18. Hansen, A. G., *Similarity Analysis of Boundary-Value Problems in Engineering*. Englewood Cliffs, NJ: Prentice-Hall, 1964.

19. Kowalski, T., "Hydrodynamics of Water-Borne Bodies," Chapter 4 in *Introduction to Ocean Engineering,* H. Schenck, ed. New York: McGraw-Hill, 1975.

PROBLEMS

7.1 At very low speeds, the drag on an object is independent of fluid density. Thus the drag force, F, on a small sphere is a function only of speed, V, fluid viscosity, μ, and sphere diameter, D. Use dimensional analysis to express the drag force as a function of these variables.

7.2 Experiments show that the pressure drop due to flow through a sudden contraction in a circular duct may be expressed as $\Delta p = p_1 - p_2 = f(\rho, \mu, \bar{V}, d, D)$. You are asked to organize some experimental data. Obtain the resulting dimensionless parameters.

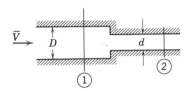

P7.2

7.3 The boundary-layer thickness, δ, on a smooth flat plate in an incompressible flow without pressure gradients depends on the freestream speed, U, the fluid density, ρ, the fluid viscosity, μ, and the distance from the leading edge of the plate, x. Express these variables in dimensionless form.

7.4 The wall shear stress, τ_w, in a boundary layer depends on distance from the leading edge of the body, x, the density, ρ, and viscosity, μ, of the fluid, and the freestream speed of the flow, U. Obtain the dimensionless groups and express the functional relationship among them.

7.5 The mean velocity, $\bar{u}$, for turbulent flow in a pipe or a boundary layer may be correlated using the wall shear stress, τ_w, distance from the wall, y, and the fluid properties, ρ and μ. Use dimensional analysis to find one dimensionless parameter containing $\bar{u}$ and one containing y that are suitable for organizing experimental data. Show that the result may be written

$$\frac{\bar{u}}{u_*} = f\left(\frac{y u_*}{\nu}\right)$$

where $u_* = (\tau_w/\rho)^{1/2}$ is the *friction velocity*.

7.6 Measurements of the liquid height upstream from an obstruction placed in an open-channel flow can be used to determine volume flow rate. (Such obstructions, designed and calibrated to measure rate of open-channel flow, are called *weirs*.) Assume the volume flow rate, Q, over a weir is a function of upstream height, h, gravity, g, and channel width, b. Use dimensional analysis to find the functional dependence of Q on the other variables.

7.7 The speed, V, of a free-surface gravity wave in deep water is a function of the wavelength, λ, depth, D, density, ρ, and acceleration of gravity, g. Use dimensional analysis to find the functional dependence of V on the other variables. Express V in the simplest form possible.

7.8 Capillary waves are formed on a liquid free surface as a result of surface tension. They have short wavelengths. The speed of a capillary wave depends on surface tension, σ, wavelength, λ, and liquid density, ρ. Use dimensional analysis to express the wave speed as a function of these variables.

7.9 The load-carrying capacity, W, of a journal bearing is known to depend on its diameter, D, length, l, and clearance, c, in addition to its angular speed, ω, and lubricant viscosity, μ. Determine the dimensionless parameters that characterize this problem.

7.10 A disk rotates near a fixed surface. The radius of the disk is R, and the space between the disk and the surface is filled with a liquid of viscosity μ. The spacing between the disk and the surface is h, and the disk rotates at angular speed ω. Find the dependence between torque on the disk, T, and the other variables.

7.11 The power per unit cross-sectional area, E, transmitted by a sound wave is a function of wave speed, V, medium density, ρ, wave amplitude, r, and wave frequency, n. Determine, by dimensional analysis, the general form of the expression for E in terms of the other variables.

7.12 The power, $\mathscr{P}$, required to drive a fan is believed to depend on fluid density, ρ, volume flow rate, Q, impeller diameter, D, and angular velocity, ω. Use dimensional analysis to determine the dependence of $\mathscr{P}$ on the other variables.

7.13 The vorticity, ζ, at a point in an axisymmetric flow field is thought to depend on initial circulation, Γ_0, radius, r, time, τ, and fluid kinematic viscosity, ν. Find a set of dimensionless parameters suitable for organizing experimental data.

7.14 You are asked to find a set of dimensionless parameters to organize data from a laboratory experiment, in which a tank is drained through an orifice from initial liquid level h_0. The time, τ, to drain the tank depends on tank diameter, D, orifice diameter, d, acceleration of gravity, g, liquid density, ρ, and liquid viscosity, μ. How many dimensionless parameters will result? How many repeating variables must be selected to determine the dimensionless parameters? Obtain the Π parameter that contains the viscosity.

7.15 In a fluid mechanics laboratory experiment a tank of water, with diameter D, is drained from initial level h_0. The smoothly rounded drain hole has diameter d. Assume the mass flow rate from the tank is a function of h, D, d, g, ρ, and μ, where g is the acceleration of gravity and ρ and μ are fluid properties. Measured data are to be correlated in dimensionless form. Determine the number of dimensionless parameters that will result. Specify the number of repeating parameters that must be selected to determine the dimensionless parameters. Obtain the Π parameter that contains the viscosity.

7.16 A continuous belt moving vertically through a bath of viscous liquid drags a layer of liquid, of thickness h, along with it. The volume flow rate of liquid, Q, is assumed to depend on μ, ρ, g, h, and V, where V is the belt speed. Apply dimensional analysis to predict the form of dependence of Q on the other variables.

7.17 Assume that the resistance force, R, of a flat plate immersed in a fluid depends on fluid density and viscosity, velocity, width, b, and height, h, of the plate. Find a convenient set of coordinates for organizing data.

7.18 Small droplets of liquid are formed when a liquid jet breaks up in spray and fuel injection processes. The resulting droplet diameter, d, is thought to depend on liquid density, viscosity, and surface tension, as well as jet speed, V, and diameter, D. How many dimensionless ratios are required to characterize this process? Determine these ratios.

7.19 The sketch shows an air jet discharging vertically. Experiments show that a ball placed in the jet is suspended in a stable position. The equilibrium height of the ball in the jet is found to depend on D, d, V, ρ, μ, and W, where W is the weight of the ball. Dimensional analysis is suggested to correlate experimental data. Find the Π parameters that characterize this phenomenon.

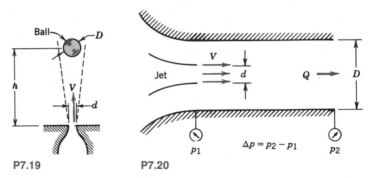

P7.19   P7.20

7.20 Consider the jet pump shown. A dimensional analysis is needed to organize performance data measured for the jet pump. The functional dependence is $\Delta p = f(\rho, V, d, D, \mu, Q)$. Find the number of dimensionless groups needed to characterize the jet pump. Obtain the nondimensional groups that contain the volume flow rate, Q, and the viscosity, μ.

7.21 A large tank of liquid under pressure is drained through a smoothly contoured nozzle of area A. The mass flow rate is thought to depend on nozzle area, A, liquid density, ρ, difference in height between the liquid surface and nozzle, h, tank gage pressure, Δp, and gravitational acceleration, g. Determine how many independent Π parameters can be formed for this problem. Find the dimensionless parameters. State the functional relationship for the mass flow rate in terms of the dimensionless parameters.

7.22 Spin plays an important role in the flight trajectory of golf, Ping-Pong, and tennis balls. Therefore, it is important to know the rate at which spin decreases for a ball in flight. The aerodynamic torque, T, acting on a ball in flight, is thought to depend on flight speed, V, air density, ρ, air viscosity, μ, ball diameter, D, spin rate (angular speed), ω, and diameter of the dimples on the ball, d. Determine the dimensionless parameters that result.

7.23 The thrust of a marine propeller is to be measured during "open-water" tests at a variety of angular speeds and forward speeds ("speeds of advance"). The thrust, F_T, is thought to depend on water density, ρ, propeller diameter, D, speed of advance, V, acceleration of gravity, g, angular speed, ω, pressure in the liquid, p, and liquid viscosity, μ. Develop a set of dimensionless parameters to characterize the performance of the propeller. (One of the resulting parameters, gD/V^2, is known as the *Froude speed of advance*.)

7.24   The power loss, $\mathcal{P}$, in a journal bearing depends on length, l, diameter, D, and clearance, c, of the bearing, in addition to its angular speed, ω. The lubricant viscosity and mean pressure are also important. Obtain the dimensionless parameters that characterize this problem. Determine the functional form of the dependence of $\mathcal{P}$ on these parameters.

7.25   The power, $\mathcal{P}$, required to drive a propeller is known to depend on the following variables: freestream speed, V, propeller diameter, D, angular speed, ω, fluid viscosity, μ, fluid density, ρ, and speed of sound in the fluid, c. How many dimensionless groups are required to characterize this situation? Obtain these dimensionless groups.

7.26   In a fan-assisted convection oven, the heat transfer rate to a roast, $\dot{Q}$ (energy per unit time), is thought to depend on the specific heat of air, c_p, temperature difference, Θ, a length scale, L, air density, ρ, air viscosity, μ, and air speed, V. How many basic dimensions are included in these variables? Determine the number of Π parameters needed to characterize the oven. Evaluate the Π parameters.

7.27   When a valve is closed suddenly in a pipe with flowing water, a *water hammer* pressure wave is set up. The very high pressures generated by such waves can damage the pipe. The maximum pressure, p_{max}, generated by water hammer is a function of liquid density, ρ, initial flow speed, U_0, and liquid bulk modulus, E_v. How many dimensionless groups are needed to characterize water hammer? Determine the functional relationship among the variables in terms of the necessary Π groups.

7.28   An airship is to operate at 20 m/sec in air at standard conditions. A model is constructed to $\frac{1}{20}$ scale and tested in a wind tunnel at the same air temperature to determine drag. What criterion should be considered to obtain dynamic similarity? If the model is tested at 75 m/sec, what pressure should be used in the wind tunnel? If the model drag force is 250 N, what will be the drag of the prototype?

7.29   Measurements of drag force are made on a model automobile in a towing tank filled with freshwater. The model length scale is $\frac{1}{5}$ that of the prototype. State the conditions required to ensure dynamic similarity between the model and prototype. Determine the fraction of the prototype speed in air at which the model test should be made in water to ensure dynamically similar conditions. Measurements made at various speeds show that the dimensionless force ratio becomes contant at model test speeds above $V_m = 4$ m/sec. The drag force measured during a test at this speed is $F_{Dm} = 182$ N. Calculate the drag force expected on the prototype vehicle operating at 90 km/hr in air.

7.30   A $\frac{1}{5}$-scale model of a torpedo is tested in a wind tunnel to determine the drag force. The prototype operates in water, has 533 mm diameter, and is 6.7 m long. The desired operating speed of the prototype is 28 m/sec. To avoid compressibility effects in the wind tunnel, the maximum speed is limited to 110 m/sec. However, the pressure in the wind tunnel can be varied while holding the temperature constant at 20 C. At what minimum pressure should the wind tunnel be operated to achieve a dynamically similar test? At dynamically similar test conditions, the drag force on the model is measured as 618 N. Evaluate the drag force expected on the full-scale torpedo.

7.31   The drag of an airfoil at zero angle of attack is a function of density, viscosity, and velocity, in addition to a length parameter. A $\frac{1}{10}$-scale model of an airfoil was tested in a wind tunnel at a Reynolds number of 5.5×10^6, based on chord length. Test conditions in the wind tunnel air stream were 15 C and 10 atm absolute pressure. The prototype airfoil has a chord length of 2 m, and it is to be flown in air at standard conditions. Determine the speed at which the wind tunnel model was tested, and the corresponding prototype speed.

7.32 An airplane wing, with chord length of 5 ft and span of 30 ft, is designed to move through standard air at a speed of 230 ft/sec. A $\frac{1}{10}$-scale model of this wing is to be tested in a water tunnel. What speed is necessary in the water tunnel to achieve dynamic similarity? What will be the ratio of forces measured in the model flow to those on the prototype wing?

7.33 Consider a smooth sphere, of diameter D, immersed in a fluid moving with speed V. The drag force on a 3 m diameter weather ballon in air moving at 1.5 m/sec is to be calculated from test data. The test is to be performed in water using a 50 mm diameter model. Under dynamically similar conditions, the model drag force is measured as 3.78 N. Evaluate the model test speed and the drag force expected on the full-scale balloon.

7.34 The fluid dynamic characteristics of a golf ball are to be tested using a model in a wind tunnel. Dependent parameters are the drag force, F_D, and lift force, F_L, on the ball. The independent parameters should include angular speed, ω, and dimple depth, d. Determine suitable dimensionless parameters and express the functional dependence among them. A golf pro can hit a ball at $V = 240$ ft/sec and $\omega = 9000$ rpm. To model these conditons in a wind tunnel with maximum speed of 80 ft/sec, what diameter model should be used? How fast must the model rotate? (The diameter of a U.S. golf ball is 1.68 in.)

7.35 A model test is performed to determine the flight characteristics of a Frisbee. Dependent parameters are drag force, F_D, and lift force, F_L. The independent parameters should include angular speed, ω, and roughness height, h. Determine suitable dimensionless parameters and express the functional dependence among them. The test (using air) on a $\frac{1}{4}$-scale model Frisbee, is to be geometrically, kinematically, and dynamically similar to the prototype. The prototype values are $V_p = 20$ ft/sec and $\omega_p = 100$ rpm. What values of V_m and ω_m should be used?

7.36 A model hydrofoil is to be tested at 1:20 scale. The test speed is chosen to duplicate the Froude number corresponding to the 60 knot prototype speed. To model cavitation correctly, the cavitation number also must be duplicated. At what ambient pressure must the test be run? Water in the model test basin can be heated to 130 F, compared to 45 F for the prototype.

7.37 SAE 10W oil at 80 F flowing in a 1 in. diameter horizontal pipe, at an average speed of 3 ft/sec, produces a pressure drop of 65.3 psig over a 500 ft length. Water at 60 F flows through the same pipe under dynamically similar conditions. Using the results of Example Problem 7.2, calculate the average speed of the water flow and the corresponding pressure drop.

7.38 The capillary rise, Δh, of a liquid in a circular tube, of diameter D, depends on surface tension, σ, and specific weight, γ, of the fluid. The significant variables found from dimensional analysis are

$$\Pi_1 = \Delta h \sqrt{\frac{\gamma}{\sigma}} \quad \text{and} \quad \Pi_2 = \frac{\sigma}{\gamma D^2}$$

Show that the Π parameters can be manipulated to yield the nondimensional parameters of Example Problem 7.3. The capillary rise for liquid A is 1.0 in. in a tube of 0.010 in. inside diameter. What will be the rise for liquid B (having the same surface tension but four times the density of A) in a tube of 0.005 in. inside diameter?

7.39 Consider water flow around a circular cylinder, of diameter D and length l. In addition to geometry, the drag force is known to depend on liquid speed, V, density, ρ, and viscosity, μ. Express drag force, F_D, in dimensionless form as a function of all relevant variables. The static pressure distribution on a circular cylinder, measured in the laboratory, can be expressed in terms of the dimensionless pressure coefficient; the lowest pressure coefficient is $C_p = -2.4$ at the location of the minimum static pressure on the cylinder surface. Estimate the maximum speed at which a cylinder could be

towed in water at atmospheric pressure, without causing cavitation, if the onset of
cavitation occurs at a cavitation number of 0.5.

7.40   A $\frac{1}{8}$-scale model of a tractor-trailer rig is tested in a pressurized wind tunnel. The
rig width, height, and length are $W = 0.305$ m, $H = 0.476$ m, and $L = 2.48$ m, respectively. At wind speed $V = 75.0$ m/sec, the model drag force is $F_D = 128$ N. (Air
density in the tunnel is $\rho = 3.23$ kg/m^3.) Calculate the aerodynamic drag coefficient
for the model. Compare the Reynolds numbers for the model test and for the prototype
vehicle at 55 mph. Calculate the aerodynamic drag force on the prototype vehicle at a
road speed of 55 mph into a headwind of 10 mph.

7.41   An automobile is to travel through standard air at 100 km/hr. To determine the pressure
distribution, a $\frac{1}{5}$-scale model is to be tested in water. What factors must be considered
to ensure kinematic similarity in the tests? Determine the water speed that should
be used. What is the corresponding ratio of drag force between prototype and model
flows? The lowest pressure coefficient is $C_p = -1.4$ at the location of the minimum
static pressure on the surface. Estimate the minimum tunnel pressure required to avoid
cavitation, if the onset of cavitation occurs at a cavitation number of 0.5.

7.42   In some speed ranges, vortices are shed from the rear of bluff cylinders placed across
a flow. The vortices alternately leave the top and bottom of the cylinder, as shown,
causing an alternating force normal to the freestream velocity. The vortex shedding
frequency, f, is thought to depend on ρ, d, and μ. Use dimensional analysis to develop
a functional relationship for f. Vortex shedding occurs in standard air on two cylinders
with a diameter ratio of 2. Determine the velocity ratio for dynamic similarity, and the
ratio of vortex shedding frequencies.

P7.42

7.43   A model test of a tractor-trailer rig is performed in a wind tunnel. The drag force, F_D,
is found to depend on frontal area, A, wind speed, V, air density, ρ, and air viscosity,
μ. The model scale is 1:4; frontal area of the model is $A = 0.625$ m^2. Obtain a set
of dimensionless parameters suitable to characterize the model test results. State the
conditions required to obtain dynamic similarity between model and prototype flows.
When tested at wind speed $V = 89.6$ m/sec, in standard air, the measured drag force
on the model was $F_D = 2.46$ kN. Estimate the aerodynamic drag force on the full-scale
vehicle at $V = 22.4$ m/sec. Calculate the power needed to overcome this drag force if
there is no wind.

7.44   A 1:50 scale model of a submarine is tested in a water tunnel. The drag force, F_D,
depends on water speed, V, density, ρ, and viscosity, μ, and on model volume, $\forall$.
Find a set of dimensionless parameters suitable to organize the resulting test data.
Estimate the drag of the full-scale submarine at 27 knots if a model test at 10 knots
yielded a measured drag force of 13 N.

7.45   Your favorite professor likes mountain climbing, so there is always a possibility that
the professor may fall into a crevasse in some glacier. If that happened today, and the
professor was trapped in a slowly moving glacier, you are curious to know whether
the professor would reappear at the downstream drop off of the glacier during this
academic year. Assuming ice is a Newtonian fluid with the density of glycerine but a
million times as viscous, you decide to build a model and use dimensional analysis
and similarity to estimate when the professor would reappear. Assume the real glacier
is 15 m deep and is on a slope that falls 1.5 m in a horizontal distance of 1850 m.
Develop the dimensionless parameters and conditions expected to govern dynamic
similarity in this problem. If the model professor reappears in the laboratory after 9.6
hours, when should you return to the end of the real glacier to provide help to your
favorite professor?

7.46 A $\frac{1}{10}$-scale model of a tractor-trailer rig is tested in a wind tunnel. The model frontal area is $A_m = 1.08$ ft^2. When tested at $V_m = 250$ ft/sec in standard air, the measured drag force is $F_D = 76.3$ lbf. Evaluate the drag coefficient for the model conditions given. Assuming that the drag coefficient is the same for model and prototype, calculate the drag force on a prototype rig at a highway speed of 55 mph. Determine the air speed at which a model *should* be tested to ensure dynamically similar results if the prototype speed is 55 mph. Is this air speed practical? Why or why not?

7.47 It is recommended in [5] that the frontal area of a model be less than 5 percent of the wind tunnel test section area and $Re = Vw/\nu > 2 \times 10^6$, where w is the model width. Further, the model height must be less than 30 percent of the test section height, and the maximum projected width of the model at maximum yaw (20°) must be less than 30 percent of the test section width. The maximum air speed should be less than 300 ft/sec to avoid compressibility effects. A model of a tractor-trailer rig is to be tested in a wind tunnel that has a test section 1.5 ft high and 2 ft wide. The height, width, and length of the full-scale rig are 13 ft 6 in., 8 ft, and 65 ft, respectively. Evaluate the scale ratio of the largest model that meets the recommended criteria. Assess whether an adequate Reynolds number can be achieved in this test facility.

7.48 A circular container, partially filled with water, is rotated about its axis at constant angular speed, ω. At any time, τ, from the start of rotation, the velocity, V_θ, at distance r from the axis of rotation, was found to be a function of τ, ω, and the properties of the liquid. Write the dimensionless parameters that characterize this problem. If, in another experiment, honey is rotated in the same cylinder at the same angular speed, determine from your dimensionless parameters whether honey will attain steady motion as quickly as water. Explain why the Reynolds number would not be an important dimensionless parameter in scaling the steady-state motion of liquid in the container.

7.49 The power, $\mathcal{P}$, required to drive a fan is assumed to depend on fluid density, ρ, volume flow rate, Q, impeller diameter, D, and angular speed, ω. If a fan with $D_1 = 8$ in. delivers $Q_1 = 800$ cfm (cubic feet per minute) of air at $\omega_1 = 2400$ rpm, what volume flow rate could be expected for a geometrically similar fan with $D_2 = 16$ in. at $\omega_2 = 1850$ rpm?

7.50 The pressure rise, Δp, of a liquid flowing steadily through a centrifugal pump depends on pump diameter, D, angular speed of the rotor, ω, volume flow rate, Q, and density, ρ. The table gives data for the prototype and for a geometrically similar model pump. For conditions corresponding to dynamic similarity between the model and prototype pumps, calculate the missing values in the table.

Variable	Prototype	Model
Δp		74.9 kPa
Q	1.25 m^3/min	
ρ	800 kg/m^3	999 kg/m^3
ω	10 rad/sec	100 rad/sec
D	60 mm	120 mm

7.51 An axial-flow pump is required to deliver 25 ft^3/sec of water at a head of 150 ft · lbf/slug. The diameter of the rotor is 1 ft, and it is to be driven at 500 rpm. The prototype is to be modeled on a small test apparatus having a 3 hp, 1000 rpm power supply. For similar performance between the prototype and the model, calculate the head, volume flow rate, and diameter of the model.

7.52 Consider again Problem 7.23. Experience shows that for ship-size propellers, viscous effects on scaling are small. Also, when cavitation is not present, the nondimensional parameter containing pressure can be ignored. Assume that torque, T, and power, $\mathcal{P}$, depend on the same parameters as thrust. For conditions under which effects of μ and p can be neglected, derive scaling "laws" for propellers, similar to the pump "laws" of Section 7-5.2, that relate thrust, torque, and power to the angular speed and diameter of the propeller.

7.53   A model propeller 2 ft in diameter is tested in a wind tunnel. Air approaches the propeller at 150 ft/sec when it rotates at 2000 rpm. The thrust and torque measured under these conditions are 25 lbf and 7.5 ft · lbf, respectively. A prototype 10 times as large as the model is to be built. At a dynamically similar operating point, the approach air speed is to be 400 ft/sec. Calculate the speed, thrust, and torque of the prototype propeller under these conditions, neglecting the effect of viscosity but including density.

7.54   The following data for a model test of a marine propeller are given in Problem 4–6 of [19]:

Parameter	Model	Prototype
Diameter	18 in.	24 ft
Angular speed	960 rpm	240 rpm
Speed of advance	20 ft/sec	
Thrust	35.1 lbf	
Torque	120 in. · lbf	
Fluid	Freshwater	Seawater
Temperature	65 F	59 F

For dynamically similar test conditions, calculate the speed of advance, thrust, and torque of the prototype propeller. (Use the results of Problem 7.52.) Evaluate the thrust power produced, required power input, and efficiency of the prototype propeller. Neglect cavitation and Reynolds number considerations.

7.55   Closed-circuit wind tunnels can produce higher speeds than open-circuit tunnels with the same power input because energy is recovered in the diffuser downstream from the test section. The *kinetic energy ratio* is a figure of merit defined as the ratio of the kinetic energy flux in the test section to the drive power. Estimate the kinetic energy ratio for the 40×80 wind tunnel at NASA–Ames.

7.56   In tests of models in wind tunnels, forces and moments are measured relative to an axis system aligned with the tunnel centerline. Forces acting on a highway vehicle are expressed most usefully in an axis system referenced to the vehicle (body-axis coordinates). Test data measured at yaw angles must be converted to body-axis coordinates for use. Develop equations to calculate body-axis drag and side force from data measured in wind tunnel coordinates. Assume that the model yaw angle is ψ.

7.57   A 1:16 model of a bus is tested in a wind tunnel in standard air. The model is 152 mm wide, 200 mm high, and 762 mm long. The measured drag force at 26.5 m/sec wind speed is 6.09 N. The longitudinal pressure gradient in the wind tunnel test section is $-11.8 \text{ N/m}^2/\text{m}$. Estimate the correction that should be made to the measured drag force to correct for horizontal buoyancy caused by the pressure gradient in the test section. Calculate the drag coefficient for the model. Evaluate the aerodynamic drag force on the prototype at 100 km/hr on a calm day.

7.58   A 1:16 model of a 20 m long truck is tested in a wind tunnel, where the axial static pressure gradient is -1.2 mm of water per meter at a test speed of 80 m/sec. The frontal area of the prototype is 10 m^2. Estimate the horizontal buoyancy correction for this situation. Express the correction as a fraction of the measured C_D, if $C_D = 0.85$.

7.59 The propagation speed of small-amplitude surface waves in a region of uniform depth is given by

$$c^2 = \left(\frac{\sigma}{\rho} \frac{2\pi}{\lambda} + \frac{g\lambda}{2\pi} \right) \tanh \frac{2\pi h}{\lambda}$$

where h is depth of the undisturbed liquid and λ is wavelength. Using L as a characteristic length and V_0 as a characteristic velocity, obtain the dimensionless groups that characterize the equation and determine the conditions for similarity.

7.60 The slope of the free surface of a steady wave in one-dimensional flow in a shallow liquid layer is described by the equation

$$\frac{\partial h}{\partial x} = -\frac{u}{g} \frac{\partial u}{\partial x}$$

Use a length scale, L, and a velocity scale, V_0, to nondimensionalize this equation. Obtain the dimensionless groups that characterize this flow. Determine the conditions for dynamic similarity.

7.61 One-dimensional unsteady flow in a thin liquid layer is described by the equation

$$\frac{\partial u}{\partial t} + u \frac{\partial u}{\partial x} = -g \frac{\partial h}{\partial x}$$

Use a length scale, L, and a velocity scale, V_0, to nondimensionalize this equation. Obtain the dimensionless groups that characterize this flow. Determine the conditions for dynamic similarity.

7.62 By using order of magnitude analysis, the continuity and Navier–Stokes equations can be simplified to the Prandtl boundary-layer equations. For steady, incompressible, and two-dimensional flow, neglecting gravity, the result is

$$\frac{\partial u}{\partial x} + \frac{\partial v}{\partial y} = 0$$

$$u \frac{\partial u}{\partial x} + v \frac{\partial u}{\partial y} = -\frac{1}{\rho} \frac{\partial p}{\partial x} + \nu \frac{\partial^2 u}{\partial y^2}$$

Use L and V_0 as characteristic length and velocity, respectively. Nondimensionalize these equations and identify the similarity parameters that result.

Chapter 8

INTERNAL INCOMPRESSIBLE VISCOUS FLOW

Flows completely bounded by solid surfaces are called internal flows. Thus internal flows include flows through pipes, ducts, nozzles, diffusers, sudden contractions and expansions, valves, and fittings.

Internal flows may be laminar or turbulent. Some laminar flow cases may be solved analytically. In the case of turbulent flow, analytical solutions are not possible, and we must rely heavily on semi-empirical theories and on experimental data. The nature of laminar and turbulent flows was discussed in Section 2-5.2. For internal flows, the flow regime (laminar or turbulent) is primarily a function of the Reynolds number.

Following a brief introductory section, we consider two cases of fully developed laminar flow of a Newtonian fluid. Although most internal flows of engineering interest are turbulent, laminar flow can be important in applications such as lubrication or chemical process flows. Our analysis begins with a differential control volume rather than with the differential equations of motion derived in Chapter 5. This analysis also provides insight into the basic nature of turbulent flows in pipes and ducts, which are considered next. The chapter concludes with a discussion of flow measurements.

8-1 INTRODUCTION

As discussed previously in Section 2-5.4, the pipe flow regime (laminar or turbulent) is determined by the Reynolds number, $Re = \rho \bar{V} D / \mu$. One can demonstrate, by the classic Reynolds experiment,[1] the qualitative difference between laminar and turbulent flows. In this experiment water flows from a large reservoir through a clear tube. A thin filament of dye injected at the entrance to the tube allows visual observation of the flow. At low flow rates (low Reynolds numbers) the dye injected into the flow remains in a single filament; there is little dispersion of dye because the flow is laminar. A laminar flow is one in which the fluid flows in laminae, or layers; there is no macroscopic mixing of adjacent fluid layers.

As the flow rate through the tube is increased, the dye filament becomes unstable and breaks up into a random motion; the line of dye is stretched and twisted into myriad entangled threads, and it quickly disperses throughout the entire flow field. This behavior of turbulent flow is due to small, high-frequency velocity fluctuations superimposed on the mean motion of a turbulent flow, as illustrated earlier in Fig. 2.14; the mixing of fluid particles from adjacent layers of fluid results in rapid dispersion of the dye.

[1] This experiment is demonstrated in the NCFMF film, *Turbulence*, R. W. Stewart, principal.

With great care to maintain the flow free from disturbances, and with smooth surfaces, experiments to date have been able to maintain laminar flow in a pipe to a Reynolds number of about 100,000! However, most engineering flow situations are not so carefully controlled. Under normal conditions, transition occurs at $Re \approx 2300$ for flow in pipes. (Transition Reynolds numbers for other flow situations are given in the Example Problems.)

Figure 8.1 illustrates laminar flow in the entrance region of a circular pipe. The flow has uniform velocity U_0 at the pipe entrance. Because of the no-slip condition at the wall, we know that the velocity at the wall must be zero along the entire length of the pipe. A boundary layer (Section 2-5.1) develops along the walls of the channel. The solid surface exerts a retarding shear force on the flow; thus the speed of the fluid in the neighborhood of the surface is reduced. At successive sections along the pipe in this entry region, the effect of the solid surface is felt farther out into the flow.

For incompressible flow, conservation of mass requires the velocity at the pipe centerline to increase with distance from the inlet. To satisfy conservation of mass for incompressible flow, the average velocity at any cross section

$$\bar{V} = \frac{1}{A} \int_{\text{Area}} u \, dA$$

must equal U_0, so

$$\bar{V} = U_0 = \text{constant}$$

Sufficiently far from the pipe entrance, the boundary layer developing on the pipe wall reaches the pipe centerline and the flow becomes entirely viscous. The velocity profile shape changes slightly after the inviscid core disappears. When the profile shape no longer changes with increasing distance, x, the flow is *fully developed*. The distance downstream from the entrance to the location at which fully developed flow begins is called the *entrance length*. The actual shape of the fully developed velocity profile depends on whether the flow is laminar or turbulent. In Fig. 8.1 the profile is shown qualitatively for a laminar flow.

For laminar flow, the entrance length, L, is a function of Reynolds number,

$$\frac{L}{D} \simeq 0.06 \frac{\rho \bar{V} D}{\mu} \tag{8.1}$$

where D is pipe diameter, $\bar{V}$ is average velocity, ρ is fluid density, and μ is fluid viscosity. Laminar flow in a pipe may be expected only for Reynolds numbers less than 2300. Thus the entrance length for laminar pipe flow may be as long as

$$L \simeq 0.06 \, Re \, D \leq (0.06)(2300) D = 138 D$$

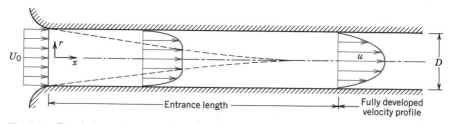

Fig. 8.1 Flow in the entrance region of a pipe.

or more than 100 pipe diameters. If the flow is turbulent, enhanced mixing among fluid layers[2] causes more rapid growth of the boundary layer. Experiments show that the mean velocity profile becomes fully developed within 25 to 40 pipe diameters from the entrance. However, the details of the turbulent motion may not be fully developed for 80 or more pipe diameters. Fully developed internal flows will be treated in Parts A and B of this chapter.

PART A   FULLY DEVELOPED LAMINAR FLOW

Although there are relatively few viscous flow problems for which we can obtain analytical solutions in closed form, the method of solution is important. In this section we consider a few classic examples of fully developed laminar flows. Our intent is to obtain detailed information about the velocity field. Knowledge of the velocity field permits calculation of shear stress, pressure drop, and flow rate.

Rather than use the complete differential equations of motion (Eqs. 5.27) for the flow of a viscous fluid, we shall derive the governing equations from first principles for each flow field of interest. Since we are interested in the details of the flow field, our aim will be to derive differential equations that describe the flow. In every case we shall begin by applying the familiar control volume formulation of Newton's second law to a suitably chosen differential control volume.

8-2   FULLY DEVELOPED LAMINAR FLOW
BETWEEN INFINITE PARALLEL PLATES

8-2.1   Both Plates Stationary

Fluid in high-pressure hydraulic systems often leaks through the annular gap between a piston and cylinder. For very small gaps (typically about 0.005 mm), this flow field may be modeled as flow between infinite parallel plates. To calculate the leakage flow rate, we must first determine the velocity field.

Let us consider the fully developed laminar flow between infinite parallel plates. The plates are separated by distance a, as shown in Fig. 8.2. The plates are considered infinite in the z direction, with no variation of any fluid property in this direction. The flow is also assumed to be steady and incompressible. Before starting our analysis,

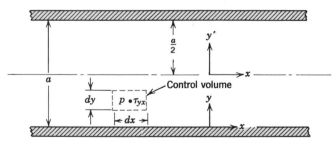

Fig. 8.2   Control volume for analysis of laminar flow between stationary infinite parallel plates.

---

[2] This mixing is illustrated extremely well in the introductory portion of the NCFMF film, *Turbulence,* R. W. Stewart, principal.

what do we know about the flow field? For one thing we know that the x component of velocity must be zero at both the upper and lower plates as a result of the no slip condition at the wall. The boundary conditions are

$$\text{at} \quad y = 0 \qquad u = 0$$

$$\text{at} \quad y = a \qquad u = 0$$

Since the flow is fully developed, the velocity cannot vary with x and, hence, depends on y only, so that $u = u(y)$. Furthermore, there is no component of velocity in either the y or z directions ($v = w = 0$).

For our analysis we select a differential control volume of size $d\Psi = dx\,dy\,dz$, and apply the x component of the momentum equation.

Basic equation:

$$\overset{= 0(3) = 0(1)}{F_{S_x} + \cancel{F_{B_x}}} = \cancel{\frac{\partial}{\partial t} \int_{CV}} u\rho\,d\Psi + \int_{CS} u\rho\vec{V} \cdot d\vec{A} \tag{4.19a}$$

Assumptions: (1)   Steady flow
                (2)   Fully developed flow
                (3) $F_{B_x} = 0$

For fully developed flow, the net momentum flux through the control surface is zero. (The momentum flux through the right face of the control surface is equal in magnitude but opposite in sign to the momentum flux through the left face; there is no momentum flux through any of the remaining faces of the control volume.) Since there are no body forces in the x direction, the momentum equation reduces to

$$F_{S_x} = 0 \tag{8.2}$$

The next step is to sum the forces acting on the control volume in the x direction. We recognize that normal forces (pressure forces) act on the left and right faces and tangential forces (shear forces) act on the top and bottom faces.

If the pressure at the center of the element is p, then the pressure force on the left face is

$$\left(p - \frac{\partial p}{\partial x} \frac{dx}{2} \right) dy\,dz$$

and the pressure force on the right face is

$$-\left(p + \frac{\partial p}{\partial x} \frac{dx}{2} \right) dy\,dz$$

If the shear stress at the center of the element is τ_{yx}, then the shear force on the bottom face is

$$-\left(\tau_{yx} - \frac{d\tau_{yx}}{dy} \frac{dy}{2} \right) dx\,dz$$

and the shear force on the top face is

$$\left(\tau_{yx} + \frac{d\tau_{yx}}{dy} \frac{dy}{2} \right) dx\,dz$$

Note that in expanding the shear stress, τ_{yx}, in a Taylor series about the center of the element, we have used the total derivative rather than a partial derivative. We did this because we recognized that τ_{yx} is only a function of y, since $u = u(y)$.

Having formulated the forces acting on each face of the control volume, we substitute them into Eq. 8.2; this equation simplifies to

$$-\frac{\partial p}{\partial x} + \frac{d\tau_{yx}}{dy} = 0$$

or

$$\frac{d\tau_{yx}}{dy} = \frac{\partial p}{\partial x} \tag{8.3}$$

Equation 8.3 must be valid for all x and y. This requires that

$$\frac{d\tau_{yx}}{dy} = \frac{\partial p}{\partial x} = \text{constant}$$

Integrating this equation, we obtain

$$\tau_{yx} = \left(\frac{\partial p}{\partial x}\right) y + c_1$$

which indicates that the shear stress varies linearly with y. Since for a Newtonian fluid

$$\tau_{yx} = \mu \frac{du}{dy} \tag{2.10}$$

then

$$\mu \frac{du}{dy} = \left(\frac{\partial p}{\partial x}\right) y + c_1$$

and

$$u = \frac{1}{2\mu}\left(\frac{\partial p}{\partial x}\right) y^2 + \frac{c_1}{\mu} y + c_2 \tag{8.4}$$

To evaluate the constants, c_1 and c_2, we must apply the boundary conditions. At $y = 0$, $u = 0$. Consequently, $c_2 = 0$. At $y = a$, $u = 0$. Hence

$$0 = \frac{1}{2\mu}\left(\frac{\partial p}{\partial x}\right) a^2 + \frac{c_1}{\mu} a$$

This gives

$$c_1 = -\frac{1}{2}\left(\frac{\partial p}{\partial x}\right) a$$

and hence

$$u = \frac{1}{2\mu}\left(\frac{\partial p}{\partial x}\right) y^2 - \frac{1}{2\mu}\left(\frac{\partial p}{\partial x}\right) a y$$

or

$$u = \frac{a^2}{2\mu}\left(\frac{\partial p}{\partial x}\right)\left[\left(\frac{y}{a}\right)^2 - \left(\frac{y}{a}\right)\right] \tag{8.5}$$

At this point we have the velocity profile. What else can we learn about the flow?

Shear Stress Distribution

The shear stress distribution is given by

$$\tau_{yx} = \left(\frac{\partial p}{\partial x}\right)y + c_1 = \left(\frac{\partial p}{\partial x}\right)y - \frac{1}{2}\left(\frac{\partial p}{\partial x}\right)a = a\left(\frac{\partial p}{\partial x}\right)\left[\frac{y}{a} - \frac{1}{2}\right] \qquad (8.6a)$$

Volume Flow Rate

The volume flow rate is given by

$$Q = \int_A \vec{V} \cdot d\vec{A}$$

For a depth l in the z direction,

$$Q = \int_0^a ul \ dy$$

or

$$\frac{Q}{l} = \int_0^a \frac{1}{2\mu}\left(\frac{\partial p}{\partial x}\right)(y^2 - ay) \ dy$$

Thus the volume flow rate per depth l is given by

$$\frac{Q}{l} = -\frac{1}{12\mu}\left(\frac{\partial p}{\partial x}\right)a^3 \qquad (8.6b)$$

Flow Rate as a Function of Pressure Drop

Since $\partial p/\partial x$ is constant, the pressure varies linearly with x and

$$\frac{\partial p}{\partial x} = \frac{p_2 - p_1}{L} = \frac{-\Delta p}{L}$$

Substituting into the expression for volume flow rate gives

$$\frac{Q}{l} = -\frac{1}{12\mu}\left[\frac{-\Delta p}{L}\right]a^3 = \frac{a^3 \Delta p}{12\mu L} \qquad (8.6c)$$

Average Velocity

The average velocity, $\bar{V}$, is given by

$$\bar{V} = \frac{Q}{A} = -\frac{1}{12\mu}\left(\frac{\partial p}{\partial x}\right)\frac{a^3 l}{l a} = -\frac{1}{12\mu}\left(\frac{\partial p}{\partial x}\right)a^2 \qquad (8.6d)$$

Point of Maximum Velocity

To find the point of maximum velocity, we set du/dy equal to zero and solve for the corresponding y. From Eq. 8.5

$$\frac{du}{dy} = \frac{a^2}{2\mu}\left(\frac{\partial p}{\partial x}\right)\left[\frac{2y}{a^2} - \frac{1}{a}\right]$$

Thus

$$\frac{du}{dy} = 0 \qquad \text{at} \qquad y = \frac{a}{2}$$

At

$$y = \frac{a}{2}, \qquad u = u_{max} = -\frac{1}{8\mu}\left(\frac{\partial p}{\partial x}\right)a^2 = \frac{3}{2}\bar{V} \qquad (8.6e)$$

Transformation of Coordinates

In deriving the above relations, the origin of coordinates, $y = 0$, was taken at the bottom plate. We could just as easily have taken the origin at the centerline of the channel. If we denote the coordinates with origin at the channel centerline as x, y', the boundary conditions are $u = 0$ at $y' = \pm a/2$.

To obtain the velocity profile in terms of x, y', we substitute $y = y' + a/2$ into Eq. 8.5. The result is

$$u = \frac{a^2}{2\mu}\left(\frac{\partial p}{\partial x}\right)\left[\left(\frac{y'}{a}\right)^2 - \frac{1}{4}\right] \qquad (8.7)$$

This equation shows that the velocity profile we have determined is parabolic, as shown in Fig. 8.3.

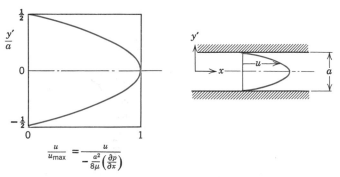

Fig. 8.3 Dimensionless velocity profile for fully developed laminar flow between infinite parallel plates.

Since all stresses were related to velocity gradients through Newton's law of viscosity, and the additional stresses that arise as a result of turbulent fluctuations have not been accounted for, all of the results in this section are valid for laminar flow only. Experiments show that this flow becomes turbulent for Reynolds numbers (defined as $Re = \rho\bar{V}a/\mu$) greater than approximately 1400. Consequently, the Reynolds number should be checked after using Eqs. 8.6 to ensure a valid solution.

EXAMPLE 8.1—Leakage Flow Past a Piston

A hydraulic system operates at a gage pressure of 20 MPa and 55 C. The hydraulic fluid is SAE 10W oil. A control valve consists of a piston 25 mm in diameter, fitted to a cylinder with a mean radial clearance of 0.005 mm. Determine the leakage flow rate if the gage pressure on the low-pressure side of the piston is 1.0 MPa. (The piston is 15 mm long.)

EXAMPLE PROBLEM 8.1

GIVEN: Flow of hydraulic oil between piston and cylinder, as shown. Fluid is SAE 10W oil at 55 C.

FIND: Leakage flow rate, Q.

$p_1 = 20$ MPa (gage)

$D = 25$ mm

$L = 15$ mm

$a = 0.005$ mm

$p_2 = 1.0$ MPa (gage)

SOLUTION:

The gap width is very small, so the flow may be modeled as flow between parallel plates. Equation 8.6c may be applied.

Computing equation:

$$\frac{Q}{l} = \frac{a^3 \Delta p}{12 \mu L} \tag{8.6c}$$

Assumptions: (1) Laminar flow
(2) Steady flow
(3) Incompressible flow
(4) Fully developed flow (note $L/a = 15/0.005 = 3000$!)

The plate width, l, is approximated as $l = \pi D$. Thus

$$Q = \frac{\pi D a^3 \Delta p}{12 \mu L}$$

For SAE 10W oil at 55 C, $\mu = 0.018$ kg/m · sec, from Fig. A.2, Appendix A. Thus

$$Q = \frac{\pi}{12} \times \frac{25 \text{ mm}}{} \times \frac{(0.005)^3 \text{ mm}^3}{} \times \frac{(20-1)10^6 \text{ N}}{\text{m}^2} \times \frac{\text{m} \cdot \text{sec}}{0.018 \text{ kg}} \times \frac{1}{15 \text{ mm}} \times \frac{\text{kg} \cdot \text{m}}{\text{N} \cdot \text{sec}^2}$$

$$Q = 57.6 \text{ mm}^3/\text{sec} \qquad\qquad\qquad Q$$

To ensure that flow is laminar, we also should check the Reynolds number.

$$\bar{V} = \frac{Q}{A} = \frac{Q}{\pi D a} = \frac{57.6 \text{ mm}^3}{\text{sec}} \times \frac{1}{\pi} \times \frac{1}{25 \text{ mm}} \times \frac{1}{0.005 \text{ mm}} \times \frac{\text{m}}{10^3 \text{ mm}} = 0.147 \text{ m/sec}$$

and

$$Re = \frac{\rho \bar{V} a}{\mu} = \frac{\text{SG} \rho_{\text{H}_2\text{O}} \bar{V} a}{\mu}$$

For SAE 10W oil, SG = 0.92, from Table A.2, Appendix A. Thus

$$Re = \frac{0.92}{} \times \frac{1000 \text{ kg}}{\text{m}^3} \times \frac{0.147 \text{ m}}{\text{sec}} \times \frac{0.005 \text{ mm}}{} \times \frac{\text{m} \cdot \text{sec}}{0.018 \text{ kg}} \times \frac{\text{m}}{10^3 \text{ mm}} = 0.0375$$

Thus flow is surely laminar, since $Re \ll 1400$.

8-2.2 Upper Plate Moving with Constant Speed, U

A second laminar flow case of practical importance is flow in a journal bearing. In such a bearing, an inner cylinder, the journal, rotates inside a stationary member. At light loads, the centers of the two members essentially coincide, and the small

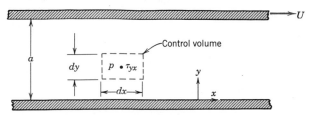

Fig. 8.4 Control volume for analysis of laminar flow between infinite parallel plates: upper plate moving with constant speed, U.

clearance gap is symmetric. Since the gap is small, it is reasonable to "unfold" the bearing and to model the flow field as flow between infinite parallel plates.

Let us now consider a case where the upper plate is moving to the right with constant speed, U, as shown in Fig. 8.4. All we have done in going from a stationary upper plate to a moving upper plate is to change one of the boundary conditions. The boundary conditions for the moving plate case are

$$u = 0 \quad \text{at} \quad y = 0$$

$$u = U \quad \text{at} \quad y = a$$

Since only the boundary conditions have changed, there is no need to repeat the entire analysis of Section 8-2.1. The analysis leading to Eq. 8.4 is equally valid for the moving plate case. Thus the velocity distribution is given by

$$u = \frac{1}{2\mu}\left(\frac{\partial p}{\partial x}\right)y^2 + \frac{c_1}{\mu}y + c_2 \tag{8.4}$$

and our only task is to evaluate constants c_1 and c_2 by using the appropriate boundary conditions.

At $y = 0$, $u = 0$. Consequently, $c_2 = 0$.

At $y = a$, $u = U$. Consequently,

$$U = \frac{1}{2\mu}\left(\frac{\partial p}{\partial x}\right)a^2 + \frac{c_1}{\mu}a$$

Thus

$$c_1 = \frac{U\mu}{a} - \frac{1}{2}\left(\frac{\partial p}{\partial x}\right)a$$

and

$$u = \frac{1}{2\mu}\left(\frac{\partial p}{\partial x}\right)y^2 + \frac{Uy}{a} - \frac{1}{2\mu}\left(\frac{\partial p}{\partial x}\right)ay$$

$$= \frac{Uy}{a} + \frac{1}{2\mu}\left(\frac{\partial p}{\partial x}\right)(y^2 - ay)$$

$$u = \frac{Uy}{a} + \frac{a^2}{2\mu}\left(\frac{\partial p}{\partial x}\right)\left[\left(\frac{y}{a}\right)^2 - \left(\frac{y}{a}\right)\right] \tag{8.8}$$

It is reassuring to note that Eq. 8.8 reduces to Eq. 8.5 for a stationary upper plate. From Eq. 8.8, for zero pressure gradient (for $\partial p/\partial x = 0$) the velocity varies linearly with y. This was the case treated earlier in Chapter 2.

From the velocity distribution of Eq. 8.8 we can obtain additional information about the flow.

Shear Stress Distribution

The shear stress distribution is given by $\tau_{yx} = \mu(du/dy)$,

$$\tau_{yx} = \mu \frac{U}{a} + \frac{a^2}{2}\left(\frac{\partial p}{\partial x}\right)\left[\frac{2y}{a^2} - \frac{1}{a}\right] = \mu \frac{U}{a} + a\left(\frac{\partial p}{\partial x}\right)\left[\frac{y}{a} - \frac{1}{2}\right] \qquad (8.9a)$$

Volume Flow Rate

The volume flow rate is given by $Q = \int_A \vec{V} \cdot d\vec{A}$. For depth l in the z direction

$$Q = \int_0^a ul\, dy$$

or

$$\frac{Q}{l} = \int_0^a \left[\frac{Uy}{a} + \frac{1}{2\mu}\left(\frac{\partial p}{\partial x}\right)(y^2 - ay)\right] dy$$

Thus the volume flow rate per depth l is given by

$$\frac{Q}{l} = \frac{Ua}{2} - \frac{1}{12\mu}\left(\frac{\partial p}{\partial x}\right)a^3 \qquad (8.9b)$$

Average Velocity

The average velocity, $\bar{V}$, is given by

$$\bar{V} = \frac{Q}{A} = l\left[\frac{Ua}{2} - \frac{1}{12\mu}\left(\frac{\partial p}{\partial x}\right)a^3\right] \Big/ la = \frac{U}{2} - \frac{1}{12\mu}\left(\frac{\partial p}{\partial x}\right)a^2 \qquad (8.9c)$$

Point of Maximum Velocity

To find the point of maximum velocity, we set du/dy equal to zero and solve for the corresponding y. From Eq. 8.8

$$\frac{du}{dy} = \frac{U}{a} + \frac{a^2}{2\mu}\left(\frac{\partial p}{\partial x}\right)\left[\frac{2y}{a^2} - \frac{1}{a}\right] = \frac{U}{a} + \frac{a}{2\mu}\left(\frac{\partial p}{\partial x}\right)\left[2\left(\frac{y}{a}\right) - 1\right]$$

Thus

$$\frac{du}{dy} = 0 \quad \text{at} \quad y = \frac{a}{2} - \frac{U/a}{(1/\mu)(\partial p/\partial x)}$$

There is no simple relation between the maximum velocity, u_{max}, and the mean velocity, $\bar{V}$, for this flow case.

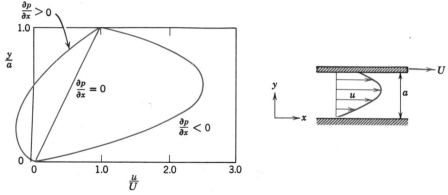

Fig. 8.5 Dimensionless velocity profile for fully developed laminar flow between infinite parallel plates: upper plate moving with constant speed, U.

Equation 8.8 suggests that the velocity profile may be treated as a combination of a linear and a parabolic velocity profile; the last term in Eq. 8.8 is identical to Eq. 8.5. The result is a family of velocity profiles, depending on U and $(1/\mu)(\partial p/\partial x)$; a few profiles are sketched in Fig. 8.5. (As shown in Fig. 8.5, some reverse flow—in the negative x direction—can occur when $\partial p/\partial x > 0$.)

Again, all of the results developed in this section are valid for laminar flow only. Experiments show that this flow becomes turbulent (for $\partial p/\partial x = 0$) at a Reynolds number of approximately 1500, where $Re = \rho U a/\mu$ for this flow case. Not much information is available for the case where the pressure gradient is not zero.

EXAMPLE 8.2—Torque and Power in a Journal Bearing

A crankshaft journal bearing in an automobile engine is lubricated by SAE 30 oil at 210 F. The bearing diameter is 3 in., the diametral clearance is 0.0025 in., and it rotates at 3600 rpm; it is 1.25 in. long. The bearing is under no load, so the clearance is symmetric. Determine the torque required to turn the journal and the power dissipated.

EXAMPLE PROBLEM 8.2

GIVEN: Journal bearing, as shown. Note that the gap width, a, is *half* the diametral clearance. Lubricant is SAE 30 oil at 210 F. Speed is 3600 rpm.

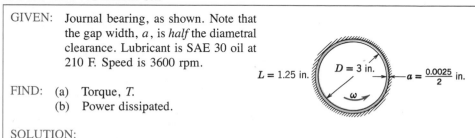

FIND: (a) Torque, T.
       (b) Power dissipated.

SOLUTION:
Torque on the journal is due to viscous shear in the oil film. The gap width is small, so the flow may be modeled as flow between infinite parallel plates:

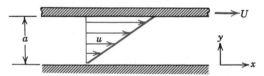

Computing equation:

$$\tau_{yx} = \mu \frac{U}{a} + a \left(\overset{=\,0(6)}{\cancel{\frac{\partial p}{\partial x}}}\right)\left[\frac{y}{a} - \frac{1}{2}\right] \tag{8.9a}$$

Assumptions: (1) Laminar flow
(2) Steady flow
(3) Incompressible flow
(4) Fully developed flow
(5) Infinite width $(L/a = 1.25/0.00125 = 1000$, so this is a reasonable assumption)
(6) $\partial p/\partial x = 0$ (flow is symmetric in the actual bearing at no load)

Then

$$\tau_{yx} = \mu \frac{U}{a} = \mu \frac{\omega R}{a} = \mu \frac{\omega D}{2a}$$

For SAE 30 oil at 210 F (99 C), $\mu = 9.6 \times 10^{-3}$ N · sec/m^2 (2.01×10^{-4} lbf · sec/ft^2), from Fig. A.2, Appendix A. Thus

$$\tau_{yx} = \frac{2.01 \times 10^{-4}\ \text{lbf} \cdot \text{sec}}{\text{ft}^2} \times 3600\ \frac{\text{rev}}{\text{min}} \times 2\pi\ \frac{\text{rad}}{\text{rev}} \times \frac{\text{min}}{60\ \text{sec}} \times 3\ \text{in.} \times \frac{1}{2} \times \frac{1}{0.00125\ \text{in.}}$$

$$\tau_{yx} = 90.9\ \text{lbf/ft}^2$$

Since $\tau_{yx} > 0$, it acts to the *left* on the upper plate, which is a minus y surface. The total shear force is given by the shear stress times the area. It is applied to the journal surface. Therefore,

$$T = FR = \tau_{yx} \pi DLR = \frac{\pi}{2} \tau_{yx} D^2 L$$

$$= \frac{\pi}{2} \times \frac{90.9\ \text{lbf}}{\text{ft}^2} \times \frac{(3)^2\ \text{in.}^2}{} \times \frac{\text{ft}^2}{144\ \text{in.}^2} \times 1.25\ \text{in.}$$

$$T = 11.2\ \text{in.} \cdot \text{lbf} \qquad\qquad\qquad\qquad\qquad T$$

The power dissipated in the bearing is

$$\dot{W} = FU = FR\omega = T\omega$$

$$= \frac{11.2\ \text{in.} \cdot \text{lbf}}{} \times 3600\ \frac{\text{rev}}{\text{min}} \times \frac{\text{min}}{60\ \text{sec}} \times 2\pi\ \frac{\text{rad}}{\text{rev}} \times \frac{\text{ft}}{12\ \text{in.}} \times \frac{\text{hp} \cdot \text{sec}}{550\ \text{ft} \cdot \text{lbf}}$$

$$\dot{W} = 0.640\ \text{hp} \qquad\qquad\qquad\qquad\qquad \dot{W}$$

To ensure laminar flow, check the Reynolds number.

$$Re = \frac{\rho U a}{\mu} = \frac{SG\rho_{H_2O} U a}{\mu} = \frac{SG\rho_{H_2O} \omega R a}{\mu}$$

Assume SG of SAE 30 oil is the same as that of SAE 10W oil. From Table A.2, Appendix A, SG = 0.92. Thus

$$Re = 0.92 \times 1.94\ \frac{\text{slug}}{\text{ft}^3} \times \frac{(3600)2\pi}{60}\ \frac{\text{rad}}{\text{sec}} \times 1.5\ \text{in.} \times 0.00125\ \text{in.}$$

$$\times \frac{ft^2}{2.01 \times 10^{-4} \ lbf \cdot sec} \times \frac{ft^2}{144 \ in.^2} \times \frac{lbf \cdot sec^2}{slug \cdot ft}$$

$$Re = 43.6$$

Therefore, the flow is laminar, since $Re \ll 1500$.

EXAMPLE 8.3—Laminar Film on a Vertical Wall

A viscous, incompressible, Newtonian liquid flows in steady, laminar flow down a vertical wall. The thickness, δ, of the liquid film is constant. Since the liquid free surface is exposed to atmospheric pressure, there is no pressure gradient. For this gravity-driven flow, apply the momentum equation to differential control volume, $dx\,dy\,dz$, to derive the velocity distribution in the liquid film.

EXAMPLE PROBLEM 8.3

GIVEN: Fully developed laminar flow of incompressible, Newtonian liquid down a vertical wall; thickness, δ, of the liquid film is constant and $\partial p/\partial x = 0$.

FIND: Expression for the velocity distribution in the film.

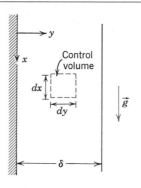

SOLUTION:

The x component of the momentum equation for a control volume is

$$F_{S_x} + F_{B_x} = \frac{\partial}{\partial t} \int_{CV} u\rho \, d\Psi + \int_{CS} u\rho \, \vec{V} \cdot d\vec{A} \qquad (4.19a)$$

Assumptions: (1) Laminar flow
              (2) Steady flow
              (3) Incompressible flow
              (4) Fully developed flow

For steady flow, $\dfrac{\partial}{\partial t} \displaystyle\int_{CV} u\rho \, d\Psi = 0$

For fully developed flow, $\displaystyle\int_{CS} u\rho \, \vec{V} \cdot d\vec{A} = 0$

Thus the momentum equation for the present case reduces to

$$F_{S_x} + F_{B_x} = 0$$

The body force, F_{B_x}, is given by $F_{B_x} = \rho g \, d\Psi = \rho g \, dx \, dy \, dz$. The only surface forces acting on the differential control volume are shear forces on the vertical surfaces. (Since $\partial p/\partial x = 0$, there is no net pressure force acting on the control volume.)

If the shear stress at the center of the differential control volume is τ_{yx}, then,

shear stress on left face is $\tau_{yx_L} = \left(\tau_{yx} - \dfrac{d\tau_{yx}}{dy}\dfrac{dy}{2}\right)$,

and

shear stress on right face is $\tau_{yx_R} = \left(\tau_{yx} + \dfrac{d\tau_{yx}}{dy}\dfrac{dy}{2}\right)$

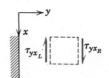

The direction of the shear stress vectors is taken consistent with the sign convention of Section 2-3. Thus on the left face, a minus y surface, τ_{yx_L} acts upward, and on the right face, a plus y surface, τ_{yx_R} acts downward.

The surface forces are obtained by multiplying each shear stress by the area over which it acts. Substituting into $F_{S_x} + F_{B_x} = 0$, we obtain

$$-\tau_{yx_L}\, dx\, dz + \tau_{yx_R}\, dx\, dz + \rho g\, dx\, dy\, dz = 0$$

or

$$-\left(\tau_{yx} - \frac{d\tau_{yx}}{dy}\frac{dy}{2}\right) dx\, dz + \left(\tau_{yx} + \frac{d\tau_{yx}}{dy}\frac{dy}{2}\right) dx\, dz + \rho g\, dx\, dy\, dz = 0$$

Simplifying gives

$$\frac{d\tau_{yx}}{dy} + \rho g = 0 \qquad \text{or} \qquad \frac{d\tau_{yx}}{dy} = -\rho g$$

Since

$$\tau_{yx} = \mu \frac{du}{dy} \qquad \text{then} \qquad \mu \frac{d^2 u}{dy^2} = -\rho g$$

and

$$\frac{d^2 u}{dy^2} = -\frac{\rho g}{\mu}$$

Integrating with respect to y gives

$$\frac{du}{dy} = -\frac{\rho g}{\mu} y + c_1$$

Integrating again, we obtain

$$u = -\frac{\rho g}{\mu}\frac{y^2}{2} + c_1 y + c_2$$

To evaluate constants c_1 and c_2, we must apply appropriate boundary conditions:

(i) $\quad y = 0, \qquad u = 0 \qquad$ (no-slip)

(ii) $\quad y = \delta, \qquad \dfrac{du}{dy} = 0 \qquad$ (neglect air resistance, i.e., assume zero shear stress at free surface)

From boundary condition (i), $c_2 = 0$

From boundary condition (ii), $\quad 0 = -\dfrac{\rho g}{\mu}\delta + c_1 \qquad$ or $\qquad c_1 = \dfrac{\rho g}{\mu}\delta$

Hence

$$u = -\frac{\rho g}{\mu}\frac{y^2}{2} + \frac{\rho g}{\mu}\delta y$$

or

$$u = \frac{\rho g}{\mu}\delta^2\left[\left(\frac{y}{\delta}\right) - \frac{1}{2}\left(\frac{y}{\delta}\right)^2\right] \qquad\qquad u(y)$$

Using the velocity profile it can be shown that

$$\text{the volume flow rate is } Q/l = \frac{\rho g}{3\mu}\delta^3$$

$$\text{the maximum velocity is } U_{\max} = \frac{\rho g}{2\mu}\delta^2$$

$$\text{the average velocity is } \bar{V} = \frac{\rho g}{3\mu}\delta^2$$

Flow in the liquid film is laminar for $Re = \bar{V}\delta/\nu \leqslant 1000$ [1].

$\begin{cases} \text{The purpose of this problem is to illustrate the application of the momentum equation} \\ \text{to a differential control volume for a fully developed laminar flow.} \end{cases}$

8-3 FULLY DEVELOPED LAMINAR FLOW IN A PIPE

As a final example of fully developed laminar flow cases, let us consider fully developed laminar flow in a pipe. Here the flow is axisymmetric. Consequently it is most convenient to work in cylindrical coordinates. We shall again use a differential control volume, but this time, since the flow is axisymmetric, the control volume will be a differential annular ring, as shown in Fig. 8.6. The annular differential control volume length is dx and its thickness is dr.

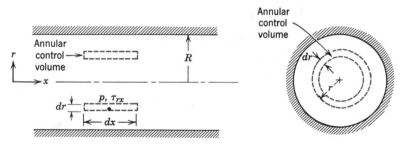

Fig. 8.6 Control volume for analysis of fully developed laminar flow in a pipe.

For a fully developed steady flow, the x component of the momentum equation (Eq. 4.19a), when applied to the differential control volume, reduces to

$$F_{S_x} = 0$$

The next step is to sum the forces acting on the control volume in the x direction. We know that normal forces (pressure forces) act on the left and right ends of the control volume, and that tangential forces (shear forces) act on the inner and outer cylindrical surfaces.

If the pressure at the center of the annular control volume is p, then the pressure force on the left end is

$$\left(p - \frac{\partial p}{\partial x}\frac{dx}{2}\right) 2\pi r \, dr$$

The pressure force on the right end is

$$-\left(p + \frac{\partial p}{\partial x}\frac{dx}{2}\right) 2\pi r \, dr$$

If the shear stress at the center of the annular control volume is τ_{rx}, then the shear force on the inner cylindrical surface is

$$-\left(\tau_{rx} - \frac{d\tau_{rx}}{dr}\frac{dr}{2}\right) 2\pi\left(r - \frac{dr}{2}\right) dx$$

The shear force on the outer cylindrical surface is

$$\left(\tau_{rx} + \frac{d\tau_{rx}}{dr}\frac{dr}{2}\right) 2\pi\left(r + \frac{dr}{2}\right) dx$$

The sum of the x components of force acting on the control volume must be zero. This leads to the condition that

$$-\frac{\partial p}{\partial x}2\pi r\,dr\,dx + \tau_{rx}\,2\pi dr\,dx + \frac{d\tau_{rx}}{dr}2\pi r\,dr\,dx = 0$$

Dividing this equation by $2\pi r\,dr\,dx$ and solving for $\partial p/\partial x$ gives

$$\frac{\partial p}{\partial x} = \frac{\tau_{rx}}{r} + \frac{d\tau_{rx}}{dr} = \frac{1}{r}\frac{d(r\tau_{rx})}{dr} \tag{8.10}$$

Since τ_{rx} is only a function of r (this is the reason for using the total rather than the partial derivative of τ_{rx} in the force components above), we recognize that Eq. 8.10 holds for all r and x only if each side of the equation is constant. Equation 8.10 can be written as

$$\frac{1}{r}\frac{d(r\tau_{rx})}{dr} = \frac{\partial p}{\partial x} = \text{constant}$$

or

$$\frac{d(r\tau_{rx})}{dr} = r\frac{\partial p}{\partial x}$$

Integrating this equation, we obtain

$$r\tau_{rx} = \frac{r^2}{2}\left(\frac{\partial p}{\partial x}\right) + c_1$$

or

$$\tau_{rx} = \frac{r}{2}\left(\frac{\partial p}{\partial x}\right) + \frac{c_1}{r}$$

Since

$$\tau_{rx} = \mu\frac{du}{dr}$$

then

$$\mu\frac{du}{dr} = \frac{r}{2}\left(\frac{\partial p}{\partial x}\right) + \frac{c_1}{r}$$

and

$$u = \frac{r^2}{4\mu}\left(\frac{\partial p}{\partial x}\right) + \frac{c_1}{\mu}\ln r + c_2 \tag{8.11}$$

We need to evaluate constants c_1 and c_2. However, we have only the one boundary condition that $u = 0$ at $r = R$. What do we do? Before throwing in the towel, let us look at the solution for the velocity profile given by Eq. 8.11. Although we do not know the velocity at the pipe centerline, we do know from physical considerations that the velocity must be finite at $r = 0$. The only way that this can be true is for c_1 to be zero. Thus, from physical considerations, we conclude that $c_1 = 0$, and hence

$$u = \frac{r^2}{4\mu} \left(\frac{\partial p}{\partial x} \right) + c_2$$

The constant, c_2, is evaluated by using the available boundary condition at the pipe wall: at $r = R$, $u = 0$. Consequently,

$$0 = \frac{R^2}{4\mu} \left(\frac{\partial p}{\partial x} \right) + c_2$$

This gives

$$c_2 = -\frac{R^2}{4\mu} \left(\frac{\partial p}{\partial x} \right)$$

and hence

$$u = \frac{r^2}{4\mu} \left(\frac{\partial p}{\partial x} \right) - \frac{R^2}{4\mu} \left(\frac{\partial p}{\partial x} \right) = \frac{1}{4\mu} \left(\frac{\partial p}{\partial x} \right)(r^2 - R^2)$$

or

$$u = -\frac{R^2}{4\mu} \left(\frac{\partial p}{\partial x} \right) \left[1 - \left(\frac{r}{R} \right)^2 \right] \tag{8.12}$$

Since we have the velocity profile, we can obtain a number of additional features of the flow.

Shear Stress Distribution

The shear stress is given by

$$\tau_{rx} = \mu \frac{du}{dr} = \frac{r}{2} \left(\frac{\partial p}{\partial x} \right) \tag{8.13a}$$

Volume Flow Rate

$$Q = \int_A \vec{V} \cdot d\vec{A} = \int_0^R u 2\pi r \, dr = \int_0^R \frac{1}{4\mu} \left(\frac{\partial p}{\partial x} \right)(r^2 - R^2) 2\pi r \, dr$$

$$Q = -\frac{\pi R^4}{8\mu} \left(\frac{\partial p}{\partial x} \right) \tag{8.13b}$$

Flow Rate as a Function of Pressure Drop

In fully developed flow, the pressure gradient, $\partial p/\partial x$, is constant. Therefore, $\partial p/\partial x = (p_2 - p_1)/L = -\Delta p/L$. Substituting into Eq. 8.13b for the volume flow rate gives

$$Q = -\frac{\pi R^4}{8\mu} \left[\frac{-\Delta p}{L} \right] = \frac{\pi \Delta p R^4}{8\mu L} = \frac{\pi \Delta p D^4}{128\mu L} \tag{8.13c}$$

for laminar flow in a horizontal pipe.

Average Velocity

The average velocity, $\bar{V}$, is given by

$$\bar{V} = \frac{Q}{A} = \frac{Q}{\pi R^2} = -\frac{R^2}{8\mu}\left(\frac{\partial p}{\partial x}\right) \tag{8.13d}$$

Point of Maximum Velocity

To find the point of maximum velocity, we set du/dr equal to zero and solve for the corresponding r. From Eq. 8.12

$$\frac{du}{dr} = \frac{1}{2\mu}\left(\frac{\partial p}{\partial x}\right)r$$

Thus

$$\frac{du}{dr} = 0 \quad \text{at} \quad r = 0$$

At $r = 0$,

$$u = u_{max} = U = -\frac{R^2}{4\mu}\left(\frac{\partial p}{\partial x}\right) = 2\bar{V} \tag{8.13e}$$

The velocity profile (Eq. 8.12) may be written in terms of the maximum (centerline) velocity as

$$\frac{u}{U} = 1 - \left(\frac{r}{R}\right)^2 \tag{8.14}$$

The parabolic velocity profile, given by Eq. 8.14 for fully developed laminar pipe flow, was sketched in Fig. 8.1.

EXAMPLE 8.4—Capillary Viscometer

A simple and accurate viscometer can be made from a length of capillary tubing. If the flow rate and pressure drop are measured, and the tube geometry is known, the viscosity of a Newtonian liquid can be computed from Eq. 8.13c. A test of a certain liquid in a capillary viscometer gave the following data:

Flow rate:	880 mm³/sec	Tube length:	1 m
Tube diameter:	0.50 mm	Pressure drop:	1.0 MPa

Determine the viscosity of the liquid.

EXAMPLE PROBLEM 8.4

GIVEN: Flow in a capillary viscometer. The flow rate is $Q = 880$ mm³/sec.

FIND: The fluid viscosity.

SOLUTION:
Equation 8.13c may be applied.

Computing equation:

$$Q = \frac{\pi \Delta p D^4}{128 \mu L}$$

(8.

Assumptions: (1) Laminar flow
(2) Steady flow
(3) Incompressible flow
(4) Fully developed flow
(5) Horizontal tube

Then

$$\mu = \frac{\pi \Delta p D^4}{128 LQ} = \frac{\pi}{128} \times \frac{1.0 \times 10^6 \ N}{m^2} \times (0.50)^4 \ mm^4 \times \frac{sec}{880 \ mm^3} \times \frac{1}{1 \ m} \times \frac{m}{10^3 \ mm}$$

$$\mu = 1.74 \times 10^{-3} \ N \cdot sec/m^2 \qquad\qquad\qquad\qquad\qquad\qquad\qquad\qquad\qquad \mu$$

Check the Reynolds number. Assume the fluid density is similar to water, 999 kg/m^3.

$$\bar{V} = \frac{Q}{A} = \frac{4Q}{\pi D^2} = \frac{4}{\pi} \times \frac{880 \ mm^3}{sec} \times \frac{1}{(0.50)^2 \ mm^2} \times \frac{m}{10^3 \ mm} = 4.48 \ m/sec$$

Then

$$Re = \frac{\rho \bar{V} D}{\mu}$$

$$= \frac{999 \ kg}{m^3} \times \frac{4.48 \ m}{sec} \times 0.50 \ mm \times \frac{m^2}{1.74 \times 10^{-3} \ N \cdot sec} \times \frac{m}{10^3 \ mm} \times \frac{N \cdot sec^2}{kg \cdot m}$$

$$Re = 1290$$

Consequently, since $Re < 2300$, the flow is laminar.

PART B   FLOW IN PIPES AND DUCTS

Our main purpose in this section is to evaluate the pressure changes that result from incompressible flow in pipes, ducts, and flow systems. The pressure changes in a flow system result from changes in elevation or flow velocity (due to area changes) and from friction. In a frictionless flow, the Bernoulli equation could be used to account for the effects of changes in elevation and flow velocity. Thus the prime concern in the analysis of real flows is to account for friction. The effect of friction is to decrease the pressure, causing a pressure "loss" compared to the ideal, frictionless flow case. To simplify analysis, the "loss" will be divided into *major losses* (due to friction in constant-area portions of the system) and *minor losses* (due to flow through valves, tees, elbows, and frictional effects in other nonconstant-area portions of the system).

To develop relations for major losses due to friction in constant-area ducts, we shall deal with fully developed flows in which the velocity profile is unvarying in the direction of flow. Our attention will focus on turbulent flows, since the pressure drop for fully developed laminar flow in a pipe can be calculated from the results of Section 8-3. The pressure drop that occurs at the entrance of a pipe will be treated as a minor loss.

Since ducts of circular cross section are most common in engineering applications, the basic analysis will be performed for circular geometries. The results can be extended to other geometries by introducing the hydraulic diameter, which is treated in Section 8-7.3. (Compressible flow in ducts will be treated in Chapter 13.)

8-4 SHEAR STRESS DISTRIBUTION IN FULLY DEVELOPED PIPE FLOW

In fully developed steady flow in a horizontal pipe, be it laminar or turbulent, the pressure drop is balanced only by shear forces at the pipe wall. This can be seen by applying the momentum equation to a cylindrical control volume in the flow, Fig. 8.7. The x component of the momentum equation is

Basic equation:

$$F_{S_x} + F_{B_x} = \frac{\partial}{\partial t} \int_{CV} u\rho \; d\Psi + \int_{CS} u\rho \vec{V} \cdot d\vec{A} \qquad (4.19\text{a})$$

$$= 0(1) \qquad = 0(2) \qquad \qquad = 0(3, 4)$$

Assumptions: (1) Horizontal pipe, $F_{B_x} = 0$
               (2) Steady flow
               (3) Incompressible flow
               (4) Fully developed flow

Then

$$F_{S_x} = 0$$

The surface forces acting on the control volume are shown in Fig. 8.7. The pressure at the center of the element is p; the pressure at each end of the element is obtained from a Taylor series expansion of p about the center of the element. The shear force acts on the circumferential surface of the element. The direction has been assumed such that the shear stress is positive. Thus

$$F_{S_x} = \left(p - \frac{\partial p}{\partial x}\frac{dx}{2}\right)\pi r^2 - \left(p + \frac{\partial p}{\partial x}\frac{dx}{2}\right)\pi r^2 + \tau_{rx} 2\pi r \; dx = 0$$

or

$$-\frac{\partial p}{\partial x} \; dx \; \pi r^2 + \tau_{rx} 2\pi r \; dx = 0$$

Therefore,

$$\tau_{rx} = \frac{r}{2}\frac{\partial p}{\partial x} \qquad (8.15)$$

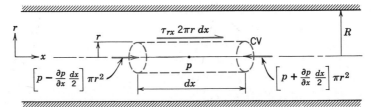

Fig. 8.7 Control volume for analysis of shear stress distribution in fully developed flow in a circular pipe.

Thus we see that the shear stress on the fluid varies linearly across the pipe, from zero at the centerline to a maximum at the pipe wall. If we denote the wall shear stress as τ_w, Eq. 8.15 shows that at the surface of the pipe

$$\tau_w = -[\tau_{rx}]_{r=R} = -\frac{R}{2}\frac{\partial p}{\partial x} \tag{8.16}$$

Equation 8.16 relates the wall shear stress to the axial pressure gradient. The momentum equation was used to derive it, but no assumption was made about a relation between the shear stress and the velocity field. Consequently, Eq. 8.16 is applicable to both laminar and turbulent fully developed pipe flow.

If we could relate the shear stress field to the mean velocity field, we could determine analytically the pressure drop over a length of pipe for fully developed flow. Such a relation between the stress field and the mean velocity field exists for laminar flow and was used in Section 8-3. The resulting equation, Eq. 8.13c, was first discovered experimentally by Jean Louis Poiseuille, a French physician, and independently by Gotthilf H. L. Hagen, a German engineer, in the 1850s [2].

In turbulent flow, no simple relation exists between the shear stress field and the mean velocity field. Velocity fluctuations in turbulent flow (discussed in Section 2-5.2) exchange momentum between adjacent layers of fluid, thereby causing apparent shear stresses that must be added to the stress caused by the mean velocity gradients. For fully developed turbulent channel flow, the total shear stress is given by

$$\tau = \mu\frac{d\bar{u}}{dy} - \rho\overline{u'v'} \tag{8.17}$$

In Eq. 8.17, y is distance from the pipe wall and $\bar{u}$ is mean velocity. As defined in Chapter 2, u' and v' are fluctuating components of velocity in the x and y directions, respectively, and $\overline{u'v'}$ is the time average of the product of u' and v'. The notion of an apparent stress was first introduced by Osborne Reynolds; the term $-\rho\overline{u'v'}$ is referred to as the *Reynolds stress*.

Dividing Eq. 8.17 by ρ gives

$$\frac{\tau}{\rho} = \nu\frac{d\bar{u}}{dy} - \overline{u'v'} \tag{8.18}$$

The term τ/ρ arises frequently in the consideration of turbulent flows; it has dimensions of velocity squared. The quantity $(\tau_w/\rho)^{1/2}$ is called the *friction velocity* and is denoted by the symbol u_*. In Fig. 8.8, experimental measurements of $\overline{u'v'}$ for fully developed turbulent pipe flow at two Reynolds numbers are presented; $Re_U = UD/\nu$, where U is the centerline velocity. The turbulent shear stress has been nondimensionalized with the square of the friction velocity. In the region very close to the wall, the *wall layer*, viscous shear is dominant. The turbulent stress goes to zero at the wall because the no-slip condition requires that the velocity at the wall be zero. Since the Reynolds stress is zero at the wall, then from Eq. 8.17, the wall shear stress is given by $\tau_w = \mu(d\bar{u}/dy)_{y=0}$. The total shear stress varies linearly across the pipe radius, so turbulent shear is dominant over the center region of the pipe. In the region between the wall layer and the central portion of the pipe both viscous and turbulent shear are important.

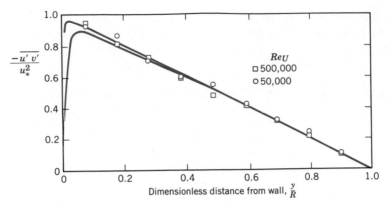

Fig. 8.8 Turbulent shear stress (Reynolds stress) for fully developed turbulent flow in a pipe. (Data from [3].)

8-5 TURBULENT VELOCITY PROFILES IN FULLY DEVELOPED PIPE FLOW

Except for flows of very viscous fluids in small diameter ducts, internal flows generally are turbulent. As noted in the discussion of shear stress distribution in fully developed pipe flow (Section 8-4), in turbulent flow there is no universal relationship between the stress field and the mean velocity field. Thus, for turbulent flows we are forced to rely on experimental data.

The velocity profile for fully developed turbulent flow through a smooth pipe is shown in Fig. 8.9. The plot is semilogarithmic; $\bar{u}/u_*$ is plotted against $\log(yu_*/\nu)$.

In the region very close to the wall where viscous shear is dominant, the mean velocity profile follows the linear viscous relation

$$u^+ = \frac{\bar{u}}{u_*} = \frac{yu_*}{\nu} = y^+ \tag{8.19}$$

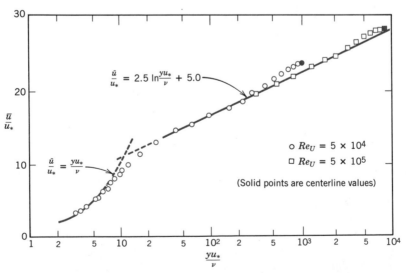

Fig. 8.9 Turbulent velocity profile for fully developed flow in a smooth pipe. (Data from [3].)

where y is distance measured from the wall ($y = R - r$; R is the pipe radius), and $\bar{u}$ is mean velocity. Equation 8.19 is valid for $0 \le y^+ \le 5$; this region is called the *viscous sublayer*.

In the region where both viscous and turbulent shear are important, the velocity profile follows the logarithmic relation

$$\frac{\bar{u}}{u_*} = 2.5 \ln \frac{y u_*}{\nu} + 5.0 \tag{8.20}$$

There is considerable scatter in the numerical constants of Eq. 8.20; the values given represent averages over many experiments [4]. From Fig. 8.9 we see that the logarithmic profile gives a reasonably good approximation of the velocity profile all the way to the pipe centerline.

In the central region, where turbulent shear is dominant, the velocity profile data are well correlated by the equation

$$\frac{U - \bar{u}}{u_*} = 2.5 \ln \frac{R}{y} \tag{8.21}$$

where U is centerline velocity. Equation 8.21 is the velocity defect law.

The velocity profile for turbulent flow through a smooth pipe can be represented by the empirical *power-law* equation

$$\frac{\bar{u}}{U} = \left(\frac{y}{R}\right)^{1/n} = \left(1 - \frac{r}{R}\right)^{1/n} \tag{8.22}$$

where the exponent, n, varies with the Reynolds number. In Fig. 8.10 the data of Laufer [3] are shown on a plot of $\ln y/R$ versus $\ln \bar{u}/U$; the slope of the straight line through the data gives n.

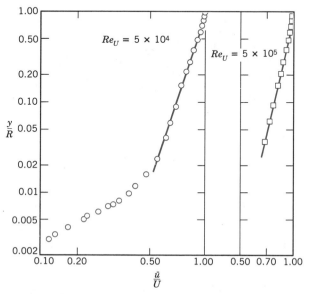

Fig. 8.10 Power-law velocity profiles for fully developed turbulent flow in a smooth pipe. (Data from [3].)

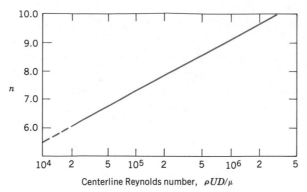

Fig. 8.11 Exponent for power-law profile. (Adapted from [5].)

The power-law profile is not applicable close to the wall ($y/R < 0.04$); the profile gives infinite velocity gradient at the wall. Although the profile fits the data close to the centerline, it fails to give zero slope at the centerline. The variation of exponent n in the power-law profile with Reynolds number (based on pipe diameter, D, and centerline velocity, U) is shown in Fig. 8.11.

Since the average velocity is $\bar{V} = Q/A$, and

$$Q = \int_A \vec{V} \cdot d\vec{A}$$

the ratio of the average velocity to the centerline velocity may be calculated for the power-law profiles of Eq. 8.22. The result is

$$\frac{\bar{V}}{U} = \frac{2n^2}{(n+1)(2n+1)} \tag{8.23}$$

From Eq. 8.23, we see that as n increases (due to increasing Reynolds number) the ratio of the average velocity to the centerline velocity increases; with increasing Reynolds number the velocity profile becomes more blunt or "fuller" (for $n = 6$, $\bar{V}/U = 0.79$; for $n = 10$, $\bar{V}/U = 0.87$). As a representative value, 7 often is used for the exponent; this gives rise to the term "a one-seventh power profile" for fully developed turbulent flow.

Velocity profiles for $n = 6$ and $n = 10$ are shown in Fig. 8.12. The parabolic profile for fully developed laminar flow has been included for comparison. It is clear that the turbulent profile has a much steeper slope near the wall.

8-6 ENERGY CONSIDERATIONS IN PIPE FLOW

Thus far in our discussion of viscous flow, we have derived all results by applying the momentum equation for a control volume. We have, of course, also used the control volume formulation of conservation of mass. Nothing has been said about conservation of energy—the first law of thermodynamics. Additional insight into the nature of the pressure losses in internal viscous flows can be obtained from the energy equation. Consider, for example, steady flow through the piping system, including

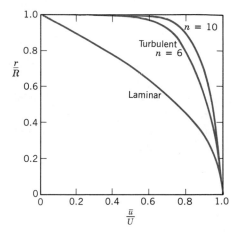

Fig. 8.12 Velocity profiles for fully developed pipe flow.

a reducing elbow, shown in Fig. 8.13. The control volume boundaries are shown as dashed lines. They are normal to the flow at sections ① and ② and coincide with the inside pipe wall elsewhere.

Basic equation:

$$\dot{Q} - \cancel{\dot{W}_s}^{=0(1)} - \cancel{\dot{W}_{shear}}^{=0(2)} - \cancel{\dot{W}_{other}}^{=0(1)} = \cancel{\frac{\partial}{\partial t}}^{=0(3)} \int_{CV} e\rho \, d\Psi + \int_{CS} (e + pv)\rho \vec{V} \cdot d\vec{A} \qquad (4.57)$$

$$e = u + \frac{V^2}{2} + gz$$

Assumptions: (1) $\dot{W}_s = 0$, $\dot{W}_{other} = 0$
(2) $\dot{W}_{shear} = 0$ (although shear stresses are present at the walls of the elbow, the velocities are zero at the walls)
(3) Steady flow
(4) Incompressible flow
(5) Internal energy and pressure uniform across sections ① and②

Under these assumptions the energy equation reduces to

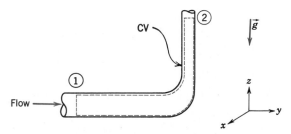

Fig. 8.13 Control volume and coordinates for energy analysis of flow through a 90° reducing elbow.

$$\dot{Q} = \dot{m}(u_2 - u_1) + \dot{m}\left(\frac{p_2}{\rho} - \frac{p_1}{\rho}\right) + \dot{m}g(z_2 - z_1)$$

$$+ \int_{A_2} \frac{V_2^2}{2}\rho V_2\, dA_2 - \int_{A_1} \frac{V_1^2}{2}\rho V_1\, dA_1 \qquad (8.24)$$

Note that we have not assumed the velocity to be uniform at sections ① and ②, since we know for viscous flows the velocity at a section cannot be uniform. However, it is convenient to introduce the average velocity into Eq. 8.24 so that we can eliminate the integrals. To do this, we define a kinetic energy coefficient.

8-6.1 Kinetic Energy Coefficient

The *kinetic energy coefficient,* α, is defined such that

$$\int_A \frac{V^2}{2}\rho V\, dA = \alpha \int_A \frac{\bar{V}^2}{2}\rho V\, dA = \alpha \dot{m}\frac{\bar{V}^2}{2} \qquad (8.25a)$$

or

$$\alpha = \frac{\int_A \rho V^3\, dA}{\dot{m}\bar{V}^2} \qquad (8.25b)$$

For laminar flow in a pipe (velocity profile given by Eq. 8.12), $\alpha = 2.0$.

In turbulent pipe flow, the velocity profile is quite flat, as shown in Fig. 8.12. We can use Eq. 8.25b together with Eqs. 8.22 and 8.23 to determine α. Substituting the power-law velocity profile of Eq. 8.22 into Eq. 8.25b, we obtain

$$\alpha = \left[\frac{U}{\bar{V}}\right]^3 \frac{2n^2}{(3+n)(3+2n)} \qquad (8.26)$$

The value of $\bar{V}/U$ is determined from Eq. 8.23. For $n = 6$, $\alpha = 1.08$; for $n = 10$, $\alpha = 1.03$. Since the exponent, n, in the power-law profile is a function of Reynolds number, α also varies with Reynolds number. Because α is reasonably close to one for large Reynolds number, unity often is assumed for pipe flow calculations. However, for developing flows at moderate Reynolds numbers the change of kinetic energy may be significant.

8-6.2 Head Loss

Using the definition of α, the energy equation (Eq. 8.24) can be written

$$\dot{Q} = \dot{m}(u_2 - u_1) + \dot{m}\left(\frac{p_2}{\rho} - \frac{p_1}{\rho}\right) + \dot{m}g(z_2 - z_1) + \dot{m}\left(\frac{\alpha_2 \bar{V}_2^2}{2} - \frac{\alpha_1 \bar{V}_1^2}{2}\right)$$

Dividing by the mass rate of flow gives

$$\frac{\delta Q}{dm} = u_2 - u_1 + \frac{p_2}{\rho} - \frac{p_1}{\rho} + gz_2 - gz_1 + \frac{\alpha_2 \bar{V}_2^2}{2} - \frac{\alpha_1 \bar{V}_1^2}{2}$$

Rearranging this equation, we write

$$\left(\frac{p_1}{\rho} + \alpha_1\frac{\bar{V}_1^2}{2} + gz_1\right) - \left(\frac{p_2}{\rho} + \alpha_2\frac{\bar{V}_2^2}{2} + gz_2\right) = (u_2 - u_1) - \frac{\delta Q}{dm} \qquad (8.27)$$

In Eq. 8.27, the term

$$\left(\frac{p}{\rho} + \alpha\frac{\bar{V}^2}{2} + gz\right)$$

represents the mechanical energy per unit mass at a cross section. The term $u_2 - u_1 - \delta Q/dm$ is equal to the difference in mechanical energy per unit mass between sections ① and ②. It represents the (irreversible) conversion of mechanical energy at section ① to unwanted thermal energy $(u_2 - u_1)$ and loss of energy via heat transfer $(-\delta Q/dm)$. We identify this group of terms as the total head loss, h_{l_T}. Then

$$\left(\frac{p_1}{\rho} + \alpha_1\frac{\bar{V}_1^2}{2} + gz_1\right) - \left(\frac{p_2}{\rho} + \alpha_2\frac{\bar{V}_2^2}{2} + gz_2\right) = h_{l_T} \qquad (8.28)$$

Head loss has dimensions of energy per unit mass $[FL/M]$; this is equivalent to dimensions of $[L^2/t^2]$.

If the flow were assumed frictionless, the velocity at a section would be uniform $(\alpha_1 = \alpha_2 = 1)$ and Bernoulli's equation would predict zero head loss.

In incompressible frictionless flow, a change in internal energy can occur only through heat transfer; there is no conversion of mechanical energy $(p/\rho + V^2/2 + gz)$ to internal energy. For viscous flow in a pipe, one effect of friction may be to increase the internal energy of the flow, as shown by Eq. 8.27.

You might wonder why the energy loss, h_{l_T}, is called a "head" loss. As the empirical science of hydraulics developed during the nineteenth century, it was common practice to express the energy balance in terms of energy per unit *weight* of flowing liquid (e.g., water) rather than energy per unit *mass*, as in Eq. 8.28. To obtain dimensions of energy per unit weight, we divide each term in Eq. 8.28 by the acceleration of gravity, g. Then the net dimensions of h_{l_T} are $[(L^2/t^2)(t^2/L)] = [L]$, or feet of flowing liquid. Since the term head loss is in common use, we shall also use it here. Remember that its physical interpretation is a loss in mechanical energy per unit mass of flowing fluid.

Equation 8.28 can be used to calculate the pressure difference between any two points in a piping system, provided the head loss, h_{l_T}, can be determined. We shall consider calculation of h_{l_T} in the next section.

8-7 CALCULATION OF HEAD LOSS

Total head loss, h_{l_T}, is regarded as the sum of major losses, h_l, due to frictional effects in fully developed flow in constant-area tubes, and minor losses, h_{l_m}, due to entrances, fittings, area changes, and so on. Consequently, we consider the major and minor losses separately.

8-7.1 Major Losses: Friction Factor

The energy balance, expressed by Eq. 8.28, can be used to evaluate the major head loss. For fully developed flow through a constant-area pipe, $h_{l_m} = 0$, and

$\alpha_1(\bar{V}_1^2/2) = \alpha_2(\bar{V}_2^2/2)$; Eq. 8.28 reduces to

$$\frac{p_1 - p_2}{\rho} = g(z_2 - z_1) + h_l \tag{8.29}$$

If the pipe is horizontal, then $z_2 = z_1$ and

$$\frac{p_1 - p_2}{\rho} = \frac{\Delta p}{\rho} = h_l \tag{8.30}$$

Thus the major head loss can be expressed as the pressure loss for fully developed flow through a horizontal pipe of constant area.

Since head loss represents the energy converted by frictional effects from mechanical to thermal energy, head loss for fully developed flow in a constant-area duct depends only on the details of the flow through the duct. Head loss is independent of pipe orientation.

a. Laminar Flow

In laminar flow, the pressure drop may be computed analytically for fully developed flow in a horizontal pipe. Thus, from Eq. 8.13c,

$$\Delta p = \frac{128\mu L Q}{\pi D^4} = \frac{128\mu L \bar{V}(\pi D^2/4)}{\pi D^4} = 32 \frac{L}{D} \frac{\mu \bar{V}}{D}$$

Substituting in Eq. 8.30 gives

$$h_l = 32\frac{L}{D}\frac{\mu\bar{V}}{\rho D} = \frac{L}{D}\frac{\bar{V}^2}{2}\left(64\frac{\mu}{\rho\bar{V}D}\right) = \left(\frac{64}{Re}\right)\frac{L}{D}\frac{\bar{V}^2}{2} \tag{8.31}$$

(We shall see the reason for writing h_l in this form shortly.)

b. Turbulent flow

In turbulent flow we cannot evaluate the pressure drop analytically; we must resort to experimental results and use dimensional analysis to correlate the experimental data. In fully developed turbulent flow, the pressure drop, Δp, due to friction in a horizontal constant-area pipe is known to depend on pipe diameter, D, pipe length, L, pipe roughness, e, average flow velocity, $\bar{V}$, fluid density, ρ, and fluid viscosity, μ. In functional form

$$\Delta p = \Delta p(D, L, e, \bar{V}, \rho, \mu)$$

We applied dimensional analysis to this problem in Example Problem 7.2. The results were a correlation of the form

$$\frac{\Delta p}{\rho\bar{V}^2} = f\left(\frac{\mu}{\rho\bar{V}D}, \frac{L}{D}, \frac{e}{D}\right)$$

We recognize that $\mu/\rho\bar{V}D = 1/Re$, so we could just as well write

$$\frac{\Delta p}{\rho\bar{V}^2} = \phi\left(Re, \frac{L}{D}, \frac{e}{D}\right)$$

Substituting from Eq. 8.30, we see that

$$\frac{h_l}{\bar{V}^2} = \phi\left(Re, \frac{L}{D}, \frac{e}{D}\right)$$

Although dimensional analysis predicts the functional relationship, we must obtain actual values experimentally.

Experiments show that the nondimensional head loss is directly proportional to L/D. Hence we can write

$$\frac{h_l}{\bar{V}^2} = \frac{L}{D}\phi_1\left(Re, \frac{e}{D}\right)$$

Since the function, ϕ_1, is still undetermined, it is permissible to introduce a constant into the left side of the above equation. The number $\frac{1}{2}$ is introduced into the denominator such that the head loss is nondimensionalized on the kinetic energy per unit mass of flow. Then

$$\frac{h_l}{\frac{1}{2}\bar{V}^2} = \frac{L}{D}\phi_2\left(Re, \frac{e}{D}\right)$$

The unknown function, $\phi_2(Re, e/D)$, is defined as the *friction factor, f,*

$$f \equiv \phi_2\left(Re, \frac{e}{D}\right)$$

and

$$h_l = f\frac{L}{D}\frac{\bar{V}^2}{2} \qquad (8.32)$$

The friction factor[3] is determined experimentally. The results, published by L. F. Moody [6], are shown in Fig. 8.14.

To determine head loss for fully developed flow with known conditions, the Reynolds number is evaluated first. Relative roughness, e/D, is obtained from Fig. 8.15. Then the friction factor, f, is read from the appropriate curve in Fig. 8.14, at the known values of Re and e/D. Finally, head loss is found using Eq. 8.32.

Several features of Fig. 8.14 require some discussion. The friction factor for laminar flow may be obtained by comparing Eqs. 8.31 and 8.32

$$h_l = \left(\frac{64}{Re}\right)\frac{L}{D}\frac{\bar{V}^2}{2} = f\frac{L}{D}\frac{\bar{V}^2}{2}$$

Consequently, for laminar flow

$$f_{\text{laminar}} = \frac{64}{Re} \qquad (8.33)$$

Thus, in laminar flow, the friction factor is a function of Reynolds number only; it is independent of roughness. Although we took no notice of roughness in deriving Eq. 8.31, experimental results verify that the friction factor is a function only of Reynolds number in laminar flow.

The Reynolds number in a pipe may be changed most easily by varying the average flow velocity. If the flow in a pipe is originally laminar, increasing the velocity until the critical Reynolds number is reached causes transition to occur; the laminar flow

---

[3] The friction factor defined by Eq. 8.32 is the *Darcy friction factor*. The *Fanning friction factor,* less frequently used, is defined in Problem 8.74.

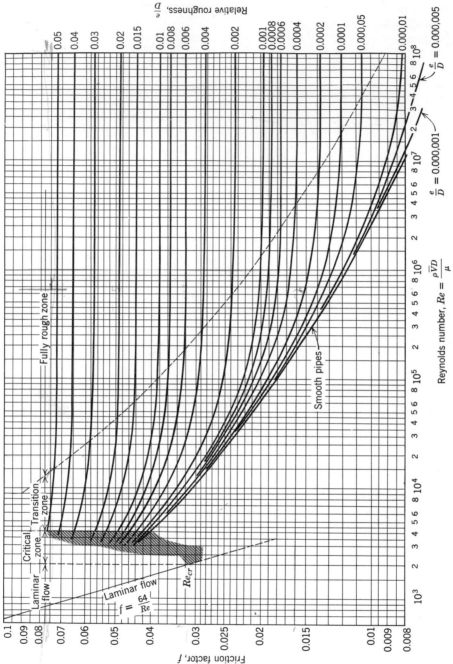

Fig. 8.14 Friction factor for fully developed flow in circular pipes. (Data from [6], used by permission.)

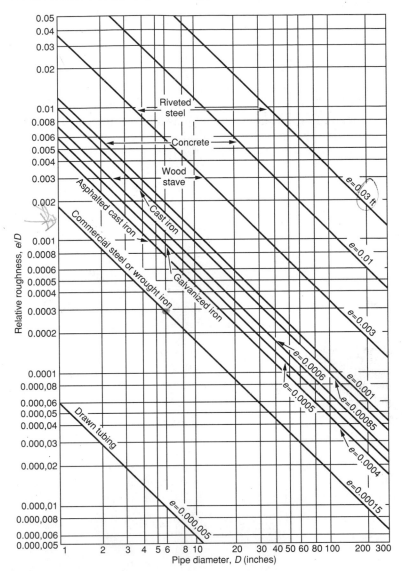

Fig. 8.15   Relative roughness for pipes of common engineering materials.
(Data from [6], used by permission.)

gives way to turbulent flow. The effect of transition on the velocity profile was discussed in Section 8-5. Figure 8.12 shows that the velocity gradient at the tube wall is much larger for turbulent flow than for laminar flow. This change in velocity profile causes the wall shear stress to increase sharply, with the same effect on the friction factor.

As the Reynolds number is increased above the transition value, the velocity profile continues to become fuller, as noted in Section 8-5; the friction factor at first tends to follow the smooth pipe curve, along which friction factor is a function of Reynolds number only. However, as the Reynolds number increases, the velocity profile becomes still fuller. The size of the thin viscous sublayer near the tube wall decreases. As roughness elements begin to poke through this layer, the effect of roughness becomes important, and the friction factor becomes a function of both the Reynolds number *and* the relative roughness.

At very large Reynolds number, most of the roughness elements on the tube wall protrude through the viscous sublayer; the drag and, hence, the pressure loss, depend only on the size of the roughness elements. This is termed the "fully rough" flow regime; the friction factor depends only on e/D in this regime.

To summarize the preceding discussion, we see that as Reynolds number is increased, the friction factor decreases as long as the flow remains laminar. At transition, f increases sharply. In the turbulent flow regime, the friction factor decreases gradually along the smooth pipe curve, and finally levels out at a constant value for extremely large Reynolds number. (The only exception to these trends is that friction factors for pipes with $e/D \gtrsim 0.001$ in turbulent flow fall above the smooth pipe curve.)

Analysis of pipe flow data [5] suggests that the friction factor, f, may be related to the exponent, n, of the power law profile for turbulent pipe flow, by the simple approximate expression

$$\frac{1}{n} = \sqrt{f} \tag{8.34}$$

for both smooth and rough pipes in the range $f < 0.1$. This relation was used in conjunction with the smooth pipe data of Fig. 8.14 to generate the curve of n versus Re_U presented in Fig. 8.11.

In order to use the computer to solve problems, it is necessary to have a mathematical formulation for the friction factor, f, in terms of Reynolds number, $Re = \bar{V}D/\nu$, and relative roughness, e/D. The Blasius correlation for turbulent flow in smooth pipes, valid for $Re \leq 10^5$, is

$$f = \frac{0.3164}{Re^{0.25}} \tag{8.35}$$

When this relation is combined with the expression for wall shear stress (Eq. 8.16), the expression for head loss (Eq. 8.30), and the definition of friction factor (Eq. 8.32), a useful expression for the wall shear stress is obtained

$$\tau_w = 0.03325 \rho \bar{V}^2 \left(\frac{\nu}{R\bar{V}} \right)^{0.25} \tag{8.36}$$

This equation will be used later in our study of turbulent boundary-layer flow over a flat plate (Chapter 9).

The most widely used formula for friction factor is from Colebrook [7],

$$\frac{1}{f^{0.5}} = -2.0 \log \left(\frac{e/D}{3.7} + \frac{2.51}{Re \, f^{0.5}} \right) \tag{8.37a}$$

Equation 8.37a is transcendental, so iteration is needed to evaluate f. Miller [8] suggests that a single iteration will produce a result within 1 percent if the initial estimate is calculated from

$$f_0 = 0.25 \left[\log \left(\frac{e/D}{3.7} + \frac{5.74}{Re^{0.9}} \right) \right]^{-2} \tag{8.37b}$$

Equation 8.37b is from [9].

Figure 8.15 also needs some explanation. All of the e/D values given are for new pipes, in relatively good condition. Over long periods of service, corrosion

Fig. 8.16 Pipe section removed after 40 years of service as a water line, showing formation of scale.

takes place and, particularly in hard water areas, lime deposits and rust scale form on pipe walls. Corrosion can weaken pipes, eventually leading to failure. Deposit formation increases wall roughness appreciably, and also decreases the effective diameter. These factors combine to cause e/D to increase by factors of 2 to 5 for old pipes. An example is shown in Fig. 8.16.

Curves presented in Figs. 8.14 and 8.15 represent average values for data obtained from numerous experiments. The curves should be considered accurate within approximately ±10 percent, which is sufficient for many engineering analyses. If more accuracy is needed, actual test data should be used.

8-7.2 Minor Losses

The flow in a piping system may be required to pass through a variety of fittings, bends, or abrupt changes in area. Additional head losses are encountered, primarily as a result of flow separation. (Energy eventually is dissipated by violent mixing in the separated zones.) These losses will be minor (hence the term *minor losses*) if the piping system includes long lengths of constant-area pipe. The minor head loss may be expressed as

$$h_{l_m} = K \frac{\bar{V}^2}{2} \qquad (8.38a)$$

where the *loss coefficient*, K, must be determined experimentally for each situation. Minor head loss also may be expressed as

$$h_{l_m} = f \frac{L_e}{D} \frac{\bar{V}^2}{2} \qquad (8.38b)$$

where L_e is an *equivalent length* of straight pipe.

For flow through pipe bends and fittings, the loss coefficient, K, is found to vary with pipe size (diameter) in much the same manner as the friction factor, f, for flow through a straight pipe. Consequently, the equivalent length, L_e/D, tends toward a constant for different sizes of a given type of fitting.

Experimental data for minor losses are plentiful, but they are scattered among a variety of sources. Different sources may give different values for the same flow configuration. The data presented here should be considered as representative for some commonly encountered situations; in each case the source of the data is identified.

a.   Inlets and Exits

A poorly designed inlet to a pipe can cause appreciable head loss. If the inlet has sharp corners, flow separation occurs at the corners, and a *vena contracta* is formed. The fluid must accelerate locally to pass through the reduced flow area at the vena contracta. Losses in mechanical energy result from the unconfined mixing as the flow stream decelerates again to fill the pipe. Three basic inlet geometries are shown in Table 8.1. From the table it is clear that the loss coefficient is reduced significantly when the inlet is rounded even slightly. For a well-rounded inlet ($r/D \geq 0.15$) the entrance loss coefficient is almost negligible. Example Problem 8.9 illustrates a procedure for experimentally determining the loss coefficient for a pipe inlet.

Table 8.1   Minor Loss Coefficients for Pipe Entrances (Data from [10].)

Entrance Type		Minor Loss Coefficient, K^a
Reentrant		0.78
Square-edged		0.5
Rounded		r/D | 0.02 | 0.06 | ≥ 0.15 K | 0.28 | 0.15 | 0.04

^{a} Based on $h_{l_m} = K(\bar{V}^2/2)$, where $\bar{V}$ is the mean velocity in the pipe.

The kinetic energy per unit mass, $\alpha \bar{V}^2/2$, is completely dissipated by mixing when flow discharges from a duct into a large reservoir or plenum chamber. The situation corresponds to flow through an abrupt expansion with $AR = 0$ (Fig. 8.17). The minor loss coefficient thus equals α. No improvement in minor loss coefficient for an exit is possible; however, addition of a diffuser can reduce $\bar{V}^2/2$ considerably (see Example Problem 8.10).

b.   Enlargements and Contractions

Minor loss coefficients for sudden expansions and contractions in circular ducts are given in Fig. 8.17. Note that both loss coefficients are based on the larger $\bar{V}^2/2$. Thus losses for a sudden expansion are based on $\bar{V}_1^2/2$, and those for a contraction are based on $\bar{V}_2^2/2$.

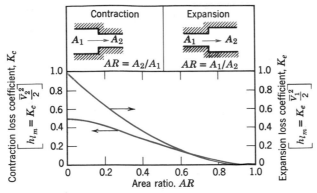

Fig. 8.17 Loss coefficients for flow through sudden area changes. (Data from [1].)

Losses due to area change can be reduced somewhat by installing a nozzle or diffuser between the two sections of straight pipe. Data for nozzles are given in Table 8.2.

Table 8.2 Loss Coefficients (K) for Gradual Contractions: Round and Rectangular Ducts (Data from [11].)

			Included Angle, θ, Degrees				
A_2/A_1	**10**	**15–40**	**50–60**	**90**	**120**	**150**	**180**
0.50	0.05	0.05	0.06	0.12	0.18	0.24	0.26
0.25	0.05	0.04	0.07	0.17	0.27	0.35	0.41
0.10	0.05	0.05	0.08	0.19	0.29	0.37	0.43

Note: Coefficients are based on $h_{l_m} = K(\bar{V}_2^2/2)$.

Losses in diffusers depend on a number of geometric and flow variables. Diffuser data most commonly are presented in terms of a pressure recovery coefficient, C_p, defined as the ratio of static pressure rise to inlet dynamic pressure,

$$C_p \equiv \frac{p_2 - p_1}{\frac{1}{2}\rho\bar{V}_1^2} \tag{8.39}$$

Data for conical diffusers with fully developed turbulent pipe flow at the inlet are presented in Fig. 8.18 as a function of geometry. From the performance map of Fig. 8.18, we see that optimum diffuser geometries may be defined. For each area ratio, AR, there is an N/R_1 above which no increase in pressure recovery would be expected; this is particularly clear for $AR < 1.40$ in Fig. 8.18. Similarly, for a given dimensionless length, N/R_1, there is an optimum area ratio for maximum pressure recovery.

Performance maps for plane wall and annular diffusers [13] and for radial diffusers [14] are available in the literature.

Diffuser pressure recovery is essentially independent of Reynolds number for inlet Reynolds numbers greater than 7.5×10^4 [15]. Diffuser pressure recovery with

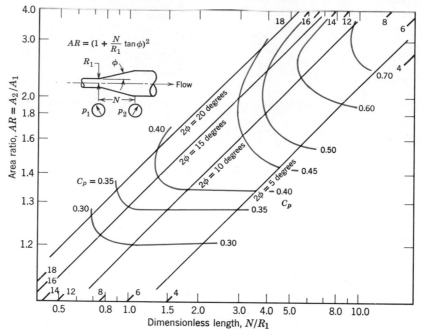

Fig. 8.18   Pressure recovery for conical diffusers with fully developed turbulent pipe flow at inlet. (Data from [12].)

uniform inlet flow is somewhat better than that for fully developed inlet flow. Performance maps for plane wall, conical, and annular diffusers for a variety of inlet flow conditions are presented in [16].

Since static pressure rises in the direction of flow in a diffuser, flow may separate from the walls. For some geometries, the outlet flow is distorted; sometimes pulsations occur. The flow regime behavior of plane wall diffusers is illustrated well in the NCFMF film, *Flow Visualization* (S. J. Kline, principal). For wide angle diffusers, vanes or splitters can be used to suppress stall and improve pressure recovery [17].

The definition of C_p may be related to head loss. If gravity is neglected, and $\alpha_1 = \alpha_2 = 1.0$, Eq. 8.28 reduces to

$$\left[\frac{p_1}{\rho} + \frac{\bar{V}_1^2}{2}\right] - \left[\frac{p_2}{\rho} + \frac{\bar{V}_2^2}{2}\right] = h_{l_T} = h_{l_m}$$

Thus

$$h_{l_m} = \frac{\bar{V}_1^2}{2} - \frac{\bar{V}_2^2}{2} - \frac{p_2 - p_1}{\rho}$$

$$h_{l_m} = \frac{\bar{V}_1^2}{2}\left[\left(1 - \frac{\bar{V}_2^2}{\bar{V}_1^2}\right) - \frac{p_2 - p_1}{\frac{1}{2}\rho\bar{V}_1^2}\right] = \frac{\bar{V}_1^2}{2}\left[\left(1 - \frac{\bar{V}_2^2}{\bar{V}_1^2}\right) - C_p\right]$$

From continuity, $A_1\bar{V}_1 = A_2\bar{V}_2$, so

$$h_{l_m} = \frac{\bar{V}_1^2}{2}\left[1 - \left(\frac{A_1}{A_2}\right)^2 - C_p\right]$$

or

$$h_{l_m} = \frac{\bar{V}_1^2}{2}\left[\left(1 - \frac{1}{(AR)^2}\right) - C_p\right]$$ (8.40)

For a frictionless flow, $h_{l_m} = 0$; for this case, Eq. 8.40 gives the ideal pressure recovery coefficient, denoted C_{p_i}, as

$$C_{p_i} = 1 - \frac{1}{(AR)^2}$$ (8.41)

This same result can be obtained by applying the Bernoulli equation, together with the continuity equation, to frictionless flow through the diffuser. Thus the head loss for flow through an actual diffuser may be written

$$h_{l_m} = (C_{p_i} - C_p)\frac{\bar{V}_1^2}{2}$$ (8.42)

c. Pipe Bends

The head loss of a bend is larger than for fully developed flow through a straight section of equal length. The additional loss is primarily the result of secondary flow,[4] and is represented most conveniently by an equivalent length of straight pipe. The equivalent length depends on the relative radius of curvature of the bend, as shown in Fig. 8.19a for 90° bends. An approximate procedure for computing the resistance of bends with other turning angles is given in [10].

Because they are simple and inexpensive to erect in the field, miter bends often are used, especially in large pipe systems. Design data for miter bends are given in Fig. 8.19b.

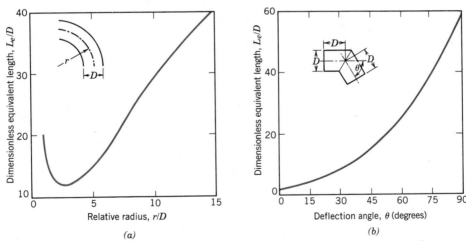

Fig. 8.19 Representative total resistance (L_e/D) for (a) 90° pipe bends and flanged elbows, and (b) miter bends. (Data from [10].)

d. Valves and Fittings

Losses for flow through valves and fittings also may be expressed in terms of an equivalent length of straight pipe. Some representative data are given in Table 8.3.

---

[4] Secondary flows are shown in the NCFMF film, *Secondary Flow*, E. S. Taylor, principal.

Table 8.3 Representative Dimensionless Equivalent Lengths (L_e/D) for Valves and Fittings (Data from [10].)

Fitting Type	Equivalent Length,[a] L_e/D
Valves (fully open)	
Gate valve	8
Globe valve	340
Angle valve	150
Ball valve	3
Lift check valve: globe lift	600
: angle lift	55
Foot valve with strainer: poppet disk	420
: hinged disk	75
Standard elbow: 90°	30
: 45°	16
Return bend, close pattern	50
Standard tee: flow through run	20
: flow through branch	60

[a] Based on $h_{l_m} = f \dfrac{L_e}{D} \dfrac{\bar{V}^2}{2}$.

All resistances are given for fully open valves; losses increase markedly when valves are partially open. Valve design varies significantly among manufacturers. Whenever possible, resistances furnished by the valve supplier should be used if accurate results are needed.

Fittings in a piping system may have threaded, flanged, or welded connections. For small diameters, threaded joints are most common; large pipe systems frequently have flanged or welded joints.

In practice, insertion losses for fittings and valves vary considerably, depending on the care used in fabricating the pipe system. If burrs from cutting pipe sections are allowed to remain, they cause local flow obstructions, which increase losses appreciably.

Although the losses discussed in this section were termed "minor losses," they can be a large fraction of the overall system loss. Thus a system for which calculations are to be made must be checked carefully to make sure all losses have been identified and their magnitudes estimated. If calculations are made carefully, the results will be of satisfactory engineering accuracy. You may expect to predict actual losses within ±10 percent.

8-7.3 Noncircular Ducts

The empirical correlations for pipe flow also may be used for computations involving noncircular ducts, provided their cross sections are not too exaggerated. Thus ducts of square or rectangular cross section may be treated if the ratio of height to width is less than about 3 or 4.

The correlations for turbulent pipe flow are extended for use with noncircular geometries by introducing the *hydraulic diameter,* defined as

$$D_h \equiv \frac{4A}{P} \qquad (8.43)$$

in place of the diameter, D. In Eq. 8.43, A is cross-sectional area, and P is *wetted perimeter,* the length of wall in contact with the flowing fluid at any cross section. The factor 4 is introduced so that the hydraulic diameter will equal the duct diameter for a circular geometry. For a circular duct, $A = \pi D^2/4$ and $P = \pi D$, so that

$$D_h = \frac{4A}{P} = \frac{4\left(\frac{\pi}{4}\right)D^2}{\pi D} = D$$

For a rectangular duct of width b and height h, $A = bh$ and $P = 2(b+h)$, so

$$D_h = \frac{4bh}{2(b+h)}$$

If the *aspect ratio, ar,* is defined as $ar = h/b$, then

$$D_h = \frac{2h}{1+ar}$$

for rectangular ducts. For a square duct, $ar = 1$ and $D_h = h$.

As noted, the hydraulic diameter concept can be applied in the approximate range $\frac{1}{4} < ar < 4$. Under these conditions, the correlations for pipe flow give acceptably accurate results for rectangular ducts; since such ducts are easy and cheap to fabricate from sheet metal, they are commonly used in air conditioning, heating, and ventilating applications. Extensive data on losses for air flow are available (e.g., see [11, 18]).

Losses due to secondary flows increase rapidly for more extreme geometries, so the correlations are not applicable to wide, flat ducts, or to ducts of triangular or other irregular shapes. Experimental data must be used when precise design information is required for specific situations.

8-8 SOLUTION OF PIPE FLOW PROBLEMS

Once the total head loss has been calculated using the methods of Section 8-7, pipe flow problems can be solved using the energy equation, Eq. 8.28. The same basic techniques are used, even for complex piping systems, but let us first consider single-path pipe flow problems.

8-8.1 Single-Path Systems

Equation 8.28 is the computing equation for pipe systems. The pressure drop across a pipe system is a function of flow rate, elevation change, and total head loss. The total head loss consists of major losses due to friction in constant-area sections (Eq. 8.32) and minor losses due to fittings, area changes, and so forth (Eqs. 8.38). The pressure drop could be written in the functional form

$$\Delta p = \phi_3(L, Q, D, e, \Delta z, \text{ system configuration, } \rho, \mu)$$

The fluid properties are constant for pipe flow of incompressible fluids. The roughness, elevation change, and system configuration depend on the pipe system layout. Once these have been fixed (for a given system and fluid), the dependence reduces to

$$\Delta p = \phi_4(L, Q, D) \tag{8.44}$$

Equation 8.44 relates four variables. Any one of these may be the unknown quantity in a practical flow situation. Thus four general cases are possible:

(a) L, Q, and D known, Δp unknown.
(b) Δp, Q, and D known, L unknown.
(c) Δp, L, and D known, Q unknown.
(d) Δp, L, and Q known, D unknown.

Cases (a) and (b) may be solved directly by applying the continuity and energy equations, and using loss data from Section 8-7. Solutions for cases (c) and (d) make use of the same equations and data, but require iteration. Each case is discussed below and illustrated by an example.

a. L, Q, and D Known, Δp Unknown

A friction factor is obtained from the Moody chart or empirical equations using Re and e/D computed from the given data. The total head loss is computed from Eqs. 8.32 and 8.38. Equation 8.28 is then used to evaluate the pressure drop, Δp. The procedure is illustrated in Example Problem 8.5.

b. Δp, Q, and D Known, L Unknown

The total head loss is calculated from Eq. 8.28. A friction factor is obtained from the Moody diagram or empirical equations using Re and e/D computed from the given data. The unknown length is determined by solving Eq. 8.32. The procedure is illustrated in Example Problem 8.6.

c. Δp, L, and D Known, Q Unknown

Equation 8.28 is combined with the defining equations for head loss; the result is an expression for $\bar{V}$ (or Q) in terms of the friction factor, f. Most pipe flows of engineering interest have relatively large Reynolds numbers. Thus, even though the Reynolds number (and hence f) cannot be calculated because Q is not known, a good first guess for the friction factor is taken from the fully rough region of Fig. 8.14. Using the assumed f, a first approximation for $\bar{V}$ is calculated. The Reynolds number is computed for this $\bar{V}$, and a new f and a second approximation for $\bar{V}$ are obtained. Since f is a rather weak function of Reynolds number, more than two iterations seldom are required for convergence. A flow of this type is evaluated in Example Problem 8.7.

d. Δp, L, and Q Known, D Unknown

When a fluid-handling device is available and the geometry of the piping system is known, the problem is to determine the smallest (and hence least costly) pipe size that can deliver the desired flow rate. Since the pipe diameter is unknown, neither the Reynolds number nor relative roughness can be computed directly, and an iterative solution is required. A flow of this type is illustrated in Example Problem 8.8.

Calculations begin by assuming a trial pipe diameter. The Reynolds number and relative roughness then are calculated using the assumed D. A friction factor is obtained from Fig. 8.14. Then head loss is computed from Eqs. 8.32 and 8.38, and Eq. 8.28 is solved for pressure drop. The resulting trial Δp is compared to the system requirement.

If the trial Δp is too large, calculations are repeated for a larger assumed D. If the trial Δp is less than the criterion, a smaller D should be checked.

In choosing a pipe size, it is logical to work with diameters that are available commercially. Pipe is manufactured in a limited number of standard sizes. Some data for standard pipe sizes are given in Table 8.4. For data on extra strong or double extra strong pipes, consult a handbook, e.g., [10]. Pipe larger than 12 in. nominal diameter is produced in multiples of 2 in. up to a nominal diameter of 36 in. and in multiples of 6 in. for still larger sizes.

Table 8.4 Standard Sizes for Carbon Steel, Alloy Steel, and Stainless Steel Pipe (Data from [10].)

Nominal Pipe Size (in.)	Inside Diameter (in.)	Nominal Pipe Size (in.)	Inside Diameter (in.)
$\frac{1}{8}$	0.269	$2\frac{1}{2}$	2.469
$\frac{1}{4}$	0.364	3	3.068
$\frac{3}{8}$	0.493	4	4.026
$\frac{1}{2}$	0.622	5	5.047
$\frac{3}{4}$	0.824	6	6.065
1	1.049	8	7.981
$1\frac{1}{2}$	1.610	10	10.020
2	2.067	12	12.000

EXAMPLE 8.5—Pipe Flow from a Reservoir: Pressure Drop Unknown

A 100 m length of smooth horizontal pipe is attached to a large reservoir. What depth, d, must be maintained in the reservoir to produce a volume flow rate of 0.0084 m³/sec of water? The inside diameter of the smooth pipe is 75 mm. The inlet is square-edged and water discharges to the atmosphere.

EXAMPLE PROBLEM 8.5

GIVEN: Water flow at 0.0084 m³/sec through 75 mm diameter pipe, with
$L = 100$ m, attached to a constant-level reservoir. Square-edged inlet.

FIND: Reservoir depth, d, to maintain the flow.

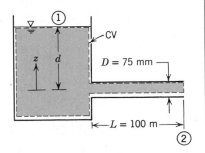

SOLUTION:
Computing equation:

$$\left(\frac{p_1}{\rho} + \alpha_1\frac{\bar{V}_1^2}{2} + gz_1\right) - \left(\frac{p_2}{\rho} + \alpha_2\frac{\bar{V}_2^2}{2} + gz_2\right) = h_{l_T} = h_l + h_{l_m} \qquad (8.28)$$

where

$$h_l = f\frac{L}{D}\frac{\bar{V}^2}{2} \qquad \text{and} \qquad h_{l_m} = K\frac{\bar{V}^2}{2}$$

For the given problem, $p_1 = p_2 = p_{atm}$, $\bar{V}_1 \simeq 0$, $\bar{V}_2 = \bar{V}$, and $\alpha_2 \simeq 1.0$. Assuming $z_2 = 0$, then $z_1 = d$. Simplifying Eq. 8.28 gives

$$g d - \frac{\bar{V}^2}{2} = f \frac{L}{D} \frac{\bar{V}^2}{2} + K \frac{\bar{V}^2}{2}$$

Then

$$d = \frac{1}{g}\left[f\frac{L}{D}\frac{\bar{V}^2}{2} + K\frac{\bar{V}^2}{2} + \frac{\bar{V}^2}{2}\right] = \frac{\bar{V}^2}{2g}\left[f\frac{L}{D} + K + 1\right]$$

Since $\bar{V} = \dfrac{Q}{A} = \dfrac{4Q}{\pi D^2}$, then

$$d = \frac{8Q^2}{\pi^2 D^4 g}\left[f\frac{L}{D} + K + 1\right]$$

Assuming water at 20 C, $\rho = 999$ kg/m^3, and $\mu = 1.0 \times 10^{-3}$ kg/m·sec. Thus

$$Re = \frac{\rho \bar{V} D}{\mu} = \frac{4\rho Q}{\pi \mu D}$$

$$Re = \frac{4}{\pi} \times \frac{999 \text{ kg}}{\text{m}^3} \times \frac{0.0084 \text{ m}^3}{\text{sec}} \times \frac{\text{m} \cdot \text{sec}}{1.0 \times 10^{-3} \text{ kg}} \times \frac{1}{0.075 \text{ m}} = 1.42 \times 10^5$$

For smooth pipe, from Fig. 8.14, $f = 0.0170$. From Table 8.1, $K = 0.5$. Then

$$d = \frac{8Q^2}{\pi^2 D^4 g}\left[f\frac{L}{D} + K + 1\right]$$

$$= \frac{8}{\pi^2} \times \frac{(0.0084)^2 \text{ m}^6}{\text{sec}^2} \times \frac{1}{(0.075)^4 \text{ m}^4} \times \frac{\text{sec}^2}{9.81 \text{ m}}\left[(0.0170)\frac{100 \text{ m}}{0.075 \text{ m}} + 0.5 + 1\right]$$

$$d = 4.45 \text{ m} \qquad\qquad\qquad\qquad d$$

{This problem illustrates the method for calculating total head loss.}

EXAMPLE 8.6—Flow in a Pipeline: Length Unknown

Crude oil flows through a level section of the Alaskan pipeline at a rate of 1.6 million barrels per day (1 barrel = 42 gal). The pipe inside diameter is 48 in.; its roughness is equivalent to galvanized iron. The maximum allowable pressure is 1200 psi; the minimum pressure required to keep dissolved gases in solution in the crude oil is 50 psi. The crude oil has SG = 0.93; its viscosity at the pumping temperature of 140 F is $\mu = 3.5 \times 10^{-4}$ lbf·sec/ft^2. For these conditions, determine the maximum possible spacing between pumping stations. If the pump efficiency is 85 percent, determine the power that must be supplied at each pumping station.

EXAMPLE PROBLEM 8.6

GIVEN: Flow of crude oil through horizontal section of Alaskan pipeline.

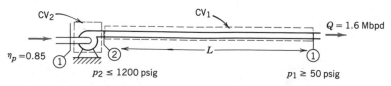

$D = 48$ in. (roughness of galvanized iron), SG = 0.93, $\mu = 3.5 \times 10^{-4}$ lbf·sec/ft^2

FIND: (a) Maximum spacing, L.

(b) Power needed at each pump station.

SOLUTION:

Apply energy equation for steady, incompressible pipe flow.

Computing equations: $\left(\dfrac{p_2}{\rho} + \alpha_2 \dfrac{\bar{V}_2^2}{2} + g z_2\right) - \left(\dfrac{p_1}{\rho} + \alpha_1 \dfrac{\bar{V}_1^2}{2} + g z_1\right) = h_{l_T}$

$$h_{l_T} = h_l + h_{l_m} = f \frac{L}{D} \frac{\bar{V}^2}{2} + K \frac{\bar{V}^2}{2} \overset{= 0(3)}{}$$

Assumptions: (1) $\alpha_1 \bar{V}_1^2 = \alpha_2 \bar{V}_2^2$

(2) Horizontal pipe; $z_1 = z_2$

(3) Neglect minor losses

(4) Constant viscosity

Then, using CV_1

$$\Delta p = p_2 - p_1 = f \frac{L}{D} \rho \frac{\bar{V}^2}{2} \quad \text{or} \quad L = \frac{2D}{f} \frac{\Delta p}{\rho \bar{V}^2} \quad \text{where } f = f(Re, e/D)$$

$$\bar{V} = \frac{Q}{A} = \frac{1.6 \times 10^6 \text{ bbl}}{\text{day}} \times \frac{4}{\pi (4)^2 \text{ ft}^2} \times \frac{42 \text{ gal}}{\text{bbl}} \times \frac{\text{ft}^3}{7.48 \text{ gal}} \times \frac{\text{day}}{24 \text{ hr}} \times \frac{\text{hr}}{3600 \text{ sec}}$$

$$\bar{V} = 8.27 \text{ ft/sec}$$

$$Re = \frac{\rho \bar{V} D}{\mu} = \frac{(0.93) \, 1.94 \text{ slug}}{\text{ft}^3} \times \frac{8.27 \text{ ft}}{\text{sec}} \times 4 \text{ ft} \times \frac{\text{ft}^2}{3.5 \times 10^{-4} \text{ lbf} \cdot \text{sec}} \times \frac{\text{lbf} \cdot \text{sec}^2}{\text{slug} \cdot \text{ft}}$$

$$Re = 1.71 \times 10^5$$

From Fig. 8.15, $e/D = 0.00012$, and from Fig. 8.14, $f \approx 0.017$. Then

$$L = \frac{2}{0.017} \times 4 \text{ ft} \times \frac{(1200 - 50) \text{ lbf}}{\text{in.}^2} \times \frac{\text{ft}^3}{(0.93) \, 1.94 \text{ slug}} \times \frac{\text{sec}^2}{(8.27)^2 \text{ ft}^2}$$

$$\times \frac{144 \text{ in.}^2}{\text{ft}^2} \times \frac{\text{slug} \cdot \text{ft}}{\text{lbf} \cdot \text{sec}^2} = 6.32 \times 10^5 \text{ ft}$$

$$L = 632,000 \text{ ft} \ (120 \text{ mi}) \underset{\longleftarrow}{\hspace{6cm}} L$$

To find the pumping power, apply the first law of thermodynamics to CV_2, across the pump between sections ① and ②.

Basic equation:

$$\dot{Q} - \dot{W}_s = \overset{= 0(1)}{\frac{\partial}{\partial t} \int_{CV} e \rho \, d\forall} + \int_{CS} \left(u + \overset{= 0(2)}{\frac{V^2}{2}} + \overset{= 0(3)}{g z} + \frac{p}{\rho}\right) \rho \vec{V} \cdot d\vec{A}$$

Assumptions: (1) Steady flow

(2) $\bar{V}_2 = \bar{V}_1$

(3) $z_2 = z_1$

(4) Uniform flow at each section

Then

$$\dot{W}_{in} = -\dot{W}_s = \left(u_2 + \frac{p_2}{\rho}\right)\{\dot{m}\} + \left(u_1 + \frac{p_1}{\rho}\right)\{-\dot{m}\} - \dot{Q}$$

$$\dot{W}_{in} = \left(\frac{p_2 - p_1}{\rho}\right)\dot{m} + (u_2 - u_1)\dot{m} - \dot{Q} = \left(\frac{p_2 - p_1}{\rho}\right)\dot{m} + \text{Losses}$$

The losses are determined in terms of the pump efficiency, η.

$$\text{Losses} = (1 - \eta)\dot{W}_{in}$$

Thus,

$$\dot{W}_{in} = \frac{1}{\eta}\frac{(p_2 - p_1)}{\rho}\dot{m} = \frac{1}{\eta}(p_2 - p_1)\bar{V}A = \frac{\bar{V}A\Delta p}{\eta}$$

$$= \frac{1}{0.85} \times 8.27\ \frac{\text{ft}}{\text{sec}} \times \frac{\pi\ (4)^2\ \text{ft}^2}{4} \times 1150\ \frac{\text{lbf}}{\text{in.}^2} \times 144\ \frac{\text{in.}^2}{\text{ft}^2} \times \frac{\text{hp}\cdot\text{sec}}{550\ \text{ft}\cdot\text{lbf}}$$

$$\dot{W}_{in} = 36,800\ \text{hp (power supplied)} \qquad\qquad \dot{W}_{needed}$$

This problem illustrates the method of solving for an unknown pipe length. Note that the theoretical pumping power, obtained from applying the first law across the pump, reduces to $\dot{W} = Q\Delta p$. The actual power needed can be determined from the definition of efficiency, $\eta = \dot{W}_{theoretical}/\dot{W}_{actual}$.

EXAMPLE 8.7—Flow from a Water Tower: Flow Rate Unknown

A fire protection system is supplied from a water tower and standpipe 80 ft tall. The longest pipe in the system is 600 ft and is made of cast iron about 20 years old. The pipe contains one gate valve; other minor losses may be neglected. The pipe diameter is 4 in. Determine the maximum rate of flow (gpm) through this pipe.

EXAMPLE PROBLEM 8.7

GIVEN: Fire protection system, as shown.

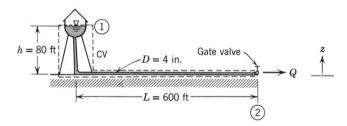

FIND: Q, gpm.

SOLUTION:
Computing equations:

$$\left(\frac{p_1}{\rho} + \alpha_1\frac{\bar{V}_1^2}{2} + gz_1\right) - \left(\frac{p_2}{\rho} + \alpha_2\frac{\bar{V}_2^2}{2} + gz_2\right) = h_{l_T} \qquad (8.28)$$

$$h_{l_T} = f\frac{L}{D}\frac{\bar{V}_2^2}{2} + h_{l_m} = f\frac{L}{D}\frac{\bar{V}_2^2}{2} + f\frac{L_e}{D}\frac{\bar{V}_2^2}{2}$$

Assumptions: (1) $p_1 = p_2 = p_{atm}$
(2) $\bar{V}_1 \approx 0$, and $\alpha_2 \approx 1.0$

Then Eq. 8.28 can be written as

$$g(z_1 - z_2) - \frac{\bar{V}_2^2}{2} = h_{l_T} = f\left(\frac{L}{D} + \frac{L_e}{D}\right)\frac{\bar{V}_2^2}{2}$$

For a fully open gate valve, from Table 8.3, $L_e/D = 8$. Thus

$$g(z_1 - z_2) = \frac{\bar{V}_2^2}{2}\left[f\left(\frac{L}{D} + 8\right) + 1\right]$$

Solving for $\bar{V}_2$, we obtain

$$\bar{V}_2 = \left[\frac{2g(z_1 - z_2)}{f(L/D + 8) + 1}\right]^{1/2}$$

To be conservative, assume the standpipe is the same diameter as the horizontal pipe. Then

$$\frac{L}{D} = \frac{600 \text{ ft} + 80 \text{ ft}}{4 \text{ in.}} \times \frac{12 \text{ in.}}{\text{ft}} = 2040$$

Also

$$z_1 - z_2 = h = 80 \text{ ft}$$

Since $\bar{V}_2$ is not known, we cannot compute Re. But we can assume a friction factor from the fully rough flow region. From Fig. 8.15, $e/D \simeq 0.0025$ for cast-iron pipe. Since the pipe is quite old, choose $e/D = 0.005$. Then, from Fig. 8.14, guess $f \simeq 0.03$. Thus a first approximation to $\bar{V}_2$ is

$$\bar{V}_2 = \left[2 \times 32.2 \frac{\text{ft}}{\text{sec}^2} \times 80 \text{ ft} \times \frac{1}{0.03(2040 + 8) + 1}\right]^{1/2} = 9.08 \text{ ft/sec}$$

Now check the assumed f.

$$Re = \frac{\rho \bar{V}D}{\mu} = \frac{\bar{V}D}{\nu} = 9.08 \frac{\text{ft}}{\text{sec}} \times \frac{\text{ft}}{3} \times \frac{\text{sec}}{1.2 \times 10^{-5} \text{ ft}^2} = 2.52 \times 10^5$$

For $e/D = 0.005$, $f = 0.031$ from Fig. 8.14. Thus we obtain

$$\bar{V}_2 = \left[2 \times 32.2 \frac{\text{ft}}{\text{sec}^2} \times 80 \text{ ft} \times \frac{1}{0.031(2040 + 8) + 1}\right]^{1/2} = 8.94 \text{ ft/sec}$$

Thus convergence is satisfactory. The volume flow rate is

$$Q = \bar{V}_2 A = \bar{V}_2 \frac{\pi D^2}{4} = 8.94 \frac{\text{ft}}{\text{sec}} \times \frac{\pi}{4}\left(\frac{1}{3}\right)^2 \text{ ft}^2 \times 7.48 \frac{\text{gal}}{\text{ft}^3} \times 60 \frac{\text{sec}}{\text{min}}$$

$$Q = 350 \text{ gpm} \hspace{7cm} Q$$

This problem illustrates the procedure for solving pipe flow problems in which the flow rate is unknown. Note that the velocity, and hence the flow rate, is essentially proportional to $1/\sqrt{f}$. Doubling e/D to account for aging reduced the flow rate by about 10 percent.

EXAMPLE 8.8—Flow in an Irrigation System: Diameter Unknown

Spray heads in an agricultural spraying system are to be supplied with water through 500 ft of drawn aluminum tubing from an engine-driven pump. In its most efficient operating range, the pump output is 1500 gpm at a discharge pressure not exceeding 65 psig. For satisfactory operation, the sprinklers must operate at 30 psig or higher pressure. Minor losses and elevation changes may be neglected. Determine the smallest standard pipe size that can be used.

EXAMPLE PROBLEM 8.8

GIVEN: Water supply system, as shown.

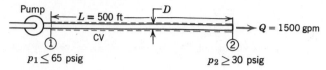

FIND: Smallest standard D.

SOLUTION:

Δp, L, and Q are known. D is unknown, so iteration is needed to determine the minimum standard diameter that satisfies the pressure drop constraint at the given flow rate. The maximum allowable pressure drop over the length, L, is

$$\Delta p_{max} = p_{1\,max} - p_{2\,min} = (65-30) \text{ psi} = 35 \text{ psi}$$

Computing equations :

$$\left(\frac{p_1}{\rho} + \alpha_1 \frac{\bar{V}_1^2}{2} + g\cancel{z_1}\right) - \left(\frac{p_2}{\rho} + \alpha_2 \frac{\bar{V}_2^2}{2} + g\cancel{z_2}\right) = h_{l_T} \qquad (8.28)$$

$$= 0(3)$$

$$h_{l_T} = h_l + \cancel{h_{l_m}} = f\frac{L}{D}\frac{\bar{V}_2^2}{2}$$

Assumptions: (1) Steady flow
(2) Incompressible flow
(3) $h_{l_T} = h_l$, i.e., $h_{l_m} = 0$
(4) $z_1 = z_2$
(5) $\bar{V}_1 = \bar{V}_2 = \bar{V}$; $\alpha_1 \simeq \alpha_2$

Then

$$\Delta p = p_1 - p_2 = f\frac{L}{D}\frac{\rho\bar{V}^2}{2}$$

Since trial diameters are to be assumed, it is convenient to substitute $\bar{V} = Q/A = 4Q/\pi D^2$, so that

$$\Delta p = f\frac{L}{D}\frac{\rho}{2}\left(\frac{4Q}{\pi D^2}\right)^2 = \frac{8fL\rho Q^2}{\pi^2 D^5} \qquad (1)$$

The Reynolds number is needed to find f. In terms of Q,

$$Re = \frac{\rho\bar{V}D}{\mu} = \frac{\bar{V}D}{\nu} = \frac{4Q}{\pi D^2}\frac{D}{\nu} = \frac{4Q}{\pi\nu D}$$

Finally, Q must be converted to cubic feet per second.

$$Q = \frac{1500 \text{ gal}}{\text{min}} \times \frac{\text{min}}{60 \text{ sec}} \times \frac{\text{ft}^3}{7.48 \text{ gal}} = 3.34 \text{ ft}^3/\text{sec}$$

For an initial guess, take nominal 4 in. (4.026 in. i.d.) pipe:

$$Re = \frac{4Q}{\pi\nu D} = \frac{4}{\pi} \times \frac{3.34 \text{ ft}^3}{\text{sec}} \times \frac{\text{sec}}{1.2\times10^{-5} \text{ ft}^2} \times \frac{1}{4.026 \text{ in.}} \times \frac{12 \text{ in.}}{\text{ft}} = 1.06\times10^6$$

For drawn tubing, $e/D = 0.000{,}016$ (Fig. 8.15), so $f \simeq 0.012$ (Fig. 8.14), and

$$\Delta p = \frac{8fL\rho Q^2}{\pi^2 D^5} = \frac{8}{\pi^2} \times 0.012 \times 500 \text{ ft} \times \frac{1.94 \text{ slug}}{\text{ft}^3} \times \frac{(3.34)^2 \text{ ft}^6}{\text{sec}^2}$$

$$\times \frac{1}{(4.026)^5 \text{ in.}^5} \times \frac{1728 \text{ in.}^3}{\text{ft}^3} \times \frac{\text{lbf} \cdot \text{sec}^2}{\text{slug} \cdot \text{ft}}$$

$$\Delta p = 172 \text{ lbf/in.}^2 > \Delta p_{max}$$

Since this pressure drop is too large, try $D = 6$ in. (actually 6.065 in. i.d.):

$$Re = \frac{4}{\pi} \times \frac{3.34 \text{ ft}^3}{\text{sec}} \times \frac{\text{sec}}{1.2 \times 10^{-5} \text{ ft}^2} \times \frac{1}{6.065 \text{ in.}} \times \frac{12 \text{ in.}}{\text{ft}} = 7.01 \times 10^5$$

For drawn tubing, $e/D = 0.000,010$ (Fig. 8.15), so $f \simeq 0.013$ (Fig. 8.14), and

$$\Delta p = \frac{8}{\pi^2} \times 0.013 \times 500 \text{ ft} \times 1.94 \frac{\text{slug}}{\text{ft}^3} \times (3.34)^2 \frac{\text{ft}^6}{\text{sec}^2}$$

$$\times \frac{1}{(6.065)^5 \text{ in.}^5} \times \frac{(12)^3 \text{ in.}^3}{\text{ft}^3} \times \frac{\text{lbf} \cdot \text{sec}^2}{\text{slug} \cdot \text{ft}}$$

$$\Delta p = 24.0 \text{ lbf/in.}^2 < \Delta p_{max}$$

Since this is less than the allowable pressure drop, we should check a 5 in. (nominal) pipe. With an actual i.d. of 5.047 in.,

$$Re = \frac{4}{\pi} \times \frac{3.34 \text{ ft}^3}{\text{sec}} \times \frac{\text{sec}}{1.2 \times 10^{-5} \text{ ft}^2} \times \frac{1}{5.047 \text{ in.}} \times \frac{12 \text{ in.}}{\text{ft}} = 8.43 \times 10^5$$

For drawn tubing, $e/D = 0.000,012$ (Fig. 8.15), so $f \simeq 0.012$ (Fig. 8.14), and

$$\Delta p = \frac{8}{\pi^2} \times 0.012 \times 500 \text{ ft} \times 1.94 \frac{\text{slug}}{\text{ft}^3} \times (3.34)^2 \frac{\text{ft}^6}{\text{sec}^2}$$

$$\times \frac{1}{(5.047)^5 \text{ in.}^5} \times \frac{(12)^3 \text{ in.}^3}{\text{ft}^3} \times \frac{\text{lbf} \cdot \text{sec}^2}{\text{slug} \cdot \text{ft}}$$

$$\Delta p = 55.5 \text{ lbf/in.}^2 > \Delta p_{max}$$

Thus the criterion for pressure drop is satisfied for a minimum nominal diameter of 6 in. pipe.

$$D$$

This problem illustrates the procedure for solving pipe flow problems when the diameter is unknown. Note from Eq. 1 that the pressure drop in turbulent pipe flow is proportional to f/D^5. The variation of f is small, so Δp at constant flow rate is approximately proportional to $1/D^5$.

We have solved each of Example Problems 8.7 and 8.8 by direct iteration. Several specialized forms of friction factor versus Reynolds number diagrams have been introduced to solve problems of this type without the need for iteration. For examples of these specialized diagrams, see [19] and [20].

Example Problems 8.9 and 8.10 illustrate the evaluation of minor loss coefficients and the application of a diffuser to reduce exit kinetic energy from a flow system.

EXAMPLE 8.9—Calculation of Entrance Loss Coefficient

Reference [21] reports results of measurements made to determine entrance losses for flow from a reservoir to a pipe with various degrees of entrance rounding. A copper pipe 10 ft long, with 1.5 in. i.d., was used for the tests. The pipe discharged to atmosphere. For a square-edged entrance, a discharge of 0.566 ft³/sec was measured when the reservoir level was 85.1 ft above the pipe centerline. From these data, evaluate the loss coefficient for a square-edged entrance.

EXAMPLE PROBLEM 8.9

GIVEN: Pipe with square-edged entrance discharging from reservoir as shown.

FIND: K_{entrance}.

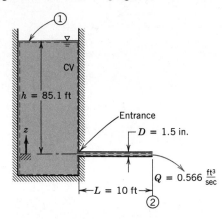

SOLUTION:
Apply the energy equation for steady, incompressible pipe flow.

Computing equations:
$$\overset{\approx 0(2)}{\cancel{\frac{p_1}{\rho}} + \alpha_1 \cancel{\frac{\bar{V}_1^2}{2}} + g z_1} = \overset{=0}{\cancel{\frac{p_2}{\rho}} + \alpha_2 \frac{\bar{V}_2^2}{2} + g\cancel{z_2} + h_{l_T}}$$

$$h_{l_T} = f \frac{L}{D} \frac{\bar{V}_2^2}{2} + K_{\text{entrance}} \frac{\bar{V}_2^2}{2}$$

Assumptions: (1) $p_1 = p_2 = p_{\text{atm}}$
(2) $\bar{V}_1 \approx 0$

Substituting for h_{l_T}, and dividing by g, gives $z_1 = h = \alpha_2 \frac{\bar{V}_2^2}{2g} + f \frac{L}{D} \frac{\bar{V}_2^2}{2g} + K_{\text{entrance}} \frac{\bar{V}_2^2}{2g}$

or

$$K_{\text{entrance}} = \frac{2gh}{\bar{V}_2^2} - f \frac{L}{D} - \alpha_2 \tag{1}$$

The average velocity is

$$\bar{V}_2 = \frac{Q}{A} = \frac{4Q}{\pi D^2} = \frac{4}{\pi} \times 0.566 \frac{\text{ft}^3}{\text{sec}} \times \frac{1}{(1.5)^2 \text{ in.}^2} \times \frac{144 \text{ in.}^2}{\text{ft}^2} = 46.1 \text{ ft/sec}$$

Assume $T \approx 75$ F (24 C), so $\nu = 8.8 \times 10^{-7}$ m²/sec (Fig. A.3). Then

$$Re = \frac{\bar{V}D}{\nu} = 46.1 \frac{\text{ft}}{\text{sec}} \times 1.5 \text{ in.} \times \frac{\text{sec}}{8.8 \times 10^{-7} \text{ m}^2} \times \frac{\text{ft}}{12 \text{ in.}} \times \frac{(0.3048)^2 \text{ m}^2}{\text{ft}^2} = 6.08 \times 10^5$$

For drawn tubing, $e/D = 0.000,04$ (Fig. 8.15), so $f = 0.013$ (Fig. 8.14). From Fig. 8.11, $n = 8.7$, so

$$\frac{\bar{V}}{U} = \frac{2n^2}{(n+1)(2n+1)} = 0.848 \tag{8.23}$$

and

$$\alpha = \left(\frac{U}{\bar{V}}\right)^3 \frac{2n^2}{(3+n)(3+2n)} = 1.04 \tag{8.26}$$

Substituting into Eq. 1, we obtain

$$K_{\text{entrance}} = 2 \times 32.2 \frac{\text{ft}}{\text{sec}^2} \times 85.1 \text{ ft} \times \frac{\text{sec}^2}{(46.1)^2 \text{ ft}^2} - (0.013) \frac{10 \text{ ft}}{1.5 \text{ in.}} \times \frac{12 \text{ in.}}{\text{ft}} - 1.04$$

$$K_{entrance} = 0.499 \qquad\qquad\qquad K_{entrance}$$

This coefficient compares favorably with that shown in Table 8.1. The hydraulic and energy grade lines are shown below. The large head loss in a square-edged entrance is due primarily to separation at the sharp inlet corner and formation of a vena contracta immediately downstream from the corner. The effective flow area reaches a minimum at the vena contracta, so the flow velocity is a maximum there. The flow expands again following the vena contracta to fill the pipe. The uncontrolled expansion following the vena contracta is responsible for most of the head loss. (See Example Problem 8.12.)

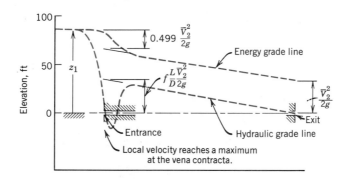

Rounding the inlet corner reduces the extent of separation significantly. This reduces the velocity increase through the vena contracta and consequently reduces the head loss due to the entrance. A "well-rounded" inlet almost eliminates flow separation; the flow pattern approaches that shown in Fig. 8.1. The added head loss in a well-rounded inlet compared to fully developed flow is the result of higher wall shear stresses in the entrance length.

EXAMPLE 8.10—Use of Diffuser to Increase Flow Rate

Water rights granted to each citizen by the Emperor of Rome gave permission to install into the public water main a calibrated, circular, tubular bronze nozzle [22]. Some citizens were clever enough to take unfair advantage of a law that regulated flow rate by such an indirect method. They installed diffusers on the outlet of the nozzles to increase their discharge. Assume the static head available from the main is $z_0 = 1.5$ m and the nozzle exit diameter is $D = 25$ mm. (The discharge is to atmospheric pressure.) Determine the increase in flow rate when a diffuser with $N/R_1 = 3.0$ and $AR = 2.0$ is attached to the end of the nozzle.

EXAMPLE PROBLEM 8.10

GIVEN: Nozzle attached to water main as shown.

FIND: Increase in discharge when diffuser with $N/R_1 = 3.0$ and $AR = 2.0$ is installed.

SOLUTION:
Apply the energy equation for steady, incompressible pipe flow.

Computing equation:
$$\frac{p_0}{\rho} + \alpha_0 \frac{\bar{V}_0^2}{2} + g z_0 = \frac{p_1}{\rho} + \alpha_1 \frac{\bar{V}_1^2}{2} + g z_1 + h_{l_T}$$

Assumptions: (1) $\bar{V}_0 \approx 0$
(2) $\alpha_1 \approx 1$

For the nozzle alone,

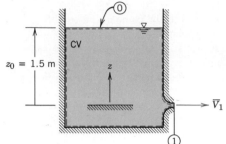

$$\frac{\overset{\approx 0(1)}{\cancel{\frac{p_0}{\rho}}} + \cancel{\alpha_0 \frac{\bar{V}_0^2}{2}} + g z_0 = \overset{}{\cancel{\frac{p_1}{\rho}}} + \overset{\approx 1(2)}{\cancel{\alpha_1}} \frac{\bar{V}_1^2}{2} + \overset{=0}{\cancel{g z_1}} + h_{l_T}$$

$$h_{l_T} = K_{\text{entrance}} \frac{\bar{V}_1^2}{2} \approx 0.04 \frac{\bar{V}_1^2}{2} \qquad \text{(from Table 8.1)}$$

Thus

$$g z_0 = \frac{\bar{V}_1^2}{2} + 0.04 \frac{\bar{V}_1^2}{2} = 1.04 \frac{\bar{V}_1^2}{2}$$

$$\bar{V}_1 = \sqrt{\frac{2 g z_0}{1.04}} = \sqrt{\frac{2}{1.04} \times 9.81 \frac{\text{m}}{\text{sec}^2} \times 1.5 \text{ m}} = 5.32 \text{ m/sec}$$

$$Q = \bar{V}_1 A_1 = \bar{V}_1 \frac{\pi D^2}{4} = 5.32 \frac{\text{m}}{\text{sec}} \times \frac{\pi}{4} \times (0.025)^2 \text{ m}^2 = 0.00261 \text{ m}^3/\text{sec} \qquad \longleftarrow Q$$

For the nozzle with diffuser attached,

$$\frac{\overset{\approx 0(1)}{\cancel{\frac{p_0}{\rho}}} + \cancel{\alpha_0 \frac{\bar{V}_0^2}{2}} + g z_0 = \overset{}{\cancel{\frac{p_2}{\rho}}} + \overset{\approx 1(2)}{\cancel{\alpha_2}} \frac{\bar{V}_2^2}{2} + \overset{=0}{\cancel{g z_2}} + h_{l_T}$$

$$h_{l_T} = K_{\text{entrance}} \frac{\bar{V}_1^2}{2} + K_{\text{diffuser}} \frac{\bar{V}_1^2}{2}$$

or

$$g z_0 = \frac{\bar{V}_2^2}{2} + (K_{\text{entrance}} + K_{\text{diffuser}}) \frac{\bar{V}_1^2}{2} \qquad (1)$$

Figure 8.18 gives data for $C_p = \dfrac{p_2 - p_1}{\frac{1}{2}\rho \bar{V}_1^2}$ for diffusers.

To obtain K_{diffuser}, apply energy equation from ① to ②.

$$\frac{p_1}{\rho} + \alpha_1 \frac{\bar{V}_1^2}{2} + \cancel{g z_1} = \frac{p_2}{\rho} + \alpha_2 \frac{\bar{V}_2^2}{2} + \cancel{g z_2} + K_{\text{diffuser}} \frac{\bar{V}_1^2}{2}$$

Solving, with $\alpha_2 \approx 1$, we obtain

$$K_{\text{diffuser}} = 1 - \frac{\bar{V}_2^2}{\bar{V}_1^2} - \frac{p_2 - p_1}{\frac{1}{2}\rho \bar{V}_1^2} = 1 - \left(\frac{A_1}{A_2}\right)^2 - C_p = 1 - \frac{1}{(AR)^2} - C_p$$

where from continuity, $\bar{V}_1 A_1 = \bar{V}_2 A_2$. From Fig. 8.18, $C_p = 0.45$, so

$$K_{\text{diffuser}} = 1 - \frac{1}{(2.0)^2} - 0.45 = 0.75 - 0.45 = 0.3$$

Thus, substituting in Eq. 1,

$$g z_0 = \frac{\bar{V}_2^2}{2} + (0.04 + 0.30)\frac{\bar{V}_1^2}{2} = \frac{\bar{V}_1^2}{2}\left[\frac{1}{(AR)^2} + 0.34\right] = 0.59\frac{\bar{V}_1^2}{2}$$

For this system

$$\bar{V}_1 = \sqrt{\frac{2g z_0}{0.59}} = \sqrt{\frac{2}{0.59} \times 9.81 \frac{m}{sec^2} \times 1.5 \, m} = 7.06 \, m/sec$$

and

$$Q_d = \bar{V}_1 A_1 = \bar{V}_1 \frac{\pi D^2}{4} = 7.06 \frac{m}{sec} \times \frac{\pi}{4} \times (0.025)^2 \, m^2 = 0.00347 \, m^3/sec \qquad \underset{\longleftarrow}{Q_d}$$

The flow rate increase that results from adding the diffuser is

$$\frac{\Delta Q}{Q} = \frac{Q_d - Q}{Q} = \frac{Q_d}{Q} - 1 = \frac{0.00347}{0.00261} - 1 = 0.330 \qquad or \qquad 33 \, percent \qquad \underset{\longleftarrow}{\frac{\Delta Q}{Q}}$$

Addition of the diffuser increases the flow rate through the nozzle significantly. The diffuser exhausts to atmospheric pressure; the pressure rise through the diffuser thus reduces the pressure in the nozzle exit plane, causing the flow rate to increase.

Energy and hydraulic grade lines are sketched in the following figures—approximately to scale—for the bare nozzle and for the nozzle-diffuser combination. The bare nozzle exhausts to atmospheric pressure, so that the hydraulic grade line drops to zero elevation at the nozzle exit. The only loss in the nozzle is 0.04 times the velocity head, so the energy grade line elevation decreases slightly as shown.

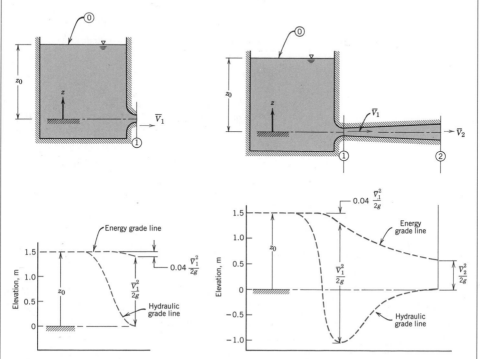

Hydrostatic pressure at the nozzle exit plane is reduced below atmospheric by the diffuser. As shown in the sketch, the pressure rises through the diffuser to atmospheric pressure at the discharge plane. With a diffuser area ratio of 2, the average velocity is reduced by a factor of 2, and the kinetic energy leaving the diffuser exit is one-fourth of the inlet value.

The energy grade line drops by 0.04 times the inlet velocity head at the nozzle exit plane. The diffuser introduces the additional loss, $K_{\text{diffuser}} = 0.3$, which further drops the level of the energy grade line.

> Water Commissioner Frontinus standardized conditions for all Romans in 97 A.D. He required that the tube attached to the nozzle of each customer's pipe be the same diameter for at least 50 lineal feet from the public water main (see Problem 8.117).

**8-8.2 Multiple-Path Systems

In many practical situations, such as water supply or fire protection systems, complex pipe networks must be analyzed. The basic techniques developed in Section 8-8.1 also may be used to analyze flow in multiple-path pipe systems. The procedure is analogous to solving direct current electric circuits, but with nonlinear elements.

The representative pipe system shown in Fig. 8.20 has two nodes, labeled A and B, and three branches. The total flow rate into the system must be distributed among the branches. Consequently, the flow rate through each branch is unknown. However, the pressure drop for each branch is the same, $p_A - p_B$. This information is sufficient to permit an iterative solution for the flow rate in each branch, as shown in Example Problem 8.11.

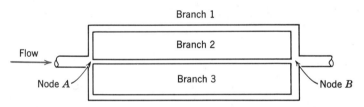

Fig. 8.20 Simple multiple-path pipe flow system.

The fluid flow rate and pressure drop are, respectively, analogous to the current and voltage in an electric circuit. However, the simple linear relation between voltage and current given by Ohm's law does not apply to the fluid flow system. Instead, the pressure drop is approximately proportional to the square of the flow rate. This nonlinearity makes iterative solutions necessary, and the resulting calculations can be quite lengthy. A number of schemes have been developed for use with digital computers. For an example, see [23].

EXAMPLE 8.11—Multiple-Path Pipe Flow System
The irrigation system of Example 8.8 is to be extended, holding the total flow rate constant at 1500 gpm, by adding three branches, each made from 3 in. pipe, as shown in the sketch. The sprinkler at the end of each branch has a nozzle of 1.5 in. minimum diameter. Minor losses at elbows should be included, but elevation terms may be neglected. Determine the flow rate in each branch, the pressure required at Ⓐ, and the pressure applied to each sprinkler nozzle.

** This section may be omitted without loss of continuity in the text material.

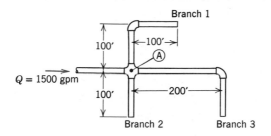

Branch 1

$Q = 1500$ gpm

100'

100'

100'

←100'→

(A)

←200'→

Branch 2

Branch 3

EXAMPLE PROBLEM 8.11

GIVEN: Pipe network shown schematically in the following figure.

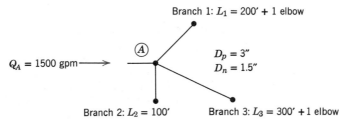

Branch 1: $L_1 = 200' + 1$ elbow

$Q_A = 1500$ gpm ⟶

(A)

$D_p = 3''$
$D_n = 1.5''$

Branch 2: $L_2 = 100'$

Branch 3: $L_3 = 300' + 1$ elbow

FIND: (a) Q_1, Q_2, Q_3.
        (b) p_A.
        (c) p_1, p_2, p_3 at nozzle inlets.

SOLUTION:

Apply Eq. 8.28 to each branch. Choose subscript n for nozzle exit, subscript p for pipe.

Computing equations:

$$\frac{p_A}{\rho} + \alpha_A \frac{\overset{\approx 0(1)}{\cancel{\bar{V}_A^2}}}{2} + g\cancel{z_A} = \cancel{\frac{p_n}{\rho}} + \alpha_n \frac{\bar{V}_n^2}{2} + g\cancel{z_n} + h_{l_T} \qquad h_{l_T} = f\frac{L}{D}\frac{\bar{V}_p^2}{2} + f\frac{L_e}{D}\frac{\bar{V}_p^2}{2}$$

Assumptions: (1) $\bar{V}_A^2 \ll \bar{V}_n^2$; $\alpha_n \approx 1.0$
               (2) $z_A \approx z_n$
               (3) $p_n = p_{\text{atm}}$
               (4)   Neglect loss due to flow split at (A)

Then, for each branch of the flow system, using gage pressure at (A), we obtain

$$\frac{p_A}{\rho} = \frac{\bar{V}_n^2}{2} + f\frac{L}{D}\frac{\bar{V}_p^2}{2} + f\frac{L_e}{D}\frac{\bar{V}_p^2}{2}$$

But for incompressible flow, $\bar{V}_p A_p = \bar{V}_n A_n$, so $\bar{V}_n^2 = \bar{V}_p^2 (A_p/A_n)^2 = \bar{V}_p^2 (D_p/D_n)^4$, and

$$\frac{p_A}{\rho} = \frac{\bar{V}_p^2}{2}\left[\left(\frac{D_p}{D_n}\right)^4 + f\left(\frac{L}{D} + \frac{L_e}{D}\right)\right]$$

or

$$\bar{V}_p = \sqrt{\frac{2p_A}{\rho}}\left[\frac{1}{\left(\dfrac{D_p}{D_n}\right)^4 + f\left(\dfrac{L}{D} + \dfrac{L_e}{D}\right)}\right]^{1/2}$$

The corresponding flow rate for each branch is

$$Q_p = \bar{V}_p A_p = A_p \sqrt{\frac{2p_A}{\rho}} \left[\frac{1}{\left(\dfrac{D_p}{D_n}\right)^4 + f\left(\dfrac{L}{D} + \dfrac{L_e}{D}\right)} \right]^{1/2} \tag{1}$$

To find f, we must determine the pipe Reynolds number. If flow is equally split as a first approximation, then $Q_p \simeq 500$ gpm per branch, and

$$Re = \frac{\bar{V}_p D_p}{\nu} = \frac{4Q_p}{\pi \nu D_p}$$

$$= \frac{4}{\pi} \times \frac{500}{} \frac{\text{gal}}{\text{min}} \times \frac{\text{sec}}{1.2 \times 10^{-5} \text{ ft}^2} \times \frac{1}{3 \text{ in.}} \times \frac{\text{min}}{60 \text{ sec}} \times \frac{\text{ft}^3}{7.48 \text{ gal}} \times \frac{12 \text{ in.}}{\text{ft}}$$

$$Re = 4.73 \times 10^5$$

From Fig. 8.14, for smooth pipe, $f \simeq 0.0133$. For an elbow, $L_e/D = 30$, from Table 8.3. We can obtain a first approximation for Q_p for each branch; substituting into Eq. 1 yields

$$Q_{p1} \simeq A_p \sqrt{\frac{2p_A}{\rho}} \left[\frac{1}{\left(\dfrac{3.0}{1.5}\right)^4 + 0.0133 \left(200 \text{ ft} \times \dfrac{1}{3 \text{ in.}} \times \dfrac{12 \text{ in.}}{\text{ft}} + 30\right)} \right]^{1/2}$$

$$Q_{p1} \simeq 0.192 A_p \sqrt{\frac{2p_A}{\rho}}$$

$$Q_{p2} \simeq A_p \sqrt{\frac{2p_A}{\rho}} \left[\frac{1}{\left(\dfrac{3.0}{1.5}\right)^4 + 0.0133 \left(100 \text{ ft} \times \dfrac{1}{3 \text{ in.}} \times \dfrac{12 \text{ in.}}{\text{ft}}\right)} \right]^{1/2}$$

$$Q_{p2} \simeq 0.217 A_p \sqrt{\frac{2p_A}{\rho}}$$

and

$$Q_{p3} \simeq A_p \sqrt{\frac{2p_A}{\rho}} \left[\frac{1}{\left(\dfrac{3.0}{1.5}\right)^4 + 0.0133 \left(300 \text{ ft} \times \dfrac{1}{3 \text{ in.}} \times \dfrac{12 \text{ in.}}{\text{ft}} + 30\right)} \right]^{1/2}$$

$$Q_{p3} \simeq 0.176 A_p \sqrt{\frac{2p_A}{\rho}}$$

From continuity,

$$Q_A = Q_{p1} + Q_{p2} + Q_{p3} \simeq (0.192 + 0.217 + 0.176) A_p \sqrt{\frac{2p_A}{\rho}} \simeq 0.585 A_p \sqrt{\frac{2p_A}{\rho}}$$

Thus

$$\frac{Q_{p1}}{Q_A} \simeq \frac{0.192 A_p \sqrt{\dfrac{2p_A}{\rho}}}{0.585 A_p \sqrt{\dfrac{2p_A}{\rho}}} = 0.328$$

or

$$Q_{p1} \simeq (0.328)1500 \text{ gpm} = 492 \text{ gpm}$$

Similarly

$$\frac{Q_{p2}}{Q_A} \simeq \frac{0.217}{0.585} = 0.371 \qquad Q_{p2} \simeq 557 \text{ gpm}$$

$$\frac{Q_{p3}}{Q_A} \simeq \frac{0.176}{0.585} = 0.301 \qquad Q_{p3} \simeq 452 \text{ gpm}$$

Better approximations for the Reynolds number and friction factor for each branch now may be calculated.

$$Re_1 = \frac{Q_{p1}}{500 \text{ gpm}} \times 4.73 \times 10^5 \simeq \frac{492}{500} \times 4.73 \times 10^5 = 4.65 \times 10^5$$

From Fig. 8.14, $f_1 \simeq 0.0133$. Similarly,

$$Re_2 \simeq \frac{557}{500} \times 4.73 \times 10^5 = 5.27 \times 10^5 \qquad f_2 \simeq 0.0130$$

and

$$Re_3 \simeq \frac{452}{500} \times 4.73 \times 10^5 = 4.28 \times 10^5 \qquad f_3 \simeq 0.0136$$

Substituting these values into Eq. 1, we obtain

$$Q_{p1} \simeq 0.192 A_p \sqrt{\frac{2 p_A}{\rho}}$$

$$Q_{p2} \simeq 0.217 A_p \sqrt{\frac{2 p_A}{\rho}}$$

and

$$Q_{p3} \simeq 0.175 A_p \sqrt{\frac{2 p_A}{\rho}}$$

From continuity,

$$Q_A = Q_{p1} + Q_{p2} + Q_{p3} \simeq (0.192 + 0.217 + 0.175) A_p \sqrt{\frac{2 p_A}{\rho}} \simeq 0.584 A_p \sqrt{\frac{2 p_A}{\rho}}$$

Solving for the individual flow rates,

$$Q_{p1} \simeq \frac{0.192}{0.584} \times 1500 \text{ gpm} = 493 \text{ gpm}$$

$$Q_{p2} \simeq \frac{0.217}{0.584} \times 1500 \text{ gpm} = 557 \text{ gpm}$$

and

$$Q_{p3} \simeq \frac{0.175}{0.584} \times 1500 \text{ gpm} = 449 \text{ gpm} \qquad\qquad Q_1, Q_2, Q_3$$

Solving Eq. 1 for p_A gives

$$p_A = \frac{\rho}{2} \left(\frac{Q_p}{A_p}\right)^2 \left[\left(\frac{D_p}{D_n}\right)^4 + f\left(\frac{L}{D} + \frac{L_e}{D}\right)\right]$$

Substituting for Branch ①,

$$p_A = \frac{1}{2} \times \frac{1.94 \text{ slug}}{\text{ft}^3} \left(493 \frac{\text{gal}}{\text{min}} \times \frac{4}{\pi} \frac{1}{(3)^2 \text{ in.}^2} \times \frac{\text{ft}^3}{7.48 \text{ gal}} \times \frac{\text{min}}{60 \text{ sec}} \times \frac{144 \text{ in.}^2}{\text{ft}^2} \right)^2$$

$$\times \left[\left(\frac{3.0}{1.5} \right)^4 + 0.0133 \left(200 \text{ ft} \times \frac{1}{3 \text{ in.}} \times \frac{12 \text{ in.}}{\text{ft}} + 30 \right) \right] \frac{\text{lbf} \cdot \text{sec}^2}{\text{slug} \cdot \text{ft}} \times \frac{\text{ft}^2}{144 \text{ in.}^2}$$

$$p_A = 91.2 \text{ lbf/in.}^2$$

Pressures for Branches ② and ③ by similar calculations are 91.4 and 90.5 psig, respectively. (The slight differences are due to rounding the flow rates.) Thus

$$p_A \approx 91.0 \text{ psig} \qquad\qquad\qquad p_A$$

The pressure at the inlet to each sprinkler nozzle may be calculated from the energy equation. The equation between Ⓐ and the nozzle inlet, section ①, is

$$\frac{p_A}{\rho} + \frac{\bar{V}_A^2}{2} + gz_A = \frac{p_i}{\rho} + \frac{\bar{V}_p^2}{2} + gz_i + h_{l_T} \qquad h_{l_T} = f \frac{L}{D} \frac{\bar{V}_p^2}{2} + f \frac{L_e}{D} \frac{\bar{V}_p^2}{2}$$

From continuity, $\bar{V}_A = Q_A/A_A$ and $\bar{V}_p = Q_p/A_p$, so

$$p_i = p_A + \frac{\rho}{2} \left[\left(\frac{Q_A}{A_A} \right)^2 - \left(f \frac{L}{D} + f \frac{L_e}{D} + 1 \right) \left(\frac{Q_p}{A_p} \right)^2 \right]$$

or

$$p_i = p_A + \frac{\rho}{2} \left(\frac{Q_A}{A_A} \right)^2 \left\{ 1 - \left[f \left(\frac{L}{D} + \frac{L_e}{D} \right) + 1 \right] \left(\frac{Q_p}{Q_A} \right)^2 \left(\frac{A_A}{A_p} \right)^2 \right\}$$

Substituting for Branch ① gives

$$p_{i_1} = \frac{91.0 \text{ lbf}}{\text{in.}^2} + \frac{1}{2} \times \frac{1.94 \text{ slug}}{\text{ft}^3}$$

$$\times \left(1500 \frac{\text{gal}}{\text{min}} \times \frac{4}{\pi} \frac{1}{(6)^2 \text{ in.}^2} \times \frac{\text{ft}^3}{7.48 \text{ gal}} \times \frac{\text{min}}{60 \text{ sec}} \times \frac{144 \text{ in.}^2}{\text{ft}^2} \right)^2$$

$$\times \left\{ 1 - \left[0.0133 \left(200 \text{ ft} \times \frac{1}{3 \text{ in.}} \times \frac{12 \text{ in.}}{\text{ft}} + 30 \right) + 1 \right] \left(\frac{493}{1500} \right)^2 \left(\frac{6}{3} \right)^4 \right\} \frac{\text{lbf} \cdot \text{sec}^2}{\text{slug} \cdot \text{ft}}$$

$$\times \frac{\text{ft}^2}{144 \text{ in.}^2}$$

$$p_{i_1} = 52.3 \text{ lbf/in.}^2$$

Similar calculations for Branches ② and ③ give

$$p_{i_2} = 66.3 \text{ lbf/in.}^2 \quad \text{and} \quad p_{i_3} = 43.3 \text{ lbf/in.}^2 \qquad\qquad p_i$$

{ This problem illustrates the general method used to solve multiple-path pipe flow problems. }

PART C   FLOW MEASUREMENT

The choice of a flow meter is influenced by the accuracy required, range, cost, complication, ease of reading or data reduction, and service life. The simplest and cheapest device that gives the desired accuracy should be chosen.

8-9   DIRECT METHODS

Tanks can be used to determine flow rate for steady liquid flows by measuring the volume or mass of liquid collected during a known time interval. If the time interval is long enough to be measured accurately, flow rates may be determined precisely in this way.

Compressibility must be considered in volume measurements for gas flows. The densities of gases generally are too small to permit accurate direct measurement of mass flow rate. However, a volume sample often can be collected by displacing a "bell," or inverted jar over water (if the pressure is held constant by counterweights). If volume or mass measurements are set up carefully, no calibration is required; this is a great advantage of direct methods.

In specialized applications, particularly for remote or recording uses, positive displacement flow meters may be specified. Common examples include household water and natural gas meters, which are calibrated to read directly in units of product, or gasoline metering pumps, which measure total flow and automatically compute the cost. Many positive-displacement meters are available commercially. Consult manufacturers' literature or References (e.g., [8]) for design and installation details.

8-10   RESTRICTION FLOW METERS FOR INTERNAL FLOWS

Most restriction flow meters for internal flow (except the laminar flow element, Section 8-10.4) are based on acceleration of a fluid stream through some form of nozzle, as shown schematically in Fig. 8.21. Flow separation at the sharp edge of the nozzle throat causes a recirculation zone to form, as shown by the dashed lines downstream from the nozzle. The mainstream flow continues to accelerate from the nozzle throat to form a *vena contracta* at section ② and then decelerates again to fill the duct. At the vena contracta, the flow area is a minimum, the flow streamlines are essentially straight, and the pressure is uniform across the channel section.

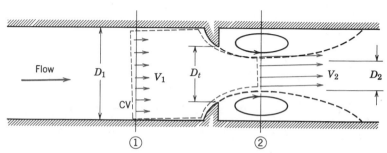

Fig. 8.21   Internal flow through a generalized nozzle, showing control volume used for analysis.

The theoretical flow rate may be related to the pressure differential between sections ① and ② by applying the continuity and Bernoulli equations. Then empirical correction factors may be applied to obtain the actual flow rate.

Basic equations:

$$= 0(1)$$

$$0 = \frac{\partial}{\partial t} \int_{CV} \rho \, d\forall + \int_{CS} \rho \vec{V} \cdot d\vec{A} \tag{4.13}$$

$$\frac{p_1}{\rho} + \frac{V_1^2}{2} + \cancel{gz_1} = \frac{p_2}{\rho} + \frac{V_2^2}{2} + \cancel{gz_2} \tag{6.9}$$

Assumptions: (1) Steady flow
(2) Incompressible flow
(3) Flow along a streamline
(4) No friction
(5) Uniform velocity at sections ① and ②
(6) No streamline curvature at sections ① or ②, so pressure is uniform across those sections
(7) $z_1 = z_2$

Then, from the Bernoulli equation,

$$p_1 - p_2 = \frac{\rho}{2}(V_2^2 - V_1^2) = \frac{\rho V_2^2}{2}\left[1 - \left(\frac{V_1}{V_2}\right)^2\right]$$

and from continuity

$$0 = \{-|\rho V_1 A_1|\} + \{|\rho V_2 A_2|\}$$

or

$$V_1 A_1 = V_2 A_2 \qquad \text{so} \qquad \left(\frac{V_1}{V_2}\right)^2 = \left(\frac{A_2}{A_1}\right)^2$$

Substituting gives

$$p_1 - p_2 = \frac{\rho V_2^2}{2}\left[1 - \left(\frac{A_2}{A_1}\right)^2\right]$$

Solving for the theoretical velocity, V_2,

$$V_2 = \sqrt{\frac{2(p_1 - p_2)}{\rho[1 - (A_2/A_1)^2]}} \tag{8.45}$$

The theoretical flow rate is then given by

$$\dot{m}_{\text{theoretical}} = \rho V_2 A_2 = \rho \sqrt{\frac{2(p_1 - p_2)}{\rho[1 - (A_2/A_1)^2]}} \, A_2$$

or

$$\dot{m}_{\text{theoretical}} = \frac{A_2}{\sqrt{1 - (A_2/A_1)^2}} \sqrt{2\rho(p_1 - p_2)} \tag{8.46}$$

Equation 8.46 shows the general relationship between mass flow rate and pressure drop for a restriction flow meter: Mass flow rate is proportional to the square root of the pressure differential across the meter taps. This relationship limits the flow rates that can be measured accurately to approximately a 4:1 range.

Several factors limit the utility of Eq. 8.46 for calculating the actual mass flow rate through a meter. The actual flow area at section ② is unknown when the vena contracta is pronounced (e.g., for orifice plates when D_t is a small fraction of D_1). The velocity profiles approach uniform flow only at very large Reynolds numbers. Frictional effects can become important (especially downstream from the meter) when the meter contours are abrupt. Finally, the location of pressure taps influences the differential pressure reading, $p_1 - p_2$.

The theoretical equation is adjusted for Reynolds number and diameter ratio by defining an empirical *discharge coefficient* as

$$C \equiv \frac{\text{actual mass flow rate}}{\text{theoretical mass flow rate}} \tag{8.47}$$

Using the discharge coefficient, the actual mass flow rate is expressed as

$$\dot{m}_{\text{actual}} = \frac{CA_t}{\sqrt{1 - (A_t/A_1)^2}} \sqrt{2\rho(p_1 - p_2)} \tag{8.48}$$

or with $\beta = D_t/D_1$, then $(A_t/A_1)^2 = (D_t/D_1)^4 = \beta^4$, so

$$\dot{m}_{\text{actual}} = \frac{CA_t}{\sqrt{1 - \beta^4}} \sqrt{2\rho(p_1 - p_2)} \tag{8.49}$$

In Eq. 8.49, $1/\sqrt{1 - \beta^4}$ is the *velocity of approach factor*. The discharge coefficient and velocity of approach factor frequently are combined into a single *flow coefficient,*

$$K \equiv \frac{C}{\sqrt{1 - \beta^4}} \tag{8.50}$$

In terms of the flow coefficient, the actual mass flow rate is expressed as

$$\dot{m}_{\text{actual}} = KA_t \sqrt{2\rho(p_1 - p_2)} \tag{8.51}$$

For standardized metering elements, test data [8, 24] have been used to develop empirical equations that predict discharge and flow coefficients from meter bore, pipe diameter, and Reynolds number. The accuracy of the equations (within specified ranges) usually is adequate so that the meter can be used without calibration. If the Reynolds number, pipe size, or bore diameter are outside the specified range of the equation, the coefficients should be measured experimentally.

For the turbulent flow regime (pipe Reynolds number greater than 4000) the discharge coefficient may be expressed by an equation of the form [8]

$$C = C_\infty + \frac{b}{Re_{D_1}^n} \tag{8.52}$$

The corresponding form for the flow coefficient equation is

$$K = K_\infty + \frac{1}{\sqrt{1 - \beta^4}} \frac{b}{Re_{D_1}^n} \tag{8.53}$$

In Eqs. 8.52 and 8.53, subscript ∞ denotes the coefficient at infinite Reynolds number; constants b and n allow for scaling to finite Reynolds numbers. Correlating equations and curves of coefficients versus Reynolds number are given for specific metering elements in the next three subsections, following a general comparison of their characteristics.

As we have noted, selection of a flow meter depends on factors such as cost, accuracy, need for calibration, and ease of installation and maintenance. Some of these factors are compared for orifice plate, flow nozzle, and venturi meters in Table 8.5.

Table 8.5 Characteristics of Orifice, Flow Nozzle, and Venturi Flow Meters

Flow Meter Type	Diagram	Head Loss	Cost
Orifice		High	Low
Flow Nozzle		Intermediate	Intermediate
Venturi		Low	High

Flow meter coefficients reported in the literature have been measured with fully developed turbulent velocity distributions at the meter inlet (Section ①). If a flow meter is to be installed downstream from a valve, elbow, or other disturbance, a straight section of pipe must be placed in front of the meter. Approximately 10 diameters of straight pipe are required for venturi meters, and up to 40 diameters for orifice plate or flow nozzle meters. When a meter has been properly installed, the flow rate may be computed from Eqs. 8.49 or 8.51, after choosing an appropriate value for the empirical discharge coefficient, C, or flow coefficient, K, defined in Eqs. 8.47 and 8.50. Some design data for incompressible flow are given in the next few sections. The same basic methods can be extended to compressible flows, but these will not be treated here. For complete details, see [8] or [24].

8-10.1 The Orifice Plate

The orifice plate (Fig. 8.22) is a thin plate that may be clamped between pipe flanges. Since its geometry is simple, it is low in cost and easy to install or replace. The sharp edge of the orifice will not foul with scale or suspended matter. However, suspended matter can build up at the inlet side of a concentric orifice in a horizontal pipe; an eccentric orifice may be placed flush with the bottom of the pipe to avoid this difficulty. The primary disadvantages of the orifice are its limited capacity and the high permanent head loss due to the uncontrolled expansion downstream from the metering element.

Pressure taps for orifices may be placed in several locations, as shown in Fig. 8.22 (see [8] or [24] for additional details). Since the location of the pressure taps

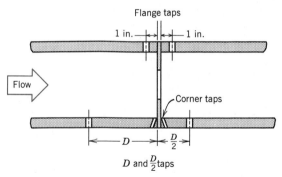

Fig. 8.22   Orifice geometry and pressure tap locations [8].

influences the empirically determined flow coefficient, one must select handbook values of C or K consistent with the location of pressure taps.

The correlating equation recommended for a concentric orifice with corner taps [8] is

$$C = 0.5959 + 0.0312\beta^{2.1} - 0.184\beta^8 + \frac{91.71\beta^{2.5}}{Re_{D_1}^{0.75}}$$ (8.54)

Equation 8.54 predicts orifice discharge coefficients within ±0.6 percent for $0.2 < \beta < 0.75$ and for $10^4 < Re_{D_1} < 10^7$. Some flow coefficients calculated from Eqs. 8.54 and 8.50 are presented in Fig. 8.23. Flow coefficients are relatively insensitive to Reynolds number for Re_{D_1} above 10^5 when $\beta > 0.5$.

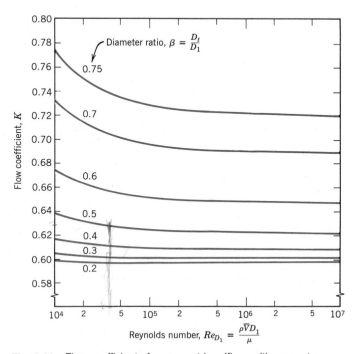

Fig. 8.23   Flow coefficients for concentric orifices with corner taps.

A similar correlating equation is available for orifice plates with D and $D/2$ taps. Flange taps require a different correlation for every line size. Pipe taps, located at $2\frac{1}{2}$ and $8\,D$, no longer are recommended for accurate work.

Example Problem 8.12, which appears later in this section, illustrates the application of flow coefficient data to orifice sizing.

8-10.2 The Flow Nozzle

Flow nozzles may be used as metering elements in either plenums or ducts, as shown in Fig. 8.24; the nozzle section is approximately a quarter ellipse. Design details and recommended locations for pressure taps are given in [24].

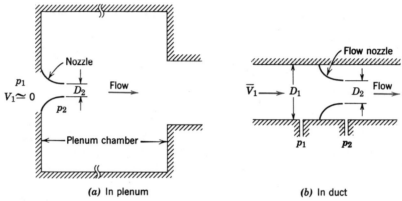

(a) In plenum   (b) In duct

Fig. 8.24 Typical installations of nozzle flow meters.

The correlating equation recommended for an ASME long-radius flow nozzle [8] is

$$C = 0.9975 - \frac{6.53\beta^{0.5}}{Re_{D_1}^{0.5}} \tag{8.55}$$

Equation 8.55 predicts discharge coefficients for flow nozzles within ± 2.0 percent for $0.25 < \beta < 0.75$ for $10^4 < Re_{D_1} < 10^7$. Some flow coefficients calculated from Eq. 8.55 and Eq. 8.50 are presented in Fig. 8.25. (K can be greater than one when the velocity of approach factor exceeds one.)

a. Pipe Installation

For pipe installation, K must be read from Fig. 8.25. Figure 8.25 shows that K is essentially independent of Reynolds number for $Re_{D_1} > 10^6$. Thus at high flow rates, the flow rate may be computed directly using Eq. 8.51. At lower flow rates, where K is a weak function of Reynolds number, iteration may be required.

b. Plenum Installation

For plenum installation, nozzles may be fabricated from spun aluminum, molded fiberglass, or other inexpensive materials. Thus they are simple and cheap to make and install. Since the plenum pressure is equal to p_2, the location of the downstream

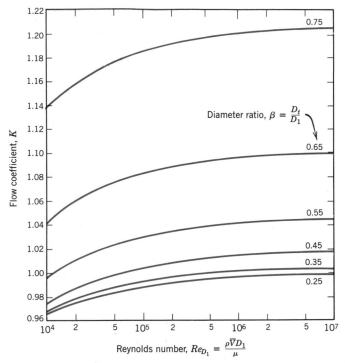

Fig. 8.25 Flow coefficients for ASME long-radius flow nozzles.

pressure tap is not critical. Meters suitable for a wide range of flow rates may be made by installing several nozzles in a plenum. At low flow rates, most of them may be plugged. For larger flow rates, more nozzles may be used.

For plenum nozzles, typical flow coefficients are in the range, $0.95 < K < 0.99$; the larger values apply at high Reynolds numbers. Thus the mass rate of flow can be computed within approximately ±2 percent using Eq. 8.51 with $K = 0.97$.

8-10.3 The Venturi

Venturi meters, as sketched in Table 8.5, are generally made from castings and machined to close tolerances to duplicate the performance of the standard design. As a result, venturi meters are heavy, bulky, and expensive. The conical diffuser section downstream from the throat gives excellent pressure recovery; therefore, overall head loss is low. Venturi meters are also self-cleaning because of their smooth internal contours.

Experimental data show that discharge coefficients for venturi meters range from 0.980 to 0.995 at high Reynolds numbers ($Re_{D_1} > 2 \times 10^5$). Thus $C = 0.99$ can be used to measure mass flow rate within about ±1 percent at high Reynolds number [8]. Manufacturers' literature should be consulted for specific information at Reynolds numbers below 10^5.

The orifice plate, flow nozzle, and venturi all produce pressure drops proportional to the square of the flow rate, according to Eq. 8.51. In practice, a meter size must be chosen to accommodate the largest flow rate expected. Because the pressure drop versus flow rate relationship is nonlinear, a limited range of flow rate can be measured accurately. Flow meters with single throats usually are considered for flow rates over a 4:1 range [8].

The unrecoverable loss in head across a metering element may be expressed as a fraction of the differential pressure, Δp, across the element. Pressure losses are displayed as functions of diameter ratio in Fig. 8.26 [8].

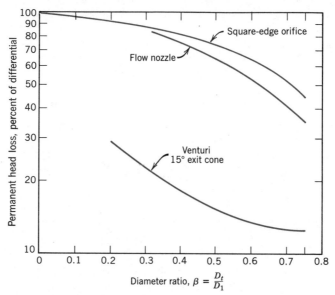

Fig. 8.26 Permanent head loss produced by various flow metering elements [8].

8-10.4 The Laminar Flow Element

The laminar flow element[5] is designed to produce a pressure differential directly proportional to flow rate. The laminar flow element (LFE) contains a metering section subdivided into many passages, each small enough in diameter to assure fully developed laminar flow. As shown in Section 8-3, the pressure drop in laminar duct flow is directly proportional to flow rate. Because the pressure drop versus flow rate relationship is linear, the LFE may be used with reasonable accuracy over a 10:1 range of flow rate. The relationship between the pressure drop and flow rate for laminar flow also depends on viscosity, which is a strong function of temperature. Therefore, the fluid temperature must be known to obtain accurate metering with a LFE.

A laminar flow element costs approximately as much as a venturi, but it is much lighter and smaller. Thus the LFE is becoming widely used in applications where compactness and extended range are important.

EXAMPLE 8.12—Flow through an Orifice Meter

An air flow rate of 1 m^3/sec at standard conditions is expected in a 0.25 m diameter duct. An orifice meter is used to measure the rate of flow. The manometer available to make the measurement has a maximum range of 300 mm of water. What diameter orifice plate should be used with corner taps? Analyze the head loss if the flow area at the vena contracta is $A_2 = 0.65 \, A_t$. Compare with data from Fig. 8.26.

---

[5] Patented and manufactured by Meriam Instrument Co., 10920 Madison Ave., Cleveland, Ohio.

EXAMPLE PROBLEM 8.12

GIVEN: Flow through duct and orifice as shown.

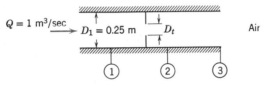

$$Q = 1 \text{ m}^3/\text{sec} \quad D_1 = 0.25 \text{ m} \quad D_t \qquad \text{Air}$$

$$(p_1 - p_2)_{max} = 300 \text{ mm H}_2\text{O}$$

FIND: (a) D_t.
(b) Head loss between sections ① and ③.
(c) Compare with data from Fig. 8.26.

SOLUTION:
The orifice plate may be designed using Eq. 8.51 and data from Fig. 8.23.

Computing equation: $\dot{m}_{actual} = KA_t\sqrt{2\rho(p_1 - p_2)}$ (8.51)

Assumptions: (1) Steady flow
(2) Incompressible flow

Since $A_t/A_1 = (D_t/D_1)^2 = \beta^2$,

$$\dot{m}_{actual} = K\beta^2 A_1\sqrt{2\rho(p_1 - p_2)}$$

or

$$K\beta^2 = \frac{\dot{m}_{actual}}{A_1\sqrt{2\rho(p_1 - p_2)}} = \frac{\rho Q}{A_1\sqrt{2\rho(p_1 - p_2)}} = \frac{Q}{A_1}\sqrt{\frac{\rho}{2(p_1 - p_2)}}$$

$$= \frac{Q}{A_1}\sqrt{\frac{\rho}{2g\rho_{H_2O}\Delta h}}$$

$$= \frac{1 \text{ m}^3}{\text{sec}} \times \frac{4}{\pi} \frac{1}{(0.25)^2 \text{ m}^2}\left[\frac{1}{2} \times \frac{1.23 \text{ kg}}{\text{m}^3} \times \frac{\text{sec}^2}{9.81 \text{ m}} \times \frac{\text{m}^3}{999 \text{ kg}} \times \frac{1}{0.30 \text{ m}}\right]^{1/2}$$

$$K\beta^2 = 0.295 \quad \text{or} \quad K = \frac{0.295}{\beta^2} \qquad (1)$$

Since K is a function of both β and Re_{D_1}, we must iterate to find β. The duct Reynolds number is

$$Re_{D_1} = \frac{\rho\bar{V}_1 D_1}{\mu} = \frac{\rho(Q/A_1)D_1}{\mu} = \frac{4Q}{\pi\nu D_1}$$

$$Re_{D_1} = \frac{4}{\pi} \times \frac{1 \text{ m}^3}{\text{sec}} \times \frac{\text{sec}}{1.45\times 10^{-5} \text{ m}^2} \times \frac{1}{0.25 \text{ m}} = 3.51\times 10^5$$

Guess $\beta = 0.75$. From Fig. 8.23, K should be 0.72. From Eq. 1,

$$K = \frac{0.295}{(0.75)^2} = 0.524$$

Thus our guess for β is too large. Guess $\beta = 0.70$. From Fig. 8.23, K should be 0.69. From Eq. 1,

$$K = \frac{0.295}{(0.70)^2} = 0.602$$

Thus our guess for β is still too large. Guess $\beta = 0.65$. From Fig. 8.23, K should be 0.67. From Eq. 1,

$$K = \frac{0.295}{(0.65)^2} = 0.698$$

There is satisfactory agreement with $\beta \simeq 0.66$ and

$$D_t = \beta D_1 = 0.66(0.25 \text{ m}) = 0.165 \text{ m} \qquad\qquad D_t$$

To evaluate the permanent head loss, apply Eq. 8.28 between sections ① and ③. Computing equation:

$$\left(\frac{p_1}{\rho} + \alpha_1 \frac{\bar{V}_1^2}{2} + g\cancel{z_1}\right) - \left(\frac{p_3}{\rho} + \alpha_3 \frac{\bar{V}_3^2}{2} + g\cancel{z_3}\right) = h_{l_T} \qquad (8.28)$$

Assumptions: (3) $\alpha_1 \bar{V}_1^2 = \alpha_3 \bar{V}_3^2$
(4) Neglect Δz

Then

$$h_{l_T} = \frac{p_1 - p_3}{\rho} = \frac{p_1 - p_2 - (p_3 - p_2)}{\rho} \qquad (2)$$

The pressure at ③ may be found by applying the x component of the momentum equation to a control volume between sections ② and ③.

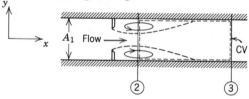

$A_2 = A_{\text{vena contracta}}$

Basic equation: $= 0(5) = 0(1)$

$$F_{S_x} + \cancel{F_{B_x}} = \cancel{\frac{\partial}{\partial t} \int_{\text{CV}}} u\rho \, d\Psi + \int_{\text{CS}} u\rho \vec{V} \cdot d\vec{A} \qquad (4.19a)$$

Assumptions: (5) $F_{B_x} = 0$
(6) Uniform flow at sections ② and ③
(7) Pressure uniform across duct at sections ② and ③
(8) Neglect friction force on CV

Then

$$(p_2 - p_3)A_1 = u_2\{-|\rho \bar{V}_2 A_2|\} + u_3\{|\rho \bar{V}_3 A_3|\} = (u_3 - u_2)\rho Q = (\bar{V}_3 - \bar{V}_2)\rho Q$$

or

$$p_3 - p_2 = (\bar{V}_2 - \bar{V}_3)\frac{\rho Q}{A_1}$$

Now $\bar{V}_3 = Q/A_1$, and

$$\bar{V}_2 = \frac{Q}{A_2} = \frac{Q}{0.65 A_t} = \frac{Q}{0.65 \beta^2 A_1}$$

Thus

$$p_3 - p_2 = \frac{\rho Q^2}{A_1^2}\left[\frac{1}{0.65\beta^2} - 1\right]$$

$$p_3 - p_2 = \frac{1.23 \text{ kg}}{\text{m}^3} \times \frac{(1)^2 \text{ m}^6}{\text{sec}^2} \times \frac{4^2}{\pi^2} \frac{1}{(0.25)^4 \text{ m}^4} \left[\frac{1}{0.65(0.66)^2} - 1 \right] \frac{\text{N} \cdot \text{sec}^2}{\text{kg} \cdot \text{m}}$$

$$p_3 - p_2 = 1290 \text{ N/m}^2$$

The diameter ratio, β, was selected to give maximum manometer deflection at maximum flow rate. Thus

$$p_1 - p_2 = \rho_{\text{H}_2\text{O}} g \, \Delta h = \frac{999 \text{ kg}}{\text{m}^3} \times \frac{9.81 \text{ m}}{\text{sec}^2} \times 0.30 \text{ m} \times \frac{\text{N} \cdot \text{sec}^2}{\text{kg} \cdot \text{m}} = 2940 \text{ N/m}^2$$

Substituting into Eq. 2 gives

$$h_{l_T} = \frac{p_1 - p_3}{\rho} = \frac{p_1 - p_2 - (p_3 - p_2)}{\rho}$$

$$h_{l_T} = \frac{(2940 - 1290) \text{ N}}{\text{m}^2} \times \frac{\text{m}^3}{1.23 \text{ kg}} = 1340 \text{ N} \cdot \text{m/kg} \qquad\qquad h_{l_T} \longleftarrow$$

To compare with Fig. 8.26, express the permanent pressure loss as a fraction of the meter differential

$$\frac{p_1 - p_3}{p_1 - p_2} = \frac{(2940 - 1290) \text{ N/m}^2}{2940 \text{ N/m}^2} = 0.561$$

The fraction from Fig. 8.26 is about 0.57. This is satisfactory agreement!

> This problem illustrates flow meter calculations and shows use of the momentum equation to compute the pressure rise in a sudden expansion.

8-11   LINEAR FLOW METERS

Several flow meter types produce outputs that are directly proportional to flow rate. These meters produce signals without the need to measure differential pressure. The most common linear flow meters are discussed briefly in the following paragraphs.

Float meters may be used to indicate flow rate directly for liquids or gases. An example is shown in Fig. 8.27. In operation, the ball or float is carried upward in the tapered clear tube by the flowing fluid until the drag force and float weight

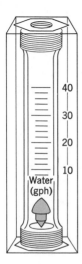

Fig. 8.27   Float-type variable-area flow meter. (Courtesy of Dwyer Instrument Co., Michigan City, Indiana.)

are in equilibrium. Such meters (often called rotameters) are available with factory calibration for a number of common fluids and flow rate ranges.

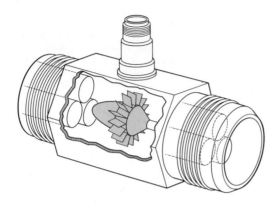

Fig. 8.28   Turbine flow meter. (Courtesy of Potter Aeronautical Corp., Union, New Jersey.)

A free-running vaned impeller may be mounted in a cylindrical section of tube (Fig. 8.28) to make a *turbine flow meter*. With proper design, the rate of rotation of the impeller may be made closely proportional to the volume flow rate over a wide range.

Rotational speed of the turbine element can be sensed using a magnetic or modulated carrier pickup external to the meter. This sensing method therefore requires no penetrations or seals in the duct. Thus turbine flow meters can be used safely to measure flow rates in corrosive or toxic fluids. The electrical signal can be displayed, recorded, or integrated to provide total flow information.

Vortex shedding from a bluff obstruction may be used to meter flow. As noted in Chapter 7, the Strouhal number, $St = fL/V$, is approximately constant ($St \approx 0.21$). Thus frequency of vortex shedding, f, is proportional to flow velocity. The vortex shedding from the obstruction results in velocity and pressure changes around and downstream from the obstruction. Pressure, thermal, or ultrasonic sensors may be used to detect the vortex shedding frequency, and thus to infer the fluid velocity. (The velocity profile does affect the constancy of the shedding frequency.) Vortex flow meters can be used over a 20:1 range of flow rates [8].

The electromagnetic flow meter uses the principle of magnetic induction. A magnetic field is created across a pipe. When a conductive fluid passes through the field, a voltage is generated at right angles to the field and velocity vectors. Electrodes placed on a pipe diameter are used to detect the resulting signal voltage. The signal voltage is proportional to the average axial velocity when the profile is axisymmetric.

Magnetic flow meters may be used with liquids that have electrical conductivities above 100 microsiemens per meter (1 siemen = 1 ampere per volt). The minimum flow speed should be above about 0.3 m/sec, but there are no restrictions on Reynolds number. The flow range normally quoted is 10:1 [8].

Ultrasonic flow meters also respond to average velocity at a pipe cross section. Two principal types of ultrasonic meters are common: propagation time is measured for clean liquids, and reflection frequency shift (Doppler effect) is measured for flows carrying particulates. The speed of an acoustic wave increases in the flow direction and decreases when transmitted against the flow. For clean liquids, an acoustic path inclined to the pipe axis is used to infer flow velocity. In practice, multiple paths are used to estimate the volume flow rate accurately.

Doppler effect ultrasonic flow meters depend on reflection of sonic waves (in the MHz range) from scattering particles in the fluid. When the particles move at flow speed, the frequency shift is proportional to flow speed; for a suitably chosen path, output is proportional to volume flow rate. One or two transducers may be used; the meter may be clamped to the outside of the pipe. Ultrasonic meters may require calibration in place. Flow rate range is 10:1 [8].

8-12 TRAVERSING METHODS

In some situations, e.g., in air handling or refrigeration equipment, it is impractical or impossible to install fixed flow meters. In such cases it may be possible to obtain flow rate data using traversing techniques.

To make a flow rate measurement by traverse, the duct cross section is subdivided into segments of equal area. The velocity is measured at the center of each area segment using a pitot tube, a total head tube, or a suitable anemometer. The volume flow rate for each segment is approximated by the product of the measured velocity and segment area. The flow rate through the entire duct is the sum of these segmental flow rates. Details of recommended procedures for flow rate measurements by the traverse method are given in [25].

Use of pitot or pitot-static tubes for traverse measurements requires direct access to the flow field. Pitot tubes give uncertain results when pressure gradients or streamline curvature are present, and their response times are slow. Two types of anemometers — thermal anemometers and laser Doppler anemometers — overcome these difficulties partially, although they introduce new complications.

Thermal anemometers use tiny elements (either hot-wire or hot-film elements) that are heated electrically. Sophisticated electronic feedback circuits are used to maintain the temperature of the element constant and to sense the input heating rate. The heating rate may be related to the local flow velocity by calibration. The primary advantage of thermal anemometers is the small size of the sensing element. Sensors as small as 0.002 mm in diameter and 0.1 mm long are available commercially. Because the thermal mass of such tiny elements is extremely small, their response to fluctuations in flow velocity is rapid. Frequency responses to the 50 kHz range have been quoted [26]. Thus thermal anemometers are ideal for measurement of turbulence quantities. Insulating coatings may be applied to permit their use in conductive or corrosive gases or liquids.

Because of their fast response and small size, thermal anemometers are used extensively for research. Numerous schemes have been published for treating the resulting data. Digital processing techniques, including fast Fourier transforms, can be applied to the signals to obtain mean values and moments, and to analyze frequency content and correlations.

Laser Doppler anemometers (LDAs) are becoming widely used for specialized applications where direct physical access to the flow field is difficult or impossible. One or more laser beams are focused to a small volume in the flow at the location of interest. Laser light is scattered from particles that are present in the flow (dust or particulates) or introduced for this purpose. A frequency shift is caused by the local flow speed (Doppler effect). Scattered light and a reference beam are collected by receiving optics. The frequency shift is proportional to the flow speed; this relationship may be calculated, so there is no need for calibration. Since velocity is measured directly, the signal is unaffected by changes in temperature, density, or composition in the flow field. The primary disadvantages of LDAs are that the optical equipment is expensive and fragile, and that extremely careful alignment is needed.

8-13 SUMMARY OBJECTIVES

After completing study of Chapter 8, you should be able to do the following:

1. Define:

internal flow   kinetic energy flux coefficient
entrance length   momentum flux coefficient
fully developed flow   head loss
friction velocity   major, minor head loss
Reynolds stress   hydraulic diameter

2. For fully developed laminar flows, apply the control volume formulation of Newton's second law to a suitably chosen differential control volume to determine: the velocity distribution, the shear stress distribution, the volume flow rate, the average velocity, and the location of the maximum velocity.

3. For fully developed flow in a pipe, determine the wall shear stress and the shear stress variation in the flow in terms of the pressure gradient.

4. For fully developed turbulent flow in a pipe, with velocity distribution represented by a power-law profile, determine $\bar{V}/U$.

5. Write the first law of thermodynamics in a form suitable for the solution of pipe flow problems.

6. Use the Moody diagram to determine the friction factor for fully developed pipe flow.

7. Solve single-path pipe flow system problems for each of the four cases discussed in Section 8-8.1; sketch pressure distribution, hydraulic, and energy grade lines.

**8. Use the basic techniques of Section 8-8.1 to analyze multiple-path pipe flow systems.

9. Determine the mass flow rate from the pressure differential measured using an orifice plate, flow nozzle, or venturi meter.

10. Solve the problems at the end of the chapter that relate to the material you have studied.

REFERENCES

1. Streeter, V. L., ed., *Handbook of Fluid Dynamics*. New York: McGraw-Hill, 1961.
2. Rouse, H., and S. Ince, *History of Hydraulics*. New York: Dover, 1957.
3. Laufer, J., "The Structure of Turbulence in Fully Developed Pipe Flow," U.S. National Advisory Committee for Aeronautics (NACA), Technical Report 1174, 1954.
4. Tennekes, H., and J. L. Lumley, *A First Course in Turbulence*. Cambridge, MA: The MIT Press, 1972.
5. Hinze, J. O., *Turbulence,* 2nd ed. New York: McGraw-Hill, 1975.
6. Moody, L. F., "Friction Factors for Pipe Flow," *Transactions of the ASME, 66,* 8, November 1944, pp. 671–684.

** This objective applies to a section that may be omitted without loss of continuity in the text material.

7. Colebrook, C. F., "Turbulent Flow in Pipes, with Particular Reference to the Transition Region between the Smooth and Rough Pipe Laws," *Journal of the Institution of Civil Engineers, London, 11,* 1938–39, pp. 133–156.

8. Miller, R. W., *Flow Measurement Engineering Handbook,* 2nd ed. New York: McGraw-Hill, 1985.

9. Swamee, P. K., and A. K. Jain, "Explicit Equations for Pipe-Flow Problems," *Proceedings of the ASCE, Journal of the Hydraulics Division, 102,* HY5, May 1976, pp. 657–664.

10. "Flow of Fluids through Valves, Fittings, and Pipe," New York: Crane Company, Technical Paper No. 410, 1982.

11. *ASHRAE Handbook—Fundamentals.* Atlanta, GA: American Society of Heating, Refrigerating, and Air Conditioning Engineers, Inc., 1981.

12. Cockrell, D. J., and C. I. Bradley, "The Response of Diffusers to Flow Conditions at Their Inlet," Paper No. 5, *Symposium on Internal Flows,* University of Salford, Salford, England, April 1971, pp. A32–A41.

13. Sovran, G., and E. D. Klomp, "Experimentally Determined Optimum Geometries for Rectilinear Diffusers with Rectangular, Conical, or Annular Cross-Sections," in *Fluid Mechanics of Internal Flows,* G. Sovran, ed. Amsterdam: Elsevier, 1967.

14. Feiereisen, W. J., R. W. Fox, and A. T. McDonald, "An Experimental Investigation of Incompressible Flow without Swirl in R-Radial Diffusers," Proceedings, Second International Japan Society of Mechanical Engineers Symposium on Fluid Machinery and Fluidics, Tokyo, Japan, September 4–9, 1972, pp. 81–90.

15. McDonald, A. T., and R. W. Fox, "An Experimental Investigation of Incompressible Flow in Conical Diffusers," *International Journal of Mechanical Sciences, 8,* 2, February 1966, pp. 125–139.

16. Runstadler, P. W., Jr., "Diffuser Data Book," Hanover, NH: Creare, Inc., Technical Note 186, 1975.

17. Reneau, L. R., J. P. Johnston, and S. J. Kline, "Performance and Design of Straight, Two-Dimensional Diffusers," *Transactions of the ASME, Journal of Basic Engineering, 89D,* 1, March 1967, pp. 141–150.

18. *Aerospace Applied Thermodynamics Manual.* New York: Society of Automotive Engineers, 1969.

19. Daily, J. W., and D. R. F. Harleman, *Fluid Dynamics.* Reading, MA: Addison-Wesley, 1966.

20. White, F. M., *Fluid Mechanics,* 2nd ed. New York: McGraw-Hill, 1986.

21. Hamilton, J. B., "The Suppression of Intake Losses by Various Degrees of Rounding," University of Washington, Seattle, WA, Experiment Station Bulletin 51, 1929.

22. Herschel, C., *The Two Books on the Water Supply of the City of Rome, from Sextus Julius Frontinus* (ca. 40–103 A.D.). Boston, 1899.

23. Lam, C. F., and M. L. Wolla, "Computer Analysis of Water Distribution Systems: Part 1, Formulation of Equations," *Proceedings of the ASCE, Journal of the Hydraulics Division, 98,* HY2, February 1972, pp. 335–344.

24. Bean, H. S., ed., *Fluid Meters, Their Theory and Application.* New York: American Society of Mechanical Engineers, 1971.

25. ISO 7145, *Determination of Flowrate of Fluids in Closed Conduits or Circular Cross Sections—Method of Velocity Determination at One Point in the Cross Section,* ISO UDC 532.57.082.25:532.542, 1st ed. Geneva: International Standards Organization, 1982.

26. Goldstein, R. J., ed., *Fluid Mechanics Measurements.* Washington, D.C.: Hemisphere, 1983.

27. Potter, M. C., and J. F. Foss, *Fluid Mechanics.* New York: Ronald, 1975.

PROBLEMS

8.1 Standard air enters a 0.3 m diameter duct. The volume flow rate is 2 m^3/min. Determine whether the flow is laminar or turbulent. Estimate the entrance length required to establish fully developed flow.

8.2 For flow in circular tubes, transition to turbulent flow for engineering systems seldom is delayed past $Re = 2300$. On a log-log plot of average velocity versus tube diameter, plot lines that correspond to $Re = 2300$ for (a) standard air and (b) water at 15 C.

8.3 For flow in circular tubes, transition to turbulent flow for engineering systems seldom is delayed past $Re = 2300$. On a log-log plot of volume flow rate versus tube diameter, plot lines that correspond to $Re = 2300$ for (a) standard air and (b) water at 15 C.

8.4 For flow in circular tubes, transition to turbulent flow for engineering systems seldom is delayed past $Re = 2300$. On a log-log plot of mass flow rate versus tube diameter, plot lines that correspond to $Re = 2300$ for (a) standard air and (b) water at 15 C.

8.5 For laminar flow in a 12.7 mm diameter pipe, determine (a) the maximum allowable volume flow rate if the fluid is water and (b) the maximum average velocity if the fluid is air. What is the corresponding entrance length?

8.6 Consider incompressible flow in a circular channel. Derive general expressions for Reynolds number in terms of (a) volume flow rate and tube diameter and (b) mass flow rate and tube diameter. The Reynolds number is 1800 in a section where the tube diameter is 10 mm. Find the Reynolds number for the same flow rate in a section where the tube diameter is 6 mm.

8.7 The velocity profile for fully developed flow between stationary parallel plates is given by $u = ay(h - y)$, where a is a constant, h is the total gap width between plates, and y is the distance measured upward from the lower plate. Determine the ratio $\bar{V}/u_{max}$.

8.8 An incompressible fluid flows between two infinite stationary parallel plates. The velocity profile is given by $u = u_{max}(Ay^2 + By + C)$, where A, B, and C are constants and y is measured from the center of the gap. The total gap width is h units. Use appropriate boundary conditions to express the magnitude and units of the constants in terms of h. Develop an expression for volume flow rate per unit depth and evaluate the ratio $\bar{V}/u_{max}$.

8.9 A viscous oil flows steadily between parallel plates. The flow is laminar and fully developed. The total gap width between the plates is $h = 3$ mm. The oil viscosity is 0.5 N·sec/m^2 and the pressure gradient is -1200 N/m^2/m. Find the magnitude and direction of the shear stress on the upper plate and the volume flow rate through the channel, per meter of width.

8.10 Viscous oil flows steadily between parallel plates. The flow is fully developed and laminar. The pressure gradient is -8 lbf/ft^2/ft and the channel half-width is $h = 0.06$ in. Calculate the magnitude and direction of the wall shear stress at the upper plate surface. Find the volume flow rate through the channel ($\mu = 0.01$ lbf·sec/ft^2).

8.11 A fluid flows steadily between two parallel plates. The flow is fully developed and laminar. The distance between the plates is h.
   (a) Derive an equation for the shear stress as a function of y. Plot this function.
   (b) For $\mu = 2.4 \times 10^{-5}$ lbf·sec/ft^2, $\partial p/\partial x = -4.0$ lbf/ft^2/ft, and $h = 0.05$ in., calculate the maximum shear stress, in lbf/ft^2.

8.12 A hydraulic jack supports a load of 20,000 pounds. The following data are given:

Diameter of piston	4.0 in.
Radial clearance between piston and cylinder	0.002 in.
Length of piston	4.8 in.

Estimate the rate of leakage of hydraulic fluid past the piston, assuming the fluid is SAE 30 oil at 80 F.

8.13 Oil is confined in a 4 in. diameter cylinder by a piston having a radial clearance of 0.001 in. and a length of 2 in. A steady force of 5000 pounds is applied to the piston. Assume the properties of SAE 30 oil at 120 F. Estimate the rate at which oil leaks past the piston.

8.14 A high pressure in a system is created by a small piston-cylinder assembly. The piston diameter is 0.25 in. and it extends 2 in. into the cylinder. The radial clearance between the piston and cylinder is 10^{-4} in. Neglect elastic deformations of the piston and cylinder due to pressure. Assume the fluid properties are those of SAE 10W oil at 100 F. When the pressure in the cylinder is 100,000 psi, estimate the leakage rate.

8.15 A hydrostatic bearing is to support a load of 3600 pounds per foot of length perpendicular to the diagram. The bearing is supplied with SAE 30 oil at 100 F and 100 psig through the central slit. Since the oil is viscous and the gap is small, the flow may be considered fully developed. Calculate (a) the required width of the bearing pad, (b) the resulting pressure gradient, dp/dx, and (c) the gap height, if $Q = 0.0006$ ft^3/min per foot of length.

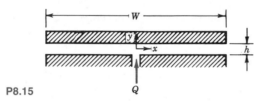

P8.15

8.16 The basic component of a pressure gage tester consists of a piston-cylinder apparatus as shown. The piston, 6 mm in diameter, is loaded to develop a pressure of known magnitude. (The piston length is 25 mm.) Calculate the mass, M, required to produce 1.5 MPa (gage) in the cylinder. Determine the leakage flow rate as a function of radial clearance, a, for this load if the liquid is SAE 30 oil at 20 C. Specify the maximum allowable radial clearance so the vertical movement of the piston due to leakage will be less than 1 mm/min.

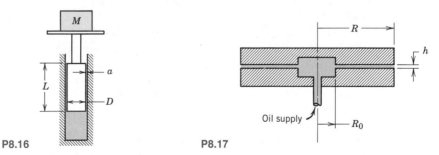

P8.16   **P8.17**

8.17 Viscous liquid, at volume flow rate Q, is pumped through the central opening into the narrow gap between the parallel disks shown. The flow rate is low, so the flow is laminar, and the pressure gradient due to convective acceleration in the gap is negligible compared to the gradient due to viscous forces (this is termed *creeping flow*). Obtain a general expression for the variation of average velocity in the gap between the disks. For creeping flow, the velocity profile at any cross section in the gap is the same as for fully developed flow between stationary parallel plates. Evaluate the pressure gradient, dp/dr, as a function of radius. Obtain an expression for $p(r)$. Show that the net force required to hold the upper plate in the position shown is

$$F = \frac{3\mu Q R^2}{h^3}\left[1-\left(\frac{R_0}{R}\right)^2\right]$$

8.18 Two immiscible layers of viscous liquid flow between large parallel plates. The space between $y = 0$ and $y = b$ is filled with liquid 1, having properties ρ_1 and μ_1. The space between $y = -b$ and $y = 0$ is filled with liquid 2, having properties ρ_2 and μ_2. Find the velocity distribution if the pressure gradient is $\partial p/\partial x = -K$.

8.19 Consider the simple power-law model for a non-Newtonian fluid given by Eq. 2.11. Extend the analysis of Section 8-2.1 to show that the velocity profile for fully developed laminar flow of a power-law fluid between parallel plates separated by distance $2h$ may be written

$$u = \left(\frac{h}{k}\frac{\Delta p}{L}\right)^{1/n}\frac{nh}{n+1}\left[1-\left(\frac{y}{h}\right)^{\frac{n+1}{n}}\right]$$

where y is the coordinate measured from the channel centerline.

8.20 Evaluate the volume flow rate for fully developed laminar flow of a power-law fluid between stationary parallel plates (Problem 8.19). Show that the velocity profile may be written

$$\frac{u}{\bar{V}} = \frac{2n+1}{n+1}\left[1-\left(\frac{y}{h}\right)^{\frac{n+1}{n}}\right]$$

where y is the coordinate measured from the channel centerline.

8.21 A sealed journal bearing is formed from concentric cylinders. The inner and outer radii are 25 and 26 mm, the journal length is 100 mm, and it turns at 2800 rpm. The gap is filled with oil in laminar motion. The velocity profile is linear across the gap. The torque needed to turn the journal is 0.2 N·m. Calculate the viscosity of the oil. Will the torque increase or decrease with time? Why?

8.22 Consider fully developed laminar flow between infinite parallel plates separated by gap width $d = 0.35$ in. The upper plate moves to the right with speed $U_2 = 2$ ft/sec; the lower plate moves to the left with speed $U_1 = 1$ ft/sec. The pressure gradient in the direction of flow is zero. Develop an expression for the velocity distribution in the gap. Find the volume flow rate per unit depth passing a given cross section.

8.23 Consider fully developed laminar flow between infinite parallel plates separated by gap width h. The upper plate moves to the right with speed U_2; the lower plate moves to the left with speed U_1. The pressure gradient in the direction of flow is zero. The net volume flow rate in the positive x direction is $Q = U_1 h/2$ ft³/sec per unit depth. Find the speed ratio, U_2/U_1.

8.24 Consider incompressible, steady, fully developed laminar flow between infinite parallel plates. The upper plate moves to the right at $U = 3$ mm/sec. There is no pressure variation in the x direction, but there is a constant body force due to an electric field, $\rho B_x = 800$ N/m³. The clearance gap between the plates is $h = 0.1$ mm and the liquid viscosity is 0.02 kg/m·sec. Evaluate the velocity profile, $u(y)$, if $y = 0$ at the stationary lower plate. Compute the volume flow rate past a vertical section.

8.25 Water at 60 C flows between two large flat plates. The lower plate moves to the left at a speed of 0.3 m/sec; the upper plate is stationary. The plate spacing is 3 mm, and the flow is laminar. Determine the pressure gradient required to produce zero net flow at a cross section.

8.26 The record-write head for a computer disk-drive memory storage system floats above the spinning disk on a very thin film of air (the film thickness is 0.5 μm). The head location is 150 mm from the disk centerline; the disk spins at 3600 rpm. The record-write head is 10 mm square. For standard air in the gap between the head and disk, determine (a) the Reynolds number of the flow, (b) the viscous shear stress, and (c) the power required to overcome viscous shear.

8.27 Consider steady, fully developed laminar flow of a viscous liquid down an inclined surface. The liquid layer is of constant thickness, h. Use a suitably chosen differential control volume to obtain the velocity profile. Develop an expression for the volume flow rate.

8.28 A film of water (at 15 C) in steady, laminar motion runs down a long slope, inclined 30° below the horizontal. The thickness of the film is 0.8 mm. Assume that flow is fully developed and at zero pressure gradient. Determine the surface shear stress and the volume flow rate per unit width.

8.29 Water at 60 F flows between parallel plates with gap width $b = 0.01$ ft. The upper plate moves with speed $U = 1$ ft/sec in the positive x direction. The pressure gradient is $\partial p/\partial x = -0.60$ lbf/ft^2/ft. Locate the point of maximum velocity and determine its magnitude (let $y = 0$ at the bottom plate). Sketch the velocity and shear stress distributions. Determine the volume of flow that passes a given cross section ($x =$ constant) in 10 sec.

8.30 Consider steady, fully developed, laminar flow between parallel plates, separated by distance h, with known imposed pressure gradient, $\partial p/\partial x$. Beginning from a differential control volume, derive an expression for the speed, U, of the upper plate when no external force is applied to it.

8.31 A continuous belt, passing upward through a chemical bath at speed U_0, picks up a liquid film of thickness h, density ρ, and viscosity μ. Gravity tends to make the liquid drain down, but the movement of the belt keeps the liquid from running off completely. Assume that the flow is fully developed and laminar with zero pressure gradient, and that the atmosphere produces no shear stress at the outer surface of the film. State clearly the boundary conditions to be satisfied by the velocity at $y = 0$ and $y = h$. Obtain an expression for the velocity profile.

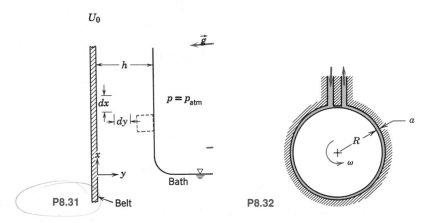

P8.31   P8.32

8.32 A viscous-shear pump is made from a stationary housing with a close-fitting rotating drum inside. The clearance is small compared to the diameter of the drum, so flow in the annular space may be treated as flow between parallel plates. Fluid is dragged around the annulus by viscous forces. Evaluate the performance characteristics of the shear pump (pressure differential, input power, and efficiency) as functions of volume flow rate. Assume that the depth normal to the diagram is b.

8.33 Consider again the viscous shear pump of Problem 8.32. Find the operating point (volume flow rate and pressure rise) that results in optimum efficiency. Express your results in terms of Q/Q_{max}, where Q is volume flow rate.

8.34 The clamping force to hold a part in a metal-turning operation is provided by high pressure oil supplied by a pump. Oil leaks axially through an annular gap with diameter D, length L, and radial clearance a. The inner member of the annulus rotates at angular speed ω. Power is required both to pump the oil and to overcome viscous dissipation in the annular gap. Develop expressions in terms of the specified geometry for the pump power, $\mathcal{P}_p$, and the viscous dissipation power, $\mathcal{P}_v$. Show that the total power requirement is minimized when the radial clearance, a, is chosen such that $\mathcal{P}_v = 3\mathcal{P}_p$.

8.35 For fully developed laminar flow in a pipe, determine the radial distance from the pipe axis at which the velocity equals the average velocity.

8.36 Consider first water and then SAE 10W lubricating oil flowing at 40 C in a 6 mm diameter tube. Determine the maximum flow rate (and the corresponding pressure gradient, $\partial p/\partial x$) for each fluid at which laminar flow would be expected.

8.37 A water injection line is made from smooth capillary tubing with inside diameter $D - 0.25$ mm. Determine the maximum volume flow rate at which flow is laminar. Evaluate the pressure drop required to produce this flow rate through a section of tubing with length $L = 0.75$ m.

8.38 A viscosity measurement setup for an undergraduate fluid mechanics laboratory is to be made from flexible plastic tubing; the fluid is to be water. Assume the tubing diameter is $D = 0.125 \pm 0.010$ in. and the length is 50 ft. Evaluate the maximum volume flow rate at which laminar flow would be expected and the corresponding pressure drop. Estimate the experimental uncertainty in viscosity measured using this apparatus. How could the setup be improved?

8.39 A hypodermic needle, with inside diameter $d = 0.1$ mm and length $L = 25$ mm, is used to inject saline solution with viscosity five times that of water. The plunger diameter is $D = 10$ mm; the maximum force that can be exerted by a thumb on the plunger is $F = 45$ N. Estimate the volume flow rate of saline that can be produced.

8.40 In a commercial viscometer, the volume flow rate is measured by timing the flow of a known volume through a vertical capillary under the influence of gravity. To reduce the uncertainty in time measurement to a negligible percentage, the size is chosen to give a flow time of about 200 seconds. A certain viscometer has a capillary diameter of 0.31 mm and a tube length of 73 mm. Estimate the maximum amount the diameter can vary to allow viscosity measurements with uncertainty less than 1 percent. (Note the precision required is the order of ± 1 *micro*meter; thus most viscometers require calibration with a liquid of known viscosity.)

8.41 Resistance to fluid flow can be defined by analogy to Ohm's law for electric current. Thus resistance to flow is given by the ratio of pressure drop (driving potential) to volume flow rate (current). Show that resistance to laminar flow is given by

$$\text{resistance} = \frac{128\mu L}{\pi D^4}$$

which is independent of flow rate. Find the maximum pressure drop for which this relation is valid for a tube 50 mm long and 0.25 mm inside diameter for both kerosine and castor oil at 40 C.

8.42 Cutting oil flows vertically downward under gravity in a long circular tube. The tube is long compared to its diameter so that end effects are negligible. The oil is SAE 10W at 100 F. The pressure is atmospheric everywhere. Determine the volume flow rate of oil that will be delivered if the tube diameter is 0.5 in. Verify that the flow is laminar.

8.43 Consider fully developed laminar flow due to gravity of a viscous liquid in a vertical circular tube. Assume that the pressure is atmospheric at both the tube inlet and outlet. Show that the relationship between tube diameter and Reynolds number may be expressed as

$$D = \left(\frac{32\, Re\, \nu^2}{g}\right)^{1/3}$$

Evaluate the maximum tube diameter for laminar flow of (a) water and (b) SAE 30 oil at 20 C.

8.44 A tube 17.6 in. long, with inside diameter of 0.030 in., is used as a capillary viscometer. Calibration tests are made using water at 60 F, and a flow rate of 1 cm³/sec is measured for an applied pressure drop of 10 psi. (Assume that the pressure drop in the entrance length is twice that for the same length of fully developed flow.) Determine the percentage error in viscosity that would result if Eq. 8.13c were used directly to compute it, without considering the entrance length.

8.45 Consider fully developed laminar flow in a circular pipe. Use a cylindrical control volume as shown. Indicate the forces acting on the control volume. Using the momentum equation, develop an expression for the velocity distribution.

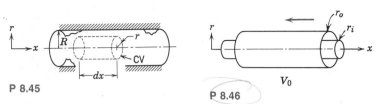

P 8.45   P 8.46

8.46 Consider fully developed laminar flow in the annulus between two concentric pipes. The inner pipe is stationary, and the outer pipe moves in the x direction with speed V_0. Assume the axial pressure gradient is zero ($\partial p/\partial x = 0$). Obtain a general expression for the shear stress, τ, as a function of the radius, r, in terms of a constant, C_1. Obtain a general expression for the velocity profile, $V(r)$, in terms of two constants, C_1 and C_2. Evaluate C_1 and C_2.

‡8.47 Consider fully developed laminar flow in the annular space formed by the two concentric cylinders shown in the diagram for Problem 8.46, but with pressure gradient, $\partial p/\partial x$, and the outer cylinder stationary. Let $r_0 = R$ and $r_i = kR$. Show that the velocity profile is given by

$$u = \frac{R^2}{4\mu}\frac{\partial p}{\partial x}\left[1 - \left(\frac{r}{R}\right)^2 + \left(\frac{1-k^2}{\ln(1/k)}\right)\ln\frac{r}{R}\right]$$

Obtain an expression for the location of the maximum velocity as a function of k. Compare the limiting case, $k \to 0$, with the corresponding expression for flow in a circular pipe.

‡8.48 For the flow of Problem 8.47 show that the volume flow rate is given by

$$Q = -\frac{\pi R^4}{8\mu}\frac{\partial p}{\partial x}\left[(1-k^4) - \frac{(1-k^2)^2}{\ln(1/k)}\right]$$

Find an expression for the average velocity. Compare the limiting case, $k \to 0$, with the corresponding expression for flow in a circular pipe.

‡8.49 It has been suggested in the design of an agricultural sprinkler that a structural member be held in place by a wire placed along the centerline of a pipe; it is

‡ You may wish to use simple computer programs to help solve problems marked with daggers.

surmised that a relatively small wire would have little effect on the pressure drop for a given flow rate. Using the result of Problem 8.48, derive an expression giving the percent change in pressure drop as a function of the ratio of wire diameter to pipe diameter for laminar flow. (You may simplify the results for small diameter wires.) Compute the percentage change in pressure drop for diameter ratios, d/D, of 0.01, 0.001, and 0.0001.

8.50 Consider the simple power-law model for a non-Newtonian fluid given by Eq. 2.11. Extend the analysis of Problem 8.45 to show that the velocity profile for fully developed laminar flow of a power-law fluid in a circular tube may be written

$$u = \left(\frac{R}{2k}\frac{\Delta p}{L}\right)^{1/n}\frac{nR}{n+1}\left[1 - \left(\frac{r}{R}\right)^{\frac{n+1}{n}}\right]$$

8.51 Evaluate the volume flow rate for fully developed laminar flow of a power-law fluid in a circular tube (Problem 8.50). Show that the velocity profile may be written

$$\frac{u}{\bar{V}} = \left(\frac{3n+1}{n+1}\right)\left[1 - \left(\frac{r}{R}\right)^{\frac{n+1}{n}}\right]$$

as suggested in Problem 2.73.

8.52 A horizontal pipe carries fluid in fully developed turbulent flow. The static pressure difference measured between two sections is 3 psi. The distance between the sections is 25 ft and the pipe diameter is 6 in. Calculate the shear stress, τ_w, that acts on the walls.

8.53 The pressure drop between two taps separated in the streamwise direction by 3 m in a horizontal, fully developed channel flow of water is 1.78 kPa. The cross section of the channel is a 30×240 mm rectangle. Calculate the average wall shear stress.

8.54 Kerosine at 70 F flows in a smooth tube with 1 in. inside diameter. The flow Reynolds number is 4000. For laminar flow, the pressure gradient is found to be $\partial p/\partial x = -0.2$ lbf/ft^2/ft, whereas for turbulent flow, $\partial p/\partial x = -0.5$ lbf/ft^2/ft. Plot the variation of shear stress, τ/τ_w, as a function of nondimensional radius, r/R, for both flow conditions.

8.55 A liquid drug, with the viscosity and density of water, is to be administered through a hypodermic needle. The inside diameter of the needle is 0.25 mm and its length is 50 mm. Determine (a) the maximum volume flow rate for which the flow will be laminar, (b) the pressure drop required to deliver the maximum flow rate, and (c) the corresponding wall shear stress.

8.56 Laufer [3] measured the following data for mean velocity in fully developed turbulent pipe flow at $Re_U = 50{,}000$:

$\bar{u}/U$	0.996	0.981	0.963	0.937	0.907	0.866	0.831
y/r	0.898	0.794	0.691	0.588	0.486	0.383	0.280

$\bar{u}/U$	0.792	0.742	0.700	0.650	0.619	0.551
y/R	0.216	0.154	0.093	0.062	0.041	0.024

Plot the data on log-log graph paper. Evaluate a power-law velocity profile exponent graphically. Compare with a value calculated by the method of least squares using the result of Problem 8.62.

8.57 Laufer [3] measured the following data for mean velocity in fully developed turbulent pipe flow at $Re_U = 500{,}000$:

| $\bar{u}/U$ | 0.997 | 0.988 | 0.975 | 0.959 | 0.934 | 0.908 |
| y/R | 0.898 | 0.794 | 0.691 | 0.588 | 0.486 | 0.383 |

| $\bar{u}/U$ | 0.874 | 0.847 | 0.818 | 0.771 | 0.736 | 0.690 |
| y/R | 0.280 | 0.216 | 0.154 | 0.093 | 0.062 | 0.037 |

Plot the data on log-log graph paper. Evaluate a power-law velocity profile exponent graphically. Compare with a value calculated by the method of least squares using the result of Problem 8.62.

8.58 Consider the velocity profile for fully developed laminar pipe flow, Eq. 8.14, and the empirical "power-law" profile for turbulent pipe flow, Eq. 8.22. Assume that $n = 7$ for the turbulent profile. Determine r/R for each profile at which u is equal to the average velocity, $\bar{V}$.

8.59 The velocity profile for turbulent flow through smooth pipes is often represented by the empirical relation of Eq. 8.22. Show that the ratio of average to centerline velocities is given by Eq. 8.23. Using n from Fig. 8.11, plot $\bar{V}/U$ as a function of Reynolds number.

8.60 Figure 8.11 is a plot of power-law velocity profile exponent, n, versus centerline Reynolds number, Re_U, for fully developed turbulent pipe flow. Equation 8.23 relates mean velocity, $\bar{V}$, to centerline velocity, U, for various values of n. Prepare plots of $\bar{V}/U$ versus $\log Re_U$ and $\log Re_{\bar{V}}$.

8.61 A momentum flux coefficient, β, is defined as

$$\int_A u\rho u\, dA = \beta \int_A \bar{V}\rho u\, dA = \beta \dot{m}\bar{V}$$

Evaluate β for a laminar velocity profile, Eq. 8.14, and for a "power-law" turbulent velocity profile, Eq. 8.22 (choose $n = 7$).

8.62 Consider the empirical power-law velocity profile for fully developed turbulent pipe flow, Eq. 8.22. Using the method of least squares, show that a best-fit n may be calculated from measured data as

$$\frac{1}{n} = \frac{\sum \ln(\bar{u}/U)\ln(y/R)}{\sum [\ln(y/R)]^2}$$

Hint: Express the profile as $\ln(\bar{u}/U) = (1/n)\ln(y/R)$.

8.63 Consider fully developed laminar flow of water between infinite parallel plates. The maximum flow speed, plate spacing, and width are 6 m/sec, 0.2 mm, and 30 mm, respectively. Find the flux of kinetic energy at a cross section.

8.64 Evaluate the kinetic energy flux coefficient, α, for the flow of Problem 8.63.

8.65 Consider fully developed laminar flow in a circular tube. Evaluate the kinetic energy flux coefficient for this flow.

8.66 Show that the kinetic energy flux coefficient, α, for the "power-law" turbulent velocity profile of Eq. 8.22 is given by Eq. 8.26. Evaluate α for $n = 7$.

8.67 Water flows in a constant-area pipeline; the pipe diameter is 50 mm and the average flow speed is 1.5 m/sec. At the pipe inlet the gage pressure is 590 kPa. The outlet of the pipe is 25 m higher than the inlet; the outlet pressure is atmospheric. Determine the head loss between the inlet and outlet of the pipe.

8.68 The pipe of Problem 8.67 is placed on a horizontal surface. The flow rate and outlet pressure are to remain the same. Compute the inlet pressure for this new condition.

8.69 Water is pumped at the rate of 2 ft^3/sec from a reservoir 20 ft above a pump to a free discharge 90 ft above the pump. The pressure on the intake side of the pump is 5 psig and the pressure on the discharge side is 50 psig. All pipes are commercial steel of 6 in. diameter. Determine (a) the head supplied by the pump and (b) the total head loss between the pump and point of free discharge.

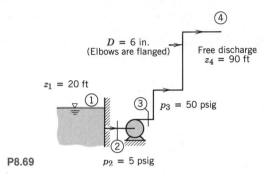

P8.69

8.70 Laufer [3] measured the following data for mean velocity near the wall in fully developed turbulent pipe flow at $Re_U = 50,000$ in air:

$\dfrac{\bar{u}}{U}$	0.343	0.318	0.300	0.264	0.228	0.221	0.179	0.152	0.140
$\dfrac{y}{R}$	0.0082	0.0075	0.0071	0.0061	0.0055	0.0051	0.0041	0.0034	0.0030

Plot the data on linear graph paper. Evaluate a best-fit value for $d\bar{u}/dy$ ($U = 9.8$ ft sec and $R = 5.9$ in.). Compare the wall shear stress evaluated from $\tau_w = \mu \, d\bar{u}/dy$ with that calculated from a friction factor taken from the Moody diagram.

8.71 A smooth, 3 in. diameter pipe carries water (150 F) horizontally at a mass flow rate of 0.006 slug/sec. The pressure drop is observed to be 0.065 lbf/ft^2 per 100 ft of pipe. From the Moody chart, the friction factor could be chosen as 0.021 or 0.042. Which is correct?

‡8.72 The curves plotted on the Moody chart are derived from the empirical correlation given by Eq. 8.37a. As noted in section 8-7.1, an initial guess for f_0, calculated from Eq. 8.37b, produces results accurate to 1 percent with a single iteration [8]. Write a subroutine for calculator or computer and validate the accuracy of this claim for $Re = 10^4$ and 10^7 for $e/D = 0$ and 0.010.

8.73 An empirical correlation for friction factor in turbulent flow in smooth pipes was developed by H. Blasius in 1911. For $Re \le 10^5$, he found that Eq. 8.35 correlated data well. Show that, in turbulent flow, the predicted pressure drop is proportional to $(\bar{V})^{7/4}$ when the Blasius correlation is used. How does pressure drop depend on tube diameter at a given flow rate?

8.74 The moody diagram gives the Darcy friction factor, f, in terms of Reynolds number and relative roughness. The *Fanning friction factor* for pipe flow is defined as

$$f_F = \frac{\tau_w}{\frac{1}{2}\rho\bar{V}^2}$$

where τ_w is the wall shear stress in the pipe. Obtain a relation between the Darcy and Fanning friction factors for fully developed pipe flow. Show that $f = 4f_F$.

‡ You may wish to use simple computer programs to help solve problems marked with daggers.

8.75 Kerosine flows in a 12 in. diameter smooth pipe at the rate of 4.5 ft³/sec. Estimate the mean velocity and shear stress at a radial location 1 in. from the pipe wall.

8.76 Water flows through a 1 in. diameter tube that suddenly enlarges to a diameter of 2 in. The flow rate through the enlargement is 20 gpm. Calculate the pressure rise across the enlargement.

8.77 Air at standard conditions flows through a sudden expansion in a circular duct. The upstream and downstream duct diameters are 3 and 9 in., respectively. The pressure downstream is 0.25 in. of water *higher* than that upstream. Determine the average speed of the air approaching the expansion and the volume flow rate.

8.78 Water flows through a 50 mm diameter tube that suddenly contracts to 25 mm diameter. The pressure drop across the contraction is 3.4 kPa. Determine the volume flow rate.

8.79 Flow through a sudden contraction is shown. The minimum flow area at the vena contracta is given in terms of the area ratio by the contraction coefficient [27],

$$C_c = \frac{A_c}{A_2} = 0.62 + 0.38 \left(\frac{A_2}{A_1}\right)^3$$

Because flow accelerates from A_1 to A_c, losses are quite small. However, losses are not negligible for the "sudden" expansion from A_c to A_2. Use these assumptions to evaluate (a) the contraction coefficient and (b) the minor loss coefficient, for a sudden contraction with $AR = A_2/A_1 = 0.5$. Compare with data from Fig. 8.17.

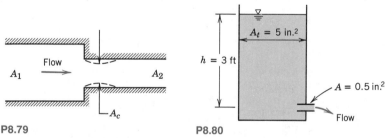

P8.79   P8.80

8.80 Water flows from the tank shown through a very short pipe. Assume the flow is quasi-steady. Estimate the flow rate at the instant shown. How could you improve the flow system if a larger flow rate were desired?

8.81 Air flows out of a clean room test chamber through a 150 mm diameter duct of length L. The original duct had a square-edged entrance, but this has been replaced with a well-rounded one. The pressure in the chamber is 2.5 mm of water above ambient. Losses due to friction are negligible compared to the entrance and exit losses. Estimate the increase in volume flow rate that results from the change in entrance contour.

8.82 Space has been found for a conical diffuser 0.45 m long in the clean room ventilation system described in Problem 8.81. The best diffuser of this size is to be used. Assume that data from Fig. 8.18 may be used. Determine the appropriate diffuser angle and area ratio for this installation and estimate the volume flow rate that will be delivered after it is installed.

8.83 Water at 20 C flows through a 0.1 m (internal diameter) concrete drainage pipe at a rate of 15 kg/sec. Determine the pressure drop per 100 m of horizontal pipe.

8.84 Air at 15 C flows through a straight, 0.3 m diameter, smooth duct 50 m long. The flow rate is 0.6 m³/sec, and the pressure is the same at both ends of the duct. Determine the change in elevation between the inlet and outlet.

8.85 Water at 78 F flows in a pipe whose inside diameter is 1.2 in. The flow rate is 0.04 ft³/sec. Determine the slope that the pipe must have to maintain constant pressure along its length. If the temperature remains constant, determine the heat transfer per 100 ft of pipe.

8.86 In a certain air-conditioning installation, a flow rate of 35 m³/min of air at standard conditions is required. A smooth sheet metal duct 0.3 m square is to be used. Determine the pressure drop for a 30 m horizontal duct run.

8.87 A smooth pipe, with inside diameter $D = 175$ mm, delivers $Q = 28$ m³/min of air at 20 C down a mine shaft. The shaft is 700 m deep and straight down. Estimate the pressure difference between the top and bottom of the pipe

8.88 An Ocean Thermal Energy Conversion (OTEC) plant draws cold seawater (at $T = 4$ C) into a cold water pipe far below the surface, as shown. The pipe inlet is located 1000 m below sea level. The hydrostatic pressure at that depth is $p_1 = 9.9$ MPa (gage). The cold water temperature stays nearly constant. The mean velocity in the cold water pipe is $\bar{V} = 1.83$ m/sec, and the pipe diameter is $D = 28.2$ m. Its effective roughness height is $e = 0.01$ m. Estimate the static pressure at sea level in the cold water pipe.

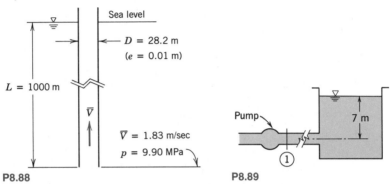

P8.88   P8.89

8.89 Water from a pump flows through a 0.25 m diameter pipe for a distance of 5 km from the pump discharge to a reservoir open to the atmosphere. The level of the water in the reservoir is 7 m above the pump discharge, and the average speed of the water in the pipe is 3 m/sec. Calculate the pressure at the pump discharge.

8.90 Water is to flow by gravity from one reservoir to a lower one through a straight, inclined pipe. The required flow rate is 0.007 m³/sec, the pipe diameter is 50 mm, and the total length is 250 m. Each reservoir is open to the atmosphere. Calculate the difference in level required to maintain this flow rate. Estimate the fraction of ΔH that is due to minor losses.

8.91 Consider flow of standard air at 35 m³/min. Compare the pressure drop per unit length of a round duct with that for rectangular ducts of aspect ratio 1, 2, and 3. Assume that all ducts are smooth, with cross-sectional areas of 0.1 m².

8.92 Water flows from a large reservoir as shown. The pipe is cast iron, with inside diameter of 0.2 m. The flow rate is 0.14 m³/sec, and the discharge is to atmospheric pressure. The mean temperature for the flow is 10 C; the entire system is insulated. Determine the gage pressure, p_1, required to produce the flow. Calculate the temperature rise between the liquid surface and the exit.

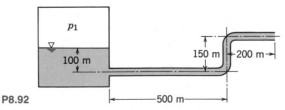

P8.92 |——————500 m——————|

8.93 Two reservoirs are connected by three clean cast-iron pipes in series, $L_1 = 600$ m, $D_1 = 0.3$ m, $L_2 = 900$ m, $D_2 = 0.4$ m, $L_3 = 1500$ m, and $D_3 = 0.45$ m. When the discharge is 0.11 m³/sec of water at 15 C, determine the difference in elevation between the reservoirs.

8.94 Water, at volume flow rate $Q = 300$ gpm, is delivered by a fire hose and nozzle assembly. The hose ($L = 200$ ft in overall length, $D = 3$ in., and $e/D = 0.004$) is made up of four 50 ft sections joined by couplings. The entrance is square-edged; the minor loss coefficient for each coupling is $K_e = 0.5$, based on mean velocity through the hose. The nozzle loss coefficient is $K_n = 0.02$, based on velocity in the exit jet, of $D_2 = 1.0$ in. diameter. Estimate the supply pressure required at this flow rate.

8.95 A 2.5 (nominal) in. pipeline conveying water contains 290 ft of straight galvanized pipe, 2 fully open gate valves, 1 fully open angle valve, 7 standard 90° elbows, 1 square-edged entrance from a reservoir, and 1 free discharge. The entrance and exit conditions are:

Location	Elevation	Pressure
Entrance	50.0 ft	20 psig
Discharge	94.0 ft	0 psig

A centrifugal pump is installed in the line to move the water. What pressure rise must the pump deliver so the volume flow rate will be $Q = 0.439$ ft^3/sec?

8.96 Data were obtained from measurements on a vertical section of old, corroded, galvanized iron pipe of 1 in. inside diameter. At one section the pressure was $p_1 = 100$ psig; at a second section, 20 ft lower, the pressure was $p_2 = 75.5$ psig. The volume flow rate of water was 0.110 ft^3/sec. Estimate the relative roughness of the pipe. What percent savings in pumping power would result if the pipe were restored to its new, clean relative roughness?

8.97 In the water flow system shown, reservoir B has variable elevation, x. Determine the water level in reservoir B so that no water flows into or out of the reservoir. The speed in the 12 in. diameter pipe is 10 ft/sec.

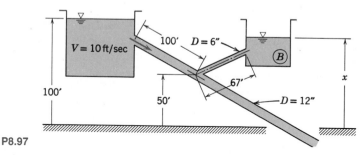

P8.97

8.98 A clean room must be supplied with 800 m^3/hr of air at standard conditions. The geometry of the supply duct is shown. Evaluate the gage pressure in the clean room. Sketch the pressure distribution along the supply duct. What improvements to the duct system could you recommend? How would these improvements reduce losses?

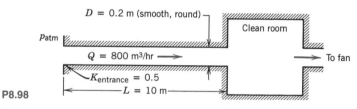

P8.98

8.99 Lightweight crude oil (SG $= 0.855$ with viscosity similar to SAE 30 oil) is pumped horizontally through a 1 mile length of 12 in. diameter pipe. The average roughness size is 0.01 in. The flow rate is 4500 gpm. Calculate the horsepower required to drive the pump, if it is 75 percent efficient.

8.100 Cooling water is pumped from a reservoir to rock drills on a construction job using the pipe system shown. The flow rate must be 600 gpm and water must leave the spray nozzle at 120 ft/sec. Calculate the minimum supply pressure needed at the pump outlet. Estimate the required power input if the pump efficiency is 70 percent.

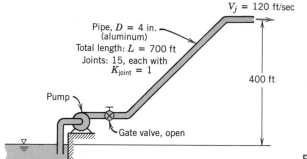

P8.100

8.101 Air conditioning for the Purdue University campus is provided by chilled water pumped through a main supply pipe. The pipe makes a loop 3 miles in length. The pipe diameter is 2 ft and the material is steel. The maximum design volume flow rate is 11,200 gpm. The circulating pump is driven by an electric motor. The efficiencies of pump and motor are $\eta_p = 0.80$ and $\eta_m = 0.90$, respectively. Electricity cost is $0.067/kW·hr. Determine (a) the pressure drop, (b) the minimum required pumping power, and (c) the annual cost of electrical energy for pumping.

8.102 The Alaskan pipeline runs from Prudhoe Bay to Valdez, a total distance of 798 mi. Both terminal points are at sea level. The pipe is 48 in. i.d. commercial steel. The capacity of the line is 2 million barrels of crude oil per day (1 barrel of oil = 42 gal). The specific gravity of the oil is 0.93, and its viscosity at the pumping temperature of 140 F is $\mu = 3.5 \times 10^{-4}$ lbf · sec/ft². Determine the total pumping power required if the pump efficiency is 85 percent. Express this result as a fraction of the chemical energy conveyed by the oil stream. (Assume that crude oil has a heating value of 18,000 Btu/lbm.)

8.103 A compressed air drill requires 0.25 kg/sec of air at 650 kPa (gage) at the drill. The hose from the air compressor to the drill is 40 mm inside diameter. The maximum compressor discharge gage pressure is 690 kPa; air leaves the compressor at 40 C. Neglect changes in density and any effects due to hose curvature. Calculate the longest hose that may be used.

8.104 Kerosine at 60 C flows through a pipe system in a refinery at the rate of 2.3 m³/min. The pipe is commercial steel, with inside diameter of 0.15 m. The gage pressure in the reactor vessel is 90 kPa. Determine the total length, L, of the straight pipe in the system.

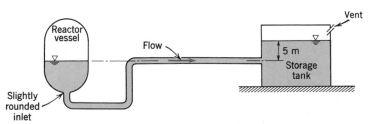

P8.104

8.105 Heavy crude oil (SG = 0.925 and $\nu = 1.1 \times 10^{-3}$ ft²/sec) is pumped through a pipeline laid on flat ground. The line is made from steel pipe with 24 in. i.d.

and has a wall thickness of $\frac{1}{2}$ in. The allowable tensile stress in the pipe wall is limited to 40,000 psi by corrosion considerations. It is important to keep the oil under pressure to ensure that gases remain in solution. The minimum recommended pressure is 75 psia. The pipeline carries a flow of 400,000 barrels (in the petroleum industry, a "barrel" is 42 gal) per day. Determine the maximum spacing between pumping stations. Compute the power added to the oil at each pumping station.

8.106 Gasoline flows in a long, underground pipeline at a constant temperature of 15 C. Two pumping stations at the same elevation are located 13 km apart. The pressure drop between the stations is 1.4 MPa. The pipeline is made from 0.6 m diameter pipe. Although made from commercial steel, age and corrosion have raised the pipe roughness to approximately that for galvanized iron. The specific gravity of the gasoline is 0.72. Compute the volume rate of flow through the pipe.

8.107 Water flows steadily in a horizontal 5 in. diameter cast-iron pipe. The pipe is 500 ft long and the pressure drop between sections ① and ② is 23 psi. Find the volume flow rate through the pipe.

8.108 Water flows steadily in a 5 in. diameter cast-iron pipe 500 ft long. The pressure drop between sections ① and ② is 23 psi. Section ② is located 30 ft above section ①. Find the volume flow rate.

8.109 A hydraulic turbine is to be supplied with water from a mountain stream through a supply pipe, as shown. The pipe diameter is 1 ft and the average roughness height is 0.05 in. Minor losses can be neglected. Flow leaves the pipe at atmospheric pressure. Calculate the discharge velocity.

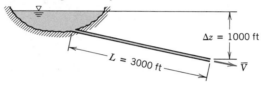

P8.109

8.110 Water flows steadily in an inclined 3 in. inside diameter cast-iron pipe 100 ft long. The pressure at section ① is 6.0 psig. At section ②, 16 ft above section ①, the pressure is 0.5 psig. Determine (a) the flow direction and (b) the volume flow rate.

8.111 A mining engineer plans to do hydraulic mining with a high-speed jet of water. A lake is located $H = 300$ m above the mine site. Water will be delivered through $L = 900$ m of fire hose; the hose has inside diameter $D = 75$ mm and relative roughness $e/D = 0.01$. Couplings, with equivalent length $L_e = 20$ D, are located every 10 m along the hose. The nozzle outlet diameter is $d = 25$ mm. Its minor loss coefficient is $K = 0.02$ based on *outlet velocity*. Estimate the maximum outlet velocity, V_o, that this system could deliver. Determine the maximum force exerted on a rock face by this water jet.

8.112 The siphon shown is fabricated from 2 in. i.d. drawn aluminum tubing. The liquid is water at 60 F. Compute the volume flow rate through the siphon. Estimate the minimum pressure inside the tube.

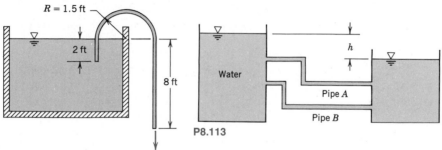

P8.112 P8.113

8.113 Two water reservoirs are connected by galvanized iron pipes, as shown on the previous page. Assume $D_A = 75$ mm, $D_B = 50$ mm, and $h = 10.5$ m. The length of both pipes is 100 m. Compare the head losses in pipes A and B. Compute the volume flow rate in each pipe.

‡**8.114** A small hydraulic turbine is supplied from a mountain reservoir that has a surface located height H above the turbine inlet. The galvanized iron supply pipe is of constant diameter, D, and length, L. Assume the power output from the turbine is proportional to the product of volume flow rate, Q, and pressure drop across the turbine, Δp. Obtain an algebraic expression for the power output of the turbine. For fixed pipe diameter, find an algebraic expression for the mean velocity, $\bar{V}$, at which turbine output power is a maximum. Plot the maximum power that theoretically could be developed by the turbine, as a function of pipe diameter, for $75 < D < 300$ mm. Assume flow in the fully rough zone; use $H = 100$, 500, and 1000 m, and $L = 2H$.

8.115 A fire nozzle is supplied through 300 ft of 1.5 in. diameter, smooth, rubber-lined hose. Water from a hydrant is supplied to a booster pump on board the pumper truck at 50 psig. At design conditions, the pressure at the nozzle inlet is 100 psig, and the pressure drop along the hose is 33 psi per 100 ft of length. Determine (a) the design flow rate, (b) the nozzle exit velocity, assuming no losses in the nozzle, and (c) the power required to drive the booster pump, if its efficiency is 70 percent.

8.116 Two reservoirs containing water are connected by a galvanized iron pipe that has one right-angle bend. The surface pressure at the upper reservoir is atmospheric, whereas the gage pressure at the lower reservoir surface is 70 kPa. The pipe diameter is 75 mm. Determine the magnitude and direction of the volume flow rate.

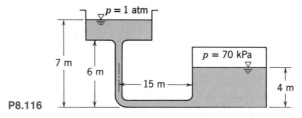

P8.116

8.117 Consider again the Roman water supply discussed in Example Problem 8.10. Assume that the 50 ft length of horizontal constant-diameter pipe required by law has been installed. The relative roughness of the pipe is 0.01. Estimate the flow rate of water delivered by the pipe under the inlet conditions of the example. What would be the effect of adding the same diffuser to the end of the 50 ft pipe?

8.118 You are watering your lawn with an *old* hose. Because lime deposits have built up over the years, the 0.75 in. i.d. hose now has an average roughness height of 0.022 in. One 50 ft length of the hose, attached to your spigot, delivers 20 gpm of water (60 F). Compute the pressure at the spigot, in psi. Estimate the delivery if two 50 ft lengths of the hose are connected. Assume that the pressure at the spigot varies with flow rate and the water main pressure remains constant at 70 psig.

‡**8.119** A circular tank 1.5 m in diameter is filled 5 m deep with water at 15 C. A smooth tube 7 m long is attached to the tank bottom. The 50 mm diameter tube discharges to atmosphere 5 m below the tank bottom. Estimate the time required for the tank level to drop 1.5 m. (The Blasius correlation for the friction factor for turbulent flow, Eq. 8.35, might be useful.)

8.120 The head versus capacity curve for a certain fan may be approximated by the equation $h = 30 - 10^{-7}Q^2$, where h is the output static head in inches of water and

‡ You may wish to use simple computer programs to help solve problems marked with daggers.

Q is the air flow rate in ft³/min. The fan outlet dimensions are 8×16 in. Determine the air flow rate delivered by the fan into a 200 ft straight length of 8×16 in. rectangular duct.

8.121 A spray system for a sewage treatment plant is shown. Water at 60 F is pumped through a spray arm. The effective flow area of each nozzle is 0.25 in.² The pipe inside diameter is 1 in., and the material is galvanized iron. Determine the flow rate of water through the spray arm.

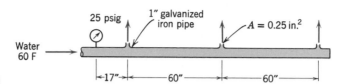

P8.121

8.122 A hydraulic press is powered by a remote high-pressure pump. The gage pressure at the pump outlet is 20 MPa, whereas the pressure required for the press is 19 MPa (gage), at a flow rate of 0.032 m³/min. The press and pump are to be connected by 50 m of smooth, drawn steel tubing. The fluid is SAE 10W oil at 40 C. Determine the minimum tubing diameter that may be used.

8.123 A pump is located 15 ft to one side of, and 12 ft above a reservoir. The pump is designed for a flow rate of 100 gpm. For satisfactory operation, the suction head at the pump inlet must not be lower than −20 ft of water gage. Determine the smallest standard commercial steel pipe that will give the required performance.

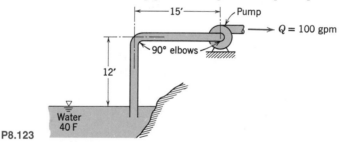

P8.123

8.124 Determine the minimum size smooth rectangular duct with an aspect ratio of 2 that will pass 80 m³/min of standard air with a head loss of 30 mm of water per 30 m of duct.

8.125 A new industrial plant requires a water flow rate of 5.7 m³/min. The gage pressure in the water main, located in the street 50 m from the plant, is 800 kPa. The supply line will require installation of 4 elbows in a total length of 65 m. The gage pressure required in the plant is 500 kPa. What size galvanized iron line should be installed?

8.126 A water jet pump is shown. The primary jet diameter is $D_1 = 10$ mm; the mixing section diameter is $D_2 = 30$ mm. A conical diffuser connects the mixing section to a horizontal pipe with $D_4 = 50$ mm and $L = 30$ m. The pipe delivers volume flow rate $Q_4 = 407$ liters per minute to atmospheric pressure. The flow rate to the primary jet is $Q_1 = 136$ liters per minute. The secondary stream (section ③) is supplied

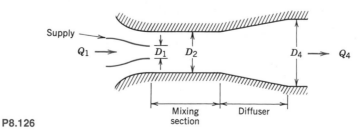

P8.126

from a reservoir. Sketch the pressure distribution from sections ① through ④. Determine the pressure needed to supply the primary jet. Estimate the minimum pool height needed to supply the secondary stream flow.

**8.127 Oil has been flowing from a large tank on a hill to a tanker at the wharf. The compartment in the tanker is nearly full and an operator is in the process of stopping the flow. A valve on the wharf is closed at a rate such that 150 psig is maintained in the line immediately upstream of the valve. Assume:

Length of line from tank to valve	10,000 ft
Inside diameter of line	8.0 in.
Elevation of oil surface in tank	200 ft
Elevation of valve on wharf	20 ft
Instantaneous flow rate	1.5 ft³/sec
Head loss in line (exclusive of valve being closed) at this rate of flow	75 ft of oil
Specific gravity of oil	0.88

Calculate the initial instantaneous rate of change of volume flow rate.

‡8.128 Incompressible liquid leaves through the wall of a horizontal perforated pipe section of length L, with no component of velocity in the axial direction. The pressure gradient along the length of perforated pipe may be positive or negative, depending on pipe diameter, D, inlet volume flow rate, Q_0, fluid properties, and the distribution of the liquid jets along the pipe. Assume flow leaves the pipe surface at a volume flow rate per unit length, q, that is a constant fraction, k, of the volume flow rate at the pipe inlet, such that $q = kQ_0/L$. Also assume the friction factor, f, is constant. Show that the axial pressure gradient is

$$\frac{dp}{dx} = \frac{\rho Q_0^2}{2A^2 L}\left[2k(1-k\lambda) - f\frac{L}{D}(1-k\lambda)^2\right]$$

where $\lambda = x/L$. Plot the results as a function λ for representative values of k.

8.129 The results of Example Problem 8.11 indicated that the pipes used were too small to produce the same pressure at each nozzle inlet. Rework the problem using the same geometry but with $3\frac{1}{2}$ in. smooth pipes.

8.130 The piping system shown has been constructed from 3 in. galvanized iron pipe. All flow rates are high enough that the flow may be considered in the fully rough zone. Minor losses may be neglected. Determine the pressure at section ②, in terms of Q_2. Find the unknown flow rates as fractions of the inlet flow rate, Q_0. The fluid is water.

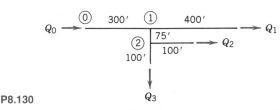

P8.130

8.131 A swimming pool has a partial-flow filtration system. Water at 75 F is pumped from the pool through the system shown. The pump delivers 30 gpm. The pipe is nominal 3/4 in. PVC (i.d. = 0.824 in.). The pressure loss through the filter is

** These problems require material from sections that may be omitted without loss of continuity in the text material.

‡ You may wish to use simple computer programs to help solve problems marked with daggers.

approximately $\Delta p = 0.6Q^2$, where Δp is in psi and Q is in gpm. Determine the pump pressure and the flow rate through each branch of the system.

P8.131

8.132 An "ergometer" is used to measure the rate at which air is consumed by a research subject running on a treadmill. The atmospheric temperature and pressure are 23 C and 752 mm of mercury, respectively. An inverted bell, which is used to measure air flow rate, is 0.45 m in diameter. During a 30 sec test run, the bell rises 43 mm. Determine the rate at which the subject consumes oxygen.

8.133 Water at 150 F flows through a 3 in. diameter orifice installed in a 6 in. i.d. pipe. The flow rate is 300 gpm. Determine the pressure difference between the corner taps.

8.134 A square-edged orifice with corner taps and a water manometer are used to meter compressed air. The following data are given:

Inside diameter of air line	6 in.
Orifice plate diameter	4 in.
Upstream pressure	90.0 psig
Temperature of air	80 F
Manometer deflection	30 in. H_2O

Calculate the volume flow rate in the line, expressed in SCFM.

8.135 A venturi meter with a 75 mm diameter throat is placed in a 150 mm diameter line carrying water at 25 C. The pressure drop between the upstream tap and the venturi throat is 300 mm of mercury. Compute the rate of flow.

8.136 Gasoline (SG = 0.73) flows through a 2×1 in. venturi meter. The differential pressure is 380 mm of mercury. Find the volume flow rate.

8.137 Consider a horizontal 2×1 in. venturi with water flow. For a differential pressure of 20 psi, calculate the volume flow rate.

8.138 Air flow rate in a test of an internal combustion engine is to be measured using a flow nozzle installed in a plenum. The engine displacement is 1.6 liters, and its maximum operating speed is 6000 rpm. To avoid loading the engine, the maximum pressure drop across the nozzle should not exceed 0.25 m of water. The manometer can be read to ± 0.5 mm of water. Determine the flow nozzle diameter that should be specified. Find the minimum rate of air flow that can be metered to ± 2 percent using this setup.

8.139 Air flows through the venturi meter described in Problem 8.135. Assume that the upstream pressure is 400 kPa, and that the temperature is everywhere constant at 20 C. Determine the maximum possible mass flow rate of air for which the assumption of incompressible flow is a valid engineering approximation. Compute the corresponding differential pressure reading on a mercury manometer.

8.140 Kerosine at 40 C flows through a 0.3 m diameter line in a refinery. The flow rate is not expected to exceed 120 kg/sec. A manometer with a range of 1 m of water is available for use with an orifice meter. Specify a recommended orifice diameter for use with this system. What minimum rate of flow could be measured within 10 percent accuracy if the manometer least count is 1 mm of water?

8.141 Water at 70 F flows steadily through a venturi. The pressure upstream from the throat is 5 psig. The throat area is 0.025 ft²; the upstream area is 0.1 ft². Estimate the maximum flow rate this device can handle without cavitation.

8.142 Drinking straws are to be used to improve the air flow in a pipe-flow experiment. Packing a section of the air pipe with drinking straws to form a "laminar flow element" might allow the air flow rate to be measured directly, and simultaneously would act as a flow straightener. To evaluate this idea, determine (a) the Reynolds number for flow in each drinking straw, (b) the friction factor for flow in each straw, and (c) the gage pressure at the exit from the drinking straws. (For laminar flow in a tube, the entrance loss coefficient is $K_{ent} \approx 1.4$ and $\alpha = 2.0$.) Comment on the utility of this idea.

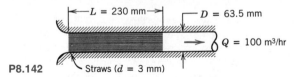

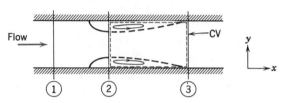

P8.142

8.143 Consider a flow nozzle installation in a pipe. Apply the basic equations to the control volume indicated, to show that the permanent head loss across the meter can be expressed, in dimensionless form, as the head loss coefficient,

$$C_l = \frac{p_1 - p_3}{p_1 - p_2} = \frac{1 - A_2/A_1}{1 + A_2/A_1}$$

P8.143

Chapter 9

EXTERNAL INCOMPRESSIBLE VISCOUS FLOW

External flows are flows over bodies immersed in an unbounded fluid. The flow over a semi-infinite flat plate (Fig. 2.11) and the flow over a cylinder (Fig. 2.12a) are examples of external flows. These were discussed qualitatively in Chapter 2. Our objective in this chapter is to quantify the behavior of viscous, incompressible fluids in external flow.

A number of phenomena that occur in external flow over a body are illustrated in the sketch of the high Reynolds number viscous flow over an airfoil (Fig. 9.1). The freestream flow divides at the stagnation point and flows around the body. Fluid at the surface takes on the velocity of the body as a result of the no-slip condition. Boundary layers form on both the upper and lower surfaces of the body. (The boundary-layer thickness on both surfaces in Fig. 9.1 is exaggerated greatly for clarity.) The flow in the boundary layers initially is laminar. Transition to turbulent flow occurs at some distance from the stagnation point, depending on freestream conditions, surface roughness, and pressure gradient. The transition points are indicated by "T" in the figure. The turbulent boundary layer following transition grows more rapidly than the laminar layer. A slight displacement of the streamlines of the external flow is caused by the thickening boundary layers on the surface. In a region of increasing pressure (an *adverse pressure gradient*) flow separation may occur. Separation points are indicated by "S" in the figure. Fluid that was in the boundary layers on the body surface forms the viscous *wake* behind the separation points.

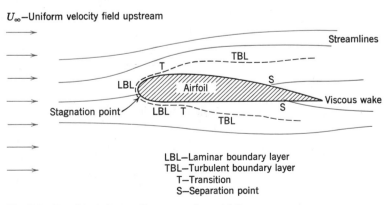

Fig. 9.1   Details of viscous flow around an airfoil.

Part A of this chapter is devoted to boundary-layer flows. Following a discussion of the boundary-layer concept, the exact solution for laminar boundary-layer flow over a flat plate (zero pressure gradient) is presented. Since exact solutions for turbulent boundary layers do not exist, approximate solutions must be used. The momentum integral equation for nonzero pressure gradient is derived from first principles as the basis for the approximate solutions; approximate solutions for both laminar and turbulent flow over flat plates are considered. Although approximate solutions for boundary-layer flows with pressure gradients are beyond the scope of this book, the effect of pressure gradients on boundary-layer flows is discussed.

The airfoil of Fig. 9.1 experiences a net force as a result of the shear and pressure forces acting on its surfaces. The component of the net force parallel to the uniform upstream flow, U_∞, is called the drag force; the component of the net force perpendicular to U_∞ is called the lift. The presence of flow separation precludes the analytical determination of lift and drag. In Part B of this chapter approximate analyses and correlations of experimental data are presented for the drag and lift on a number of bodies of interest.

PART A   BOUNDARY LAYERS

9-1   THE BOUNDARY-LAYER CONCEPT

The concept of a boundary layer was first introduced by Ludwig Prandtl [1], a German aerodynamicist, in 1904.

Prior to Prandtl's historic breakthrough, the science of fluid mechanics had been developing in two rather different directions. Theoretical hydrodynamics evolved from Euler's equation (Eq. 6.2, published by Leonhard Euler in 1755) of motion for a nonviscous fluid. Since the results of hydrodynamics contradicted many experimental observations, practicing engineers developed their own empirical art of hydraulics. This was based on experimental data and differed significantly from the purely mathematical approach of theoretical hydrodynamics.

Although the complete equations describing the motion of a viscous fluid (the Navier–Stokes equations, Eqs. 5.26, developed by Navier, 1827, and independently by Stokes, 1845) were known prior to Prandtl, the mathematical difficulties in solving these equations (except for a few simple cases) prohibited a theoretical treatment of viscous flows. Prandtl showed [1] that many viscous flows can be analyzed by dividing the flow into two regions, one close to solid boundaries, the other covering the rest of the flow. Only in the thin region adjacent to a solid boundary (the boundary layer) is the effect of viscosity important. In the region outside of the boundary layer, the effect of viscosity is negligible and the fluid may be treated as inviscid.

The boundary-layer concept provided the link that had been missing between theory and practice. Furthermore, the boundary-layer concept permitted the solution of viscous flow problems that would have been impossible through application of the Navier–Stokes equations to the complete flow field.[1] Thus the introduction of the boundary-layer concept marked the beginning of the modern era of fluid mechanics.

The development of a boundary layer on a solid surface was discussed in Section 2-5.1. The development of a laminar boundary layer on a flat plate was illustrated in Fig. 2.11. In the boundary layer both viscous and inertia forces are important.

---

[1] Today, computer solutions of the Navier–Stokes equations are common.

Consequently, it is not surprising that the Reynolds number (which represents the ratio of inertia to viscous forces) is significant in characterizing boundary-layer flows. The characteristic length used in the Reynolds number is either the length in the flow direction over which the boundary layer has developed or some measure of the boundary-layer thickness.

As for flow in a duct, flow in a boundary layer may be laminar or turbulent. There is no unique value of the Reynolds number at which transition from laminar to turbulent flow occurs in a boundary layer. Among the factors that affect boundary-layer transition are pressure gradient, surface roughness, heat transfer, body forces, and freestream disturbances. Detailed consideration of these effects is beyond the scope of this book.

In many real flow situations, a boundary layer develops over a long, essentially flat surface. Examples include flow over ship and submarine hulls, aircraft wings, and atmospheric motions over flat terrain. Since the basic features of all these flows are illustrated in the simpler case of flow over a flat plate, let us consider this first.

A qualitative picture of the boundary-layer growth over a flat plate is shown in Fig. 9.2. The boundary layer is laminar for a short distance downstream from the leading edge; transition occurs over a region of the plate rather than at a single line across the plate. The transition region extends downstream to the location where the boundary-layer flow becomes completely turbulent.

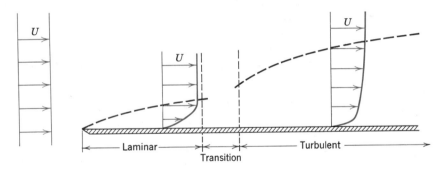

Fig. 9.2 Boundary layer on a flat plate (vertical thickness exaggerated greatly).

For incompressible flow over a smooth flat plate (zero pressure gradient), in the absence of heat transfer, transition from laminar to turbulent flow in the boundary layer can be delayed to a Reynolds number, $Re_x = \rho U x/\mu$, greater than one million if external disturbances are minimized. For calculation purposes, under typical flow conditions, transition usually is considered to occur at a length Reynolds number of 500,000. For air at standard conditions, with freestream velocity $U = 30$ m/sec, this corresponds to length, x, along the plate of $x \approx 0.24$ m. In the qualitative picture of Fig. 9.2, we have shown the turbulent boundary layer growing at a faster rate than the laminar layer. In later sections of this chapter we shall show that this is indeed true.

9-2 BOUNDARY-LAYER THICKNESSES

The boundary layer is the region adjacent to a solid surface in which viscous forces are important. The boundary-layer *disturbance thickness*, δ, usually is defined as the distance from the surface to the point where the velocity is within 1 percent of the freestream velocity. Since the velocity profile merges smoothly and asymptotically into the freestream, the boundary-layer thickness, δ, is difficult to measure.

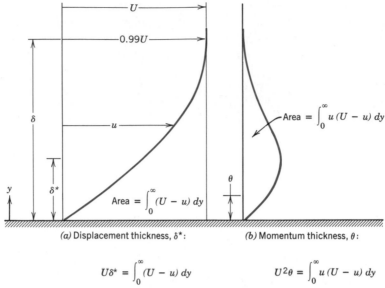

(a) Displacement thickness, δ^*:

$$U\delta^* = \int_0^\infty (U - u)\,dy$$

(b) Momentum thickness, θ:

$$U^2\theta = \int_0^\infty u\,(U - u)\,dy$$

Fig. 9.3 Boundary-layer thickness definitions.

The effect of viscous forces in the boundary layer is to retard the flow. The mass flow rate adjacent to a solid surface is less than the mass flow rate that would pass through the same region in the absence of a boundary layer. The decrease in mass flow rate due to the influence of viscous forces is $\int_0^\infty \rho(U - u)w\,dy$, where w is the width of the surface in the direction perpendicular to the flow. If viscous forces were absent, the velocity at a section would be U. The *displacement thickness*, δ^*, is the distance by which the solid boundary would have to be displaced in a frictionless flow to give the same mass flow rate deficit as exists in the boundary layer. Displacing the boundary by a distance δ^* would result in a mass flow deficiency of $\rho U\delta^* w$. Thus, as illustrated in Fig. 9.3a,

$$\rho U\delta^* w = \int_0^\infty \rho(U - u)w\,dy$$

For incompressible flow, $\rho = $ constant, and

$$\delta^* = \int_0^\infty \left(1 - \frac{u}{U}\right) dy \approx \int_0^\delta \left(1 - \frac{u}{U}\right) dy \tag{9.1}$$

Since $u = U$ at $y = \delta$, the integrand is essentially zero for $y \geq \delta$. Application of the displacement-thickness concept is illustrated in Example Problem 9.1.

Flow retardation within the boundary layer also results in a reduction in momentum flux at a section compared to inviscid flow. The momentum deficiency of the actual mass flow rate, $\int_0^\infty \rho u w\,dy$, through the boundary layer is $\int_0^\infty \rho u(U - u)w\,dy$. If viscous forces were absent, it would be necessary to move the solid boundary outward to obtain a momentum deficiency; denoting this distance (the momentum thickness) as θ, the momentum deficiency would be $\rho U^2 \theta w$. The *momentum thickness*, θ, is defined as the thickness of a layer of fluid, of velocity U, for which the momentum flux is equal to the deficit of momentum flux through the boundary layer. Thus, as illustrated in Fig. 9.3b,

$$\rho U^2 \theta = \int_0^\infty \rho u(U - u)\, dy$$

For incompressible flow, $\rho = $ constant, and

$$\theta = \int_0^\infty \frac{u}{U}\left(1 - \frac{u}{U}\right) dy \approx \int_0^\delta \frac{u}{U}\left(1 - \frac{u}{U}\right) dy \qquad (9.2)$$

Again, the integrand is essentially zero for $y \geq \delta$.

The displacement and momentum thicknesses, δ^* and θ, are *integral* thicknesses, because their definitions, Eqs. 9.1 and 9.2, are in terms of integrals across the boundary layer. Because they are defined in terms of integrals for which the integrand vanishes in the freestream, they are appreciably easier to evaluate accurately from experimental data than the boundary-layer disturbance thickness, δ. This fact, coupled with their physical significance, accounts for their common use in specifying boundary-layer thickness.

EXAMPLE 9.1—Boundary Layer in Channel Flow
A laboratory wind tunnel has a test section that is 305 mm square. Boundary-layer velocity profiles are measured at two cross sections and displacement thicknesses are evaluated from the measured profiles. At section ①, where the freestream speed is $U_1 = 26$ m/sec, the displacement thickness is $\delta_1^* = 1.5$ mm. At section ②, located downstream from section ①, $\delta_2^* = 2.1$ mm. Calculate the change in static pressure between sections ① and ②. Express the result as a fraction of the freestream dynamic pressure at section ①. Assume standard atmosphere conditions.

EXAMPLE PROBLEM 9.1

GIVEN: Flow of standard air in laboratory wind tunnel. Test section is $L = 305$ mm square. Displacement thicknesses are $\delta_1^* = 1.5$ mm and $\delta_2^* = 2.1$ mm. Freestream speed is $U_1 = 26$ m/sec.

FIND: Change in static pressure between sections ① and ②. (Express as a fraction of freestream dynamic pressure at section ①.)

SOLUTION:
Use the displacement thickness concept to find the effective flow area for the freestream flow outside the thin wall boundary layers. Replace the actual boundary-layer velocity profiles with uniform velocity profiles as sketched in the following figures.

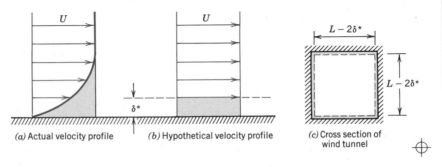

(a) Actual velocity profile (b) Hypothetical velocity profile (c) Cross section of wind tunnel

Apply the continuity and Bernoulli equations to freestream flow outside the boundary-layer displacement thickness, where viscous effects are negligible.

Basic equations :

$$= 0(1)$$

$$0 = \frac{\partial}{\partial t} \int_{CV} \rho \, d\forall + \int_{CS} \rho \vec{V} \cdot d\vec{A}$$

$$\frac{p_1}{\rho} + \frac{V_1^2}{2} + g z_1 = \frac{p_2}{\rho} + \frac{V_2^2}{2} + g z_2$$

Assumptions:
(1) Steady flow
(2) Incompressible flow
(3) Flow uniform at each section outside δ^*
(4) Flow along a streamline from sections ① to ②
(5) No frictional effects in freestream
(6) Neglect elevation changes

From the Bernoulli equation we obtain

$$p_1 - p_2 = \frac{1}{2}\rho\left(V_2^2 - V_1^2\right) = \frac{1}{2}\rho\left(U_2^2 - U_1^2\right) = \frac{1}{2}\rho U_1^2\left[\left(\frac{U_2}{U_1}\right)^2 - 1\right]$$

or

$$\frac{p_1 - p_2}{\frac{1}{2}\rho U_1^2} = \left(\frac{U_2}{U_1}\right)^2 - 1$$

From continuity, $V_1 A_1 = U_1 A_1 = V_2 A_2 = U_2 A_2$, so $\dfrac{U_2}{U_1} = \dfrac{A_1}{A_2}$, where $A = (L - 2\delta^*)^2$ is the effective flow area. Substituting gives

$$\frac{p_1 - p_2}{\frac{1}{2}\rho U_1^2} = \left(\frac{A_1}{A_2}\right)^2 - 1 = \left[\frac{(L - 2\delta_1^*)^2}{(L - 2\delta_2^*)^2}\right]^2 - 1$$

$$\frac{p_1 - p_2}{\frac{1}{2}\rho U_1^2} = \left[\frac{305 - 2(1.5)}{305 - 2(2.1)}\right]^4 - 1 = 0.0161 \qquad \text{or} \qquad 1.61 \text{ percent} \qquad\qquad \frac{p_1 - p_2}{\frac{1}{2}\rho U_1^2}$$

$\left\{\begin{array}{l}\text{This problem illustrates the application of the displacement-thickness concept. The} \\ \text{real, viscous boundary-layer flow is modeled as a uniform, inviscid flow displaced} \\ \text{from the boundary through distance } \delta^*.\end{array}\right\}$

**9-3 LAMINAR FLAT-PLATE BOUNDARY LAYER: EXACT SOLUTION

The solution for the laminar boundary layer on a horizontal flat plate was obtained by H. Blasius [2], one of Prandtl's students, in 1908. For two-dimensional, steady, incompressible flow with zero pressure gradient, the governing equations of motion reduce to [3]

$$\frac{\partial u}{\partial x} + \frac{\partial v}{\partial y} = 0 \tag{9.3}$$

$$u\frac{\partial u}{\partial x} + v\frac{\partial u}{\partial y} = \nu\frac{\partial^2 u}{\partial y^2} \tag{9.4}$$

** This section may be omitted without loss of continuity in the text material.

with boundary conditions

$$\text{at} \quad y = 0, \qquad u = 0$$

$$\text{at} \quad y = \infty, \qquad u = U, \qquad \frac{du}{dy} = 0 \qquad (9.5)$$

Blasius reasoned that the velocity profile, u/U, should be similar for all values of x when plotted versus a nondimensional distance from the wall; the boundary-layer thickness, δ, was a natural choice for nondimensionalizing the distance from the wall. Thus the solution is of the form

$$\frac{u}{U} = g(\eta) \qquad \text{where} \quad \eta \propto \frac{y}{\delta} \qquad (9.6)$$

Based on the solution of Stokes [4], Blasius reasoned that $\delta \propto \sqrt{\nu x/U}$ and set

$$\eta = y \sqrt{\frac{U}{\nu x}} \qquad (9.7)$$

Introducing the stream function, ψ, where

$$u = \frac{\partial \psi}{\partial y} \qquad \text{and} \qquad v = -\frac{\partial \psi}{\partial x} \qquad (5.4)$$

satisfies the continuity equation (Eq. 9.3) identically; substituting for u and v into Eq. 9.4 reduces the equation to one in which ψ is the single dependent variable. Defining a dimensionless stream function as

$$f(\eta) = \frac{\psi}{\sqrt{\nu x U}} \qquad (9.8)$$

makes $f(\eta)$ the dependent variable and η the independent variable in Eq. 9.4. With ψ defined by Eq. 9.8 and η defined by Eq. 9.7, we can evaluate each of the terms in Eq. 9.4.

The velocity components are given by

$$u = \frac{\partial \psi}{\partial y} = \frac{\partial \psi}{\partial \eta} \frac{\partial \eta}{\partial y} = \sqrt{\nu x U} \frac{df}{d\eta} \sqrt{\frac{U}{\nu x}} = U \frac{df}{d\eta} \qquad (9.9)$$

and

$$v = -\frac{\partial \psi}{\partial x} = -\left[\sqrt{\nu x U} \frac{\partial f}{\partial x} + \frac{1}{2} \sqrt{\frac{\nu U}{x}} f \right]$$

$$= -\left[\sqrt{\nu x U} \frac{df}{d\eta} \left(-\frac{1}{2} \eta \frac{1}{x} \right) + \frac{1}{2} \sqrt{\frac{\nu U}{x}} f \right]$$

$$v = \frac{1}{2} \sqrt{\frac{\nu U}{x}} \left[\eta \frac{df}{d\eta} - f \right] \qquad (9.10)$$

By differentiating the velocity components, it also may be shown that

$$\frac{\partial u}{\partial x} = -\frac{U}{2x} \eta \frac{d^2 f}{d\eta^2}$$

$$\frac{\partial u}{\partial y} = U \sqrt{U/\nu x} \frac{d^2 f}{d\eta^2}$$

and

$$\frac{\partial^2 u}{\partial y^2} = \frac{U^2}{\nu x}\frac{d^3 f}{d\eta^3}$$

Substituting these expressions into Eq. 9.4, we obtain

$$2\frac{d^3 f}{d\eta^3} + f\frac{d^2 f}{d\eta^2} = 0 \qquad (9.11)$$

with boundary conditions:

$$\text{at} \quad \eta = 0, \quad f = \frac{df}{d\eta} = 0$$

$$\text{at} \quad \eta \to \infty, \quad \frac{df}{d\eta} = 1 \qquad (9.12)$$

The second-order partial differential equations governing the growth of the laminar boundary layer on a flat plate (Eqs. 9.3 and 9.4) have been transformed to a nonlinear, third-order ordinary differential equation (Eq. 9.11) with boundary conditions given by Eq. 9.12. It is not possible to solve Eq. 9.11 in closed form; Blasius solved it using a power series expansion about $\eta = 0$ matched to an asymptotic expansion for $\eta \to \infty$. The same equation later was solved more precisely—again using numerical methods—by Howarth [5], who reported results to 5 decimal places. The numerical values of f, $df/d\eta$, and $d^2 f/d\eta^2$ in Table 9.1 were calculated with a personal computer using 4th-order Runge-Kutta numerical integration.

The velocity profile is obtained in dimensionless form by plotting u/U versus η, using values from Table 9.1. The resulting profile is plotted in Fig. 9.3a. Velocity profiles measured experimentally are in excellent agreement with the analytical so-

Table 9.1 The Function $f(n)$ for the Laminar Boundary Layer along a Flat Plate at Zero Incidence

$\eta = y\sqrt{\dfrac{U_\infty}{\nu x}}$	f	$f' = \dfrac{u}{U_\infty}$	f''
0	0	0	0.3321
0.5	0.0415	0.1659	0.3309
1.0	0.1656	0.3298	0.3230
1.5	0.3701	0.4868	0.3026
2.0	0.6500	0.6298	0.2668
2.5	0.9963	0.7513	0.2174
3.0	1.3968	0.8460	0.1614
3.5	1.8377	0.9130	0.1078
4.0	2.3057	0.9555	0.0642
4.5	2.7901	0.9795	0.0340
5.0	3.2833	0.9915	0.0159
5.5	3.7806	0.9969	0.0066
6.0	4.2796	0.9990	0.0024
6.5	4.7793	0.9997	0.0008
7.0	5.2792	0.9999	0.0002
7.5	5.7792	1.0000	0.0001
8.0	6.2792	1.0000	0.0000

lution. Profiles from all locations on a flat plate are *similar;* they collapse to a single profile when plotted in nondimensional coordinates.

From Table 9.1, we see that at $\eta = 5.0$, $u/U = 0.992$. Defining the boundary layer thickness, δ, as the value of y for which $u/U = 0.99$, then from Eq. 9.7,

$$\delta \approx \frac{5.0}{\sqrt{U/\nu x}} = \frac{5.0x}{\sqrt{Re_x}} \qquad (9.13)$$

The boundary-layer thickness, δ, is indicated on the velocity profile plot of Fig. 9.3a.

The wall shear stress may be expressed as

$$\tau_w = \mu \frac{\partial u}{\partial y}\bigg]_{y=0} = \mu U \sqrt{U/\nu x} \frac{d^2 f}{d\eta^2}\bigg]_{\eta=0}$$

Then

$$\tau_w = 0.332 U \sqrt{\rho \mu U/x} = \frac{0.332 \rho U^2}{\sqrt{Re_x}} \qquad (9.14)$$

and the wall shear stress coefficient, C_f, is given by

$$C_f = \frac{\tau_w}{\frac{1}{2}\rho U^2} = \frac{0.664}{\sqrt{Re_x}} \qquad (9.15)$$

Each of the results for boundary-layer thicknesses, δ, wall shear stress, τ_w, and skin friction coefficient, C_f, Eqs. 9.13 through 9.15, depends on the length Reynolds number, Re_x, to the one-half power. The boundary-layer thickness increases as $x^{1/2}$, and the wall shear stress and skin friction coefficient vary as $1/x^{1/2}$. These results characterize the behavior of the laminar boundary layer on a flat plate.

EXAMPLE 9.2—Laminar Boundary Layer on a Flat Plate: Exact Solution

Use the numerical results presented in Table 9.1 to evaluate the following quantities for laminar boundary-layer flow on a flat plate:

   (a) δ^*/δ (evaluate for $\eta = 5$ and as $\eta \to \infty$).
   (b) v/U at the boundary-layer edge.
   (c) Compare the slope of a streamline at the boundary-layer edge with the slope of δ versus x.

EXAMPLE PROBLEM 9.2

GIVEN: Numerical solution for laminar flat-plate boundary layer, Table 9.1.

FIND: (a) δ^*/δ (evaluate for $\eta = 5$ and as $\eta \to \infty$).
       (b) v/U at boundary-layer edge.
       (c) Compare the slope of a streamline at the boundary-layer edge with the slope of δ versus x.

SOLUTION:
The displacement thickness is defined by Eq. 9.1 as

$$\delta^* = \int_0^\infty \left(1 - \frac{u}{U}\right) dy \approx \int_0^\delta \left(1 - \frac{u}{U}\right) dy$$

From Eq. 9.7, $\eta = y\sqrt{\dfrac{U}{\nu x}}$, so $y = \eta\sqrt{\dfrac{\nu x}{U}}$ and $dy = d\eta\sqrt{\dfrac{\nu x}{U}}$

Thus

$$\delta^* = \int_0^{\eta_{max}}\left(1 - \frac{u}{U}\right)\sqrt{\frac{\nu x}{U}}\,d\eta = \sqrt{\frac{\nu x}{U}}\int_0^{\eta_{max}}\left(1 - \frac{u}{U}\right)d\eta$$

Note: Corresponding to the upper limit on y in Eq. 9.1, $\eta_{max} = \infty$, or $\eta_{max} \approx 5$.
From Eq. 9.13,

$$\delta \approx \frac{5}{\sqrt{U/\nu x}}, \qquad \text{so} \qquad \sqrt{\frac{\nu x}{U}} = \frac{\delta}{5}$$

Thus

$$\frac{\delta^*}{\delta} = \frac{1}{5}\int_0^{\eta_{max}}\left(1 - \frac{u}{U}\right)d\eta$$

Substituting from Eq. 9.9, we obtain

$$\frac{\delta^*}{\delta} = \frac{1}{5}\int_0^{\eta_{max}}\left(1 - \frac{df}{d\eta}\right)d\eta$$

Integrating gives

$$\frac{\delta^*}{\delta} = \frac{1}{5}\Big[\eta - f(\eta)\Big]_0^{\eta_{max}}$$

Evaluating at $\eta_{max} = 5$, we obtain

$$\frac{\delta^*}{\delta} = \frac{1}{5}(5.0 - 3.2833) = 0.343 \qquad\qquad \frac{\delta^*}{\delta}(\eta = 5)$$

The quantity $\eta - f(\eta)$ becomes constant for $\eta > 8$. Evaluating at $\eta_{max} = 8$ gives

$$\frac{\delta^*}{\delta} = \frac{1}{5}(8.0 - 6.2792) = 0.344 \qquad\qquad \frac{\delta^*}{\delta}(\eta \to \infty)$$

Thus $\delta^*_{\eta\to\infty}$ is 0.29 percent larger than $\delta^*_{\eta=5}$.
From Eq. 9.10,

$$v = \frac{1}{2}\sqrt{\frac{\nu U}{x}}\left(\eta\frac{df}{d\eta} - f\right), \quad \text{so} \quad \frac{v}{U} = \frac{1}{2}\sqrt{\frac{\nu}{Ux}}\left(\eta\frac{df}{d\eta} - f\right) = \frac{1}{2\sqrt{Re_x}}\left(\eta\frac{df}{d\eta} - f\right)$$

Evaluating at the boundary-layer edge ($\eta = 5$), we obtain

$$\frac{v}{U} = \frac{1}{2\sqrt{Re_x}}[5(0.9915) - 3.2833] = \frac{0.837}{\sqrt{Re_x}} \approx \frac{0.84}{\sqrt{Re_x}} \qquad \frac{v}{U}(\eta = 5)$$

Thus v is only 0.84 percent of U at $Re_x = 10^4$, and only about 0.12 percent of U at $Re_x = 5\times 10^5$.

The slope of a streamline at the boundary-layer edge is

$$\left.\frac{dy}{dx}\right)_{streamline} = \frac{v}{u} = \frac{v}{U} \approx \frac{0.84}{\sqrt{Re_x}}$$

The slope of the boundary-layer edge may be obtained from Eq. 9.13.

$$\delta \approx \frac{5}{\sqrt{U/\nu x}} = 5\sqrt{\frac{\nu x}{U}}, \qquad \text{so} \qquad \frac{d\delta}{dx} = 5\sqrt{\frac{\nu}{U}}\frac{1}{2}x^{-1/2} = 2.5\sqrt{\frac{\nu}{Ux}} = \frac{2.5}{\sqrt{Re_x}}$$

Thus $\left. \dfrac{dy}{dx} \right)_{\text{streamline}} = \dfrac{0.84}{2.5}\dfrac{d\delta}{dx} = 0.336\dfrac{d\delta}{dx}$

This result shows that streamlines penetrate the boundary-layer edge, as sketched:

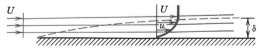

{ This problem illustrates application of results from the numerical solution of the laminar boundary layer on a flat plate. }

9-4 MOMENTUM INTEGRAL EQUATION

The exact solution of Blasius provided an expression for the boundary-layer thickness, $\delta(x)$, and the wall shear stress, $\tau_w(x)$. The velocity profiles were found to be similar when plotted nondimensionally as u/U versus y/δ. A closed-form solution for the velocity profile was not possible; a numerical solution was necessary.

Approximate methods may be used to obtain closed-form solutions for laminar boundary-layer flow on a flat plate. The same approximate methods may be used to solve for the characteristics of turbulent boundary-layer development. Since exact solutions for turbulent boundary layers do not exist, approximate solution techniques are necessary in this case. In this section we shall develop an analysis that will enable us to closely approximate the thickness of a laminar or turbulent boundary layer as a function of distance along a body. We shall again apply the integral equations to a differential control volume. Our aim is to develop an equation that will enable us to predict (at least approximately) the manner in which the boundary layer grows as a function of distance along a body. We shall derive a relation that may be applied to both laminar and turbulent flow; the relation will not be restricted to zero pressure gradient flows.

Consider the incompressible, steady, two-dimensional flow over a solid surface. The boundary-layer thickness, δ, grows in some manner with increasing distance, x. For our analysis we choose a differential control volume, of length dx, width w, and height $\delta(x)$, as shown in Fig. 9.4.

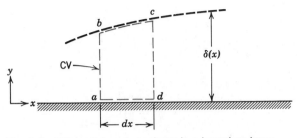

Fig. 9.4   Differential control volume in a boundary layer.

We wish to determine the boundary-layer thickness, δ, as a function of x. There will be mass flow across surfaces ab and cd of differential control volume $abcd$. What about surface bc? Will there be a mass flow across this surface? In our earlier discussion of boundary layers (Chapter 2), and in Example Problem 9.2, we found that the edge of the boundary layer is not a streamline. Thus there will be mass flow across surface bc. Since control surface ad is adjacent to a solid boundary, there will not be flow across ad. Before considering the forces acting on the control volume and the momentum fluxes through the control surface, let us apply the continuity equation to determine the mass flux through each portion of the control surface.

a. Continuity Equation

Basic equation:

$$= 0(1)$$

$$0 = \frac{\partial}{\partial t}\int_{CV} \rho \, d\forall + \int_{CS} \rho \vec{V} \cdot d\vec{A} \tag{4.13}$$

Assumptions: (1) Steady flow
(2) Two-dimensional flow

Then

$$0 = \int_{CS} \rho \vec{V} \cdot d\vec{A} = \dot{m}_{ab} + \dot{m}_{bc} + \dot{m}_{cd}$$

or

$$\dot{m}_{bc} = -\dot{m}_{ab} - \dot{m}_{cd}$$

Now let us evaluate these terms for the differential control volume of width w:

Surface   *Mass Flux*

ab Surface ab is located at x. Since the flow is two-dimensional (no variation with z), the mass flux through ab is

$$\dot{m}_{ab} = -\left\{ \int_0^\delta \rho u \, dy \right\} w$$

cd Surface cd is located at $x + dx$. Expanding $\dot{m}$ in a Taylor series about location x, we obtain

$$\dot{m}_{x+dx} = \dot{m}_x + \frac{\partial \dot{m}}{\partial x}\bigg]_x dx$$

and hence

$$\dot{m}_{cd} = \left\{ \int_0^\delta \rho u \, dy + \frac{\partial}{\partial x}\left[\int_0^\delta \rho u \, dy \right] dx \right\} w$$

bc Thus for surface bc we obtain

$$\dot{m}_{bc} = -\left\{ \frac{\partial}{\partial x}\left[\int_0^\delta \rho u \, dy \right] dx \right\} w$$

Now let us consider the momentum fluxes and forces associated with control volume $abcd$. These are related by the momentum equation.

b. Momentum Equation

Apply the x component of the momentum equation to control volume $abcd$:

Basic equation:

$$= 0(3) = 0(1)$$

$$F_{S_x} + \cancel{F_{B_x}} = \cancel{\frac{\partial}{\partial t}} \int_{CV} u \rho \, d\Psi + \int_{CS} u \rho \vec{V} \cdot d\vec{A} \qquad (4.19a)$$

Assumption: (3) $F_{B_x} = 0$

Then

$$F_{S_x} = \mathrm{mf}_{ab} + \mathrm{mf}_{bc} + \mathrm{mf}_{cd}$$

where mf represents the x component of momentum flux.

To apply this equation to differential control volume $abcd$, we must obtain expressions for the x momentum flux through the control surface and also the surface forces acting on the control volume in the x direction. Let us consider the momentum flux first and again consider each segment of the control surface.

Surface	*Momentum Flux (mf)*

ab   Surface ab is located at x. Since the flow is two-dimensional, the x momentum flux through ab is

$$\mathrm{mf}_{ab} = -\left\{ \int_0^{\delta} u \rho u \, dy \right\} w$$

cd   Surface cd is located at $x + dx$. Expanding the x momentum flux (mf) in a Taylor series about location x, we obtain

$$\mathrm{mf}_{x+dx} = \mathrm{mf}_x + \frac{\partial \mathrm{mf}}{\partial x}\bigg]_x dx$$

or

$$\mathrm{mf}_{cd} = \left\{ \int_0^{\delta} u \rho u \, dy + \frac{\partial}{\partial x}\left[\int_0^{\delta} u \rho u \, dy \right] dx \right\} w$$

bc   Since the mass crossing surface bc has velocity component U in the x direction, the x momentum flux across bc is given by

$$\mathrm{mf}_{bc} = U \dot{m}_{bc}$$

$$\mathrm{mf}_{bc} = -U \left\{ \frac{\partial}{\partial x}\left[\int_0^{\delta} \rho u \, dy \right] dx \right\} w$$

From the above we can evaluate the net x momentum flux through the control surface as

$$\int_{CS} u \rho \vec{V} \cdot d\vec{A} = -\left\{ \int_0^{\delta} u \rho u \, dy \right\} w + \left\{ \int_0^{\delta} u \rho u \, dy \right\} w$$

$$+ \left\{ \frac{\partial}{\partial x}\left[\int_0^{\delta} u \rho u \, dy \right] dx \right\} w - U \left\{ \frac{\partial}{\partial x}\left[\int_0^{\delta} \rho u \, dy \right] dx \right\} w$$

Collecting terms, we find that

$$\int_{CS} u\rho\vec{V}\cdot d\vec{A} = \left\{ \frac{\partial}{\partial x}\left[\int_0^\delta u\rho u\, dy\right]dx - U\frac{\partial}{\partial x}\left[\int_0^\delta \rho u\, dy\right]dx\right\} w$$

Now that we have a suitable expression for the x momentum flux through the control surface, let us consider the surface forces acting on the control volume in the x direction. (For convenience the differential control volume has been redrawn in Fig. 9.5.) We recognize that normal forces act on three surfaces of the control surface. In addition, a shear force acts on surface ad. Since the velocity gradient goes to zero at the edge of the boundary layer, no shear force acts along surface bc.

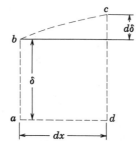

Fig. 9.5 Differential control volume.

Surface	*Force*

ab If the pressure at x is p, then the force acting on surface ab is given by

$$F_{ab} = pw\delta$$

(The boundary layer is very thin; its thickness has been greatly exaggerated in all the sketches we have made. Because it is thin, pressure variations in the y direction may be neglected, and we assume that within the boundary layer, $p = p(x)$.)

cd Expanding in a Taylor series, the pressure at $x + dx$ is given by

$$p_{x+dx} = p + \frac{dp}{dx}\bigg]_x dx$$

The force on surface cd is then given by

$$F_{cd} = -\left(p + \frac{dp}{dx}\bigg]_x dx\right)w(\delta + d\delta)$$

bc The average pressure acting over surface bc is

$$p + \frac{1}{2}\frac{dp}{dx}\bigg]_x dx$$

Then the x component of the normal force acting over bc is given by

$$F_{bc} = \left(p + \frac{1}{2}\frac{dp}{dx}\bigg]_x dx\right)w\, d\delta$$

ad   The shear force acting on *ad* is given by

$$F_{ad} = -\tau_w w \, dx$$

Summing the x component of each force acting on the control volume, we obtain

$$F_{S_x} = \left\{ -\frac{dp}{dx} \delta \, dx - \overset{\simeq 0}{\overbrace{\frac{1}{2} \frac{dp}{dx} dx \, d\delta}} - \tau_w \, dx \right\} w$$

where we note that $dx \, d\delta \ll \delta \, dx$, and so neglect the second term.

Substituting the expressions for $\int_{CS} u\rho \vec{V} \cdot d\vec{A}$ and F_{S_x} into the x momentum equation, we obtain

$$\left\{ -\frac{dp}{dx} \delta \, dx - \tau_w \, dx \right\} w = \left\{ \frac{\partial}{\partial x} \left[\int_0^\delta u\rho u \, dy \right] dx - U \frac{\partial}{\partial x} \left[\int_0^\delta \rho u \, dy \right] dx \right\} w$$

Dividing this equation by $w \, dx$ gives

$$-\delta \frac{dp}{dx} - \tau_w = \frac{\partial}{\partial x} \int_0^\delta u\rho u \, dy - U \frac{\partial}{\partial x} \int_0^\delta \rho u \, dy \qquad (9.16)$$

Equation 9.16 is a "momentum integral" equation that gives a relation between the x components of the forces acting in a boundary layer and the x momentum flux.

The pressure gradient, dp/dx, can be determined by applying the Bernoulli equation to the inviscid flow outside the boundary layer; $dp/dx = -\rho U \, dU/dx$. If we recognize that $\delta = \int_0^\delta dy$, then Eq. 9.16 can be written as

$$\tau_w = -\frac{\partial}{\partial x} \int_0^\delta u\rho u \, dy + U \frac{\partial}{\partial x} \int_0^\delta \rho u \, dy + \frac{dU}{dx} \int_0^\delta \rho U \, dy$$

Since

$$U \frac{\partial}{\partial x} \int_0^\delta \rho u \, dy = \frac{\partial}{\partial x} \int_0^\delta \rho u U \, dy - \frac{dU}{dx} \int_0^\delta \rho u \, dy$$

then

$$\tau_w = \frac{\partial}{\partial x} \int_0^\delta \rho u (U - u) \, dy + \frac{dU}{dx} \int_0^\delta \rho (U - u) \, dy$$

and

$$\tau_w = \frac{\partial}{\partial x} U^2 \int_0^\delta \rho \frac{u}{U} \left(1 - \frac{u}{U} \right) dy + U \frac{dU}{dx} \int_0^\delta \rho \left(1 - \frac{u}{U} \right) dy$$

Using the definitions of displacement thickness, δ^*, (Eq. 9.1) and momentum thickness, θ, (Eq. 9.2), then

$$\frac{\tau_w}{\rho} = \frac{d}{dx} (U^2 \theta) + \delta^* U \frac{dU}{dx} \qquad (9.17)$$

Equation 9.17 is the momentum integral equation; this equation will yield an ordinary differential equation for boundary-layer thickness, provided that a suitable form is assumed for the velocity profile and that the wall shear stress can be related to other variables. Once the boundary-layer thickness is determined, the momentum thickness, the displacement thickness, and the wall shear stress can be calculated.

Equation 9.17 was obtained by applying the basic equations (continuity and x momentum) to a differential control volume. Reviewing the assumptions we made in the derivation, we see that the equation is restricted to steady, incompressible, two-dimensional flow with no body forces.

We have not made any specific assumption relating the wall shear stress, τ_w, to the velocity field. Thus Eq. 9.17 is valid for either a laminar or turbulent boundary-layer flow. In order to use this equation to estimate the boundary-layer thickness as a function of x, we must:

1. Obtain a first approximation to the velocity distribution, $U(x)$. This is determined from inviscid flow theory (the velocity that would exist in the absence of a boundary layer). The pressure in the boundary layer is related to the freestream velocity, U, using the Bernoulli equation.
2. Assume a reasonable velocity profile shape inside the boundary layer.
3. Relate the wall shear stress to the velocity field.

To illustrate the application of Eq. 9.17 to boundary-layer flows, we consider first the zero pressure gradient case of flow over a flat plate (Section 9-5). The effects of pressure gradients in boundary-layer flow are then discussed in Section 9-6.

9-5 USE OF THE MOMENTUM INTEGRAL EQUATION FOR ZERO PRESSURE GRADIENT FLOW

For the special case of flow over a flat plate, $U = $ constant. From Bernoulli's equation, we see that for this case, $p = $ constant, and thus $dp/dx = 0$.

The momentum integral equation then reduces to

$$\tau_w = \rho U^2 \frac{d\theta}{dx} = \rho U^2 \frac{d}{dx} \int_0^\delta \frac{u}{U}\left(1 - \frac{u}{U}\right) dy \qquad (9.18)$$

The velocity distribution, u/U, in the boundary layer normally is specified as a function of y/δ. (Note that u/U is dimensionless and δ is a function of x only.) Consequently, it is convenient to change the variable of integration from y to y/δ. Defining

$$\eta = \frac{y}{\delta}$$

then

$$dy = \delta \, d\eta$$

and the momentum integral equation for zero pressure gradient is written

$$\tau_w = \rho U^2 \frac{d\theta}{dx} = \rho U^2 \frac{d\delta}{dx} \int_0^1 \frac{u}{U}\left(1 - \frac{u}{U}\right) d\eta \qquad (9.19)$$

We wish to solve this equation for the boundary-layer thickness as a function of x. To do this, we must:

1. Assume a velocity distribution in the boundary layer—a functional relationship of the form

$$\frac{u}{U} = f\left(\frac{y}{\delta}\right)$$

(a) The assumed velocity distribution should satisfy certain physical boundary conditions:

$$\text{at} \quad y = 0, \qquad u = 0$$
$$\text{at} \quad y = \delta, \qquad u = U$$
$$\text{at} \quad y = \delta, \qquad \frac{\partial u}{\partial y} = 0$$

(b) Note that once the velocity distribution has been assumed, then the numerical value of the integral in Eq. 9.19 is simply

$$\int_0^1 \frac{u}{U}\left(1 - \frac{u}{U}\right) d\eta = \frac{\theta}{\delta} = \text{constant} = \beta$$

and the momentum integral equation becomes

$$\tau_w = \rho U^2 \frac{d\delta}{dx}\beta$$

2. Obtain an expression for τ_w in terms of δ. This will then permit us to solve for $\delta(x)$, as illustrated below.

9-5.1 Laminar Flow

For laminar flow over a flat plate, a reasonable assumption for the velocity profile is a polynomial in y:

$$u = a + by + cy^2$$

The physical boundary conditions are:

$$\text{at} \quad y = 0, \qquad u = 0$$
$$\text{at} \quad y = \delta, \qquad u = U$$
$$\text{at} \quad y = \delta, \qquad \frac{\partial u}{\partial y} = 0$$

Evaluating the constants, a, b, and c, gives

$$\frac{u}{U} = 2\left(\frac{y}{\delta}\right) - \left(\frac{y}{\delta}\right)^2 = 2\eta - \eta^2 \tag{9.20}$$

The wall shear stress is given by

$$\tau_w = \mu \frac{\partial u}{\partial y}\bigg)_{y=0}$$

Substituting the assumed velocity profile, Eq. 9.20, into this expression for τ_w gives

$$\tau_w = \mu \frac{\partial u}{\partial y}\bigg]_{y=0} = \mu \frac{U\partial(u/U)}{\delta\partial(y/\delta)}\bigg]_{y/\delta=0} = \frac{\mu U}{\delta}\frac{d(u/U)}{d\eta}\bigg]_{\eta=0}$$

or

$$\tau_w = \frac{\mu U}{\delta}\frac{d}{d\eta}(2\eta - \eta^2)\bigg]_{\eta=0} = \frac{\mu U}{\delta}(2 - 2\eta)\bigg]_{\eta=0} = \frac{2\mu U}{\delta}$$

We are now in a position to apply the momentum integral equation

$$\tau_w = \rho U^2 \frac{d\delta}{dx}\int_0^1 \frac{u}{U}\left(1 - \frac{u}{U}\right) d\eta \tag{9.19}$$

Substituting for τ_w and u/U, we obtain

$$\frac{2\mu U}{\delta} = \rho U^2 \frac{d\delta}{dx} \int_0^1 (2\eta - \eta^2)(1 - 2\eta + \eta^2)\, d\eta$$

or

$$\frac{2\mu U}{\delta \rho U^2} = \frac{d\delta}{dx} \int_0^1 (2\eta - 5\eta^2 + 4\eta^3 - \eta^4)\, d\eta$$

Integrating and substituting limits yields

$$\frac{2\mu}{\delta \rho U} = \frac{2}{15}\frac{d\delta}{dx}$$

or

$$\delta\, d\delta = \frac{15\mu}{\rho U}\, dx$$

which is a differential equation for δ. Integrating again gives

$$\frac{\delta^2}{2} = \frac{15\mu}{\rho U}x + c$$

If it is assumed that $\delta = 0$ at $x = 0$, then $c = 0$ and thus

$$\delta = \sqrt{\frac{30\mu x}{\rho U}}$$

or

$$\frac{\delta}{x} = \sqrt{\frac{30\mu}{\rho U x}} = \frac{5.48}{\sqrt{Re_x}} \tag{9.21}$$

Equation 9.21 shows that the ratio of laminar boundary-layer thickness to distance along a flat plate varies inversely with the square root of length Reynolds number. It has the same form as the exact solution derived from the complete differential equations of motion by H. Blasius in 1908. Remarkably, Eq. 9.21 is only in error (the constant is too large) by about 10 percent compared to the exact solution (Section 9-3). Table 9.2 summarizes corresponding results calculated using approximate velocity profiles and lists results obtained from the exact solution. The shapes of the approximate profiles may be compared readily by plotting u/U versus y/δ (see Problem 9.9).

Once we know the boundary-layer thickness, all details of the flow may be determined. The wall shear stress, or "skin friction," coefficient is defined as

$$C_f \equiv \frac{\tau_w}{\frac{1}{2}\rho U^2} \tag{9.22}$$

Substituting from the velocity profile and Eq. 9.21 gives

$$C_f = \frac{\tau_w}{\frac{1}{2}\rho U^2} = \frac{2\mu(U/\delta)}{\frac{1}{2}\rho U^2} = \frac{4\mu}{\rho U \delta} = 4\frac{\mu}{\rho U x}\frac{x}{\delta} = 4\frac{1}{Re_x}\frac{\sqrt{Re_x}}{5.48}$$

Finally,

$$C_f = \frac{0.730}{\sqrt{Re_x}} \tag{9.23}$$

Table 9.2 Results of the Calculation of the Laminar Boundary-Layer Flow over a Flat Plate at Zero Incidence Based on Approximate Velocity Profiles

Velocity Distribution

$\dfrac{u}{U}=f\left(\dfrac{y}{\delta}\right)=f(\eta)$	$\dfrac{\theta}{\delta}$	$\dfrac{\delta^*}{\delta}$	$H=\dfrac{\delta^*}{\theta}$	$a=\dfrac{\delta}{x}\sqrt{Re_x}$	$b=C_f\sqrt{Re_x}$
$f(\eta)=\eta$	$\frac{1}{6}$	$\frac{1}{2}$	3.00	3.46	0.577
$f(\eta)=2\eta-\eta^2$	$\frac{2}{15}$	$\frac{1}{3}$	2.50	5.48	0.730
$f(\eta)=\dfrac{3}{2}\eta-\dfrac{1}{2}\eta^3$	$\frac{39}{280}$	$\frac{3}{8}$	2.69	4.64	0.647
$f(\eta)=2\eta-2\eta^3+\eta^4$	$\frac{37}{315}$	$\frac{3}{10}$	2.55	5.84	0.685
$f(\eta)=\sin\left(\dfrac{\pi}{2}\eta\right)$	$\dfrac{4-\pi}{2\pi}$	$\dfrac{\pi-2}{\pi}$	2.66	4.80	0.654
Exact	—	—	2.59	5.0	0.664

Once the variation of τ_w is known, the viscous drag on the surface can be evaluated by integration over the area of the flat plate, as illustrated in Example Problem 9.3.

Equation 9.21 can be used to calculate the thickness of the laminar boundary layer at transition. At $Re_x = 5 \times 10^5$, with $U = 30$ m/sec, $x = 0.24$ m for air at standard conditions. Thus

$$\frac{\delta}{x} = \frac{5.48}{\sqrt{Re_x}} = \frac{5.48}{\sqrt{5 \times 10^5}} = 0.00775$$

and the boundary-layer thickness is

$$\delta = 0.00775x = 0.00775(0.24 \text{ m}) = 1.86 \text{ mm}$$

The boundary-layer thickness at transition is less than 1 percent of the development length, x. These calculations confirm that viscous effects are confined to a very thin layer near the surface of a body.

The results in Table 9.2 indicate that reasonable results may be obtained with a variety of approximate velocity profiles.

EXAMPLE 9.3—Laminar Boundary Layer on a Flat Plate: Approximate Solution Using Sinusoidal Velocity Profile

Consider two-dimensional laminar boundary-layer flow along a flat plate. Assume the velocity profile in the boundary layer is sinusoidal,

$$\frac{u}{U} = \sin\left(\frac{\pi}{2}\frac{y}{\delta}\right)$$

Find expressions for:

(a) the rate of growth of δ as a function of x.
(b) the displacement thickness, δ^*, as a function of x.
(c) the total friction force on a plate of length L and width b.

EXAMPLE PROBLEM 9.3

GIVEN: Two-dimensional, laminar boundary-layer flow along a flat plate. The boundary-layer velocity profile is

$$\frac{u}{U} = \sin\left(\frac{\pi}{2}\frac{y}{\delta}\right) \qquad \text{for } 0 \le y \le \delta$$

and

$$\frac{u}{U} = 1 \qquad \text{for } y > \delta$$

FIND: (a) $\delta(x)$.
(b) $\delta^*(x)$.
(c) Total friction force on a plate of length L and width b.

SOLUTION:
For flat plate flow, $U = $ constant, $dp/dx = 0$, and

$$\tau_w = \rho U^2 \frac{d\theta}{dx} = \rho U^2 \frac{d\delta}{dx} \int_0^1 \frac{u}{U}\left(1 - \frac{u}{U}\right) d\eta \qquad (9.19)$$

Assumptions: (1) Steady flow
(2) Incompressible flow

Substituting $\dfrac{u}{U} = \sin\dfrac{\pi}{2}\eta$ into Eq. 9.19, we obtain

$$\tau_w = \rho U^2 \frac{d\delta}{dx} \int_0^1 \sin\frac{\pi}{2}\eta\left(1 - \sin\frac{\pi}{2}\eta\right) d\eta$$

$$= \rho U^2 \frac{d\delta}{dx} \int_0^1 \left(\sin\frac{\pi}{2}\eta - \sin^2\frac{\pi}{2}\eta\right) d\eta$$

$$= \rho U^2 \frac{d\delta}{dx} \frac{2}{\pi}\left[-\cos\frac{\pi}{2}\eta - \frac{1}{2}\frac{\pi}{2}\eta + \frac{1}{4}\sin\pi\eta\right]_0^1$$

$$= \rho U^2 \frac{d\delta}{dx} \frac{2}{\pi}\left[0 + 1 - \frac{\pi}{4} + 0 + 0 - 0\right]$$

$$\tau_w = 0.137\rho U^2 \frac{d\delta}{dx} = \beta\rho U^2 \frac{d\delta}{dx}; \quad \beta = 0.137$$

Now

$$\tau_w = \mu \frac{\partial u}{\partial y}\bigg]_{y=0} = \mu \frac{U}{\delta}\frac{\partial(u/U)}{\partial(y/\delta)}\bigg]_{y=0} = \mu \frac{U}{\delta}\frac{\pi}{2}\cos\frac{\pi}{2}\eta\bigg]_{\eta=0} = \frac{\pi\mu U}{2\delta}$$

Therefore,

$$\tau_w = \frac{\pi\mu U}{2\delta} = 0.137\rho U^2 \frac{d\delta}{dx}$$

Separating variables gives

$$\delta\, d\delta = 11.5\frac{\mu}{\rho U}dx$$

Integrating, we obtain

$$\frac{\delta^2}{2} = 11.5\frac{\mu}{\rho U}x + c$$

But $c = 0$, since $\delta = 0$ at $x = 0$, so

$$\delta = \sqrt{23.0\frac{x\mu}{\rho U}}$$

or

$$\frac{\delta}{x} = 4.80 \sqrt{\frac{\mu}{\rho U x}} = \frac{4.80}{\sqrt{Re_x}} \qquad\qquad\qquad \delta(x)$$

The displacement thickness, δ^*, is given by

$$\delta^* = \delta \int_0^1 \left(1 - \frac{u}{U}\right) d\eta$$

$$= \delta \int_0^1 \left(1 - \sin\frac{\pi}{2}\eta\right) d\eta = \delta \left[\eta + \frac{2}{\pi}\cos\frac{\pi}{2}\eta\right]_0^1$$

$$\delta^* = \delta \left[1 - 0 + 0 - \frac{2}{\pi}\right] = \delta \left[1 - \frac{2}{\pi}\right]$$

Since, from part (a),

$$\frac{\delta}{x} = \frac{4.80}{\sqrt{Re_x}}$$

then

$$\frac{\delta^*}{x} = \left(1 - \frac{2}{\pi}\right)\frac{4.80}{\sqrt{Re_x}} = \frac{1.74}{\sqrt{Re_x}} \qquad\qquad \delta^*(x)$$

The total friction force on one side of the plate is given by

$$F = \int_{A_p} \tau_w \, dA$$

Since $dA = b \, dx$ and $0 \le x \le L$, then

$$F = \int_0^L \tau_w b \, dx = \int_0^L \rho U^2 \frac{d\theta}{dx} b \, dx = \rho U^2 b \int_0^{\theta_L} d\theta = \rho U^2 b \theta_L$$

$$\theta_L = \int_0^{\delta_L} \frac{u}{U}\left(1 - \frac{u}{U}\right) dy = \delta_L \int_0^1 \frac{u}{U}\left(1 - \frac{u}{U}\right) d\eta = \beta \delta_L$$

From part (a), $\beta = 0.137$ and $\delta_L = \dfrac{4.80L}{\sqrt{Re_L}}$, so

$$F = \frac{0.658 \rho U^2 b L}{\sqrt{Re_L}} \qquad\qquad\qquad F$$

$$\left\{\begin{array}{l}\text{This problem illustrates the application of the momentum integral equation to a flat} \\ \text{plate, laminar, boundary-layer flow.}\end{array}\right\}$$

9-5.2 Turbulent Flow

Details of the turbulent velocity profile for boundary layers at zero pressure gradient are very similar to those for turbulent flow in pipes and channels. Data for turbulent boundary layers plot on the universal velocity profile using coordinates of $\bar{u}/u_*$ versus y/δ. However, the universal velocity profile is too complex mathematically for easy use with the momentum integral equation. The momentum integral equation is approximate; a suitable velocity profile for turbulent boundary layers on smooth flat plates is the empirical power-law profile. An exponent of $\frac{1}{7}$ is typically used to model the velocity profile,

$$\frac{u}{U} = \left(\frac{y}{\delta}\right)^{1/7} = \eta^{1/7} \tag{9.24}$$

However, this profile does not hold in the immediate vicinity of the wall, since at the wall it predicts $du/dy = \infty$. Consequently, we cannot use this profile in the definition of τ_w to obtain an expression for τ_w in terms of δ as we did for laminar boundary-layer flow. For turbulent boundary-layer flow we adapt the expression developed for pipe flow,

$$\tau_w = 0.03325 \rho \bar{V}^2 \left[\frac{\nu}{R\bar{V}}\right]^{0.25} \tag{8.35}$$

For a $\frac{1}{7}$-power profile in a pipe, Eq. 8.23 gives $\bar{V}/U = 0.817$. Substituting $\bar{V} = 0.817U$ and $R = \delta$ into Eq. 8.35, we obtain

$$\tau_w = 0.0233 \rho U^2 \left(\frac{\nu}{U\delta}\right)^{1/4} \tag{9.25}$$

We are now in a position to apply the momentum integral equation

$$\tau_w = \rho U^2 \frac{d\delta}{dx} \int_0^1 \frac{u}{U}\left(1 - \frac{u}{U}\right) d\eta \tag{9.19}$$

Substituting for τ_w and u/U and integrating, we obtain

$$0.0233 \left(\frac{\nu}{U\delta}\right)^{1/4} = \frac{d\delta}{dx} \int_0^1 \eta^{1/7}(1 - \eta^{1/7}) d\eta = \frac{7}{72}\frac{d\delta}{dx}$$

Thus we obtain a differential equation for δ:

$$\delta^{1/4} d\delta = 0.240\left(\frac{\nu}{U}\right)^{1/4} dx$$

Integrating gives

$$\frac{4}{5}\delta^{5/4} = 0.240 \left(\frac{\nu}{U}\right)^{1/4} x + c$$

If it is assumed that $\delta \simeq 0$ at $x = 0$ (this is equivalent to assuming turbulent flow from the leading edge), then $c = 0$ and

$$\delta = 0.382 \left(\frac{\nu}{U}\right)^{1/5} x^{4/5}$$

or

$$\frac{\delta}{x} = 0.382 \left(\frac{\nu}{Ux}\right)^{1/5} = \frac{0.382}{Re_x^{1/5}} \tag{9.26}$$

Using Eq. 9.25, we obtain the skin friction coefficient in terms of δ:

$$C_f = \frac{\tau_w}{\frac{1}{2}\rho U^2} = 0.0466 \left(\frac{\nu}{U\delta}\right)^{1/4}$$

Substituting for δ, we obtain

$$C_f = \frac{\tau_w}{\frac{1}{2}\rho U^2} = \frac{0.0594}{Re_x^{1/5}} \tag{9.27}$$

Experiments show that Eq. 9.27 predicts turbulent skin friction on a flat plate very well for $5\times10^5 < Re_x < 10^7$. This agreement is remarkable in view of the approximate nature of our analysis.

Application of the momentum integral equation for turbulent boundary-layer flow is illustrated in Example Problem 9.4.

Use of the momentum integral equation is an approximate technique to predict boundary-layer development; the equation predicts trends correctly. Parameters of the laminar boundary layer depend on $Re_x^{1/2}$; those for the turbulent boundary layer depend on $Re_x^{1/5}$. The turbulent boundary layer develops more rapidly than the laminar boundary layer. The agreement we have obtained with experimental results shows that use of the momentum integral equation is an effective method that gives us considerable insight into the general behavior of boundary layers.

EXAMPLE 9.4—Turbulent Boundary Layer on a Flat Plate: Approximate Solution Using $\frac{1}{7}$-power Velocity Profile

Water flows at $U = 1$ m/sec past a flat plate with $L = 1$ m in the flow direction. The boundary layer is tripped so it becomes turbulent at the leading edge. Evaluate the disturbance thickness, δ, displacement thickness, δ^*, and wall shear stress at $x = L$. Compare with laminar flow maintained to the same position. Assume a $\frac{1}{7}$-power turbulent velocity profile.

EXAMPLE PROBLEM 9.4

GIVEN: Flat-plate boundary-layer flow; turbulent flow from the leading edge. Assume $\frac{1}{7}$-power velocity profile.

FIND: (a) Disturbance thickness, δ.
  (b) Displacement thickness, δ^*.
  (c) Wall shear stress, τ_w.
  (d) Compare with results for laminar flow from the leading edge.

SOLUTION:

Apply results from the momentum integral equation.

Computing equations:

$$\frac{\delta}{x} = \frac{0.382}{Re_x^{1/5}} \tag{9.26}$$

$$\delta^* = \int_0^\infty \left(1 - \frac{u}{U}\right) dy \tag{9.1}$$

$$C_f = \frac{\tau_w}{\frac{1}{2}\rho U^2} = \frac{0.0594}{Re_x^{1/5}} \tag{9.27}$$

At $x = L$, with $\nu = 1.00\times10^{-6}$ m²/sec for water,

$$Re_L = \frac{UL}{\nu} = 1\ \frac{m}{sec} \times 1\ m \times \frac{sec}{10^{-6}\ m^2} = 10^6$$

From Eq. 9.26,

$$\delta_L = \frac{0.382}{Re_L^{1/5}} L = \frac{0.382}{(10^6)^{1/5}}\ 1\ m = 0.0241\ m \qquad or \qquad \delta_L = 24.1\ mm \qquad\qquad \underleftarrow{\delta_L}$$

Using Eq. 9.1, with $u/U = (y/\delta)^{1/7} = \eta^{1/7}$, we obtain

$$\delta_L^* = \int_0^\infty \left(1 - \frac{u}{U}\right) dy = \delta_L \int_0^1 \left(1 - \frac{u}{U}\right) d\left(\frac{y}{\delta}\right) = \delta_L \int_0^1 (1 - \eta^{1/7})\, d\eta = \delta_L \left[\eta - \frac{7}{8}\eta^{8/7}\right]_0^1$$

$$\delta_L^* = \frac{\delta_L}{8} = \frac{24.1 \text{ mm}}{8} = 3.01 \text{ mm} \qquad\qquad\qquad \delta_L^*$$

From Eq. 9.27,

$$C_f = \frac{0.0594}{(10^6)^{1/5}} = 0.00375$$

$$\tau_w = C_f \frac{1}{2}\rho U^2 = \frac{0.00375}{} \times \frac{1}{2} \times 999 \ \frac{\text{kg}}{\text{m}^3} \times (1)^2 \ \frac{\text{m}^2}{\text{sec}^2} \times \frac{\text{N} \cdot \text{sec}^2}{\text{kg} \cdot \text{m}}$$

$$\tau_w = 1.87 \text{ N/m}^2 \qquad\qquad\qquad \tau_w(L)$$

For laminar flow, use Blasius solution values. From Eq. 9.13,

$$\delta_L = \frac{5.0}{\sqrt{Re_L}} \ L = \frac{5.0}{(10^6)^{1/2}} \times \frac{1 \text{ m}}{} = 0.005 \text{ m} \qquad \text{or} \qquad 5.00 \text{ mm}$$

From Example 9.2, $\delta^*/\delta = 0.344$, so

$$\delta^* = 0.344 \ \delta = 0.344 \times 5.0 \text{ mm} - 1.72 \text{ mm}$$

From Eq. 9.15, $C_f = \dfrac{0.664}{\sqrt{Re_x}}$, so

$$\tau_w = C_f \frac{1}{2}\rho U^2 = \frac{0.664}{\sqrt{10^6}} \times \frac{1}{2} \times 999 \ \frac{\text{kg}}{\text{m}^3} \times (1)^2 \ \frac{\text{m}^2}{\text{sec}^2} \times \frac{\text{N} \cdot \text{sec}^2}{\text{kg} \cdot \text{m}} = 0.332 \text{ N/m}^2$$

Comparing values, we obtain

$$\text{Disturbance thickness,} \quad \frac{\delta_{\text{turbulent}}}{\delta_{\text{laminar}}} = \frac{24.1 \text{ mm}}{5.00 \text{ mm}} = 4.82$$

$$\text{Displacement thickness,} \quad \frac{\delta_{\text{turbulent}}^*}{\delta_{\text{laminar}}^*} = \frac{3.01 \text{ mm}}{1.72 \text{ mm}} = 1.75$$

$$\text{Wall shear stress,} \quad \frac{\tau_{w,\text{turbulent}}}{\tau_{w,\text{laminar}}} = \frac{1.88 \text{ N/m}^2}{0.332 \text{ N/m}^2} = 5.66$$

This problem illustrates use of the momentum integral equation for turbulent boundary layers. The results, when compared to those for laminar flow, indicate much more rapid growth due to the higher wall shear stress for the turbulent boundary layer.

9-6 PRESSURE GRADIENTS IN BOUNDARY-LAYER FLOW

We have restricted our analyses of boundary-layer flows to flow over a flat plate for which the pressure gradient is zero. The momentum integral equation for this case was given as

$$\tau_w = \rho U^2 \frac{d\theta}{dx} = \rho U^2 \frac{d}{dx} \int_0^\delta \frac{u}{U}\left(1 - \frac{u}{U}\right) dy \qquad (9.18)$$

Recall that in deriving this equation, no assumption was made regarding the boundary-layer flow regime; the equation is valid for both laminar and turbulent boundary layers. Equation 9.18 indicates that the wall shear stress is balanced by a decrease in fluid momentum. Thus the velocity profiles change as we move along the plate. The boundary-layer thickness continues to increase and the fluid close to the wall is continually being slowed down (losing momentum). One question of interest is, "Will the fluid close to the wall ever be brought to rest?" Rephrasing the question, "For the case of $dp/dx = 0$, is it possible that $\partial u/\partial y)_{y=0} = 0$?"[2]

In considering the wall shear stress distributions for flat plates, we found that for laminar flow

$$\frac{\tau_w(x)}{\rho U^2} = \frac{\text{constant}}{\sqrt{Re_x}}$$

and for turbulent flow

$$\frac{\tau_w(x)}{\rho U^2} = \frac{\text{constant}}{Re_x^{1/5}}$$

Recalling that $\tau_w = \mu \, \partial u/\partial y)_{y=0}$, we can then say that for any finite length plate, $\partial u/\partial y)_{y=0}$ will never be zero. The point on a solid boundary at which $\partial u/\partial y = 0$ is defined as the point of *separation*. Consequently, we conclude that for $dp/dx = 0$, the flow will not separate; the fluid layer in the neighborhood of a solid surface cannot be brought to zero velocity.

The pressure gradient is said to be adverse if the pressure increases in the direction of flow (if $\partial p/\partial x > 0$). When $\partial p/\partial x < 0$ (when the pressure decreases in the direction of flow), the pressure gradient is said to be favorable.

Consider the flow through a channel of variable cross section shown in Fig. 9.6. To simplify the discussion, consider the flow along the straight wall.

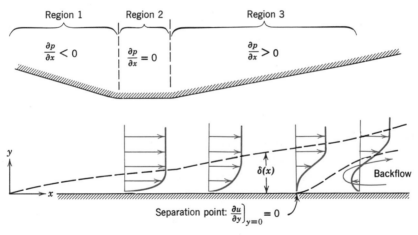

Fig. 9.6   Boundary-layer flow with pressure gradient (boundary-layer thickness exaggerated for clarity).

If we consider the forces acting on a fluid particle close to the solid boundary, we see that there is a net retarding shear force on the particle no matter what the sign of the pressure gradient. For $\partial p/\partial x = 0$, the result is a decrease in momentum,

---

[2] Note that if $\partial u/\partial y)_{y=0} = 0$, then the fluid layer near the wall will have zero velocity, since

$$u_{0+dy} = u_0 + \frac{\partial u}{\partial y}\bigg)_{y=0} dy$$

and $u_0 = 0$ from the no-slip condition.

but as we have already shown, it is not sufficient to bring the particle to rest. Since $\partial p/\partial x < 0$ in region 1, the pressure behind the particle (aiding its motion) is greater than that opposing its motion; the particle is "sliding down a pressure hill," without danger of being slowed to zero velocity. However, in attempting to flow through region 3, the particle encounters an adverse pressure gradient, $\partial p/\partial x > 0$, and the particle must "climb a pressure hill." The fluid particle could be brought to rest, thus causing the neighboring fluid to be deflected away from the boundary; when this occurs, the flow is said to separate from the surface. Just downstream from the point of separation, the flow direction in the separated region is opposite to the main flow direction. The low energy fluid in the separated region is forced back upstream by the increased pressure downstream.

Thus we see that an adverse pressure gradient, $\partial p/\partial x > 0$, is a necessary condition for separation. Does this mean that if $\partial p/\partial x > 0$, we will have separation? No, it does not. We have not shown that $\partial p/\partial x > 0$ will always lead to separation, but rather we have reasoned that separation cannot occur unless $\partial p/\partial x > 0$. This conclusion can be shown rigorously using the complete differential equations of motion for boundary-layer flow ([3], p. 132).

The nondimensional velocity profiles for laminar and turbulent boundary-layer flow over a flat plate are shown in Fig. 9.7a. The turbulent profile is much fuller (more blunt) than the laminar profile. At the same freestream speed, the momentum flux within the turbulent boundary layer is greater than within the laminar layer (Fig. 9.7b). Separation occurs when the momentum of fluid layers near the surface is reduced to zero by the combined action of pressure and viscous forces. As shown in Fig. 9.7b, the momentum of the fluid near the surface is significantly greater for the turbulent profile. Consequently, the turbulent layer is better able to resist separation in an adverse pressure gradient. We shall discuss some consequences of this behavior in Section 9-7.3.

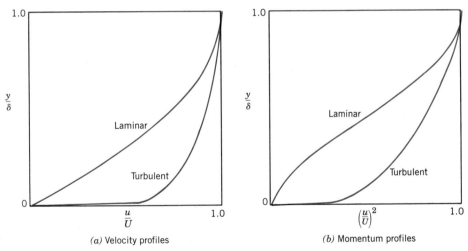

(a) Velocity profiles   (b) Momentum profiles

Fig. 9.7 Nondimensional profiles for flat plate boundary-layer flow.

Adverse pressure gradients cause significant changes in velocity profiles for both laminar and turbulent boundary-layer flows. Approximate solutions for nonzero pressure gradient flow may be obtained from the momentum integral equation

$$\frac{\tau_w}{\rho} = \frac{d}{dx}(U^2\theta) + \delta^* U \frac{dU}{dx} \qquad (9.17)$$

Expanding the first term, we can write

$$\frac{\tau_w}{\rho} = U^2 \frac{d\theta}{dx} + (\delta^* + 2\theta)U\frac{dU}{dx}$$

or

$$\frac{\tau_w}{\rho U^2} = \frac{C_f}{2} = \frac{d\theta}{dx} + (H+2)\frac{\theta}{U}\frac{dU}{dx} \tag{9.28}$$

where $H = \delta^*/\theta$ is a velocity-profile "shape factor." The shape factor increases in an adverse pressure gradient. For turbulent boundary-layer flow, H increases from 1.3 for a zero pressure gradient to approximately 2.5 at separation. For laminar flow with zero pressure gradient, $H = 2.6$; at separation $H = 3.5$.

The freestream velocity distribution, $U(x)$, must be known before Eq. 9.28 can be applied. Since $dp/dx = -\rho U \, dU/dx$, specifying $U(x)$ is equivalent to specifying the pressure gradient. We can obtain a first approximation for $U(x)$ from ideal flow theory for an inviscid flow under the same conditions. As pointed out in Chapter 6, for frictionless irrotational flow (potential flow), the stream function, ψ, and the velocity potential, ϕ, satisfy Laplace's equation. These can be used to determine $U(x)$ over the body surface.

Much effort has been devoted to calculation of velocity distributions over bodies of known shape (the "direct" problem) and to the determination of body shapes to produce a desired pressure distribution (the "inverse" problem). Smith and co-workers [6] have developed calculation methods that use singularities distributed over the body surface to solve the direct problem for two-dimensional or axisymmetric body shapes. A type of finite-element method that uses singularities defined on discrete surface panels (the "panel" method [7]) recently has gained increased popularity for application to three-dimensional flows.

Once the velocity distribution, $U(x)$, is known, Eq. 9.28 can be integrated to determine $\theta(x)$ if H and C_f can be correlated with θ. A detailed discussion of various calculation methods for nonzero pressure gradient flows is beyond the scope of this book. Numerous solutions for laminar flows are given in [8]. Calculation methods for turbulent boundary-layer flow based on the momentum integral equation are reviewed in [9].

Because of the importance of turbulent boundary layers in engineering flow situations, the state-of-the-art of calculation schemes is advancing rapidly. Numerous calculation schemes have been proposed [10,11]; most such schemes for turbulent flow use models to predict turbulent shear stress and then solve the boundary-layer equations numerically [12,13]. Continuing improvements in size and speed of digital computers is beginning to make possible the solution of the full Navier–Stokes equations using numerical methods.

PART B FLUID FLOW ABOUT IMMERSED BODIES

Whenever there is relative motion between a solid body and the fluid in which it is immersed, the body experiences a net force, $\vec{F}$, due to the action of the fluid. In general, the infinitesimal force, $d\vec{F}$, acting on an element of surface area, will be neither normal nor parallel to the element. This can be seen clearly when one considers the nature of the surface forces that contribute to the net force, $\vec{F}$. If the body is moving through a viscous fluid, then both shear and pressure forces act on the body,

$$\vec{F} = \int_{\text{body surface}} d\vec{F} = \int_{\text{body surface}} d\vec{F}_{\text{shear}} + \int_{\text{body surface}} d\vec{F}_{\text{pressure}}$$

The resultant force, $\vec{F}$, can be resolved into components parallel and perpendicular to the direction of motion. The component of force parallel to the direction of motion is the drag force, F_D, and the force component perpendicular to the direction of motion is the lift force, F_L.

Recognizing that

$$d\vec{F}_{\text{shear}} = \vec{\tau}_w \, dA$$

and

$$d\vec{F}_{\text{pressure}} = -p \, d\vec{A}$$

one might be inclined to think that drag and lift could be evaluated analytically. This proves not to be so; there are very few cases in which the lift and drag can be determined without recourse to experimental results. As we have seen, the presence of an adverse pressure gradient often leads to separation; flow separation prohibits the analytical determination of the force acting on a body. Therefore, for most shapes of interest, we must resort to the use of experimentally measured coefficients to compute lift and drag.

9-7 DRAG

Drag is the component of force on a body acting parallel to the direction of motion. In discussing the need for experimental results in fluid mechanics (Chapter 7), we considered the problem of determining the drag force, F_D, on a smooth sphere of diameter d, moving through a viscous, incompressible fluid with speed V; the fluid density and viscosity were ρ and μ, respectively. The drag force, F_D, was written in the functional form

$$F_D = f_1(d, V, \mu, \rho)$$

Application of the Buckingham Pi theorem resulted in two dimensionless Π parameters that were written in functional form as

$$\frac{F_D}{\rho V^2 d^2} = f_2\left(\frac{\rho V d}{\mu}\right)$$

Note that d^2 is proportional to the cross-sectional area ($A = \pi d^2/4$) and, hence, we could write

$$\frac{F_D}{\rho V^2 A} = f_3\left(\frac{\rho V d}{\mu}\right) = f_3(Re) \tag{9.29}$$

Although Eq. 9.29 was obtained for a sphere, the form of the equation is valid for incompressible flow over any body; the characteristic length used in the Reynolds number depends on the body shape.

The *drag coefficient*, C_D, is defined as

$$C_D \equiv \frac{F_D}{\frac{1}{2}\rho V^2 A} \tag{9.30}$$

The number $\frac{1}{2}$ has been inserted (as was done in the defining equation for the friction factor) to form the familiar dynamic pressure. Then Eq. 9.29 can be written as

$$C_D = f(Re) \tag{9.31}$$

We have not considered compressibility or free-surface effects in this discussion of the drag force. Had these been included, we would have obtained the functional form

$$C_D = f(Re, Fr, M)$$

At this point we shall consider the drag force on several bodies for which Eq. 9.31 is valid. The total drag force is the sum of friction drag and pressure drag. However, the drag coefficient is a function only of the Reynolds number.

9-7.1   Flow over a Flat Plate Parallel to the Flow: Friction Drag

This flow situation has been considered in detail in Section 9-5. Since the pressure gradient is zero, the total drag is equal to the friction drag. Thus

$$F_D = \int_{\text{plate surface}} \tau_w \, dA$$

and

$$C_D = \frac{F_D}{\frac{1}{2}\rho V^2 A} = \frac{\int_{PS} \tau_w \, dA}{\frac{1}{2}\rho V^2 A} \tag{9.32}$$

where A is the total surface area in contact with the fluid (i.e., the *wetted area*). The drag coefficient for a flat plate parallel to the flow depends on the shear stress distribution along the plate.

For laminar flow over a flat plate, the shear stress coefficient was given by

$$C_f = \frac{\tau_w}{\frac{1}{2}\rho U^2} = \frac{0.664}{\sqrt{Re_x}} \tag{9.15}$$

The drag coefficient for flow with freestream velocity V, over a flat plate of length L, and width b, is obtained by substituting for τ_w from Eq. 9.15 into Eq. 9.32. Thus

$$C_D = \frac{1}{A}\int_A 0.664 \, Re_x^{-0.5} \, dA = \frac{1}{bL}\int_0^L 0.664 \left(\frac{V}{\nu}\right)^{-0.5} x^{-0.5} b \, dx$$

$$= \frac{0.664}{L}\left(\frac{\nu}{V}\right)^{0.5}\left[\frac{x^{0.5}}{0.5}\right]_0^L = 1.328\left(\frac{\nu}{VL}\right)^{0.5}$$

$$C_D = \frac{1.328}{\sqrt{Re_L}} \tag{9.33}$$

Assuming the boundary layer is turbulent from the leading edge, the shear stress coefficient, based on the approximate analysis of Section 9-5.2, is given by

$$C_f = \frac{\tau_w}{\frac{1}{2}\rho U^2} = \frac{0.0594}{Re_x^{1/5}} \tag{9.27}$$

Substituting for τ_w from Eq. 9.27 into Eq. 9.32, we obtain

$$C_D = \frac{1}{A}\int_A 0.0594 \, Re_x^{-0.2} \, dA = \frac{1}{bL}\int_0^L 0.0594 \left(\frac{V}{\nu}\right)^{-0.2} x^{-0.2} b \, dx$$

$$= \frac{0.0594}{L}\left(\frac{\nu}{V}\right)^{0.2}\left[\frac{x^{0.8}}{0.8}\right]_0^L = 0.074\left(\frac{\nu}{VL}\right)^{0.2}$$

$$C_D = \frac{0.074}{Re_L^{1/5}} \tag{9.34}$$

Equation 9.34 is valid for $5 \times 10^5 < Re_L < 10^7$.

For $Re_L < 10^9$ the empirical equation given by Schlichting [3]

$$C_D = \frac{0.455}{(\log Re_L)^{2.58}} \tag{9.35}$$

fits experimental data very well.

For a boundary layer that is initially laminar and undergoes transition at some location on the plate, the turbulent drag coefficient must be adjusted to account for the laminar flow over the initial length. The adjustment is made by subtracting the quantity B/Re_L from the C_D determined for completely turbulent flow. The value of B depends on the Reynolds number at transition; B is given by

$$B = Re_{\text{tr}}(C_{D_{\text{turbulent}}} - C_{D_{\text{laminar}}}) \tag{9.36}$$

For a transition Reynolds number of 5×10^5, the drag coefficient may be calculated by making the adjustment to Eq. 9.34, in which case

$$C_D = \frac{0.074}{Re_L^{1/5}} - \frac{1740}{Re_L} \qquad (5 \times 10^5 < Re_L < 10^7) \tag{9.37a}$$

or to Eq. 9.35, in which case

$$C_D = \frac{0.455}{(\log Re_L)^{2.58}} - \frac{1610}{Re_L} \qquad (5 \times 10^5 < Re_L < 10^9) \tag{9.37b}$$

The variation in drag coefficient for a flat plate parallel to the flow is shown in Fig. 9.8.

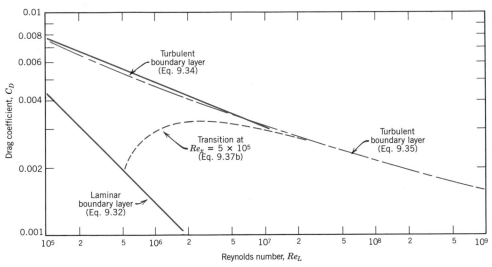

Fig. 9.8 Variation of drag coefficient with Reynolds number for a smooth flat plate parallel to the flow.

In the plot of Fig. 9.8, transition was assumed to occur at $Re_x = 5 \times 10^5$ for flows in which the boundary layer was initially laminar. The Reynolds number at which transition occurs depends on a combination of factors, such as surface roughness and freestream disturbances. Transition tends to occur earlier (at lower Reynolds number) as surface roughness or freestream turbulence is increased. For transition at other Re, the constant in the second term of Eqs. 9.37 is modified using Eq. 9.36. Figure 9.8 shows that the drag coefficient is less, for a given length of plate, when laminar flow is maintained over the longest possible distance. However, at large $Re_L(>10^7)$ the contribution of the laminar drag is negligible.

EXAMPLE 9.5—Skin Friction Drag on a Supertanker

A supertanker is 360 m long, has a beam width of 70 m, and a draft of 25 m. Estimate the force and power required to overcome skin friction drag at a cruising speed of 13 kt in seawater at 10 C.

EXAMPLE PROBLEM 9.5

GIVEN: Supertanker cruising at $U = 13$ kt.

FIND: (a)   Force, and
       (b)   Power
           required
           to overcome
           skin friction
           drag.

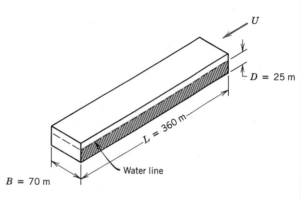

SOLUTION:
Model the tanker hull as a flat plate, of length L and width $b = B + 2D$, in contact with water. Estimate skin friction drag from the drag coefficient.

Computing equations:
$$C_D = \frac{F_D}{\frac{1}{2}\rho U^2 A} \tag{9.32}$$

$$C_D = \frac{0.455}{(\log Re_L)^{2.58}} - \frac{1610}{Re_L} \tag{9.37b}$$

The ship speed is 13 kt (nautical miles per hour), so

$$U = \frac{13 \text{ nm}}{\text{hr}} \times \frac{6076 \text{ ft}}{\text{nm}} \times \frac{0.3048 \text{ m}}{\text{ft}} \times \frac{\text{hr}}{3600 \text{ sec}} = 6.69 \text{ m/sec}$$

From Appendix A, at 10 C, $\nu \approx 1.4 \times 10^{-6} \text{ m}^2/\text{sec}$ for seawater. Then

$$Re_L = \frac{UL}{\nu} = \frac{6.69 \text{ m}}{\text{sec}} \times 360 \text{ m} \times \frac{\text{sec}}{1.4 \times 10^{-6} \text{ m}^2} = 1.72 \times 10^9$$

Assuming Eq. 9.37b is valid,

$$C_D = \frac{0.455}{(\log 1.72 \times 10^9)^{2.58}} - \frac{1610}{1.72 \times 10^9} = 0.00147$$

and from Eq. 9.32,

$$F_D = C_D A \frac{1}{2}\rho U^2$$

$$= 0.00147 \times (360 \text{ m})(70 + 50) \text{ m} \times \frac{1}{2} \times \frac{1020 \text{ kg}}{\text{m}^3} \times \frac{(6.69)^2 \text{ m}^2}{\text{sec}^2} \times \frac{\text{N} \cdot \text{sec}^2}{\text{kg} \cdot \text{m}}$$

$$F_D = 1.45 \text{ MN} \qquad\qquad\qquad\qquad\qquad\qquad\qquad\qquad\qquad\qquad\qquad F_D$$

The corresponding power is

$$\mathcal{P} = F_D U = \frac{1.45 \times 10^6 \text{ N}}{} \times 6.69 \frac{\text{m}}{\text{sec}} \times \frac{\text{W} \cdot \text{sec}}{\text{N} \cdot \text{m}}$$

$$\mathcal{P} = 9.70 \text{ MW} \qquad \qquad \mathcal{P}$$

This power requirement ($\sim$ 13,000 hp) is substantial. Although the drag coefficient is very low, the wetted surface area is large ($\sim$ 10.7 acres). Because the Reynolds number is large, the effect of laminar flow is negligible; transition occurs at $x \approx 0.1$ m.

9-7.2 Flow over a Flat Plate Normal to the Flow: Pressure Drag

In flow over a flat plate normal to the flow (Fig. 9.9), the wall shear stress does not contribute to the drag force. The drag is given by

$$F_D = \int_{\text{surface}} p \, dA$$

For this geometry the flow separates from the edges of the plate; there is backflow in the low energy wake of the plate. Although the pressure over the rear surface of the plate is essentially constant, its magnitude cannot be determined analytically. Consequently, we must resort to experiments to determine the drag force.

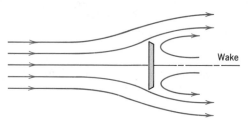

Fig. 9.9 Flow over a flat plate normal to the flow.

The *drag coefficient* for flow over an immersed object usually is based on the *frontal area* (or *projected area*) of the object. (For airfoils and wings, the planform area is used, see Section 9-8.)

The drag coefficient for a finite plate normal to the flow depends on the ratio of plate width to height and on the Reynolds number. For Re (based on height) greater than about 1000, the drag coefficient is essentially independent of Reynolds number. The variation of C_D with the ratio of plate width to height (b/h) is shown in Fig. 9.10. (The ratio b/h is defined as the *aspect ratio* of the plate.) For $b/h = 1.0$, the drag coefficient is a minimum at $C_D = 1.18$; this is just slightly higher than for a circular disk ($C_D = 1.17$) at large Reynolds number.

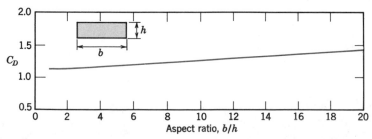

Fig. 9.10 Variation of drag coefficient with aspect ratio for a flat plate of finite width normal to the flow with $Re_h > 1000$ [14].

The drag coefficient for all objects with sharp edges is essentially independent of Reynolds number (for $Re \gtrsim 1000$) because the separation points are fixed by the geometry of the object. Drag coefficients for a few selected objects are given in Table 9.3.

Table 9.3 Drag Coefficient Data for Selected Objects ($Re \gtrsim 10^3$)[a]

Object	Diagram		$C_D (Re \gtrsim 10^3)$
Square cylinder		$b/h = \infty$	2.05
		$b/h = 1$	1.05
Disk			1.17
Ring			1.20[b]
Hemisphere (open end facing flow)			1.42
Hemisphere (open end facing downstream)			0.38
C-section (open side facing flow)			2.30
C-section (open side facing downstream)			1.20

[a] Data from [14].
[b] Based on ring area.

9-7.3   Flow over a Sphere and Cylinder: Friction and Pressure Drag

We have looked at two special flow cases in which either friction or pressure drag was the sole form of drag present. In the former case, the drag coefficient was a strong function of Reynolds number, while in the latter case, C_D was essentially independent of Reynolds number for $Re \gtrsim 1000$.

In the case of flow over a sphere, both friction drag and pressure drag contribute to total drag. The drag coefficient for flow over a smooth sphere is shown in Fig. 9.11 as a function of Reynolds number.[3]

At very low Reynolds number,[4] $Re \lesssim 1$, there is no flow separation from a sphere; the wake is laminar and the drag is predominantly friction drag. Stokes has shown analytically, for very low Reynolds number flows where inertia forces may be ne-

---

[3] An approximate curve fit to the data of Fig. 9.11 is presented in Problem 9.88.

[4] See the NCFMF film, *The Fluid Dynamics of Drag*, A. H. Shapiro, principal, or [15] for a good discussion of drag on spheres and other shapes. Another excellent (NCFMF) film is *Low Reynolds Number Flows*, Sir G. I. Taylor, principal. See also [16].

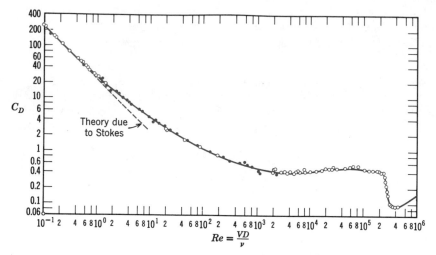

Fig. 9.11 Drag coefficient of a smooth sphere as a function of Reynolds number [3].

glected, that drag force on a sphere of diameter d, moving at speed V, through a fluid of viscosity μ, is given by

$$F_D = 3\pi\mu V d$$

The drag coefficient, C_D, defined by Eq. 9.30, is then

$$C_D = \frac{24}{Re}$$

As shown in Fig. 9.11, this expression agrees with experimental values at low Reynolds number but begins to deviate significantly from the experimental data for $Re > 1.0$.

As the Reynolds number is increased up to about 1000, the drag coefficient drops continuously. As a result of flow separation, the drag is a combination of friction and pressure drag. The relative contribution of friction drag decreases with increasing Reynolds number; at $Re \simeq 1000$, the friction drag is approximately 5 percent of the total drag.

In the range $10^3 < Re < 3 \times 10^5$, the drag coefficient curve is relatively flat. The drag coefficient undergoes a rather sharp drop at a *critical* Reynolds number of approximately 3×10^5. Experiments show that for $Re < 3 \times 10^5$ the boundary layer on the forward portion of the sphere is laminar. Separation of the boundary layer occurs just upstream of the sphere midsection; a relatively wide turbulent wake is present downstream from the sphere. In the separated region behind the sphere, the pressure is essentially constant and lower than the pressure over the forward portion of the sphere (Fig. 9.12). This pressure difference is the main contributor to the drag.

For Reynolds numbers larger than about 3×10^5, transition occurs and the boundary layer on the forward portion of the sphere becomes turbulent. The point of separation then moves downstream from the sphere midsection, and the size of the wake is decreased. The net pressure force on the sphere is reduced (Fig. 9.12), and the drag coefficient decreases abruptly.

A turbulent boundary layer, since it has more momentum than a laminar boundary layer, can better resist an adverse pressure gradient, as discussed in Section 9-6. Consequently, turbulent boundary-layer flow is desirable on a blunt body because it delays separation and thus reduces the pressure drag.

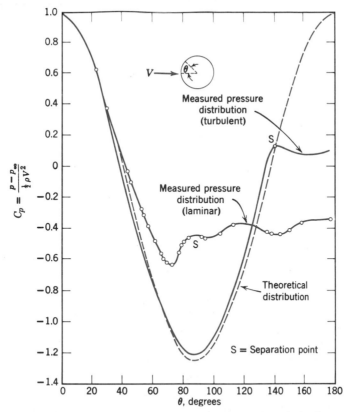

Fig. 9.12   Pressure distribution around a smooth sphere for laminar and turbulent boundary-layer flow, compared to inviscid flow [16].

Transition in the boundary layer is affected by roughness of the sphere surface and turbulence in the flow stream. Therefore, the reduction in drag associated with a turbulent boundary layer does not occur at a unique value of Reynolds number. Experiments with smooth spheres in a flow with low turbulence level show that transition may be delayed to a critical Reynolds number, Re_D, of about 4×10^5. For rough surfaces and/or highly turbulent freestream flow, transition can occur at a critical Reynolds number as low as 50,000.

The drag coefficient of a sphere with turbulent boundary-layer flow is about 5 times less than that for laminar flow near the critical Reynolds number. The corresponding reduction in drag force can affect the range of a sphere (e.g., a golf ball) appreciably. The "dimples" on a golf ball are designed to "trip" the boundary layer and, thus, to guarantee turbulent boundary-layer flow and minimum drag. To illustrate this effect more graphically, we obtained samples of golf balls without dimples some years ago. One of our students volunteered to hit some drives with the smooth balls. In 50 tries with each type of ball, the average distance with the standard balls was 215 yards; the average with the smooth balls was only 125 yards!

Adding roughness elements to a sphere also can suppress local oscillations in location of the transition between laminar and turbulent flow in the boundary layer. These oscillations can lead to variations in drag and to random fluctuations in lift (see Section 9-8). In baseball, the "knuckle ball" pitch is intended to behave erratically to confuse the batter. By throwing the ball with almost no spin, the pitcher re-

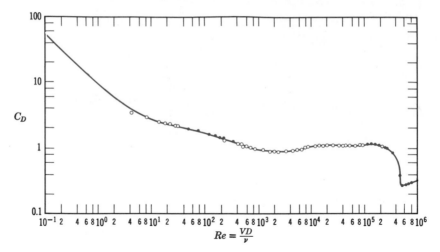

Fig. 9.13 Drag coefficient for a smooth circular cylinder as a function of Reynolds number [3].

lies on the seams to cause transition in an unpredictable fashion as the ball moves on its way to the batter. This causes the desired variation in the flight path of the ball.

Figure 9.13 shows the drag coefficient for flow over a smooth cylinder. The variation of C_D with Reynolds number shows the same characteristics as observed in the flow over a smooth sphere, but the values of C_D are about twice as large.

Flow about a smooth circular cylinder may develop a regular pattern of alternating vortices downstream. The *vortex trail*[5] causes an oscillatory lift force on the cylinder perpendicular to the stream motion. Vortex shedding excites oscillations that cause telegraph wires to "sing" and ropes on flag poles to "slap" annoyingly. Sometimes structural oscillations can reach dangerous magnitudes and cause high stresses; they can be reduced or eliminated by applying roughness elements or fins— either axial or helical—that destroy the symmetry of the cylinder and stabilize the flow.

Experimental data show that regular vortex shedding occurs most strongly in the range of Reynolds number from about 60 to 5000. For $Re > 1000$ the dimensionless frequency of vortex shedding, expressed as a Strouhal number, $St = nD/V$, is approximately equal to 0.21 [3].

Roughness affects drag of cylinders and spheres similarly: the critical Reynolds number is reduced by the rough surface and transition from laminar to turbulent flow in the boundary layers occurs earlier. The drag coefficient is reduced by a factor of about 4 when the boundary layer on the cylinder becomes turbulent.

EXAMPLE 9.6—Aerodynamic Drag and Moment on a Chimney

A cylindrical chimney 1 m in diameter and 25 m tall is exposed to a uniform 50 km/hr wind at standard atmospheric conditions. End effects and gusts may be neglected. Estimate the bending moment at the base of the chimney due to wind forces.

---

[5] The regular pattern of vortices in the wake of a cylinder sometimes is called a *Karman vortex street* in honor of the prominent fluid mechanician, Theodor von Karman, who was first to predict the stable spacing of the vortex trail on theoretical grounds in 1911 [17].

EXAMPLE PROBLEM 9.6

GIVEN: Cylindrical chimney, $D = 1$ m, $L = 25$ m, in uniform flow with

$$V = 50 \text{ km/hr} \qquad p = 101 \text{ kPa} \qquad T = 15 \text{ C}$$

Neglect end effects.

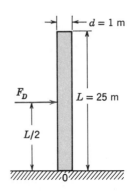

FIND: Bending moment at bottom of chimney.

SOLUTION:

The drag coefficient is given by $\quad C_D = \dfrac{F_D}{\frac{1}{2}\rho V^2 A}$,

and thus $\quad F_D = C_D A \dfrac{1}{2}\rho V^2$.

Since the force per unit length is uniform over the entire length, the resultant force, F_D, will act at the midpoint of the chimney. Hence the moment about the chimney base is

$$M_0 = F_D \frac{L}{2} = C_D A \frac{1}{2}\rho V^2 \frac{L}{2} = C_D A \frac{L}{4}\rho V^2$$

$$V = \frac{50 \text{ km}}{\text{hr}} \times \frac{10^3 \text{ m}}{\text{km}} \times \frac{\text{hr}}{3600 \text{ sec}} = 13.9 \text{ m/sec}$$

For air at standard conditions, $\rho = 1.23 \text{ kg/m}^3$, and $\mu = 1.78 \times 10^{-5} \text{ kg/m} \cdot \text{sec}$. Thus

$$Re = \frac{\rho V D}{\mu} = \frac{1.23 \text{ kg}}{\text{m}^3} \times \frac{13.9 \text{ m}}{\text{sec}} \times \frac{1 \text{ m}}{} \times \frac{\text{m} \cdot \text{sec}}{1.78 \times 10^{-5} \text{ kg}} = 9.61 \times 10^5$$

From Fig. 9.13, $C_D \approx 0.35$. For a cylinder, $A = DL$, so

$$M_0 = C_D A \frac{L}{4}\rho V^2 = C_D D L \frac{L}{4}\rho V^2 = C_D D \frac{L^2}{4}\rho V^2$$

$$= \frac{1}{4} \times \frac{0.35}{} \times \frac{1 \text{ m}}{} \times \frac{(25)^2 \text{ m}^2}{} \times \frac{1.23 \text{ kg}}{\text{m}^3} \times \frac{(13.9)^2 \text{ m}^2}{\text{sec}^2} \times \frac{\text{N} \cdot \text{sec}^2}{\text{kg} \cdot \text{m}}$$

$$M_0 = 13.0 \text{ kN} \cdot \text{m} \qquad\qquad\qquad\qquad\qquad\qquad M_0$$

{ This problem illustrates application of drag coefficient data to calculate the moment due to wind force on a structure. Actually, the velocity profile for wind over the ground is not uniform. Wind speed in the atmospheric boundary layer frequently is modeled using the power-law profile. It can be shown (see Problem 9.130) that the moment due to a power-law profile is $n/(n+1)$ times the moment for a uniform flow. Thus little error is introduced by assuming uniform flow. }

EXAMPLE 9.7—Deceleration of an Automobile by a Drag Parachute

A dragster weighing 1600 lbf attains a speed of 270 mph in the quarter mile. Immediately after passing through the timing lights, the driver opens the drag chute, of area $A = 25$ ft^2. Air and rolling resistance of the car may be neglected. Find the time required for the machine to decelerate to 100 mph in standard air.

EXAMPLE PROBLEM 9.7

GIVEN: Dragster weighing 1600 lbf, moving with speed $V = 270$ mph, is slowed by the drag force on a chute of area $A = 25$ ft^2. Neglect air and rolling resistance of the car. Assume standard air.

FIND: Time required for the machine to decelerate to 100 mph.

SOLUTION:
Taking the car as a system and writing Newton's second law in the direction of motion gives

$$-F_D = ma = m\frac{dV}{dt}$$

$V_0 = 270$ mph
$V_f = 100$ mph
$\rho = 0.00238$ slug/ft^3

Since $\quad C_D = \dfrac{F_D}{\frac{1}{2}\rho V^2 A}, \quad$ then $\quad F_D = \frac{1}{2}C_D\rho V^2 A.$

Substituting into Newton's second law gives

$$-\frac{1}{2}C_D\rho V^2 A = m\frac{dV}{dt}$$

Separating variables and integrating, we obtain

$$-\frac{1}{2}C_D\rho\frac{A}{m}\int_0^t dt = \int_{V_0}^{V_f}\frac{dV}{V^2}$$

$$-\frac{1}{2}C_D\rho\frac{A}{m}t = -\frac{1}{V}\bigg]_{V_0}^{V_f} = -\frac{1}{V_f} + \frac{1}{V_0} = -\frac{(V_0 - V_f)}{V_f V_0}$$

Finally,

$$t = \frac{(V_0 - V_f)}{V_f V_0}\frac{2m}{C_D\rho A} = \frac{(V_0 - V_f)}{V_f V_0}\frac{2W}{C_D\rho Ag}$$

Model the drag chute as a hemisphere (with open end facing flow). From Table 9.3, $C_D = 1.42$ (assuming $Re > 10^3$). Then, substituting numerical values,

$$t = \frac{(270 - 100)\text{ mph}}{} \times 2 \times 1600\text{ lbf} \times \frac{1}{100\text{ mph}} \times \frac{\text{hr}}{270\text{ mi}} \times \frac{1}{1.2} \times \frac{\text{ft}^3}{0.00238\text{ slug}}$$

$$\times \frac{1}{25\text{ ft}^2} \times \frac{\text{sec}^2}{32.2\text{ ft}} \times \frac{\text{slug}\cdot\text{ft}}{\text{lbf}\cdot\text{sec}^2} \times \frac{\text{mi}}{5280\text{ ft}} \times \frac{3600\text{ sec}}{\text{hr}}$$

$t = 5.05$ sec $\xleftarrow{\hspace{5cm}} t$

Check the assumption on Re:

$$Re = \frac{DV}{\nu} = \left[\frac{4A}{\pi}\right]^{1/2}\frac{V}{\nu}$$

$$= \left[\frac{4}{\pi}\times 25\text{ ft}^2\right]^{1/2} \times \frac{100\text{ mi}}{\text{hr}} \times \frac{\text{hr}}{3600\text{ sec}} \times \frac{5280\text{ ft}}{\text{mi}} \times \frac{\text{sec}}{1.56\times 10^{-4}\text{ ft}^2}$$

$Re = 5.30\times 10^6$

Hence the assumption is valid.

All experimental data presented in this section are for single objects immersed in an unbounded fluid stream. The objective of wind tunnel tests is to simulate the conditions of an unbounded flow. Limitations on equipment size make this goal unreachable in practice. Frequently it is necessary to apply corrections to measured data to obtain results applicable to unbounded flow conditions.

In numerous realistic flow situations, interactions occur with nearby objects or surfaces. Drag can be reduced significantly when two or more objects, moving in tandem, interact. This phenomenon is well known to bicycle riders and those interested in automobile racing, where "drafting" is a common practice. Drag reductions of 80 percent may be achieved with optimum spacing [18]. Drag also can be increased significantly when spacing is not optimum.

Drag can be affected by adjacent neighbors as well. Small particles falling under gravity travel more slowly when they have neighbors than when they are isolated. This phenomenon, which is illustrated in the NCFMF film, *Low Reynolds Number Flows,* has important applications to mixing and sedimentation processes.

Experimental data for drag coefficients on objects must be selected and applied carefully. Due regard must be given to the differences between the actual conditions and the more controlled conditions under which measurements were made.

9-7.4 Streamlining

The extent of the separated flow region behind many of the objects discussed in the previous section can be reduced or eliminated by streamlining, or fairing, the body shape. The objective of streamlining is to reduce the adverse pressure gradient that occurs behind the point of maximum thickness on the body. This delays boundary-layer separation and thus reduces the pressure drag. However, addition of a faired tail section increases the surface area of the body; this causes skin friction drag to increase. The optimum streamlined shape is thus the shape that gives minimum total drag. These effects are discussed at length in the NCFMF film series, *The Fluid Dynamics of Drag.*

The pressure gradient around a "teardrop" shape (a "streamlined" cylinder) is less severe than that around a cylinder of circular section. The trade-off between pressure and friction drag for this case is illustrated by the results presented in Fig. 9.14,

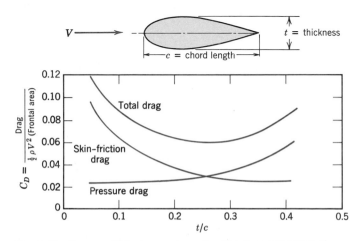

Fig. 9.14   Drag coefficient on a streamlined strut as a function of thickness ratio, showing contributions of skin friction and pressure to total drag [18].

for tests at $Re_c = 4 \times 10^5$. (This Reynolds number is typical of that for a strut on an early aircraft.) From the figure, the minimum drag coefficient is $C_D \simeq 0.06$, which occurs when the thickness to chord ratio is $t/c \simeq 0.25$. This value is approximately 20 percent of the minimum drag coefficient for a circular cylinder of the same thickness! Consequently, a streamlined strut about 5 times the thickness of a cylindrical strut could be used, with no penalty in aerodynamic drag.

The maximum thickness (and therefore minimum pressure) for the shapes shown in Fig. 9.14 is located approximately 25 percent of the chord distance from the leading edge. Most of the drag on these shapes is due to skin friction in the turbulent boundary layers on the tapered rear sections. Interest in low-drag airfoils increased during the 1930s. The National Advisory Committee for Aeronautics (NACA) developed several series of "laminar-flow" airfoils for which transition was postponed to 60 or 65 percent chord length aft from the airfoil nose.

Pressure distribution and drag data[6] for two symmetric airfoils of infinite span and 15 percent thickness at zero angle of attack are presented in Fig. 9.15. Transition on the conventional (NACA 0015) airfoil takes place where the pressure gradient becomes adverse, at $x/c = 0.13$, near the point of maximum thickness. Thus most of the airfoil surface is covered with a turbulent boundary layer; the drag coefficient

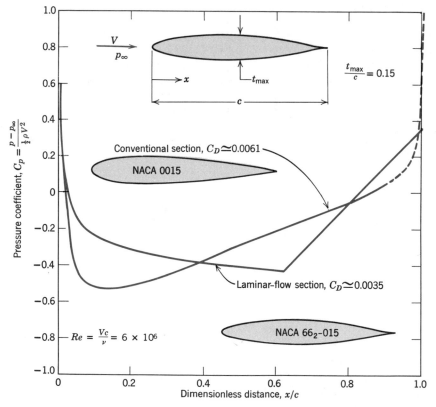

Fig. 9.15 Theoretical pressure distributions at zero angle of attack for two symmetric airfoil sections of 15 percent thickness ratio. (Data from [19].)

---

[6] Note that drag coefficients for airfoils are based on the planform area, i.e., $C_D = F_D / \frac{1}{2} \rho V^2 A_p$, where A_p is the maximum projected wing area.

is $C_D \simeq 0.0061$. The point of maximum thickness has been moved aft on the airfoil (NACA 66_2–015) designed for laminar flow. The boundary layer is maintained in the laminar regime by the favorable pressure gradient to $x/c = 0.63$. Thus the bulk of the flow is laminar; $C_D \simeq 0.0035$ for this section, based on planform area. The drag coefficient based on frontal area is $C_{D_f} = C_D/0.15 = 0.0233$, or about 40 percent of that for the shapes shown in Fig. 9.14.

Tests in special wind tunnels have shown that laminar flow can be maintained up to length Reynolds numbers as high as 30 million by appropriate profile shaping. Because they have favorable drag characteristics, laminar-flow airfoils are used in the design of most modern subsonic aircraft.

Recent advances have made possible development of low-drag shapes even better than the NACA 60-series shapes. Experiments [20] led to the development of a pressure distribution that prevented separation while maintaining the turbulent boundary layer in a condition that produces negligible skin friction. Improved methods for calculating body shapes that produced a desired pressure distribution [21, 22] led to development of nearly optimum shapes for thick struts with low drag. Figure 9.16 shows an example of the results.

Reduction of aerodynamic drag also is important for road vehicle applications. Interest in fuel economy has provided significant incentive to balance efficient aerodynamic performance with attractive design for automobiles. Drag reduction also has become important for buses and trucks.

Practical considerations limit the overall length of road vehicles. Fully streamlined tails are impractical for all but land-speed record cars. Consequently, it is not possible to achieve results comparable to those for optimum airfoil shapes. However, it is possible to optimize both front and rear contours within given constraints on overall length [23-25].

Much attention has been focused on front contours. Studies on buses have shown that drag reductions up to 30 percent are possible with careful attention to front contour [25]. Thus it is possible to reduce the drag coefficient of a bus from about 0.65 to less than 0.5 with practical designs. Highway tractor-trailer rigs have higher drag coefficients—C_Ds from 0.90 to 1.1 have been reported. Commercially available add-on devices offer improvements in drag of up to 15 percent, particularly for windy conditions where yaw angles are nonzero. The typical fuel saving is half the percentage by which aerodynamic drag is reduced.

Front contours and details are important for automobiles. A low nose and smoothly rounded contours are the primary features that promote low drag. Radii of "A-pillar" and windshield header, and blending of accessories to reduce parasite and interference drag have received increased attention. As a result, drag coefficients have been reduced from about 0.55 to 0.30 or less for recent production vehicles.

Fig. 9.16 Nearly optimum shape
for low-drag strut [22].

Recent advances in computational methods have led to development of computer generated optimum shapes. A number of designs have been proposed, with claims of C_Ds below 0.2 for vehicles complete with running gear.

9-8 LIFT

Lift is the component of resultant aerodynamic force perpendicular to the fluid motion. A common example of dynamic lift is flow over an airfoil.[7] The *lift coefficient, C_L*, is defined as

$$C_L \equiv \frac{F_L}{\frac{1}{2}\rho V^2 A_p} \tag{9.38}$$

The lift and drag coefficients for an airfoil are functions of both Reynolds number and angle of attack; the angle of attack, α, is the angle between the airfoil chord and the freestream velocity vector. The *chord* of an airfoil is the straight line joining the mean thickness line between the airfoil leading edge and the trailing edge. When the airfoil has a symmetric section, the *mean line* and the chord line both are straight lines, and they coincide. An airfoil with a curved mean line is said to be *cambered*.

The area at right angles to the flow changes with angle of attack. Consequently, the planform area, A_p (the maximum projected area of the wing), is used to define lift and drag coefficients for an airfoil.

Lift and drag coefficient data for typical conventional and laminar-flow profiles are plotted in Fig. 9.17, for a Reynolds number of 9×10^6, based on chord length. The section shapes in Fig. 9.17 are designated as follows:

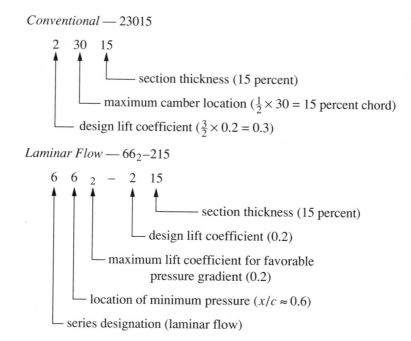

Conventional — 23015

2   30   15

—— section thickness (15 percent)

—— maximum camber location ($\frac{1}{2} \times 30 = 15$ percent chord)

—— design lift coefficient ($\frac{3}{2} \times 0.2 = 0.3$)

Laminar Flow — 66_2–215

6   6   2   –   2   15

—— section thickness (15 percent)

—— design lift coefficient (0.2)

—— maximum lift coefficient for favorable pressure gradient (0.2)

—— location of minimum pressure ($x/c \approx 0.6$)

—— series designation (laminar flow)

Both sections are cambered to give lift at zero angle of attack. As the angle of attack is increased, the lift coefficients increase smoothly until a maximum is reached. Further increases in angle of attack produce a sudden decrease in C_L. The airfoil is said to have *stalled* when C_L drops in this fashion.

---

[7] Flow over an airfoil is shown in the NCFMF film, *Boundary Layer Control*, D. C. Hazen, principal.

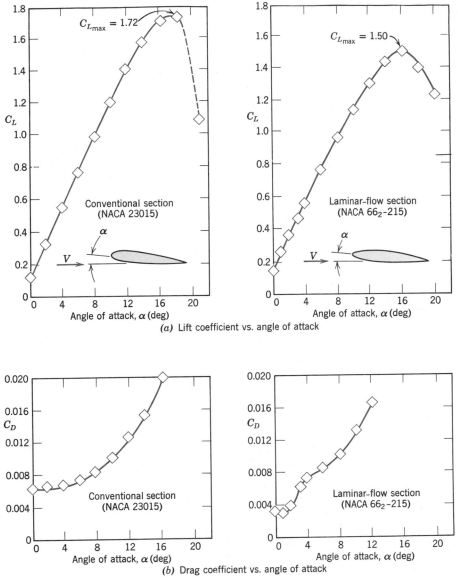

Fig. 9.17 Lift and drag coefficients versus angle of attack for two airfoil sections of 15 percent thickness ratio. (Data from [19].)

Airfoil stall results when flow separation occurs over a major portion of the upper surface of the airfoil. As the angle of attack is increased, the stagnation point moves back along the lower surface of the airfoil, as shown schematically in Fig. 9.18. Flow on the upper surface then must accelerate sharply to round the nose of the airfoil.[8] The minimum pressure becomes lower, and its location moves forward on the upper surface. A severe adverse pressure gradient appears following the point of minimum pressure; finally, the adverse pressure gradient causes the flow to separate completely from the upper surface and the airfoil stalls.

---

[8] Flow patterns and pressure distributions for airfoil sections are shown in the NCFMF film, *Boundary Layer Control*, D. C. Hazen, principal.

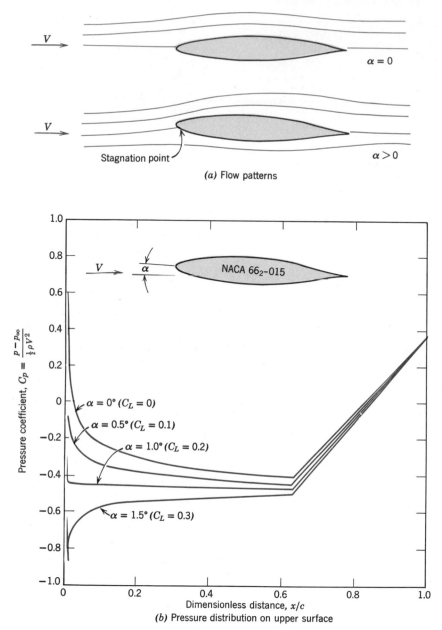

Fig. 9.18 Effect of angle of attack on flow pattern and theoretical pressure distribution for a symmetric laminar-flow airfoil of 15 percent thickness ratio. (Data from [19].)

Movement of the minimum pressure point and accentuation of the adverse pressure gradient are responsible for the sudden increase in C_D for the laminar-flow section, which is apparent in Fig 9.17. The sudden rise in C_D is due to early transition from laminar to turbulent boundary-layer flow on the upper surface. Aircraft with laminar-flow sections are designed to cruise in the low drag region.

Because laminar-flow sections have very sharp leading edges, all of the effects we have described are exaggerated, and they stall at lower angles of attack than

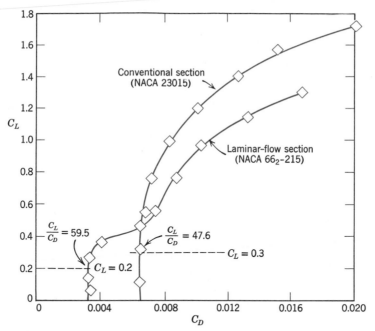

Fig. 9.19 Lift-drag polars for two airfoil sections of 15 percent thickness ratio. (Data from [19].)

conventional sections, as shown in Fig. 9.17. The maximum possible lift coefficient, $C_{L_{\max}}$, also is less for laminar-flow sections.

Plots of C_L versus C_D (called lift-drag polars) often are used to present airfoil data. A polar plot is given in Fig. 9.19 for the two sections we have discussed. The lift/drag ratio, C_L/C_D, is shown at the design lift coefficient for both sections. For an aircraft of given mass at fixed speed, the power required for level flight is inversely proportional to the lift/drag ratio. The advantage of the laminar-flow section is clear.

Recent improvements in modeling and computational capabilities have made it possible to design airfoil sections that develop high lift while maintaining very low drag [21, 22]. Boundary-layer calculation codes are used with inverse methods for calculating potential flow to develop pressure distributions and the resulting body shapes that postpone transition to the most rearward location possible. The turbulent boundary layer following transition is maintained in a state of incipient separation with nearly zero skin friction by appropriate shaping of the pressure distribution.

Such computer-designed airfoils have been used on racing cars to develop very high negative lift (downforce) to improve high-speed stability and cornering performance [21]. Airfoil sections especially designed for operation at low Reynolds number were used for the wings and propeller on the Kremer prize-winning man-powered "Gossamer Condor" [26], which now hangs in the National Air and Space Museum in Washington, D.C.

All of the data presented so far have been for *sections,* slices from airfoils of infinite span. End effects on *wings* of finite span reduce lift and increase drag. Thus the lift/drag ratios that can be achieved in practice are less than those obtained from tests of airfoil sections.

Finite-span effects can be correlated using the *aspect ratio,* defined as

$$ar \equiv \frac{b^2}{A_p} \qquad (9.39)$$

where A_p is planform area and b is wingspan. For a rectangular planform of span b and chord c,

$$ar = \frac{b^2}{A_p} = \frac{b^2}{bc} = \frac{b}{c}$$

The maximum lift/drag ratio ($L/D = C_L/C_D$) for a modern low-drag section may be as high as 400 for infinite aspect ratio. A high-performance sailplane (glider) with $ar = 40$ might have $L/D = 40$, and a typical light plane ($ar \sim 12$) might have $L/D \sim 20$ or so. Two examples of rather poor shapes are lifting bodies used for reentry from the upper atmosphere, and water skis, which are *hydrofoils* of low aspect ratio. For both of these shapes, L/D typically is less than unity.

Mother nature is well aware of the effects of aspect ratio on aerodynamic performance. Soaring birds, such as the albatross or California condor, have thin wings of long span. Birds that must maneuver quickly to catch their prey, such as owls, have wings of relatively short span, but large area, which gives low *wing loading* (ratio of weight to planform area) and thus high maneuverability.

A wing of finite span carries with it a system of *trailing vortices,* as shown schematically in Fig. 9.20, whenever it generates lift. Trailing vortices result from leakage flows around the wing tips, from the high pressure below to the low pressure above the wing. Trailing vortices may be very strong and persistent, and they may present a hazard to light planes for 5 to 10 miles behind a large plane. Air speeds exceeding 200 mph have been measured in trailing vortices from large, heavy aircraft.[9]

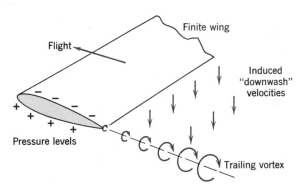

Fig. 9.20 Schematic representation of the trailing vortex system of a finite wing.

It is possible to increase the *effective* aspect ratio for a wing of given geometric aspect ratio by adding an *endplate* or *winglet* to the wing tip. An endplate may be a simple plate attached at the tip, perpendicular to the wing span, as on the rear-mounted wing of the racing car, shown (later) in Fig. 9.26. An endplate functions by blocking the flow that tends to migrate from the high-pressure region below the wing tip to the low-pressure region above the tip when the wing is producing lift. When the endplate is added, the strength of the trailing vortex and the induced drag are reduced.

Winglets are short, aerodynamically contoured wings set perpendicular to the wingtip. Like the endplate, the winglet reduces the strength of the trailing vortex

[9] Sforza, P. M., "Aircraft Vortices: Benign or Baleful?" *Space/Aeronautics*, *53*, 4, April 1970, pp. 42–49. See also the University of Iowa film, *Form Drag, Lift, and Propulsion,* H. Rouse, principal.

system and the induced drag. The winglet also produces a small component of force in the flight direction, which has the effect of further reducing the overall drag of the aircraft. The contour and angle of attack of the winglet are adjusted to provide optimum results based on wind tunnel tests.

Induced downwash velocities on a lifting wing reduce the effective angle of attack, reducing lift. (At fixed geometric angle of attack, the wing "sees" a flow at approximately the mean of the upstream and downstream flow directions.) To maintain the same lift force, the geometric angle of attack must be increased. This causes drag to increase compared to the infinite aspect ratio case. These effects are illustrated schematically in Fig. 9.21.

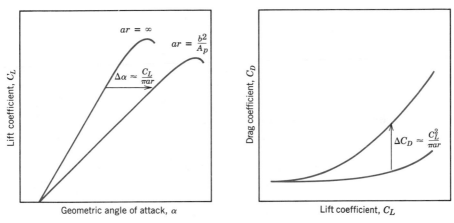

Fig. 9.21   Effect of finite aspect ratio on lift and drag coefficients for a wing.

Theory and experiment show that downwash velocities reduce the effective angle of attack in proportion to the lift coefficient. Compared to an airfoil section with $ar = \infty$, the geometric angle of attack of a wing must be increased by

$$\Delta\alpha \approx \frac{C_L}{\pi ar} \tag{9.40}$$

to achieve the lift coefficient of the section. This causes an increase in drag coefficient for the wing given by

$$\Delta C_D \approx C_L \, \Delta\alpha \approx \frac{C_L^2}{\pi ar} \tag{9.41}$$

This increase in drag due to lift is termed *induced drag*.

The *effective* aspect ratio includes the effect of planform shape. For most planform shapes the effective aspect ratio is within 15 percent of the geometric aspect ratio. When written in terms of effective aspect ratio, the drag of a wing of finite span becomes [19]

$$C_D = C_{D,\infty} + C_{D,i} = C_{D,\infty} + \frac{C_L^2}{\pi ar} \tag{9.42}$$

where $C_{D,\infty}$ is the section drag coefficient at C_L, $C_{D,i}$ is the induced drag coefficient at C_L, and ar is the effective aspect ratio of the finite-span wing.

Drag on airfoils arises from viscous and pressure forces. Viscous drag changes with Reynolds number but only slightly with angle of attack. These relationships and some commonly used terminology are illustrated in Fig. 9.22.

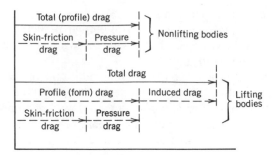

Fig. 9.22 Drag breakdown on nonlifting and lifting bodies.

A useful approximation to the drag polar for a complete aircraft may be obtained by adding the induced drag to the drag at zero lift. The drag at any lift coefficient is obtained from

$$C_D = C_{D,0} + C_{D,i} = C_{D,0} + \frac{C_L^2}{\pi ar} \qquad (9.43)$$

where $C_{D,0}$ is the drag coefficient at zero lift and ar is the effective aspect ratio.

As we have seen, aircraft can be fitted with low-drag airfoils to give excellent performance at cruise conditions. However, since the maximum lift coefficient is low for thin airfoils, additional effort must be expended to obtain acceptably low landing speeds. In steady-state flight conditions, the lift must equal the aircraft weight. Thus, from Eq. 9.38,

$$W = F_L = C_L \tfrac{1}{2}\rho V^2 A$$

Minimum flight speed is obtained when $C_L = C_{L_{\max}}$. Solving for $V_{\min}$,

$$V_{\min} = \sqrt{\frac{2W}{\rho C_{L_{\max}} A}} \qquad (9.44)$$

According to Eq. 9.44, the minimum landing speed can be reduced by increasing either $C_{L_{\max}}$ or wing area. Two basic techniques are available to control these variables: variable-geometry wing sections (e.g., through the use of flaps) or boundary-layer control techniques.

Flaps are movable portions of a wing trailing edge that may be extended during landing and takeoff to increase effective wing area. The effects on lift and drag of two typical flap configurations are shown in Fig. 9.23, as applied to the NACA 23012 airfoil section. The maximum lift coefficient for this section is increased from 1.52 in the "clean" condition to 3.48 with double-slotted flaps. From Eq. 9.44, the corresponding reduction in landing speed would be 34 percent.

Figure 9.23 shows that section drag is increased substantially by high-lift devices. From Fig. 9.23b, section drag at $C_{L_{\max}}(C_D \simeq 0.28)$ with double-slotted flaps is about 5 times larger than section drag at $C_{L_{\max}}(C_D \simeq 0.055)$ for the clean airfoil. Induced drag due to lift must be added to section drag to obtain total drag. Because induced drag is proportional to C_L^2, Eq. 9.41, total drag rises sharply at low aircraft speeds. At speeds near stall, drag may increase enough to exceed the thrust available from the engines. To avoid this dangerous region of unstable operation, the Federal Aviation Administration (FAA) limits operation of commercial aircraft to speeds above 1.2 times stall speed.

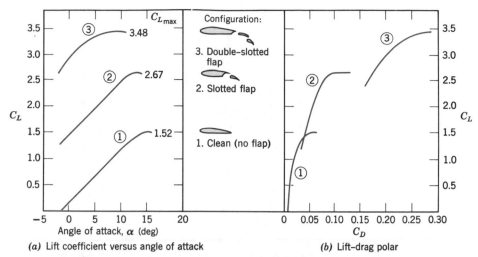

(a) Lift coefficient versus angle of attack

(b) Lift–drag polar

Fig. 9.23 Effect of flaps on aerodynamic characteristics of NACA 23012 airfoil section. (Data from [19].)

Fig. 9.24(a) Application of high-lift boundary-layer control devices to jet transport aircraft for reduction of landing speed. The Boeing 727 wing is highly mechanized. During the approach to landing, triple-slotted, trailing edge flaps roll out from under the wing and deflect downward to increase lift. A section of the leading edge near the outboard end of the wing slides forward to open a slot that keeps the air flow close to the wing's upper surface. At the leading edge near the root, a Kruger flap drops down from under the wing, increasing the effective radius of the leading edge to prevent flow separation. After touchdown, spoilers (not shown) pop up in front of each flap to kill the lift and ensure that the plane remains on the ground, despite the lift-augmenting devices. (Photograph courtesy of Boeing Airplane Company.)

Fig. 9.24(*b***)** On the Boeing 707 both the inboard and outboard trailing edge flaps (shown here fully deflected) are equipped with double slots near the pivot. The slots duct air from the under surface to the upper surface of the flap, creating a streamwise jet that causes the air flow to cling more closely to the upper surface. Two rows of short blades in front of the flaps are vortex generators whose purpose is to encourage the slow-moving layers of air next to the surface to follow the wing surface more closely, preventing separation. The raised panels just above the flaps are lift-destroying devices called spoilers. Normally, they are used to control lift for landing and maneuvering. Here they are raised to help the plane descend from high altitude. (Photograph courtesy of Boeing Airplane Company.)

Although details of boundary-layer control techniques[10] are beyond the scope of this book, the basic purpose of all of them is to delay separation or reduce drag, by adding momentum to the boundary layer through blowing, or by removing low-momentum boundary-layer fluid by suction. Many examples of practical boundary-layer control systems may be seen on commercial transport aircraft at your local airport. One of the more sophisticated systems, on the Boeing 727 transport, is shown in Fig. 9.24. On this aircraft, leading edge devices are used in conjunction with triple-slotted trailing edge flaps, to achieve a $C_{L_{max}}$ in excess of 3.6.

EXAMPLE 9.8—Optimum Cruise Performance of a Jet Transport

Jet engines burn fuel at a rate proportional to thrust delivered. The optimum cruise condition for a jet aircraft is at maximum speed for a given thrust. In steady level flight, thrust and drag are equal. Optimum cruise occurs at the speed when the ratio of drag force to air speed is minimized.

---

[10] See the excellent film, *Boundary Layer Control*, D. C. Hazen, principal, for a review of these techniques.

A Boeing 727-200 jet transport has wing planform area $A_p = 1600$ square feet, and effective aspect ratio $ar = 6.5$. Stall speed at sea level for this aircraft with flaps up and a gross weight of 150,000 pounds is 175 mph. Below $M = 0.6$, drag due to compressibility effects is negligible, so Eq. 9.43 may be used to estimate total drag on the aircraft. $C_{D,0}$ for the aircraft is constant at 0.0182. Assume sonic speed at sea level is $c = 759$ mph.

Evaluate the performance envelope for this aircraft at sea level by plotting drag force versus speed, between stall and $M = 0.6$. Use this graph to estimate optimum cruise speed for the aircraft at sea-level conditions. Comment on stall speed and optimum cruise speed for the aircraft at an altitude of 30,000 ft on a standard day.

EXAMPLE PROBLEM 9.8

GIVEN: Boeing 727-200 jet transport at sea-level conditions.

$$W = 150,000 \text{ lbf}, \ A = 1600 \text{ ft}^2, \ ar = 6.5, \text{ and } C_{D,0} = 0.0182$$

Stall speed is $V_{\text{stall}} = 175$ mph, and compressibility effects on drag are negligible for $M \le 0.6$ (sonic speed at sea level is $c = 759$ mph).

FIND: (a) Evaluate and plot drag force versus speed from V_{stall} to $M = 0.6$.
  (b) Estimate optimum cruise speed at sea level.
  (c) Comment on performance at altitude of 30,000 ft.

SOLUTION:
For steady, level flight, weight equals lift and thrust equals drag.

Computing equations: $\quad F_L = C_L A \tfrac{1}{2}\rho V^2 = W \qquad C_D = C_{D,0} + \dfrac{C_L^2}{\pi \, ar}$

$$F_D = C_D A \tfrac{1}{2}\rho V^2 = T \qquad M = \dfrac{V}{c}$$

At sea level, $\rho = 0.00238 \text{ slug/ft}^3$, and $c = 759$ mph.
Since $F_L = W$ for level flight at any speed, then

$$C_L = \frac{W}{\tfrac{1}{2}\rho V^2 A} = \frac{2W}{\rho V^2 A}$$

At stall speed, $V = 175$ mph, so

$$C_L = \frac{2}{} \times \frac{150,000 \text{ lbf}}{} \times \frac{\text{ft}^3}{0.00238 \text{ slug}} \left[\frac{\text{hr}}{175 \text{ mi}} \times \frac{\text{mi}}{5280 \text{ ft}} \times \frac{3600 \text{ sec}}{\text{hr}} \right]^2 \frac{1}{1600 \text{ ft}^2}$$

$$\times \frac{\text{slug} \cdot \text{ft}}{\text{lbf} \cdot \text{sec}^2}$$

$$C_L = \frac{3.65 \times 10^4}{[V(\text{mph})]^2} = \frac{3.65 \times 10^4}{(175)^2} = 1.19, \text{ and}$$

$$C_D = C_{D,0} + \frac{C_L^2}{\pi \, ar} = 0.0182 + \frac{(1.19)^2}{\pi (6.5)} = 0.0875$$

Then

$$F_D = W \frac{C_D}{C_L} = 150,000 \text{ lbf} \left(\frac{0.0875}{1.19} \right) = 11,000 \text{ lbf}$$

At $M = 0.6$, $V = Mc = (0.6)759$ mph $= 455$ mph, so $C_L = 0.176$ and

$$C_D = 0.0182 + \frac{(0.176)^2}{\pi (6.5)} = 0.0197; \quad F_D = 150,000 \text{ lbf} \left(\frac{0.0197}{0.176} \right) = 16,800 \text{ lbf}$$

Similar calculations lead to the following table:

V(mph)	175	200	300	400	455
C_L	1.19	0.913	0.406	0.228	0.176
C_D	0.0875	0.0590	0.0263	0.0207	0.0197
F_D(lbf)	11,000	9690	9720	13,600	16,800

These data may be plotted as:

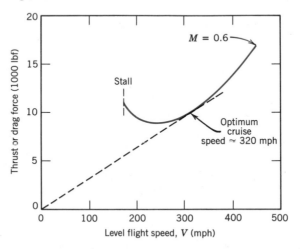

From the plot, the optimum cruise speed at sea-level is estimated as 320 mph.

At 30,000 ft (9140 m) altitude, the density is only about 0.375 times sea level density, from Table A.3. The speeds for corresponding forces are calculated from

$$F_L = C_L A \tfrac{1}{2} \rho V^2 \quad \text{or} \quad V = \sqrt{\frac{2F_L}{C_L \rho A}} \quad \text{or} \quad \frac{V_{30}}{V_{SL}} = \sqrt{\frac{\rho_{SL}}{\rho_{30}}} = \sqrt{\frac{1}{0.375}} = 1.63$$

Thus speeds increase 63 percent at 30,000 ft altitude: $V_{stall} \approx 285$ mph

$$V_{cruise} \approx 522 \text{ mph}$$

A primary reason to operate jet transport aircraft at high altitude is the increase in optimum cruising speed compared to sea level; in this example, range is 63 percent longer at altitude. (Secondary reasons are improved engine operation and less weather disturbance.) Optimum cruise altitude depends on aircraft configuration, gross weight, segment length, and winds aloft.

Aerodynamic lift is an important consideration in the design of high-speed land vehicles such as racing cars and land-speed record machines. A road vehicle generates lift by virtue of its shape [27]. A representative centerline pressure distribution measured in the wind tunnel for an automobile is shown in Fig. 9.25 [28].

The pressure is low around the nose due to streamline curvature as the flow rounds the nose. The pressure reaches a maximum at the base of the windshield, again due to streamline curvature. Low-pressure regions also occur at the windshield header and over the top of the automobile. The air speed across the top is approximately 30 percent higher than the freestream air speed. The same effect occurs around the "A-pillars" at the windshield edges. The drag increase due to an added object, such as an antenna, spotlight, or mirror at that location, thus would be $(1.3)^2 \approx 1.7$ times the drag the object would experience in an undisturbed flow field. Thus the *parasite drag*

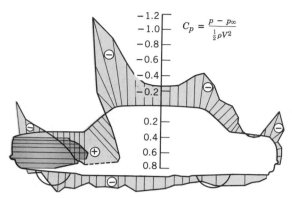

$$C_p = \frac{p - p_\infty}{\frac{1}{2}\rho V^2}$$

Fig. 9.25 Pressure distribution along the centerline of an automobile [28].

of an added component can be much higher than predicted from its drag coefficient measured in free flow.

At high speeds, aerodynamic lift forces can unload tires, causing serious reductions in steering control and reducing stability to a dangerous extent. Lift forces on early racing cars were counteracted somewhat by "spoilers," at considerable penalty in drag. In 1965 Jim Hall introduced the use of movable inverted airfoils on his Chaparral sports cars to develop aerodynamic downforce and provide aerodynamic braking [29]. Since then the developments in application of aerodynamic devices have been rapid. Aerodynamic design is used to reduce lift on all modern racing cars, as exemplified in Fig. 9.26. Liebeck airfoils [21] are used frequently for high-speed automobiles. Their high lift coefficients and relatively low drag allow downforce equal to or greater than the car weight to be developed at racing speeds. "Ground effect" cars use venturi-shaped ducts under the car and side skirts to seal leakage flows. The result of aerodynamic downforce is significantly higher cornering speeds and lower lap times.

Another method of boundary-layer control is to use moving surfaces to reduce skin friction effects on the boundary layer [30]. This method is hard to apply to practical devices, because of geometric and weight complications, but it is very important in

Fig. 9.26 Contemporary sports-racing car, showing aerodynamic design features. To achieve 200+ mph performance requires careful attention to aerodynamic design for low drag and to aerodynamic downforce for stability and high cornering speeds. The photo shows the carefully streamlined mirrors, flush inlet ducts, and other details needed to achieve low drag. The low front contour, underbody shape, and rear wing create downforce for stability and high-speed cornering performance. (Photo courtesy of Goodyear Tire & Rubber Co., Inc.)

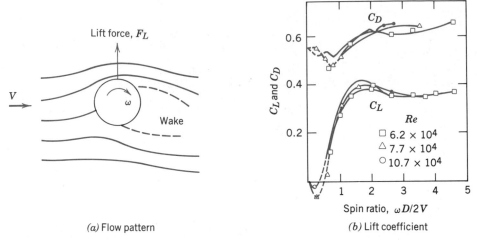

(a) Flow pattern   (b) Lift coefficient

Fig. 9.27 Flow pattern, lift, and drag coefficients for a smooth spinning sphere in uniform flow. (Data from [17].)

recreation. Most golfers, tennis players, Ping-Pong enthusiasts, and baseball pitchers can attest to this! Tennis and Ping-Pong players use spin to control the trajectory and bounce of a shot. In golf, a drive can leave the tee at 275 ft/sec or more, with backspin of 9000 rpm! Spin provides significant aerodynamic lift that substantially increases the carry of a drive. Spin is also largely responsible for hooking and slicing, when shots are not hit squarely. The baseball pitcher uses spin to throw a curve ball.

Flow about a spinning sphere is shown in Fig. 9.27a. Spin alters the pressure distribution and also affects the location of boundary-layer separation. Separation is delayed on the upper surface of the sphere in Fig. 9.27a, and it occurs earlier on the lower surface. Pressure is reduced on the upper surface and increased on the lower surface; the wake is deflected downward as shown. Pressure forces cause a lift in the direction shown; spin in the opposite direction would produce negative lift—a downward force. The force is directed perpendicular to both V and the spin axis.

Lift and drag data for spinning smooth spheres are presented in Fig. 9.27b. The most important parameter is the *spin ratio*, $\omega D/2V$, the ratio of surface speed to freestream flow speed; Reynolds number plays a secondary role. At low spin ratio, lift is negative for clockwise spin. Only above $\omega D/2V \approx 0.5$ does lift become positive and continue to increase as spin ratio increases. Lift coefficient levels out at about 0.35. Spin has little effect on sphere drag coefficient, which varies from about 0.5 to about 0.65 over the range of spin ratio shown.

Earlier we mentioned the effect of dimples on the drag of a golf ball. Experimental data for lift and drag coefficients for spinning golf balls are presented in Fig. 9.28 for subcritical Reynolds numbers between 126,000 and 238,000. Again the independent variable is spin ratio; a much smaller range of spin ratio, typical of golf balls, is presented in Fig. 9.28.

A clear trend is evident: the lift coefficient increases consistently with spin ratio for both hexagonal and "conventional" (round) dimples. The lift coefficient on a golf ball with hexagonal dimples is significantly—as much as 15 percent—larger than on a ball with round dimples. The advantage for hexagonal dimples continues to the largest spin ratios that were measured. The drag coefficient for a ball with hexagonal dimples is consistently 5 to 7 percent lower than the drag coefficient for a ball with round dimples at low spin ratios, but the difference becomes less pronounced as spin ratio increases.

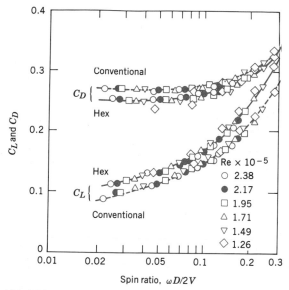

Fig. 9.28 Comparison of conventional and hex-dimpled golf balls [31].

The combination of higher lift and lower drag increases the carry of a golf shot. Some years ago Royal introduced the "Plus 6" ball—with hexagonal dimples—in the U.S. Their advertising campaign assured golfers that drives consistently would be six yards longer than when using balls with conventional round dimples!

EXAMPLE 9.9—Lift of a Spinning Ball
A smooth tennis ball, with 57 g mass and 64 mm diameter, is hit at 25 m/sec with topspin of 7500 rpm. Calculate the aerodynamic lift acting on the ball. Evaluate the radius of curvature of its path in a vertical plane. Compare with the radius for no spin.

EXAMPLE PROBLEM 9.9

GIVEN: Tennis ball in flight, with $m = 57$ g and $D = 64$ mm, hit with $V = 25$ m/sec and topspin of 7500 rpm.

FIND: (a) Aerodynamic lift acting on ball.
       (b) Radius of curvature of path in vertical plane.
       (c) Compare with radius for no spin.

SOLUTION:
Assume ball is smooth.

Use data from Fig. 9.27 to find lift: $C_L = f\left(\dfrac{\omega D}{2V}, Re_D\right)$.

From given data (for standard air, $\nu = 1.45 \times 10^{-5}$ m²/sec),

$$\frac{\omega D}{2V} = \frac{1}{2} \times 7500 \frac{\text{rev}}{\text{min}} \times 0.064 \text{ m} \times \frac{\text{sec}}{25 \text{ m}} \times \frac{2\pi \text{ rad}}{\text{rev}} \times \frac{\text{min}}{60 \text{ sec}} = 1.01$$

$$Re_D = \frac{VD}{\nu} = 25 \frac{\text{m}}{\text{sec}} \times 0.064 \text{ m} \times \frac{\text{sec}}{1.45 \times 10^{-5} \text{ m}^2} = 1.1 \times 10^5$$

From Fig. 9.27, $C_L \approx 0.3$, so

$$F_L = C_L A \frac{1}{2}\rho V^2 = C_L \frac{\pi D^2}{4}\frac{1}{2}\rho V^2 = \frac{\pi}{8}C_L D^2 \rho V^2$$

$$F_L = \frac{\pi}{8} \times 0.3 \times (0.064)^2 \text{ m}^2 \times 1.23 \frac{\text{kg}}{\text{m}^3} \times (25)^2 \frac{\text{m}^2}{\text{sec}^2} \times \frac{\text{N} \cdot \text{sec}^2}{\text{kg} \cdot \text{m}} = 0.371 \text{ N} \qquad \underleftarrow{} F_L$$

Because the ball is hit with topspin, this force acts down.

Use Newton's second law to evaluate curvature of path. In the vertical plane,

$$\sum F_z = -F_L - mg = ma_z = -m\frac{V^2}{R} \qquad \text{or} \qquad R = \frac{V^2}{g + F_L/m}$$

$$R = \frac{(25)^2 \text{ m}^2}{\text{sec}^2} \left[\cfrac{1}{9.81 \dfrac{\text{m}}{\text{sec}^2} + 0.371 \text{ N} \times \dfrac{1}{0.057 \text{ kg}} \times \dfrac{\text{kg} \cdot \text{m}}{\text{N} \cdot \text{sec}^2}} \right]$$

$$R = 38.3 \text{ m (with spin)} \qquad \underleftarrow{} R$$

$$R = \frac{(25)^2 \text{ m}^2}{\text{sec}^2} \times \frac{\text{sec}^2}{9.81 \text{ m}} = 63.7 \text{ m (without spin)} \qquad \underleftarrow{} R$$

Thus topspin has a significant effect on trajectory of the shot!

It has long been known that a spinning projectile in flight is affected by a force perpendicular to the direction of motion and to the spin axis. This effect, known as the *Magnus effect,* is responsible for the systematic drift of artillery shells.

Cross flow about a rotating circular cylinder is qualitatively similar to flow about the spinning sphere shown schematically in Fig. 9.27a. With clockwise rotation of

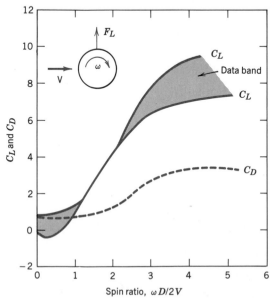

Fig. 9.29 Lift and drag of a rotating cylinder as a function of relative rotational speed; Magnus force. (Data from [32].)

a cylinder, separation is delayed on the upper surface; it occurs earlier on the lower surface. Thus the wake is deflected and the pressure distribution on the cylinder surface is altered when rotation is present. Pressure is reduced on the upper surface and increased on the lower surface, causing a net lift force acting upward. Spin in the opposite direction reverses these effects and causes a downward lift force.

Lift and drag coefficients for the rotating cylinder are based on projected area, LD. Experimentally measured lift and drag coefficients for subcritical Reynolds numbers between 40,000 and 660,000 are shown as functions of spin ratio in Fig. 9.29. When surface speed exceeds flow speed, the lift coefficient increases to surprisingly large values, while in two-dimensional flow, drag is affected only moderately. Induced drag, which must be considered for finite cylinders, can be reduced by using end disks larger in diameter than the body of the cylinder.

The power required to rotate a cylinder may be estimated from the skin friction drag of the cylinder surface. Hoerner [32] suggests basing the skin friction drag estimate on the tangential surface speed and surface area. Goldstein [17] suggests that the power required to spin the cylinder, when expressed as an equivalent drag coefficient, may represent 20 percent or more of the aerodynamic C_D of a stationary cylinder.

9-9 SUMMARY OBJECTIVES

After completing study of Chapter 9, you should be able to do the following:

1. Define:

external flow	drag
boundary-layer disturbance thickness	lift
displacement thickness	angle of attack
momentum thickness	chord
pressure gradient (favorable, adverse)	wing span
separation	aspect ratio
skin friction coefficient	induced drag

2. Beginning with the momentum integral equation for zero pressure gradient flow, develop expressions for $\delta(x)$, $\delta^*(x)$, $\tau_w(x)$, $C_f(x)$; determine the total friction force on a flat plate placed parallel to the flow.
3. Use the displacement-thickness concept to approximate the pressure drop in the entrance region of a channel.
4. Determine the drag and lift forces for bodies in external flow.
5. Calculate the moment on an object due to aerodynamic drag forces.
6. Solve the problems at the end of the chapter that relate to the material you have studied.

REFERENCES

1. Prandtl, L., "Fluid Motion with Very Small Friction (in German)," Proceedings of the Third International Congress on Mathematics, Heidelberg, 1904; English translation available as NACA TM 452, March 1928.
2. Blasius, H., "The Boundary Layers in Fluids with Little Friction (in German)," *Zeitschrift für Mathematik und Physik, 56,* 1, 1908, pp. 1–37; English translation available as NACA TM 1256, February 1950.
3. Schlichting, H., *Boundary-Layer Theory,* 7th ed. New York: McGraw-Hill, 1979.
4. Stokes, G. G., "On the Effect of the Internal Friction of Fluids on the Motion of Pendulums," *Cambridge Philosophical Transactions, IX,* 8, 1851.

5. Howarth, L., "On the Solution of the Laminar Boundary-Layer Equations," *Proceedings of the Royal Society of London, A164*, 1938, pp. 547–579.
6. Hess, J. L., and A. M. O. Smith, "Calculation of Potential Flow about Arbitrary Bodies," in *Progress in Aeronautical Sciences*, Vol. 8, D. Kuchemann, et al., eds. Elmsford, NY: Pergamon Press, 1966.
7. Kraus, W., "Panel Methods in Aerodynamics," in *Numerical Methods in Fluid Dynamics*, H. J. Wirz and J. J. Smolderen, eds. Washington, D.C.: Hemisphere, 1978.
8. Rosenhead, L., ed., *Laminar Boundary Layers*. London: Oxford University Press, 1963.
9. Rotta, J. C., "Turbulent Boundary Layers in Incompressible Flow," in *Progress in Aeronautical Sciences*, A. Ferri, et al., eds. New York: Pergamon Press, 1960, pp. 1–220.
10. Kline, S. J., et al., eds., *Proceedings, Computation of Turbulent Boundary Layers— 1968 AFOSR-IFP-Stanford Conference*, Vol. I: Methods, Predictions, Evaluation, and Flow Structure, and Vol. II: Compiled Data. Stanford, CA: Thermosciences Division, Department of Mechanical Engineering, Stanford University, 1969.
11. Kline, S. J., et al., eds., *Proceedings, 1980–81 AFOSR-HTTM-Stanford Conference on Complex Turbulent Flows: Comparison of Computation and Experiment*, three volumes. Stanford, CA: Thermosciences Division, Department of Mechanical Engineering, Stanford University, 1982.
12. Cebeci, T., and P. Bradshaw, *Momentum Transfer in Boundary Layers*. Washington, D.C.: Hemisphere, 1977.
13. Bradshaw, P., T. Cebeci, and J. H. Whitelaw, *Engineering Calculation Methods for Turbulent Flow*. New York: Academic Press, 1981.
14. Hoerner, S. F., *Fluid-Dynamic Drag*, 2nd ed. Midland Park, NJ: Published by the author, 1965.
15. Shapiro, A. H., *Shape and Flow, The Fluid Dynamics of Drag*. New York: Anchor, 1961 (paperback).
16. Fage, A., "Experiments on a Sphere at Critical Reynolds Numbers," Great Britain, *Aeronautical Research Council, Reports and Memoranda*, No. 1766, 1937.
17. Goldstein, S., ed., *Modern Developments in Fluid Dynamics*, Vols. I and II. Oxford: Clarendon Press, 1938. (Reprinted in paperback by Dover, New York, 1967.)
18. Morel, T., and M. Bohn, "Flow over Two Circular Disks in Tandem," *Transactions of the ASME, Journal of Fluids Engineering, 102*, 1, March 1980, pp. 104–111.
19. Abbott, I. H., and A. E. von Doenhoff, *Theory of Wing Sections, Including a Summary of Airfoil Data*. New York: Dover, 1959 (paperback).
20. Stratford, B. S., "An Experimental Flow with Zero Skin Friction," *Journal of Fluid Mechanics, 5*, Pt. 1, January 1959, pp. 17–35.
21. Liebeck, R. H., "Design of Subsonic Airfoils for High Lift," *AIAA Journal of Aircraft, 15*, 9, September 1978, pp. 547–561.
22. Smith, A. M. O., "Aerodynamics of High-Lift Airfoil Systems," in *Fluid Dynamics of Aircraft Stalling*, AGARD CP-102, 1973, pp. 10-1 through 10-26.
23. Morel, T., "Effect of Base Slant on Flow in the Near Wake of an Axisymmetric Cylinder," *Aeronautical Quarterly, XXXI*, Pt. 2, May 1980, pp. 132–147.
24. Hucho, W. H., "The Aerodynamic Drag of Cars—Current Understanding, Unresolved Problems, and Future Prospects," in *Aerodynamic Drag Mechanisms of Bluff Bodies and Road Vehicles*, G. Sovran, T. Morel, and W. T. Mason, eds. New York: Plenum, 1978.
25. McDonald, A. T., and G. M. Palmer, "Aerodynamic Drag Reduction of Intercity Buses," *Transactions, Society of Automotive Engineers, 89*, Section 4, 1980, pp. 4469–4484 (SAE Paper No. 801404).
26. Grosser, M., *Gossamer Odyssey*. Boston: Houghton Mifflin, 1981.

27. Carr, G. W., "The Aerodynamics of Basic Shapes for Road Vehicles. Part 3: Streamlined Bodies," The Motor Industry Research Association, Warwickshire, England, Report No. 107/4, 1969.

28. Goetz, H., "The Influence of Wind Tunnel Tests on Body Design, Ventilation, and Surface Deposits of Sedans and Sports Cars," SAE Paper No. 710212, 1971.

29. Hall, J., "What's Jim Hall Really Like?" *Automobile Quarterly, VIII,* 3, Spring 1970, pp. 282–293.

30. Moktarian, F., and V. J. Modi, "Fluid Dynamics of Airfoils with Moving Surface Boundary-Layer Control," *AIAA Journal of Aircraft, 25,* 2, February 1988, pp. 163–169.

31. Mehta, R. D., "Aerodynamics of Sports Balls," in *Annual Review of Fluid Mechanics,* ed. by M. van Dyke, et al. Palo Alto, CA: Annual Reviews, 1985, *17,* pp. 151–189.

32. Hoerner, S. F., and H. V. Borst, *Fluid-Dynamic Lift.* Bricktown, NJ: Hoerner Fluid Dynamics, 1975.

33. Chow, C.-Y., *An Introduction to Computational Fluid Mechanics.* New York: Wiley, 1980.

34. Carr, G. W., "The Aerodynamics of Basic Shapes for Road Vehicles, Part 1: Simple Rectangular Bodies," The Motor Industry Research Association, Warwickshire, England, Report No. 1968/2, 1967.

PROBLEMS

9.1 A submarine moves at 20 knots through seawater at 7 C. Assume that near the bow the boundary layer behaves as though on a flat plate. Estimate the distance from the bow where transition from laminar to turbulent flow in the boundary layer might be expected.

9.2 A model of a river towboat is to be tested at 1:13.5 scale. The boat is designed to travel at 8 mph in fresh water at 10 C. Estimate the distance from the bow where transition occurs. Where should transition be stimulated on the model towboat?

9.3 An airplane cruises at 300 knots at 10 km altitude on a standard day. Assume the boundary layers on the wing surfaces behave as on a flat plate. Estimate the expected extent of laminar flow in the wing boundary layers.

9.4 The Reynolds number is $Re = \rho V L/\mu$, where L is a characteristic dimension for the flow field. Consider boundary-layer development in parallel flow past a flat plate, for which the characteristic dimension is x, the distance measured from the leading edge. Prepare a log-log plot of velocity versus distance for $0.01 \le x \le 10$ m. Show lines for which $Re_x = 5 \times 10^5$ and 1×10^6 if the fluid is water.

9.5 The Reynolds number is $Re = \rho V L/\mu$, where L is a characteristic dimension for the flow field. Consider boundary-layer development in parallel flow past a flat plate, for which the characteristic dimension is x, the distance measured from the leading edge. Prepare a log-log plot of velocity versus distance for $0.01 \le x \le 10$ m. Show lines for which $Re_x = 5 \times 10^5$ and 1×10^6 for standard air at (a) sea level conditions and (b) 10 km altitude.

9.6 Aircraft and missiles flying at high altitude may have regions of laminar flow that become turbulent at lower altitudes at the same speed. Explain a possible mechanism for this effect. Support your answer with calculations based on standard atmosphere data.

9.7 The most general sinusoidal velocity profile for laminar boundary-layer flow on a flat plate is $u = A \sin (By) + C$. State three boundary conditions applicable to the laminar boundary-layer velocity profile. Evaluate constants A, B, and C.

**9.8 The velocity profile for a laminar boundary layer is to be approximated by the expression

$$\frac{u}{U} = a + b\left(\frac{y}{\delta}\right) + c\left(\frac{y}{\delta}\right)^m$$

** These problems require material from sections that may be omitted without loss of continuity in the text material.

Evaluate a, b, and c for the situation where $m = 1.5$. Compare the resulting profile with the Blasius profile, Table 9.1.

9.9 Velocity profiles in laminar boundary layers often are approximated by the equations

Linear: $\quad \dfrac{u}{U} = \dfrac{y}{\delta}$ $\qquad\qquad$ Cubic: $\quad \dfrac{u}{U} = \dfrac{3}{2}\left(\dfrac{y}{\delta}\right) - \dfrac{1}{2}\left(\dfrac{y}{\delta}\right)^3$

Parabolic: $\quad \dfrac{u}{U} = 2\left(\dfrac{y}{\delta}\right) - \left(\dfrac{y}{\delta}\right)^2$ $\qquad$ Sinusoidal: $\quad \dfrac{u}{U} = \sin\!\left(\dfrac{\pi}{2}\dfrac{y}{\delta}\right)$

Compare the shapes of these velocity profiles by plotting y/δ (on the ordinate) versus u/U (on the abscissa).

9.10 The velocity profile in a turbulent boundary layer often is approximated by the "$\frac{1}{7}$-power-law" equation

$$\frac{u}{U} = \left(\frac{y}{\delta}\right)^{1/7}$$

Compare the shape of this profile with the parabolic laminar boundary-layer velocity profile (Problem 9.9) by plotting y/δ (on the ordinate) versus u/U (on the abscissa) for both profiles.

9.11 Transition from laminar to turbulent boundary-layer flow actually occurs over a finite length of surface, during which the velocity profile and wall shear stress adjust from laminar to turbulent forms. A useful approximation during transition is that the momentum thickness of the boundary layer remains constant. Assuming constant momentum thickness, find the ratio $\delta_{\text{turbulent}}/\delta_{\text{laminar}}$ for transition from a parabolic laminar velocity profile to a "$\frac{1}{7}$-power" turbulent velocity profile.

9.12 Evaluate δ^*/δ for each of the laminar boundary-layer velocity profiles given in Problem 9.9.

9.13 Evaluate δ^*/δ and θ/δ for the turbulent $\frac{1}{7}$-power-law velocity profile given in Problem 9.10. Compare with ratios for the cubic laminar boundary-layer velocity profile given in Problem 9.9.

9.14 Consider a laminar boundary layer on a flat plate with velocity profile given by the cubic expression of Problem 9.9. For this profile

$$\frac{\delta}{x} = \frac{4.64}{\sqrt{Re_x}}$$

Find expressions for δ^*/x and θ/x.

9.15 Air at standard conditions flows over a flat plate. The flow is uniform at the leading edge of the plate. The velocity profile in the boundary layer is of the form

$$\frac{u}{U} = 2\eta - \eta^2 + C_1, \qquad \text{where } \eta \equiv y/\delta$$

At a section, $U = 20$ m/sec, $L = 0.20$ m, and $\delta = 5.7$ mm. Is the flow at this section laminar or turbulent? Why? What boundary conditions must the equation for u/U satisfy, and what is C_1? Determine δ^*, θ, and τ_w at the given section.

9.16 Evaluate θ/δ for each of the laminar boundary-layer velocity profiles given in Problem 9.9.

9.17 Evaluate $H \equiv \delta^*/\theta$ for each of the laminar boundary-layer velocity profiles given in Problem 9.9.

9.18 Evaluate $H \equiv \delta^*/\theta$ for the power-law profile used to represent the turbulent velocity profile (Problem 9.10). Compare with H for the cubic, laminar boundary-layer velocity profile given in Problem 9.9.

‡9.19 Water flows over a flat plate at freestream speed 0.5 ft/sec. There is no pressure gradient, and the laminar boundary layer is 0.25 in. thick. Assume a sinusoidal velocity profile as given in Problem 9.9. Derive an equation for the shear stress at any location within the boundary layer. For the flow conditions given, compute the local wall shear stress. Find θ by numerical or graphical integration, and compare with the analytical result.

9.20 Air at standard conditions flows over a thin flat plate 1 m long and 0.3 m wide. The flow is uniform at the leading edge of the plate. Assume the velocity profile in the boundary layer is linear, and the freestream velocity is $U = 2.7$ m/sec. Treat the flow as two-dimensional; assume that flow conditions are independent of z. Using control volume $abcd$, shown by the dashed lines, compute the mass flow rate across surface ab. Determine the magnitude and direction of the x component of force required to hold the plate stationary.

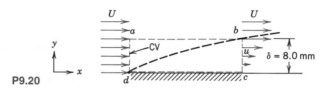

P9.20

9.21 Rework Problem 9.20 with the velocity profile at section bc given by the parabolic expression from Problem 9.9 and $\delta = 12.7$ mm.

9.22 A viscous boundary-layer velocity profile and the equivalent inviscid velocity profile with the same mass flow rate are sketched in Example Problem 9.1. Consider the defect of kinetic energy flux between the equivalent inviscid flow and the viscous flow. Derive an expression for the "energy defect thickness," δ^{**}. Sketch a diagram showing the physical interpretation of this thickness.

9.23 Air flows in the entrance region of a square duct, as shown. The velocity is uniform, $U_0 = 30$ m/sec, and the duct is 80 mm square. At a section 0.3 m downstream from the entrance, the displacement thickness, δ^*, on each wall measures 1.0 mm. Determine the pressure change between sections ① and ②.

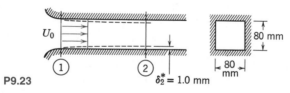

P9.23

9.24 Laboratory wind tunnels have test sections 1 ft square and 2 ft long. With nominal air speed $U_1 = 80$ ft/sec at the test section inlet, turbulent boundary layers form on the top, bottom, and side walls of the tunnel. The boundary-layer thickness is $\delta_1 = 0.8$ in. at the inlet and $\delta_2 = 1.2$ in. at the outlet from the test section. The boundary-layer velocity profiles are of power-law form, with $u/U = (y/\delta)^{1/7}$. Evaluate the freestream velocity, U_2, at the exit from the wind-tunnel test section. Determine the change in static pressure along the test section.

9.25 The square test section of a small laboratory wind tunnel has sides of width $W = 305$ mm. At one measurement location, the turbulent boundary layers on the tunnel walls are $\delta_1 = 9.5$ mm thick. The velocity profile is approximated well by the "$\frac{1}{7}$-power" expression. At this location the freestream air speed is $U_1 = 18.3$ m/sec, and the static pressure is $p_1 = -22.9$ mm H_2O (gage). At a second measurement location downstream, the boundary-layer thickness is $\delta_2 = 12.7$ mm. Evaluate the air speed in the freestream at the second section. Calculate the difference in static pressure from section ① to section ②.

‡You may wish to use simple computer programs to help solve problems marked with daggers.

9.26 Standard air flows from the atmosphere into the wide flat channel shown. Laminar boundary layers form on the top and bottom walls of the channel (ignore boundary-layer effects on the side walls). Assume the boundary layers behave as on a flat plate, with linear velocity profiles. Evaluate the displacement thickness at section ②. Determine the static pressure at section ①.

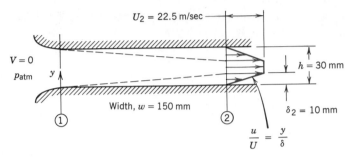

P9.26

9.27 Flow of air develops in a flat horizontal duct following a well-rounded entrance section. The duct height is $H = 300$ mm. Turbulent boundary layers grow on the duct walls, but the flow is not yet fully developed. Assume that the velocity profile in each boundary layer is $u/U = (y/\delta)^{1/7}$. The inlet flow is uniform at $\bar{V} = 10$ m/sec at section ①. At section ②, the boundary-layer thickness on each wall of the channel is $\delta_2 = 100$ mm. Show that for this flow, $\delta^* = \delta/8$. Evaluate the static gage pressure at section ②. Find the average wall shear stress between the entrance and section ②, located at $L = 5$ m.

9.28 A laboratory wind tunnel has a square test section, with sides of width $W = 305$ mm and length $L = 610$ mm. When the freestream air speed at the test section entrance is $U_1 = 24.4$ m/sec, the head loss from the atmosphere is 6.5 mm H_2O. Turbulent boundary layers form on the top, bottom, and side walls of the test section. Measurements show the boundary-layer thicknesses are $\delta_1 = 20.3$ mm at the entrance and $\delta_2 = 25.4$ mm at the outlet from the test section. The velocity profiles are of $\frac{1}{7}$-power form. Evaluate the freestream air speed at the outlet from the test section. Determine the static pressures at the test section inlet and outlet.

9.29 Air enters a 3 in. diameter circular duct through a smoothly contoured inlet. The flow is steady, and the duct area is constant. The velocity is uniform at section ①, where the static pressure is -0.567 in. of water (gage). At section ②, the velocity varies linearly, from 0 at the wall to V_2 at a distance of 0.15 in. from the wall. Determine (a) the volume flow rate of air through the duct, (b) the core velocity at section ②, and (c) the displacement thickness at section ②.

9.30 For the conditions of Problem 9.24, compute (a) the total wall shear force and (b) the average wall shear stress on the test section wall.

**9.31 Using numerical results for the Blasius exact solution for laminar boundary-layer flow on a flat plate, plot the dimensionless velocity profile, u/U (on the abscissa), versus dimensionless distance from the surface, y/δ (on the ordinate). Compare with the approximate parabolic velocity profile of Problem 9.9.

‡**9.32 Using graphical or numerical integration, evaluate θ/δ for the Blasius exact solution for the laminar boundary layer on a flat plate (Table 9.1). Show that the result is $\theta/x = 0.664/\sqrt{Re_x}$.

**9.33 Using numerical results obtained by Blasius (Table 9.1), evaluate the distribution of shear stress in a laminar boundary layer on a flat plate. Plot τ/τ_w versus y/δ. Compare

---

** These problems require material from sections that may be omitted without loss of continuity in the text material.

‡ You may wish to use simple computer programs to help solve problems marked with daggers.

with results derived from the approximate parabolic velocity profile given in Problem 9.9.

9.34 Using numerical results obtained by Blasius (Table 9.1), evaluate the distribution of shear stress in a laminar boundary layer on a flat plate. Plot τ/τ_w versus y/δ. Compare with results derived from the approximate sinusoidal velocity profile given in Problem 9.9.

9.35 Using numerical results obtained by Blasius (Table 9.1), evaluate the distribution of shear stress in a laminar boundary layer on a flat plate. Plot τ/τ_w versus y/δ. Compare with results derived from the approximate cubic velocity profile given in Problem 9.9.

9.36 Using numerical results obtained by Blasius (Table 9.1), evaluate the vertical component of velocity in a laminar boundary layer on a flat plate. Plot v/U versus y/δ for $Re_x = 10^5$.

9.37 The Prandtl boundary-layer equations may be written as an ordinary differential equation by introducing the stream function defined by Eq. 9.8. By direct substitution, express the derivative $\partial v/\partial x$, in terms of $f(\eta)$ and its derivatives. Evaluate at the wall and explain the physical significance of this result.

9.38 Verify that the y component of velocity for the Blasius solution to the Prandtl boundary-layer equations is given by Eq. 9.10. Obtain an algebraic expression for the x component of the acceleration of a fluid particle in the laminar boundary layer. Estimate the maximum x component of acceleration at a given x.

9.39 Numerical results of the Blasius solution to the Prandtl boundary-layer equations are presented in Table 9.1. Consider steady, incompressible flow of standard air over a flat plate at freestream speed $U = 4.3$ m/sec. At $x = 0.2$ m, estimate the distance from the surface at which $u = 0.95\ U$. Evaluate the slope of the streamline through this point. Obtain an algebraic expression for the local skin friction, $\tau_w(x)$. Obtain an algebraic expression for the total skin friction drag force on the plate. Evaluate the momentum thickness at $L = 0.8$ m.

9.40 Consider horizontal, steady, incompressible flow in a boundary layer with distributed wall suction. The wall suction velocity is constant with $v = -v_0$ at $y = 0$. There is no pressure gradient. Use a differential control volume to show that

$$\frac{d\theta}{dx} = \frac{\tau_w}{\rho U^2} - \frac{v_0}{U}$$

where U is the freestream speed and τ_w is the wall shear stress.

9.41 Solve Problem 9.40 for a boundary-layer flow with distributed wall *blowing* rather than suction. Assume the wall blowing velocity is constant with $v = v_0$ at $y = 0$.

9.42 A thin flat plate, $L = 0.3$ m long and $b = 1$ m wide, is installed in a water tunnel as a splitter. The freestream speed is $U = 2$ m/sec and the velocity profile in the boundary layer is approximated as parabolic. For this profile, $\delta/x = 5.48/\sqrt{Re_x}$. Plot δ, δ^*, and τ_w versus x/L for the plate.

9.43 A thin flat plate is installed in a water tunnel as a splitter. The plate is 0.3 m long and 1 m wide. The freestream speed is 1.6 m/sec. Laminar boundary layers form on both sides of the plate. The boundary-layer velocity profile is approximated as parabolic. Determine the total viscous drag force on the plate assuming that pressure drag is negligible.

9.44 Consider flow over the splitter plate of Problem 9.42. Show algebraically that the total drag force on one side of the splitter plate may be written $F_D = \rho U^2 \theta_L b$. Evaluate θ_L and the total drag for the given conditions.

9.45 Calculate the drag force on a flat plate with dimensions of 0.75 m × 0.75 m when it is aligned in a flow of standard air where the freestream speed is 1.8 m/sec.

** These problems require material from sections that may be omitted without loss of continuity in the text material.

9.46 A horizontal surface, with length $L = 1.8$ m and width $b = 0.9$ m, is immersed in a stream of standard air flowing at $U = 3.2$ m/sec. Assume a laminar boundary layer forms and approximate the velocity profile as sinusoidal. Plot δ, δ^*, and τ_w versus x/L for the plate.

9.47 For the flow conditions of Problem 9.46, develop an algebraic expression for the variation of wall shear stress with distance along the surface. Integrate to obtain an algebraic expression for the total skin friction drag on the surface. Evaluate the drag for the given conditions.

9.48 Consider again the flow conditions of Problem 9.46. Show algebraically that the total drag force on one side of the plate may be written $F_D = \rho U^2 \theta_L b$. Evaluate θ_L and the total drag for the given conditions.

9.49 The velocity profile in a laminar boundary-layer flow at zero pressure gradient is approximated by the linear expression given in Problem 9.9. Use the momentum integral equation with this profile to obtain expressions for δ/x and C_f.

9.50 A horizontal surface, with length $L = 0.8$ m and width $b = 1.9$ m, is immersed in a stream of standard air flowing at $U = 5.3$ m/sec. Assume a laminar boundary layer forms and approximate the velocity profile as linear. Plot δ, δ^*, and τ_w versus x/L for the plate.

9.51 For the flow conditions of Problem 9.50, develop an algebraic expression for the variation of wall shear stress with distance along the surface. Integrate to obtain an algebraic expression for the total skin-friction drag on the surface. Evaluate the drag for the given conditions.

9.52 Consider again the flow conditions of Problem 9.50. Show algebraically that the total drag force on one side of the plate may be written $F_D = \rho U^2 \theta_L b$. Evaluate θ_L and the total drag for the given conditions.

9.53 Water at 15 C flows over a flat plate at a speed of 1 m/sec. The plate is 0.4 m long and 1 m wide. The boundary layer on each surface of the plate is laminar. Assume that the velocity profile is approximated as linear. Determine the drag force on the plate.

9.54 The velocity profile in a laminar boundary-layer flow at zero pressure gradient is to be approximated by the cubic expression given in Problem 9.9. Use the momentum integral equation with this profile to obtain expressions for the ratio δ/x and the skin friction coefficient, C_f.

9.55 Water flows over a flat plate, with length $L = 1.2$ ft and width $b = 3$ ft, at freestream speed $U = 4$ ft/sec. Assume a laminar boundary layer forms and approximate the velocity profile using the cubic expression given in Problem 9.9. Plot δ, δ^*, and τ_w versus x/L for the plate.

9.56 Water flows over the top surface of a flat plate, forming a laminar boundary layer. The boundary-layer velocity profile is approximated by the cubic expression given in Problem 9.9. The plate is 0.5 ft long and 3 ft wide. The freestream flow speed is 4 ft/sec. Determine the maximum δ for the plate. Where does the minimum wall shear stress occur? Illustrate with a sketch of τ_w versus x. Calculate the minimum wall shear stress. Determine the drag force on the plate.

9.57 Consider again the flow conditions of Problem 9.55. Show algebraically that the total drag force on one side of the plate may be written $F_D = \rho U^2 \theta_L b$. Evaluate θ_L and the total drag for the given conditions.

9.58 Assume the flow conditions given in Example Problem 9.4. Plot δ, δ^*, and τ_w versus x/L for the plate.

9.59 For the flow conditions of Example Problem 9.4, develop an algebraic expression for the variation of wall shear stress with distance along the surface. Integrate to obtain an algebraic expression for the total skin friction drag on the surface. Evaluate the drag for the given conditions.

9.60 Consider again the flow conditions of Example Problem 9.4. Show algebraically that the total drag force on one side of the plate may be written $F_D = \rho U^2 \theta_L b$. Evaluate θ_L and the total drag for the given conditions.

9.61 The velocity profile in a turbulent boundary-layer flow at zero pressure gradient is approximated by the "$\frac{1}{6}$-power" profile expression,

$$\frac{u}{U} = \eta^{1/6}, \quad \text{where} \quad \eta = \frac{y}{\delta}$$

Use the momentum integral equation with this profile to obtain expressions for δ/x and C_f. Compare with results obtained in Section 9-5.2 for the "$\frac{1}{7}$-power" profile.

9.62 For the flow conditions of Example Problem 9.4, but using the "$\frac{1}{6}$-power" velocity profile of Problem 9.61, develop an algebraic expression for the variation of wall shear stress with distance along the surface. Integrate to obtain an algebraic expression for the total skin friction drag on the surface. Evaluate the drag for the given conditions.

9.63 Repeat Problem 9.61, using the "$\frac{1}{8}$-power" profile expression.

9.64 For the flow conditions of Example Problem 9.4, but using the "$\frac{1}{8}$-power" velocity profile, develop an algebraic expression for the variation of wall shear stress with distance along the surface. Integrate to obtain an algebraic expression for the total skin friction drag on the surface. Evaluate the drag for the given conditions.

9.65 Air at standard conditions flows over a flat plate. The freestream speed is 15 m/sec. Find δ and τ_w at $x = 1$ m from the leading edge for (a) completely laminar flow (assume a parabolic velocity profile) and (b) completely turbulent flow (assume a "$\frac{1}{7}$-power" velocity profile).

9.66 Standard air flows over a horizontal smooth flat plate at freestream speed $U = 14.5$ m/sec. The plate length is $L = 1.5$ m and its width is $b = 0.8$ m. The pressure gradient is zero. The boundary layer is tripped so that it is turbulent from the leading edge; the velocity profile is well represented by the "$\frac{1}{7}$-power" expression. Evaluate the boundary-layer thickness, δ, at the trailing edge of the plate. Calculate the wall shear stress at the trailing edge of the plate. Estimate the skin friction drag on the portion of the plate between $x = 0.5$ m and the trailing edge.

9.67 A uniform flow of standard air at 60 m/sec enters a plane-wall diffuser with negligible boundary-layer thickness. The inlet width is 75 mm. The diffuser walls diverge slightly to accommodate the boundary-layer growth so that the pressure gradient is negligible. Assume flat-plate boundary-layer behavior. Explain why the Bernoulli equation is applicable to this flow. Estimate the diffuser width 1.2 m downstream from the entrance.

9.68 A laboratory wind tunnel has a *flexible upper wall* that can be adjusted to compensate for boundary-layer growth, giving zero pressure gradient along the test section. The wall boundary layers are well represented by the "$\frac{1}{7}$-power" velocity profile. At the inlet the tunnel cross section is square, with height H_1 and width W_1, each equal to 305 mm. With freestream speed $U_1 = 26.5$ m/sec, measurements show that $\delta_1 = 12.2$ mm and downstream, $\delta_6 = 16.6$ mm. Calculate the height of the tunnel walls at section ⑥. Determine the equivalent length of flat plate that would produce the inlet boundary layer thickness. Estimate the streamwise distance between sections ① and ⑥ in the tunnel.

9.69 We wish to compare the flow of an ideal fluid ($\mu = 0$) and a real fluid in a plane-wall diffuser, as shown on the next page. Consider first the straight-channel case where $\phi = 0$. What can be said of the pressure gradient for the real and ideal fluids? Which fluid gives the higher p_2? Now consider a case where ϕ is not equal to zero, but is small enough to avoid separation. Again, what can be said of the pressure gradient for real and ideal fluids? Which case results in the highest exit pressure?

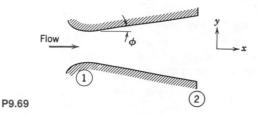

P9.69

9.70 Two hypothetical boundary-layer velocity profiles are shown. Obtain an expression for the momentum flux of each profile. If the two profiles were subjected to the same pressure gradient conditions, which would be most likely to separate first? Why?

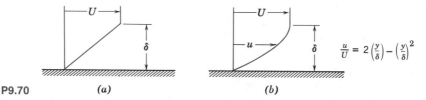

P9.70   *(a)*   *(b)*

9.71 Boundary-layer separation occurs when the shear stress at the surface becomes zero. Assume a polynomial representation for the laminar boundary layer of the form, $u/U = a + b\lambda + c\lambda^2 + d\lambda^3$, where $\lambda = y/\delta$. Specify boundary conditions on the velocity profile at separation. Find appropriate constants, a, b, c, and d, for the separation profile. Plot the profile and compare with the parabolic approximate profile. Calculate H at separation.

9.72 By suitable choice of adverse pressure gradient, it is possible to maintain a turbulent boundary layer over appreciable distance with nearly zero wall shear stress. For such flows, experiments show that shape factor H remains nearly constant and equal to 2. Consider turbulent flow of standard air through a flat diffuser with upstream width $W_1 = 3.0$ in. Assume the freestream flow speed at the diffuser inlet is $U_1 = 200$ ft/sec and that the corresponding momentum thickness is $\theta_1 = 0.10$ in. The freestream flow speed at section ② downstream is $U_2 = 100$ ft/sec. Plot the outlet width, W_2, for shape factors in the range from 1.29 to 2.6.

9.73 Standard air flows between two parallel flat plates as shown. The upper plate is porous from B to C and additional air is injected through this surface. As a result, the freestream speed, $U(x)$, varies as $U(x) = U_0 + C_1 x$, where $U_0 = 5$ m/sec, $C_1 = 4$ sec^{-1}, and x is the distance in meters measured from B. A laminar boundary layer develops along the lower surface; at $x = 0.05$ m, $\delta = 3.5$ mm. Assume the velocity distribution in this boundary layer is linear. Estimate the rate of growth of the boundary layer, $d\delta/dx$, at $x = 0.05$ m. Comment on this result in relation to the nature of the pressure gradient for $x > 0$.

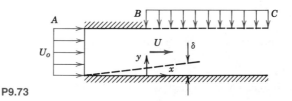

P9.73

9.74 Standard air flows between two parallel flat plates as shown. The upper plate is parallel to the lower plate from A to B, then sloped from B to C. As a result of the slope, the channel height, $h(x)$, varies as $h(x) = h_0 - C_1 x$, where $h_0 = 50$ mm, $C_1 = 0.005$, and x is the distance in meters measured from B. A laminar boundary layer develops along both surfaces; at $x = 0.05$ m, $\delta = 3.5$ mm. Assume the velocity distribution in the laminar boundary layer is parabolic. Estimate the rate of growth of the boundary layer, $d\delta/dx$, at $x = 0.05$ m.

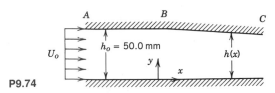

P9.74

9.75 Consider again the flow geometry and initial conditions of Problem 9.73. Assume a laminar boundary layer, with the approximate *linear* velocity profile, develops along the bottom surface. Neglect any boundary-layer development along surface AC. Evaluate the freestream velocity distribution, $U(x)$, needed to maintain the laminar boundary layer at *constant thickness* from $x = 50$ mm to $x = 100$ mm.

‡9.76 Prepare a computer program to solve for the distribution of boundary-layer thickness, $\delta(x)$, between $x = 50$ mm and $x = 100$ mm in Problem 9.73.

9.77 Cooling air is supplied through the wide, flat channel shown. For minimum noise and disturbance of the outlet flow, laminar boundary layers should be maintained on the channel walls. Estimate the maximum inlet flow speed at which the outlet flow will be laminar. Assuming parabolic velocity profiles in the laminar boundary layers, evaluate the pressure drop, $p_1 - p_2$. Express your answer in inches of water.

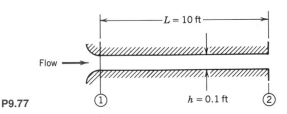

P9.77

9.78 A laboratory wind tunnel has a test section that is square in cross section, with inlet width, W_1, and height, H_1, each equal to 305 mm. At freestream speed $U_1 = 24.5$ m/sec, measurements show the boundary-layer thickness is $\delta_1 = 9.75$ mm with a "$\frac{1}{7}$-power" turbulent velocity profile. The pressure gradient in this region is given approximately by $dp/dx = -0.035$ mm H_2O/mm. Evaluate the reduction in effective flow area caused by the boundary layers on the tunnel bottom, top, and walls at section ①. Calculate the rate of change of boundary-layer momentum thickness, $d\theta/dx$, at section ①. Estimate the momentum thickness at the end of the test section, located $L = 254$ mm downstream.

‡9.79 The variable-wall concept is proposed to maintain constant boundary-layer thickness in the wind tunnel of Problem 9.78. Beginning with the initial conditions of Problem 9.78, evaluate the pressure gradient needed to maintain constant boundary-layer thickness. Assume constant width, W_1. Estimate the top-height settings along the test section from $x = 0$ at section ① to $x = 254$ mm at section ② downstream.

9.80 A flat-bottomed barge, 25 m long and 10 m wide, submerged to a depth of 1.5 m, is to be pushed up a river at the rate of 8 km/hr. Estimate the power required to overcome skin friction if the water temperature is 15 C.

9.81 A vertical stabilizing fin on a land-speed record car is $L = 1.65$ m long and $H = 0.785$ m tall. The automobile is to be driven at the Bonneville Salt Flats in Utah, where the elevation is 1340 m and the summer temperature reaches 50 C. The car speed is 560 km/hr. Evaluate the length Reynolds number of the fin. Estimate the location of transition from laminar to turbulent flow in the boundary layers. Calculate the power required to overcome skin friction drag on the fin.

‡ You may wish to use simple computer programs to help solve problems marked with daggers.

9.82 A jet transport aircraft cruises at 12 km altitude in steady level flight at 820 km/hr. Model the aircraft fuselage as a circular cylinder with $D = 4$ m diameter and $L = 40$ m length. Neglecting compressibility effects, estimate the skin friction drag force on the fuselage. Evaluate the power needed to overcome this force.

9.83 A towboat for river barges is tested in a towing tank. The towboat model is built at a scale ratio of 1:13.5. Dimensions of the model are overall length 11.1 ft, beam 3.11 ft, and draft 0.62 ft. (The model displacement in fresh water is 1200 lb.) Estimate the average length of wetted surface on the hull. Calculate the skin friction drag force on the prototype at a speed of 8 mph relative to the water.

9.84 Resistance of a barge is to be determined from model test data. The model is constructed to a scale ratio of 1:13.5, and has length, beam, and draft of 22.0, 4.00, and 0.667 ft, respectively. The test is to simulate performance of the prototype at 8 mph. At what speed should the model be tested? Are boundary layers on the prototype laminar or turbulent? Where should boundary-layer trips be placed on the model? Estimate the skin friction drag forces for the model and prototype barges.

9.85 A sheet of plastic material $\frac{3}{8}$ in. thick, with specific gravity SG = 1.5, is dropped into a large tank containing water. The sheet is 2 ft high and 3 ft wide. It falls vertically. Estimate the terminal speed of the sheet, assuming that the only drag is due to skin friction, and that the boundary layers are turbulent from the leading edge.

9.86 A nuclear submarine cruises fully submerged at 27 knots. The hull is approximately a circular cylinder with $D = 11.0$ m diameter and $L = 107$ m length. Estimate the percentage of the hull length for which the boundary layer is laminar. Calculate the skin friction drag on the hull. Estimate the approximate rate of deceleration of the submarine if all propulsion power were removed suddenly.

9.87 Consider the ship model test data presented in Figs. 7.1 and 7.2. Calculate the skin friction drag coefficients and corresponding forces for model and prototype at $Fr = 0.5$. For the prototype, $L = 409$ ft and $A = 19{,}500$ ft^2.

9.88 A supertanker displacement is approximately 600,000 metric tons. This ship has length $L = 300$ m, beam (width) $b = 80$ m, and draft (depth) $D = 25$ m. The ship steams at 14 knots through seawater at 4 C. For these conditions, estimate (a) the thickness of the boundary layer at the stern of the ship, (b) the total skin friction drag acting on the ship, (c) the power required to overcome the drag force, (d) the kinetic energy contained in the boundary layers on the ship, and (e) the minimum distance required to bring the ship to a stop.

9.89 As a part of the 1976 bicentennial celebration, an enterprising group hung a giant American flag (59 m high and 112 m wide) from the suspension cables of the Verrazano Narrows Bridge. They apparently were reluctant to make holes in the flag to alleviate the wind force, and hence they effectively had a flat plate normal to the flow. The flag tore loose from its mountings when the wind speed reached 16 km/hr. Estimate the wind force acting on the flag at this wind speed. Should they have been surprised that the flag blew down?

9.90 A rotary mixer is constructed from two circular disks as shown. The mixer is rotated at 60 rpm in a large vessel containing a brine solution (SG = 1.1). Neglect the drag on the rods and the motion induced in the liquid. Estimate the minimum torque and power required to drive the mixer.

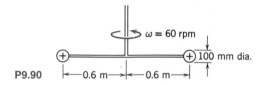

9.91 The vertical component of the landing speed of a parachute is to be less than 6 m/sec. The total mass of chute and jumper is 120 kg. Determine the minimum diameter of the open parachute.

9.92 Ballistic data obtained on a firing range show that aerodynamic drag reduces the speed of a 44 magnum revolver bullet from 250 m/sec to 210 m/sec as it travels over a horizontal distance of 150 m. The diameter and mass of the bullet are 11.2 mm and 15.6 g, respectively. Evaluate the average drag coefficient for the bullet.

9.93 The resistance to motion of a good bicycle on smooth pavement is nearly all due to aerodynamic drag. Assume that the total mass of rider and bike is $M = 100$ kg. The frontal area measured from a photograph is $A = 0.46$ m^2. Experiments on a hill, where the road grade is 8 percent, show that terminal speed is $V_t = 15$ m/sec. From these data, the drag coefficient is estimated as $C_D = 1.2$. Verify this calculation of drag coefficient. Estimate the distance needed for the bike and rider to decelerate from 15 to 10 m/sec while coasting after reaching level road.

9.94 A circular disk is hung in an air stream from a pivoted strut as shown. In a wind-tunnel experiment, performed in air at 50 ft/sec with a 1 in. diameter disk, α was measured at 10°. For these conditions determine the mass of the disk. Assume drag on the strut and friction in the pivot are negligible. Plot a theoretical curve of α as a function of air speed.

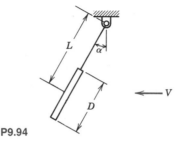

P9.94

9.95 It has been proposed to use surplus 55 gal oil drums to make simple windmills for underdeveloped countries. Two possible configurations are shown. Estimate which would be better. Why, and by how much? The diameter and length of a 55 gal drum are $D = 24$ in. and $H = 29$ in.

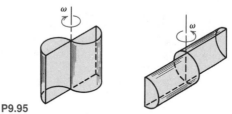

P9.95

9.96 A simple but effective anemometer to measure wind speed can be made from a thin plate hinged to deflect in the wind. Consider a thin plate made from brass that is 20 mm high and 10 mm wide. Derive a relationship for wind speed as a function of deflection angle, θ. What thickness of brass should be used to give $\theta = 30°$ at 10 m/sec?

9.97 A top-notch athlete can ride a bicycle at a sustained 37 km/hr on a calm day at maximum exertion. (The total mass of rider and bike is $M = 80$ kg. The rolling resistance force from the tires is $F_R = 4$ N. The drag coefficient and frontal area of the bike and rider are $C_D = 1.2$ and $A = 0.25$ m^2.) The athlete has made a bet that he can ride at a ground speed of 30 km/hr into a headwind that blows at 10 km/hr. Determine the maximum power output that the athlete can sustain. Evaluate the athlete's prospects to win this bet.

9.98 Supports for traffic lights in downtown Detroit are shown. Each member of the support structure is made from 6 in. × 6 in. square steel tubing. Estimate the maximum forces and moments at the base in a 90 mph wind.

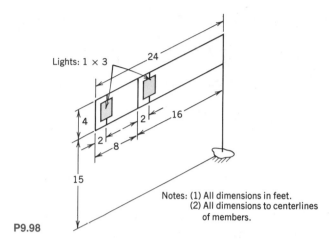

Lights: 1 × 3

24

4

2

16

2

8

15

Notes: (1) All dimensions in feet.
(2) All dimensions to centerlines
of members.

P9.98

9.99 An F-4 aircraft is slowed after landing by dual parachutes deployed from the rear. Each parachute is 12 ft in diameter. The F-4 weighs 32,000 lbf and lands at 160 knots. Estimate the time and distance required to decelerate the aircraft to 100 knots, assuming that the brakes are not used and the drag of the aircraft is negligible.

9.100 Experimental data [14] suggest that the maximum and minimum drag area $(C_D A)$ for a skydiver varies from about 0.85 m^2 for a prone, spread-eagle position to about 0.11 m^2 for vertical fall. Estimate the terminal speeds for a 75 kg skydiver in each position. Calculate the time and distance needed for the skydiver to reach 95 percent of terminal speed at an altitude of 3000 m on a standard day.

9.101 A vehicle is built to try for the land-speed record at the Bonneville Salt Flats, elevation 4400 ft. The engine delivers 500 hp to the rear wheels, and careful streamlining has resulted in a drag coefficient of 0.15, based on a 15 ft^2 frontal area. Compute the theoretical maximum ground speed of the car (a) in still air and (b) with a 20 mph headwind.

‡9.102 In the early 1970s a typical large American sedan had a frontal area of 23.4 ft^2 and a drag coefficient of 0.5. Plot a curve of horsepower required to overcome aerodynamic drag versus road speed in standard air. If rolling resistance is 1.5 percent of curb weight (4500 lbf), determine the speed at which the aerodynamic force exceeds frictional resistance. How much power is required to cruise at 55 mph and at 70 mph on level road with no wind?

9.103 A tractor-trailer rig has frontal area $A = 102$ ft^2 and drag coefficient $C_D = 0.9$. Rolling resistance is 6 lbf per 1000 lbf of vehicle weight. The specific fuel consumption of the diesel engine is 0.34 lbm of fuel per horsepower hour, and drivetrain efficiency is 92 percent. The density of diesel fuel is 6.9 lbm/gal. Estimate the fuel economy of the rig at 55 mph if its gross weight is 72,000 lbf. An air fairing system reduces aerodynamic drag 15 percent. The truck travels 120,000 miles per year. Calculate the fuel saved per year by the roof fairing.

9.104 According to an advertisement, the Porsche 944 has the following characteristics: $C_D = 0.35$, $A = 1.83$ m^2, and maximum power $\mathscr{P} = 143$ bhp. The ad further states that the vehicle requires 13.9 hp to cruise at 55 mph. Use these data to estimate (a) the maximum acceleration capability at 55 mph and (b) the top speed of the car. (Assume that rolling resistance is 1 percent of car weight.)

‡ You may wish to use simple computer programs to help solve problems marked with daggers.

9.105   A Ford "Probe GT" automobile is driven along level highway at 100 km/hr in standard air. The frontal area of the vehicle is 1.8 m² and the drag coefficient is 0.31. How much power is required to overcome aerodynamic drag? Estimate the maximum speed of the car if the engine is rated at 145 hp.

9.106   A round thin disk of radius R is oriented perpendicular to a fluid stream. The pressure distributions on the front and back surfaces are measured and presented in the form of pressure coefficients. The data are modeled with the following expressions for the front and back surfaces, respectively:

$$\text{Front Surface} \qquad C_p = 1 - \left(\frac{r}{R}\right)^6$$

$$\text{Rear Surface} \qquad C_p = -0.42$$

Calculate the drag coefficient for the disk.

9.107   An anemometer to measure wind speed is made from four hemispherical cups of 50 mm diameter, as shown. The center of each cup is placed at $R = 75$ mm from the pivot. The anemometer is to start rotating when the wind speed is above 1 km/hr. Determine the relation between the rotational speed, ω, and the wind speed, V, if friction in the pivot is neglected. Calculate the maximum frictional torque that can be present in the pivot. Estimate the error in speed measurement caused by the pivot frictional torque at a wind speed of 10 km/hr.

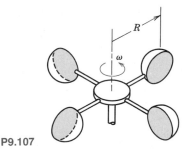

P9.107

9.108   An object falls in air down a long vertical chute. The speed of the object is constant at 3 m/sec. The flow pattern around the object is shown. The static pressure is uniform across sections ① and ②; pressure is atmospheric at section ①. The effective flow area at section ② is 20 percent of the chute area. Frictional effects between sections ① and ② are negligible. Evaluate the flow speed relative to the object at section ②. Calculate the static pressure at section ②. Determine the mass of the object.

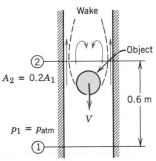

P9.108 $A_1 = 0.09$ m²

9.109   An object of mass m, with cross-sectional area equal to half the size of the chute, falls down a mail chute. The motion is steady. The wake area is $\frac{3}{4}$ the size of the chute at its maximum area. Use the assumption of constant pressure in the wake. Apply the continuity, Bernoulli, and momentum equations to develop an expression for terminal speed of the object in terms of its mass and other quantities.

9.110 A large paddle wheel is immersed in the current of a river to generate power. Each paddle has area A and drag coefficient C_D; the center of each paddle is located at radius R from the centerline of the paddle wheel. Assume the equivalent of one paddle is submerged continuously in the flowing stream. Obtain an expression for the drag force on a single paddle in terms of geometric variables, current speed, V, and linear speed of the paddle center, $U = R\omega$. Develop expressions for the torque and power produced by the paddle wheel. Find the speed at which the paddle wheel should rotate to obtain maximum power output from the wheel in a given current.

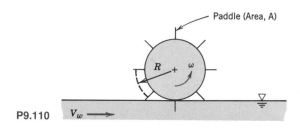

P9.110

9.111 A pitot-static probe with stem diameter $d = 6$ mm is inserted $L = 300$ mm into a wind-tunnel air flow, where the uniform speed is 25 m/sec. Calculate the drag force and bending moment acting on the probe.

9.112 The CB antenna on a car is 8 mm in diameter and 2 m long. Estimate the torque that tends to snap it off if the car is driven at 125 km/hr on a standard day.

9.113 Consider small oil droplets (SG = 0.85) rising in water. Develop a relation for calculating terminal speed of a droplet (in m/sec) as a function of droplet diameter (in mm) assuming Stokes flow. For what range of droplet diameter is Stokes flow a reasonable assumption?

9.114 A dust particle falling in air is observed to settle at 2 mm/sec. The specific gravity of the particle is 4.5. Estimate its size.

9.115 A spherical hydrogen-filled balloon, 0.6 m in diameter, exerts an upward force of 1.3 N on a restraining string when held stationary in standard air with no wind. With a wind speed of 3 m/sec, the string holding the balloon makes an angle of 60° with the horizontal. Calculate the drag coefficient of the balloon under these conditions, neglecting the weight of the string.

9.116 An antique airplane carries 60 m of external guy wires stretched normal to the direction of motion. The wire diameter is 6 mm. Basing calculations on two-dimensional flow around the wires, what power saving may be effected by removing the wires if the speed is 150 km/hr in standard air at sea level?

9.117 A field hockey ball has diameter $D = 73$ mm and mass $m = 160$ g. When struck well, it leaves the stick with initial speed $U_0 = 50$ m/sec. The ball is essentially smooth. Estimate the distance traveled in horizontal flight before the speed of the ball is reduced 10 percent by aerodynamic drag.

9.118 Compute the terminal speed of 10 mm diameter hailstones (assume spherical) in standard air.

9.119 Compute the terminal speed of a $\frac{1}{8}$ in. diameter raindrop (assume spherical) in standard air.

9.120 Determine the terminal speed of a smooth tennis ball in still air. Its weight is 2 oz and its diameter is 2.5 in.

9.121 A light plane tows an advertising banner over a football stadium on a Saturday afternoon. The banner is 1 m tall and 12 m long. According to Hoerner [14], the drag coefficient based on area (Lh) for such a banner is approximated by $C_D = 0.05$ L/h, where L is the banner length and h is the banner height. Estimate the power required to tow the banner at $V = 90$ km/hr. Compare with the drag of a rigid flat plate. Why is the drag larger for the banner?

9.122 A tennis ball with mass of 57 g and diameter of 64 mm is dropped in standard sea-level air. Calculate the terminal speed of the ball. Estimate the time and distance required for the ball to reach 95 percent of its terminal speed.

9.123 A small sphere ($D = 6$ mm) is observed to fall through castor oil at a terminal speed of 60 mm/sec. The temperature is 20 C. Compute the drag coefficient for the sphere. Determine the density of the sphere. If dropped in water, would the sphere fall slower or faster? Why?

9.124 In the ink-jet printing process, small spherical droplets of ink (SG = 1.2) are sprayed from a nozzle. Each droplet is electrically charged, and its trajectory is controlled by an electric field. The diameter of a typical droplet is $D = 60$ μm and its initial speed is $V_0 = 17.5$ m/sec. The drag coefficient for a single spherical droplet in this speed range is given approximately by

$$C_D \approx \frac{10.4}{\sqrt{Re_D}}$$

Consider a single spherical droplet of ink moving horizontally in otherwise undisturbed air. Evaluate the droplet speed when it reaches the paper, located $L = 30$ mm from the nozzle.

9.125 The following curve-fit for the drag coefficient of a smooth sphere as a function of Reynolds number has been proposed by Chow [33]:

$$C_D = 24/Re \qquad\qquad\qquad Re \leq 1$$
$$C_D = 24/Re^{0.646} \qquad\qquad 1 < Re \leq 400$$
$$C_D = 0.5 \qquad\qquad\qquad 400 < Re \leq 3 \times 10^5$$
$$C_D = 0.000366\ Re^{0.4275} \qquad 3 \times 10^5 < Re \leq 2 \times 10^6$$
$$C_D = 0.18 \qquad\qquad\qquad Re > 2 \times 10^6$$

Use data from Fig. 9.11 to estimate the magnitude and location of the maximum error between the curve fit and data.

9.126 Problem 9.94 showed a circular disk hung in an air stream from a cylindrical strut. Assume the strut is $L = 40$ mm long and $d = 3$ mm in diameter. Solve Problem 9.94 including the effect of drag on the support. Assume the drag coefficients for the cylinder and disk apply when the component of wind speed perpendicular to the object is used.

9.127 A water tower consists of a 12 m diameter sphere on top of a vertical tower 30 m tall. Estimate the bending moment exerted on the base of the tower due to the aerodynamic force imposed by a 100 km/hr wind on a standard day. Neglect interference at the joint between the sphere and tower.

9.128 A spherical balloon contains helium and ascends through standard air. The mass of the balloon and its payload is 150 kg. Determine the required diameter if it is to ascend at 3 m/sec.

9.129 A cast iron "12-pounder" cannon ball rolls off the deck of a ship and falls into the ocean at a location where the depth is 1000 m. Estimate the time that elapses before the cannonball hits the sea bottom.

9.130 Consider a cylindrical flag pole of height H. For constant drag coefficient, evaluate the drag force and bending moment on the pole if wind speed varies as $u/U = (y/H)^{1/7}$, where y is distance measured from the ground. Compare with drag and moment for a uniform wind profile with constant speed U.

9.131 The Stokes drag law for smooth spheres is to be verified experimentally by dropping steel ball bearings in glycerin. Evaluate the largest diameter steel ball for which $Re < 1$

at terminal speed. Calculate the height of glycerin column needed for a bearing to reach 95 percent of terminal speed.

9.132 Hydrogen bubbles frequently are used as markers for flow visualization. Consider bubbles with diameters from 0.01 to 0.1 mm. Estimate their terminal speeds in water. (Be sure to consider the effect of surface tension on the pressure of hydrogen within the bubble.)

‡**9.133** The plot shows pressure difference versus angle, measured for air flow around a circular cylinder at $Re = 80,000$. Use these data to estimate C_D for this flow. Compare with data from Fig. 9.13. How can you explain the difference?

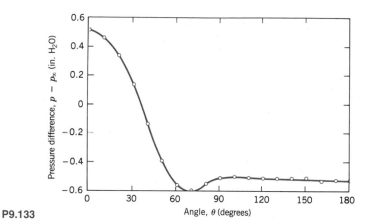

P9.133

9.134 A thin-walled plastic sphere 10 mm in diameter with mass of 0.05 g, immersed in a glycerin bath at a depth of 1 m, is released and begins to rise to the surface. Calculate the time required for the sphere to reach the surface.

9.135 A falling raindrop behaves essentially as a sphere moving through an infinite medium. Obtain an algebraic relationship for the terminal speed of a raindrop for a constant drag coefficient of 0.5. Evaluate numerically to obtain the form

$$V(\text{m/sec}) = K\sqrt{D(\text{mm})}$$

where K is a dimensional constant and D is drop diameter. For what range of drop diameter would you expect this equation to (a) over-predict or (b) under-predict the raindrop speed?

‡**9.136** Air bubbles rise from the regulator of a scuba diver who swims at a depth of 10 m in seawater. Consider a bubble of 10 mm diameter at this depth. Evaluate the terminal speed of this bubble as a function of depth. Estimate the time needed for the bubble to rise to the surface. Use numerical or graphical integration if necessary.

9.137 Coastdown tests, performed on a level road on a calm day, can be used to measure aerodynamic drag and rolling resistance coefficients for a full-scale vehicle. Rolling resistance is estimated from dV/dt measured at low speed, where aerodynamic drag is small. Rolling resistance then is deducted from dV/dt measured at high speed to determine the aerodynamic drag. The following data were obtained during a test with a vehicle, of weight $W = 25,000$ lbf and frontal area $A = 79$ ft²:

V(mph)	5	55
$\dfrac{dV}{dt}\left(\dfrac{\text{mph}}{\text{sec}}\right)$	-0.150	-0.475

Evaluate the aerodynamic drag coefficient for this vehicle.

‡ You may wish to use simple computer programs to help solve problems marked with daggers.

9.138 Approximate dimensions of a rented rooftop carrier are shown. Estimate the drag force on the carrier ($r = 4$ in.) at 65 mph. If the drivetrain efficiency of the vehicle is 0.85 and the brake specific fuel consumption of its engine is 0.46 lbm/hp·hr, estimate the additional rate of fuel consumption due to the carrier. Compute the effect on fuel economy if the auto achieves 30 mpg without the carrier.

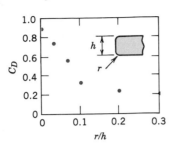

Drag coefficient v. radius ratio [34]

P9.138

‡9.139 Motion of a small rocket was analyzed in Example Problem 4.12 assuming negligible aerodynamic drag. This was not realistic at the final calculated speed of 369 m/sec. Write a simple computer or calculator program to evaluate rocket speed as a function of time, assuming $C_D = 0.3$ and a rocket diameter of 700 mm. Compare with results from Example Problem 4.12.

9.140 The guy wires on the antique plane in Problem 9.116 are to be streamlined rather than removed. Use data from Fig. 9.14 to select an optimum faired shape for the guys. Evaluate the maximum power saving that results from fairing the wires.

9.141 The bicycle of Problem 9.97 is equipped with a fairing to reduce aerodynamic drag. The fairing reduces C_D to 0.90 but increases the frontal area to 0.30 m². Estimate the rider's top speed in still air with the fairing installed.

9.142 An aircraft is in level flight at 250 km/hr through air at standard conditions. The lift coefficient at this speed is 0.4 and the drag coefficient is 0.065. The mass of the aircraft is 850 kg. Calculate the effective lift area for the craft.

9.143 An airplane with an effective lift area of 25 m² is fitted with airfoils of NACA 23012 section (Fig. 9.23). The maximum flap setting that can be used at takeoff corresponds to configuration ② in Fig. 9.23. Determine the maximum gross mass possible for the airplane if its takeoff speed is 150 km/hr (neglect added lift due to ground effect).

9.144 Consider a kite of mass 0.2 kg as a flat plate with an area of 1 m², flown in standard air moving horizontally at 10 m/sec. The kite makes a 5° angle with the horizontal. Assume the lift coefficient is given by the equation $C_L = 2\pi \sin\alpha$, where α is the angle of attack, and the lift-drag ratio is 4.0. If the string makes an angle θ with the horizontal, determine θ and the tension in the string.

9.145 A flat plate, $L = 3$ ft long and $b = 4$ ft wide, moves through still air at 12 degrees angle of attack and 44.5 ft/sec. At this angle of attack, the lift and drag coefficients for a plate with this aspect ratio are $C_L = 0.72$ and $C_D = 0.17$, based on plate area. Evaluate the resultant force vector that acts on the plate. Determine the power required to propel the plate at this speed.

9.146 The foils of a surface-piercing hydrofoil watercraft have a total effective area of 0.7 m². Their coefficients of lift and drag are 1.6 and 0.5, respectively. The total mass of the craft in running trim is 1800 kg. Determine the minimum speed at which the craft is supported by the hydrofoils. At this speed, find the power required to overcome water resistance. If the craft is fitted with a 110 kW engine, estimate its top speed.

9.147 A light airplane, with mass $M = 2000$ lbm, has wing planform area $A = 100$ ft². Its top speed in level flight is $V_{max} = 180$ mph. The airplane may be used safely for

‡ You may wish to use simple computer programs to help solve problems marked with daggers.

aerobatic maneuvers provided the maximum vertical acceleration is less than $5g$. The airfoil used for the craft has a NACA 23015 section. Is it possible for the pilot to exceed an instantaneous vertical acceleration of $5g$ with this aircraft from level flight? (In level flight, the pilot and aircraft are considered subject to $1g$.)

9.148 The U.S. Air Force F-16 fighter aircraft has wing planform area $A = 27.9$ m²; it can achieve a maximum lift coefficient of $C_L = 1.6$. When fully loaded its maximum mass is $M = 11,600$ kg. The airframe is capable of maneuvers that produce $9g$ vertical accelerations. However, student pilots are restricted to $5g$ maneuvers during training. Consider a turn flown in level flight with the aircraft banked. Find the minimum speed in standard air at which the pilot can produce a $5g$ total acceleration. Calculate the corresponding flight radius. Discuss the effect of altitude on these results.

9.149 A light plane has 10 m effective wingspan and 1.8 m chord. It was originally designed to use a conventional (NACA 23015) airfoil section. With this airfoil, its cruising speed on a standard day near sea level is 225 km/hr. A conversion to a laminar-flow (NACA 66_2-215) section airfoil is proposed. Determine the cruising speed that could be achieved with the new airfoil section for the same power.

9.150 Assume the Boeing 727 aircraft has wings with NACA 23012 section, planform area of 1600 ft², double-slotted flaps, and effective aspect ratio of 6.5. If the aircraft flies at 150 knots in standard air at 175,000 lb gross weight, estimate the thrust required to maintain level flight.

9.151 An airplane, with mass of 4500 kg, is flown at constant elevation and speed on a circular path at 250 km/hr. The flight circle has a radius of 1000 m. The plane has lifting area of 22 m² and is fitted with NACA 23015 section airfoils with effective aspect ratio of 7. Determine the drag on the aircraft and the power required.

9.152 Jim Hall's Chaparral 2F sports-racing cars in the 1960s pioneered use of airfoils mounted above the rear suspension to enhance stability and improve braking performance. The airfoil was effectively 6 ft wide (span) and had a 1 ft chord. Its angle of attack was variable between 0 and minus 12 degrees. Assume lift and drag coefficient data are given by curves (for conventional section) in Fig. 9.17. Consider a car speed of 120 mph on a calm day. For an airfoil deflection of 12° down, calculate (a) the maximum downward force and (b) the maximum increase in deceleration force produced by the airfoil.

9.153 The glide angle for unpowered flight is such that lift, drag, and weight are in equilibrium. Show that the glide slope angle, θ, is such that $\tan \theta = C_D/C_L$. The minimum glide slope occurs at the speed where C_L/C_D is a maximum. For the conditions of Example Problem 9.8, evaluate the minimum glide slope angle for a Boeing 727-200. How far could this aircraft glide from an initial altitude of 10 km on a standard day?

9.154 Performance of jet aircraft is considered in Example Problem 9.8. Show analytically that, at the optimum cruise speed, $C_{D,i} = \frac{1}{3}C_{D,0}$.

9.155 Performance of jet aircraft is considered in Example Problem 9.8. Show analytically that, at the speed for maximum endurance, $C_{D,i} = C_{D,0}$.

9.156 The wing loading of the Gossamer Condor is 0.4 lbf/ft² of wing area. Crude measurements showed drag was approximately 6 lbf at 12 mph. The total weight of the Condor was 200 lbf. The effective aspect ratio of the Condor is 17. Estimate the minimum power required to fly the aircraft. Compare to the 0.39 hp that pilot Brian Allen could sustain for 2 hr.

9.157 Air moving over an automobile is accelerated to speeds higher than the travel speed, as shown in Fig. 9.25. This causes changes in interior pressure when windows are opened or closed. Use the data of Fig. 9.25 to estimate the pressure reduction when a window is opened slightly at a speed of 100 km/hr. What is the air speed in the freestream near the window opening?

9.158 A class demonstration showed that lift is present when a cylinder rotates in an air stream. A string wrapped around a paper cylinder and pulled causes the cylinder to spin and move forward simultaneously. Assume a cylinder of 2 in. diameter and 10 in. length is given a rotational speed of 300 rpm and a forward speed of 4 ft/sec. Estimate the approximate lift force that acts on the cylinder.

9.159 Rotating cylinders were proposed as a means of ship propulsion in 1924 by the German engineer, Flettner. The original Flettner rotor ship had two rotors, each about 3 m in diameter and 15 m high, rotating at up to 750 rpm. Calculate the maximum lift and drag forces that act on each rotor in a 50 km/hr wind. Compare the total force to that produced at the optimum L/D at the same wind speed. Estimate the power needed to spin the rotor at 750 rpm.

9.160 A golf ball (diameter $D = 43$ mm) with circular dimples is hit from a sand trap at 20 m/sec with backspin of 2000 rpm. The mass of the ball is 48 g. Evaluate the lift and drag forces acting on the ball. Express your results as fractions of the body force due to gravity acting on the ball.

9.161 American and British golf balls have slightly different diameters but the same mass (see Problems 1.32 and 1.33). Assume a professional golfer hits each type of ball from a tee at 85 m/sec with backspin of 9000 rpm. Evaluate the lift and drag forces on each ball. Express your answers as fractions of the body force on each ball. Estimate the radius of curvature of the trajectory of each ball. Which ball should have the longer range for these conditions?

9.162 A baseball pitcher throws a ball at 90 km/hr. Home plate is 18 m away from the pitcher's mound. What spin should be placed on the ball for maximum horizontal deviation from a straight path? (A baseball has $m = 145$ g and $D = 74$ mm.) How far will the ball deviate from a straight line?

Chapter 10

FLOW IN OPEN CHANNELS

Many flows in nature occur with a *free surface*. Because free surface flows differ in several important respects from flows in closed conduits, they are treated separately in this chapter. Rainwater runoff, and flows in rivers, aqueducts, irrigation canals, and drainage ditches are familiar examples where the free surface is at atmospheric pressure. Geometric properties of common open-channel shapes are presented in Section 10-1.

Surface waves can form in flows with a free surface. The propagation speed of a single, or solitary, wave is analogous in many respects to the propagation of a sound wave in a compressible fluid medium. The propagation rate of a disturbance in open-channel flow depends on the Froude number of the flow (Section 10-2). Changes in channel cross section or depth, and their effects on the mean flow velocity, also are distinguishing features of free surface flows.

In contrast to flow in a closed conduit, where the flow is sustained by a pressure difference, the driving force for open-channel flow is gravity. The gravity force is opposed by a friction force on the solid boundaries of the channel.

Most flows of interest are large in physical scale, so the Reynolds numbers generally are large. Consequently, open-channel flow seldom is laminar. As in the case of turbulent flow in pipes, we must rely on empirical correlations to relate frictional effects to the average velocity of flow. The empirical correlation is included through a head loss term in the energy equation (Section 10-3). Additional complications in many practical cases include the presence of sediment, or other particulate matter, in the flow, and the erosion of earthen channels or structures by water action.

In this chapter we shall analyze several aspects of steady open-channel flow using the basic control volume equations of Chapter 4. In Section 10-4 we consider flows in which the effects of area change predominate and frictional forces may be neglected. When the flow cross section does not vary in the flow direction, the flow is said to be at normal depth, or uniform (Section 10-5). For flow at normal depth, the liquid surface is parallel to the channel bed. This is analogous to fully developed flow in a pipe. When the liquid depth is not constant, we have varied flow. In Section 10-6 we consider gradually varied flow. The major objective in the analysis of gradually varied flow is to predict the shape of the free surface. When conditions require the flow to change in depth abruptly, this is accomplished through a hydraulic jump (Section 10-7). The chapter concludes with a brief discussion of flow measurement techniques for use in open channels (Section 10-8).

10-1 CHARACTERISTICS OF OPEN CHANNELS

Any conduit with a liquid free surface is classified as an open channel; examples of natural channels are abundant. Man-made channels are given many different names, including canal, flume, or culvert; these are defined and applied rather loosely. A *canal* usually is excavated below ground level, and may be unlined or lined. Canals

488

Table 10.1 Geometric Properties of Common Open-Channel Shapes

Shape	Section	Flow Area, A	Wetted Perimeter, P	Hydraulic Radius, R_h
Trapezoidal		$y(b + y \cot \alpha)$	$b + \dfrac{2y}{\sin \alpha}$	$\dfrac{y(b + y \cot \alpha)}{b + \dfrac{2y}{\sin \alpha}}$
Triangular		$y^2 \cot \alpha$	$\dfrac{2y}{\sin \alpha}$	$\dfrac{y \cos \alpha}{2}$
Rectangular		by	$b + 2y$	$\dfrac{by}{b + 2y}$
Wide Flat		by	b	y
Circular		$(\alpha - \sin \alpha)\dfrac{D^2}{8}$	$\dfrac{\alpha D}{2}$	$\dfrac{D}{4}\left(1 - \dfrac{\sin \alpha}{\alpha}\right)$

generally are long and of very mild slope; they are used to carry irrigation or storm water or for navigation. A *flume* usually is built above ground level to carry water across a depression. A *culvert*—which usually is designed to flow only part full—is a short covered channel used to drain water under a highway or railroad embankment.

Channels may be constructed in a variety of cross-sectional shapes; usually regular geometric shapes are used. A channel with constant slope and cross section is termed *prismatic*. Lined canals often are built with rectangular or trapezoidal sections; smaller troughs or ditches sometimes are triangular. Culverts and tunnels generally are circular or elliptical in section. Natural channels are highly irregular and nonprismatic, but often they are approximated by trapezoid or paraboloid sections. Geometric properties of common open-channel shapes are summarized in Table 10.1. The depth of flow, y, is the perpendicular distance measured from the channel bed to the free surface. The flow area, A, is the cross section of the flow perpendicular

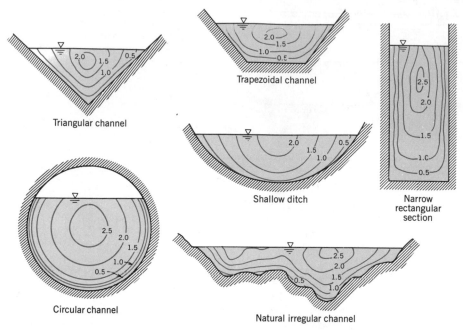

Triangular channel

Trapezoidal channel

Shallow ditch

Narrow rectangular section

Circular channel

Natural irregular channel

Fig. 10.1 Typical contours of equal velocity in open-channel sections. (From [1], used by permission.)

to the flow direction. The wetted perimeter, P, is the length of the solid channel surface in contact with the liquid. The hydraulic radius, R_h, is defined as

$$R_h = \frac{A}{P} \tag{10.1}$$

Note that, for flow in noncircular closed conduits (Section 8-7.3), the hydraulic diameter was defined as

$$D_h = 4\frac{A}{P} \tag{8.43}$$

Thus, for a circular pipe, the hydraulic diameter, from Eq. 8.43, is equal to the pipe diameter. From Eq. 10.1, the hydraulic radius is half the pipe radius. The hydraulic radius, as defined by Eq. 10.1, is commonly used in the analysis of open-channel flows, so it will be used throughout this chapter.

For nonrectangular channels, the *hydraulic depth* is defined as $y_h = A/b_s$, where b_s is the surface width. The hydraulic depth represents the average depth of the channel at any cross section. It gives the depth of an equivalent rectangular channel.

Typical contours of streamwise velocity for a number of open-channel sections are shown in Fig. 10.1. Although the profiles are not uniform, the approach followed is to assume uniform flow at a section. The kinetic energy coefficient, α, is taken as unity.

Note from Fig. 10.1 that the measured maximum velocity occurs below the free surface. Since there is negligible shear stress due to air drag, one would expect the maximum velocity to occur at the free surface. Secondary flows are responsible for distorting the axial velocity profile.[1]

---

[1] The NCFMF film, *Secondary Flow*, E. S. Taylor, principal, illustrates several examples of secondary flow phenomena.

The nonuniform profile of axial velocity causes strong secondary flows at the base of an obstruction, such as a bridge pier. The location of the maximum velocity is well above the channel floor; the high stagnation pressure at this location causes a recirculation zone to form at the base in front of the obstruction. As this swirling flow moves around the sides of the obstruction, it forms a horseshoe vortex. The vortex core is stretched and the swirl velocities are increased along the sides of the obstruction.[2] The high velocities present in the vortex can seriously erode the bottom of a natural channel along the sides of a pier.

10-2 PROPAGATION OF SURFACE WAVES

Consider an open channel with movable end wall, containing a liquid initially at rest. If the end wall is given a sudden displacement, as in Fig. 10.2a, a small wave forms and travels down the channel of width b. The speed of wave propagation, the wave *celerity*, is denoted by c.

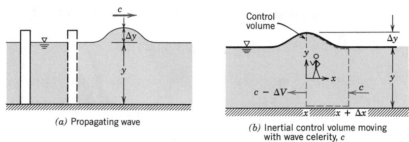

(a) Propagating wave

(b) Inertial control volume moving with wave celerity, c

Fig. 10.2   Generation of a small solitary wave in quiescent liquid.

10-2.1 Wave Speed

The wave speed may be computed by applying the basic equations. As seen by a fixed observer, the wave propagation is unsteady. However, the flow appears steady to an observer *on* a differential control volume that moves with the wave, Fig. 10.2b.

a. Continuity Equation

Basic equation:
$$0 = \underbrace{\frac{\partial}{\partial t} \int_{CV} \rho \, d\forall}_{= 0(1)} + \int_{CS} \rho \vec{V} \cdot d\vec{A} \tag{4.13}$$

Assumptions: (1) Steady flow
(2) Uniform flow at a section
(3) Incompressible, ρ = constant

Then
$$0 = \{+|\rho b(y + \Delta y)(c - \Delta V)|\} + \{-|\rho b y c|\} \tag{10.2a}$$
$$0 = yc - y\,\Delta V + c\,\Delta y - \Delta y\,\Delta V - yc$$

Solving for ΔV,
$$\Delta V = c\,\frac{\Delta y}{y + \Delta y} \tag{10.2b}$$

---

[2] Formation and stretching of vortices are shown in the NCFMF film, *Vorticity*, A. H. Shapiro, principal.

b. Momentum Equation

$$= 0(6) = 0(1)$$

Basic equation: $\quad F_{S_x} + F_{B_x} = \dfrac{\partial}{\partial t} \displaystyle\int_{CV} u\rho \, d\mathbb{V} + \int_{CS} u\rho \vec{V} \cdot d\vec{A} \qquad (4.19a)$

Assumptions: (4) Hydrostatic pressure variation (this will be exactly true if stream-line curvature effects are negligible, i.e., for small Δy), so

$$\frac{dp}{dy} = -\rho g$$

(5) No viscous or surface tension effects
(6) $F_{B_x} = 0$

In the absence of viscous and surface tension effects, F_{S_x} will be due to pressure forces only. Since the pressure variation is hydrostatic at both vertical faces of the control volume,

$$\frac{dp}{dy} = -\rho g \quad \text{and} \quad p = \rho g(y_s - y)$$

where y_s is the distance to the free surface. The magnitude of the pressure force is

$$F_{S_x} = \int p \, dA = \int_0^{y_s} pb \, dy = \int_0^{y_s} \rho g(y_s - y) b \, dy$$

$$F_{S_x} = \rho g b \left[y y_s - \frac{y^2}{2} \right]_0^{y_s} = \frac{\rho g b y_s^2}{2}$$

Then

$$F_{S_x}]_{x+\Delta x} = -\frac{\rho g b}{2} y^2$$

and

$$F_{S_x}]_x = \frac{\rho g b}{2} (y + \Delta y)^2$$

Substituting for the pressure forces into the momentum equation, we obtain

$$\frac{\rho g b}{2}(y + \Delta y)^2 - \frac{\rho g b}{2} y^2 = -|c - \Delta V|\{+|\rho b(y + \Delta y)(c - \Delta V)|\} - |c|\{-|\rho b y c|\}$$

The two terms { } in this equation are equal by continuity (Eq. 10.2a) so the momentum equation reduces to

$$g y \, \Delta y + g \frac{(\Delta y)^2}{2} = \Delta V y c$$

or

$$g \left(1 + \frac{\Delta y}{2y} \right) \Delta y = \Delta V c \qquad (10.3)$$

Combining Eqs. 10.2b and 10.3, we obtain

$$c^2 = g \left(1 + \frac{\Delta y}{2y} \right)(y + \Delta y)$$

or

$$c^2 = g y \left(1 + \frac{\Delta y}{2y} \right)\left(1 + \frac{\Delta y}{y} \right) \qquad (10.4)$$

In our development of Eq. 10.4, we assumed hydrostatic pressure variation in the liquid. A careful study of the details of wave motions [2] indicates that this is a good assumption, provided the wavelength, λ, is long compared to the liquid depth. Such waves are called *shallow water waves*.

Thus, for the case $\Delta y \ll y$, there will be negligible variation in propagation speed across the wave, and our assumption of hydrostatic pressure variation is reasonable. We may use

$$c = \pm \sqrt{gy} \qquad (10.5)$$

to represent the speed of such waves.

The celerity, c, depends on local depth, y. Consequently, c will be larger at the peak of the wave than at the leading or trailing edge. Thus real waves of finite amplitude get steeper as they travel. This causes the "breaking" of waves on a beach.[3]

EXAMPLE 10.1—Propagation Speed of an Isolated Free-Surface Wave

Calculate and plot the propagation speed of isolated waves for depths from 10 mm to 10 km. Comment on the significance of the propagation speed for the average ocean depth of 4 km [3].

EXAMPLE PROBLEM 10.1

GIVEN: Isolated wave propagation.

FIND: Evaluate and plot propagation speed for depths from 10 mm to 10 km. Comment on speed for average ocean depth of 4 km.

SOLUTION:
From Eq. 10.5, $c = \pm \sqrt{gy}$, where y is water depth. The equation will plot as a straight line on log-log coordinates. At $y = 4$ km,

$$c = \sqrt{gy} = \left[9.81 \frac{m}{sec^2} \times 4 \text{ km} \times 1000 \frac{m}{km} \right]^{1/2} = 198 \text{ m/sec} \qquad \leftarrow c$$

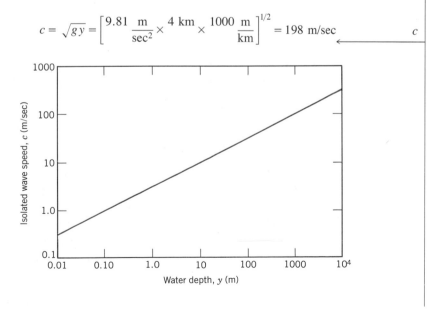

[3] These and other surface wave phenomena are illustrated in the NCFMF film, *Waves in Fluids*, A. E. Bryson, principal.

At the average ocean depth of 4 km, wave speed is 198 m/sec (713 km/hr). Tidal waves (*tsunamis*) generated by underwater earthquakes propagate rapidly; they can do tremendous damage near a coastline [4, 5].

10-2.2 The Froude Number

Most phenomena of interest take place in moving streams. The analysis of Section 10-2.1 would apply equally well to a small wave propagating on the surface of a stream moving with speed V. The wave speed seen by a fixed observer would be

$$V_w = V \pm c = V \pm \sqrt{gy}$$

Consequently, V_w can take on any value depending on the magnitudes of V and $\sqrt{gy}$. The wave speed, V_w, can be negative—a wave can move upstream—only when $V < \sqrt{gy}$. Thus we see that the character of the flow changes at the condition when $V = \sqrt{gy}$, or when

$$\frac{V}{\sqrt{gy}} = Fr = 1 \qquad (10.6)$$

where Fr is the Froude number introduced in Chapter 7. When $Fr = 1$, the character of the wave motion changes.

Open-channel flows[4] may be classified on the basis of Froude number:

$Fr < 1$ Flow is *subcritical, tranquil,* or *streaming*. Disturbances can travel upstream; downstream conditions can affect the flow upstream.

$Fr = 1$ Flow is *critical*.

$Fr > 1$ Flow is *supercritical, rapid,* or *shooting*. No disturbance can travel upstream; downstream conditions cannot be felt upstream.

These regimes of flow behavior are qualitatively analogous to the subsonic, sonic, and supersonic regimes of gas flow discussed in Chapter 12.

10-3 ENERGY EQUATION FOR OPEN-CHANNEL FLOW

As in the case of pipe flow, friction in open-channel flows results in a loss of mechanical energy; this can be characterized by a head loss. The effect of friction is particularly important in long channels.

Consider flow through a long rectangular channel, of width b and bed slope $S_b = \tan\theta$, where S_b is small. The flow depth may vary. To derive a suitable form of the energy equation for open-channel flow, we assume uniform flow at each section. The control volume used for the analysis is shown in Fig. 10.3. Coordinate z indicates distances measured in the vertical direction; distances measured normal to the bed are denoted by y.

---

[4] The Froude number for nonrectangular channels must be based on the hydraulic depth, so that

$$Fr = \frac{V}{\sqrt{gy_h}}$$

For a rectangular channel, $y_h = y$, so this equation reduces to the expression given above.

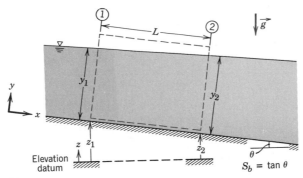

Fig. 10.3 Control volume and coordinates for energy analysis of open-channel flow.

The energy equation for a control volume is

$$\dot{Q} - \cancel{\dot{W}_s} - \cancel{\dot{W}_{\text{shear}}} - \cancel{\dot{W}_{\text{other}}} = \overset{=0(1)}{\cancel{\frac{\partial}{\partial t} \int_{\text{CV}} e\rho\, d\Psi}}$$

$$+ \int_{\text{CS}} \left(u + p\upsilon + \frac{V^2}{2} + gz \right) \rho \vec{V} \cdot d\vec{A} \qquad (4.57)$$

The equation is to be applied to the control volume of Fig. 10.3 under the assumptions:

(1)   Steady flow
(2)   Incompressible flow
(3)   Uniform flow at a section
(4)   Depth varies gradually so that pressure distribution is hydrostatic
(5)   Bed slope is small, $\theta \sim \sin\theta \sim \tan\theta = S_b$, and $\cos\theta \sim 1$
(6) $\dot{W}_s = \dot{W}_{\text{shear}} = \dot{W}_{\text{other}} = 0$

Under these assumptions, the energy equation reduces to

$$\dot{Q} = \int_{\text{CS}} \left(u + p\upsilon + \frac{V^2}{2} + gz \right) \rho \vec{V} \cdot d\vec{A} \qquad (10.7)$$

As a result of friction, mechanical energy will be dissipated between sections ① and ②. As in the case of pipe flow, the dissipation is characterized by a head loss. To demonstrate this, Eq. 10.7 is rewritten as

$$\int_{\text{CS}} \left(p\upsilon + \frac{V^2}{2} + gz \right) \rho \vec{V} \cdot d\vec{A} + \int_{\text{CS}} u\rho \vec{V} \cdot d\vec{A} - \dot{Q} = 0$$

$$\int_{\text{CS}} \left(p\upsilon + \frac{V^2}{2} + gz \right) \rho \vec{V} \cdot d\vec{A} + (u_2 - u_1)\dot{m} - \frac{\delta Q}{dm} \dot{m} = 0$$

$$\int_{\text{CS}} \left(p\upsilon + \frac{V^2}{2} + gz \right) \rho \vec{V} \cdot d\vec{A} + \dot{m} h_l = 0 \qquad (10.8)$$

The integral of Eq. 10.8 must be evaluated at sections ① and ②. At section ①, $dA = b\, dy$, $z = z_1 + y\cos\theta$, and the velocity is uniform over the area. The pressure variation is hydrostatic; $dp = -\rho g\, dz = -\rho g \cos\theta\, dy$, and the hydrostatic pressure distribution is given by $p = \rho g \cos \theta (y_1 - y)$. Thus

$$\int_{CS_1} \left(pv + \frac{V^2}{2} + gz \right) \rho \vec{V} \cdot d\vec{A}$$

$$= \int_0^{y_1} \left(g \cos\theta(y_1 - y) + \frac{V_1^2}{2} + g(z_1 + y \cos\theta) \right) \{-|\rho V_1 b \, dy|\}$$

$$= \int_0^{y_1} \left(g \cos\theta y_1 + \frac{V_1^2}{2} + g z_1 \right) \{-|\rho V_1 b \, dy|\}$$

Since all of the terms under the integral sign are independent of y,

$$\int_{CS_1} \left(pv + \frac{V^2}{2} + gz \right) \rho \vec{V} \cdot d\vec{A} = -\dot{m}_1 \left(\frac{V_1^2}{2} + g z_1 + g \cos\theta y_1 \right)$$

The form of the integral at section ② is identical to that at section ① (except the sign will be positive). Further, $\dot{m}_1 = \dot{m}_2 = \dot{m}$. Substituting into Eq. 10.8 gives

$$\left(\frac{V_2^2}{2} + g z_2 + g \cos\theta y_2 \right) - \left(\frac{V_1^2}{2} + g z_1 + g \cos\theta y_1 \right) + h_l = 0 \qquad (10.9)$$

For small slopes, $\cos\theta \approx 1$. To obtain head loss in dimensions of length (head loss per unit weight rather than per unit mass), Eq. 10.9 is divided by g. Thus

$$\frac{V_1^2}{2g} + y_1 + z_1 = \frac{V_2^2}{2g} + y_2 + z_2 + h_l \qquad (10.10)$$

For flow without friction there is no head loss and the energy equation (Eq. 10.10) becomes

$$\frac{V_1^2}{2g} + y_1 + z_1 = \frac{V_2^2}{2g} + y_2 + z_2$$

10-3.1 Specific Energy

The energy equation for open-channel flow (Eq. 10.10) contained the sum of terms, $V^2/2g + y$, on both sides. This sum is defined as the *specific energy* (or *specific head*) and denoted by the symbol E,

$$E = \frac{V^2}{2g} + y \qquad (10.11)$$

where the flow depth, y, is measured normal to the channel bed.

For uniform flow at a section, the velocity can be written in terms of the volume flow rate. From continuity, $V = Q/A$, so

$$E = \frac{Q^2}{2g A^2} + y \qquad (10.12)$$

For a given flow rate, specific energy is a function of depth. Depth also appears in the expression for area (Table 10.1). To illustrate the relation between depth and specific energy at constant flow rate, consider flow in a rectangular channel of width b. Then $A = by$, and

$$E = \frac{Q^2}{2g b^2 y^2} + y \qquad (10.13)$$

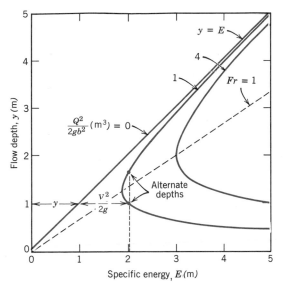

Fig. 10.4   Specific energy diagram.

The variation of depth as a function of specific energy for a given flow rate is plotted in Fig. 10.4; curves for several values of Q (actually $Q^2/2gb^2$) are shown.

When $Q = 0$, then $y = E$; this limiting case is a 45° line on the plot. For a given flow rate $(Q > 0)$ and specific energy, there are two possible depths. These two depths are called *alternate depths*. The curve for constant Q gives the locus of all possible depths and the corresponding specific energies that satisfy Eq. 10.13. As Q increases, the curves are displaced to the right.

For any E, the horizontal distance, from the vertical axis $(E = 0)$ to the line $y = E$, gives the depth y. The distance from the line $y = E$ to the Q curve is then equal to the kinetic energy, $V^2/2g$; this is shown in Fig. 10.4.

For each curve representing a given flow rate in Fig. 10.4, there is a depth that gives a minimum E. We may determine this depth, y, by differentiating Eq. 10.13; E will be a minimum when

$$\frac{dE}{dy} = -\frac{Q^2}{gb^2y^3} + 1 = 0 \tag{10.14}$$

Solving for y, we obtain

$$\frac{Q^2}{gb^2} = y^3 \qquad \text{or} \qquad y = \left[\frac{Q^2}{gb^2}\right]^{1/3}$$

Substituting this result into Eq. 10.13 gives

$$E_{\min} = \tfrac{1}{2}y + y = \tfrac{3}{2}y$$

The locus of minimum values of E is thus a straight line with $y = \tfrac{2}{3}E_{\min}$.

We can solve for the velocity at $E_{\min}$ using Eq. 10.14,

$$V^2 = \frac{Q^2}{A^2} = \frac{Q^2}{b^2y^2} = gy$$

Thus, at the minimum value of E, $Fr = V/\sqrt{gy} = 1.0$, and the condition corresponds to critical flow. The depth at E_{min} is termed the *critical depth*, y_c. Thus, for flow in a rectangular channel of width b,

$$y_c = \left[\frac{Q^2}{gb^2}\right]^{1/3} \qquad \text{and} \qquad E_{min} = \frac{3}{2}y_c$$

From Eq. 10.11

$$E_{min} = \frac{3}{2}y_c = \frac{V_c^2}{2g} + y_c$$

so

$$\frac{V_c^2}{2g} = \frac{y_c}{2} \tag{10.15}$$

We can investigate the nature of the flow on the branches of the curve above and below the critical depth by writing an expression for the Froude number. From continuity

$$Q = V_c b y_c = V b y \tag{10.16}$$

Then, using Eqs. 10.15 and 10.16,

$$Fr = \frac{V}{\sqrt{gy}} = \frac{V_c y_c}{y\sqrt{gy}} = \frac{\sqrt{gy_c} y_c}{y\sqrt{gy}} = \left[\frac{y_c}{y}\right]^{3/2} \tag{10.17}$$

On the upper branch of the curve,

$$y > y_c, \quad \text{so } Fr < 1; \qquad \text{flow is subcritical}$$

On the lower branch of the curve,

$$y < y_c, \quad \text{so } Fr > 1; \qquad \text{flow is supercritical}$$

For nonrectangular channels, the channel depth varies across the width. For these cases, evaluation of the critical depth (and the corresponding critical speed and area) usually requires a trial and error solution. At the minimum specific energy, differentiating Eq. 10.12 with respect to y at constant Q gives

$$\frac{dE}{dy} = 0 = -\frac{Q^2}{gA^3}\frac{dA}{dy} + 1 \tag{10.18}$$

Since $dA = b_s\,dy$, where b_s is the channel width at the free surface, then

$$\frac{gA_c^3}{b_{sc}Q^2} = 1 \qquad \text{at} \qquad E = E_{min} \tag{10.19}$$

Both A_c and b_{sc} are functions of critical depth, y_c. For a specific cross-sectional shape, Eq. 10.19 can be solved numerically for the critical depth.

The critical area, A_c, corresponding to $Fr = 1$ is

$$A_c = \left[\frac{b_{sc}Q^2}{g}\right]^{1/3} \tag{10.20}$$

From continuity, $V_c = Q/A_c$. From Eq. 10.19, $Q^2/A_c^2 = gA_c/b_{sc}$. Thus

$$V_c = \frac{Q}{A_c} = \left[\frac{Q^2}{A_c^2}\right]^{1/2} = \left[\frac{gA_c}{b_{sc}}\right]^{1/2} = [gy_{hc}]^{1/2} \tag{10.21}$$

where y_{hc} is the hydraulic depth at critical conditions. Thus the Froude number is unity for flow at critical conditions that correspond to minimum specific energy.

For a rectangular channel, $A_c = by_c$ and $b_{sc} = b$; Eq. 10.21 gives $V_c = (gy_c)^{1/2}$. (This is the same result obtained from Eq. 10.15.)

Near the minimum E, the rate of change of y with E is nearly infinite. Even small changes in E, due to channel irregularities or disturbances, can cause pronounced changes in fluid depth. Thus, surface waves usually form when a flow is near critical conditions. Long runs of near-critical flow consequently are avoided in practice.

10-4   FRICTIONLESS FLOW: EFFECT OF AREA CHANGE

We begin by considering two simple flow cases in which the channel bed is horizontal, the effects of channel cross section (area change) predominate, and the effect of friction may be neglected. Since the flow is assumed to be frictionless, the energy equation (Eq. 10.8) reduces to a form of the Bernoulli equation. These flow cases are analyzed using the Bernoulli and continuity equations.

10-4.1   Flow over a Bump

Consider frictionless flow in a horizontal rectangular channel of constant width, b, with a bump in the channel bed, as illustrated in Fig. 10.5. The bump height above the horizontal bed of the channel is $h(x)$; the water depth, $y(x)$, is measured from the local channel bottom surface. The flow is assumed to be uniform at each section. We are interested in investigating the shape of the free surface as the flow passes over the bump.

Since the flow is steady, incompressible, and frictionless, we may apply the Bernoulli equation along a streamline,

$$\frac{p}{\rho} + \frac{V^2}{2} + gz = \text{constant} \tag{6.9}$$

Applying the equation along the free surface streamline between upstream location ① and a point above the bump, we obtain

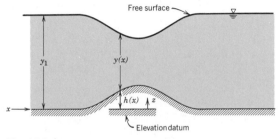

Fig. 10.5   Open-channel flow over a bump in a horizontal channel bed.

$$\frac{p_1}{\rho g} + \frac{V_1^2}{2g} + y_1 = \frac{p}{\rho g} + \frac{V^2}{2g} + y + h$$

Since the pressure is atmospheric along the free surface, $p_1 = p = p_{atm}$, and

$$\frac{V_1^2}{2g} + y_1 = \frac{V^2}{2g} + y + h = \text{constant} \qquad (10.22)$$

The volume flow rate is a constant. Thus, from continuity for steady, incompressible, uniform flow at a section,

$$\frac{Q}{b} = V_1 y_1 = V y$$

Substituting for V_1 and V into Eq. 10.22 yields

$$\frac{Q^2}{2g\,b^2 y_1^2} + y_1 = \frac{Q^2}{2g\,b^2 y^2} + y + h = \text{constant} \qquad (10.23)$$

We can obtain an expression for the variation of the free surface depth by differentiating Eq. 10.23

$$\frac{-Q^2}{g\,b^2 y^3}\frac{dy}{dx} + \frac{dy}{dx} + \frac{dh}{dx} = 0$$

Solving for the slope of the free surface, we obtain

$$\frac{dy}{dx} = \frac{dh/dx}{\left[\dfrac{Q^2}{g\,b^2 y^3} - 1\right]} = \frac{dh/dx}{\left[\dfrac{V^2}{g y} - 1\right]} = \frac{dh/dx}{Fr^2 - 1} \qquad (10.24)$$

From Eq. 10.24 we see that the slope of the free surface depends on the local Froude number. For $Fr < 1$, an increase in bed elevation causes a decrease in water depth; a decrease in bed elevation causes water depth to increase. For $Fr > 1$, an increase in bed elevation causes an increase in water depth; a decrease in bed elevation causes water depth to decrease. These results are summarized in Fig. 10.6.

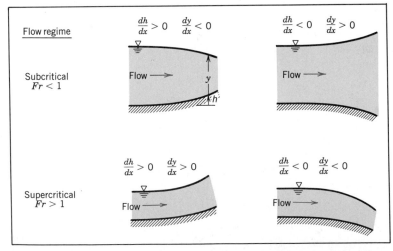

Fig. 10.6 Effect of bed elevation changes on water depth in open-channel flow.

When the Froude number is unity, Eq. 10.24 predicts an infinite water surface slope, unless dh/dx equals zero. Since the free surface slope cannot be infinite, then dh/dx must be zero when $Fr = 1$; a Froude number of unity can only exist at the location where $dh/dx = 0$. If critical flow is attained, then downstream of the critical flow location the flow may be subcritical or supercritical, depending on downstream conditions. If critical flow does not occur where $dh/dx = 0$, then flow downstream from this location will be the same type as the flow upstream from the location.

EXAMPLE 10.2—Flow over a Bump on the Bed of a Horizontal Channel

Water flows in a horizontal rectangular channel. The flow speed and depth at section ① are 0.5 m/sec and 0.3 m, respectively. The flow passes over a smooth bump on the channel floor. Evaluate the flow speed and depth directly over the peak of the bump. Assume that the peak height is 0.03 m and neglect friction.

EXAMPLE PROBLEM 10.2

GIVEN: Water flow in a horizontal rectangular channel. Neglect frictional effects.

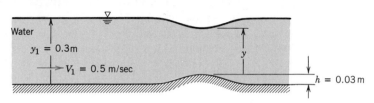

FIND: Flow speed and depth above the peak of the bump.

SOLUTION:
For steady, incompressible flow along a streamline without friction, the Bernoulli equation may be applied. Along the free surface the pressure is constant, so

$$\frac{\cancel{p_{atm}}}{\rho g} + \frac{V_1^2}{2g} + y_1 = \frac{\cancel{p_{atm}}}{\rho g} + \frac{V^2}{2g} + y + h$$

For uniform flow at each section, the continuity equation reduces to $V_1 y_1 = V y$. These two equations permit us to solve for V and y; the solution process may be visualized using the specific energy diagram as follows. From Eq. 10.22,

$$E_1 = \frac{V_1^2}{2g} + y_1 = E + h = \frac{V^2}{2g} + y + h \qquad \text{or} \qquad E = E_1 - h$$

Evaluating, we obtain

$$E_1 = \frac{1}{2} \times \frac{(0.5)^2 \; \text{m}^2}{\text{sec}^2} \times \frac{\text{sec}^2}{9.807 \; \text{m}} + 0.3 \; \text{m}$$

$$E_1 = 0.3127 \; \text{m}$$

and

$$E = E_1 - h = (0.3127 - 0.03) \; \text{m} = 0.2827 \; \text{m}$$

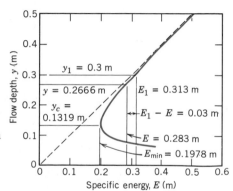

These points are plotted on the specific energy diagram. By iteration it may be shown that the flow depth for this E is

$$y = 0.2666 \text{ m} \qquad\qquad\qquad\qquad\qquad y$$

The change in surface level is

$$\Delta = y + h - y_1 = (0.2666 + 0.0300 - 0.3000) \text{ m} = -3.4 \text{ mm}$$

The flow speed is

$$V = V_1 \frac{y_1}{y} = \frac{0.5}{\text{sec}} \frac{\text{m}}{\text{sec}} \times \frac{0.3000 \text{ m}}{0.2666 \text{ m}} = 0.563 \text{ m/sec} \qquad V$$

$\left\{ \begin{array}{l} \text{The specific energy diagram suggests that the flow could traverse a bump approximately} \\ E_1 - E_{\min} = 0.313 \text{ m} - 0.1978 \text{ m} = 0.115 \text{ m in height without reaching critical speed.} \end{array} \right\}$

10-4.2 Flow through a Sluice Gate

As a second example of open-channel flow where friction may be neglected, consider flow through a sluice gate. A sluice gate is a form of control structure often used to regulate discharge. Flow beneath a sluice gate is shown in Fig. 10.7. Far upstream from the gate, the water depth, y_0, is constant and the flow speed is negligible. This is equivalent to considering flow from a large reservoir. Section ① is chosen so flow is uniform across depth y_1.

Since flow is steady, incompressible, and frictionless, we may apply the Bernoulli equation along a streamline. Applying the equation along the surface streamline between sections ⓪ and ①, then

$$y_0 = \frac{V_1^2}{2g} + y_1 \tag{10.25}$$

We can express V in terms of the volume flow rate. Since $V = Q/A$ and $A = by$, where b is the width of gate, then

$$y_0 = \frac{Q^2}{2g\, b^2 y_1^2} + y_1 \tag{10.26}$$

Equation 10.26 gives a relation between depth y_1 and volume flow rate,

$$y_1^2(y_0 - y_1) = \frac{Q^2}{2g\, b^2} \tag{10.27}$$

This relation is shown in Fig. 10.8, where depth y_1 has been nondimensionalized by reservoir depth, y_0. Figure 10.8 shows that for a given flow rate there are two

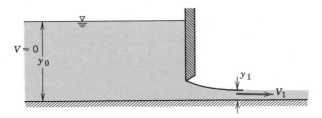

Fig. 10.7 Flow through a sluice gate.

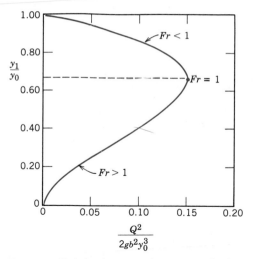

Fig. 10.8 Relation between downstream depth and flow rate for flow through a sluice gate.

possible depths for any flow rate less than maximum. At maximum flow rate there is a single depth. We can evaluate the depth at maximum flow rate by differentiating Eq. 10.27 with respect to y_1,

$$\frac{d}{dy_1}\left[\frac{Q^2}{2gb^2}\right] = 0 = 2y_1 y_0 - 3y_1^2$$

At maximum flow rate

$$y_1 = \frac{2}{3}y_0$$

and from Eq. 10.27

$$\frac{Q_{max}^2}{gb^2} = 2\left[\frac{2}{3}y_0\right]^2\left[y_0 - \frac{2}{3}y_0\right] = \frac{8}{27}y_0^3$$

The Froude number at maximum flow rate is given by

$$Fr^2 = \frac{V_{max}^2}{gy_1} = \frac{Q_{max}^2}{gb^2 y_1^3} = \frac{8}{27}y_0^3\left[\frac{3}{2y_0}\right]^3 = 1$$

Thus, in Fig. 10.8, maximum flow rate occurs at $y_1/y_0 = \frac{2}{3}$. The upper portion of the curve corresponds to subcritical flow ($Fr < 1$) and the lower portion of the curve corresponds to supercritical flow ($Fr > 1$).

Flow leaving a sluice gate passes through a vena contracta, as shown in Fig. 10.7. It is not possible analytically to relate depth y_1 to the gate opening. Experimental data must be used to determine the gate setting needed to pass a desired flow rate, as discussed in Section 10-8.

EXAMPLE 10.3—Flow under a Sluice Gate

Water flow under a sluice gate was considered in Example Problem 4.6, at the following conditions:

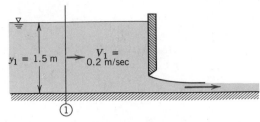

Evaluate the water depth, y_0, at a location upstream from the gate where the flow speed is negligible. Find the exact values of downstream depth and speed at this flow rate required by the condition that the specific energy remain constant. Find the depth and speed that would provide maximum flow rate. Calculate this flow rate.

EXAMPLE PROBLEM 10.3

GIVEN: Water flow under a sluice gate at the conditions shown.

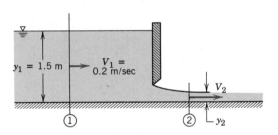

FIND: (a) Water depth, y_0, where flow speed is negligible.
       (b) Water depth, y_2, and speed, V_2, downstream at this flow rate.
       (c) Water depth, y, and speed, V, at downstream conditions that produce the maximum flow rate possible.
       (d) Maximum flow rate.

SOLUTION:

Assumptions: (1) Steady flow
               (2) Incompressible flow
               (3) Uniform flow at sections ① and ②

The specific energy upstream is

$$E_1 = \frac{V_1^2}{2g} + y_1 = \frac{1}{2} \times (0.2)^2 \, \frac{m^2}{sec^2} \times \frac{sec^2}{9.807 \, m} + 1.5 \, m = 1.502 \, m$$

For constant specific energy, $E_0 = \overset{\approx 0}{\frac{\cancel{V_0^2}}{\cancel{2g}}} + y_0 = E_1$, so $y_0 = 1.502$ m $\overset{\displaystyle \longleftarrow}{\underset{}{}}$ y_0

For constant specific energy, downstream conditions must be at the alternate depth corresponding to the rate of flow specified, as shown on the specific energy diagram,

$$E_2 = y_0 = 1.502 \, m = \frac{Q^2}{2g \, b^2 y_2^2} + y_2$$

By iteration, for $Q/b = 0.3$ m²/sec,

$y_2 = 0.05632$ m

The corresponding speed is

$$V_2 = V_1 \frac{y_1}{y_2}$$

$$= \frac{0.2}{\text{sec}} \frac{\text{m}}{\text{sec}} \times \frac{1.5 \text{ m}}{0.05632 \text{ m}}$$

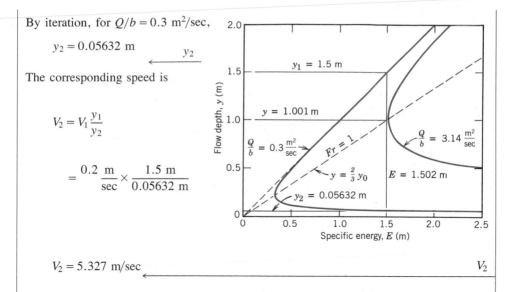

$V_2 = 5.327$ m/sec

These conditions define the alternate depth, as shown above. (To three significant figures, these results agree with the data given in Example Problem 4.6.)

From the solution to Eq. 10.27, at maximum flow rate,

$$y = \tfrac{2}{3} y_0 = \tfrac{2}{3} \times 1.502 \text{ m} = 1.001 \text{ m}$$

and

$$V = \sqrt{g y} = \left[9.807 \frac{\text{m}}{\text{sec}^2} \times 1.001 \text{ m} \right]^{1/2} = 3.133 \text{ m/sec}$$

The maximum flow rate is

$$\frac{Q}{b} = Vy = 3.133 \frac{\text{m}}{\text{sec}} \times 1.001 \text{ m} = 3.14 \text{ m}^3/\text{sec/m (m}^2/\text{sec)}$$

The corresponding curve of y versus E also is shown above.

$\begin{cases} \text{This is the maximum rate of flow possible from the specified reservoir conditions,} \\ V_0 \approx 0 \text{ and } y_0 = 1.502 \text{ m. Conditions at section } \textcircled{1} \text{ would change when the sluice gate} \\ \text{is opened to the position for maximum flow.} \end{cases}$

10-5   FLOW AT NORMAL DEPTH: UNIFORM FLOW

Fully developed flow through a prismatic channel (a channel with constant slope and cross section) at constant depth, y_n, is termed flow at *normal depth* or *uniform flow*. The slope of the channel bottom, or bed, is denoted by S_b and is assumed to be small. To analyze the flow, we assume uniform flow at each cross section and apply the basic equations to the control volume of Fig. 10.9.

10-5.1   Basic Equations

a.   Continuity Equation

Basic equation:

$$0 = \overset{= 0(1)}{\cancel{\frac{\partial}{\partial t} \int_{CV} \rho \, d\forall}} + \int_{CS} \rho \vec{V} \cdot d\vec{A} \tag{4.13}$$

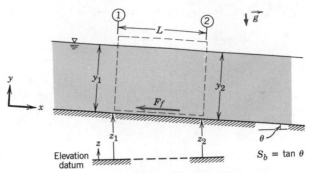

Control volume and coordinates used for analysis
of flow at normal depth.

Assumptions: (1) Steady flow
(2) Incompressible flow
(3) Uniform flow at a section
(4) Normal depth, $y_1 = y_2 = y_n$

The continuity equation gives

$$0 = -|\rho V_1 A_1| + |\rho V_2 A_2|$$

Since $A_1 = A_2$, then

$$V_1 = V_2$$

b.   Momentum Equation

Basic equation:
$$F_{S_x} + F_{B_x} = \overset{=0(1)}{\cancel{\frac{\partial}{\partial t} \int_{CV} u\rho\,d\forall}} + \int_{CS} u\rho\vec{V}\cdot d\vec{A} \qquad (4.19a)$$

Assumptions: (5) Hydrostatic pressure distribution
(6) Small slope, $\theta \sim \sin\theta \sim \tan\theta = S_b$

Since the flow has constant depth and uniform velocity, the momentum flux through
the control surface is zero. The net pressure force on the control volume in the x
direction is zero because the pressure distribution is hydrostatic. The x component
of body force is the component of the weight of liquid in the control volume in the
x direction. Thus the momentum equation reduces to

$$-F_f + W\sin\theta = 0$$

or

$$F_f = W\sin\theta \qquad (10.28)$$

The friction force may be expressed as the product of wall shear stress, τ_w, and
the channel surface area on which the stress acts. Equation 10.28 shows that for flow
at normal depth, the component of the gravity force driving the flow is balanced by
the friction force acting on the channel walls.

c. Energy Equation

The energy equation for open-channel flow was derived in Section 10-3 as

$$\frac{V_1^2}{2g} + y_1 + z_1 = \frac{V_2^2}{2g} + y_2 + z_2 + h_l \tag{10.10}$$

For flow at normal depth, $y_1 = y_2 = y_n$, and $V_1 = V_2$. Thus, from Eq. 10.10,

$$h_l = z_1 - z_2 = LS_b \tag{10.29}$$

For flow at normal depth, the head loss due to friction is equal to the change in elevation of the bed. The specific energy, E, is constant at each section normal to the bed. The energy grade line, the hydraulic grade line, and the channel bed are all parallel.

10-5.2 The Manning Correlation for Velocity

Open-channel flows in practice invariably are turbulent. Since no simple constitutive relation is available to relate shear stress and velocity gradient, we must rely on empirical correlations. As in the case of fully developed turbulent pipe flow, the head loss may be written in terms of a friction factor (Eq. 8.32). For application to open-channel flow, the pipe diameter D is replaced by hydraulic radius ($D = 2R = 4R_h$) and head loss is expressed as head loss per unit weight (h_l has units of length). Thus, for open-channel flow,

$$h_l = f \frac{L}{4R_h} \frac{V^2}{2g} \tag{10.30}$$

From Eq. 10.29, $h_l/L = S_b$, and thus the velocity for flow at normal depth is

$$V = \left[\frac{8g}{f}\right]^{1/2} \sqrt{R_h S_b} \tag{10.31}$$

For most open-channel flows, friction factor is only a function of surface roughness; it is independent of Reynolds number. This is analogous to flow in the fully rough regime for turbulent pipe flow. For a given roughness, Eq. 10.31 can be written as

$$V = C\sqrt{R_h S_b} \tag{10.32}$$

Equation 10.32 is known as the Chezy equation. Empirical values of C were determined by Manning. He suggested that

$$C = \left[\frac{8g}{f}\right]^{1/2} = \frac{R_h^{1/6}}{n} \tag{10.33}$$

where n is a roughness coefficient having different values for different types of boundary roughness. With this expression for C, the velocity for flow at normal depth becomes

$$V = \frac{R_h^{2/3} S_b^{1/2}}{n} \tag{10.34}$$

and the volume flow rate is written

$$Q = VA = \frac{R_h^{2/3} S_b^{1/2}}{n} A \tag{10.35}$$

Table 10.2 Manning Roughness Coefficient, n, for Representative Surfaces

Channel Surface	Representative[a] n-value
Lucite, glass, or plastic film	0.010
Wood or finished concrete	0.013
Unfinished concrete, well-laid brickwork, concrete or cast-iron pipe	0.015
Riveted or spiral steel pipe	0.017
Smooth, uniform earth channel	0.022
Corrugated metal flumes, typical canals, rivers free from large stones and heavy weeds	0.025
Canals and rivers with many stones and weeds	0.035

[a] See [1] for a more complete table.

Values of n for some typical surfaces are given in Table 10.2. These values of n, taken to be dimensionless (which they are not; from Eq. 10.34 we see that n has dimensions of $L^{-1/3}t$), were based on measurements in SI units.[5]

The relationship among variables in Eq. 10.35 can be viewed in a number of ways. The volume flow rate through a prismatic channel of given slope and roughness is a function of both channel size and channel shape. This is illustrated in Example Problem 10.4. For a specified flow rate through a prismatic channel of given slope and roughness, the depth of uniform flow is a function of both channel size and channel shape. There is only one depth for uniform flow at a given flow rate; it may be greater than, less than, or equal to the critical depth. This is illustrated in Example Problem 10.5.

EXAMPLE 10.4—Flow vs. Area through Two Channel Shapes

Open channels, of square and semicircular shapes, are considered to carry flow on a slope of $S_b = 0.001$; the channel walls are to be poured concrete with $n = 0.015$. Evaluate the flow rate delivered by the channels for maximum dimensions between 0.5 and 2.0 m. Compare the channels on the basis of volume flow rate for given cross-sectional area.

EXAMPLE PROBLEM 10.4

GIVEN: Square and semicircular channels; $S_b = 0.001$ and $n = 0.015$. Sizes between 0.5 and 2.0 m across.

FIND: (a) Evaluate flow rate as a function of size.
(b) Compare channels on the basis of volume flow rate, Q, versus cross-sectional area, A.

---

[5] Equations 10.34 and 10.35 are valid for SI units. In British units the numerator of these equations must be multiplied by $(0.3048)^{-1/3} \approx 1.49$. With this modification n can be considered to have the same numerical value in both systems of units.

SOLUTION:
Apply Eq. 10.35 for flow at normal depth in a long channel.

Computing equation:

$$Q = \frac{R_h^{2/3} S_b^{1/2}}{n} A$$
(10.35)

Assumptions: (1) Steady flow
               (2) Incompressible flow
               (3) Uniform flow at a section
               (4) Flow at normal depth

For the square channel,

$$P = 3L \quad \text{and} \quad A = L^2, \quad \text{so } R_h = \frac{L}{3}$$

Substituting into Eq. 10.35, we obtain

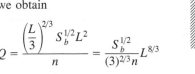

$$Q = \frac{\left(\dfrac{L}{3}\right)^{2/3} S_b^{1/2} L^2}{n} = \frac{S_b^{1/2}}{(3)^{2/3} n} L^{8/3}$$

For $L = 1$ m,

$$Q = \frac{(0.001)^{1/2}}{(3)^{2/3}(0.015)}(1)^{8/3} = 1.01 \text{ m}^3/\text{sec} \qquad\qquad Q_\square$$

Tabulating for a range of sizes yields

L(m)	0.5	1.0	1.5	2.0
A(m^2)	0.25	1.00	2.25	4.00
Q(m^3/sec)	0.160	1.01	2.99	6.44

For the semicircular channel,

$$P = \frac{\pi D}{2} \quad \text{and} \quad A = \frac{\pi D^2}{8}, \quad \text{so } R_h = \frac{\pi D^2}{8}\frac{2}{\pi D} = \frac{D}{4}$$

Substituting into Eq. 10.35, we obtain

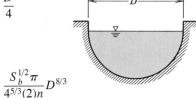

$$Q = \frac{\left(\dfrac{D}{4}\right)^{2/3} S_b^{1/2}}{n}\frac{\pi D^2}{8} = \frac{S_b^{1/2}\pi}{4^{5/3}(2)n} D^{8/3}$$

For $D = 1$ m,

$$Q = \frac{(0.001)^{1/2}\pi}{(2)(4)^{5/3}(0.015)}(1)^{8/3} = 0.329 \text{ m}^3/\text{sec} \qquad\qquad Q_\smile$$

Tabulating for a range of sizes yields

D(m)	0.5	1.0	1.5	2.0
A(m^2)	0.0982	0.393	0.884	1.57
Q(m^3/sec)	0.0517	0.329	0.969	2.09

For both channels, volume flow rate varies as

$$Q \sim L^{8/3} \quad \text{or} \quad Q \sim A^{4/3}$$

since $A \sim L^2$. The plot of flow rate versus cross-sectional area shows that the semicircular channel is more "efficient."

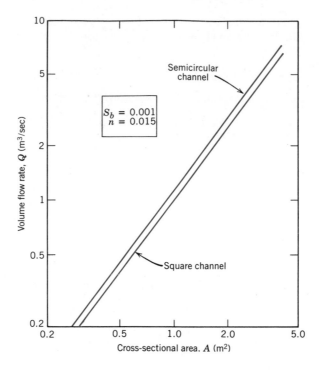

Performance of the two channels may be compared at any specified area. At $A = 1$ m^2, $Q/A = 1.01$ m/sec for the square channel. For the semicircular channel with $A = 1$ m^2, then $D = 1.60$ m, and $Q = 1.15$ m^3/sec, so $Q/A = 1.15$ m/sec. Thus the semicircular channel carries approximately 14 percent more flow per unit area than the square channel.

$$\left\{ \begin{array}{l} \text{The comparison on cross-sectional area relates to the excavation required to build the} \\ \text{channel. The channel shapes also could be compared on the basis of perimeter, which} \\ \text{would relate to the amount of concrete needed to finish the channel.} \end{array} \right\}$$

EXAMPLE 10.5—Determination of Flume Size

An above-ground flume, built from timber, is to convey water from a mountain lake to a small hydroelectric plant. The flume is to deliver water at $Q = 2$ m^3/sec; the slope is $S_b = 0.002$ and $n = 0.013$. Evaluate the required flume size for (a) a rectangular section with $y/b = 0.5$ and (b) an equilateral triangular section.

EXAMPLE PROBLEM 10.5

GIVEN: Flume to be built from timber, with $S_b = 0.002$, $n = 0.013$, and $Q = 2.00$ m^3/sec.

FIND: Required flume size for
(a) Rectangular section with $y/b = 0.5$.
(b) Equilateral triangular section.

SOLUTION:
Assume flume is long so flow is uniform (at normal depth). Then Eq. 10.35 applies.

Computing equation:
$$Q = \frac{R_h^{2/3} S_b^{1/2}}{n} A \tag{10.35}$$

The choice of channel shape fixes the relationship between R_h and A, so Eq. 10.35 may be solved for normal depth, y_n, thus determining the channel size required.

Assumptions: (1) Steady flow
(2) Incompressible flow
(3) Uniform flow at a section
(4) Flow at normal depth

(a) Rectangular section

$P = 2y_n + b; \quad y_n/b = 0.5, \quad \text{so } b = 2y_n$

$\left. \begin{array}{l} P = 2y_n + 2y_n = 4y_n \\[2mm] A = y_n b = y_n(2y_n) = 2y_n^2 \end{array} \right\}$ $R_h = \dfrac{A}{P} = \dfrac{2y_n^2}{4y_n} = 0.5y_n$

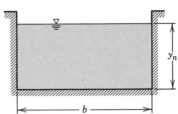

Substituting into Eq. 10.35, we obtain

$$Q = \frac{R_h^{2/3} S_b^{1/2}}{n} A = \frac{(0.5y_n)^{2/3} S_b^{1/2}(2y_n^2)}{n} = \frac{2(0.5)^{2/3} S_b^{1/2} y_n^{8/3}}{n}$$

$$y_n = \left[\frac{nQ}{2(0.5)^{2/3} S_b^{1/2}} \right]^{3/8} = \left[\frac{0.013(2.00)}{2(0.5)^{2/3}(0.002)^{1/2}} \right]^{3/8} = 0.748 \text{ m}$$

The required dimensions for the rectangular channel are

$$y_n = 0.748 \text{ m} \qquad A = 1.12 \text{ m}^2$$
$$b = 1.50 \text{ m} \qquad P = 3.00 \text{ m} \qquad\qquad\qquad \text{Flume size}$$

Also $Fr = \dfrac{V}{\sqrt{g y_n}} = \dfrac{Q}{A \sqrt{g y_n}}$

$$Fr = \frac{2.00 \text{ m}^3}{\text{sec}} \times \frac{1}{1.12 \text{ m}^2} \times \frac{1}{\left[9.81 \dfrac{\text{m}}{\text{sec}^2} \times 0.748 \text{ m}\right]^{1/2}} = 0.659 \qquad\qquad Fr$$

(b) Equilateral triangular section

$\left. \begin{array}{l} P = 2s = \dfrac{2y_n}{\cos 30°} \\[4mm] A = \dfrac{y_n s}{2} = \dfrac{y_n^2}{2\cos 30°} \end{array} \right\}$ $R_h = \dfrac{A}{P} = \dfrac{y_n}{4}$

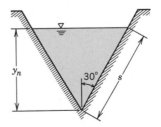

Substituting into Eq. 10.35 yields

$$Q = \frac{R_h^{2/3} S_b^{1/2}}{n} A = \frac{\left(\dfrac{y_n}{4}\right)^{2/3} S_b^{1/2} \left(\dfrac{y_n^2}{2\cos 30°}\right)}{n} = \frac{S_b^{1/2} y_n^{8/3}}{2(4)^{2/3} \cos 30° \, n}$$

$$y_n = \left[\frac{2(4)^{2/3} \cos 30° \, nQ}{S_b^{1/2}} \right]^{3/8} = \left[\frac{2(4)^{2/3} \cos 30°(0.013)(2.00)}{(0.002)^{1/2}} \right]^{3/8} = 1.42 \text{ m}$$

The required dimensions for the triangular channel are

$$y_n = 1.42 \text{ m} \qquad P = 3.28 \text{ m}$$
$$A = 1.16 \text{ m}^2 \qquad b_s = 1.64 \text{ m}$$

Flume size

Also $V = \dfrac{Q}{A} = \dfrac{2.0 \text{ m}^3}{\text{sec}} \times \dfrac{1}{1.16 \text{ m}^2} = 1.72 \text{ m/sec}$

and $Fr = \dfrac{V}{\sqrt{g y_h}} = \dfrac{V}{\sqrt{g A / b_s}}$

$$Fr = \frac{1.72}{\text{sec}} \frac{\text{m}}{\left[9.81 \dfrac{\text{m}}{\text{sec}^2} \times 1.16 \text{ m}^2 \times \dfrac{1}{1.64 \text{ m}}\right]^{1/2}} = 0.653$$

Fr

Comparing results, we see that the rectangular flume would be cheaper to build; its perimeter is about 8.5 percent less than that of the triangular flume.

This example shows the effect of channel shape on the size required to deliver a given flow rate at a specified bed slope and roughness coefficient. At specified S_b and n, flow may be subcritical, critical, or supercritical, depending on Q.

10-5.3   Optimum Channel Cross Section

For given slope and roughness, the optimum channel cross section requires a minimum flow area for a given flow rate. From Eq. 10.35

$$\frac{Q}{A} = \frac{R_h^{2/3} S_b^{1/2}}{n} \tag{10.36}$$

Thus the optimum cross section has maximum hydraulic radius, R_h. Since $R_h = A/P$, R_h is maximum when the wetted perimeter is minimum. Solving Eq. 10.36 for A (with $R_h = A/P$) then yields

$$A = \left[\frac{nQ}{S_b^{1/2}}\right]^{3/5} P^{2/5} \tag{10.37}$$

From Eq. 10.37, the flow area will be a minimum when the wetted perimeter is a minimum.

Wetted perimeter, P, is a function of channel shape. For any given prismatic channel shape (rectangular, trapezoidal, triangular, circular, etc.), the channel cross section can be optimized. Optimum cross sections for common channel shapes are given in Table 10.3. The determination of the optimum cross section for a trapezoidal channel is illustrated in Example Problem 10.6.

Once the optimum cross section for a given channel shape has been determined, expressions for normal depth, y_n, and area, A, as functions of flow rate can be obtained from Eq. 10.35. These expressions are included in Table 10.3.

Table 10.3 Properties of Optimum Open-Channel Sections

Shape	Section	Optimum Geometry	Normal Depth, y_n	Cross-Sectional Area, A
Trapezoidal		$\alpha = 60°$ $b = \dfrac{2}{\sqrt{3}} y_n$	$0.968 \left[\dfrac{Qn}{S_b^{1/2}}\right]^{3/8}$	$1.622 \left[\dfrac{Qn}{S_b^{1/2}}\right]^{3/4}$
Rectangular		$b = 2y_n$	$0.917 \left[\dfrac{Qn}{S_b^{1/2}}\right]^{3/8}$	$1.682 \left[\dfrac{Qn}{S_b^{1/2}}\right]^{3/4}$
Triangular		$\alpha = 45°$	$1.297 \left[\dfrac{Qn}{S_b^{1/2}}\right]^{3/8}$	$1.682 \left[\dfrac{Qn}{S_b^{1/2}}\right]^{3/4}$
Wide Flat		None	$1.00 \left[\dfrac{(Q/b)n}{S_b^{1/2}}\right]^{3/8}$	— — —
Circular		$D = 2y_n$	$1.00 \left[\dfrac{Qn}{S_b^{1/2}}\right]^{3/8}$	$1.583 \left[\dfrac{Qn}{S_b^{1/2}}\right]^{3/4}$

EXAMPLE 10.6—Optimization of Trapezoidal Channel Cross Section

A trapezoidal channel is to be optimized so that excavation is minimized for a given discharge at normal depth. Find the optimum ratio of channel side length to bottom width and the optimum side slope angle.

EXAMPLE PROBLEM 10.6

GIVEN: Trapezoidal channel section.

FIND: (a) Optimum ratio of channel side length to bottom width.
(b) Optimum side slope angle.

SOLUTION:

Assumptions: (1) Steady flow   (3) Uniform flow at a section
(2) Incompressible flow   (4) Flow at normal depth

From the discussion of Section 10-5.3, optimum channel shape is obtained when R_h is maximized, or when P is minimized for a given flow area. The area for the section is

$$A = by_n + L\cos\theta\, L\sin\theta$$

But $L = \dfrac{y_n}{\sin\theta}$, so

$$A = by_n + y_n^2\frac{\cos\theta}{\sin\theta} = by_n + y_n^2\cot\theta \tag{1}$$

The wetted perimeter is

$$P = b + 2L = b + 2\frac{y_n}{\sin\theta}$$

The bottom width may be eliminated using Eq. 1 to obtain

$$P = \frac{A}{y_n} - y_n\cot\theta + \frac{2y_n}{\sin\theta} \tag{2}$$

For any side slope angle, the minimum perimeter can be determined by differentiating Eq. 2 with respect to y_n. Thus

$$\frac{dP}{dy_n} = -\frac{A}{y_n^2} - \cot\theta + \frac{2}{\sin\theta} = 0$$

and

$$y_n^2 = \frac{A\sin\theta}{2 - \cos\theta} \tag{3}$$

(For a rectangular channel, $\theta = \pi/2$ and $A = by_n$. Equation 3 becomes

$$y_n^2 = \frac{by_n}{2} \qquad\text{or}\qquad y_n = \frac{b}{2}$$

as shown in Table 10.3.)

To find the optimum side slope angle, differentiate Eq. 3 with respect to θ. Thus

$$2y_n\frac{dy_n}{d\theta} = \frac{A\cos\theta}{2-\cos\theta} + \frac{A\sin^2\theta}{(2-\cos\theta)^2} = A\left[\frac{\cos\theta(2-\cos\theta) - \sin^2\theta}{(2-\cos\theta)^2}\right] = 0$$

which reduces to $2\cos\theta - 1 = 0$, $\cos\theta = \frac{1}{2}$, or $\theta = 60°$ ⟵ $\qquad\qquad \theta$

Substituting into Eq. 3 gives

$$y_n^2 = \frac{A(\sqrt{3}/2)}{2 - \frac{1}{2}} = A\frac{\sqrt{3}}{3} = \frac{A}{\sqrt{3}} \qquad\text{or}\qquad A = \sqrt{3}y_n^2$$

Finally, substituting into Eq. 1 gives

$$\sqrt{3}y_n^2 = by_n + y_n^2\frac{1}{\sqrt{3}} = by_n + \frac{\sqrt{3}}{3}y_n^2 \qquad\text{or}\qquad b = \frac{2}{3}\sqrt{3}y_n = \frac{2y_n}{\sqrt{3}}$$

so that

$$L = \frac{y_n}{\sin\theta} = \frac{2y_n}{\sqrt{3}} = b \qquad\qquad L$$

Thus the optimum trapezoidal channel has sides and bottom of equal length and a side slope angle of 60°; it is half a hexagon.

10-5.4 Critical Normal Flow

Equation 10.35 indicates that for a given volume flow rate through a prismatic channel of fixed roughness, there is only one bed slope for flow at normal depth. Solving Eq. 10.35 for S_b,

$$S_b = \left[\frac{nQ}{R_h^{2/3}A}\right]^2 \tag{10.38}$$

If the channel bed slope is such that normal depth for a given flow rate is exactly equal to critical depth, that slope is called the critical slope. Denoting the critical bed slope as S_c, then

$$S_c = \left[\frac{nQ}{R_{hc}^{2/3} A_c} \right]^2 = \frac{n^2 Q^2}{R_{hc}^{4/3} A_c^2} \tag{10.39}$$

From Eq. 10.21, for critical flow,

$$\frac{Q^2}{A_c^2} = \frac{gA_c}{b_s}$$

Thus we can write

$$S_c = \frac{n^2 g A_c}{b_s R_{hc}^{4/3}} \tag{10.40}$$

For a wide rectangular channel of width b, with $b \gg y_c$, then $b_s = b$, $A_c = by_c$, and $R_{hc} = y_c$. Substituting into Eq. 10.40, we obtain

$$S_c = \frac{n^2 g}{y_c^{1/3}} \tag{10.41}$$

In terms of critical depth and critical slope,

if $y_n > y_c$, then $S_b < S_c$:   slope is mild (subcritical)

if $y_n = y_c$, then $S_b = S_c$:   slope is critical

if $y_n < y_c$, then $S_b > S_c$:   slope is steep (supercritical)

EXAMPLE 10.7—Critical Normal Flow through Two Channel Shapes

For critical normal flow in a wide flat channel made from unfinished concrete, plot bed slope as a function of flow depth for $0.01 \text{ m} < y < 10 \text{ m}$. Compare with the corresponding curve for critical normal flow in an optimum rectangular channel.

EXAMPLE PROBLEM 10.7

GIVEN: Critical normal flow in (a) a wide flat channel and (b) an optimum rectangular channel, made from unfinished concrete.

FIND: Plot S_c versus y_c for both cases.

SOLUTION:

Assumptions: (1)   Steady flow
                     (2)   Incompressible flow
                     (3)   Uniform flow at a section
                     (4)   Flow at critical depth

For unfinished concrete, Table 10.2 gives $n = 0.015$. For the wide channel, apply Eq. 10.41 directly.

Computing equation: $$S_c = \frac{n^2 g}{y_c^{1/3}} \tag{10.41}$$

Thus

$$S_c = (0.015)^2(9.81)y_c^{-1/3} = 0.00221y_c^{-1/3} \tag{1}$$

For the optimum rectangular channel, start with Eq. 10.34, and use $y_n/b = 0.5$ for optimum conditions.

Computing equation:
$$V = \frac{R_h^{2/3} S_b^{1/2}}{n} \tag{10.34}$$

For optimum conditions, $P = b + 2y_n = 4y_n$, $A = by_n = 2y_n^2$, and

$$R_h = \frac{A}{P} = \frac{2y_n^2}{4y_n} = 0.5y_n$$

For critical normal flow, $y_n = y_c$, and $V = V_c = \sqrt{gy_c}$. Substituting into Eq. 10.34,

$$V_c = \sqrt{gy_c} = g^{1/2}y_c^{1/2} = \frac{(0.5y_c)^{2/3}S_c^{1/2}}{n}$$

Thus

$$S_c^{1/2} = \frac{ng^{1/2}}{(0.5)^{2/3}y_c^{1/6}} \qquad \text{or} \qquad S_c = \frac{n^2 g}{(0.5)^{4/3}y_c^{1/3}}$$

and

$$S_c = \frac{(0.015)^2(9.81)}{(0.5)^{4/3}}y_c^{-1/3} = 0.00556y_c^{-1/3} \tag{2}$$

Slopes calculated from Eqs. 1 and 2 are shown in the following figure:

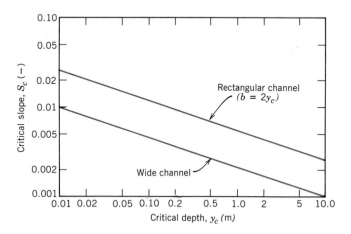

There is only one bed slope for flow at normal depth for given flow rate, channel geometry, and roughness coefficient. If the normal depth is the critical depth, then the slope is as given above. Comparing Eqs. 1 and 2 shows that for given depth (and hence the same flow rate per unit width, since $V_c = \sqrt{gy_c}$) the critical slope for the rectangular channel is 2.52 times larger than for the flat channel, due to the extra friction from the channel sides.

10-6 FLOW WITH GRADUALLY VARYING DEPTH

When open-channel flow encounters a change in bed slope or is approaching normal depth, flow depth changes gradually. Flow with gradually varying depth must be analyzed by applying the energy equation to a differential control volume; the result is a differential equation that relates changes in depth to distance along the flow. The resulting equation may be solved numerically if one assumes the head loss at each section is the same as for flow at normal depth, with the velocity and hydraulic radius of the section. Water depth and channel bed height are assumed to change slowly. As in the case of flow at normal depth, velocity is assumed uniform, and the pressure distribution is assumed hydrostatic at each section.

The energy equation (Eq. 10.10) for open-channel flow was applied to a finite control volume in Section 10-3. We shall adapt this equation to the differential control volume, of length dx, shown in Fig. 10.10. Thus

$$\frac{V^2}{2g} + y + z = \frac{V^2}{2g} + d\left[\frac{V^2}{2g}\right] + y + dy + z + dz + dh_l$$

The change in bed elevation, dz, can be written in terms of the slope as $dz = -S_b dx$. The head loss, dh_l, can be written in terms of the slope of the energy grade line as $dh_l = S\,dx$, where S is positive because the EGL drops in the direction of flow.

Thus the differential energy equation can be written as

$$d\left[\frac{V^2}{2g}\right] + dy = (S_b - S)dx$$

or

$$\frac{dy}{dx} + \frac{d}{dx}\left[\frac{V^2}{2g}\right] = S_b - S \qquad (10.42)$$

For flow in a rectangular channel of width b, $Q = Vby$, and $V = Q/by$, so

$$\frac{d}{dx}\left[\frac{V^2}{2g}\right] = \frac{d}{dx}\left[\frac{Q^2}{2gb^2y^2}\right] = -2\frac{Q^2}{2gb^2y^3}\frac{dy}{dx} = -\frac{V^2}{gy}\frac{dy}{dx}$$

Since $Fr = V/\sqrt{gy}$, then

$$\frac{d}{dx}\left[\frac{V^2}{2g}\right] = -Fr^2\frac{dy}{dx}$$

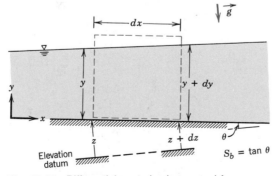

Fig. 10.10 Differential control volume used for analysis of gradually varied flow.

Substituting into Eq. 10.42, we obtain

$$\frac{dy}{dx} = \frac{S_b - S}{1 - Fr^2} \tag{10.43}$$

The slope, S, of the energy grade line is determined by assuming that the rate of head loss at a section is the same as for a flow at normal depth, having the velocity and hydraulic radius of the section.

From Eq. 10.34, for flow at normal depth,

$$V = \frac{R_h^{2/3} S^{1/2}}{n}$$

Solving for the slope of the energy grade line yields

$$S = \frac{n^2 V^2}{R_h^{4/3}} \tag{10.44}$$

Equation 10.43, with S given by Eq. 10.44, is a first-order, nonlinear, ordinary differential equation that describes the variation of the water surface profile for gradually varying flows. The sign of the slope of the water surface profile depends on whether the flow is subcritical or supercritical ($Fr < 1$ or $Fr > 1$) and on the relative magnitudes of S and S_b.

10-6.1 Classification of Surface Profiles

The general behavior of surface profiles can be studied most easily for wide rectangular channels; for such channels $b \gg y$ and $R_h \approx y$. For this case, the surface profile is governed by

$$\frac{dy}{dx} = \frac{S_b - S}{1 - Fr^2} = \frac{S_b \left[1 - \left(\dfrac{S}{S_b} \right) \right]}{1 - Fr^2}$$

where

$$S = \frac{n^2 V^2}{y^{4/3}} = \frac{n^2 Q^2}{b^2 y^{10/3}} \tag{10.45}$$

For a given flow rate through a wide rectangular channel, of width b and specified roughness, channel bed slope is uniquely related to normal depth. From Eq. 10.35

$$Q = \frac{R_h^{2/3} S_b^{1/2}}{n} A = \frac{y_n^{2/3} S_b^{1/2}}{n} b y_n$$

Solving for S_b, we obtain

$$S_b = \frac{n^2 Q^2}{b^2 y_n^{10/3}} \tag{10.46}$$

Thus

$$\frac{S}{S_b} = \left[\frac{y_n}{y} \right]^{10/3} \tag{10.47}$$

Froude number can be written in terms of critical depth. From Eq. 10.17,

$$Fr = \left[\frac{y_c}{y}\right]^{3/2} \tag{10.17}$$

Then the slope of the water surface for flow in a wide rectangular channel can be written as

$$\frac{dy}{dx} = S_b \frac{[1 - (y_n/y)^{10/3}]}{[1 - (y_c/y)^3]} \tag{10.48}$$

Equation 10.48 shows that the sign of dy/dx depends only on the relation among depths y, y_n, and y_c. In discussing critical normal flow (Section 10-5.4), we noted that

if $y_n > y_c$, then $S_b < S_c$: slope is mild (M)

if $y_n = y_c$, then $S_b = S_c$: slope is critical (C)

if $y_n < y_c$, then $S_b > S_c$: slope is steep (S)

In addition, the channel bed may be horizontal (H), in which case $S_b = 0$, or bed slope may be adverse (A), with $S_b < 0$.

For given bed slope (designated by the letter M, C, S, H, or A), the shape of the surface profile depends on the actual depth, y, relative to y_n and y_c. Numbers used to designate the three possibilities are

Curve 1: $y > y_n$ and $y > y_c$

Curve 2: $y_n > y > y_c$ or $y_c > y > y_n$

Curve 3: $y < y_n$ and $y < y_c$

The possible surface profiles and their properties for each of the five possible channel bed slopes are summarized in Table 10.4.

a. Surface Profiles in Channels with Mild Slope ($S_b < S_c$)

For mild channel slopes, $y_n > y_c$. For the $M1$ profile, $y > y_n$ and hence both the numerator and denominator of Eq. 10.48 are positive; thus $dy/dx > 0$. As y increases, dy/dx approaches S_b and the free surface approaches the horizontal. The $M1$ backwater curve typically is found upstream from a dam or control structure. Since the actual water surface lies above normal depth, flooding may occur near the dam if this is not accounted for in the design.

For the $M2$ profile, $y_n > y > y_c$ and the flow is subcritical. The numerator of Eq. 10.48 is negative and the denominator is positive; thus $dy/dx < 0$ and depth decreases in the direction of flow. As y approaches y_c, Eq. 10.48 indicates that $dy/dx \rightarrow \infty$, which is not possible. With strong surface curvature, the assumptions of straight streamlines and hydrostatic pressure variation (inherent in Eq. 10.48) no longer are valid. Consequently, the profile is shown dashed as flow depth approaches the critical depth. The $M2$ drawdown curve can occur upstream from a section where the channel slope changes from mild to critical or supercritical. An example is the flow over a spillway crest.

For the $M3$ profile, $y_n > y_c > y$ and the flow is supercritical. Both the numerator and denominator of Eq. 10.48 are negative, so dy/dx is positive and depth increases

Table 10.4 Surface Profiles for Gradually Varied Flow

Surface Profiles	Curve	Depth	Flow	Surface Slope
Mild slope, $S_b < S_c$	$M1$ $M2$ $M3$	$y > y_n > y_c$ $y_n > y > y_c$ $y_n > y_c > y$	Subcritical Subcritical Supercritical	Positive Negative Positive
Steep slope, $S_b > s_c$	$S1$ $S2$ $S3$	$y > y_c > y_n$ $y_c > y > y_n$ $y_c > y_n > y$	Subcritical Supercritical Supercritical	Positive Negative Positive
Critical slope, $S_b = S_c$	$C1$ $C3$	$y > y_c = y_n$ $y < y_c = y_n$	Subcritical Supercritical	Positive Positive
Horizontal slope, $S_b = 0$	$H2$ $H3$	$y > y_c$ $y < y_c$	Subcritical Supercritical	Negative Positive
Adverse slope, $S_b < 0$	$A2$ $A3$	$y > y_c$ $y < y_c$	Subcritical Supercritical	Negative Positive

in the flow direction. The $M3$ profile, also a backwater curve, occurs in supercritical flow, e.g., downstream from a spillway or sluice gate. As critical flow depth is approached, a sudden transition from supercritical to subcritical flow occurs. This sudden transition is a *hydraulic jump*, which will be analyzed and discussed in detail in Section 10-7.

b. Surface Profiles in Channels with Steep Slope ($S_b > S_c$)

For steep channel slopes, $y_n < y_c$. For the $S1$ profile, $y > y_c$ and hence both numerator and denominator of Eq. 10.48 are positive; thus $dy/dx > 0$. Furthermore, as y becomes much larger than y_c, then $dy/dx \rightarrow S_b$. Consequently, the $S1$ curve approaches the horizontal.

The $S2$ profile with $y_c > y > y_n$ is a drawdown curve with $dy/dx < 0$; the flow is supercritical. As y approaches y_n, the slope of the curve approaches zero; the depth approaches the normal depth. The $S2$ profile may occur downstream from a transition in channel slope from mild to steep.

For the $S3$ profile, $y < y_n < y_c$; the flow is supercritical, and the slope of the surface profile is positive. The depth increases in the direction of flow. As y approaches y_n, the surface slope approaches zero. This backwater curve may result downstream of a sluice gate if the flow depth is less than the normal depth on a steep slope.

c. Surface Profiles in Channels with Critical Slope ($S_b = S_c$)

For critical channel slopes, $y_n = y_c$. The slope of the surface profile is positive for both the $C1$ and $C3$ curves. For the $C1$ curve, $y > y_c$; the flow is subcritical, and the slope approaches S_b for large y. Consequently, the $C1$ curve approaches the horizontal. When $y < y_c$, the flow is supercritical. As y approaches y_c, by application of L'Hospital's rule to Eq. 10.48, it can be shown that the slope of the curve approaches $\frac{10}{9}$.

There is no $C2$ curve for gradually varying flow. When $y = y_n = y_c$, we have critical flow at normal depth.

d. Surface Profiles in Horizontal Channels ($S_b = 0$)

For flow in a horizontal channel, $S_b = 0$ and the normal depth is infinite. Consequently, an $H1$ profile cannot exist. The $H2$ and $H3$ profiles correspond to the $M2$ and $M3$ curves for $S_b = 0$. Equation 10.48 becomes indeterminate for $S_b = 0$ and $y_n \rightarrow \infty$. The profile slope can be determined from Eq. 10.43. For the $H2$ profile, $y > y_c$, the flow is subcritical, and the slope is negative. For the $H3$ profile, $y < y_c$, the flow is supercritical, and the slope is positive. For both profiles, Eq. 10.43 predicts infinite slope as $y \rightarrow y_c (Fr \rightarrow 1)$.

Since flow cannot continue indefinitely on a horizontal bed, these profiles can be found only in horizontal sections of more complex channels. The drawdown profile, $H2$, might be found across the top of a broad-crested weir (Section 10-8.2), and the profile $H3$ might be found on the flat floor of a stilling basin downstream from a spillway or below a sluice gate.

e. Surface Profiles in Channels with Adverse Slope ($S_b < 0$)

When the channel bed has adverse slope, S_b is negative; there is no real value of y_n. From Eq. 10.43, we conclude that the $A2$ and $A3$ profiles are similar to the $H2$ and $H3$ profiles.

As in the case of a horizontal channel, sustained flow is not possible over a long reach of adverse slope. A hydraulic jump may occur as profile $A3$ develops.

10-6.2 Calculation of Surface Profiles

Table 10.4 illustrated the general nature of surface profiles. To determine the actual profile for a given flow rate and initial conditions, we must integrate the governing differential equation numerically,

$$\frac{dy}{dx} + \frac{d}{dx}\left[\frac{V^2}{2g}\right] = S_b - S \tag{10.42}$$

This equation can be written in terms of the specific energy, E, as

$$\frac{dE}{dx} = S_b - S$$

Writing the equation in finite difference form, we obtain

$$\Delta x = \frac{\Delta E}{(S_b - S)_m} = \frac{E(y + \Delta y) - E(y)}{(S_b - S)_m} \tag{10.49}$$

where the subscript m denotes the mean properties over a channel reach of length Δx. Equation 10.49 can be used to solve for the channel section length, Δx, which corresponds to a given or assumed Δy, provided that S can be determined. If we assume that the local energy loss is equal to that for flow at normal depth, we can obtain S from the Manning correlation, since S_b for flow at normal depth equals S. From Eq. 10.44,

$$S_m = \frac{n^2 V_m^2}{R_{h_m}^{4/3}} \tag{10.50}$$

where $V_m = \frac{1}{2}[V(y + \Delta y) + V(y)]$ and $R_{h_m} = \frac{1}{2}[R_h(y + \Delta y) + R_h(y)]$.

Computation of the surface profile must begin at a point where the coordinates are known. Calculations may proceed upstream or downstream. To use Eq. 10.49, a depth y_2 is assumed at an adjacent point, Δx away from the initial point. For depth y_2, velocity V_2 is calculated from continuity. The average velocity, V_m, and hydraulic radius are calculated, and the mean slope of the energy grade line is determined from Eq. 10.50. The corresponding distance, Δx, is then computed directly from Eq. 10.49.

Obviously, hand calculations using Eq. 10.49 would be laborious, since small increments of y must be chosen to permit use of mean properties. The calculations are easy to program for computer solution; this permits evaluation for small steps. The denominator of Eq. 10.49 vanishes when flow at normal depth is approached (at normal depth, $S = S_b$). Consequently, the program should include a test of the denominator.

Calculation of surface profiles is illustrated in Example Problem 10.8.

EXAMPLE 10.8—Calculation of Free-Surface Profile

Water flows in a 5 m wide rectangular channel made from unfinished concrete with $n = 0.015$. The channel contains a long reach on which S_b is constant at $S_b = 0.020$. At one section, flow is at depth $y_1 = 1.5$ m, with speed $V_1 = 4.0$ m/sec. Estimate the channel location where the flow reaches a depth of 0.9 m.

EXAMPLE PROBLEM 10.8

GIVEN: Water flow in a long rectangular channel, with $S_b = 0.020$, $n = 0.015$, and $b = 5.0$ m. At section ①, $y_1 = 1.5$ m, and $V_1 = 4.0$ m/sec.

FIND: Location in channel where flow depth is $y_2 = 0.9$ m.

SOLUTION:

Assumptions: (1) Steady flow
(2) Incompressible flow
(3) Uniform flow at a section

The location where $y_2 = 0.9$ m may be upstream or downstream from section ①, depending on y_1, y_c, and y_n.

The flow rate is

$$Q = V_1 b y_1 = \frac{4 \text{ m}}{\text{sec}} \times 5 \text{ m} \times 1.5 \text{ m} = 30 \text{ m}^3/\text{sec}$$

At critical conditions, $V = V_c = \sqrt{g y_c}$, so

$$Q = V_1 b y_1 = V_c b y_c = \sqrt{g y_c} b y_c = b \sqrt{g} y_c^{3/2}$$

Thus

$$y_c = \left(\frac{Q}{b\sqrt{g}}\right)^{2/3} = \left[\frac{30 \text{ m}^3}{\text{sec}} \times \frac{1}{5 \text{ m}} \times \frac{\text{sec}}{\sqrt{9.81} \text{ m}^{1/2}}\right]^{2/3} = 1.54 \text{ m}$$

Since $y < y_c$, flow is supercritical.

$$Fr_1 = \frac{V_1}{\sqrt{g y_1}} = \frac{4 \text{ m}}{\text{sec}} \times \frac{1}{\left[9.81 \frac{\text{m}}{\text{sec}^2} \times 1.5 \text{ m}\right]^{1/2}} = 1.04 > 1$$

The normal depth at this slope, roughness, and discharge may be determined from

$$Q = \frac{R_h^{2/3} S_b^{1/2}}{n} A = \frac{\left[\dfrac{y_n}{1 + 2 y_n/b}\right]^{2/3} S_b^{1/2}}{n} b y_n = 47.14 y_n \left(\frac{y_n}{1 + 2 y_n/b}\right)^{2/3}$$

By iteration it may be found that $y_n = 0.858$ m.

Thus $y_c > y > y_n$, so flow is on an $S2$ curve. We expect the depth to *decrease* in the direction of flow; y should approach $y_n = 0.858$ m asymptotically.

Equation 10.49 may be used to solve numerically for the channel reach, Δx, required to change the surface level by an assumed amount, Δy. Thus, since $S_b = $ constant, $(S_b - S)_m = S_b - S_m$, and Eq. 10.49 becomes

Computing equation:
$$\Delta x = \frac{E(y + \Delta y) - E(y)}{S_b - S_m}$$

To apply this equation, we must express E as a function of y and we must evaluate S_m. Since $V = Q/by$, then

$$E = \frac{V^2}{2g} + y = \frac{Q^2}{2g b^2 y^2} + y$$

and

$$S_m = \frac{n^2 V_m^2}{R_{h_m}^{4/3}} \tag{10.50}$$

where $V_m = \frac{1}{2}[V(y + \Delta y) + V(y)]$, and $R_{h_m} = \frac{1}{2}[R_h(y + \Delta y) + R_h(y)]$.

Calculating equations for quantities in terms of y are

$$V = \frac{Q}{by} = \frac{V_1 b y_1}{by} = V_1 \frac{y_1}{y} = \frac{4 \text{ m}}{\text{sec}} \times \frac{1.5 \text{ m}}{1} \times \frac{1}{y} = \frac{6.000}{y}$$

$$E = \frac{Q^2}{2g\,b^2 y^2} + y = \frac{1}{2} \times \frac{(30)^2 \text{ m}^6}{\text{sec}^2} \times \frac{\text{sec}^2}{9.807 \text{ m}} \times \frac{1}{(5)^2 \text{ m}} \times \frac{1}{y^2} + y = \frac{1.835}{y^2} + y$$

$$R_h = \frac{A}{P} = \frac{by}{b+2y} = \frac{y}{1+2y/b} = \frac{y}{1+2y \times \dfrac{1}{5 \text{ m}}} = \frac{y}{1+0.4y}$$

Calculations may be tabulated as shown in the following table. (Four significant figures are shown to reduce roundoff errors.)

y (m)	V (m/sec)	E (m)	R_h (m)	R_{h_m} (m)	V_m (m/sec)	S_m (−)	Δx (m)	x (m)
1.5	4.000	2.316	0.9375					0.0
				0.9177	4.143	0.004330	1.321	
1.4	4.286	2.336	0.8974					1.32
				0.8766	4.451	0.005132	3.378	
1.3	4.615	2.386	0.8553					4.70
				0.8333	4.808	0.006632	6.624	
1.2	5.000	2.475	0.8108					11.3
				0.7877	5.227	0.008452	12.32	
1.1	5.455	2.617	0.7639					23.6
				0.7394	5.727	0.01104	24.39	
1.0	6.000	2.835	0.7143					48.0
				0.6884	6.333	0.01485	64.15	
0.9	6.667	3.166	0.6618					112.2

A graph of the results follows. The plot shows the water depth dropping rapidly at first, then approaching $y_n = 0.858$ m asymptotically farther downstream.

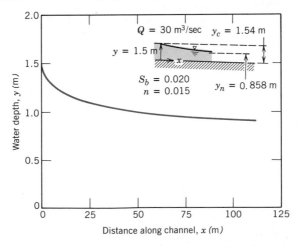

Accuracy of the results depends on (a) the accuracy of the mathematical model and (b) the technique used for numerical integration. The assumptions of negligible streamline curvature (and hydrostatic pressure distribution) are questionable from $y = 1.5$ to $y = 1.3$

m, where the depth changes rapidly. However, the channel distance during this interval is only 4.7 m, which is a small part of the total reach. Repeating the calculations with half the depth interval ($\Delta y = 0.05$ m) increased total reach to 117.7 m, a 5 percent increase. See [6] and [7] for more details on numerical methods.

10-7 THE HYDRAULIC JUMP

We have shown that open-channel flow may be subcritical ($Fr < 1$) or supercritical ($Fr > 1$). For subcritical flow, disturbances caused by a change in bed slope or flow cross section may move upstream and downstream; the result is a smooth adjustment of the flow. When flow at a section is supercritical, and downstream conditions require a change to subcritical flow, the need for the change cannot be communicated upstream. Thus a gradual change with a smooth transition through the critical point is not possible. The transition from supercritical to subcritical flow occurs abruptly through a hydraulic jump. The abrupt change in depth involves a significant loss of mechanical energy through turbulent mixing.

The general features of a hydraulic jump are sketched in Fig. 10.11. We shall analyze the jump phenomenon by applying the basic equations to the control volume shown in the sketch. Experiments show that the jump occurs over a relatively short distance; the maximum nondimensional jump length, L/y_2, is approximately 6.1 [8]. In view of this short length, it is reasonable to assume that friction forces acting on the control volume are negligible compared to pressure forces. Although hydraulic jumps can occur on inclined surfaces, for simplicity we assume a horizontal bed and select a control volume of width b.

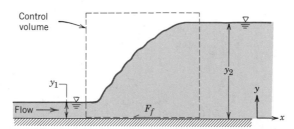

Fig. 10.11 Schematic of hydraulic jump, showing control volume used for analysis.

10-7.1 Basic Equations

a. Continuity Equation

Basic equation:
$$0 = \frac{\partial}{\partial t}\int_{CV} \rho \, d\Psi + \int_{CS} \rho \vec{V} \cdot d\vec{A} \qquad (4.13)$$

$$= 0(1)$$

Assumptions: (1) Steady flow
(2) Uniform flow at a section
(3) Incompressible flow

Then

$$0 = \{-|\rho V_1 b y_1|\} + \{|\rho V_2 b y_2|\} \quad \text{or} \quad V_1 y_1 = V_2 y_2 \qquad (10.51)$$

b. Momentum Equation

Basic equation:
$$F_{S_x} + \overset{=0(4)}{\cancel{F_{B_x}}} = \overset{=0(1)}{\cancel{\frac{\partial}{\partial t}}} \int_{CV} u\rho\, d\Psi + \int_{CS} u\rho\vec{V}\cdot d\vec{A} \tag{4.19a}$$

Assumptions: (4) $F_{B_x} = 0$

(5) Hydrostatic pressure distribution, so that

$$\frac{dp}{dy} = -\rho g, \text{ and } p = p_{\text{atm}} + \rho g(y_s - y)$$

(6) Neglect friction force, F_f

The momentum equation becomes

$$\int_0^{y_1} [p_{\text{atm}} + \rho g(y_1 - y)]b\, dy - \int_0^{y_2} [p_{\text{atm}} + \rho g(y_2 - y)]b\, dy$$

$$+ p_{\text{atm}}b(y_2 - y_1) - \overset{\approx 0(6)}{\cancel{F_f}} = V_1\{-|\rho V_1 b y_1|\} + V_2\{|\rho V_2 b y_2|\}$$

or

$$p_{\text{atm}}by_1 + \rho g\left(y_1^2 - \frac{y_1^2}{2}\right)b - p_{\text{atm}}by_2 - \rho g\left(y_2^2 - \frac{y_2^2}{2}\right)b$$

$$+ p_{\text{atm}}b(y_2 - y_1) = -\rho V_1^2 b y_1 + \rho V_2^2 b y_2$$

Finally, after simplifying and dividing by $\rho g b$,

$$\frac{V_1^2}{g}y_1 + \frac{y_1^2}{2} = \frac{V_2^2}{g}y_2 + \frac{y_2^2}{2} \tag{10.52}$$

c. Energy Equation

The energy equation for open-channel flow was derived in Section 10-3 as

$$\frac{V_1^2}{2g} + y_1 + z_1 = \frac{V_2^2}{2g} + y_2 + z_2 + h_l \tag{10.10}$$

For the hydraulic jump of Fig. 10.11, the channel bed is horizontal, and the energy equation becomes

$$\frac{V_1^2}{2g} + y_1 = \frac{V_2^2}{2g} + y_2 + h_l \tag{10.53}$$

In terms of the specific energy, Eq. 10.53 becomes

$$E_1 = E_2 + h_l \tag{10.54}$$

Since h_l is positive, Eq. 10.54 shows that the flow leaving a hydraulic jump has a lower specific energy than the entering flow.

The continuity and momentum equations, Eqs. 10.51 and 10.52, can be solved for y_2 in terms of upstream conditions (Section 10-7.2). Then the head loss across the jump also can be computed in terms of upstream conditions (Section 10-7.3).

10-7.2 Depth Increase across a Hydraulic Jump

To determine the downstream or *sequent* depth in terms of conditions upstream from the hydraulic jump, we begin by eliminating V_2 from the momentum equation. From continuity, $V_2 = V_1 y_1/y_2$; hence, Eq. 10.52 can be written as

$$\frac{V_1^2 y_1}{g} + \frac{y_1^2}{2} = \frac{V_1^2 y_1}{g}\left(\frac{y_1}{y_2}\right) + \frac{y_2^2}{2}$$

Rearranging this equation gives

$$y_2^2 - y_1^2 = \frac{2V_1^2 y_1}{g}\left(1 - \frac{y_1}{y_2}\right) = \frac{2V_1^2 y_1}{g}\left(\frac{y_2 - y_1}{y_2}\right)$$

Dividing both sides by the common factor $(y_2 - y_1)$, we obtain

$$y_2 + y_1 = \frac{2V_1^2 y_1}{g y_2}$$

Multiplying through by y_2/y_1^2 gives

$$\left(\frac{y_2}{y_1}\right)^2 + \left(\frac{y_2}{y_1}\right) = \frac{2V_1^2}{g y_1} = 2\,Fr_1^2$$

Solving for y_2/y_1 using the quadratic formula (and choosing the positive root, since y_2/y_1 must be positive), we find that

$$\frac{y_2}{y_1} = \frac{1}{2}\left(\sqrt{1 + 8\,Fr_1^2} - 1\right) \tag{10.55}$$

Thus, the ratio of downstream to upstream depths across a hydraulic jump is only a function of the upstream Froude number. Depths y_1 and y_2 are referred to as *conjugate* depths. From Eq. 10.55, we see that an increase in depth requires an upstream Froude number greater than one. Experimental confirmation of Eq. 10.55 is shown in Fig. 10.12.

Hydraulic jumps often are utilized to dissipate energy below spillways as a means of preventing erosion of artificial or natural channel bottom or sides. The head loss due to a hydraulic jump is treated in the next section.

10-7.3 Head Loss across a Hydraulic Jump

The head loss across a hydraulic jump may be calculated from the energy equation

$$h_l = E_1 - E_2 = \frac{V_1^2}{2g} + y_1 - \left[\frac{V_2^2}{2g} + y_2\right] \tag{10.54}$$

From continuity, $V_2 = V_1 y_1/y_2$; hence

$$h_l = \frac{V_1^2}{2g}\left[1 - \left(\frac{y_1}{y_2}\right)^2\right] + (y_1 - y_2)$$

or

$$\frac{h_l}{y_1} = \frac{Fr_1^2}{2}\left[1 - \left(\frac{y_1}{y_2}\right)^2\right] + \left[1 - \frac{y_2}{y_1}\right] \tag{10.56}$$

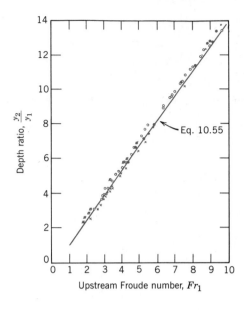

Fig. 10.12 Comparison of theory and experimental results for depth ratio across a hydraulic jump. (Data from [8], used by permission.)

Solving Eq. 10.55 for Fr_1, in terms of y_2/y_1, and substituting into Eq. 10.56, we obtain (after considerable algebraic manipulation),

$$\frac{h_l}{y_1} = \frac{1}{4} \frac{\left[\frac{y_2}{y_1} - 1\right]^3}{\frac{y_2}{y_1}} \tag{10.57}$$

Since h_l is positive, Eq. 10.57 shows that y_2/y_1 must be greater than one.

The specific energy, E_1, can be written as

$$E_1 = \frac{V_1^2}{2g} + y_1 = y_1\left[\frac{V_1^2}{2gy_1} + 1\right] = y_1 \frac{(Fr_1^2 + 2)}{2}$$

Nondimensionalizing h_l on E_1, then

$$\frac{h_l}{E_1} = \frac{1}{2} \frac{\left[\frac{y_2}{y_1} - 1\right]^3}{\frac{y_2}{y_1}[Fr_1^2 + 2]} \tag{10.58}$$

The depth ratio is given in terms of Fr_1 by Eq. 10.55. Thus h_l/E_1 can be written as a function of entering Froude number. The result is

$$\frac{h_l}{E_1} = \frac{\left[\sqrt{1 + 8\,Fr_1^2} - 3\right]^3}{8\left[\sqrt{1 + 8\,Fr_1^2} - 1\right][Fr_1^2 + 2]} \tag{10.59}$$

The effect of the entering Froude number on head loss is shown in dimensionless form in Fig. 10.13. Experimental data correlate well with the prediction of Eq. 10.59; Fig. 10.13 shows that more than 70 percent of the mechanical energy of the entering stream is dissipated in jumps with $Fr_1 > 9$.

Inspection of Eq. 10.59 shows that if $Fr_1 = 1$, then $h_l = 0$. Imaginary values are predicted for $Fr_1 < 1$. Since h_l must be positive in any real flow, a hydraulic jump can occur only in supercritical flow with $Fr > 1$. Flow downstream from a jump always is subcritical.

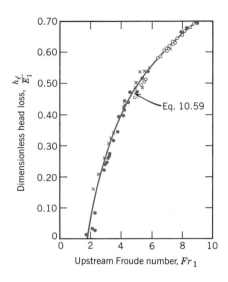

Fig. 10.13 Head loss across a hydraulic jump as a function of approach Froude number. (Data from [8], used by permission.)

EXAMPLE 10.9—Analysis of Hydraulic Jump

A hydraulic jump occurs on a horizontal bed immediately downstream from the sluice gate of Example Problem 10.3. Calculate the sequent depth following the hydraulic jump. Evaluate the head loss across the jump.

EXAMPLE PROBLEM 10.9

GIVEN: Flow under a sluice gate at the conditions shown.

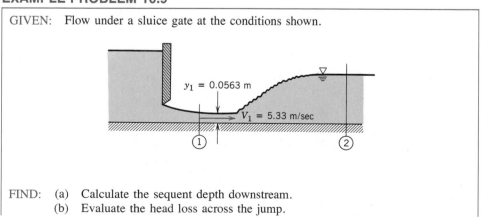

FIND: (a) Calculate the sequent depth downstream.
  (b) Evaluate the head loss across the jump.

SOLUTION:

Assumptions: (1) Steady flow
(2) Incompressible flow
(3) Uniform flow at sections ① and ②

Apply Eqs. 10.55 and 10.59 directly.

Computing equations:
$$\frac{y_2}{y_1} = \frac{1}{2}\left[\sqrt{1 + 8\,Fr_1^2} - 1\right] \tag{10.55}$$

$$\frac{h_l}{E_1} = \frac{\left[\sqrt{1 + 8\,Fr_1^2} - 3\right]^3}{8\left[\sqrt{1 + 8\,Fr_1^2} - 1\right][Fr_1^2 + 2]} \tag{10.59}$$

The Froude number of the flow entering the jump is

$$Fr_1 = \frac{V_1}{\sqrt{g\,y_1}} = 5.33\ \frac{m}{sec} \times \frac{1}{\left[9.81\ \frac{m}{sec^2} \times 0.0563\ m\right]^{1/2}} = 7.17$$

Then

$$\frac{y_2}{y_1} = \frac{1}{2}\left[\sqrt{1 + 8(7.17)^2} - 1\right] = 9.65$$

and

$$y_2 = 9.65\,y_1 = 9.65 \times 0.0563\ m = 0.543\ m \qquad\qquad \underleftarrow{\hspace{2cm}}\ y_2$$

Also

$$\frac{h_l}{E_1} = \frac{\left[\sqrt{1 + 8(7.17)^2} - 3\right]^3}{8\left[\sqrt{1 + 8(7.17)^2} - 1\right][(7.17)^2 + 2]} = 0.629$$

Since

$$E_1 = \frac{V_1^2}{2g} + y_1 = \frac{1}{2} \times \frac{(5.33)^2\ m^2}{sec^2} \times \frac{sec^2}{9.81\ m} + 0.0563\ m = 1.50\ m$$

then

$$h_l = 0.629\,E_1 = (0.629)1.50\ m = 0.944\ m \qquad\qquad \underleftarrow{\hspace{2cm}}\ h_l$$

This hydraulic jump dissipates nearly 63 percent of the specific energy of the incoming flow.

10-8 MEASUREMENTS IN OPEN-CHANNEL FLOW

Two basic methods are available to measure flow rates in open channels. At a critical section, the flow speed is equal to the critical speed, so the flow rate may be calculated from a depth measurement. At an obstruction in a channel (a *weir*), flow depth correlates with the rate of flow.

10-8.1 Sharp-Crested Weirs

Flow over a sharp-crested weir is sketched in Fig. 10.14. It is tempting to apply the Bernoulli equation from far upstream to flow at the weir *nappe*. Near the weir crest, the streamlines are highly curved, making the assumptions of uniform flow and hydrostatic pressure variation poor. Consequently, calculation of flow rate over a sharp-crested weir requires use of an empirically determined discharge coefficient. Many experiments have been run on a variety of weir configurations. Figure 10.15 illustrates three principal types of sharp-crested weirs.

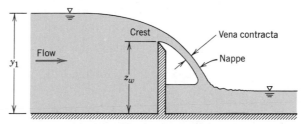

Fig. 10.14   Section view of flow over a sharp-crested weir.

The flow area across a horizontal weir is proportional to the height difference, $y_1 - z_w$. Thus

$$A \sim b\left(\frac{L}{b}\right)(y_1 - z_w)$$

If we neglect the upstream velocity, the velocity across the weir may be determined approximately from the Bernoulli equation as

$$V \approx \sqrt{2g(y_1 - z_w)} \sim \sqrt{g(y_1 - z_w)}$$

Combining these equations yields

$$Q \sim b\left(\frac{L}{b}\right)\sqrt{g}(y_1 - z_w)^{3/2}$$

Introducing the empirically determined discharge coefficient, C_d, to account for velocity nonuniformity, stream contraction, and frictional effects, then

$$Q = C_d b\left(\frac{L}{b}\right)\sqrt{g}(y_1 - z_w)^{3/2} \tag{10.60}$$

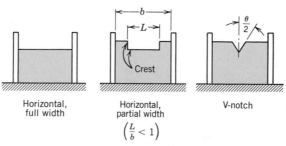

Fig. 10.15   Principal types of sharp-crested weirs.

Many correlations have been suggested for the discharge coefficient. A simple equation of the form

$$C_d = 0.59 + 0.08\left[\frac{y_1}{z_w} - 1\right] \tag{10.61}$$

gives satisfactory results for many cases [9, 10].

The nappe (Fig. 10.14) of a full-width horizontal weir must be aerated (ventilated to the atmosphere). If the nappe is not aerated, entrainment forms a low-pressure region, causing the discharge to increase for a given head.

Aeration is automatic when the weir spans only part of the channel. However, additional contraction of the stream is caused by the sharp vertical edges of the crest. For approximate calculations [11], the length, L, used in Eq. 10.60 may be reduced by $0.2\,(y_1 - z_w)$.

The flow area over a V-notch crest is proportional to the square of the height difference

$$A \sim (y_1 - z_w)\tan\frac{\theta}{2}(y_1 - z_w) = \tan\frac{\theta}{2}(y_1 - z_w)^2$$

The velocity is related to $(y_1 - z_w)$ by

$$V \sim \sqrt{g(y_1 - z_w)}$$

which suggests

$$Q = C_d \tan\frac{\theta}{2}\sqrt{g}(y_1 - z_w)^{5/2} \tag{10.62}$$

Experiments show that $C_d \approx 0.44$ for a V-notch weir [9, 10].

Selection of a weir for a given situation depends upon the range of flow rate to be measured, the accuracy desired, and whether the weir can be calibrated after installation. Equations 10.60 and 10.62 are applicable only for $y_1 - z_w$ greater than about 0.06 m ($\sim$ 0.2 ft). This suggests that a V-notch weir should be chosen for low flow rates. For repeatability, the weir crest must be kept sharp. Periodic removal of scale or rust is simple in the laboratory but may be more difficult in the field.

When calibrated in place using volume measurements or weigh tanks, the accuracy of a weir in use is limited only by the accuracy of measuring head. Errors in head measurement can be reduced using a stilling well and hook gage, as shown in Fig. 10.16. The surface can be located with extreme precision by sighting across it nearly horizontally and noting the reflection of the hook point in the surface.

Measurement uncertainties for weirs depend on many factors. Flow rates can be measured to ± 5 percent. Consult [9] through [11] if more precise results are needed.

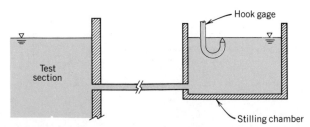

Fig. 10.16 Use of a stilling chamber and hook gage for precise liquid level measurements.

EXAMPLE 10.10—Flow over a Sharp-Crested Weir

A horizontal, full width, sharp-crested weir is to be used in a laboratory setup to control and measure water flow rate, which is expected to vary between 200 and 300 gpm. The width, b, is to be such that the level change between these flow rates is held to 0.10 ft.

EXAMPLE PROBLEM 10.10

GIVEN: Horizontal (full width) weir, water flow rate between 200 and 300 gpm.

FIND: Width, b, such that $\Delta z_{hi} - \Delta z_{lo} < 0.10$ ft.

SOLUTION:
From Section 10-8.1,

$$Q = C_d b \left(\frac{L}{b}\right) \sqrt{g}(y_1 - z_w)^{3/2} \qquad (10.60)$$

and

$$C_d = 0.59 + 0.08\left[\frac{y_1}{z_w} - 1\right] \qquad (10.61)$$

Assumptions: (1)   Steady flow
(2)   Incompressible flow
(3) $L/b = 1$
(4)   Neglect the second term in Eq. 10.61, as a first approximation, by assuming $\Delta z = y_1 - z_w \ll z_w$

Then $C_d = 0.59$, and from Eq. 10.60,

$$y_1 - z_w = \Delta z = \left(\frac{Q}{C_d L \sqrt{g}}\right)^{2/3}$$

Hence,

$$\Delta z_{hi} - \Delta z_{lo} = \left[\frac{1}{0.59 b \sqrt{g}}\right]^{2/3} [Q_{hi}^{2/3} - Q_{lo}^{2/3}]$$

Finally,

$$b = \frac{1}{0.59 \sqrt{g}} \frac{\left[Q_{hi}^{2/3} - Q_{lo}^{2/3}\right]^{3/2}}{\left[\Delta z_{hi} - \Delta z_{lo}\right]^{3/2}}$$

The flow rates are

$$Q_{hi} = \frac{300 \text{ gal}}{\text{min}} \times \frac{\text{ft}^3}{7.48 \text{ gal}} \times \frac{\text{min}}{60 \text{ sec}} = 0.668 \text{ ft}^3/\text{sec}$$

and

$$Q_{lo} = \frac{200 \text{ gal}}{\text{min}} \times \frac{\text{ft}^3}{7.48 \text{ gal}} \times \frac{\text{min}}{60 \text{ sec}} = 0.446 \text{ ft}^3/\text{sec}$$

Substituting,

$$b = \frac{\sec}{0.59(32.2)^{1/2} \text{ ft}^{1/2}} \frac{[(0.668)^{2/3} - (0.446)^{2/3}]^{3/2}}{(0.10)^{3/2} \text{ ft}^{3/2}} \frac{\text{ft}^3}{\sec} = 0.724 \text{ ft} \qquad \longleftarrow \qquad b$$

Equation 10.60 is valid if the elevation change, Δz, is greater than 0.2 ft. Solving for Δz at the lowest flow rate, we obtain

$$\Delta z = \left[\frac{Q}{C_d b \sqrt{g}} \right]^{2/3} = \left[0.446 \frac{\text{ft}^3}{\sec} \times \frac{1}{0.59} \times \frac{1}{0.724 \text{ ft}} \times \frac{\sec}{(32.2)^{1/2} \text{ ft}^{1/2}} \right]^{2/3}$$

$$\Delta z = 0.324 \text{ ft}$$

which satisfies the criterion.

For the second term in Eq. 10.61 to be negligible,

$$0.08 \left[\frac{y_1}{z_w} - 1 \right] = (0.08) \frac{y_1 - z_w}{z_w} = (0.08) \frac{\Delta z}{z_w} < 0.01$$

so z_w must be

$$z_w > 8 \Delta z = 8 \times 0.324 \text{ ft} = 2.59 \text{ ft}$$

10-8.2 Broad-Crested Weirs

Flow over a broad-crested weir is sketched in Fig. 10.17. When the tailwater level downstream is sufficiently low, critical depth, y_c, is reached at some point on top of the weir. There $Fr = 1$, so

$$V_c = \sqrt{g y_c}$$

For a wide channel, the discharge is given by

$$Q \approx VA = \sqrt{g y_c} b y_c = b \sqrt{g} y_c^{3/2}$$

or

$$Q = C_d b \sqrt{g} y_c^{3/2} \qquad (10.63)$$

where C_d is an empirical discharge coefficient. Experimental data [9, 10] suggest that when $(y_1 - z_w)/L > 0.2$, the discharge coefficient is between 0.95 and 0.98.

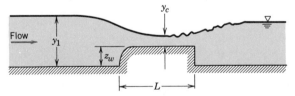

Fig. 10.17 Flow over a broad-crested weir.

When a broad-crested weir is long and the tailwater is low enough, a *free overfall* occurs [11]. As shown in Fig. 10.18, water depth at the brink, y_b, is less than critical depth, y_c. Also, y_c is located some distance, L_c, upstream from the brink. Experiments [11] show that

$$y_b \approx 0.72 y_c \qquad \text{and} \qquad L_c \approx 3.5 y_c$$

Thus, if the brink depth, y_b, is known, then y_c and the flow rate may be computed.

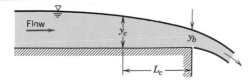

Fig. 10.18 Flow over a broad-crested weir with a free overfall.

10-8.3 Sluice Gates

A sluice gate often is used to regulate discharge. Flow beneath a typical sluice gate is shown in Fig. 10.19 for both low and high tailwater conditions.

For low tailwater, as shown in Fig. 10.19a, the vena contracta is the section of minimum area in the discharge stream. There streamlines are straight and parallel, so the pressure distribution is hydrostatic. If we assume frictionless flow, the Bernoulli equation may be applied between sections ① and ②. By introducing an empirical discharge coefficient, C_d, the flow rate may be written

$$Q = C_d b z_g \sqrt{2 g y_1} \qquad (10.64)$$

Experiments [11] show that $0.6 < C_d < 0.9$.

For high tailwater, as shown in Fig. 10.19b, analysis of flow under a sluice gate is impossible. The general form of Eq. 10.64 may be used with an appropriate value of C_d, e.g., from [11].

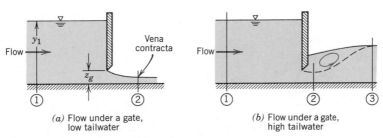

(a) Flow under a gate, low tailwater

(b) Flow under a gate, high tailwater

Fig. 10.19 Flow under a sluice gate.

10-8.4 Critical Flumes

Accurate flow measurements can be made using weirs. However, practical problems that limit their field use include:

- Fouling by silt or debris
- Deterioration of sharp edges
- Large head loss

These problems can be overcome by using a critical flow meter, such as the Parshall flume, shown in Fig. 10.20.

The Parshall flume is widely used to measure irrigation water flows because it is self-cleaning, requires only a small head, and gives reasonably accurate results over a wide range of flow rate [12]. A Parshall flume is built in three sections: the upstream section has a flat floor and converging walls, the center (or throat) section has parallel walls and a floor that slopes downward, and the outlet section has diverging walls

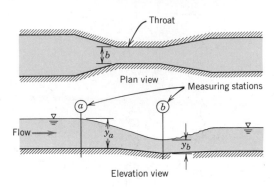

Fig. 10.20 Parshall flume for measuring open-channel flow rate.

and a floor that slopes upward. Actual dimensions and the location of measuring sections are based on throat width, which may range from 75 mm to many meters.

When the downstream level is low, critical flow occurs. The flow rate may be determined from tables if y_a is known. When $y_b > 0.7 y_a$, both depths must be measured. Tables and charts are available for many sizes of Parshall flumes [12].

10-9 SUMMARY OBJECTIVES

After completing study of Chapter 10, you should be able to do the following:

1. Define the terms:

Froude number	energy grade line
prismatic channel	hydraulic grade line
wetted perimeter	Manning roughness coefficient
hydraulic radius	optimum channel cross section
secondary flow	critical depth
wave celerity	mild slope
subcritical flow	critical slope
critical flow	steep slope
supercritical flow	hydraulic jump
specific energy	sequent depth
alternate depths	conjugate depths
uniform flow	free overfall
normal depth	

2. Calculate cross-sectional area, wetted perimeter, and hydraulic radius for common open-channel shapes.

3. Derive an equation for the propagation speed of an isolated free-surface wave and show that for small amplitudes $c = \sqrt{gy}$.

4. Apply the control volume form of the energy equation to open-channel flow. Discuss the variation in specific energy for frictionless flow and for flow at normal depth.

5. Evaluate the effect of changes in channel bottom contour on frictionless flow. Sketch changes in surface profile for subcritical and supercritical flows over a bump and a depression in the channel bottom.

6. Derive the conditions required to maximize flow rate through a sluice gate assuming frictionless flow.

7. Use the Manning correlation for velocity to evaluate normal depth for uniform flow on a bed of constant slope. Calculate the critical slope for a given flow rate.

8. Define and calculate the optimum channel cross section or flow depth for flow at normal depth.

9. Apply the basic equations to a differential control volume to evaluate the rate of change of surface profile for gradually varied flow. Classify and calculate surface profiles for various bed slopes in subcritical and supercritical flows.

10. Analyze flow through a hydraulic jump to calculate the downstream surface level and head loss across the jump.

11. Determine the volume flow rate from elevation measurements for sharp-edged and broad-crested weirs and sluice gates.

12. Solve the problems at the end of the chapter that relate to the material you have studied.

REFERENCES

1. Chow, V. T., *Open Channel Hydraulics*. New York: McGraw-Hill, 1959.

2. White, F. M., "Hydrodynamics of Coastal Areas," in *Introduction to Ocean Engineering*, H. Schenck, ed. New York: McGraw-Hill, 1975, pp. 8–42.

3. Todd, D. K., ed., *The Water Encyclopedia*. Port Washington, NY: Water Information Center, 1970.

4. Volt, S. S., "Tsunamis," in *Annual Review of Fluid Mechanics, 19*, 1987, pp. 217–236.

5. Varley, E., R. Venkataraman, and E. Cumberbatch, "The Propagation of Large Amplitude Tsunamis across a Basin of Varying Depth. Part I, Off-Shore Behavior," *J. Fluid Mechanics, 49*, Pt. 4, 29 October 1971, pp. 771–802.

6. Smith, P. D., *BASIC Hydraulics*. London: Butterworth Scientific, 1982.

7. Computer Program HEC2, "Water Surface Profiles," The Hydrologic Engineering Center, Water Resources Support Center, U. S. Army Corps of Engineers, 609 Second Street, Davis, CA 95616-4687.

8. Peterka, A. J., "Hydraulic Design of Stilling Basins and Energy Dissipators," U. S. Department of the Interior, Bureau of Reclamation, Engineering Monograph No. 25, (Revised) July 1963.

9. Ackers, P., et al., *Weirs and Flumes for Flow Measurement*. New York: Wiley, 1978.

10. Bos, M. G., J. A. Replogle, and A. J. Clemmens, *Flow Measuring Flumes for Open Channel Systems*. New York: Wiley-Interscience, 1984.

11. King, H. W., and E. F. Brater, *Handbook of Hydraulics*, 5th ed. New York: McGraw-Hill, 1963.

12. Scott, V. L., and C. E. Houston, "Measuring Irrigation Water," Agricultural Extension Service, University of California, Davis, Circular 473, January 1959.

13. Hoerner, S. F., *Fluid Dynamic Drag*, 2nd ed. Midland Park, NJ: Published by the author, 1965.

14. Lynch, D. K., "Tidal Bores," *Scientific American, 247*, 4, October 1982, pp. 146–156.

PROBLEMS

10.1 Experiments show that transition to turbulent flow, in a broad inclined open channel, occurs at $Re = yV_{max}/\nu \approx 1200$. Derive an expression for the velocity profile, assuming fully developed laminar flow. Show that to maintain laminar flow down an inclined surface the flow rate per unit width, Q/b, must be less than 800ν.

‡10.2 Derive an expression for the hydraulic radius of a trapezoidal channel with bottom width b, liquid depth y, and side slope angle θ. Verify the equation given in Table

‡ You may wish to use simple computer programs to help solve problems marked with daggers.

10.1. Plot the ratio R_h/y for $b = 2$ m with side slope angles of $30°$ and $60°$ for $0.5 < y < 3$ m.

‡10.3 Verify the equation given in Table 10.1 for the hydraulic radius of a circular channel. Evaluate and plot the ratio R_h/D, for liquid depths between 0 and D.

10.4 A wave from a passing boat in a lake is observed to travel at 10 mph. Determine the approximate water depth at this location.

‡10.5 Solution of the complete differential equations for wave motion without surface tension shows that wave speed is given by

$$V_w = \sqrt{\frac{g\lambda}{2\pi} \tanh\left(\frac{2\pi y}{\lambda}\right)}$$

where λ is wavelength and y is liquid depth. Show that when $\lambda/y \ll 1$, wave speed becomes proportional to $\sqrt{\lambda}$. In the limit as $\lambda/y \to \infty$, $V_w = \sqrt{gy}$. Determine the value of λ/y above which $V_w > 0.99\sqrt{gy}$.

‡10.6 Solution of the complete differential equations for wave motion in quiescent liquid, including the effect of surface tension, shows that wave speed is given by

$$V_w = \sqrt{\left(\frac{g\lambda}{2\pi} + \frac{2\pi\sigma}{\rho\lambda}\right)\tanh\left(\frac{2\pi y}{\lambda}\right)}$$

where λ is wavelength, y is liquid depth, and σ is surface tension. Prepare a plot of wave speed versus wavelength for the range $1 < \lambda < 100$ mm for (a) water and (b) mercury. Assume that $y = 7$ mm for both liquids.

10.7 Water flows in a rectangular channel at a depth of 2 ft. If the flow speed is (a) 4 ft/sec and (b) 12 ft/sec, compute the corresponding Froude numbers.

10.8 A pebble is dropped into a stream of water that flows in a rectangular channel at 8 ft depth. In one second, a ripple caused by the stone is carried 20 ft downstream. What is the speed of the flowing water?

10.9 A pebble is dropped into a stream of water of uniform depth. A wave is observed to travel upstream 5 ft in 1 sec, and 13 ft downstream in the same time. Determine the flow speed and depth.

10.10 Surface waves are caused by a sharp object that just touches the free surface of a stream of flowing water, forming the wave pattern shown. The stream depth is 6 in. Determine the flow speed and Froude number.

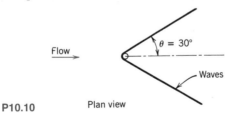

P10.10   Plan view

10.11 The Froude number characterizes flow with a free surface. On a log-log plot of speed versus depth for $0.1 < V < 3$ m/sec and $0.001 < y < 1$ m, plot the line $Fr = 1$, and indicate regions that correspond to tranquil and rapid flow.

10.12 A submerged body traveling horizontally beneath a liquid surface at Froude number (based on body length) near 0.5 produces a strong surface wave pattern if submerged less than half its length [13]. (The wave pattern of a surface ship also is pronounced at this Froude number.) On a log-log plot of speed versus body (or ship) length for $1 < V < 30$ m/sec and $1 < x < 300$ m, plot the line $Fr = 0.5$.

‡ You may wish to use simple computer programs to help solve problems marked with daggers.

10.13 A rectangular channel carries a discharge of 10 ft^3/sec per foot of width. Determine the minimum specific energy possible for this flow. Compute the corresponding flow depth and speed.

10.14 Flow in the channel of Problem 10.13 ($E_{min} = 2.19$ ft) is to be at twice the minimum specific energy. Compute the alternate depths for this E.

10.15 Water flows in a wide channel of uniform slope as shown. Determine the elevations of the hydraulic and energy grade lines above the datum for sections ① and ②.

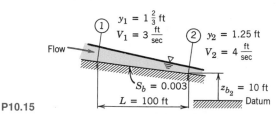

P10.15

10.16 Consider flow in a channel with arbitrary cross section. For uniform velocity show that the flux of energy across each section is given by

$$ef = \left(\frac{p_{atm}}{\rho} + \frac{V^2}{2} + gy + gz_b \right) \dot{m}$$

where y is the flow depth.

10.17 A long rectangular channel 10 ft wide is observed to have a wavy surface at a depth of about 6 ft. Estimate the rate of discharge.

10.18 Water flows at 300 ft^3/sec in a trapezoidal channel with bottom width of 8 ft. The sides are sloped at 2:1. Find the critical depth for this channel.

10.19 For a channel of nonrectangular cross section, specific energy can be written in terms of area as Eq. 10.12. Critical depth occurs at minimum specific energy. Obtain a general equation for critical depth in a triangular channel in terms of Q, g, and α.

10.20 For a channel of nonrectangular cross section, specific energy can be written in terms of area as Eq. 10.12. Critical depth occurs at minimum specific energy. Obtain a general equation for critical depth in a channel of trapezoidal section in terms of Q, g, b, and α.

10.21 Consider the Venturi flume shown. The bed is horizontal and flow may be considered frictionless. The upstream depth is 1 ft and the downstream depth is 0.75 ft. The upstream breadth is 2 ft and the breadth of the throat is 1 ft. Estimate the flow rate through the flume.

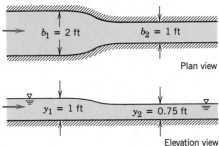

P10.21

10.22 A rectangular channel 10 ft wide carries 100 cfs on a horizontal bed at 1.0 ft depth. A smooth bump across the channel rises 4 in. above the channel bottom. Find the elevation of the liquid free surface above the bump.

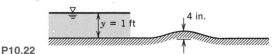

P10.22

10.23 A rectangular channel 10 ft wide carries a discharge of 20 ft³/sec at 0.9 ft depth. A smooth bump 0.2 ft high is placed on the floor of the channel. Estimate the local change in flow depth caused by the bump.

10.24 At a section of a 10 ft wide rectangular channel, the depth is 0.3 ft for a discharge of 20 ft³/sec. A smooth bump 0.1 ft high is placed on the floor of the channel. Determine the local change in flow depth caused by the bump.

10.25 Water, at 3 ft/sec and 2 ft depth, approaches a smooth rise in a wide channel. Estimate the stream depth after the 0.5 ft rise.

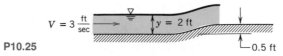

P10.25

10.26 Water issues from a sluice gate at 0.6 m depth. The discharge per unit width is 6.0 m³/sec/m. Estimate the water level far upstream where the flow speed is negligible. Calculate the maximum rate of flow per unit width that could be delivered through the sluice gate.

10.27 A horizontal rectangular channel 3 ft wide contains a sluice gate. Upstream of the gate the depth is 6 ft; the depth downstream is 0.9 ft. Estimate the volume flow rate in the channel.

10.28 A partially open sluice gate in a 3 m wide rectangular channel carries water at 8.5 m³/sec. The upstream depth is 2 m. Find the downstream depth and Froude number.

10.29 A rectangular flume built of timber, with 1 ft per 1000 ft slope, is 6 ft wide. Water flows at a normal depth of 3 ft. Compute the discharge.

10.30 A rectangular flume built of timber is 3 ft wide. The flume is to handle a flow of 90 ft³/sec at a normal depth of 6 ft. Determine the slope required.

10.31 A channel with square cross section is to carry 20 m³/sec of water at normal depth on a slope of 0.003. Compare the dimensions of the channel required for (a) unfinished and (b) finished concrete.

10.32 Water flows in a trapezoidal channel at a normal depth of 1.2 m. The bottom width is 2.4 m and the sides slope at 1:1 (45°). The flow rate is 7.1 m³/sec. The channel is excavated from firm, smooth earth. Find the bed slope.

10.33 A triangular channel with side angles of 45° is to carry 10 m³/sec at a slope of 0.001. The channel is unfinished concrete. Find the required dimensions.

10.34 A semicircular trough of corrugated steel, with diameter $D = 1$ m, carries water at depth $y_n = 0.25$ m. The slope is 0.01. Find the discharge.

10.35 Find the discharge at which the channel of Problem 10.34 flows full.

10.36 The flume of Problem 10.29 is fitted with a new plastic film liner. Find the new depth of flow if the discharge remains constant at 85.5 ft³/sec.

10.37 Discharge through the channel of Problem 10.32 is increased to 15 m³/sec. Find the corresponding normal depth, if the bed slope is 0.00193.

10.38 The channel of Problem 10.32 has 0.00193 bed slope. Find the normal depth for the given discharge after a new plastic liner is installed.

10.39 Consider again the semicircular channel of Problem 10.34. Find the normal depth that corresponds to a discharge of 0.3 m³/sec.

10.40 Consider a symmetric open channel of triangular cross section. Show that for a given flow area, the wetted perimeter is minimized when the sides meet at a right angle.

10.41 An above-ground rectangular flume is to be constructed of timber. For a drop of 10 ft/mile, what will be the depth and width for the most economical flume if it is to discharge 40 cfs?

10.42 An irrigation canal is to be designed to carry water at 250 m³/sec on a bed slope of 0.0004. Calculate the required dimensions of the best trapezoidal channel with side slopes of 45° in smooth earth. Estimate the reduction in excavation that would be possible if the canal were lined with plastic film.

‡10.43 Consider flow in a triangular channel at depth y_n. Obtain expressions for A, P, and R_h, in terms of y_n and side angle α. By plotting P versus α for a given A, confirm that the optimum channel configuration has $\alpha = 45°$. How far off optimum are channels with side angles of 60°?

10.44 A wide flat unfinished concrete channel discharges water at 20 ft³/sec per foot of width. Find the critical slope.

10.45 An optimum rectangular storm sewer channel made of unfinished concrete is to be designed to carry a maximum flow rate of 100 ft³/sec, at critical normal flow conditions. Determine the channel width and slope.

10.46 Consider flow in a rectangular channel. Show that for flow at critical depth and optimum aspect ratio $(b = 2y)$, the volume flow rate and bed slope are given by the expressions:

$$Q = 6.26 y_c^{5/2} \qquad \text{and} \qquad S_c = 24.7 \frac{n^2}{y_c^{1/3}}$$

10.47 A trapezoidal canal lined with brick has side slopes of 2:1 and bottom width of 10 ft. It carries 600 ft³/sec at critical speed. Determine the critical slope.

10.48 Water flows at 1400 ft³/sec in a 15 ft wide rectangular channel. The bottom slope is constant and the channel roughness is such that normal depth is 6.0 ft. At a certain section the depth is 2.8 ft. How will the depth change with distance downstream?

10.49 Classify the water surface profile of Problem 10.52. Show all necessary calculations.

10.50 A rectangular flume of timber is 5 ft wide and carries 60 cfs of water. The bed slope is 0.0006, and at a certain section the depth is 3 ft. Find the distance (in one reach) to the section where the depth is 2.5 ft. Is this section upstream or downstream?

10.51 Suppose that the slope of the flume in Problem 10.50 is changed so that, with the same flow rate, the depth varies from 3 ft at one section to 4 ft at a section 1000 ft downstream. Using a single reach, estimate the new bed slope of the flume. Sketch the flume and the water surface to assure that the answer is reasonable.

10.52 Water discharges under a gate onto a wide channel of adverse slope. The depth at the vena contracta is 0.6 m and the speed is 12 m/sec. For a channel roughness of 0.015, estimate the depth at the downstream end of the adverse slope. Use a single reach.

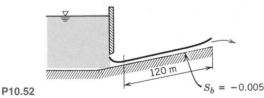

P10.52 — 120 m — $S_b = -0.005$

‡10.53 Consider a rectangular channel 15 ft wide, made from unfinished concrete. The normal depth of water flow in the channel is 3 ft at a Froude number of 2. At one section in the channel, the depth is 4.5 ft. Estimate the channel length required for the flow depth to change to 3.1 ft. Does the change occur upstream or downstream from the section with 4.5 ft depth?

10.54 A horizontal wooden flume is 5 m wide and 300 m in length. The flow depth at the flume exit is 2.0 m. If the discharge is 40 m³/sec, estimate the flow depth at the flume inlet.

‡You may wish to use simple computer programs to help solve problems marked with daggers.

‡10.55 A dam partially restricts the flow of a small river. Assume that the river can be modeled as a wide channel with $S_b = 0.001$ and $n = 0.025$. The flow rate per unit width is 3 m³/sec/m and the water depth immediately upstream from the dam is 5 m. Find the upstream distance to the location where $y = 2$ m.

‡10.56 Water from a sluice gate issues onto a wide horizontal floor of unfinished concrete. The depth is 0.6 m and the discharge per unit width is 6 m²/sec. Estimate the distance to the downstream location where the flow depth reaches 1.2 m. (Assume that no hydraulic jump occurs.)

‡10.57 Water flows at 40 m³/sec in a 5 m wide rectangular channel. The bottom slope is constant and the roughness is $n = 0.017$; normal depth is $y_n = 2.0$ m. At a certain section, the depth is 0.9 m. Evaluate S_b. Estimate the channel length needed to raise the water depth to 1.6 m.

10.58 Consider the flow conditions of the hydraulic jump of Example Problem 10.9. Calculate the increase in water temperature that would result if all the head loss due to the jump were converted to thermal energy. Would it be practical to measure the jump head loss with a thermometer?

10.59 A wide channel carries 20 ft³/sec per foot of width at a depth of 1 ft at the toe of a hydraulic jump. Determine the sequent depth of the jump and the head loss across it.

10.60 A hydraulic jump occurs in a wide horizontal channel. The discharge is 30 ft³/sec per foot of width. The upstream depth is 1.3 ft. Determine the sequent depth for the jump.

10.61 A hydraulic jump occurs in a rectangular channel. The flow rate is 200 ft³/sec, and the depth before the jump is 1.2 ft. Determine the depth behind the jump and the head loss, if the channel is 10 ft wide.

10.62 The hydraulic jump may be used as a crude flow meter. Suppose that in a horizontal rectangular channel 5 ft wide the observed depths before and after a hydraulic jump are 0.66 and 3.0 ft. Find the rate of flow and the head loss.

10.63 A hydraulic jump occurs on a horizontal apron downstream from a wide spillway, at a location where depth is 0.9 m and speed is 25 m/sec. Estimate the depth and speed downstream from the jump. Compare the specific energy downstream from the jump to that upstream.

10.64 A hydraulic jump occurs in a rectangular channel. The flow rate is 6.5 m³/sec and the depth before the jump is 0.4 m. Determine the depth behind the jump and the head loss, if the channel is 1 m wide.

10.65 A positive surge wave, or moving hydraulic jump, can be produced in the laboratory by suddenly opening a sluice gate. Consider a surge of depth y_2 advancing into a quiescent channel of depth y_1. Obtain an expression for surge speed in terms of y_1 and y_2.

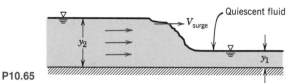

P10.65

10.66 A tidal bore—an abrupt translating wave or moving hydraulic jump—often forms when the tide flows into the wide estuary of a river [14]. In one case, a bore is observed to have a height of 12 ft above the undisturbed level of the river that is 8 ft deep. The bore travels upstream at 18 mph. Determine the approximate speed of the current of the undisturbed river.

‡You may wish to use simple computer programs to help solve problems marked with daggers.

10.67 The crest of a broad-crested weir is 1 ft below the level of an upstream reservoir, where the water depth is 8 ft. What is the maximum flow rate per unit width that could pass over the weir?

10.68 Water flows over a horizontal sharp-crested weir 5 ft wide in a channel of 10 ft width. The height of the weir is 3 ft at the crest. Upstream from the weir, the flow depth is 4 ft. Determine the flow rate.

10.69 Water flows over a 60° V-notch weir. The height differential is $y_1 - z_w = 0.6$ ft. Determine the discharge.

10.70 Water flows under a sluice gate from a reservoir, where the depth is 8 ft. The gate is raised 1.5 ft above the channel floor. Estimate the flow rate per unit width if $C_d = 0.8$.

Chapter 11

FLUID MACHINERY

Man has sought to control nature since antiquity. Early man carried water by the bucket; as larger groups formed, this process was mechanized. Thus the first fluid machines developed as bucket wheels and screw pumps to lift water. The Romans introduced paddle wheels around 70 B.C. to obtain energy from streams [1]. Later, windmills were developed to harness wind power, but the low power density of the wind limited output to a few hundred horsepower. Development of water wheels made possible the extraction of thousands of horsepower at a single site.

Today we take for granted many fluid machines. On a typical day we draw pressurized water from the tap, use a blower to dry our hair, drive a car in which fluid machines operate the lubrication, cooling, and power steering systems, and work in a comfortable environment provided by air circulation. The list could be extended indefinitely.

The purpose of this chapter is to introduce the concepts needed to analyze, design, and apply fluid machines. Our treatment deals almost exclusively with incompressible flows. The major emphasis is on work-absorbing devices (pumps and fans), since these devices are most frequently encountered in fluid systems.

First the terminology of the field is introduced and machines are classified by operating principle and physical characteristics. Rather than attempt a treatment of the entire field, the focus is on machines in which energy transfer to or from the fluid is through a rotating element. Basic equations are reviewed and then simplified to forms useful for analysis of fluid machines. Performance characteristics of typical machines are considered. Examples are given of pump and turbine applications in typical systems. Problems ranging from simple applications to system design conclude the chapter.

11-1 INTRODUCTION AND CLASSIFICATION OF FLUID MACHINES

Fluid machines may be broadly classified as either *positive displacement* or *dynamic*. In positive displacement machines, energy transfer is accomplished by volume changes that occur while the fluid is confined completely within a chamber or passage. Fluid handling devices that direct the flow with blades or vanes attached to a rotating member are termed *turbomachines*. In contrast to positive displacement machinery, fluid never is confined completely in a turbomachine. All work interactions in a turbomachine result from dynamic effects of the rotor on the fluid stream. The emphasis in this chapter will be on dynamic machines.

A further distinction among types of turbomachines is based on the geometry of the flow path. In *radial flow* machines, the flow path is essentially radial, with significant changes in radius from inlet to outlet. (Such machines sometimes are called *centrifugal* machines.) In *axial flow* machines, the flow path is nearly parallel to the machine centerline, and the radius of the flow path does not vary significantly.

544

In *mixed flow* machines the flow path radius changes only moderately. Schematic diagrams of some typical turbomachines are shown in Figs. 11.1 through 11.3.

Machines that add energy to a fluid stream are called *pumps* when the flow is liquid or slurry, and *fans, blowers,* or *compressors* for gas or vapor handling units, depending on pressure rise. Fans usually have small pressure rise (less than 1 inch of water) and blowers have moderate pressure rise (perhaps 1 inch of mercury); pumps and compressors may have very high pressure rises. Current industrial systems operate at pressures up to 150,000 psi (10^4 atmospheres).

The rotating element of a pump frequently is called the *impeller*; the impeller is contained within the pump *housing* or *casing*. The shaft that transfers mechanical energy to the impeller usually must penetrate the casing; a system of bearings and seals is required to complete the mechanical design of the unit.

Three typical centrifugal machines are shown schematically in Fig. 11.1. Flow enters each machine nearly axially at small radius through the *eye* of the rotor, diagram (*a*), at radius r_1. Flow is turned and leaves through the impeller discharge at radius r_2, where the width is b_2. Flow leaving the impeller is collected in the *scroll* or *volute*, which gradually increases in area as it nears the outlet of the machine, diagram (*b*). The impeller usually has vanes; it may be *shrouded* (enclosed) as shown in diagram (*a*), or *open* as shown in diagram (*c*). The impeller vanes may be relatively straight, or they may curve to become non-radial at the outlet. Diagram (*c*) shows that there may be a diffuser between the impeller discharge and the volute. This *radial* diffuser may be *vaneless* or it may have vanes.

Typical axial-flow and mixed-flow turbomachines are shown schematically in Fig. 11.2. Diagram (*a*) shows a typical axial-flow compressor *stage*.[1] Flow enters nearly parallel to the rotor axis and maintains nearly the same radius through the stage. The mixed-flow pump in diagram (*b*) shows the flow being turned radially and moving to larger radius as it passes through the stage.

The pressure rise that can be achieved efficiently in a single stage is limited, depending on the type of machine. However, stages may be combined to produce multi-stage machines, virtually without limit on pressure rise. Axial-flow compressors, as typically found in turbojet engines, are examples of multi-stage compressors. Centrifugal pumps frequently are built with multiple stages in a single housing.

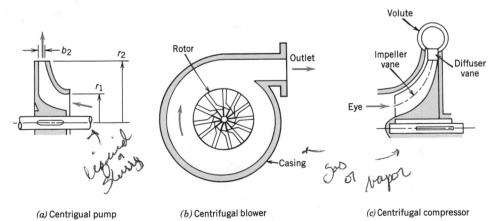

(*a*) Centrigual pump (*b*) Centrifugal blower (*c*) Centrifugal compressor

Fig. 11.1 Schematic diagrams of typical centrifugal-flow turbomachines, adapted from [2].

[1] The combination of a stationary blade row and a moving blade row is called a *stage*. (The stationary blades may be guide vanes placed before the rotor; more commonly they are *antiswirl vanes* placed after the rotor.)

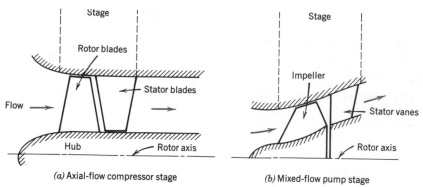

Fig. 11.2   Schematic diagrams of typical axial-flow and mixed-flow turbomachines, adapted from [2].

Fans, blowers, compressors, and pumps are found in many sizes and types, ranging from simple household units to complex industrial units of large capacity. Torque and power requirements for idealized pumps and turboblowers can be analyzed by applying the angular momentum principle using a suitable control volume.

Propellers are essentially axial-flow devices that operate without an outer housing. Propellers may be designed to operate in gases or liquids. As you might expect, propellers designed for these very different applications are quite distinct. Marine propellers tend to have wide blades compared to their radius, giving high *solidity*. Aircraft propellers tend to have long, thin blades with relatively low solidity.

Machines that extract energy from a fluid stream are called *turbines*. The assembly of *vanes, blades,* or *buckets* attached to the turbine shaft is called the *rotor, wheel,* or *runner*. In *hydraulic turbines* the working fluid is water, so the flow is incompressible. In *gas turbines* and *steam turbines* the density of the working fluid may change significantly.

The two most general classifications of turbines are *impulse* and *reaction* turbines. Impulse turbines are driven by one or more high-speed free jets. Each jet is accelerated in a nozzle external to the turbine wheel. If friction and gravity are neglected, neither the fluid pressure nor its speed relative to the runner change as it passes over the turbine buckets. Thus for an impulse turbine, the fluid expansion from high to low pressure takes place in nozzles external to the blades and the runner does not flow full of fluid.

Several typical hydraulic turbines are shown schematically in Fig. 11.3. Diagram (*a*) shows an impulse turbine driven by a single jet, which lies in the plane of the

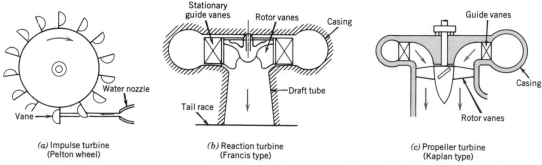

Fig. 11.3   Schematic diagrams of typical hydraulic turbines, adapted from [2].

turbine runner. Water from the jet strikes each bucket in succession, is turned, and leaves the bucket with relative velocity nearly opposite to that with which it entered the bucket. Spent water falls into the *tailrace* (not shown).

In reaction turbines, part of the fluid pressure change takes place externally and part takes place within the moving blades. External acceleration occurs and the flow is turned to enter the runner in the proper direction, as it passes through nozzles or stationary blades called *guide vanes* or *wicket gates*. Additional fluid acceleration relative to the rotor occurs within the moving blades, so both the relative velocity and the pressure of the stream change across the runner. Because reaction turbines flow full of fluid, they generally can produce more power for a given overall size than impulse turbines.

A reaction turbine of the Francis type is shown in Fig. 11.3b. Incoming water flows circumferentially through the turbine casing. It enters the outside periphery of the stationary guide vanes and flows toward the runner. In cross-section, it enters the runner nearly radially and is turned downward to leave nearly axially; the flow pattern may be thought of as a centrifugal pump in reverse. Water leaving the runner flows through a diffuser known as a *draft tube* before entering the tailrace.

Diagram (c) shows a propeller turbine of the Kaplan type. The water entry is similar to the Francis turbine just described. However, it is turned to flow nearly axially before encountering the turbine runner. Flow leaving the runner may pass through a draft tube.

Thus turbines range from simple windmills to complex gas and steam turbines with many stages of carefully designed blading. These devices also can be analyzed in idealized form by applying the angular momentum principle.

Dimensionless parameters such as *specific speed, flow coefficient, torque coefficient, power coefficient*, and *pressure ratio* frequently are used to characterize the performance of turbomachines. These parameters were introduced in Chapter 7; their development and use will be considered in more detail later in this chapter.

11-2   SCOPE OF COVERAGE

Approximately a third of the energy consumed in the U.S. is used by industry. It is estimated that 40-50 percent of industrial energy is used to drive pumps and compressors [3]. Proper design, construction, selection, and application of pumps and compressors are economically significant.

Design of actual machines involves diverse technical knowledge, including fluid mechanics, materials, bearings, seals, and vibrations. These topics are covered in numerous specialized texts. Our objective here is to present only enough detail to illustrate the analytical basis of fluid flow design and to discuss briefly the limitations on results obtained from simple analytical methods. For more detailed design information consult the references.

Applications or "system" engineering requires a wealth of experience. Much of this experience must be gained by working with other engineers in the field. Our coverage is not intended to be comprehensive; instead we discuss only the most important considerations for successful system application of pumps, compressors, and turbines.

As stated earlier in this chapter, our treatment deals almost exclusively with incompressible flows. Even with this limitation, the material presented in Chapter 11 is intrinsically more difficult than the fundamental topics considered in earlier chapters, because so much of it involves integration of empirical information with theory.

Consequently, no presentation of fluid machinery can be as clear or straightforward as the earlier chapters dealing with fundamentals.

11-3 TURBOMACHINERY ANALYSIS

The analysis method used for turbomachinery is chosen according to the information sought. If overall information on flow rate, pressure change, torque, and power is desired, then a finite control volume analysis may be used. If detailed information is desired about blade angles or velocity profiles, then individual blade elements must be analyzed using an infinitesimal control volume or other detailed procedure. We consider only idealized flow processes in this book, so we concentrate on the finite control volume approach, applying the angular momentum principle.

11-3.1 The Angular Momentum Principle

The angular momentum principle was applied to finite control volumes in Chapter 4. The result was Eq. 4.47,

$$\vec{r} \times \vec{F}_S + \int_{CV} \vec{r} \times \vec{g} \rho \, d\Psi + \vec{T}_{shaft} = \frac{\partial}{\partial t} \int_{CV} \vec{r} \times \vec{V} \rho \, d\Psi + \int_{CS} \vec{r} \times \vec{V} \rho \vec{V} \cdot d\vec{A} \qquad (4.47)$$

In the next section Eq. 4.47 is simplified for analysis of turbomachinery.

11-3.2 Euler Turbomachine Equation

For turbomachinery analysis, it is convenient to choose a fixed control volume enclosing the rotor to evaluate shaft torque. As a first approximation, torques due to surface forces may be ignored. The body force contribution may be neglected by symmetry. Then, for steady flow, Eq. 4.47 becomes

$$\vec{T}_{shaft} = \int_{CS} \vec{r} \times \vec{V} \rho \vec{V} \cdot d\vec{A} \qquad (11.1a)$$

Let us write this equation in scalar form and illustrate its application to axial and radial flow machines.

As shown in Fig. 11.4, we select a *fixed* control volume enclosing a generalized turbomachine rotor. The fixed coordinate system is chosen with the z axis aligned with the axis of rotation of the machine. The idealized velocity components are shown in the figure. The fluid is assumed to enter the rotor at radial location r_1, with uniform absolute velocity $\vec{V}_1$; the fluid leaves the rotor at radial location r_2, with uniform absolute velocity $\vec{V}_2$.

The term on the right side of Eq. 11.1a is the product of $\vec{r} \times \vec{V}$ with the mass flow rate at each section. For uniform flow into the rotor at section ①, and out of the rotor at section ②, Eq. 11.1a becomes

$$T_{shaft}\hat{k} = (r_2 V_{t_2} - r_1 V_{t_1})\dot{m}\hat{k} \qquad (11.1b)$$

or in scalar form,

$$T_{shaft} = (r_2 V_{t_2} - r_1 V_{t_1})\dot{m} \qquad (11.1c)$$

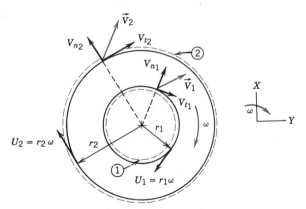

Fig. 11.4 Finite control volume and absolute velocity components for angular momentum analysis.

Equation 11.1c is the basic relationship between torque and angular momentum for all turbomachines. It often is called the *Euler turbomachine equation.*

The velocities that appear in Eq. 11.1c are the tangential components of the absolute velocity of the fluid crossing the control surface. The tangential velocities are chosen positive when in the direction as the blade speed, U. This sign convention gives $T_{\text{shaft}} > 0$ for pumps, fans, blowers, and compressors and $T_{\text{shaft}} < 0$ for turbines.

The rate of work done on a turbomachine rotor (the *mechanical power,* $\dot{W}_m$) is given by the dot product of rotor angular velocity, $\vec{\omega}$, and applied torque, $\vec{T}_{\text{shaft}}$. Using Eq. 11.1b, we obtain

$$\dot{W}_m = \vec{\omega} \cdot \vec{T}_{\text{shaft}} = \omega\hat{k} \cdot T_{\text{shaft}}\hat{k} = \omega\hat{k} \cdot (r_2 V_{t_2} - r_1 V_{t_1})\dot{m}\hat{k}$$

or

$$\dot{W}_m = \omega T_{\text{shaft}} = \omega(r_2 V_{t_2} - r_1 V_{t_1})\dot{m} \tag{11.2a}$$

According to Eq. 11.2a, the angular momentum of the fluid is increased by the addition of shaft work. For a pump, $\dot{W}_m > 0$ and the angular momentum of the fluid must increase. For a turbine, $\dot{W}_m < 0$ and the angular momentum of the fluid must decrease.

Equation 11.2a may be written in two other useful forms. Introducing $U = r\omega$, where U is the tangential speed of the rotor at radius r, then

$$\dot{W}_m = (U_2 V_{t_2} - U_1 V_{t_2})\dot{m} \tag{11.2b}$$

Dividing Eq. 11.2b by $\dot{m}g$, we obtain a quantity with the dimensions of length, often termed the *head* added to the flow.[2]

$$H = \frac{\dot{W}_m}{\dot{m}g} = \frac{1}{g}(U_2 V_{t_2} - U_1 V_{t_1}) \tag{11.2c}$$

Equations 11.1 and 11.2 are simplified forms of the angular momentum equation for a control volume. They all are written for a fixed control volume under the assumptions of steady, uniform flow at each section. The equations show that only

---

[2] Since $\dot{W}_m$ is energy per unit time and $\dot{m}g$ is weight flow per unit time, *head* is actually energy per unit weight of flowing fluid.

the difference in the product rV_t or UV_t, between the outlet and inlet sections, is important in determining the torque applied to the rotor or the energy transfer to the fluid. No restriction has been made on geometry; the fluid may enter and leave at the same or different radii.

If one applies the first law of thermodynamics for incompressible flow across a pump (or turbine), the rate of energy addition to (or extraction from) the fluid stream in the ideal case is $\dot{W}_h = \dot{m}\,\Delta p/\rho = \dot{m}\,gH = \rho Q\,gH$; $\dot{W}_h$ is referred to as the *hydraulic power*. When friction and flow losses are neglected, the hydraulic power is equal to the mechanical power, $\dot{W}_m = T\omega$. When friction and flow losses are included, the hydraulic power is related to the mechanical power using an *efficiency*.

For pumps, the efficiency is defined as $\eta_p = \dot{W}_h/\dot{W}_m$. Hence,

$$\eta_p = \frac{\dot{W}_h}{\dot{W}_m} = \frac{\rho Q\,gH}{\omega T} \tag{11.3a}$$

For hydraulic turbines, the efficiency is defined as $\eta_t = \dot{W}_m/\dot{W}_h$. Hence,

$$\eta_t = \frac{\dot{W}_m}{\dot{W}_h} = \frac{\omega T}{\rho Q\,gH} \tag{11.3b}$$

11-3.3 Velocity Polygon Analysis

The equations that we have derived also suggest the importance of clearly defining the velocity components of the fluid and rotor at the inlet and outlet sections. For this purpose, it is useful to develop *velocity polygons* for the inlet and outlet flows. Figure 11.5 shows the velocity polygons and introduces the notation for blade and flow angles.

In the idealized situation at the design point, flow relative to the rotor is assumed to enter and leave tangent to the blade profile at each section. (This idealized inlet condition is sometimes called *shockless* entry flow.) Blade angles, β, are measured relative to the circumferential direction, as shown in Fig. 11.5a. The inlet blade angle, β_1, fixes the direction of the relative inlet velocity at design conditions.

The runner speed at inlet is $U_1 = \omega R_1$, and therefore it is specified by the impeller geometry and the machine operating speed. The absolute fluid velocity is the vector sum of the impeller velocity and the flow velocity relative to the blade. The absolute inlet velocity may be determined graphically, as shown in Fig. 11.5b. The angle of

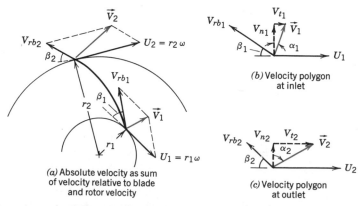

(a) Absolute velocity as sum of velocity relative to blade and rotor velocity

(b) Velocity polygon at inlet

(c) Velocity polygon at outlet

Fig. 11.5 Geometry and notation used to develop velocity polygons for typical radial-flow machines.

the absolute fluid velocity, α_1, is measured from the normal direction, as shown.[3] The tangential component of the absolute velocity, V_{t_1}, and the component normal to the flow area, V_{n_1}, also are shown in Fig. 11.5b. Note from the geometry of the figure that at each section the normal component of the *absolute* velocity, V_n, and the normal component of the velocity *relative to the blade,* $V_{r b_n}$, are equal.

When the inlet flow is swirl-free, the absolute inlet velocity will be purely radial. The inlet blade angle may be specified for the design flow rate and pump speed to provide shockless entry flow. Pre-swirl, which may be present in the inlet flow, or introduced by *inlet guide vanes,* causes the absolute inlet flow direction to differ from radial.

Velocity polygons are constructed similarly at the outlet section. The runner speed at the outlet is $U_2 = \omega R_2$, which again is known from the geometry and operating speed of the turbomachine. The relative flow is assumed to leave the impeller tangent to the blades, as shown in Fig. 11.5c. This idealizing assumption of perfect guidance fixes the direction of the relative outlet flow at design conditions.

For a centrifugal pump or reaction turbine, the velocity relative to the blade generally changes in magnitude from inlet to outlet. The continuity equation must be applied, using the impeller geometry, to determine the normal component of velocity at each section. The normal component, together with the outlet blade angle, is sufficient to establish the velocity relative to the blade at the impeller outlet for a radial-flow machine. The velocity polygon is completed by the vector addition of the velocity relative to the blade and the wheel velocity, as shown in Fig. 11.5c.

The inlet and outlet velocity polygons provide all the information needed to calculate the ideal torque or power, absorbed or delivered by the impeller, using Eqs. 11.1 or 11.2. The results represent the performance of a turbomachine under idealized conditions at the design operating point, since we have assumed that all flows are uniform at each section and that they enter and leave the rotor tangent to the blades. These idealized results represent the upper limits of performance for a turbomachine.

Performance of an actual machine may be estimated using the same basic approach, but accounting for variations in flow properties across the blade span at the inlet and outlet sections, and for deviations between the blade angles and the flow directions. Such detailed calculations are beyond the scope of this book.

The alternative is to measure the overall performance of a machine on a suitable test stand. Manufacturers' data are examples of measured performance information.

In Example Problem 11.1, we apply the idealized results that we have developed to an axial-flow fan. The subject of Example Problem 11.2 is an idealized centrifugal pump.

EXAMPLE 11.1—Idealized Axial-Flow Fan
An axial-flow fan operates at 1200 rpm. The blade tip diameter is 1.1 m and the hub diameter is 0.8 m. The blade inlet and exit angles are 30° and 60°, respectively. Inlet guide vanes give the absolute flow entering the first stage an angle of 30°. The fluid is air at standard conditions and the flow may be considered incompressible. There is no change in axial component of velocity across the rotor. Assume the relative flow enters and leaves the rotor at the geometric blade angles and use properties at the mean blade radius for calculations. For these idealized conditions, draw the inlet velocity polygon, determine the volume flow rate of the fan, and sketch the rotor blade shapes. Using the data so obtained, draw the outlet velocity polygon and calculate the minimum torque and power needed to drive the fan.

[3] The notation varies from book to book, so be careful when comparing references.

EXAMPLE PROBLEM 11.1

GIVEN: Flow through rotor of axial-flow fan.

Tip diameter: 1.1 m
Hub diameter: 0.8 m
Operating speed: 1200 rpm
Absolute inlet angle: 30°
Blade inlet angle: 30°
Blade outlet angle: 60°

Fluid is air at standard conditions.
Use properties at mean diameter of blades.

FIND: (a) Inlet velocity polygon.
(b) Volume flow rate.
(c) Rotor blade shape.

(d) Outlet velocity polygon.
(e) Rotor torque.
(f) Power required.

SOLUTION:
Apply the angular momentum equation to a fixed control volume.

Computing equations:
$$\vec{T}_{\text{shaft}} = \int_{CS} \vec{r} \times \vec{V} \rho \vec{V} \cdot d\vec{A} \tag{11.1a}$$

$$0 = \overset{=0(2)}{\cancel{\frac{\partial}{\partial t} \int_{CV} \rho \, d\Psi}} + \int_{CS} \rho \vec{V} \cdot d\vec{A}$$

Assumptions: (1) Neglect torques due to body or surface forces
(2) Steady flow
(3) Uniform flow at inlet and outlet sections
(4) Incompressible flow
(5) No change in axial flow area
(6) Use mean radius of rotor blades, R_m

The blade shapes are

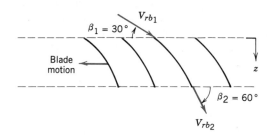

The inlet velocity polygon is

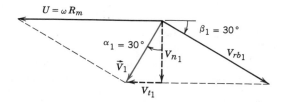

From continuity

$$0 = \{-|\rho V_{n_1} A_1|\} + \{|\rho V_{n_2} A_2|\}$$

or

$$Q = V_{n_1} A_1 = V_{n_2} A_2$$

Since $A_1 = A_2$, then $V_{n_1} = V_{n_2}$, and the outlet velocity polygon is as shown in the following figure:

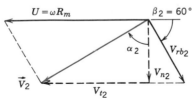

At the mean blade radius,

$$U = R_m \omega = \frac{D_m}{2}\omega$$

$$U = \frac{\frac{1}{2}(1.1+0.8)\,\text{m}}{2} \times \frac{1200\ \text{rev}}{\text{min}} \times \frac{2\pi\ \text{rad}}{\text{rev}} \times \frac{\text{min}}{60\ \text{sec}} = 59.7\ \text{m/sec}$$

From the geometry of the inlet velocity polygon,

$$U = V_{n_1}(\tan\alpha_1 + \cot\beta_1)$$

so that

$$V_{n_1} = \frac{U}{\tan\alpha_1 + \cot\beta_1} = \frac{59.7\ \text{m}}{\text{sec}} \times \frac{1}{\tan 30° + \cot 30°} = 25.9\ \text{m/sec}$$

Consequently,

$$V_1 = \frac{V_{n_1}}{\cos\alpha_1} = \frac{25.9\ \text{m}}{\text{sec}} \times \frac{1}{\cos 30°} = 29.9\ \text{m/sec}$$

$$V_{t_1} = V_1 \sin\alpha_1 = \frac{29.9\ \text{m}}{\text{sec}} \times \sin 30° = 15.0\ \text{m/sec}$$

and

$$V_{rb_1} = \frac{V_{n_1}}{\sin\beta_1} = \frac{25.9\ \text{m}}{\text{sec}} \times \frac{1}{\sin 30°} = 51.8\ \text{m/sec}$$

The volume flow rate is

$$Q = V_{n_1} A_1 = \frac{\pi}{4} V_{n_1}(D_t^2 - D_h^2) = \frac{\pi}{4} \times \frac{25.9\ \text{m}}{\text{sec}}[(1.1)^2 - (0.8)^2]\ \text{m}^2$$

$$Q = 11.6\ \text{m}^3/\text{sec} \qquad\qquad Q$$

From the geometry of the outlet velocity polygon,

$$\tan\alpha_2 = \frac{V_{t_2}}{V_{n_2}} = \frac{U - V_{n_2}\cot\beta_2}{V_{n_2}} = \frac{U - V_{n_1}\cot\beta_2}{V_{n_1}}$$

or

$$\alpha_2 = \tan^{-1}\left[\frac{\dfrac{59.7\ \text{m}}{\text{sec}} - \dfrac{25.9\ \text{m}}{\text{sec}} \times \cot 60°}{\dfrac{25.9\ \text{m}}{\text{sec}}}\right] = 59.9°$$

and

$$V_2 = \frac{V_{n_2}}{\cos\alpha_2} = \frac{V_{n_1}}{\cos\alpha_2} = \frac{25.9 \text{ m}}{\text{sec}} \times \frac{1}{\cos 59.9°} = 51.6 \text{ m/sec}$$

Finally,

$$V_{t_2} = V_2 \sin\alpha_2 = \frac{51.6 \text{ m}}{\text{sec}} \times \sin 59.9° = 44.6 \text{ m/sec}$$

The angular momentum equation becomes

$$\vec{T} = T_z\hat{k} = \int_{CS} \vec{R}_m \times \vec{V}\rho\vec{V} \cdot d\vec{A} = \hat{k}\int_{CS} R_m V_t \rho\vec{V} \cdot d\vec{A}$$

so that for uniform flow

$$T_z = R_m V_{t_1}\{-|\rho V_{n_1} A_1|\} + R_m V_{t_2}\{|\rho V_{n_2} A_2|\} = \rho Q R_m(V_{t_2} - V_{t_1})$$

$$= \frac{1.23 \text{ kg}}{\text{m}^3} \times \frac{11.6 \text{ m}^3}{\text{sec}} \times \frac{0.95 \text{ m}}{2} \times \frac{(44.6 - 15.0) \text{ m}}{\text{sec}} \times \frac{\text{N} \cdot \text{sec}^2}{\text{kg} \cdot \text{m}}$$

$$T_z = 201 \text{ N} \cdot \text{m} \qquad\qquad\qquad\qquad\qquad\qquad\qquad\qquad\qquad\qquad T_z$$

Thus the torque *on* the CV is in the same sense as $\vec{\omega}$. The power required is

$$\dot{W}_m = \vec{\omega} \cdot \vec{T} = \omega_z T_z = \frac{1200 \text{ rev}}{\text{min}} \times \frac{2\pi \text{ rad}}{\text{rev}} \times \frac{\text{min}}{60 \text{ sec}} \times \frac{201 \text{ N} \cdot \text{m}}{} \times \frac{\text{W} \cdot \text{sec}}{\text{N} \cdot \text{m}}$$

$$\dot{W}_m = 25.3 \text{ kW} \qquad\qquad\qquad\qquad\qquad\qquad\qquad\qquad\qquad\qquad \dot{W}_m$$

{ This problem illustrates construction of velocity polygons and application of the angular momentum equation for a fixed control volume to an axial-flow machine under idealized conditions. }

EXAMPLE 11.2—Idealized Centrifugal Pump

Water at 150 gpm enters a centrifugal pump impeller axially through a 1.25 in. diameter inlet. The inlet velocity is axial and uniform. The impeller outlet diameter is 4 in. Flow leaves the impeller at 10 ft/sec relative to the radial blades. The impeller speed is 3450 rpm. Determine the impeller exit width, b_2, the minimum torque input to the impeller, and the minimum power required.

EXAMPLE PROBLEM 11.2

GIVEN: Flow as shown in the following figure: $V_{rb_2} = 10$ ft/sec, $Q = 150$ gpm.

FIND: (a) b_2.
(b) T_{shaft}.
(c) $\dot{W}_m$.

SOLUTION:
Apply the angular momentum equation to a fixed control volume.

Computing equations:
$$\vec{T}_{shaft} = \int_{CS} \vec{r} \times \vec{V} \rho \vec{V} \cdot d\vec{A} \tag{11.1a}$$

$$0 = \cancelto{0(2)}{\frac{\partial}{\partial t}} \int_{CV} \rho \, d\Psi + \int_{CS} \rho \vec{V} \cdot d\vec{A}$$

Assumptions: (1) Neglect torques due to body and surface forces
(2) Steady flow
(3) Uniform flow at inlet and outlet sections
(4) Incompressible flow

Then, from continuity,

$$0 = \{-|\rho V_1 \pi R_1^2|\} + \{|\rho V_{rb_2} 2\pi R_2 b_2|\}$$

or

$$\rho Q = \rho V_{rb_2} 2\pi R_2 b_2$$

so that

$$b_2 = \frac{Q}{2\pi R_2 V_{rb_2}} = \frac{1}{2\pi} \times \frac{150 \text{ gal}}{\text{min}} \times \frac{1}{2 \text{ in.}} \times \frac{\text{sec}}{10 \text{ ft}} \times \frac{\text{ft}^3}{7.48 \text{ gal}} \times \frac{\text{min}}{60 \text{ sec}} \times \frac{12 \text{ in.}}{\text{ft}}$$

$$b_2 = 0.0319 \text{ ft or } 0.383 \text{ in.} \qquad\qquad\qquad \underset{\longleftarrow}{b_2}$$

The axial inlet flow has no z component of angular momentum. From the angular momentum equation with uniform exit flow,

$$\hat{k}_2 T_{shaft} = \vec{r}_2 \times \vec{V}_2 \{|\rho Q|\}$$

At section ②,

$$\vec{r}_2 = R_2 \hat{e}_r$$

$$\vec{V}_2 = V_{rb_2} \hat{e}_r + \omega R_2 \hat{e}_\theta$$

so

$$\vec{r}_2 \times \vec{V}_2 = R_2(\omega R_2)\hat{k} = \omega R_2^2 \hat{k}$$

Thus,

$$T_{shaft} = \omega R_2^2 \rho Q = \frac{3450 \text{ rev}}{\text{min}} \times \frac{(2)^2 \text{ in.}^2}{} \times \frac{1.94 \text{ slug}}{\text{ft}^3} \times \frac{150 \text{ gal}}{\text{min}}$$

$$\times \frac{2\pi \text{ rad}}{\text{rev}} \times \frac{\text{min}^2}{3600 \text{ sec}^2} \times \frac{\text{ft}^3}{7.48 \text{ gal}} \times \frac{\text{ft}^2}{144 \text{ in.}^2} \times \frac{\text{lbf} \cdot \text{sec}^2}{\text{slug} \cdot \text{ft}}$$

$$T_{shaft} = 6.51 \text{ ft} \cdot \text{lbf} \qquad\qquad\qquad \underset{\longleftarrow}{T_{shaft}}$$

and

$$\dot{W}_m = \omega T_{shaft} = \frac{3450 \text{ rev}}{\text{min}} \times \frac{6.51 \text{ ft} \cdot \text{lbf}}{} \times \frac{2\pi \text{ rad}}{\text{rev}} \times \frac{\text{min}}{60 \text{ sec}} \times \frac{\text{hp} \cdot \text{sec}}{550 \text{ ft} \cdot \text{lbf}}$$

$$\dot{W}_m = 4.28 \text{ hp} \qquad\qquad\qquad \underset{\longleftarrow}{\dot{W}_m}$$

This problem illustrates application of the angular momentum equation for a fixed control volume to a centrifugal flow machine under idealized conditions and assuming perfect guidance. The actual input torque and power requirements may be larger than predicted by the theoretical analysis. However, the predicted head rise may not be achieved because of imperfect guidance of the flow leaving the impeller.

Figure 11.5 represents the flow through a simple centrifugal pump impeller. If the fluid enters the impeller with a purely radial absolute velocity, then the fluid entering the impeller has no angular momentum and V_{t_1} is identically zero.

With $V_{t_1} = 0$, the increase in head (from Eq. 11.2c) is given by

$$H = \frac{U_2 V_{t_2}}{g} \tag{11.4}$$

From the exit velocity triangle of Fig. 11.5c,

$$V_{t_2} = U_2 - V_{rb_2}\cos\beta_2 = U_2 - \frac{V_{n_2}}{\sin\beta_2}\cos\beta_2 = U_2 - V_{n_2}\cot\beta_2 \tag{11.5}$$

Then

$$H = \frac{U_2^2 - U_2 V_{n_2}\cot\beta_2}{g} \tag{11.6}$$

For an impeller of width w, the volume flow rate is

$$Q = \pi D_2 w V_{n_2} \tag{11.7}$$

To express the increase in head in terms of volume flow rate, we substitute for V_{n_2} in terms of Q from Eq. 11.7. Thus

$$H = \frac{U_2^2}{g} - \frac{U_2\cot\beta_2}{\pi D_2 w g}Q \tag{11.8a}$$

Equation 11.8a is of the form

$$H = C_1 - C_2 Q \tag{11.8b}$$

where constants C_1 and C_2 are functions of machine geometry and speed,

$$C_1 = \frac{U_2^2}{g} \quad \text{and} \quad C_2 = \frac{U_2\cot\beta}{\pi D_2 w g}$$

Thus Eq. 11.8a predicts a linear variation of head, H, with volume flow rate, Q.

Constant $C_1 = U_2^2/g$ represents the ideal head developed by the pump for zero flow rate; this is referred to as the *shutoff head*. The slope of the curve of head versus flow rate (the $H-Q$ curve) depends on the sign and magnitude of C_2.

For radial outlet vanes, $\beta_2 = 90°$ and $C_2 = 0$. The tangential component of the absolute velocity at the outlet is equal to the wheel speed and is independent of flow rate. From Eq. 11.8a, the ideal head is independent of flow rate. This characteristic $H-Q$ curve is plotted in Fig. 11.6.

If the vanes are *backward curved* (as shown in Fig. 11.5a), $\beta_2 < 90°$ and $C_2 > 0$. Then the absolute outlet speed is less than the wheel speed and it decreases in proportion to the flow rate. From Eq. 11.8a, the ideal head decreases linearly with increasing flow rate. The corresponding $H-Q$ curve is plotted in Fig. 11.6.

If the vanes are *forward curved*, then $\beta_2 > 90°$ and $C_2 < 0$. The absolute fluid velocity at the outlet is greater than the wheel speed and it increases as the flow rate increases. From Eq. 11.8a, the ideal head increases linearly with increasing flow rate. The corresponding $H-Q$ curve is plotted in Fig. 11.6.

The characteristics of a radial-flow machine can be altered by changing the outlet vane angle; the idealized model predicts the trends as the outlet vane angle is changed.

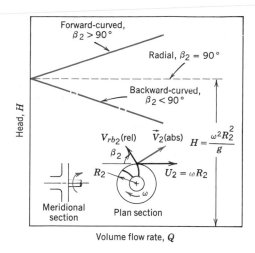

Fig. 11.6 Idealized relationship between head and volume flow rate for centrifugal pump with forward-curved, radial, and backward-curved impeller blades.

The predictions of the idealized angular momentum theory for a centrifugal pump are summarized in Fig. 11.6. Internal flow losses and deviations from perfect guidance that occur in a real machine reduce the actual head compared to the ideal predictions. The trends predicted by the idealized theory are compared with experimental results in the next section.

11-4 PERFORMANCE CHARACTERISTICS

To specify fluid machines for flow systems, the designer must know the pressure rise (or head), torque, power requirement, and efficiency of a machine. For a given machine each of these characteristics is a function of flow rate; the characteristics for similar machines depend upon size and operating speed. In this section we define performance characteristics for pumps and turbines and review experimentally measured trends for typical machines. We discuss dimensionless parameters to illustrate the similarities among families of machines and the trends in design features as functions of flow rate and head rise. We review scaling laws and present examples to illustrate their use. The section concludes with a discussion of cavitation and the net head that must be available at the inlet of a pump to assure satisfactory cavitation-free operation.

11-4.1 Performance Parameters

The idealized analyses presented in Section 11-3 are useful to predict trends and to approximate the design point performance of an energy absorbing or energy producing machine. However, the complete performance of a real machine, including operation at off-design conditions, must be determined experimentally.

To determine performance, a pump, fan, blower, or compressor must be set up on an instrumented test stand with the capability of measuring flow rate, speed, input torque, and pressure rise. The test must be performed according to a standardized procedure corresponding to the machine being tested [4, 5]. Measurements are made as flow rate is varied from shutoff (zero flow) to maximum delivery. Power input to the machine is determined from a calibrated motor or calculated from measured speed and torque, then efficiency is computed as illustrated in Example Problem 11.3. Finally, the calculated characteristics are plotted in the desired engineering units or nondimensionally. If appropriate, smooth curves may be faired through the plotted points or curve-fits may be made to the results, as illustrated in Example Problem 11.4.

EXAMPLE 11.3—Calculation of Pump Characteristics from Test Data

The flow system used to test a centrifugal pump at a nominal speed of 1750 rpm is shown. The liquid is water at 80 F, and the suction and discharge pipe diameters are 6 in. Data measured during the test are given in the table. The motor is supplied at 460 V, 3-phase, has a power factor of 0.875, and a constant efficiency of 90 percent.

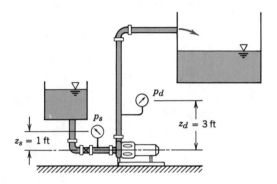

Rate of Flow (gpm)	Suction Pressure (psig)	Discharge Pressure (psig)	Motor Current (amp)	Pump Speed (rpm)
0	−3.7	53.3	18.0	1750
500	−4.2	48.3	26.2	1745
800	−4.7	42.3	31.0	1749
1000	−5.7	34.3	36.0	1750
1100	−6.2	31.3	37.0	1747
1200	−6.7	27.3	37.3	1752
1400	−7.7	15.3	39.0	1750
1500	−8.4	7.3	41.5	1753

Calculate the net head delivered and the pump efficiency at a volume flow rate of 1000 gpm. Plot the pump head, power, and efficiency as functions of volume flow rate.

EXAMPLE PROBLEM 11.3

GIVEN: Pump test flow system and data shown.

FIND: (a) Pump head and efficiency at $Q = 1000$ gpm.
      (b) Plot pump head, power input, and efficiency versus volume flow rate.

SOLUTION: Apply the energy equation to determine the power delivered to the fluid. Consider the ideal case ($\dot{Q} = 0$, $u_2 = u_1$).

$$\underset{\cancel{}}{\dot{Q}} - \dot{W}_s = \underset{=0(1)}{\cancel{\frac{\partial}{\partial t}\int_{CV} e\, \rho\, d\Psi}} + \int_{CS}\left(e + \frac{p}{\rho}\right)\rho\vec{V}\cdot d\vec{A}; \quad e = u + \frac{V^2}{2} + gz$$

Basic equation: (with $=0(2)$ marked over integral)

Assumptions: (1) $\dot{Q} = 0$
            (2) Steady flow
            (3) Uniform flow at each section
            (4) $u_2 = u_1$
            (5) $\bar{V}_2 = \bar{V}_1$
            (6) Correct all heads to the same elevation

Then

$$-\dot{W}_s = \dot{m}\left[\left(\frac{\bar{V}_2^2}{2} + gz_2 + \frac{p_2}{\rho}\right) - \left(\frac{\bar{V}_1^2}{2} + gz_1 + \frac{p_1}{\rho}\right)\right] = \dot{m}\frac{\Delta p}{\rho} = \rho Q\frac{\Delta p}{\rho} = Q\,\Delta p$$

Correct measured static pressures to the pump centerline (let subscript s represent the pump suction and subscript d represent the pump discharge):

$$p_1 = p_s + \rho g\, z_s$$

$$p_1 = \frac{-5.70\ \text{lbf}}{\text{in.}^2} + 1.94\ \frac{\text{slug}}{\text{ft}^3} \times 32.2\ \frac{\text{ft}}{\text{sec}^2} \times 1.0\ \text{ft} \times \frac{\text{lbf} \cdot \text{sec}^2}{\text{slug} \cdot \text{ft}} \times \frac{\text{ft}^2}{144\ \text{in.}^2} = -5.27\ \text{psig}$$

and

$$p_2 = p_d + \rho g\, z_d$$

$$p_2 = \frac{34.3\ \text{lbf}}{\text{in.}^2} + 1.94\ \frac{\text{slug}}{\text{ft}^3} \times 32.2\ \frac{\text{ft}}{\text{sec}^2} \times 3.0\ \text{ft} \times \frac{\text{lbf} \cdot \text{sec}^2}{\text{slug} \cdot \text{ft}} \times \frac{\text{ft}^2}{144\ \text{in.}^2} = 35.6\ \text{psig}$$

Compute the mechanical power delivered to the fluid:

$$\dot{W}_{in} = -\dot{W}_s = Q\, \Delta p$$

$$= \frac{1000\ \frac{\text{gal}}{\text{min}}} {} \times \frac{[35.6 - (-5.27)]\ \text{lbf}}{\text{in.}^2} \times \frac{\text{ft}^3}{7.48\ \text{gal}} \times \frac{\text{min}}{60\ \text{sec}} \times \frac{144\ \text{in.}^2}{\text{ft}^2} \times \frac{\text{hp} \cdot \text{sec}}{550\ \text{ft} \cdot \text{bf}}$$

$$\dot{W}_{in} = 23.8\ \text{hp}$$

Calculate the motor power output (pump input power) from electrical information:

$$\mathscr{P}_{in} = \eta\, \sqrt{3}(PF)EI$$

$$\mathscr{P}_{in} = 0.90 \times \sqrt{3} \times 0.875 \times 460\ \text{V} \times 36.0\ \text{A} \times \frac{\text{W}}{\text{VA}} \times \frac{\text{hp}}{746\ \text{W}} = 30.3\ \text{hp}$$

The corresponding pump efficiency is

$$\eta_p = \frac{\dot{W}_{in}}{\mathscr{P}_{in}} = \frac{23.8\ \text{hp}}{30.3\ \text{hp}} = 0.785 \quad \text{or} \quad 78.5\ \text{percent}$$

Results from similar calculations at the other volume flow rates are plotted below:

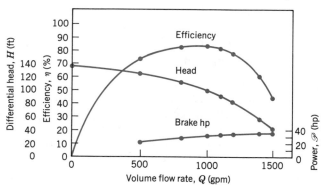

At $Q = 1000$ gpm the net positive suction head (see Section 11-4.4) is calculated as the difference between the absolute stagnation pressure at the inlet and the vapor pressure of the liquid being pumped.

The absolute stagnation pressure at the inlet is

$$p_{01} = p_{atm} + p_{1,gage} + \frac{1}{2}\rho \bar{V}_1^2$$

If we assume $D = 6$ in., then

$$\bar{V} = \frac{Q}{A} = \frac{1000\ \text{gal}}{\text{min}} \times \frac{\text{ft}^3}{7.48\ \text{gal}} \times \frac{4}{\pi(6)^2\ \text{in.}^2} \times \frac{144\ \text{in.}^2}{\text{ft}^2} \times \frac{\text{min}}{60\ \text{sec}} = 11.3\ \text{ft/sec}$$

and

$$\frac{1}{2}\rho \bar{V}_1^2 = \frac{1}{2} \times 1.94\ \frac{\text{slug}}{\text{ft}^3} \times (11.3)^2\ \frac{\text{ft}^2}{\text{sec}^2} \times \frac{\text{lbf} \cdot \text{sec}^2}{\text{slug} \cdot \text{ft}} \times \frac{\text{ft}^2}{144\ \text{in.}^2} = 0.860\ \text{lbf/in.}^2$$

Thus, upon substituting,

$$p_{01} = \frac{(14.7 - 5.27 + 0.860)}{\text{in.}^2} \frac{\text{lbf}}{} = 10.3 \text{ psia}$$

From steam tables, at 80 F, $p_v = 0.947$ psia, so

$$NPSH = \frac{p_{01} - p_v}{\rho g} = \frac{(10.3 - 0.947)}{\text{in.}^2} \frac{\text{lbf}}{} \times \frac{\text{ft}^3}{1.94 \text{ slug}} \times \frac{\text{sec}^2}{32.2 \text{ ft}} \times \frac{\text{slug} \cdot \text{ft}}{\text{lbf} \cdot \text{sec}^2} \times \frac{144 \text{ in.}^2}{\text{ft}^2}$$

$$NPSH = 21.6 \text{ ft} \qquad\qquad\qquad\qquad\qquad\qquad\qquad\qquad\qquad NPSH$$

$\left\{\begin{array}{l}\text{This problem illustrates the reduction of measured data to obtain a performance curve} \\ \text{for a pump. Peak efficiency for this pump occurs between 675 and 1150 gpm.}\end{array}\right\}$

The basic procedure used to calculate machine performance was illustrated for a centrifugal pump in Example Problem 11.3. The difference in static pressures between the pump suction and discharge was used to calculate the head rise produced by the pump. For pumps, dynamic pressure typically is a small fraction of the head rise developed by the pump, so it may be neglected compared to the head rise.

The test and data reduction procedures for fans, blowers, and compressors are basically the same as for centrifugal pumps. However, blowers, and especially fans, add relatively small amounts of static head to gas or vapor flows. For these machines, the dynamic head may increase from inlet to discharge, and it may be appreciable compared to the static head rise. For these reasons, it is important to state clearly the basis on which performance calculations are made. Standard definitions are available for machine efficiency based on either the static-to-static pressure rise or the static-to-total pressure rise [6].

EXAMPLE 11.4—Curve-Fit to Pump Performance Data

Pump test data were given and performance was calculated in Example Problem 11.3. Fit a parabolic curve, $H = H_0 - AQ^2$, to these calculated pump performance results and compare the fitted curve to the measured data.

EXAMPLE PROBLEM 11.4

GIVEN: Pump test data and performance calculated in Example Problem 11.3.

FIND: (a)   Fit a parabolic curve, $H = H_0 - AQ^2$, to the pump performance data.
　　　　 (b)   Compare the curve fit to the calculated performance.

SOLUTION:
The curve fit may be obtained by fitting a linear curve to H versus Q^2. Tabulating,

From calculated performance:			From the curve fit:	
Q (gpm)	Q^2 (gpm^2)	H (ft)	H (ft)	Error (%)
0	0	133	133	0
500	25×10^4	123	125	1.2
800	64×10^4	110	109	−1.4
1000	100×10^4	94.2	93.7	−0.5
1200	144×10^4	80.4	75.6	−5.9
1400	196×10^4	55.0	54.3	−1.3
1500	225×10^4	38.2	42.4	10.9

Using the method of least squares, the equation for the fitted curve is obtained as

$$H\,[\text{ft}] = 135 - 4.11 \times 10^5 \, [Q\,(\text{gpm})]^2$$

with coefficient of determination $r^2 = 0.993$.

> The coefficient of determination, $r^2 = 0.993$, indicates a very good fit to the measured data. As shown in the table, the percentage errors are small, except at the 1200 and 1500 gpm points. The differences between the fitted curve and calculated values would not be noticeable on a plot.

Typical characteristic curves for a centrifugal pump tested at constant speed were shown qualitatively in Fig. 7.4;[4] the head versus capacity curve is reproduced in Fig. 11.7 to compare with characteristics predicted by the idealized analysis. Figure 11.7 shows that the head at any flow rate in the real machine may be significantly lower than predicted by the idealized analysis. Some of the causes are:

1. At very low flow rate some fluid recirculates in the impeller.
2. Friction loss and leakage loss both increase with flow rate.
3. "Shock loss" results from mismatch between relative velocity and the impeller blade at the inlet.[5]

Curves such as those in Figs. 7.4 and 11.7 are measured at constant speed with a single impeller diameter. It is common practice to vary pump capacity by changing the impeller size in a given casing. To present information compactly, it is desirable to include data from tests of several impeller diameters on a single graph, as shown in Fig. 11.8. As before, for each diameter, head is plotted versus flow rate; each curve is labeled with the corresponding diameter. Efficiency contours are plotted by joining points having the same constant efficiency. Power requirement contours also are plotted. Finally, the *NPSH* requirements are shown for the extreme diameters; in Fig. 11.8, the 8 in. impeller curve would lie between the 6 in. and 10 in. impeller curves.

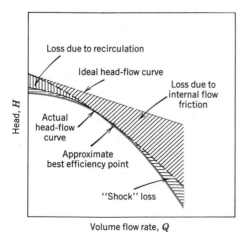

Volume flow rate, Q

Fig. 11.7 Comparison of ideal and actual head-flow curves for a centrifugal pump with backward-curved impeller blades [7].

---

[4] The only important pump characteristic not shown in Fig. 7.4 is the net positive suction head (*NPSH*) required to prevent cavitation. Cavitation and *NPSH* will be treated in Section 11-4.4.

[5] This loss is largest at high and low flow rates; it decreases essentially to zero as optimum operating conditions are approached [8].

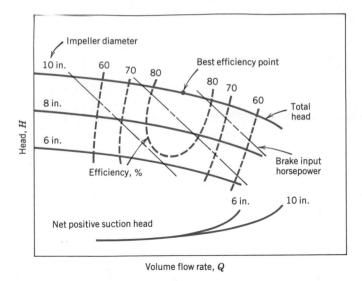

Fig. 11.8 Typical pump performance curves from tests with three impeller diameters at constant speed [7].

For this typical machine, head is a maximum at shutoff and decreases continuously as flow rate increases. Input power is minimum at shutoff and increases as delivery is increased. Consequently, to minimize the starting load, it may be advisable to start the pump with the outlet valve closed. (However, the valve may not be left closed for long, lest the pump overheat as energy dissipated by friction is transferred to the water in the housing.) Pump efficiency increases with capacity until the *best efficiency point* (BEP) is reached, then decreases as flow rate is increased. For minimum energy consumption, it is desirable to operate as close to BEP as possible.

Centrifugal pumps may be combined in parallel to deliver greater flow or in series to deliver greater head. A number of manufacturers build multi-stage pumps, which essentially are several pumps arranged in series within a single casing. Pumps and blowers usually are tested at several constant speeds. Common practice is to drive machines with electric motors at nearly constant speed, but in some system applications impressive energy savings can result from variable-speed operation. These pump application topics are discussed in Section 11-5.

The test procedure for turbines is similar to that for pumps, except that a dynamometer is used to absorb the turbine power output while speed and torque are measured. Turbines usually are intended to operate at a constant speed that is a fraction or multiple of the electric power frequency. Therefore turbine tests are run at constant speed under varying load, while water usage is measured and efficiency is calculated.

The impulse turbine is a relatively simple turbomachine, so we use it to illustrate typical test results. Impulse turbines are chosen when the head available exceeds about 300 m. Most impulse turbines used today are improved versions of the *Pelton wheel* developed in the 1880s by American mining engineer Lester Pelton [9]. An impulse turbine is supplied with water under high head through a long conduit called a *penstock*. The water is accelerated through a nozzle and discharges as a high-speed free jet at atmospheric pressure. The jet strikes deflecting buckets attached to the rim of a rotating wheel (Fig. 11.3*a*). Its kinetic energy is given up as it is turned by the buckets. Turbine output is controlled at essentially constant jet speed by changing the

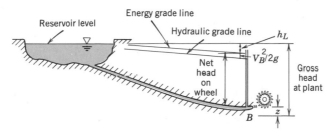

Fig. 11.9   Schematic of impulse turbine installation, showing definitions of gross and net heads [8].

flow rate of water striking the buckets. A variable-area nozzle may be used to make small and gradual changes in turbine output. Larger or more rapid changes must be accomplished by means of jet deflectors, or auxiliary nozzles, to avoid sudden changes in flow speed and the resulting high pressures in the long water column in the penstock. Water discharged from the wheel at relatively low speed falls into the tailrace. The tailrace level is set to avoid submerging the wheel during flooded conditions. When large amounts of water are available, additional power can be obtained by connecting two wheels to a single shaft or by arranging two or more jets to strike a single wheel.

Figure 11.9 illustrates an impulse turbine installation and the definitions of gross and net head [8]. The *gross head* available is the difference between the levels in the supply reservoir and the tailrace. The effective or *net head, H*, used to calculate efficiency, is taken as the difference between the total head at the nozzle entrance and the elevation of the nozzle centerline [8]. Thus, the impulse turbine is charged with the loss in the nozzle and with any residual kinetic energy of the water that falls into the tailrace. However, it is not charged with the elevation of the nozzle above the tailrace. (For high head installations this correction is a small percentage of the gross head.) In practice, the penstock usually is sized so that at rated power the net head is 85-95 percent of the gross head.

In addition to nozzle loss, windage, bearing friction, and surface friction between the jet and bucket reduce performance compared to the ideal, frictionless case. Figure 11.10 shows typical results from tests performed at constant head.

The peak efficiency of the impulse turbine corresponds to the peak power, since the tests are performed at constant head and flow rate. For the ideal turbine, as

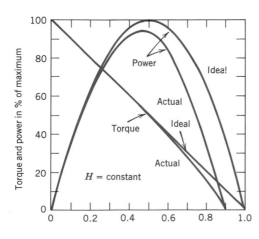

Fig. 11.10   Ideal and actual variable-speed performance for an impulse turbine [10].

shown in Example Problem 11.5, this occurs when the wheel speed is half the jet speed. In actual installations, peak efficiency occurs at a wheel speed only slightly less than half the jet speed. This condition fixes the wheel speed once the jet speed is determined for a given installation. For large units, overall efficiency may be as high as 88 percent [11].

EXAMPLE 11.5—Optimum Speed for Impulse Turbine

A Pelton wheel is a form of impluse turbine well adapted to situations of high head and low flow rate. Consider the Pelton wheel and single jet arrangement shown, in which the jet stream strikes the bucket tangentially and is turned through angle θ. Obtain an expression for the torque exerted by the water stream on the wheel and the corresponding power output. Show that the power is a maximum when the bucket speed, $U = R\omega$, is half the jet speed, V.

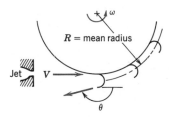

EXAMPLE PROBLEM 11.5

GIVEN: Pelton wheel and single jet shown.

FIND: (a) Expression for torque exerted on the wheel.
    (b) Expression for power output.
    (c) Show that power is a maximum when $U = V/2$.

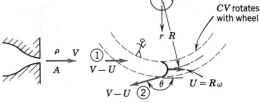

SOLUTION:
Apply the angular momentum equation to the rotating CV shown.

Basic Equation:

$$\overset{=0(1)}{\cancel{\vec{r}\times\vec{F}_s}} + \overset{=0(2)}{\cancel{\int_{CV}\vec{r}\times\vec{g}\,\rho\,d\forall}} + \vec{T}_{\text{shaft}} - \int_{CV}\vec{r}\times[2\vec{\omega}\times\vec{V}_{xyz}+\vec{\omega}\times(\vec{\omega}\times\vec{r})+\overset{\approx0(3)}{\cancel{\dot{\vec{\omega}}\times\vec{r}}}]\rho\,d\forall$$

$$= \overset{=0(4)}{\cancel{\frac{\partial}{\partial t}\int_{CV}\vec{r}\times\vec{V}_{xyz}\,\rho\,d\forall}} + \int_{CS}\vec{r}\times\vec{V}_{xyz}\,\rho\vec{V}_{xyz}\cdot d\vec{A} \qquad (4.53)$$

Assumptions: (1) Neglect torque due to surface forces
    (2) Neglect torque due to body forces
    (3) Neglect mass of water on wheel
    (4) Steady flow with respect to wheel
    (5) All water that issues from the nozzle acts upon the buckets
    (6) Bucket height is small compared to R, hence $r_1 \approx r_2 \approx R$

(7) Uniform flow at each section
(8) No change in jet speed relative to bucket

Then, since all water from the jet crosses the buckets,

$$\vec{T}_{shaft} = \vec{r}_1 \times \vec{V}_1 \{-|\rho V A|\} + \vec{r}_2 \times \vec{V}_2 \{+|\rho V A|\}$$

$$\vec{r}_1 = R\hat{e}_r \qquad\qquad \vec{r}_2 = R\hat{e}_r$$

$$\vec{V}_1 = (V-U)\hat{e}_\theta \qquad\qquad \vec{V}_2 = (V-U)\cos\theta\,\hat{e}_\theta$$

$$T_{shaft}\hat{k} = R(V-U)\hat{k}(-\rho V A) + R(V-U)\cos\theta\,\hat{k}(\rho V A)$$

so that finally

$$T_{shaft}\hat{k} = -R(1-\cos\theta)\rho V A(V-U)\hat{k}$$

This is the external torque of the shaft on the control volume, i.e., on the wheel. The torque exerted by the water on the wheel is equal and opposite,

$$\vec{T}_{out} = -\vec{T}_{shaft} = R(1-\cos\theta)\rho V A(V-U)\hat{k} = \rho Q R(V-U)(1-\cos\theta)\hat{k} \qquad \underleftarrow{\vec{T}_{out}}$$

The corresponding power output is

$$\dot{W}_{out} = \vec{\omega}\cdot\vec{T}_{out} = R\omega(1-\cos\theta)\rho V A(V-U) = \rho Q U(V-U)(1-\cos\theta) \qquad \underleftarrow{\dot{W}_{out}}$$

To find the condition for maximum power, differentiate the expression for power with respect to wheel speed, U, and set the result equal to zero. Thus

$$\frac{d\dot{W}}{dU} = \rho Q(V-U)(1-\cos\theta) + \rho Q U(-1)(1-\cos\theta) = 0$$

$$\therefore (V-U) - U = V - 2U = 0$$

Thus for maximum power, $U = V/2$.

Note: Turning the flow through $\theta = 180°$ would give maximum power with $U = V/2$. In practice, it is possible to deflect the jet stream through angles up to 165°. With $\theta = 165°$, $1-\cos\theta \approx 1.97$, or about 1.5 percent below the value for maximum power.

> The purpose of this problem was to apply the angular momentum principle to a rotating control volume and to show that, for an ideal impulse turbine, peak power occurs when the wheel speed is half the jet speed.

In practice, hydraulic turbines usually are run at constant speed, and output is varied by changing the opening area of the needle valve jet nozzle. Nozzle loss increases slightly and mechanical losses become a larger fraction of output as the valve is closed, so efficiency drops sharply at low load, as shown in Fig. 11.11; for this Pelton wheel, efficiency remains above 85 percent from 40 to 110 percent of full load.

At lower heads, reaction turbines provide better efficiency than impulse turbines. In contrast to flow in a centrifugal pump, flow in a reaction turbine enters the rotor at the largest (outer) radial section and discharges at the smaller (inner) radial section after transferring most of its energy to the rotor. Reaction turbines tend to be high flow, low head machines. A typical reaction turbine installation is shown schematically in Fig. 11.12, where the terminology used to define the heads is indicated.

Reaction turbines flow full of water. Consequently, it is possible to use a diffuser or draft tube to regain a fraction of the kinetic energy that remains in water leaving the

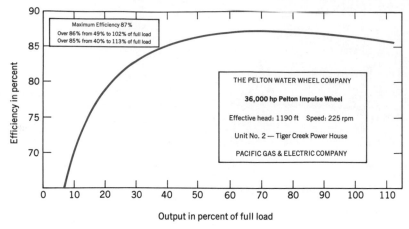

Fig. 11.11 Relation between efficiency and output for a typical Pelton water turbine (adapted from [11]).

rotor. The draft tube forms an integral part of the installation design. As shown in Fig. 11.12, the *gross head* available is the difference between the supply reservoir head and the tailrace head. The *effective head* or *net head*, H, used to calculate efficiency, is the difference between the elevation of the energy grade line just upstream of the turbine and that of the tailrace; the benefit to be gained from the draft tube in reducing discharge velocity is apparent. Viewed in another way, the effect of the draft tube is to lower the pressure at the turbine discharge; this increases the change in pressure head across the turbine over what it would be without the draft tube.

An efficient mixed-flow turbine runner was developed by James B. Francis during a careful series of experiments at Lowell, Massachusetts during the 1840s [9]. An efficient axial-flow propeller turbine, with adjustable blades, was developed by German professor Victor Kaplan between 1910 and 1924. The *Francis Turbine* (Fig. 11.3*b*) usually is chosen when $15 \le H \le 300$ m, and the *Kaplan turbine* (Fig. 11.3*c*) usually is chosen for heads of 15 m or less. Performance of reaction turbines may be measured in the same manner as performance of the impulse turbine. However, because the gross heads are less, any change in water level during operation is more significant. Consequently, measurements must be made at a series of heads to completely define the performance of a reaction turbine.

An example of the data presentation for a reaction turbine is given in Fig 11.13, where efficiency is shown at various output powers for a series of constant heads

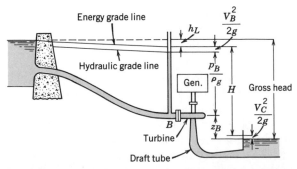

Fig. 11.12 Schematic of typical reaction turbine installation, showing definitions of head terminology [8].

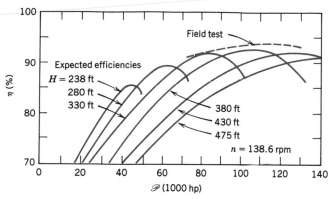

Fig. 11.13   Performance of typical reaction turbine as predicted by model tests (expected efficiencies) and confirmed by field test [10].

[10]. The reaction turbine has higher maximum efficiency than the impulse turbine, but efficiency varies more sharply with load.

As a sharp-eyed reader, you will have noticed that Fig. 11.13 contained both model-test (expected efficiencies) and full-scale results. You should be asking, "How are model tests designed and conducted, and how are model test results scaled to predict prototype performance?" To answer these questions, read on.

11-4.2   Dimensional Analysis and Specific Speed

Dimensional analysis for turbomachines was introduced in Chapter 7, where dimensionless flow, head, and power coefficients were derived in generalized form. It was shown that the independent parameters were the flow coefficient and a form of Reynolds number. The dependent parameters were the head and power coefficients.

Our objective here is to develop the forms of dimensionless coefficients in common use, and to give examples illustrating their use in selecting a machine type, designing model tests, and scaling results. Since we developed an idealized theory for turbomachines in Section 11-3, we can gain additional physical insight by developing dimensionless coefficients directly from the resulting computing equations.

The *flow coefficient*, Φ, is defined by normalizing the volume flow rate using the exit area and the wheel speed at the outlet. Thus

$$\Phi = \frac{Q}{A_2 U_2} = \frac{V_{n_2}}{U_2} \tag{11.9}$$

where V_{n_2} is the velocity component perpendicular to the exit area. This component is also referred to as the *meridional velocity* at the wheel exit plane. It appears in true projection in the *meridional plane*, which is any radial cross-section through the centerline of a machine.

A dimensionless *head coefficient*, Ψ, may be obtained by normalizing the head, H, (Eq. 11.2c) with U_2^2/g. Thus

$$\Psi = \frac{gH}{U_2^2} \tag{11.10}$$

A dimensionless *torque coefficient*, τ, may be obtained by normalizing the torque, T, (Eq. 11.1c) with $\rho A_2 U_2^2 R_2$. Thus

$$\tau = \frac{T}{\rho A_2 U_2^2 R_2} \tag{11.11}$$

Finally, the dimensionless *power coefficient*, Π, is obtained by normalizing the power $\dot{W}$, (Eq. 11.2b) with $\dot{m} U_2^2 = \rho Q U_2^2$. Thus

$$\Pi = \frac{\dot{W}}{\rho Q U_2^2} = \frac{\dot{W}}{\rho \omega^2 Q R_2^2} \tag{11.12}$$

For pumps, mechanical input power exceeds hydraulic power, and the efficiency is defined as $\eta_p = \dot{W}_h / \dot{W}_m$ (Eq. 11.3a). Hence

$$\dot{W}_m = T\omega = \frac{1}{\eta_p} \dot{W}_h = \frac{\rho Q g H}{\eta_p} \tag{11.13}$$

Introducing dimensionless coefficients Φ (Eq. 11.9), Ψ (Eq. 11.10), and τ (Eq. 11.11) into Eq. 11.13, we obtain an analogous relation among the dimensionless coefficients as

$$\tau = \frac{\Psi \Phi}{\eta_p} \tag{11.14}$$

For turbines, mechanical output power is less than hydraulic power, and the efficiency is defined as $\eta_t = \dot{W}_m / \dot{W}_h$ (Eq. 11.3b). Hence,

$$\dot{W}_m = T\omega = \eta_t \dot{W}_h = \eta_t \rho Q g H \tag{11.15}$$

Introducing dimensionless coefficients Φ, Ψ, and τ into Eq. 11.15, we obtain an analogous relation among the dimensionless coefficients as

$$\tau = \eta_t \Psi \Phi \tag{11.16}$$

The dimensionless coefficients form the basis for designing model tests and scaling the results. As shown in Chapter 7, the flow coefficient is treated as the independent parameter. Then, if viscous effects are neglected, the head, torque, and power coefficients are treated as multiple dependent parameters. Under these assumptions, dynamic similarity is achieved when the flow coefficient is matched between model and prototype machines.

As discussed in Chapter 7, a useful parameter called *specific speed* can be obtained by combining the flow and head coefficients and eliminating the machine size (Eq. 7.6). The result was

$$N_s = \frac{\omega Q^{1/2}}{H^{3/4}} \tag{11.17}$$

When head is expressed as energy per unit mass (i.e., with dimensions equivalent to L^2/t^2, or g times head in height of liquid), and ω is expressed in radians per second, the specific speed defined by Eq. 11.17 is dimensionless.

Although specific speed is a dimensionless parameter, it is common practice to use a convenient but inconsistent set of units to specify the variables, ω, Q, and H. When this is done, the specific speed is not a unitless parameter and the magnitude of the specific speed depends on the units used to calculate it. Typical units used in U.S. engineering practice for pumps are rpm for ω, gpm for Q, and feet (energy per unit weight) for H. Values of the dimensionless specific speed, N_s (Eq. 11.17), must be multiplied by 2733 to obtain the values of specific speed corresponding to this commonly used but inconsistent set of units (see Example Problem 11.6).

Power output from a hydraulic turbine is proportional to the product of volume flow rate and head, $\mathscr{P} \propto QH$; thus the dimensions of Q are $\mathscr{P}/H$. When written in terms of turbine power output, the specific speed becomes

$$N_s = \frac{\omega \mathscr{P}^{1/2}}{H^{5/4}}$$

Typical units used in U.S. engineering practice for hydraulic turbines are rpm for ω, horsepower for $\mathscr{P}$, and feet for H.

Specific speed may be thought of as the operating speed at which a machine produces unit head at unit volume flow rate. Holding specific speed constant describes all operating conditions of geometrically similar machines with similar flow conditions.

It is customary to characterize a machine by its specific speed at the design point. This specific speed has been found to characterize the hydraulic design features of a machine. Low specific speeds are produced efficiently by radial-flow machines. High specific speeds are produced efficiently by axial-flow machines. For a specified head and flow rate, one can choose either a low specific speed machine (which operates at low speed) or a high specific speed machine (which operates at higher speed).

Typical proportions for commercial pump designs and their variation with dimensionless specific speed are shown in Fig. 11.14. In this figure, the size of each machine has been adjusted to give the same head and flowrate when rotating at a speed corresponding to the specific speed. Thus it can be seen that if the machine's size and weight are critical, one should choose a higher specific speed.

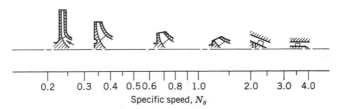

Fig. 11.14 Typical geometric proportions of commercial pumps as they vary with dimensionless specific speed [12].

The corresponding efficiency trends for typical pumps are shown in Fig. 11.15. Figure 11.14 shows the trend from radial (purely centrifugal pumps), through mixed-flow, to axial-flow geometries as specific speed increases. Figure 11.15 shows that pump capacity generally increases as specific speed increases; the figure also shows that at any given specific speed, efficiency is greater for large pumps than for small ones. Physically this *scale effect* means that viscous losses become less important as the pump size is increased.

Characteristic proportions of hydraulic turbines also are correlated by specific speed, as shown in Fig. 11.16. As in Fig. 11.14, the machine size has been scaled in this illustration to deliver approximately the same power at unit head when rotating at a speed equal to the specific speed. The corresponding efficiency trends for typical turbine types are shown in Fig. 11.17.

Several variations of specific speed, calculated directly from engineering units, are widely used in practice. The most commonly used forms of specific speed for pumps are defined and compared in Example Problem 11.6.

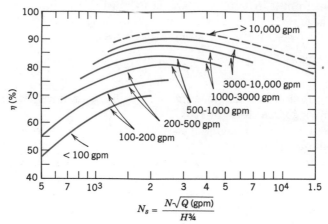

Fig. 11.15 Average efficiencies of commercial pumps as they vary with specific speed and pump size [10]. (Note the specific speed parameter is calculated in U.S. customary units of rpm, gpm, and ft, so it is *not* dimensionless.)

$$N_s = \frac{N\sqrt{Q\,(\text{gpm})}}{H^{3/4}}$$

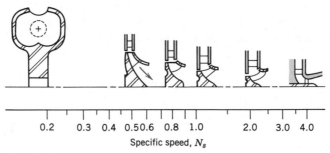

Fig. 11.16 Typical geometric proportions of commercial hydraulic turbines as they vary with dimensionless specific speed [12].

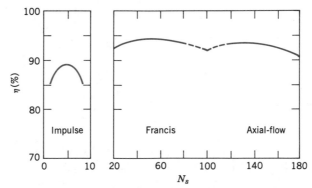

Fig. 11.17 Average efficiencies of commercial hydraulic turbines as they vary with specific speed [10]. (Note the specific speed parameter is calculated in U. S. Customary units of rpm, hp, and ft, so it is *not* dimensionless.)

EXAMPLE 11.6—Comparison of Specific Speed Definitions

At the best efficiency point, a centrifugal pump, with impeller diameter $D = 8$ in., produces $H = 21.9$ ft at $Q = 300$ gpm with $\omega = 1170$ rpm. Compute the corresponding specific speeds using: (a) U.S. customary units, (b) SI units (rad/sec, m³/sec, m²/sec²), and (c) European units (rev/sec, m³/sec, m²/sec²). Develop conversion factors to relate the specific speeds.

EXAMPLE PROBLEM 11.6

GIVEN: Centrifugal pump at best efficiency point (BEP). Assume the pump characteristics are $H = 21.9$ ft, $Q = 300$ gpm, and $\omega = 1170$ rpm.

FIND: (a) Evaluate the specific speed in U.S. customary units.
(b) Compute the specific speed in SI units.
(c) Calculate the specific speed in European units.
(d) Develop conversion factors to relate the specific speeds.

SOLUTION:

Computing equation: $\quad N_s = \dfrac{\omega Q^{1/2}}{H^{3/4}}$

From the given information, the specific speed in U.S. customary units is

$$N_s(\text{US}) = \frac{1170 \text{ rpm}}{} \times (300)^{1/2} \text{ gpm}^{1/2} \times \frac{1}{(21.9)^{3/4} \text{ ft}^{3/4}} = 2000 \qquad\qquad N_s(\text{US})$$

Convert information to SI units:

$$\omega = \frac{1170 \text{ rev}}{\text{min}} \times \frac{2\pi \text{ rad}}{\text{rev}} \times \frac{\text{min}}{60 \text{ sec}} = 123 \text{ rad/sec}$$

$$Q = \frac{300 \text{ gal}}{\text{min}} \times \frac{\text{ft}^3}{7.48 \text{ gal}} \times \frac{\text{min}}{60 \text{ sec}} \times \frac{(0.305)^3 \text{ m}^3}{\text{ft}^3} = 0.0190 \text{ m}^3/\text{sec}$$

$$H = \frac{21.9 \text{ ft}}{} \times \frac{0.305 \text{ m}}{\text{ft}} = 6.68 \text{ m}$$

The energy per unit mass is

$$gH = \frac{9.81 \text{ m}}{\text{sec}^2} \times \frac{6.68 \text{ m}}{} = 65.5 \text{ m}^2/\text{sec}^2$$

The specific speed in SI units is

$$N_s(\text{SI}) = \frac{123 \text{ rad}}{\text{sec}} \times \frac{(0.0190)^{1/2} \text{ m}^{3/2}}{\text{sec}^{1/2}} \times \frac{(\text{sec}^2)^{3/4}}{(65.5)^{3/4} (\text{m}^2)^{3/4}} = 0.736 \qquad\qquad N_s(\text{SI})$$

Convert the operating speed to hertz:

$$\omega = \frac{1170 \text{ rev}}{\text{min}} \times \frac{\text{min}}{60 \text{ sec}} \times \frac{\text{Hz} \cdot \text{sec}}{\text{rev}} = 19.5 \text{ Hz}$$

Finally, the specific speed in European units is

$$N_s(\text{Eur}) = \frac{19.5 \text{ Hz}}{} \times \frac{(0.0190)^{1/2} \text{ m}^{3/2}}{\text{sec}^{1/2}} \times \frac{(\text{sec}^2)^{3/4}}{(65.5)^{3/4} (\text{m}^2)^{3/4}} = 0.117 \qquad\qquad N_s(\text{Eur})$$

To relate the specific speeds, form ratios:

$$\frac{N_s(\text{US})}{N_s(\text{Eur})} = \frac{2000}{0.117} = 17,100$$

$$\frac{N_s(\text{US})}{N_s(\text{SI})} = \frac{2000}{0.736} = 2720$$

> The purpose of this problem is to demonstrate the method used to calculate specific speed for pumps from each of three commonly used definitions and to compare the results. (Three significant figures have been used for all calculations in this example. Slightly different numerical values might be obtained if more significant figures were carried in intermediate calculations.)

11-4.3 Similarity Rules

Pump manufacturers offer a limited number of casing sizes and designs. Frequently, casings of different sizes are developed from a common design by increasing or decreasing all dimensions by the same scale ratio. Additional variation in characteristic curves may be obtained by varying the operating speed or by changing the impeller size within a given pump housing. The dimensionless parameters developed in Chapter 7 form the basis for predicting changes in performance that result from changes in pump size, operating speed, or impeller diameter.

To achieve dynamic similarity requires geometric and kinematic similarity. Assuming similar pumps and flow fields, and neglecting viscous effects, as shown in Chapter 7, dynamic similarity is obtained when the dimensionless flow coefficient is held constant. Dynamically similar operation is assured when two flow conditions satisfy the relation

$$\frac{Q_1}{\omega_1 D_1^3} = \frac{Q_2}{\omega_2 D_2^3} \tag{11.18a}$$

The dimensionless head and power coefficients depend only on the flow coefficient, i.e.,

$$\frac{H}{\omega^2 D^2} = f_1\left(\frac{Q}{\omega D^3}\right) \quad \text{and} \quad \frac{\mathscr{P}}{\rho \omega^3 D^5} = f_2\left(\frac{Q}{\omega D^3}\right)$$

When this is true, as shown in Example Problem 7.6, pump characteristics at a new condition (subscript 2) may be related to those at an old condition (subscript 1) by

$$\frac{H_1}{\omega_1^2 D_1^2} = \frac{H_2}{\omega_2^2 D_2^2} \tag{11.18b}$$

and

$$\frac{\mathscr{P}_1}{\rho_1 \omega_1^3 D_1^5} = \frac{\mathscr{P}_2}{\rho_2 \omega_2^3 D_2^5} \tag{11.18c}$$

These scaling relationships may be used to predict the effects of changes in pump operating speed, pump size, or impeller diameter within a given housing.

The simplest situation is when only the pump speed changes. Then geometric similarity is assured. Kinematic similarity holds if there is no cavitation; flows are then dynamically similar when the flow coefficients are matched. For this case of speed change with fixed diameter, Eqs. 11.18 become

$$\frac{Q_2}{Q_1} = \frac{\omega_2}{\omega_1} \tag{11.19a}$$

$$\frac{H_2}{H_1} = \left(\frac{\omega_2}{\omega_1}\right)^2 \tag{11.19b}$$

$$\frac{\mathscr{P}_2}{\mathscr{P}_1} = \left(\frac{\omega_2}{\omega_1}\right)^3 \tag{11.19c}$$

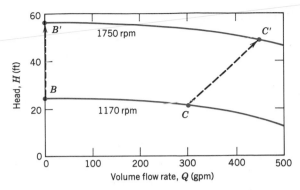

Fig. 11.18 Schematic of a pump performance curve, illustrating the effect of a change in pump operating speed.

In Example Problem 11.4, it was shown that a pump performance curve can be modeled within engineering accuracy by the parabolic relationship, $H = H_0 - AQ^2$. Since this representation contains two parameters, the pump curve for the new operating condition may be derived by scaling any two points from the performance curve measured at the original operating condition. Usually, the shutoff condition and the best efficiency point are chosen for scaling. These points are represented by points B and C in Fig. 11.18.

As shown by Eq. 11.19a, the flow rate increases by the ratio of operating speeds, so

$$Q_{B'} = \frac{\omega_2}{\omega_1} Q_B = 0 \quad \text{and} \quad Q_{C'} = \frac{\omega_2}{\omega_1} Q_C$$

Thus point B' is located directly above point B and point C' moves to the right of point C.

The head increases by the square of the speed ratio, so

$$H_{B'} = H_B \left(\frac{\omega_2}{\omega_1} \right)^2 \quad \text{and} \quad H_{C'} = H_C \left(\frac{\omega_2}{\omega_1} \right)^2$$

Points C and C', where dynamically similar flow conditions are present, are termed *homologous* points for the pump.

It is easy to show (see Example Problem 11.7) that the resulting performance curve is given by

$$H' = H_0' - AQ^2 \tag{11.20}$$

where A is the constant from the original curve.

Efficiency remains relatively constant between dynamically similar operating points when only the pump operating speed is changed. Application of these ideas is illustrated in Example Problem 11.7.

EXAMPLE 11.7—Scaling Pump Performance Curves

When operated at $\omega = 1170$ rpm, a centrifugal pump, with impeller diameter $D = 8$ in., has shutoff head $H_0 = 25.0$ ft of water. At the same operating speed, best efficiency occurs at $Q = 300$ gpm, where the head is $H = 21.9$ ft of water. Develop a curve-fit for the pump curve at 1170 rpm. Scale the results to a new operating speed of 1750 rpm. Plot and compare the curve-fit results.

EXAMPLE PROBLEM 11.7

GIVEN: Centrifugal pump (with $D = 8$ in. impeller) operated at $\omega = 1170$ rpm.

Q (gpm)	0	300
H (ft of water)	25.0	21.9

FIND: (a) Develop a curve-fit for the pump characteristics at 1170 rpm.
(b) Scale the curve-fit to a new operating speed of 1750 rpm.
(c) Plot and compare the curve-fit results.

SOLUTION:
Assume a parabolic variation in pump head of the form, $H = H_0 - AQ^2$. Solving for A,

$$A = \frac{H_0 - H}{Q^2} = \frac{(25.0 - 2.19)\ \text{ft}}{} \times \frac{1}{(300)^2\ (\text{gpm})^2} = 3.44 \times 10^{-5}\ \text{ft/(gpm)}^2$$

The curve-fit equation is

$$H\ (\text{ft}) = 25.0 - 3.44 \times 10^{-5}\ [Q\ (\text{gpm})]^2$$

The pump remains the same, so the two flow conditions are geometrically similar. Assuming no cavitation occurs, the two flows also will be kinematically similar. Then dynamic similarity will be obtained when the two flow coefficients are matched. Denoting the 1170 rpm condition by subscript 1 and the 1750 rpm condition by subscript 2, then

$$\frac{Q_2}{\omega_2 D_2^3} = \frac{Q_1}{\omega_1 D_1^3} \quad \text{or} \quad \frac{Q_2}{Q_1} = \frac{\omega_2}{\omega_1}$$

since $D_2 = D_1$. For the shutoff condition,

$$Q_2 = \frac{\omega_2}{\omega_1} Q_1 = \frac{1750\ \text{rpm}}{1170\ \text{rpm}} \times 0\ \text{gpm} = 0\ \text{gpm}$$

From the best efficiency point, the new flow rate is

$$Q_2 = \frac{\omega_2}{\omega_1} Q_1 = \frac{1750\ \text{rpm}}{1170\ \text{rpm}} \times 300\ \text{gpm} = 449\ \text{gpm}$$

The pump heads are related by

$$\frac{H_2}{H_1} = \frac{\omega_2^2 D_2^2}{\omega_1^2 D_1^2} \quad \text{or} \quad \frac{H_2}{H_1} = \frac{\omega_2^2}{\omega_1^2} = \left(\frac{\omega_2}{\omega_1}\right)^2$$

since $D_2 = D_1$. For the shutoff condition,

$$H_2 = \left(\frac{\omega_2}{\omega_1}\right)^2 H_1 = \left(\frac{1750\ \text{rpm}}{1170\ \text{rpm}}\right)^2 25.0\ \text{ft} = 55.9\ \text{ft}$$

At the best efficiency point,

$$H_2 = \left(\frac{\omega_2}{\omega_1}\right)^2 H_1 = \left(\frac{1750\ \text{rpm}}{1170\ \text{rpm}}\right)^2 21.9\ \text{ft} = 49.0\ \text{ft}$$

The curve-fit parameters at 1750 rpm may now be found. Solving for A,

$$A_2 = \frac{H_{02} - H_2}{Q_2^2} = \frac{(55.9 - 49.0)\ \text{ft}}{} \times \frac{1}{(449)^2\ (\text{gpm})^2} = 3.44 \times 10^{-5}\ \text{ft/(gpm)}^2$$

Note that A_2 at 1750 rpm is the same as A_1 at 1170 rpm. Thus coefficient A in the curve-fit equation does not change when the pump speed is changed. The two curve-fit equations are

$$H_1 = 25.0 - 3.44 \times 10^{-5} [Q \text{ (gpm)}]^2 \qquad \text{(at 1170 rpm)}$$

and

$$H_2 = 55.9 - 3.44 \times 10^{-5} [Q \text{ (gpm)}]^2 \qquad \text{(at 1750 rpm)}$$

The pump curves are compared in the following plot:

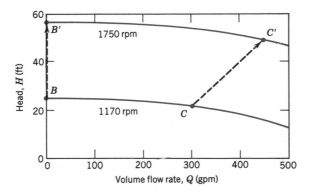

The purpose of this problem was to illustrate the procedure for scaling pump curves from one speed to another.

In principle, geometric similarity would be maintained when pumps of the same geometry, differing in size only by a scale ratio, were tested at the same operating speed. The flow and power would be predicted to vary with pump size as

$$Q_2 = Q_1 \left(\frac{D_2}{D_1} \right)^3, \quad H_2 = H_1 \left(\frac{D_2}{D_1} \right)^2, \quad \text{and} \quad \mathscr{P}_2 = \mathscr{P}_1 \left(\frac{D_2}{D_1} \right)^5 \qquad (11.21)$$

It is not possible to compare the efficiencies at the two operating conditions directly. However, viscous effects should become relatively less important as the pump size increases. Thus efficiency should improve slightly as diameter is increased.

Moody [13] suggested an empirical equation that may be used to estimate the maximum efficiency of a prototype pump based on test data from a geometrically similar model of the prototype pump. His equation is written

$$\frac{1 - \eta_p}{1 - \eta_m} = \left(\frac{D_m}{D_p} \right)^{\frac{1}{5}} \qquad (11.22)$$

To develop Eq. 11.22, Moody assumed that only the surface resistance changes with model scale so that losses in passages of the same roughness vary as $1/D^5$. Unfortunately, it is difficult to maintain the same relative roughness between the model and prototype pumps. Further, the Moody model does not account for any difference in mechanical losses between the model and prototype, nor does it allow determination of off-peak efficiencies. Nevertheless, scaling of the maximum efficiency point is useful to obtain a general knowledge of the efficiency curve for the prototype pump.

The final case is when the impeller diameter is changed inside a casing of fixed geometry. Geometric similarity is not strictly preserved in this case, because the clearance dimensions between the impeller and the pump housing change. However,

analysis based on a simple scale change may still give a useful estimate
1ance if the change in impeller size is not too drastic [14].

Cavitation and Net Positive Suction Head

ıon can occur in any machine handling liquid whenever the local static
pressu.e falls below the vapor pressure of the liquid. When this occurs, the liquid can
flash to vapor locally, forming a vapor cavity and changing the flow pattern from the
non-cavitating condition. The vapor cavity changes the effective shape of the flow passage, thus altering the local pressure field. Since the size and shape of the vapor cavity
are influenced by the local pressure field, the flow may become unsteady. The unsteadi-
ness may cause the entire flow to oscillate and the machine to vibrate.

As cavitation commences, the effect is to reduce the performance of a pump or
turbine rapidly. Thus cavitation must be avoided to maintain stable and efficient operation. In addition, local surface pressures may become high when the vapor cavity
collapses, causing erosion damage or surface pitting. The damage may be severe
enough to destroy a machine made from a brittle low-strength material. Obviously
cavitation also must be avoided to assure long machine life.

In a pump, cavitation tends to begin at the section where the flow is accelerated into
the impeller. Cavitation in a turbine begins where pressure is lowest. The tendency
to cavitate increases as local flow speeds increase; this occurs whenever the flow rate
or the machine operating speed is increased.

Cavitation can be avoided if the pressure everywhere in the machine is kept above
the vapor pressure of the operating liquid. At constant speed, this requires that a
positive pressure — greater than the vapor pressure of the liquid — be maintained at a
pump inlet (the *suction*). Because of pressure losses in the inlet piping, the suction
pressure may be sub-atmospheric. Therefore it is important to carefully limit the
pressure drop in the inlet piping system.

Net positive suction head (NPSH) is defined as the difference between the absolute
stagnation pressure in the flow at the pump suction and the liquid vapor pressure,
expressed as head of flowing liquid [17].[6] The *net positive suction head required
(NPSHR)* by a specific pump to suppress cavitation varies with the liquid pumped,
and with the liquid temperature and pump condition (e.g., as critical geometric features of the pump are affected by wear). *NPSHR* may be measured in a pump test
facility by controlling the input pressure. The results are plotted on the pump performance curve. Typical pump characteristic curves for three impellers tested in the
same housing were shown in Fig. 11.8. Experimentally determined *NPSHR* curves
for the largest and smallest diameter impellers are plotted near the bottom of the
figure.

The *net positive suction head available (NPSHA)* at the pump inlet must be greater
than the *NPSHR* to suppress cavitation. Pressure drops in the inlet piping and pump
entrance increase as volume flow rate increases. Thus for any system, the *NPSHA*
decreases as flow rate is raised. The *NPSHR* of the pump increases as the flow rate
is raised. Therefore, as the system flow rate is increased, the curves for *NPSHA* and
NPSHR versus flow rate ultimately cross. For any inlet system, there is a flow rate
that cannot be exceeded if flow through the pump is to remain free from cavitation.
Inlet pressure losses may be reduced by increasing the diameter of the inlet piping;

---

[6] *NPSH* may be expressed in any convenient units of measure, such as height of the flowing liquid,
e.g., ft of water (hence the term *suction head*), psia, or kPa (abs). When expressed as *head, NPSH* is
measured relative to the pump impeller centerline.

for this reason many centrifugal pumps have larger flanges or couplings at the inlet than at the outlet.

11-5 APPLICATIONS TO FLUID SYSTEMS

We define a *fluid system* as the combination of a fluid machine and a network of pipes or channels that convey fluid. The engineering application of fluid machines in an actual system requires matching the machine and system characteristics, while satisfying constraints of energy efficiency, capital economy, and durability. We have alluded to the vast assortment of hardware offered by competing suppliers; this variety verifies the commercial importance of fluid machinery in modern engineering systems.

Usually it is more economical to specify a production machine rather than a custom unit, because products of established vendors have known, published performance characteristics, and they must be durable to survive in the marketplace. Application engineering consists of making the best selection from catalogs of available products. In addition to machine characteristic curves, all manufacturers provide a wealth of dimensional data, alternative configuration and mounting schemes, and technical information bulletins to guide intelligent application of their products.

This section consists of a brief review of relevant theory, followed by example applications using data taken from manufacturer literature. Two subsections treat work-absorbing machines (pumps, fans, blowers, compressors, and propellers) and work-producing machines (turbines and windmills). Selected performance curves for centrifugal pumps and fans are presented in Appendix D. These may be studied as typical examples of performance data supplied by manufacturers. The curves also may be used to help solve the equipment selection and system design problems at the end of the chapter.

11-5.1 Work-Absorbing Machines

The system pressure-flow curve represents the relation between the pressure applied to the system and the flow rate. The system will operate at the combination of head and flow rate at which the machine performance exactly matches the system requirement. Graphically, the system pressure-flow curve and the machine pressure-flow curve may be superimposed.[7] Then the intersection of the system pressure-flow curve with the machine pressure-capacity curve defines the operating point of the machine and the system.

The system pressure requirement at a given flow rate is composed of frictional pressure drop (major loss due to friction in straight sections of constant area and minor loss due to entrances, fittings, valves, and exits) and pressure changes due to gravity (static lift may be positive or negative). It is useful to discuss the two limiting cases of pure friction and pure lift before considering their combination.

The *all friction* system pressure-flow curve, with no static lift, starts at zero flow and head, as shown in Fig. 11.19a. Frictional pressure drop in turbulent flow (the usual flow regime in engineering systems) varies as the flow rate raised to a power between 1.75 and 2, so this system curve is nearly parabolic. The system curve with pure friction becomes steeper as flow rate increases. To develop the friction curve, losses are computed at various flow rates and then plotted.

---

[7] For illustrative purposes, we use the notion of graphically superimposing the curves. System matching also may be done analytically or numerically, using curve-fits to the system pressure-flow curve and to the machine characteristic curve.

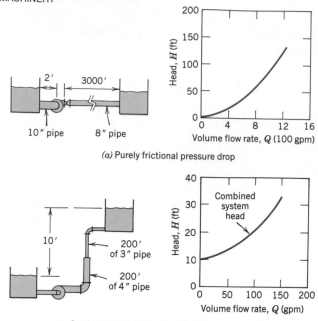

(a) Purely frictional pressure drop

(b) Combination of frictional and gravity pressure changes

Fig. 11.19 Schematic diagrams illustrating basic types of system head-flow curves (adapted from [7]).

Pressure change due to elevation difference is independent of flow rate. Thus the *pure lift* system pressure-flow curve is a horizontal straight line. The gravity pressure is evaluated from the change in elevation of the system.

All actual flow systems have some frictional pressure drop and some elevation change. Thus all system pressure-flow curves may be treated as the sum of a frictional component and a static lift component. The pressure drop for the complete system at any flow rate is the sum of the frictional and lift pressures. The system pressure-flow curve is plotted in Fig. 11.19b.

Whether the resulting system curve is *steep* or *flat* depends on the relative importance of friction and gravity. Friction drop may be relatively unimportant in the water supply to a high-rise building (e.g., the Sears Tower in Chicago, which is nearly 400 m tall), and gravity lift may be negligible in an air-handling system for a one-story building.

a. Pumps

The pump operating point is defined by superimposing the system curve and the pump performance curve, as shown in Fig. 11.20. The point of intersection is the

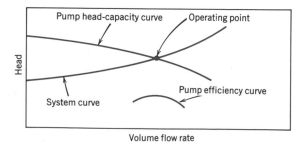

Fig. 11.20 Superimposed system head-flow and pump head-capacity curves.

only condition where the pump and system flow rates are equal and the pump and system pressures are equal simultaneously. The procedure used to determine the match point for a pumping system is illustrated in Example Problem 11.8.

EXAMPLE 11.8—Finding the Operating Point for a Pumping System

The pump of Example Problem 11.7, operating at 1750 rpm, is used to pump water through the pipe system of Fig. 11.19a. Develop an algebraic expression for the general shape of the system resistance curve. Calculate and plot the system resistance curve. Solve graphically for the system operating point. Obtain an approximate analytical expression for the system resistance curve. Solve analytically for the system operating point.

EXAMPLE PROBLEM 11.8

GIVEN: Pump of Example Problem 11.7, operating at 1750 rpm, with $H = H_0 - AQ^2$, where $H_0 = 55.9$ ft and $A = 3.44 \times 10^{-5}$ ft/(gpm)^2. System of Fig. 11.19a, where $L_1 = 2$ ft of $D_1 = 10$ in. pipe and $L_2 = 3000$ ft of $D_2 = 8$ in. pipe, conveying water between two large reservoirs whose surfaces are at the same level.

FIND: (a) Develop a general algebraic expression for the system resistance curve.
       (b) Calculate and plot the system resistance curve.
       (c) Solve graphically for the system operating point.
       (d) Obtain an approximate analytical expression for the system resistance curve.
       (e) Solve analytically for the system operating point.

SOLUTION:
Apply the energy equation for steady, fully developed pipe flow.
Computing equations:

$$\frac{\cancel{p_0}}{\rho} + \alpha_0 \frac{\cancel{\bar{V}_0^2}}{2} + g z_0 - \left(\frac{p_1}{\rho} + \alpha_1 \frac{\bar{V}_1^2}{2} + \cancel{g z_1} \right) = h_{l_{T_{01}}}$$

$$\frac{p_2}{\rho} + \alpha_2 \frac{\bar{V}_2^2}{2} + \cancel{g z_2} - \left(\frac{\cancel{p_3}}{\rho} + \alpha_3 \frac{\cancel{\bar{V}_3^2}}{2} + g z_3 \right) = h_{l_{T_{23}}}$$

where points ① and ② are located just upstream and downstream from the pump, respectively, and z_0 and z_3 are the surface elevations of the supply and discharge reservoirs, respectively.

Assumptions: (1) $p_0 = p_3 = p_{atm}$
           (2) $V_0 = V_3 = 0$
           (3) Uniform flow at each section

Adding, we obtain

$$\frac{p_2}{\rho} + \alpha_2 \frac{\bar{V}_2^2}{2} + g z_0 - \frac{p_1}{\rho} - \alpha_1 \frac{\bar{V}_1^2}{2} - g z_3 = h_{l_{T_{01}}} + h_{l_{T_{23}}}$$

The pressure rise across the pump is then

$$\frac{\Delta p}{\rho} = \frac{p_2 - p_1}{\rho} = \alpha_1 \frac{\bar{V}_1^2}{2} - \alpha_2 \frac{\bar{V}_2^2}{2} + g(z_3 - z_0) + h_{l_{T_{01}}} + h_{l_{T_{23}}}$$

Convert to head rise by dividing through by g

$$H = \frac{\Delta p}{\rho g} = \frac{\alpha_1 \bar{V}_1^2 - \alpha_2 \bar{V}_2^2}{2g} + z_3 - z_0 + \frac{h_{l_{T_{01}}}}{g} + \frac{h_{l_{T_{23}}}}{g}$$

The total head losses are the sum of the major and minor losses, so

$$h_{lT_{01}} = K_{ent}\frac{\bar{V}_1^2}{2} + f_1\frac{L_1}{D_1}\frac{\bar{V}_1^2}{2} = \left(K_{ent} + f_1\frac{L_1}{D_1}\right)\frac{\bar{V}_1^2}{2}$$

$$h_{lT_{23}} = f_2\frac{L_2}{D_2}\frac{\bar{V}_2^2}{2} + K_{exit}\frac{\bar{V}_2^2}{2} = \left(f_2\frac{L_2}{D_2} + K_{exit}\right)\frac{\bar{V}_2^2}{2}$$

From continuity, $\bar{V}_1 A_1 = \bar{V}_2 A_2$, so $\bar{V}_1 = \bar{V}_2\dfrac{A_2}{A_1} = \bar{V}_2\left(\dfrac{D_2}{D_1}\right)^2$.

Substituting

$$H = \frac{\alpha_1\bar{V}_2^2\left(\dfrac{D_2}{D_1}\right)^4 - \alpha_2\bar{V}_2^2}{2g} + \left(K_{ent} + f_1\frac{L_1}{D_1}\right)\frac{\bar{V}_2^2}{2g}\left(\frac{D_2}{D_1}\right)^4 + \left(f_2\frac{L_2}{D_2} + K_{exit}\right)\frac{\bar{V}_2^2}{2g} + z_3 - z_0$$

or, upon simplifying

$$H = \left[\alpha_1\left(\frac{D_2}{D_1}\right)^4 - \alpha_2 + \left(K_{ent} + f_1\frac{L_1}{D_1}\right)\left(\frac{D_2}{D_1}\right)^4 + f_2\frac{L_2}{D_2} + K_{exit}\right]\frac{\bar{V}_2^2}{2g} + z_3 - z_0 \qquad H \leftarrow$$

This is the general relationship for the system resistance curve. Calculate and tabulate at various volume flow rates (with $z_3 = z_0$):

Q (gpm)	$\bar{V}_1$ (ft/sec)	Re_1 (1000)	f_1 (—)	$\bar{V}_2$ (ft/sec)	Re_2 (1000)	f_2 (—)	H (ft)
0	0.00	0	0.000	0.00	0	0.000	0
100	0.41	32	0.026	0.64	40	0.026	1
200	0.82	63	0.023	1.28	79	0.024	3
300	1.23	95	0.022	1.91	119	0.023	6
400	1.63	127	0.022	2.55	158	0.022	10
500	2.04	158	0.021	3.19	198	0.022	16
600	2.45	190	0.021	3.83	238	0.022	23
700	2.86	222	0.021	4.47	277	0.022	31
800	3.27	253	0.021	5.11	317	0.022	40
900	3.68	285	0.021	5.74	356	0.022	51
1000	4.09	317	0.021	6.38	396	0.022	62
1100	4.49	348	0.021	7.02	435	0.021	75
1200	4.90	380	0.021	7.66	475	0.021	89
1300	5.31	412	0.020	8.30	515	0.021	105
1400	5.72	443	0.020	8.94	554	0.021	121
1500	6.13	475	0.020	9.57	594	0.021	139

The pump curve and the system resistance curve are plotted below:

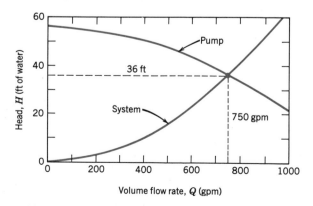

The graphical solution is shown on the plot. At the operating point, $H \approx 36$ ft and $Q \approx 750$ gpm (the graphical results are accurate to two significant figures at best).

The approximate analytical expression for the system resistance curve is developed as follows. If flow is at high Reynolds number it will be in the fully rough flow regime. Then the friction factors will be constant and

$$H = Z_0 + C Q^2$$

where $Z_0 = z_3 - z_0$ and $C = 2/\pi D_2^2 g$ times the term in square brackets in the general equation for H above. For this flow, $Z_0 = 0$ and $C = 6.17 \times 10^{-5}$ ft/(gpm)^2. Thus the analytical expression for the system curve is

$$H = 6.17 \times 10^{-5} \text{ ft/(gpm)}^2 [Q \text{ (gpm)}]^2 \qquad\qquad\qquad\qquad\qquad H$$

The pump curve equation and the system resistance curve equation are solved simultaneously to find the system operating point. Thus

$$H_p = H_0 - A\, Q^2 = Z_0 + C\, Q^2$$

Solving for Q, the volume flow rate at the operating point,

$$Q = \left[\frac{H_0 - Z_0}{A + C}\right]^{1/2}$$

For this case, $Z_0 = 0$ and

$$Q = \left[55.9 \text{ ft} \times \frac{\text{(gpm)}^2}{(3.44 \times 10^{-5} + 6.17 \times 10^{-5}) \text{ ft}}\right]^{1/2} = 763 \text{ gpm} \qquad\qquad Q$$

The volume flow rate may be substituted into either expression for head to calculate the head at the operating point as

$$H = C\, Q^2 = \frac{6.17 \times 10^{-5}}{\text{(gpm)}^2} \text{ ft } \frac{(763)^2 \text{ (gpm)}^2}{} = 35.9 \text{ ft} \qquad\qquad\qquad H$$

Comparing results shows that the graphical solution was of acceptable accuracy. The graphical solution was about 0.3 percent high in head and about 1.7 percent high in flow rate. When we recall that the system resistance calculation depended on pipe flow coefficients accurate to within approximately ± 10 percent, this is acceptable.

> The purpose of this problem was to illustrate the procedures used to find the operating point of a pump and flow system using both graphical and analytical solution methods. The example shows that either method may be used to obtain results of engineering accuracy.

The shapes of both the pump curve and the system curve can be important to system stability in certain applications. The pump curve shown in Fig. 11.20 is typical of the curve for a new centrifugal pump of intermediate specific speed, for which the head (or pressure) decreases smoothly and monotonically as the flow rate increases from shutoff. Two effects take place gradually as the system ages: the pump wears and its performance decreases (the pump curve gradually moves downward toward lower pressure at each flow rate) and the system resistance increases (the system curve gradually moves toward higher pressure at each flow rate because of pipe aging[8]). The effect of these changes is to move the operating point toward

---

[8] As the pipe ages, mineral deposits form on the wall (see Fig. 8.16), raising the relative roughness and reducing the pipe diameter compared to the as-new condition. See Problem 11.45 for typical friction factor data.

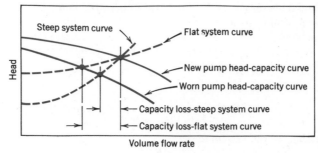

Fig. 11.21 Effect of pump wear on flow delivery to system.

lower flow rates over time. The magnitude of the change in flow rate depends on the shapes of the pump and system curves.

The capacity losses, as pump wear occurs, are compared for steep and flat system curves in Fig. 11.21. The loss in capacity is greater for the flat system curve than for the steep system curve.

The pump efficiency curve is also plotted in Fig. 11.20. The original system operating point usually is chosen to coincide with the maximum efficiency by careful choice of pump size and operating speed. Pump wear increases internal leakage, thus reducing delivery and lowering peak efficiency. In addition, as shown in Fig. 11.21, the operating point moves toward lower flow rate, away from the best efficiency point. Thus the reduced system performance may not be accompanied by reduced energy usage.

Sometimes it is necessary to satisfy a high-head, low-flow requirement; this forces selection of a pump with low specific speed. Such a pump may have a performance curve with a flat or slightly rising head near shutoff, as shown in Fig. 11.22. When used with a steep system curve, the operating point is well defined and no problems with system operation should result. However, use of the pump with a flat system curve easily could cause problems, especially if the actual system curve were slightly above the computed curve or the pump delivery below the charted head-capacity performance.

If there are two points of intersection between the pump and system curves, the system may operate at either, depending on conditions at startup; a disturbance could cause the system operating point to shift to the second point of intersection. Under certain conditions, the system operating point can alternate between the two points of intersection, causing unsteady flow and unsatisfactory performance.

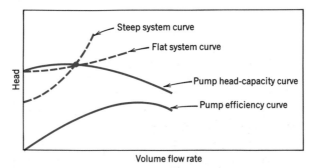

Fig. 11.22 Operation of low specific speed pump near shutoff.

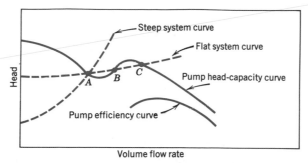

Fig. 11.23 Operation of high specific speed pump near the dip.

Instead of a single pump of low specific speed, a multi-stage pump may be used in this situation. Since the flow rate through each stage is the same, but the head per stage is less than in the single-stage unit, the specific speed of the multi-stage pump is higher (see Eq. 11.17).

The head-flow characteristic curve of some high specific speed pumps shows a dip at capacities below the peak efficiency point, as shown in Fig. 11.23. Caution is needed in applying such pumps if ever it is necessary to operate the pump at or near the dip in the head-flow curve. No trouble should occur if the system characteristic is steep, for there will be only one point of intersection with the pump curve. Unless this intersection is near point B, the system should return to stable, steady-state operation following any transient disturbance.

Operation with a flat system curve may be more problematic. It is possible to have one, two, or three points of intersection between the pump and system curves, as suggested in the figure. With the flat system curve, the pump may "hunt" or oscillate (periodically or aperiodically) between the two pairs of adjacent points or among all three points.

Several other factors can adversely influence pump performance: pumping hot liquid, pumping liquid with entrained vapor, and pumping liquid with high viscosity. According to [5], the presence of small amounts of entrained gas can drastically reduce performance. As little as 4 percent vapor can reduce pump capacity by more than 40 percent. Air can enter the suction side of the pumping circuit where pressure is below atmospheric if any leaks are present.

Adequate submergence of the suction pipe is necessary to prevent air entrainment. Insufficient submergence can cause a vortex to form at the pipe inlet. If the vortex is strong, air can enter the suction pipe. References [15] and [16] give guidelines for adequate suction basin design to eliminate the likelihood of vortexing.

Increased fluid viscosity may dramatically reduce the performance of a centrifugal pump [17]. Typical experimental test results are plotted in Fig. 11.24. In the figure, pump performance with water ($\nu = 1$ cP) is compared to performance with a more viscous liquid ($\nu = 220$ cP). The increased viscosity reduces the head produced by the pump. At the same time the input power requirement is increased. The result is a dramatic drop in pump efficiency at all flow rates.

Heating a liquid raises its vapor pressure. Thus to pump a hot liquid requires additional pressure at the pump inlet to prevent cavitation. The *NPSHA* must be raised, or the pump delivery must be reduced, to maintain the same cavitation number as when the *NPSHR* curve was measured in the laboratory.

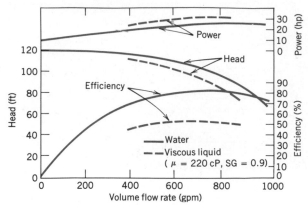

Fig. 11.24 Effect of liquid viscosity on performance of a centrifugal pump [5].

In some systems, such as city water supply or chilled water circulation, there may be a wide range in demand with a relatively constant system resistance. In these cases, it may be possible to operate constant-speed pumps in series or parallel to supply the system requirements without excessive energy dissipation due to outlet throttling. Two or more pumps may be operated in parallel or series to supply flow at high demand conditions, and fewer units can be used when demand is low.

For pumps in *series*, the combined performance curve is derived by adding the pressure rises at each flow rate (Fig. 11.25). The increase in flow rate gained by operating pumps in series depends upon the resistance of the system being supplied. For two pumps in series, delivery will increase at any system pressure. The characteristic curves for one pump and for two identical pumps in series are

$$H_1 = H_0 - A\,Q^2$$

and

$$H_{2s} = 2(H_0 - A\,Q^2) = 2H_0 - 2A\,Q^2$$

Figure 11.25 is a schematic illustrating the application of two identical pumps in series. A reasonable match to the system requirement is possible—while keeping efficiency high—if the system curve is relatively steep.

In an actual system, it is not appropriate simply to connect two pumps in series. If only one pump were powered, flow through the second, unpowered pump would

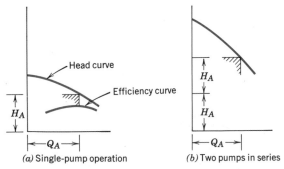

Fig. 11.25 Operation of two centrifugal pumps in series.

cause additional losses, raising the system resistance. It also is desirable to arrange the pumps and piping so that each pump can be taken out of the pumping circuit for maintenance, repair, or replacement when needed. Thus a system of bypasses, valves, and check valves may be necessary in an actual installation [14, 16].

Pumps also may be combined in *parallel*. The resulting performance curve, shown in Fig. 11.26, is obtained by adding the pump capacities at each head. The characteristic curves for one pump and for two identical pumps in parallel are

$$H_1 = H_0 - A\,Q^2$$

and

$$H_{2p} = H_0 - A\left(\frac{Q}{2}\right)^2 = H_0 - \frac{1}{4}A\,Q^2$$

The schematic in Fig. 11.26 shows that the parallel combination may be used most effectively to increase system capacity when the system curve is relatively flat.

An actual system installation with parallel pumps also requires more thought to allow satisfactory operation with only one pump powered. It is necessary to prevent backflow through the pump that is not powered. To prevent backflow, and to permit pump removal, a more complex and expensive piping setup is needed.

Many other piping arrangements and pump combinations are possible. Pumps of different sizes, heads, and capacities may be combined in series, parallel, or series-parallel arrangements. Obviously the complexity of the piping and control system increases rapidly. No matter what the arrangement or number of pumps, a system with discrete pumps operated at constant speed is not capable of continuously varied flow rate without use of throttling valves. Control without valves is possible only in discrete steps according to the specific combination of pumps powered at any time.

Use of variable-speed operation allows infinitely variable control of system flow rate with high energy efficiency and without extra plumbing complexity. A further advantage is that a variable-speed drive system offers much simplified control of system flow rate. The cost of efficient variable-speed drive systems continues to decrease because of advances in power electronic components and circuits. The system flow rate can be controlled by varying pump operating speed with impressive savings in pumping power and energy usage. The input power reduction afforded by variable-speed drive is illustrated in Table 11.1 [18]. At 1100 gpm the power input is cut almost 54 percent for the variable-speed system; the reduction at 600 gpm is more than 75 percent.

The reduction in input power requirement at reduced flow with the variable-speed drive is impressive. The energy savings, and therefore the cost savings, depend on the specific duty cycle on which the machine operates. Reference [18] presents

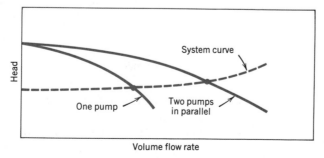

Fig. 11.26 Operation of two centrifugal pumps in parallel.

Table 11.1 Power Requirements for Constant- and Variable-Speed Drive Pumps [18]

Throttle Valve Control with Constant-Speed (1750 rpm) Motor								
Flow Rate (gpm)	System Head (ft)	Valve[a] Efficiency (%)	Pump Head (ft)	Pump Efficiency (%)	Pump Power (bhp)	Motor Efficiency (%)	Motor Input (hp)	Power Input[b] (hp)
1700	180	100.0	180	80.0	96.7	90.8	106.5	106.7
1500	150	78.1	192	78.4	92.9	90.7	102.4	102.6
1360	131	66.2	198	76.8	88.6	90.7	97.7	97.9
1100	102	49.5	206	72.4	79.1	90.6	87.3	87.5
900	83	39.5	210	67.0	71.3	90.3	79.0	79.1
600	62	29.0	214	54.0	60.1	90.0	66.8	66.9

Variable-Speed Drive with Energy-Efficient Motor								
Flow Rate (gpm)	Pump/System Head (ft)	Pump Efficiency (%)	Pump Power (bhp)	Motor Speed (rpm)	Motor Efficiency (%)	Motor Input (hp)	Control Efficiency (%)	Power Input (hp)
1700	180	80.0	96.7	1750	93.7	103.2	97.0	106.4
1500	150	79.6	71.5	1580	94.0	76.0	96.1	79.1
1360	131	78.8	57.2	1470	93.9	60.9	95.0	64.1
1100	102	78.4	36.2	1275	93.8	38.6	94.8	40.7
900	83	77.1	24.5	1140	92.3	26.5	92.8	28.6
600	62	72.0	13.1	960	90.0	14.5	89.1	16.3

[a] Valve efficiency is the ratio of system pressure to pump pressure.
[b] Power input is motor input divided by 0.998 starter efficiency.

information on mean duty cycles for centrifugal pumps used in the chemical process industry; Fig. 11.27 is a plot showing the histogram of these data. The plot shows that although the system must be designed and installed to deliver full rated capacity, this seldom occurs. Instead, more than half the time, the system operates at 70 percent capacity or below. The energy saving that results from use of a variable-speed drive for this duty cycle is computed in Example Problem 11.9.

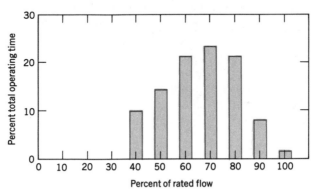

Fig. 11.27 Mean duty cycle for centrifugal pumps in the chemical and petroleum industries [18].

EXAMPLE 11.9—Energy Savings with Variable-Speed Centrifugal Pump Drive

Combine the information on mean duty cycle for centrifugal pumps given in Fig. 11.27 with the drive data in Table 11.1. Estimate the annual savings in pumping energy and cost that could be achieved by implementing a variable-speed drive system.

EXAMPLE PROBLEM 11.9

GIVEN: Consider the variable-flow, variable-pressure pumping system of Table 11.1. Assume the system operates on the typical duty cycle shown in Fig. 11.27, 24 hours per day, year round.

FIND: (a) Estimate the reduction in annual energy usage obtained with this variable-speed drive.
(b) Compute the energy costs and evaluate the cost saving due to variable-speed operation.

SOLUTION:
Full-time operation involves 365 days × 24 hours per day, or 8760 hours per year. Thus the percentages in Fig. 11.27 may be multiplied by 8760 to give annual hours of operation.

First plot the pump input power versus flow rate using data from Table 11.1 to allow interpolation, as shown below.

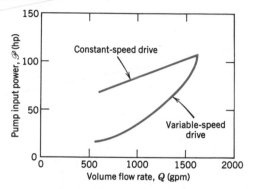

Illustrate the procedure using operation at 70 percent flow as a sample calculation. At 70 percent flow, the pump delivery is 0.7 × 1700 gpm = 1190 gpm. From the plot, the pump input power requirement at this flow rate is 91 hp for the constant-speed drive. At this flow rate, the pump operates 23 percent of the time, or 0.23 × 8760 = 2010 hours per year. The total energy consumed at this duty point is 91 hp × 2010 hr = 1.91×10⁵ hp · hr. The electrical energy consumed is

$$E = \frac{1.91 \times 10^5 \text{ hp} \cdot \text{hr}}{} \times \frac{0.746 \text{ kW} \cdot \text{hr}}{\text{hp} \cdot \text{hr}} = 1.42 \times 10^5 \text{ kW} \cdot \text{hr}$$

The corresponding cost of electricity (at $0.08/kW · hr) is

$$C = \frac{1.42 \times 10^5 \text{ kW} \cdot \text{hr}}{} \times \frac{\$0.08}{\text{kW} \cdot \text{hr}} = \$11,400$$

The following tables were prepared using similar calculations:

	Constant-Speed Drive, 8760 hr/yr				
Flow (%)	Flow (gpm)	Time (%)	Time (hr)	Power (hp)	Energy (hp·hr)
100	1700	2	175	106	1.86×10⁴
90	1530	8	701	102	7.15×10⁴
80	1360	21	1840	98	18.0×10⁴
70	1190	23	2101	91	19.1×10⁴
60	1020	21	1840	83	15.3×10⁴
50	850	15	1310	77	10.1×10⁴
40	680	10	876	70	6.13×10⁴
				Total:	77.6×10⁴

Summing the last column of the table shows that for the constant-speed drive system the annual energy consumption is 7.76×10^5 hp · hr. The electrical energy consumption is

$$E = \frac{7.76 \times 10^5 \text{ hp} \cdot \text{hr}}{} \times \frac{0.746 \text{ kW} \cdot \text{hr}}{\text{hp} \cdot \text{hr}} = 579{,}000 \text{ kW} \cdot \text{hr} \qquad\qquad E_{CSD}$$

At $0.08 per kilowatt hour, the energy cost for the constant-speed drive system is

$$C = \frac{579{,}000 \text{ kW} \cdot \text{hr}}{} \times \frac{\$0.08}{\text{kW} \cdot \text{hr}} = \$46{,}300 \qquad\qquad C_{CSD}$$

		Variable-Speed Drive, 8760 hr/yr			
Flow %	Flow (gpm)	Time %	Time (hr)	Power (hp)	Energy (hp · hr)
100	1700	2	175	106	1.86×10^4
90	1530	8	701	83	5.82×10^4
80	1360	21	1840	64	11.8×10^4
70	1190	23	2010	49	9.85×10^4
60	1020	21	1840	35	6.44×10^4
50	850	15	1310	25	3.28×10^4
40	680	10	876	19	1.66×10^4
				Total:	40.7×10^4

Summing the last column of the table shows that for the variable-speed drive system, the annual energy consumption is 4.07×10^5 hp · hr. The electrical energy consumption is

$$E = \frac{4.07 \times 10^5 \text{ hp} \cdot \text{hr}}{} \times \frac{0.746 \text{ kW} \cdot \text{hr}}{\text{hp} \cdot \text{hr}} = 304{,}000 \text{ kW} \cdot \text{hr} \qquad\qquad E_{VSD}$$

At $0.08 per kilowatt hour, the energy cost for the variable-speed drive system is only

$$C = \frac{304{,}000 \text{ kW} \cdot \text{hr} \times}{} \frac{\$0.08}{\text{kW} \cdot \text{hr}} = \$24{,}300 \qquad\qquad C_{VSD}$$

Thus, in this application, the variable-speed drive reduces energy consumption by 275,000 kW · hr (47 percent). The cost saving is an impressive $22,000 annually. One could afford to install a variable-speed drive even at considerable cost penalty. The savings in energy cost are appreciable each year and continue throughout the life of the system.

The purpose of this problem was to illustrate the energy and cost savings that can be gained by using variable-speed pump drives. The specific benefits depend on the system and on its operating duty cycle.

b. Fans, Blowers, and Compressors

Fans are designed to handle air or vapor. Fan sizes range from the cooling fan on a piece of electronic equipment, that moves a cubic meter of air per hour and requires a few watts of power, to a mine ventilation fan, that moves thousands of cubic meters of air per minute and requires many hundreds of kilowatts of power. Fans are produced in varieties similar to those of pumps: they range from radial-flow (centrifugal) to axial-flow devices. As with pumps, the characteristic curve shapes for fans depend on the fan type. Some typical performance curves for centrifugal fans are presented in Appendix D. The curves may be used to choose fans to solve some of the equipment selection and system design problems at the end of the chapter.

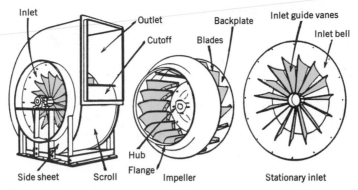

Fig. 11.28   Exploded view of typical centrifugal fan [19].

An exploded view of a medium-size centrifugal fan is shown in Fig. 11.28. Some terminology in common use is shown on the figure.

The pressure rise produced by fans is several orders of magnitude less than for pumps. Another difference between fans and pumps is that measurement of flow rate is more difficult in gases and vapors than in liquids. There is no convenient analog to the "catch the flow in a bucket" method of measuring liquid flow rates! Consequently, fan testing requires special facilities and procedures [6, 20]. Because the pressure rise produced by a fan is small, usually it is impractical to measure flow rate with a restriction flow meter such as an orifice, flow nozzle, or venturi. It may be necessary to use an auxiliary fan to develop enough pressure rise to permit measurement of flow rate with acceptable accuracy using a restriction flow meter. An alternative is to use an instrumented duct in which the flow rate is calculated from a pitot traverse. Appropriate standards may be consulted to obtain complete information on specific fan test methods and data reduction procedures for each application [6, 20].

Because the pressure change across a fan is small, the dynamic pressure at the fan exit may be an appreciable fraction of the pressure rise. Consequently, it is necessary to carefully specify the basis on which pressure measurements are made. Data for both static and total pressure rise and for efficiency, based on both pressure rises, frequently are plotted on the same characteristic graph (Fig. 11.29).

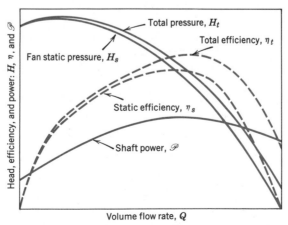

Fig. 11.29   Typical characteristic curves for fan with backward-curved blades [21].

The coordinates may be plotted in physical units (e.g., inches of water, cubic feet per minute, and horsepower) or as dimensionless flow and pressure coefficients. The difference between the total and static pressures is the dynamic pressure, so the vertical distance between these two curves is proportional to Q^2.

Centrifugal fans are used frequently; we will use them as examples. The centrifugal fan developed from simple paddle wheel designs, in which the wheel was a disk carrying radial flat plate segments. (This primitive form still is used in non-clogging fans such as in commercial clothes dryers.) Refinements have led to the three general types shown in Figs. 11.30a through c, with backward-curved, radial, and forward-curved blades. The fans illustrated all have blades that are curved at their inlet edges to approximate shockless flow between the blade and the inlet flow direction. These three designs are typical of fans with sheet metal blades, which are relatively simple to manufacture and thus relatively inexpensive. The forward-curved design illustrated in the figure has very closely spaced blades; it frequently is called a *squirrel cage* fan because of its resemblance to the exercise wheels found in animal cages.

As fans become larger in size and power demand, efficiency becomes more important. The streamlined *airfoil blades* shown in Fig. 11.30d are much less sensitive to inlet flow direction and improve efficiency markedly compared to the thin blades shown in diagrams a through c. The added expense of airfoil blades for large metal fans may be life-cycle cost effective. Airfoil blades are used frequently on small fans as impellers molded from plastic become common.

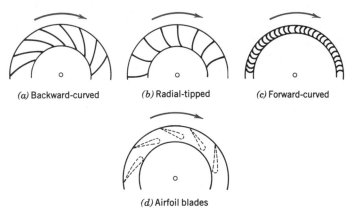

(a) Backward-curved   (b) Radial-tipped   (c) Forward-curved

(d) Airfoil blades

Fig. 11.30 Typical types of blading used for centrifugal fan wheels [21].

As for pumps, the total pressure rise across a fan is approximately proportional to the absolute velocity of the fluid at the exit from the wheel. Therefore the characteristic curves produced by the basic blade shapes tend to differ. The typical curve shapes are shown in Fig. 11.31, where both pressure rise and power requirements are sketched. Fans with backward-curved blade tips typically have a power curve that reaches a maximum and then decreases as flow rate increases. If the fan drive is sized properly to handle the peak power, it is impossible to overload the drive with this type of fan.

The power curves for fans with radial and forward-curved blades rise as flow rate increases. If the fan operating point is higher than the design flow rate, the motor may be overloaded. Such fans cannot be run for long periods at low back pressures.

Fans with backward-curved blades are best for installations with large power demand and continuous operation. The forward-curved blade fan is preferred where low first cost and small size are important and where service is intermittent. Forward-curved blades require lower tip speed to produce a specified head; lower blade tip

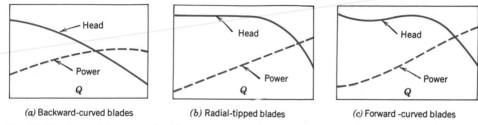

(a) Backward-curved blades   (b) Radial-tipped blades   (c) Forward -curved blades

Fig. 11.31   General features of performance curves for centrifugal fans with backward-, radial-, and forward-curved blades [21].

speed means reduced noise. Thus forward-curved blades may be specified for heating and air conditioning applications to minimize noise.

Characteristic curves for axial-flow (*propeller*) fans differ markedly from those for centrifugal fans. The power curve, Fig. 11.32, is especially different, as it tends to decrease continuously as flow rate increases. Thus it is impossible to overload a properly sized drive for an axial-flow fan.

The simple propeller fan often is used for ventilation; it may be free-standing or mounted in an opening, as a window fan, with no inlet or outlet duct work. Ducted axial-flow fans have been studied extensively and developed to high efficiency [22]. Modern designs, with airfoil blades, mounted in ducts and often fitted with guide vanes, can deliver large volumes against high resistances with high efficiency. The primary deficiency of the axial-flow fan is the non-monotonic shape of the pressure characteristic: in certain ranges of flow rate the fan may pulsate. Because axial-flow fans tend to have high rotational speeds, they can be noisy.

Selection and installation of a fan always requires compromise. To minimize energy consumption, it is desirable to operate a fan at its highest efficiency point. To reduce the fan size for a given capacity, it is tempting to operate at higher flow rate than at maximum efficiency. In an actual installation, this tradeoff must be made considering such factors as available space, initial cost, and annual hours of operation. It is not wise to operate a fan at a flow rate below maximum efficiency. Such a fan would be larger than necessary and some designs can be unstable and noisy when operated in this region.

It is necessary to consider the duct system at both the fan inlet and outlet to develop a satisfactory installation. Anything that disrupts the uniform flow at the fan inlet is likely to impair performance. Nonuniform flow at the inlet causes the wheel to operate unsymmetrically and may decrease capacity dramatically. Swirling flow also adversely affects fan performance. Swirl in the direction of rotation reduces the

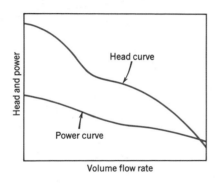

Fig. 11.32   Characteristic curves for a typical axial-flow fan [21].

pressure developed; swirl in the opposite direction can increase the power required to drive the fan.

The fan specialist may not be allowed total freedom in designing the best flow system for the fan. Sometimes a poor flow system can be improved without too much effort by adding splitters or straightening vanes to the inlet. Some fan manufacturers offer guide vanes that can be installed for this purpose.

Flow conditions at the fan discharge also affect installed performance. Every fan produces nonuniform outlet flow. When connected to a length of straight duct, the flow becomes more uniform and some excess kinetic energy is transformed to static pressure. If the fan discharges directly into a large space with no duct, the excess kinetic energy of the nonuniform flow is dissipated. A fan in a flow system with no discharge ducting may fall considerably short of the performance measured in a laboratory test setup.

The flow pattern at the fan outlet may be affected by the amount of resistance present downstream. The effect of the system on fan performance may be different at different points along the fan pressure-flow curve. Thus, it may not be possible to accurately predict the performance of a fan, *as installed*, on the basis of curves measured in the laboratory.

Fans may be scaled up or down in size or speed using the basic laws developed for fluid machines in Chapter 7. It is possible for two fans to operate with fluids of significantly different density,[9] so pressure must replace head as a dependent parameter and density must be retained in the dimensionless groups. The dimensionless groups appropriate for fan analysis are

$$\Pi_1 = \frac{Q}{\omega D^3}, \quad \Pi_2 = \frac{p}{\rho \omega^2 D^2}, \quad \text{and} \quad \Pi_3 = \frac{\mathcal{P}}{\rho \omega^3 D^5} \tag{11.23}$$

Once again dynamic similarity is assured when the flow coefficients are matched. Thus when

$$Q' = Q \left(\frac{\omega'}{\omega}\right)\left(\frac{D'}{D}\right)^3 \tag{11.24a}$$

then

$$p' = p \left(\frac{\rho'}{\rho}\right)\left(\frac{\omega'}{\omega}\right)^2 \left(\frac{D'}{D}\right)^2 \tag{11.24b}$$

and

$$\mathcal{P}' = \mathcal{P} \left(\frac{\rho'}{\rho}\right)\left(\frac{\omega'}{\omega}\right)^3 \left(\frac{D'}{D}\right)^5 \tag{11.24c}$$

As a first approximation, the efficiency of the scaled fan is assumed to remain constant, so

$$\eta' = \eta \tag{11.24d}$$

When head is replaced by pressure, and density is included, the expression defining the specific speed of a fan becomes

$$N_s = \frac{\omega Q^{1/2} \rho^{3/4}}{\mathcal{P}^{3/4}} \tag{11.25}$$

A fan scale-up with density variation is the subject of Example Problem 11.10.

---

[9] Density of the flue gas handled by an induced draft fan on a steam powerplant may be 40 percent less than the density of the air handled by the forced draft fan in the same plant.

EXAMPLE 11.10—Scaling of Fan Performance

Given below are performance curves [21] for a centrifugal fan with $D = 36$ in. and $\omega = 600$ rpm, as measured on a test stand using air at standard density ($\rho = 0.075$ lbm/ft^3). Scale the data to predict the performance of a similar fan with $D' = 42$ in., $\omega' = 1150$ rpm, and $\rho' = 0.045$ lbm/ft^3. Estimate the delivery and power of the larger fan when it operates at a system pressure equivalent to 7.4 in. of H$_2$O. Check the specific speed of the fan at the new operating point.

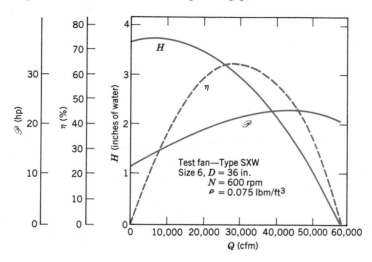

EXAMPLE PROBLEM 11.10

GIVEN: Performance data as shown for centrifugal fan with $D = 36$ in., $\omega = 600$ rpm, and $\rho = 0.075$ lbm/ft^3.

FIND: (a) Scale these data to predict the performance of a geometrically similar fan with $D' = 42$ in., at $\omega' = 1150$ rpm, with $\rho' = 0.045$ lbm/ft^3.
 (b) Estimate the delivery and input power requirement if the larger fan operates against a system resistance of 7.4 in. H$_2$O.
 (c) Check the specific speed of the larger fan at this operating point.

SOLUTION:
Develop the performance curves at the new operating condition by scaling the test data point-by-point. Using Eqs. 11.24 and the data from the curves at $Q = 30,000$ cfm, the new volume flow rate is

$$Q' = Q\left(\frac{\omega'}{\omega}\right)\left(\frac{D'}{D}\right)^3 = 30,000 \text{ cfm}\left(\frac{1150}{600}\right)\left(\frac{42}{36}\right)^3 = 91,300 \text{ cfm}$$

The fan pressure rise is

$$p' = p\,\frac{\rho'}{\rho}\left(\frac{\omega'}{\omega}\right)^2\left(\frac{D'}{D}\right)^2 = 2.96 \text{ in. H}_2\text{O}\left(\frac{0.045}{0.075}\right)\left(\frac{1150}{600}\right)^2\left(\frac{42}{36}\right)^2 = 8.88 \text{ in. H}_2\text{O}$$

and the new power input is

$$\mathscr{P}' = \mathscr{P}\left(\frac{\rho'}{\rho}\right)\left(\frac{\omega'}{\omega}\right)^3\left(\frac{D'}{D}\right)^5 = 21.4 \text{ hp}\left(\frac{0.045}{0.075}\right)\left(\frac{1150}{600}\right)^3\left(\frac{42}{36}\right)^5 = 195 \text{ hp}$$

We assume the efficiency remains constant between the two scaled points, so

$$\eta' = \eta = 0.64$$

Similar calculations at other operating points give the results tabulated below:

Q (cfm)	p (in. H_2O)	$\mathscr{P}$ (hp)	η (%)	Q' (cfm)	p' (in. H_2O)	$\mathscr{P}'$ (hp)
0	3.68	11.1	0	0	11.0	101
10,000	3.75	15.1	37.4	30,400	11.2	138
20,000	3.50	18.6	59.2	60,900	10.5	170
30,000	2.96	21.4	64.8	91,300	8.88	195
40,000	2.12	23.1	57.4	122,000	6.36	211
50,000	1.02	23.1	34.5	152,000	3.06	211
60,000	0	21.0	0	177,000	0	192

To allow interpolation among the datum points, it is convenient to plot the results:

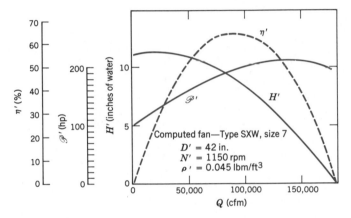

From the pressure versus capacity curve, the larger fan should deliver 110,000 cfm at 7.4 in. H_2O system pressure, with an efficiency of approximately 61 percent.

This operating point is slightly to the right of peak efficiency for this fan, so it is a reasonable point at which to operate the fan. The specific speed of the fan at this operating point (in U.S. customary units) is

$$N_s = \frac{(1150 \text{ rpm})(110{,}000 \text{ cfm})^{1/2}(0.045 \text{ lbm/ft}^3)^{3/4}}{(7.4 \text{ in.}H_2O)^{3/4}} = 8310 \qquad \longleftarrow \quad N_s$$

In nondimensional (SI) units,

$$N_s = \frac{(120 \text{ rad/sec})(3110 \text{ m}^3/\text{sec})^{1/2}(0.721 \text{ kg/m}^3)^{3/4}}{(1.84\times10^3 \text{ N/m}^2)^{3/4}} = 18.6 \qquad \longleftarrow \quad N_s$$

$\left\{\begin{array}{l}\text{The purpose of this problem was to illustrate the procedure for scaling performance of}\\ \text{fans operating on gases with two different densities.}\end{array}\right\}$

Three methods are available to control fan delivery: motor speed control, inlet dampers, and outlet throttling. Speed control was treated thoroughly in the section on pumps. The same benefits of reduced energy usage and noise are obtained with fans and the cost of variable-speed drive systems continues to drop.

Inlet dampers may be used effectively on some large centrifugal fans. However, they penalize efficiency and cannot be used to reduce the fan flow rate below about 40 percent of rated capacity. Outlet throttling is cheap but wasteful of energy. For further details, consult References [19] or [21], which are particularly comprehensive. Osborne [23] also treats noise, vibration, and the mechanical design of fans.

Fans also may be combined in series, parallel, or more complex arrangements to match varying system resistance and flow needs. These combinations may be analyzed using the methods described for pumps. References [24] and [25] are excellent sources for loss data on air flow systems.

Blowers have performance characteristics similar to fans, but they operate (typically) at higher speeds and increase the fluid pressure more than do fans. Jorgensen [19] divides the territory between fans and compressors at an arbitrary pressure level that changes the air density by 5 percent; he does not demarcate between fans and blowers.

Compressors may be centrifugal or axial, depending on specific speed. Automotive turbochargers, small gas turbine engines, and natural gas pipeline boosters usually are centrifugal. Large gas turbines and jet aircraft engines frequently are axial-flow machines.

Compressor performance depends on operating speed, mass flow rate, and density of the working fluid. It is common practice to present compressor performance data on the coordinates shown in Fig. 11.33, as *pressure ratio* versus *corrected mass flow rate*, with *corrected speed* as a parameter. Normalizing the mass flow rate by $\sqrt{T}/p$, where T and p are absolute temperature and pressure, removes the effects of density variations. Normalizing the compressor operating speed with $1/\sqrt{T}$ relates the tip speed of the compressor wheel to the speed of sound (this forms a Mach number—see Chapter 13).

At first glance, the performance curves in Fig. 11.33 appear only partially complete. Two phenomena limit the range of mass flow rate at which a compressor can be operated at any given speed. The maximum mass flow rate is limited by choking— the approach to $M = 1$ at some point in the machine—see Chapter 13. Performance deteriorates rapidly as the choking limit is approached.

The minimum mass flow rate is limited by either *rotating stall* or *surge* in the compressor. Rotating stall occurs when cells of separated flow form and block a segment of the compressor rotor. The effect is to reduce performance and unbalance the rotor; this causes severe vibration and can lead quickly to damage. Thus it is impossible to operate a compressor with rotating stall and it must be avoided.

Centrifugal and axial compressors also may be limited by surge, a cyclic pulsation phenomenon that causes the mass flow rate through the machine to vary, and can even reverse it! Surge is accompanied by loud noises and can damage the compressor or related components; it too must be avoided.

In general, the higher the performance, the more narrow the range in which the compressor may be operated successfully. Thus a compressor must be carefully

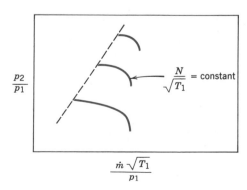

Fig. 11.33 Typical performance map for a centrifugal compressor [2].

matched to its flow system to assure satisfactory operation. Compressor matching in natural gas pipeline applications is discussed by Vincent-Genot [26]. Perhaps the most common application of high-speed fluid machinery today is in automotive turbochargers (approximately 3 million units worldwide are sold each year with turbochargers). Automotive turbocharger matching is described in manufacturers' literature [27].

c. Positive-Displacement Pumps

Positive-displacement pumps develop pressure by reducing the size of a volume in which liquid is completely confined. In contrast to turbomachines, high pressures may be developed at relatively low speeds, because the pumping effect depends on volume change instead of dynamic action.

Positive-displacement pumps frequently are used in hydraulic systems at pressures ranging up to 35 MPa (5000 psi). A principal advantage of hydraulic power is the high *power density* (power per unit weight or unit size) that can be achieved: for a given power output, a hydraulic system can be lighter and smaller than a typical electric or internal combustion engine drive system.

Numerous types of positive-displacement pumps have been developed. A few examples include piston pumps, vane pumps, and gear pumps. Within each type, pumps may be fixed- or variable-displacement. A comprehensive classification of pump types is given in [15].

The performance characteristics of most positive-displacement pumps are similar; in this section we shall focus on gear pumps. This pump type typically is used, for example, to supply pressurized lubricating oil in internal combustion engines. Figure 11.34 is a schematic diagram of a typical gear pump. Oil enters the space between the gears at the bottom of the pump cavity. Oil is carried outward and upward by the teeth of the rotating gears and exits through the outlet port at the top of the cavity. Pressure is generated as the oil is forced toward the pump outlet; leakage and backflow are prevented by the closely fitting gear teeth at the center of the pump, and by the small clearances maintained between the side faces of the gears and the pump housing. The close clearances require the hydraulic fluid to be kept extremely clean by full-flow filtration.

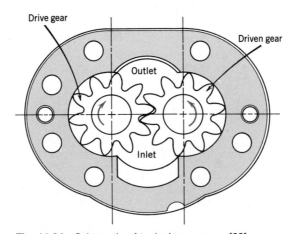

Fig. 11.34 Schematic of typical gear pump [28].

Fig. 11.35   Illustration of gear pump with pressure-loaded side plates
[28]. (Photo courtesy Sauer Sundstrand Company.)

Figure 11.35 is a photo showing the parts of an actual gear pump; it gives a good idea of the robust housing and bearings needed to withstand the large pressure forces developed within the pump. It also shows pressure-loaded side plates designed to "float"—to allow thermal expansion—while maintaining the smallest possible side clearance between gears and housing. Many ingenious designs have been developed for pumps; details are beyond the scope of our treatment here, which will focus on performance characteristics. For more details consult References [28] or [29].

Typical performance curves of pressure versus delivery for a medium-duty gear pump are shown in Fig. 11.36. The pump size is specified by its displacement per revolution and the working fluid is characterized by its viscosity and temperature. Curves for tests at three constant speeds are presented in the diagram. At each speed, delivery decreases slightly as pressure is raised. The pump displaces the same volume, but as pressure is raised, both leakage and backflow increase, so delivery decreases slightly. Leakage fluid ends up in the pump housing, so a case drain must be provided to return this fluid to the system reservoir.

Volumetric efficiency—shown by the dashed curves—is defined as actual delivery divided by pump displacement. Volumetric efficiency decreases as pressure is raised or pump speed is reduced. *Overall efficiency*—shown by the solid curves—is defined as power delivered to the fluid divided by power input to the pump. Overall efficiency tends to rise (and reaches a maximum at intermediate pressure) as pump speed increases.

Thus far we have shown pumps of fixed displacement only. The extra cost and complication of variable-displacement pumps is motivated by the energy saving they permit during partial-flow operation. In a variable-displacement pump, delivery can be varied to accommodate the load. Load sensing can be used to reduce still further the delivery pressure and thus the energy expenditure during part-load operation. Some pump designs allow pressure relief to further reduce power loss during standby operation.

Figure 11.37 illustrates system losses with a fixed-displacement pump compared to losses for variable-displacement and variable-pressure pumps. Assume the pressure

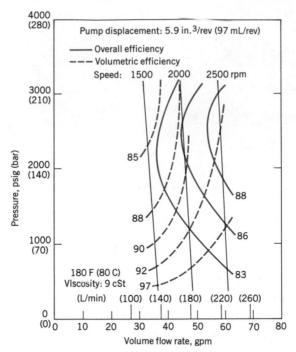

Fig. 11.36 Performance characteristics of typical gear pump [28].

and flow required by the load at partial-flow operation correspond to point L on the diagram. A fixed-displacement pump will operate along curve CD; its delivery will be at point A. Since the load requires only the flow at L, the remaining flow (between L and A) must be bypassed back to the reservoir. Its pressure is dissipated by throttling. Consequently the system power loss will be the area beneath line LA.

A variable-displacement pump operating at constant pressure will deliver just enough flow to supply the load, but at a pressure represented by point B. The system power loss will be proportional to the area to the left of line BL. Control of delivery pressure using load sensing can be used to reduce power loss. With a load sensing pump of variable displacement, the pressure supplied is only slightly higher than needed to move the load. A pump with load sensing would operate at the flow

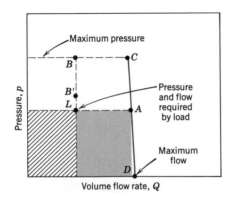

Fig. 11.37 Pressure-flow diagram illustrating system losses at part load [28].

and pressure of point B'. The system loss would be reduced significantly to the area to the left of line $B'L$.

The best system choice depends on the operating duty cycle. Complete details of these and other hydraulic power systems are presented in [28].

EXAMPLE 11.11—Performance of a Positive-Displacement Pump

A hydraulic pump, with the performance characteristics of Fig. 11.36, operates at 2000 rpm in a system that requires $Q = 20$ gpm at $p = 1500$ psig to the load at one operating condition. Check the volume of oil per revolution delivered by this pump. Compute the required pump power input, the power delivered to the load, and the power dissipated by throttling at this condition. Compare with the power dissipated by using (i) a variable-displacement pump at 3000 psig and (ii) a pump with load sensing that operates at 100 psig above the load requirement.

EXAMPLE PROBLEM 11.11

GIVEN: Hydraulic pump, with performance characteristics of Fig. 11.36, operating at 2000 rpm. System requires $Q = 20$ gpm at $p = 1500$ psig.

FIND: (a) Check the volume of oil per revolution delivered by this pump.
(b) Compute the required pump power input.
(c) Calculate the power delivered to the load.
(d) Calculate the power dissipated by throttling at this condition.
(e) Compare with the power dissipated using:
(i) a variable-displacement pump at 3000 psig, and
(ii) a pump with load sensing that operates at 100 psig above the load pressure requirement.

SOLUTION:
To estimate the maximum delivery, extrapolate the curve of pressure versus flow rate to zero pressure. Under these conditions, $Q = 48.5$ gpm at $N = 2000$ rpm with negligible Δp. Thus

$$\Psi = \frac{Q}{N} = \frac{48.5 \text{ gal}}{\text{min}} \times \frac{\text{min}}{2000 \text{ rev}} \times \frac{231 \text{ in.}^3}{\text{gal}} = 5.60 \text{ in.}^3/\text{rev} \qquad \Psi$$

The volumetric efficiency of the pump at maximum flow is

$$\eta_V = \frac{\Psi_{calc}}{\Psi_{pump}} = \frac{5.60}{5.9} = 0.949$$

The operating point of the pump may be found from Fig. 11.36. At 1500 psig, it operates at $Q \approx 46.5$ gpm. Applying the first law of thermodynamics to a CV enclosing the pump gives the power delivered to the fluid as

$$\mathscr{P}_{fluid} = Q\Delta p$$

$$= \frac{46.5 \text{ gal}}{\text{min}} \times \frac{2000 \text{ lbf}}{\text{in.}^2} \times \frac{\text{ft}^3}{7.48 \text{ gal}} \times \frac{\text{min}}{60 \text{ sec}} \times \frac{144 \text{ in.}^2}{\text{ft}^2} \times \frac{\text{hp} \cdot \text{sec}}{550 \text{ ft} \cdot \text{lbf}}$$

$$\mathscr{P}_{fluid} = 54.3 \text{ hp}$$

From the graph, at this operating point, the pump efficiency is approximately $\eta = 0.84$. Therefore the required input power is

$$\mathscr{P}_{input} = \frac{\mathscr{P}_{fluid}}{\eta} = \frac{54.3 \text{ hp}}{0.84} = 64.6 \text{ hp} \qquad \mathscr{P}_{input}$$

The power delivered to the load is

$$\mathcal{P}_{load} = Q_{load}\Delta p$$

$$= \frac{20.0 \text{ gal}}{\text{min}} \times 2000 \frac{\text{lbf}}{\text{in.}^2} \times \frac{\text{ft}^3}{7.48 \text{ gal}} \times \frac{\text{min}}{60 \text{ sec}} \times \frac{144 \text{ in.}^2}{\text{ft}^2} \times \frac{\text{hp}\cdot\text{sec}}{550 \text{ ft}\cdot\text{lbf}}$$

$\mathcal{P}_{load} = 23.3 \text{ hp}$ ⟵ $\mathcal{P}_{load}$

The power dissipated by throttling is

$$\mathcal{P}_{dissipated} = \mathcal{P}_{fluid} - \mathcal{P}_{load} = 54.3 - 23.3 = 31.0 \text{ hp} \quad\quad\quad\quad\quad \mathcal{P}_{dissipated}$$

The dissipation with the variable-displacement pump is

$$\mathcal{P}_{var\text{-}disp} = Q\Delta p$$

$$= \frac{20.0 \text{ gal}}{\text{min}} \times \frac{(3000 - 1500) \text{ lbf}}{\text{in.}^2} \times \frac{\text{ft}^3}{7.48 \text{ gal}} \times \frac{\text{min}}{60 \text{ sec}} \times \frac{144 \text{ in.}^2}{\text{ft}^2} \times \frac{\text{hp}\cdot\text{sec}}{550 \text{ ft}\cdot\text{lbf}}$$

$\mathcal{P}_{var\text{-}disp} = 17.5 \text{ hp}$ ⟵ $\mathcal{P}_{var\text{-}disp}$

The dissipation with the variable-displacement pump is therefore less than the 31.0 hp dissipated with the constant-displacement pump and throttle. The saving is approximately 14 hp.

The final computation is for the load-sensing pump. If the pump pressure is 100 psig above that required by the load, the excess energy dissipation is

$$\mathcal{P}_{load\text{-}sense} = Q\Delta p$$

$$= \frac{20.0 \text{ gal}}{\text{min}} \times \frac{100 \text{ lbf}}{\text{in.}^2} \times \frac{\text{ft}^3}{7.48 \text{ gal}} \times \frac{\text{min}}{60 \text{ sec}} \times \frac{144 \text{ in.}^2}{\text{ft}^2} \times \frac{\text{hp}\cdot\text{sec}}{550 \text{ ft}\cdot\text{lbf}}$$

$\mathcal{P}_{load\text{-}sense} = 1.17 \text{ hp}$ ⟵ $\mathcal{P}_{load\text{-}sense}$

> The purpose of this problem was to contrast the performance of a system with a pump of constant displacement to that of systems with variable-displacement and load-sensing pumps. The specific savings depend on the system operating point and on the duty cycle of the system.

d. Propellers

It has been suggested that a propeller may be considered an axial-flow machine without a housing [10]. In common with other propulsion devices, a propeller produces thrust by imparting linear momentum to a fluid. Thrust production always leaves the stream with some kinetic energy and angular momentum that are not recoverable, so the process is never 100 percent efficient.

The one-dimensional flow model shown schematically in Fig. 11.38 is drawn as seen by an observer moving with the propeller, so the flow is steady. The actual propeller is replaced by a thin *actuator disk*, across which flow speed is continuous but pressure rises abruptly. Relative to the propeller, the upstream flow is at speed V and ambient pressure. The axial speed at the actuator disk is $V + \Delta V/2$, with a corresponding reduction in pressure. Downstream, the speed is $V + \Delta V$ and the pressure returns to ambient. (It is shown in Example Problem 11.12 that half the speed increase occurs before and half after the actuator disk.) The contraction of the slipstream area to satisfy continuity and the pressure rise across the propeller disk are shown in the figure.

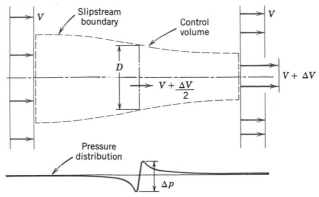

Fig. 11.38 One-dimensional flow model and control volume used to analyze an idealized propeller [10].

Not shown in the figure are the swirl velocities that result from the torque required to turn the propeller. The kinetic energy due to swirl in the slipstream also is lost unless it is removed by a counter-rotating propeller or partially recovered in stationary guide vanes.

As for all turbomachinery, propellers may be analyzed in two ways. Application of linear momentum in the axial direction, using a finite control volume, provides overall relations among slipstream speed, thrust, useful power output, and minimum residual kinetic energy in the slipstream. A more detailed *blade element* theory is needed to calculate the interaction between a propeller blade and the stream. A general relation for ideal propulsive efficiency can be derived using the control volume approach, as shown following Example 11.12.

EXAMPLE 11.12—Control Volume Analysis of Idealized Flow through a Propeller

Consider the one-dimensional model shown in Fig. 11.38 for the idealized flow through a propeller. The propeller advances into still air at steady speed V_1. Obtain expressions for the pressure immediately upstream and the pressure immediately downstream from the actuator disk. Write the thrust on the propeller as the product of this pressure difference times the disk area. Equate this expression for thrust to one obtained by applying the linear momentum equation to the control volume. Show that half the velocity increase occurs ahead of and half behind the propeller disk.

EXAMPLE PROBLEM 11.12

GIVEN: Propeller advancing into still air at speed V_1, as shown in Fig. 11.38.

FIND: (a) Expressions for the pressures immediately upstream and immediately downstream from the actuator disk.

         (b) Show that half the velocity increase occurs ahead of the actuator disk and half occurs behind the actuator disk.

SOLUTION:
Apply the Bernoulli equation and the x component of linear momentum using the CV shown.

$$\approx 0(5)$$

Basic equations: $\qquad\qquad \dfrac{p}{\rho} + \dfrac{V^2}{2} + g\cancel{z} = \text{constant}$

$$= 0(5) \quad = 0(1)$$

$$F_{S_x} + F_{B_x} = \frac{\partial}{\partial t} \int_{CV} u_{xyz}\rho \, d\mathbf{V} + \int_{CS} u_{xyz}\rho \vec{V} \cdot d\vec{A}$$

Assumptions:
(1) Steady flow relative to the CV
(2) Incompressible flow
(3) Flow along a streamline
(4) Frictionless flow
(5) Horizontal: neglect changes in z; $F_{B_x} = 0$
(6) Uniform flow at each section
(7) p_{atm} surrounds the CV

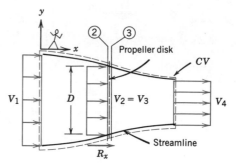

Applying the Bernoulli equation from section ① to section ② gives

$$\frac{p_{atm}}{\rho} + \frac{V_1^2}{2} = \frac{p_2}{\rho} + \frac{V_2^2}{2}; \quad p_{2(gage)} = \tfrac{1}{2}\rho(V_1^2 - V_2^2)$$

Applying Bernoulli from section ③ to section ④ gives

$$\frac{p_3}{\rho} + \frac{V_3^2}{2} = \frac{p_{atm}}{\rho} + \frac{V_4^2}{2}; \quad p_{3(gage)} = \tfrac{1}{2}\rho(V_4^2 - V_3^2)$$

The thrust on the propeller is given by

$$T = (p_3 - p_2)A = \tfrac{1}{2}\rho A(V_4^2 - V_1^2) \qquad (V_3 = V_2 = V)$$

From the momentum equation, using *relative* velocities,

$$R_x = T = u_1\{-\dot{m}\} + u_4\{+\dot{m}\} = \rho V A(V_4 - V_1) \qquad \{u_1 = V_1, \quad u_4 = V_4\}$$

$$T = \rho V A(V_4 - V_1)$$

Equating these two expressions for T,

$$T = \tfrac{1}{2}\rho A(V_4^2 - V_1^2) = \rho V A(V_4 - V_1) \quad \text{or} \quad \tfrac{1}{2}(V_4 + V_1)(V_4 - V_1) = V(V_4 - V_1)$$

Thus $V = \tfrac{1}{2}(V_1 + V_4)$, so

$$\Delta V_{12} = V - V_1 = \tfrac{1}{2}(V_1 + V_4) - V_1 = \tfrac{1}{2}(V_4 - V_1) = \frac{\Delta V}{2}$$

$$\Delta V_{34} = V_4 - V = V_4 - \tfrac{1}{2}(V_1 + V_4) = \tfrac{1}{2}(V_4 - V_1) = \frac{\Delta V}{2}$$

Velocity Increase

$$\left\{ \begin{array}{l} \text{The purpose of this problem was to apply the continuity, momentum, and Bernoulli} \\ \text{equations to an idealized flow model of a propeller, and to verify the Rankine theory} \\ \text{of 1885 that half the velocity change occurs on either side of the propeller disk.} \end{array} \right\}$$

The control volume forms of the continuity and momentum equations were applied in Example Problem 11.12 to the propeller flow shown in Fig. 11.38. The results obtained are discussed further below. The thrust produced is

$$F_T = \dot{m} \, \Delta V \tag{11.26}$$

For incompressible flow, in the absence of friction and heat transfer, the energy equation indicates that the minimum required input to the propeller is the power required to increase the kinetic energy of the flow, which may be expressed as

$$\mathscr{P}_{\text{input}} = \dot{m}\left[\frac{(V+\Delta V)^2}{2} - \frac{V^2}{2}\right] = \dot{m}\left[\frac{2V\Delta V + (\Delta V)^2}{2}\right] = \dot{m}V\Delta V\left[1 + \frac{\Delta V}{2V}\right] \quad (11.27)$$

The useful power produced is the product of thrust and speed of advance, V; using Eq. 11.26, this may be written as

$$\mathscr{P}_{\text{useful}} = F_T V = \dot{m}V\Delta V \quad (11.28)$$

Combining Eqs. 11.27 and 11.28, and simplifying, gives the propulsive efficiency as

$$\eta = \frac{\mathscr{P}_{\text{useful}}}{\mathscr{P}_{\text{input}}} = \frac{1}{1 + \dfrac{\Delta V}{2V}} \quad (11.29)$$

Equations 11.26 through 11.29 are applicable to any device that creates thrust by increasing the speed of a fluid stream. Thus they apply equally well to propeller-driven or jet-propelled aircraft, boats, or ships.

Equation 11.29 for propulsive efficiency is of fundamental importance. It indicates that propulsive efficiency can be increased by reducing ΔV or by increasing V. At constant thrust, as shown by Eq. 11.26, ΔV can be reduced if $\dot{m}$ is increased, i.e., if more fluid is accelerated over a smaller speed increase. More mass flow can be handled if propeller diameter is increased, but overall size and tip speed ultimately limit this approach. The same principle is used to increase the propulsive efficiency of a *turbofan* engine by using a large fan to move additional air flow outside the engine core.

Propulsive efficiency also can be increased by increasing the speed of motion relative to the fluid. Speed of advance may be limited by cavitation in marine applications. Flight speed is limited for propeller-driven aircraft by compressibility effects at the propeller tips, but progress is being made in the design of propellers to maintain high efficiency with low noise levels while operating with transonic flow at the blade tips. Jet-propelled aircraft can fly much faster than propeller-driven craft, giving them superior propulsive efficiency.

A more detailed *blade element* theory may be used to calculate the interaction between a propeller blade and the stream. If the blade spacing is large and the *disk loading*[10] is light, blades can be considered independent and relations can be derived for the torque required and the thrust produced by a propeller. These approximate relations are most accurate for low solidity propellers.[11] Aircraft propellers typically are of fairly low solidity, having long, thin blades.

A schematic of an element of a rotating propeller blade is shown in Fig. 11.39. The blade is set at angle θ to the plane of the propeller disk.[12] Flow is shown as it would be seen by an observer *on* the propeller blade.

The relative speed of flow, V_r, passing over the blade element depends on both the blade peripheral speed, ωr, and the *speed of advance*, V. Consequently, for a

---

[10] *Disk loading* is the propeller thrust divided by the swept area of the actuator disk.

[11] *Solidity* is defined as the ratio of projected blade area to the swept area of the actuator disk.

[12] *Pitch* is defined as the distance a propeller would travel in still fluid per revolution if it advanced along the blade setting angle θ. The pitch, H, of this blade element is equal to $2\pi r \tan\theta$. To obtain constant pitch along the blade, θ must follow the relation, $\tan\theta = H/2\pi r$, from hub to tip. Thus the geometric blade angle is smallest at the tip and increases steadily toward the root.

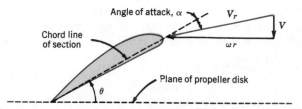

Fig. 11.39 Diagram of blade element and relative flow vector.

given blade setting, the angle of attack, α, depends on *both* V and ωr. Thus, the performance of a propeller is influenced by both ω and V.

Even if the geometry of the propeller is adjusted to give constant geometric pitch, the flow field in which it operates may not be uniform. Thus, the angle of attack across the blade elements may vary from the ideal, and can be calculated only with the aid of a comprehensive computer code that can predict local flow directions and speeds.

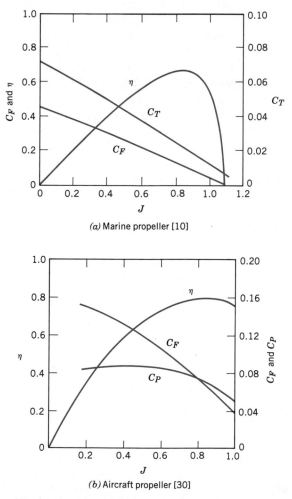

(a) Marine propeller [10]

(b) Aircraft propeller [30]

Fig. 11.40 Typical measured characteristics of two propellers.

Propeller performance characteristics usually are measured experimentally. Figure 11.40 shows typical measured characteristics for a marine propeller [10], and for an aircraft propeller [30]. The variables used to plot the characteristics are almost dimensionless: by convention, the rotational speed, n, is expressed in revolutions per second (rather than as ω, in radians per second). The independent variable is the *speed of advance coefficient, J,*

$$J \equiv \frac{V}{nD} \tag{11.30}$$

Dependent variables are the *thrust coefficient, C_F,* the *torque coefficient, C_T,* the *power coefficient, C_P,* and the *propeller efficiency, η,* defined as

$$C_F = \frac{F_T}{\rho n^2 D^4}, \quad C_T = \frac{T}{\rho n^2 D^5}, \quad C_P = \frac{\mathcal{P}}{\rho n^3 d^5}, \quad \text{and} \quad \eta = \frac{F_T V}{\omega T} \tag{11.31}$$

The performance curves for both propellers show similar trends. Both thrust and torque coefficients are highest, and efficiency is zero, at zero speed of advance. This corresponds to the largest angle of attack for each blade element. Efficiency is zero because no useful work is being done by the stationary propeller. As advance speed increases, thrust and torque decrease smoothly. Efficiency increases to a maximum at an optimum advance speed and then decreases to zero as thrust tends to zero.

EXAMPLE 11.13—Sizing a Marine Propeller

Consider the supertanker of Example Problem 9.5. Assume the total power required to overcome viscous resistance and wave drag is 11.4 MW. Use the performance characteristics of the marine propeller shown in Fig. 11.40a to estimate the diameter and operating speed required to propel the supertanker using a single propeller.

EXAMPLE PROBLEM 11.13

GIVEN: Supertanker of Example Problem 9.5, with total propulsion power requirement of 11.4 MW to overcome viscous and wave drag, and performance data for the marine propeller shown in Fig. 11.40a.

FIND: (a) Estimate the diameter of a single propeller needed to power the ship.
(b) Compute the operating speed of this propeller.

SOLUTION:
From the curves in Fig. 11.40a, at optimum propeller efficiency, the coefficients are

$$J = 0.85, \quad C_F = 0.10, \quad C_T = 0.020, \quad \text{and} \quad \eta = 0.66$$

The ship steams at $V = 6.69$ m/sec and requires $\mathcal{P}_{\text{useful}} = 11.4$ MW. Therefore, the propeller thrust must be

$$F_T = \frac{\mathcal{P}_{\text{useful}}}{V} = \frac{11.4 \times 10^6 \text{ W}}{} \times \frac{\text{sec}}{6.69 \text{ m}} \times \frac{\text{N} \cdot \text{m}}{\text{W} \cdot \text{sec}} = 1.70 \text{ MN}$$

The required power input to the propeller is

$$\mathcal{P}_{\text{input}} = \frac{\mathcal{P}_{\text{useful}}}{\eta} = \frac{11.4 \text{ MW}}{0.66} = 17.3 \text{ MW}$$

From $J = \dfrac{V}{nD} = 0.85$, then

$$nD = \frac{V}{J} = \frac{6.69 \text{ m}}{\text{sec}} \times \frac{1}{0.85} = 7.87 \text{ m/sec}$$

Since

$$C_F = \frac{F_T}{\rho n^2 D^4} = 0.10 = \frac{F_T}{\rho (n^2 D^2) D^2} = \frac{F_T}{\rho (nD)^2 D^2}$$

then, solving for D,

$$D = \left[\frac{F_T}{\rho (nD)^2 C_F} \right]^{1/2} = \left[1.70 \times 10^6 \text{ N} \times \frac{\text{m}^3}{1025 \text{ kg}} \times \frac{\text{sec}^2}{(7.87)^2 \text{ m}^2} \times \frac{1}{0.10} \times \frac{\text{kg} \cdot \text{m}}{\text{N} \cdot \text{sec}^2} \right]^{1/2}$$

$$D = 16.4 \text{ m} \hspace{8cm} D$$

From $nD = \dfrac{V}{J} = 7.87$ m/sec, $n = \dfrac{nD}{D} = 7.87 \dfrac{\text{m}}{\text{sec}} \times \dfrac{1}{16.4 \text{ m}} = 0.480 \text{ sec}^{-1}$

so that

$$n = \frac{0.480}{\text{sec}} \times \frac{60 \text{ sec}}{\text{min}} = 28.8 \text{ rev/min} \hspace{5cm} n$$

The required propeller is quite large, but still smaller than the 25 m draft of the supertanker. The ship would need to take on seawater for ballast to keep the propeller submerged when not carrying a full load of petroleum.

$$\left\{ \begin{array}{l} \text{The purpose of this problem was to illustrate the use of dimensionless coefficient data} \\ \text{for the preliminary sizing of a marine propeller. This preliminary design process would} \\ \text{be repeated, using data for other propeller types, to find the optimum combination of} \\ \text{propeller size, speed, and efficiency.} \end{array} \right.$$

Marine propellers tend to have high solidity. This packs a lot of lifting surface within the swept area of the disk to keep the pressure difference small across the propeller and to avoid cavitation. Cavitation tends to unload the blades of a marine propeller, reducing both the torque required and the thrust produced [10]. Cavitation becomes more prevalent along the blades as the cavitation number,

$$Ca = \frac{p - p_v}{\frac{1}{2} \rho V^2} \tag{11.32}$$

is reduced. Inspection of Eq. 11.32 shows that Ca decreases when p is reduced by operating near the free surface or by increasing V. Those who have operated motor boats also are aware that local cavitation can be caused by distorted flow approaching the propeller, e.g., from turning sharply.

Compressibility affects aircraft propellers when tip speeds approach the *critical Mach number* at which the local Mach number approaches $M = 1$ at some point on the blade. Under these conditions, torque increases due to increased drag, thrust drops due to reduced section lift, and efficiency drops drastically.

If a propeller operates within the boundary layer of a propelled body, where the relative flow is slowed, its apparent thrust and torque may increase compared to those in a uniform freestream at the same rate of advance. The residual kinetic energy in the slipstream also may be reduced. The combination of these effects may increase the overall propulsive efficiency of the combined body and propeller. Advanced computer codes are used in the design of modern ships (and submarines, where noise may be an overriding consideration) to optimize performance of each propeller/hull combination.

For certain special applications, a propeller may be placed within a *shroud* or *duct*. Such configurations may be integrated into a hull, e.g., as a bow thruster to increase maneuverability, built into the wing of an aircraft, or placed on the deck of a hover-craft. Thrust may be improved by the favorable pressure forces on the duct lip, but efficiency may be reduced by the added skin friction losses encountered in the duct.

11-5.2 Work-Producing Machines

a. Hydraulic Turbines

Hydraulic turbines convert the potential energy of stored water to mechanical work. To maximize turbine efficiency, it is always a design goal to discharge water from a turbine at ambient pressure, as close to the tailwater elevation as possible, and with the minimum possible residual kinetic energy.

Conveying water flow into the turbine with minimum energy loss also is important. Numerous design details must be considered, such as inlet geometry, trash racks, etc. [31]. References [1, 8, 10, and 31-37] contain a wealth of information about turbine siting, selection, hydraulic design, and optimization of hydropower plants. The number of large manufacturers has dwindled to just a few, but small-scale units are becoming plentiful [34]. The enormous cost of a commercial-scale hydro plant justifies the use of comprehensive scale-model testing to finalize design details. See [31] for detailed coverage of hydraulic power generation.

Hydraulic losses in long supply pipes (known as *penstocks*) must be considered when designing the installation for high-head machines such as impulse turbines; an optimum diameter for the inlet pipe that maximizes turbine output power can be determined for these units, as shown in Example Problem 11.14.

Turbine power output is proportional to volume flow rate times the pressure difference across the turbine. At zero flow, the full hydrostatic head is available but power is zero. As flow rate increases, the net head at the turbine inlet decreases. Power first increases, reaches a maximum, then decreases again as flow rate increases. For a given penstock diameter, the theoretical maximum power is obtained when one third of the gross head is dissipated by friction losses in the penstock. In practice, penstock diameter is chosen larger than the theoretical minimum, and only 10-15 percent of the gross head is dissipated by friction [8].

A certain minimum penstock diameter is required to produce a given power output. The minimum diameter depends on the desired power output, the available head, and the penstock material and length. Some representative values are shown in Fig. 11.41.

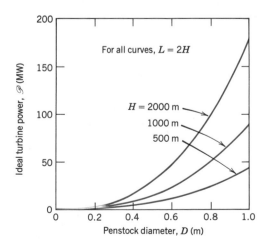

Fig. 11.41 Maximum hydraulic impulse turbine power output versus penstock diameter.

EXAMPLE 11.14—Performance and Optimization of an Impulse Turbine

Consider the hypothetical impulse turbine installation shown. Analyze flow in the penstock to develop an expression for turbine output power as a function of jet velocity. Obtain an expression for the volume flow rate at which output power is maximized. Develop a parametric equation for the minimum penstock diameter needed to produce a specified power output, using gross head and penstock length as parameters.

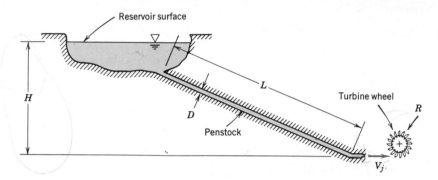

EXAMPLE PROBLEM 11.14

GIVEN: Impulse turbine installation shown.

FIND: (a) Expression for turbine output power as a function of jet velocity.
(b) Flow rate at which output power is maximized.
(c) Parametric equation for the minimum penstock diameter needed to produce a specified output power, using gross head and penstock length as parameters.

SOLUTION:
According to the results of Example 11.5, the output power of an idealized impulse turbine is given by $\mathcal{P}_{\text{out}} = \rho Q U (V - U)(1 - \cos\theta)$. For optimum power output, $U = V/2 = V_j/2$, and

$$\mathcal{P}_{\text{out}} = \rho Q \frac{V}{2}(V - \frac{V}{2})(1 - \cos\theta) = \rho A_j V_j \frac{V_j}{2} \frac{V_j}{2}(1 - \cos\theta)$$

$$\mathcal{P}_{\text{out}} = \rho A_j \frac{V_j^3}{4}(1 - \cos\theta)$$

Thus output power is proportional to V_j^3.

Apply the energy equation for steady incompressible pipe flow through the penstock to analyze V_j^2 at the nozzle outlet. The free surface of the reservoir is designated as section ①; there $\bar{V}_1 \approx 0$.

Computing equation:

$$\frac{\cancel{p_1}}{\rho} + \alpha_1 \overset{\approx 0}{\cancel{\frac{\bar{V}_1^2}{2}}} + g z_1 - \left(\frac{\cancel{p_j}}{\rho} + \alpha_j \frac{\bar{V}_j^2}{2} + g z_j\right) = h_{lT} = \left(K_{\text{ent}} + f \frac{L}{D}\right) \frac{\bar{V}_p^2}{2} + K_{\text{nozzle}} \frac{\bar{V}_j^2}{2}$$

Assumptions: (1) Steady flow
(2) Incompressible flow
(3) Fully developed flow
(4) Atmospheric pressure at jet exit
(5) $\alpha_j = 1$, so $\bar{V}_j = V_j$
(6) $K_{\text{ent}} \ll f \dfrac{L}{D}$
(7) $K_{\text{nozzle}} = 0$

Then

$$g(z_1 - z_j) = gH = f\frac{L}{D}\frac{\bar{V}_p^2}{2} + \frac{V_j^2}{2} \quad \text{or} \quad V_j^2 = 2gH - f\frac{L}{D_p}\bar{V}_p^2$$

The turbine power can be written as

$$\mathcal{P} = \rho A_j \frac{V_j^3}{4}(1 - \cos\theta) = \rho A_j(1 - \cos\theta)V_j^2\frac{V_j}{4} = C_1 V_j^2\frac{V_j}{2}$$

where $C_1 = \frac{1}{2}\rho A_j(1 - \cos\theta) = $ constant. Hence

$$\mathcal{P} = C_1\left[2gH - f\frac{L}{D_p}\bar{V}_p^2\right]\frac{V_j}{2}$$

From continuity, $\bar{V}_p A_p = V_j A_j$, so $\bar{V}_p = V_j\dfrac{A_j}{A_p} = V_j\left(\dfrac{D_j}{D_p}\right)^2$. Substituting,

$$\mathcal{P} = C_1\left[2gH - f\frac{L}{D_p}\left(\frac{D_j}{D_p}\right)^4 V_j^2\right]\frac{V_j}{2} = C_1\left[gHV_j - f\frac{L}{D_p}\left(\frac{D_j}{D_p}\right)^4\frac{V_j^3}{2}\right]$$

To find the condition for maximum power output, at fixed penstock diameter, D_p, differentiate with respect to V_j and set equal to zero,

$$\frac{d\mathcal{P}}{dV_j} = C_1\left[gH - 3f\frac{L_j}{D_p}\left(\frac{D_j}{D_p}\right)^4\frac{V_j^2}{2}\right] = C_1\left[gH - 3f\frac{L}{D_p}\frac{V_p^2}{2}\right] = C_1\left[gH - 3h_{lT}\right] = 0$$

or

$$h_{lT} = \frac{1}{3}gH$$

For fixed penstock diameter, the turbine power is maximized when the frictional head loss, h_{lT}, is equal to 1/3 of the gross head available at the site. The net head is then 2/3 of the gross head.

The jet and pipe diameters for optimum operation also may be found. The jet speed at optimum conditions is

$$V_j^2 = 2gH - f\frac{L}{D_p}\bar{V}_p^2 = 2gH - \frac{1}{3}(2gH) = \frac{4}{3}gH \quad \text{or} \quad V_j = \sqrt{\frac{4}{3}gH} \qquad\qquad V_j$$

The pipe flow speed may be obtained from

$$f\frac{L}{D_p}\frac{\bar{V}_p^2}{2} = \frac{1}{3}gH \quad \text{or} \quad \bar{V}_p = \left[\frac{2}{3}\frac{gH}{f\dfrac{L}{D_p}}\right]^{\frac{1}{2}} \qquad\qquad \bar{V}_p$$

Using the continuity equation to relate the flow velocities and diameters leads to the relation,

$$\bar{V}_p^2 = \frac{2}{3}\frac{gHD_p}{fL} = V_j^2\left(\frac{D_j}{D_p}\right)^4 = \frac{4}{3}gh\left(\frac{D_j}{D_p}\right)^4$$

Solving for D_j,

$$D_j = \left[\frac{D_p^5}{2fL}\right]^{\frac{1}{4}}$$

Substituting into the equation for ideal turbine ouptut power,

$$\mathcal{P}_{optimum} = \rho V_j^3\frac{A_j}{4}(1 - \cos\theta) = \rho\left(\frac{4}{3}gH\right)^{\frac{3}{2}}\frac{\pi}{16}\left[\frac{D_p^5}{2fL}\right]^{\frac{1}{2}}(1 - \cos\theta)$$

Finally, solving for penstock diameter, the equation may be written in the form

$$D_p \propto \left(\frac{L}{H}\right)^{\frac{1}{5}} \left(\frac{\mathscr{P}}{H}\right)^{\frac{2}{5}}$$

D_p

{ The purpose of this problem was to illustrate the optimization of an idealized impulse turbine. The relations derived above define the minimum penstock size needed to obtain a specified power output. Actual practice is to use larger penstocks, thus reducing the frictional head loss compared to the case analyzed here. }

b. Wind-Power Machines

Windmills (or more properly, wind turbines) have been used for centuries to harness the power of natural winds. Two well-known examples are shown in Fig. 11.42.

(a) Traditional Dutch mill

(b) American farm windmill

Fig. 11.42 Examples of well-known windmills [38]. (Photo courtesy (a) Netherlands Board of Tourism, (b) U.S. Department of Agriculture.)

Dutch windmills (Fig. 11.42a) turned slowly so the power could be used to turn stone wheels for milling grain, hence the name "windmill." They evolved into large structures; the practical maximum size was limited by the materials of the day. Calvert [39] reports that, based on his laboratory-scale tests, a traditional Dutch windmill of 26 m diameter produced 41 kW in a wind of 36 km/hr at an angular speed of 20 rpm.

American multi-blade windmills (Fig. 11.42b) were found on many American farms between about 1850 and 1950. They performed valuable service in powering water pumps before rural electrification.

The recent emphasis on renewable resources has revived interest in windmill design and optimization. Horizontal-axis wind turbine (HAWT) and vertical-axis wind turbine (VAWT) configurations have been studied extensively. Most HAWT designs feature 2- or 3-bladed propellers turning at high speed. The large modern HAWT, shown in Fig. 11.43a, is capable of producing power in any wind above a light breeze.

The final example (Fig. 11.43b) is a *Darrieus* VAWT. This device uses a modern symmetric airfoil section for the rotor, which is formed into a *troposkien* shape.[13] In

[13] This shape (which would be assumed by a flexible cord whirled about a vertical axis) minimizes bending stresses in the Darrieus turbine rotor.

(a) Horizontal-axis wind turbine [38]   (b) Vertical-axis wind turbine [40]

Fig. 11.43 Examples of modern wind turbine designs. (Photo courtesy U.S. Department of Energy.)

contrast to the other designs, the Darrieus VAWT is not capable of starting from rest; it can only produce usable power above a certain minimum angular speed. It may be combined with a self-starting turbine, such as a *Savonius rotor*, (see illustration for Problem 9.95 or [47]) to provide starting torque.

A horizontal-axis wind turbine may be analyzed as a propeller operated in reverse. The one-dimensional flow, idealized actuator disk model of Rankine (Fig. 11.44) is used. The simplified notation of the figure is frequently used to analyze wind turbines.

The wind speed far upstream is V. The stream is decelerated to $V(1-a)$ at the turbine disk and to $V(1-2a)$ in the wake of the turbine (a is called the *interference factor*). Thus the streamtube of air captured by the windmill is small upstream and its diameter increases as it moves downstream.

Straightforward application of linear momentum (see Example Problem 11.15) predicts the axial thrust on a turbine of radius R to be

$$F_T = 2\pi R^2 \rho V^2 a (1-a) \tag{11.33}$$

Application of the energy equation, assuming no losses (no change in internal energy or heat transfer), gives the power taken from the fluid stream as

$$\mathcal{P} = 2\pi R^2 \rho V^3 a (1-a)^2 \tag{11.34}$$

The efficiency of a windmill is most conveniently defined with reference to the kinetic energy flux contained within a streamtube the size of the actuator disk. This

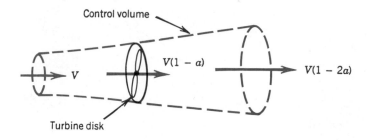

Control volume

Turbine disk

V $V(1-a)$ $V(1-2a)$

Fig. 11.44 Control volume and simplified notation used to analyze wind turbine performance.

kinetic energy flux is

$$KEF = \frac{1}{2}\rho V^3 \pi R^2 \qquad (11.35)$$

Combining Eqs. 11.34 and 11.35 gives the efficiency (or alternatively, the *power coefficient* [40]) as

$$\eta = \frac{\wp}{KEF} = 4a(1-a)^2 \qquad (11.36)$$

Betz [40] was the first to derive this result and to show that the theoretical efficiency is maximized when $a = 1/3$. The maximum theoretical efficiency is $\eta = 0.593$.

If the windmill is lightly loaded (a is small), it will affect a large mass flow rate of air, but the energy extracted per unit mass will be small and the efficiency low. Most of the kinetic energy in the initial air stream will be left in the wake and wasted. If the windmill is heavily loaded ($a \approx 1/2$), it will affect a much smaller mass flow rate. The energy removed per unit mass will be large, but the power produced will be small compared to the kinetic energy flux through the undisturbed area of the actuator disk. Thus a peak efficiency occurs at intermediate disk loadings.

The Rankine analysis includes some important assumptions that limit its applicability [40]. First, the wind turbine is assumed to affect only the air contained within the streamtube defined in Fig. 11.44. Second, the kinetic energy produced as swirl behind the turbine is not accounted for. Third, any radial pressure gradient is ignored. Glauert [30] partially accounted for the wake swirl to predict the dependence of ideal efficiency on *tip speed ratio*, $X = \omega R/V$, as shown in Fig. 11.45 (ω is the angular velocity of the turbine).

As the tip speed ratio increases, ideal efficiency increases, approaching the peak value ($\eta = 0.593$) asymptotically. (Physically, the swirl left in the wake is reduced as the tip speed ratio increases.) Reference [40] presents a summary of the detailed blade element theory used to develop the limiting efficiency curve shown in Fig. 11.45.

Each type of wind turbine has its most favorable range of application. The traditional American multi-bladed windmill has a large number of blades and operates at relatively slow speed. Its solidity, σ (the ratio of blade area to the swept area of the turbine disk, πR^2) is high. Because of its relatively slow operating speed, its tip speed ratio and theoretical performance limit are low. Its relatively poor performance, compared to its theoretical limit, is largely due to use of unsophisticated blades, which are simple bent sheet metal surfaces rather than airfoil shapes.

It is necessary to increase the tip speed ratio considerably to reach a more favorable operating range. Modern high-speed wind turbine designs use carefully shaped airfoils and operate at tip speed ratios up to 7 [48].

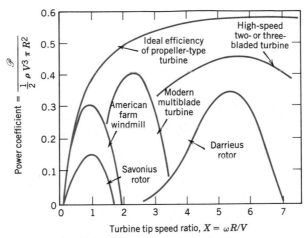

Fig. 11.45   Efficiency trends of wind turbine types versus tip speed ratio [40].

EXAMPLE 11.15—Performance of an Idealized Windmill

Develop general expressions for thrust, power output, and efficiency of an idealized windmill, as shown in Fig. 11.44. Calculate the thrust, ideal efficiency, and actual efficiency for the Dutch windmill tested by Calvert ($D = 26$ m, $\omega = 20$ rpm, $V = 36$ km/hr, and $\mathscr{P}_{output} = 41$ kW).

EXAMPLE PROBLEM 11.15

GIVEN: Idealized windmill, as shown in Fig. 11.44, and Dutch windmill tested by Calvert:

$$D = 26 \text{ m} \qquad \omega = 20 \text{ rpm} \qquad V = 36 \text{ km/hr} \qquad \mathscr{P}_{output} = 41 \text{ kW}$$

FIND: (a) General expressions for the ideal thrust, power output, and efficiency.
(b) Calculate the thrust, power output, and the ideal and actual efficiencies for the Dutch windmill tested by Calvert.

SOLUTION:
Apply the continuity, x component of momentum, and energy equations, using the CV and coordinates shown.

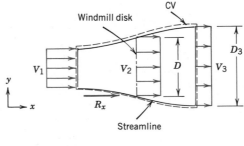

Basic equations: $0 = \dfrac{\partial}{\partial t}\!\!\!\diagup\!\!\!\int_{CV} \rho \, d\forall + \int_{CS} \rho \vec{V} \cdot d\vec{A}$
$$\qquad\qquad\qquad\quad = 0(3)$$

$$F_{S_x} + \cancelto{=0(2)}{F_{B_x}} = \cancelto{=0(1)}{\frac{\partial}{\partial t}} \int_{CS} u\, \rho d\Psi + \int_{CS} u\, \rho \vec{V} \cdot d\vec{A}$$

$$\cancelto{=0(7)}{\dot{Q}} - \dot{W}_s = \cancelto{=0(3)}{\frac{\partial}{\partial t}} \int_{CV} e\, \rho d\Psi + \int_{CS} \left(e + \frac{p}{\rho}\right)\rho \vec{V} \cdot d\vec{A}$$

Assumptions: (1) Atmospheric pressure acts on CV; $F_{S_x} = R_x$
(2) $F_{B_x} = 0$
(3) Steady flow
(4) Uniform flow at each section
(5) Incompressible flow of standard air
(6) $V_1 - V_2 = V_2 - V_3 = \frac{1}{2}(V_1 - V_3)$, as shown by Rankine
(7) $\dot{Q} = 0$
(8) $u_1 = u_2 = u_3$, for frictionless incompressible flow

In terms of the interference factor, a, $V_1 = V$, $V_2 = (1-a)V$, and $V_3 = (1-2a)V$.

From continuity, for uniform flow at each cross section, $V_1 A_1 = V_2 A_2 = V_3 A_3$.

From momentum,

$$R_x = u_1\{-|\rho V_1 A_1|\} + u_3\{+|\rho V_3 A_3|\} = (V_3 - V_1)\rho V_2 A_2 \quad \{u_1 = V_1, \quad u_3 = V_3\}$$

R_x is the external force acting *on* the control volume. The thrust force exerted *by* the CV *on* the surroundings is

$$K_x = -R_x = (V_1 - V_3)\rho V_2 A_2$$

In terms of the interference factor, the equation for thrust may be written in the general form,

$$K_x = \rho V^2 \pi R^2 2a(1-a) \qquad\qquad\qquad\qquad \overset{\longleftarrow}{} \qquad K_x$$

The energy equation becomes

$$-\dot{W}_s = \frac{V_1^2}{2}\{-|\rho V_1 A_1|\} + \frac{V_3^2}{2}\{+|\rho V_3 A_3|\} = \rho V_2 \pi R^2 \frac{1}{2}(V_3^2 - V_1^2)$$

The output power, $\mathscr{P}$, is equal to $\dot{W}_s$. In terms of the interference factor,

$$\mathscr{P} = \dot{W}_s = \rho V(1-a)\pi R^2\left[\frac{V^2}{2} - \frac{V^2}{2}(1-2a)^2\right] = \rho V^3(1-a)\frac{\pi R^2}{2}\left[1 - (1-2a)^2\right]$$

After simpifying algebraically,

$$\mathscr{P}_{\text{ideal}} = 2\rho V^3 \pi R^2 a(1-a)^2 \qquad\qquad \overset{\longleftarrow}{} \qquad \mathscr{P}_{\text{ideal}}$$

The kinetic energy flux through a streamtube of undisturbed flow, equal in area to the actuator disk, is

$$KEF = \rho V \pi R^2 \frac{V^2}{2} = \frac{1}{2}\rho V^3 \pi R^2$$

Thus the ideal efficiency may be written

$$\eta = \frac{\mathscr{P}_{\text{ideal}}}{KEF} = \frac{2\rho V^3 \pi R^2 a(1-a)^2}{\frac{1}{2}\rho V^3 \pi R^2} = 4a(1-a)^2 \qquad \overset{\longleftarrow}{} \qquad \eta$$

To find the condition for maximum possible efficiency, set $d\eta/da$ equal to zero. The maximum efficiency is $\eta = 0.593$, which occurs when $a = 1/3$.

The Dutch windmill tested by Calvert had a tip speed ratio of

$$X = \frac{\omega R}{V} = \frac{20 \text{ rev}}{\text{min}} \times \frac{2\pi \text{ rad}}{\text{rev}} \times \frac{\text{min}}{60 \text{ sec}} \times \frac{13 \text{ m}}{} \times \frac{\text{sec}}{10 \text{ m}} = 2.72 \qquad\qquad X$$

The maximum theoretically attainable efficiency at this tip speed ratio, accounting for swirl (Fig. 11.45), would be about 0.53.

The actual efficiency of the Dutch windmill is

$$\eta_{actual} = \frac{\mathcal{P}_{actual}}{KEF}$$

Based on Calvert's test data, the kinetic energy flux is

$$KEF = \frac{1}{2}\rho V^3 \pi R^2$$

$$= \frac{1}{2} \times \frac{1.23 \text{ kg}}{m^3} \times \frac{(10)^3 \text{ m}^3}{\text{sec}^3} \times \pi(13)^2 \text{ m}^2 \times \frac{N \cdot \text{sec}^2}{\text{kg} \cdot m} \times \frac{W \cdot \text{sec}}{N \cdot m}$$

$$KEF = 3.27 \times 10^5 \text{ W} \quad\text{or}\quad 327 \text{ kW}$$

Substituting into the definition of actual efficiency,

$$\eta_{actual} = \frac{41 \text{ kW}}{327 \text{ kW}} = 0.125 \qquad\qquad \eta_{actual}$$

Thus the actual efficiency of the Dutch windmill is about 24 percent of the maximum efficiency theoretically attainable at this tip speed ratio.

The actual thrust on the Dutch windmill can only be estimated, because the interference factor, a, is not known. The maximum possible thrust would occur at $a = 1/4$, in which case,

$$K_x = \rho V^2 \pi R^2 2a(1-a)$$

$$= \frac{1.23 \text{ kg}}{m^3} \times \frac{(10)^2 \text{ m}^2}{\text{sec}^2} \times \pi(13)^2 \text{ m}^2 \times 2\left(\frac{1}{4}\right)\left(1 - \frac{1}{4}\right)$$

$$K_x = 2.45 \times 10^4 \text{ N} \quad\text{or}\quad 24.5 \text{ kN} \qquad\qquad K_x$$

This does not sound like too large a thrust force, considering the size ($D = 26$ m) of the windmill. However, $V = 36$ km/hr is only a moderate wind. The actual machine would have to withstand much more severe wind conditions during storms.

The purpose of this problem was to illustrate an application of the concepts of ideal thrust, power, and efficiency for a windmill, and to calculate the corresponding values for an actual machine.

11-6 SUMMARY OBJECTIVES

After completing study of Chapter 11, you should be able to do the following:

1. Give operational definitions of:
    positive displacement machine   compressor stage
    turbomachine   solidity

radial flow	turbine
axial flow	impulse turbine
mixed flow	reaction turbine
pump	draft tube
fan	hydraulic power
blower	mechanical power
compressor	turbine efficiency
impeller, rotor, wheel, or runner	shutoff head
pump efficiency	shock loss
scroll or volute	specific speed

2. Give a word statement of the angular momentum principle and write the control volume form of the basic equation.

3. Draw velocity polygons and apply the Euler turbomachine equation to calculate the ideal torque, head, and power developed by a fluid machine rotor.

4. Evaluate performance (head developed, power input, and efficiency) of a fluid machine from measured data.

5. Calculate and use dimensionless parameters to scale the performance of a fluid machine as the size, operating speed, or operating conditions are changed.

6. Use the net positive suction head concept to define and calculate the *NPSH* available and required by a fluid machine.

7. Develop and apply system resistance curves to predict the performance of a fluid machine as part of a flow system.

8. Predict performance of fluid machines installed in parallel or series combinations in a flow system.

9. Use manufacturers' literature to specify pumps and fans appropriate for use in defined flow systems.

10. Perform preliminary analyses and sizing for hydraulic and wind turbine installations.

11. Solve the problems at the end of the chapter that relate to the material you have studied.

REFERENCES

1. Wilson, D. G., "Turbomachinery—From Paddle Wheels to Turbojets," Mechanical Engineering, *104*, 10, October 1982, pp. 28–40.

2. Logan, E. S., Jr., *Turbomachinery: Basic Theory and Applications.* New York: Dekker, 1981.

3. Private communication, Rocky Mountain Institute, 1990.

4. American Society of Mechanical Engineers, *Performance Test Codes: Centrifugal Pumps,* ASME PTC 8.2-1990. New York: ASME, 1990.

5. American Institute of Chemical Engineers, *Equipment Testing Procedure: Centrifugal Pumps (Newtonian Liquids).* New York: AIChE, 1984.

6. Air Movement and Control Association, *Laboratory Methods of Testing Fans for Rating.* AMCA Standard 210-74, ASHRAE Standard 51-75. Atlanta, GA: ASHRAE, 1975.

7. Peerless Pump,* Brochure B-4003, "System Analysis for Pumping Equipment Selection," 1979.

8. Daugherty, R. L., J. B. Franzini, and E. J. Finnemore, *Fluid Mechanics with Engineering Applications,* 8th ed. New York: McGraw-Hill, 1985.

9. Rouse, H., and S. Ince, *History of Hydraulics.* Iowa City, IA: Iowa University Press, 1957.

10. Daily, J. W., "Hydraulic Machinery," in Rouse, H., ed., *Engineering Hydraulics.* New York: Wiley, 1950.

* Peerless Pump Division, FMC Corporation, P.O. Box 7026, Indianapolis, IN 46206-7026.

11. Russell, G. E., *Hydraulics*, 5th ed. New York: Henry Holt, 1942.

12. Sabersky, R. H., A. J. Acosta, and E. G. Hauptmann, *Fluid Flow: A First Course in Fluid Mechanics*, 3rd ed. New York: Macmillan, 1989.

13. Moody, L. F., "Hydraulic Machinery," in *Handbook of Applied Hydraulics,* ed. by C. V. Davis. New York: McGraw-Hill, 1942.

14. Hodge, B. K., *Analysis and Design of Energy Systems*, 2nd ed. Englewood Cliffs, NJ: Prentice-Hall, 1990.

15. Dickinson, C., *Pumping Manual,* 8th ed. Surrey, England: Trade & Technical Press, Ltd., 1988.

16. Hicks, T. G., and T. W. Edwards, *Pump Application Engineering.* New York: McGraw-Hill, 1971.

17. Hydraulic Institute, *Hydraulic Institute Standards.* New York: Hydraulic Institute, 1969.

18. Armintor, J. K., and D. P. Conners, "Pumping Applications in the Petroleum and Chemical Industries," IEEE Transactions on Industry Applications, *IA-23*, 1, January 1987.

19. Jorgensen, R., ed., *Fan Engineering,* 8th ed. Buffalo, NY: Buffalo Forge, 1983.

20. American Society of Mechanical Engineers, *Power Test Code for Fans.* New York: ASME, Power Test Codes, PTC 11-1946.

21. Berry, C. H., *Flow and Fan: Principles of Moving Air through Ducts*, 2nd ed. New York: Industrial Press, 1963.

22. Wallis, R. A., *Axial Flow Fans and Ducts.* New York: Wiley, 1983.

23. Osborne, W. C., *Fans,* 2nd ed. London: Pergamon Press, 1977.

24. American Society of Heating, Refrigeration, and Air Conditioning Engineers, *Handbook of Fundamentals.* Atlanta, GA: ASHRAE, 1980.

25. Idelchik, I. E., *Handbook of Hydraulic Resistance*, 2nd ed. New York: Hemisphere, 1986.

26. Vincent-Genod, J., *Fundamentals of Pipeline Engineering.* Houston: Gulf Publishing Co., 1984.

27. Warner-Ishi Turbocharger brochure.**

28. Lambeck, R. R., *Hydraulic Pumps and Motors: Selection and Application for Hydraulic Power Control Systems.* New York: Dekker, 1983.

29. Warring, R. H., ed., *Hydraulic Handbook,* 8th ed. Houston: Gulf Publishing Co., 1983.

30. Durand, W. F., ed., *Aerodynamic Theory,* 6 Volumes. New York: Dover, 1963.

31. Gulliver, J. S., and R. E. A. Arndt, *Hydropower Engineering Handbook.* New York: McGraw-Hill, 1990.

32. Fritz, J. J., *Small and Mini Hydropower Systems: Resource Assessment and Project Feasibility.* New York: McGraw-Hill, 1984.

33. Gladwell, J. S., *Small Hydro: Some Practical Planning and Design Considerations.* Idaho Water Resources Institute. Moscow, ID: University of Idaho, April 1980.

34. McGuigan, D., *Small Scale Water Power.* Dorchester: Prism Press, 1978.

35. Olson, R. M., and S. J. Wright, *Essentials of Engineering Fluid Mechanics,* 5th ed. New York: Harper & Row, 1990.

36. Quick, R. S., "Problems Encountered in the Design and Operation of Impulse Turbines," Transactions of the ASME, *62*, 1940, pp. 15–27.

37. Warnick, C. C., *Hydropower Engineering.* Englewood Cliffs, NJ: Prentice-Hall, 1984.

38. Putnam, P. C., *Power from the Wind.* New York: Van Nostrand, 1948.

39. Calvert, N. G., *Windpower Principles: Their Application on the Small Scale.* London: Griffin, 1978.

** Warner-Ishi, P.O. Box 580, Shelbyville, IL 62565-0580.

40. Baumeister, T., E. A. Avallone, and T. Baumeister, III, eds., *Marks' Standard Handbook for Mechanical Engineers,* 8th ed. New York: McGraw-Hill, 1978.

41. White, F. M., *Fluid Mechanics,* 2nd ed. New York: McGraw-Hill, 1986.

42. Sovern, D. T., and G. J. Poole, "Column Separation in Pumped Pipelines," in Kienow, K. K., ed., *Pipeline Design and Installation,* Proceedings of the International Conference on Pipeline Design and Installation, Las Vegas, Nevada, March 25–27, 1990. New York: American Society of Civil Engineers, 1990, pp. 230–243.

43. Stepanoff, A. J., *Pumps and Blowers, Two-Phase Flow.* New York: Wiley, 1965.

44. Drella, M., "Aerodynamics of Human-Powered Flight," in *Annual Review of Fluid Mechanics, 22,* pp. 93–110. Palo Alto, CA: Annual Reviews, 1990.

45. Shepherd, D. G., *Principles of Turbomachinery.* New York: Macmillan, 1956.

46. U.S. Department of the Interior, "Selecting Hydraulic Reaction Turbines," A Water Resources Technical Publication, *Engineering Monograph No. 20.* Denver, CO: U.S. Department of the Interior, Bureau of Reclamation, 1976.

47. Eldridge, F. R., *Wind Machines,* 2nd ed. New York: Van Nostrand Reinhold, 1980.

48. Migliore, P. G., "Comparison of NACA 6-Series and 4-Digit Airfoils for Darrieus Wind Turbines," Journal of Energy, *7,* 4, Jul–Aug 1983, pp. 291–292.

49. Abbott, I. H., and A. E. von Doenhoff, *Theory of Wing Sections, Including a Summary of Airfoil Data.* New York: Dover, 1959 (paperback).

50. Heald, C. C., ed., *Cameron Hydraulic Data,* 17th ed. Woodcliff Lake, NJ: Ingersoll-Rand, 1988.

PROBLEMS

11.1 Gasoline is pumped by a centrifugal pump. When the flow rate is 375 gpm, the pump requires 19.3 hp input, and its efficiency is 81.2 percent. Calculate the pressure rise produced by the pump. Express this result as (a) ft of water and (b) ft of gasoline.

‡11.2 Dimensions of a centrifugal pump impeller are

Parameter	Inlet, Section ①	Outlet, Section ②
Radius, r (mm)	100	300
Blade width, b (mm)	50	40
Blade angle, β (deg)	70	80

Assume the pump is driven at 1150 rpm while pumping water. Calculate the theoretical head and hydraulic power delivered to the fluid, and plot versus volume flow rate.

‡11.3 Dimensions of a centrifugal pump impeller are

Parameter	Inlet, Section ①	Outlet, Section ②
Radius, r (mm)	200	600
Blade width, b (mm)	60	40
Blade angle, β (deg)	50	70

The pump handles water and is driven at 850 rpm. Calculate the theoretical head, torque, and power input, and plot versus volume flow rate.

---

‡ You may wish to use simple computer programs to help solve problems marked with daggers.

‡11.4 Dimensions of a centrifugal pump impeller are

Parameter	Inlet, Section ①	Outlet, Section ②
Radius, r (mm)	400	1200
Blade width, b (mm)	120	80
Blade angle, β (deg)	40	60

The pump is driven at 575 rpm and the fluid is water. Calculate the theoretical head, torque, and power input, and plot versus volume flow rate.

11.5 The theoretical head delivered by a centrifugal pump at shutoff depends on the discharge radius and angular speed of the impeller. For preliminary design, it is useful to have a plot showing the theoretical shutoff characteristics and approximating the actual performance. Prepare a log-log plot of impeller radius versus theoretical head rise at shutoff with standard motor speeds as parameters. Assume the fluid is water and the actual head at the design flow rate is 30 percent less than the theoretical shutoff head (show these as dashed lines on the plot). Explain how this plot might be used for preliminary design.

‡11.6 Consider the centrifugal pump impeller dimensions given in Example Problem 11.2. Construct the velocity polygon for shockless flow at the impeller inlet, if b = constant. Calculate the effective flow angle with respect to the radial impeller blades for the case of no inlet swirl. Investigate the effects on flow angle of (a) variations in impeller width and (b) inlet swirl velocities.

‡11.7 Consider the centrifugal pump impeller dimensions given in Example Problem 11.2. Construct the velocity polygon for flow leaving the impeller tangent to the vanes, at outlet angles in the range $60 < \beta < 90$ degrees. Estimate the ideal pump torque, power input, and head rise at the given flow rate, for flow with zero inlet swirl. Plot the results versus β.

11.8 Consider a centrifugal pump whose geometry and flow conditions are

Impeller inlet radius, R_1	37.5 mm
Impeller outlet radius, R_2	150 mm
Impeller outlet width, b_2	12.7 mm
Design speed, ω	1750 rpm
Design flow rate, Q	4.25 L/sec
Backward-curved vanes (outlet blade angle), β_2	60°
Axial velocity at inlet, V_1	<3.05 m/sec
Required flow rate range	50–100% of design

Assume ideal pump behavior with 100 percent efficiency. Find the shutoff head. Calculate the absolute and relative discharge velocities, the total head, and the theoretical power required at the design flow rate.

‡11.9 Consider the geometry of the idealized centrifugal pump described in Problem 11.8. Draw inlet and outlet velocity polygons. Calculate the inlet vane angles required for "shockless" entry flow at the design flow rate. Estimate the difference between the inlet vane angles and the flow directions at 50 and 150 percent of design flow rate. Evaluate and plot the theoretical power input to the pump over this range of flow rate. Determine the flow rate at which the power delivered to the fluid is maximum.

11.10 A centrifugal water pump, with 6 in. diameter impeller and axial inlet flow, is driven at 1750 rpm. The impeller vanes are backward-curved ($\beta_2 = 65°$) and have

‡ You may wish to use simple computer programs to help solve problems marked with daggers.

axial width $b_2 = 0.75$ in. The pump casing acts as a diffuser, which converts 60 percent of the absolute velocity head at the impeller outlet to static pressure rise. The head loss through the pump suction and discharge channels is 0.75 times the *radial component* of velocity head leaving the impeller. Estimate the volume flow rate, head rise, power input, and pump efficiency at the maximum efficiency point.

11.11   Use data from Appendix D to choose points from the performance curves for a Peerless horizontal split case Type 4AE12 pump at 1750 and 3550 nominal rpm. Obtain and plot curve-fits for total head versus delivery at each speed for this pump.

11.12   Use data from Appendix D to choose points from the performance curves for a Peerless horizontal split case Type 10AE12 pump at 1150 and 1750 nominal rpm. Obtain and plot curve-fits of total head versus delivery for this pump.

11.13   Data from tests of a Peerless end suction Type 1430 pump operated at 1750 rpm with a 14.0 in. diameter impeller are

Flow rate, Q (gpm)	270	420	610	720	1000
Total head, H (ft)	198	195	178	165	123
Power input, $\mathcal{P}$ (hp)	25	30	35	40	45

Plot the performance curves for this pump; include a curve of efficiency versus volume flow rate. Locate the best efficiency point and specify the pump rating at this point.

11.14   Data from tests of a Peerless end suction Type 1440 pump operated at 1750 rpm with a 14.0 in. diameter impeller are

Flow rate, Q (gpm)	290	440	550	790	920	1280
Total head, H (ft)	204	203	200	187	175	135
Power input, $\mathcal{P}$ (hp)	30	35	40	45	50	60

Plot the performance curves for this pump; include a curve of efficiency versus volume flow rate. Locate the best efficiency point and specify the pump rating at this point.

11.15   Data measured during tests of a centrifugal pump at 3500 rpm are

Parameter	Inlet, Section ①	Outlet, Section ②
Gage pressure, p (kPa)	95.2	412
Elevation above datum, z (m)	1.25	2.75
Average speed of flow, $\bar{V}$ (m /sec)	2.35	3.62

The flow rate is 11.5 m^3/hr and the torque applied to the pump shaft is 3.68 N·m. Evaluate the total dynamic heads at the pump inlet and outlet, the hydraulic power input to the fluid, and the pump efficiency. Specify the electric motor size needed to drive the pump. If the electric motor efficiency is 85 percent, calculate the electric power requirement.

11.16   The *kilogram force* (kgf), defined as the force exerted by a kilogram mass in standard gravity, is commonly used in European practice. The *metric horsepower* (hpm) is defined as 1 hpm ≡ 75 m·kgf/sec. Develop a conversion relating metric horsepower to U.S. horsepower. Relate the specific speed for a hydraulic turbine—calculated in units of rpm, metric horsepower, and meters—to the specific speed calculated in U.S. customary units.

11.17   A small centrifugal pump, when tested at $N = 2875$ rpm with water, delivered $Q = 252$ gpm and $H = 138$ ft at its best efficiency point ($\eta = 0.76$). Determine the specific speed of the pump at this test condition. Sketch the impeller shape you expect. Compute the required power input to the pump.

11.18   Typical performance curves for a centrifugal pump, tested with three different impeller diameters in a single casing, are shown. Specify the flow rate and head produced by the pump at its best efficiency point with a 12 in. diameter impeller. Scale these data to predict the performance of this pump when tested with 11 in. and 13 in. impellers. Comment on the accuracy of the scaling procedure.

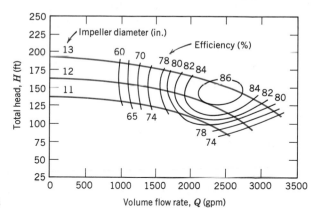

P11.18

11.19   At its best efficiency point ($\eta = 0.82$), a mixed-flow pump, with $D = 500$ mm, delivers $Q = 0.75$ m³/sec of water at $H = 39$ m when operating at $N = 1170$ rpm. Calculate the specific speed of this pump. Evaluate the required power input. Determine the curve-fit parameters of the pump performance curve based on the shutoff point and the best efficiency point. Scale the performance curve to estimate the flow, head, efficiency, and power input required to run the same pump at 860 rpm.

11.20   A pumping system must be specified for a lift station at a wastewater treatment facility. The average flow rate is 30 million gallons per day and the required lift is 30 ft. Non-clogging impellers must be used; about 65 percent efficiency is expected. For convenient installation, electric motors of 50 hp or less are desired. Determine the number of motor/pump units needed and recommend an appropriate operating speed.

11.21   A centrifugal water pump operates at 1750 rpm; the impeller has backward-curved vanes with $\beta_2 = 60°$ and $b_2 = 0.50$ in. At a flow rate of 350 gpm, the radial outlet velocity is $V_{n_2} = 11.7$ ft/sec. Estimate the head this pump could deliver at 1150 rpm.

11.22   Appendix D contains area bound curves for pump model selection and performance curves for individual pump models. Use these data to verify the similarity rules for a Peerless Type 4AE12 pump operated at 1750 and 3550 nominal rpm.

11.23   Use data from Appendix D to verify the similarity rules for the effect of changing the impeller diameter of a Peerless Type 10AE12 pump operated at 1150 nominal rpm. Verify the similarity rules for the effect of speed change for the same pump, using data for 1150 and 1750 nominal rpm.

11.24   Consider the Peerless Type 16A18B horizontal split case centrifugal pump (Appendix D). Use these performance data to verify the similarity rules for (a) impeller diameter change and (b) operating speeds of 705 and 880 rpm (note the scale change between speeds).

11.25   Performance curves for Peerless horizontal split case pumps are presented in Appendix D. Develop and plot curve-fits for a Type 4AE12 pump driven at 1750 and 3550 nominal rpm. Verify the effects of pump speed on scaling pump curves using the procedure described in Example Problem 11.7.

11.26   Appendix D contains area bound curves for pump model selection and performance curves for individual pump models. Use these data to verify the similarity rules for a Peerless Type 10AE12 pump operated at 1150 and 1750 nominal rpm.

11.27   Performance curves for Peerless horizontal split case pumps are presented in Appendix D. Develop and plot curve-fits for a Type 10AE12 pump driven at 1150 and

1750 nominal rpm. Verify the effects of pump speed on scaling pump curves using the procedure described in Example Problem 11.7.

11.28   Performance curves for Peerless horizontal split case pumps are presented in Appendix D. Develop and plot curve-fits for a Type 16A18B pump driven at 705 and 880 rpm. Verify the effects of pump speed on scaling pump curves using the procedure described in Example Problem 11.7.

11.29   A centrifugal pump is to operate at $Q = 250$ cfs, $H = 400$ ft, and $N = 870$ rpm. A model test is planned in a facility where the maximum water flow rate is 5 cfs and a 300 hp dynamometer is available. Assume the model and prototype efficiencies are comparable. Determine the appropriate model test speed and scale ratio.

11.30   Catalog data for a centrifugal water pump at design conditions are $Q = 250$ gpm and $\Delta p = 18.6$ psi at 1750 rpm. A laboratory flume requires 200 gpm at 32 ft of head. The only motor available develops 3 hp at 1750 rpm. Is this motor suitable for the laboratory flume? How might the pump/motor match be improved?

11.31   A reaction turbine is designed to produce 25,000 hp at 90 rpm under 150 ft of head. Laboratory facilities are available to provide 25 ft of head and to absorb 50 hp from the model turbine. Assume comparable efficiencies for the model and prototype turbines. Determine the appropriate model test speed, scale ratio, and volume flow rate.

11.32   White [41] suggests modeling the efficiency for a centrifugal pump using the curve-fit, $\eta = aQ - bQ^3$, where a and b are constants. Describe a procedure to evaluate a and b from experimental data. Evaluate a and b using data for the Peerless Type 10AE12 pump at 1750 rpm (Appendix D). Plot and illustrate the accuracy of the curve-fit by comparing measured and predicted efficiencies for this pump.

11.33   Sometimes the variation of water viscosity with temperature can be used to achieve dynamic similarity. A model pump delivers 1.25 L/sec of water at 15 C against a head of 18.6 m, when operating at 3500 rpm. Determine the water temperature that must be used to obtain dynamically similar operation at 1750 rpm. Estimate the volume flow rate and head produced by the pump at the lower-speed test condition. Comment on the *NPSH* requirements for the two tests.

11.34   A four-stage boiler feed pump has suction and discharge lines of 4 in. and 3 in. inside diameter. At 3500 rpm, the pump is rated at 400 gpm against a head of 400 ft while handling water at 240 F. The inlet pressure gage, located 1.5 ft below the impeller centerline, reads 21.5 psig. The pump is to be factory certified by tests at the same flow rate, head rise, and speed, but using water at 80 F. Calculate the *NPSHA* at the pump inlet in the field installation. Evaluate the suction head that must be used in the factory test to duplicate field suction conditions.

11.35   Data from tests of a Peerless end suction Type 1430 pump operated at 1750 rpm, with a 14.0 in. diameter impeller, are

Flow Rate, Q (gpm)	200	400	600	800	1000
Net positive suction head required, *NPSHR* (ft)	7.8	9.8	13.6	19.2	28.7

Develop and plot a curve-fit equation for *NPSHR* versus volume flow rate in the form, $NPSHR = a + bQ^2$, where a and b are constants. Compare the results with the measured data.

11.36   Data from tests of a Peerless end suction Type 1440 pump operated at 1750 rpm, with a 14.0 in. diameter impeller, are

Flow Rate, Q (gpm)	200	400	600	800	1000	1200	1300
Net positive suction head required, *NPSHR* (ft)	5.5	6.3	8.5	11.3	15.8	22.0	27.0

Develop and plot a curve-fit equation for *NPSHR* versus volume flow rate in the form, $NPSHR = a + bQ^2$, where a and b are constants. Compare the results with the measured data.

‡11.37 The net positive suction head required (*NPSHR*) by a pump may be expressed approximately as a parabolic function of volume flow rate. The *NPSHR* for a particular pump operating at 1750 rpm on cold water is given as $H_r = H_0 + AQ^2$, where $H_0 = 10$ ft of water and $A = 4.1 \times 10^{-5}$ ft/(gpm)^2. Assume the pipe system supplying the pump suction consists of a reservoir, whose surface is 20 ft above the pump centerline, a square entrance, 20 ft of 6 in. cast-iron pipe, and a 90° elbow. Calculate the maximum volume flow rate at which the suction head is sufficient to operate this pump on cold water without cavitation. Repeat this calculation for hot water at various temperatures and plot versus water temperature. (Be sure to consider the density variation as water temperature increases.)

‡11.38 A pump lifts water between two reservoirs connected by 200 ft of 4 in. and 200 ft of 3 in. cast-iron pipe in series. The gravity lift is 10 ft. Calculate and plot the system head curve. Estimate the head requirement, power needed, and annual cost of electrical energy to pump water at 150 gpm to the higher reservoir.

‡11.39 A centrifugal pump is installed in a piping system with $L = 1000$ ft of $D = 16$ in. cast-iron pipe. The downstream reservoir surface is 50 ft lower than the upstream reservoir. Determine and plot the system head curve. Find the volume flow rate (magnitude and direction) through the system when the pump is not operating. Estimate the friction loss, power requirement, and annual energy cost to pump water at 14,600 gpm through this system.

‡11.40 A pumping system with two different static lifts is shown. Each reservoir is supplied by a line consisting of 1000 ft of 8 in. cast-iron pipe. Evaluate and plot the system head versus flow curve. Explain what happens when the pump head is less than the height of the upper reservoir. Calculate the flow rate delivered at a pump head of 88 ft.

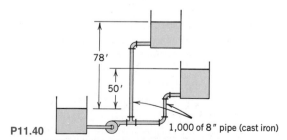

P11.40

78'

50'

1,000 of 8″ pipe (cast iron)

‡11.41 A pump transfers water from one reservoir to another through two cast-iron pipes in series. The first is 1500 ft of 10 in. pipe and the second is 500 ft of 6 in. pipe. A constant flow rate of 300 gpm is tapped off at the junction between the two pipes. Obtain and plot the system head versus flow rate curve. Find the delivery if the system is supplied by the pump of Fig. 11.18, operating at 1750 rpm.

‡11.42 Performance data for a pump are

H (ft)	148	140	130	115	100	75	50
Q (gpm)	0	800	1200	1600	2000	2400	2800

Estimate the delivery when the pump is used to move water between two open reservoirs, through 1200 ft of 12 in. commercial steel pipe containing two 90° elbows and an open gate valve, if the elevation increase is 50 ft. Determine the gate valve loss coefficient needed to reduce the volume flow rate by half.

‡ You may wish to use simple computer programs to help solve problems marked with daggers.

‡**11.43** Consider again the pump and piping system of Problem 11.42. Determine the volume flow rate and gate valve loss coefficient for the case of two identical pumps installed in *parallel*.

‡**11.44** Consider again the pump and piping system of Problem 11.42. Determine the volume flow rate and gate valve loss coefficient for the case of two identical pumps installed in *series*.

‡**11.45** The resistance of a given pipe increases with age as deposits form, increasing the roughness and reducing the pipe diameter (see Fig. 8.16). Typical multipliers to be applied to the friction factor are given in [16]:

Pipe age, years	Small pipes, 4–10 in.	Large pipes, 12–60 in.
New	1.00	1.00
10	2.20	1.60
20	5.00	2.00
30	7.25	2.20
40	8.75	2.40
50	9.60	2.86
60	10.0	3.70
70	10.1	4.70

Consider again the pump and piping system of Problem 11.42. Estimate the percentage reductions in volume flow rate that occur after (a) 20 years and (b) 40 years of use, if the pump characteristics remain constant. Repeat the calculation if the pump head is reduced 10 percent after 20 years of use and 25 percent after 40 years.

‡**11.46** Consider again the pump and piping system of Problem 11.43. Estimate the percentage reductions in volume flow rate that occur after (a) 20 years and (b) 40 years of use, if the pump characteristics remain constant. Repeat the calculation if the pump head is reduced 10 percent after 20 years of use and 25 percent after 40 years. (Use the data of Problem 11.45 for increase in pipe friction factor with age.)

‡**11.47** Consider again the pump and piping system of Problem 11.44. Estimate the percentage reductions in volume flow rate that occur after (a) 20 years and (b) 40 years of use, if the pump characteristics remain constant. Repeat the calculation if the pump head is reduced 10 percent after 20 years of use and 25 percent after 40 years. (Use the data of Problem 11.45 for increase in pipe friction factor with age.)

11.48 Part of the water supply for the South Rim of Grand Canyon National Park is taken from the Colorado River [42]. A flow rate of 600 gpm, taken from the river at Elevation 3734 ft, is pumped to a storage tank atop the South Rim at 7022 ft elevation. Part of the pipeline is above ground and part is in a hole directionally drilled at angles up to 70° from the vertical; the total pipe length is approximately 13,200 ft. Under steady flow operating conditions, the head loss due to friction is 290 ft of water in addition to the static lift. Estimate the diameter of the commercial steel pipe in the system. Compute the pumping power requirement if the pump efficiency is 61 percent. Discuss the tradeoff between impeller diameter and number of pump stages for operation at 3500 rpm.

‡**11.49** The city of Englewood, Colorado diverts water for municipal use from the South Platte River at Elevation 5280 ft [42]. The water is pumped to storage reservoirs at 5310 ft elevation. The inside diameter of the steel water line is 27 in.; its length is 5800 ft. The facility is designed for an initial capacity (flow rate) of 31 cfs, with an ultimate capacity of 38 cfs. Calculate and plot the system resistance curve.

‡ You may wish to use simple computer programs to help solve problems marked with daggers.

Specify an appropriate pumping system. Estimate the pumping power required for steady-state operation, at both the initial and ultimate flow rates.

11.50 A pump in the system shown draws water from a sump and delivers it to an open tank through 1250 ft of new 4 in. nominal diameter, schedule 40 steel pipe. The vertical suction pipe is 5 ft long and includes a foot valve with hinged disk and a 90° standard elbow. The discharge line includes two 90° standard elbows, an angle lift check valve, and a fully open gate valve. The design flow rate is 200 gpm. Find the head losses in the suction and discharge lines. Calculate the *NPSHA*. Select a pump suitable for this application.

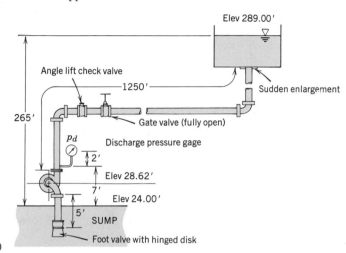

P11.50

‡**11.51** Consider the flow system and data of Problem 11.50 and the data for pipe aging given in Problem 11.45. Select pump(s) that will maintain the system flow at the desired rate for (a) 10 years and (b) 20 years. Compare the delivery with these pumps to the delivery with the pump sized for new pipes only.

‡**11.52** Consider the flow system shown in Problem 8.69. Evaluate the *NPSHA* at the pump inlet. Select a pump appropriate for this application. Use the data on pipe aging from Problem 11.45 to estimate the reduction in flow rate after 10 years of operation.

‡**11.53** Consider the flow system shown in Problem 8.89. Assume the minimum *NPSHR* at the pump inlet is 4.5 m of water. Select a pump appropriate for this application. Use the data for increase in friction factor with pipe age given in Problem 11.45 to determine and compare the system flow rate after 10 years of operation.

‡**11.54** Consider the flow system shown in Problem 8.95. Select a pump appropriate for this application. Check the *NPSHR* versus the *NPSHA* for this system.

‡**11.55** Consider the flow system shown in Problem 8.100. Select an appropriate pump for this application. Check the pump efficiency and power requirement compared to those in the problem statement.

‡**11.56** Consider the chilled water circulation system of Problem 8.101. Select pumps that may be combined in parallel to supply the total flow requirement. Calculate the power required for 3 pumps in parallel. Also calculate the volume flow rates and power required when only 1 or 2 of these pumps operates.

‡**11.57** Consider the gasoline pipeline flow of Problem 8.106. Select pumps that, combined in parallel, supply the total flow requirement. Calculate the power required for 4 pumps in parallel. Also calculate the volume flow rates and power required when only 1, 2, or 3 of these pumps operates.

‡ You may wish to use simple computer programs to help solve problems marked with daggers.

11.58 Consider the pipe network of Problem 8.129. Assume the system is to be supplied by a small diesel engine-driven pump. Specify a pump, impeller diameter, and operating speed suitable for this application. If the optimum brake specific fuel consumption (BSFC) of the engine is 0.38 lbm/hp·hr between 1500 and 1700 rpm, estimate the amount of diesel fuel consumed during each 24 hr period.

11.59 Consider the pipe network of Problem 8.130. Select a pump suitable to deliver a total flow rate of 300 gpm through the pipe network.

‡11.60 Manufacturer's data for the "Little Giant Water Wizard" submersible utility pump are

Discharge Height		1	2	5	10	15	20	26.3
Water Flow Rate (gpm)	20.4	20	19	16	13	8	0	

The Owners Manual also states, "Note: These ratings are based on discharge into 1 in. pipe with friction loss neglected. Using 3/4 in. garden hose adaptor, performance will be reduced approximately 15%." Plot a performance curve for the pump. Develop a curve-fit equation for the performance curve; show the curve-fit on the plot. Calculate and plot the pump delivery versus discharge height through a 50 ft length of smooth 3/4 in. garden hose. Compare to the curve for delivery into 1 in. pipe.

11.61 Consider the fire hose and nozzle of Problem 8.115. Discuss pump selection for this system. Would a bypass to increase flow through the pump improve operation of this system? Specify an appropriate pump and impeller diameter for this application. Calculate the power input to the pump. Estimate the power delivered to the hose and the power dissipated in the bypass.

11.62 Consider the swimming pool filtration system of Problem 8.131. Assume the pipe used is 3/4 in. nominal PVC (smooth plastic). Specify the speed and impeller diameter and estimate the efficiency of a suitable pump.

11.63 According to Stepanoff [43], a 6 in. boiler feed pump, operating at 3500 rpm, delivered 2200 gpm at a total head rise of 6900 ft, while absorbing 4700 hp. The inlet conditions were $p_1 = 132.7$ psia, $T_1 = 331$ F, and $NPSH = 72$ ft. Estimate the number of pump stages required for this application. Estimate the specific speed per stage and the overall efficiency for this pump.

11.64 A centrifugal pump, with characteristic curve given by $H_p = H_0 - AQ^2$, supplies a system with pure frictional loss, $H_s = CQ^2$. Obtain general expressions for the operating point flow rate, Q_1, with a single pump. Explore the effects of combining (a) n pumps in series and (b) n pumps in parallel. Express the operating point flow rates, Q_n, as ratios to Q_1. Plot in nondimensional form and interpret the results.

11.65 Consider the fan and duct system of Problem 8.120. Specify a fan appropriate for this system. Estimate the actual fan efficiency and input power requirement.

11.66 Performance data for a Buffalo Forge Type BL centrifugal fan of 36.5 in. diameter, tested at 600 rpm, are

Volume flow rate, Q (cfm)	6000	8000	10,000	12,000	14,000	16,000
Static pressure rise, Δp (in. H_2O)	2.10	2.00	1.76	1.37	0.92	0.42
Power input, $\mathscr{P}$ (hp)	2.75	3.18	3.48	3.51	3.50	3.22

‡ You may wish to use simple computer programs to help solve problems marked with daggers.

Plot the performance data versus volume flow rate. Calculate static efficiency and show the curve on the plot. Find the best efficiency point and specify the fan rating at this point.

11.67 Consider the fan and performance data of Problem 11.66. At $Q = 12,000$ cfm, the dynamic pressure is equivalent to 0.16 in. of water. Evaluate the fan outlet area. Plot total pressure rise and input horsepower for this fan versus volume flow rate. Calculate the fan total efficiency and show the curve on the plot. Find the best efficiency point and specify the fan rating at this point.

11.68 The performance data of Problem 11.66 are for a 36.5 in. diameter fan wheel. This fan also is manufactured with 40.3, 44.5, 49.0, and 54.3 in. diameter wheels. Pick a standard fan to deliver 30,000 cfm against a 5 in. H_2O static pressure rise. Assume standard air at the fan inlet. Determine the required fan speed and the input power needed.

11.69 In U.S. customary units, the specific speed for fans is given by

$$N_s = \frac{N \text{ (rpm)}[Q \text{ (cfm)}]^{\frac{1}{2}}}{[\Delta p \text{ (in. } H_2O)]^{\frac{3}{4}}}$$

Evaluate and plot specific speed versus volume flow rate for the fan data of Problem 11.66. Use the specific speed concept to determine the best fan size and motor speed for the application of Problem 11.68. Use standard motor speeds (fractions of 3550 nominal rpm).

‡11.70 Performance characteristics of a Buffalo Forge axial flow fan are presented below. The fan is used to power a wind tunnel with 1 ft square test section. The tunnel consists of a smooth inlet contraction, two screens (each with loss coefficient $K = 0.12$), the test section, and a diffuser where the cross section is expanded to 24 in. diameter at the fan inlet. Flow from the fan is discharged back to the room. Calculate and plot the system characteristic curve of pressure loss versus volume flow rate. Estimate the maximum air flow speed available in this wind tunnel test section.

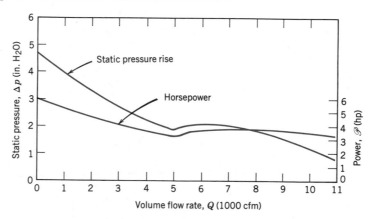

P11.70

‡11.71 Consider again the axial-flow fan and wind tunnel of Problem 11.70. Scale the performance of the fan as it varies with operating speed. Develop and plot a "calibration curve" showing test section flow speed (in m/sec) versus fan speed (in rpm).

11.72 Experimental test data for an aircraft engine fuel pump are presented below. This gear pump is required to supply jet fuel at 450 pounds per hour and 150 psig to the engine fuel controller. Tests were conducted at 10, 96, and 100 percent of the rated pump speed of 4536 rpm. At each constant speed, the back pressure on the pump was set, and the flow rate measured. On one graph, plot curves of pressure versus delivery at the three constant speeds. Estimate the pump displacement volume per revolution. Calculate the volumetric efficiency at each test point and sketch contours

‡ You may wish to use simple computer programs to help solve problems marked with daggers.

of constant η_v. Evaluate the energy loss due to valve throttling at 100 percent speed and full delivery to the engine.

Pump Speed (rpm)	Back Pressure (psig)	Fuel Flow (pph*)	Pump Speed (rpm)	Back Pressure (psig)	Fuel Flow (pph)	Pump Speed (rpm)	Back Pressure (psig)	Fuel Flow (pph)
	200	1810		200	1730		200	89
4536	300	1810	4355	300	1750	453	250	73
(100%)	400	1810	(96%)	400	1735	(10%)	300	58.5
	500	1790		500	1720		350	45
	900	1720		900	1635		400	30

* Fuel flow rate measured in pounds per hour (pph).

11.73 Consider the hydraulic pump and system of Problem 8.122. Assume the fixed displacement pump operates at 3550 rpm, has a maximum pressure of 3200 psig, and that the pump characteristic curve slope is similar to that of Fig. 11.36. Determine the minimum pump displacement needed to deliver the required flow rate. Estimate the power input, the power delivered to the fluid, and the power delivered to the load.

11.74 An air boat in the Florida Everglades is powered by a propeller, with $D = 6$ ft, driven at maximum speed, $N = 1800$ rpm, by a 60 hp engine. Estimate the maximum thrust produced by the propeller at (a) standstill and (b) $V = 30$ mph.

11.75 Drag data for model and prototype guided missile frigates are presented in Figs. 7.1 and 7.2. Dimensions of the prototype vessel are given in Problem 9.87. Use these data, with the propeller performance characteristics of Fig. 11.40, to size a single propeller to power the full-scale vessel. Calculate the propeller size, operating speed, and power input, if the propeller operates at maximum efficiency when the vessel travels at its maximum speed, $V = 37.6$ knots.

11.76 The propeller for the Gossamer Condor human-powered aircraft has $D = 12$ ft and rotates at $N = 107$ rpm. Additional details on the aircraft are given in Problem 9.156. Estimate the dimensionless performance characteristics and efficiency of this propeller at cruise conditions. (See Reference 44 for more information on human-powered flight.)

‡11.77 A *variable-pitch* propeller—equipped with a mechanism to change blade angles— can be controlled in flight to operate the engine at constant speed (for maximum

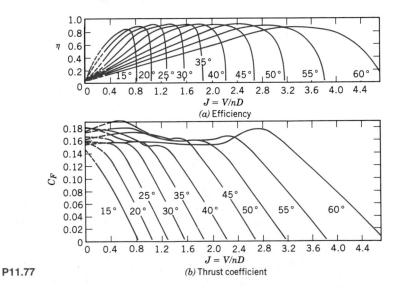

P11.77

(a) Efficiency

(b) Thrust coefficient

‡ You may wish to use simple computer programs to help solve problems marked with daggers.

power or minimum fuel consumption). Characteristics of a NACA 5868-9 variable-pitch propeller are presented in Fig. P11.77; the numbers on each curve refer to blade angles at 0.75R. Consider an aircraft equipped with this propeller, having $D = 15$ ft, set to maintain constant propeller speed of 1200 rpm. Assume the engine delivers 2400 shaft horsepower. Calculate and plot thrust horsepower and blade angle versus flight speed for the range $0 < V < 500$ mph.

11.78 Equations for the thrust, power, and efficiency of propulsion devices were derived in Section 11-5.1d. Show that these equations may be combined for the condition of constant thrust to obtain

$$\eta = \frac{2}{1 + \left(1 + \dfrac{F_T}{\dfrac{\rho V^2}{2} \dfrac{\pi D^2}{4}}\right)^{\frac{1}{2}}}$$

Interpret this result physically.

11.79 Preliminary calculations for a hydroelectric power generation site show a net head of 2350 ft is available at a water flow rate of 75 ft³/sec. Compare the geometry and efficiency of Pelton wheels designed to run at (a) 450 rpm and (b) 600 rpm.

11.80 Conditions at the inlet to the nozzle of a Pelton wheel are $p = 4.81$ MPa (gage) and $V = 6.10$ m/sec. The jet diameter is $d = 200$ mm and the nozzle loss coefficient is $K_{nozzle} = 0.04$. The wheel diameter is $D = 2.45$ m. At this operating condition, $\eta = 0.86$. Calculate (a) the power output, (b) the normal operating speed, (c) the approximate runaway speed, (d) the torque at normal operating, and (e) the approximate torque at zero speed.

11.81 The reaction turbines at Niagara Falls are of the Francis type. The impeller outside diameter is 176 in. Each turbine produces 72,500 hp at 107 rpm, with 93.8 percent efficiency under 214 ft of net head. Calculate the specific speed of these units. Evaluate the volume flow rate to each turbine. Estimate the penstock size, if it is 1300 ft long and the net head is 85 percent of the gross head.

11.82 The cross-flow (Banki) hydraulic turbine illustrated below is simple to build because the blades curve in only one direction, no casing is needed, and the area of the rectangular nozzle can be controlled easily. Shepherd [45] quotes results from a model test of a Banki turbine at its best efficiency point as

$Q = 3$ cfs
$H = 16$ ft
$N = 263$ rpm
$D_o = 13$ in.
$D_i = 8.56$ in.
$L = 12$ in.
$\mathcal{P} = 2.75$ hp (maximum)

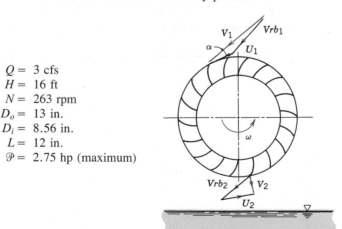

P11.82

Calculate the specific speed for this operating condition and compare to the optimum specific speeds of other hydraulic turbines. Evaluate the efficiency at this condition.

11.83   Figure 11.11 contains data for the efficiency of a large Pelton water wheel installed in the Tiger Creek Power House of Pacific Gas & Electric Company near Jackson, California. This unit is rated at 36,000 hp when operated at 225 rpm under a net head of 1190 ft of water. Assume reasonable flow angles and nozzle loss coefficient. Determine the rotor diameter and estimate the jet diameter. Compute the rate at which water is consumed during full-load operation.

11.84   Measured data for performance of the reaction turbines at Shasta Dam near Redding, California are shown in Fig. 11.13. Each turbine is rated at 103,000 hp when operating at 138.6 rpm under a net head of 380 ft. Evaluate the specific speed and compute the shaft torque developed by each turbine at rated operating conditions. Calculate and plot the water flow rate per turbine required to produce rated output power as a function of head.

11.85   A typical Pelton wheel, with a pitch diameter of 3.6 m, is to be installed in a location where a single jet is supplied at a net head of 490 m. Calculate and plot the expected turbine output power as a function of jet diameter between 0.1 and 0.5 m. Determine the best speed of operation for the wheel under these operating conditions. Calculate the torque delivered to the shaft for a 0.2 m diameter jet.

11.86   An impulse turbine is to develop 20,000 hp from a single wheel at a location where the net head is 1120 ft. Determine the appropriate speed, wheel diameter, and jet diameter for single- and multiple-jet operation. Compare with a double-overhung wheel installation. Estimate the required water consumption.

11.87   Tests of a model impulse turbine under a net head of 65.5 ft produced the following results:

Wheel Speed (rpm)	No-Load Discharge (cfs)	Net Brake Scale Reading (lbf) ($R = 5.25$ ft)					
275	0.110	6.8	14.9	22.0	28.9	40.0	48.0
300	0.125	5.9	12.9	19.8	25.5	36.0	43.8
Discharge (cfs)		0.397	0.773	1.114	1.414	1.896	2.315

Calculate and plot the machine power output and efficiency versus water flow rate.

11.88   Francis turbine Units 19, 20, and 21, installed at the Grand Coulee Dam on the Columbia River, are *very* large [46]. Each runner is 32.6 ft in diameter and contains 550 tons of cast steel. At rated conditions, each turbine develops 820,000 hp at 72 rpm under 285 ft of head. Efficiency is nearly 95 percent at rated conditions. The turbines operate at heads from 220 to 355 ft. Calculate the specific speed at rated operating conditions. Estimate the maximum water flow rate through each turbine.

11.89   In English units, the common definition of specific speed for a hydraulic turbine is

$$N_s = \frac{N \text{ (rpm)}[\mathscr{P} \text{ (hp)}]^{1/2}}{[h \text{ (ft)}]^{5/4}}$$

Develop a conversion between this definition and a truly dimensionless one in SI units. Evaluate the specific speed of an impulse turbine, operating at 400 rpm under a net head of 1190 ft with 86 percent efficiency, when supplied by a single 6 in. diameter jet. Use both English and SI units. Estimate the wheel diameter.

‡11.90   A small hydraulic impulse turbine is supplied with water through a penstock made from commercially rough pipe with diameter D and length L. The elevation difference between the reservoir surface and nozzle centerline is H. The nozzle head loss coefficient is K_{nozzle} and the loss coefficient from the reservoir to the penstock en-

‡ You may wish to use simple computer programs to help solve problems marked with daggers.

trance is $K_{entrance}$. Prepare an interactive computer program to calculate and tabulate the water jet speed, the volume flow rate, and the hydraulic power of the jet, for any input value of jet diameter, d. Use the program to find the optimum jet diameter and the resulting hydraulic power of the jet, for the case where $H = 300$ ft, $L = 1000$ ft, $D = 6$ in., $K_{entrance} = 0.5$, and $K_{nozzle} = 0.04$, if the pipe is made from welded steel. Explore the effects of varying the loss coefficients and pipe roughness.

11.91 Problem 15.17 of Reference 8 describes a laboratory test of a 24 in. Pelton wheel under 65.5 ft of head. At 275 rpm, the brake load was 40 lbf, applied at 5.25 ft radius; the discharge was 1.90 ft^3/sec. At this speed, the bearing friction and windage losses totaled 0.21 hp. Reduce these data to obtain the net power output, turbine efficiency, and specific speed. Express the mechanical losses as a fraction of the output power and calculate the mechanical efficiency. Estimate the hydraulic efficiency of this machine (assume $C_v = 0.98$ for the nozzle, i.e., $V_n = 0.98$ of the theoretical velocity if friction were neglected).

‡11.92 Prepare a set of design charts for a single-nozzle impulse turbine. On log-log coordinates, plot power output versus net head. Plot lines corresponding to various numbers of generator poles (rotating speeds).

11.93 Two or more identical hydraulic turbines are to be chosen for a location where the net head is 110 m and the available water flow rate is 17.0 m^3/sec. If the initial estimate of turbine efficiency is 90 percent, discuss the possible turbine choices.

11.94 According to a spokesperson for Pacific Gas & Electric Company, the Tiger Creek plant, located east of Jackson, California, is one of 71 PG&E hydroelectric power-plants. The plant has 1219 ft of gross head, consumes 750 cfs of water, is rated at 60 MW, and operates at 58 MW. The plant is claimed to produce 968 kW · hr/acre · ft of water and 336.4×10^6 kW · hr/yr of operation. Estimate the net head at the site, the turbine specific speed, and its efficiency. Comment on the internal consistency of these data.

‡11.95 Prepare an interactive computer program to design an impulse turbine installation. Assume the given input data are the net head available and the desired power output. Set up the program to choose the optimum number of generator poles (turbine operating speed) with interaction from the user. After the user achieves a satisfactory design, have the program calculate the required turbine wheel diameter, nozzle diameter, and water consumption.

11.96 The National Aeronautics & Space Administration (NASA) and the U.S. Department of Energy (DOE) co-sponsor a large demonstration wind turbine generator at Plum Brook, near Sandusky, Ohio [47]. The turbine has two blades, with $D = 38$ m, and delivers maximum power when the wind speed is above $V = 29$ km/hr. It is designed to produce 100 kW with powertrain efficiency of 0.75. The rotor is designed to operate at a constant speed of 40 rpm in winds over 6 mph by controlling system load and adjusting blade angles. For the maximum power condition, estimate the rotor tip speed and power coefficient.

11.97 A model of an American multiblade farm windmill is to be built for display. The model, with $D = 2$ ft, is to develop full power at $V = 22$ mph wind speed. Calculate the angular speed of the model for optimum power generation. Estimate the power output.

11.98 The largest known Darrieus vertical axis wind turbine was built by the U.S. Department of Energy near Sandia, New Mexico [48]. This machine is 60 ft tall and 30 ft in diameter; the area swept by the rotor is almost 1200 ft^2. Estimate the maximum power this windmill can produce in a 20 mph wind.

‡11.99 A typical American multiblade farm windmill has $D = 7$ ft and is designed to produce maximum power in winds with $V = 15$ mph. Estimate the rate of water delivery, as a function of the height to which the water is pumped, for this windmill.

‡ You may wish to use simple computer programs to help solve problems marked with daggers.

‡**11.100** Develop an equation for maximum power output available from a horizontal-axis wind turbine. Write the equation in a form to obtain power, $\mathcal{P}$ (kW), by substituting wind speed, V (km/hr), and blade diameter, D (m). Plot the results for the ranges $1 < D < 30$ m and $10 < V < 100$ km/hr.

‡**11.101** Aluminum extrusions, patterned after NACA symmetric airfoil sections, frequently are used to form Darrieus wind turbine "blades." Below are section lift and drag coefficient data [49] for a NACA 0012 section, tested at $\mathrm{Re} = 6 \times 10^6$ with standard roughness (the section stalled for $\alpha > 12°$):

Angle of attack, α (deg)	0	2	4	6	8	10	12
Lift coefficient, C_L (—)	0	0.23	0.45	0.68	0.82	0.94	1.02
Drag coefficient, C_D (—)	0.0098	0.0100	0.0119	0.0147	0.0194	—	—

Analyze the air flow relative to a blade element of a Darrieus wind turbine rotating about its troposkein axis. Develop a numerical model for the blade element. Calculate the power coefficient developed by the blade element as a function of tip speed ratio. Compare your result with the general trend of power output for Darrieus rotors shown in Fig. 11.45.

‡**11.102** Lift and drag data for the NACA 23015 airfoil section are presented in Fig. 9.17. Consider a two-blade horizontal-axis propeller wind turbine with NACA 23015 blade section. Analyze the air flow relative to a blade element of the rotating wind turbine. Develop a numerical model for the blade element. Calculate the power coefficient developed by the blade element as a function of tip speed ratio. Compare your result with the general trend of power output for high-speed two-bladed turbine rotors shown in Fig. 11.45.

‡ You may wish to use simple computer programs to help solve problems marked with daggers.

Chapter 12

INTRODUCTION TO COMPRESSIBLE FLOW

In Chapter 4 we developed control volume formulations of the basic equations. For incompressible flow, the two variables of principal interest were pressure and velocity. The continuity and momentum equations provided the two independent relations needed to solve for these variables. In Chapter 8, the energy equation was used to identify the losses in mechanical energy due to friction in duct flows.

"Compressible" flow implies appreciable variations in density throughout a flow field. Compressibility becomes important at high flow speeds or for large temperature changes. Large changes in velocity involve large pressure changes; for gas flows, these pressure changes are accompanied by significant variations in both density and temperature. Since two additional variables are encountered in treating compressible flow, two additional equations are needed. Both the energy equation and an equation of state must be applied to solve compressible flow problems.

In our study of compressible fluid flow, we shall deal primarily with steady, one-dimensional flow of an ideal gas. Although many real flows of interest are more complex, these restrictions allow us to concentrate on the effects of basic flow processes.

The thermodynamics necessary for the study of compressible flows, including an equation of state, and $T\,ds$ equations, is reviewed in the next section.

12-1 REVIEW OF THERMODYNAMICS

The pressure, density, and temperature of a substance may be related by an equation of state. Although many substances are complex in behavior, experience shows that most gases of engineering interest, at moderate pressure and temperature, are well represented by the ideal gas equation of state,

$$p = \rho RT \tag{12.1}$$

where R is a constant for each gas;[1] R is given by

$$R = \frac{R_u}{M_m}$$

where R_u is the universal gas constant, $R_u = 8314$ N $\cdot$ m/kgmole $\cdot$ K(1544 ft $\cdot$ lbf/lbmole $\cdot$ R) and M_m is the molecular mass of the gas. Although no real substance

---

[1] For air, $R = 287$ N $\cdot$ m/kg $\cdot$ K (53.3 ft $\cdot$ lbf/lbm $\cdot$ R).

behaves exactly as an ideal gas,[2] Eq. 12.1 is in error by less than 1 percent for air at room temperature for pressures as high as 30 atm. For air at 1 atm, the equation is less than 1 percent in error for temperatures as low as 140 K.

The ideal gas has other features that are simple and useful. In general, the internal energy of a substance may be expressed as $u = u(v, T)$. Then

$$du = \left(\frac{\partial u}{\partial T}\right)_v dT + \left(\frac{\partial u}{\partial v}\right)_T dv$$

where $v \equiv 1/\rho$ is the specific volume. The specific heat at constant volume is defined as $c_v \equiv (\partial u/\partial T)_v$, so that

$$du = c_v dT + \left(\frac{\partial u}{\partial v}\right)_T dv$$

For any substance that follows the ideal gas equation of state, $p = \rho RT$, then $(\partial u/\partial v)_T = 0$, and hence $u = u(T)$. Consequently,

$$du = c_v dT \tag{12.2}$$

for an ideal gas; this means that internal energy and temperature changes may be related if c_v is known. Furthermore, since $u = u(T)$, then $c_v = c_v(T)$.

The enthalpy of a substance is defined as $h \equiv u + p/\rho$. For an ideal gas, $p = \rho RT$, and hence, $h = u + RT$. Since $u = u(T)$ for an ideal gas, then h also must be a function of temperature alone.

To obtain a relation between h and T, we express h in its most general form as

$$h = h(p, T)$$

Then

$$dh = \left(\frac{\partial h}{\partial T}\right)_p dT + \left(\frac{\partial h}{\partial p}\right)_T dp$$

Since the specific heat at constant pressure is defined as $c_p \equiv (\partial h/\partial T)_p$, then

$$dh = c_p dT + \left(\frac{\partial h}{\partial p}\right)_T dp$$

We have shown that for an ideal gas h is a function of T only. Consequently, $(\partial h/\partial p)_T = 0$ and

$$dh = c_p dT \tag{12.3}$$

Again, since h is a function of T alone, Eq. 12.3 requires that c_p be a function of T only, for an ideal gas.

The specific heats for an ideal gas have been shown to be functions of temperature only. Their difference is a constant for each gas. From

$$h = u + RT$$

we can write

$$dh = du + R dT$$

[2] See, e.g., [1].

Combining this with Eq. 12.3, and using Eq. 12.2, we can write

$$dh = c_p \, dT = du + R \, dT = c_v \, dT + R \, dT$$

Then

$$c_p - c_v = R \tag{12.4}$$

The ratio of specific heats is defined as

$$k \equiv \frac{c_p}{c_v} \tag{12.5}$$

By using the definition of k, Eq. 12.4 can be solved for either c_p or c_v in terms of k and R. Thus

$$c_p = \frac{kR}{k-1} \tag{12.6a}$$

and

$$c_v = \frac{R}{k-1} \tag{12.6b}$$

For an ideal gas, the specific heats are functions of temperature only. Within reasonable temperature ranges, the specific heats of an ideal gas may be treated as constants for calculations of engineering accuracy. Under these conditions

$$u_2 - u_1 = \int_{u_1}^{u_2} du = \int_{T_1}^{T_2} c_v \, dT = c_v(T_2 - T_1) \tag{12.7a}$$

$$h_2 - h_1 = \int_{h_1}^{h_2} dh = \int_{T_1}^{T_2} c_p \, dT = c_p(T_2 - T_1) \tag{12.7b}$$

These equations obviously may be used to advantage in simplifying analyses.

Data for M_m, c_p, c_v, R, and k for common gases are given in Appendix A, Table A.6.

The property entropy is extremely useful in analyzing compressible flows. State diagrams, particularly the temperature-entropy (Ts) diagram, are valuable aids in the physical interpretation of analytical results. Since we shall make extensive use of Ts diagrams in solving compressible flow problems, let us review briefly some useful relationships involving the property entropy [2, 3].

Entropy is defined by the equation

$$\Delta S \equiv \int_{\text{rev}} \frac{\delta Q}{T} \qquad \text{or} \qquad dS = \left(\frac{\delta Q}{T} \right)_{\text{rev}} \tag{12.8}$$

The inequality of Clausius, deduced from the second law, states that

$$\oint \frac{\delta Q}{T} \leq 0$$

As a consequence of the second law, these results can be extended to

$$dS \geq \frac{\delta Q}{T} \qquad \text{or} \qquad T \, dS \geq \delta Q \tag{12.9a}$$

For *reversible* processes, the equality holds, and

$$T\,ds = \frac{\delta Q}{dm} \qquad \text{(reversible process)} \qquad (12.9\text{b})$$

The inequality holds for *irreversible* processes, and

$$T\,ds > \frac{\delta Q}{dm} \qquad \text{(irreversible process)} \qquad (12.9\text{c})$$

For an *adiabatic* process, $\delta Q/dm \equiv 0$. Thus

$$ds = 0 \qquad \text{(reversible adiabatic process)} \qquad (12.9\text{d})$$

and

$$ds > 0 \qquad \text{(irreversible adiabatic process)} \qquad (12.9\text{e})$$

Thus a process that is reversible and adiabatic also is *isentropic;* the entropy remains constant during the process. Equation 12.9e shows that entropy must *increase* for an adiabatic process that is irreversible.

Close study of Eqs. 12.9 shows that any two of the restrictions—reversible, adiabatic, or isentropic—must imply the third. For example, a process that is isentropic and reversible also must be adiabatic.

A useful relationship among properties (p, v, T, s, u) can be obtained by considering the first and second laws together. The result is the Gibbs, or $T\,ds$ equation,

$$T\,ds = du + p\,dv \qquad (12.10\text{a})$$

This is a relationship among properties, valid for all processes between equilibrium states. Although it is derived from the first and second laws, in itself, it is a statement of neither.

An alternate form of Eq. 12.10a can be obtained by substituting

$$du = d(h - pv) = dh - p\,dv - v\,dp$$

to obtain

$$T\,ds = dh - v\,dp \qquad (12.10\text{b})$$

For an ideal gas, entropy change can be evaluated from the $T\,ds$ equations as

$$ds = \frac{du}{T} + \frac{p}{T}\,dv = c_v \frac{dT}{T} + R \frac{dv}{v}$$

$$ds = \frac{dh}{T} - \frac{v}{T}\,dp = c_p \frac{dT}{T} - R \frac{dp}{p}$$

For constant specific heats, these equations may be integrated to yield

$$s_2 - s_1 = c_v \ln \frac{T_2}{T_1} + R \ln \frac{v_2}{v_1}$$

$$s_2 - s_1 = c_p \ln \frac{T_2}{T_1} - R \ln \frac{p_2}{p_1}$$

Example Problem 12.1 illustrates use of the ideal gas relations and the $T\,ds$ equations to evaluate thermodynamic properties and the entropy change for a process.

For the special case of an isentropic process, $ds = 0$, and the $T\,ds$ equations reduce to

$$0 = du + p\,dv$$

$$0 = dh - v\,dp$$

For an ideal gas, we have

$$0 = c_v\,dT + p\,dv$$

$$0 = c_p\,dT - v\,dp$$

Solving for dT gives

$$dT = \frac{v\,dp}{c_p} = -\frac{p\,dv}{c_v}$$

or

$$\frac{dp}{p} + \frac{c_p}{c_v}\frac{dv}{v} = \frac{dp}{p} + k\frac{dv}{v} = 0$$

Integrating (for $k =$ constant) gives

$$\ln p + k \ln v = \ln c$$

or

$$\ln p + \ln v^k = \ln c$$

Taking antilogarithms, this equation reduces to

$$pv^k = \text{constant} \tag{12.11a}$$

or

$$\frac{p}{\rho^k} = \text{constant} \tag{12.11b}$$

Equations 12.11 are property relations for an ideal gas undergoing an isentropic process.

Qualitative information that is useful in drawing state diagrams also can be obtained from the $T\,ds$ equations. To complete our review of the thermodynamic fundamentals, the slopes of lines of constant pressure and of constant volume on the Ts diagram are evaluated in Example Problem 12.2.

EXAMPLE 12.1—Property Changes in Compressible Duct Flow

Air flows through a long duct of constant area at 0.15 kg/sec. A short section of the duct is cooled by liquid nitrogen that surrounds the duct. The rate of heat loss in this section is 15.0 kJ/sec from the air. The absolute pressure, temperature, and velocity entering the cooled section are 188 kPa, 440 K, and 210 m/sec, respectively. At the outlet, the absolute pressure and temperature are 213 kPa and 351 K. Compute the duct cross-sectional area and the changes in enthalpy, internal energy, and entropy for this flow.

EXAMPLE PROBLEM 12.1

GIVEN: Air flows steadily through a short section of constant-area duct that is cooled by liquid nitrogen.

$T_1 = 440$ K

$p_1 = 188$ kPa (abs)

$V_1 = 210$ m/sec

$T_2 = 351$ K

$p_2 = 213$ kPa (abs)

Flow →   ← CV

① $\dot{Q} < 0$ ②

FIND: (a) Duct area. (b) Δh. (c) Δu. (d) Δs.

SOLUTION:
The duct area may be found from the continuity equation.

Basic equation:

$$0 = \overset{= 0(1)}{\cancel{\frac{\partial}{\partial t} \int_{CV} \rho\, d\forall}} + \int_{CS} \rho \vec{V} \cdot d\vec{A}$$

(4.13)

Assumptions: (1) Steady flow
(2) Uniform flow at each section
(3) Ideal gas

Then

$$0 = \{-|\rho_1 V_1 A_1|\} + \{|\rho_2 V_2 A_2|\}$$

or

$$\dot{m} = \rho_1 V_1 A = \rho_2 V_2 A$$

since $A = A_1 = A_2 = $ constant. Using the ideal gas relation, $p = \rho R T$, we find

$$\rho_1 = \frac{p_1}{RT_1} = \frac{1.88 \times 10^5}{} \frac{\text{N}}{\text{m}^2} \times \frac{\text{kg} \cdot \text{K}}{287\ \text{N} \cdot \text{m}} \times \frac{1}{440\ \text{K}} = 1.49\ \text{kg/m}^3$$

From continuity,

$$A = \frac{\dot{m}}{\rho_1 V_1} = \frac{0.15}{} \frac{\text{kg}}{\text{sec}} \times \frac{\text{m}^3}{1.49\ \text{kg}} \times \frac{\text{sec}}{210\ \text{m}} = 4.79 \times 10^{-4}\ \text{m}^2 \qquad \underleftarrow{\hspace{2cm}} \quad A$$

For an ideal gas, $dh = c_p\, dT$, so

$$\Delta h = h_2 - h_1 = \int_{T_1}^{T_2} c_p\, dT = c_p(T_2 - T_1)$$

$$\Delta h = \frac{1.00}{} \frac{\text{kJ}}{\text{kg} \cdot \text{K}} \times (351 - 440)\ \text{K} = -89.0\ \text{kJ/kg} \qquad \underleftarrow{\hspace{2cm}} \quad \Delta h$$

Also $du = c_v\, dT$, so

$$\Delta u = u_2 - u_1 = \int_{T_1}^{T_2} c_v\, dT = c_v(T_2 - T_1)$$

$$\Delta u = \frac{0.717}{} \frac{\text{kJ}}{\text{kg} \cdot \text{K}} \times (351 - 440)\ \text{K} = -63.8\ \text{kJ/kg} \qquad \underleftarrow{\hspace{2cm}} \quad \Delta u$$

The entropy change may be obtained from the $T\,ds$ equation,

$$T\,ds = dh - v\,dp$$

$$ds = \frac{dh}{T} - \frac{v\,dp}{T} = c_p\frac{dT}{T} - R\frac{dp}{p}$$

or

$$\Delta s = s_2 - s_1 = \int_{T_1}^{T_2} c_p\frac{dT}{T} - \int_{p_1}^{p_2} R\frac{dp}{p} = c_p\ln\frac{T_2}{T_1} - R\ln\frac{p_2}{p_1}$$

$$= \frac{1.00}{kg\cdot K}\,\frac{kJ}{\,}\ln\!\left(\frac{351}{440}\right) - \frac{0.287}{kg\cdot K}\,\frac{kJ}{\,}\ln\!\left(\frac{2.13\times10^5}{1.88\times10^5}\right)$$

$$\Delta s = -0.262\ \text{kJ/kg}\cdot\text{K} \qquad\qquad\qquad\qquad \Delta s$$

$\left\{\begin{array}{l}\text{The purpose of this problem was to review the calculation of thermodynamic properties}\\ \text{and the entropy change for a process of an ideal gas.}\end{array}\right\}$

EXAMPLE 12.2—Constant Property Lines on Ts Diagram

For an ideal gas, determine the slope of (a) a constant volume line and (b) a constant pressure line in the Ts plane.

EXAMPLE PROBLEM 12.2

GIVEN: An ideal gas.

FIND: (a) Slope of constant volume line in Ts plane.
        (b) Slope of constant pressure line in Ts plane.

SOLUTION:
The $T\,ds$ equations may be applied.

$$T\,ds = du + p\,dv \qquad\qquad\qquad (12.10a)$$

$$T\,ds = dh - v\,dp \qquad\qquad\qquad (12.10b)$$

Substituting for an ideal gas, $du = c_v dT$ and $dh = c_p dT$, we obtain

$$T\,ds = c_v\,dT + p\,dv$$

$$T\,ds = c_p\,dT - v\,dp$$

For a constant volume process, $dv = 0$. From the first equation,

$$\left.\frac{dT}{ds}\right)_{\substack{\text{constant}\\\text{volume}}} = \left.\frac{\partial T}{\partial s}\right)_v = \frac{T}{c_v} \qquad\qquad \text{constant volume slope}$$

For a constant pressure process, $dp = 0$. From the second equation,

$$\left.\frac{dT}{ds}\right)_{\substack{\text{constant}\\\text{pressure}}} = \left.\frac{\partial T}{\partial s}\right)_p = \frac{T}{c_p} \qquad\qquad \text{constant pressure slope}$$

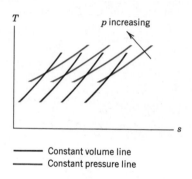

Constant volume line
Constant pressure line

Note that the slope of each line is proportional at any point to the absolute temperature. Furthermore, at any point, a constant volume line slope is $c_p/c_v = k$ times larger than a constant pressure line slope.

12-2 PROPAGATION OF SOUND WAVES

12-2.1 Speed of Sound

Supersonic and *subsonic* are familiar terms; they refer to speeds that are, respectively, greater than and less than the speed of sound. The speed of sound (a pressure wave of infinitesimal strength) is an important characteristic parameter for compressible flow. We have previously (Chapters 2 and 7) introduced the Mach number, $M = V/c$, the ratio of the local flow speed to the local speed of sound, as an important nondimensional parameter characterizing compressible flows. Before studying compressible flows, we obtain a general relation for calculating sonic speed.

Consider propagation of a sound wave of infinitesimal strength into an undisturbed medium, as shown in Fig. 12.1a. We are interested in relating the speed of wave propagation, c, to fluid property changes across the wave. If pressure and density in the undisturbed medium ahead of the wave are denoted by p and ρ, passage of the wave will cause them to undergo infinitesimal changes and to become $p+dp$ and

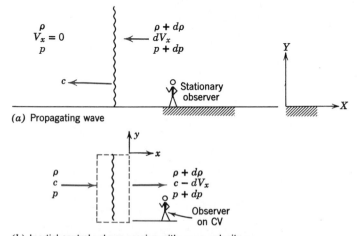

(a) Propagating wave

(b) Inertial control volume moving with wave, velocity c

Fig. 12.1 Propagating sound wave showing control volume chosen for analysis.

$\rho + d\rho$. Since the wave propagates into a stationary fluid, the velocity ahead of the wave, V_x, is zero. The magnitude of the velocity behind the wave, $V_x + dV_x$, then will be simply dV_x; in Figure 12.1a, the direction of the motion behind the wave has been assumed to the left.[3]

The flow of Fig. 12.1a appears unsteady to a stationary observer, viewing the wave motion from a fixed point on the ground. However, the flow appears steady to an observer located *on* an inertial control volume moving with a segment of the wave, as shown in Fig. 12.1b. The velocity approaching the control volume is c, and the velocity leaving is $c - dV_x$.

The basic equations may be applied to the differential control volume shown in Fig. 12.1b (we use V_x for the x component of velocity to avoid confusion with internal energy, u).

a. Continuity Equation

Basic equation:

$$0 = \overset{= 0(1)}{\cancel{\frac{\partial}{\partial t} \int_{CV} \rho \, d\Psi}} + \int_{CS} \rho \vec{V} \cdot d\vec{A} \tag{4.13}$$

Assumptions: (1) Steady flow
(2) Uniform flow at each section

Then

$$0 = \{-|\rho c A|\} + \{|(\rho + d\rho)(c - dV_x)A|\} \tag{12.12a}$$

or

$$0 = -\rho \cancel{c A} + \rho \cancel{c A} - \rho \, dV_x A + d\rho c A - \overset{\approx 0}{\cancel{d\rho \, dV_x A}}$$

Neglecting the product of differentials, $d\rho \, dV_x$, compared to either $d\rho$ or dV_x, we obtain

$$0 = -\rho A \, dV_x + c A \, d\rho$$

or

$$dV_x = \frac{c}{\rho} \, d\rho \tag{12.12b}$$

b. Momentum Equation

Basic equation:

$$F_{S_x} + \overset{= 0(3)}{\cancel{F_{B_x}}} - \overset{= 0(4)}{\cancel{\int_{CV} a_{rf_x} \rho \, d\Psi}} = \overset{= 0(1)}{\cancel{\frac{\partial}{\partial t} \int_{CV} V_x \rho \, d\Psi}} + \int_{CS} V_x \rho \vec{V}_{xyz} \cdot d\vec{A} \tag{4.35a}$$

Assumptions: (3) $F_{B_x} = 0$
(4) $a_{rf_x} = 0$

---

[3] The same final result is obtained regardless of the direction initially assumed for motion behind the wave.

Surface forces acting on the infinitesimal control volume are

$$F_{S_x} = dR_x + pA - (p + dp)A$$

where dR_x represents all forces applied to the horizontal portions of the control surface shown in Fig. 12.1b. We consider only a portion of a moving sound wave, so $dR_x = 0$ because there is no relative motion along the wave. Thus the surface force simplifies to

$$F_{S_x} = -A\,dp$$

Substituting into the basic equation gives

$$-A\,dp = c\{-|\rho cA|\} + (c - dV_x)\{|(\rho + d\rho)(c - dV_x)A|\}$$

Using the continuity equation, in the form of Eq. 12.12a, reduces this to

$$-A\,dp = c\{-|\rho cA|\} + (c - dV_x)\{|\rho cA|\} = (-c + c - dV_x)\{|\rho cA|\}$$

$$-A\,dp = -\rho cA\,dV_x$$

or

$$dV_x = \frac{1}{\rho c}\,dp \qquad\qquad (12.12c)$$

Combining Eqs. 12.12b and 12.12c, we obtain

$$dV_x = \frac{c}{\rho}\,d\rho = \frac{1}{\rho c}\,dp$$

from which

$$dp = c^2\,d\rho$$

or

$$c^2 = \frac{dp}{d\rho}$$

To evaluate the derivative of a thermodynamic property, we must specify the property to be held constant during differentiation. For the present case, the limit at vanishing strength of a sound wave will be

$$\lim_{\text{strength}\to 0} \frac{dp}{d\rho} = \frac{\partial p}{\partial \rho}\bigg)_{s=\text{constant}}$$

Perhaps a more physical justification for the assumption of isentropic propagation is that an infinitesimal pressure change is reversible. Hence, since there is too little time for heat transfer, the process is reversible and adiabatic; a reversible adiabatic process must be isentropic. Thus the speed of the propagation of a sound wave is given by

$$c = \sqrt{\frac{\partial p}{\partial \rho}\bigg)_s}$$

Data for solid and liquid media usually are reported as the bulk modulus,

$$E_v = \frac{dp}{d\rho/\rho} = \rho\frac{dp}{d\rho} \qquad\qquad (3.8)$$

For these media

$$c = \sqrt{E_v/\rho} \qquad (12.13)$$

For an ideal gas, the pressure and density in isentropic flow are related by

$$\frac{p}{\rho^k} = \text{constant} \qquad (12.11b)$$

as shown previously. Taking logarithms, we obtain

$$\ln p - k \ln \rho = \ln c$$

and differentiating

$$\frac{dp}{p} - k \frac{d\rho}{\rho} = 0$$

Therefore,

$$\left. \frac{\partial p}{\partial \rho} \right)_s = k \frac{p}{\rho}$$

But $p/\rho = RT$, so finally

$$c = \sqrt{kRT} \qquad (12.14)$$

for an ideal gas. The speed of sound in air has been measured precisely by numerous investigators [4]. The results agree closely with the theoretical prediction of Eq. 12.14.

The important feature of sound propagation in an ideal gas, as shown by Eq. 12.14, is that the speed of sound is a function of temperature only. The variation in atmospheric temperature with altitude on a standard day was discussed in Chapter 3; the properties are summarized in Table A.3. The corresponding variation in c is computed in Example Problem 12.3 and plotted as a function of altitude.

EXAMPLE 12.3—Speed of Sound in the Standard Atmosphere

Compute the speed of sound at sea level in standard air. Evaluate the speed of sound and plot for altitudes to 15 km.

EXAMPLE PROBLEM 12.3

GIVEN: Air under standard atmospheric conditions.

FIND: (a)   The speed of sound at sea level.
         (b)   Plot the speed of sound for altitudes to 15 km.

SOLUTION:
Assume an ideal gas.

Computing equation: $\qquad\qquad c = \sqrt{kRT}$

From Table 3.1, the temperature at sea level on a standard day is 288 K. Thus

$$c = \left(1.4 \times 287 \frac{\text{N} \cdot \text{m}}{\text{kg} \cdot \text{K}} \times 288\ \text{K} \times \frac{\text{kg} \cdot \text{m}}{\text{N} \cdot \text{sec}^2} \right)^{1/2} = 340\ \text{m/sec}$$

Temperatures at various altitudes may be found from Table A.3. The resulting sound speeds are plotted in the following figure.

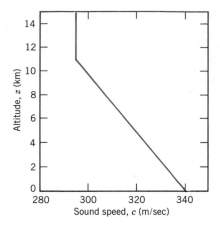

From the plot we see that the speed of sound in air on a standard day varies from 340 m/sec at sea level to 295 m/sec at 11 km altitude.

> Air temperature varies linearly from sea level to about 11 km altitude. Sonic speed is proportional to the square root of absolute temperature, so its variation with altitude is nonlinear. The nonlinearity is not obvious at the scale of this plot.

12-2.2 Types of Flow—The Mach Cone

Flows for which $M < 1$ are *subsonic,* while those with $M > 1$ are *supersonic.* Flow fields that have both subsonic and supersonic regions are termed *transonic.* (The transonic regime occurs for Mach numbers between about 0.9 and 1.2.) Although most flows within our experience are subsonic, there are important practical cases where $M \geq 1$ occurs in a flow field. Perhaps the most obvious are supersonic aircraft and transonic flows in aircraft compressors and fans. Yet another flow regime, *hypersonic* flow ($M \gtrsim 5$), is of interest in missile and reentry vehicle design. The National Aerospace Plane (NASP), currently under study, would cruise at Mach numbers approaching 20. Some important qualitative differences between subsonic and supersonic flows can be deduced from the properties of a simple moving sound source.

Consider a point source that emits instantaneous infinitesimal disturbances, which propagate in all directions with speed c. At any time t, the location of the wave front from the disturbance emitted at time t_0, will be represented by a sphere, with radius $c(t - t_0)$, whose center coincides with the location of the disturbance at time t_0.

We are interested in determining the nature of the disturbance propagation for different speeds of the moving source. Four cases are shown in Fig. 12.2:

1. $V = 0$. The sound pattern propagates uniformly in all directions. At an instant Δt after emission, any given sound pulse is located at radius $c\Delta t$ from the source. At instant $2\Delta t$, the radius is $c(2\Delta t)$. Each wave front is spherical; all wave fronts are concentric spheres.

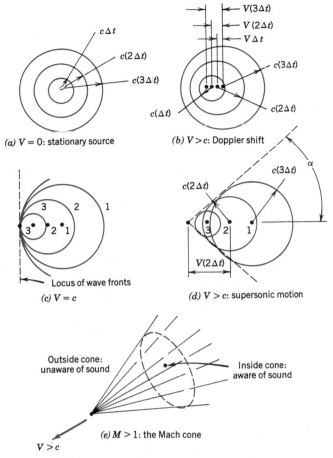

Fig. 12.2 Propagation of sound waves from a moving source: The Mach cone.

2. $0 < V < c$. The concentricity of the wave pattern is lost. Individual wave fronts are spherical, but each successive sound is emitted from a different position, $V \Delta t$ distant from the previous position.

If you imagine the circles shown to be the amplitude peaks of a sinusoidal tone, the same qualitative picture holds for a moving source of continuous sound. If this source moves with constant speed, V, the whole pattern shown in Fig. 12.2b is carried along with the emitter. Thus a stationary observer hears more peaks per unit time as the source approaches than after it passes. This is known as the "Doppler effect." (Have you ever heard a fast-moving train whistle through a crossing?)

3. $V = c$. The locus of leading surfaces of all waves will be a plane at the source, perpendicular to the path of motion. No sound wave can travel in front of the source. Consequently, an observer in front of the source will not hear it approaching.

4. $V > c$. In this case, the locus of leading surfaces of the sound waves will be a cone. Again, no sound will be heard in front of the cone.

The cone angle can be related to the Mach number at which the source moves. From the geometry of Fig. 12.2d,

$$\sin\alpha = \frac{c}{V} = \frac{1}{M}$$

or

$$\alpha = \sin^{-1}\left(\frac{1}{M}\right) \qquad (12.15)$$

The cone depicted in Fig. 12.2e is termed the *Mach cone*; α is the *Mach angle*. The regions inside and outside the cone are sometimes called the *zone of action* and the *zone of silence,* respectively.

12-3 REFERENCE STATE: LOCAL ISENTROPIC STAGNATION PROPERTIES

If we wish to describe the state of a fluid at any point in a flow field, we must specify two independent intensive thermodynamic properties (usually pressure and temperature), plus the velocity at the point.[4]

In our discussion of compressible flow, we shall find it convenient to use the stagnation state as a reference state. The stagnation state is characterized by zero velocity; stagnation properties at any point in a flow field are those properties that would exist at that point if the velocity were reduced to zero. Consider a point in a flow field having temperature T, pressure p, and velocity V. The stagnation state at that point in the flow field would be characterized by stagnation pressure p_0, stagnation temperature T_0, and zero velocity. Before we can calculate the stagnation properties, we must specify the process by which the fluid is imagined to decelerate to zero velocity.

For incompressible flow (Chapter 6), integration of Euler's equation for frictionless, incompressible flow along a streamline led to the Bernoulli equation

$$\frac{p}{\rho} + \frac{V^2}{2} + gz = \text{constant} \qquad (6.9)$$

For incompressible flow, a frictionless deceleration process to zero velocity led to the stagnation pressure, p_0, given by

$$p_0 = p + \tfrac{1}{2}\rho V^2 \qquad (6.12)$$

For compressible flow we again use a frictionless deceleration process; in addition, we specify that the deceleration process be adiabatic. In short, we specify an isentropic deceleration process to define local isentropic stagnation properties:

Local isentropic stagnation properties are those properties that would be obtained at any point in a flow field if the fluid at that point were decelerated from local conditions to zero velocity following a frictionless, adiabatic (an isentropic) process.

---

[4] The state of a pure substance, in the absence of motion, gravity, and surface, magnetic, or electrical effects, is defined by two independent intensive thermodynamic properties.

Isentropic stagnation properties are reference properties that may be evaluated at any point in a flow field. Variations in these properties from point to point in a flow field give information about the flow process between points. This will become apparent in our discussion of one-dimensional flow cases. To calculate the local isentropic stagnation properties, we must imagine a hypothetical deceleration process to zero velocity. At the beginning of the process, conditions correspond to the actual flow at the point (velocity V, pressure p, temperature T, etc.); at the end of the process the speed is zero, and conditions are the corresponding local isentropic stagnation properties (stagnation pressure p_0, stagnation temperature T_0, etc.).

12-3.1 Local Isentropic Stagnation Properties for the Flow of an Ideal Gas

We need to develop an expression describing the relationship among fluid properties during the process. Since both initial and final properties are specified, we develop the relationship among properties in differential form. Then we can integrate to obtain expressions for stagnation conditions in terms of initial conditions corresponding to the actual flow at the point.

The hypothetical deceleration process is shown schematically in Fig. 12.3. We are interested in finding the stagnation properties for flow at point ①. To find a relationship among fluid properties during the deceleration process, we apply the continuity and momentum equations to the stationary differential streamtube control volume shown.

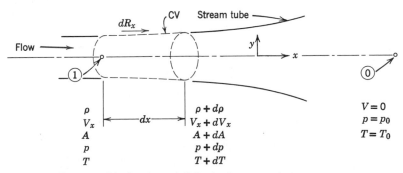

Fig. 12.3 Compressible flow in an infinitesimal stream tube.

a. Continuity Equation

Basic equation:
$$= 0(1)$$

$$0 = \frac{\partial}{\partial t} \int_{cv} \rho \, d\mathbf{V} + \int_{cs} \rho \vec{V} \cdot d\vec{A} \tag{4.13}$$

Assumptions: (1) Steady flow
(2) Uniform flow at each section

Then

$$0 = \{-|\rho V_x A|\} + \{|(\rho + d\rho)(V_x + dV_x)(A + dA)|\}$$

or

$$\rho V_x A = (\rho + d\rho)(V_x + dV_x)(A + dA) \tag{12.16a}$$

b. Momentum Equation

Basic equation:

$$\overset{=0(3)}{\cancel{F_{S_x}} + \cancel{F_{B_x}}} - \int_{cv} \cancel{a_{rf_x}} \rho \, d\Psi = \frac{\partial}{\partial t} \int_{cv} V_x \rho \, d\Psi + \int_{cs} V_x \rho \vec{V}_{xyz} \cdot d\vec{A} \qquad (4.35a)$$

Assumptions: (3) $F_{B_x} = 0$
(4) $a_{rf_x} = 0$
(5) Frictionless flow

The surface forces acting on the infinitesimal control volume are

$$F_{S_x} = dR_x + pA - (p + dp)(A + dA)$$

The force dR_x is applied along the streamtube boundary, as shown in Fig. 12.3. There the average pressure is $p + dp/2$, and the area component in the x direction is dA. There is no friction. Thus

$$F_{S_x} = \left(p + \frac{dp}{2}\right) dA + pA - (p + dp)(A + dA)$$

or

$$F_{S_x} = p \, \cancel{dA} + \overset{\approx 0}{\cancel{\frac{dp \, dA}{2}}} + \cancel{pA} - \cancel{pA} - dp \, A - p \, \cancel{dA} - \overset{\approx 0}{\cancel{d \, p \, dA}}$$

Substituting this result into the momentum equation,

$$-dp \, A = V_x\{-|\rho V_x A|\} + (V_x + d V_x)\{|(\rho + d\rho)(V_x + d V_x)(A + dA)|\}$$

which may be simplified using Eq. 12.16a to obtain

$$-dp \, A = (-V_x + V_x + d V_x)(\rho V_x A)$$

Finally,

$$dp = -\rho V_x \, d V_x = -\rho d\left(\frac{V_x^2}{2}\right)$$

or

$$\frac{dp}{\rho} + d\left(\frac{V_x^2}{2}\right) = 0 \qquad (12.16b)$$

Equation 12.16b is a relation among properties during the deceleration process. In developing this relation, we have specified a frictionless deceleration process. Before we can integrate between the initial and final (stagnation) states, we must specify the relation that exists between pressure, p, and density, ρ, along the process path.

Since the deceleration process is isentropic, then p and ρ for an ideal gas are related by the expression

$$\frac{p}{\rho^k} = \text{constant} \qquad (12.11b)$$

Our task now is to integrate Eq. 12.16b subject to this relation. Along the stagnation streamline there is only a single component of velocity; V_x is the magnitude of the velocity. Hence we can drop the subscript and write

$$\frac{dp}{\rho} + d\left(\frac{V^2}{2}\right) = 0 \tag{12.16c}$$

From $p/\rho^k = \text{constant} = C$, we can write

$$p - C\rho^k \qquad \text{and} \qquad \rho = p^{1/k} C^{-1/k}$$

Then, from Eq. 12.16c,

$$-d\left(\frac{V^2}{2}\right) = \frac{dp}{\rho} = p^{-1/k} C^{1/k} dp$$

We can integrate this equation between the initial state and the corresponding stagnation state

$$-\int_V^0 d\left(\frac{V^2}{2}\right) = C^{1/k} \int_p^{p_0} p^{-1/k} dp$$

to obtain

$$\frac{V^2}{2} = C^{1/k} \frac{k}{k-1} \left[p^{(k-1)/k} \right]_p^{p_0} = C^{1/k} \frac{k}{k-1} \left[p_0^{(k-1)/k} - p^{(k-1)/k} \right]$$

$$\frac{V^2}{2} = C^{1/k} \frac{k}{k-1} p^{(k-1)/k} \left[\left(\frac{p_0}{p} \right)^{(k-1)/k} - 1 \right]$$

Since $C^{1/k} = p^{1/k}/\rho$, then

$$\frac{V^2}{2} = \frac{k}{k-1} \frac{p^{1/k}}{\rho} p^{(k-1)/k} \left[\left(\frac{p_0}{p} \right)^{(k-1)/k} - 1 \right] = \frac{k}{k-1} \frac{p}{\rho} \left[\left(\frac{p_0}{p} \right)^{(k-1)/k} - 1 \right]$$

Since we seek an expression for stagnation pressure, we can rewrite this equation as

$$\left(\frac{p_0}{p} \right)^{(k-1)/k} = 1 + \frac{k-1}{k} \frac{\rho}{p} \frac{V^2}{2}$$

and

$$\frac{p_0}{p} = \left[1 + \frac{k-1}{k} \frac{\rho V^2}{2p} \right]^{k/(k-1)}$$

For an ideal gas, $p = \rho RT$, and hence

$$\frac{p_0}{p} = \left[1 + \frac{k-1}{2} \frac{V^2}{kRT} \right]^{k/(k-1)}$$

Also, for an ideal gas the sonic speed is $c = \sqrt{kRT}$, and thus

$$\frac{p_0}{p} = \left[1 + \frac{k-1}{2} \frac{V^2}{c^2} \right]^{k/(k-1)}$$

$$\frac{p_0}{p} = \left[1 + \frac{k-1}{2} M^2 \right]^{k/(k-1)} \tag{12.17a}$$

Equation 12.17a enables us to calculate the isentropic stagnation pressure at any point in a flow field of an ideal gas, provided that we know the static pressure and Mach number at that point.

We can readily obtain expressions for other isentropic stagnation properties by applying the relation

$$\frac{p}{\rho^k} = \text{constant}$$

between end states of the process. Thus

$$\frac{p_0}{p} = \left(\frac{\rho_0}{\rho}\right)^k$$

and

$$\frac{\rho_0}{\rho} = \left(\frac{p_0}{p}\right)^{1/k}$$

From the ideal gas equation of state, $p = \rho RT$. Then

$$\frac{T_0}{T} = \frac{p_0}{p}\frac{\rho}{\rho_0} = \frac{p_0}{p}\left(\frac{p_0}{p}\right)^{-1/k} = \left(\frac{p_0}{p}\right)^{(k-1)/k}$$

Using Eq. 12.17a, we can summarize the equations for determining local isentropic stagnation properties of an ideal gas as

$$\frac{p_0}{p} = \left[1 + \frac{k-1}{2}M^2\right]^{k/(k-1)} \tag{12.17a}$$

$$\frac{T_0}{T} = 1 + \frac{k-1}{2}M^2 \tag{12.17b}$$

$$\frac{\rho_0}{\rho} = \left[1 + \frac{k-1}{2}M^2\right]^{1/(k-1)} \tag{12.17c}$$

From Eqs. 12.17, the ratio of each local isentropic stagnation property to the corresponding static property at any point in a flow field for an ideal gas can be found, if the local Mach number is known. In fact, these ratios could be calculated once and for all and tabulated. The calculation procedure is illustrated in Example Problem 12.4; ratios of local isentropic stagnation properties to the corresponding static properties for an ideal gas are tabulated in Table E.1 of Appendix E.

The Mach number range for validity of the assumption of incompressible flow is investigated in Example Problem 12.5.

EXAMPLE 12.4—Local Isentropic Stagnation Conditions in Channel Flow
Air flows steadily through the duct shown from 350 kPa(abs), 60 C, and 183 m/sec at the initial state to $M = 1.3$ at the outlet, where local isentropic stagnation conditions are known to be 385 kPa(abs) and 350 K. Compute the local isentropic stagnation pressure and temperature at the inlet and the static pressure and temperature at the duct outlet. Locate the inlet and outlet static state points on a Ts diagram, and indicate the stagnation processes.

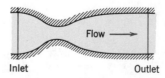

Inlet   Outlet

EXAMPLE PROBLEM 12.4

GIVEN: Steady flow of air through a duct as shown in the sketch.

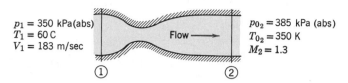

$p_1 = 350$ kPa(abs)
$T_1 = 60$ C
$V_1 = 183$ m/sec

Flow →

$p_{0_2} = 385$ kPa (abs)
$T_{0_2} = 350$ K
$M_2 = 1.3$

①   ②

FIND: (a) p_{0_1}. (b) T_{0_1}. (c) p_2. (d) T_2.
(e) Locate state points ① and ② on a Ts diagram and indicate the stagnation processes.

SOLUTION:
To evaluate local isentropic stagnation conditions at section ①, we must calculate the Mach number, $M_1 = V_1/c_1$. For an ideal gas, $c = \sqrt{kRT}$. Then

$$c_1 = \sqrt{kRT_1} = \left[1.4 \times 287 \frac{\text{N} \cdot \text{m}}{\text{kg} \cdot \text{K}} \times (273 + 60) \text{ K} \times \frac{\text{kg} \cdot \text{m}}{\text{N} \cdot \text{sec}^2} \right]^{1/2} = 366 \text{ m/sec}$$

and

$$M_1 = \frac{V_1}{c_1} = \frac{183}{366} = 0.5$$

Local isentropic stagnation properties may be evaluated from Eqs. 12.17. Thus

$$\frac{p_{0_1}}{p_1} = \left[1 + \frac{k-1}{2} M_1^2 \right]^{k/(k-1)} = [1 + 0.2(0.5)^2]^{3.5} = 1.186$$

and

$$p_{0_1} = 1.186 p_1 = (1.186)(350 \text{ kPa}) = 415 \text{ kPa (abs)} \qquad\qquad p_{0_1}$$

$$\frac{T_{0_1}}{T_1} = 1 + \frac{k-1}{2} M_1^2 = 1.05$$

and

$$T_{0_1} = 1.05 T_1 = (1.05)(333 \text{ K}) = 350 \text{ K} \qquad\qquad T_{0_1}$$

At section ②, Eqs. 12.17 may be applied again. Thus from Eq. 12.17a,

$$\frac{p_{0_2}}{p_2} = \left[1 + \frac{k-1}{2} M_2^2 \right]^{k/(k-1)} = [1 + 0.2(1.3)^2]^{3.5} = 2.77$$

and

$$p_2 = \frac{p_{0_2}}{2.77} = \frac{385 \text{ kPa}}{2.77} = 139 \text{ kPa (abs)} \qquad\qquad p_2$$

From Eq. 12.17b,

$$\frac{T_{0_2}}{T_2} = 1 + \frac{k-1}{2} M_2^2 = 1.338$$

and

$$T_2 = \frac{T_{0_2}}{1.338} = \frac{350 \text{ K}}{1.338} = 262 \text{ K} \qquad\qquad T_2$$

The entropy change must be evaluated to locate state ② with respect to state ①. Using the $T\,ds$ equation,

$$T\,ds = dh - v\,dp$$

or

$$ds = \frac{dh}{T} - \frac{v\,dp}{T} = c_p\frac{dT}{T} - R\frac{dp}{p}$$

Integrating gives

$$s_2 - s_1 = \int_{T_1}^{T_2} c_p\frac{dT}{T} - \int_{p_1}^{p_2} R\frac{dp}{p} = c_p\,\ln\frac{T_2}{T_1} - R\,\ln\frac{p_2}{p_1}$$

$$s_2 - s_1 = \frac{1.00}{\text{kg}\cdot\text{K}}\,\frac{\text{kJ}}{}\times\ln\!\left(\frac{262}{333}\right) - \frac{0.287}{\text{kg}\cdot\text{K}}\,\frac{\text{kJ}}{}\times\ln\!\left(\frac{1.39}{3.50}\right) = 0.0252\ \text{kJ/kg}\cdot\text{K}$$

Therefore, state ② lies to the right of state ① on the Ts plane, as shown in the following sketch:

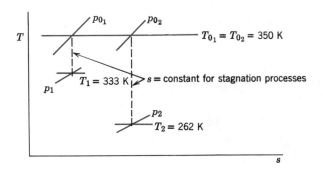

The process the fluid follows between states ① and ② is not specified. However, it need not be known. A unique isentropic stagnation process is defined at each state point. Note that $s_{0_2} - s_{0_1} = s_2 - s_1$.

EXAMPLE 12.5—Mach Number Limit for Incompressible Flow

We have derived equations for p_0/p for both compressible and "incompressible" flow. By writing both equations in terms of Mach number, compare their behavior. Find the Mach number below which the two equations agree within engineering accuracy.

EXAMPLE PROBLEM 12.5

GIVEN: The incompressible and compressible forms of the equations for stagnation pressure, p_0.

Incompressible $\qquad p_0 = p + \tfrac{1}{2}\rho V^2$ $\qquad\qquad\qquad\qquad$ (6.12)

Compressible $\qquad \dfrac{p_0}{p} = \left[1 + \dfrac{k-1}{2}M^2\right]^{k/(k-1)}$ $\qquad\qquad$ (12.17a)

FIND: (a) Behavior of both equations as a function of Mach number.

(b) Mach number below which calculated values of $(p_0 - p)/p_0$ agree within engineering accuracy.

SOLUTION:
First, let us write Eq. 6.12 in terms of Mach number. Using the ideal gas equation of state and $c^2 = kRT$,

$$\frac{p_0}{p} = 1 + \frac{\rho V^2}{2p} = 1 + \frac{V^2}{2RT} = 1 + \frac{kV^2}{2kRT} = 1 + \frac{kV^2}{2c^2}$$

Thus

$$\frac{p_0}{p} = 1 + \frac{k}{2}M^2 \tag{1}$$

for "incompressible" flow.

Equation 12.17a may be expanded using the binomial theorem,

$$(1+x)^n = 1 + nx + \frac{n(n-1)}{2!}x^2 + \cdots, \ |x| < 1$$

For Eq. 12.17a, $x = [(k-1)/2]M^2$, and $n = k/(k-1)$. Thus the series converges for $[(k-1)/2]M^2 < 1$, and

$$\frac{p_0}{p} = 1 + \left(\frac{k}{k-1}\right)\left[\frac{k-1}{2}M^2\right] + \left(\frac{k}{k-1}\right)\left(\frac{k}{k-1}-1\right)\frac{1}{2!}\left[\frac{k-1}{2}M^2\right]^2$$

$$+ \left(\frac{k}{k-1}\right)\left(\frac{k}{k-1}-1\right)\left(\frac{k}{k-1}-2\right)\frac{1}{3!}\left[\frac{k-1}{2}M^2\right]^3 + \cdots$$

$$\frac{p_0}{p} = 1 + \frac{k}{2}M^2 + \frac{k}{8}M^4 + \frac{k(2-k)}{48}M^6 + \cdots$$

$$\frac{p_0}{p} = 1 + \frac{k}{2}M^2\left[1 + \frac{1}{4}M^2 + \frac{(2-k)}{24}M^4 + \cdots\right] \tag{2}$$

for compressible flow.

In the limit, as $M \to 0$, the term in brackets in Eq. 2 approaches 1.0. Thus, for flow at low Mach number, the incompressible and compressible equations give the same result. The variation of p_0/p with Mach number is shown in the sketch below. As Mach number is increased, the compressible equation gives a larger ratio, p_0/p.

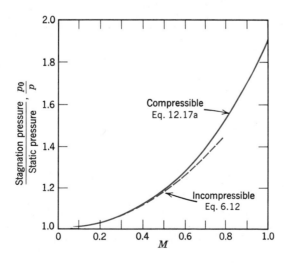

Equations 1 and 2 may be compared quantitatively most simply by writing

$$\frac{p_0}{p} - 1 = \frac{k}{2}M^2 \quad \text{("incompressible")}$$

$$\frac{p_0}{p} - 1 = \frac{k}{2}M^2\left[1 + \frac{1}{4}M^2 + \frac{(2-k)}{24}M^4 + \cdots\right] \quad \text{(compressible)}$$

The term in brackets is approximately equal to 1.02 at $M = 0.3$, and to 1.04 at $M = 0.4$. Thus, for calculations of engineering accuracy, flow may be considered incompressible if $M < 0.3$. The two agree within 5 percent for $M \leqslant 0.45$.

12-4 CRITICAL CONDITIONS

Stagnation conditions are extremely useful as reference conditions for thermodynamic properties; this is not true for velocity, since $V = 0$ by definition at stagnation. A useful reference value for velocity is the *critical speed*—the speed at a Mach number of unity. Even if there is no point in a given flow field where the Mach number is equal to unity, such a hypothetical condition still is useful as a reference condition.

Using asterisks to denote conditions at $M = 1$, then by definition

$$V^* \equiv c^* \tag{12.18}$$

At critical conditions, Eqs. 12.17 for isentropic stagnation properties become (for $k = 1.4$)

$$\frac{p_0^*}{p^*} = \left[1 + \frac{k-1}{2}\right]^{k/(k-1)} = 1.893$$

$$\frac{T_0^*}{T^*} = 1 + \frac{k-1}{2} = 1.200$$

$$\frac{\rho_0^*}{\rho^*} = \left[1 + \frac{k-1}{2}\right]^{1/(k-1)} = 1.577$$

The critical speed may be written in terms of either critical temperature, T^*, or critical stagnation temperature, T_0^*.

For an ideal gas, $c^* = \sqrt{kRT^*}$, and thus $V^* = \sqrt{kRT^*}$. Since

$$T^* = \frac{T_0^*}{1 + (k-1)/2} = \frac{2}{k+1}T_0^*$$

then

$$V^* = c^* = \sqrt{\frac{2k}{k+1}RT_0^*} \tag{12.19}$$

We shall use both stagnation conditions and critical conditions as reference conditions in the next chapter when we consider a variety of one-dimensional compressible flows.

12-5 SUMMARY OBJECTIVES

After completing study of Chapter 12, you should be able to do the following:

1. Define:

Mach number	Mach angle
subsonic flow	zone of action
supersonic flow	zone of silence
transonic flow	local isentropic stagnation properties
hypersonic flow	critical conditions
Mach cone	

2. For an ideal gas, write expressions for
    (a) the change in internal energy and enthalpy.
    (b) the $T\,ds$ equations.
    (c) the relation between pressure and density for an isentropic process.

3. Derive an equation for the speed of sound in a medium and show that, for an ideal gas, $c = \sqrt{kRT}$.

4. Write (and derive) expressions for local isentropic stagnation properties (temperature, pressure, and density) for flow of an ideal gas.

5. Solve the problems at the end of the chapter that relate to the material you have studied.

REFERENCES

1. Zucrow, M. J., and J. D. Hoffman, *Gas Dynamics, Vol. 1.* New York: Wiley, 1976, Chapter 1.

2. Van Wylen, G. J., and R. E. Sonntag, *Fundamentals of Classical Thermodynamics (English/SI Version),* 3rd ed. New York: Wiley, 1986, Chapter 7.

3. Moran, M. J., and H. N. Shapiro, *Fundamentals of Engineering Thermodynamics.* New York: Wiley, 1988, Chapter 6.

4. Wong, G. S. K., "Speed of Sound in Standard Air," *J. Acoustical Society of America, 79,* 5, May 1986, pp. 1359–1366.

PROBLEMS

12.1 Five kg of air in a closed system expands reversibly with constant entropy from 300 kPa (abs), 60 C, to 150 kPa (abs). Calculate the air temperature after expansion. Show the process state points on a *Ts* diagram.

12.2 Air is compressed irreversibly from 100 kPa (abs), 5 C, to 200 kPa(abs), 115 C. Calculate the change in entropy of the air. Show the process state points on a *Ts* diagram.

12.3 Air is cooled at constant pressure from 858 K and 4.5 MPa (gage) to 15 C. Show the process on a *Ts* diagram. Calculate the change in specific entropy for the air if it behaves as an ideal gas. Evaluate the heat transferred per unit mass if the process is reversible.

12.4 Air is compressed isothermally from standard conditions to 4.5 MPa (gage). Show the process on a *Ts* diagram. Calculate the change in specific entropy for the process if the air behaves as an ideal gas. Evaluate the heat transferred per unit mass if the process is reversible.

12.5 Air is expanded in a steady flow process through a turbine. Initial conditions are 1300 C and 2.0 MPa (abs). Final conditions are 500 C and atmospheric pressure. Show this process on a *Ts* diagram. Evaluate the changes in internal energy, enthalpy, and specific entropy for this process.

12.6 Is an adiabatic expansion of air from 300 kPa (abs), 60 C, to 150 kPa (abs), 27 C, possible? Justify your answer. Show the process state points on a Ts diagram.

12.7 Ten lbm of air is cooled in a closed tank from 500 to 100 F. The initial pressure is 400 psia. Compute the changes in entropy, internal energy, and enthalpy. Show the process state points on a Ts diagram.

12.8 In a closed system, a gas undergoes a cycle made up of the following processes: 1–2 reversible isothermal compression, 2–3 reversible constant volume heating, 3–4 reversible constant pressure expansion, and 4–1 reversible adiabatic expansion.
(a) Sketch pv and Ts diagrams of the cycle.
(b) State whether each of the following quantities is positive, zero, negative, or indeterminate in sign:

$$\oint \delta W, \quad \oint \delta Q, \quad \oint dS, \quad \oint dU, \quad \oint dH$$

12.9 An ideal gas is heated at constant volume from state ① to state ②, expanded isothermally to state ③, expanded adiabatically to state ④, which is at the same pressure as state ①, and then restored to state ① by a constant pressure process. All four processes are reversible.
(a) Sketch pv and Ts diagrams of the cycle.
(b) State whether each of the following quantities is positive, zero, negative, or indeterminate in sign:

$$\oint \delta Q, \quad \oint \delta W, \quad \oint du, \quad \oint dh, \quad \oint ds$$

12.10 Air enters a turbine in steady flow at 0.5 kg/sec with negligible velocity. Inlet conditions are 1300 C and 2.0 MPa (abs). The air is expanded through the turbine to atmospheric pressure. If the actual temperature and velocity at the turbine exit are 500 C and 200 m/sec, determine the power produced by the turbine. Label state points on a Ts diagram for this process.

12.11 A tank contains 10 m³ of compressed air at 15 C. The gage pressure in the tank is 4.50 MPa. Evaluate the work required to fill the tank by compressing air from standard atmosphere conditions for (a) isothermal compression and (b) isentropic compression followed by cooling at constant pressure. What is the peak temperature of the isentropic compression process? Calculate the energy removed during cooling for process (b). Assume ideal gas behavior and reversible processes. Label state points on a Ts diagram for each process.

12.12 Natural gas, with the thermodynamic properties of methane, flows in an underground pipeline of 0.6 m diameter. The gage pressure at the inlet to a compressor station is 0.5 MPa; outlet pressure is 8.0 MPa (gage). The gas temperature and speed at inlet are 13 C and 32 m/sec, respectively. The compressor efficiency is $\eta = 0.85$. Calculate the mass flow rate of natural gas through the pipeline. Label state points on a Ts diagram for compressor inlet and outlet. Evaluate the gas temperature and speed at the compressor outlet and the power required to drive the compressor.

12.13 In an isothermal process, 0.1 cubic feet of standard air per minute (SCFM) is pumped into a balloon. Tension in the rubber skin of the balloon is given by $\sigma = kA$, where $k = 200$ lbf/ft³, and A is the surface area of the balloon in ft². Compute the time required to increase the balloon radius from 5 to 7 inches.

12.14 An aircraft flies at 960 km/hr through air at 82 kPa and 0 C. Calculate the Mach number of the craft.

12.15 An airplane flies at 180 m/sec at 500 m altitude on a standard day. The plane climbs to 15 km and flies at 320 m/sec. Calculate the Mach number of flight in both cases.

12.16 The Boeing 727 aircraft of Example Problem 9.8 cruises at 520 mph at 33,000 ft altitude on a standard day. Calculate the cruise Mach number of the aircraft. If the maximum allowable operating Mach number for the aircraft is 0.9, what is the corresponding flight speed?

12.17 Actual performance characteristics of the Lockheed SR-71 "Blackbird" reconnaissance aircraft never were released. However, it was thought to cruise at $M = 3.3$ at 85,000 ft altitude. Evaluate the speed of sound and flight speed for these conditions. Compare to the muzzle speed of a 30–06 rifle bullet (700 m/sec).

12.18 What is the speed of sound in carbon dioxide at 150 C? What is the acoustic speed in water at 20 C?

12.19 The speed of sound in steel is observed to be about 5.3 km/sec. Estimate the bulk modulus for steel. Compare your estimate with the bulk modulus for mercury. Compute the speed of sound in mercury.

12.20 Heptane and octane are used to define the 0 and 100 octane rating points for gasoline. Use data from Appendix A to estimate the bulk modulus of gasoline. Compare the speeds of sound in gasoline and in water.

‡12.21 Use data for specific volume from a steam table to calculate and plot the speed of sound in saturated liquid water over the temperature range from 32 to 400 F.

12.22 The transonic flow regime begins at about $M = 0.9$. Compare the flow speeds for which the Mach number equals 0.9 in (a) air and (b) low-pressure steam at 75 C (assuming that the steam behaves as an ideal gas), and check using steam table data.

12.23 Published data indicate the F-5G aircraft can accelerate from $M = 0.9$ to $M = 1.2$ in 30 sec, and from $M = 0.9$ to $M = 1.6$ in 80 sec at 10 km altitude on a standard day. Express the average rates of acceleration of the aircraft in g s.

12.24 Published data indicate that the F-5G aircraft is capable of making sustained horizontal turns at a rate of 6°/sec at $M = 0.7$ at 30,000 ft altitude. At $M = 1.6$ the aircraft can sustain a turn at 3.5°/sec. Calculate the radius of curvature and normal acceleration produced by these turns.

12.25 The temperature varies linearly from sea level to approximately 11 km altitude in the standard atmosphere. Evaluate the *lapse rate*—the rate of decrease of temperature with altitude—in the standard atmosphere. Derive an expression for the rate of change of sonic speed with altitude in an ideal gas under standard atmospheric conditions. Evaluate at sea level and at 10 km altitude.

12.26 Mixtures of helium and oxygen are used instead of air for long-term undersea work to minimize the risk of causing the "bends" from release of nitrogen bubbles in the bloodstream upon decompression. The speech of "aquanauts" who breathe this mixture is strange, partly because the speed of sound is changed in the mixture. Estimate the gas constant, specific heat ratio, and speed of sound at standard conditions in a mixture that contains 20 percent oxygen and 80 percent helium by volume.

12.27 Air at 20 C flows at 150 m/sec. A bullet is fired into the air stream at 800 m/sec. (The direction of the bullet is opposite to the air flow.) Calculate (a) the Mach number of the air flow, (b) the Mach number of the bullet if it were fired in still air, and (c) the Mach number of the bullet with respect to the moving air.

12.28 A photograph of a bullet shows a Mach angle of 28°. Determine the speed of the bullet for standard air.

12.29 Air at 25 C flows at $M = 1.9$. Determine the air speed and the Mach angle.

12.30 Measurements in very high-speed flow often are made by firing projectiles upstream into a high-speed gas flow. A projectile is fired at 4500 m/sec into a stream of helium at $M = 3.5$ and $T = -20$ C. Calculate the Mach number of the projectile with respect to the gas stream. What Mach angle would be expected on a photograph?

‡ You may wish to use simple computer programs to help solve problems marked with double daggers.

12.31 A projectile is fired into a gas in which the pressure is 50 psia and the density is 0.27 lbm/ft^3. It is observed experimentally that a Mach cone emanates from the projectile with 20° total angle. What is the speed of the projectile with respect to the gas?

12.32 An F-4 aircraft makes a high-speed pass over an airfield on a day when $T = 35$ C. The aircraft flies at $M = 1.4$ and 200 m altitude. Calculate the speed of the aircraft. How long after it passes directly over point A on the ground does its Mach cone pass over point A?

12.33 The National Transonic Facility (NTF) is a high-speed wind tunnel designed to operate with air at cryogenic temperatures to reduce viscosity, thus raising the unit Reynolds number (Re/x) and reducing pumping power requirements. Operation is envisioned at temperatures of −270 F and below. A schlieren photograph taken in the NTF shows a Mach angle of 57° where $T = -270$ F and $p = 1.3$ psia. Evaluate the local Mach number and flow speed. Calculate the unit Reynolds number for the flow.

12.34 An F-5G aircraft passes overhead at 3 km altitude. The aircraft flies at $M = 1.35$; assume air temperature is constant at 30 C. Find the air speed of the aircraft. A headwind blows at 10 m/sec. How long after the aircraft passes directly overhead does its sound reach a point on the ground?

12.35 A supersonic aircraft flies at 10,000 ft altitude at a speed of 3000 ft/sec on a standard day. How long after passing directly above a ground observer is the sound of the aircraft heard by the ground observer?

12.36 For the conditions of Problem 12.34, find the location at which the sound wave that first strikes the point on the ground was emitted.

12.37 The Concorde supersonic transport cruises at $M = 2.2$ at 17 km altitude on a standard day. How long after the aircraft passes directly above a ground observer is the sound of the aircraft heard?

12.38 Air flows in a duct at $T = 50$ F and $p = 10$ psia. Calculate the local isentropic stagnation pressure if the air speed is (a) 200 ft/sec and (b) 2000 ft/sec.

12.39 A plane is flying at 1500 m altitude. At a point on the plane where the relative air speed is zero, the temperature is 49 C. Determine the Mach number and speed of the plane.

12.40 The maximum density in a compressible flow field occurs at stagnation conditions. Evaluate the Mach numbers for flow of air at which ρ_0 and ρ differ by 2 percent and 5 percent. Comment on the significance of these results.

12.41 The pressure at the nose of an aircraft in flight is 44.3 kPa (abs). (The air speed relative to the craft is zero at this point.) Estimate the Mach number, speed, and altitude of the craft, if the undisturbed air is at 27.6 kPa (abs) and −50 C.

12.42 Consider flow of standard air at 600 m/sec. What is the local isentropic stagnation pressure? The stagnation enthalpy? The stagnation temperature?

12.43 A body moves through standard air at 200 m/sec. What is the pressure at a point on the body where the air speed relative to the body is zero? Assume (a) compressible flow and (b) incompressible flow.

12.44 A DC-10 aircraft cruises at 12 km altitude on a standard day. A pitot-static tube on the nose of the aircraft measures stagnation and static pressures of 29.6 kPa and 19.4 kPa. Calculate (a) the flight Mach number of the aircraft, (b) the speed of the aircraft, and (c) the stagnation temperature that would be sensed by a probe on the aircraft.

12.45 The Anglo-French "Concorde" supersonic transport cruises at $M = 2.2$ at 20 km altitude. Evaluate the speed of sound, aircraft flight speed, and Mach angle. Compare the aircraft speed to the muzzle speed of a .22-caliber rifle bullet (460 m/sec). What is the maximum air temperature at stagnation points on the aircraft structure?

12.46 A jet transport aircraft cruises at $M = 0.85$ at 12.5 km altitude on a standard day. Evaluate the stagnation pressure sensed by a probe on the aircraft. What speed would be calculated from the incompressible Bernoulli equation? By what percentage is this speed in error compared to the true speed of the aircraft?

12.47 A smoothbore "12-pounder" cannon used on a sailing ship fires a spherical cast-iron shot, with diameter $D = 110$ mm and mass $m - 5.44$ kg, horizontally at sea level on a standard day. Initially the shot travels at supersonic speed, but it is slowed rapidly by aerodynamic drag. At the instant when the speed is sonic, estimate (a) the horizontal acceleration of the shot (assume the drag coefficient for a sphere at sonic speed is $C_D = 1.3$), (b) the maximum pressure on the surface of the shot, and (c) the maximum air temperature near the surface of the shot.

12.48 A supersonic wind tunnel test section is designed to have $M = 3.0$ at 60 F and 5 psia. The fluid is air. Determine the required inlet (stagnation) conditions, T_0 and p_0. Calculate the required mass flow rate for a test section area of 2.0 ft^2.

12.49 Air flows steadily through a length (① denotes inlet and ② denotes exit) of insulated constant-area duct. Properties change along the duct as a result of friction.
(a) Beginning with the control volume form of the first law of thermodynamics, show that the equation can be reduced to

$$h_1 + \frac{V_1^2}{2} = h_2 + \frac{V_2^2}{2} = \text{constant}$$

(b) Denoting the constant by h_0 (the stagnation enthalpy), show that for adiabatic flow of an idea gas with friction

$$\frac{T_0}{T} = 1 + \frac{k-1}{2}M^2$$

(c) For this flow does $T_{0_1} = T_{0_2}$? $p_{0_1} = p_{0_2}$?

12.50 According to *Popular Science* magazine, the maximum air speed in Ford's Variable Venturi carburetor is approximately 430 ft/sec. For standard day atmospheric conditions, evaluate the minimum static pressure in the carburetor venturi. What percentage error in calculated static pressure would result if this calculation were made assuming incompressible flow?

12.51 For aircraft flying at supersonic speeds, lift and drag coefficients are functions of Mach number only. A supersonic transport with wingspan of 75 m is to fly at 780 m/sec at 20 km altitude on a standard day. Performance of the aircraft is to be measured from tests of a model with 0.9 m wingspan in a supersonic wind tunnel. The wind tunnel is to be supplied from a large reservoir of compressed air, which can be heated if desired. The static temperature of air in the test section is to be 10 C to avoid freezing of moisture. At what air speed should the wind tunnel tests be run to duplicate the Mach number of the prototype? What must be the stagnation temperature in the reservoir? What pressure is required in the reservoir if the test section pressure is to be 10 kPa (abs)?

12.52 A transonic wind tunnel test section operates at $M = 1.2$, with $T = -34$ C and $p = 41.8$ kPa (abs). Calculate the local isentropic stagnation conditions for this flow. A normal shock forms ahead of a pitot-static tube placed in the flow stream. The total-head tube measures a stagnation pressure of 100.6 kPa (abs). Determine the loss in stagnation pressure across the normal shock. By what percentage is the local isentropic stagnation pressure reduced compared to the upstream value?

12.53 A supersonic wind tunnel test section operates at $M = 1.8$, with $T = -80$ C and $p = 258$ kPa (gage). A normal shock that stands in front of a total-head tube reduces the stagnation pressure by 18.7 percent. Calculate (a) the local isentropic stagnation conditions for the test section conditions and (b) the pressure sensed by the total-head tube.

12.54 Published data indicate the F-5G aircraft has a maximum speed of $M = 2.1$ at 36,000 ft altitude on a standard day. Calculate the aircraft flight speed. Determine the local isentropic stagnation pressure that corresponds to these flight conditions. Because the aircraft speed is supersonic, a normal shock occurs in front of a total-head tube. The stagnation pressure decreases by 32.6 percent across the shock. Evaluate the stagnation pressure sensed by a probe on the aircraft. What is the maximum air temperature at stagnation points on the aircraft structure?

12.55 Actual performance characteristics of the Lockheed SR-71 "Blackbird" reconnaissance aircraft were classified. However, it was thought to cruise at $M = 3.3$ at 26 km altitude. Calculate the aircraft flight speed for these conditions. Determine the local isentropic stagnation pressure. Because the aircraft speed is supersonic, a normal shock occurs in front of a total-head tube. The stagnation pressure decreases by 74.7 percent across the shock. Evaluate the stagnation pressure sensed by a probe on the aircraft. What is the maximum air temperature at stagnation points on the aircraft structure?

12.56 Air enters a long, insulated duct at $M_1 = 0.2$, $T_1 = 286$ K, and $p_1 = 98.5$ kPa (abs). Downstream the properties are $M_2 = 0.6$, $T_2 = 268.9$ K, and $p_2 = 31.3$ kPa (abs). (Four significant figures are given to minimize roundoff errors.) Evaluate local isentropic stagnation conditions (a) at the inlet section and (b) at the outlet section. Calculate the change in specific entropy along the duct. Plot static and stagnation state points on a *Ts* diagram.

12.57 Air enters a combustion chamber at $M_1 = 0.2$, $T_1 = 580$ K, and $p_1 = 1.0$ MPa (abs). Heat addition causes the exit properties to be $M_2 = 0.4$, $T_2 = 1727$ K, and $p_2 = 862.7$ kPa (abs). (Four significant figures are given to minimize roundoff errors.) Evaluate local isentropic stagnation conditions (a) at the inlet to and (b) at the outlet from the combustion chamber. Calculate the change in specific entropy across the combustor. Plot static and stagnation state points on a *Ts* diagram.

12.58 Air passes through a normal shock in a supersonic wind tunnel. Upstream conditions are $M_1 = 1.8$, $T_1 = 270$ K, and $p_1 = 10.0$ kPa (abs). Downstream conditions are $M_2 = 0.6165$, $T_2 = 413.6$ K, and $p_2 = 36.13$ kPa (abs). (Four significant figures are given to minimize roundoff errors.) Evaluate local isentropic stagnation conditions (a) upstream from and (b) downstream from the normal shock. Calculate the change in specific entropy across the shock. Plot static and stagnation state points on a *Ts* diagram.

12.59 Air enters a turbine at $M_1 = 0.4$, $T_1 = 2350$ F, and $p_1 = 90.0$ psia. Conditions leaving the turbine are $M_2 = 0.8$, $T_2 = 1200$ F, and $p_2 = 3.00$ psia. (Four significant figures are given to minimize roundoff errors.) Evaluate local isentropic stagnation conditions (a) at the turbine inlet and (b) at the turbine outlet. Calculate the change in specific entropy across the turbine. Plot static and stagnation state points on a *Ts* diagram.

12.60 A Boeing 747 cruises at $M = 0.87$ at an altitude of 13 km on a standard day. A window in the cockpit is located where the external flow Mach number is 0.2 relative to the plane surface. The cabin is pressurized to an equivalent altitude of 2500 m in a standard atmosphere. Estimate the pressure difference across the window. Be sure to specify the direction of the net pressure force.

12.61 Consider again the gas pipeline compressor of Problem 12.12. Calculate local isentropic stagnation conditions for flow of gas (a) at the compressor inlet and (b) at the compressor outlet. Calculate the change in specific entropy across the compressor. Show the static and stagnation state points on a *Ts* diagram. (The compressor outlet temperature and Mach number are 572 K and 0.0762, respectively.)

12.62 Air flows steadily through a constant-area duct. At section ①, the air is at 60 psia, 600 R, and 500 ft/sec. As a result of heat transfer and friction, the air at section

② downstream is at 40 psia, 800 R. Calculate the heat transfer per pound of air between sections ① and ②, and the stagnation pressure at section ②.

12.63 Air flows through a constant-area duct. At section ①, conditions are 550 K, 973 kPa (abs), and $M_1 = 0.20$. At section ②, flow is at 1100 K, 910 kPa (abs), and $M_2 = 0.30$. Calculate stagnation enthalpy, stagnation temperature, and stagnation pressure at sections ① and ②. Is the flow isentropic? Justify your answer.

12.64 Consider steady, adiabatic flow of air through a long straight pipe with $A = 0.05$ m². At the inlet (section ①) the air is at 200 kPa (abs), 60 C, and 146 m/sec. Downstream at section ②, the air is at 95.6 kPa (abs) and 280 m/sec. Determine p_{0_1}, p_{0_2}, T_{0_1}, T_{0_2}, and the entropy change for the flow. Show static and stagnation state points on a Ts diagram.

12.65 Air flows steadily through a constant-area duct. At section ①, the air is at 60 psia, 600 R, and 500 ft/sec. As a result of heat transfer and friction, the air at section ② downstream is at 40 psia, 800 R. Determine p_{0_1}, p_{0_2}, T_{0_1}, T_{0_2}, and the entropy change for the flow. Show static and stagnation state points on a Ts diagram.

12.66 All modern high-speed aircraft use "air data computers" to calculate air speed from measured dynamic pressure. Evaluate the subsonic Mach number above which the incompressible Bernoulli equation predicts a speed error of 2 percent compared to true air speed calculated including compressibility effects. Assume flight at 10 km altitude on a standard day. Is the result you obtain independent of freestream conditions?

12.67 The *critical pressure ratio*, p^*/p_0^*, is the ratio of static pressure to local isentropic stagnation pressure at the condition where $V = V^* = c^*$. For an ideal gas, the critical pressure ratio is a function only of specific heat ratio, k. Show that for air, $p^*/p_0^* = 0.5283$. For comparison, evaluate the critical pressure ratios for gases with the maximum and minimum ks given in Table A.6.

12.68 The tires on a modern lightweight bicycle are inflated to 800 kPa (gage). Consider a day when a tire is in equilibrium with the ambient at 37 C. Calculate the critical conditions (temperature, pressure, and flow speed) that correspond to these stagnation conditions.

12.69 A service station tank holds air at 800 kPa (gage) and 30 C. Calculate the critical conditions (temperature, pressure, and flow speed) that correspond to these stagnation conditions.

12.70 A CO_2 cartridge is used to propel a toy rocket. Gas in the cartridge is pressurized to 45 MPa (gage) and is at 25 C. Calculate the critical conditions (temperature, pressure, and flow speed) that correspond to these stagnation conditions.

12.71 A fire extinguisher, filled with carbon dioxide gas, is pressurized to 35 MPa (gage) and stored at 30 C. Calculate the critical conditions (temperature, pressure, and flow speed) that correspond to these stagnation conditions.

12.72 The gas storage reservoir for a high-speed wind tunnel contains helium at 2500 K and 6.0 MPa (gage). Calculate the critical conditions (temperature, pressure, and flow speed) that correspond to these stagnation conditions.

12.73 Stagnation conditions in a solid propellant rocket motor are $T_0 = 3500$ K and $p_0 = 40$ MPa (gage). Critical conditions occur in the throat of the rocket nozzle where the Mach number is equal to one. Evaluate the temperature, pressure, and flow speed at the throat. Assume ideal gas behavior with $R = 323$ J/kg·K and $k = 1.2$.

12.74 Consider the natural gas pipeline of Problem 12.12. Calculate the critical conditions (temperature, pressure, and flow speed) that correspond to the compressor inlet conditions.

12.75 The hot gas stream at the turbine inlet of a JT9-D jet engine is at 2350 F, 140 kPa (abs), and $M = 0.32$. Calculate the critical conditions (temperature, pressure, and flow speed) that correspond to these conditions. Assume the fluid properties of pure air.

Chapter 13

STEADY ONE-DIMENSIONAL COMPRESSIBLE FLOW

Fluid properties in compressible flow are affected by area change, friction, heat transfer, and normal shocks. In this chapter each of these effects is considered separately for steady, one-dimensional, compressible flow.

Isentropic flow, in which area is the independent variable (friction and heat transfer are neglected), is considered first for a general fluid. Then isentropic flow of an ideal gas and applications to nozzles are considered in greater detail.

Following isentropic flow, adiabatic flow in a constant-area duct with friction and frictionless flow in a constant-area duct with heat transfer are considered. A discussion of normal shocks and supersonic channel flows with shocks concludes the chapter.

13-1 BASIC EQUATIONS FOR ISENTROPIC FLOW

Consider steady, one-dimensional, isentropic flow of any compressible fluid through a channel of arbitrary cross section; a portion of such a duct is shown in Fig. 13.1. To develop the governing equations for this flow, we apply the basic equations, derived in Chapter 4, to the finite, fixed control volume of Fig. 13.1. Properties at sections ① and ② are labeled with appropriate subscripts; R_x is the x component of the surface force acting on the control volume.

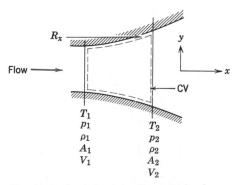

Fig. 13.1 Control volume for analysis of a general isentropic flow.

a. Continuity Equation

Basic equation:

$$= 0(1)$$

$$0 = \frac{\partial}{\partial t} \int_{CV} \rho \, d\mathcal{V} + \int_{CS} \rho \vec{V} \cdot d\vec{A} \tag{4.13}$$

Assumptions: (1) Steady flow
(2) One-dimensional flow

Then

$$0 = \{-|\rho_1 V_1 A_1|\} + \{|\rho_2 V_2 A_2|\}$$

Using scalar magnitudes and dropping absolute value signs gives the familiar form

$$\rho_1 V_1 A_1 = \rho_2 V_2 A_2 = \rho V A = \dot{m} = \text{ constant} \tag{13.1a}$$

b. Momentum Equation

Basic equation:

$$= 0(3) = 0(1)$$

$$F_{S_x} + F_{B_x} = \frac{\partial}{\partial t} \int_{CV} V_x \rho \, d\mathcal{V} + \int_{CS} V_x \rho \vec{V} \cdot d\vec{A} \tag{4.19a}$$

Assumptions: (3) $F_{B_x} = 0$

The surface force will be due to pressure forces at the surfaces ① and ②, and to the distributed pressure force, R_x, along the channel walls. Substituting gives

$$R_x + p_1 A_1 - p_2 A_2 = V_1\{-|\rho_1 V_1 A_1|\} + V_2\{|\rho_2 V_2 A_2|\}$$

Using scalar magnitudes and dropping absolute value signs, we obtain

$$R_x + p_1 A_1 - p_2 A_2 = \dot{m} V_2 - \dot{m} V_1 \tag{13.1b}$$

c. First Law of Thermodynamics

Basic equation:

$$= 0(4) = 0(5) = 0(6) \quad = 0(6) \quad = 0(1)$$

$$\dot{Q} - \dot{W}_s - \dot{W}_{shear} - \dot{W}_{other} = \frac{\partial}{\partial t} \int_{CV} e\rho \, d\mathcal{V} + \int_{CS} (e + pv)\rho \vec{V} \cdot d\vec{A} \tag{4.57}$$

where

$$\simeq 0(7)$$

$$e = u + \frac{V^2}{2} + gz$$

Assumptions: (4) $\dot{Q} = 0$ (isentropic, i.e., frictionless, adiabatic flow)
(5) $\dot{W}_s = 0$
(6) $\dot{W}_{shear} = \dot{W}_{other} = 0$
(7) Effects of gravity are negligible

Under these assumptions, the first law reduces to

$$0 = \left(u_1 + p_1 v_1 + \frac{V_1^2}{2}\right)\{-|\rho_1 V_1 A_1|\} + \left(u_2 + p_2 v_2 + \frac{V_2^2}{2}\right)\{|\rho_2 V_2 A_2|\}$$

But we know from continuity that the mass flow rate terms in brackets are equal, so they may be canceled. We may also substitute $h \equiv u + pv$, to obtain

$$h_1 + \frac{V_1^2}{2} = h_2 + \frac{V_2^2}{2} = h + \frac{V^2}{2} = \text{constant} \tag{13.1c}$$

The combination $h + V^2/2$ occurs often in compressible flow problems. It is convenient to define the stagnation enthalpy, h_0, as

$$h_0 \equiv h + \frac{V^2}{2}$$

Physically, stagnation enthalpy is the enthalpy that would be reached if the fluid were decelerated adiabatically to zero velocity. We note, from Eq. 13.1c, that stagnation enthalpy is constant throughout an adiabatic flow field.

d. Second Law of Thermodynamics

Basic equation:

$$\int_{CS} \frac{1}{T} \overset{=0(4)}{\cancel{\dot{Q}}}{dA} \le \overset{=0(1)}{\cancel{\frac{\partial}{\partial t}}} \int_{CV} s\rho \, d\Psi + \int_{CS} s\rho \vec{V} \cdot d\vec{A} \tag{4.59}$$

Then, for a reversible adiabatic process, the equality holds and

$$0 = s_1\{-|\rho_1 V_1 A_1|\} + s_2\{|\rho_2 V_2 A_2|\}$$

Since the mass flow rate terms { } are equal by continuity,

$$s_1 = s_2 = s = \text{constant} \tag{13.1d}$$

e. Equation of State

Equations of state are relations among intensive thermodynamic properties. These relations may be in the form of tables, charts, or algebraic equations. For a pure substance, it is possible to specify any intensive thermodynamic property in terms of any other two intensive thermodynamic properties; we can write

$$h = h(s, p) \tag{13.1e}$$

and

$$\rho = \rho(s, p) \tag{13.1f}$$

as equations of state.

Before summarizing the simplified forms of the basic equations for steady, one-dimensional, isentropic flow of any compressible fluid, let us turn to a representation of isentropic flow on an hs diagram. At some point in the isentropic flow field (call it state ①), flow properties are h_1, s_1, p_1, V_1, and so on. Clearly, in isentropic flow, the flow may proceed to state ② (with properties h_2, $s_2 = s_1, p_2, V_2$, etc.), or to state ③ (with properties h_3, $s_3 = s_1$, p_3, V_3, etc.), as shown in Fig. 13.2.

How are isentropic stagnation properties at states ①, ②, and ③ related? To answer this question, consider results obtained from the first law analysis for isentropic flow. From Eq. 13.1c and the definition of stagnation enthalpy, we have

$$h_1 + \frac{V_1^2}{2} = h_2 + \frac{V_2^2}{2} = h + \frac{V^2}{2} = h_0 = \text{constant}$$

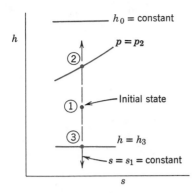

Fig. 13.2 Representation of isentropic flow in the hs plane.

Thus all states in an isentropic flow have the same stagnation enthalpy. In addition, by definition, all states in an isentropic flow (including the stagnation state) have the same entropy. Thus all stagnation states in an isentropic flow have the same stagnation enthalpy and stagnation entropy. Since the velocity is zero at the stagnation state, it follows that stagnation properties are constant for all points in an isentropic flow.

For isentropic flow, the first law, in the form

$$h + \frac{V^2}{2} = h_0 = \text{constant}$$

suggests another interpretation for stagnation enthalpy: stagnation enthalpy, h_0, represents the total energy per unit mass of flowing fluid. The kinetic energy per unit mass is represented by the enthalpy difference, $h_0 - h$, as illustrated graphically in Fig. 13.3.

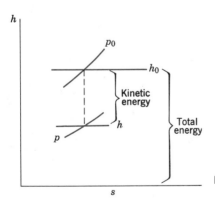

Fig. 13.3 Schematic hs diagram illustrating interpretation of energy per unit mass in a flow.

Equations 13.1a through 13.1f are the simplified forms of the basic equations that describe steady, one-dimensional, isentropic flow of any compressible fluid. There are six independent equations. If all properties of the flow are known at state ①, then we have a total of seven unknowns ($p_2, A_2, V_2, \rho_2, h_2, s_2,$ and R_x) in these six equations. Consequently, an isentropic flow can proceed to a variety of states from state ①. Each such state must have $s_2 = s_1$, satisfying Eq. 13.1d. This leaves, in effect, six unknowns and five equations, and the problem is indeterminate. To determine conditions at state ②, one of the six unknowns must be specified.

What causes fluid property changes in isentropic flow? You undoubtedly recognize that it is area variation. In the next section, we analyze the effect of area variation on properties in isentropic flow.

13-2 EFFECT OF AREA VARIATION ON PROPERTIES IN ISENTROPIC FLOW

In considering the effect of area variation on fluid properties in isentropic flow, we shall concern ourselves primarily with velocity and pressure. We wish to determine the effect of a change in area, A, on velocity, V, and pressure, p; i.e., for a change dA in area, are dV and dp positive or negative?

To answer these questions, it is convenient to work with the differential forms of the governing equations. These were derived for the differential control volume of Fig. 12.3 (Section 12-3.1). The differential momentum equation for isentropic flow reduces to

$$\frac{dp}{\rho} + d\left(\frac{V^2}{2}\right) = 0 \qquad (12.16c)$$

or

$$dp = -\rho V \, dV$$

Dividing by ρV^2, we obtain

$$\frac{dp}{\rho V^2} = -\frac{dV}{V} \qquad (13.2)$$

A convenient differential form of the continuity equation can be obtained from Eq. 13.1a,

$$\rho A V = \text{constant} \qquad (13.1a)$$

Taking the natural logarithm of both sides yields

$$\ln \rho + \ln A + \ln V = \ln C$$

Differentiating,

$$\frac{d\rho}{\rho} + \frac{dA}{A} + \frac{dV}{V} = 0 \qquad (13.3)$$

Solving Eq. 13.3 for dA/A gives

$$\frac{dA}{A} = -\frac{dV}{V} - \frac{d\rho}{\rho}$$

Substituting from Eq. 13.2,

$$\frac{dA}{A} = \frac{dp}{\rho V^2} - \frac{d\rho}{\rho}$$

or

$$\frac{dA}{A} = \frac{dp}{\rho V^2}\left[1 - \frac{V^2}{dp/d\rho}\right]$$

Now recall that for an isentropic process, $dp/d\rho = \partial p/\partial \rho)_s = c^2$, so

$$\frac{dA}{A} = \frac{dp}{\rho V^2}\left[1 - \frac{V^2}{c^2}\right] = \frac{dp}{\rho V^2}[1 - M^2] \qquad (13.4)$$

From Eq. 13.4, we see that for $M < 1$, an area change causes a pressure change of the same sign (positive dA means positive dp for $M < 1$); for $M > 1$, an area change causes a pressure change of opposite sign.

Substituting from Eq. 13.2 into Eq. 13.4, we obtain

$$\frac{dA}{A} = \frac{-dV}{V}[1 - M^2]$$

(13.5)

From Eq. 13.5, we see that for $M < 1$ an area change causes a velocity change of opposite sign (positive dA means negative dV for $M < 1$); for $M > 1$ an area change causes a velocity change of the same sign.

These results are summarized in Fig. 13.4. For subsonic flows ($M < 1$) flow acceleration in a *nozzle* requires a passage of diminishing cross section; area must decrease to cause a velocity increase. This produces a passage shaped like that shown in the upper left of Fig. 13.4, and this result is in accord with our experience. A subsonic *diffuser* requires that the passage area increase to cause a velocity decrease. Again this result agrees with our experience.

In supersonic flows ($M > 1$), the effects of area change are different. According to Eq. 13.5, a *supersonic nozzle* must be built with an area increase in the flow direction. A *supersonic diffuser* must be a converging channel. Although these predictions may be contrary to our experience, laboratory experiments show that they are valid. We also can recall seeing divergent nozzles designed to produce supersonic flow on missiles and launch vehicles.

What of the remaining case, $M = 1$? Further inspection of Eq. 13.5 shows that at $M = 1$, $dA/dV = 0$; this means the channel area must pass through a minimum or maximum at $M = 1$. Inspection of Fig. 13.4 shows that $M = 1$ can be reached only in a *throat*, or section of minimum area.

To accelerate flow from rest to supersonic speed ($M > 1$) requires first a subsonic *converging nozzle*. Under proper conditions, the flow will be at $M = 1$ at the throat, where the area is a minimum. Further acceleration is possible if a supersonic *diverging nozzle* segment is added downstream from a throat. Isentropic flow in converging

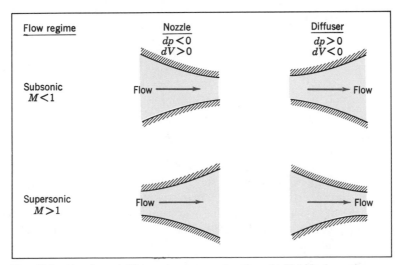

Fig. 13.4 Nozzle and diffuser shapes as a function of initial Mach number.

nozzles will be treated in Section 13-3.4, and isentropic flow in converging-diverging nozzles will be covered in Section 13-3.5.

To decelerate flow from supersonic ($M > 1$) to subsonic speed requires first a supersonic (converging) diffuser. In theory, the flow speed could be reduced isentropically to $M = 1$ at a throat where the area is a minimum, and further isentropic deceleration could take place in a diverging subsonic diffuser section. In practice, supersonic flow cannot be decelerated to exactly $M = 1$ at a throat because sonic flow near a throat is unstable in a rising (adverse) pressure gradient. (Disturbances that always are present in a real subsonic flow propagate upstream, disturbing the sonic flow at the throat, causing shock waves to form and travel upstream, where they are disgorged from the inlet of the supersonic diffuser.)

The throat area of a real supersonic diffuser must be slightly larger than required to reduce the flow to $M = 1$. Under the proper downstream conditions, a weak normal shock forms in the diverging channel just downstream from the throat. Flow leaving the shock is subsonic and decelerates in the diverging channel. Thus deceleration from supersonic to subsonic flow cannot occur isentropically in practice, since the weak normal shock causes an entropy increase. Normal shocks will be analyzed in Section 13-6.

For accelerating flows (favorable pressure gradients) the idealization of isentropic flow is generally a realistic model of the actual flow behavior. For decelerating flows, the idealization of isentropic flow may not be a realistic model because of the adverse pressure gradients and the attendant possibility of flow separation, as discussed for incompressible boundary-layer flow in Chapter 9.

13-3 ISENTROPIC FLOW OF AN IDEAL GAS

13-3.1 Basic Equations

In Section 13-1, we applied the basic equations to a finite control volume for steady, one-dimensional, isentropic flow of any compressible fluid. To restrict our discussion to an ideal gas, we only need to modify the equation of state. For an ideal gas, the equation of state is $p = \rho RT$. In addition, for isentropic flow of an ideal gas, we have the process equation, $p/\rho^k = \text{constant}$. Then, for isentropic flow of an ideal gas, we can summarize the basic equations as follows:

Continuity: $$\rho_1 V_1 A_1 = \rho_2 V_2 A_2 = \rho V A = \dot{m} \tag{13.1a}$$

Momentum: $$R_x + p_1 A_1 - p_2 A_2 = \dot{m} V_2 - \dot{m} V_1 \tag{13.1b}$$

First law: $$h_1 + \frac{V_1^2}{2} = h_2 + \frac{V_2^2}{2} = h + \frac{V^2}{2} \tag{13.1c}$$

Second law: $$s_1 = s_2 = s \tag{13.1d}$$

Equation of state: $$p = \rho RT \tag{12.1}$$

Process equation: $$p/\rho^k = \text{constant} \tag{12.11b}$$

These are the governing equations for steady, one-dimensional, isentropic flow of an ideal gas. If all properties at state ① are known, then we have eight unknowns ($\rho_2, A_2, V_2, p_2, h_2, s_2, T_2$, and R_x) in these six equations. However, we have the known relationship between h and T for an ideal gas, $dh = c_p \, dT$. For an ideal gas with constant specific heats,

$$\Delta h = h_2 - h_1 = c_p \, \Delta T = c_p \, (T_2 - T_1) \tag{12.7b}$$

Thus, as in the general case (Section 13-1), the problem is indeterminate. One condition (other than s_2) must be specified at state ② before conditions at state ② can be completely determined.

13-3.2 Reference Conditions for Isentropic Flow of an Ideal Gas

Expressions for local isentropic stagnation properties for an ideal gas were developed in Chapter 12 (Section 12-3.1). For completeness these expressions are repeated here.

Stagnation pressure:
$$\frac{p_0}{p} = \left[1 + \frac{k-1}{2}M^2\right]^{k/(k-1)} \qquad (12.17a)$$

Stagnation temperature:
$$\frac{T_0}{T} = 1 + \frac{k-1}{2}M^2 \qquad (12.17b)$$

Stagnation density:
$$\frac{\rho_0}{\rho} = \left[1 + \frac{k-1}{2}M^2\right]^{1/(k-1)} \qquad (12.17c)$$

As shown in Section 13-1, stagnation properties are constant throughout a steady, isentropic flow field.

Critical conditions—flow properties at which the Mach number is unity—were introduced in Section 12-3.2. Since stagnation properties are constant in an isentropic flow, then from Eqs. 12.17 we can write, for $k = 1.4$,

$$\frac{p_0}{p^*} = \left[1 + \frac{k-1}{2}\right]^{k/(k-1)} = 1.893; \qquad \frac{p^*}{p_0} = 0.5283$$

$$\frac{T_0}{T^*} = 1 + \frac{k-1}{2} = 1.200; \qquad \frac{T^*}{T_0} = 0.8333$$

$$\frac{\rho_0}{\rho^*} = \left[1 + \frac{k-1}{2}\right]^{1/(k-1)} = 1.577$$

In addition, from Eq. 12.19, we have

$$V^* = c^* = \sqrt{\frac{2k}{k+1}RT_0}$$

In Section 13-2, we saw that it was necessary for a passage to have a section of minimum area (a throat) to accelerate a flow isentropically from rest to a Mach number greater than unity. Furthermore, in such a flow $M = 1$ at the throat. If the area at which the Mach number is unity is designated by A^*, then it is possible to express the contour of a passage in terms of area ratio A/A^*.

For steady, one-dimensional flow, the continuity equation can be written

$$\rho A V = \text{constant} = \rho^* A^* V^*$$

Then

$$\frac{A}{A^*} = \frac{\rho^*}{\rho}\frac{V^*}{V} = \frac{\rho^*}{\rho}\frac{c^*}{Mc} = \frac{1}{M}\frac{\rho^*}{\rho}\sqrt{\frac{T^*}{T}}$$

$$\frac{A}{A^*} = \frac{1}{M}\frac{\rho^*}{\rho_0}\frac{\rho_0}{\rho}\sqrt{\frac{T^*/T_0}{T/T_0}}$$

$$\frac{A}{A^*} = \frac{1}{M} \frac{\left[1 + \frac{k-1}{2}M^2\right]^{1/(k-1)}}{\left[1 + \frac{k-1}{2}\right]^{1/(k-1)}} \left[\frac{1 + \frac{k-1}{2}M^2}{1 + \frac{k-1}{2}}\right]^{1/2}$$

$$\frac{A}{A^*} = \frac{1}{M} \left[\frac{1 + \frac{k-1}{2}M^2}{1 + \frac{k-1}{2}}\right]^{(k+1)/2(k-1)} \tag{13.6}$$

From Eq. 13.6, we see that a choice of M gives a unique value of A/A^*. The variation of A/A^* with M is shown in Fig. 13.5. The curve is double-valued; for any A/A^* other than unity, there are two possible values of Mach number. This is consistent with the results of Section 13-2 (see Fig. 13.4), where it was found that a converging-diverging passage with a section of minimum area is required to accelerate from subsonic to supersonic speed.

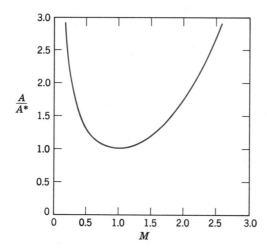

Fig. 13.5 Variation of A/A^* with Mach number in isentropic flow of an ideal gas with $k = 1.4$.

EXAMPLE 13.1—Isentropic Flow in a Converging Channel

Air flows isentropically in a channel. At section ①, the Mach number is 0.3, the area is 0.001 m², and the absolute pressure and the temperature are 650 kPa and 62 C, respectively. At section ②, the Mach number is 0.8. Sketch the channel shape, plot a Ts diagram for the process, and evaluate properties at section ②.

EXAMPLE PROBLEM 13.1

GIVEN: Isentropic flow of air in a channel. At sections ① and ②, the following data are given: $M_1 = 0.3$, $T_1 = 62$ C, $p_1 = 650$ kPa (abs), $A_1 = 0.001$ m², and $M_2 = 0.8$.

FIND: (a) Sketch the channel shape.
       (b) Plot a Ts diagram for the process.
       (c) Properties at section ②.

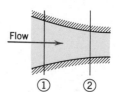

SOLUTION:
To accelerate a subsonic flow requires a converging nozzle.
The channel shape must be as shown.

On the Ts plane, the process follows a line, $s = $ constant.
Stagnation conditions remain fixed for isentropic flow.

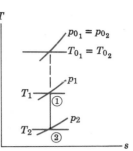

Consequently, the stagnation temperature at section ② can be calculated (for air, $k = 1.4$) from

$$T_{0_2} = T_{0_1} = T_1 \left[1 + \frac{k-1}{2} M_1^2 \right]$$

$$= (62 + 273) \text{ K} \left[1 + \frac{1.4-1}{2}(0.3)^2 \right]$$

$$T_{0_2} = T_{0_1} = 341 \text{ K} \qquad\qquad\qquad\qquad\qquad\qquad T_{0_1}, T_{0_2}$$

and

$$T_2 = \frac{T_{0_2}}{\left[1 + \frac{k-1}{2} M_2^2 \right]} = \frac{341 \text{ K}}{[1 + 0.2(0.8)^2]} = 302 \text{ K or } 29 \text{ C} \qquad\qquad T_2$$

Also, for an ideal gas,

$$c_2 = \sqrt{kRT_2} = \left[1.4 \times \frac{287 \text{ N} \cdot \text{m}}{\text{kg} \cdot \text{K}} \times 302 \text{ K} \times \frac{\text{kg} \cdot \text{m}}{\text{N} \cdot \text{sec}^2} \right]^{1/2} = 348 \text{ m/sec} \qquad c_2$$

From the definition of Mach number,

$$V_2 = M_2 c_2 = (0.8)348 \text{ m/sec} = 278 \text{ m/sec} \qquad\qquad\qquad\qquad V_2$$

Using the isentropic relation, $p/\rho^k = $ constant, and the ideal gas equation of state, we obtain

$$\frac{p_2}{p_1} = \left(\frac{T_2}{T_1} \right)^{k/(k-1)} = \left(\frac{302 \text{ K}}{335 \text{ K}} \right)^{3.5} = 0.696$$

and

$$p_2 = 0.696 \, p_1 = (0.696)650 \text{ kPa (abs)} = 452 \text{ kPa (abs)} \qquad\qquad p_2$$

Also,

$$\rho_2 = \frac{p_2}{RT_2} = \frac{4.52 \times 10^5 \text{ N}}{\text{m}^2} \times \frac{\text{kg} \cdot \text{K}}{287 \text{ N} \cdot \text{m}} \times \frac{1}{302 \text{ K}} = 5.21 \text{ kg/m}^3 \qquad \rho_2$$

From the continuity equation,

$$\dot{m} = \rho_1 V_1 A_1 = \rho_2 V_2 A_2 = \text{constant}$$

so that

$$A_2 = A_1 \frac{\rho_1}{\rho_2} \frac{V_1}{V_2} = A_1 \left(\frac{T_1}{T_2} \right)^{1/(k-1)} \frac{M_1 c_1}{M_2 c_2} = A_1 \left(\frac{T_1}{T_2} \right)^{1/(k-1)} \frac{M_1}{M_2} \sqrt{\frac{kRT_1}{kRT_2}}$$

or

$$A_2 = A_1 \frac{M_1}{M_2} \left(\frac{T_1}{T_2} \right)^{(k+1)/2(k-1)} = 0.001 \text{ m}^2 \times \frac{0.3}{0.8} \left(\frac{335}{302} \right)^3 = 5.12 \times 10^{-4} \text{ m}^2 \qquad A_2$$

Thus $A_2 < A_1$, as expected. Finally,

$$p_{0_2} = p_2 \left[1 + \frac{k-1}{2} M_2^2 \right]^{k/(k-1)} = 452 \text{ kPa}[1 + 0.2(0.8)^2]^{3.5}$$

$$p_{0_2} = 689 \text{ kPa (abs)} \qquad\qquad\qquad\qquad\qquad\qquad\qquad\qquad p_{0_2}$$

The stagnation pressure should be constant for isentropic flow. Checking gives

$$p_{0_1} = p_1 \left[1 + \frac{k-1}{2}M_1^2\right]^{k/(k-1)} = 650 \text{ kPa}[1 + 0.2(0.3)^2]^{3.5}$$

$$p_{0_1} = 692 \text{ kPa (abs)}$$

The discrepancy between the calculated p_{0_1} and p_{0_2} is due to use of temperatures rounded to three significant figures to calculate p_{0_2}. Thus, we note again for isentropic flow that $T_{0_1} = T_{0_2}$ and $p_{0_1} = p_{0_2}$.

**13-3.3 Tables for Computation of Isentropic Flow of an Ideal Gas

In the previous section, we saw (Eqs. 12.17a, 12.17b, 12.17c, and 13.6) that the properties at a point in a compressible flow of an ideal gas may be related to appropriate reference conditions by functions of the local Mach number. This makes it possible to tabulate or plot them as functions of Mach number for a given k.

Table E.1 of Appendix E lists values of T/T_0, p/p_0, ρ/ρ_0, and A/A^* as functions of M for isentropic flow of an ideal gas, with $k = 1.4$.[1] The use of tables can reduce the labor of calculations significantly.

Since the reference conditions remain constant in isentropic flow, the ratio of properties at two points in a flow may readily be found from the tables. Use of the tables is illustrated in Example Problem 13.2.

EXAMPLE 13.2—Isentropic Flow in a Converging Channel: Table Solution

Air flows isentropically in a channel. At section ①, the Mach number is 0.3, the area is 0.001 m², and the absolute pressure and the temperature are 650 kPa and 62 C, respectively. Evaluate properties at section ②, where the Mach number is 0.8, using the isentropic flow tables. (Note these are the data of Example 13.1.)

EXAMPLE PROBLEM 13.2

GIVEN: Isentropic flow of air in a channel. At sections ① and ②, the following data are given: $M_1 = 0.3$, $T_1 = 62$ C, $p_1 = 650$ kPa (abs), $A_1 = 0.001$ m², and $M_2 = 0.8$.

FIND: Properties at section ②, using isentropic flow tables.

SOLUTION:

The Ts diagram was sketched in Example Problem 13.1. From Table E.1, Appendix E, we find

M	T/T_0	p/p_0	ρ/ρ_0	A/A^*
0.3	0.9823	0.9395	0.9564	2.035
0.8	0.8865	0.6560	0.7400	1.038

** This section may be omitted without loss of continuity in the text material.

[1] It is easy to program Eqs. 12.17a, 12.17b, 12.17c, and 13.6 for a calculator or personal computer. Thus tables may be generated for any desired value of k. Olfe [2] has written a commercial software package that includes table solutions.

For isentropic flow, $T_{0_1} = T_{0_2} = T_0$. Thus

$$\frac{T_2}{T_1} = \frac{T_2}{T_0}\frac{T_0}{T_1} = \frac{(T/T_0)_2}{(T/T_0)_1} = \frac{0.8865}{0.9823} = 0.9025$$

$$T_2 = 0.9025T_1 = 0.9025(273+62)\text{ K} = 302\text{ K} \qquad\qquad T_2$$

Also $p_{0_2} = p_{0_1} = p_0$, so

$$\frac{p_2}{p_1} = \frac{p_2}{p_0}\frac{p_0}{p_1} = \frac{(p/p_0)_2}{(p/p_0)_1} = \frac{0.6560}{0.9395} = 0.6982$$

$$p_2 = 0.6982\,p_1 = 0.6982(650\text{ kPa}) = 454\text{ kPa (abs)} \qquad\qquad p_2$$

and

$$\rho_2 = \frac{p_2}{RT_2} = \frac{4.54\times10^5\text{ N}}{\text{m}^2} \times \frac{\text{kg}\cdot\text{K}}{287\text{ N}\cdot\text{m}} \times \frac{1}{302\text{ K}} = 5.24\text{ kg/m}^3 \qquad\qquad \rho_2$$

The stagnation properties are

$$T_{0_2} = T_{0_1} = \frac{T_1}{(T/T_0)_1} = \frac{(273+62)\text{ K}}{0.9823} = 341\text{ K} \qquad\qquad T_{0_2}$$

and

$$p_{0_2} = p_{0_1} = \frac{p_1}{(p/p_0)_1} = \frac{650\text{ kPa}}{0.9395} = 692\text{ kPa (abs)} \qquad\qquad p_{0_2}$$

The area may be computed using A/A^*. Thus, since $A^* = $ constant,

$$\frac{A_2}{A_1} = \frac{A_2}{A^*}\frac{A^*}{A_1} = \frac{(A/A^*)_2}{(A/A^*)_1} = \frac{1.038}{2.035} = 0.5101$$

$$A_2 = 0.5101A_1 = 0.5101(0.001\text{ m}^2) = 5.10\times10^{-4}\text{ m}^2 \qquad\qquad A_2$$

The speed at section ② may be calculated from $V_2 = M_2 c_2$.

13-3.4 Isentropic Flow in a Converging Nozzle

In this section we investigate the operation of a converging nozzle under various back pressures. Flow through the converging nozzle shown in Fig. 13.6 is supplied from a large plenum chamber, where stagnation conditions are assumed; flow is induced by a vacuum pump downstream and is controlled by the valve shown.

The back pressure, p_b, to which the nozzle discharges, is controlled by the valve. The upstream stagnation conditions (T_0, p_0, etc.) are maintained constant. The pressure in the exit plane of the nozzle is p_e. We wish to investigate the effect of variations in back pressure on the pressure distribution through the nozzle, on mass flow rate, and on exit plane pressure. Results are illustrated graphically in Fig. 13.6. Let us look at each of the cases shown.

When the valve is closed, there is no flow through the nozzle. The pressure is p_0 throughout, as shown by condition (i) in Fig. 13.6a.

If the back pressure, p_b, is now reduced to slightly less than p_0, there will be flow through the nozzle with a decrease in pressure in the direction of flow, as shown by condition (ii). Flow at the exit plane will be subsonic with the exit plane pressure equal to the back pressure.

What happens as we continue to decrease back pressure? The flow rate will continue to increase, and the exit plane pressure will continue to decrease, as shown

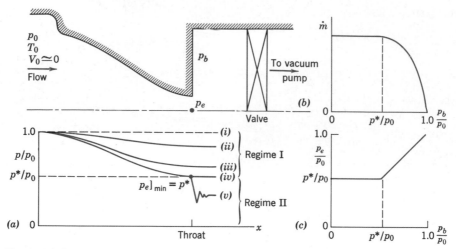

Fig. 13.6 Converging nozzle operating at various back pressures.

by condition (*iii*) in Fig. 13.6. Will these trends continue indefinitely as the back pressure is lowered?

Recall from our previous discussion that, in a converging channel, the Mach number cannot increase beyond unity in isentropic flow. Thus, with continued decrease of back pressure, flow at the exit plane of the nozzle eventually will reach a Mach number of unity. The corresponding pressure is the critical pressure, p^*. Condition (*iv*) illustrates the condition where M_e equals unity and p_b/p_0 equals p^*/p_0.

From Eq. 12.17a, with $M = 1$, the critical pressure ratio for an ideal gas is given by

$$\frac{p^*}{p_0} = \left(\frac{2}{k+1}\right)^{k/(k-1)}$$

For $k = 1.4$, $p^*/p_0 = 0.528$.

What happens when back pressure is reduced further, below p^*, such as condition (*v*)? Since the throat Mach number is unity ($V_e = c_e$), information about conditions in the exhaust duct cannot be transmitted upstream. Consequently, reductions in p_b below p^* have no effect on flow conditions in the nozzle; thus, neither the pressure distribution through the nozzle, nozzle exit pressure, nor mass flow rate are affected by lowering p_b below p^*. When p_b is less than or equal to p^*, the nozzle is said to be *choked*.

For p_b less than p^*, flow leaving the nozzle will expand to match the lower back pressure, as shown for condition (*v*) in Fig. 13.6a. This unconfined expansion process is three-dimensional; the pressure distribution cannot be predicted by one-dimensional theory. Experiments show that a series of shocks form in the exit stream, resulting in an increase in entropy.

Flow through a converging nozzle may be divided into two regimes:

1. In Regime I, $1 \geq p_b/p_0 \geq p^*/p_0$. Flow to the throat is isentropic; $p_e = p_b$.
2. In Regime II, $p_b/p_0 < p^*/p_0$. Flow to the throat is isentropic, but a nonisentropic expansion occurs in the flow leaving the nozzle; $p_e = p^* > p_b$.

The flow processes corresponding to Regime II are shown on a *Ts* diagram in Fig. 13.7. Two problems involving converging nozzles are solved in Example Problems 13.3 and 13.4.

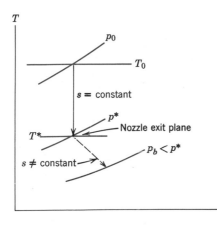

Fig. 13.7 Schematic Ts diagram for choked flow through a converging nozzle.

Although isentropic flow is an idealization, it often is a very good approximation for the actual behavior of nozzles. Since a nozzle is a device that accelerates a flow, the internal pressure gradient is favorable. This tends to keep the wall boundary layers thin and to minimize the effects of friction.

EXAMPLE 13.3—Isentropic Flow in a Converging Nozzle

A converging nozzle, with a throat area of 0.001 m², is operated with air at a back pressure of 591 kPa (abs). The nozzle is fed from a large plenum chamber where the absolute stagnation pressure and temperature are 1.0 MPa and 60 C. The exit Mach number and mass flow rate are to be determined using (*i*) isentropic flow relations and (*ii*) tables for isentropic flow.

EXAMPLE PROBLEM 13.3

GIVEN: Air flow through a converging nozzle at the conditions shown:
Flow is isentropic.

FIND:
(a) M_e.   (b) $\dot{m}$.

SOLUTION:
The first step is to check for choking. The pressure ratio is

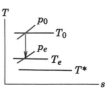

$$\frac{p_b}{p_0} = \frac{5.91 \times 10^5}{1.0 \times 10^6} = 0.591 > 0.528$$

so the flow is not choked. Thus $p_b = p_e$, and the flow is isentropic, as sketched on the Ts diagram.

(*i*) *Isentropic Flow Relations*

Since $p_0 = $ constant, M_e may be found from the pressure ratio,

$$\frac{p_0}{p_e} = \left[1 + \frac{k-1}{2}M_e^2\right]^{k/(k-1)}$$

Solving for M_e, since $p_e = p_b$, we obtain

$$1 + \frac{k-1}{2}M_e^2 = \left(\frac{p_0}{p_b}\right)^{(k-1)/k}$$

and

$$M_e = \left\{\left[\left(\frac{p_0}{p_b}\right)^{(k-1)/k} - 1\right]\frac{2}{k-1}\right\}^{1/2} = \left\{\left[\left(\frac{1.0\times10^6}{5.91\times10^5}\right)^{0.286} - 1\right]\frac{2}{1.4-1}\right\}^{1/2} = 0.9 \quad\longleftarrow\quad M_e$$

The mass flow rate will be given by

$$\dot{m} = \rho_e V_e A_e = \rho_e M_e c_e A_e$$

Thus we need T_e to find ρ_e and c_e. Since $T_0 = $ constant,

$$\frac{T_0}{T_e} = 1 + \frac{k-1}{2}M_e^2$$

or

$$T_e = \frac{T_0}{1 + \dfrac{k-1}{2}M_e^2} = \frac{(273+60)\text{ K}}{1+0.2(0.9)^2} = 287\text{ K}$$

$$c_e = \sqrt{kRT_e} = \left[1.4 \times 287\,\frac{\text{N}\cdot\text{m}}{\text{kg}\cdot\text{K}} \times 287\text{ K} \times \frac{\text{kg}\cdot\text{m}}{\text{N}\cdot\text{sec}^2}\right]^{1/2} = 340\text{ m/sec}$$

and

$$\rho_e = \frac{p_e}{RT_e} = 5.91\times10^5\,\frac{\text{N}}{\text{m}^2} \times \frac{\text{kg}\cdot\text{K}}{287\text{ N}\cdot\text{m}} \times \frac{1}{287\text{ K}} = 7.18\text{ kg/m}^3$$

Finally,

$$\dot{m} = \rho_e M_e c_e A_e = 7.18\,\frac{\text{kg}}{\text{m}^3} \times 0.9 \times 340\,\frac{\text{m}}{\text{sec}} \times 0.001\text{ m}^2 = 2.20\text{ kg/sec} \quad\longleftarrow\quad \dot{m}$$

(ii)   **Tables for Isentropic Flow*

The pressure ratio is

$$\frac{p_b}{p_0} = \frac{5.91\times10^5}{1.0\times10^6} = 0.591 > 0.528$$

so the flow is not choked. Thus $p_e = p_b$, and the flow is isentropic. From Table E.1, Appendix E, $p/p_0 = 0.591$ at $M = 0.90$. Thus

$$M_e = 0.90 \quad\longleftarrow\quad M_e$$

The simplest procedure in using the tables is to express the desired result in terms of property ratios, which then may be found from the tables. Thus

$$\dot{m} = \rho_e V_e A_e = \rho_e M_e c_e A_e = \rho_e M_e \sqrt{kRT_e}A_e$$

$$\dot{m} = \rho_0 \frac{\rho_e}{\rho_0}M_e\sqrt{kRT_0}\sqrt{\frac{T_e}{T_0}}A_e = \frac{p_0}{RT_0}\frac{\rho_e}{\rho_0}M_e\sqrt{kRT_0}\sqrt{\frac{T_e}{T_0}}A_e$$

or finally,

$$\dot{m} = \frac{\rho_e}{\rho_0}\sqrt{\frac{T_e}{T_0}}p_0 M_e\sqrt{\frac{k}{RT_0}}A_e$$

Using property ratios for $M_e = 0.9$ gives

$$\dot{m} = (0.6870)(0.8606)^{1/2} \left(1.0 \times 10^6 \, \frac{N}{m^2}\right)(0.9)$$

$$\times \left[1.4 \times \frac{kg \cdot K}{287 \, N \cdot m} \times \frac{1}{333 \, K} \times \frac{kg \cdot m}{N \cdot sec^2}\right]^{1/2} (0.001 \, m^2)$$

$$\dot{m} = 2.20 \, kg/sec \qquad\qquad\qquad\qquad\qquad\qquad \dot{m}$$

EXAMPLE 13.4—Choked Flow in a Converging Nozzle

Air flows isentropically through a converging nozzle. At a section where the nozzle area is 0.013 ft², the local pressure, temperature, and Mach number are 60 psia, 40 F, and 0.52, respectively. The back pressure is 30 psia. The Mach number at the throat, the mass flow rate, and the throat area are to be determined, using (*i*) isentropic flow relations and (*ii*) tables for isentropic flow.

EXAMPLE PROBLEM 13.4

GIVEN: Air flow through a converging nozzle at the conditions shown:

$M_1 = 0.52$

$T_1 = 40 \, F$

$p_1 = 60 \, psia$

$A_1 = 0.013 \, ft^2$

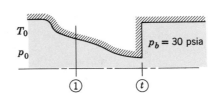

FIND: (a) M_t.   (b) $\dot{m}$.   (c) A_t.

SOLUTION:

(*i*) *Isentropic Flow Relations*

First we check for choking, to determine if flow is isentropic down to p_b. To check, we evaluate the stagnation conditions.

$$\frac{p_0}{p_1} = \left[1 + \frac{k-1}{2}M_1^2\right]^{k/(k-1)} = [1 + 0.2(0.52)^2]^{3.5} = 1.20$$

and

$$p_0 = 1.20 \, p_1 = (1.2)60 \, psia = 72.0 \, psia$$

The back pressure ratio is

$$\frac{p_b}{p_0} = \frac{30.0}{72.0} = 0.417 < 0.528$$

so the flow is choked! For choked flow,

$$M_t = 1.0 \qquad\qquad\qquad\qquad\qquad\qquad M_t$$

The Ts diagram is

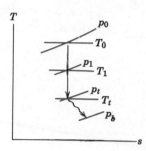

The mass flow rate may be found from conditions at section ①, using $\dot{m} = \rho_1 V_1 A_1$.

$$V_1 = M_1 c_1 = M_1 \sqrt{kRT_1}$$

$$= 0.52 \left[1.4 \times \frac{53.3 \text{ ft} \cdot \text{lbf}}{\text{lbm} \cdot \text{R}} \times (460+40) \text{ R} \times \frac{32.2 \text{ lbm}}{\text{slug}} \times \frac{\text{slug} \cdot \text{ft}}{\text{lbf} \cdot \text{sec}^2} \right]^{1/2}$$

$$V_1 = 570 \text{ ft/sec}$$

$$\rho_1 = \frac{p_1}{RT_1} = \frac{60 \text{ lbf}}{\text{in.}^2} \times \frac{\text{lbm} \cdot \text{R}}{53.3 \text{ ft} \cdot \text{lbf}} \times \frac{1}{500 \text{ R}} \times \frac{144 \text{ in.}^2}{\text{ft}^2} = 0.324 \text{ lbm/ft}^3$$

$$\dot{m} = \rho_1 V_1 A_1 = \frac{0.324 \text{ lbm}}{\text{ft}^3} \times \frac{570 \text{ ft}}{\text{sec}} \times 0.013 \text{ ft}^2 = 2.40 \text{ lbm/sec} \qquad \overset{\longleftarrow}{} \quad \dot{m}$$

The throat area may be computed by applying continuity between section ① and the throat, i.e., $\dot{m} = \rho_1 V_1 A_1 = \rho_t V_t A_t$, so

$$A_t = A_1 \frac{\rho_1}{\rho_t} \frac{V_1}{V_t} = A_1 \frac{p_1}{RT_1} \frac{RT_t}{p_t} \frac{M_1 \sqrt{kRT_1}}{M_t \sqrt{kRT_t}} = A_1 \frac{p_1}{p_t} \frac{M_1}{M_t} \sqrt{\frac{T_t}{T_1}}$$

For isentropic flow, $T_0 = \text{constant}$, so

$$\frac{T_t}{T_1} = \frac{T_t}{T_0} \frac{T_0}{T_1} = \frac{1 + \dfrac{k-1}{2} M_1^2}{1 + \dfrac{k-1}{2} M_t^2} = \frac{1 + (0.2)(0.52)^2}{1.2} = 0.878$$

Also,

$$\frac{p_0}{p_t} = \left[1 + \frac{k-1}{2} M_t^2 \right]^{k/(k-1)} = (1.2)^{3.5} = 1.89$$

so that

$$p_t = \frac{p_0}{1.89} = \frac{72.0 \text{ psia}}{1.89} = 38.1 \text{ psia}$$

Substituting gives

$$A_t = A_1 \frac{p_1}{p_t} \frac{M_1}{M_t} \sqrt{\frac{T_t}{T_1}} = 0.013 \text{ ft}^2 \times \frac{60.0 \text{ psia}}{38.1 \text{ psia}} \times \frac{0.52}{1.0} \sqrt{0.878} = 9.98 \times 10^{-3} \text{ ft}^2 \qquad \overset{\longleftarrow}{} \quad A_t$$

(ii)   **Tables for Isentropic Flow*

First we check for choking. From Table E.1, Appendix E, at $M_1 = 0.52$

$$\frac{p_1}{p_0} = 0.8317; \qquad p_0 = \frac{p_1}{0.8317} = \frac{60 \text{ psia}}{0.8317} = 72.1 \text{ psia}$$

From the table at $M = 1.0$, the minimum isentropic pressure ratio in a converging nozzle is

$$\frac{p}{p_0} = 0.5283$$

From the conditions given,

$$\frac{p_b}{p_0} = \frac{30}{72.1} = 0.416 < 0.5283$$

so the flow is choked! For choked flow,

$$M_t = 1.0 \qquad\qquad M_t$$

The Ts diagram was given above. The mass flow rate calculation is the same as in the previous solution. From Table E.1, at $M_1 = 0.52$, $A_1/A^* = 1.303$. For choked flow, $A_t = A^*$. Thus

$$A_t = A^* = \frac{A_1}{1.303} = \frac{0.013 \text{ ft}^2}{1.303} = 9.98 \times 10^{-3} \text{ ft}^2 \qquad\qquad A_t$$

13-3.5 Isentropic Flow in a Converging-Diverging Nozzle

Having considered isentropic flow in a converging nozzle, we turn now to isentropic flow in a converging-diverging (C-D) nozzle. As in the previous case, flow through the converging-diverging passage of Fig. 13.8 is induced by a vacuum pump downstream, and is controlled by the valve shown. Upstream stagnation conditions are assumed constant. Pressure in the exit plane of the nozzle is p_e; the nozzle discharges to back pressure p_b. We wish to investigate the effect of variations in back pressure on the pressure distribution through the nozzle. The results are illustrated graphically in Fig. 13.8. Let us consider each of the cases shown.

With the valve initially closed, there is no flow through the nozzle; the pressure is constant at p_0. Opening the valve slightly (p_b slightly less than p_0) produces the pressure distribution curve (*i*). If the flow rate is low enough, at all points on this curve the flow will be subsonic and essentially incompressible. Under these conditions, the C-D nozzle will behave as a venturi, with flow accelerating in the converging portion, until a point of maximum velocity and minimum pressure is reached at the throat, then decelerating in the diverging portion to the nozzle exit.

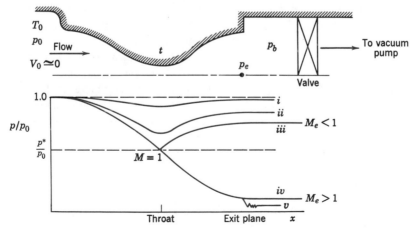

Pressure distributions for isentropic flow in a converging-diverging nozzle.

As the valve is opened farther and the flow rate is increased, a more sharply defined pressure minimum occurs, as shown by curve (*ii*). Although compressibility effects become important, the flow is still subsonic everywhere, and deceleration takes place in the diverging section. Finally, as the valve is opened farther, curve (*iii*) results. At the section of minimum area the flow finally reaches $M = 1$, and the nozzle is choked—the flow rate is the maximum possible for the given nozzle and stagnation conditions.

All flows with pressure distributions (*i*), (*ii*), and (*iii*) are isentropic; each curve is associated with a unique rate of mass flow. Finally, when curve (*iii*) is reached, critical conditions are present at the throat. For this flow rate, the flow is choked, and

$$\dot{m} = \rho^* V^* A^*$$

where $A^* = A_t$.

In our discussion of the effect of area variation on isentropic flow, we noted that a diverging section was required to accelerate a flow to supersonic speed from $M = 1$ at a throat. At this point, then, we ask the question, "What back pressure, p_b, is necessary to accelerate the flow isentropically in the diverging portion of the nozzle?"

To accelerate flow in the diverging section requires a pressure decrease. This condition is illustrated by curve (*iv*) in Fig. 13.8. The flow will accelerate isentropically in the nozzle provided the exit pressure is set at p_{iv}. Thus, we see that with a throat Mach number of unity, there are two possible isentropic flow conditions in the converging-diverging nozzle. This is consistent with the results of Fig. 13.5, where we found two Mach numbers for each A/A^* in isentropic flow.

Lowering the back pressure below condition (*iv*), say to condition (*v*), has no effect on flow in the nozzle. The flow is isentropic from the plenum chamber to the nozzle exit [same as condition (*iv*)] and then undergoes a three-dimensional irreversible expansion to the lower back pressure. A nozzle operating under these conditions is said to be *underexpanded*, since additional expansion takes place outside the nozzle.

A converging-diverging nozzle generally is intended to produce supersonic flow at the exit plane. If the back pressure is set at p_{iv}, flow will be isentropic through the nozzle, and supersonic at the nozzle exit. Nozzles operating at $p_b = p_{iv}$ [corresponding to curve (*iv*) in Fig. 13.8] are said to be at *design conditions*.

Flow leaving a C-D nozzle is supersonic when the back pressure is at or below nozzle design pressure. The exit Mach number is fixed once the area ratio, A_e/A^*, is specified. All other exit plane properties (for isentropic flow) are uniquely related to stagnation properties by the fixed exit plane Mach number.

The assumption of isentropic flow for a real nozzle at design conditions is a reasonable one. However, the one-dimensional flow model is inadequate for the design of relatively short nozzles to produce uniform supersonic exit flow.

Rocket-propelled vehicles use C-D nozzles to accelerate the exhaust gases to the maximum possible speed to produce high thrust. A propulsion nozzle is subject to varying ambient conditions during flight through the atmosphere, so it is impossible to attain the maximum theoretical thrust over the complete operating range. Because only a single supersonic Mach number can be obtained for each area ratio, nozzles for supersonic wind tunnels often are built with interchangeable sections, or with variable geometry.

You undoubtedly have noticed that nothing has been said about the operation of converging-diverging nozzles with back pressure in the range $p_{iii} > p_b > p_{iv}$.

For such cases the flow cannot expand isentropically to p_b. Under these conditions a shock (which may be treated as an irreversible discontinuity involving entropy increase) occurs somewhere within the flow. Following a discussion of normal shocks in Section 13-6, we shall return to complete the discussion of converging-diverging nozzle flows.

Nozzles operating with $p_{iii} > p_b > p_{iv}$ are said to be *overexpanded* because the pressure at some point in the nozzle is less than the back pressure. Obviously, an overexpanded nozzle could be made to operate at a new design condition by cutting off a portion of the diverging section.

One other comment should be made at this point. Real compressible fluid flows are affected by friction, heating or cooling, and the possible presence (in supersonic flow) of shock waves. We have treated isentropic flow first because it is a useful idealized model for many real flow processes and because it gives us valuable insight into the behavior of fluids in compressible flow. In the next two sections we consider the effects of friction and heat transfer separately to gain insight into the effect of each factor on flow behavior. Following this, we return to a discussion of the normal shock and shocks in channels, to complete our study of converging-diverging nozzle flows. In later courses it will be possible to explore real flows and the results of combining several of these effects.

EXAMPLE 13.5—Isentropic Flow in a Converging-Diverging Nozzle

Air flows isentropically in a converging-diverging nozzle, with exit area of 0.001 m². The nozzle is fed from a large plenum where the stagnation conditions are 350 K and 1.0 MPa (abs). The exit pressure is 954 kPa (abs) and the Mach number at the throat is 0.68. Flow conditions at the throat and the exit Mach number are to be determined.

EXAMPLE PROBLEM 13.5

GIVEN: Isentropic flow of air in C-D nozzle as shown:

$T_0 = 350$ K

$p_0 = 1.0$ MPa (abs)

$p_b = 954$ kPa (abs)

$M_t = 0.68$ $A_e = 0.001$ m²

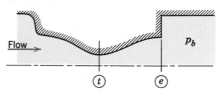

FIND: (a) Properties at nozzle throat.   (b) M_e.

SOLUTION:

(i) *Isentropic Flow Relations*

Stagnation temperature is constant for isentropic flow. Thus, since

$$\frac{T_0}{T} = 1 + \frac{k-1}{2}M^2$$

then

$$T_t = \frac{T_0}{1 + \dfrac{k-1}{2}M_t^2} = \frac{350 \text{ K}}{1 + 0.2(0.68)^2} = 320 \text{ K} \qquad\qquad \overset{T_t}{\longleftarrow}$$

Also, since p_0 is constant for isentropic flow, then

$$p_t = p_0 \left(\frac{T_t}{T_0}\right)^{k/(k-1)} = p_0 \left[\frac{1}{1+\dfrac{k-1}{2}M_t^2}\right]^{k/(k-1)}$$

$$p_t = 1.0 \times 10^6 \text{ Pa} \left[\frac{1}{1+0.2(0.68)^2}\right]^{3.5} = 734 \text{ kPa (abs)} \hspace{2cm} \longleftarrow \quad p_t$$

so

$$\rho_t = \frac{p_t}{RT_t} = \frac{7.34 \times 10^5 \text{ N}}{\text{m}^2} \times \frac{\text{kg} \cdot \text{K}}{287 \text{ N} \cdot \text{m}} \times \frac{1}{320 \text{ K}} = 7.99 \text{ kg/m}^3 \hspace{1cm} \longleftarrow \quad \rho_t$$

and

$$V_t = M_t c_t = M_t \sqrt{kRT_t}$$

$$V_t = 0.68 \left[1.4 \times \frac{287 \text{ N} \cdot \text{m}}{\text{kg} \cdot \text{K}} \times 320 \text{ K} \times \frac{\text{kg} \cdot \text{m}}{\text{N} \cdot \text{sec}^2}\right]^{1/2} = 244 \text{ m/sec} \hspace{1cm} \longleftarrow \quad V_t$$

Since $M_t < 1$, flow at the exit must be subsonic. Therefore, $p_e = p_b$. Stagnation properties are constant, so

$$\frac{p_0}{p_e} = \left[1 + \frac{k-1}{2}M_e^2\right]^{k/(k-1)}$$

Solving for M_e gives

$$M_e = \left\{\left[\left(\frac{p_0}{p_e}\right)^{(k-1)/k} - 1\right]\frac{2}{k-1}\right\}^{1/2}$$

$$M_e = \left\{\left[\left(\frac{1.0 \times 10^6}{9.54 \times 10^5}\right)^{0.286} - 1\right](5)\right\}^{1/2} = 0.26 \hspace{1cm} \longleftarrow \quad M_e$$

The Ts diagram for this flow is

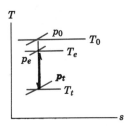

(ii) **Tables for Isentropic Flow*

Stagnation properties are constant for isentropic flow. From Table E.1, Appendix E, at $M = 0.68$,

$$\frac{T}{T_0} = 0.9154; \qquad T_t = 0.9154T_0 = (0.9154)(350 \text{ K}) = 320 \text{ K} \hspace{1cm} \longleftarrow \quad T_t$$

$$\frac{p}{p_0} = 0.7338; \qquad p_t = 0.7338 p_0 = (0.7338)(1.0 \times 10^6 \text{ Pa}) = 734 \text{ kPa (abs)} \hspace{0.3cm} \longleftarrow \quad p_t$$

$$\frac{\rho}{\rho_0} = 0.8016; \qquad \rho_t = 0.8016\rho_0 = 0.8016\frac{p_0}{RT_0}$$

$$\rho_t = 0.8016 \times \frac{1.0 \times 10^6 \text{ N}}{\text{m}^2} \times \frac{\text{kg} \cdot \text{K}}{287 \text{ N} \cdot \text{m}} \times \frac{1}{350 \text{ K}} = 7.98 \text{ kg/m}^3 \qquad\qquad \underleftarrow{\rho_t}$$

and

$$\frac{A}{A^*} = 1.110; \qquad A_t = 1.110A^*$$

but at this point A^* is not known.
  At the exit, $p_e = 954$ kPa (abs). Thus $p_e/p_0 = 0.954$, and from Table E.1,

$$M_e = 0.26 \qquad\qquad\qquad M_e$$

  Since A_e is known, we can compute A^*. From Table E.1, at $M = 0.26$, $A/A^* = 2.317$.
Thus

$$A^* = \frac{A_e}{2.317} = \frac{0.001 \text{ m}^2}{2.317} = 4.32 \times 10^{-4} \text{ m}^2$$

and

$$A_t = 1.110A^* = (1.110)(4.32 \times 10^{-4}\text{m}^2) = 4.80 \times 10^{-4} \text{ m}^2 \qquad\qquad \underleftarrow{A_t}$$

Note that the solution for A_t using the tables was relatively effortless. To find A_t using
the isentropic relations, we could have applied continuity between the throat and exit
planes. Since this would have required calculation of all properties at the exit, it would
have been a lengthy process.

EXAMPLE 13.6—Isentropic Flow in a Converging-Diverging Nozzle: Choked Flow

The nozzle of Example 13.5 has a design back pressure of 87.5 kPa (abs) but is
operated at a back pressure of 50.0 kPa (abs). Assume flow within the nozzle is
isentropic. Determine the exit Mach number and mass flow rate. Use (*i*) isentropic
flow relations and (*ii*) tables for isentropic flow.

EXAMPLE PROBLEM 13.6

GIVEN: Air flow through C-D nozzle as shown:

  $T_0 = 350$ K

  $p_0 = 1.0$ MPa (abs)

p_e(design) $= 87.5$ kPa (abs)

  $p_b = 50.0$ kPa (abs)

  $A_e = 0.001$ m^2

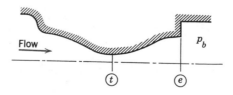

FIND: (a) M_e.   (b) $\dot{m}$.

SOLUTION:

The operating back pressure is *below* the design pressure. Consequently, the nozzle is underexpanded, and the Ts diagram and pressure distribution will be as shown:

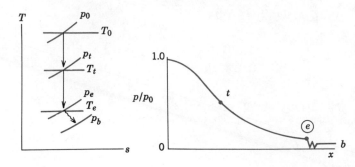

Flow *within* the nozzle will be isentropic, but the irreversible expansion from p_e to p_b will cause an entropy increase; $p_e = p_e(\text{design}) = 87.5$ kPa (abs).

(i) Isentropic Flow Relations

Since stagnation properties are constant for isentropic flow, the exit Mach number can be computed from the pressure ratio. Thus

$$\frac{p_0}{p_e} = \left[1 + \frac{k-1}{2}M_e^2\right]^{k/(k-1)}$$

or

$$M_e = \left\{\left[\left(\frac{p_0}{p_e}\right)^{(k-1)/k} - 1\right]\frac{2}{k-1}\right\}^{1/2} = \left\{\left[\left(\frac{1.0 \times 10^6}{8.75 \times 10^4}\right)^{0.286} - 1\right]\frac{2}{0.4}\right\}^{1/2} = 2.24 \qquad\qquad M_e$$

The mass flow rate is given by

$$\dot{m} = \rho_e V_e A_e = \rho_e M_e c_e A_e = \rho_e M_e \sqrt{kRT_e}A_e = \frac{p_e}{RT_e}M_e\sqrt{kRT_e}A_e$$

or

$$\dot{m} = p_e M_e \sqrt{\frac{k}{RT_e}}A_e$$

Since T_0 is constant,

$$\frac{T_0}{T_e} = 1 + \frac{k-1}{2}M_e^2; \qquad T_e = \frac{T_0}{1 + \frac{k-1}{2}M_e^2} = \frac{350 \text{ K}}{1 + 0.2(2.24)^2} = 175 \text{ K}$$

Then

$$\dot{m} = p_e M_e \sqrt{\frac{k}{RT_e}}A_e$$

$$= \frac{8.75 \times 10^4 \text{ N}}{\text{m}^2} \times 2.24\left[1.4 \times \frac{\text{kg}\cdot\text{K}}{287 \text{ N}\cdot\text{m}} \times \frac{1}{175 \text{ K}} \times \frac{\text{kg}\cdot\text{m}}{\text{N}\cdot\text{sec}^2}\right]^{1/2} 0.001 \text{ m}^2$$

$$\dot{m} = 1.03 \text{ kg/sec} \qquad\qquad\qquad\qquad\qquad\qquad \dot{m}$$

(ii)   ***Tables for Isentropic Flow*

From Table E.1, Appendix E, at $p/p_0 = p_e/p_0 = 0.0875$,

$$M_e \simeq 2.24$$

Also,

$$\frac{T_e}{T_0} = 0.4991; \qquad T_e = 0.4991 T_0 = 0.4991(350 \text{ K}) = 175 \text{ K}$$

$\Big\{$This temperature is the same as obtained above. The mass flow rate calculation also would be as above.$\Big\}$

13-4   FLOW IN A CONSTANT-AREA DUCT WITH FRICTION

Gas flow in constant-area ducts is commonly encountered in a variety of engineering applications. In this section we consider flows in which wall friction is responsible for changes in fluid properties.

Sections 13-4.1 through 13-4.3 consider adiabatic flow in a constant-area duct with friction. The assumption of adiabatic flow is appropriate for flows in which the duct length is reasonably short. When ducts are long, as in uninsulated natural gas pipe lines, significant surface area is available for heat transfer and flow is approximately isothermal. Isothermal flow is discussed in Section 13-4.4.

To analyze compressible flow in constant-area ducts with friction, the wall friction force may be related to the flow properties through the friction factor, using the methods developed for incompressible flow in Chapter 8.

13-4.1   Basic Equations for Adiabatic Flow

To obtain an overall view of the problem of frictional adiabatic flow, apply the basic equations to steady uniform flow of an ideal gas, with constant specific heats, through the finite control volume shown in Fig. 13.9.

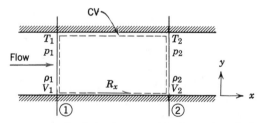

Fig. 13.9   Control volume used for integral analysis of frictional adiabatic flow.

a.   Continuity Equation

Basic equation:

$$0 = \frac{\cancel{\partial}}{\cancel{\partial t}} \overset{= 0(1)}{\int_{CV}} \rho \, d\forall + \int_{CS} \rho \vec{V} \cdot d\vec{A} \tag{4.13}$$

Assumptions: (1)   Steady flow
(2)   Uniform flow at each section

Then

$$0 = \{-|\rho_1 V_1 A_1|\} + \{|\rho_2 V_2 A_2|\}$$

The area is constant, so

$$\rho_1 V_1 = \rho_2 V_2 = G = \frac{\dot{m}}{A} \qquad (13.7a)$$

b. Momentum Equation

Basic equation:

$$= 0(3) \quad = 0(1)$$

$$F_{S_x} + \cancel{F_{B_x}} = \cancel{\frac{\partial}{\partial t} \int_{CV} V_x \rho \, d\forall} + \int_{CS} V_x \rho \vec{V} \cdot d\vec{A} \qquad (4.19a)$$

Assumption: (3) $F_{B_x} = 0$

The surface force is due to pressure forces at sections ① and ②, and to the friction force, R_x, of the duct wall *on* the flow. Substituting, and recognizing that $A_2 = A_1 = A$,

$$R_x + p_1 A - p_2 A = V_1\{-|\rho_1 V_1 A|\} + V_2\{|\rho_2 V_2 A|\}$$

Using scalar magnitudes and dropping absolute value signs gives

$$R_x + p_1 A - p_2 A = \dot{m} V_2 - \dot{m} V_1 \qquad (13.7b)$$

c. First Law of Thermodynamics

Basic equation:

$$= 0(4) = 0(5) = 0(6) \quad = 0(6) \quad = 0(1)$$

$$\cancel{\dot{Q}} - \cancel{\dot{W}_s} - \cancel{\dot{W}_{shear}} - \cancel{\dot{W}_{other}} = \cancel{\frac{\partial}{\partial t} \int_{CV} e\rho \, d\forall} + \int_{CS} (e + pv)\rho \vec{V} \cdot d\vec{A} \qquad (4.57)$$

where

$$\approx 0(7)$$

$$e = u + \frac{V^2}{2} + \cancel{gz}$$

Assumptions: (4) $\dot{Q} = 0$ (adiabatic flow)
(5) $\dot{W}_s = 0$
(6) $\dot{W}_{shear} = \dot{W}_{other} = 0$
(7) Effects of gravity are negligible

With these restrictions, the equation becomes

$$0 = \left(u_1 + \frac{V_1^2}{2} + p_1 v_1\right)\{-|\rho_1 V_1 A|\} + \left(u_2 + \frac{V_2^2}{2} + p_2 v_2\right)\{|\rho_2 V_2 A|\}$$

Since the mass flow rate terms in { } are identical by continuity, then

$$u_1 + \frac{V_1^2}{2} + p_1 v_1 = u_2 + \frac{V_2^2}{2} + p_2 v_2$$

or

$$h_1 + \frac{V_1^2}{2} = h_2 + \frac{V_2^2}{2} \qquad (13.7c)$$

We also could write

$$h_{0_1} = h_{0_2}$$

which is a physical consequence of our assumption of adiabatic flow.

d. Second Law of Thermodynamics

Basic equation:

$$\int_{CS} \frac{1}{T} \overset{= 0(4)}{\frac{\dot{Q}}{A}} dA \le \overset{= 0(1)}{\frac{\partial}{\partial t} \int_{CV}} s\rho \, d\forall + \int_{CS} s\rho \, \vec{V} \cdot d\vec{A} \qquad (4.59)$$

Then because the flow is frictional, and hence irreversible,

$$0 < s_1\{-|\rho_1 V_1 A|\} + s_2\{|\rho_2 V_2 A|\} = \dot{m}(s_2 - s_1)$$

The control volume form of the second law tells us that $s_2 - s_1 > 0$.

This fact is of little help in calculating the actual entropy change between any two sections in an adiabatic frictional flow. To calculate the entropy change, we rely on the $T \, ds$ equations. Since

$$T \, ds = dh - v \, dp$$

for an ideal gas, we can write

$$ds = c_p \frac{dT}{T} - R \frac{dp}{p}$$

For constant specific heats, the equation can be integrated to give

$$s_2 - s_1 = c_p \ln \frac{T_2}{T_1} - R \ln \frac{p_2}{p_1} \qquad (13.7d)$$

e. Equations of State

For an ideal gas, the equation of state is given by

$$p = \rho R T \qquad (13.7e)$$

Equations 13.7a through 13.7e are the governing equations for steady, one-dimensional, adiabatic, frictional flow of an ideal gas in a constant-area duct. If all properties at state ① are known, then we have seven unknowns $(T_2, p_2, \rho_2, V_2, h_2, s_2,$ and $R_x)$ in these five equations. However, we also have the known relationship between h and T for an ideal gas, $dh = c_p \, dT$. For an ideal gas with constant specific heats,

$$\Delta h = h_2 - h_1 = c_p \, \Delta T = c_p(T_2 - T_1) \qquad (13.7f)$$

We thus have the situation of six equations and seven unknowns.

If all conditions at state ① are known, how many possible states ② are there? The mathematics of the situation (six equations and seven unknowns) indicate an infinite number of possible states ②.

With an infinite number of possible states ② for a given state ①, what is to be expected if all possible states ② are plotted on a Ts diagram? It follows that the locus of all possible states ②, reachable from state ①, is a continuous curve passing through state ①.

How might we determine this curve? Perhaps the simplest way is to assume different values of T_2. For each assumed T_2, we could calculate all other properties at state ② and also R_x.

13-4.2 Adiabatic Flow: The Fanno Line

The results of these calculations are shown qualitatively on the Ts plane of Fig. 13.10. The locus of all possible downstream states is referred to as the *Fanno line*.[2] Detailed calculations show some interesting features of Fanno line flow. At the point of maximum entropy, the Mach number is unity. On the upper branch of the curve, the Mach number is always less than unity, and it increases monotonically as we proceed to the right along the curve. At every point on the lower portion of the curve, the Mach number is greater than unity; the Mach number decreases monotonically as we move to the right along the curve.

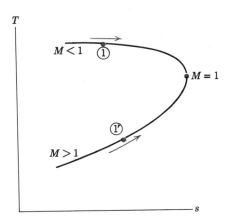

Fig. 13.10 Schematic Ts diagram for frictional adiabatic (Fanno line) flow in a constant-area duct.

For any initial state on a Fanno line, each point on the Fanno line represents a mathematically possible downstream state. Indeed, we determined the locus of all possible downstream states by assuming T_2 and calculating the corresponding properties. Although the Fanno line represents all mathematically possible downstream states, are they all physically attainable downstream states? A moment's reflection will indicate they are not. Why not? The second law requires that entropy must increase. Consequently, flow states must always move to the right along the Fanno line of Fig. 13.10. Indeed, it is the effect of friction that causes flow properties to change from the initial state. Referring again to Fig. 13.10, we see that for an initially subsonic flow (state ①), the effect of friction is to increase the Mach number toward unity. For a flow that is initially supersonic (state ①'), the effect of friction is to decrease the Mach number toward unity.

In developing the simplified form of the first law for Fanno line flow, we found that stagnation enthalpy remains constant. Consequently, when the fluid is an ideal gas with constant specific heats, stagnation temperature also must remain constant. What happens to stagnation pressure? Friction causes the local isentropic stagnation pressure to decrease for all Fanno line flows, as shown in Fig. 13.11. Recall that the second law requires $s_2 - s_1 > 0$. Since entropy must increase in the direction of

---

[2] A convenient way to remember this name is to think of Frictional Adiabatic flow.

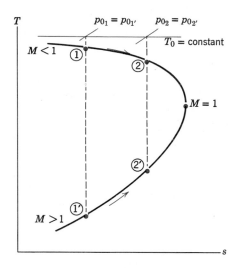

Fig. 13.11 Schematic of Fanno line flow on Ts plane, showing reduction in local isentropic stagnation pressure caused by friction.

flow, the flow process must proceed to the right on the Ts diagram. In Fig. 13.11, a path from state ① to state ② is shown on the subsonic portion of the curve. The corresponding local isentropic stagnation pressures, p_{0_1} and p_{0_2}, clearly show that $p_{0_2} < p_{0_1}$. An identical result is obtained for flow on the supersonic branch of the curve from state ①' to state ②'. Again $p_{0_{2'}} < p_{0_{1'}}$. Thus p_0 decreases for any Fanno line flow.

At this point we summarize the effects of friction on flow properties in Fanno line flow. The summary is presented in Table 13.1.

Table 13.1 Summary of Effects of Friction on Properties in Fanno Line Flow

Property	Subsonic $M<1$	Supersonic $M>1$	Obtained from:
Stagnation temperature, T_0	Constant	Constant	Energy equation
Entropy, s	Increases	Increases	Second law
Stagnation pressure, p_0	Decreases	Decreases	$T_0 = $ constant; s increases
Temperature, T	Decreases	Increases	Shape of Fanno line
Velocity, V	Increases	Decreases	Energy equation, and trend of T
Mach number, M	Increases	Decreases	Trends of V, T, and definition of M
Density, ρ	Decreases	Increases	Continuity equation, and effect on V
Pressure, p	Decreases	Increases	Equation of state, and effects on ρ, T

In deducing the effect of friction on flow properties for Fanno line flow, we used the shape of the Fanno line on the Ts diagram and the basic governing equations (Eqs. 13.7a through 13.7f). You should follow through the logic indicated in the right-hand column of the table.

We have noted that entropy must increase in the direction of flow: it is the effect of friction that causes the change in flow properties along the Fanno line curve. From Fig. 13.11, we see that there is a maximum entropy point corresponding to $M = 1$ for each Fanno line. The maximum entropy point is reached by increasing the amount of friction (through addition of duct length), just enough to produce a Mach number of unity (choked flow) at the exit. How do we compute this critical length of duct?

To compute the critical duct length, we must analyze the flow in detail, accounting for friction. This analysis requires that we begin with a differential control volume, develop expressions in terms of Mach number, and integrate along the duct to the section where $M = 1$. This analysis is completed in Section 13-4.3, where tables for Fanno line flow are developed. The algebra required for the detailed analysis tends to obscure the physics of the flow; the general trends in properties caused by friction can be demonstrated using finite control volumes and the basic governing equations. This approach is illustrated in Example Problem 13.7.

EXAMPLE 13.7—Frictional Adiabatic Flow in a Constant-Area Channel

Air flow is induced in an insulated tube of 7.16 mm diameter by a vacuum pump. The air is drawn from a room, where $p_0 = 101$ kPa (abs) and $T_0 = 23$ C, through a smoothly contoured converging nozzle. At section ①, where the nozzle joins the constant-area tube, the static pressure is 98.5 kPa (abs). At section ②, located some distance downstream in the constant-area tube, the air temperature is 14 C. Determine the mass flow rate, the local isentropic stagnation pressure at section ②, and the friction force on the duct wall between sections ① and ②.

EXAMPLE PROBLEM 13.7

GIVEN: Air flow in insulated tube.

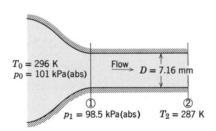

$T_0 = 296$ K
$p_0 = 101$ kPa(abs)

Flow

$D = 7.16$ mm

① $p_1 = 98.5$ kPa(abs)   ② $T_2 = 287$ K

FIND: (a) $\dot{m}$.
(b) Stagnation pressure at section ②.
(c) Force on duct wall.

SOLUTION:
The mass flow rate can be obtained from properties at section ①. For isentropic flow through the converging nozzle, local isentropic stagnation properties remain constant. Thus

$$\frac{p_{0_1}}{p_1} = \left(1 + \frac{k-1}{2}M_1^2\right)^{k/(k-1)}$$

and

$$M_1 = \left\{ \frac{2}{k-1} \left[\left(\frac{p_{0_1}}{p_1} \right)^{(k-1)/k} - 1 \right] \right\}^{1/2} = \left\{ \frac{2}{0.4} \left[\left(\frac{1.01 \times 10^5}{9.85 \times 10^4} \right)^{0.286} - 1 \right] \right\}^{1/2} = 0.190$$

$$T_1 = \frac{T_{0_1}}{1 + \frac{k-1}{2} M_1^2} = \frac{296 \text{ K}}{1 + 0.2(0.190)^2} = 294 \text{ K}$$

For an ideal gas,

$$\rho_1 = \frac{p_1}{RT_1} = \frac{9.85 \times 10^4 \text{ N}}{\text{m}^2} \times \frac{\text{kg} \cdot \text{K}}{287 \text{ N} \cdot \text{m}} \times \frac{1}{294 \text{ K}} = 1.17 \text{ kg/m}^3$$

$$V_1 = M_1 c_1 = M_1 \sqrt{kRT_1} = (0.19) \left[1.4 \times \frac{287 \text{ N} \cdot \text{m}}{\text{kg} \cdot \text{K}} \times 294 \text{ K} \times \frac{\text{kg} \cdot \text{m}}{\text{N} \cdot \text{sec}^2} \right]^{1/2}$$

$$V_1 = 65.3 \text{ m/sec}$$

The area, A_1, is

$$A_1 = A = \frac{\pi D^2}{4} = \frac{\pi}{4} (7.16 \times 10^{-3})^2 \text{ m}^2 = 4.03 \times 10^{-5} \text{ m}^2$$

From continuity,

$$\dot{m} = \rho_1 V_1 A_1 = \frac{1.17 \text{ kg}}{\text{m}^3} \times \frac{65.3 \text{ m}}{\text{sec}} \times 4.03 \times 10^{-5} \text{ m}^2$$

$$\dot{m} = 3.08 \times 10^{-3} \text{ kg/sec} \qquad\qquad\qquad\qquad \overleftarrow{} \quad \dot{m}$$

Flow is adiabatic, so T_0 is constant, and

$$T_{0_2} = T_{0_1} = 296 \text{ K} \qquad\qquad\qquad\qquad \overleftarrow{} \quad T_{0_2}$$

Then

$$\frac{T_{0_2}}{T_2} = 1 + \frac{k-1}{2} M_2^2$$

Solving for M_2 gives

$$M_2 = \left[\frac{2}{k-1} \left(\frac{T_{0_2}}{T_2} - 1 \right) \right]^{1/2} = \left[\frac{2}{0.4} \left(\frac{296}{287} - 1 \right) \right]^{1/2} = 0.396 \quad \overleftarrow{} \quad M_2$$

$$V_2 = M_2 c_2 = M_2 \sqrt{kRT_2} = (0.396) \left[1.4 \times \frac{287 \text{ N} \cdot \text{m}}{\text{kg} \cdot \text{K}} \times 287 \text{ K} \times \frac{\text{kg} \cdot \text{m}}{\text{N} \cdot \text{sec}^2} \right]^{1/2}$$

$$V_2 = 134 \text{ m/sec} \qquad\qquad\qquad\qquad \overleftarrow{} \quad V_2$$

From continuity, $\rho_1 V_1 = \rho_2 V_2$, so

$$\rho_2 = \rho_1 \frac{V_1}{V_2} = \frac{1.17 \text{ kg}}{\text{m}^3} \times \frac{65.3}{134} = 0.570 \text{ kg/m}^3 \qquad \overleftarrow{} \quad \rho_2$$

and

$$p_2 = \rho_2 R T_2 = \frac{0.570 \text{ kg}}{\text{m}^3} \times \frac{287 \text{ N} \cdot \text{m}}{\text{kg} \cdot \text{K}} \times 287 \text{ K} = 47.0 \text{ kPa (abs)} \quad \overleftarrow{} \quad p_2$$

The local isentropic stagnation pressure is

$$p_{0_2} = p_2 \left(1 + \frac{k-1}{2} M_2^2 \right)^{k/(k-1)} = 4.70 \times 10^4 \text{ Pa}[1 + 0.2(0.396)^2]^{3.5}$$

$p_{0_2} = 52.4$ kPa (abs) p_{0_2}

The friction force may be found by applying the momentum equation to the control volume shown:

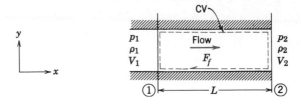

Basic equation:

$$F_{S_x} + \cancelto{0(1)}{F_{B_x}} = \cancelto{0(2)}{\frac{\partial}{\partial t}} \int_{CV} V_x \rho\, d\forall + \int_{CS} V_x \rho \vec{V} \cdot d\vec{A} \qquad (4.19a)$$

Assumptions: (1) $F_{B_x} = 0$
             (2) Steady flow
             (3) Uniform flow at each section

Then

$$-F_f + p_1 A - p_2 A = V_1\{-|\rho_1 V_1 A|\} + V_2\{|\rho_2 V_2 A|\} = \dot{m}(V_2 - V_1)$$

and

$$-F_f = (p_2 - p_1)A + \dot{m}(V_2 - V_1)$$

$$-F_f = (4.70 - 9.85)10^4 \frac{\text{N}}{\text{m}^2} \times 4.03 \times 10^{-5} \text{ m}^2$$

$$+ \; 3.08 \times 10^{-3} \frac{\text{kg}}{\text{sec}} \times (134 - 65.3) \frac{\text{m}}{\text{sec}} \times \frac{\text{N} \cdot \text{sec}^2}{\text{kg} \cdot \text{m}}$$

or

$$F_f = 1.86 \text{ N} \qquad \text{(to the left, as shown)}$$

This is the force exerted on the control volume by the duct wall. The force of the *fluid* on the *duct* is

$$K_x = F_f = 1.86 \text{ N} \qquad \text{(to the right)} \qquad\qquad K_x$$

**13-4.3 Tables for Computation of Fanno Line Flow of an Ideal Gas

The primary independent variable in Fanno line flow is the friction force, F_f. Knowledge of the total friction force between any two points in a Fanno line flow would enable us to predict downstream conditions from known upstream conditions. The total friction force is the integral of the wall shear stress over the duct surface area. Since wall shear stress varies along the duct, we must develop a differential equation and then integrate to find property variations. To set up the differential equation, we use the differential control volume shown in Fig. 13.12 for our analysis.

** This section may be omitted without loss of continuity in the text material.

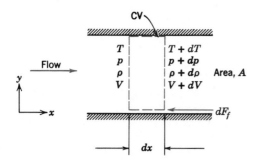

Fig. 13.12   Differential control volume used for analysis of Fanno line flow.

a.   Continuity Equation

Basic equation:

$$0 = \frac{\cancelto{0(1)}{\partial}}{\partial t} \int_{CV} \rho \, d\forall + \int_{CS} \rho \vec{V} \cdot d\vec{A} \qquad (4.13)$$

Assumptions: (1)   Steady flow
                     (2)   Uniform flow at each section

Then

$$0 = \{-|\rho V A|\} + \{|(\rho + d\rho)(V + dV)A|\}$$

or

$$\rho V A = (\rho + d\rho)(V + dV)A$$

Simplifying and canceling A gives

$$0 = -\rho \cancel{V} + \rho \cancel{V} + \rho \, dV + V \, d\rho + \cancelto{\approx 0}{d\rho \, dV}$$

which reduces to

$$\rho \, dV + V \, d\rho = 0 \qquad (13.8a)$$

since products of differentials are negligible.

b.   Momentum Equation

Basic equation:

$$F_{S_x} + \cancelto{0(3)}{F_{B_x}} = \frac{\cancelto{0(1)}{\partial}}{\partial t} \int_{CV} V_x \rho \, d\forall + \int_{CS} V_x \rho \vec{V} \cdot d\vec{A} \qquad (4.19a)$$

Assumption: (3) $F_{B_x} = 0$

The momentum equation becomes

$$-dF_f + pA - (p + dp)A = V\{-|\rho V A|\} + (V + dV)\{|(\rho + d\rho)(V + dV)A|\}$$

which simplifies, using continuity, to give

$$-\frac{dF_f}{A} - dp = \rho V \, dV \qquad (13.8b)$$

c. First Law of Thermodynamics

Basic equation:

$$= 0(4) = 0(5) = 0(6) \quad = 0(6) \quad = 0(1)$$

$$\dot{Q} - \dot{W}_s - \dot{W}_{shear} - \dot{W}_{other} = \frac{\partial}{\partial t} \int_{CV} e\rho \, d\Psi + \int_{CS} (e + pv)\rho\vec{V} \cdot d\vec{A} \qquad (4.57)$$

where

$$\simeq 0(7)$$

$$e = u + \frac{V^2}{2} + gz$$

Assumptions: (4) Adiabatic flow, $\dot{Q} = 0$
(5) $\dot{W}_s = 0$
(6) $\dot{W}_{shear} = \dot{W}_{other} = 0$
(7) Effects of gravity are negligible

Under these restrictions, we obtain

$$0 = \left(u + \frac{V^2}{2} + pv\right)\{-|\rho VA|\}$$

$$+ \left[u + du + \frac{V^2}{2} + d\left(\frac{V^2}{2}\right) + pv + d(pv)\right]\{|(\rho + d\rho)(V + dV)A|\}$$

Noting from continuity that the flow rate terms { } are equal, and substituting $h = u + pv$, we obtain

$$dh + d\left(\frac{V^2}{2}\right) = 0 \qquad (13.8c)$$

To complete our formulation, we must relate the friction force, dF_f, to flow variables at each cross section. We note that

$$dF_f = \tau_w \, dA_w = \tau_w P \, dx \qquad (13.9)$$

where P is the wetted perimeter of the duct. To obtain an expression for τ_w, in terms of flow variables at each cross section, we assume changes in flow variables with x are gradual and use correlations developed in Chapter 8 for fully developed, incompressible duct flow. For incompressible flow, the local wall shear stress can be written in terms of flow properties and friction factor. From Eqs. 8.16, 8.30, and 8.32 we have, for incompressible flow,

$$\tau_w = -\frac{R}{2}\frac{dp}{dx} = \frac{\rho R}{2}\frac{dh_l}{dx} = \frac{f\rho V^2}{8} \qquad (13.10)$$

where f is the friction factor for pipe flow, Fig. 8.14. We assume that this correlation of experimental data also applies to compressible flow. This assumption, when checked against experimental data, shows surprisingly good agreement for subsonic flows; data for supersonic flow are sparse.

Ducts of other than circular shape can be included in our analysis by introducing the hydraulic diameter

$$D_h = \frac{4A}{P} \qquad (8.43)$$

(The factor of 4 was included in Eq. 8.43 so that D_h would reduce to diameter D for circular ducts.)

Combining Eqs. 8.43, 13.9, and 13.10, we obtain

$$dF_f = \tau_w P \, dx = f \frac{\rho V^2}{8} \frac{4A}{D_h} dx$$

or

$$dF_f = \frac{fA}{D_h} \frac{\rho V^2}{2} dx \qquad (13.11)$$

Substituting this result into the momentum equation (Eq. 13.8b), we obtain

$$-\frac{f}{D_h} \frac{\rho V^2}{2} dx - dp = \rho V \, dV$$

or, after dividing by p,

$$\frac{dp}{p} = -\frac{f}{D_h} \frac{\rho V^2}{2p} dx - \frac{\rho V \, dV}{p}$$

Noting that $p/\rho = RT = c^2/k$, and $V \, dV = d(V^2/2)$, we obtain

$$\frac{dp}{p} = -\frac{f}{D_h} \frac{kM^2}{2} dx - \frac{k}{c^2} d\left(\frac{V^2}{2}\right)$$

and finally,

$$\frac{dp}{p} = -\frac{f}{D_h} \frac{kM^2}{2} dx - \frac{kM^2}{2} \frac{d(V^2)}{V^2} \qquad (13.12)$$

To relate M and x, we must eliminate dp/p and $d(V^2)/V^2$ from Eq. 13.12. From the definition of Mach number, $M = V/c$, then $V^2 = M^2 c^2 = M^2 kRT$ and

$$\frac{d(V^2)}{V^2} = \frac{dT}{T} + \frac{d(M^2)}{M^2} \qquad (13.13a)$$

From the continuity equation, $d\rho/\rho = -dV/V$ and

$$\frac{d\rho}{\rho} = -\frac{1}{2} \frac{d(V^2)}{V^2}$$

From the ideal gas equation of state, $p = \rho RT$,

$$\frac{dp}{p} = \frac{d\rho}{\rho} + \frac{dT}{T}$$

Combining these three equations, we obtain

$$\frac{dp}{p} = \frac{1}{2} \frac{dT}{T} - \frac{1}{2} \frac{d(M^2)}{M^2} \qquad (13.13b)$$

Substituting Eqs. 13.13 into Eq. 13.12 gives

$$\frac{1}{2} \frac{dT}{T} - \frac{1}{2} \frac{d(M^2)}{M^2} = -\frac{f}{D_h} \frac{kM^2}{2} dx - \frac{kM^2}{2} \frac{dT}{T} - \frac{kM^2}{2} \frac{d(M^2)}{M^2}$$

This equation can be simplified to

$$\left(\frac{1+kM^2}{2}\right)\frac{dT}{T} = -\frac{f}{D_h}\frac{kM^2}{2}\,dx + \left(\frac{1-kM^2}{2}\right)\frac{d(M^2)}{M^2} \tag{13.14}$$

We have been successful in reducing the number of variables somewhat. However, to relate M and x, we must obtain an expression for dT/T in terms of M. Such an expression can be obtained most readily from the stagnation temperature equation

$$\frac{T_0}{T} = 1 + \frac{k-1}{2}M^2 \tag{12.17b}$$

Since stagnation temperature is constant for Fanno line flow,

$$T\left(1 + \frac{k-1}{2}M^2\right) = \text{constant}$$

and

$$\frac{dT}{T} + \frac{M^2\frac{(k-1)}{2}}{\left(1 + \frac{k-1}{2}M^2\right)}\frac{d(M^2)}{M^2} = 0$$

Substituting for dT/T into Eq. 13.14 yields

$$\frac{M^2\frac{(k-1)}{2}\left(\frac{1+kM^2}{2}\right)}{\left(1 + \frac{k-1}{2}M^2\right)}\frac{d(M^2)}{M^2} = \frac{f}{D_h}\frac{kM^2}{2}\,dx - \left(\frac{1-kM^2}{2}\right)\frac{d(M^2)}{M^2}$$

Combining terms, we obtain

$$\frac{(1-M^2)}{\left(1 + \frac{k-1}{2}M^2\right)}\frac{d(M^2)}{kM^4} = \frac{f}{D_h}dx \tag{13.15}$$

We have obtained a differential equation that relates changes in M with x. Now we must integrate the equation to find M as a function of x.

Integration of Eq. 13.15 between states ① and ② would produce a complicated function of both M_1 and M_2. The function would have to be evaluated numerically for each new combination of M_1 and M_2 encountered in a problem. Calculations can be simplified considerably using critical conditions (where, by definition, $M = 1$). All Fanno line flows tend toward $M = 1$, so integration is between a section where the Mach number is M and the section where sonic conditions occur (the critical conditions). Mach number will reach unity when the maximum possible length of duct is used, as shown schematically in Fig. 13.13.

The task is to perform the integration

$$\int_M^1 \frac{(1-M^2)}{kM^4\left(1 + \frac{k-1}{2}M^2\right)}d(M^2) = \int_0^{L_{max}}\frac{f}{D_h}dx \tag{13.16}$$

The left side may be integrated by parts. On the right side, the friction factor, f, may vary with x, since Reynolds number will vary along the duct. Note, however,

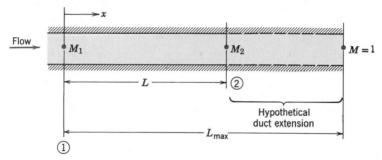

Fig. 13.13 Coordinates and notation used for analysis of Fanno line flow.

that since ρV is constant along the duct (from continuity), the variation in Reynolds number is caused solely by variations in fluid viscosity.

Defining a mean friction factor, $\bar{f}$, over the duct length as

$$\bar{f} = \frac{1}{L_{max}} \int_0^{L_{max}} f \, dx$$

then integration of Eq. 13.16 leads to

$$\frac{1-M^2}{kM^2} + \frac{k+1}{2k} \ln \left[\frac{(k+1)M^2}{2\left(1+\frac{k-1}{2}M^2\right)} \right] = \frac{\bar{f}L_{max}}{D_h} \tag{13.17}$$

Equation 13.17 gives the maximum $\bar{f}L/D_h$ corresponding to any initial Mach number. These values are tabulated in Table E.2 of Appendix E.

Since $\bar{f}L_{max}/D_h$ is a function of M, the duct length, L, required for the Mach number to change from M_1 to M_2 (as illustrated in Fig. 13.13) may be found from

$$\frac{\bar{f}L}{D_h} = \left(\frac{\bar{f}L_{max}}{D_h}\right)_{M_1} - \left(\frac{\bar{f}L_{max}}{D_h}\right)_{M_2}$$

Critical conditions are appropriate reference conditions to use in tabulating fluid properties as a function of local Mach number. Thus, for example, since T_0 is constant, we can write

$$\frac{T}{T^*} = \frac{T/T_0}{T^*/T_0} = \frac{1}{1+\frac{k-1}{2}M^2} \bigg/ \frac{1}{1+\frac{k-1}{2}} = \frac{\left(\frac{k+1}{2}\right)}{\left(1+\frac{k-1}{2}M^2\right)} \tag{13.18a}$$

Similarly,

$$\frac{V}{V^*} = \frac{M\sqrt{kRT}}{\sqrt{kRT^*}} = M\sqrt{\frac{T}{T^*}} = \left[\frac{\left(\frac{k+1}{2}\right)M^2}{1+\frac{k-1}{2}M^2} \right]^{1/2} \tag{13.18b}$$

From continuity,

$$\frac{\rho}{\rho^*} = \frac{V^*}{V} = \left[\frac{1+\dfrac{k-1}{2}M^2}{\left(\dfrac{k+1}{2}\right)M^2}\right]^{1/2} \qquad (13.18c)$$

From the ideal gas equation of state,

$$\frac{p}{p^*} = \frac{\rho}{\rho^*}\frac{T}{T^*} = \frac{1}{M}\left[\frac{\left(\dfrac{k+1}{2}\right)}{1+\dfrac{k-1}{2}M^2}\right]^{1/2} \qquad (13.18d)$$

The ratio of local stagnation pressure to the reference stagnation pressure is given by

$$\frac{p_0}{p_0^*} = \frac{p_0}{p}\frac{p}{p^*}\frac{p^*}{p_0^*}$$

$$= \left(1+\frac{k-1}{2}M^2\right)^{k/(k-1)}\frac{1}{M}\left[\frac{\left(\dfrac{k+1}{2}\right)}{1+\dfrac{k-1}{2}M^2}\right]^{1/2}\frac{1}{\left(1+\dfrac{k-1}{2}\right)^{k/(k-1)}}$$

or

$$\frac{p_0}{p_0^*} = \frac{1}{M}\left[\left(\frac{2}{k+1}\right)\left(1+\frac{k-1}{2}M^2\right)\right]^{(k+1)/2(k-1)} \qquad (13.18e)$$

The ratios in Eqs. 13.18 are tabulated as functions of Mach number for an ideal gas with $k = 1.4$ in Table E.2 of Appendix E.

EXAMPLE 13.8—Frictional Adiabatic Flow in a Constant-Area Channel: Table Solution

Air flow is induced in a smooth insulated tube of 7.16 mm diameter by a vacuum pump. Air is drawn from a room, where $p_0 = 760$ mm Hg (abs) and $T_0 = 23$ C, through a smoothly contoured converging nozzle. At section ①, where the nozzle joins the constant-area tube, the static pressure is -18.9 mm Hg (gage). At section ②, located some distance downstream in the constant-area tube, the static pressure is -412 mm Hg (gage). The duct walls are smooth; assume the average friction factor, $\bar{f}$, is the value at section ①. Determine the length of duct required for choking from section ①, the Mach number at section ②, and the duct length, L_{12}, between sections ① and ②.

EXAMPLE PROBLEM 13.8

GIVEN: Air flow (with friction) in an insulated constant-area tube.

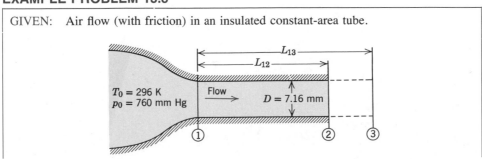

Gage pressures: $p_1 = -18.9$ mm Hg, $p_2 = -412$ mm Hg. $M_3 = 1.0$

FIND: (a) L_{13}.   (b) M_2.   (c) L_{12}.

SOLUTION:
Flow in the constant-area tube is frictional and adiabatic, a Fanno line flow. To find the friction factor, we need to know the flow conditions at section ①. If we assume flow in the nozzle is isentropic, local properties at the nozzle exit may be computed using isentropic relations. Thus

$$\frac{p_{0_1}}{p_1} = \left(1 + \frac{k-1}{2} M_1^2\right)^{k/(k-1)}$$

Solving for M_1, we obtain

$$M_1 = \left\{\frac{2}{k-1}\left[\left(\frac{p_{0_1}}{p_1}\right)^{(k-1)/k} - 1\right]\right\}^{1/2} = \left\{\frac{2}{0.4}\left[\left(\frac{760}{760-18.9}\right)^{0.286} - 1\right]\right\}^{1/2} = 0.190$$

$$T_1 = \frac{T_{0_1}}{1 + \frac{k-1}{2}M_1^2} = \frac{296 \text{ K}}{1 + 0.2(0.190)^2} = 294 \text{ K}$$

$$V_1 = M_1 c_1 = M_1 \sqrt{kRT_1} = 0.190\left[1.4 \times 287 \frac{\text{N} \cdot \text{m}}{\text{kg} \cdot \text{K}} \times 294 \text{ K} \times \frac{\text{kg} \cdot \text{m}}{\text{N} \cdot \text{sec}^2}\right]^{1/2}$$

$$V_1 = 65.3 \text{ m/sec}$$

Using the density of mercury at room temperature (23 C),

$$p_1 = g\rho_{Hg}h_1 = g\text{SG}\rho_{H_2O}h_1$$

$$= \frac{9.81 \text{ m}}{\text{sec}^2} \times 13.5 \times \frac{1000 \text{ kg}}{\text{m}^3} \times (760 - 18.9)10^{-3} \text{ m} \times \frac{\text{N} \cdot \text{sec}^2}{\text{kg} \cdot \text{m}}$$

$$p_1 = 98.1 \text{ kPa (abs)}$$

$$\rho_1 = \frac{p_1}{RT_1} = \frac{9.81 \times 10^4 \text{ N}}{\text{m}^2} \times \frac{\text{kg} \cdot \text{K}}{287 \text{ N} \cdot \text{m}} \times \frac{1}{294 \text{ K}} = 1.16 \text{ kg/m}^3$$

At $T = 294$ K (21 C), $\mu = 1.9 \times 10^{-5}$ kg/m · sec from Fig. A.2, Appendix A. Thus

$$Re_1 = \frac{\rho_1 V_1 D_1}{\mu_1} = \frac{1.16 \text{ kg}}{\text{m}^3} \times \frac{65.3 \text{ m}}{\text{sec}} \times 0.00716 \text{ m} \times \frac{\text{m} \cdot \text{sec}}{1.9 \times 10^{-5} \text{ kg}} = 2.85 \times 10^4$$

From Fig. 8.14, for smooth pipe, $f = 0.0235$. From Table E.2 at $M_1 = 0.19$, $p/p^* = 5.745$, and $\bar{f}L_{max}/D_h = 16.38$. Thus, assuming $\bar{f} = f_1$,

$$L_{13} = (L_{max})_1 = \left(\frac{\bar{f}L_{max}}{D_h}\right)_1 \frac{D_h}{f_1} = 16.38 \times \frac{0.00716 \text{ m}}{0.0235} = 4.99 \text{ m} \qquad \underleftarrow{L_{13}}$$

Since p^* is constant for Fanno line flow, conditions at section ② can be determined from the pressure ratio, $(p/p^*)_2$. Thus

$$\left(\frac{p}{p^*}\right)_2 = \frac{p_2}{p^*} = \frac{p_2}{p_1}\frac{p_1}{p^*} = \frac{p_2}{p_1}\left(\frac{p}{p^*}\right)_1 = \left(\frac{760-412}{760-18.9}\right)5.745 = 2.698$$

From Table E.2, at $(p/p^*)_2 = 2.698$, $M_2 \simeq 0.40$ \qquad \underleftarrow{M_2}$

At $M_2 = 0.40$, $\bar{f}L_{max}/D_h = 2.309$ ('Table E.2). Thus

$$L_{23} = (L_{max})_2 = \left(\frac{\bar{f}L_{max}}{D_h}\right)_2 \frac{D_h}{f_1} = 2.309 \times \frac{0.00716 \text{ m}}{0.0235} = 0.704 \text{ m}$$

Finally,

$$L_{12} = L_{13} - L_{23} = (4.99 - 0.704) \text{ m} = 4.29 \text{ m} \qquad\qquad L_{12}$$

$\left\{\begin{array}{l}\text{This is the same physical system analyzed in Example Problem 13.7. Use of tables}\\ \text{simplifies calculations and makes it possible to determine the duct length.}\end{array}\right\}$

**13-4.4 Isothermal Flow

As noted earlier, there are cases of gas flow in constant-area ducts in which the flow is essentially isothermal. Mach numbers in such flows are generally low, but significant pressure changes can occur as a result of frictional effects acting over long duct lengths. Hence, such flows cannot be treated as incompressible. The assumption of isothermal flow is much more appropriate.

Analysis of isothermal flow is similar to that for adiabatic flow, with one major change. For adiabatic flow, the heat exchange, $\delta Q/dm$, was zero; for isothermal flow, temperature is constant, and hence $dT = 0$. For isothermal flow, the first law statement for the finite control volume of Fig. 13.9 is written as

$$h_1 + \frac{V_1^2}{2} + \frac{\delta Q}{dm} = h_2 + \frac{V_2^2}{2} \tag{13.19a}$$

or

$$q = \frac{\delta Q}{dm} = h_{0_2} - h_{0_1} = \frac{V_2^2 - V_1^2}{2} \tag{13.19b}$$

The continuity and momentum equations are given by Eqs. 13.7a and 13.7b, respectively,

$$\rho_1 V_1 = \rho_2 V_2 \equiv G = \frac{\dot{m}}{A} \tag{13.7a}$$

$$R_x + p_1 A - p_2 A = \dot{m}V_2 - \dot{m}V_1 \tag{13.7b}$$

For an ideal gas,

$$p = \rho R T \tag{13.7e}$$

Equations 13.7a, 13.7b, 13.7e, and 13.19a are the governing equations for steady, one-dimensional, isothermal flow of an ideal gas in a constant-area duct. If all conditions at state ① are known, we have five unknowns in these four equations. Thus there are an infinite number of possible states ②. The locus of these possible downstream states is a horizontal line through state ① on the Ts diagram. For isothermal flow, the entropy change may be calculated from Eq. 13.7d as

$$s_2 - s_1 = -R \ln \frac{p_2}{p_1} \tag{13.20}$$

** This section may be omitted without loss of continuity in the text material.

As in Fanno line flow, the primary independent variable in isothermal flow is the friction force, R_x. Knowledge of the friction force between any two sections in an isothermal flow would enable us to predict downstream conditions when conditions upstream were known. Downstream properties can be readily determined once the downstream Mach number is known. For isothermal flow, $c =$ constant, so $V_2/V_1 = M_2/M_1$, and from Eq. 13.7a we have

$$\frac{\rho_2}{\rho_1} = \frac{V_1}{V_2} = \frac{M_2}{M_1}$$
(13.21)

Combining the ideal gas equation of state with Eq. 13.21, we obtain

$$\frac{p_2}{p_1} = \frac{\rho_2}{\rho_1} = \frac{M_2}{M_1}$$
(13.22)

The stagnation temperature ratio is given by

$$\frac{T_{0_2}}{T_{0_1}} = \frac{1 + \dfrac{k-1}{2}M_2^2}{1 + \dfrac{k-1}{2}M_1^2}$$
(13.23)

To determine the variation in Mach number along the duct length, it is necessary to consider the differential momentum equation for flow with friction. The analysis leading to Eq. 13.14 is valid for isothermal flow. Since $T =$ constant for isothermal flow, then from Eq. 13.14, with $dT = 0$,

$$\frac{f}{D_h}\frac{kM^2}{2}dx = \left(\frac{1 - kM^2}{2}\right)\frac{d(M^2)}{M^2}$$
(13.24)

and

$$\frac{f}{D_h}dx = \frac{(1 - kM^2)d(M^2)}{kM^4}$$
(13.25)

Equation 13.25 shows that the limiting Mach number for isothermal flow is $M = 1/\sqrt{k}$. Since T is constant, then the friction factor, $f = f(Re)$, is also constant. Integration of Eq. 13.25 between the limits of $M = M$ at $x = 0$ and $M = 1/\sqrt{k}$ at $x = L_{max}$, where L_{max} is the distance beyond which the isothermal flow may not proceed, gives

$$\frac{fL_{max}}{D_h} = \frac{1 - kM^2}{kM^2} + \ln kM^2$$
(13.26)

The duct length, L, required for the flow Mach number to change from M_1 to M_2 can be obtained from

$$f\frac{L}{D_h} = f\frac{L_{max_1} - L_{max_2}}{D_h} = \frac{1 - kM_1^2}{kM_1^2} - \frac{1 - kM_2^2}{kM_2^2} + \ln\frac{M_1^2}{M_2^2}$$
(13.27)

The distribution of heat exchange along the duct required to maintain isothermal flow can be determined from the differential form of Eq. 13.19b, as

$$dq = dh_0 = c_p dT_0 = c_p d\left[T(1 + \frac{k-1}{2}M^2)\right]$$

or, since $T = \text{constant}$,

$$dq = c_p T \left(\frac{k-1}{2}\right) dM^2 = \frac{c_p T_0(k-1)}{2(1 + \frac{k-1}{2}M^2)} dM^2$$

Substituting for dM^2 from Eq. 13.25,

$$dq = \frac{c_p T_0(k-1)kM^4}{2(1 + \frac{k-1}{2}M^2)(1-kM^2)} \frac{f}{D_h} dx \qquad (13.28)$$

From Eq. 13.28 we note that as $M \to 1/\sqrt{k}$, then $dq/dx \to \infty$. Thus, an infinite rate of heat exchange is required to maintain isothermal flow as the Mach number approaches the limiting value. Hence, we conclude that isothermal acceleration of flow in a constant-area duct is only physically possible for low Mach number flow.

13-5   FRICTIONLESS FLOW IN A CONSTANT-AREA DUCT WITH HEAT EXCHANGE

To explore the effects of heat exchange on a compressible flow, we apply the basic equations to steady, one-dimensional, frictionless flow of an ideal gas with constant specific heats through the finite control volume shown in Fig. 13.14.

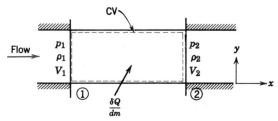

Fig. 13.14   Control volume used for integral analysis of frictionless flow with heat exchange.

13-5.1   Basic Equations

a.   Continuity Equation

Basic equation:
$$= 0(1)$$
$$0 = \frac{\partial}{\partial t} \int_{CV} \rho \, d\mathcal{V} + \int_{CS} \rho \vec{V} \cdot d\vec{A} \qquad (4.13)$$

Assumptions: (1)   Steady flow
(2)   Uniform flow at each section

Then

$$0 = \{-|\rho_1 V_1 A_1|\} + \{|\rho_2 V_2 A_2|\}$$

The area is constant, so

$$\rho_1 V_1 = \rho_2 V_2 \equiv G = \frac{\dot{m}}{A} \qquad (13.29a)$$

b. Momentum Equation

Basic equation:

$$\overset{=\,0(3)}{\cancel{F_{S_x}}} + \overset{=\,0(1)}{\cancel{F_{B_x}}} = \cancel{\frac{\partial}{\partial t} \int_{CV} V_x \rho \, d\Psi} + \int_{CS} V_x \rho \vec{V} \cdot d\vec{A} \qquad (4.19a)$$

Assumption: (3) $F_{B_x} = 0$

Since there is no friction between the duct walls and the flow, and $A_2 = A_1 = A$, then

$$p_1 A - p_2 A = V_1\{-|\rho_1 V_1 A|\} + V_2\{|\rho_2 V_2 A|\}$$

This can be written as

$$p_1 A - p_2 A = \dot{m} V_2 - \dot{m} V_1 \qquad (13.29b)$$

or

$$p_1 + \rho_1 V_1^2 = p_2 + \rho_2 V_2^2$$

c. First Law of Thermodynamics

Basic equation: $= 0(4) \quad = 0(5) \quad = 0(5) \quad = 0(1)$

$$\dot{Q} - \overset{}{\cancel{\dot{W}_s}} - \overset{}{\cancel{\dot{W}_{shear}}} - \overset{}{\cancel{\dot{W}_{other}}} = \cancel{\frac{\partial}{\partial t} \int_{CV} e\rho \, d\Psi} + \int_{CS} (e + pv)\rho\vec{V} \cdot d\vec{A} \qquad (4.57)$$

where

$$e = u + \frac{V^2}{2} + \overset{\approx\,0(6)}{\cancel{gz}}$$

Assumptions: (4) $\dot{W}_s = 0$
  (5) $\dot{W}_{shear} = \dot{W}_{other} = 0$
  (6) Effects of gravity are negligible

With these restrictions,

$$\dot{Q} = \left(u_1 + \frac{V_1^2}{2} + p_1 v_1\right)\{-|\rho_1 V_1 A|\} + \left(u_2 + \frac{V_2^2}{2} + p_2 v_2\right)\{|\rho_2 V_2 A|\}$$

or

$$\dot{Q} = \dot{m}\left(h_2 + \frac{V_2^2}{2} - h_1 - \frac{V_1^2}{2}\right)$$

But

$$\frac{\delta Q}{dm} = \frac{1}{\dot{m}}\dot{Q}$$

so

$$\frac{\delta Q}{dm} + h_1 + \frac{V_1^2}{2} = h_2 + \frac{V_2^2}{2} \tag{13.29c}$$

or

$$\frac{\delta Q}{dm} = h_{0_2} - h_{0_1}$$

We see that heat exchange causes the stagnation enthalpy, and hence the stagnation temperature, to change.

d. Second Law of Thermodynamics

Basic equation:

$$\int_{CS} \frac{1}{T} \frac{\dot{Q}}{A} dA \leq \overset{= 0(1)}{\cancel{\frac{\partial}{\partial t}} \int_{CV} s\rho \, d\forall} + \int_{CS} s\rho \vec{V} \cdot d\vec{A} \tag{4.59}$$

or

$$\int_{CS} \frac{1}{T} \frac{\dot{Q}}{A} dA \leq \dot{m}(s_2 - s_1)$$

Since the flow is frictionless, the equality in Eq. 4.59 would hold if the process of heat exchange were considered to be reversible. However, even if this dubious assumption were made, the integral on the left side of the equation cannot be evaluated. Consequently, the control volume form of the second law does not enable us to calculate the actual entropy change between any two locations in the flow. The rate of heat exchange, $\dot{Q}$, may be positive (heat addition to the flow) or negative (heat rejection from the flow). Therefore, the entropy change in a frictionless flow with heat exchange may be either positive or negative.

To calculate the entropy change, we rely on the $T\,ds$ equations. Since

$$T\,ds = dh - v\,dp$$

for an ideal gas, we can write

$$ds = c_p \frac{dT}{T} - R\frac{dp}{p}$$

For constant specific heats, the equation can be integrated to give

$$s_2 - s_1 = c_p \ln \frac{T_2}{T_1} - R \ln \frac{p_2}{p_1} \tag{13.29d}$$

e. Equations of State

For an ideal gas, the equation of state is

$$p = \rho RT \tag{13.29e}$$

Equations 13.29a through 13.29e are the governing equations for the steady, one-dimensional, frictionless flow of an ideal gas in a constant-area duct with heat exchange. If all properties at state ① are known, then we have seven unknowns

(p_2, V_2, p_2, h_2, s_2, T_2, and $\delta Q/dm$) in these five equations. However, we have the known relationship between h and T for an ideal gas, $dh = c_p dT$. For an ideal gas with constant specific heats,

$$\Delta h = h_2 - h_1 = c_p \Delta T = c_p(T_2 - T_1) \qquad (13.29f)$$

We thus have the situation of six equations and seven unknowns.

13-5.2 The Rayleigh Line

If all conditions at state ① are known, how many possible states ② are there? The mathematics of the situation (six equations and seven unknowns) indicate an infinite number of possible states ②.

With an infinite number of possible states ② for a given state ①, what is to be expected if all possible states ② are plotted on a Ts diagram? It follows that the locus of all possible states ②, reachable from state ①, is a continuous curve passing through state ①.

How can we determine this curve? Perhaps the simplest way is to assume different values of T_2. For each T_2 we calculate all other properties at state ② and also $\delta Q/dm$.

The results of these calculations are shown qualitatively on the Ts plane in Fig. 13.15. The locus of all possible downstream states is called the *Rayleigh line*. The calculations show some interesting features of Rayleigh line flow. At the point of maximum temperature (point a of Fig. 13.15), the Mach number for an ideal gas is $1/\sqrt{k}$. At the point of maximum entropy (point b of Fig. 13.15), $M = 1$. On the upper branch of the curve, Mach number is always less than unity, and it increases monotonically as we proceed to the right along the curve. At every point on the lower portion of the curve, Mach number is greater than unity, and it decreases monotonically as we move to the right along the curve. Regardless of the initial

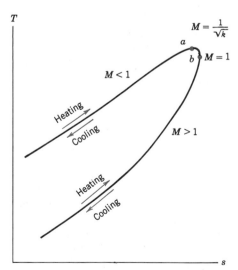

Fig. 13.15 Schematic Ts diagram for frictionless flow in a constant-area duct with heat exchange (Rayleigh line flow).

Mach number, with heat addition the flow state proceeds to the right, and with heat rejection the flow state proceeds to the left along the Rayleigh line.

For any initial state in a Rayleigh line flow, any point on the Rayleigh line represents a mathematically possible downstream state. Indeed, we determined the locus of all possible downstream states by assuming values of T_2 and calculating the corresponding properties. Although the Rayleigh line represents all mathematically possible states, are they all physically attainable downstream states? A moment's reflection will indicate that they are. Since we are considering a flow with heat exchange, the second law does not impose any restrictions on the sign of the entropy change.

The effects of heat exchange on properties in steady, frictionless, compressible flow of an ideal gas may be found from the basic equations, Eqs. 13.29a through 13.29f, and the Rayleigh line, Fig. 13.15. These effects are summarized in Table 13.2; the basis of the indicated trends is discussed in the next few paragraphs.

The direction of entropy change is always determined by the heat exchange; entropy increases with heating and decreases with cooling. Similarly, the first law, Eq. 13.29c, shows that heating increases the stagnation enthalpy and cooling decreases it; since $\Delta h_0 = c_p \Delta T_0$, the effect on stagnation temperature is the same.

The effect on temperature of heating and cooling may be deduced from the shape of the Rayleigh line in Fig. 13.15. We see that for $M < 1/\sqrt{k}$ (for air, $1/\sqrt{k} \approx 0.85$) or for $M > 1$, heating causes T to increase, and in the same regions cooling causes T to decrease. However, we also see the unexpected result that for $1/\sqrt{k} < M < 1$, *heat addition* causes the stream temperature to *decrease*, and *heat rejection* causes the stream temperature to *increase!*

For subsonic flow, the Mach number increases monotonically with heating, until $M = 1$ is reached. For given inlet conditions, all possible downstream states lie on a single Rayleigh line. Therefore, the point $M = 1$ determines the maximum possible heat addition without choking. If the flow is initially supersonic, heating will reduce the Mach number. Again, the maximum possible heat addition without choking is that which reduces the Mach number to $M = 1.0$.

The effect of heat exchange on static pressure is obtained from the shapes of the Rayleigh line and of constant pressure lines on the Ts plane (see Fig. 13.16). For $M < 1$, pressure falls with heating and for $M > 1$, pressure increases with heating, as shown by the shapes of the constant-pressure lines. Once the pressure variation has been found, the effect on velocity may be found from the momentum equation,

$$p_1 A - p_2 A = \dot{m} V_2 - \dot{m} V_1 \qquad (13.29b)$$

or

$$p + \left(\frac{\dot{m}}{A}\right) V = \text{constant}$$

Thus, since $\dot{m}/A = \text{constant}$, trends in p and V must be opposite. From the continuity equation, the trend in ρ is opposite to that in V.

Local isentropic stagnation pressure always decreases with heating. This is illustrated schematically in Fig. 13.16. A reduction in stagnation pressure has obvious practical implications for heating processes, such as combustion chambers. Adding the same amount of energy per unit mass (same change in T_0) causes a larger change in p_0 for supersonic flow; because heating occurs at a lower temperature in supersonic flow, the entropy increase is larger.

Table 13.2 Summary of Heat Exchange Effects on Fluid Properties

Property	Heating		Cooling		Obtained from:
	$M < 1$	$M > 1$	$M < 1$	$M > 1$	
Entropy, s	Increase	Increase	Decrease	Decrease	Second law
Stagnation temperature, T_0	Increase	Increase	Decrease	Decrease	First law, and $\Delta h_0 = c_p \Delta T_0$
Temperature, T	$\left(M < \dfrac{1}{\sqrt{k}}\right)$ Increase	Increase	$\left(M < \dfrac{1}{\sqrt{k}}\right)$ Decrease	Decrease	Shape of Rayleigh line
	$\left(\dfrac{1}{\sqrt{k}} < M < 1\right)$ Decrease		$\left(\dfrac{1}{\sqrt{k}} < M < 1\right)$ Increase		
Mach number, M	Increase	Decrease	Decrease	Increase	Trend on Rayleigh line
Pressure, p	Decrease	Increase	Increase	Decrease	Trend on Rayleigh line
Velocity, V	Increase	Decrease	Decrease	Increase	Momentum equation, and effect on p
Density, ρ	Decrease	Increase	Increase	Decrease	Continuity equation, and effect on V
Stagnation pressure, p_0	Decrease	Decrease	Increase	Increase	Fig. 13.16

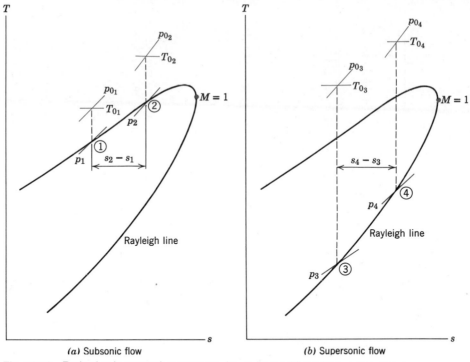

Fig. 13.16 Reduction in stagnation pressure due to heat addition for two flow cases.

EXAMPLE 13.9—Frictionless Flow in a Constant-Area Duct with Heat Addition

Air flows with negligible friction through a duct of area $A = 0.25 \text{ ft}^2$. At section ①, flow properties are $T_1 = 600$ R, $p_1 = 20$ psia, and $V_1 = 360$ ft/sec. At section ②, $p_2 = 10$ psia. The flow is heated between sections ① and ②. Determine the properties at section ②, the energy added, and the entropy change. Finally, plot the process on a Ts diagram.

EXAMPLE PROBLEM 13.9

GIVEN: Frictionless flow of air in duct shown:

$T_1 = 600$ R

$p_1 = 20$ psia $p_2 = 10$ psia

$V_1 = 360$ ft/sec $A_1 = A_2 = A = 0.25 \text{ ft}^2$

FIND: (a) Properties at section ②.   (b) $\delta Q/dm$.
       (c) $s_2 - s_1$.   (d) Sketch on Ts diagram.

SOLUTION:
Apply the x component of the momentum equation, using the coordinates and control volume shown.

Basic equation:

$$F_{S_x} + \cancel{F_{B_x}}^{= 0(1)} = \cancel{\frac{\partial}{\partial t} \int_{CV} V_x \rho \, d\forall}^{= 0(2)} + \int_{CS} V_x \rho \vec{V} \cdot d\vec{A} \qquad (4.19a)$$

Assumptions: (1) $F_{B_x} = 0$
(2) Steady flow
(3) Uniform flow at each section

Then

$$p_1 A - p_2 A = V_1\{-|\rho_1 V_1 A|\} + V_2\{|\rho_2 V_2 A|\} = \dot{m}(V_2 - V_1)$$

or

$$p_1 - p_2 = \frac{\dot{m}}{A}(V_2 - V_1) = \rho_1 V_1 (V_2 - V_1)$$

Solving for V_2 gives

$$V_2 = \frac{p_1 - p_2}{\rho_1 V_1} + V_1$$

For an ideal gas,

$$\rho_1 = \frac{p_1}{RT_1} = \frac{20 \text{ lbf}}{\text{in.}^2} \times \frac{144 \text{ in.}^2}{\text{ft}^2} \times \frac{\text{lbm} \cdot \text{R}}{53.3 \text{ ft} \cdot \text{lbf}} \times \frac{1}{600 \text{ R}} = 0.0901 \text{ lbm/ft}^3$$

$$V_2 = \frac{(20-10) \text{ lbf}}{\text{in.}^2} \times \frac{144 \text{ in.}^2}{\text{ft}^2} \times \frac{\text{ft}^3}{0.0901 \text{ lbm}} \times \frac{\text{sec}}{360 \text{ ft}} \times \frac{32.2 \text{ lbm}}{\text{slug}}$$

$$\times \frac{\text{slug} \cdot \text{ft}}{\text{lbf} \cdot \text{sec}^2} + \frac{360 \text{ ft}}{\text{sec}}$$

$\underline{V_2 = 1790 \text{ ft/sec}}$ $\qquad\qquad\qquad\qquad\qquad\qquad\qquad\qquad\qquad\qquad V_2$

From continuity, $G = \rho_1 V_1 = \rho_2 V_2$, so

$$\rho_2 = \rho_1 \frac{V_1}{V_2} = \frac{0.0901 \text{ lbm}}{\text{ft}^3} \left(\frac{360}{1790}\right) = 0.0181 \text{ lbm/ft}^3 \qquad\qquad \rho_2$$

Solving for T_2, we obtain

$$T_2 = \frac{p_2}{\rho_2 R} = \frac{10 \text{ lbf}}{\text{in.}^2} \times \frac{144 \text{ in.}^2}{\text{ft}^2} \times \frac{\text{ft}^3}{0.0181 \text{ lbm}} \times \frac{\text{lbm} \cdot \text{R}}{53.3 \text{ ft} \cdot \text{lbf}} = 1490 \text{ R} \qquad T_2$$

The local isentropic stagnation temperature is given by

$$T_{0_2} = T_2\left(1 + \frac{k-1}{2}M_2^2\right)$$

$$c_2 = \sqrt{kRT_2} = 1890 \text{ ft/sec}; \qquad M_2 = \frac{V_2}{c_2} = \frac{1790}{1890} = 0.947$$

$$T_{0_2} = 1490 \text{ R}[1 + 0.2(0.947)^2] = 1760 \text{ R} \qquad\qquad\qquad\qquad\qquad T_{0_2}$$

and

$$p_{0_2} = p_2\left(\frac{T_{0_2}}{T_2}\right)^{k/(k-1)} = 10 \text{ psia}\left(\frac{1760}{1490}\right)^{3.5} = 17.9 \text{ psia} \qquad\qquad p_{0_2}$$

The heat addition is determined from the energy equation.
Basic equation:

$$\overset{=0(4)}{} \quad \overset{=0(5)}{} \quad \overset{=0(5)}{} \quad \overset{=0(2)}{}$$

$$\dot{Q} - \cancel{\dot{W}_s} - \cancel{\dot{W}_{\text{shear}}} - \cancel{\dot{W}_{\text{other}}} = \frac{\cancel{\partial}}{\partial t}\int_{CV} e\rho \, d\Psi + \int_{CS}(e + pv)\rho\vec{V}\cdot d\vec{A} \qquad (4.57)$$

where

$$\overset{\approx\, 0(6)}{e = u + \frac{V^2}{2} + g\!\!\nearrow\!\!z}$$

Assumptions: (4) $\dot{W}_s = 0$

(5) $\dot{W}_{\text{shear}} = \dot{W}_{\text{other}} = 0$

(6) Neglect changes in z

Then

$$\dot{Q} = \left(u_1 + p_1 v_1 + \frac{V_1^2}{2}\right)\{-|\rho_1 V_1 A|\} + \left(u_2 + p_2 v_2 + \frac{V_2^2}{2}\right)\{|\rho_2 V_2 A|\}$$

$$\dot{Q} = \dot{m}\left(h_2 + \frac{V_2^2}{2} - h_1 - \frac{V_1^2}{2}\right) = \dot{m}(h_{0_2} - h_{0_1}) = \dot{m}c_p(T_{0_2} - T_{0_1})$$

and

$$\frac{\delta Q}{dm} = \frac{1}{\dot{m}}\dot{Q} = c_p(T_{0_2} - T_{0_1})$$

$$T_{0_1} = T_1\left(1 + \frac{k-1}{2}M_1^2\right)$$

$$c_1 = \sqrt{kRT_1} = 1200 \text{ ft/sec}; \qquad M_1 = \frac{V_1}{c_1} = \frac{360}{1200} = 0.3$$

$$T_{0_1} = 600 \text{ R}[1 + 0.2(0.3)^2] = 611 \text{ R}$$

so

$$\frac{\delta Q}{dm} = 0.240 \frac{\text{Btu}}{\text{lbm}\cdot\text{R}}(1760 - 611) \text{ R} = 276 \text{ Btu/lbm} \qquad \underleftarrow{\delta Q/dm}$$

Using the $T\,ds$ equation, $T\,ds = dh - v\,dp$, we obtain, for an ideal gas with constant specific heats,

$$s_2 - s_1 = c_p \ln\frac{T_2}{T_1} - R \ln\frac{p_2}{p_1} = c_p \ln\frac{T_2}{T_1} - (c_p - c_v)\ln\frac{p_2}{p_1}$$

Then

$$s_2 - s_1 = \frac{0.240}{\;} \frac{\text{Btu}}{\text{lbm}\cdot\text{R}} \times \ln\left(\frac{1490}{600}\right) - \frac{(0.240 - 0.171)}{\;} \frac{\text{Btu}}{\text{lbm}\cdot\text{R}} \times \ln\left(\frac{10}{20}\right)$$

$$s_2 - s_1 = 0.266 \text{ Btu/lbm}\cdot\text{R} \qquad \underleftarrow{s_2 - s_1}$$

The process follows a Rayleigh line:

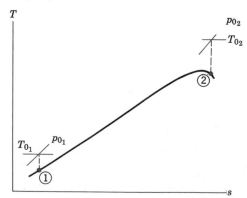

To complete our analysis, let us examine the change in p_0 by comparing p_{0_2} with p_{0_1}.

$$p_{0_1} = p_1 \left(\frac{T_{0_1}}{T_1} \right)^{k/(k-1)} = 20.0 \text{ psia} \left(\frac{611}{600} \right)^{3.5} = 21.3 \text{ psia} \qquad\qquad\qquad p_{0_1}$$

Comparing, we see that p_{0_2} is less than p_{0_1}.

$$\left\{ \begin{array}{l} \text{In general, stagnation pressure is decreased by heating and increased by cooling in} \\ \text{Rayleigh line flow.} \end{array} \right\}$$

**13-5.3 Tables for Computation of Rayleigh Line Flow of an Ideal Gas

In Section 13-5.1 we wrote the basic equations for Rayleigh line flow between two arbitrary states ① and ② in the flow. To facilitate solving problems, it is convenient to tabulate dimensionless properties in terms of local Mach number as we did for Fanno line flow. The reference state is again taken as the critical condition where $M = 1$; properties at the critical condition are denoted by (*).

Dimensionless properties (such as T/T^* and p/p^*) may be obtained by writing the basic equations between a point in the flow where properties are M, T, p, etc. (unsubscripted), and the critical state ($M = 1$, with properties denoted as T^*, p^*, etc.).

The pressure ratio, p/p^*, may be obtained from the momentum equation

$$pA - p^*A = \dot{m}V^* - \dot{m}V \qquad (13.29b)$$

or

$$p + \rho V^2 = p^* + \rho^* V^{*2}$$

Substituting $\rho = p/RT$, and factoring out pressures, yields

$$p \left[1 + \frac{V^2}{RT} \right] = p^* \left[1 + \frac{V^{*2}}{RT^*} \right]$$

Noting that $V^2/RT = k(V^2/kRT) = kM^2$, then we find

$$p[1 + kM^2] = p^*[1 + k]$$

and finally,

$$\frac{p}{p^*} = \frac{1+k}{1+kM^2} \qquad (13.30a)$$

From the ideal gas equation of state,

$$\frac{T}{T^*} = \frac{p}{p^*} \frac{\rho^*}{\rho}$$

From the continuity equation,

$$\frac{\rho^*}{\rho} = \frac{V}{V^*} = M \frac{c}{c^*} = M \sqrt{\frac{T}{T^*}}$$

** This section may be omitted without loss of continuity in the text material.

Then, substituting for ρ^*/ρ, we obtain

$$\frac{T}{T^*} = \frac{p}{p^*} M \sqrt{\frac{T}{T^*}}$$

Squaring and substituting from Eq. 13.30a gives

$$\frac{T}{T^*} = \left[\frac{p}{p^*} M\right]^2 = \left[M\left(\frac{1+k}{1+kM^2}\right)\right]^2 \tag{13.30b}$$

From continuity, using Eq. 13.30b,

$$\frac{\rho^*}{\rho} = \frac{V}{V^*} = \frac{M^2(1+k)}{1+kM^2} \tag{13.30c}$$

The dimensionless stagnation temperature, T_0/T_0^*, can be determined from

$$\frac{T_0}{T_0^*} = \frac{T_0}{T}\frac{T}{T^*}\frac{T^*}{T_0^*} = \left(1 + \frac{k-1}{2}M^2\right)\left[M\left(\frac{1+k}{1+kM^2}\right)\right]^2 \frac{1}{\left(1 + \frac{k-1}{2}\right)}$$

$$\frac{T_0}{T_0^*} = \frac{2(k+1)M^2\left(1 + \frac{k-1}{2}M^2\right)}{(1+kM^2)^2} \tag{13.30d}$$

Similarly,

$$\frac{p_0}{p_0^*} = \frac{p_0}{p}\frac{p}{p^*}\frac{p^*}{p_0^*} = \left(1 + \frac{k-1}{2}M^2\right)^{k/(k-1)}\left(\frac{1+k}{1+kM^2}\right)\frac{1}{\left(1 + \frac{k-1}{2}\right)^{k/(k-1)}}$$

$$\frac{p_0}{p_0^*} = \frac{1+k}{1+kM^2}\left[\left(\frac{2}{k+1}\right)\left(1 + \frac{k-1}{2}M^2\right)\right]^{k/(k-1)} \tag{13.30e}$$

The ratios in Eqs. 13.30a through 13.30e are functions of Mach number only; they are tabulated as functions of Mach number, for an ideal gas with $k = 1.4$, in Table E.3 of Appendix E.

EXAMPLE 13.10—Frictionless Flow in a Constant-Area Duct with Heat Addition: Table Solution

Air flows with negligible friction in a constant-area duct. At section ①, properties are $T_1 = 60$ C, $p_1 = 135$ kPa (abs), and $V_1 = 732$ m/sec. Heat is added between section ① and section ②, where $M = 1.2$. Determine the properties at section ②, the heat exchange per unit mass, the entropy change, and sketch the process on a Ts diagram. Use tables.

EXAMPLE PROBLEM 13.10

GIVEN: Frictionless flow of air as shown:

$T_1 = 333$ K $M_2 = 1.2$

$p_1 = 135$ kPa (abs)

$V_1 = 732$ m/sec

FIND: (a) Properties at section ②. (b) $\delta Q/dm$.
(c) $s_2 - s_1$. (d) Sketch on a Ts diagram.

SOLUTION:
To obtain property ratios from tables, we need both Mach numbers.

$$c_1 = \sqrt{kRT_1} = \left[1.4 \times \frac{287 \text{ N} \cdot \text{m}}{\text{kg} \cdot \text{K}} \times 333 \text{ K} \times \frac{\text{kg} \cdot \text{m}}{\text{N} \cdot \text{sec}^2} \right]^{1/2} = 366 \text{ m/sec}$$

$$M_1 = \frac{V_1}{c_1} - \frac{732 \text{ m}}{\text{sec}} \times \frac{\text{sec}}{366 \text{ m}} = 2.00$$

From Table E.3, Appendix E,

M	T_0/T_0^*	p_0/p_0^*	T/T^*	p/p^*	V/V^*
2.00	0.7934	1.503	0.5289	0.3636	1.455
1.20	0.9787	1.019	0.9119	0.7958	1.146

Using these data and recognizing that critical properties are constant, we obtain

$$\frac{T_2}{T_1} = \frac{T_2/T^*}{T_1/T^*} = \frac{0.9119}{0.5289} = 1.72; \qquad T_2 = 1.72 T_1 = (1.72)333 \text{ K} = 573 \text{ K} \qquad \underleftarrow{T_2}$$

$$\frac{p_2}{p_1} = \frac{p_2/p^*}{p_1/p^*} = \frac{0.7958}{0.3636} = 2.19; \qquad p_2 = 2.19 p_1 = (2.19)135 \text{ kPa}$$
$$p_2 = 296 \text{ kPa (abs)} \qquad \underleftarrow{p_2}$$

$$\frac{V_2}{V_1} = \frac{V_2/V^*}{V_1/V^*} = \frac{1.146}{1.455} = 0.788; \qquad V_2 = 0.788 V_1 = (0.788)732 \text{m/sec}$$
$$V_2 = 577 \text{ m/sec} \qquad \underleftarrow{V_2}$$

$$\rho_2 = \frac{p_2}{RT_2} = \frac{2.96 \times 10^5 \text{ N}}{\text{m}^2} \times \frac{\text{kg} \cdot \text{K}}{287 \text{ N} \cdot \text{m}} \times \frac{1}{573 \text{ K}} = 1.80 \text{ kg/m}^3 \qquad \underleftarrow{\rho_2}$$

The heat addition may be determined from the energy equation, which reduces to (see Example Problem 13.9)

$$\frac{\delta Q}{dm} = h_{0_2} - h_{0_1} = c_p(T_{0_2} - T_{0_1})$$

From the isentropic flow tables (Table E.1) at $M = 2.0$,

$$\frac{T}{T_0} = \frac{T_1}{T_{0_1}} = 0.5556; \qquad T_{0_1} = \frac{T_1}{0.5556} = \frac{333 \text{ K}}{0.5556} = 599 \text{ K}$$

and at $M = 1.2$

$$\frac{T}{T_0} = \frac{T_2}{T_{0_2}} = 0.7764; \qquad T_{0_2} = \frac{T_2}{0.7764} = \frac{573 \text{ K}}{0.7764} = 738 \text{ K} \qquad \underleftarrow{T_{0_2}}$$

Substituting gives

$$\frac{\delta Q}{dm} = c_p(T_{0_2} - T_{0_1}) = \frac{1.00}{\text{kg} \cdot \text{K}}(738 - 599) \text{ K} = 139 \text{ kJ/kg} \qquad \underleftarrow{\delta Q/dm}$$

The entropy change may be found from the $T\,ds$ equation, $T\,ds = dh - v\,dp$. For an ideal gas with constant specific heats,

$$s_2 - s_1 = c_p \ln \frac{T_2}{T_1} - R \ln \frac{p_2}{p_1}$$

$$s_2 - s_1 = \frac{1.00}{\text{kg} \cdot \text{K}} \frac{\text{kJ}}{\text{kg} \cdot \text{K}} \ln\left(\frac{573}{333}\right) - \frac{287}{\text{kg} \cdot \text{K}} \frac{\text{N} \cdot \text{m}}{\text{kg} \cdot \text{K}} \ln\left(\frac{2.96 \times 10^5}{1.35 \times 10^5}\right) \frac{\text{kJ}}{1000 \text{ N} \cdot \text{m}}$$

$$s_2 - s_1 = 0.317 \text{ kJ/kg} \cdot \text{K} \qquad\qquad\qquad s_2 - s_1$$

Finally, check the effect on p_0. From Table E.1, at $M = 2.0$,

$$\frac{p}{p_0} = \frac{p_1}{p_{0_1}} = 0.1278; \qquad p_{0_1} = \frac{p_1}{0.1278} = \frac{135 \text{ kPa}}{0.1278} = 1.06 \text{ MPa (abs)}$$

and at $M = 1.2$,

$$\frac{p}{p_0} = \frac{p_2}{p_{0_2}} = 0.4124; \qquad p_{0_2} = \frac{p_2}{0.4124} = \frac{196 \text{ kPa}}{0.4124} = 718 \text{ kPa (abs)} \qquad p_{0_2}$$

Thus $p_{0_2} < p_{0_1}$, as expected for a heating process.
The process follows the supersonic branch of a Rayleigh line:

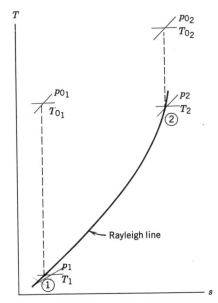

13-6 NORMAL SHOCKS

We have previously mentioned normal shocks in the section on nozzle flow. In practice, these irreversible discontinuities can occur in any supersonic flow field, in either internal flow or external flow.[2] Knowledge of property changes across shocks and of shock behavior is important in understanding the design of supersonic diffusers, e.g., for inlets on high performance aircraft, and supersonic wind tunnels. Accordingly, the purpose of this section is to analyze the normal shock process.

Before applying the basic equations to normal shocks, it is important to form a clear physical picture of the shock itself. Although it is physically impossible to have discontinuities in fluid properties, the normal shock is nearly discontinuous. The thickness of a shock is about 0.2 micrometers (10^{-5} in.), or roughly 4 times the mean free path of the gas molecules [1]. Large changes in pressure, temperature, and other properties occur across this small distance. Local fluid decelerations reach

---

[2] The NCFMF film, *Channel Flow of a Compressible Fluid*, D. Coles, principal, shows several examples of shock formation in internal flow.

tens of millions of *g*s! These considerations justify treating the normal shock as an abrupt discontinuity; we are interested in changes occurring across the shock rather than in the details of its structure.

13-6.1 Basic Equations

To begin our analysis, apply the basic equations to the thin control volume shown in Fig. 13.17, where, for generality, we have depicted a shock standing in a passage of arbitrary shape.

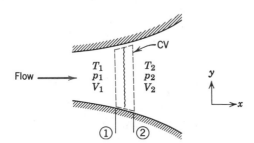

ig. 13.17 Control volume used for analysis of normal shock.

a. Continuity Equation

Basic equation:

$$= 0(1)$$

$$0 = \frac{\partial}{\partial t} \int_{CV} \rho \, d\Psi + \int_{CS} \rho \vec{V} \cdot d\vec{A} \tag{4.13}$$

Assumptions: (1) Steady flow
(2) Uniform flow at each section
(3) $A_1 = A_2 = A$, because the shock is so extraordinarily thin

Then

$$0 = \{-|\rho_1 V_1 A|\} + \{|\rho_2 V_2 A|\}$$

Writing the result in terms of scalar magnitudes, we obtain

$$\rho_1 V_1 = \rho_2 V_2 = G = \frac{\dot{m}}{A} \tag{13.31a}$$

b. Momentum Equation

Basic equation:

$$= 0(5) \ = 0(1)$$

$$F_{S_x} + F_{B_x} = \frac{\partial}{\partial t} \int_{CV} V_x \rho \, d\Psi + \int_{CS} V_x \rho \vec{V} \cdot d\vec{A} \tag{4.19a}$$

Assumptions: (4) Negligible friction at duct walls because the shock is so thin
(5) $F_{B_x} = 0$

Under these conditions,

$$F_{S_x} = p_1 A - p_2 A = V_1 \{-|\rho_1 V_1 A|\} + V_2 \{|\rho_2 V_2 A|\}$$

Using scalar magnitudes and dropping absolute value signs, we obtain

$$p_1 A - p_2 A = \dot{m} V_2 - \dot{m} V_1 \tag{13.31b}$$

or

$$p_1 + \rho_1 V_1^2 = p_2 + \rho_2 V_2^2$$

c. First Law of Thermodynamics

Basic equation:

$$= 0(6) \quad = 0(7) \quad = 0(8) \quad = 0(8) \quad = 0(1)$$

$$\cancel{\dot{Q}} - \cancel{\dot{W}_s} - \cancel{\dot{W}_{shear}} - \cancel{\dot{W}_{other}} = \cancel{\frac{\partial}{\partial t}} \int_{CV} e\rho \, d\Psi + \int_{CS} (e + pv)\rho \vec{V} \cdot d\vec{A} \qquad (4.57)$$

where

$$\approx 0(9)$$

$$e = u + \frac{V^2}{2} + \cancel{gz}$$

Assumptions:

(6) $\dot{Q} = 0$ (adiabatic flow)
(7) $\dot{W}_s = 0$
(8) $\dot{W}_{shear} = \dot{W}_{other} = 0$
(9) Effects of gravity are negligible

Then

$$0 = \left(u_1 + p_1 v_1 + \frac{V_1^2}{2}\right)\{-|\rho_1 V_1 A|\} + \left(u_2 + p_2 v_2 + \frac{V_2^2}{2}\right)\{|\rho_2 V_2 A|\}$$

However, from continuity, the mass flow rate terms in braces are equal. We may also substitute $h = u + pv$, to obtain

$$h_1 + \frac{V_1^2}{2} = h_2 + \frac{V_2^2}{2} \qquad (13.31c)$$

or, in terms of stagnation enthalpy,

$$h_{0_1} = h_{0_2}$$

Physically, we should expect the total energy of the flow to remain constant, since there is no energy addition.

d. Second Law of Thermodynamics

Basic equation:

$$= 0(6) \quad = 0(1)$$

$$\int_{CS} \frac{1}{T} \cancel{\frac{\dot{Q}}{A}} dA \le \cancel{\frac{\partial}{\partial t}} \int_{CV} s\rho \, d\Psi + \int_{CS} s\rho \vec{V} \cdot d\vec{A} \qquad (4.59)$$

Then

$$0 \le s_1\{-|\rho_1 V_1 A|\} + s_2\{|\rho_2 V_2 A|\}$$

Flow through the normal shock is irreversible because of the almost discontinuous property changes across the shock. Consequently, the inequality in the above equation holds. The control volume form of the second law then tells us that $s_2 > s_1$.

This fact is of little help in calculating the actual entropy change across the shock. To calculate the entropy change, we rely on the $T\,ds$ equations. Since

$$T\,ds = dh - v\,dp$$

then, for an ideal gas, we can write

$$ds = c_p \frac{dT}{T} - R \frac{dp}{p}$$

For constant specific heats, this equation can be integrated to give

$$s_2 - s_1 = c_p \ln \frac{T_2}{T_1} - R \ln \frac{p_2}{p_1} \qquad (13.31\text{d})$$

e. Equation of State

For an ideal gas, the equation of state is

$$p = \rho RT \qquad (13.31\text{e})$$

Equations 13.31a through 13.31e are the governing equations for the flow of an ideal gas through a normal shock. If all properties at state ① (immediately upstream of the shock) are known, then we have six unknowns (T_2, p_2, ρ_2, V_2, h_2, s_2) in these five equations. However, we have the known relationship between h and T for an ideal gas, $dh = c_p dT$. For an ideal gas with constant specific heats,

$$\Delta h = h_2 - h_1 = c_p \Delta T = c_p(T_2 - T_1) \qquad (13.31\text{f})$$

We thus have the situation of six equations and six unknowns.

Then, if all conditions at state ① (immediately ahead of the shock) are known, how many possible states ② (immediately behind the shock) are there? The mathematics of the situation (six equations and six unknowns) shows there is a unique state ② for a given state ①.

We can obtain a physical picture of the flow through a normal shock by using some of the notions developed in analyzing the Fanno line and Rayleigh line flows. For convenience let us first rewrite the governing equations for a normal shock:

$$\rho_1 V_1 = \rho_2 V_2 = G = \frac{\dot{m}}{A} \qquad (13.31\text{a})$$

$$p_1 A - p_2 A = \dot{m} V_2 - \dot{m} V_1 \qquad (13.31\text{b})$$

$$h_1 + \frac{V_1^2}{2} = h_2 + \frac{V_2^2}{2} \qquad (13.31\text{c})$$

$$s_2 - s_1 = c_p \ln \frac{T_2}{T_1} - R \ln \frac{p_2}{p_1} \qquad (13.31\text{d})$$

$$p = \rho RT \qquad (13.31\text{e})$$

$$h_2 - h_1 = c_p(T_2 - T_1) \qquad (13.31\text{f})$$

Flow through a normal shock must satisfy Eqs. 13.31a through 13.31f. Since all conditions at state ① are known, we can locate state ① on a Ts diagram. If we were to draw a Fanno line curve through state ①, we would have a locus of mathematical states that satisfy Eqs. 13.31a and 13.31c through 13.31f. (The Fanno line curve does not satisfy Eq. 13.31b.) Drawing a Rayleigh line curve through state ① gives a locus of mathematical states that satisfy Eqs. 13.31a, 13.31b, and 13.31d through 13.31f. (The Rayleigh line curve does not satisfy Eq. 13.31c.) These curves are shown in Fig. 13.18.

The normal shock must satisfy all six of Eqs. 13.31a through 13.31f. Consequently, for a given state ①, the end state (state ②) of the normal shock must lie

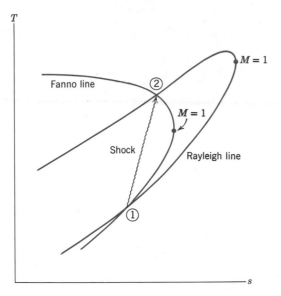

Fig. 13.18 Intersection of Fanno line and Rayleigh line as a solution of the normal shock equations.

on both the Fanno line and the Rayleigh line passing through state ①. Hence, the intersection of the two lines at state ② represents conditions downstream from the shock, corresponding to upstream conditions at state ①. In Fig. 13.18, flow through the shock has been indicated as occurring from state ① to state ②. This is the only possible direction of the shock process as dictated by the second law ($s_2 > s_1$).

From Fig. 13.18 we note also that flow through a normal shock involves a change from supersonic to subsonic speeds. Normal shocks can occur only in flow that is initially supersonic.

As an aid in summarizing the effects of a normal shock on the flow properties, a schematic of the normal shock process is illustrated on the Ts plane of Fig. 13.19. This figure, together with the governing basic equations, is the basis for Table 13.3. You should follow through the logic indicated in the table.

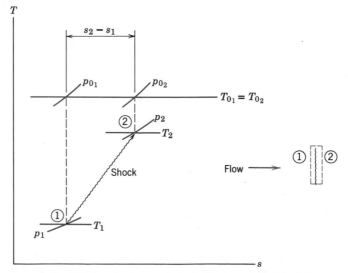

Fig. 13.19 Schematic of normal shock process on the Ts plane.

Table 13.3 Summary of Property Changes across a Normal Shock

Property	Effect	Obtained from:
Stagnation temperature, T_0	Constant	Energy equation
Entropy, s	Increase	Second law
Stagnation pressure, p_0	Decrease	Ts diagram
Temperature, T	Increase	Ts diagram
Velocity, V	Decrease	Energy equation, and effect on T
Density, ρ	Increase	Continuity equation, and effect on V
Pressure, p	Increase	Momentum equation, and effect on V
Mach number, M	Decrease	$M = V/c$, and effect on V and T

It is of interest to note the parallel between normal shocks (Table 13.3) and supersonic flow with friction (Table 13.1). Both represent irreversible processes in supersonic flow, and all properties change in the same direction.

In theory, the solution of six equations in six unknowns poses no difficulty, but in practice the algebra becomes involved. It makes sense to recast the equations to tabulate property ratios across the shock; this we shall do in the next section. To illustrate the direct application of the control volume equations to a normal shock, consider Example Problem 13.11, in which the problem is overspecified (one property downstream from the shock is given).

EXAMPLE 13.11—Normal Shock in a Duct

A normal shock stands in a duct. The fluid is air, which may be considered an ideal gas. Properties upstream from the shock are $T_1 = 5$ C, $p_1 = 65.0$ kPa (abs), and $V_1 = 668$ m/sec. The temperature at section ②, downstream from the shock, is $T_2 = 469$ K. Determine properties at section ② and compare them with upstream properties. Sketch the process on a Ts diagram.

EXAMPLE PROBLEM 13.11

GIVEN: Normal shock in a duct as shown:

$T_1 = 5$ C $T_2 = 469$ K

$p_1 = 65.0$ kPa (abs)

$V_1 = 668$ m/sec

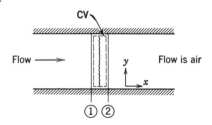

FIND: (a) Properties at section ②.
　　　　(b) Ts diagram.

SOLUTION:
First compute the remaining properties at section ①. For an ideal gas,

$$\rho_1 = \frac{p_1}{RT_1} = \frac{6.5 \times 10^4 \text{ N}}{\text{m}^2} \times \frac{\text{kg} \cdot \text{K}}{287 \text{ N} \cdot \text{m}} \times \frac{1}{278 \text{ K}} = 0.815 \text{ kg/m}^3$$

$$c_1 = \sqrt{kRT_1} = \left[1.4 \times \frac{287 \text{ N} \cdot \text{m}}{\text{kg} \cdot \text{K}} \times 278 \text{ K} \times \frac{\text{kg} \cdot \text{m}}{\text{N} \cdot \text{sec}^2} \right]^{1/2} = 334 \text{ m/sec}$$

Then $\quad M_1 = \dfrac{V_1}{c_1} = \dfrac{668}{334} = 2.00,\ \text{and}$

$$T_{0_1} = T_1\left(1 + \frac{k-1}{2}M_1^2\right) = 278\ \text{K}\left[1 + 0.2(2.0)^2\right] = 500\ \text{K}$$

$$p_{0_1} = p_1\left(1 + \frac{k-1}{2}M_1^2\right)^{k/(k-1)} = 65.0\ \text{kPa}[1 + 0.2(2.0)^2]^{3.5} = 509\ \text{kPa (abs)}$$

V_2 may be evaluated by applying the energy equation to the control volume shown.

Basic equation:

$$\overset{= 0(1)}{\cancel{\dot Q}} - \overset{= 0(2)}{\cancel{\dot W_s}} - \overset{= 0(3)}{\cancel{\dot W_{\text{shear}}}} - \overset{= 0(3)}{\cancel{\dot W_{\text{other}}}} = \overset{= 0(4)}{\cancel{\frac{\partial}{\partial t}\int_{CV} e\rho\, d\forall}} + \int_{CS}(e + pv)\rho\vec V\cdot d\vec A \qquad (4.57)$$

where

$$e = u + \frac{V^2}{2} + \overset{\approx 0(6)}{\cancel{gz}}$$

Assumptions: (1) $\dot Q = 0$ (5) Uniform flow at each section
                      (2) $\dot W_s = 0$ (6) Neglect gravity term
                      (3) $\dot W_{\text{shear}} = \dot W_{\text{other}} = 0$ (7) $A_1 = A_2 = A$
                      (4) Steady flow

Then

$$0 = \left(u_1 + p_1 v_1 + \frac{V_1^2}{2}\right)\{-|\rho_1 V_1 A|\} + \left(u_2 + p_2 v_2 + \frac{V_2^2}{2}\right)\{|\rho_2 V_2 A|\}$$

$$0 = \dot m\left(u_2 + p_2 v_2 + \frac{V_2^2}{2} - u_1 - p_1 v_1 - \frac{V_1^2}{2}\right)$$

or

$$h_1 + \frac{V_1^2}{2} = h_2 + \frac{V_2^2}{2}$$

Solving for V_2 gives

$$V_2 = [V_1^2 + 2(h_1 - h_2)]^{1/2} = [V_1^2 + 2c_p(T_1 - T_2)]^{1/2}$$

$$= \left[(668)^2\,\frac{\text{m}^2}{\text{sec}^2} + 2 \times 1000\,\frac{\text{N}\cdot\text{m}}{\text{kg}\cdot\text{K}} \times (278 - 469)\ \text{K} \times \frac{\text{kg}\cdot\text{m}}{\text{N}\cdot\text{sec}^2}\right]^{1/2}$$

$$V_2 = 253\ \text{m/sec} \qquad\qquad\qquad\qquad\qquad\qquad\qquad\qquad\qquad\qquad V_2$$
←

Continuity reduces to $G = \rho_1 V_1 = \rho_2 V_2$, so

$$\rho_2 = \rho_1\frac{V_1}{V_2} = 0.815\frac{\text{kg}}{\text{m}^3}\left(\frac{668}{253}\right) = 2.15\ \text{kg/m}^3 \qquad\qquad\qquad\qquad \rho_2$$
←

The pressure can be obtained two ways:
(1) From the ideal gas equation of state

$$p_2 = \rho_2 R T_2 = \frac{2.15\ \text{kg}}{\text{m}^3} \times \frac{287\ \text{N}\cdot\text{m}}{\text{kg}\cdot\text{K}} \times 469\ \text{K} = 289\ \text{kPa (abs)} \qquad\qquad p_2$$
←

(2)   From the momentum equation

Basic equation:

$$F_{S_x} + \cancel{F_{B_x}}^{=0(8)} = \cancelto{=0(4)}{\frac{\partial}{\partial t} \int_{CV} V_x \rho \, d\Psi} + \int_{CS} V_x \rho \vec{V} \cdot d\vec{A} \tag{4.19a}$$

Assumption: (8) $F_{B_x} = 0$

Then

$$p_1 A - p_2 A = V_1\{-|\rho_1 V_1 A|\} + V_2\{|\rho_2 V_2 A|\} = \dot{m}(V_2 - V_1)$$

or

$$p_1 - p_2 = \rho_1 V_1 (V_2 - V_1)$$

Solving gives

$$p_2 = p_1 - \rho_1 V_1 (V_2 - V_1)$$

$$= \frac{6.5 \times 10^4 \ \text{N}}{\text{m}^2} - 0.815 \frac{\text{kg}}{\text{m}^3} \times 668 \frac{\text{m}}{\text{sec}} \times (253 - 668) \frac{\text{m}}{\text{sec}} \times \frac{\text{N} \cdot \text{sec}^2}{\text{kg} \cdot \text{m}}$$

$$p_2 = 291 \ \text{kPa (abs)}$$

(The two calculated pressures are in close, but not exact, agreement due to roundoff.) For adiabatic flow, $T_0 = $ constant. Thus

$$T_{0_2} = T_{0_1} = 500 \ \text{K} \hspace{4cm} T_{0_2}$$

The local isentropic stagnation pressure at section ② is

$$p_{0_2} = p_2 \left(\frac{T_{0_2}}{T_2}\right)^{k/(k-1)} = 289 \ \text{kPa} \left(\frac{500}{469}\right)^{3.5} = 362 \ \text{kPa (abs)} \hspace{2cm} p_{0_2}$$

Comparing conditions downstream and upstream of the shock, we see that $T_{0_2} = T_{0_1}$, $p_{0_2} < p_{0_1}$, $T_2 > T_1$, $p_2 > p_1$, and $V_2 < V_1$, in accord with Table 13.3.

The entropy change may be computed from the $T\,ds$ equation

$$T\,ds = dh - v\,dp$$

For an ideal gas

$$ds = c_p \frac{dT}{T} - R \frac{dp}{p}$$

Integrating, for constant specific heats, yields

$$s_2 - s_1 = c_p \ln \frac{T_2}{T_1} - R \ln \frac{p_2}{p_1}$$

Since $s_{0_2} - s_{0_1} = s_2 - s_1$, the entropy change is easiest to evaluate at stagnation conditions because $T_{0_2} = T_{0_1}$. At stagnation conditions we have

$$s_2 - s_1 = s_{0_2} - s_{0_1} = c_p \cancelto{=0}{\ln \frac{T_{0_2}}{T_{0_1}}} - R \ln \frac{p_{0_2}}{p_{0_1}} = -\frac{287 \ \text{N} \cdot \text{m}}{\text{kg} \cdot \text{K}} \ln \left(\frac{3.62 \times 10^5}{5.09 \times 10^5}\right)$$

$$s_2 - s_1 = 0.0978 \ \text{kJ/kg} \cdot \text{K} \hspace{4cm} s_2 - s_1$$

Finally, the Ts diagram may be sketched:

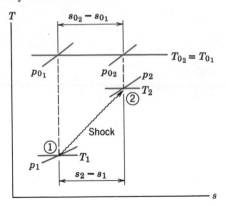

$$\{ \text{Note once again that } s_{0_2} - s_{0_1} = s_2 - s_1, \text{ since } s_{0_1} = s_1 \text{ and } s_{0_2} = s_2. \}$$

**13-6.2   Tables for Computation of Normal Shocks in an Ideal Gas

The basic equations for the flow through a normal shock indicate that for given conditions ahead of the shock there is a unique downstream state. Therefore, it is possible to develop expressions for property ratios across the shock in terms of M_1 ahead of the shock; these results can be tabulated as functions of M_1.

To obtain the results, we proceed in three steps. First, we obtain property ratios (e.g., T_2/T_1 and p_2/p_1) in terms of M_1 and M_2. Then we develop a relation between M_1 and M_2. Finally, we use this relation to obtain expressions for property ratios in terms of upstream Mach number, M_1.

The temperature ratio can be expressed as

$$\frac{T_2}{T_1} = \frac{T_2}{T_{0_2}} \frac{T_{0_2}}{T_{0_1}} \frac{T_{0_1}}{T_1}$$

Since stagnation temperature is constant across the shock, we have

$$\frac{T_2}{T_1} = \frac{1 + \dfrac{k-1}{2}M_1^2}{1 + \dfrac{k-1}{2}M_2^2} \tag{13.32a}$$

A velocity ratio may be obtained by using

$$\frac{V_2}{V_1} = \frac{M_2 c_2}{M_1 c_1} = \frac{M_2}{M_1} \frac{\sqrt{kRT_2}}{\sqrt{kRT_1}} = \frac{M_2}{M_1} \sqrt{\frac{T_2}{T_1}}$$

or

$$\frac{V_2}{V_1} = \frac{M_2}{M_1} \left[\frac{1 + \dfrac{k-1}{2}M_1^2}{1 + \dfrac{k-1}{2}M_2^2} \right]^{1/2} \tag{13.32b}$$

A ratio of densities may be obtained from the continuity equation

$$\rho_1 V_1 = \rho_2 V_2 \tag{13.31a}$$

** This section may be omitted without loss of continuity in the text material.

Substituting from Eq. 13.32b gives

$$\frac{\rho_2}{\rho_1} = \frac{V_1}{V_2} = \frac{M_1}{M_2} \left[\frac{1 + \frac{k-1}{2}M_2^2}{1 + \frac{k-1}{2}M_1^2} \right]^{1/2} \tag{13.32c}$$

Finally, we can obtain a pressure ratio from the momentum equation

$$p_1 A - p_2 A = \dot{m}V_2 - \dot{m}V_1 \tag{13.31b}$$

or

$$p_1 + \rho_1 V_1^2 = p_2 + \rho_2 V_2^2$$

Substituting $\rho = p/RT$, and factoring out pressures, gives

$$p_1 \left[1 + \frac{V_1^2}{RT_1} \right] = p_2 \left[1 + \frac{V_2^2}{RT_2} \right]$$

Since

$$\frac{V^2}{RT} = k\frac{V^2}{kRT} = kM^2$$

then

$$p_1[1 + kM_1^2] = p_2[1 + kM_2^2]$$

Finally,

$$\frac{p_2}{p_1} = \frac{1 + kM_1^2}{1 + kM_2^2} \tag{13.32d}$$

To solve for M_2 in terms of M_1, we must obtain another expression for one of the property ratios given by Eqs. 13.32a through 13.32d.

From the ideal gas equation of state, the temperature ratio may be written

$$\frac{T_2}{T_1} = \frac{p_2/\rho_2 R}{p_1/\rho_1 R} = \frac{p_2}{p_1}\frac{\rho_1}{\rho_2}$$

Substituting from Eqs. 13.32c and 13.32d yields

$$\frac{T_2}{T_1} = \left[\frac{1 + kM_1^2}{1 + kM_2^2} \right] \frac{M_2}{M_1} \left[\frac{1 + \frac{k-1}{2}M_1^2}{1 + \frac{k-1}{2}M_2^2} \right]^{1/2} \tag{13.33}$$

Equations 13.32a and 13.33 are two equations for T_2/T_1. We can combine them and solve for M_2 in terms of M_1. Combining and canceling gives

$$\left[\frac{1 + \frac{k-1}{2}M_1^2}{1 + \frac{k-1}{2}M_2^2} \right]^{1/2} = \frac{M_2}{M_1} \left[\frac{1 + kM_1^2}{1 + kM_2^2} \right]$$

Squaring, we obtain

$$\frac{1 + \frac{k-1}{2}M_1^2}{1 + \frac{k-1}{2}M_2^2} = \frac{M_2^2}{M_1^2} \left[\frac{1 + 2kM_1^2 + k^2M_1^4}{1 + 2kM_2^2 + k^2M_2^4} \right]$$

which may be solved explicitly for M_2^2. Two solutions are obtained:

$$M_2^2 = M_1^2 \qquad (13.34a)$$

and

$$M_2^2 = \frac{M_1^2 + \dfrac{2}{k-1}}{\dfrac{2k}{k-1}M_1^2 - 1} \qquad (13.34b)$$

Obviously, the first of these is trivial. The second expresses the unique dependence of M_2 on M_1. This relation is tabulated for an ideal gas with $k = 1.4$ in Appendix E, Table E.4.

Now, having a relationship between M_2 and M_1, we can solve for property ratios across a shock. Knowing M_1, M_2 is obtained from Eq. 13.34b; the property ratios subsequently can be determined from Eqs. 13.32a through 13.32d. These property ratios are tabulated as functions of M_1, for an ideal gas with $k = 1.4$, in Table E.4 of Appendix E.

Since the stagnation temperature remains constant, the stagnation temperature ratio across the shock is unity. The ratio of stagnation pressures is evaluated as

$$\frac{p_{0_2}}{p_{0_1}} = \frac{p_{0_2}}{p_2}\frac{p_2}{p_1}\frac{p_1}{p_{0_1}} = \frac{p_2}{p_1}\left[\frac{1 + \dfrac{k-1}{2}M_2^2}{1 + \dfrac{k-1}{2}M_1^2}\right]^{k/(k-1)} \qquad (13.35)$$

Combining Eqs. 13.32d and 13.34b, we obtain (after considerable algebra)

$$\frac{p_2}{p_1} = \frac{1 + kM_1^2}{1 + kM_2^2} = \frac{2k}{k+1}M_1^2 - \frac{k-1}{k+1} \qquad (13.36)$$

Using Eqs. 13.34b and 13.36, we find that Eq. 13.35 becomes

$$\frac{p_{0_2}}{p_{0_1}} = \frac{\left[\dfrac{\dfrac{k+1}{2}M_1^2}{1 + \dfrac{k-1}{2}M_1^2}\right]^{k/(k-1)}}{\left[\dfrac{2k}{k+1}M_1^2 - \dfrac{k-1}{k+1}\right]^{1/(k-1)}} \qquad (13.37)$$

Stagnation pressure ratio as a function of upstream Mach number is tabulated in Table E.4 of Appendix E.

Use of the tables in the solution of a problem involving a normal shock is illustrated in Example Problem 13.12.

EXAMPLE 13.12—Normal Shock in a Duct: Table Solution

A normal shock stands in a duct. The fluid is air, which can be considered an ideal gas. Properties upstream from the shock are $T_1 = 5$ C, $p_1 = 65.0$ kPa (abs), and $V_1 = 668$ m/sec. Determine properties downstream and $s_2 - s_1$. Include a Ts plot.

(These upstream conditions are the same as in Example Problem 13.11. However, with the tables available, we do not need to know any property downstream to complete a solution.)

EXAMPLE PROBLEM 13.12

GIVEN: Normal shock in a duct as shown:

$T_1 = 278$ K

$p_1 = 65.0$ kPa (abs)

$V_1 = 668$ m/sec

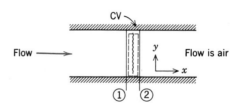

FIND: (a) Properties at section ②.
(b) $s_2 - s_1$.
(c) Ts plot.

SOLUTION:
To use the shock tables, we need to know M_1. For an ideal gas,

$$c_1 = \sqrt{kRT_1} = \left[1.4 \times \frac{287 \text{ N} \cdot \text{m}}{\text{kg} \cdot \text{K}} \times 278 \text{ K} \times \frac{\text{kg} \cdot \text{m}}{\text{N} \cdot \text{sec}^2} \right]^{1/2} = 334 \text{ m/sec}$$

Then

$$M_1 = \frac{V_1}{c_1} = \frac{668}{334} = 2.0$$

Normal shock property ratios are given in Table E.4, Appendix E. At $M_1 = 2.0$,

M_1	M_2	p_{0_2}/p_{0_1}	T_2/T_1	p_2/p_1	ρ_2/ρ_1
2.00	0.5774	0.7209	1.688	4.500	2.667

From these data

$T_2 = 1.688T_1 = (1.688)278 \text{ K} = 469 \text{ K}$ ←——————————————— T_2

$p_2 = 4.500\,p_1 = (4.500)65.0 \text{ kPa} = 293 \text{ kPa (abs)}$ ←——————————————— p_2

For an ideal gas,

$$\rho_2 = \frac{p_2}{RT_2} = \frac{2.93 \times 10^5 \text{ N}}{\text{m}^2} \times \frac{\text{kg} \cdot \text{K}}{287 \text{ N} \cdot \text{m}} \times \frac{1}{469 \text{ K}} = 2.18 \text{ kg/m}^3$$ ←————— ρ_2

and

$$V_2 = M_2 c_2 = M_2 \sqrt{kRT_2} = (0.5774) \left[1.4 \times \frac{287 \text{ N} \cdot \text{m}}{\text{kg} \cdot \text{K}} \times 469 \text{ K} \times \frac{\text{kg} \cdot \text{m}}{\text{N} \cdot \text{sec}^2} \right]^{1/2}$$

$V_2 = 251$ m/sec ←——————————————— V_2

Local isentropic stagnation properties at section ① may be evaluated using tables for isentropic flow; from Table E.1, Appendix E, at $M = 2.0$,

$$\frac{T}{T_0} = \frac{T_1}{T_{0_1}} = 0.5556; \qquad T_{0_1} = \frac{T_1}{0.5556} = \frac{278 \text{ K}}{0.5556} = 500 \text{ K}$$

$$\frac{p}{p_0} = \frac{p_1}{p_{0_1}} = 0.1278; \qquad p_{0_1} = \frac{p_1}{0.1278} = \frac{65.0 \text{ kPa}}{0.1278} = 509 \text{ kPa (abs)}$$

Stagnation temperature is constant in adiabatic flow. Thus

$$T_{0_2} = T_{0_1} = 500 \text{ K}$$ ←——————————————— T_{0_2}

Using the property ratios for a normal shock, we obtain

$$p_{0_2} = p_{0_1} \frac{p_{0_2}}{p_{0_1}} = 509 \text{ kPa}\,(0.7209) = 367 \text{ kPa (abs)} \qquad\qquad p_{0_2}$$

The entropy change across the shock may be found from the $T\,ds$ equation

$$T\,ds = dh - v\,dp$$

For an ideal gas,

$$ds = c_p \frac{dT}{T} - R\frac{dp}{p}$$

Integrating, for constant specific heats, gives

$$s_2 - s_1 = c_p \ln \frac{T_2}{T_1} - R \ln \frac{p_2}{p_1}$$

But $s_{0_2} - s_{0_1} = s_2 - s_1$, so

$$s_{0_2} - s_{0_1} = s_2 - s_1 = c_p \ln \frac{T_{0_2}}{T_{0_1}}^{=0} - R \ln \frac{p_{0_2}}{p_{0_1}} = -\frac{0.287}{} \frac{\text{kJ}}{\text{kg}\cdot\text{K}} \ln(0.7209)$$

$$s_2 - s_1 = 0.0939 \text{ kJ/kg}\cdot\text{K} \qquad\qquad s_2 - s_1$$

The Ts diagram is

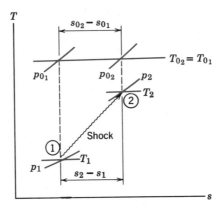

Comparing the solutions presented in Example Problems 13.11 and 13.12, we see that using the tables simplifies computations appreciably. In addition, the consistency of calculated values is improved as a result of using four significant figures in all calculations. You must be careful when using the tables to check your work as you proceed. Examine each calculated result to be sure both its trend and magnitude are reasonable. This process is simplified and made almost automatic when you draw a Ts diagram for each problem.

13-7 SUPERSONIC CHANNEL FLOW WITH SHOCKS

Supersonic flow is a necessary condition for a normal shock to occur. The possibility of a normal shock must be considered in any supersonic flow. Sometimes a shock *must* occur to match a downstream pressure condition; it is desirable to determine if a shock will occur and the shock location when it does occur.

In Section 13-6.1 we showed that stagnation pressure decreases dramatically across a shock: the stronger the shock, the larger the decrease in stagnation pressure. It is necessary to control shock position to obtain acceptable performance from a supersonic diffuser or supersonic wind tunnel.

The purpose of this section is to consider supersonic channel flows with shocks. Flows with area change, friction, and heat addition are considered. Operation of supersonic diffusers and supersonic wind tunnels is described.

13-7.1 Flow in a Converging-Diverging Nozzle

Since we have considered normal shocks, we now can complete our discussion of flow in a converging diverging nozzle operating under varying back pressures, begun in Section 13-3.5. The pressure distribution through the nozzle for different back pressures is shown in Fig. 13.20.

Four flow regimes are possible. In Regime I the flow is subsonic throughout. The flow rate increases with decreasing back pressure. At condition (*iii*), which forms the dividing line between Regimes I and II, flow at the throat is sonic, $M_t = 1$.

As the back pressure is lowered below condition (*iii*), a normal shock appears downstream from the throat. There is a pressure rise across the shock. Since the flow is subsonic ($M < 1$) behind the shock, the flow decelerates, with an accompanying increase in pressure, through the diverging channel. As the back pressure is lowered further, the shock moves downstream until it appears at the exit plane (condition *vii*). In Regime II, as in Regime I, the exit flow is subsonic, and consequently $p_e = p_b$. Since flow properties at the throat are constant for all conditions in Regime II, the flow rate in Regime II does not vary with back pressure.

In Regime III, as exemplified by condition (*viii*), the back pressure is higher than the exit pressure, but not sufficiently high to sustain a normal shock in the exit plane. The flow adjusts to the back pressure through a series of oblique compression shocks outside the nozzle; these oblique shocks cannot be treated by one-dimensional theory.

As previously noted in Section 13-3.5, condition (*iv*) represents the design condition. In Regime IV the flow adjusts to the lower back pressure through a series of oblique expansion waves outside the nozzle; these oblique expansion waves cannot be treated by one-dimensional theory.

The *Ts* diagram for converging-diverging nozzle flow with a normal shock is shown in Fig. 13.21; state ① is located immediately upstream from the shock and state ② is immediately downstream. The entropy increase across the shock moves the subsonic

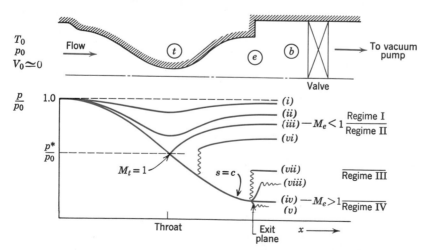

Fig. 13.20 Pressure distributions for flow in a converging-diverging nozzle as a function of back pressure.

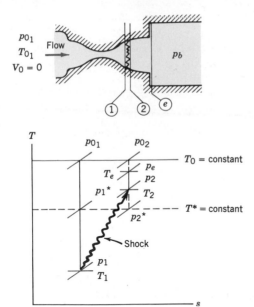

Fig. 13.21 Schematic Ts diagram for flow in a converging-diverging nozzle with a normal shock.

downstream flow to a new isentropic line. The critical temperature is constant, so p_2^* is lower than p_1^*. Since $\rho^* = p^*/RT^*$, the critical density downstream also is reduced. To carry the same mass flow rate, the downstream flow must have a larger critical area. From continuity (and the equation of state), the critical area ratio is the inverse of the critical pressure ratio, i.e., across a shock, $p^*A^* = $ constant.

If the Mach number (or position) of the normal shock in the nozzle is known, the exit plane pressure can be calculated directly. In the more realistic situation, the exit plane pressure is specified, and the position and strength of the shock are unknown. This problem can be solved by iteration [1]. The subsonic flow downstream must leave the nozzle at back pressure, so $p_b = p_e$. Then

$$\frac{p_b}{p_{0_1}} = \frac{p_e}{p_{0_1}} = \frac{p_e}{p_{0_2}} \frac{p_{0_2}}{p_{0_1}} = \frac{p_e}{p_{0_2}} \frac{A_1^*}{A_2^*} = \frac{p_e}{p_{0_2}} \frac{A_t}{A_e} \frac{A_e}{A_2^*} \qquad (13.38)$$

The left side of Eq. 13.38 is given. The right side is a function of exit Mach number only: p_e/p_{0_2} is obtained from the local isentropic stagnation relation, Eq. 12.17a, A_t/A_e is known from the geometry of the nozzle, and A_e/A_2^* is obtained from Eq. 13.6. Iteration may be needed to solve for M_e. A personal computer program to calculate and plot the distributions of pressure and Mach number in a C-D nozzle with shock is given in [2].

**13-7.2 Supersonic Diffuser

Analysis of the effects of area change in isentropic flow (Section 13-2) showed that a converging channel reduces the speed of a supersonic stream; a converging channel is a *supersonic diffuser*. Because flow speed decreases, pressure rises in the flow direction, creating an adverse pressure gradient. Isentropic flow is not a completely accurate model for flow with an adverse pressure gradient,[3] but the isentropic

---

** This section may be omitted without loss of continuity in the text material.

[3] Boundary layers develop rapidly in adverse pressure gradients, so viscous effects may be important or even dominant. In the presence of thick boundary layers, supersonic flows in diffusers may form complicated systems of oblique and normal shocks.

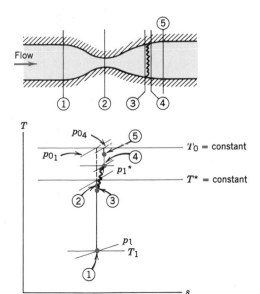

Fig. 13.22 Schematic *Ts* diagram for flow in a supersonic diffuser with a normal shock.

flow model with a normal shock may be used to demonstrate the basic features of supersonic diffusion.

For isentropic flow, a shock cannot stand in a stable position in a converging passage; a shock may stand stably only in a diverging passage. Real flow near $M = 1$ is unstable, so it is not possible to reduce a supersonic flow exactly to sonic speed. The minimum Mach number that can be reached at a throat is 1.2 to 1.3.

Thus in real supersonic diffusers, flow is decelerated to $M \approx 1.3$ in a converging passage. Downstream from the throat section of minimum area, the flow is allowed to accelerate to $M \approx 1.4$, where a normal shock takes place. At this Mach number, the stagnation pressure loss is only about 4.2 percent (Table E.4, Appendix E). This small loss is an acceptable compromise in exchange for flow stability.

Figure 13.22 shows the idealized process of supersonic diffusion, in which flow is isentropic except across a normal shock. The slight reduction in stagnation pressure all takes place across the shock.

In the actual flow, additional losses in stagnation pressure occur during the supersonic and subsonic diffusion processes before and after the shock. Experimental data must be used to predict the actual losses in supersonic and subsonic diffusers [3, 4].

Supersonic diffusion also is important for high-speed aircraft, where a supersonic external freestream flow must be decelerated efficiently to subsonic speed. Some diffusion can occur outside the inlet by means of a weak oblique shock system [5]. Variable geometry may be needed to accomplish efficient supersonic diffusion within the inlet as the flight Mach number varies. Multi-dimensional compressible flows, although beyond the scope of this text, are treated in detail elsewhere [6, 7].

**13-7.3 Supersonic Wind Tunnel Operation

To build an efficient supersonic wind tunnel, it is necessary to understand shock behavior and to control shock location. The basic physical phenomena are described by Coles in the NCFMF film, *Channel Flow of a Compressible Fluid* [8]. In addition to *choking* — sonic flow at a throat, with upstream flow independent of downstream conditions — Coles discusses blocking and starting conditions for supersonic wind tunnels.

** This section may be omitted without loss of continuity in the text material.

A closed-circuit supersonic wind tunnel must have a converging-diverging nozzle to accelerate flow to supersonic speed, followed by a test section of nearly constant area, and then a supersonic diffuser with a second throat. The circuit must be completed by compression machinery, coolers, and flow-control devices, as shown in Fig. 13.23 [9].

Consider the process of accelerating flow from rest to supersonic speed in the test section. Soon after flow at the nozzle throat becomes sonic, a shock wave forms in the divergence. The shock attains its maximum strength when it reaches the nozzle exit plane. Consequently, to *start* the tunnel and achieve steady supersonic flow in the test section, the shock must move through the second throat and into the subsonic diffuser. When this occurs, we say the shock has been *swallowed* by the second throat. Consequently, to start the tunnel, the supersonic diffuser throat must be larger than the nozzle throat. The second throat must be large enough to exceed the critical area for flow downstream from the strongest possible shock.

Blocking occurs when the second throat is not large enough to swallow the shock. When the channel is blocked, flow is sonic at both throats and flow in the test section is subsonic; flow in the test section cannot be controlled by varying conditions downstream from the supersonic diffuser.

When the tunnel is *running* there is no shock in the nozzle or test section, so the energy dissipation is much reduced. The second throat area may be reduced slightly during running to improve the diffuser efficiency. The compressor pressure ratio may be adjusted to move the shock in the subsonic diffuser to a lower Mach number. A combination of adjustable second throat and pressure ratio control may be used to achieve optimum running conditions for the tunnel. Small differences in efficiency are important when the tunnel drive system may consume more than half a million kilowatts [10]!

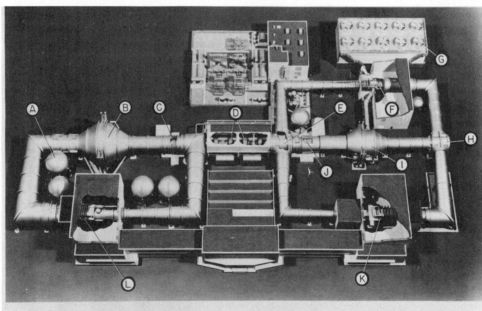

A. Dry Air Storage Spheres	G. Cooling Tower
B. Aftercooler	H. Flow Diversion Valve
C. 3-Stage Axial Flow Fan	I. Aftercooler
D. Drive Motors	J. 11-Stage Axial Flow Compressor
E. Flow Diversion Valve	K. 9- by 7-Foot Supersonic Test Section
F. 8- by 7-Foot Supersonic Test Section	L. 11- by 11-Foot Transonic Test Section

Fig. 13.23 Schematic view of NASA-Ames closed-circuit, high-speed wind tunnel with supporting facilities [9]. (Photo courtesy of NASA.)

**13-7.4 Constant-Area Channel with Friction

Flow in a constant-area channel with friction is dominated by viscous effects. Even when the main flow is supersonic, the no-slip condition at the channel wall guarantees subsonic flow near the wall. Consequently, supersonic flow in constant-area channels may form complicated systems of oblique and normal shocks. However, the basic behavior of adiabatic supersonic flow with friction in a constant-area channel is revealed by considering the simpler case of normal shock formation in Fanno line flow.

Supersonic flow along the Fanno line becomes choked after only a short length of duct, because at high speed the effects of friction are pronounced. Table E.2 (Appendix E) shows that the limiting value of $\bar{f}L_{max}/D_h$ is less than one; subsonic flows can have much longer runs. Thus when choking results from friction and duct length is increased further, the supersonic flow shocks down to subsonic to match downstream conditions.

The Ts diagrams in Figs. 13.24a through 13.24d illustrate what happens when the length of constant-area duct, fed by a converging-diverging nozzle supplied from a reservoir with constant stagnation conditions, is increased. Supersonic flow on the Fanno line of Fig. 13.24a is choked by friction when the duct length is L_a. When additional duct is added to produce $L_b > L_a$, Fig. 13.24b, a normal shock appears. Flow upstream from the shock does not change, because it is supersonic (no change in downstream condition can affect the supersonic flow before the shock).

In Fig. 13.24b the shock is shown in an arbitrary position. The shock moves toward the entrance of the constant-area channel (toward higher initial Mach number) as more duct is added.

Flow remains on the same Fanno line as the shock is driven upstream to state ① by adding duct length; thus the mass flow rate remains unchanged. The duct length, L_c, which moves the shock into the channel entrance plane, Fig. 13.24c, may be calculated directly using the methods of Section 13-4.3.

When duct length L_c is exceeded, the shock is driven back into the C-D nozzle, Fig. 13.24d. The mass flow rate remains constant until the shock reaches the nozzle throat. Only when more duct is added after the shock reaches the throat does the mass flow rate decrease.

If the shock position is known, flow properties at each section and duct length can be calculated directly. When length is specified and shock location is to be determined, iteration is necessary.

**13-7.5 Constant-Area Channel with Heat Addition

Supersonic flow with heat addition in a frictionless channel of constant area is shown in Fig. 13.25a. Assume the channel is fed by a converging-diverging nozzle, supplied from a reservoir with constant stagnation conditions, and flow is supersonic at state ①. Heat addition causes state points to move up and to the right along the Rayleigh line. Fig. 13.25a illustrates the condition in which the heat addition is just sufficient to choke the flow. Flow is sonic at the exit, so $p_e = p^*$ and $T_e = T^*$; the heat addition per unit mass, $\int_{s_1}^{s_e} T\,ds$, is represented by the shaded area beneath the Rayleigh line.

A normal shock involves no heat addition, so T_0 is constant across a shock. Consequently, a shock in the constant-area channel would not change the heat addition required to change the flow state from the inlet condition to choking. When the shock stands at the channel inlet, Fig. 13.25b, the heat addition needed to reach Mach one at the exit is the same as in Fig. 13.25a; the shaded areas also must be identical.

** This section may be omitted without loss of continuity in the text material.

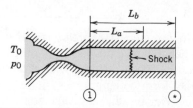

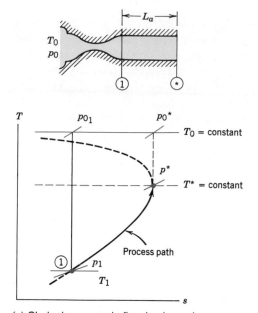

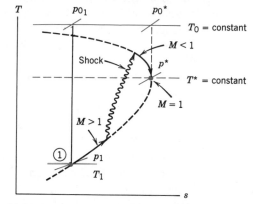

(a) Choked supersonic flow in channel.

(b) Choked flow in channel with shock.

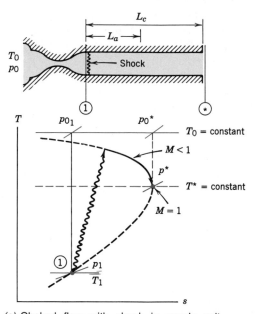

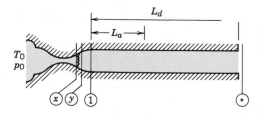

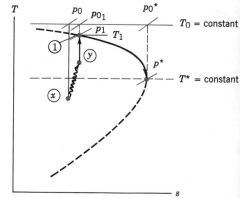

(c) Choked flow with shock in nozzle exit plane.

(d) Choked flow with shock in nozzle; subsonic flow in channel.

Fig. 13.24 Schematic *Ts* diagrams for supersonic Fanno line flows with normal shocks.

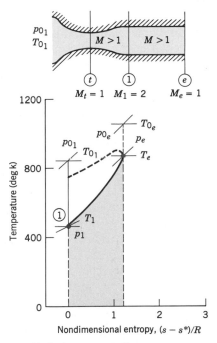

(a) Choked supersonic flow.

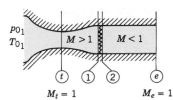
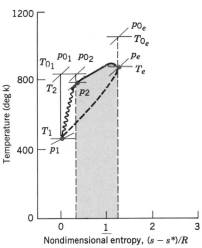

(b) Choked flow with shock at nozzle exit plane.

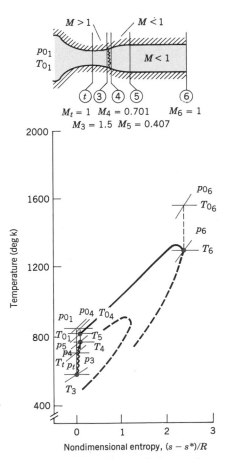

(c) Choked flow with shock in nozzle; same mass flow rate, but flow shifts to a new Rayleigh line.

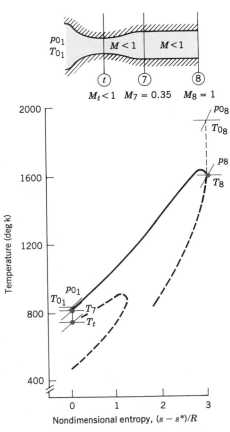

(d) Subsonic flow throughout; decreased mass flow rate and flow shifted to another new Rayleigh line.

Fig. 13.25 Schematic Ts diagrams for supersonic Rayleigh line flows with normal shocks.

If more thermal energy is added to flow at the conditions shown in Fig. 13.25b, the shock will be pushed from the entrance of the constant-area duct back into the diverging portion of the nozzle, where the Mach number is lower.

With a shock in the nozzle, conditions at the duct entrance are changed, and heat addition occurs along a different Rayleigh line, as shown in Fig. 13.25c. There is no change in T_0 or T^* across the shock (thus $T_{0_3} = T_{0_4}$ and $T_3^* = T_4^*$), but the Mach number downstream changes. Additional subsonic diffusion occurs from state ④ to the nozzle exit, state ⑤, thus moving the choked condition upward on the Ts plane, allowing for increased heat addition on the new Rayleigh line. All of these changes occur at the same mass flow rate, because nozzle throat conditions remain unchanged.

The Mach number immediately upstream from the shock, state ③, is less than M_1 of Fig. 13.25b; the corresponding temperature, T_3, is higher than T_1. Since the shock strength is reduced, the entropy rise across the shock is less, $(s_4 - s_3) < (s_2 - s_1)$. The subsonic diffusion following the shock results in a lower Mach number and higher temperature at the duct entrance. Thus $M_5 < M_2$ and $T_5 > T_2$.

When the heat addition rate is increased enough to drive the shock to the nozzle throat, a further increase in heat addition will result in a decrease in the mass flow rate. The Mach number at the channel inlet is reduced, $M_7 < M_5$, and the channel flow shifts to another new Rayleigh line, as shown in Fig. 13.25d.

Thus for specified mass flow rate, there is a maximum rate of heat addition for supersonic flow throughout. For larger heat addition rates, a shock occurs in the nozzle and flow is subsonic in the constant-area channel, but the exit flow remains sonic. If the shock position is specified, the heat addition along the Rayleigh line may be calculated directly. If the heat addition is specified but the shock position or mass flow rate are unknown, iteration is required to obtain a solution.

Additional consideration of flow with shock waves is given in [11].

13-8 SUMMARY OBJECTIVES

After completing study of Chapter 13 you should be able to do the following:

1. Write the basic equations for the steady, one-dimensional isentropic flow of (a) any compressible fluid and (b) an ideal gas, through a channel of arbitrary cross section. Use these equations, together with the appropriate Ts plot, to solve isentropic flow problems.

2. Determine the effect of area change on fluid properties for isentropic flow. Sketch flow passages for the following: subsonic nozzle, subsonic diffuser, supersonic nozzle, and supersonic diffuser.

3. For flow in (a) a converging nozzle and (b) a converging-diverging nozzle, plot the pressure distribution through the nozzle as a function of back pressure. State the conditions under which the nozzle is choked.

4. Write the basic equations for steady, one-dimensional, adiabatic flow of an ideal gas with constant specific heats through a constant-area duct. Use these equations, together with the appropriate Ts plot, to solve Fanno line flow problems.

5. Write the basic equations for steady, one-dimensional, frictionless flow of an ideal gas with heat exchange through a constant-area duct. Use these equations, together with the appropriate Ts plot, to solve Rayleigh line flow problems.

6. Write the basic equations for steady, one-dimensional flow of an ideal gas through a normal shock. Use these equations, together with the appropriate Ts plot, to solve normal shock problems.

**7. Use the tables for computation of (a) isentropic, (b) Fanno line, (c) Rayleigh line, and (d) normal shock flow of an ideal gas.

** This objective applies to sections that may be omitted without loss of continuity in the text material.

8. Describe the principal features of supersonic channel flows with normal shocks.

9. Solve the problems at the end of the chapter that relate to the material you have studied.

REFERENCES

1. Chapman, A. J., and W. F. Walker, *Introductory Gas Dynamics*. New York: Holt, Rinehart & Winston, 1971.

2. Olfe, D. B., *Fluid Mechanics Programs for the IBM PC*. New York: McGraw-Hill, 1987.

3. Hermann, R., *Supersonic Inlet Diffusers*. Minneapolis, MN: Minneapolis-Honeywell Regulator Co., Aeronautical Division, 1956.

4. Runstadler, P.W., Jr., "Diffuser Data Book," Creare, Inc., Hanover, NH, Technical Note 186, 1975.

5. Seddon, J., and E. L. Goldsmith, *Intake Aerodynamics*. New York: American Institute of Aeronautics and Astronautics, 1985.

6. Shapiro, A. H., *The Dynamics and Thermodynamics of Compressible Fluid Flow*, Vol. 1. New York: Ronald Press, 1953.

7. Zucrow, M. J., and J. D. Hoffman, *Compressible Flow*, Vol. 1. New York: Wiley, 1976.

8. Coles, D., *Channel Flow of a Compressible Fluid*, NCFMF Film.

9. Baals, D. W., and W. R. Corliss, *Wind Tunnels of NASA*. Washington, D. C.: National Aeronautics and Space Administration, SP-440, 1981.

10. Pope, A., and K. L. Goin, *High-Speed Wind Tunnel Testing*. New York: Krieger, 1978.

11. Glass, I.I., "Some Aspects of Shock-Wave Research," *AIAA J., 25*, 2, February 1987, pp. 214–229.

PROBLEMS

13.1 Steam flows steadily and isentropically through a nozzle. At one section the steam is at 700 F and 240 psia, and the speed is 600 ft/sec. The steam is to be accelerated to the maximum possible speed such that the exit quality is 1.0. Determine the corresponding exit speed.

13.2 Steam flows steadily and isentropically through a nozzle. At one section the steam is at 700 F and 240 psia, and the speed is 600 ft/sec. The steam is to be accelerated to the maximum possible speed such that the exit quality is 0.9. Determine the corresponding exit speed.

13.3 Steam flows steadily and isentropically through a nozzle. At one section the steam is at 350 C and an absolute pressure of 1.0 MPa, and the speed is 200 m/sec. The steam is to be accelerated to the maximum possible speed such that the exit quality is 1.0. Determine the corresponding exit speed.

13.4 Steam flows steadily and isentropically through a nozzle. At one section the steam is at 350 C and an absolute pressure of 1.0 MPa, and the speed is 200 m/sec. The steam is to be accelerated to the maximum possible speed such that the exit quality is 0.9. Determine the corresponding exit speed.

13.5 Steam flows steadily and isentropically through a nozzle. At an upstream section where the speed is negligible, the temperature and pressure are 900 F and 900 psia. At a section where the nozzle diameter is 0.188 in., the steam pressure is 600 psia. Determine the speed and Mach number at this section and the mass flow rate of steam. Sketch the passage shape.

13.6 Steam flows steadily and isentropically through a nozzle. At an upstream section where the speed is negligible, the temperature and pressure are 880 F and 875 psia. At a section where the nozzle diameter is 0.50 in., the steam pressure is 290 psia. Determine the speed and Mach number at this section and the mass flow rate of steam. Sketch the passage shape.

13.7 Steam flows steadily and iscntropically through a nozzlc. At an upstream section where the speed is negligible, the temperature and absolute pressure are 475 C and 6.0 MPa. At a section where the nozzle diameter is 6 mm, the absolute pressure of the steam is 4.0 MPa. Determine the speed and Mach number at this section and the mass flow rate of steam. Sketch the passage shape.

13.8 Steam flows steadily and isentropically through a nozzle. At an upstream section where the speed is negligible, the temperature and absolute pressure are 475 C and 6.0 MPa. At a section where the nozzle diameter is 12 mm, the absolute pressure is 2.0 MPa. Determine the speed and Mach number at this section and the mass flow rate of steam. Sketch the passage shape.

13.9 An F-4 aircraft makes a low-level pass over an airfield at sea level on a standard day. A pitot tube on the aircraft senses a stagnation pressure of 23.6 psia. Determine the Mach number at which the aircraft flies. Evaluate the speed of the aircraft.

13.10 At a section in a passage, the pressure is 30 psia, the temperature is 90 F, and the speed is 575 ft/sec. For isentropic flow of air, determine the Mach number at the point where the pressure is 12 psia. Sketch the passage shape.

13.11 Air flows steadily and isentropically through a passage. At section ①, where the cross-sectional area is 0.02 m², the air is at 40.0 kPa (abs), 60 C, and $M = 2.0$. At section ② downstream, the speed is 519 m/sec. Calculate the Mach number at section ②. Sketch the shape of the passage between sections ① and ②.

13.12 Air, at an absolute pressure of 60.0 kPa and 27 C, enters a passage at 486 m/sec, where $A = 0.02$ m². At section ② downstream, $p = 78.8$ kPa (abs). Assuming isentropic flow, calculate the Mach number at section ②. Sketch the flow passage.

13.13 Air flows steadily and isentropically through a passage at 100 kg/sec. At the section where $A = 0.464$ m², $M = 3$, $T = -60$ C, and $p = 15.0$ kPa (abs). Determine the speed and cross-sectional area downstream where $T = 138$ C. Sketch the flow passage.

13.14 A passage is designed to expand air isentropically to atmospheric pressure from a large tank in which properties are held constant at 5 C and 304 kPa (abs). The desired flow rate is 1 kg/sec. Determine the exit area of the passage. Sketch the Mach number and pressure as functions of distance along the passage.

13.15 The aircraft of Problem 13.9 flies at $M = 0.851$. Air is slowed in the engine inlet system to 475 ft/sec relative to the aircraft. Determine the temperature of the air at this location. If the deceleration process were modeled as isentropic, what would be the static pressure at this section?

13.16 A supersonic diffuser decelerates air isentropically from $M_1 = 3.0$ to $M_2 = 1.4$. If the static pressure at the diffuser inlet is 30.0 kPa (abs), calculate the static pressure rise in the diffuser and the ratio of inlet to outlet area of the diffuser.

13.17 Air with $T_{0_1} = 600$ R flows isentropically through a converging nozzle. At the point in the flow where the temperature is 571 R, the pressure is 100 psia. Determine the speed and the stagnation pressure at the downstream location where $T = 532$ R.

13.18 Air flows isentropically through a converging nozzle into a receiver in which the absolute pressure is 240 kPa. The air enters the nozzle with negligible speed at a pressure of 406 kPa (abs) and a temperature of 95 C. Determine the flow rate through the nozzle for a throat area of 0.01 m².

13.19 Air flows isentropically through a converging nozzle into a receiver where the pressure is 33 psia. If the pressure is 50 psia and the speed is 500 ft/sec, at the nozzle location where the Mach number is 0.4, determine the pressure, speed, and Mach number at the nozzle throat.

13.20 Air flowing isentropically through a converging nozzle discharges to the atmosphere. At the section where the absolute pressure is 179 kPa, the temperature is 39 C and the air speed is 177 m/sec. Determine the nozzle throat pressure.

13.21 Air flows from a large tank ($p = 650$ kPa (abs), $T = 550$ C) through a converging nozzle, with a throat area of 600 mm^2, and discharges to the atmosphere. Determine the mass rate of flow for isentropic flow through the nozzle.

13.22 Helium in a very large tank is maintained at 800 kPa (abs) and 250 C. Helium leaves the tank steadily and isentropically through a converging nozzle that discharges to the atmosphere. The nozzle throat area is 0.002 m^2. Determine the helium density at the nozzle throat.

13.23 Air, with $p_0 = 650$ kPa (abs) and $T_0 = 350$ K, flows isentropically through a converging nozzle. At the section in the nozzle where the area is 2.6×10^{-3} m^2, the Mach number is 0.5. The nozzle discharges to a back pressure of 270 kPa (abs). Determine the exit area of the nozzle.

13.24 A small spark-ignition engine is tested under standard atmosphere conditions. At idle it consumes air at 0.35 kg/min; the pressure in the inlet manifold is 635 mm of mercury (vacuum). Find the minimum flow area of the carburetor.

13.25 A converging nozzle is connected to a large tank that contains compressed air at 75 F. The nozzle exit area is 1.5 in.^{2} The exhaust is discharged to the atmosphere. To obtain a satisfactory shadow photograph of the flow pattern leaving the nozzle exit, the pressure in the exit plane must be greater than 45 psig. What pressure is required in the tank? What mass flow rate of air must be supplied if the system is to run continuously? Show static and stagnation state points on a Ts diagram.

13.26 A converging nozzle discharges helium to atmosphere. Conditions at section ① in the nozzle result in $p_1 = 350$ kPa (abs), $T_1 = 20$ C, and $V_1 = 201$ m/sec. The mass flow rate is to be 0.15 kg/sec. Assume that the flow is frictionless and adiabatic. Determine the Mach number at the nozzle exit and the exit area of the nozzle.

13.27 Carbon dioxide, at 50 MPa (abs) and 150 C, discharges isentropically from a large tank to atmosphere through a converging nozzle whose throat area is 1.0 mm^2. Find the temperature at the exit, the pressure difference between nozzle exit and atmosphere, and the mass flow rate.

13.28 A large tank supplies air to a convergent nozzle that discharges to atmospheric pressure. Assume the flow is reversible and adiabatic. For what range of tank pressures will the flow at the nozzle exit be sonic ($M = 1$)? If the tank pressure is 600 kPa (abs) and the temperature is 600 K, determine the mass flow rate through the nozzle, if the exit area is 1.29×10^{-3} m^2.

13.29 A large tank initially is evacuated to 27 in. Hg (vacuum). (Ambient conditions are 29.4 in. Hg at 70 F.) At $t = 0$, an orifice of 0.25 in. diameter is opened in the tank wall; the vena contracta area is 65 percent of the geometric area. Calculate the mass flow rate at which air initially enters the tank. Show the process on a Ts diagram. Make a schematic plot of mass flow rate as a function of time. Explain why the plot is nonlinear.

13.30 An 18 in. diameter spherical cavity initially is evacuated. The cavity is to be filled with air for a combustion experiment. The pressure is to be 5 psia, measured after its temperature reaches T_{atm}. Assume the valve on the cavity is a converging nozzle with throat diameter of 0.05 in., and the surrounding air is at standard conditions. For how long should the valve be opened to achieve the desired final pressure in the cavity? Calculate the entropy change for the air in the cavity.

13.31 Air flows isentropically through a converging nozzle attached to a large tank, where the absolute pressure is 171 kPa and the temperature is 27 C. At the inlet section the Mach number is 0.2. The nozzle discharges to the atmosphere; the discharge area is 0.015 m². Determine the magnitude and direction of the force that must be applied to hold the nozzle in place.

13.32 A stream of air flowing in a duct (area = 1 in.²) is at $p = 20$ psia, has $M = 0.6$, and flows at $\dot{m} = 0.5$ lbm/sec. Determine the local isentropic stagnation temperature. If the cross-sectional area of the passage were reduced downstream, determine the maximum percentage reduction of area allowable without reducing the flow rate (assume isentropic flow). Determine the speed and pressure at the minimum area location.

13.33 Consider a "rocket cart" propelled by a jet supplied from a tank of compressed air on the cart. Initially, air in the tank is at 1.52 MPa (abs) and 27 C, and the mass of the cart and tank is $M_0 = 23.2$ kg. The air exhausts through a converging nozzle with exit area $A_e = 31.7$ mm². Rolling resistance of the cart is $F_R = 5.2$ N; aerodynamic resistance is negligible. At the instant after air begins to flow through the nozzle: (a) compute the pressure in the nozzle exit plane, (b) evaluate the mass flow rate of air through the nozzle, and (c) calculate the acceleration of the tank and cart assembly.

13.34 A cylinder of gas used for welding contains helium at 3000 psig and room temperature. The cylinder is knocked over, its valve is broken off, and gas escapes through a converging passage. The minimum flow area is 0.10 in.² at the outlet section where the gas flow is uniform. Find (a) the mass flow rate at which gas leaves the cylinder and (b) the instantaneous acceleration of the cylinder (assume the cylinder axis is horizontal and its mass is 125 lb). Show static and stagnation states and the process path on a Ts diagram.

13.35 A converging nozzle is bolted to the side of a large tank. Air inside the tank is maintained at a constant 50 psia and 100 F. The inlet area of the nozzle is 10 in.² and the exit area is 1 in.² The nozzle discharges to the atmosphere. For isentropic flow in the nozzle, determine the total force on the bolts, and indicate whether the bolts are in tension or compression.

13.36 An ideal gas, with $k = 1.4$, flows isentropically through the converging nozzle shown and discharges into a large duct where the pressure is $p_2 = 125$ kPa (abs). The gas is *not* air and the gas constant, R, is unknown. Flow is steady and uniform at all cross sections. Find the exit area of the nozzle, A_2, and the exit speed, V_2.

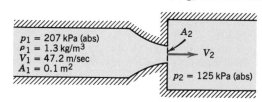

$p_1 = 207$ kPa (abs)
$\rho_1 = 1.3$ kg/m³
$V_1 = 47.2$ m/sec
$A_1 = 0.1$ m²

A_2

V_2

$p_2 = 125$ kPa (abs)

P13.36

13.37 An insulated air tank with $V = 107$ ft³ is used in a blowdown installation. Initially the tank is charged to 400 psia at 800 R. The mass flow rate of air from the tank is a function of time; during the first 30 sec of blowdown 64.5 lbm of air leaves the tank. Determine the air temperature in the tank after 30 sec of blowdown. Estimate the nozzle throat area.

13.38 A jet transport aircraft, with pressurized cabin, cruises at 11 km altitude. The cabin temperature and pressure initially are at 25 C and 2.5 km altitude. The interior volume of the cabin is 25 m³. Air escapes through a small hole with effective flow area of 0.002 m². Calculate the time required for the cabin pressure to decrease by 40 percent.

13.39 Air escapes from a high-pressure bicycle tire through a hole with diameter $d = 0.254$ mm. The initial pressure in the tire is $p_1 = 620$ kPa (gage). (Assume the temperature remains constant at 27 C.) The internal volume of the tire is approximately 4.26×10^{-4} m^3, and is constant. Estimate the time needed for the pressure in the tire to drop to 310 kPa (gage). Compute the change in specific entropy of the air in the tire during this process.

13.40 A converging-diverging nozzle is attached to a very large tank of air in which the pressure is 20 psia and the temperature is 40 F. The nozzle exhausts to the atmosphere where the pressure is 14.7 psia. The exit area of the nozzle is 2 in.^{2} What is the flow rate through the nozzle? Assume the flow is isentropic.

13.41 Consider the isentropic flow of helium through a converging-diverging wind-tunnel nozzle. The stagnation pressure at the nozzle entrance is 700 kPa (abs) and the stagnation temperature is 60 C. At a section downstream from the throat, the pressure is 528 kPa (abs) and the area is 1.2×10^{-3} m^2. At this section, determine the Mach number, the temperature, the stagnation pressure, and the mass flow rate.

13.42 A converging-diverging nozzle, designed to expand air to $M = 3.0$, has 250 mm^2 exit area. The nozzle is bolted to the side of a large tank and discharges to standard atmosphere. Air in the tank is pressurized to 4.5 MPa (gage) at 750 K. Assume flow within the nozzle is isentropic. Evaluate the pressure in the nozzle exit plane. Calculate the mass flow rate of air through the nozzle.

13.43 A large tank supplies helium through a converging-diverging nozzle to the atmosphere. Pressure in the tank remains constant at 8.00 MPa (abs) and temperature remains constant at 1000 K. Flow throughout the nozzle is isentropic. The nozzle is designed to discharge at exit Mach number $M = 3.5$. The exit area of the nozzle is 100 mm^2. What is the mass flow rate through the nozzle?

13.44 Air, at a stagnation pressure of 7.20 MPa (abs) and a stagnation temperature of 1100 K, flows isentropically through a converging-diverging nozzle having a throat area of 0.01 m^2. Determine the speed and the mass flow rate at the downstream section where the Mach number is 4.0.

13.45 A converging-diverging nozzle, with a throat area of 2 in.^{2}, is connected to a large tank in which air is kept at a pressure of 80 psia and a temperature of 60 F. If the nozzle is to operate at design conditions (flow is isentropic) and the ambient pressure outside the nozzle is 12.9 psia, calculate the exit area of the nozzle and the mass flow rate.

13.46 At a point upstream of the throat in a converging-diverging nozzle, the air speed is 172 m/sec; $p = 200$ kPa (abs) and $T = 22$ C. The flow is isentropic and is supersonic at the nozzle exit. If the nozzle throat area is 0.01 m^2, determine the flow rate.

13.47 Air flows isentropically through a converging-diverging nozzle attached to a large tank, in which the pressure is 100 psia and the temperature is 500 R. The nozzle is operating at design conditions for which the nozzle exit pressure, p_e, is equal to the surrounding atmospheric pressure, p_a. The exit area of the nozzle is $A_e = 4.0$ in.^{2} Calculate the flow rate through the nozzle. If the temperature of the air in the tank is increased to 2000 R (all pressures remaining the same), how will the flow rate be affected?

13.48 Air is to be expanded through a converging-diverging nozzle by a frictionless adiabatic process, from a pressure of 1.10 MPa (abs) and a temperature of 115 C, to a pressure of 141 kPa (abs). Determine the throat and exit areas for a well-designed shockless nozzle, if the mass flow rate is 2 kg/sec.

13.49 Air flows steadily and isentropically through a converging-diverging nozzle. At the throat the air is at 140 kPa (abs) and 60 C. The throat cross-sectional area is 0.05 m^2. At a certain section in the diverging part of the nozzle the pressure is 70.0 kPa (abs). Calculate the speed and the area at this section.

13.50 Nitrogen, at a pressure and temperature of 371 kPa (abs) and 400 K, enters a nozzle with negligible speed. The exhaust jet is directed against a large flat plate that is perpendicular to the jet axis. The flow leaves the nozzle at atmospheric pressure. The exit area is 0.003 m². Find the force required to hold the plate.

13.51 A small, solid fuel rocket motor is tested on a thrust stand. The chamber pressure and temperature are 600 psia and 6000 R. The propulsion nozzle is designed to expand the exhaust gases isentropically to a back pressure of 10.0 psia. The nozzle exit area is 0.60 ft². Treat the gas as ideal with $k = 1.2$ and $R = 60.0$ ft·lbf/lbm·R. Determine the mass flow rate of propellant gas and the thrust force exerted against the test stand.

13.52 A liquid rocket motor is fueled with hydrogen and oxygen. The chamber temperature and absolute pressure are 3300 K and 6.90 MPa. The nozzle is designed to expand the exhaust gases isentropically to a design back pressure corresponding to an altitude of 10 km on a standard day. The thrust produced by the motor is to be 100 kN at the design conditions. Treat the exhaust gases as water vapor and assume ideal gas behavior. Determine the propellant mass flow rate needed to produce the desired thrust, the nozzle exit area, and the area ratio, A_e/A_t.

13.53 A small rocket motor, fueled with hydrogen and oxygen, is tested on a thrust stand at a simulated altitude of 10 km. The motor is operated at chamber stagnation conditions of 1500 K and 8.0 MPa (gage). The combustion product is water vapor, which may be treated as an ideal gas. Expansion occurs through a converging-diverging nozzle with design Mach number of 3.5 and exit area of 700 mm². Evaluate the pressure at the nozzle exit plane. Calculate the mass flow rate of exhaust gas. Determine the force exerted by the rocket motor on the thrust stand.

13.54 Air flows isentropically in a converging-diverging nozzle. At section ① in the converging portion where the area is 1250 mm², the pressure is 600 kPa (abs), the temperature is 22 C, and the Mach number is 0.50. Develop a general expression for the area ratio, A_2/A_1, in terms of M_1 and M_2. Determine the area, A_2, in the diverging section where the Mach number is 2.

13.55 A CO_2 cartridge is used to propel a small rocket cart. Compressed gas, stored at 6000 psig and 70 F, is expanded through a smoothly contoured converging nozzle with 0.020 in. throat diameter. The back pressure is atmospheric. Calculate the pressure at the nozzle throat. Evaluate the mass flow rate of carbon dioxide through the nozzle. Determine the thrust available to propel the cart. How much would the thrust increase if a diverging section were added to the nozzle to expand the gas to atmospheric pressure? Show stagnation states, static states, and the processes on a Ts diagram.

13.56 A converging-diverging nozzle has a throat diameter of 10 mm. Measurements show that a turbulent boundary layer forms on the nozzle walls. The velocity profile in the boundary layer is modeled closely by the $\frac{1}{7}$th power-law expression. The boundary-layer thickness is 1.0 mm at the throat where $M = 1.0$, and 2.5 mm at the exit plane where $M = 2.0$. The nozzle is supplied with helium from a tank at 25 C and 1.0 MPa (abs). Calculate (a) the pressure at the nozzle throat, (b) the mass flow rate through the nozzle, and (c) the diameter at the nozzle exit plane. Show static and stagnation state points for the nozzle throat and exit plane on a Ts diagram.

*13.57 A converging-diverging nozzle has a throat diameter of 10 mm and an exit plane diameter of 20 mm. Measurements show that a turbulent boundary layer forms on the nozzle walls. The velocity profile in the boundary layer is modeled closely by the $\frac{1}{7}$th power-law expression. The boundary-layer thickness is 1.0 mm at the throat where $M = 1.0$, and 2.5 mm at the exit plane. Evaluate the effective flow

* Problems marked with an asterisk are designed to be solved using tables.

areas at the nozzle throat and exit plane. For flow of an ideal gas with $k = 1.4$, estimate (a) the subsonic and (b) the supersonic Mach numbers in the nozzle exit plane. How much do the boundary layers alter the mass flow rate and the Mach number in the exit plane? Show static and stagnation state points for the nozzle throat and exit plane on a Ts diagram.

13.58 Consider compressible flow in a long straight duct. The inlet to the duct is from the atmosphere where $T = 25$ C. Consider the flow to be adiabatic and the pipe long enough that the flow is choked. Find the speed and temperature at the pipe exit.

13.59 Air flows adiabatically with friction through a duct of 0.3 m square cross section. At one section, the pressure, temperature, and speed are 60.0 kPa (abs), 50 C, and 180 m/sec. At the duct exit, the pressure and temperature are 31.7 kPa (abs) and 19 C. Calculate the exit Mach number and the stagnation temperature midway between the first section and the exit.

13.60 Air flows steadily and adiabatically through a constant-area duct. The density at the inlet is 3.51 kg/m³. If the exit Mach number is unity and the exit temperature and pressure are 227 C and 110 kPa (abs), determine the inlet Mach number and the entropy change from inlet to exit.

13.61 Room air is drawn into an insulated duct of constant area through a smoothly contoured converging nozzle. Room conditions are $T = 27$ C and $p = 101$ kPa (abs). The duct diameter is $D = 25$ mm. The Mach number at the duct inlet (nozzle outlet) is $M_1 = 0.40$. Find (a) the mass flow rate in the duct and (b) the range of exit pressures for which the duct exit flow is choked.

13.62 A Fanno line flow apparatus in an undergraduate fluid mechanics laboratory consists of a smooth brass tube of 7.16 mm inside diameter, fed by a converging nozzle. The lab temperature and uncorrected barometer reading are 23.5 C and 755.1 mm of mercury. The pressure at the exit from the converging nozzle (entrance to the constant-area duct) is −20.8 mm of mercury (gage). Compute the Mach number at the entrance to the constant-area tube. Calculate the mass flow rate in the tube. Evaluate the pressure at the location in the tube where the Mach number is 0.4.

13.63 Air flows through a well-insulated, constant-area channel. At a section where the speed is 144 m/sec, $T = 50$ C and $p = 600$ kPa (abs). Determine (a) the temperature and stagnation pressure at the downstream section where the density is 3.05 kg/m³ and (b) the entropy increase.

13.64 Air flows steadily and adiabatically from a large tank through a converging nozzle connected to an insulated constant-area duct. The nozzle may be considered frictionless. Air in the tank is at $p = 1.00$ MPa (abs) and $T = 125$ C. The absolute pressure at the nozzle exit (duct inlet) is 784 kPa. Determine the pressure at the end of the duct, if the temperature there is 65 C. Find the entropy increase.

13.65 Air flows through a smooth well-insulated 4 in. diameter pipe at 600 lbm/min. At one section the air is at 100 psia and 80 F. Determine the minimum pressure and the maximum speed that can occur in the pipe.

13.66 Measurements are made of compressible flow in a long smooth 7.16 mm i.d. tube. Air is drawn from the surroundings (20 C and 101 kPa) by a vacuum pump downstream. Pressure readings along the tube become steady when the downstream pressure is reduced to 626 mm Hg (vacuum) or below. For these conditions, determine (a) the maximum mass flow rate possible through the tube, (b) the stagnation pressure of the air leaving the tube, and (c) the entropy change of the air in the tube. Show static and stagnation state points and the process on a Ts diagram.

13.67 Air is drawn from the atmosphere (20 C and 101 kPa) through a converging nozzle into a long insulated 20 mm diameter tube of constant area. Flow in the nozzle is isentropic. The Mach number at the inlet to the constant-area tube is 0.15. Evaluate the mass flow rate through the tube. Calculate T^* and p^* for the isentropic process. Calculate T^* and p^* for flow through the constant-area tube. Show the corresponding static and stagnation state points on a Ts diagram.

13.68 Consider adiabatic flow of air in a constant-area pipe with friction. At one section of the pipe, $p_0 = 100$ psia, $T_0 = 500$ R, and $M = 0.70$. If the cross-sectional area is 1 ft^2 and the Mach number at the exit is $M_2 = 1$, find the friction force exerted on the fluid by the pipe.

13.69 Air flows through an insulated duct with constant area of 0.03 m^2. At section ①, the static temperature, static pressure, and Mach number are 277 K, 690 kPa (abs), and 0.60. At section ② downstream, the temperature is 260 K. Draw a Ts diagram for the flow between states ① and ②, showing both the static and stagnation states. Calculate M_2 and p_2. Evaluate the frictional force exerted by the air on the duct wall.

13.70 A dental drill is powered by a miniature air turbine. Air is supplied by a compressor at a stagnation pressure of 50 psia. Air flows to the turbine through a long 0.060 in. i.d. tube. Air enters the tube through a smooth nozzle; flow within the tube is adiabatic. The Mach number at the turbine inlet is 0.608. Model flow through the turbine as isentropic. The pressure drop across the turbine is 10 psi. Air leaves the turbine at atmospheric pressure. Because the exhaust stream is used to cool the cutting area, its temperature must be held at 30 F. Determine the air temperature at the turbine inlet. Calculate the stagnation temperature at the compressor outlet (nozzle inlet). Evaluate the loss in stagnation pressure and the change in entropy through the supply tube. Show static and stagnation state points and the process path on a Ts diagram.

13.71 Room air (75 F and 14.7 psia) is to be drawn through a converging nozzle into a Fanno line demonstration apparatus. Flow in the nozzle may be modeled as isentropic. Air from the nozzle enters a 0.50 in. diameter constant-area duct at $M_1 = 0.3$. Flow in the duct is adiabatic but frictional. Evaluate properties T_{0_1}, p_{0_1}, T_1, p_1, ρ_1, V_1, and $\dot{m}$. Assuming that flow is choked at section ②, evaluate T_2, V_2, ρ_2, p_2, T_{0_2}, and p_{0_2}. Compute the change in specific entropy. Show all static and stagnation state points on a Ts diagram.

13.72 Air flows through a converging nozzle and then a length of insulated duct. The air is supplied from a tank where the temperature is constant at 15 C and the pressure is variable. The outlet end of the duct exhausts to atmosphere. When the exit flow is just choked, pressure measurements show the duct inlet pressure and Mach number are 53.2 psia and 0.30. Determine the pressure in the tank and the temperature, stagnation pressure, and mass flow rate of the outlet flow, if the tube diameter is 0.249 in. Show on a Ts diagram the effect of raising the tank pressure to 100 psia. Plot the pressure distribution versus distance along the channel for this new flow condition.

13.73 Air flows steadily and adiabatically in a horizontal 50 mm diameter pipe. At section ①, the pressure is 340 kPa (abs) and the temperature is 53 C. At section ② (downstream), the temperature is 114 C and the speed is 591 m/sec. Determine the speed and Mach number at section ①.

13.74 A converging-diverging nozzle supplies air to a well-insulated, constant-area duct. At the duct inlet, $M = 2.0$, $p = 19.4$ psia, and $T = 278$ R. At the duct exit, the Mach number is unity and the stagnation pressure is 90 psia. Determine the pressure and temperature at the duct exit and the entropy change.

13.75 A converging-diverging nozzle discharges air into an insulated pipe with cross-sectional area of 650 mm^2. At the pipe inlet, $T_1 = 39$ C, $p_1 = 128$ kPa (abs), and

$M_1 = 2.0$. Assume flow is shockless and it leaves the pipe at $M_2 = 1.0$. Determine the temperature at the pipe exit. Calculate the static pressure at the pipe exit.

13.76 A converging-diverging nozzle discharges air into an insulated pipe with area $A = 650$ mm^2. At the pipe inlet, $p = 128$ kPa (abs), $T = 39$ C, and $M = 2.0$. For shockless flow to a Mach number of unity at the pipe exit, calculate the exit temperature, the net force of the fluid on the pipe, and the entropy change.

13.77 Air flows steadily and adiabatically in a horizontal tube of 50 mm diameter. Measurements at section ① show that the temperature and pressure are 53 C and 340 kPa (abs). At section ② downstream, the temperature is 114 C and the flow speed is 591 m/sec. Find (a) the flow speed and Mach number at section ① and (b) the magnitude and direction of the friction force exerted by the air on the duct wall. Draw a Ts diagram for the process, showing all static and stagnation state points.

13.78 We wish to build a supersonic wind tunnel using an insulated nozzle and constant-area duct assembly. Shock-free operation is desired, with $M_1 = 2.1$ at the test section inlet and $M_2 = 1.1$ at the test section outlet. Stagnation conditions are $T_0 = 295$ K and $p_0 = 101$ kPa (abs). Calculate the outlet pressure and temperature and the entropy change through the test section.

*13.79 Solve Problem 13.58 for duct length, L, if the diameter is 12 mm and the surface is smooth. At the duct inlet, $p = 94.1$ kPa (abs).

*13.80 Consider the laboratory Fanno line flow channel of Problem 13.62. Assume laboratory conditions are 22.5 C and 760 mm of mercury (uncorrected). The manometer reading at a pressure tap at the end of the converging nozzle is -11.8 mm of mercury (gage). Calculate the Mach number at this location. Determine the duct length required to attain choked flow. Calculate the temperature and stagnation pressure at the choked state in the constant-area duct.

*13.81 For the conditions of Problem 13.64, find the length, L, of commercial steel pipe of 50 mm diameter between sections ① and ②.

*13.82 Air flows in an insulated duct. At one section, the temperature, absolute pressure and velocity are 200 C, 2.00 MPa and 140 m/sec. Find the temperature in this duct where the pressure has dropped to 1.26 MPa (abs) as a result of friction. If the duct ($e/D = 0.0003$) has 150 mm diameter, find the distance between the two sections.

13.83 Hydrogen flows in a horizontal insulated pipe 900 ft long and 2.5 in. diameter. The initial temperature and pressure are 140 F and 60.5 psia. The average friction factor is 0.015. Determine the maximum possible flow rate. Calculate the pressure drop required to maintain this flow rate.

*13.84 For the conditions of Problem 13.68, determine the duct length. Assume the duct is circular and made from commercial steel. Plot the variations of pressure and Mach number versus distance along the duct.

*13.85 Consider the flow described in Example Problem 13.8. Using the tables for Fanno line flow of an ideal gas, plot static pressure, temperature, and Mach number versus L/D measured from the tube inlet; continue until the choked state is reached.

*13.86 Using coordinates T/T_0 and $(s - s^*)/c_p$, where s^* is the entropy at $M = 1$, plot the Fanno line starting from the inlet conditions specified in Example Problem 13.8. Use tables for Fanno line flow, and proceed to $M = 1$.

*13.87 Using coordinates T/T^* and $(s - s^*)/c_p$, where s^* is the entropy at $M = 1$, plot the Fanno line for air flow ($k = 1.4$) for $0.1 < M < 3.0$.

*13.88 Air flows through a 40 ft length of insulated constant-area duct with $D = 2.12$ ft. The relative roughness is $e/D = 0.002$. At the duct inlet, $T_1 = 100$ F and $p_1 = 17.0$

* Problems marked with an asterisk are designed to be solved using tables.

psia. At a location downstream, $p_2 = 14.7$ psia and the flow is subsonic. Is sufficient information given to solve for M_1 and M_2? Prove your answer graphically. Find the mass flow rate in the duct and T_2.

*13.89 Air brought into a tube through a converging-diverging nozzle initially has stagnation temperature and pressure of 1000 R and 200 psia. Flow in the nozzle is isentropic; flow in the tube is adiabatic. At the junction between the nozzle and tube the pressure is 2.62 psia. The tube is 4 ft long and 1 in. in diameter. If the outlet Mach number is unity, find the average friction factor over the tube length. Calculate the change in pressure between the tube inlet and discharge.

*13.90 The duct of Problem 13.74 has relative roughness $e/D = 0.0002$ and hydraulic diameter of 0.5 ft. Determine the duct length, L.

*13.91 For the conditions of Problem 13.76, determine the duct length. Assume the duct is circular and made from commercial steel. Plot the variations of pressure and Mach number versus distance along the duct.

*13.92 Beginning with the inlet conditions of Problem 13.76, and using coordinates T/T_0 and $(s - s^*)/c_p$, plot the supersonic and subsonic branches of the Fanno line.

*13.93 A smooth constant-area duct assembly ($D = 150$ mm) is to be fed by a converging-diverging nozzle from a tank containing air at 295 K and 1.0 MPa (abs). Shock-free operation is desired. The Mach number at the duct inlet is to be 2.1 and the Mach number at the duct outlet is to be 1.4. The entire assembly will be insulated. Find (a) the pressure required at the duct outlet, (b) the duct length required, and (c) the change in specific entropy along the duct. Show the static and stagnation state points and the process line on a Ts diagram.

**13.94 Consider compressible flow through a duct of constant area. The flow is adiabatic and the Mach number is held constant at $M = 1/\sqrt{k}$ by bleeding gas through one porous wall. The bleed flow leaves in a direction normal to the wall. The flow is frictional with friction factor, f, and the duct equivalent diameter is D. Derive an expression for the pressure change in a duct section of length L, in terms of f, k, L, and D.

**13.95 In long, constant-area pipelines, as used for natural gas, temperature may be considered constant. Assume gas leaves a pumping station at 50 psia and 70 F at $M = 0.10$. At the section along the pipe where the pressure has dropped to 20 psia, calculate the Mach number of the flow. Is heat added to or removed from the gas over the length between the pressure taps? Justify your answer. Sketch the process on a Ts diagram. Indicate (qualitatively) T_{0_1}, T_{0_2}, and p_{0_2}.

**13.96 Air flows through a channel, where friction and heat transfer combine to keep the temperature constant and cause the flow to be choked at the channel exit. Inlet conditions are 500 R, 150 psia, and $M = 0.5$. The channel diameter is $D = 1.25$ ft. Calculate the flow speed at the exit section. Evaluate the magnitude and direction of the heat exchange for the process. Determine the pressure in the exit plane.

**13.97 Air enters a horizontal channel of constant area at 200 F, 600 psia, and 350 ft/sec. Determine the limiting pressure for isothermal flow. Compare with the limiting pressure for frictional adiabatic flow.

**13.98 Natural gas, with the thermodynamic properties of pure methane, enters a horizontal tube at 200 F, 600 psia, and 325 ft/sec. Determine the limiting pressure for isothermal flow. Compare with the limiting pressure for frictional adiabatic flow.

**13.99 A clean steel pipe is 950 ft long and 5.25 in. inside diameter. Air at 80 F, 120 psia, and 80 ft/sec enters the pipe. Calculate and compare the pressure drops through the pipe for (a) incompressible, (b) isothermal, and (c) adiabatic flow.

---

 * Problems marked with an asterisk are designed to be solved using tables.
** These problems require material from sections that may be omitted without loss of continuity in the text material.

**13.100 Air enters a 6 in. diameter pipe at 60 F, 200 psia, and 190 ft/sec. The average friction factor is 0.016. Flow is isothermal. Calculate the local Mach number and the distance from the entrance of the channel, at the point where the pressure reaches 75 psia.

**13.101 The following data apply to a cross-country pipeline for natural gas:

Inside diameter:	2.00 ft	Temperature:	520 R
Maximum pressure:	45,000 psfa	Mass flow rate:	5.15 slug/sec
Minimum pressure:	15,000 psfa	Gas constant:	3000 ft·lbf/slug·R
Average friction factor:	0.012		

Assuming steady isothermal flow, calculate maximum and minimum gas velocities. Estimate the maximum distance between pumping stations. What minimum power is required to operate each pipeline compressor?

**13.102 Natural gas (molecular mass $M_m = 18$ and $k = 1.3$) is to be pumped through a 36 in. i.d. pipe connecting two compressor stations 40 miles apart. At the upstream station the pressure is not to exceed 90 psig, and at the downstream station it is to be at least 10 psig. Calculate the maximum allowable rate of flow (ft³/day at 70 F and 1 atm) assuming sufficient heat exchange through the pipe to maintain the gas at 70 F.

13.103 Consider frictionless flow of air in a constant-area duct. At section ①, $M_1 = 0.50$, $p_1 = 1.10$ MPa (abs), and $T_{0_1} = 333$ K. Through the effect of heat exchange, the Mach number at section ② is $M_2 = 0.90$ and the stagnation temperature is $T_{0_2} = 478$ K. Determine the amount of heat exchange per unit mass to or from the fluid between sections ① and ② and the pressure difference, $p_1 - p_2$.

13.104 Air flows through a 2 in. inside diameter pipe with negligible friction. Inlet conditions are $T_1 = 60$ F, $p_1 = 150$ psia, and $M_1 = 0.30$. Determine the heat exchange per pound of air required to produce $M_2 = 1.0$ at the pipe exit, where $p_2 = 72.0$ psia.

13.105 Consider steady, one-dimensional, frictionless flow of air in a constant-area duct ($A = 0.0231$ m²) with heat addition. Flow properties are: $T_{0_1} = 900$ K, $p_{0_1} = 400$ kPa (abs), $\rho_1 = 1.30$ kg/m³, $T_2 = 929$ K, and $M_2 = 0.75$. Calculate the Mach number at section ①. Find the rate of heat addition to the flow. Evaluate the pressure at section ②. Compute the entropy change, $s_2 - s_1$.

13.106 Air flows without friction through a short duct of constant area. At the duct entrance, $M_1 = 0.30$, $T_1 = 50$ C, and $\rho_1 = 2.16$ kg/m³. As a result of heating, the Mach number and density at the tube outlet are $M_2 = 0.60$ and $\rho_2 = 0.721$ kg/m³. Determine the heat addition per unit mass and the entropy change for the process.

13.107 A fuel-air mixture, with the thermodynamic properties of pure air, enters a combustor of constant area. The stagnation temperature of the inlet stream is constant at 400 K. Friction is negligible. When heat is added to the flow at the rate of 1.4 MJ/kg, flow at the duct exit is choked. At this condition, the static pressure at the inlet is 144.6 kPa (abs) and the pressure at the duct exit is 64.7 kPa (abs). Calculate (a) the combustor outlet temperature, (b) the combustor inlet Mach number, and (c) the loss in stagnation pressure through the combustor. Show static and stagnation state points on a Ts diagram; indicate the process direction.

13.108 Air flows without friction through a 75 mm diameter duct. At the inlet, the Mach number, stagnation temperature, and absolute pressure are $M_1 = 0.2$, $T_{0_1} = 278$ K, and $p_1 = 275$ kPa. Heat is added at the rate of 219 kJ/kg of flowing fluid, and the outlet temperature is 489 K. Determine the stagnation temperature and Mach number at the duct outlet, and the entropy change for the process.

** These problems require material from sections that may be omitted without loss of continuity in the text material.

13.109   Liquid Freon, used to cool electronic components, flows steadily into a horizontal tube of constant diameter, $D = 15.9$ mm. Heat is transferred to the flow, and the liquid boils and leaves the tube as vapor. The effects of friction are negligible compared to the effects of heat addition. Flow conditions are shown. Find (a) the rate of heat transfer and (b) the pressure difference, $p_1 - p_2$.

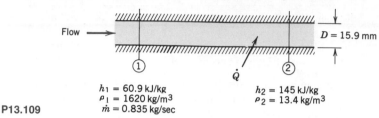

Flow →

$D = 15.9$ mm

① $\dot{Q}$ ②

$h_1 = 60.9$ kJ/kg
$\rho_1 = 1620$ kg/m³
$\dot{m} = 0.835$ kg/sec

$h_2 = 145$ kJ/kg
$\rho_2 = 13.4$ kg/m³

P13.109

13.110   Air flows at 1.42 kg/sec through a 100 mm diameter duct. At the inlet section, the temperature and absolute pressure are 52 C and 60.0 kPa. At the section downstream where the flow is choked, $T_2 = 45$ C. Determine the heat addition per unit mass, the entropy change, and the change in stagnation pressure for the process, assuming frictionless flow.

13.111   Air flows steadily and without friction in a duct of 0.5 ft² cross-sectional area. The inlet conditions are $M_1 = 0.30$, $T_{0_1} = 400$ R, and $p_{0_1} = 12.0$ psia. The exit stagnation temperature and static pressure are 797 R and 9.43 psia. Determine the heat exchange rate and the exit Mach number.

13.112   Air flows in a constant-area duct without friction. At section ①, $p_1 = 97.3$ psia, $T_1 = 992$ R, and $V_1 = 309$ ft/sec. Through the effect of heat addition, downstream $V_2 = 652$ ft/sec. Determine the pressure, temperature, stagnation pressure, and stagnation temperature at section ②, and the heat addition per unit mass between sections ① and ②.

13.113   Gasoline is added to the air stream in the carburetor of a spark-ignition engine to form a nearly stoichiometric mixture (1 lbm gasoline per 16 lbm air). The fuel evaporates in a short section of constant-area duct where friction can be neglected. (The latent heat of vaporization of gasoline is 140 Btu/lbm.) As a first approximation, neglect changes in the chemical composition and mass of the air stream, which leaves the evaporation section at 9.6 F. Determine the stagnation temperature and pressure of the flow leaving the duct, if the temperature at the duct inlet is 43.6 F (assume standard atmosphere conditions and isentropic flow from atmosphere to the duct inlet). Calculate the change in stagnation pressure caused by the evaporation process.

13.114   Air flows without friction through a section of constant-area duct. Inlet temperature, pressure, and Mach number are 818 R, 200 psia, and 0.3. The speed at the duct outlet is to be $V_2 \leq 2000$ ft/sec. Calculate the flow properties at the duct outlet and the maximum allowable rate of heat addition.

13.115   At 3 m before the exit of a constant-area duct (area = 0.02 m²), air is at 126 kPa (abs) and 260 C. The air flows steadily through the duct (without friction) at 1.83 kg/sec. The air leaves the duct subsonically at atmospheric pressure. Determine the Mach number, temperature, and stagnation temperature at the duct exit and the heat addition over the 3 m of duct length.

13.116   Consider frictionless flow of air in a duct of constant area, $A = 0.087$ ft². At one section, the static properties are 500 R and 15.0 psia and the Mach number is 0.2. At a section downstream, the static pressure is 10.0 psia. Draw a Ts diagram showing the static and stagnation states. Calculate the flow speed and temperature at the downstream location. Evaluate the rate of heat exchange for the process.

13.117 A combustor from a JT8D jet engine (as used on the Douglas DC-9 aircraft) has an air flow rate of 15 lbm/sec. The area is constant and frictional effects are negligible. Properties at the combustor inlet are 1260 R, 235 psia, and 609 ft/sec. At the combustor outlet, $T = 1850$ R and $M = 0.476$. The heating value of the fuel is 18,000 Btu/lbm; the air-fuel ratio is large enough so properties are those of air. Calculate the pressure at the combustor outlet. Determine the rate of energy addition to the air stream. Find the mass flow rate of fuel required; compare it to the air flow rate. Show the process on a Ts diagram, indicating static and stagnation states and the process direction.

13.118 Consider frictionless flow of air in a duct with $D = 4$ in. At section ①, the temperature and pressure are 30 F and 10 psia; the mass flow rate is 1.2 lbm/sec. How much heat may be added without choking the flow? Evaluate the resulting change in stagnation pressure.

13.119 A constant-area duct is fed with air from a converging-diverging nozzle. At the entrance to the duct, the following properties are known: $p_{0_1} = 800$ kPa (abs), $T_{0_1} = 700$ K, and $M_1 = 3.0$. A short distance down the duct (at section ②) $\rho_2 = 0.334$ kg/m^3. Assuming frictionless flow, determine the speed, pressure, and Mach number at section ②, and the heat exchange between the inlet and section ②.

13.120 Consider steady, one-dimensional flow of air in a constant-area duct with *both* friction and heat exchange. At section ①, the temperature, pressure, and speed are 600 R, 60 psia, and 500 ft/sec. At section ② downstream, the temperature and pressure are 800 R and 40 psia. Evaluate the heat exchange per unit mass and the stagnation pressure at section ②.

13.121 Frictionless flow of air in a constant-area duct discharges to atmospheric pressure at section ②. Upstream at section ①, $M_1 = 3.0$, $T_1 = 215$ R, and $p_1 = 1.73$ psia. Between sections ① and ②, 48.5 Btu per pound mass of air is added to the flow. Is p_2 greater than, equal to, or less than atmospheric pressure? Is M_2 supersonic, sonic, or subsonic? In addition to a Ts diagram, sketch the pressure distribution versus distance along the channel, labeling sections ① and ②.

*13.122 Air flows without friction in a constant-area duct. At section ①, $M_1 = 0.50$, $p_1 = 1.10$ MPa (abs), and $T_{0_1} = 335$ K. Through the effect of heat exchange, the Mach number is raised to $M_2 = 0.9$ at section ②. Determine the heat addition per unit mass between sections ① and ② and the pressure p_2.

*13.123 Air is drawn from the atmosphere (288 K and 101 kPa) through a smooth converging nozzle into the inlet manifold of a racing engine. Flow is isentropic through the nozzle. Air leaves the nozzle and enters the manifold at $M = 0.30$. In the manifold, alcohol fuel is evaporated in a short section of constant-area duct where friction is negligible. The effect is to remove 60.3 kJ/kg of thermal energy from the air stream. Assume the fluid properties are those of pure air. Find (a) the Mach number and (b) the pressure at the outlet of the evaporation section. Show static and stagnation state points and the process on a Ts diagram.

*13.124 Air flows without friction in a constant-area duct. At section ①, $T_1 = 992$ R, $p_1 = 97.3$ psia, and $V_1 = 309$ ft/sec. Heat addition causes the speed to increase to 652 ft/sec at section ②. Determine the Mach number, temperature, pressure, stagnation temperature, and stagnation pressure at section ②.

*13.125 Air flows steadily and without friction at 1.83 kg/sec through a duct with cross-sectional area of 0.02 m^2. At the duct inlet, the temperature and absolute pressure are 260 C and 126 kPa. The exit flow discharges subsonically to atmospheric pressure. Determine the Mach number, temperature, and stagnation temperature at the duct outlet and the heat exchange rate.

---

* Problems marked with an asterisk are designed to be solved using tables.

*13.126 A converging-diverging nozzle feeds a short duct of constant area. At the duct inlet, local isentropic stagnation conditions are $T_{0_1} = 700$ K and $p_{0_1} = 800$ kPa (abs), and $M_1 = 3.0$. A short distance down the duct at section ②, $\rho_2 = 0.334$ kg/m³. Assuming frictionless flow in the duct, determine the speed, pressure, and Mach number at section ②, and the heat exchange between the inlet and section ②.

*13.127 Air flows without friction in a short section of constant-area duct. At the duct inlet, $M_1 = 0.30$, $T_1 = 50$ C, and $\rho_1 = 2.16$ kg/m³. At the duct outlet, $M_2 = 0.60$. Determine the heat addition per unit mass, the entropy change, and the change in stagnation pressure for the process.

*13.128 In the frictionless flow of air through a 100 mm diameter duct, 1.42 kg/sec enters at 52 C and 60.0 kPa (abs). Determine the amount of heat that must be added to choke the flow, and the fluid properties at the choked state.

*13.129 Air flows without friction through a duct of 75 mm diameter. At the inlet, $M_1 = 0.20$, $T_{0_1} = 278$ K, and $p_1 = 275$ kPa. Heat is added at 219 kJ/kg of flowing air. Determine the stagnation temperature and Mach number at the duct outlet, and the changes in entropy and stagnation pressure for the process.

*13.130 Air flows steadily in a constant-area, frictionless duct of 0.5 ft² area. The inlet conditions are $M_1 = 0.3$, $T_{0_1} = 400$ R, and $p_{0_1} = 10$ psia; the exit stagnation temperature is $T_{0_2} = 797$ R. Determine the amount of heat addition and the exit Mach number and pressure.

*13.131 Air, from an aircraft inlet system, enters the engine combustion chamber where heat is added during a frictionless process in a tube with constant area of 0.01 m². The local isentropic stagnation temperature and Mach number entering the combustor are 427 K and 0.3. The mass flow rate is 0.5 kg/sec. When the rate of heat addition is set at 404 kW, flow leaves the combustor at 1026 K and 22.9 kPa (abs). Determine for this process (a) the Mach number at the combustor outlet, (b) the static pressure at the combustor inlet, and (c) the change in local isentropic stagnation pressure during the heat addition process. Show static and stagnation state points and indicate the process direction on a Ts diagram.

*13.132 Air enters the inlet manifold of a spark-ignition engine at 0.308 kg/sec with $T_{0_1} = 300$ K. At a certain location, $M_1 = 0.30$ and $p_1 = 94.0$ kPa (abs). The air stream is cooled by the evaporation of gasoline. After the evaporation section, $M_2 = 0.28$. The effects of friction and area change are negligible. Treat the fluid as pure air and find the flow area. Calculate the heat exchange per unit mass of air, the change in stagnation pressure for the process, and the entropy change. On a Ts diagram show the static and stagnation states and the process direction.

*13.133 Consider steady, one-dimensional flow of air in a combustor with constant area of 0.5 ft², where hydrocarbon fuel, added to the air stream, burns. The process is equivalent to simple heating because the amount of fuel is small compared to the amount of air; heating occurs over a short distance so that friction is negligible. Properties at the combustor inlet are 818 R, 200 psia, and $M = 0.3$. The speed at the combustor outlet must not exceed 2000 ft/sec. Find the properties at the combustor outlet and the heat addition rate. Show the process on a Ts diagram, indicating static and stagnation state points before and after the heat addition.

*13.134 Flow in a gas turbine combustor is modeled as steady, one-dimensional, frictionless heating of air in a channel of constant area. For a certain process, the inlet conditions are 960 F, 225 psia, and $M = 0.4$. Calculate the maximum possible heat addition. Find all fluid properties at the outlet section and the reduction in stagnation pressure. Show the process on a Ts diagram, indicating all static and stagnation state points.

---

* Problems marked with an asterisk are designed to be solved using tables.

*13.135 A supersonic wind tunnel is supplied from a high-pressure tank of air at 25 C. The test section temperature is to be maintained above 0 C to prevent formation of ice particles. To accomplish this, air from the tank is heated before it flows into a converging-diverging nozzle which feeds the test section. The heating is done in a short section with constant area. The heater output is $\dot{Q} = 10$ kW. The design Mach number in the wind tunnel test section is to be 3.0. Evaluate the stagnation temperature required at the heater exit. Calculate the maximum mass flow rate at which air can be supplied to the wind tunnel test section. Determine the area ratio, A_e/A_t.

*13.136 Consider steady flow of air in a combustor where thermal energy is added by burning fuel. Neglect friction. Assume thermodynamic properties are constant and equal to those of pure air. Calculate the stagnation temperature at the burner exit. Compute the Mach number at the burner exit. Evaluate the heat addition per unit mass and the heat exchange rate. Express the rate of heat addition as a fraction of the maximum rate of heat addition possible with this inlet Mach number.

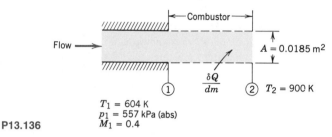

P13.136

*13.137 A jet aircraft cruises at $M = 0.85$ at 10 km altitude on a standard day. Air is slowed by the inlet system to $M_1 = 0.3$ relative to the aircraft. Flow in the inlet system is adiabatic, but friction causes the absolute stagnation pressure to drop by 5 percent. Air from the inlet is compressed to $p_2 = 30\,p_1$ in the engine's compressor section. Flow in the compressor is adiabatic, but the stagnation temperature of the stream increases by 750 K due to work addition. The Mach number leaving the compressor and entering the combustor is $M_2 = 0.4$. Heat is added in the combustor, where the maximum temperature must be held to $T_3 = 1630$ K or less. Evaluate for these conditions (a) the maximum possible heat addition in the combustor, (b) the combustor exit Mach number for part (a), and (c) the change in stagnation pressure across the combustor. On a Ts diagram for the process, show static and stagnation states and indicate the process direction.

*13.138 Using coordinates T/T^* and $(s - s^*)/c_p$, where s^* is the entropy at $M = 1$, plot the Rayleigh line for air flow ($k = 1.4$) for $0.4 < M < 3.0$.

*13.139 Beginning with the inlet conditions of Problem 13.76, and using coordinates T/T_{0_1} and $(s - s^*)/c_p$, plot the supersonic and subsonic branches of the Rayleigh line for the flow.

13.140 Air flows steadily through a round tube with $D = 50$ mm. The process is unspecified. Conditions at section ① are $T_1 = 260$ K, $p_1 = 95.0$ kPa (abs), and $M_1 = 0.6$; at section ② downstream, $p_2 = 64.9$ kPa (abs) and $M_2 = 0.85$. Calculate the heat added per unit mass, the friction force exerted by the gas on the tube, and the entropy change.

13.141 A small hand-held hair styling dryer with 1200 W heating element is shown. Heat losses to the surroundings are negligible. The inlet temperature is 70 F; the outlet temperature must be held below 140 F to avoid damaging hair. Neglect friction,

---

* Problems marked with an asterisk are designed to be solved using tables.

but assume the entrance loss coefficient is $K_{ent} = 0.50$. Find the minimum volume flow rate of air that must be provided. Estimate the pressure rise across the fan (express your answer in inches of water).

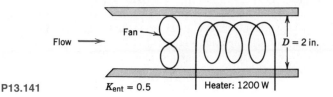

P13.141 $K_{ent} = 0.5$ Heater: 1200 W

Flow   Fan $D = 2$ in.

13.142   A jet transport aircraft cruises at $M = 0.85$ at an altitude of 40,000 ft. Air for the cabin pressurization system is taken aboard through an inlet duct and slowed isentropically to 100 ft/sec relative to the aircraft. Then it enters a compressor where its pressure is raised adiabatically to provide a cabin pressure equivalent to 8000 ft altitude. The air temperature increase across the compressor is 170 F. Finally, the air is cooled to 70 F (in a heat exchanger with negligible friction) before it is added to the cabin air. Sketch a diagram of the system, labeling all components and numbering appropriate cross sections. Determine the stagnation and static temperature and pressure at each cross section. Sketch to scale and label a Ts diagram showing the static and stagnation state points and indicating the process paths. Evaluate the work added in the compressor and the energy rejected in the heat exchanger.

13.143   In a scheme proposed by an inventor for a rocket engine, air and hydrogen will be stored in separate containers at 293 K and 305 kPa (abs). The gases will flow separately and isentropically into a constant-area, frictionless duct, where the hydrogen will be burned with a small percentage of the air. The Mach number at the duct entrance is 0.3; at the duct exit the stagnation temperature is 559 K and the static pressure is 244 kPa (abs). The gases—which are essentially heated air—then flow through a converging nozzle to the atmosphere. The area of the duct is 0.04 m^2 and the nozzle exit area is 0.0285 m^2. Sketch the duct and nozzle and draw a Ts diagram for the air, showing all static and stagnation states. Calculate the static thrust of this engine at sea level. Sketch a nozzle shape that would provide shock-free flow at sea level and calculate the exit area for this new nozzle, assuming conditions upstream from the nozzle entrance are unchanged.

13.144   Air, with $T_{0_1} = 333$ K and $p_{0_1} = 600$ kPa (abs), approaches a normal shock at $M_1 = 2.0$. The speed downstream from the shock is $V_2 = 204$ m/sec. Determine the static pressure downstream from the shock.

13.145   A normal shock stands in a constant-area duct. Air approaches the shock with $T_{0_1} = 1000$ R, $p_{0_1} = 100$ psia, and $M_1 = 3.0$. The Mach number downstream from the shock is $M_2 = 0.475$. Determine the static pressure downstream from the shock.

13.146   Air approaches a normal shock at $V_1 = 951$ m/sec, with $T_{0_1} = 700$ K and $p_1 = 125$ kPa (abs); $M_1 = 3.0$. Downstream from the shock $p_2 = 1.29$ MPa (abs). Determine the downstream speed and temperature.

13.147   A normal shock occurs in air at a section where the flow speed is 924 m/sec. At this section, $T_1 = 10$ C and $p_1 = 35.0$ kPa (abs). Downstream from the shock $p_2 = 301$ kPa (abs). Determine the speed and Mach number downstream from the shock.

13.148   An air stream, with $T_1 = 0$ C, $p_1 = 60.0$ kPa (abs), and $V_1 = 497$ m/sec, undergoes a normal shock. The temperature downstream from the shock is 87 C. Determine the Mach number, speed, and stagnation pressure downstream from the shock.

13.149 Air undergoes a normal shock. Upstream, $T_1 = 35$ C, $p_1 = 229$ kPa, and $V_1 = 704$ m/sec. Downstream from the shock, $\rho_2 = 6.91$ kg/m^3. Determine the temperature and stagnation pressure of the air stream leaving the shock.

13.150 An air stream approaches a normal shock at $M_1 = 2.64$. Upstream, $p_{0_1} = 3.00$ MPa (abs) and $\rho_1 = 1.65$ kg/m^3. The ratio of static pressure to stagnation pressure immediately behind the shock is 0.843. Determine the downstream Mach number and temperature.

13.151 A supersonic aircraft cruises at $M = 2.2$ at 12 km altitude. A pitot tube is used to sense stagnation pressure for calculating air speed. A normal shock stands in front of the pitot tube; the Mach number following the shock is 0.547. Evaluate local isentropic stagnation conditions in front of the shock. Estimate the pressure sensed by the pitot tube. Show all static and stagnation state points and the process path on a Ts diagram.

13.152 A normal shock occurs when a pitot-static tube is inserted into a supersonic wind tunnel. Pressures measured by the tube are $p_{0_2} = 68.1$ kPa (abs) and $p_2 = 54.8$ kPa (abs). Before the shock, $T_1 = 160$ K and $p_1 = 11.0$ kPa (abs). Calculate the air speed in the wind tunnel.

13.153 A total-pressure probe is placed in a supersonic wind tunnel where $T = 530$ R and $M = 2.0$. A normal shock stands in front of the probe. Behind the shock, $M_2 = 0.577$ and $p_2 = 5.76$ psia. Find (a) the downstream stagnation pressure and stagnation temperature and (b) all fluid properties upstream from the shock. Show static and stagnation state points and the process on a Ts diagram.

13.154 Air flows steadily through a long, insulated constant-area pipe. At section ①, $M_1 = 2.0$, $T_1 = 140$ F, and $p_1 = 35.9$ psia. At section ②, downstream from a normal shock, $V_2 = 1080$ ft/sec. Determine the density and Mach number at section ②. Make a qualitative sketch of the pressure distribution along the pipe.

13.155 Air approaches a normal shock at 0 C, 60 kPa (abs), and 497 m/sec. The air speed immediately downstream from the shock is 267 m/sec. Evaluate all flow properties downstream from the shock and the entropy change across the shock. Show static and stagnation states and the process direction on a Ts diagram.

13.156 Air approaches a normal shock with $T_1 = 18$ C, $p_1 = 101$ kPa, and $V_1 = 766$ m/sec. The temperature immediately downstream from the shock is $T_2 = 551$ K. Determine the velocity immediately downstream from the shock and the pressure change across the shock. Calculate the corresponding pressure change for a frictionless, shockless deceleration between the same speeds.

13.157 A supersonic wind tunnel is to be operated at $M = 2.2$ in the test section. Upstream from the test section, the nozzle throat area is 0.07 m^2. Air is supplied at stagnation conditions of 500 K and 1.0 MPa (abs). At one flow condition, while the tunnel is being brought up to speed, a normal shock stands at the nozzle exit plane. The flow is steady and the temperature downstream from the shock is 472 K. For this *starting* condition, immediately downstream from the shock find (a) the Mach number, (b) the static pressure, (c) the stagnation pressure, and (d) the minimum area theoretically possible for the second throat downstream from the test section. On a Ts diagram show static and stagnation state points and the process direction.

13.158 A supersonic aircraft flies at $M_1 = 2.7$ at 20 km altitude on a standard day. Air enters the engine inlet system where it is slowed isentropically to $M_2 = 1.3$. A normal shock occurs at that location. (Immediately downstream from the shock $M_3 = 0.786$.) The resulting subsonic flow is decelerated further to $M_4 = 0.40$. The subsonic diffusion is adiabatic but not isentropic; the final pressure is 104 kPa (abs). Evaluate (a) the stagnation temperature for the flow, (b) the pressure change across the shock, (c) the entropy change, $s_4 - s_1$, and (d) the final stagnation pressure. Show the process on a Ts diagram, indicating all static and stagnation states.

13.159 A blast wave propagates outward from an explosion. At large radii, curvature is small and the wave may be treated as a strong normal shock. (The pressure and temperature rise associated with the blast wave decrease slowly as the wave travels outward.) At one instant, a blast wave front travels at $M = 1.60$ with respect to undisturbed air at standard conditions. The static pressure immediately behind the wave is 286 kPa (abs). Find (a) the speed of the air behind the blast wave with respect to the wave and (b) the speed of the air behind the blast wave as seen by an observer on the ground. Draw a Ts diagram for the process as seen by an observer on the wave, indicating static and stagnation state points and property values.

13.160 An air stream, with $T_1 = 0$ C, $p_1 = 60.0$ kPa (abs), and $V_1 = 497$ m/sec, undergoes a normal shock. Determine the Mach number, speed, and stagnation pressure downstream from the shock.

*13.161 Air, with $T_{0_1} = 333$ K and $p_{0_1} = 600$ kPa (abs), approaches a normal shock at $M_1 = 2.0$. Determine the static pressure downstream from the shock and the decrease in stagnation pressure across the shock.

*13.162 Air approaches a normal shock at $V_1 = 951$ m/sec, with $T_{0_1} = 700$ K and $p_1 = 125$ kPa (abs); $M_1 = 3.0$. Determine the speed and temperature of the air leaving the shock and the entropy change across the shock.

*13.163 A normal shock stands in a constant-area duct. Air approaches the shock with $T_{0_1} = 1000$ R, $p_{0_1} = 100$ psia, and $M_1 = 3.0$. Determine the static pressure downstream from the shock. Compare the downstream pressure with that reached by decelerating isentropically to the same subsonic Mach number.

*13.164 Air undergoes a normal shock. Upstream, $T_1 = 35$ C, $p_1 = 229$ kPa (abs), and $V_1 = 704$ m/sec. Determine the temperature and stagnation pressure of the air stream leaving the shock.

*13.165 An air stream approaches a normal shock at $M_1 = 2.64$. Upstream, $p_{0_1} = 3.00$ MPa (abs) and $\rho_1 = 1.65$ kg/m³. Determine the downstream Mach number and temperature, and the entropy change across the shock.

*13.166 A normal shock occurs in air at a section where $V_1 = 924$ m/sec, $T_1 = 10$ C, and $p_1 = 35.0$ kPa (abs). Determine the speed and Mach number downstream from the shock, and the change in stagnation pressure across the shock.

*13.167 Air approaches a normal shock with $T_1 = 18$ C, $p_1 = 101$ kPa (abs), and $V_1 = 766$ m/sec. Determine the speed immediately downstream from the shock and the pressure change across the shock. Calculate the corresponding pressure change for a frictionless, shockless deceleration between the same speeds.

*13.168 A supersonic aircraft cruises at $M = 2.2$ at 12 km altitude. A pitot tube is used to sense pressure for calculating air speed. A normal shock stands in front of the tube. Evaluate the local isentropic stagnation conditions in front of the shock. Estimate the stagnation pressure sensed by the pitot tube. Show static and stagnation state points and the process path on a Ts diagram.

*13.169 A supersonic aircraft flies at $M = 2.7$ at 20 km altitude on a standard day. Air approaching a forward-facing total-head (pitot) tube passes through a normal shock and then is brought isentropically to rest relative to the aircraft. Determine the final temperature of the air and the pressure sensed by the total-head tube.

*13.170 Stagnation pressure and temperature probes are located on the nose of a supersonic aircraft at 35,000 ft altitude. A normal shock stands in front of the probes. The temperature probe indicates $T_0 = 420$ F behind the shock. Calculate the Mach number and air speed of the plane. Find the static and stagnation pressures behind the shock. Show the process and the static and stagnation state points on a Ts diagram.

* Problems marked with an asterisk are designed to be solved using tables.

*13.171 The Concorde supersonic transport flies at $M = 2.2$ at 20 km altitude. Air is decelerated isentropically by the engine inlet system to a local Mach number of 1.3. The air passes through a normal shock and is decelerated further to $M = 0.4$ at the engine compressor section. Assume, as a first approximation, that this subsonic diffusion process is isentropic and use standard atmosphere data for freestream conditions. Determine the temperature, pressure, and stagnation pressure of the air entering the engine compressor.

*13.172 The subsonic portion of the inlet diffuser described in Problem 13.171 has an isentropic efficiency of 0.97; the actual static pressure rise is 97 percent of the value that could be obtained from an isentropic deceleration between the same initial and final speeds. The supersonic diffusion may be assumed to be isentropic. The entire flow is adiabatic. Determine the overall reduction in stagnation pressure for the inlet flow and the static pressure of the air stream entering the engine compressor at $M = 0.4$.

*13.173 Consider a supersonic wind tunnel starting as shown. The nozzle throat area is 1.25 ft^2, and the test section design Mach number is 2.50. As the tunnel starts, a normal shock stands in the divergence of the nozzle where the area is 3.05 ft^2. Upstream stagnation conditions are $T_0 = 1080$ R and $p_0 = 115$ psia. Find the minimum possible diffuser throat area at this instant. Calculate the entropy increase across the shock.

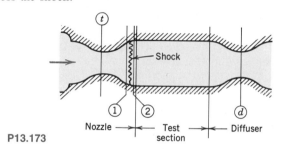

P13.173

*13.174 A supersonic aircraft cruises at $M = 2.7$ at 60,000 ft altitude. A normal shock stands in front of a pitot tube on the aircraft; the tube senses a stagnation pressure of 10.4 psia. Calculate the static pressure and temperature behind the shock. Evaluate the loss in stagnation pressure through the shock. Determine the change in specific entropy across the shock. Show static and stagnation states and the process path on a Ts diagram.

*13.175 An aircraft flies supersonically at 10 km altitude on a standard day. The true air speed of the plane is 659 m/sec. Calculate the flight Mach number of the aircraft. A total-head tube attached to the plane is used to sense stagnation pressure which is converted to flight Mach number by an on-board computer. However, the computer programmer has ignored the normal shock that stands in front of the total-head tube and has assumed isentropic flow. Evaluate the pressure sensed by the total head tube. Determine the erroneous air speed calculated by the computer program.

*13.176 A supersonic aircraft flies at 22 km altitude on a standard day. A forward-facing stagnation tube senses a total pressure of 55.3 kPa (abs). Determine the Mach number and speed of the aircraft.

*13.177 A total-pressure probe is used to measure the speed of supersonic air flow. A normal shock stands in front of the tube. Upstream flow conditions are $T_1 = 40$ F

* Problems marked with an asterisk are designed to be solved using tables.

and $p_1 = 4.15$ psia. Downstream from the shock, $p_{0_2} = 22.6$ psia. Calculate the air speed approaching the shock.

*13.178 Air flows steadily through a long insulated pipe of 2 in. diameter. At section ①, $M_1 = 2.0$, $T_1 = 140$ F, and $p_1 = 35.9$ psia. A normal shock occurs at section ② where $M_2 = 1.88$. At section ④, some distance downstream from the shock, the speed is 1080 ft/sec. Determine the flow properties at sections ② and ③, immediately upstream and downstream from the shock, and the Mach number and pressure at section ④. Determine the length of the pipe and plot the pressure distribution along the pipe.

13.179 Air flows through a converging-diverging nozzle designed to give exit Mach number $M = 2.80$. The upstream stagnation conditions are atmospheric; the back pressure is maintained by a vacuum pump. Determine the back pressure required to cause a normal shock to stand in the exit plane and the flow speed after the shock. (The static pressure ratio across a normal shock occurring at $M = 2.80$ is 8.98.)

13.180 A normal shock occurs in the diverging section of a converging-diverging nozzle where $A = 4.0$ in.^{2} and $M = 2.00$. (The Mach number immediately downstream from the shock is 0.577.) Upstream, $T_{0_1} = 1000$ R and $p_{0_1} = 100$ psia. The nozzle exit area is 6.0 in.^{2} and the exit temperature is 977 R. Assume that flow is isentropic except across the shock. Find the nozzle exit pressure. Show the processes on a Ts diagram, and indicate the static and stagnation state points.

13.181 A supersonic diffuser for an aircraft is designed to operate with a flight speed of $M = 3.0$ and inlet area of 3.0 m^2 at 30 km altitude. A normal shock occurs downstream from the diffuser throat at a section where the Mach number is 1.4. (The Mach number immediately behind the shock is 0.740.) The final Mach number leaving the diffuser is to be 0.4. Sketch the passage shape required. Compute T_0, T, p_0, and p at the inlet, the throat, before and after the shock, and the diffuser outlet, assuming all processes are isentropic except across the shock. Show the process paths and the static and stagnation states on a Ts diagram.

13.182 A supersonic diffuser for an aircraft is designed to operate with a flight speed of $M = 2.2$ and inlet area of 1.5 m^2 at 20 km altitude. A normal shock occurs downstream from the diffuser throat at a section where the Mach number is 1.3. (The Mach number immediately behind the shock is 0.786.) The final Mach number leaving the diffuser is to be 0.3. Sketch the passage shape required. Compute T_0, T, p_0, and p at the inlet, the throat, before and after the shock, and the diffuser outlet, assuming all processes are isentropic except across the shock. Show the process paths and the static and stagnation states on a Ts diagram.

13.183 A missile powered by a ramjet engine is to fly at $M = 3.5$ at 30 km altitude. Air can be decelerated isentropically to $M = 1.4$ in the inlet system before a normal shock occurs. Calculate the pressure rise that theoretically could be obtained in this process. If the efficiency of the diffusion process actually is 85 percent, calculate the loss in stagnation pressure for the process. (Diffuser efficiency is defined as $\eta = \Delta h(\text{isentropic})/\Delta h(\text{actual})$ between the same pressure levels.) Show static and stagnation states for the theoretical and actual processes on a Ts diagram.

13.184 Air flows adiabatically from a reservoir, where $T_{0_1} = 60$ C and $p_{0_1} = 600$ kPa (abs), through a converging-diverging nozzle. The design Mach number of the nozzle is 2.94. A normal shock occurs at the location in the nozzle where $M = 2.42$. (The Mach number immediately behind the shock is $M = 0.521$.) Assuming isentropic flow before and after the shock, determine the back pressure downstream from the nozzle, if the temperature there is 54.6 C. Sketch the pressure distribution.

* Problems marked with an asterisk are designed to be solved using tables.

13.185 A normal shock occurs in the diverging section of a converging-diverging nozzle where $A = 4.0$ in.^{2} and $M = 2.50$. (The Mach number immediately downstream from the shock is 0.513.) The stagnation conditions for the upstream flow are $T_0 = 1000$ R and $p_0 = 100$ psia. The nozzle exit area is 6.0 in.^{2} and the exit temperature is 981 R. Assume the flow is isentropic except across the shock. Determine the nozzle exit pressure, throat area, and mass flow rate.

13.186 A converging-diverging nozzle is designed to expand air isentropically to atmospheric pressure from a large tank where $T_{0_1} = 150$ C and $p_{0_1} = 790$ kPa (abs). A normal shock stands in the diverging section at a location where $p_{0_2} = 160$ kPa (abs) and $A = 600$ mm^2. The Mach number immediately behind the shock is 0.641. Determine the nozzle back pressure, exit area, and throat area.

13.187 Consider flow of air through a converging-diverging nozzle. Sketch the approximate behavior of the mass flow rate versus back pressure ratio, p_b/p_0. Sketch the variation of pressure with distance along the nozzle, and the Ts diagram for the nozzle flow, when the back pressure is p^*.

13.188 Because of friction, real nozzle performance differs slightly from that computed using the isentropic flow relations. In both cases it is reasonable to consider the flow adiabatic. The *nozzle efficiency*, η, is defined as the ratio of the actual exit kinetic energy to the exit kinetic energy that would be obtained in a frictionless nozzle expanding the gas to the same final pressure. Derive an expression for nozzle efficiency in terms of stagnation temperature and the actual and ideal temperatures at the nozzle exit plane. Assume in both cases that flow up to the nozzle throat is isentropic and the exit flow is supersonic. Consider a nozzle designed to produce $M_e = 2.2$ for isentropic flow of air. Evaluate the actual exit plane Mach number for $0.8 < \eta < 1.0$.

*13.189 Air flows through a converging-diverging nozzle with $A_e/A_t = 3.5$. The upstream stagnation conditions are atmospheric; the back pressure is maintained by a vacuum pump. Determine the back pressure required to cause a normal shock to stand in the nozzle exit plane and the flow speed leaving the shock.

*13.190 A converging-diverging nozzle expands air from 250 F and 50.5 psia to 14.7 psia. The throat and exit plane areas are 0.801 and 0.917 in.^{2}, respectively. Calculate the exit Mach number. Evaluate the mass flow rate through the nozzle.

*13.191 A converging-diverging nozzle is attached to a large tank of air, in which $T_{0_1} = 300$ K and $p_{0_1} = 250$ kPa (abs). At the nozzle throat (section of minimum area) the pressure is 132 kPa (abs). In the diverging section, the pressure falls to 68.1 kPa before rising suddenly across a normal shock. At the nozzle exit the pressure is 180 kPa. Find the Mach number immediately behind the shock. Determine the pressure immediately downstream from the shock. Calculate the entropy change across the shock. Sketch the Ts diagram for this flow, indicating static and stagnation state points for conditions at the nozzle throat, both sides of the shock, and the exit plane.

*13.192 A converging-diverging nozzle, with throat area $A_t = 1.0$ in.^{2}, is attached to a large tank in which the pressure and temperature are maintained at 100 psia and 600 R. The nozzle exit area is 1.58 in.^{2} Determine the exit Mach number at design conditions. Referring to Fig. 13.20, determine the back pressures corresponding to the boundaries of Regimes I, II, III, and IV. Sketch the corresponding plot for this nozzle.

*13.193 A converging-diverging nozzle, with $A_e/A_t = 4.0$, is designed to expand air isentropically to atmospheric pressure. Determine the exit Mach number at design conditions and the required inlet stagnation pressure. Referring to Fig. 13.20,

---

* Problems marked with an asterisk are designed to be solved using tables.

determine the back pressures that correspond to the boundaries of Regimes I, II, III, and IV. Sketch the plot of pressure ratio versus axial distance for this nozzle.

*13.194 Air flows adiabatically from a reservoir, where $T = 60$ C and $p = 600$ kPa (abs), through a converging-diverging nozzle with $A_e/A_t = 4.0$. A normal shock occurs where $M = 2.42$. Assuming isentropic flow before and after the shock, determine the back pressure downstream from the nozzle. Sketch the pressure distribution.

*13.195 A normal shock occurs in the diverging section of a converging-diverging nozzle where $A = 4.0$ in.^{2} and $M = 2.50$. Upstream, $T_0 = 1000$ R and $p_0 = 100$ psia. The nozzle exit area is 6.0 in.^{2} Assume the flow is isentropic except across the shock. Determine the nozzle exit pressure, throat area, and mass flow rate.

*13.196 A converging-diverging nozzle is designed to expand air isentropically to atmospheric pressure from a large tank, where $T_0 = 150$ C and $p_0 = 790$ kPa (abs). A normal shock stands in the diverging section, where $p = 160$ kPa (abs) and $A = 600$ mm^2. Determine the nozzle back pressure, exit area, and throat area.

*13.197 A converging-diverging nozzle, with design pressure ratio $p_e/p_0 = 0.128$, is operated with a back pressure condition such that $p_b/p_0 = 0.830$, causing a normal shock to stand in the diverging section. Determine the Mach number at which the shock occurs.

*13.198 Air flows through a converging-diverging nozzle, with $A_e/A_t = 3.5$. The upstream stagnation conditions are atmospheric; the back pressure is maintained by a vacuum system. Determine the range of back pressures for which a normal shock will occur within the nozzle and the corresponding mass flow rate, if $A_t = 500$ mm^2.

*13.199 Air flows through a converging-diverging nozzle with $A_e/A_t = 1.87$. Upstream, $T_{0_1} = 240$ F and $p_{0_1} = 100$ psia. The back pressure is maintained at 40 psia. Determine the Mach number and flow speed in the nozzle exit plane.

*13.200 A converging-diverging nozzle, with $A_e/A_t = 1.633$, is designed to operate with atmospheric pressure at the exit plane. Determine the range(s) of stagnation pressures for which the nozzle will be free from normal shocks.

*13.201 A normal shock stands in a section of insulated constant-area duct. The flow is frictional. At section ①, some distance upstream from the shock, $T_1 = 470$ R. At section ④, some distance downstream from the shock, $T_4 = 750$ R and $M_4 = 1.0$. Denote conditions immediately upstream and downstream from the shock by subscripts ② and ③, respectively. Sketch the pressure distribution along the duct, indicating clearly the locations of sections ① through ④. Sketch a Ts diagram for the flow. Determine the Mach number at section ①.

*13.202 A normal shock stands in a section of insulated constant-area duct. The flow is frictional. At section ①, some distance upstream from the shock, $T_1 = 668$ R, $p_{0_1} = 78.2$ psia, and $M_1 = 2.05$. At section ④, some distance downstream from the shock, $M_4 = 1.00$. Calculate the air speed, V_2, immediately ahead of the shock, where $T_2 = 388$ F. Evaluate the entropy change, $s_4 - s_1$.

* Problems marked with an asterisk are designed to be solved using tables.

Appendix A

FLUID PROPERTY DATA

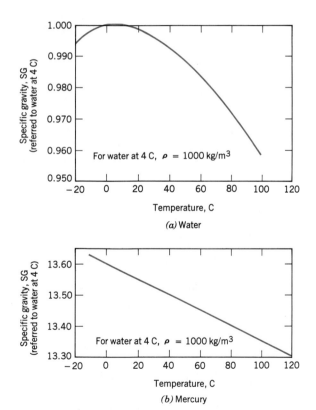

For water at 4 C, $\rho = 1000$ kg/m³

Temperature, C

(a) Water

For water at 4 C, $\rho = 1000$ kg/m³

Temperature, C

(b) Mercury

Fig. A.1 Specific gravity of water and mercury as functions of
temperature. (Data from [1].)]
(The specific gravity of mercury varies linearly with
temperature. The variation is given by SG $= 13.60 - 0.00240T$
when T is measured in degrees C.)

Table A.1 Specific Gravities of Selected Engineering Materials

(a) Common Manometer Liquids at 20 C (Data from [1, 2, 3].)

Liquid	Specific Gravity[a]
E.V. Hill blue oil	0.797
Meriam red oil	0.827
Benzene	0.879
Dibutyl phthalate	1.04
Monochloronaphthalene	1.20
Carbon tetrachloride	1.595
Bromoethylbenzene (Meriam blue)	1.75
Tetrabromoethane	2.95
Mercury	13.55

[a] Specific gravity, $SG = \rho/\rho_{H_2O}$ (at 4 C); ρ_{H_2O} (at 4 C) = 1000 kg/m^3 (1.94 slug/ft^3).

(b) Common Materials (Data from [4].)

Material	Specific Gravity (—)
Aluminum	2.64
Balsa wood	0.14
Brass	8.55
Cast Iron	7.08
Concrete (cured)	2.4*
Concrete (liquid)	2.5*
Copper	8.91
Ice (0 C)	0.917
Lead	11.4
Oak	0.77
Steel	7.83
Styrofoam (1 pcf**)	0.0160
Styrofoam (3 pcf)	0.0481
Uranium (depleted)	18.7
White pine	0.43

* depending on aggregate
** pounds per cubic foot

Table A.2   Physical Properties of Common Liquids at 20 C (Data
from [1, 5, 6].)

Liquid	Isentropic Bulk Modulus[a] (GN/m^2)	Specific Gravity (—)
Benzene	1.48	0.879
Carbon tetrachloride	1.36	1.595
Castor oil	2.11	0.969
Crude oil	—	0.82–0.92
Gasoline	—	0.72
Glycerin	4.59	1.26
Heptane	0.886	0.684
Kerosine	1.43	0.82
Lubricating oil	1.44	0.88
Mercury	28.5	13.55
Octane	0.963	0.702
Seawater[b]	2.42	1.025
SAE 10W oil	—	0.92
Water	2.24	0.998

[a] Calculated from speed of sound; 1 GN/m^2 = 10^9 N/m^2 (1 N/m^2 = 1.45×10^{-4} lbf/in.^{2}).
[b] Dynamic viscosity of seawater at 20 C is $\mu = 1.08 \times 10^{-3}$ N·sec/m^2.

Table A.3 Properties of the U.S. Standard Atmosphere (Data from [7].)

Geometric Altitude (m)	Temperature (K)	p/p_{SL} (—)	ρ/ρ_{SL} (—)
−500	291.4	1.061	1.049
0	288.2	1.000[a]	1.000[b]
500	284.9	0.9421	0.9529
1,000	281.7	0.8870	0.9075
1,500	278.4	0.8345	0.8638
2,000	275.2	0.7846	0.8217
2,500	271.9	0.7372	0.7812
3,000	268.7	0.6920	0.7423
3,500	265.4	0.6492	0.7048
4,000	262.2	0.6085	0.6689
4,500	258.9	0.5700	0.6343
5,000	255.7	0.5334	0.6012
6,000	249.2	0.4660	0.5389
7,000	242.7	0.4057	0.4817
8,000	236.2	0.3519	0.4292
9,000	229.7	0.3040	0.3813
10,000	223.3	0.2615	0.3376
11,000	216.8	0.2240	0.2978
12,000	216.7	0.1915	0.2546
13,000	216.7	0.1636	0.2176
14,000	216.7	0.1399	0.1860
15,000	216.7	0.1195	0.1590
16,000	216.7	0.1022	0.1359
17,000	216.7	0.08734	0.1162
18,000	216.7	0.07466	0.09930
19,000	216.7	0.06383	0.08489
20,000	216.7	0.05457	0.07258
22,000	218.6	0.03995	0.05266
24,000	220.6	0.02933	0.03832
26,000	222.5	0.02160	0.02797
28,000	224.5	0.01595	0.02047
30,000	226.5	0.01181	0.01503
40,000	250.4	0.002834	0.003262
50,000	270.7	0.0007874	0.0008383
60,000	255.8	0.0002217	0.0002497
70,000	219.7	0.00005448	0.00007146
80,000	180.7	0.00001023	0.00001632
90,000	180.7	0.000001622	0.000002588

[a] $p_{SL} = 1.01325 \times 10^5$ N/m^2 abs (= 14.696 psia)
[b] $\rho_{SL} = 1.2250$ kg/m^3 (= 0.002377 slug/ft^3)

A-1 SURFACE TENSION

The values of surface tension, σ, for most organic compounds are remarkably similar at room temperature; the typical range is 25 to 40 mN/m. Water is higher, at about 73 mN/m at 20 C. Liquid metals have values in the range between 300 and 600 mN/m; liquid mercury has a value of about 480 mN/m at 20 C. Surface tension decreases with temperature; the decrease is nearly linear with absolute temperature. Surface tension at the critical temperature is zero.

Values of σ are usually reported for surfaces in contact with the pure vapor of the liquid being studied or with air. At low pressures both values are about the same.

Table A.4 Surface Tension of Common Liquids at 20 C (Data from [1, 5, 8, 9].)

Liquid	Surface Tension, σ (mN/m)[a]	Contact Angle, θ (degrees)
(a) In contact with air		
Benzene	28.9	
Carbon tetrachloride	27.0	
Glycerin	63.0	
Hexane	18.4	
Kerosine	26.8	
Lube oil	25–35	
Mercury	484	140
Methanol	22.6	
Octane	21.8	
Water	72.8	~ 0
(b) In contact with water		
Benzene	35.0	
Carbon tetrachloride	45.0	
Hexane	51.1	
Mercury	375	140
Methanol	22.7	
Octane	50.8	

[a] 1 mN/m = 10^{-3} N/m.

A-2 THE PHYSICAL NATURE OF VISCOSITY

Viscosity is a measure of internal fluid friction, i.e., resistance to deformation. The mechanism of gas viscosity is reasonably well understood, but the theory is poorly developed for liquids. We can gain some insight into the physical nature of viscous flow by discussing these mechanisms briefly.

The viscosity of a Newtonian fluid is fixed by the state of the material. Thus $\mu = \mu(T, p)$. Temperature is the more important variable, so let us consider it first. Excellent empirical equations for viscosity as a function of temperature are available.

A-2.1 Effect of Temperature on Viscosity

a. Gases

All gas molecules are in continuous random motion. When there is bulk motion due to flow, the bulk motion is superimposed on the random motions. It is then distributed throughout the fluid by molecular collisions. Analyses based on kinetic theory predict

$$\mu \propto \sqrt{T}$$

The kinetic theory prediction is in fair agreement with experimental trends, but the constant of proportionality and one or more correction factors must be determined; this limits practical application of this simple equation.

If two or more experimental points are available, the data may be correlated using the empirical Sutherland correlation

$$\mu = \frac{bT^{1/2}}{1 + S/T} \tag{A.1}$$

Constants b and S may be determined most simply by writing

$$\mu = \frac{bT^{3/2}}{S + T}$$

or

$$\frac{T^{3/2}}{\mu} = \left(\frac{1}{b}\right)T + \frac{S}{b}$$

(Compare this with $y = mx + c$.) From a plot of $T^{3/2}/\mu$ versus T, one obtains the slope, $1/b$, and the intercept, S/b. For air,

$$b = 1.458 \times 10^{-6} \frac{\text{kg}}{\text{m} \cdot \text{sec} \cdot \text{K}^{1/2}}$$

$$S = 110.4 \text{ K}$$

These constants were used with Eq. A.1 to compute viscosities for the standard atmosphere in [7].

b. Liquids

Viscosities for liquids cannot be estimated well theoretically. The phenomenon of momentum transfer by molecular collisions seems overshadowed in liquids by the effects of interacting force fields among the closely packed liquid molecules.

Liquid viscosities are affected drastically by temperature. This dependence on absolute temperature is well represented by the empirical equation

$$\mu = Ae^{B/T} \tag{A.2}$$

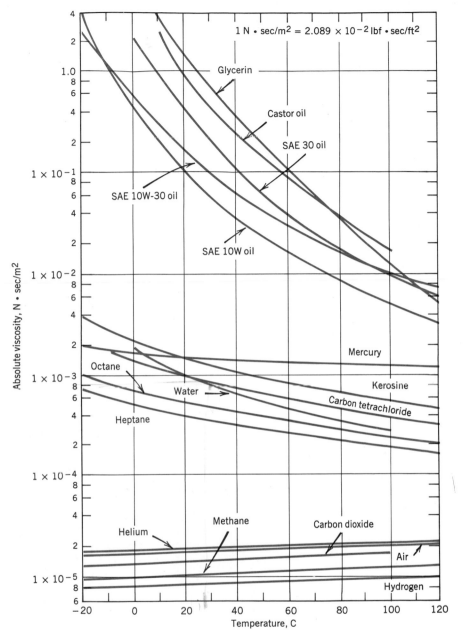

Fig. A.2 Dynamic (absolute) viscosity of common fluids as a function of temperature. (Data from [1, 6, and 10].)

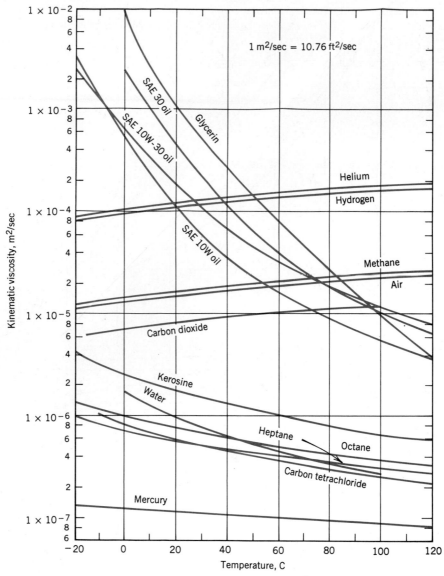

Fig. A.3 Kinematic viscosity of common fluids (at atmospheric pressure) as a function of temperature. (Data from [1, 6, and 10].)

Equation A.2 requires two points to fit A and B. This may be done easily by plotting $\log \mu$ versus $1/T$ in the form

$$\log \mu = \log A + \frac{B}{T}\log e = \log A + \frac{0.434\,B}{T}$$

Log A is the intercept and $0.434B$ the slope of the resulting plot.

Note that the viscosity of a liquid decreases with temperature, while that of a gas increases with temperature.

A-2.2   Effect of Pressure on Viscosity

a.   Gases

The viscosity of gases is essentially independent of pressure between a few hundredths of an atmosphere and a few atmospheres. However, viscosity at high pressures increases with pressure (or density).

b.   Liquids

The viscosities of most liquids are not affected by moderate pressures, but large increases have been found at very high pressures. For example, the viscosity of water at 10,000 atm is twice that at 1 atm. More complex compounds show a viscosity increase of several orders of magnitude over the same pressure range.

More information may be found in [11].

Table A.5   Allowable Viscosity Ranges for SAE Lubricant Classifications (Data from [12].)

| Lubricant Type | SAE Viscosity Number | Viscosity Range (centistokes)[a] | | | |
| | | At 0 F | | At 210 F | |
		Minimum	Maximum	Minimum	Maximum
Crankcase	5 W		1300	3.9	
	10 W	1300	2600	3.9	
	20 W	2600	10,500	3.9	
	20			5.7	9.6
	30			9.6	12.9
	40			12.9	16.8
	50			16.8	22.7
Transmission and axle	75		15,000		
	80	15,000	100,000		
	90			75	120
	140			120	200
	250			200	
Automatic transmission fluid	Type A	39[b]	43[b]	7	8.5

[a] centistoke = 1 cSt = 10^{-6} m^2/sec (= 1.08×10^{-5} ft^2/sec).
[b] At 100 F.

Table A.6 Thermodynamic Properties of Common Gases at STP[a] (Data from [7, 14, 15].)

Gas	Chemical Symbol	Molecular Mass, M_m	R^b $\left(\dfrac{\text{J}}{\text{kg} \cdot \text{K}}\right)$	c_p $\left(\dfrac{\text{J}}{\text{kg} \cdot \text{K}}\right)$	c_v $\left(\dfrac{\text{J}}{\text{kg} \cdot \text{K}}\right)$	$k = \dfrac{c_p}{c_v}$ (—)	R^b $\left(\dfrac{\text{ft} \cdot \text{lbf}}{\text{lbm} \cdot \text{R}}\right)$	c_p $\left(\dfrac{\text{Btu}}{\text{lbm} \cdot \text{R}}\right)$	c_v $\left(\dfrac{\text{Btu}}{\text{lbm} \cdot \text{R}}\right)$
Air	——	28.98	286.9	1004	717.4	1.40	53.33	0.2399	0.1713
Carbon dioxide	CO_2	44.01	188.9	840.4	651.4	1.29	35.11	0.2007	0.1556
Carbon monoxide	CO	28.01	296.8	1039	742.1	1.40	55.17	0.2481	0.1772
Helium	He	4.003	2077	5225	3147	1.66	386.1	1.248	0.7517
Hydrogen	H_2	2.016	4124	14,180	10,060	1.41	766.5	3.388	2.402
Methane	CH_4	16.04	518.3	2190	1672	1.31	96.32	0.5231	0.3993
Nitrogen	N_2	28.01	296.8	1039	742.0	1.40	55.16	0.2481	0.1772
Oxygen	O_2	32.00	259.8	909.4	649.6	1.40	48.29	0.2172	0.1551
Steam[c]	H_2O	18.02	461.4	~2000	~1540	~1.30	85.78	~0.478	~0.368

[a] STP = standard temperature and pressure, T = 15 C (59 F) and p = 101.325 kPa abs (14.696 psia).

[b] $R \equiv R_u/M_m$; R_u = 8314.3 J/kgmol·K (1545.3 ft·lbf/lbmol·R); 1Btu = 778.2 ft·lbf.

[c] Water vapor behaves as an ideal gas when superheated by 55 C (100 F) or more.

A-3 LUBRICATING OILS

Engine and transmission lubricating oils are classified by viscosity according to standards established by the Society of Automotive Engineers [12]. The allowable viscosity ranges for several grades are given in Table A.5, page 765.

Viscosity numbers with W (e.g., 20W) are classified by viscosity at 0 F. Those without W are classified by viscosity at 210 F.

Multigrade oils (e.g., 10W-40) are formulated to minimize viscosity variation with temperature. High polymer "viscosity index" improvers are used in blending these multigrade oils. Such additives are highly non-Newtonian; they may suffer permanent viscosity loss due to shearing.

Special charts used to estimate the viscosity of petroleum products as a function of temperature are available. The charts were used to develop the data for typical lubricating oils plotted in Figs. A.2 and A.3. For details, see [13].

REFERENCES

1. *Handbook of Chemistry and Physics,* 62nd ed. Cleveland, OH: Chemical Rubber Publishing Co., 1981–1982.

2. "Meriam Standard Indicating Fluids," Pamphlet No. 920GEN: 430-1, The Meriam Instrument Co., 10920 Madison Avenue, Cleveland, OH 44102.

3. E. Vernon Hill, Inc., P.O. Box 7053, Corte Madera, CA 94925.

4. Avallone, E. A., and T. Baumeister, III, eds., *Marks' Standard Handbook for Mechanical Engineers,* 9th ed. New York: McGraw-Hill, 1987.

5. *Handbook of Tables for Applied Engineering Science.* Cleveland, OH: Chemical Rubber Publishing Co., 1970.

6. Vargaftik, N. B., *Tables on the Thermophysical Properties of Liquids and Gases,* 2nd ed. Washington, D.C.: Hemisphere Publishing Corp., 1975.

7. *The U.S. Standard Atmosphere (1962).* Washington, D.C.: U.S. Government Printing Office, 1962.

8. Trefethen, L., "Surface Tension in Fluid Mechanics," in *Illustrated Experiments in Fluid Mechanics.* Cambridge, MA: The M.I.T. Press, 1972.

9. Streeter, V. L., ed., *Handbook of Fluid Dynamics.* New York: McGraw-Hill, 1961.

10. Touloukian, Y. S., S. C. Saxena, and P. Hestermans, *Thermophysical Properties of Matter, the TPRC Data Series, Vol. 11—Viscosity.* New York: Plenum Publishing Corp., 1975.

11. Reid, R. C., and T. K. Sherwood, *The Properties of Gases and Liquids,* 2nd ed. New York: McGraw-Hill, 1966.

12. "Crankcase Oil Viscosity Classification, Recommended Practice SAE J300b," *SAE Handbook,* 1976 ed. Warrendale, PA: Society of Automotive Engineers, 1976.

13. ASTM Standard D 341–77, "Viscosity-Temperature Charts for Liquid Petroleum Products," American Society for Testing and Materials, 1916 Race Street, Philadelphia, PA 19103.

14. NASA, *Compressed Gas Handbook* (Revised). Washington, D.C.: National Aeronautics and Space Administration, SP-3045, 1970.

15. ASME, *Thermodynamic and Transport Properties of Steam.* New York: American Society of Mechanical Engineers, 1967.

EQUATIONS OF MOTION IN CYLINDRICAL COORDINATES

The continuity equation in cylindrical coordinates for constant density is

$$\frac{1}{r}\frac{\partial}{\partial r}(rv_r) + \frac{1}{r}\frac{\partial}{\partial \theta}(v_\theta) + \frac{\partial}{\partial z}(v_z) = 0 \tag{B.1}$$

Normal and shear stresses in cylindrical coordinates for constant density and viscosity are

$$\sigma_{rr} = -p + 2\mu\frac{\partial v_r}{\partial r} \qquad\qquad \tau_{r\theta} = \mu\left[r\frac{\partial}{\partial r}\left(\frac{v_\theta}{r}\right) + \frac{1}{r}\frac{\partial v_r}{\partial \theta}\right]$$

$$\sigma_{\theta\theta} = -p + 2\mu\left(\frac{1}{r}\frac{\partial v_\theta}{\partial \theta} + \frac{v_r}{r}\right) \qquad \tau_{\theta z} = \mu\left(\frac{\partial v_\theta}{\partial z} + \frac{1}{r}\frac{\partial v_z}{\partial \theta}\right)$$

$$\sigma_{zz} = -p + 2\mu\frac{\partial v_z}{\partial z} \qquad\qquad \tau_{zr} = \mu\left(\frac{\partial v_r}{\partial z} + \frac{\partial v_z}{\partial r}\right) \tag{B.2}$$

The Navier–Stokes equations in cylindrical coordinates for constant density and viscosity are

r component:

$$\rho\left(\frac{\partial v_r}{\partial t} + v_r\frac{\partial v_r}{\partial r} + \frac{v_\theta}{r}\frac{\partial v_r}{\partial \theta} - \frac{v_\theta^2}{r} + v_z\frac{\partial v_r}{\partial z}\right)$$

$$= \rho g_r - \frac{\partial p}{\partial r} + \mu\left\{\frac{\partial}{\partial r}\left(\frac{1}{r}\frac{\partial}{\partial r}[rv_r]\right) + \frac{1}{r^2}\frac{\partial^2 v_r}{\partial \theta^2} - \frac{2}{r^2}\frac{\partial v_\theta}{\partial \theta} + \frac{\partial^2 v_r}{\partial z^2}\right\} \tag{B.3a}$$

θ component:

$$\rho\left(\frac{\partial v_\theta}{\partial t} + v_r\frac{\partial v_\theta}{\partial r} + \frac{v_\theta}{r}\frac{\partial v_\theta}{\partial \theta} + \frac{v_r v_\theta}{r} + v_z\frac{\partial v_\theta}{\partial z}\right)$$

$$= \rho g_\theta - \frac{1}{r}\frac{\partial p}{\partial \theta} + \mu\left\{\frac{\partial}{\partial r}\left(\frac{1}{r}\frac{\partial}{\partial r}[rv_\theta]\right) + \frac{1}{r^2}\frac{\partial^2 v_\theta}{\partial \theta^2} + \frac{2}{r^2}\frac{\partial v_r}{\partial \theta} + \frac{\partial^2 v_\theta}{\partial z^2}\right\} \tag{B.3b}$$

z component:

$$\rho\left(\frac{\partial v_z}{\partial t} + v_r\frac{\partial v_z}{\partial r} + \frac{v_\theta}{r}\frac{\partial v_z}{\partial \theta} + v_z\frac{\partial v_z}{\partial z}\right)$$

$$= \rho g_z - \frac{\partial p}{\partial z} + \mu\left\{\frac{1}{r}\frac{\partial}{\partial r}\left(r\frac{\partial v_z}{\partial r}\right) + \frac{1}{r^2}\frac{\partial^2 v_z}{\partial \theta^2} + \frac{\partial^2 v_z}{\partial z^2}\right\} \tag{B.3c}$$

Appendix C

VIDEOTAPES AND FILMS FOR FLUID MECHANICS

Listed below by supplier are titles of videotapes and 16-mm films on fluid mechanics.

1. Encyclopaedia Britannica Educational Corporation
   310 South Michigan Avenue
   Chicago, Illinois 60604

   The following twenty two films developed by the National Committee for Fluid Mechanics Films (NCFMF)[1] are available on videotape (length as noted):

   Aerodynamic Generation of Sound (44 min, principals: M. J. Lighthill, J. E. Ffowcs-Williams)

   Boundary Layer Control (25 min, principal: D. C. Hazen)

   Cavitation (31 min, principal: P. Eisenberg)

   Channel Flow of a Compressible Fluid (29 min, principal: D. E. Coles)

   Deformation of Continuous Media (38 min, principal: J. L. Lumley)

   Eulerian and Lagrangian Descriptions in Fluid Mechanics (27 min, principal: J. L. Lumley)

   Flow Instabilities (27 min, principal: E. L. Mollo-Christensen)

   Flow Visualization (31 min, principal: S. J. Kline)

   The Fluid Dynamics of Drag[2] (4 parts, 120 min, principal: A. H. Shapiro)

   Fundamentals of Boundary Layers (24 min, principal: F. H. Abernathy)

   Low-Reynolds-Number Flows (33 min, principal: Sir G. I. Taylor)

   Magnetohydrodynamics (27 min, principal: J. A. Shercliff)

   Pressure Fields and Fluid Acceleration (30 min, principal: A. H. Shapiro)

   Rarefied Gas Dynamics (33 min, principals: F. C. Hurlbut, F. S. Sherman)

   Rheological Behavior of Fluids (22 min, principal: H. Markovitz)

   Rotating Flows (29 min, principal: D. Fultz)

   Secondary Flow (30 min, principal: E. S. Taylor)

   Stratified Flow (26 min, principal: R. R. Long)

   Surface Tension in Fluid Mechanics (29 min, principal: L. M. Trefethen)

   Turbulence (29 min, principal: R. W. Stewart)

   Vorticity (2 parts, 44 min, principal: A. H. Shapiro)

   Waves in Fluids (33 min, principal: A. E. Bryson)

---

[1] Detailed summaries of the NCFMF films are contained in *Illustrated Experiments in Fluid Mechanics* (Cambridge, MA: The M.I.T. Press, 1972).

[2] The contents of this film are summarized and illustrated in *Shape and Flow: The Fluid Dynamics of Drag*, by Ascher H. Shapiro (New York: Anchor Books, 1961).

2. University of Iowa
   The Audiovisual Center
   C4 Seashore Hall
   Iowa City, Iowa 52240

   The following six films were prepared as a series, in the order listed. They can be viewed individually without serious loss of continuity.

   Introduction to the Study of Fluid Motion (24 min, principal: H. Rouse). This orientation film shows a variety of familiar flow phenomena. Use of scale models for empirical study of complex phenomena is illustrated and the significance of the Euler, Froude, Mach, and Reynolds numbers as similarity parameters is shown using several sequences of model and prototype flows.
   Fundamental Principles of Flow (23 min, principal: H. Rouse). The basic concepts and physical relationships needed to analyze fluid motions are developed in this film. The continuity, momentum, and energy equations are derived and used to analyze a jet propulsion device.
   Fluid Motion in a Gravitational Field (23 min, principal: H. Rouse). Buoyancy effects and free-surface flows are illustrated in this film. The Froude number is shown to be a fundamental parameter for flows with a free surface. Wave motions are shown for open-channel and density-stratified flows.
   Characteristics of Laminar and Turbulent Flow (26 min, principal: H. Rouse). Dye, smoke, suspended particles, and hydrogen bubbles are used to visualize laminar and turbulent flows. Instabilities that lead to turbulence are shown; production and decay of turbulence and mixing are described.
   Form Drag, Lift, and Propulsion (24 min, principal: H. Rouse). The effects of boundary-layer separation on flow patterns and pressure distributions are shown for several body shapes. The basic characteristics of lifting shapes, including effects of aspect ratio, are discussed, and the results are applied to analysis of the performance of propellers and torque converters.
   Effects of Fluid Compressibility (17 min, principal: H. Rouse). The hydraulic analogy between open-channel liquid flow and compressible gas flow is used to show representative wave patterns. Schlieren optical flow visualization is used in a supersonic wind tunnel to show patterns of flow past several bodies at subsonic and supersonic speeds.

3. University of Minnesota
   Saint Anthony Falls Hydraulics Laboratory
   Mississippi River and 3rd Avenue, SE
   Minneapolis, Minnesota 55414

   Some Phenomena of Open-Channel Flow (33 min, silent). Many features of open-channel flow are demonstrated. Applications of the specific energy and pressure-momentum curves are illustrated.
   Fluid Mechanics—The Boundary Layer (30 min, sound). The film contains a series of demonstrations of physical principles. It is planned for use as a summary after regular classroom discussion of boundary-layer phenomena.

4. American Institute of Aeronautics and Astronautics
   370 L'Enfant Promenade, S.W.
   Washington, D.C. 20024-2518

   America's Wings (29 min). Individuals who made significant contributions to development of aircraft for high-speed flight are interviewed; they discuss and explain their contributions. This is an effective film for a relatively sophisticated audience.

5. Purdue University
    Center for Instructional Services
    Film Booking
    Stewart Center
    West Lafayette, Indiana 47907

    Tacoma Narrows Bridge Collapse (3 min, color, silent; available as 16-mm film or VHS videocassette). This brief film contains spectacular original footage from the spontaneous collapse in a light breeze of the 2800 ft suspension bridge over the Tacoma Narrows, which occurred November 7, 1940.

Appendix D

SELECTED PERFORMANCE CURVES FOR PUMPS AND FANS

D-1 INTRODUCTION

Many firms, worldwide, manufacture fluid machines in numerous standard types and sizes. Each manufacturer publishes complete performance data to allow application of their machines in systems. This Appendix contains selected performance data for use in solving pump and fan system problems. Two pump types and one fan type are included.

Choice of a manufacturer may be based on established practice, location, or cost. Once a manufacturer is chosen, machine selection is a three-step process:

1. Select a machine type, suited to the application, from a manufacturer's full-line catalog, which gives the ranges of pressure rise (head) and flow rate for each machine type.
2. Choose an appropriate machine model and driver speed from a master selector chart, which superposes the head and flow rate ranges of a series of machines on one graph.
3. Verify that the candidate machine is satisfactory for the intended application, using a detailed performance curve for the specific machine.

It is wise to consult with experienced system engineers, either employed by the machine manufacturer, or in your own organization, before making a final purchase decision.

Many manufacturers currently use computerized procedures to select a machine that is most suitable for each given application. Such procedures are simply automated versions of the traditional selection method. Use of the master selector chart and the detailed performance curves is illustrated below for pumps and fans, using data from one manufacturer of each type of machine. Literature of other manufacturers differs in detail, but contains the necessary information for machine selection.

D-2 PUMP SELECTION

Representative data are shown in Figs. D.1 through D.10 for Peerless[1] horizontal split case single-stage (series AE) pumps and in Figs. D.11 and D.12 for Peerless multi-stage (series TU and TUT) pumps.

Figures D.1 and D.2 are master pump selector charts for series AE pumps at 3500 and 1750 nominal rpm. On these charts, the model number (e.g., 6AE14) indicates

---

[1] Peerless Pump Division, FMC Corporation, P.O. Box 7026, Indianapolis, IN 46207-7026.

the discharge line size (6 in. nominal pipe), the pump series (AE), and the maximum impeller diameter (approximately 14 in.).

Figures D.3 through D.10 are detailed performance charts for individual pump models in the AE series.

Figures D.11 and D.12 are master pump selector charts for series TU and TUT pumps at 1750 nominal rpm. Data for two-stage pumps are presented in Fig. D.11, while Fig. D.12 contains data for pumps with three, four, and five stages.

Each pump performance chart contains curves of total head versus volume flow rate; curves for several impeller diameters—tested in the same casing—are presented on a single graph. Each performance chart also contains contours showing pump efficiency and driver power; the *NPSH* requirement, as it varies with flow rate, is shown by the curve at the bottom of each chart. The best efficiency point (BEP) for each impeller may be found using the efficiency contours.

Use of the master pump selector chart and detailed performance curves is illustrated in Example Problem D.1.

EXAMPLE D.1—Pump Selection Procedure

Select a pump to deliver 1750 gpm of water at 120 ft total head. Choose the appropriate pump model and driver speed. Specify the pump efficiency, driver power, and *NPSH* requirement.

EXAMPLE PROBLEM D.1

GIVEN: Select a pump to deliver 1750 gpm of water at 120 ft total head.

FIND: (a) Pump model and driver speed.   (b) Pump efficiency.
      (c) Driver power.   (d) *NPSH* requirement.

SOLUTION:

Use the pump selection procedure described in Section D-1. (The numbers below correspond to the numbered steps given in the procedure.)

1. First, select a machine type suited to the application. (This step actually requires a manufacturer's full-line catalog, which is not reproduced here. The Peerless product line catalog specifies a maximum delivery and head of 2500 gpm and 660 ft for series AE pumps. Therefore the required performance can be obtained; assume the selection is to be made from this series.)

2. Second, consult the master pump selector chart. The desired operating point is not within any pump contour on the 3500 rpm selector chart (Fig. D.1). From the 1750 rpm chart (Fig. D.2), select a model 6AE14 pump. From the performance curve for the 6AE14 pump (Fig. D.6), choose a 13 in. impeller.

3. Third, verify the performance of the machine using the detailed performance chart. On the performance chart for the 6AE14 pump, project up from the abscissa at $Q = 1750$ gpm. Project across from $H = 120$ ft on the ordinate. The intersection is the pump performance at the desired operating point:

$$\eta \approx 85.8 \text{ percent} \qquad \mathcal{P} \approx 64 \text{ hp}$$

From the operating point, project down to the *NPSH* requirement curve. At the intersection, read $NPSH \approx 17$ ft.

This completes the selection process for this pump. One should consult with experienced system engineers to verify that the system operating condition has been predicted accurately and the pump has been selected correctly.

D-3 FAN SELECTION

Fan selection is similar to pump selection. A representative master fan selection chart is shown in Fig. D.13 for a series of Buffalo Forge[2] axial-flow fans. The chart shows the performance of the entire series of fans, as a function of fan size and driver speed.

The master fan selector chart is used to select a fan size and driver speed for detailed consideration. Final evaluation of suitability of the fan model for the application is done using detailed performance charts for the specific model. A sample performance chart for a Buffalo Forge Size 48, Vaneaxial fan, is presented in Fig. D.14.

The performance chart is plotted as total pressure rise versus volume flow rate. Figure D.14 contains curves for HB, LB, and MB wheels, operating at various constant speeds; the shaded bands represent measured total efficiency for the fans.

EXAMPLE D.2—Fan Selection Procedure

Select an axial-flow fan to deliver 40,000 cfm of standard air at 1.25 in. H_2O total pressure. Choose the appropriate fan model and driver speed. Specify the fan efficiency and driver power.

EXAMPLE PROBLEM D.2

GIVEN: Select an axial fan to deliver 40,000 cfm of standard air at 1.25 in. H_2O total head.

FIND: (a) Fan size and driver speed.   (b) Fan efficiency.
       (c) Driver power.

SOLUTION:
Use the fan selection procedure described in Section D-1. (The numbers below correspond to the numbered steps given in the procedure.)

1. First, select a machine type suited to the application. (This step actually requires a manufacturer's full-line catalog, which is not reproduced here. Assume the fan selection is to be made from the axial fan data presented in Fig. D.13.)
2. Second, consult the master fan selector chart. The desired operating point is within the contour for the Size 48 fan on the selector chart (Fig. D.13). To achieve the desired performance requires driving the fan at 870 rpm.
3. Third, verify the performance of the machine using the detailed performance chart (Fig. D.14). On the detailed performance chart, project up from the abscissa at $Q = 40,000$ cfm. Project across from $p = 1.25$ in. H_2O on the ordinate. The intersection is the desired operating point.

    These operating conditions cannot be delivered by a Type LB wheel; however, they are close to peak efficiency for either HB or MB wheels. The operating conditions can be delivered at about 72 percent total efficiency using an HB wheel. With an MB wheel, slightly in excess of 75 percent total efficiency may be expected. From the chart, the "efficiency factor" is 4780 at $\eta = 0.75$, and the fan driver power requirement is

$$\mathscr{P} = \frac{\text{Total Pressure} \times \text{Capacity}}{\text{Efficiency Factor}} = \frac{1.25 \text{ in. } H_2O \times 40,000 \text{ cfm}}{4780} = 10.5 \text{ hp}$$

$\left[\begin{array}{l}\text{This completes the fan selection process. Again, one should consult with experienced} \\ \text{system engineers to verify that the system operating condition has been predicted} \\ \text{accurately and the fan has been selected correctly.}\end{array}\right\}$

---

[2] Buffalo Forge, 465 Broadway, Buffalo, 14240 NY.

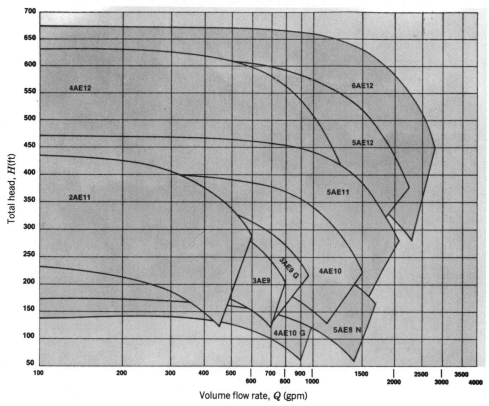

Fig. D.1   Selector chart for Peerless horizontal split case (series AE) pumps at 3500 nominal rpm.

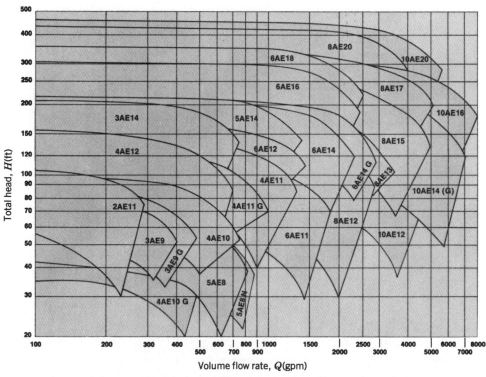

Fig. D.2   Selector chart for Peerless horizontal split case (series AE) pumps at 1750 nominal rpm.

775

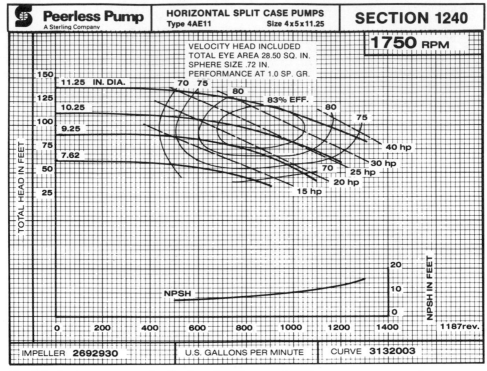

Fig. D.3   Performance curve for Peerless 4AE11 pump at 1750 rpm.

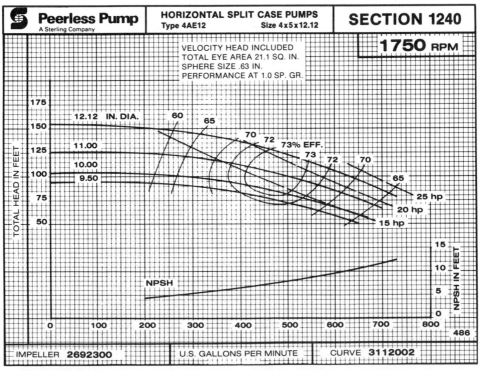

Fig. D.4   Performance curve for Peerless 4AE12 pump at 1750 rpm.

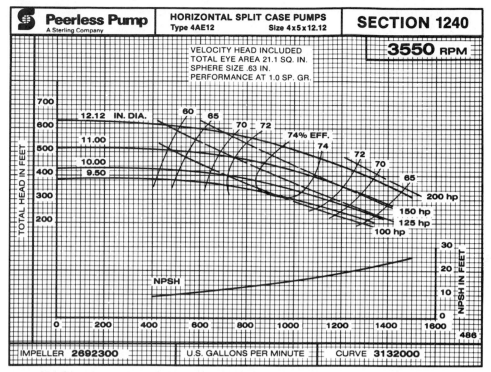

Fig. D.5 Performance curve for Peerless 4AE12 pump at 3500 rpm.

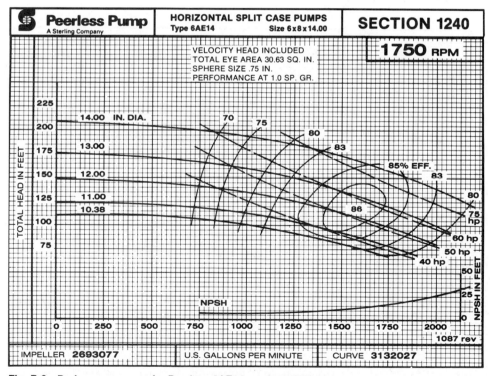

Fig. D.6 Performance curve for Peerless 6AE14 pump at 1750 rpm.

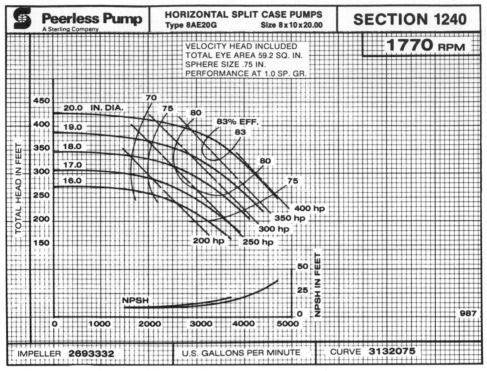

Fig. D.7 Performance curve for Peerless 8AE20G pump at 1770 rpm.

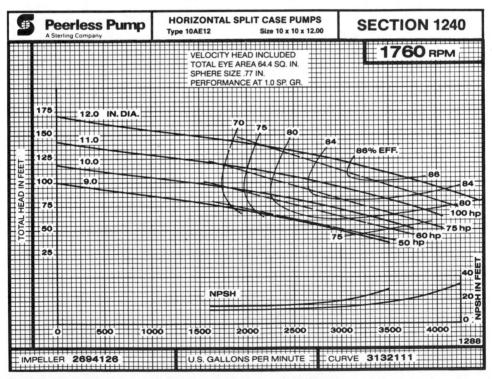

Fig. D.8 Performance curve for Peerless 10AE12 pump at 1760 rpm.

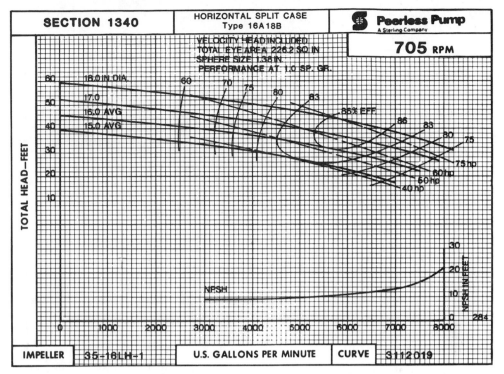

Fig. D.9   Performance curve for Peerless 16AE18B pump at 705 rpm.

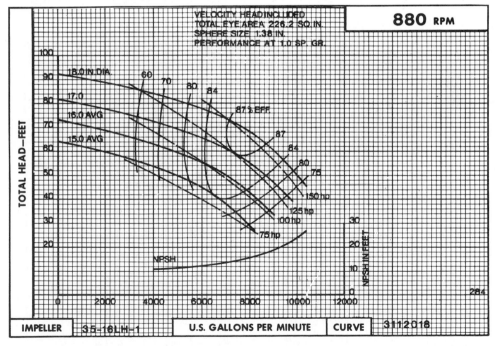

Fig. D.10   Performance curve for Peerless 16AE18B pump at 880 rpm.

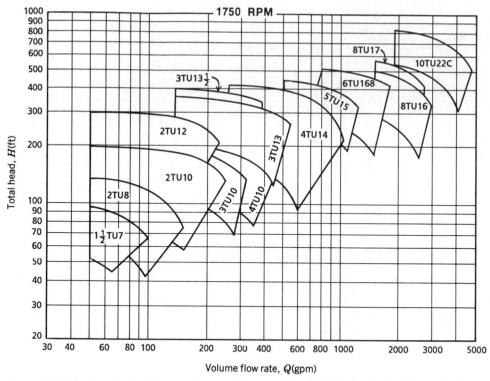

Fig. D.11 Selector chart for Peerless two-stage (series TU and TUT) pumps at 1750 nominal rpm.

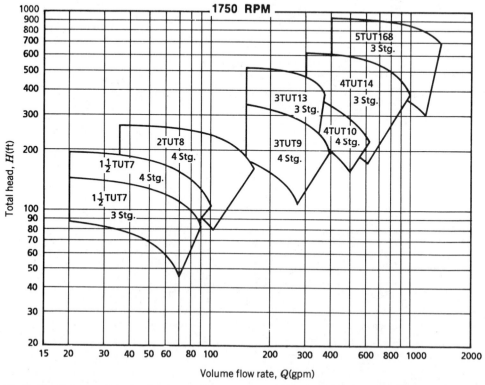

Fig. D.12 Selector chart for Peerless multi-stage (series TU and TUT) pumps at 1750 nominal rpm.

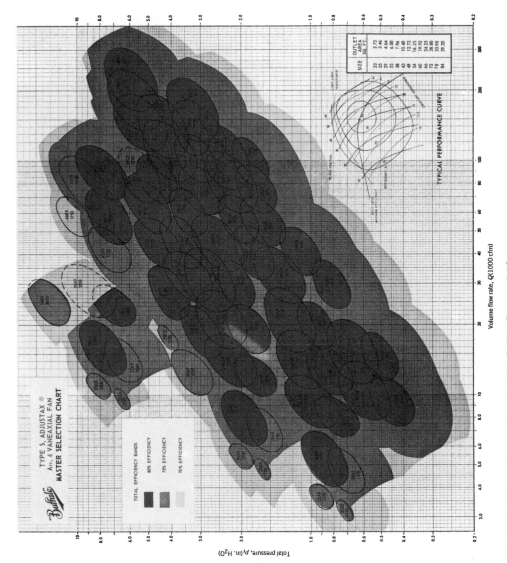

Fig. D.13 Master fan selection chart for Buffalo Forge axial fans.

SIZE 48

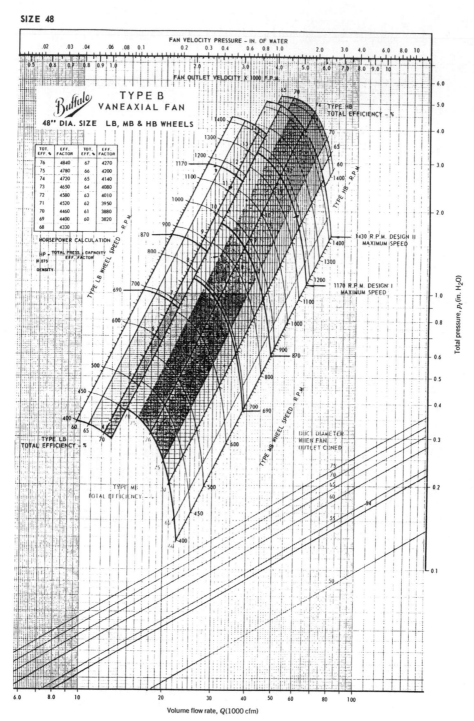

Fig. D.14 Performance chart for Buffalo Forge axial fans.

REFERENCES

1. Peerless Pump literature:
   - Horizontal Split Case Single Stage Double Suction Pumps, Series AE, Brochure B-1200, 1987.
   - Horizontal Split Case Series AE Pumps, 60 Hertz, Performance Curves, Brochure B-1240, n.d.
   - Horizontal Split Case, Multistage Single Suction Pumps, Types TU, TUT, 60 Hertz, Performance Curves, Brochure B-1440, n.d.
2. Buffalo Forge literature:
   - Axial Flow Fans, Bulletin F-305A, 1968.

Appendix E

TABLES FOR COMPUTATION OF COMPRESSIBLE FLOW

Table E.1 Isentropic Flow Functions (one-dimensional flow, ideal gas, $k = 1.4$)

M	T/T_0	p/p_0	ρ/ρ_0	A/A^*
0.00	1.0000	1.0000	1.0000	∞
0.02	0.9999	0.9997	0.9998	28.94
0.04	0.9997	0.9989	0.9992	14.48
0.06	0.9993	0.9975	0.9982	9.666
0.08	0.9987	0.9955	0.9968	7.262
0.10	0.9980	0.9930	0.9950	5.822
0.12	0.9971	0.9900	0.9928	4.864
0.14	0.9961	0.9864	0.9903	4.182
0.16	0.9949	0.9823	0.9873	3.673
0.18	0.9936	0.9777	0.9840	3.278
0.20	0.9921	0.9725	0.9803	2.964
0.22	0.9904	0.9669	0.9762	2.708
0.24	0.9886	0.9607	0.9718	2.496
0.26	0.9867	0.9541	0.9670	2.317
0.28	0.9846	0.9470	0.9619	2.166
0.30	0.9823	0.9395	0.9564	2.035
0.32	0.9799	0.9315	0.9506	1.922
0.34	0.9774	0.9231	0.9445	1.823
0.36	0.9747	0.9143	0.9380	1.736
0.38	0.9719	0.9052	0.9313	1.659
0.40	0.9690	0.8956	0.9243	1.590
0.42	0.9659	0.8857	0.9170	1.529
0.44	0.9627	0.8755	0.9094	1.474
0.46	0.9594	0.8650	0.9016	1.425
0.48	0.9560	0.8541	0.8935	1.380
0.50	0.9524	0.8430	0.8852	1.340

Table E.1 Isentropic Flow Functions (*Continued*)

M	T/T_0	p/p_0	ρ/ρ_0	A/A^*
0.50	0.9524	0.8430	0.8852	1.340
0.52	0.9487	0.8317	0.8766	1.303
0.54	0.9449	0.8201	0.8679	1.270
0.56	0.9410	0.8082	0.8589	1.240
0.58	0.9370	0.7962	0.8498	1.213
0.60	0.9328	0.7840	0.8405	1.188
0.62	0.9286	0.7716	0.8310	1.166
0.64	0.9243	0.7591	0.8213	1.145
0.66	0.9199	0.7465	0.8115	1.127
0.68	0.9154	0.7338	0.8016	1.110
0.70	0.9108	0.7209	0.7916	1.094
0.72	0.9061	0.7080	0.7814	1.081
0.74	0.9013	0.6951	0.7712	1.068
0.76	0.8964	0.6821	0.7609	1.057
0.78	0.8915	0.6691	0.7505	1.047
0.80	0.8865	0.6560	0.7400	1.038
0.82	0.8815	0.6430	0.7295	1.030
0.84	0.8763	0.6300	0.7189	1.024
0.86	0.8711	0.6170	0.7083	1.018
0.88	0.8659	0.6041	0.6977	1.013
0.90	0.8606	0.5913	0.6870	1.009
0.92	0.8552	0.5785	0.6764	1.006
0.94	0.8498	0.5658	0.6658	1.003
0.96	0.8444	0.5532	0.6551	1.001
0.98	0.8389	0.5407	0.6445	1.000
1.00	0.8333	0.5283	0.6339	1.000
1.02	0.8278	0.5160	0.6234	1.000
1.04	0.8222	0.5039	0.6129	1.001
1.06	0.8165	0.4919	0.6024	1.003
1.08	0.8108	0.4801	0.5920	1.005
1.10	0.8052	0.4684	0.5817	1.008
1.12	0.7994	0.4568	0.5714	1.011
1.14	0.7937	0.4455	0.5612	1.015
1.16	0.7880	0.4343	0.5511	1.020
1.18	0.7822	0.4232	0.5411	1.025
1.20	0.7764	0.4124	0.5311	1.030
1.22	0.7706	0.4017	0.5213	1.037
1.24	0.7648	0.3912	0.5115	1.043
1.26	0.7590	0.3809	0.5019	1.050
1.28	0.7532	0.3708	0.4923	1.058
1.30	0.7474	0.3609	0.4829	1.066
1.32	0.7416	0.3512	0.4736	1.075
1.34	0.7358	0.3417	0.4644	1.084
1.36	0.7300	0.3323	0.4553	1.094
1.38	0.7242	0.3232	0.4463	1.104
1.40	0.7184	0.3142	0.4374	1.115

Table E.1 Isentropic Flow Functions (*Continued*)

M	T/T_0	p/p_0	ρ/ρ_0	A/A^*
1.40	0.7184	0.3142	0.4374	1.115
1.42	0.7126	0.3055	0.4287	1.126
1.44	0.7069	0.2969	0.4201	1.138
1.46	0.7011	0.2886	0.4116	1.150
1.48	0.6954	0.2804	0.4032	1.163
1.50	0.6897	0.2724	0.3950	1.176
1.52	0.6840	0.2646	0.3869	1.190
1.54	0.6783	0.2570	0.3789	1.204
1.56	0.6726	0.2496	0.3711	1.219
1.58	0.6670	0.2423	0.3633	1.234
1.60	0.6614	0.2353	0.3557	1.250
1.62	0.6558	0.2284	0.3483	1.267
1.64	0.6502	0.2217	0.3409	1.284
1.66	0.6447	0.2152	0.3337	1.301
1.68	0.6392	0.2088	0.3266	1.319
1.70	0.6337	0.2026	0.3197	1.338
1.72	0.6283	0.1966	0.3129	1.357
1.74	0.6229	0.1907	0.3062	1.376
1.76	0.6175	0.1850	0.2996	1.397
1.78	0.6121	0.1794	0.2931	1.418
1.80	0.6068	0.1740	0.2868	1.439
1.82	0.6015	0.1688	0.2806	1.461
1.84	0.5963	0.1637	0.2745	1.484
1.86	0.5911	0.1587	0.2686	1.507
1.88	0.5859	0.1539	0.2627	1.531
1.90	0.5807	0.1492	0.2570	1.555
1.92	0.5756	0.1447	0.2514	1.580
1.94	0.5705	0.1403	0.2459	1.606
1.96	0.5655	0.1360	0.2405	1.633
1.98	0.5605	0.1318	0.2352	1.660
2.00	0.5556	0.1278	0.2301	1.688
2.02	0.5506	0.1239	0.2250	1.716
2.04	0.5458	0.1201	0.2200	1.745
2.06	0.5409	0.1164	0.2152	1.775
2.08	0.5361	0.1128	0.2105	1.806
2.10	0.5314	0.1094	0.2058	1.837
2.12	0.5266	0.1060	0.2013	1.869
2.14	0.5219	0.1027	0.1968	1.902
2.16	0.5173	0.09956	0.1925	1.935
2.18	0.5127	0.09650	0.1882	1.970
2.20	0.5081	0.09352	0.1841	2.005
2.22	0.5036	0.09064	0.1800	2.041
2.24	0.4991	0.08784	0.1760	2.078
2.26	0.4947	0.08514	0.1721	2.115
2.28	0.4903	0.08252	0.1683	2.154
2.30	0.4859	0.07997	0.1646	2.193

Table E.1   Isentropic Flow Functions (*Continued*)

M	T/T_0	p/p_0	ρ/ρ_0	A/A^*
2.30	0.4859	0.07997	0.1646	2.193
2.32	0.4816	0.07751	0.1610	2.233
2.34	0.4773	0.07513	0.1574	2.274
2.36	0.4731	0.07281	0.1539	2.316
2.38	0.4689	0.07057	0.1505	2.359
2.40	0.4647	0.06840	0.1472	2.403
2.42	0.4606	0.06630	0.1440	2.448
2.44	0.4565	0.06426	0.1408	2.494
2.46	0.4524	0.06229	0.1377	2.540
2.48	0.4484	0.06038	0.1347	2.588
2.50	0.4444	0.05853	0.1317	2.637
2.52	0.4405	0.05674	0.1288	2.687
2.54	0.4366	0.05500	0.1260	2.737
2.56	0.4328	0.05332	0.1232	2.789
2.58	0.4289	0.05169	0.1205	2.842
2.60	0.4252	0.05012	0.1179	2.896
2.62	0.4214	0.04859	0.1153	2.951
2.64	0.4177	0.04711	0.1128	3.007
2.66	0.4141	0.04568	0.1103	3.065
2.68	0.4104	0.04429	0.1079	3.123
2.70	0.4068	0.04295	0.1056	3.183
2.72	0.4033	0.04166	0.1033	3.244
2.74	0.3998	0.04039	0.1010	3.306
2.76	0.3963	0.03917	0.09885	3.370
2.78	0.3928	0.03800	0.09671	3.434
2.80	0.3894	0.03685	0.09462	3.500
2.82	0.3860	0.03574	0.09259	3.567
2.84	0.3827	0.03467	0.09059	3.636
2.86	0.3794	0.03363	0.08865	3.706
2.88	0.3761	0.03262	0.08674	3.777
2.90	0.3729	0.03165	0.08489	3.850
2.92	0.3697	0.03071	0.08308	3.924
2.94	0.3665	0.02980	0.08130	3.999
2.96	0.3633	0.02891	0.07957	4.076
2.98	0.3602	0.02805	0.07788	4.155
3.00	0.3571	0.02722	0.07623	4.235
3.10	0.3422	0.02345	0.06852	4.657
3.20	0.3281	0.02023	0.06165	5.121
3.30	0.3147	0.01748	0.05554	5.629
3.40	0.3019	0.01512	0.05009	6.184
3.50	0.2899	0.01311	0.04523	6.790
3.60	0.2784	0.01138	0.04089	7.450
3.70	0.2675	0.009903	0.03702	8.169
3.80	0.2572	0.008629	0.03355	8.951
3.90	0.2474	0.007532	0.03044	9.799
4.00	0.2381	0.006586	0.02766	10.72

Table E.1 Isentropic Flow Functions (*Continued*)

M	T/T_0	p/p_0	ρ/ρ_0	A/A^*
4.00	0.2381	0.006586	0.02766	10.72
4.10	0.2293	0.005769	0.02516	11.71
4.20	0.2208	0.005062	0.02292	12.79
4.30	0.2129	0.004449	0.02090	13.95
4.40	0.2053	0.003918	0.01909	15.21
4.50	0.1980	0.003455	0.01745	16.56
4.60	0.1911	0.003053	0.01597	18.02
4.70	0.1846	0.002701	0.01463	19.58
4.80	0.1783	0.002394	0.01343	21.26
4.90	0.1724	0.002126	0.01233	23.07
5.00	0.1667	0.001890	0.01134	25.00

Table E.2 Fanno Line Flow Functions (one-dimensional flow, ideal gas, $k = 1.4$)

M	p_0/p_0^*	T/T^*	p/p^*	V/V^*	$\bar{f}L_{max}/D_h$
0.00	∞	1.200	∞	0.0000	∞
0.02	28.94	1.200	54.77	0.02191	1778
0.04	14.48	1.200	27.38	0.04381	440.5
0.06	9.666	1.199	18.25	0.06570	193.0
0.08	7.262	1.199	13.68	0.08758	106.7
0.10	5.822	1.198	10.94	0.1094	66.92
0.12	4.864	1.197	9.116	0.1313	45.41
0.14	4.182	1.195	7.809	0.1531	32.51
0.16	3.673	1.194	6.829	0.1748	24.20
0.18	3.278	1.192	6.066	0.1965	18.54
0.20	2.964	1.191	5.456	0.2182	14.53
0.22	2.708	1.189	4.955	0.2398	11.60
0.24	2.496	1.186	4.538	0.2614	9.387
0.26	2.317	1.184	4.185	0.2829	7.688
0.28	2.166	1.182	3.882	0.3044	6.357
0.30	2.035	1.179	3.619	0.3257	5.299
0.32	1.922	1.176	3.389	0.3470	4.447
0.34	1.823	1.173	3.185	0.3682	3.752
0.36	1.736	1.170	3.004	0.3894	3.180
0.38	1.659	1.166	2.842	0.4104	2.706
0.40	1.590	1.163	2.696	0.4313	2.309
0.42	1.529	1.159	2.563	0.4522	1.974
0.44	1.474	1.155	2.443	0.4729	1.692
0.46	1.425	1.151	2.333	0.4936	1.451
0.48	1.380	1.147	2.231	0.5141	1.245
0.50	1.340	1.143	2.138	0.5345	1.069
0.52	1.303	1.138	2.052	0.5548	0.9174
0.54	1.270	1.134	1.972	0.5750	0.7866
0.56	1.240	1.129	1.898	0.5951	0.6736
0.58	1.213	1.124	1.828	0.6150	0.5757
0.60	1.188	1.119	1.763	0.6348	0.4908
0.62	1.166	1.114	1.703	0.6545	0.4172
0.64	1.145	1.109	1.646	0.6740	0.3533
0.66	1.127	1.104	1.592	0.6934	0.2979
0.68	1.110	1.098	1.541	0.7127	0.2498
0.70	1.094	1.093	1.493	0.7318	0.2081
0.72	1.081	1.087	1.448	0.7508	0.1722
0.74	1.068	1.082	1.405	0.7696	0.1411
0.76	1.057	1.076	1.365	0.7883	0.1145
0.78	1.047	1.070	1.326	0.8068	0.09167
0.80	1.038	1.064	1.289	0.8251	0.07229
0.82	1.030	1.058	1.254	0.8433	0.05593
0.84	1.024	1.052	1.221	0.8614	0.04226
0.86	1.018	1.045	1.189	0.8793	0.03097
0.88	1.013	1.039	1.158	0.8970	0.02180
0.90	1.0089	1.033	1.129	0.9146	0.01451

Table E.2 Fanno Line Flow Functions (*Continued*)

M	p_0/p_0^*	T/T^*	p/p^*	V/V^*	$\bar{f}L_{max}/D_h$
0.90	1.0089	1.033	1.129	0.9146	0.01451
0.92	1.0056	1.026	1.101	0.9320	0.008913
0.94	1.0031	1.020	1.074	0.9493	0.004815
0.96	1.0014	1.013	1.049	0.9663	0.002057
0.98	1.0003	1.007	1.024	0.9832	0.0004947
1.00	1.0000	1.000	1.000	1.000	0.0000
1.02	1.0003	0.9933	0.9771	1.017	0.0004587
1.04	1.0013	0.9866	0.9551	1.033	0.001769
1.06	1.0029	0.9798	0.9338	1.049	0.003838
1.08	1.0051	0.9730	0.9134	1.065	0.006585
1.10	1.0079	0.9662	0.8936	1.081	0.009935
1.12	1.011	0.9593	0.8745	1.097	0.01382
1.14	1.015	0.9524	0.8561	1.113	0.01819
1.16	1.020	0.9455	0.8383	1.128	0.02298
1.18	1.025	0.9386	0.8210	1.143	0.02814
1.20	1.030	0.9317	0.8044	1.158	0.03364
1.22	1.037	0.9247	0.7882	1.173	0.03942
1.24	1.043	0.9178	0.7726	1.188	0.04547
1.26	1.050	0.9108	0.7574	1.203	0.05174
1.28	1.058	0.9038	0.7427	1.217	0.05820
1.30	1.066	0.8969	0.7285	1.231	0.06483
1.32	1.075	0.8899	0.7147	1.245	0.07161
1.34	1.084	0.8829	0.7012	1.259	0.07850
1.36	1.094	0.8760	0.6882	1.273	0.08550
1.38	1.104	0.8690	0.6755	1.286	0.09259
1.40	1.115	0.8621	0.6632	1.300	0.09974
1.42	1.126	0.8551	0.6512	1.313	0.1069
1.44	1.138	0.8482	0.6396	1.326	0.1142
1.46	1.150	0.8413	0.6282	1.339	0.1215
1.48	1.163	0.8345	0.6172	1.352	0.1288
1.50	1.176	0.8276	0.6065	1.365	0.1361
1.52	1.190	0.8208	0.5960	1.377	0.1434
1.54	1.204	0.8139	0.5858	1.389	0.1506
1.56	1.219	0.8072	0.5759	1.402	0.1579
1.58	1.234	0.8004	0.5662	1.414	0.1651
1.60	1.250	0.7937	0.5568	1.425	0.1724
1.62	1.267	0.7870	0.5476	1.437	0.1795
1.64	1.284	0.7803	0.5386	1.449	0.1867
1.66	1.301	0.7736	0.5299	1.460	0.1938
1.68	1.319	0.7670	0.5213	1.471	0.2008
1.70	1.338	0.7605	0.5130	1.482	0.2078
1.72	1.357	0.7539	0.5048	1.494	0.2147
1.74	1.376	0.7474	0.4969	1.504	0.2216
1.76	1.397	0.7410	0.4891	1.515	0.2284
1.78	1.418	0.7345	0.4815	1.526	0.2352
1.80	1.439	0.7282	0.4741	1.536	0.2419

Table E.2 Fanno Line Flow Functions (*Continued*)

M	p_0/p_0^*	T/T^*	p/p^*	V/V^*	$\bar{f}L_{max}/D_h$
1.80	1.439	0.7282	0.4741	1.536	0.2419
1.82	1.461	0.7218	0.4668	1.546	0.2485
1.84	1.484	0.7155	0.4597	1.556	0.2551
1.86	1.507	0.7093	0.4528	1.566	0.2616
1.88	1.531	0.7030	0.4460	1.576	0.2680
1.90	1.555	0.6969	0.4394	1.586	0.2743
1.92	1.580	0.6907	0.4329	1.596	0.2806
1.94	1.606	0.6847	0.4265	1.605	0.2868
1.96	1.633	0.6786	0.4203	1.615	0.2930
1.98	1.660	0.6726	0.4142	1.624	0.2990
2.00	1.688	0.6667	0.4083	1.633	0.3050
2.02	1.716	0.6608	0.4024	1.642	0.3109
2.04	1.745	0.6549	0.3967	1.651	0.3168
2.06	1.775	0.6491	0.3911	1.660	0.3225
2.08	1.806	0.6433	0.3856	1.668	0.3282
2.10	1.837	0.6376	0.3802	1.677	0.3339
2.12	1.869	0.6320	0.3750	1.685	0.3394
2.14	1.902	0.6263	0.3698	1.694	0.3449
2.16	1.935	0.6208	0.3648	1.702	0.3503
2.18	1.970	0.6152	0.3598	1.710	0.3556
2.20	2.005	0.6098	0.3549	1.718	0.3609
2.22	2.041	0.6043	0.3502	1.726	0.3661
2.24	2.078	0.5990	0.3455	1.734	0.3712
2.26	2.115	0.5936	0.3409	1.741	0.3763
2.28	2.154	0.5883	0.3364	1.749	0.3813
2.30	2.193	0.5831	0.3320	1.756	0.3862
2.32	2.233	0.5779	0.3277	1.764	0.3911
2.34	2.274	0.5728	0.3234	1.771	0.3959
2.36	2.316	0.5677	0.3193	1.778	0.4006
2.38	2.359	0.5626	0.3152	1.785	0.4053
2.40	2.403	0.5576	0.3111	1.792	0.4099
2.42	2.448	0.5527	0.3072	1.799	0.4144
2.44	2.494	0.5478	0.3033	1.806	0.4189
2.46	2.540	0.5429	0.2995	1.813	0.4233
2.48	2.588	0.5381	0.2958	1.819	0.4277
2.50	2.637	0.5333	0.2921	1.826	0.4320
2.52	2.687	0.5286	0.2885	1.832	0.4362
2.54	2.737	0.5239	0.2850	1.839	0.4404
2.56	2.789	0.5193	0.2815	1.845	0.4445
2.58	2.842	0.5147	0.2781	1.851	0.4486
2.60	2.896	0.5102	0.2747	1.857	0.4526
2.62	2.951	0.5057	0.2714	1.863	0.4565
2.64	3.007	0.5013	0.2682	1.869	0.4604
2.66	3.065	0.4969	0.2650	1.875	0.4643
2.68	3.123	0.4925	0.2619	1.881	0.4681
2.70	3.183	0.4882	0.2588	1.887	0.4718

Table E.2 Fanno Line Flow Functions (*Continued*)

M	p_0/p_0^*	T/T^*	p/p^*	V/V^*	$\bar{f}L_{max}/D_h$
2.70	3.183	0.4882	0.2588	1.887	0.4718
2.72	3.244	0.4839	0.2558	1.892	0.4755
2.74	3.306	0.4797	0.2528	1.898	0.4792
2.76	3.370	0.4755	0.2499	1.903	0.4827
2.78	3.434	0.4714	0.2470	1.909	0.4863
2.80	3.500	0.4673	0.2441	1.914	0.4898
2.82	3.567	0.4632	0.2414	1.919	0.4932
2.84	3.636	0.4592	0.2386	1.925	0.4966
2.86	3.706	0.4553	0.2359	1.930	0.5000
2.88	3.777	0.4513	0.2333	1.935	0.5033
2.90	3.850	0.4474	0.2307	1.940	0.5065
2.92	3.924	0.4436	0.2281	1.945	0.5097
2.94	3.999	0.4398	0.2256	1.950	0.5129
2.96	4.076	0.4360	0.2231	1.954	0.5160
2.98	4.155	0.4323	0.2206	1.959	0.5191
3.00	4.235	0.4286	0.2182	1.964	0.5222
3.50	6.79	0.3478	0.1685	2.064	0.5864
4.00	10.72	0.2857	0.1336	2.138	0.6331
4.50	16.56	0.2376	0.1083	2.194	0.6676
5.00	25.00	0.2000	0.08944	2.236	0.6938

Table E.3 Rayleigh Line Flow Functions (one-dimensional flow, ideal gas, $k = 1.4$)

M	T_0/T_0^*	p_0/p_0^*	T/T^*	p/p^*	V/V^*
0.00	0.0000	1.268	0.0000	2.400	0.0000
0.02	0.001918	1.268	0.002301	2.399	0.0009595
0.04	0.007648	1.267	0.009175	2.395	0.003831
0.06	0.01712	1.265	0.02053	2.388	0.008597
0.08	0.03021	1.262	0.03621	2.379	0.01522
0.10	0.04678	1.259	0.05602	2.367	0.02367
0.12	0.06661	1.255	0.07970	2.353	0.03388
0.14	0.08947	1.251	0.1070	2.336	0.04578
0.16	0.1151	1.246	0.1374	2.317	0.05931
0.18	0.1432	1.241	0.1708	2.296	0.07438
0.20	0.1736	1.235	0.2066	2.273	0.09091
0.22	0.2057	1.228	0.2445	2.248	0.1088
0.24	0.2395	1.221	0.2841	2.221	0.1279
0.26	0.2745	1.214	0.3250	2.193	0.1482
0.28	0.3104	1.206	0.3667	2.163	0.1696
0.30	0.3469	1.199	0.4089	2.131	0.1918
0.32	0.3837	1.190	0.4512	2.099	0.2149
0.34	0.4206	1.182	0.4933	2.066	0.2388
0.36	0.4572	1.174	0.5348	2.031	0.2633
0.38	0.4935	1.165	0.5755	1.996	0.2883
0.40	0.5290	1.157	0.6152	1.961	0.3137
0.42	0.5638	1.148	0.6535	1.925	0.3395
0.44	0.5975	1.139	0.6903	1.888	0.3656
0.46	0.6301	1.131	0.7254	1.852	0.3918
0.48	0.6614	1.122	0.7587	1.815	0.4181
0.50	0.6914	1.114	0.7901	1.778	0.4445
0.52	0.7199	1.106	0.8196	1.741	0.4708
0.54	0.7470	1.098	0.8470	1.704	0.4970
0.56	0.7725	1.090	0.8723	1.668	0.5230
0.58	0.7965	1.083	0.8955	1.632	0.5489
0.60	0.8189	1.075	0.9167	1.596	0.5745
0.62	0.8398	1.068	0.9359	1.560	0.5998
0.64	0.8592	1.061	0.9530	1.525	0.6248
0.66	0.8771	1.055	0.9682	1.491	0.6494
0.68	0.8935	1.049	0.9814	1.457	0.6737
0.70	0.9085	1.043	0.9929	1.424	0.6975
0.72	0.9221	1.038	1.003	1.391	0.7209
0.74	0.9344	1.033	1.011	1.359	0.7439
0.76	0.9455	1.028	1.017	1.327	0.7665
0.78	0.9553	1.023	1.022	1.296	0.7885
0.80	0.9639	1.019	1.025	1.266	0.8101
0.82	0.9715	1.016	1.028	1.236	0.8313
0.84	0.9781	1.012	1.029	1.207	0.8519
0.86	0.9836	1.010	1.028	1.179	0.8721
0.88	0.9883	1.007	1.027	1.152	0.8918
0.90	0.9921	1.005	1.025	1.125	0.9110

Table E.3 Rayleigh Line Flow Functions (*Continued*)

M	T_0/T_0^*	p_0/p_0^*	T/T^*	p/p^*	V/V^*
0.90	0.9921	1.005	1.025	1.125	0.9110
0.92	0.9951	1.003	1.021	1.098	0.9297
0.94	0.9973	1.002	1.017	1.073	0.9480
0.96	0.9988	1.001	1.012	1.048	0.9658
0.98	0.9997	1.000	1.006	1.024	0.9831
1.00	1.000	1.000	1.000	1.000	1.000
1.02	0.9997	1.000	0.9930	0.9770	1.016
1.04	0.9990	1.001	0.9855	0.9546	1.032
1.06	0.9977	1.002	0.9776	0.9328	1.048
1.08	0.9960	1.003	0.9691	0.9115	1.063
1.10	0.9939	1.005	0.9603	0.8909	1.078
1.12	0.9915	1.007	0.9512	0.8708	1.092
1.14	0.9887	1.010	0.9417	0.8512	1.106
1.16	0.9856	1.012	0.9320	0.8322	1.120
1.18	0.9823	1.016	0.9220	0.8137	1.133
1.20	0.9787	1.019	0.9119	0.7958	1.146
1.22	0.9749	1.023	0.9015	0.7783	1.158
1.24	0.9709	1.028	0.8911	0.7613	1.171
1.26	0.9668	1.033	0.8805	0.7447	1.182
1.28	0.9624	1.038	0.8699	0.7287	1.194
1.30	0.9580	1.044	0.8592	0.7130	1.205
1.32	0.9534	1.050	0.8484	0.6978	1.216
1.34	0.9487	1.056	0.8377	0.6830	1.226
1.36	0.9440	1.063	0.8270	0.6686	1.237
1.38	0.9392	1.070	0.8161	0.6546	1.247
1.40	0.9343	1.078	0.8054	0.6410	1.256
1.42	0.9293	1.086	0.7947	0.6278	1.266
1.44	0.9243	1.094	0.7841	0.6149	1.275
1.46	0.9193	1.103	0.7735	0.6024	1.284
1.48	0.9143	1.112	0.7629	0.5902	1.293
1.50	0.9093	1.122	0.7525	0.5783	1.301
1.52	0.9042	1.132	0.7422	0.5668	1.310
1.54	0.8992	1.142	0.7319	0.5555	1.318
1.56	0.8942	1.153	0.7217	0.5446	1.325
1.58	0.8892	1.164	0.7117	0.5339	1.333
1.60	0.8842	1.176	0.7017	0.5236	1.340
1.62	0.8792	1.188	0.6919	0.5135	1.348
1.64	0.8743	1.200	0.6822	0.5036	1.355
1.66	0.8694	1.213	0.6726	0.4941	1.361
1.68	0.8645	1.226	0.6631	0.4847	1.368
1.70	0.8597	1.240	0.6538	0.4756	1.375
1.72	0.8549	1.255	0.6446	0.4668	1.381
1.74	0.8502	1.269	0.6355	0.4581	1.387
1.76	0.8455	1.284	0.6265	0.4497	1.393
1.78	0.8409	1.300	0.6177	0.4415	1.399
1.80	0.8363	1.316	0.6089	0.4335	1.405

Table E.3   Rayleigh Line Flow Functions (*Continued*)

M	T_0/T_0^*	p_0/p_0^*	T/T^*	p/p^*	V/V^*
1.80	0.8363	1.316	0.6089	0.4335	1.405
1.82	0.8317	1.332	0.6004	0.4257	1.410
1.84	0.8273	1.349	0.5919	0.4181	1.416
1.86	0.8228	1.367	0.5836	0.4107	1.421
1.88	0.8185	1.385	0.5754	0.4035	1.426
1.90	0.8141	1.403	0.5673	0.3964	1.431
1.92	0.8099	1.422	0.5594	0.3896	1.436
1.94	0.8057	1.442	0.5516	0.3828	1.441
1.96	0.8015	1.462	0.5439	0.3763	1.446
1.98	0.7974	1.482	0.5364	0.3699	1.450
2.00	0.7934	1.503	0.5289	0.3636	1.455
2.02	0.7894	1.525	0.5216	0.3575	1.459
2.04	0.7855	1.547	0.5144	0.3516	1.463
2.06	0.7816	1.569	0.5074	0.3458	1.467
2.08	0.7778	1.592	0.5004	0.3401	1.471
2.10	0.7741	1.616	0.4936	0.3345	1.475
2.12	0.7704	1.640	0.4868	0.3291	1.479
2.14	0.7667	1.665	0.4802	0.3238	1.483
2.16	0.7631	1.691	0.4737	0.3186	1.487
2.18	0.7596	1.717	0.4673	0.3136	1.490
2.20	0.7561	1.743	0.4611	0.3086	1.494
2.22	0.7527	1.771	0.4549	0.3038	1.497
2.24	0.7493	1.799	0.4488	0.2991	1.501
2.26	0.7460	1.827	0.4429	0.2945	1.504
2.28	0.7428	1.856	0.4370	0.2899	1.507
2.30	0.7395	1.886	0.4312	0.2855	1.510
2.32	0.7364	1.917	0.4256	0.2812	1.513
2.34	0.7333	1.948	0.4200	0.2770	1.517
2.36	0.7302	1.979	0.4145	0.2728	1.520
2.38	0.7272	2.012	0.4091	0.2688	1.522
2.40	0.7242	2.045	0.4038	0.2648	1.525
2.42	0.7213	2.079	0.3986	0.2609	1.528
2.44	0.7184	2.114	0.3935	0.2571	1.531
2.46	0.7156	2.149	0.3885	0.2534	1.533
2.48	0.7128	2.185	0.3836	0.2497	1.536
2.50	0.7101	2.222	0.3787	0.2462	1.539
2.52	0.7074	2.259	0.3739	0.2427	1.541
2.54	0.7047	2.298	0.3692	0.2392	1.543
2.56	0.7021	2.337	0.3646	0.2359	1.546
2.58	0.6995	2.377	0.3601	0.2326	1.548
2.60	0.6970	2.418	0.3556	0.2294	1.551
2.62	0.6945	2.459	0.3512	0.2262	1.553
2.64	0.6921	2.502	0.3469	0.2231	1.555
2.66	0.6896	2.545	0.3427	0.2201	1.557
2.68	0.6873	2.589	0.3385	0.2171	1.559
2.70	0.6849	2.634	0.3344	0.2142	1.561

Table E.3 Rayleigh Line Flow Functions (*Continued*)

M	T_0/T_0^*	p_0/p_0^*	T/T^*	p/p^*	V/V^*
2.70	0.6849	2.634	0.3344	0.2142	1.561
2.72	0.6826	2.680	0.3304	0.2113	1.563
2.74	0.6804	2.727	0.3264	0.2085	1.565
2.76	0.6782	2.775	0.3225	0.2058	1.567
2.78	0.6760	2.824	0.3186	0.2031	1.569
2.80	0.6738	2.873	0.3149	0.2004	1.571
2.82	0.6717	2.924	0.3111	0.1978	1.573
2.84	0.6696	2.975	0.3075	0.1953	1.575
2.86	0.6675	3.028	0.3039	0.1927	1.577
2.88	0.6655	3.081	0.3004	0.1903	1.578
2.90	0.6635	3.136	0.2969	0.1879	1.580
2.92	0.6615	3.191	0.2934	0.1855	1.582
2.94	0.6596	3.248	0.2901	0.1832	1.583
2.96	0.6577	3.306	0.2868	0.1809	1.585
2.98	0.6558	3.365	0.2835	0.1787	1.587
3.00	0.6540	3.424	0.2803	0.1765	1.588
3.50	0.6158	5.328	0.2142	0.1322	1.620
4.00	0.5891	8.227	0.1683	0.1026	1.641
4.50	0.5698	12.50	0.1354	0.08177	1.656
5.00	0.5556	18.63	0.1111	0.06667	1.667

Table E.4 Normal Shock Flow Functions (one-dimensional flow, ideal gas, $k = 1.4$)

M_1	M_2	p_{0_2}/p_{0_1}	T_2/T_1	p_2/p_1	ρ_2/ρ_1
1.00	1.000	1.000	1.000	1.000	1.000
1.02	0.9805	1.000	1.013	1.047	1.033
1.04	0.9620	0.9999	1.026	1.095	1.067
1.06	0.9444	0.9998	1.039	1.144	1.101
1.08	0.9277	0.9994	1.052	1.194	1.135
1.10	0.9118	0.9989	1.065	1.245	1.169
1.12	0.8966	0.9982	1.078	1.297	1.203
1.14	0.8820	0.9973	1.090	1.350	1.238
1.16	0.8682	0.9961	1.103	1.403	1.272
1.18	0.8549	0.9946	1.115	1.458	1.307
1.20	0.8422	0.9928	1.128	1.513	1.342
1.22	0.8300	0.9907	1.141	1.570	1.376
1.24	0.8183	0.9884	1.153	1.627	1.411
1.26	0.8071	0.9857	1.166	1.686	1.446
1.28	0.7963	0.9827	1.178	1.745	1.481
1.30	0.7860	0.9794	1.191	1.805	1.516
1.32	0.7760	0.9757	1.204	1.866	1.551
1.34	0.7664	0.9718	1.216	1.928	1.585
1.36	0.7572	0.9676	1.229	1.991	1.620
1.38	0.7483	0.9630	1.242	2.055	1.655
1.40	0.7397	0.9582	1.255	2.120	1.690
1.42	0.7314	0.9531	1.268	2.186	1.724
1.44	0.7235	0.9477	1.281	2.253	1.759
1.46	0.7157	0.9420	1.294	2.320	1.793
1.48	0.7083	0.9360	1.307	2.389	1.828
1.50	0.7011	0.9298	1.320	2.458	1.862
1.52	0.6941	0.9233	1.334	2.529	1.896
1.54	0.6874	0.9166	1.347	2.600	1.930
1.56	0.6809	0.9097	1.361	2.673	1.964
1.58	0.6746	0.9026	1.374	2.746	1.998
1.60	0.6684	0.8952	1.388	2.820	2.032
1.62	0.6625	0.8876	1.402	2.895	2.065
1.64	0.6568	0.8799	1.416	2.971	2.099
1.66	0.6512	0.8720	1.430	3.048	2.132
1.68	0.6458	0.8640	1.444	3.126	2.165
1.70	0.6406	0.8557	1.458	3.205	2.198
1.72	0.6355	0.8474	1.473	3.285	2.230
1.74	0.6305	0.8389	1.487	3.366	2.263
1.76	0.6257	0.8302	1.502	3.447	2.295
1.78	0.6210	0.8215	1.517	3.530	2.327
1.80	0.6165	0.8127	1.532	3.613	2.359
1.82	0.6121	0.8038	1.547	3.698	2.391
1.84	0.6078	0.7947	1.562	3.783	2.422
1.86	0.6036	0.7857	1.577	3.870	2.454
1.88	0.5996	0.7766	1.592	3.957	2.485
1.90	0.5956	0.7674	1.608	4.045	2.516

Table E.4 Normal Shock Flow Functions (*Continued*)

M_1	M_2	p_{0_2}/p_{0_1}	T_2/T_1	p_2/p_1	ρ_2/ρ_1
1.90	0.5956	0.7674	1.608	4.045	2.516
1.92	0.5918	0.7581	1.624	4.134	2.546
1.94	0.5880	0.7488	1.639	4.224	2.577
1.96	0.5844	0.7395	1.655	4.315	2.607
1.98	0.5808	0.7302	1.671	4.407	2.637
2.00	0.5774	0.7209	1.687	4.500	2.667
2.02	0.5740	0.7115	1.704	4.594	2.696
2.04	0.5707	0.7022	1.720	4.689	2.725
2.06	0.5675	0.6928	1.737	4.784	2.755
2.08	0.5643	0.6835	1.754	4.881	2.783
2.10	0.5613	0.6742	1.770	4.978	2.812
2.12	0.5583	0.6649	1.787	5.077	2.840
2.14	0.5554	0.6557	1.805	5.176	2.868
2.16	0.5525	0.6464	1.822	5.277	2.896
2.18	0.5498	0.6373	1.839	5.378	2.924
2.20	0.5471	0.6281	1.857	5.480	2.951
2.22	0.5444	0.6191	1.875	5.583	2.978
2.24	0.5418	0.6100	1.892	5.687	3.005
2.26	0.5393	0.6011	1.910	5.792	3.032
2.28	0.5368	0.5921	1.929	5.898	3.058
2.30	0.5344	0.5833	1.947	6.005	3.085
2.32	0.5321	0.5745	1.965	6.113	3.110
2.34	0.5297	0.5658	1.984	6.222	3.136
2.36	0.5275	0.5572	2.002	6.331	3.162
2.38	0.5253	0.5486	2.021	6.442	3.187
2.40	0.5231	0.5402	2.040	6.553	3.212
2.42	0.5210	0.5318	2.059	6.666	3.237
2.44	0.5189	0.5234	2.079	6.779	3.261
2.46	0.5169	0.5152	2.098	6.894	3.285
2.48	0.5149	0.5071	2.118	7.009	3.310
2.50	0.5130	0.4990	2.137	7.125	3.333
2.52	0.5111	0.4910	2.157	7.242	3.357
2.54	0.5092	0.4832	2.177	7.360	3.380
2.56	0.5074	0.4754	2.198	7.479	3.403
2.58	0.5056	0.4677	2.218	7.599	3.426
2.60	0.5039	0.4601	2.238	7.720	3.449
2.62	0.5022	0.4526	2.259	7.842	3.471
2.64	0.5005	0.4452	2.280	7.965	3.494
2.66	0.4988	0.4379	2.301	8.088	3.516
2.68	0.4972	0.4307	2.322	8.213	3.537
2.70	0.4956	0.4236	2.343	8.338	3.559
2.72	0.4941	0.4166	2.364	8.465	3.580
2.74	0.4926	0.4097	2.386	8.592	3.601
2.76	0.4911	0.4028	2.407	8.721	3.622
2.78	0.4897	0.3961	2.429	8.850	3.643
2.80	0.4882	0.3895	2.451	8.980	3.664

Table E.4 Normal Shock Flow Functions (*Continued*)

M_1	M_2	p_{0_2}/p_{0_1}	T_2/T_1	p_2/p_1	ρ_2/ρ_1
2.80	0.4882	0.3895	2.451	8.980	3.664
2.82	0.4868	0.3829	2.473	9.111	3.684
2.84	0.4854	0.3765	2.496	9.243	3.704
2.86	0.4840	0.3701	2.518	9.376	3.724
2.88	0.4827	0.3639	2.540	9.510	3.743
2.90	0.4814	0.3577	2.563	9.645	3.763
2.92	0.4801	0.3517	2.586	9.781	3.782
2.94	0.4788	0.3457	2.609	9.918	3.801
2.96	0.4776	0.3398	2.632	10.06	3.820
2.98	0.4764	0.3340	2.656	10.19	3.839
3.00	0.4752	0.3283	2.679	10.33	3.857
3.10	0.4695	0.3012	2.799	11.05	3.947
3.20	0.4644	0.2762	2.922	11.78	4.031
3.30	0.4596	0.2533	3.049	12.54	4.112
3.40	0.4552	0.2322	3.180	13.32	4.188
3.50	0.4512	0.2130	3.315	14.13	4.261
3.60	0.4474	0.1953	3.454	14.95	4.330
3.70	0.4440	0.1792	3.596	15.81	4.395
3.80	0.4407	0.1645	3.743	16.68	4.457
3.90	0.4377	0.1510	3.893	17.58	4.516
4.00	0.4350	0.1388	4.047	18.50	4.571
4.10	0.4324	0.1276	4.205	19.45	4.624
4.20	0.4299	0.1173	4.367	20.41	4.675
4.30	0.4277	0.1080	4.532	21.41	4.723
4.40	0.4255	0.09948	4.702	22.42	4.768
4.50	0.4236	0.09170	4.875	23.46	4.812
4.60	0.4217	0.08459	5.052	24.52	4.853
4.70	0.4199	0.07809	.5.233	25.61	4.893
4.80	0.4183	0.07214	5.418	26.71	4.930
4.90	0.4167	0.06670	5.607	27.85	4.966
5.00	0.4152	0.06172	5.800	29.00	5.000

Appendix F

ANALYSIS OF EXPERIMENTAL UNCERTAINTY

F-1 INTRODUCTION

Results of experimental tests often are used for engineering analysis and design. Not all data are equally good; the validity of data should be documented before test results are used for design. Uncertainty analysis is the procedure used to quantify data validity and accuracy.

Analysis of uncertainty also is useful during experiment design. Careful study may indicate potential sources of unacceptable error and suggest improved measurement methods.

F-2 TYPES OF ERROR

Errors always are present when experimental measurements are made. Aside from gross blunders by the experimenter, experimental error may be of two types. Fixed (or systematic) error causes repeated measurements to be in error by the same amount for each trial. Fixed error is the same for each reading and can be removed by proper calibration or correction. Random error (nonrepeatability) is different for every reading and hence cannot be removed. The factors that introduce random error are uncertain by their nature. The objective of uncertainty analysis is to estimate the probable random error in experimental results.

We assume that equipment has been constructed correctly and calibrated properly to eliminate fixed errors. We assume that instrumentation has adequate resolution and that fluctuations in readings are not excessive. We assume also that care is used in making and recording observations so that only random errors remain.

F-3 ESTIMATION OF UNCERTAINTY

Our goal is to estimate the uncertainty of experimental measurements and calculated results due to random errors. The procedure has three steps:

1. Estimate the uncertainty interval for each measured quantity.
2. State the confidence limit on each measurement.
3. Analyze the propagation of uncertainty into results calculated from experimental data.

800

Below we outline the procedure for each step and illustrate applications with examples.

Step 1. *Estimate the measurement uncertainty interval.* Designate the measured variables in an experiment as $x_1, x_2, \ldots, x_n$. One possible way to find the uncertainty interval for each variable would be to repeat each measurement many times. The result would be a distribution of data for each variable. Random errors in measurement usually produce a *normal (Gaussian)* frequency distribution of measured values. The data scatter for a normal distribution is characterized by the standard deviation, σ. The uncertainty interval for each measured variable, x_i, may be stated as $\pm n\sigma_i$, where $n = 1, 2,$ or 3.

For normally distributed data, over 99 percent of measured values of x_i lie within $\pm 3\sigma_i$ of the mean value, 95 percent lie within $\pm 2\sigma_i$, and 50 percent lie within $\pm \sigma_i$ of the mean value of the data set [1]. Thus it would be possible to quantify expected errors within any desired *confidence limit* if a statistically significant set of data were available.

The method of repeated measurements usually is impractical. In most applications it is impossible to obtain enough data for a statistically significant sample owing to the excessive time and cost involved. However, the normal distribution suggests several important concepts:

1. Small errors are more likely than large ones.
2. Plus and minus errors are about equally likely.
3. No finite maximum error can be specified.

A more typical situation in engineering work is a "single-sample" experiment, where only one measurement is made for each point [2]. A reasonable estimate of the measurement uncertainty due to random error in a single-sample experiment usually is plus or minus half the smallest scale division (the *least count*) of the instrument. However, this approach also must be used with caution, as illustrated in the following example.

EXAMPLE F.1—Uncertainty in Barometer Reading

The observed height of the mercury barometer column is $h = 752.6$ mm. The least count on the vernier scale is 0.1 mm, so one might estimate the probable measurement error as ± 0.05 mm.

A measurement probably could not be made this precisely. The barometer sliders and meniscus must be aligned by eye. The slider has a least count of 1 mm. As a conservative estimate, a measurement could be made to the nearest millimeter. The probable value of a single measurement then would be expressed as 752.6 ± 0.5 mm. The relative uncertainty in barometric height would be stated as

$$u_h = \pm \frac{0.5 \text{ mm}}{752.6 \text{ mm}} = \pm 0.000664 \quad \text{or} \quad \pm 0.0664 \text{ percent}$$

Comments:

1. An uncertainty interval of ± 0.1 percent corresponds to a result specified to three significant figures; this precision is sufficient for most engineering work.
2. The measurement of barometer height was precise, as shown by the uncertainty estimate. But was it accurate? At typical room temperatures, the observed barometer reading must be reduced by a temperature correction of nearly 3 mm! This is an example of a fixed error that requires a correction factor.

Step 2. *State the confidence limit on each measurement.* The uncertainty interval of a measurement should be stated at specified odds. For example, one may write $h = 752.6 \pm 0.5$ mm (20 to 1). This means that one is willing to bet 20 to 1 that the height of the mercury column actually is within ± 0.5 mm of the stated value. It should be obvious [3] that "... the specification of such odds can only be made by the experimenter based on ... total laboratory experience. There is no substitute for sound engineering judgment in estimating the uncertainty of a measured variable."

The confidence interval statement is based on the concept of standard deviation for a normal distribution. Odds of 100 to 1 correspond to $\pm 3\sigma$; 99 percent of all future readings are expected to fall within the interval. Odds of 20 to 1 correspond to $\pm 2\sigma$ and odds of 2 to 1 correspond to $\pm \sigma$ confidence limits. Odds of 20 to 1 typically are used for engineering work.

Step 3. *Analyze the propagation of uncertainty in calculations.* Suppose that measurements of independent variables, $x_1, x_2, \ldots, x_n$, are made in the laboratory. The relative uncertainty of each independently measured quantity is estimated as u_i. The measurements are used to calculate some result, R, for the experiment. We wish to analyze how errors in the x_is *propagate* into the calculation of R from measured values.

In general, R may be expressed mathematically as $R = R(x_1, x_2, \ldots, x_n)$. The effect on R of an error in measuring an individual x_i may be estimated by analogy to the derivative of a function [4]. A variation, δx_i, in x_i would cause R to vary according to

$$\delta R_i = \frac{\partial R}{\partial x_i} \delta x_i$$

For applications, it is convenient to normalize this equation by dividing by R to obtain

$$\frac{\delta R_i}{R} = \frac{1}{R}\frac{\partial R}{\partial x_i}\delta x_i = \frac{x_i}{R}\frac{\partial R}{\partial x_i}\frac{\delta x_i}{x_i} \tag{F.1}$$

Equation F.1 might be used to estimate the uncertainty interval in the result due to variations in x_i. To do this, substitute the uncertainty interval for x_i

$$u_{R_i} = \frac{x_i}{R}\frac{\partial R}{\partial x_i}u_{x_i} \tag{F.2}$$

How do we estimate the uncertainty in R due to the combined effects of uncertainty intervals in all the x_is? The uncertainty interval in each variable has a range of values. It is unlikely that all will have adverse values at the same time. It can be shown [2] that the best representation is

$$u_R = \pm\left[\left(\frac{x_1}{R}\frac{\partial R}{\partial x_1}u_1\right)^2 + \left(\frac{x_2}{R}\frac{\partial R}{\partial x_2}u_2\right)^2 + \cdots + \left(\frac{x_n}{R}\frac{\partial R}{\partial x_n}u_n\right)^2\right]^{1/2} \tag{F.3}$$

EXAMPLE F.2—Uncertainty in Volume of Cylinder

Obtain an expression for the uncertainty in determining the volume of a cylinder from measurements of its radius and height. The volume of a cylinder in terms of radius and height is

$$\Psi = \Psi(r, h) = \pi r^2 h$$

Differentiating, we obtain

$$d\Psi = \frac{\partial \Psi}{\partial r} dr + \frac{\partial \Psi}{\partial h} dh = 2\pi r h \; dr + \pi r^2 dh$$

since

$$\frac{\partial \Psi}{\partial r} = 2\pi r h \qquad \text{and} \qquad \frac{\partial \Psi}{\partial h} = \pi r^2$$

From Eq. F.2, the fractional uncertainty due to radius is

$$u_{\Psi,r} = \frac{\delta \Psi_r}{\Psi} = \frac{r}{\Psi} \frac{\partial \Psi}{\partial r} u_r = \frac{r}{\pi r^2 h}(2\pi r h) u_r = 2u_r$$

and the uncertainty due to height is

$$u_{\Psi,h} = \frac{\delta \Psi_h}{\Psi} = \frac{h}{\Psi} \frac{\partial \Psi}{\partial h} u_h = \frac{h}{\pi r^2 h}(\pi r^2) u_h = u_h$$

The combined uncertainty in volume is then

$$u_\Psi = \pm \; [(2u_r)^2 + (u_h)^2]^{1/2} \tag{F.4}$$

Comment: The coefficient 2, in Eq. F.4, shows that the uncertainty in measuring cylinder radius has a larger effect than the uncertainty in measuring height. This is true because the radius is squared in the equation for volume.

F-4 APPLICATIONS TO DATA

Applications to data obtained from laboratory measurements are illustrated in the following examples.

EXAMPLE F.3—Uncertainty in Liquid Mass Flow Rate

The mass flow rate of water through a tube is to be determined by collecting water in a beaker. The mass flow rate is calculated from the net mass of water collected divided by the time interval,

$$\dot{m} = \frac{\Delta m}{\Delta t} \tag{F.5}$$

where $\Delta m = m_f - m_e$. Error estimates for the measured quantities are

$$\text{Mass of full beaker, } m_f = 400 \pm 2 \text{ g (20 to 1)}$$

$$\text{Mass of empty beaker, } m_e = 200 \pm 2 \text{ g (20 to 1)}$$

$$\text{Collection time interval, } \Delta t = 10 \pm 0.2 \text{ sec (20 to 1)}$$

The relative uncertainties in measured quantities are

$$u_{m_f} = \pm \frac{2 \text{ g}}{400 \text{ g}} = \pm 0.005$$

$$u_{m_e} = \pm \frac{2 \text{ g}}{200 \text{ g}} = \pm 0.01$$

$$u_{\Delta t} = \pm \frac{0.2 \text{ sec}}{10 \text{ sec}} = \pm 0.02$$

The relative uncertainty in the measured value of net mass is calculated from Eq. F.3 as

$$u_{\Delta m} = \pm \left[\left(\frac{m_f}{\Delta m} \frac{\partial \Delta m}{\partial m_f} u_{m_f} \right)^2 + \left(\frac{m_e}{\Delta m} \frac{\partial \Delta m}{\partial m_e} u_{m_e} \right)^2 \right]^{1/2}$$

$$= \pm \{ [(2)(1)(\pm 0.005)]^2 + [(1)(-1)(\pm 0.01)]^2 \}^{1/2}$$

$$u_{\Delta m} = \pm 0.0141$$

Because $\dot{m} = \dot{m}(\Delta m, \Delta t)$, we may write Eq. F.3 as

$$u_{\dot{m}} = \pm \left[\left(\frac{\Delta m}{\dot{m}} \frac{\partial \dot{m}}{\partial \Delta m} u_{\Delta m} \right)^2 + \left(\frac{\Delta t}{\dot{m}} \frac{\partial \dot{m}}{\partial \Delta t} u_{\Delta t} \right)^2 \right]^{1/2} \tag{F.6}$$

The required partial derivative terms are

$$\frac{\Delta m}{\dot{m}} \frac{\partial \dot{m}}{\partial \Delta m} = 1 \qquad \text{and} \qquad \frac{\Delta t}{\dot{m}} \frac{\partial \dot{m}}{\partial \Delta t} = -1$$

Substituting into Eq. F.6 gives

$$u_{\dot{m}} = \pm \{ [(1)(\pm 0.0141)]^2 + [(-1)(\pm 0.02)]^2 \}^{1/2}$$

$$u_{\dot{m}} = \pm 0.0245 \qquad \text{or} \qquad \pm 2.45 \text{ percent (20 to 1)}$$

Comment: The 2 percent uncertainty interval in time measurement makes the most important contribution to the uncertainty interval in the result.

EXAMPLE F.4—Uncertainty in Reynolds Number for Water Flow

The Reynolds number is to be calculated for flow of water in a tube. The computing equation for the Reynolds number is

$$Re = \frac{4\dot{m}}{\pi \mu D} = Re(\dot{m}, D, \mu) \tag{F.7}$$

We have considered the uncertainty interval in calculating the mass flow rate. What about uncertainties in μ and D? The tube diameter is given as $D = 6.35$ mm. Do we assume that it is exact? The diameter might be measured to the nearest 0.1 mm. If so, the relative uncertainty in diameter would be estimated as

$$u_D = \pm \frac{0.05 \text{ mm}}{6.35 \text{ mm}} = \pm 0.00787 \qquad \text{or} \qquad \pm 0.787 \text{ percent}$$

The viscosity of water depends on temperature. The temperature is estimated as $T = 24 \pm 0.5$ C. How will the uncertainty in temperature affect the uncertainty in μ? One way to estimate this is to write

$$u_{\mu(T)} = \pm \frac{\delta \mu}{\mu} = \frac{1}{\mu} \frac{d\mu}{dT} (\pm \delta T) \tag{F.8}$$

The derivative can be estimated from tabulated viscosity data near the nominal temperature of 24 C. Thus

$$\frac{d\mu}{dT} \approx \frac{\Delta \mu}{\Delta T} = \frac{\mu(25 \text{ C}) - \mu(23 \text{ C})}{(25 - 23) \text{ C}}$$

$$= \frac{(0.000890 - 0.000933) \, \text{N} \cdot \text{sec}}{\text{m}^2} \times \frac{1}{2 \, \text{C}}$$

$$\frac{d\mu}{dT} = -2.15 \times 10^{-5} \, \text{N} \cdot \text{sec/m}^2 \cdot \text{C}$$

It follows from Eq. F.8 that the uncertainty in viscosity due to temperature is

$$u_{\mu(T)} = \frac{1}{0.000911 \, \text{N} \cdot \text{sec}} \frac{\text{m}^2}{\text{}} \times \frac{-2.15 \times 10^{-5} \, \text{N} \cdot \text{sec}}{\text{m}^2 \cdot \text{C}} \times (\pm 0.5 \, \text{C})$$

$$u_{\mu(T)} = \pm 0.0118 \quad \text{or} \quad \pm 1.18 \text{ percent}$$

Tabulated viscosity data themselves also have some uncertainty. If this is ± 1.0 percent, an estimate for the overall uncertainty in viscosity is

$$u_\mu = \pm [(0.01)^2 + (\pm 0.0118)^2]^{1/2} = \pm 0.0155 \quad \text{or} \quad \pm 1.55 \text{ percent}$$

The uncertainties in mass flow rate, tube diameter, and viscosity, needed to compute the uncertainty interval for the calculated Reynolds number, now are known. The required partial derivatives, determined from Eq. F.7, are

$$\frac{\dot{m}}{Re} \frac{\partial Re}{\partial \dot{m}} = \frac{\dot{m}}{Re} \frac{4}{\pi \mu D} = \frac{Re}{Re} = 1$$

$$\frac{\mu}{Re} \frac{\partial Re}{\partial \mu} = \frac{\mu}{Re}(-1)\frac{4\dot{m}}{\pi \mu^2 D} = -\frac{Re}{Re} = -1$$

$$\frac{D}{Re} \frac{\partial Re}{\partial D} = \frac{D}{Re}(-1)\frac{4\dot{m}}{\pi \mu D^2} = -\frac{Re}{Re} = -1$$

Substituting into Eq. F.3 gives

$$u_{Re} = \pm \{[(1)(\pm 0.0245)]^2 + [(-1)(\pm 0.0155)]^2 + [(-1)(\pm 0.00787)]^2\}^{1/2}$$

$$u_{Re} = \pm 0.0300 \quad \text{or} \quad \pm 3.00 \text{ percent}$$

Comment: Examples F.3 and F.4 illustrate two points important for experiment design. First, the mass of water collected, Δm, is calculated from two measured quantities, m_f and m_e. For any stated uncertainty interval in the measurements of m_f and m_e, the *relative* uncertainty in Δm can be decreased by making Δm larger. This might be accomplished by using larger containers or a longer measuring interval, Δt, which also would reduce the relative uncertainty in the measured Δt. Second, the uncertainty in tabulated property data may be significant. The data uncertainty also is increased by the uncertainty in measurement of fluid temperature.

EXAMPLE F.5—Uncertainty in Air Speed

Air speed is calculated from pitot tube measurements in a wind tunnel. From the Bernoulli equation,

$$V = \left(\frac{2gh\rho_{\text{water}}}{\rho_{\text{air}}}\right)^{1/2} \tag{F.9}$$

where h is the observed height of the manometer column.

The only new element in this example is the square root. The variation in V due to the uncertainty interval in h is

$$\frac{h}{V}\frac{\partial V}{\partial h} = \frac{h}{V}\frac{1}{2}\left(\frac{2gh\rho_{\text{water}}}{\rho_{\text{air}}}\right)^{-1/2}\frac{2g\rho_{\text{water}}}{\rho_{\text{air}}}$$

$$\frac{h}{V}\frac{\partial V}{\partial h} = \frac{h}{V}\frac{1}{2}\frac{1}{V}\frac{2g\rho_{\text{water}}}{\rho_{\text{air}}} = \frac{1}{2}\frac{V^2}{V^2} = \frac{1}{2}$$

Using Eq. F.3, we calculate the uncertainty in V as

$$u_V = \pm\left[\left(\frac{1}{2}u_h\right)^2 + \left(\frac{1}{2}u_{\rho_{\text{water}}}\right)^2 + \left(-\frac{1}{2}u_{\rho_{\text{air}}}\right)^2\right]^{1/2}$$

If $u_h = \pm 0.01$ and the other uncertainties are negligible,

$$u_V = \pm\left\{\left[\frac{1}{2}(\pm 0.01)\right]^2\right\}^{1/2}$$

$$u_V = \pm 0.00500 \qquad \text{or} \qquad \pm 0.500 \text{ percent}$$

Comment: The square root reduces the uncertainty interval in the calculated velocity to half that of u_h.

F-5 SUMMARY

A statement of the probable uncertainty of data is an important part of reporting experimental results completely and clearly. Estimating uncertainty in experimental results requires care, experience, and judgment, in common with many endeavors in engineering. We have emphasized the need to quantify the uncertainty of measurements, but space allows including only a few examples. Much more information is available in the references that follow (e.g., [5]). We urge you to consult them when designing experiments or analyzing data.

F-6 REFERENCES

1. Pugh, E. M., and G. H. Winslow, *The Analysis of Physical Measurements*. Reading, MA: Addison-Wesley, 1966.
2. Kline, S. J., and F. A. McClintock, "Describing Uncertainties in Single-Sample Experiments," *Mechanical Engineering, 75,* 1, January 1953, pp. 3-9.
3. Doebelin, E. O., *Measurement Systems,* 4th ed. New York: McGraw-Hill, 1990.
4. Young, H. D., *Statistical Treatment of Experimental Data*. New York: McGraw-Hill, 1962.
5. Holman, J. P., *Experimental Methods for Engineers,* 5th ed. New York: McGraw-Hill, 1989.

Appendix G

SI UNITS, PREFIXES, AND CONVERSION FACTORS

Table G.1 SI Units and Prefixes[a]

SI Units	Quantity	Unit	SI Symbol	Formula
SI base units:	Length	meter	m	—
	Mass	kilogram	kg	—
	Time	second	sec	—
	Temperature	kelvin	K	—
SI supplementary unit:	Plane angle	radian	rad	—
SI derived units:	Energy	joule	J	$N \cdot m$
	Force	newton	N	$kg \cdot m/sec^2$
	Power	watt	W	J/sec
	Pressure	pascal	Pa	N/m^2
	Work	joule	J	$N \cdot m$

SI prefixes	Multiplication Factor		Prefix	SI Symbol
	1 000 000 000 000=10^{12}		tera	T
	1 000 000 000=10^{9}		giga	G
	1 000 000=10^{6}		mega	M
	1 000=10^{3}		kilo	k
	0.01=10^{-2}		centi[b]	c
	0.001=10^{-3}		milli	m
	0.000 001=10^{-6}		micro	μ
	0.000 000 001=10^{-9}		nano	n
	0.000 000 000 001=10^{-12}		pico	p

[a] Source: ASTM standard for Metric Practice E 380–89, 1989.
[b] To be avoided where possible.

G-1 UNIT CONVERSIONS

The data needed to solve problems are not always available in consistent units. Thus it often is necessary to convert from one system of units to another.

In principle, all derived units can be expressed in terms of basic units. Then, only conversion factors for basic units would be required.

In practice, many engineering quantities are expressed in terms of defined units, for example, the horsepower, British thermal unit (Btu), quart, or nautical mile. Definitions for such quantities are necessary, and additional conversion factors are useful in calculations.

Basic SI units and necessary conversion factors, plus a few definitions and convenient conversion factors are given in Table G.2.

Table G.2 Conversion Factors and Definitions

Fundamental Conversion Factor	English Unit	Exact SI Value	Approximate SI Value
Length	1 in.	0.0254 m	—
Mass	1 lbm	0.453 592 37 kg	0.4536 kg
Temperature	1 F	5/9 K	—

Definitions:

Acceleration of gravity: $g = 9.8066$ m/sec^2 ($= 32.174$ ft/sec^2)

Energy: Btu (British thermal unit) $\equiv$ amount of energy required to raise the temperature of 1 lbm of water 1 F (1 Btu $= 778.2$ ft·lbf)

kilocalorie $\equiv$ amount of energy required to raise 1 kg of water 1 K (1 kcal $= 4187$ J)

Length: 1 mile $= 5280$ ft; 1 nautical mile $= 6076.1$ ft

Power: 1 horsepower $\equiv 550$ ft·lbf/sec

Pressure: 1 bar $\equiv 10^5$ Pa

Temperature: degree Fahrenheit, $T_F = \frac{9}{5}T_C + 32$ (where T_C is degrees Celsius)

degree Rankine, $T_R = T_F + 459.67$

Kelvin, $T_K = T_C + 273.15$ (exact)

Viscosity: 1 Poise $\equiv 0.1$ kg/m·sec

1 Stoke $\equiv 0.0001$ m^2/sec

Volume: 1 gal $\equiv 231$ in.^{3} (1 ft$^3 = 7.48$ gal)

Useful Conversion Factors:

1 lbf $= 4.448$ N

1 lbf/in.$^2 = 6895$ Pa

1 Btu $= 1055$ J

1 hp $= 746$ W $= 2545$ Btu/hr

1 kW $= 3413$ Btu/hr

1 quart $= 0.000946$ m$^3 = 0.946$ liter

1 kcal $= 3.968$ Btu

ANSWERS TO SELECTED PROBLEMS

Chapter 1

1.2 $\mathbf{V} = 0.0567$ m^3, $d = 0.477$ m

1.3 KE $= 5 \times 10^3$ ft $\cdot$ lbf/slug; 31.7 ft $\cdot$ lbf/ft^3

1.4 KE $= 450$ J/kg; 1.61 kJ/m^3

1.6 $\rho_m = 1.218$ kg/m^3, $\rho_d = 1.225$ kg/m^3

1.7 $\rho = 930 \pm 27.2$ kg/m^3 (20 to 1)

1.8 $\rho = 1130 \pm 21.4$ kg/m^3,
    SG $= 1.13 \pm 0.0214$ (20 to 1)

1.9 $\rho = 1260 \pm 24.6$ kg/m^3,
    SG $= 1.26 \pm 0.0246$ (20 to 1)

1.11 $\rho = 0.0765$ lbm/ft^3, $u_\rho = \pm 0.348\%$

1.12 $\rho = 1.39$ kg/m^3, $u_\rho = \pm 0.238\%$

1.13 $u_{\dot{m}} = \pm 1.30\%$, $\pm 0.217\%$

1.16 $u_a = \pm 3.47\%$

1.18 $\delta D = \pm 0.00441$ in.

1.19 $u_8 = \pm 1.95\%$, $u_{12} = \pm 0.122\%$

1.22 $\mathbf{V} = 0.254$ ft^3, $L = 1.29$ ft

1.23 $W = 394$ kJ

1.24 $T_2 = 285$ C

1.25 $T_2 \approx 269$ F, $p_2 \approx 41$ psia, $W_{\text{out}} = 5390$ Btu

1.27 $F_D = 878$ lbf, $\Delta \mathcal{P} = -7.73$ hp

1.28 $R = 2V_0^2 \sin\theta \, \cos\theta/g$, $\theta_{\text{max}} = 45°$

1.29 $t = 3W/gk$

1.30 $s = 2.05 \, W^2/gk^2$

1.31 $V_t = 56.8$ m/sec, $V_{100} = 38.3$ m/sec

1.32 $s \approx 10$ m

1.33 $V_{\text{up}} \approx 0.7$ m/sec, $V_{\text{down}} \approx 2.3$ m/sec

1.34 $V_0 \doteq 37.7$ m/sec, $\theta_0 = 21.8°$

1.35 $V_t = 53.9$ m/sec, $s = 345$ m, $s = 134$ m

1.36 1 Pa $= 1.45 \times 10^{-4}$ lbf/in.^{2}

1.41 1 ft$^3 = 7.48$ gal, 1 gal $= 3.79$ L

1.43 $\mathbf{V} = 15,600$ gal

1.45 SG $= 13.6$, $v = 7.37 \times 10^{-5}$ m^3/kg,
    $\gamma_E = 847$ lbf/ft^3, $\gamma_M = 144$ lbf/ft^3

1.46 1 m^3/sec $= 35.3$ ft^3/sec;
    1 ft^3/sec $= 449$ gpm;
    1 gal/hr $= 0.631$ kg/min;
    1 ft^3/min $= 4.59$ lbm/hr

1.48 1 home run $= 117$ m; 1 fastball $= 40.2$ m/sec;
    1 baseball $= 1.41$ N
    mass $=$ baseball $\cdot$ homerun $\cdot$ fastball^{-2}
    $= 0.103$ kg

1.49 $W = 59.9$ lbf; $\mathbf{V} = 0.964$ ft^3

1.50 $\Delta \gamma = -1.75$ lbf/ft^3

Chapter 2

2.7 $(x, y) = (2.72, 2.94)$ to $(8.15, 2.94)$

2.8 $y = cx^{-b/a}$

2.9 $xy = c$

2.11 $\vec{V} = 8\,\hat{i} - 4\hat{j}$ m/sec; $y = 2x^{-2}$

2.13 $(x - B/A)y = $ constant

2.14 $x = y/3$; $t = 2$ sec

2.16 $xy = 2$ m^2

2.17 $y - y_0 = (B/2A^2)(x - x_0)^2$

2.20 $x = y$, $x = \exp(\ln y - A(\ln y)^2/2)$

2.31 $\vec{F}_B = 937 \,\hat{j}$ lbf

2.32 $\vec{F}_B = 16.8\,\hat{j} + 88.0\,\hat{k}$ N

2.33 $\vec{F}_B = 261\,\hat{i} + 312\,\hat{k}$ N

2.34 $\vec{F}_B = 144\,\hat{i} + 192\,\hat{j}$ N

2.35 $\sigma = 51.0$ psi; $\tau = 115$ psi

2.36 $A = 5.92 \times 10^7$ kg/m $\cdot$ sec; $B = 2180$ K

2.41 $\tau_{yx} = -2.64$ N/m^2; Plus x

2.42 $F = 0.0144$ N; Right

2.44 $F = 8.07$ mN

2.47 $F = 37.4$ lbf

2.48 $V_t = 0.0123$ m/sec

2.50 $V = 34.3$ ft/sec

2.51 $t = 2.30$ sec

2.52 $F_v = \mu V A/h$, $V_{\text{max}} = mgh/\mu A$

2.54 $U = 9.84$ m/sec

2.55 $F = 1.01$ N

2.56 $T = 15.3$ ft $\cdot$ lbf

2.57 $\mu = 8.07 \times 10^{-4}$ N $\cdot$ sec/m^2

2.58 $\mu = 0.0208$ N $\cdot$ sec/m^2

2.60 $\mu = 0.0781$ N $\cdot$ sec/m^2

2.65 $T = \pi \mu \omega R^4/2h$

2.66 $\dot{\gamma} = \omega/\theta$; $T = 2\pi R^3 \tau_{yx}/3$

2.67 $T = \pi \mu \Delta \omega R^4/2a$; $\mathcal{P} = \pi \mu \omega_0 \Delta \omega R^4/2a$;
    $s = 2Ta/\pi \mu R^4 \omega_i$; $\eta = 1 - s$

2.69 $\tau = \mu \omega z \, \tan\theta/a$; $T = 0.0206$ N $\cdot$ m

2.70 $T = \dfrac{2\pi \mu \omega R^4}{h} \left(\dfrac{\cos^3 \alpha}{3} - \cos\alpha + \dfrac{2}{3} \right)$

2.72 $\tau_{yx} = 4.6$ N/m^2; $\tau_{\text{slurry}}/\tau_{\text{H}_2\text{O}} = 9.2$

2.75 $n = 0.661$; $k = 0.121$ N $\cdot$ sec$^{0.661}$/m^2;
    $\tau_{yx} = 0.0264$ N/m^2; $\eta_{\text{cream}}/\mu_{\text{H}_2\text{O}} = 264$

2.76 $n = 1.227$; $k = 0.00251$ N $\cdot$ sec$^{1.227}$/m^2

2.78 Laminar

2.79 $\mathbf{V}_{\text{max}} = 5.07 \times 10^{-2}$ m/sec

2.80 $Re = 1440$; Laminar

Chapter 3

3.1 $D = 6.0$ mm

3.2 (a) 0.3 percent; (b) -0.2 percent

3.3 $D = 0.254$ m; $p = 154$ kPa (gage)

3.4 $\sigma = 432$ MPa; Circumferential

3.5 $m = 14.7$ kg; $t = 11.9$ mm

3.6 $p = 51.0$ MPa; $\sigma_a = 421$ MPa;
 $\sigma_c = 842$ MPa; $\tau = 211$ MPa

3.7 $p_{abs} = 927$ kPa; $p_{gage} = 824$ kPa

3.8 $p_{abs} = 79.3$ kPa

3.9 $\delta h = 5.08 \times 10^{-5}$ in.

3.10 $p_{atm} = 2080$ psf; $h_0 = 746$ mm Hg

3.11 $\rho_s = 5.09$ kg/m^2

3.12 $F = 21.9$ N

3.13 $F = 2620$ lbf; $T = 62.4$ lbf

3.15 $p_{gas} = -365$ Pa (gage);
 $\Delta h = -37.2$ mm H$_2$O

3.16 $p = 34.7$ psig; $F = 19.3$ lbf; $s = 3.81$ in.

3.17 $\Delta z = 89.1$ m

3.18 $\Delta z = 1390$ m

3.20 No

3.22 $\rho = 0.00332$ kg/m^3

3.23 $p = 57.5$ kPa; $p = 60.2$ kPa

3.27 $p/p_0 = [(1 + mz)/(1 + mz_0)]^{-g/mRT_0}$

3.29 $p = 6.72$ psig

3.30 $H = 20.9$ mm

3.31 $H = 30.0$ mm

3.33 $p_1 - p_2 = 58.8$ Pa

3.34 $d = 75.0$ mm

3.35 $p_a = 1.18$ psig

3.36 SG $= 0.900$

3.38 $l = 1.60$ m

2.29 $l = 0.546$ m

3.40 $h = 1.11$ in.

3.41 $h = 7.85$ mm; $s = 0.308$

3.43 $\Delta p = 1.05$ kPa; $s = 2.15$

3.44 $\Delta p = 98.0$ Pa; $s = 3.27$

3.45 $p = 2.45$ kPa (gage); $s = 2.40$

3.46 $\theta = 12.5°$; $s = 5.0$

3.47 $\theta = 11.1°$

3.48 $L = 56.1$ mm; $s = 5.6$

3.49 $l = 0.316$ m

3.52 $\Delta h = 75.6$ mm

3.53 $T = 96.6$ C; $T = 93.2$ C

3.54 $F = 552$ kN; $x' = 2.50$ m, $y' = 2$ m

3.57 $R = 14.7$ kN

3.58 $R = 52.6$ kN

3.59 $R = 1.74$ kN, to the right

3.60 $F_R = 61,700$ lbf

3.61 $F_R = 376$ N; $y' = 0.3$ m

3.62 $F = 33.3$ kN; $D \approx 7.3$ mm

3.63 $F_A = 32.7$ kN

3.64 $d = 2.66$ m

3.65 $F/b = 174$ lbf/ft

3.66 $b = 6.15$ m

3.67 $W = 15,800$ lbf

3.68 $M = 344$ slug

3.69 $F_A/b = 15.6$ kN/m

3.70 $F = 1800$ lbf

3.71 $D = 2.60$ m

3.72 SG $= 0.542$

3.73 $F = 2\rho gR^3/3$; $y' = 3\pi R/16$

3.74 $F = 63,800$ lbf

3.75 $F = 31.6$ N

3.76 $F = 1.82 \times 10^6$ lbf;
 $R_x = 1.76 \times 10^6$ lbf;
 $R_y = 3.04 \times 10^6$ lbf

3.77 $F_{R_y} = -73.9$ kN; $x' = 1.06$ m

3.78 $F_{R_y} = 7.63$ kN; $M = 3.76$ kN·m

3.80 $F_{R_y} = \rho g w \pi R^2/4$; $x' = 4R/3\pi$

3.81 $F_{R_y} = 17,100$ lbf; $x' = 2.14$ ft

3.82 $F_{R_y} = 582$ lbf; $x' = 0.858$ ft

3.83 $F_V = 14,200$ lbf; $z' = 1.91$ ft

3.84 $F_V = 12,500$ lbf; $z' = 1.93$ ft

3.86 $F_A = 30.2$ kN

3.88 $F_A = 5.71$ kN

3.90 $M = 7.68$ kN·m(ccw)

3.93 $F_R = 1.83 \times 10^7$ N; $\alpha = 20.2°$

3.95 $F_R = 370$ kN; $\alpha = 57.6°$

3.96 $M/L = \rho R^2[1 + 3\pi/4]$, $F/L = \rho g R^2/2$

3.97 $F_V = 1.55$ kN; $x' = 0.120$ m

3.98 $F_V = 2.48$ kN; $x' = 0.642$ m;
 $F_H = 7.35$ kN; $y' = 0.217$ m

3.100 SG $= \left(\cos^{-1}(1 - \alpha) + (\alpha - 1)\sqrt{2 - \alpha} \right)/\pi$

3.101 $\gamma_s = 51.2$ lbf/ft^3; No

3.103 $h = 177$ mm

3.104 SG $=$ SG$_{H_2O}W_{air}/(W_{air} - W_{net})$

3.106 $F_B = 8.02 \times 10^{-11}$ N;
 $V = 0.341$ mm/sec

3.108 $D = 27.4$ m; $\mathrm{V}_{sw}/\mathrm{V}_{oil} = 0.507$

3.109 Claims are valid; Lift is increased
 45 percent

3.110 $\theta = 23.8°$

3.111 Tank is not neutrally buoyant

3.112 $D = 82.7$ m; $M = 637$ kg

3.113 $D = 116$ m ; $M = 703$ kg

3.115 SG > 0.70

3.116 $R = 8.79$ mN

3.118 $M = \rho g L a^3 \Delta \theta/12$; stable for $0 \le l \le$ a/3

3.119 $\omega = 1.81$ rad/sec

3.120 $\omega = 13.1$ rad/sec; No

3.121 $a = gh/L$

3.122 Slope $= 0.22$

3.123 $a_r = -1.61$ m/sec^2; $\alpha = 9.3°$

3.125 $a_r = -r\omega^2$; $\partial p/\partial r = \rho r \omega^2$; $p = 7.19$ MPa

3.126 $\Delta p = \rho \omega^2 R^2/2$; $\omega = 7.16$ rad/sec

3.127 $\omega = 188$ rad/sec

3.128 Slope $= -0.20$;
 $p(x, 0) = 106 - 1.57x(m)$ kPa

3.129 $\alpha = 13.3°$

3.130 $\alpha = 30°$; Slope $= 0.346$

3.131 Slope $= 0.540$; $\omega = 3.48$ rad/sec

3.132 $T = 47.6$ lbf; $p = 55.3$ lbf/ft^2 (gage)

3.133 $p_2/p_1 = 24.2$

3.134 $F_{rear} = 107$ N; $F_{front} = 54.1$ N;
 $F_{bottom} = 156$ N

3.135 $\theta = 14.3°$ (toward center of curvature)

3.136 $\omega = 31.3$ rad/sec;
 $p_{max} = 51.5$ kPa (gage);
 $p_{min} = 43.9$ kPa (gage)

Chapter 4

4.1 $F = 961$ N; $F = 586$ N
4.2 $W = 469$ kJ
4.3 $Q = 9.57$ kJ
4.4 $s_1 - s_1 = -0.291$ kJ/kg·K
4.5 $t = 16.7$ sec
4.6 $V = 87.5$ km/hr
4.7 $s = 2680$ ft; $t = 26.1$ sec
4.8 $k = 2.50 \times 10^{-5}$ N·sec²/m²
4.9 $\theta = 48.2°$
4.11 $\Delta u = 77.5$ kJ/kg
4.12 $t = 419$ sec
4.13 $t = 4.18$ hr
4.14 $t = 9.38$ sec; $s = 156$ m
4.15 $a_r = -31.7$ m/sec²; $V = 89.0$ m/sec
4.16 $h = 21.2$ mm; $\mu = 0.604$
4.17 $\vec{T} = -780\hat{k}$ N·m
4.18 $\vec{T}_{avg} = -3.25 \times 10^{-4}$ $\hat{k}$N·m
4.20 $Q = 1.50$ ft³/sec
4.22 $mf_x = -\rho V^2 wh/3$
4.23 (a) $u_{max}\pi R^2/2$; (b) $\hat{i}u_{max}^2\pi R^2/3$
4.25 $Q_3 = -5.00$ ft³/sec (into CV)
4.26 $D = 55.6$ mm
4.27 $\dot{m}_3 = 2500 + 999\cos(4\pi t)$ kg/sec
4.28 $\vec{V}_3 = 4.04\,\hat{i} - 2.34\,\hat{j}$ m/sec
4.29 $\dot{m}/w = \rho^2 g h^3 \sin\theta/6\mu$
4.30 $k = 2$ sec⁻¹
4.31 $u_{max} = 7.50$ m/sec
4.32 $V_m = \pi V_1/4$
4.33 $U = 5.00$ ft/sec
4.34 $V_3 = 3.33$ ft/sec (into CV)
4.36 $\dot{m}_2 = 16.2$ kg/sec
4.37 $\partial V/\partial t = -0.181$ gal/sec
4.38 $dh/dt = -0.50$ m/sec (falling)
4.39 $dh/dt = -0.289$ mm/sec (falling)
4.41 $\partial\rho/\partial t = -0.369$ kg/m³·sec
4.42 $\partial\rho_0/\partial t = 2.50 \times 10^{-3}$ slug/ft³·sec
4.43 $dh/dt = -56.6$ mm/sec
4.44 $y = 0.134$ m
4.45 $t = 28.5$ sec
4.46 $dy/dt = -9.01$ mm/sec
4.47 $Q_0 = 3.61 \times 10^{-5}$ m³/sec; $dh/dt = -0.0532$ m/sec
4.49 $\dot{m}_{bc} = 1.42$ kg/sec (out)
4.50 $t = 6V_0/5Q_0$
4.54 $mf = 349\,\hat{i} - 13.5\,\hat{j}$ N
4.55 Ratio = 1.2
4.56 Ratio = 1.33
4.57 $T = 59.9$ N
4.58 $M = 409$ kg
4.60 $F_x = 0.0230$ lbf
4.61 $F_x = 184$ N
4.62 $F_{max} = 97.0$ lbf
4.63 $M = 671$ kg
4.64 Scale reads 213 lbf; $W = 203$ lbf
4.65 $F = 1.81$ kN, tension
4.66 $F = 496$ N, tension
4.67 $F = 321$ N
4.68 $F = 0.446$ lbf

4.69 $F = 370$ N
4.70 $F = 206$ lbf, tension
4.71 $F = 8.32$ kN
4.72 $F = 1.70$ lbf
4.73 $Q = 0.424$ m³/sec; $F_y = 4.05$ kN
4.76 $\dot{m}_2 = 0.760\,\dot{m}_1\sqrt{A_2/A_1}$
4.78 $T = 30.0$ N; $\vec{F} = 30\,\hat{i} - 51.9\,\hat{j}$ N
4.80 $t = 1.19$ mm; $F_x = 3.63$ kN
4.81 $\vec{F} = -26.7\,\hat{i} - 139\,\hat{j}$ lbf
4.82 $\vec{F} = -4.68\,\hat{i} + 1.66\hat{j}$ kN
4.83 $V_2 = 6.60$ m/sec; $p_2 - p_1 = 84.2$ kPa
4.84 $\vec{F} = -139\,\hat{i} - 740\,\hat{j}$ N
4.85 $Q_2 = 1.43 \times 10^{-3}$ m³/sec (out); $p_3 = 146$ kPa (abs)
4.86 $F = 837$ lbf
4.87 $F = 4.77$ lbf
4.88 $F = 5.11$ kN
4.89 $Q = 0.141$ m³/sec; $\vec{F} = -1.65\,\hat{i} - 1.34\,\hat{j}$ kN
4.92 $\dot{m} = 9.67$ kg/sec; $V_{2,max} = 15.0$ m/sec; $F_D = 65$ N
4.93 $u_{max} = 30$ ft/sec; $p_1 - p_2 = 0.190$ lbf/ft²
4.94 $u_{max} = 60$ ft/sec; $p_1 - p_2 = 0.699$ lbf/ft²
4.96 $F = 7.90 \times 10^{-4}$ N
4.97 $u_{max} = 31.6$ ft/sec; $p_1 - p_2 = 0.0762$ lbf/ft²
4.98 $F/w = 0.277$ N/m
4.99 $D = 0.446$ N
4.100 $F/w = 0.0393$ N/m
4.101 $F/w = 2.94$ N/m
4.102 $F = 760$ N
4.103 $F = 15.6$ kN
4.105 $h_2/h_1 = 0.5(1 + \sin\theta)$
4.106 $F_D/w = 54.1$ N/m
4.108 Error = 1.73 percent
4.110 $h = H/2$
4.111 $h = 89.6$ mm; $F = 0.242$ N
4.112 $\mu_s < 0.202$
4.113 $h = 3.26$ m
4.115 $V = [V_0^2 + 2gh]^{1/2}$; $F = 1.49$ N
4.116 $V = [V_0^2 - 2gh]^{1/2}$; $h = 4.15$ ft
4.118 $t \approx 730$ sec
4.120 $z = 3V_0^2/2g$
4.121 $z = V_0^2/2g$
4.122 $p(x) = p(0) - \rho(Qx/whL)^2$
4.124 $V(x) = V_0 - qx/A$; $p(x) = p_0 + (2\rho qV_0x/A)(1 - qx/2V_0A)$
4.125 $V(x) = V_0 - (q_{max}x^2/2AL)$; $p(x) = p_0 + (\rho q_{max}V_0x^2/AL)\times(1 - q_{max}x^2/4ALV_0)$
4.128 $c = [gh]^{1/2}$
4.130 $h_1 = [h_2^2 + 2Q^2/gb^2h_2]^{1/2}$
4.131 $D/D_0 = [1 - x/L]^{1/2}$
4.132 $V(x) = V_0x/h$
4.133 $V(r) = V_0r/2h$
4.134 $\vec{F} = -822\,\hat{i} + 220\,\hat{j}$ N
4.135 $\vec{F} = -570\,\hat{i} + 329\,\hat{j}$ lbf
4.136 $\vec{F} = -135\,\hat{i} + 135\,\hat{j}$ N
4.137 $F = 1.73$ kN
4.138 $F = 3210$ lbf
4.139 $V = 95.4$ ft/sec
4.140 $F = 167$ N

4.141 $F = 3840$ lbf at $U = 75$ mph

4.142 $F = 15.5$ kN

4.144 $\dot{W} = \rho(V - U)^2 UA(1 - \cos\theta)$

4.145 $t = 4.17$ mm; $F = 4240$ N

4.146 $t = 6.25$ mm; $F = 7940$ N

4.147 $\alpha = 30°$; $F = 10.3$ kN

4.148 $U = V/2$

4.149 $\dot{m}_2/\dot{m}_3 = 0.5$; $F = 7.46$ kN

4.151 $D = 1.44$ m; $T = 1.60$ kN; $T = 800$ N

4.152 $F = 187$ N

4.153 $a_{rf} = 13.5$ m/sec^2

4.154 $U = 83.9$ ft/sec

4.155 $dU/dt = 13.7$ m/sec^2; $U_t = 15.8$ m/sec

4.156 $U/V = \ln[M_0/(M_0 - \rho VAt)]$

4.157 $\theta = 19.7°$

4.158 $A = 111$ mm^2

4.159 $A = 0.900$ in.^{2}; $t = 3$ sec

4.160 $t = 22.6$ sec

4.161 $h = 17.9$ mm

4.162 $U = 22.5$ m/sec

4.163 $a_{rf} = 5.99$ m/ sec^2; $U/U_t = 0.667$

4.164 $t = 1.71$ sec; $s = 7.47$ m

4.165 $t = 0.0668$ sec; $x = 0.177$ m

4.166 $dU/dt = 14.2$ m/sec^2; $U_t = 15.2$ m/sec

4.167 $U_t = 15.8$ m/sec

4.169 $U/U_0 = e^{-4\rho VAt/M}$

4.170 $t = 0.750\, M/\rho VA$; $x = 0.238\, MU_0/\rho VA$

4.172 $a_y = -16.5$ ft/sec^2

4.175 $t = 129$ sec

4.176 $U = 834$ m/sec; $a_{max} = 96.7$ m/sec^2

4.177 $a = 33.3$ m/sec^2; $U = 288$ m/sec

4.178 $M_f = 186$ lbm

4.179 $U = 344$ m/sec

4.180 Mass fraction $= 0.393$

4.181 $a_0 = 17.3\, g$

4.182 $V = 1910$ m/sec

4.183 $V = 3860$ ft/sec; $Y = 33,500$ ft

4.184 $F = aM_0 - \rho V_j^2 A + 3\rho V_j(at)A - 1.5\rho(at)^2 A$

4.187 $t = 3$ sec

4.188 $\theta = 18.9°$

4.189 $U = 16.5$ m/sec

4.190 $t = M/2\rho VA$

4.192 $\dot{m} = Mg/V_e$; $t = 110$ sec

4.193 $I = 2.21$ lbf $\cdot$ sec; $I_{sp} = 80.3$ sec; $V_{max} = 139$ m/sec; $Y_{max} = 1100$ m

4.194 $\dot{m} = 186$ lbm/sec

4.199 $\vec{a}_{XYZ} = -\hat{e}_r(V_0^2 r_0^2/R^3 + \omega^2 R) + \hat{e}_\theta 2\omega V_0 r_0/R$

4.204 $\dot{\omega} = -2.75$ rad/sec^2

4.205 $k = 0.004$ N $\cdot$ sec^2/m; $t = 21.5$ sec

4.206 $M = 8.14$ kN $\cdot$ m

4.207 $F = 44.4$ kN; $T = 920$ kN $\cdot$ m

4.212 $T = 0.0161$ N $\cdot$ m

4.213 $T = 0.0722$ N $\cdot$ m

4.214 $\dot{\omega} = 0.161$ rad/sec^2

4.215 $\dot{W} = R\omega(1 - \cos\theta)\rho VA(V - U)$; $U/V = 1/2$

4.216 $\vec{F} = 969\,\hat{i} - 2\,\hat{j} + 492\,\hat{k}$ lbf

4.217 $\omega = 6.04$ rad/sec; $A = 1720$ m^2

4.218 $T = 29.4$ N $\cdot$ m; $\vec{M} = 51.0\,\hat{i} + 1.40\,\hat{j}$ N $\cdot$ m

4.222 $\dot{W} = -80.0$ kW

4.223 $\dot{Q} = -146$ Btu/sec

4.224 $\dot{W} = -96.0$ kW

4.225 $p_1 - p_2 = 75.4$ kPa

4.226 $\eta = 0.348$

4.227 $\partial T/\partial t = -0.177$ R/sec

4.228 $\dot{W} = -3.41$ kW

4.231 $\eta = 0.571$

4.232 $V = 133$ ft/sec; $z_0 = 248$ ft; $\dot{W} = -5.69$ hp; $\Delta z = 30.1$ ft

4.233 $Q = 0.0166$ m^3/sec; $z_{max} = 61.4$ m; $F = 561$ N

4.234 $\eta = 0.631$

4.235 $\eta = 0.500$; $\eta = 0$

4.236 $\eta = 0.554$

4.238 $V = 94.5$ m/sec; $\dot{W} = -739$ kW

4.239 $\Delta m.e. = -1.88$ N $\cdot$ m/kg; $\Delta T = 4.49 \times 10^{-4}$ K

Chapter 5

5.1 (a)

5.2 (b), (c)

5.3 (a), (b)

5.4 $A + E + J = 0$

5.5 $u = -2yx - 2x + f(y)$

5.6 $v = Ay/x^2$

5.7 $(v/U)_{max} = 0.0025$

5.8 $(v/U)_{max} = 0.00167$

5.9 $(v/U)_{max} = 0.00182$

5.11 (a), (b), (c)

5.12 $r = q/2\pi$, $\theta = \pm\pi$

5.13 $V_\theta = -\Lambda \sin\theta/r^2 + f(r)$

5.14 $\vec{V} = \hat{e}_\theta \omega rz/h$

5.15 Yes; $r = e^\theta$

5.16 $\Psi = Vy \cos\alpha - Vx \sin\alpha$

5.17 $\vec{V} = Ax\,\hat{i} - Ay\hat{j}$

5.18 $\Psi = Uy^2/2h$; $y = h/\sqrt{2}$

5.19 $\Psi = x^2 y - y^3/3$

5.20 $\Psi = A\theta - A\ln r$

5.21 $\vec{V} = (-U\cos\theta + q/2\pi r)\hat{e}_r + U\sin\theta\,\hat{e}_\theta$

5.22 $|\vec{V}| = U$ at $\theta = \pm 30°, \pm 150°$

5.23 $\Psi = -y^3 z - 2z^2$

5.24 $Q = 4$ m^3/ sec/m

5.25 $\Psi = Uy^2/2h$; $y = 2.83$ ft

5.28 $y/\delta = 0.460, 0.667$

5.29 $\Psi = -\omega r^2/2$; $Q/b = 6 \times 10^{-4}$ m^3/ sec/m

5.30 $\Psi = -c\ln r$; $Q/b = 0.168$ m^3/ sec/m

5.32 Yes; $\vec{a}_p = 0.625\,\hat{i} + 0.313\,\hat{j}$ m/sec^2

5.33 $\vec{a}_p = (16\,\hat{i} + 32\,\hat{j} + 16\,\hat{k})/3$ m/sec^2

5.34 3-D; No; $\vec{a}_p = 27\,\hat{i} + 9\,\hat{j} + 64\,\hat{k}$ m/sec^2

5.36 $u = Ax^2/2$; $\vec{a}_p = A^2(0.5\,\hat{i} + \hat{j})$

5.37 $\vec{a}_p = -4\,\hat{i} - 12\,\hat{j}$ m/sec^2

5.38 $\vec{a}_p = -\hat{i}aV_1^2 e^{-2ax}$

5.39 $\vec{a}_p = -(Q/2\pi h)^2 r^{-3}\hat{e}_r$

5.40 $a_r = -81.0$ km/sec^2; $a_r = -3.0$ km/sec^2

5.41 $V_r = qr/2h$; $a_r = 60.0$ m/sec^2

5.42 $\vec{a}_p = -5.56\hat{e}_r + 1.36\hat{e}_\theta$ ft/sec^2; $\vec{a}_p = -22.2\hat{e}_r$ ft/sec^2

5.45 $\partial T/\partial x = -0.0873$ F/mi

5.47 $\vec{a}_p = x\hat{i} + y\hat{j}$

5.49 $c = -2\,\text{sec}^{-1}$; $\vec{a}_p = 4\,\hat{i} + 8\,\hat{j} + 5\,\hat{k}$ m/sec^2

5.50 Ratio = 100

5.51 $y/\delta = 0.634$

5.53 $v = v_0(1 - y/h)$;
$\vec{a}_p = \hat{i}v_0^2 x/h^2 - \hat{j}(v_0^2/h)(1 - y/h)$

5.54 $\vec{a}_p = \hat{e}_r(v_0/2h)^2 r - \hat{k}(v_0^2/h)(1 - z/h)$

5.56 $\vec{a}_p = 44.5\,\hat{i}$ ft/sec^2; $\vec{a}_p = 103\,\hat{i}$ ft/sec^2

5.58 $v = -bU_0 y(1 - bx^2)$; $a_{px,\max} = 201$ m/sec^2
$$\vec{a}_p = \frac{U_0^2 b}{(1 - bx)^2}\left(\hat{i} - \hat{j}\,\frac{by}{(1 - bx)}\right)$$

5.59 $a_x = 4(20 + 2\sin(\omega t))^2 + 1.2\cos(\omega t)$ m/sec^2

5.60 $f_1 = x_0 e^{At}$, $f_2 = y_0 e^{-At}$;
$t(1, 1) = 0.693$ sec, $t(2, 0.5) = 1.39$ sec;
$\vec{a}_p(1, 1) = \hat{i} + \hat{j}$ m/sec^2,
$\vec{a}_p(2, 0.5) = 2\,\hat{i} + 0.5\hat{j}$ m/sec^2

5.63 Yes; Yes

5.64 No; Yes

5.66 $\Gamma = -0.100$ m^2/sec

5.67 $\Gamma = 0$

5.69 Yes; Yes

5.70 $\vec{\omega} = -0.05\,\text{sec}^{-1}\,\hat{k}$; $\Psi = Ay^2/2 + c$

5.73 $\omega = -U/2h$

5.75 $\vec{V} = -2y\,\hat{i} - 2x\hat{j}$

5.76 Yes; $\Psi = -(q\theta + K\ln r)/2\pi$

5.83 $df/d\mathbb{V} = -356$ N/m^3

5.84 $df/d\mathbb{V} = -1.85$ kN/m^3

5.85 $df/d\mathbb{V} = -0.0134$ lbf/ft^3

Chapter 6

6.1 $\vec{a}_p = 2\,\hat{i} + 2\hat{j}$ ft/sec^2;
$\nabla p = -(4\,\hat{i} + 68.4\hat{j})$ lbf/ft^2/ft

6.2 $a = 2.24$ m/sec^2 at $\theta = 63.4°$ above x axis;
$\nabla p = -(1.0\,\hat{i} + 11.8\hat{j})$ kN/m^2/m

6.3 $\vec{a}_p = 605\,\hat{i} + 5\hat{j}$ ft/sec^2;
$\partial p/\partial x = -1170$ lbf/ft^2/ft

6.4 $\nabla p = -(3.0\,\hat{i} + 9.0\hat{j})$ kN/m^2/m

6.5 $v = -Ay$; $\vec{a}_p = 8\,\hat{i} + 4\hat{j}$ m/sec^2;
$\nabla p = -12\,\hat{i} - 6\,\hat{j} - 14.7\,\hat{k}$ N/m^3;
$p(x) = 190 - 3x^2$ Pa (gage)

6.7 Yes; $(x, y) = (2.5, 1.5)$;
$\nabla p = -\rho[(4x - 10)\,\hat{i} + (4y - 6)\hat{j} + g\hat{k}]$;
$\Delta p = 9.6$ N/m^2

6.10 $a_x = 16v_0^2 x/D^2$; $p(0) = 8\rho v_0^2(L/D)^2$

6.12 $F = 1.56$ N, down

6.14 $\nabla p = -4.23\,\hat{i} - 12.1\hat{j}$ N/m^3;
$(x/h) = [1 - y/h] = $ constant

6.18 $p_2 - p_1 = 150$ kN/m^2

6.19 $p_2 - p_1 = 37.5$ kN/m^2

6.20 $a_r = -2800g$; $\partial p/\partial r = 270$ lbf/ft^2/ft

6.22 $p_{L/2} - p_0 = -30.6$ N/m^2

6.25 $V = 20.1$ m/sec

6.26 $\Delta h = 47.7$ mm water

6.29 $\Delta h = 628$ mm water

6.30 $V = 89.5$ ft/sec

6.32 $V = 27.5$ m/sec

6.33 $p_{\text{dyn}} = 296$ N/m^2; $p = -355$ N/m^2 (gage)

6.34 $p = 227$ kPa (gage)

6.35 $p = 291$ kPa (gage)

6.36 $h = 7.72$ m

6.37 $V = 21.5$ ft/sec; $Q = 0.469$ ft^3/sec

6.39 $p = -0.404$ kPa (gage)

6.40 $p = 52.6$ kPa (gage)

6.42 $V = 330$ ft/sec

6.44 $V = 44.2$ m/sec

6.45 $V = 101$ m/sec

6.46 $\Delta h = 1.77$ in. water; $\Delta h = 3.15$ in. water

6.47 $\Delta p = 5.54$ kPa; $\Delta p/q = 0.933$

6.49 $Q = 301$ gpm; $F_x = 565$ lbf; Tension

6.51 $Q = 2.55 \times 10^{-3}$ m^3/sec

6.52 $p_{1g} = 49.2$ kPa; $K_x = 57.5$ N

6.53 $Q = 1.25 \times 10^{-2}$ m^3/sec; $p_c = 71.6$ kPa

6.55 $p = 1.35$ psig; $p_{\max} = 1.79$ psig;
$F = 4.76$ lbf

6.56 $p_1 = 249$ kPa (abs);
$p_2 = 365$ kPa (abs); $\mathscr{P} = 17.4$ MW

6.57 $h/h_0 = \left[1 - \sqrt{\dfrac{g}{2h_0(AR^2 - 1)}}\,t\right]^2$

6.58 $h = H/2$; $R = H$

6.59 $\Delta h = 89.6$ mm; $K_x = 0.242$ N

6.61 $F_v = 532$ kN

6.62 $p = p_\infty + \frac{1}{2}\rho U^2(1 - 4\sin^2\theta)$;
$\theta = 30°, 150°, 210°, 330°$; $F = 0$

6.63 $F_v = 83.3$ kN

6.64 $p = 164$ kPa (gage); $F = 152$ N

6.65 $p = -85.4$ psfg; $p = 73.8$ in. Hg

6.67 $C_c = 0.5$

6.68 $p = 12.3$ kN/m^2 (gage)

6.69 $p_1 - p_2 = 1.73\cos(\omega t)$ kN/m^2

6.71 $dQ/dt = 0.0516$ m^3/sec/sec

6.72 $d^2l/dt^2 = 2gl/L$

6.73 $p_{\text{gage}} = 3\rho V^2 R^2/8b^2$

6.74 $D/d = 0.32$

6.75 $\Delta p = -615$ N/m^2

6.76 $\Delta p = -18.8$ kN/m^2

6.77 $\vec{\omega} = -\hat{k}A(x^2 + y^2)/2$

6.78 No; Yes

6.80 $\vec{\omega} = \hat{k}$

6.82 $\phi = xy^2 - x^3/3$

6.83 $\vec{V} = -2(y\,\hat{i} + x\,\hat{j})$; $\phi = 2xy$

6.84 $\psi = -2xy$

6.85 $\psi = B(x^2 - y^2)/2 - 2Axy$

6.86 $\vec{V} = -(2x + 1)\,\hat{i} + 2y\hat{j}$; $\psi = 2xy + y$;
$\Delta p = 23.3$ psf

6.88 $\phi = 3A(x^2 y - x^3)$

6.91 $\psi = Ax^2 y^3/3$; $2\vec{\omega} = -B(y^3 + 3x^2 y)\hat{k}$; No

6.96 $C_{p,\max} = 1$, $C_{p,\min} = -3$; $\theta = 30°, 150°$,
$210°, 270°$; $\partial C_p/\partial\theta = -3.46$

6.98 $r > 10a$

6.100 $0 \le K < 4\pi Ua$

6.102 $r > 9.77$ m, $p = -6.37$ kPa (gage)

6.104 $y = \pm a$

6.105 $\psi = 0.324\,q$; $Q/b = 0.176\,q$;
$p_{\text{dyn}} = 0.8\rho q^2/4\pi^2 a^2$

6.107 Stagnation at $r = 0.367m$, $\theta = 0, \pi$

6.108 $h = 0.162$ m; $V = 44.3\,\hat{i}$ m/sec,
$p = -957$ N/m^2 (gage)

6.110 $q = 25\pi$ m^2/sec; $y = \pm\,\pi/2$

6.111 $r = 1.82$ m, $\theta = 63°$;
$\rho = -317$ N/m^2 (gage)
6.112 $R_x/b = 5.51$ kN/m
6.113 $V_\infty = 4K/5\pi a$

Chapter 7

7.1 $F/\mu VD = $ constant
7.2 $\Delta p/\rho V^2 = f(\mu/\rho VD, d/D)$
7.3 $\delta/x = f(\rho U x/\mu)$
7.4 $\tau_w/\rho U^2 = f(\mu/\rho UL)$
7.6 $Q = h^2(gh)^{1/2} f(b/h)$
7.7 $V = \sqrt{gD} f(\lambda/d)$
7.8 $V(\rho\lambda/\sigma)^{1/2} = $ constant
7.9 $W/D^2 \omega \mu = f(l/D, c/D)$
7.10 $T = R^3 \mu \omega f(h/R)$
7.11 $E = \rho V^3 f(nr/V)$
7.12 $\mathcal{P} = \rho D^5 \omega^3 f(Q/D^3 \omega)$
7.13 $\zeta r^2/\Gamma_0, \Gamma_0 \tau/r^2, \nu/\Gamma_0$
7.14 4; 3; $\mu/\rho d^{3/2} g^{1/2}$
7.15 4; 3; $\mu/\rho d^{3/2} g^{1/2}$
7.16 $Q = V h^2 f(\rho V h/\mu, V^2 g/h)$
7.17 $R/\rho V^2 b^2, \rho V b/\mu, h/b$
7.18 3; $\rho VD/\mu, d/D, \sigma/\rho DV^2$
7.19 $\rho VD/\mu, h/d, D/d$
7.20 4; $Q/V d^2$; $\mu/\rho VD$
7.21 3; $\dot{m} = \rho A^{5/4} g^{1/2} f(h/A^{1/2}, \Delta p/\rho A^{1/2} g)$
7.22 $T/\rho V^2 D^3, \mu/\rho VD, \omega D/V, d/D$
7.23 $F_T/\rho V^2 D^2, gD/V^2, \omega D/V, p/\rho V^2, \mu/\rho VD$
7.24 $\mathcal{P} = p\omega D^3 f(\mu\omega/p, c/D, \ell/D)$
7.25 $\mathcal{P}/\rho D^2 V^3, \omega D/V, \mu/\rho VD, c/V$
7.26 $\dot{Q}/\rho V^3 L^2, c_p \Theta/V^2, \mu/\rho VL$
7.27 $p_{max}/\rho U_0^2 = f(E_v/\rho U_0^2)$
7.28 $p = 539$ kPa; $F = 1.34$ kN
7.29 $V_m/V_p = 0.345$; $F_D = 219$ N
7.30 $p = 1.93$ MPa (abs); $F_D = 43.7$ kN
7.31 $V_m = 39.2$ m/sec; $V_p = 39.2$ m/sec
7.32 $V = 158$ ft/sec; $F_m/F_p = 3.85$
7.33 $V_m = 6.21$ m/sec; $F_D = 0.978$ N
7.34 $D_m = 5.04$ in.; $\omega_m = 1000$ rpm
7.35 $V_m = 80$ ft/sec; $\omega_m = 1600$ rpm
7.36 $p_m = 2.95$ psia
7.37 $\bar{V} = 0.051$ ft/sec; $\Delta p = 0.021$ psig
7.38 $\Delta h_B = 0.50$ in.
7.39 $V_{max} = 27.1$ ft/sec
7.40 $C_{D,m} = 0.0972$; $F_D = 470$ N
7.42 $V_1/V_2 = 1/2$; $f_1/f_2 = 1/4$
7.43 $F_D = 2.46$ kN; $\mathcal{P} = 55.1$ kW
7.44 $F_D = 237$ kN
7.45 $\tau = 1070$ hr
7.46 $C_D = 0.951$; $F_D = 794$ lbf; $V = 807$ ft/sec
7.47 Scale ratio = 1/50; Not possible
7.49 $Q = 4930$ cfm
7.51 $H = 145$ ft · lbf/slug; $Q = 5.92$ ft^3/sec;
$D = 0.491$ ft
7.52 $F_t/\rho\omega^2 D^4 = f_1(g/\omega^2 D, \omega D/V)$;
$T/\rho\omega^2 D^5 = f_2(g/\omega^2 D, \omega D/V)$;
$\mathcal{P}/\rho\omega^3 D^5 = f_3(g/\omega^2 D, \omega D/V)$
7.53 $\omega = 533$ rpm; $F_t = 17,800$ lbf;
$T = 53,000$ ft · lbf

7.54 $V = 80$ ft/sec; $F_t = 1.47 \times 10^5$ lbf;
$T = 6.72 \times 10^5$ ft · lbf; $\mathcal{P}_t = 2.14 \times 10^4$ hp;
$\mathcal{P}_{in} = 3.07 \times 10^4$ hp; $\eta = 0.697$
7.55 KE ratio = 7.38
7.57 $C_D = 0.464$; $F_B \approx 0.273$ N; $F_D = 5.82$ N
7.58 $F_B = 0.574$; 0.44%
7.59 $\sigma/\rho LV_0^2, gL/V_0^2$
7.60 V_0^2/gL
7.61 gL/V_0^2
7.62 $\nu/V_0 L$

Chapter 8

8.1 Turbulent; $7.5 < L < 12.0$ m
8.5 $Q = 2.29 \times 10^{-5}$ m^3/sec; $\bar{V} = 2.63$ m/sec;
$L = 1.75$ m
8.6 $Re = 4Q/\pi D\mu$; $Re = 3000$
8.7 $\bar{V}/u_{max} = 2/3$
8.8 $Q/b = 2h/3$; $\bar{V}/u_{max} = 2/3$
8.9 $\tau_{yx} = -1.80$ N/m^2 (to right);
$Q/b = 5.40 \times 10^{-6}$ m^3/sec/m
8.10 $\tau_{yx} = -0.040$ lbf/ft^2 (to right);
$Q/b = 6.67 \times 10^{-5}$ ft^3/sec/ft
8.11 $\tau_{yx} = y\partial p/\partial x$; $\tau_{max} = -0.00835$ lbf/ft^2
8.12 $Q = 6.32 \times 10^{-2}$ in.^{3}/sec
8.13 $Q = 2.44 \times 10^{-2}$ in.^{3}/sec
8.14 $Q = 5.93 \times 10^{-4}$ in.^{3}/sec
8.15 $w = 0.50$ ft; $dp/dx = -400$ psi/ft;
$h = 1.96 \times 10^{-3}$ in.
8.16 $M = 4.32$ kg; $a = 1.28 \times 10^{-5}$ m
8.21 $\mu = 0.0695$ N · sec/m^2
8.22 $Q/b = 0.0146$ ft^3/sec/ft
8.23 $U_2/U_1 = 2$
8.24 $Q/b = 1.53 \times 10^{-7}$ m^3/sec/m
8.25 $\partial p/\partial x = -94.0$ N/m^2/m
8.26 $Re = 1.95$; $\tau = 2.01$ kN/m^2; $\mathcal{P} = 11.4$ W
8.28 $\tau = 3.92$ N/m^2; $Q/w = 7.60 \times 10^{-4}$ m^3/sec/m
8.29 $y/b = 0.892$; $u_{max}/U = 1.02$;
$\Psi = 7.13 \times 10^{-2}$ ft^3/ft
8.30 $U = -(h^2/2\mu)\partial p/\partial x$
8.31 $u = U_0 + (\rho g/\mu)(y^2/2 - hy)$
8.35 $r = 0.707R$
8.37 $\Delta p = 3.53 \times 10^6$ Pa
8.38 $\Delta p = 5.10$ lbf/in.^{2}; $u_\mu = 32\%$
8.39 $Q = 11.3$ mm^3/sec
8.40 $\delta D = \pm 0.775 \mu m$
8.41 $\Delta p = 349$ kPa, 14.0 GPa
8.42 $Q = 0.00203$ ft^3/sec
8.43 $D = 1.96$ mm, 115 mm
8.44 Error = 15.5 percent
8.46 $u = (c_1/\mu)\ln r + c_2$; $c_1 = \mu V_0/\ln(r_o/r_i)$;
$c_2 = -V_0 \ln r_i/\ln(r_o/r_i)$
8.52 $\tau_w = 2.16$ lbf/ft^2
8.53 $\bar{\tau}_w = -7.91$ N/m^2
8.55 $Q = 4.52 \times 10^{-7}$ m^3/sec; $\Delta p = 235$ kPa;
$\tau_w = 294$ N/m^2
8.56 $n = 6.17$ (graphical); 6.50 (least squares)
8.57 $n = 8.34$ (graphical); 9.16 (least squares)
8.58 $r/R = 0.707$ (laminar); 0.757 (turbulent)
8.61 $\beta = 4/3$ (laminar); 1.02 (turbulent)

8.63 KE flux $= 0.296$ N $\cdot$ m/sec

8.64 $\alpha = 1.54$

8.65 $\alpha = 2.0$

8.66 $\alpha = 1.055$

8.67 $h_l = 345$ J/kg

8.68 $p = 345$ kPa (gage)

8.69 $H = 104$ ft, $h_l = 25.2$ ft

8.71 $f = 0.042$

8.73 $\Delta p \propto D^{-5}$

8.75 $\tau_w = 0.0790$ lbf/ft^2; $u = 5.48$ ft/sec

8.76 $p_2 - p_1 = 24.4$ lbf/ft^2

8.77 $\bar{V} = 76.2$ ft/sec; $Q = 224$ ft^3/min

8.78 $Q = 1.10 \times 10^{-3}$ m^3/sec

8.79 $C_c = 0.715$; $K = 0.196$

8.80 $Q = 0.0361$ ft^3/sec

8.81 $\Delta Q = 0.0184$ m^3/sec

8.82 $AR = 2.7$, $2\phi = 12°$; $Q = 0.172$ m^3/sec

8.83 $\Delta p = 67.4$ kPa

8.84 $\Delta z = 9.80$ m

8.85 $S = -4.85°$; $\dot{Q} = -0.0271$ Btu/sec

8.86 $\Delta p = 43.9$ N/m^2

8.87 $\Delta p = 5.76$ kN/m^2

8.89 $p = 1.15$ MPa (gage)

8.90 $\Delta z = 52.8$ m; Fraction $= 1.8$ percent

8.92 $p_1 = 1.40$ MPa (gage); $\Delta T = 0.215$ K

8.93 $\Delta z = 8.13$ m

8.94 $p = 135$ psig

8.96 $e/D = 0.021$; Saving $= 48.2$ percent

8.97 $x = 95.8$ ft

8.98 $p = -73.8$ Pa (gage)

8.99 $\mathscr{P} = 664$ hp

8.100 $p = 336$ psig; $\mathscr{P} = 169$ hp

8.101 $\Delta p = 43.7$ psi; $\mathscr{P} = 286$ hp; Cost $= \$174,000$/yr

8.102 $\mathscr{P} = 4.47 \times 10^5$ hp; Fraction $= 0.2$ percent

8.103 $L = 53.1$ m

8.104 $L = 211$ m

8.105 $L = 51.1$ miles

8.107 $Q = 1.49$ ft^3/sec

8.108 $Q = 0.980$ ft^3/sec

8.109 $V = 27.5$ ft/sec

8.110 $Q = 0.227$ ft^3/sec

8.111 $V_0 = 28.0$ m/sec; $F = 365$ N

8.113 $Q_A = 0.0108$ m^3/sec, $Q_B = 0.00373$ m^3/sec

8.115 $Q = 108$ gpm; $V = 124$ ft/sec; $\mathscr{P} = 13.4$ hp

8.116 $Q = 0.0136$ m^3/sec

8.117 $Q = 5.30 \times 10^{-4}$ m^3/sec

8.120 $Q \approx 10^4$ ft^3/min

8.121 $Q = 106$ gpm

8.122 $D \geq 14$ mm

8.123 $D = 2.5$ in. (nominal)

8.124 $h = 0.194$ m, $b = 0.388$ m

8.125 $D = 6$ in.

8.130 p_2 (gage) $= 26.3 Q_2^2$; $Q_1 = Q_2 = Q_3 = Q_0/3$

8.132 $\dot{m} = 0.0592$ g/sec

8.133 $\Delta p = 462$ lbf/ft^2

8.134 $Q = 8.15$ ft^3/sec

8.135 $Q = 0.0404$ m^3/sec

8.136 $Q = 92.4$ gpm

8.137 $Q = 136$ gpm

8.138 $D = 40.8$ mm; $\dot{m} = 0.0220$ kg/sec

8.139 $\dot{m} = 2.10$ kg/sec; $\Delta h = 170$ mm Hg

8.141 $Q = 1.37$ ft^3/sec

8.142 $Re = 1810$; $f = 0.0354$; $p = -289$ N/m^2 (gage)

Chapter 9

9.1 $x = 71.6$ mm

9.2 $x = 196$ mm; $x_m = 14.5$ mm

9.3 $x = 0.114$ m

9.7 $A = U$; $B = \pi/2\delta$, $c = 0$

9.8 $a = 0$, $b = 3$, $c = -2$

9.11 $\delta_t/\delta_l = 144/105$

9.12 $\delta^*/\delta = 1/2, 1/3, 3/8, 0.363$

9.13 $\delta^*/\delta = 0.125, \theta/\delta = 0.0972$

9.15 $\delta^* = 1.9$ mm; $\theta = 0.76$ mm; $\tau_w = 0.125$ N/m^2

9.16 $\theta/\delta = 0.167, 0.133, 0.139, 0.137$

9.17 $H = 3, 2.50, 2.69, 2.65$

9.18 $H_{turbulent} = 1.29$; $H_{laminar} = 2.69$

9.19 $\theta = 0.0342$ in.

9.23 $\Delta p = 59.0$ Pa

9.24 $U_2 = 81.4$ ft/sec; $p_1 - p_2 = 0.264$ lbf/ft^2

9.25 $U_2 = 18.4$ m/sec; $\Delta p = 2.19$ Pa

9.26 $\delta_2^* = 5.00$ mm; $p_1 = -138$ Pa (gage)

9.27 $p_2 = -73.1$ Pa (gage); $\tau = 0.300$ N/m^2

9.28 $U_2 = 24.6$ m/sec; $p_1 = -43.9$ mm H_2O; $p_2 = -44.5$ mm H_2O

9.29 $Q = 2.44$ ft^3/sec; $V_2 = 55.1$ ft/sec; $\delta_2^* = 0.075$ in.

9.30 $F = 0.150$ lbf; $\bar{\tau}_w = 0.0188$ lbf/ft^2

9.39 $y = 3.28$ mm; Slope $= 0.00327$; $\theta = 1.09$ mm

9.43 $F = 1.62$ N

9.44 $\theta = 0.317$ mm; $F = 0.810$ N

9.45 $F = 10.7$ mN

9.48 $\theta = 0.0343$ in.; $F = 0.00416$ lbf

9.53 $F = 0.767$ N

9.57 $\theta = 0.00986$ in.; $F = 0.0765$ lbf

9.59 $F = 2.27$ N/m

9.62 $F = 2.32$ N

9.65 $\delta_l = 5.48$ mm, $\tau_w = 0.101$ N/m^2; $\delta_t = 23.7$ mm, $\tau_w = 0.502$ N/m^2

9.66 $\delta = 32.3$ mm; $\tau_w = 0.433$ N/m^2; $F = 0.380$ N

9.67 $W = 80.1$ mm

9.70 $mf_a = \rho U^2 w\delta/3$; $mf_b = 8\rho U^2 w\delta/15$

9.71 $a = b = 0, c = 3, d = -2$; $H = 3.89$

9.72 $W_2 = 12.4 - 0.4H_1$

9.73 $d\delta/dx = -0.00865$

9.74 $d\delta/dx = -0.00828$

9.75 $U = 5.11 + 1.42x$

9.77 $U_{max} = 7.82$ ft/sec; $\Delta h = 0.00340$ in. H_2O

9.80 $\mathscr{P} = 3.86$ kW

9.81 $Re_L = 1.54 \times 10^7$; $x_t = 53.6$ mm; $\mathscr{P} = 15.3$ kW

9.82 $F = 7.87$ kN; $\mathscr{P} = 1.79$ MW

9.83 $\bar{L} = 9.96$ ft; $F = 2250$ lbf

9.84 $V = 2.18$ mph; $x_t = 0.0339$ ft; $F_m = 3.65$ lbf, $F_p = 4110$ lbf

9.85 $V = 11.0$ ft/sec

9.86 $F = 5.49 \times 10^5$ N; $a_x = -0.0528$ m/sec^2

9.88 $\delta = 1.71$ m; $F = 1.59$ MN; $\mathscr{P} = 11.4$ MW; KE $= 2.73 \times 10^7$ W $\cdot$ sec; $\Delta t = 23.4$ min

9.89 $F = 92.3$ kN

9.90 $T = 86.2$ N $\cdot$ m; $\mathscr{P} = 542$ W

9.91 $D = 6.90$ m

9.92 $\bar{C}_D = 0.299$

9.93 $s = 117$ m

9.95 Horizontal is 20 percent better

9.96 $t = 1.33$ mm

9.97 $P_{max} = 243$ W

9.100 $V_t = 43.5$, 121 m/sec; $t = 8.11$, 22.6 sec; $y = 224$, 1730 m

9.104 $a_{max} = 2.54$ m/sec^2; $V_{max} = 60.3$ m/sec

9.106 $C_D = 1.17$

9.108 $V_{2(rel)} = 15$ m/sec; $p_2 = -133$ N/m^2 (gage); $m = 0.814$ kg

9.110 $\omega_{opt} = V/3R$

9.111 $F = 0.830$ N; $M = 0.125$ N $\cdot$ m

9.112 $T = 11.9$ N $\cdot$ m

9.113 $D < 0.231$ mm

9.114 $D = 3.81 \mu$m

9.115 $C_D = 0.479$

9.116 $\mathscr{P} = 19.2$ kW

9.117 $x = 13.9$ m

9.118 $V \approx 15.4$ m/sec

9.119 $V \approx 29.8$ ft/sec

9.120 $V \approx 85$ ft/sec

9.122 $V = 25.9$ m/sec; $t = 4.83$ sec; $y = 79.6$ m

9.123 $C_D = 61.9$; $\rho = 3720$ kg/m^3; $V = 0.731$ m/sec

9.124 $V = 10.3$ m/sec

9.127 $M = 386$ kN $\cdot$ m

9.128 $D \approx 6.2$ m

9.133 $C_D = 1.08$

9.137 $C_D = 0.606$

9.140 $\Delta\mathscr{P} = 18.2$ kW

9.141 $V \approx 10.7$ m/sec

9.142 $A_p = 7.03$ m^2

9.143 $m = 7260$ kg

9.145 $\vec{F} = 4.79\,\hat{i} + 20.3\,\hat{j}$ lbf; $\mathscr{P} = 0.388$ hp

9.146 $V = 5.62$ m/sec; $\mathscr{P} = 13.0$ kW; $V = 19.9$ m/sec

9.148 $V = 140$ m/sec; $R = 408$ m

9.150 $T = 14{,}800$ lbf

9.151 $F_D = 2.15$ kN; $\mathscr{P} = 149$ kW

9.153 $\theta = 3.42°$; $L = 168$ km

9.156 $\mathscr{P} = 0.190$ hp

9.157 $p = -190$ N/m^2 (gage); $V = 149$ km/hr

9.158 $F_L = 5.91$ mN

9.162 $\omega = 11{,}600$ rpm; $s = 1.19$ m

Chapter 10

10.1 $u = 2V_{max}y(1 - y/2D)/D$

10.4 $y = 6.68$ ft

10.5 $\lambda/y = 31.3$

10.7 $Fr = 0.498, 1.50$

10.8 $V = 3.95$ ft/sec

10.9 $V = 4$ ft/sec; $y = 2.52$ ft

10.10 $V = 8.02$ ft/sec; $Fr = 2$

10.13 $E = 2.19$ ft; $y = 1.46$ ft; $V = 6.85$ ft/sec

10.14 $y = 0.645$ ft, 4.30 ft

10.15 HGL$_1 = 11.97$ ft, EGL$_1 = 12.11$ ft; HGL$_2 = 11.25$ ft, EGL$_2 = 11.50$ ft

10.17 $Q = 834$ ft^3/sec

10.18 $y_c = 3.28$ ft

10.19 $y_c = (Q^2/g \; \cot^2\alpha)^{1/5}$

10.20 $(b + 2y \; \cot\alpha)/(by + y^2\cot^2\alpha)^3 = g/Q^2$

10.21 $Q = 3.25$ ft^3/sec

10.22 $y = 1.30$ ft

10.23 $y = 0.61$ ft (32 percent drop)

10.24 $y = 0.334$ ft (11.3 percent increase)

10.25 $y = 1.32$ ft

10.26 $y_0 = 5.70$ m; $Q/b = 23.2$ m^2/sec

10.27 $Q = 49.5$ ft^3/sec

10.28 $y = 0.507$ m; $Fr = 2.51$

10.29 $Q = 85.5$ ft^3/sec

10.30 $S_b = 1.49 \times 10^{-3}$

10.31 $b = 2.36$ m; $b = 2.49$ m

10.32 $S_b = 1.93 \times 10^{-3}$

10.33 $y_n = 2.32$ m

10.34 $Q = 0.171$ m^3/sec

10.35 $Q = 0.623$ m^3/sec

10.36 $y_n = 2.47$ ft

10.37 $y_n = 1.79$ m

10.38 $y_n = 0.773$ m

10.39 $y_n = 0.334$ m

10.41 $y_n = 2.00$ ft; $b = 4.00$ ft

10.42 $y_n = 7.79$ m, $b = 6.46$ m; 44.6 percent

10.44 $S_c = 2.48 \times 10^{-3}$

10.45 $b = 4.78$ ft; $S_c = 6.15 \times 10^{-3}$

10.47 $S_c = 3.82 \times 10^{-3}$

10.48 Depth increases (S3 curve)

10.49 A3 profile

10.50 $\Delta x = 892$ ft, downstream

10.51 $S_b = 0.00146$

10.52 $y = 0.966$ m

10.53 $\Delta x = 240$ ft (downstream)

10.54 $y = 2.86$ m (one reach)

10.55 $x = 3450$ m (upstream)

10.56 $x = 202$ m (221 m in one reach)

10.57 $x = 161$ m

10.58 $\Delta T = 0.00221$ K

10.59 $y = 4.51$ ft; $h_l = 2.40$ ft

10.60 $y = 5.95$ ft

10.61 $y = 3.99$ ft; $h_l = 1.13$ ft

10.62 $Q = 54.0$ ft^3/sec; $h_l = 1.62$ ft

10.63 $y = 10.3$ m; $V = 2.18$ m/sec; $E_d/E_u = 0.321$

10.64 $y = 4.44$ m; $h_l = 9.30$ m

10.65 $V = (g y_2 (1 + y_2/y_1)/2)^{1/2}$

10.66 $V = 4.89$ mph

10.67 $Q/b = 3.41$ ft^3/sec/ft

10.68 $Q = 16.8$ ft^3/sec

10.69 $Q = 0.402$ ft^3/sec

10.70 $Q/b = 27.2$ ft^3/sec/ft

Chapter 11

11.1 $\Delta p = 71.6$ psi; $H = 165$ ft (H_2O), 230 ft (gasoline)

11.6 $\theta_{eff} = 30.4°$

11.8 $H_0 = 77.0$ m; $V_2 = 25.7$ m/sec; $V_{rb_2} = 4.10$ m/sec; $H = 72.1$ m; $\mathcal{P} = 30.0$ kW

11.9 $\beta_1 = 64.2°; \theta_{eff} = +18.2°$ at $Q/2$; $\theta_{eff} = -7.95°$ at $3Q/2$;

11.11 $H(\text{ft}) = 156 - 1.36 \times 10^{-4}[Q(\text{gpm})]^2$

11.12 $H(\text{ft}) = 91.5 - 4.01 \times 10^{-7}[Q(\text{gpm})]^2$

11.13 $\eta \approx 0.79; H \approx 196$ ft at $Q \approx 630$ gpm

11.14 $\eta \approx 0.83; H \approx 184$ ft at $Q \approx 820$ gpm

11.15 $H_{1t} = 11.2$ m; $H_{2t} = 45.5$ m; $\mathcal{P}_h = 1.07$ kW; $\eta = 0.80; \mathcal{P}_m = 2$ hp; $\mathcal{P}_{in} = 2.11$ kW

11.16 1 hp (US) = 1.01 hpm; N_s (mhp) = $4.44 N_s$ (US)

11.17 $N_s = 1130; \mathcal{P}_{in} = 11.6$ hp

11.18 $\eta \approx 0.86$ at $Q = 2220$ gpm, $H = 127$ ft; $D = 13$ in.: $Q = 2400$ gpm, $H = 149$ ft; $D = 11$ in.: $Q = 2040$ gpm, $H = 107$ ft

11.19 $N_s = 1.21; \mathcal{P}_{in} = 350$ kW; $H(\text{m}) = 95.6 - 101 [Q (\text{m}^3/\text{sec})]^2$; $Q = 0.551$ m^3/sec; $H = 21.2$ m; $\eta = 0.82; \mathcal{P}_{in} = 139$ kW

11.20 $n = 4$ pump/motor units

11.21 $H_{1150} \approx 25.6$ ft

11.29 $N_m = 6730$ rpm; $D_m/D_p = 0.137$

11.30 Yes; Operate at flow rate below *BEP*, lower speed.

11.31 $N_m = 214$ rpm; $D_m/D_p = 0.172$

11.32 $a = 0.0426$ (gpm)$^{-1}$, $b = -1.56 \times 10^{-9}$(gpm)$^{-3}; r^2 = 0.998$

11.33 $T = 47$ C; $Q = 0.625$ L/sec; $H \approx 4.65$ m

11.34 $NPSHA = 23.0$ ft; $H = 36.5$ ft ($p = 15.8$ psig)

11.35 $a = 6.29$ ft, $b = 2.17 \times 10^{-5}$ ft/(gpm)^2; $r^2 = 0.996$

11.36 $a = 3.55$ ft, $b = 1.31 \times 10^{-5}$ ft/(gpm)^2; $r^2 = 0.990$

11.37 $Q_{max} = 470$ gpm

11.38 $H = 30.8$ ft; $\mathcal{P}_h = 1.17$ hp

11.39 $H = 10,300$ ft; $\mathcal{P}_h = 38,100$ hp

11.40 $Q = 2020$ gpm

11.41 $Q = 650$ gpm

11.42 $Q = 2540$ gpm; $L_e/D = 26,300$

11.43 $Q = 4190$ gpm; $L_e/D = 9620$

11.44 $Q = 3040$ gpm; $L_e/D = 42,000$

11.48 $D = 6$ in. (nominal); $\mathcal{P}_m = 890$ hp

11.49 With 3 pumps, $\eta \approx 0.90; \mathcal{P}_m = 241, 368$ hp

11.50 $H_p = 296$ ft

11.52 $Q = 898$ gpm, $H = 104$ ft; Type 4AE11, 11 in. impeller, 1750 rpm

11.53 $Q = 2330$ gpm, $H = 374$ ft; Type 8AE20G, 19 in. impeller, 1770 rpm

11.54 $Q = 197$ gpm, $H = 116$ ft; Type 4AE12, 11 in. impeller, 1750 rpm

11.55 $Q = 600$ gpm, $H = 775$ ft; Type 5TUT-16B, 5-stage, 1750 rpm

11.56 $Q = 11,200$ gpm, $H = 101$ ft; 3 Type 10AE12, 12 in. impeller, 1750 rpm

11.57 $Q = 16,500$ gpm, $H = 692$ ft (gasoline); 4 Type 10TU-22C, 2-stage, 1750 rpm

11.58 $Q = 1500$ gpm, $H = 170$ ft; Type 6AE14, 13 in. impeller, 1750 rpm

11.60 $Q_{max} = 11.2$ gpm at $\Delta z = 0$

11.62 $N = 3500$ rpm; $D = 4.3$ in.; $\eta \approx 0.53$

11.63 $n = 38$ stages; $N_s = 2530$ per stage; $\eta \approx 0.74$

11.64 $Q_1 = \sqrt{H_0/(A+C)}; Q_2 = \sqrt{H_0/(A+C/2)}$, etc.

11.66 $\eta \approx 0.8$ at $Q = 9200$ cfm

11.67 $A_{outlet} = 7.56$ ft^2; $\eta \approx 0.89$

11.72 $\mathcal{P}_{loss} = 0.290$ hp

11.74 $F_T = 442, 387$ lbf

11.76 $V = 16.0$ ft/sec; $J = 0.748; C_F = 0.0415$

11.80 $\mathcal{P}_{out} = 16, 200$ hp; $N = 353$ rpm; $T = 2.41 \times 10^5$ ft · lbf

11.81 $N_s = 35.1; Q = 884$ m^3/sec; $D = 8.91$ m

11.82 $N_s = 13.6; \eta = 0.505$

11.83 $D = 3.30$ m; $D_j = 0.518$ m; $Q = 17.4$ m^3/sec

11.84 $N_s = 26.5; T = 3.90 \times 10^6$ ft · lbf, $Q = 2590$ cfs at H = 380 ft

11.85 $N = 257$ rpm; $T = 4.68 \times 10^5$ N · m

11.86 For one jet, $N = 229$ rpm; $D = 10.5$ ft

11.88 $N_s = 55.0; Q = 876$ m^3/sec

11.89 $N_s(\text{US}) = 4.55; D = 6.20$ ft

11.90 $D_j \approx 2$ in.; $\mathcal{P} \approx 30$ hp

11.91 $\eta \approx 0.78$

11.94 $H_{net} \approx 1100$ ft; $N_s \approx 5$

11.96 $V_t \approx 167$ km/hr; $\mathcal{P}_{max} \approx 190$ kW

11.97 $\omega \approx 26$ sec^{-1}; $\mathcal{P}_{model} \approx 0.034$ hp

11.98 $\mathcal{P} \approx 22$ hp

11.99 $Q \approx 8.8$ gpm at $h = 30$ ft with $\eta_{pump} = 0.5$

11.100 $\mathcal{P}_{out}$ (W) = 0.285 $C_P[V (\text{m/sec})]^3[D (\text{m})]^2$

Chapter 12

12.1 $T = 273$ K

12.2 $s_2 - s_1 = 134$ J/kg · K

12.3 $s_2 - s_1 = -1.10$ kJ/kg · K; $Q/m = -572$ kJ/kg

12.4 $s_2 - s_1 = -1.10$ kJ/kg · K; $Q/m = -317$ kJ/kg

12.5 $\Delta u = -574$ kJ/kg; $\Delta h = -803$ kJ/kg; $\Delta s = 143$ J/kg · K

12.6 Yes

12.7 $\Delta S = -0.923$ Btu/R; $\Delta U = -684$ Btu; $\Delta H = -960$ Btu

12.8 $< 0; > 0; 0; 0$

12.9 $> 0; < 0; 0; 0$

12.10 $\dot{W} = 392$ kW

12.11 $W = 176$ MJ; $W = 228$ MJ; $T = 858$ K; $Q = -317$ MJ

12.13 $\Delta t = 828$ sec

12.14 $M = 0.806$

12.15 $M = 0.525, 1.08$

12.16 $M = 0.776; V = 269$ m/sec

12.17 $c = 299$ m/sec; $V = 987$ m/sec; $V/V_b = 1.41$

12.18 $c = 321, 1500$ m/sec

12.19 $E_v = 219$ GPa; $c = 1450$ m/sec

12.20 $c = 1150$, 1500 m/sec

12.22 $V = 306$, 415 m/sec

12.23 $a = 0.305$, $0.267g$

12.24 $r = 2030$, 7940 m; $a = 22.2$, 29.6 m/sec^2

12.25 $dT/dz = -6.49 \times 10^{-3}$ K/m;
$dc/dz = -3.83 \times 10^{-3}$, -4.36×10^{-3} sec^{-1}

12.27 $M = 0.437$, 2.33, 2.77

12.28 $V = 725$ m/sec

12.29 $V = 657$ m/sec; $\alpha = 31.8°$

12.30 $M = 8.32$; $\alpha = 6.90°$

12.31 $V = 6320$ ft/sec

12.32 $V = 493$ m/sec; $\Delta t = 0.398$ sec

12.33 $M = 1.19$; $V = 804$ ft/sec;
$Re/x = 9.84 \times 10^6$ m^{-1}

12.34 $V = 471$ m/sec; $t = 5.90$ sec

12.35 $t = 8.51$ sec

12.36 $\Delta x = 3.18$ km

12.37 $t \approx 48.5$ sec

12.38 $p_0 = 10.2$ psia; 57.5 psia

12.39 $M = 0.885$; $V = 296$ m/sec

12.40 $M = 0.199$, 0.314

12.41 $M = 0.851$; $V = 255$ m/sec; $z \approx 10$ km

12.42 $p_0 = 546$ kPa; $T_0 = 466$ K;
$h_0 - h = 178$ kJ/kg

12.43 $p_0 = 125$, 127.6 kPa

12.44 $M = 0.801$; $V = 236$ m/sec; $T_0 = 245$ K

12.45 $c = 295$ m/sec; $V = 649$ m/sec;
$\alpha = 27.0°$; $V/V_b = 1.41$; $T_0 = 426$ K

12.46 $p_0 = 28.85$ kPa; $V = 274.1$ m/sec;
9.3 percent

12.47 $a_x = -161$ m/sec^2; $p_0 = 191$ kPa (abs);
$T_0 = 346$ K

12.48 $T_0 = 1460$ R; $p_0 = 184$ psia;
$\dot{m} = 174$ lbm/sec

12.49 Yes; No

12.50 $p = 13.23$ psia; -0.454 percent

12.51 $V = 890$ m/sec; $T_0 = 677$ K; $p_0 = 212$ kPa

12.52 $T_0 = 308$ K; $p_0 = 101.4$ kPa; $\Delta p_0 = 800$ Pa;
0.79 percent

12.53 $T_0 = 318$ K; $p_0 = 2.06$ MPa (abs);
$p = 1.67$ MPa (abs)

12.54 $V = 620$ m/sec; $p_0 = 208$ kPa;
$p_0 = 140$ kPa; $T_0 = 408$ K

12.55 $V = 987$ m/sec; $p_0 = 125$ kPa;
$p_0 = 31.6$ kPa; $T_0 = 707$ K

12.56 $T_0 = 288$ K; $p_0 = 101$, 39.9 kPa;
$s_2 - s_1 = 267$ J/kg · K

12.57 $T_0 = 585$, 1782 K; $p_0 = 1.03$ MPa,
963 kPa; $s_2 - s_1 = 1140$ J/kg · K

12.58 $T_0 = 445$ K; $p_0 = 57.5$, 46.7 kPa (abs);
$s_2 - s_1 = 59.6$ J/kg · K

12.59 $T_0 = 2900$, 1870 R; $p_0 = 101$, 4.57 psia;
$s_2 - s_1 = 0.107$ Btu/lbm · R

12.60 $\Delta p = 48.2$ kPa

12.61 $T_0 = 309$, 573 K; $p_0 = 838$ kPa (abs),
8.13 MPa (abs); $s_2 - s_1 = 170$ J/kg · K

12.62 $\delta Q/dm = 63.0$ Btu/lbm; $p_{0_2} = 56.5$ psia

12.64 $T_0 = 344$ K; $p_0 = 223$, 145 kPa;
$s_2 - s_1 = 0.124$ kJ/kg · K

12.65 $T_0 = 621$, 883 R; $p_0 = 67.6$, 56.6 psia;
$s_2 - s_1 = 0.0968$ Btu/lbm · R

12.67 $p^*/p_0^* = 0.488$, 0.547

12.68 $T^* = 258$ K, $p^* = 476$ kPa (abs);
$V^* = 322$ m/sec

12.69 $T^* = 253$ K, $p^* = 476$ kPa (abs);
$V^* = 319$ m/sec

12.70 $T^* = 260$ K, $p^* = 24.7$ MPa (abs);
$V^* = 252$ m/sec

12.71 $T^* = 265$ K, $p^* = 19.2$ MPa (abs);
$V^* = 254$ m/sec

12.72 $T^* = 1880$ K, $p^* = 2.98$ MPa (abs);
$V^* = 2550$ m/sec

12.73 $T^* = 3180$ K, $p^* = 22.6$ MPa (abs);
$V^* = 1110$ m/sec

12.74 $T^* = 268$ K, $p^* = 456$ kPa (abs);
$V^* = 426$ m/sec

12.75 $T^* = 2390$ R, $p^* = 79.2$ kPa (abs);
$V^* = 2400$ ft/sec

Chapter 13

13.1 $V = 3280$ ft/sec

13.2 $V = 4210$ ft/sec

13.3 $V = 996$ m/sec

13.4 $V = 1280$ m/sec

13.5 $V = 1660$ ft/sec; $M = 0.787$;
$\dot{m} = 0.274$ lbm/sec

13.6 $V = 2620$ ft/sec; $M = 1.36$;
$\dot{m} = 1.76$ lbm/sec

13.7 $V = 502$ m/sec; $M = 0.784$;
$\dot{m} = 0.190$ kg/sec

13.8 $V = 797$ m/sec; $M = 1.35$;
$\dot{m} = 0.706$ kg/sec

13.9 $M = 0.851$; $V = 648$ mph

13.10 $M = 1.35$

13.11 $M_2 = 1.20$

13.12 $M_2 = 1.20$

13.13 $V = 610$ m/sec; $A = 0.129$ m^2

13.14 $A = 1.49 \times 10^{-3}$ m^2

13.15 $T = 575$ R; $p = 21.0$ psia

13.16 $\Delta p = 315$ kPa; $A_1/A_2 = 3.79$

13.17 $V = 903$ ft/sec; $p_0 = 119$ psia

13.18 $\dot{m} = 8.50$ kg/sec

13.19 $p_t = 33$ psia; $M_t = 0.90$; $V_t = 1060$ ft/sec

13.20 $p_t = 112$ kPa

13.21 $\dot{m} = 0.548$ kg/sec

13.22 $\rho_t = 0.478$ kg/m^3

13.23 $A = 1.92 \times 10^{-3}$ m^2

13.24 $A_{min} = 24.1$ mm^2

13.25 $p_0 = 113$ psia; $\dot{m} = 3.25$ lbm/sec

13.26 $M = 1.0$; $A = 450$ mm^2

13.27 $T = 369$ K; $\Delta p = 2.64$ MPa;
$\dot{m} = 1.14 \times 10^{-2}$ kg/sec

13.28 $p_0 \geq 191$ kPa; $\dot{m} = 1.28$ kg/sec

13.29 $\dot{m} = 0.0107$ lbm/sec

13.30 $t = 68.4$ sec; $\Delta s = 0.0739$ Btu/lbm · R

13.31 $R_x = 1560$ N (to the left)

13.32 $T_0 = 522$ R; $\Delta A = -15.9$ percent;
$p = 13.5$ psia; $V = 1020$ ft/sec

13.33 $p = 803$ kPa (abs); $\dot{m} = 0.113$ kg/sec; $a_{rf_x} = 2.28$ m/sec^2

13.34 $\dot{m} = 2.73$ lbm/sec; $a_{rf_x} = 99.8$ ft/sec^2

13.35 $R_x = 304$ lbf, tension

13.36 $A = 0.0173$ m^2; $V = 390$ m/sec

13.38 $t = 23.6$ sec

13.39 $t = 23.5$ sec; $\Delta s = 161$ J/kg $\cdot$ K

13.40 $\dot{m} = 0.856$ lbm/sec

13.41 $M = 0.60$; $T = 298$ K; $p_0 = 700$ kPa; $\dot{m} = 0.622$ kg/sec

13.42 $p = 125$ kPa (abs); $\dot{m} = 0.401$ kg/sec

13.43 $\dot{m} = 0.0957$ kg/sec

13.44 $V = 1300$ m/sec; $\dot{m} = 87.4$ kg/sec

13.45 $A = 2.99$ in.^{2}; $\dot{m} = 3.74$ lbm/sec

13.46 $\dot{m} = 5.44$ kg/sec

13.47 $\dot{m} = 6.05$ lbm/sec; Decrease by a factor of 2

13.48 $A = 8.86 \times 10^{-4}$ m^2, 1.50×10^{-3} m^2

13.49 $V = 504$ m/sec; $A = 0.0596$ m^2

13.50 $R_x = 550$ N

13.51 $\dot{m} = 39.4$ lbm/sec; $F_x = 9750$ lbf

13.52 $\dot{m} = 33.0$ kg/sec; $A_e = 0.158$ m^2; $A_e/A_t = 18.0$

13.53 $p = 88.3$ kPa (abs); $\dot{m} = 0.499$ kg/sec; $K_x = 1030$ N

13.54 $A = 1560$ mm^2

13.55 $p_t = 3290$ psia; $\dot{m} = 0.0524$ lbm/sec; Thrust = 2.37 lbf; 36.3 percent

13.56 $p = 488$ kPa (abs); $\dot{m} = 0.0688$ kg/sec; D = 12.7 mm

13.57 $A_{\text{eff},t} = 74.7$ mm^2; $A_{\text{eff},e} = 295$ mm^2; $M = 0.149$, 2.93; 4.9 percent decrease

13.58 $V = 316$ m/sec; $T = 248$ K

13.59 $M = 0.897$; $T_0 = 339$ K

13.60 $M = 0.20$; $\Delta s = 311$ J/kg $\cdot$ K

13.61 $\dot{m} = 0.0726$ kg/sec; $p \leq 33.5$ kPa (abs)

13.62 $M = 0.20$; $\dot{m} = 3.19 \times 10^{-3}$ kg/sec; $p = 47.9$ kPa (abs)

13.63 $T = 287$ K; $p_{0_2} = 423$ kPa; $\Delta s = 132$ J/kg $\cdot$ K

13.64 $p = 477$ kPa (abs); $\Delta s = 49.5$ J/kg $\cdot$ K

13.65 $p = 18.5$ psia; $V = 1040$ ft/sec

13.66 $\dot{m} = 0.00321$ kg/sec; $p_0 = 33.8$ kPa (abs); $\Delta s = 314$ J/kg $\cdot$ K

13.67 $\dot{m} = 0.0192$ kg/sec; $T^* = 244$ K; $p^* = 53.4$, 13.6 kPa (abs)

13.68 $F = 822$ lbf

13.69 $M = 0.844$; $p_2 = 475$ kPa; $F = 2650$ N

13.70 $T = 568$ R; $T_0 = 610$ R; $\Delta p_0 = -18.3$ psi; $\Delta s = 0.0312$ Btu/lbm $\cdot$ R

13.72 $p_t = 56.6$ psia; $T = 433$ R; $p_0 = 27.8$ psia; $\dot{m} = 0.0316$ lbm/sec

13.73 $V = 686$ m/sec; $M = 1.90$

13.74 $p = 47.5$ psia; $T = 417$ R; $\Delta s = 0.0360$ Btu/lbm $\cdot$ R

13.75 $T = 468$ K; $p = 313$ kPa (abs)

13.76 $T = 468$ K; $F = 60$ N; $\Delta s = 149$ J/kg $\cdot$ K

13.77 $V_1 = 688$ m/sec; $M_1 = 1.90$; $F = 220$ N (to the right)

13.78 $T = 238$ K; $p = 26.1$ kPa (abs); $\Delta s = 172$ J/kg $\cdot$ K

13.79 $L = 2.88$ m

13.81 $L = 1.27$ m

13.82 $T = 459$ K; $L = 34.5$ m

13.84 $L = 18.8$ ft

13.90 $L = 11.1$ ft

13.91 $L = 0.405$ m

13.93 $p = 191$ kPa (abs); $L = 5.02$ m; $\Delta s = 326$ J/kg $\cdot$ K

13.95 $M = 0.25$; Added

13.102 $Q = 1.84 \times 10^8$ ft^3/day

13.103 $\delta Q/dm = 145$ kJ/kg; $\Delta p = 405$ kPa

13.104 $\delta Q/dm = 243$ Btu/lbm

13.105 $M_1 = 0.604$; $\dot{Q} = 1.37$ MW; $p_2 = 265$ kPa (abs); $\Delta s = 150$ J/kg $\cdot$ K

13.106 $\delta Q/dm = 449$ kJ/kg; $\Delta s = 0.892$ kJ/kg $\cdot$ K

13.107 $T = 1490$ K; $M = 0.230$; $\Delta p_0 = -28.0$ kPa

13.108 $T_0 = 497$ K; $M = 0.286$; $\Delta s = 593$ J/kg $\cdot$ K

13.109 $\dot{Q} = 111$ kW; $p_1 - p_2 = 1.30$ MPa

13.110 $\delta Q/dm = 18$ kJ/kg; $\Delta s = 53.2$ J/kg $\cdot$ K; $\Delta p_0 = 2.0$ kPa

13.111 $\dot{Q} = 1080$ Btu/sec; $M = 0.501$

13.112 $\delta Q/dm = 240$ Btu/lbm; $p_{0_2} = 97.1$ psia

13.113 $T_0 = 483$ R; $p = 13.4$ psia; $\Delta p_0 = 0.1$ psia

13.114 $p = 105$ psia; $T_0 = 2380$ R; $M = 0.901$; $\delta Q/dm = 371$ Btu/lbm

13.115 $M = 0.498$; $T = 1480$ K; $T_0 = 1550$ K; $\dot{Q} = 1.85$ MJ/sec

13.116 $V = 1520$ ft/sec; $T = 2310$ R; $\dot{Q} = 740$ Btu/sec

13.117 $p = 209$ psia; $\dot{Q} = 2300$ Btu/sec; $\dot{m}_f = 0.128$ lbm/sec

13.118 $\delta Q/dm = 330$ Btu/lbm; $\Delta p_0 = -2.45$ psia

13.119 $V = 866$ m/sec; $p = 46.4$ kPa; $M = 1.96$; $\delta Q/dm = 156$ kJ/kg

13.120 $\delta Q/dm = 62.9$ Btu/lbm; $p_0 = 56.6$ psia

13.121 $p_2 < p_{\text{atm}}$; $M_2 > 1.0$

13.122 $\delta Q/dm = 145$ kJ/kg; $p = 696$ kPa

13.123 $M = 0.26$; $p = 97.9$ kPa

13.124 $M = 0.30$; $T_0 = 2000$ R; $p_0 = 97.1$ psia

13.125 $M = 0.50$; $T_0 = 1560$ R; $\dot{Q} = 1.86$ MJ/sec

13.126 $V = 865$ m/sec; $p = 46.7$ kPa; $M = 1.96$; $\delta Q/dm = 162$ kJ/kg

13.127 $\Delta p_0 = -22$ kPa; $\delta Q/dm = 447$ kJ/kg; $\Delta s = 889$ J/kg $\cdot$ K

13.128 $\delta Q/dm = 17.0$ kJ/kg; $T = 318$ K; $p = 46.3$ kPa; $p_0 = 87.7$ kPa

13.129 $T_0 = 497$ K; $M = 0.28$; $\Delta p_0 = -7.0$ kPa; $s_2 - s_1 = 588$ J/kg $\cdot$ K

13.130 $\dot{Q} = 898$ Btu/sec; $M = 0.50$; $p = 7.84$ psia

13.131 $M = 1.0$; $p = 48.8$ kPa; $\Delta p_0 = -8.60$ kPa

13.132 $A = 0.00269$ m^2; $\delta Q/dm = -26.4$ kJ/kg; $\Delta p_0 = 0.60$ kPa; $\Delta s = -95.6$ J/kg $\cdot$ K

13.133 $\dot{Q} = 5.16 \times 10^4$ Btu/sec

13.134 $\delta Q/dm = 313$ Btu/lbm; $\Delta p_0 = -34$ psia

13.135 $T_0 = 764$ K; $\dot{m} = 0.0215$ kg/sec; $A_e/A_t = 4.23$

13.136 $T_0 = 623$ K; $M = 0.601$;
$\delta Q/dm = 342$ kJ/kg; Fraction $= 0.614$

13.137 $\delta Q/dm = 899$ kJ/kg; $M = 1.0$;
$\Delta p_0 = -173$ kPa

13.140 $\delta Q/dm = 0$; $F = 24.3$ N;
$\Delta s = 45.6$ J/kg $\cdot$ K

13.141 $Q = 53.1$ ft^3/min; $\Delta h = 0.665$ in. H_2O

13.144 $p = 344$ kPa

13.145 $p = 28.1$ psia

13.146 $V = 247$ m/sec; $T = 671$ K

13.147 $V = 256$ m/sec; $M = 0.490$

13.148 $M = 0.707$; $V = 269$ m/sec; $p_0 = 205$ kPa

13.149 $T = 521$ K; $p_0 = 1.29$ MPa

13.150 $M = 0.50$; $T = 679$ K

13.151 $T_0 = 426$ K; $p_0 = 207, 130$ kPa

13.152 $V = 536$ m/sec

13.153 $p_0 = 7.22$ psia; $T_0 = 954$ R

13.154 $\rho = 0.359$ lbm/ft^3; $M = 0.701$

13.156 $V = 258$ m/sec; $\Delta p = 471, 842$ kPa

13.157 $M = 0.545$; $p = 514$ kPa; $p_0 = 629$ kPa;
$A = 0.111$ m^2

13.158 $T_0 = 533$ K; $\Delta p = 37.4$ kPa;
$\Delta s = 30.0$ J/kg $\cdot$ K; $p_0 = 116$ kPa

13.159 $V = 265, 279$ m/sec

13.160 $M = 0.701$; $V = 267$ m/sec; $p_0 = 205$ kPa

13.161 $p = 345$ kPa; $\Delta p_0 = -167$ kPa

13.162 $V = 247$ m/sec; $T = 670$ K;
$\Delta s = 315$ J/kg $\cdot$ K

13.163 $p = 28.1, 85.7$ psia

13.164 $T = 520$ K; $p_0 = 1.29$ MPa

13.165 $M = 0.501$; $T = 680$ K; $\Delta s = 229$ J/kg $\cdot$ K

13.166 $V = 257$ m/sec; $M = 0.493$;
$\Delta p_0 = -512$ kPa

13.167 $V = 255$ m/sec; $\Delta p = 473, 824$ kPa

13.168 $T_0 = 426$ K; $p_0 = 207, 130$ kPa

13.169 $T = 533$ K; $p_0 = 54.6$ kPa

13.170 $M = 2.48$; $V = 2420$ ft/sec; $p = 24.3$ psia;
$p_0 = 29.1$ psia

13.171 $T = 414$ K; $p = 51.9$ kPa; $p_0 = 57.9$ kPa

13.172 $\Delta p_0 = 1.6$ kPa; $p = 51.5$ kPa

13.173 $A = 2.32$ ft^2; $\Delta s = 0.0423$ Btu/lbm $\cdot$ R

13.174 $\Delta p_0 = -14.1$ psi;
$\Delta s = 0.0591$ Btu/lbm $\cdot$ R

13.176 $M = 3.20$; $V = 949$ m/sec

13.177 $V = 2170$ ft/sec

13.178 $M_4 = 0.70$; $p_4 = 131$ psia; $L = 5.93$ ft

13.179 $p = 33.4$ kPa; $V = 162$ m/sec

13.180 $p = 66.6$ psia

13.183 $\Delta p = 27.5$ kPa; $\Delta p_0 = -26.9$ kPa

13.184 $p = 301$ kPa

13.185 $p = 46.7$ psia; $A = 1.52$ in.^{2};
$\dot{m} = 2.55$ lbm/sec

13.186 $p = 587$ kPa; $A_e = 755$ mm^2; $A = 448$ mm^2

13.189 $p = 33.4$ kPa; $V = 162$ m/sec

13.190 $M = 1.45$; $\dot{m} = 0.808$ lbm/sec

13.191 $M = 0.701$; $p = 167$ kPa;
$\Delta s = 20.9$ J/kg $\cdot$ K

13.192 $M = 1.92$; $p = 89.4, 58.6, 14.5$ psia

13.193 $M = 2.94$; $p_0 = 3.39$ MPa;
$p = 3.35, 1.00$ MPa, 101 kPa

13.194 $p = 301$ kPa

13.195 $p = 46.7$ psia; $A = 1.52$ in.^{2};
$\dot{m} = 2.55$ lbm/sec

13.196 $p = 587$ kPa; $A_e = 756$ mm^2; $A = 448$ m^2

13.197 $M = 1.50$

13.198 $33.4 < p_b < 99.6$ kPa; $\dot{m} = 0.121$ kg/sec

13.199 $M = 2.12$; $V = 2000$ ft/sec

13.200 $p_{atm} < p_0 < 112$ kPa and $p_0 > 743$ kPa

13.201 $M = 2.14$

13.202 $V = 2140$ ft/sec; $\Delta s = 0.0388$ Btu/lbm $\cdot$ R

INDEX

Conversion Factors and Definitions

Fundamental Conversion Factor	English Unit	Exact SI Value	Approximate SI Value
Length	1 in.	0.0254 m	—
Mass	1 lbm	0.453 592 37 kg	0.4536 kg
Temperature	1 F	5/9 K	—

Definitions:

Acceleration of gravity: $g = 9.8066$ m/sec^2 ($= 32.174$ ft/sec^2)

Energy: Btu (British thermal unit) $\equiv$ amount of energy required to raise the temperature of 1 lbm of water 1 F (1 Btu $=$ 778.2 ft · lbf)

kilocalorie $\equiv$ amount of energy required to raise 1 kg of water 1 K (1 kcal $=$ 4187 J)

Length: 1 mile $=$ 5280 ft; 1 nautical mile $=$ 6076.1 ft

Power: 1 horsepower $\equiv$ 550 ft · lbf/sec

Pressure: 1 bar $\equiv$ 10^5 Pa

Temperature: degree Fahrenheit, $T_F = \frac{9}{5}T_C + 32$ (where T_C is degrees Celsius)

degree Rankine, $T_R = T_F + 459.67$

Kelvin, $T_K = T_C + 273.15$ (exact)

Viscosity: 1 Poise $\equiv$ 0.1 kg/m · sec

1 Stoke $\equiv$ 0.0001 m^2/sec

Volume: 1 gal $\equiv$ 231 in.^{3} (1 ft^3 $=$ 7.48 gal)

Useful Conversion Factors:

1 lbf $=$ 4.448 N

1 lbf/in.^{2} $=$ 6895 Pa

1 Btu $=$ 1055 J

1 hp $=$ 746 W $=$ 2545 Btu/hr

1 kW $=$ 3413 Btu/hr

1 quart $=$ 0.000946 m^3 $=$ 0.946 liter

1 kcal $=$ 3.968 Btu